Lecture Notes in Computer Science 5555

Commenced Publication in 1973
Founding and Former Series Editors:
Gerhard Goos, Juris Hartmanis, and Jan van Leeuwen

Advanced Research in Computing and Software Science

Subline of Lectures Notes in Computer Science

Susanne Albers
Alberto Marchetti-Spaccamela
Yossi Matias
Sotiris Nikoletseas
Wolfgang Thomas (Eds.)

Automata, Languages and Programming

36th International Colloquium, ICALP 2009
Rhodes, Greece, July 5-12, 2009
Proceedings, Part I

Springer

Volume Editors

Susanne Albers
University of Freiburg, Department of Computer Science
Georges Köhler Allee 79, 79110, Freiburg, Germany
E-mail: salbers@informatik.uni-freiburg.de

Alberto Marchetti-Spaccamela
Sapienza University of Rome
Department of Computer and Systems Sciences
Via Ariosto 25, 00184 Roma, Italy
E-mail: alberto@dis.uniroma1.it

Yossi Matias
Tel Aviv University, School of Computer Science
Google R&D Center, Tel Aviv 69978, Israel
E-mail: matias@cs.tau.ac.il

Sotiris Nikoletseas
University of Patras and CTI
N. Kazantzaki Street 1, 26504 Rion, Patras, Greece
E-mail: nikole@cti.gr

Wolfgang Thomas
RWTH Aachen, Lehrstuhl Informatik 7
Ahornstraße 55, 52074 Aachen, Germany
E-mail: thomas@informatik.rwth-aachen.de

Library of Congress Control Number: 2009929832

CR Subject Classification (1998): F.2, F.3, I.2.1, G.1, G.2, I.1

LNCS Sublibrary: SL 1 – Theoretical Computer Science and General Issues

ISSN 0302-9743
ISBN-10 3-642-02926-4 Springer Berlin Heidelberg New York
ISBN-13 978-3-642-02926-4 Springer Berlin Heidelberg New York

springer.com

Printed in Germany

Typesetting: Camera-ready by author, data conversion by Scientific Publishing Services, Chennai, India
Printed on acid-free paper SPIN: 12715473 06/3180 5 4 3 2 1 0

Preface

ICALP 2009, the 36th edition of the International Colloquium on Automata, Languages and Programming, was held on the island of Rhodes, July 6–10, 2009. ICALP is a series of annual conferences of the European Association for Theoretical Computer Science (EATCS) which first took place in 1972. This year, the ICALP program consisted of the established track A (focusing on algorithms, complexity and games) and track B (focusing on logic, automata, semantics and theory of programming), and of the recently introduced track C (in 2009 focusing on foundations of networked computation).

In response to the call for papers, the Program Committee received 370 submissions: 223 for track A, 84 for track B and 63 for track C. Out of these, 108 papers were selected for inclusion in the scientific program: 62 papers for track A, 24 for track B and 22 for track C. The selection was made by the Program Committees based on originality, quality, and relevance to theoretical computer science. The quality of the manuscripts was very high indeed, and many deserving papers could not be selected.

ICALP 2009 consisted of five invited lectures and the contributed papers. This volume of the proceedings contains all contributed papers presented in track A together with the papers by the invited speakers Kurt Mehlhorn (Max-Planck-Institut für Informatik, Saarbrücken) and Christos Papadimitriou (University of California at Berkeley). A companion volume contains all contributed papers presented at the conference in track B and track C, together with the papers by the invited speakers Georg Gottlob (University of Oxford), Tom Henzinger (École Polytechnique Fédérale de Lausanne), and Noam Nisan (Google, Tel Aviv, and Hebrew University).

The following workshops were held as satellite events of ICALP 2009:

ALGOSENSORS 2009—5th International Workshop on Algorithmic Aspects of Wireless Sensor Networks
DCM 2009—5th International Workshop on Developments in Computational Models
FOCLASA 2009—8th International Workshop on Foundations of Coordination Languages and Software Architectures
QUANTLOG 2009—Workshop on Quantitative Logics 2009

We wish to thank all authors who submitted extended abstracts for consideration, the Program Committees for their scholarly effort, and all referees who assisted the Program Committees in the evaluation process.

Thanks are due to the sponsors (Ministry of National Education and Religious Affairs of Greece, Research Academic Computer Technology Institute (CTI), Piraeus Bank) for their support, and to the Research Academic

Computer Technology Institute (CTI) for the local organization. We are also grateful to all members of the Organizing Committee.

Thanks also to Andrei Voronkov for his help with the conference management system EasyChair, which was used in handling the submissions and the electronic PC meeting as well as in assisting in the assembly of the proceedings.

April 2009

Susanne Albers
Alberto Marchetti Spaccamela
Yossi Matias
Paul G. Spirakis
Wolfgang Thomas

Organization

Program Committee

Track A

Susanne Albers	University of Freiburg, Germany (Chair)
Gerth Brodal	University of Aarhus, Denmark
Martin Dyer	University of Leeds, UK
Irene Finocchi	University of Rome "La Sapienza", Italy
Anna Gal	University of Texas at Austin, USA
Naveen Garg	IIT Delhi, India
Raffaele Giancarlo	University of Palermo, Italy
Andrew Goldberg	Microsoft Research, Silicon Valley, USA
Mordecai Golin	Hong Kong University
Michel Habib	LIAFA, Paris 7, France
Thore Husfeldt	Lund University, Sweden
Kazuo Iwama	University of Kyoto, Japan
Howard Karloff	AT&T Labs, USA
Yishay Mansour	Tel Aviv University and Google, Israel
Jiri Matoušek	Charles University, Prague, Czech Republic
Marios Mavronicolas	University of Cyprus, Cyprus
Piotr Sankowski	University of Warsaw and ETH Zurich, Switzerland
Raimund Seidel	University of Saarbrücken, Germany
Paul Spirakis	CTI and University of Patras, Greece
Dorothea Wagner	University of Karlsruhe, Germany
Peter Widmayer	ETH Zurich, Switzerland
Ronald de Wolf	CWI Amsterdam, The Netherlands

Track B

Albert Atserias	Universitat Politecnica de Catalunya, Barcelona, Spain
Jos Baeten	Eindhoven University of Technology, The Netherlands
Gilles Barthe	IMDEA Software, Madrid, Spain
Mikolaj Bojanczyk	Warsaw University, Poland
Christian Choffrut	University Denis Diderot, Paris, France
Thierry Coquand	Göteborg University, Sweden
Roberto di Cosmo	University Denis Diderot, Paris, France
Kousha Etessami	University of Edinburgh, Scotland, UK
Dexter Kozen	Cornell University, USA

Stephan Kreutzer	Oxford University, UK
Orna Kupferman	Hebrew University, Israel
Kim Guldstrand Larsen	Aalborg University, Denmark
Dale Miller	Ecole Polytechnique, Palaiseau, France
Markus Müller-Olm	University of Münster, Germany
Anca Muscholl	University Bordeaux 1, France
R. Ramanujam	Institute of Mathematical Sciences, Chennai, India
Simona Ronchi Della Rocca	University of Turin, Italy
Jan Rutten	CWI, Amsterdam, The Netherlands
Vladimiro Sassone	University of Southampton, UK
Peter Sewell	University of Cambridge, UK
Howard Straubing	Boston College, USA
Wolfgang Thomas	RWTH Aachen University, Germany (Chair)

Track C

Hagit Attiya	Technion, Israel
Andrei Broder	Yahoo, USA
Xiaotie Deng	City University of Hong Kong
Danny Dolev	Hebrew University, Israel
Michele Flammini	University of L'Aquila, Italy
Pierre Fraigniaud	CNRS, Paris, France
Ashish Goel	University of Stanford, USA
Matthew Hennessy	Trinity College Dublin, Ireland
Kohei Honda	University of London, UK
Robert Kleinberg	Cornell University, USA
Elias Koutsoupias	University of Athens, Greece
Alberto Marchetti Spaccamela	University of Rome "La Sapienza", Italy (Co-chair)
Yossi Matias	Google and Tel Aviv University, Israel (Co-chair)
Silvio Micali	MIT, USA
Muthu Muthukrishnan	Google, NY, USA
Moni Naor	Weizmann Institute, Israel
Mogens Nielsen	University of Aarhus, Denmark
Harald Raecke	University of Warwick, UK
Jose Rolim	University of Geneva, Switzerland
Christian Schindelhauer	University of Freiburg, Germany
Roger Wattenhofer	ETH Zurich, Switzerland
Martin Wirsing	University of Munich, Germany

Organizing Committee

- Paul G. Spirakis, Research Academic Computer Technology Institute and University of Patras, Greece (Conference Chair)
- Elias Koutsoupias, Department of Informatics and Telecommunications, University of Athens, Greece (Conference Chair)
- Christos Kaklamanis, Research Academic Computer Technology Institute and University of Patras, Greece (Conference Chair)
- Christos D. Zaroliagis, Research Academic Computer Technology Institute and University of Patras, Greece (Workshops Chair)
- Sotiris Nikoletseas, Research Academic Computer Technology Institute and University of Patras, Greece (Proceedings Chair)
- Ioannis Chatzigiannakis, Research Academic Computer Technology Institute and University of Patras, Greece (Publicity Chair)
- Rozina Efstathiadou, Research Academic Computer Technology Institute, Greece (Finance Chair)
- Lena Gourdoupi, Research Academic Computer Technology Institute, Greece (Conference Secretariat)

Referees (Track A)

Scott Aaronson
Peyman Afshani
Noga Alon
Carme Alvarez
Kazuyuki Amano
Andris Ambainis
Yossi Azar
Maxim Babenko
Eric Bach
Michael Backes
Nikhil Bansal
Jérémy Barbay
Reinhard Bauer
Michael Baur
Amos Beimel
Boaz Benmoshe
Stéphane Bessy
Philip Bille
Davide Bilò
Andreas Björklund
Markus Bläser
Liad Blumrosen
Hans Bodlaender
Andrej Bogdanov
Jan Bouda
Tomáš Brázdil
Mark Braverman
Hajo Broersma
Niv Buchbinder
Leizhen Cai
Alberto Caprara
Giusi Castiglione
Ho-Leung Chan
Pierre Charbit
Hong-Bin Chen
Jianer Chen
Zhi-Zhong Chen
Flavio Chierichetti
Marek Chrobak
Amin Coja-Oghlan
Colin Cooper
Joshua Cooper
Graham Cormode
Artur Czumaj
Pooya Daboodi
Ivan Damgård
Stefan Dantchev
Shantanu Das
Daniel Delling
Erik Demaine
Josep Diaz
Florian Diedrich
Yann Disser
Hristo Djidjev
Feodor Dragan
Klaus Dräger
Arnaud Durand
Stefan Dziembowski
Khaled Elbassioni
Michael Elkin
David Eppstein
Leah Epstein
Rolf Fagerberg
Andreas Emil Feldmann
Amos Fiat
Eldar Fischer
Johannes Fischer
Abie Flaxman
Lisa Fleischer
Rudolf Fleischer
Holger Flier
Fedor Fomin

Dimitris Fotakis
Alan Frieze
Keith Frikken
Saturo Fujishige
Toshihiro Fujito
Emanuele Guido Fusco
Peter Gacs
Marco Gaertler
Francois Le Gall
Michael Gatto
Cyril Gavoile
Beat Gfeller
Paun Gheorghe
Rodolphe Giroudeau
Paul Goldberg
Navin Goyal
Fabrizio Grandoni
Mark Greve
Michelangelo Grigni
Elena Grigorescu
Martin Grohe
Sudipto Guha
Anupam Gupta
Robert Görke
Inge Li Gørtz
MohammadTaghi Hajiaghayi
Magnus Halldorsson
Xin Han
Kristoffer Arnsfelt Hansen
Nick Harvey
Andreas Hinz
Martin Holzer
Florian Horn
Tomas Hruz
John Iacono
Gabor Ivanyos
Maurice Jansen
Mark Jerrum
David Johnson
Natasha Jonoska
Stasys Jukna
Allan Grønlund Jørgensen
Christos Kaklamanis
Ming-Yang Kao
Haim Kaplan
Anna Karlin
Petteri Kaski
Bastian Katz
Telikepalli Kavitha
Rohit Khandekar
Lefteris Kirousis
Bjorn Kjos-Hanssen
Rolf Klein
Mikko Koivisto
Jean-Claude König
Spyros Kontogiannis
Takeshi Koshiba
Lukasz Kowalik
Mark Krentel
Marcus Krug
Antonín Kučera
Amit Kumar
Stefan Langerman
James Lee
Troy Lee
Stefano Leonardi
Leonid Levin
Minming Li
Mathieu Liedloff
Vincent Limouzy
Andrzej Lingas
Zhenming Liu
Bruno Loff
Zvi Lotker
Mark Manasse
Bodo Manthey
Conrado Martinez
Claire Mathieu
Arie Matsliah
Tomomi Matsui
Jens Maue
Frédéric Mazoit
Andrew McGregor
Colin McDiarmid
Steffen Mecke
Sascha Meinert
Mark Mercer
Wolfgang Merkle
Othon Michail
Matus Mihalak
Shuichi Miyazaki
Bojan Mohar
Burkhard Monien
Fabien de Montgolfier
Cris Moore
Pat Morin
Hiroki Morizumi
Anthony Morphett
Marcin Mucha
Haiko Müller
Filip Murlak
Muthu Muthukrishnan
Masaki Nakanishi
Satyadev Nandakumar
Gonzalo Navarro
Phuong Nguyen
Rolf Niedermeier
Jesper Buus Nielsen
Sotiris Nikoletseas
Evdokia Nikolova
Harumichi Nishimura
Martin Nöllenburg
Mitsunori Ogohara
Yoshio Okamoto
Alexander Okhotin
Alessio Orlandi
Rasmus Pagh
Panagiota Panagopoulou
Rina Panigrahy
Christophe Paul
Andrzej Pelc
Fernando Pereira
Seth Pettie
Wojciech Plandowski
Kirk Pruhs
Maurice Queyranne
Yuval Rabani
Michael Rao
Christoforos Raptopoulos
Saurabh Ray
Rudy Raymond

Andrea Ribichini
David Richerby
Dana Ron
Peter Rossmanith
Aaron Roth
Tim Roughgarden
Bjarke Hammersholt Roune
Ignaz Rutter
Milan Ružić
Heiko Röglin
Anand S.
Srinivasa Rao Satti
Christian Schaffner
Marcel Schöngens
Florian Schoppmann
Robert Schweller
Marinella Sciortino
Allan Scott
Robert Sedgewick
Jean-Sebastien Sereni
Olivier Serre
Hadas Shachnai
Natalia Shakhlevich
David Shmoys
Anastasios Sidiropoulos
Riccardo Silvestri
Stephen Simpson
Man-Cho So
David Garcia Soriano
Robert Spalek
Rastislav Sramek
Juraj Stacho
Rob van Stee
Vitaly Strusevich
Maxim Sviridenko
Mario Szegedy
Tadao Takaoka
Eiji Takimoto
Suguru Tamaki
Seiichiro Tani
Jun Tarui
Pascal Tesson
Dimitrios Thilikos
Mikkel Thorup
Takeshi Tokuyama
Szymon Torunczyk
Geza Toth
Jakob Truelsen
Ryuhei Uehara
Ugo Vaccaro
Moshe Vardi
Elad Verbin
Cristian Versari
Jan Vondrak
Tomasz Walen
John Watrous
Michael Weiss
Emo Welzl
Cenny Wenner
Renato Werneck
Udi Wieder
Ryan Williams
Karl Wimmer
Andreas Winter
Prudence Wong
David Woodruff
Jian Xia
Binh-Minh Bui-Xua
Masaki Yamamoto
Shigeru Yamashita
Mihalis Yannakakis
Ke Yi
Martin Zachariasen
Christos Zaroliagis
Guochuan Zhang
Shengyu Zhang
Qin Zhang
Michal Ziv-Ukelson
Uri Zwick
Anna Zych

Sponsoring Organizations

- Ministry of National Education and Religious Affairs of Greece
- Research Academic Computer Technology Institute (CTI)
- Piraeus Bank
- Rigorous Mathematical Connections between the Theory of Computation and Statistical Physics (RIMACO), European Research Council (ERC) Starting Grant
- Algorithmic Principles for Building Efficient Overlay Computers (AEOLUS) Project, EU/FET/Global Computing
- Papasotiriou Books and more
- Google

Table of Contents – Part I

Invited Lectures

Contributed Papers

Table of Contents – Part II

Track B: Invited Lectures

Track B: Contributed Papers

Track C: Invited Lecture

Track C: Contributed Papers

Assigning Papers to Referees

Kurt Mehlhorn

Max-Planck-Institut für Informatik
and
Department of Computer Science, Saarland University

Refereed conferences require every submission to be reviewed by members of a program committee (PC) in charge of selecting the conference program. A main responsibility of the PC chair is to organize the review process, in particular, to decide which papers are assigned to which member of the PC. The PC chair typically bases her decision on input from the PC, her knowledge of submissions and PC members, or scores that are computed automatically from keywords provided by authors and PC members. From now on, we call PC members *reviewers* or *referees*.

There are many software systems available that support the PC chair in her task; for example, EasyChair [8], HotCRP [7], Softconf [2], Linklings [1], CMT [4], and Websubrev [6]. Used in more than 1300 conferences in 2008 alone [9], EasyChair is currently the most popular conference management software. The system asks the reviewers to declare conflicts of interests and to rank the papers (for which the reviewer has no conflict of interest) into three classes: high interest, medium interest, and low interest. This process is called bidding. Based on this information, the system automatically computes an assignment that the PC chair can later review and modify accordingly. Creating an assignment from scratch by hand is normally not feasible since many conferences get in excess of 500 submissions [3].

The talk will be based on the paper *Assigning Papers to Referees* [5] by Naveen Garg, Telikepalli Kavitha, Amit Kumar, Kurt Mehlhorn, and Julián Mestre. The paper is available at

`http://www.mpi-inf.mpg.de/~mehlhorn/ftp/RefereeAssignment.pdf`

In this paper, we propose to optimize a number of criteria that aim at achieving fairness among referees/papers. Some of these variants can be solved optimally in polynomial time, while others are NP-hard, in which case we design approximation algorithms. Experimental results strongly suggest that the assignments computed by our algorithms are considerably better than those computed by popular conference management software.

References

1. Linklings, `http://www.linklings.com/`
2. Sofconf, `http://www.softconf.com/`
3. Apers, P.: Acceptance rates major database conferences, `http://wwwhome.cs.utwente.nl/~apers/rates.html`

S. Albers et al. (Eds.): ICALP 2009, Part I, LNCS 5555, pp. 1–2, 2009.

4. Chaudhuri, S.: Microsoft's academic conference management service, http://cmt.research.microsoft.com/cmt/
5. Garg, N., Kavitha, T., Kumar, A., Mehlhorn, K., Mestre, J.: Assigning Papers to Referees (2008), http://www.mpi-inf.mpg.de/~mehlhorn/ftp/RefereeAssignment.pdf
6. Halevi, S.: Websubrev, http://people.csail.mit.edu/shaih/websubrev/
7. Kohler, E.: HotCRP, http://www.cs.ucla.edu/~kohler/hotcrp/
8. Voronkov, A.: EasyChair, http://www.easychair.org
9. Voronkov, A.: EasyChair - users, http://www.easychair.org/users.cgi

Algorithmic Game Theory: A Snapshot

Christos H. Papadimitriou*

Division of Computer Science, U.C. Berkeley
christos@cs.berkeley.edu

Abstract. Algorithmic game theory is the research area in the interface between the theories of algorithms, networks, and games, which emerged more than a decade ago motivated by the advent of the Internet. "Snapshot" means several things: very personal point of view, of topical and possibly ephemeral interest, and put together in a hurry.

1 Introduction

Algorithmic game theory is arguably at a watershed point of its development. The field has grown tremendously in community size and stature. The basic problem areas have been defined and progress has been made — enough to justify a field-crystallizing edited book [36]. The emergence, and persistence, of important questions still outpace the exciting answers (see below for an idiosyncratic collection of both), but now it is a race. Economists are starting to pay attention, so we better have something meaningful — to them — to say.

2 Computing an Equilibrium

Games are thought experiments for understanding and predicting the behavior of rational strategic agents. The predictions of the theory are called *equilibria*, of which the Nash equilibrium is perhaps the most famous. One of the earliest goals of algorithmic game theory was to understand the complexity of computing equilibria. This was quite predictable, given our field's obsessions, but it also entails an important contribution to the other side, as algorithmic issues have influenced and shaped the debate on equilibrium concepts. We now know that computing a Nash equilibrium is PPAD-complete ([13], see [14] for a high-level exposition), even for 2-player games [8]; as a result, the most important question in this realm now is: *Is there a polynomial-time approximation scheme (PTAS) for computing a Nash equilibrium?* The standard approximation concept used in the literature is additive with normalized positive payoffs; for relative (multiplicative) approximation when negative payoffs are allowed, a negative answer is now known [11]. A quasipolynomial-time approximation scheme for this problem had been known for some time [31]; in fact the algorithm in [31] is of a special

* Supported by NSF grants CCF-0635319, CCF-0515259, a gift from Yahoo! Research, and a MICRO grant.

S. Albers et al. (Eds.): ICALP 2009, Part I, LNCS 5555, pp. 3–11, 2009.

interesting kind called *oblivious,* in that it examines possible solutions without looking at the game except to check the quality of the approximation. It can be shown [15] that the algorithm in [31] is nearly optimal among oblivious ones.

But how about special cases for which exact polynomial algorithms exist? It is quite remarkable that zero-sum games are essentially the only special case of the problem that we know how to solve — for 80 years now [48]. This can be generalized slightly to bimatrix games whose matrices, or their sum, are of low rank [25,31]. In another page of this volume [16] we show an interesting generalization to *networks* whose edges are zero-sum games (see below for further discussion). *Strictly competitive games*, apparently defined by Aumann, is an intriguing class of games generalizing the zero-sum ones; these are the games which have, along with zero-sum games, the following property: if both players switch from a pair of mixed strategies to another, then one is winning and the other is losing. Unfortunately, and rather astonishingly, it was recently shown [1] that this well studied and often mentioned "generalization" is *void:* zero-sum games (and their trivial affine variants) are the only examples! Finally on the subject of polynomially solvable special cases, recall that *correlated equilibria* are efficiently computable in general games via linear programming (essentially by design); for a nontrivial extension to succinctly representable games see [37,38].

Symmetry. Nash points out in his 1951 paper [34] that a modification of his proof establishes that a symmetric game (i.e., a game in which all players are identical) has a symmetric equilibrium (all identical players do the same thing). But are symmetric games easier? A beautiful reduction due to Gale, Kuhn and Tucker [20], which actually predated Nash's result, establishes that finding an equilibrium in a symmetric 2-player game cannot be easier than finding a Nash equilibrium in a general game (which we now know, is not easy at all). The reduction creates a game whose strategy space is the union of the two strategy spaces. Interestingly, in that same volume with [20], there is an independent proof of the same fact by Brown and von Neumann [5] based on the *product* of the strategy spaces. Somewhat surprisingly, neither reduction, or straightforward modification, works for three players, so it is currently *open* whether finding an equilibrium in a symmetric three-player game is easier than finding an equilibrium in a general game. We conjecture that symmetric games are no easier than general ones, for any number of players. Note also that it is not known how hard it is to find a *non-symmetric* Nash equilibrium in a symmetric game — it is at least PPAD-hard.

But symmetry does have definite dividends in multiplayer games. There is a polynomial-time algorithm (albeit, one relying on decision algorithms for real closed fields and therefore not very efficient with current technology) for finding symmetric Nash equilibria in symmetric games with n players and few (about $\log n$) strategies [38]. A very important class of games results from a particular relaxed kind of symmetry called *anonymity:* the players have different utility functions, but these depend on the strategy the player chooses, and on *how many* of the other players choose each strategy — not on the *identities* of the players who choose them (think of the game "shall I drive or take the bus?"). That is, the utilities are symmetric functions of the other players' choices. There is now a host of positive algorithmic results for such games culminating in an $n^{O(\log^2 \frac{1}{\epsilon})}$ PTAS [15]. These algorithms rely on results approximating, with more and more sophisticated techniques, the distribution of sums of binomial/multinomial variables, via other such sums whose probabilities are multiples of ϵ. There is no known limit to possible improvements: for all we know, this class of games can be solved exactly in polynomial time — I do believe that they are PPAD-complete.

Price Equilibria. Nash's theorem not only launched Game Theory, but also inspired the price equilibrium theorems of Arrow and Debreu [2], a very important realm of positive results in Microeconomics. Algorithms for finding such price equilibria had been an open question for some time. We now have polynomial algorithms for certain special cases (all falling within the subclass of "markets with gross substitutability," that is, markets in which increasing the price of a good never decreases the demand of other goods, a case long known to have positive algorithmic properties akin to convexity, see Chapters 5 and 6 of [36]). One particular algorithm [10], again for such a special case, is of an interesting and realistic sort that can be called "a price adjustment mechanism:" we look at the prices and the excess demands, and possibly also the history of both, and, based on this information, we adjust the prices. In contrast, all other known algorithms for price equilibria zero in to the equilibrium via much less natural primitives such as pivoting or hyperplane separation. We also have, at long last, some intractability results for classes of markets (consumer utilities) that do not fall in the gross substitutability category [9,7], as well as an exponential lower bound (without complexity assumptions) for any price adjustment mechanism that works in general markets [42].

3 Choosing an Equilibrium

The multiplicity of Nash equilibria has always been a tension between the algorithmic and game-theoretic perspective. Somewhat forgotten is an answer provided decades ago by two great game theorists [22], page 144: Players can be assumed to have prior ideas about how their opponents will play, and to start by best-responding to those. If now the players' beliefs evolve from that prior to the realities of the actual response, this creates a *linear tracing procedure* which, in the absence of degeneracies, points to one equilibrium. It would be interesting to revisit this idea from the complexity-theoretic point of view. Is the linear tracing procedure, for example, PSPACE-complete? Incidentally, another important idea in [22], risk dominance of equilibria, was recently taken on in [32].

But the ultimate conceptual contribution to game theory is the introduction of a meaningful, compelling, and influential equilibrium concept. I believe that the novel point of view of algorithmic game theory (with its computational angle and Internet zeitgeist) is capable of producing very interesting ideas here — besides, the recent negative complexity results for Nash equilibria provide new motivation toward this goal. The territory explored so far extends almost exclusively in the direction of *learning* and *repeated games,* see [18] for an early discussion and [21,17] for some later attempts.

Repeated Games. The study of repeated games is, of course, a time honored branch of classical game theory, the source of powerful models, as well as the justification for concepts such as the mixed Nash equilibrium (and the arena of early interactions between game theory and Computer Science, recall for example [41]). Part of the conventional wisdom in repeated games is that the equilibrium space is much richer and better behaved than that of one-shot games by dint of an important cluster of insights and results known as *the Folk Theorem:* any reasonable combination of payoffs can be realized by a Nash equilibrium of the repeated game (for example, in the repeated prisoner's dilemma nearly-always-collaboration is possible). It was recently pointed out in [4] that this fundamental result comes with serious, if somewhat covert, computational difficulties: Nash equilibria in repeated games are no more easily accessible (for three or more players) than those of one-shot games.

Learning. The point of repetition in games is to help players adjust to the game and to each other — that is to say, to *learn.* Learning in games has a long history, see Chapter 4 of [36]; it had been known for some time that a particular flavor of learning known as *no-regret learning* — which in a very real sense *is* a novel equilibrium concept — converges to the Nash equilibrium in zero-sum games [19], and, in a different variant, to correlated equilibria in general games [24]. More recently, important connections between no-regret learning and the *price of anarchy* (see the next section) has been brought to the fore, see for example [3,47,28].

But of course nobody believes seriously that learning can fathom the intractable, converge fast to a Nash equilibrium in general games; surprisingly, we have very little concrete information on this. It can be shown through communication complexity arguments [23] that in n-player games exponential time, in n, is needed to converge to a Nash equilibrium (pure or mixed) by learning algorithms — but then in such games the input would be exponential in n, and the proof entails pointing out that all this information must be exchanged for coordinated convergence to equilibrium. It was recently pointed out that a class of learning-type algorithms [12] fail to converge in polynomial time (in the accuracy) to a Nash equilibrium even for two-person games with three strategies per player — they do in the case of zero-sum games, and in the case of two strategies. Can we come up with exponential lower bounds (independent of any complexity assumptions) on restricted classes of algorithms for finding Nash equilibria in bimatrix games? Since our ambition here falls short of proving that $P \neq NP$, we better restrict ourselves to algorithms that are incapable of identifying — or approximating — the payoff matrices.

Incidentally, a very early instance of the learning approach to games is Julia Robinson's 1950 proof [43] that a particular algorithm called *fictitious play* (both players know all payoffs, assume that the other player's mixed strategy is captured by the histogram of her past plays, and best-respond to that) converges in the case of zero-sum games — but the best known upper bound on the number of iterations needed to converge is exponential in the number of strategies (it is implicit in Robinson's inductive proof [43]). Karlin[1] conjectured fifty years ago that the true convergence rate is in fact *quadratic.*

4 Networks and Games

Scott Shenker's famous quote *"The Internet is an equilibrium — we just have to identify the game"* captures perfectly the complex of reasons that made game-theoretic thought relevant to Computer Science ca. the late 1990s. The Internet is a computational artifact that was *not* designed (except in the loosest and broadest sense) by an entity but emerged from the unstructured interaction of many. The area of the "price of anarchy" (see [30] and Chapter 17 of [36]) seeks to gauge the loss of performance inherent in such process — the paradigmatic work in this area is Roughgarden's thesis [46,45] on the price of anarchy in routing.

The model of [46], in which routing decisions are made by flows, is an adaptation of classical models from transportation theory, and has little relevance to the Internet. It was recently shown [40] that, if routing decisions are made by each *edge* of the network so as to minimize downstream congestion experienced by the flows *through the edge*, then the price of anarchy becomes unbounded. However, if, instead, the edges charge per-unit-of-flow *prices* to their upstream neighbors and maximize revenue minus cost, it is shown (under assumptions) that the price of anarchy becomes *one.* In other words, this is another instance in which prices, magically, usher in efficiency. One important cluster of questions is, how do the conventions and protocols of today's Internet, such as the BGP protocol governing the interactions between autonomous systems, limit this ideal efficiency? Examples of features of BGP that may be sources of inefficiency are: Long-term agreements that are oblivious to fluctuations in downstream congestion conditions; and selection of a single downstream routing path per destination.

But networks are changing. *Social networks* such as Facebook are only the latest and most explicit examples of the important networks of interactions between entities that had always been Internet's *raison d' être* — enabled by it and embedded in it in a variety of ways. I believe that it is productive to understand such networks as graphs whose edges are *games* played by the nodes — the so called *graphical polymatrix games*, in which each node chooses a common strategy to play in all incident games, and receives the sum of the resulting payoffs. In fact, such modeling is not new: past work on the spreading of technologies and ideas in a social network [29], for example, falls squarely into this framework, with coordination games at the edges (see [32] for recent work on the convergence of such games). In a paper in this volume [16], a simple but rather surprising result is shown: If the edges of the graph are zero-sum games, then *the whole game has the minmax property and is easily solvable.* In fact,

[1] Sam Karlin (1924–2007), a giant of early game theory among other fields, whose work is not as broadly known to our community as one would have expected.

the nodes/players can converge to the equilibrium via distributed learning, and each of them has a *value* capturing the sum total of the advantages in both her position in the network and in the structure of the incident games. Incidentally, economists are getting increasingly interested in the effects of network structure on markets and other economic interactions, and the point of view in this paragraph may be an important opportunity for our field to contribute to modern economic thought.

5 Mechanism Design

Mechanism design [44] seeks to create games in which the desired (socially optimum or, more generally, beneficial to the designer) behavior emerges as an equilibrium of selfish participants, *independently of the participants's unknown true preferences;* this is done, of course, by providing appropriate incentives that will make "gaming the system" unpalatable. It is a mature and much lauded (see the above reference) area of economic theory, which, however, had been perceived as a little too rich in sweeping positive results. The area of *algorithmic mechanism design* (see [35] and Chapter 9 of [36]) addresses this concern by fathoming the intriguing tradeoffs between the effectiveness of the designed mechanism and the computational complexity of its implementation.

Perhaps the most classical sweeping positive result in mechanism design is the so-called *VCG mechanism* (named after the economists Vickrey, Clarke and Groves who devised it): If monetary incentives are possible (and inconsequential, which is another important aspect where computer scientists beg to differ, see for example [26]), then essentially anything goes: there are appropriate incentives that will elicit essentially any desired behavior of the agents. But algorithmic game theory researchers have pointed out that the VCG approach to mechanism design is paved with computational obstacles, as the computation of such payments is often an intractable problem. Now in computer science we know how to deal with such intractability: We *approximate*. Unfortunately, incentives are fragile, and work only if computed exactly. This three-way tension between complexity, approximability, and incentive compatibility, has created an exciting and deep research area. A recent result has connected, for the first time, this tradeoff with classical complexity theory: It was shown that there is an NP-complete optimization problem that can be approximated well (within a small constant factor) in polynomial time but which, unless NP = BPP, cannot be approximated well (better than the square root of the problem's natural parameter) in an incentive-compatible way [39]. And in more recent work [33,6] steps have been made towards extending this result to the central problem of *combinatorial auctions* (see Chapter 11 of [36]).

Mechanism design is about important realities which lie at the roots of the remarkable convergence of computational and economic thought over the past decade or so: computational systems are no longer entities whose only job is to produce correct outputs for the data on their input tape. They must encapsulate incentives so they are invoked with the right inputs — let alone invoked at all... Turing's question "what can be computed?" is being revisited once more (as it has been revisited several times in the past century to accommodate considerations of complexity, distributed computation and on-line computation, for example): *Which functions can be computed, and with what accuracy, when the inputs are owned by entities that are keenly interested in the outcome of the computation?*

Acknowledgment. Many thanks to Costis Daskalakis and Tim Roughgarden for feedback on an earlier version.

References

1. Adler, I., Daskalakis, C., Papadimitriou, C.H.: Manuscript (2009)
2. Arrow, K.J., Debreu, G.: Existence of equilibria for a competitive economy. Econometrica 22(3), 265–290 (1954)
3. Blum, A., Hajiaghayi, M., Ligett, K., Roth, A.: Regret minimization and the price of total anarchy. In: STOC (2008)
4. Borgs, C., Chayes, J.T., Immorlica, N., Kalai, A.T., Mirrokni, V.S., Papadimitriou, C.H.: The myth of the folk theorem. In: STOC 2008, pp. 365–372 (2008)
5. Brown, G.W., von Neumann, J.: Solutions of games by differential equations. In: Kuhn, H.W., Tucker, A.W. (eds.) Contributions to the Theory of Games, Princeton University Press, Princeton (1950)
6. Buchfuhrer, D., Umans, C.: Limits on the social welfare of maximal-in-range auction mechanisms (manuscript, 2009)
7. Chen, X., Dai, D., Du, Y., Teng, S.: Settling the complexity of Arrow-Debreu equilibria in markets with linearly separable utilities (manuscript, 2009)
8. Chen, X., Deng, X.: Settling the complexity of two-player Nash equilibrium. In: FOCS (2006)
9. Codenotti, B., Saberi, A., Varadarajan, K., Ye, Y.: Leontief economies encode nonzero sum two-player games. In: Proceedings of the seventeenth annual ACM-SIAM Symposium on Discrete Algorithms, SODA (2006)
10. Cole, R., Fleischer, L.: Fast-converging tatonnement algorithms for one-time and ongoing market problems. In: Proceedings of the 40th annual ACM Symposium on Theory of Computing, STOC (2008)
11. Daskalakis, C.: The complexity of approximating a Nash equilibrium (submitted, 2009)
12. Daskalakis, C., Frongillo, R., Pierrakos, G., Papadimitriou, C.H., Valiant, G.: In preparation (2009)
13. Daskalakis, C., Goldberg, P.W., Papadimitriou, C.H.: The complexity of computing a Nash equilibrium. In: STOC 2006 (2006); SIAM Journal on Computing, special issue for 2006 STOC (to appear)
14. Daskalakis, C., Goldberg, P.W., Papadimitriou, C.H.: The complexity of computing a Nash equilibrium. CACM 58(2), 89–97 (2009)
15. Daskalakis, C., Papadimitriou, C.H.: On oblivious PTAS's for Nash equilibrium. In: STOC (2009)
16. Daskalakis, C., Papadimitriou, C.H.: On a network generalization of the minmax theorem. In: Marchetti-Spaccamela, A., Matias, Y., Nikoletseas, S., Thomas, W. (eds.) ICALP 2009. LNCS, vol. 5555, Springer, Heidelberg (2009)

17. Fabrikant, A., Papadimitriou, C.H.: The complexity of game dynamics: BGP oscillations, sink equilibria, and beyond. In: SODA 2008, pp. 844–853 (2008)
18. Friedmand, E.J., Shenker, S.J.: Learning and implementation on the Internet, Working paper (1997)
19. Freund, Y., Schapire, R.E.: Adaptive game playing using multiplicative weights. Games and Economic Behavior 29, 79–103 (1999)
20. Gale, D., Kuhn, H.W., Tucker, A.W.: On symmetric games. In: Kuhn, H.W., Tucker, A.W. (eds.) Contributions to the Theory of Games, Princeton University Press, Princeton (1950)
21. Goemans, M.X., Mirrokni, V.S., Vetta, A.: Sink Equilibria and Convergence. In: FOCS 2005, pp. 142–154 (2005)
22. Harsanyi, J.C., Selten, R.: A General Theory of Equilibrium Selection in Games. MIT Press Classis, Cambridge (1982)
23. Hart, S., Mansour, Y.: The communication complexity of uncoupled Nash equilibrium procedures. In: STOC 2007 (2007)
24. Hart, S., Mas-Colell, A.: A simple adaptive procedure leading to correlated equilibrium. Econometrica 68(5), 1127–1150 (2000)
25. Kannan, R., Theobald, T.: Games of fixed rank: A hierarchy of bimatrix games. In: SODA (2007)
26. Karlin, A.R., Kempe, D., Tamir, T.: Frugality of truthful mechanisms. In: FOCS (2005)
27. Kearns, M.J., Littman, M.L., Singh, S.P.: Graphical Models for Game Theory. In: UAI (2001)
28. Kleinberg, R., Piliouras, G., Tardos, É.: Multiplicative updates outperform generic no-regret learning in congestion games. In: STOC (2009)
29. Kempe, D., Kleinberg, J., Tardos, É.: Maximizing the spread of influence through a social network. In: SIGKDD (2003)
30. Koutsoupias, E., Papadimitriou, C.: Worst-case equilibria. In: Meinel, C., Tison, S. (eds.) STACS 1999, vol. 1563, pp. 404–413. Springer, Heidelberg (1999)
31. Lipton, R., Markakis, E., Mehta, A.: Playing large games using simple strategies. In: ACM EC (2003)
32. Montanari, A., Saberi, A.: Convergence to Equilibrium in Local Interaction Games (2008)
33. Mossel, E., Papadimitriou, C.H., Schapira, M., Singer, Y.: Combinatorial Auctions: VC v. VCG (submitted, 2009)
34. Nash, J.: Noncooperative Games. Annals of Mathematics 54, 289–295 (1951)
35. Nisan, N., Ronen, A.: Algorithmic Mechanism Design. In: STOC (1999)
36. Nisan, N., Roughgarden, T., Tardos, É., Vazirani, V.V.: Algorithmic Game Theory. Cambridge University Press, New York (2007)
37. Papadimitriou, C.H.: Computing correlated equilibria in multiplayer games. In: STOC (2005)
38. Papadimitriou, C.H., Roughgarden, T.: Computing equilibria in multi-player games. In: SODA 2005 (2005); J.ACM (full version, 2008)
39. Papadimitriou, C.H., Schapira, M., Singer, Y.: On the hardness of being truthful. In: FOCS (2008)
40. Papadimitriou, C.H., Valiant, G.: Selfish routers and the price of anarchy (submitted, 2009)
41. Papadimitriou, C.H., Yannakakis, M.: On complexity as bounded rationality (extended abstract). In: STOC 1994, pp. 726–733 (1994)
42. Papadimitriou, C.H., Yannakakis, M.: An impossibility theorem for price adjustment mechanisms (manuscript, 2009)

43. Robinson, J.: An iterative method of solving a game. Annals of Mathematics (1951)
44. Royal Swedish Academy of Sciences Mechanism Design Theory (2007), `http://nobelprize.org/nobelprizes/economics/laureates/2007/ecoadv07.pdf`
45. Roughgarden, T.: Selfish Routing. MIT Press, Cambridge (2002)
46. Roughgarden, T., Tardos, É.: How bad is selfish routing? JACM (2002)
47. Roughgarden, T.: Intrinsic robustness of the price of anarchy. In: STOC (2009)
48. von Neumann, J.: Zur Theorie der Gesellschaftsspiele. Math. Annalen 100, 295–320 (1928)

SDP-Based Algorithms for Maximum Independent Set Problems on Hypergraphs

Geir Agnarsson[1], Magnús M. Halldórsson[2], and Elena Losievskaja[3]

[1] Dept. of Mathematical Sciences, George Mason University, Fairfax, VA, USA
geir@math.gmu.edu

[2] School of Computer Science, Reykjavík University, IS-103 Reykjavik, Iceland
mmh@ru.is

[3] Faculty of Industrial, Mechanical Engineering and Computer Science, University of Iceland, IS-107 Reykjavik, Iceland
elenal@hi.is

Abstract. This paper deals with approximations of maximum independent sets in non-uniform hypergraphs of low degree. We obtain the first performance ratio that is sublinear in terms of the maximum or average degree of the hypergraph. We extend this to the weighted case and give a $O(\bar{D} \log\log \bar{D}/\log \bar{D})$ bound, where $\bar{D}$ is the average weighted degree in a hypergraph, matching the best bounds known for the special case of graphs. Our approach is to use an semi-definite technique to sparsify a given hypergraph and then apply combinatorial algorithms to find a large independent set in the resulting sparser instance.

1 Introduction

This paper deals with approximations of maximum independent sets in hypergraphs of low degree. Recall that a hypergraph (set system) $H = (V, E)$ has a vertex set V and a collection E of *(hyper)edges* that are arbitrary subsets of V. A hypergraph is weighted if vertices in V are assigned weights. It has *rank* r if all edges are of size at most r, and is *r-uniform* if all are of size exactly r. A set of vertices is *independent* if it does not properly contain any edge in E. The *degree* of a vertex is its number of incident edges. We consider approximation algorithms for the maximum independent set (MIS) problem in sparse non-uniform hypergraphs.

The MIS problem is of fundamental interest, capturing conflict-free sets in a very general way. It generalizes the classic independent set problem in graphs, and thus inherits all its hardness properties. The vertices not in an independent set form a hitting set of the hypergraph. Algorithms for MIS can therefore be viewed as set covering algorithms with a differential measure, which lends it an additional interest.

Hypergraph problems tend to be more difficult to resolve than the corresponding graph problems, with the MIS problem a typical case. The best performance ratio known for MIS in general hypergraphs, in terms of the number n of vertices, is only $O(n/\log n)$, which has a rather trivial argument [7]. For the graph case,

S. Albers et al. (Eds.): ICALP 2009, Part I, LNCS 5555, pp. 12–23, 2009.

for comparison, the ratio is $O(n(\log\log n)^2/\log^3 n)$ [4]. In terms of the maximum degree Δ, a ratio of Δ is trivial, while previous work on MIS in hypergraphs has improved only the constant term [2,8]. More specifically, a $\Delta/1.365$-upper bound was obtained for a greedy algorithm and a tight $(\Delta+1)/2$-ratio for a local search method, while in [8] a tight bound of $(\Delta+1)/2$ was obtained for the greedy algorithm as well as the best previously known bound of $(\Delta+3)/5$. The main sign of success has been on sparse hypergraphs, where Turan-like bounds have been proven [3,16,15]. Unlike graphs, however, the exact constant in the bounds is not known.

The most powerful approach for the approximation of challenging optimization problems has involved the use of semi-definite programming (SDP). It is responsible for the best ratio known for IS in graphs of $O(\Delta\log\log\Delta/\log\Delta)$ [6]. It is also involved in the complementary vertex cover problem [9], both in graphs and in hypergraphs. Yet, it has failed to yield much success for MIS in hypergraphs, except for some special cases. One intuition may be that hyperedges result in significantly weaker constraints in the semi-definite relaxation than the graph edges. The special cases where it has been successful — 2-colorable k-uniform hypergraphs [6] and 3-uniform hypergraphs with a huge independence number [11] — have properties that result in strengthened constraints. The usefulness of SDP for general MIS has remained open.

This state-of-the-art suggests several directions and research questions. A key question is to what extent approximation ratios for IS in graphs can be matched in hypergraphs. This can be asked in terms of different degree parameters, as well as extensions. Given that graphs are 2-uniform hypergraphs and k-uniform hypergraphs have certain nice properties, the question is also how well we can handle non-uniform hypergraphs.

Our results. We derive the first $o(\Delta)$-approximation for IS in hypergraphs matching the $O(\Delta\log\log\Delta/\log\Delta)$-approximation for the special case of graphs. Our approach is to use an SDP formulation to sparsify the part of the instance formed by 2-edges (edges of size 2), followed by a combinatorial algorithm on the resulting sparser instance. This is extended to obtain an identical bound in terms of the average degree $\bar{d}$ of an unweighted hypergraph. As part of the method, we also obtain a $k^{5/2-1/k}\bar{d}^{1-1/k+o(1)}$-approximation for hypergraphs with independence number at least n/k.

We generalize the results to the vertex-weighted problem. In that case, no non-trivial bound is possible in terms of the average degree alone, so we turn our attention to a weighted version. The *average weighted degree* $\bar{D}$ is the node-weighted average of the vertex degrees. We give a $O(\bar{D}\log\log\bar{D}/\log\bar{D})$-approximation for MIS.

We apply two combinatorial algorithms to hypergraphs with few 2-edges. One is a greedy algorithm analyzed by Caro and Tuza [3] for the k-uniform case and Thiele [16] for the non-uniform case. The bound obtained in [16] is in general unwieldy, but we can show that it gives a good approximation when the number of 2-edges has been reduced. The other is a simple randomized algorithm analyzed by Shachnai and Srinivasan [15].

Organization. The paper is organized in the following way. In Sect. 2 we describe how to find a large sparse hypergraph in a given hypergraph H using SDP technique. In Sect. 3 we give analysis of the greedy algorithm for MIS on hypergraphs of rank 3 with small 2-degree, and then show how to apply this greedy algorithm together with SDP to find a large hypergraph in H. In Sect. 4 we describe how to use a randomized algorithm together with SDP to find a good approximation of a weighted MIS in hypergraphs.

2 Definitions

Given a hypergraph $H = (V, E)$, let n and m be the number of vertices and edges in H, respectively. We assume that H is a *simple* hypergraph, i.e. no edge is a proper subset of another edge. An edge of size t is a *t-edge*. A hypergraph is *r-uniform* if all edges are r-edges. A *graph* is a 2-uniform hypergraph. The *rank* r of a hypergraph H is the maximum edge size in H.

Let $d_t(v)$ be *t-degree* of a vertex v, or the number of t-edges incident on v. We denote by Δ_t and $\bar{d}_t$ the maximum and the average t-degree in a hypergraph, respectively. The *degree* $d(v)$ of a vertex v is the total number of edges incident on v, i.e. $d(v) = \sum_{t=2}^{r} d_t(v)$. We denote by Δ and $\bar{d}$ the maximum and the average degree in a hypergraph, respectively.

Given a function $f : V \to R$ that assigns weights to the vertices of H, let $w(H) = w(V) = \sum_{v \in V} w(v)$. We define $D(v) = w(v)d(v)$ and $\bar{D} = \sum_{v \in V} w(v)d(v) / \sum_{v \in V} w(v)$ to be the *weighted degree* of a vertex v and the *average weighted degree* in H, respectively.

By *deleting a vertex v from a hypergraph H* we mean the operation of deleting v and all incident edges from H. By *deleting a vertex v from an edge e* we mean the operation of replacing e by $e \setminus \{v\}$.

A (weak) *independent set* in H is a subset of V that doesn't properly contain any edge of H. Let $\alpha(H, w)$ be the weight of a maximum independent set in H. If H is unweighted, then it is denoted as $\alpha(H)$.

3 Semidefinite Programming

We use semidefinite programming to find large subgraphs with few 2-edges. More generally, we find subgraphs of large weight and small weighted average degree. This is obtained by rounding the vector representation of a suitable subgraph; such a subgraph is found by a result of Alon-Kahale. Along the way, we twice eliminate vertices of high-degree to ensure degree properties.

Let us recall the definition of a vector coloring of a graph [10].

Definition 3.1 ([10]). *Given a graph G and a real number $h \geq 1$, a* vector h-coloring *of G is an assignment of a $|V(G)|$-dimensional unit vector $\boldsymbol{v_i}$ to each vertex v_i of G so that for any pair v_i, v_j of adjacent vertices the inner product of their vectors satisfies*

$$\boldsymbol{v}_i \cdot \boldsymbol{v}_j \leq -\frac{1}{h-1} \ . \tag{1}$$

The *vector chromatic number* $\overline{\chi}(G)$ is the smallest positive number h, such that there exists a feasible vector h-coloring of G.

A vector representation given by a vector coloring is used to find a sparse subgraph by the means of *vector rounding* [10]: choose a random vector $\boldsymbol{b}$, and retain all vertex vectors whose inner product with $\boldsymbol{b}$ is above a certain threshold. The quality (i.e. sparsity) of the rounded subgraph depends on the vector chromatic number of the graph. In order to approximate independent sets we need to use this on graphs that do not necessarily have a small vector chromatic number but have a large independent set.

A graph with a large independent set contains a large subgraph with a small vector chromatic number, and there is a polynomial time algorithm to find it. This comes from the following variation of a result due to Alon and Kahale [1]:

Theorem 3.2 ([7]). *Let $G = (V, E, w)$ be a weighted graph and ℓ, p be numbers such that $\alpha(G, w) \geq w(G)/\ell + p$. Then, there is a polynomial time algorithm that gives an induced subgraph G_1 in G with $w(G_1) \geq p$ and $\overline{\chi}(G_1) \leq l$.*

Let us now present our algorithm for finding a large-weight low 2-degree hypergraph. It assumes that it is given the size α of the maximum independent set in the graph. We can sidestep that by trying all possible values for α, up to a sufficient precision (say, factor 2).

Algorithm SparseHypergraph
Input: Hypergraph $H(V, E)$, and its independence number α
Output: Induced hypergraph $\hat{H}$ in H of maximum degree $2k\bar{D}$ and maximum 2-degree $\sqrt{2k\bar{D}}$

Let $k = w(H)/\alpha$ and $a = 1 + \frac{1}{\ln \ln \bar{D} - 1}$.
Let G be the graph induced by the 2-edges of H.
Let G_0 be the subgraph of G induced by nodes of degree at most $2k\bar{D}$ in H.
Find an induced subgraph G_1 in G_0 with $w(G_1) \geq \frac{(a-1)w(G)}{2ak}$ with a vector $2ak$-coloring.
Choose a random $|V(G_1)|$-dimensional vector $\boldsymbol{b}$.
Let G_2 be the subgraph of G_1 induced by vertices $\{v \in V(G_1) : \boldsymbol{v} \cdot \boldsymbol{b} \geq c\}$, where $c = \sqrt{\frac{ak-2}{ak} \ln (2k\bar{D})}$.
Let $\hat{V}$ be the set of vertices of degree at most $\sqrt{2k\bar{D}}$ in G_2.
Output $\hat{H}$, the subhypergraph in H induced by $\hat{V}$.

Fig. 1. The sparsifying algorithm

The algorithm SparseHypergraph can be implemented to run in polynomial time. The subgraph G_1 in G_0 with small vector chromatic number and large independent set can be found in polynomial time [1]. A vector representation can be found within an additive error of ϵ in time polynomial in $\ln 1/\epsilon$ and n using the ellipsoid method [5] and Choleski decomposition.

Analysis

Lemma 3.3. *The graph G_0 has weight at least $w(H)(1-1/2k)$ and independence number at least $\alpha(H)/2$.*

Proof. The graph G has the same weight as H, or $w(V)$. The independence number of G is also at least that of H, since it contains only a subset of the edges. Let $X = V(G) - V(G_0)$ be the high-degree vertices deleted to obtain G_0. Then,

$$\sum_{v \in X} w(v)d(v) \geq \sum_{v \in X} w(v) \cdot 2k\bar{D} = 2k\bar{D}w(X) \ . \tag{2}$$

Since

$$\bar{D} \cdot w(V) = \sum_{v \in V} w(v)d(v) \geq \sum_{v \in X} w(v)d(v) \ , \tag{3}$$

we get from combining (3) with (2) that the weight $w(X)$ of the deleted vertices is at most $w(V)/2k$. Thus, $w(G) \geq (1-1/2k)w(H)$. Also, G_0 has a maximum independent set of weight at least $\alpha(G_0, w) \geq \alpha(G, w) - w(X) \geq \alpha(H, w) - w(X) \geq w(H)/2k$.

Observe that $\alpha(G_0, w) \geq w(H)/2k \geq w(G_0)/2k = w(G_0)/2ak + w(G_0)(a-1)/2ak$. Then, Theorem 3.2 ensures that a subgraph G_1 can be found with $w(G_1) \geq w(G_0)(a-1)/2ak$ and $\overline{\chi}(G_1) \leq 2ak$. From that, a vector $2ak$-coloring of G_1 can be found.

The main task is to bound the properties of the rounded subgraph G_2. Karger et al. [10] estimated the probability that G_2 contains a given vertex or an edge. Let $N(x)$ denote the tail of the standard normal distribution: $N(x) = \int_x^\infty \phi(z)dz$, where $\phi(z) = \frac{1}{\sqrt{2\pi}} \exp\left(-\frac{x^2}{2}\right)$ is the density function. Let $\tau = \sqrt{\frac{2(ak-1)}{ak-2}}$.

Lemma 3.4 ([10]). *A graph G_2 induced in $G_1(V_1, E_1)$ after vector-rounding contains a given vertex in V_1 with probability $N(c)$ and a given edge in E_1 with probability $N(c\tau)$.*

The following lemma states well-known bounds on the tail of the normal distribution.

Lemma 3.5 ([14]). *For every $x > 0$, $\phi(x)\left(\frac{1}{x} - \frac{1}{x^3}\right) < N(x) < \phi(x)\frac{1}{x}$.*

We can now bound the weight of the subgraph found.

Lemma 3.6. *$\hat{V}$ has expected weight $\Omega\left(\frac{w(G_1)}{(k\bar{D})^{\frac{1}{2}-\frac{1}{k}}\sqrt{\ln(k\bar{D})}}\right)$. This can be derandomized to obtain an induced subgraph $\hat{V}$ with this much weight and maximum 2-degree at most $\sqrt{2k\bar{D}}$.*

Proof. First, for any edge (u, v) in G_1 and G_2 we define a weight function $w(u, v) = w(u) + w(v)$. Let $w(V_1) = \sum_{v \in V(G_1)} w(v)$ and $w(E_1) = \sum_{(v,u) \in E(G_1)} (w(v) + w(u))$ be the weight of vertices and edges in G_1. Similarly, let $w(V_2)$

and $w(E_2)$ be the weight of vertices and edges in G_2. Let X_i be an indicator random variable with $X_i = 1$, if V_2 contains $v_i \in V_1$ and $X_i = 0$ otherwise. Then, $w(V_2) = \sum_{v_i \in V_1} w(v_i)X_i$. Using Lemma 3.4 and linearity of expectation we bound $E[w(V_2)]$ by

$$E[w(V_2)] = w(V_1)N(c) \ . \tag{4}$$

Similarly, we bound $E[w(E_2)]$ by

$$E[w(E_2)] = w(E_1)N(c\tau) \leq 2k\bar{D}w(V_1)N(c\tau) \ , \tag{5}$$

where in the last inequality we use the fact that maximum degree in G_1 is bounded by $2k\bar{D}$ (since we deleted the high-degree vertices from G and G_1 is an induced subgraph in G). Combining (4) and (5), we get that

$$E\left[w(V_2) - \frac{w(E_2)}{\sqrt{2k\bar{D}}}\right] = w(V_1)N(c) - \sqrt{2k\bar{D}}w(V_1)N(c\tau) \ . \tag{6}$$

Observe that

$$c\tau = \sqrt{\frac{2(ak-1)}{ak}\ln(2k\bar{D})} = \sqrt{2\left(1 - \frac{1}{ak}\right)\ln(2k\bar{D})}$$

and

$$exp(-(c\tau)^2/2) = (2k\bar{D})^{-1+1/ak} \ .$$

Then, by Lemma 3.5

$$N(c\tau) < \phi(c\tau)\frac{1}{c\tau} = \frac{(2k\bar{D})^{-1+1/ak}}{\sqrt{2\pi}\cdot\sqrt{\frac{2(ak-1)}{ak}\ln(2k\bar{D})}} \tag{7}$$

and

$$N(c) > \phi(c)\frac{1}{c}\left(1 - \frac{1}{c^2}\right) = \frac{(2k\bar{D})^{-1/2+1/ak}}{\sqrt{2\pi}\cdot\sqrt{\frac{ak-2}{ak}\ln(2k\bar{D})}}\left(1 - \frac{ak}{(ak-2)\ln(2k\bar{D})}\right) \ . \tag{8}$$

Combining (6), (7) and (8), we deduce that

$$\begin{aligned} E\left[w(V_2) - \frac{w(E_2)}{\sqrt{2k\bar{D}}}\right] &> w(V_1)\frac{(2k\bar{D})^{-1/2+1/ak}}{\sqrt{2\pi}\cdot\sqrt{\frac{ak-2}{ak}\ln(2k\bar{D})}} \\ &\quad \left(1 - \frac{ak}{(ak-2)\ln(2k\bar{D})} - \sqrt{\frac{ak-2}{2(ak-1)}}\right) \\ &= \Omega\left(\frac{w(V_1)}{(k\bar{D})^{1/2-1/k}\sqrt{\ln(k\bar{D})}}\right) \ , \end{aligned} \tag{9}$$

where in the last line we use $a = 1 + \frac{1}{\ln\ln\bar{D}-1}$.

The weight of vertices with degree greater than $\sqrt{2k\bar{D}}$ is at most $\sum_{v_i \in V_2} w(v_i)d(v_i)/\sqrt{2k\bar{D}} = w(E_2)/\sqrt{2k\bar{D}}$. After deleting all such vertices from G_2, the expected weight of $\hat{V}$ is $E\left[w(V_2) - \frac{w(E_2)}{\sqrt{2k\bar{D}}}\right]$ which is bounded by (10).

Finally, we can apply derandomization technique from [13] to derandomize the vector rounding in polynomial time. In our algorithm an elementary event corresponds to an edge in G_2 and involves only two vectors corresponding to the endpoints of the edge. This completes the proof.

We can bound the weight of the resulting hypergraph $\hat{H}$ in terms of the original hypergraph. We have that $w(\hat{V}) = \Omega\left(\frac{w(G_1)}{(k\bar{D})^{\frac{1}{2}-\frac{1}{k}}\sqrt{\ln k\bar{D}}}\right)$, while using $a = 1 + \frac{1}{\ln\ln\bar{D}-1}$, we have that

$$w(G_1) = \frac{(a-1)w(G)}{2ak} = \frac{w(G)}{2k\ln\ln\bar{D}} = \frac{w(H)\left(1-\frac{1}{2k}\right)}{2k\ln\ln\bar{D}} = \Omega\left(\frac{w(H)}{k\ln\ln\bar{D}}\right) .$$

Theorem 3.7. *Let H be a hypergraph with average weighted degree $\bar{D}$. The* SparseHypergraph *algorithm finds an induced hypergraph in H of weight $\Omega\left(\frac{w(H)}{k^{3/2-1/k}\bar{D}^{1/2-1/k}\ln\ln\bar{D}\sqrt{\ln(k\bar{D})}}\right)$, maximum 2-degree at most $\sqrt{2k\bar{D}}$, and maximum degree at most $2k\bar{D}$.*

4 Greedy Algorithm

Given a hypergraph H on n vertices with average degree $\bar{d}$, our GreedySDP algorithm first finds a sparse induced hypergraph H' in H using the Sparse-Hypergraph algorithm and then uses the Greedy algorithm to find an independent set in H'.

The Greedy algorithm is a natural extension of the max-degree greedy algorithm on graphs and uniform hypergraphs and was analyzed by Thiele [16]. Given a hypergraph $H(V,E)$ with rank r, for any vertex $v \in V$ let $\bar{d}(v) = (d_1(v), \ldots, d_r(v))$ be the *degree vector* of v, where $d_i(v)$ is the number of edges of size i incident on v. Then, for any vertex $v \in V$ let

$$f(\bar{d}(v)) = \sum_{i_1}^{d_1(v)} \sum_{i_2}^{d_2(v)} \cdots \sum_{i_r}^{d_r(v)} \prod \binom{d_1}{i_1} \prod \binom{d_2}{i_2} \cdots \prod \binom{d_r}{i_r} \frac{(-1)^{\sum_{j=1}^{r} i_j}}{\sum_{j=1}^{r}(j-1)i_j + 1}$$

and let $F(H) = \sum_{v \in V} f(\bar{d}(v))$. The Greedy algorithm iteratively chooses a vertex $v \in V$ with $F(H \backslash v) \geq F(H)$ and deletes v with all incident edges from H until the edge set is empty. The remaining vertices form an independent set in H.

Caro and Tuza [3] showed that in r-uniform hypergraphs the Greedy algorithm always finds a weak independent set of size at least $\Theta\left(n/\Delta^{\frac{1}{r-1}}\right)$. Thiele

[16] extended their result to non-uniform hypergraphs and gave a lower bound on the independence number as a complicated function of the degree vectors of the vertices in a hypergraph. Using these two bounds, we prove the following lemma.

Lemma 4.1. *Given a hypergraph H on n vertices with maximum 2-degree $\sqrt{d}$ and maximum degree d, the* Greedy *algorithm finds an independent set of size $\Omega(n/\sqrt{d})$.*

Proof. First, we truncate edges in H to a maximum size three by arbitrarily deleting excess vertices. The resulting hypergraph H' still has maximum 3-degree d and maximum 2-degree $\sqrt{d}$, and is now of rank 3. Moreover, an independent set in H' is also independent in H. Thus, to prove the claim it is sufficient to bound from below the size of an independent set found by the greedy algorithm in H'.

As shown in [16], Greedy finds an independent set in a rank-3 hypergraph of size at least

$$\alpha(H') \geq n \sum_{j=0}^{d} \sum_{i=0}^{\sqrt{d}} \binom{d}{j} \binom{\sqrt{d}}{i} \frac{(-1)^{(j+i)}}{i+2j+1} \ . \tag{10}$$

By using the equality $\sum_{k} \binom{n}{k} \frac{(-1)^k}{x+k} = x^{-1} \binom{x+n}{n}^{-1}$ we can simplify (10) as:

$$\begin{aligned} \alpha(H') &\geq n \sum_{i=0}^{\sqrt{d}} (-1)^i \binom{\sqrt{d}}{i} \frac{1}{2} \left(\sum_{j=0}^{d} \binom{d}{j} \frac{(-1)^j}{j+(i+1)/2} \right) \\ &= \frac{n}{2(d+1)} \sum_{i=0}^{\sqrt{d}} (-1)^i \binom{\sqrt{d}}{i} \binom{(i+1)/2+d}{d+1}^{-1} \ . \end{aligned} \tag{11}$$

We show that for any value of d

$$F_d = \sum_{i=0}^{\sqrt{d}} (-1)^i \binom{\sqrt{d}}{i} \binom{(i+1)/2+d}{d+1}^{-1} \tag{12}$$

is lower bounded by $x\sqrt{d}$ for some $x > 0$. Then, from (11) the Greedy algorithm finds an independent set of size at least $\Omega(n/\sqrt{d})$.

Let $f_d(i) = \binom{\sqrt{d}}{i} \binom{(i+1)/2+d}{d+1}^{-1}$. Abusing binomial notation, we assume that $\binom{\sqrt{d}}{i} = 0$, for any $i > \sqrt{d}$ and $\sqrt{d}$ integral. Then,

$$F_d = \sum_{i=0}^{\sqrt{d}} (-1)^i f_d(i) \ . \tag{13}$$

We define

$$q_d(i) = \frac{(i+2)(i+4)\cdots(i+2d+2)}{(i+3)(i+5)\cdots(i+2d+1)} \tag{14}$$

for any $i \geq 0$. Using Stirling's approximation for the factorial function[1] we obtain

$$q_d(0) = \frac{2^{2d+1}(d+1)d}{(2d+1)} = \sqrt{\pi d}\left(1 + O\left(\frac{1}{d}\right)\right)$$

and

$$q_d(1) = \frac{(2d+3)}{2^{2d+1}(d+1)(d+1)} = \sqrt{\frac{d}{\pi}}\left(1 + O\left(\frac{1}{d}\right)\right) .$$

Note that $q_d(i+2) = \frac{(i+3)(i+2d+4)}{(i+2)(i+2d+3)} q_d(i) > q_d(i)$, and so $q_d(i) > \sqrt{d}$ for any $i \geq 0$. Then, from the definition of $f_d(i)$ and (14) we have that $\frac{f_d(i+1)}{f_d(i)} = \frac{\sqrt{d}-i}{q_d(i)} < 1$. From (12) and (13) it follows that $F_d > f_d(0) - f_d(1)$ and $f_d(0) = q_d(0)$, then

$$\begin{aligned} F_d &> f_d(0) - f_d(1) \\ &= f_d(0)\left(1 - \frac{\sqrt{d}}{q_d(0)}\right) \\ &= q_d(0) - \sqrt{d} \\ &= (\sqrt{\pi} - 1)\sqrt{d}\left(1 + O\left(\frac{1}{d}\right)\right) . \end{aligned} \quad (15)$$

Thus, from (11), (12) and (15) the Greedy algorithm finds an independent set of size at least $\Omega(n/\sqrt{d})$.

The bound on the performance ratio of GreedySDP then follows from Lemma 4.1, Theorem 3.7 and the fact that truncating edges in SparseHypergraph doesn't increase the weight of a maximal independent set in a hypergraph.

Theorem 4.2. *Given a hypergraph H on n vertices with average degree $\bar{d}$ and the independence number $\alpha(H) = n/k$, the* GreedySDP *algorithm finds an independent set of size at least* $\Omega\left(\frac{n}{k^{5/2-1/k}\bar{d}^{1-1/k}\ln\ln\bar{d}\sqrt{\ln(k\bar{d})}}\right)$.

From Theorem 4.2 it is easy to see that if the maximum independent set in H is relatively big, say $\Omega\left(\frac{n\ln\ln\bar{d}}{\ln d}\right)$, i.e. $k = O\left(\frac{\ln\bar{d}}{\ln\ln d}\right)$, then GreedySDP obtains an approximation ratio of $O\left(\frac{\bar{d}}{\ln d}\right)$. However, if the maximum independent set is at most $\Omega\left(\frac{n\ln\ln\bar{d}}{\ln d}\right)$, then Greedy alone is within a factor of $O\left(\frac{\bar{d}\ln\ln\bar{d}}{\ln d}\right)$. Therefore, we run both Greedy and GreedySDP and output the larger independent set found.

Theorem 4.3. *Given a hypergraph H with average degree $\bar{d}$, the* GreedySDP-MIS *approximates the maximum independent set within a factor of* $O\left(\frac{\bar{d}\ln\ln\bar{d}}{\ln d}\right)$.

Corollary 4.4. *For small k the approximation factor of* GreedySDP-MIS *is* $O\left(\bar{d}^{1-\frac{1}{k}+o(1)}\right)$.

---

[1] Stirling's approximation: $N! = \sqrt{2\pi N}\left(\frac{N}{e}\right)^N\left(1 + O\left(\frac{1}{N}\right)\right)$.

The implication is the first approximation factor for the independent set problem in hypergraphs that is sublinear in the average degree.

Corollary 4.5. *The independent set problem in hypergraphs is $o(\Delta)$-approximable.*

5 Randomized Algorithm

The RANDOMIS algorithm extends the randomized version of Turán bound on graphs and was analyzed by Shachnai and Srinivasan in [15]. Given a hypergraph $H(V, E)$, the algorithm creates a random permutation π of V and adds a vertex v to the independent set I, if there is no edge e such that e contains v and v appears last in π among the vertices of e. Clearly, RANDOMIS outputs a feasible independent set I, since it never contains the last vertex in any edge under the permutation π.

Shachnai and Srinivasan [15] analyzed RANDOMIS on weighted hypergraphs. They gave a lower bound on the probability that a vertex $v \in H$ is added by the algorithm to the independent set, using conditional probabilities and the FKG inequality. In uniform hypergraphs the lower bound on the size of a independent set found by RANDOMIS follows by summing the probabilities over the vertices and applying linearity of expectation, giving a bound identical to that of Caro and Tuza [3].

Theorem 5.1 ([15], Theorem 2). *For any $k \geq 2$ and any k-uniform hypergraph H,* RANDOMIS *finds an independent set of size at least* $\sum_{v \in V} \binom{d(v)+1/(k-1)}{d(v)}^{-1}$ $= \Omega\left(\sum_{v \in V} \frac{w(v)}{(d(v))^{\frac{1}{k-1}}}\right)$.

To extend the bound to non-uniform weighted hypergraphs, Shachnai and Srinivasan introduced the following potential function on a vertex v:

$$f(v) = \min_{j=1,2,\cdots,a(v)} (d_j(v))^{-\frac{1}{k_j(v)-1}} ,$$

where a vertex v lies in edges of $a(v)$ different sizes: $k_j(v)$, for $j = 1, 2, \cdots, a(v)$, and $d_j(v)$ is the number of edges of size $k_j(v)$. Using similar analysis as in Theorem 5.1, they proved the following bound:

Theorem 5.2 ([15], Theorem 3). *Given a weighted hypergraph $H(V, E)$, the expected weight of the independent set produced by* RANDOMIS *is at least* $\Omega\left(\sum_{v \in V} \frac{w(v)}{a(v)^{1/b(v)}} f(v)\right)$, *where* $b(v) = (\min_j (k_j(v) - 1))$.

Shachnai and Srinivasan also show in [15] how to derandomize RANDOMIS for hypergraphs with bounded maximum degree, or logarithmic degree and sparse neighborhoods.

Our algorithm RANDOMSDP first uses SPARSEHYPERGRAPH to find an induced hypergraph H' in H with maximum 2-degree at most $\sqrt{2k\bar{D}}$ and maximum degree at most $2k\bar{D}$; and then uses RANDOMIS to find an independent set in H'.

The bound on the performance ratio of RANDOMSDP follows from Theorem 3.7 and Theorem 5.2, using that $\Omega\left(\sum_{v\in V} \frac{w(v)}{a(v)^{1/b(v)}} f(v)\right) = \Omega\left(\sum_{v\in V} \frac{w(v)}{\sum_{i=2}^{r} d_i^{\frac{1}{i-1}}}\right)$ by the definitions of $a(v)$, $b(v)$ and $f(v)$.

Theorem 5.3. *Given a weighted hypergraph H with average weighted degree $\bar{D}$, the* RANDOMSDP *algorithm finds an independent set of weight at least* $\Omega\left(\frac{w(H)}{k^{2-1/k}\bar{D}^{1/2-1/k}\ln\ln\bar{D}\sqrt{\ln(k\bar{D})}}\right)$.

From Theorem 5.3 it follows that the RANDOMSDP algorithm approximates MIS within a factor of $O\left(\frac{\bar{D}}{\ln \bar{D}}\right)$ if $\alpha(H,w) = \Omega\left(\frac{w(V)\ln\ln\bar{D}}{\ln\bar{D}}\right)$, whereas RANDOMIS alone finds an approximation within a factor of $O\left(\frac{\bar{D}\ln\ln\bar{D}}{\ln\bar{D}}\right)$ if $\alpha(H,w) = O\left(\frac{w(V)\ln\ln\bar{D}}{\ln\bar{D}}\right)$. Therefore, given a hypergraph H, we run both RANDOMIS and RANDOMSDP on H and output the larger of the independent sets.

Theorem 5.4. *Given a hypergraph $H(V,E)$ with average weighted degree $\bar{D}$, the* RANDOMSDP-MIS *approximates the weight of a maximum independent set in H within a factor of* $O\left(\frac{\bar{D}\ln\ln\bar{D}}{\ln\bar{D}}\right)$.

6 Conclusions

In this paper we propose a new approach to the Maximum Independent Set problem in weighted non-uniform hypergraphs. Our approach is to use SDP techniques to sparsify a given hypergraph and then apply a combinatorial algorithm to find a large independent set. Using this approach we derive $o(\bar{d})$-approximation for IS in unweighted hypergraphs, matching the best known ratio for IS in graphs, both in terms of maximum and average degree. We generalize the results to weighted hypergraphs, proving similar bounds in terms of the average weighted degree $\bar{D}$.

For further work one possible direction is to extend the result on the GREEDYSDP-MIS to weighted hypergraphs. Another (and perhaps more interesting) open question is to prove similar bounds in terms of the maximum and average weighted *hyperdegree*, where the *hyperdegree* $d^*(v)$ of a vertex v is defined as $d^*(v) = \sum_{t=2}^{r} d_t(v)^{\frac{1}{t-1}}$. The hyperdegree is a generalization of a vertex degree in a graph.

References

1. Alon, N., Kahale, N.: Approximating the independence number via the θ-function. Math. Prog. 80(3), 253–264 (1998)
2. Bazgan, C., Monnot, J., Paschos, V., Serrière, F.: On the differential approximation of MIN SET COVER. Theor. Comp. Science 332, 497–513 (2005)

3. Caro, Y., Tuza, Z.: Improved lower bounds on k-independence. J. Graph Theory 15, 99–107 (1991)
4. Feige, U.: Approximating maximum clique by removing subgraphs. SIAM J. Disc. Math. 18(2), 219–225 (2005)
5. Grötschel, M., Lovász, L., Schrijver, A.: The ellipsoid method and its consequences in combinatorial optimization. Combinatorica 1(2), 169–197 (1981)
6. Halldórsson, M.M.: Approximations of independent sets in graphs. In: Jansen, K., Rolim, J.D.P. (eds.) APPROX 1998. LNCS, vol. 1444, pp. 1–13. Springer, Heidelberg (1998)
7. Halldórsson, M.M.: Approximations of Weighted Independent Set and hereditary subset problems. J. Graph Algorithms and Applications 4(1), 1–16 (2000)
8. Halldórsson, M.M., Losievskaja, E.: Independent sets in bounded-degree hypergraphs. Disc. Appl. Math. 157, 1773–1786 (2009)
9. Halperin, E.: Improved approximation algorithms for the vertex cover problem in graphs and hypergraphs. SIAM J. Computing 31(5), 1608–1623 (2002)
10. Karger, D., Motwani, R., Sudan, M.: Approximate graph coloring by semidefinite programming. J. ACM 45(2), 246–265 (1998)
11. Krivelevich, M., Nathaniel, R., Sudakov, B.: Approximating coloring and maximum independent sets in 3-uniform hypergraphs. J. Algorithms 41(1), 99–113 (2001)
12. Lovász, L.: On Shannon capacity of a graph. IEEE Trans. Inform. Theory 25(1), 1–7 (1979)
13. Mahajan, S., Ramesh, H.: Derandomizing semidefinite programming based approximation algorithms. SIAM J. Computing 28(5), 1641–1663 (1999)
14. Rényi, A.: Probability theory. Elsevier, New York (1970)
15. Shachnai, H., Srinivasan, A.: Finding large independent sets of hypergraphs in parallel. SIAM J. Disc. Math. 18(3), 488–500 (2005)
16. Thiele, T.: A lower bound on the independence number of arbitrary hypergraphs. J. Graph Theory 32, 241–249 (1999)

Correlation Clustering Revisited: The "True" Cost of Error Minimization Problems*

Nir Ailon[1] and Edo Liberty[2]

[1] Google Research, New York, NY
nailon@gmail.com
[2] Yale University, New Haven, CT
edo.liberty@yale.edu

Abstract. Correlation Clustering was defined by Bansal, Blum, and Chawla as the problem of clustering a set of elements based on a, possibly inconsistent, binary similarity function between element pairs. Their setting is agnostic in the sense that a ground truth clustering is not assumed to exist, and the cost of a solution is computed against the input similarity function. This problem has been studied in theory and in practice and has been subsequently proven to be APX-Hard.

In this work we assume that there does exist an unknown correct clustering of the data. In this setting, we argue that it is more reasonable to measure the output clustering's accuracy against the unknown underlying true clustering.

We present two main results. The first is a novel method for continuously morphing a general (non-metric) function into a pseudometric. This technique may be useful for other metric embedding and clustering problems. The second is a simple algorithm for randomly rounding a pseudometric into a clustering. Combining the two, we obtain a certificate for the possibility of getting a solution of factor strictly less than 2 for our problem. This approximation coefficient could not have been achieved by considering the agnostic version of the problem unless $P = NP$.

1 Introduction

Correlation Clustering was defined by Bansal, Blum and Chawla [1] as the problem of producing a clustering x of data points based on a binary function, h, which tells us, for each pair, whether they are similar or not. The objective is to find the clustering x that minimizes $f(x, h)$, the number of disagreements between h and x. The problem is *agnostic* in the sense that the clustering of the data is not taken into account or even assumed to exist. This gives rise to an APX-hard optimization problem which is studied in their paper and in consequent work [1,2,3,4,5,6,7,8,9,10]. In this paper we assume a setting in which *there*

* Work done while second author was visiting Google Research as a summer intern.

S. Albers et al. (Eds.): ICALP 2009, Part I, LNCS 5555, pp. 24–36, 2009.

is an (unknown) correct way to cluster the data, τ. Such a scenario arises, for example, in duplicate detection and elimination in large data (also known as the record linkage). In this setting we argue that one should try to minimize $f(x, \tau)$, the number of disagreements between the output clustering and the ground truth clustering.

A related problem that has been studied in the literature [11] is *planted clustering*. In this model, the observation h is given by random noise applied to the ground truth clustering τ. Solving the traditional Correlation Clustering problem on h, thus obtained, gives precisely a *maximum likelihood* configuration for τ. It is not clear, however, why this random noise model should be at all realistic. If for instance h is obtained as an output of a machine learned hypothesis, then it is very reasonable to assume that the error will be highly structured and correlated. Also, it is often the case that h is obtained as a robust version (using e.g. spectral techniques [11] or dot product techniques [12][1]) of some raw input. In these cases, it is clear that any independence assumption that we may have had on the raw input would be lost in the process of obtaining h. Our approach is *adversarial*, and the practitioner may use it given h that is obtained using any preprocessing, even if heavy dependencies are introduced. The advantage of our work is that the practitioner need not worry about transitivity issues when preprocessing the data, and that unlike other techniques for obtaining a final (transitive) clustering (e.g. k-means over h obtained as a low dimensional Euclidean approximation of raw data using spectral techniques), we provide provable approximation guarantees.

In our case, the ground truth clustering τ is not only unknown but can also be arbitrarily different from the similarity function h. Since the algorithm can only access h, we can expect the output x to respect the ground truth only insofar as the input h does. We thus try to minimize C such that $f(x, \tau) \leq Cf(h, \tau)$ where f measures the distance between the different objects. In other words, the more h disagrees with the ground truth τ (larger $f(h, \tau)$) the weaker the requirements from the output x.

Traditional optimization gives the following indirect solution to our problem: Find x which approximately minimizes $f(x, h)$ so $f(x, h) \leq C^* f(h, \tau)$ for some $C^* \geq 1$ and for all possible clusterings τ. By the triangle inequality $f(\tau, x) \leq f(\tau, h) + f(h, x) \leq (C^* + 1)f(\tau, h)$. Hence an approximation factor of C^* for the traditional corresponding combinatorial optimization problem gives an upper bound of $C = C^* + 1$ for our problem. Since $C^* > 1$ (Correlation Clustering is APX-hard) this approach would yield $C > 2$. A similar argument can be made for randomized combinatorial optimization and an expected approximation ratio. This immediately raises the interesting question of whether we can go below 2 and shortcut the traditional optimization detour (often an obstruction under complexity theoretical assumptions).

Our main result, Section 3, is a morphing process which proves the existence of a good relaxed solution to our problem, which we name CorrelationClusteringX. More precisely, one can continuously change the values of the input h into

[1] In [12], a Gaussian random noise model is assumed.

a "soft" clustering x_{dif} which is $[0,1]$ valued and a metric (satisfies all triangle inequalities). More importantly, we show that $f(x_{dif}, \tau) \leq 4/3 f(h, \tau)$. The relaxation x_{dif} is obtained as the limit at infinity of a solution to a piecewise linear differential equation. This algorithm, which we also refer to as a *morphing process*, is interesting in its own right and may be useful for other problems on metric spaces. The intuitive idea behind the differential equation is a physical system in which edges "exert forces" on each other proportional to the size of triangle inequality violations. The main technical lemma shows that all triangle inequality violations decay exponentially in time. This fast decay allows us to bound the loss with respect to the ground truth from above.

In Section 4 we show how to randomly convert the relaxed solution x_{dif} into an integer solution x to CorrelationClusteringX such that $f(x, \tau) \leq 3/2 f(x_{dif}, \tau)$. Applying the rounding algorithm to x_{dif} gives a $C \leq 2$ approximation algorithm. As a side effect, our algorithm allows computing an invariant $C' = C'(h) \leq 4/3$ which serves as a witness for getting a solution to CorrelationClusteringX with $C = 3C'/2$. In particular, if $C' < 4/3$ then we get $C < 2$.

Our work is related to recent work by Ailon and Mehryar [13] on Machine Learning reductions for ranking. Balcan et al. [14] also consider clustering problems in which a ground truth is assumed to exist. However, there are two main differences. First, they consider objective functions in which the cost is computed pointwise (here we consider pairwise costs). A second and more fundamental difference is that they make strong assumptions about the (input observation, ground truth) pair. Their assumptions, in some sense, exactly state that an approximation to the traditional optimization problem is "good" for the problem in which errors are computed against the truth. In our case, we make no assumptions about the input or the ground truth. Further investigation of the connection between the two results is an interesting research direction.

We begin in Section 2 by formally defining our notation and stating our results. Some proofs (Theorems 4, 5 and 6) are entirely omitted due to space limitations and can be found in [15]. The interested reader is also referred to the last reference for further discussion.

2 Definitions and Statement of Results

We are given a set V of n elements to cluster together with a symmetric distance function h serving as clustering information. We use the convention that $h(u, v) = h(v, u) = 1$ if u, v are believed to belong to separate clusters, and 0 otherwise.[2]

Let $\mathcal{K}$ denote the set of $[0,1]$-valued symmetric functions on $V \times V$ (with a null diagonal). Let $\mathcal{I} \subseteq \mathcal{K}$ denote the subset of $\{0,1\}$ valued functions in $\mathcal{K}$. Let $\Delta \subseteq \mathcal{K}$ denote the set of functions $k \in \mathcal{K}$ satisfying the triangle inequality $k(u, v) \leq k(v, w) + k(w, u)$ for all $u, v, w \in V$. Let $\mathcal{C}$ denote $\mathcal{I} \cap \Delta$. Clearly $c \in \mathcal{C}$

[2] In other literature, h is a similarity measures, with higher values corresponding to higher belief in co-clustering. We find our convention easier to work with because a clustering is equivalently a pseudometric over the values {0,1}.

is an encoding of a clustering of V, with $c(u, v) = 1$ if u, v are separated and $c(u, v) = 0$ if they are co-clustered.

Our input h lives in $\mathcal{I}$ but not in Δ, hence the function h encodes possibly inconsistent $\{0, 1\}$ clustering information. Indeed, it may tell us that $h(u, v) = h(v, w) = 0$ but $h(u, w) = 1$, hence violating transitivity. For a number $a \in [0, 1]$ let $\overline{a}$ denote $1-a$. Define the Correlation Clustering cost function [1] $f : \mathcal{K} \times \mathcal{K} \to \mathbb{R}^+$ as $f(k_1, k_2) = \sum_{u<v}(k_1(u, v)\overline{k_2(u, v)} + \overline{k_1(u, v)}k_2(u, v))$. For integer valued k_1, k_2 this is the Hamming distance.

The problem of CorrelationClusteringX is given in the following:

Definition 1. *Given $h \in \mathcal{I}$ and $C \geq 1$ output $x \in \mathcal{C}$ such that for all $\tau \in \mathcal{C}$, $f(\tau, x) \leq Cf(\tau, h)$ (assuming such an x exists). In the randomized setting, the goal is to output a sample x from a distribution $\mathcal{D}$ on $\mathcal{C}$, such that $E_{x\sim\mathcal{D}}[f(\tau, x)] \leq Cf(\tau, h)$ (assuming such $\mathcal{D}$ exists). An algorithm outputting x in the deterministic case or drawing it from $\mathcal{D}$ in the randomized case is called a C-approximation algorithm to CorrelationClusteringX.*

Deterministic CorrelationClusteringX has a corresponding integer program over the $\binom{n}{2}$ variables of $x \in \mathcal{C}$ with an exponential number of constraints:

$$\text{IP: minimize } C \text{ s.t. } f(x, \tau) \leq Cf(h, \tau) \text{ for all } \tau \in \mathcal{C}$$
$$x \in \mathcal{C}, C \geq 0$$

Note that in traditional correlation clustering, we would have used the constraint $f(x, h) \leq Cf(h, \tau)$ for all $\tau \in \mathcal{C}$ instead. IP can be relaxed by allowing $x \in \Delta$ and adding a constraint for each $\tau \in \Delta$.

$$\text{LP: minimize } C \text{ s.t. } f(\tau, x) \leq Cf(\tau, h) \text{ for all } \tau \in \Delta$$
$$x \in \Delta, C \geq 1$$

Clearly, an equivalent program can be obtained by using only constraints that correspond to vertices of Δ, of which there are exponentially many. Let (x_{LP}, C_{LP}) denote the minimizer of LP.

Observation 1. *LP has a separation oracle and can therefore be solved optimally in polynomial time.*

To see Observation 1, note that given a candidate solution (x, C) it is possible to find $\tau \in \Delta$ satisfying $f(\tau, x) > Cf(\tau, h)$ (if one exists) using another simple standard linear program with $\tau \in \Delta$ as variable. Note that unlike in the usual case of combinatorial optimization LP relaxations, it is not immediate to compare between the values of IP and LP, because the relaxation is obtained by both adding constraints and removing others. The reason we enlarged the collection of constraints $\{f(\tau, x) \leq Cf(\tau, h)\}_\tau$ in LP is to give rise to an efficient separation oracle.

Our first result states that the optimal solution to LP is a (deterministic) fractional solution x_{LP} for CorrelationClusteringX with approximation factor

C_{LP} of at most 4/3 (in the sense that $f(x_{LP}, \tau) \leq C_{LP} f(h, \tau)$ for all $\tau \in \Delta$). The proof of the theorem is constructive. It is shown that the limit at infinity of a solution to a certain differential equation is a feasible solution to LP.

Theorem 1. *For any $h \in I$, the value of LP is at most* 4/3.

In the proof of Theorem 1 we will point to one particular solution (x_{dif}, C_{dif}) which is a limit at infinity of a solution to a piecewise linear differential equation. Finding this limit may be done exactly, but we omit the details because together with Observation 1, general purpose convex optimization may be used instead.

Our next theorems refer to the QuickCluster algorithm which is defined in Section 4. It is a tweaked version of the algorithm in [2] (also with a different analysis). QuickCluster takes as input $h \in \mathcal{K}$ and outputs $x \in \mathcal{C}$. Let $QC(h)$ denote the distribution over outputs $x \in \mathcal{C}$ of QuickCluster for input h.

Theorem 2. *For any $\hat{h} \in \Delta$ and $\tau \in \mathcal{C}$ we have $E_{x \sim QC(\hat{h})}[f(x, \tau)] \leq \frac{3}{2} f(\hat{h}, \tau)$.*

Combining Theorems 1 and Theorem 2 we get a randomized solution with $C = \frac{3}{2} C_{LP} \leq 2$ for CorrelationClusteringX. If $C_{LP} < 4/3$, we get a witness for achieving C strictly less than 2.

Theorem 3. *For any $h \in \mathcal{I}$ and $\tau \in \mathcal{C}$ we have $E_{x \sim QC(h)}[f(\tau, x)] \leq 2 f(\tau, h)$.*

The running time of QuickCluster is analyzed for two representation dependent regimes. In the pairwise-queries model, only pairwise queries to h are allowed, i.e, evaluating $h(u, v)$ for a pair $\{u, v\}$. In the neighborhood-queries regime, the algorithm is allowed neighborhood queries, returning for a query u its neighborhood $N(u) = \{u\} \cup \{v \in V \mid h(u, v) = 0\}$ as a linked list. We obtain the following bounds.

Theorem 4. *In the pairwise-queries model, any constant factor randomized approximation algorithm for CorrelationClusteringX performs $\Omega(n^2)$ queries to h in expectation for some input h.*

Trivially, QuickCluster performs $O(n^2)$ queries to h for some input h and thus has an optimal worst case running time with respect to this model.

Theorem 5. *In the neighborhood-queries model, the expected running time of QuickCluster is $O(n + \min_{\tau \in C} f(\tau, h))$.*

The following is a lower bound on the approximation guarantee any deterministic algorithm can achieve.

Theorem 6. *There exists an input h for which any deterministic algorithm for CorrelationClusteringX incurs an approximation factor of at least 2 for some ground truth $\tau \in \mathcal{C}$. For the same input, a randomized algorithm can obtain a factor of at most 4/3.*

3 Morphing h into a Metric: A Differential Program

In this section we prove Theorem 1. The idea is to "morph" $h \in \mathcal{K}$, which is not necessarily a metric, into a pseudometric. The solution $x_{dif} \in \Delta$ is obtained by theoretically running a differential equation to infinity. More precisely, we define a differential morphing process such that $h_t(u,v)$ is the changed value of $h(u,v)$ at time t and $h_0(u,v) = h(u,v)$ for all u and v. The solution is given by $x_{dif} = \lim_{t\to\infty} h_t$.

We look at a triangle created by the triplet $\{u,v,w\}$. For ease of notation we set $a = h(u,v)$, $b = h(v,w)$, and $c = h(w,u)$. First, we define the gap g_{uvw} of the triangle $\{u,v,w\}$ away from satisfying the triangle inequality as:

$$g_{uvw} = \max\{0,\ a-(b+c),\ b-(c+a),\ c-(a+b)\} \tag{1}$$

We define the *force* that triangle $\{u,v,w\}$ exerts on a as follows:

$$F(a;b,c) = \begin{cases} -g_{uvw} & \text{if } a > b+c \\ g_{uvw} & \text{otherwise.} \end{cases} \tag{2}$$

The morphing process is such that the contribution of the triangle $\{u,v,w\}$ to the change in a, $\frac{da}{dt}$, is the force $F(a;b,c)$. Intuitively, the force serves to reduce the gap. If a, b, and c satisfy the triangle inequality then no force is applied. If $a > b+c$ then $\frac{da}{dt}$ is negative and a is reduced. If $b > c+a$ or $c > a+b$ then $\frac{da}{dt}$ is positive and a is increased. Summing over all triangles containing u and v gives our differential equation in Figure 1. with the starting boundary condition $h_0(u,v) = h(u,v)\ \ \forall u,v \in V$. Similarly to our previous notation, let $a(t)$, $b(t)$ and $c(t)$ denote $h_t(u,v), h_t(v,w)$ and $h_t(w,u)$ throughout.

$$\frac{dh_t(u,v)}{dt} = \sum_{w\in V\setminus\{u,v\}} F(h_t(u,v); h_t(v,w), h_t(w,u)).$$

Fig. 1. The morphed input h_t is given by the solution to the above differential equation at time t. The initial starting point is the input $h_0 = h$. The solution x_{dif} is given by $x_{dif} = \lim_{t\to\infty} h_t$.

The following is the main technical lemma of the proof. It asserts that the external forces applied to a triangle $\{u,v,w\}$ by other triangles only contribute to reducing the gap g_{uvw}. It implies both the exponential decay of all positive gaps and the stability of null gap.

Lemma 1. *Let $g_{uvw}(t)$ denote the gap of h_t on the triplet $\{u,v,w\}$ at time t, as defined in (1). Then $\frac{dg_{uvw}(t)}{dt} \le -3g_{uvw}(t)$ for all t.*

Note: Clearly the lemma implies that $g_{uvw}(t) \le g_{uvw}(t_0)e^{-3(t-t_0)}$ for any $t_0 \le t$. The lemma is easy to prove if $|V| = 3$. For larger V, the difficulty is in showing that the interference between triangles is constructive.

Proof. It is enough to prove the lemma for the case $\{a(t) \geq b(t)+c(t)\} \cup \{b(t) \geq c(t)+a(t)\} \cup \{c(t) \geq a(t)+b(t)\}$. Indeed, in the open set $\{a(t) < b(t)+c(t)\} \cap \{b(t) < c(t)+a(t)\} \cap \{c(t) < a(t)+b(t)\}$ the value of g is 0 identically. Assume w.l.o.g. therefore that $a(t) \geq b(t)+c(t)$ (hence $g_{uvw}(t) = a(t)-b(t)-c(t)$).

$$\frac{d\ g_{uvw}(t)}{dt} = \frac{d\ (a(t)-b(t)-c(t))}{dt}$$
$$= F(a(t);b(t),c(t)) - F(b(t);c(t),a(t)) - F(c(t);a(t),b(t))$$
$$+ \sum_{s \in V \setminus \{u,v,w\}} F(a(t);x_s(t),y_s(t)) - F(b(t);z_s(t),y_s(t)) - F(c(t);x_s(t),z_s(t))$$

where $x_s(t) = h_t(u,s)$, $y_s(t) = h_t(v,s)$, and $z_s(t) = h_t(w,s)$. The first line of the RHS is exactly $-3g_{uvw}$. It suffices to prove that for any $s \in V \setminus \{u,v,w\}$, $F(a(t);x_s(t),y_s(t)) - F(b(t);z_s(t),y_s(t)) - F(c(t);x_s(t),z_s(t)) \leq 0$. This is proved by enumerating over all possible configurations of the three triangles $\{u,v,s\}$, $\{v,w,s\}$ and $\{w,u,s\}$ and is omitted from this abstract. The details can be found in [15].

The following lemma tells us that if $a(0)$, $b(0)$, and $c(0)$ violate the triangle inequality then at each moment $t > 0$ they either violate the same inequality or the violation is resolved.

Lemma 2. *Let $a(t)$, $b(t)$, and $c(t)$ denote $h_t(u,v)$, $h_t(v,w)$, and $h_t(w,u)$ respectively. If $a(0) \geq b(0)+c(0)$ then for all $t \geq 0$ either $a(t) \geq b(t)+c(t)$ or $a(t)$, $b(t)$, and $c(t)$ satisfy the triangle inequality.*

Proof. First note that if for some time t_0 the triplet $\{a(t_0), b(t_0), c(t_0)\}$ satisfies the triangle inequality, then this will continue to hold for all $t \geq t_0$ in virtue of the note following Lemma 1. Also note that $a(t) > b(t)+c(t)$ and $(b(t) > c(t)+a(t)$ or $c(t) > a(t)+b(t))$ cannot hold simultaneously. Let t' be the infimum of t such that $a(t) \leq b(t)+c(t)$, or ∞ if no such t exists. If $t' = \infty$ then the lemma is proved. Otherwise by continuity and the first note above, $a(t') = b(t')+c(t')$, $b(t') \leq a(t')+c(t')$ and $c(t') \leq a(t')+b(t')$, hence $a(t')$, $b(t')$, and $c(t')$ satisfy the triangle inequality and thus continue to do so for all $t > t'$, completing the proof of the lemma.

Now fix a ground truth clustering $\tau \in \Delta$. Consider the cost $f(\tau, h_t)$ as a function of t. Letting $L_t(u,v) = h_t(u,v)\overline{\tau(u,v)} + \overline{h_t(u,v)}\tau(u,v)$, we get $f(\tau,h_t) = \sum_{u<v} L_t(u,v) = \frac{1}{n-2}\sum_{u<v<w} C_{uvw}(t)$, where $C_{uvw}(t) := L_t(u,v) + L_t(v,w) + L_t(w,u)$. The derivative of the cost is $\frac{d\ f(\tau,h_t)}{dt} = \frac{1}{n-2}\sum_{uvw} G_{uvw}(t)$, where[3]

$$G_{uvw}(t) := (1-2\tau(u,v))F(h_t(u,v);h_t(v,w),h_t(w,u))$$
$$+(1-2\tau(v,w))F(h_t(v,w);h_t(w,u),h_t(u,v))$$
$$+(1-2\tau(w,u))F(h_t(w,u);h_t(u,v),h_t(v,w)).$$

[3] Note that G_{uvw} is *not* the derivative of C_{uvw}, but the sum $\sum_{uvw} G_{uvw}$ is the derivative of $\sum_{uvw} C_{uvw}$.

The cost at time t is $f(\tau, h_t) = \frac{1}{n-2}\sum_{uvw} C_{uvw}(0) + \frac{1}{n-2}\sum_{uvw}\int_0^t G_{uvw}(s)ds$. We concentrate on the contribution of one triangle to this sum: $H_{uvw}(t) = C_{uvw}(0) + \int_0^t G_{uvw}(s)ds$.

Let us consider the possible values of the term $G_{uvw}(t)$. If the values $h_t(u,v)$, $h_t(v,w)$, and $h_t(w,u)$ satisfy the triangle inequality then $G_{uvw}(t) = 0$ since the forces F are all zero. Assume then w.l.o.g. that $h_t(u,v) \geq h_t(v,w) + h_t(w,u)$ and so by the definition of F, $G_{uvw}(t) = [2(\tau(u,v) - \tau(v,w) - \tau(w,u)) + 1]g_{uvw}(t)$. Notice that $G_{uvw}(t) \leq g_{uvw}(t)$ since $\tau \in \Delta$. Therefore $G_{uvw}(t) \leq g_{uvw}(t)$ and by Lemma 1 $G_{uvw}(t) \leq g_{uvw}(0)e^{-3t}$.

Lemma 3. *Set $\tau \in \Delta$. Given the above process, let $x_{dif} = \lim_{t\to\infty} h_t$. Then $f(x_{dif}, \tau) \leq \frac{4}{3}f(h,\tau)$. Additionally, $x_{dif} \in \Delta$.*

Proof. In what follows we use the facts that $f(x_{dif}, \tau) = \lim_{t\to\infty}\frac{1}{n-2}\sum_{uvw} H_{uvw}(t)$ and that $\int_0^\infty G_{uvw}(t)dt \leq \int_0^\infty g_{uvw}(0)e^{-3t}dt \leq \frac{1}{3}g_{uvw}(0)$.

$$\begin{aligned} f(x_{dif}, \tau) &= \frac{1}{n-2}\lim_{t\to\infty}\sum_{u<v<w} H_{uvw}(t) = \frac{1}{n-2}\sum_{u<v<w} C_{uvw}(0) + \int_0^\infty G_{uvw}(t) \\ &\leq \frac{1}{n-2}\sum_{u<v<w} C_{uvw}(0) + \frac{1}{3}g_{uvw}(0) \\ &\leq \frac{1}{n-2}\sum_{u<v<w}\frac{4}{3}C_{uvw}(0) \leq \frac{4}{3}f(h,\tau) \end{aligned}$$

The last equation relies on the fact that $C_{uvw}(0) \geq g_{uvw}(0)$. Indeed, that would imply $C_{uvw}(0) + \frac{1}{3}g_{uvw}(0) \leq \frac{4}{3}C_{uvw}(0)$. To see that, it suffices to check that $C_{uvw}(0) \geq 0$ and $C_{uvw}(0) \geq h(u,v) - h(v,w) - h(w,u)$ for any $h(u,v)$, $h(v,w)$ and $h(w,u)$ in $[0,1]$. Notice that $C_{uvw}(0) - [h(u,v) - h(v,w) - h(w,u)]$ is a linear function in h defined on the convex set $[0,1]^3$ and thus attains its maximal values at its extreme points, i.e. integer values of h. Enumerating these cases and validating the statement is straightforward.

Lemma 3 immediately implies that $(x_{dif} = \lim_{t\to\infty} h_t, C_{dif} = 4/3)$ is a feasible solution to LP.

4 QuickCluster

We prove Theorem 2 and Theorem 3. The QuickCluster algorithm described here is very similar to the one used in [2] but the new analysis provides a shortcut that allows us to directly argue about the cost of the algorithm against an unknown truth $\tau \in \mathcal{C}$ which we hold fixed. To describe our algorithm we need to define a piecewise linear tweaking function $\psi : [0,1] \to [0,1]$ as follows: $\psi(a) = 0$ for $a \leq 1/6$, $\psi(a) = 1$ for $a \geq 5/6$, and in the middle section $a \in [1/6, 5/6]$ ψ is obtained by linear interpolation as $\psi(a) = (6a - 1)/4$. Moreover, for convenience we overload the definition of ψ such that $\psi(u,v) \equiv \psi(h(u,v))$. The algorithm begins by setting all nodes $u \in V$ as *free*. In each iteration one node is chosen uniformly at random from all *free* nodes, say u, to serve as a cluster center. Then, each node $v \neq u$ is added to the cluster centered at u with probability $\overline{\psi(u,v)}$ (and set as not-*free*). The algorithm terminates when there are no *free* nodes left. Note that QuickCluster is defined for all $h \in \mathcal{K}$ and that for $h \in \mathcal{I}$ QuickCluster is identical to the algorithm in [2]. Also, Ailon [16] used a similar tweaking idea to improve rounding of a ranking LP in a traditional combinatorial optimization setting.

4.1 The Expected Cost of QuickCluster

Let $QC(h)$ be the distribution over outputs produced by QuickCluster for input h. By definition of f and the fact that τ is fixed we have that $E_{x\sim QC(h)}[f(x,\tau)] = \sum_{u<v} E_{x\sim QC(h)}[x(u,v)]\overline{\tau(u,v)} + E_{x\sim QC(h)}[\overline{x(u,v)}]\tau(u,v)$. Since each $x(u,v)$ is a binary random variable its expectation is equal to the probability of it being equal 1 which is equal to the probability of QuickCluster separating (cross-clustering) u and v. This happens if either u or v are chosen as centers and then not co-clustered (w.p. $\psi(u,v)$). This also happens if a third node w is chosen as a center and and it co-clusters either u or v but not both. Similarly $E_{x\sim QC(h)}[\overline{x(u,v)}]$ is equal to the co-clustering probability of u and v which occurs if either u or v are chosen as centers and joined or if a third node, w, co-clusters both of them. Define p_{uv} as the probability that during the execution of the algorithm v and u are both free and one of them is chosen as a center. Define p_{uvw} as the probability that during the execution of QuickCluster, u, v and w are all free and one of them is chosen as center. Also note that the relation of u and v in the output of QuickCluster is determined exactly once. In what follows, $\binom{V}{b}$ denotes the collection of unordered b-tuples of the set V. When it is clear from the context, the notation (u,v) means an unordered tuple $\{u,v\} \in \binom{V}{2}$ and similarly (u,v,w) means an unordered tuple $\{u,v,w\} \in \binom{V}{3}$.

Lemma 4. *Fix $\tau \in \mathcal{C}$. Let $L_\psi : \binom{V}{2} \to \mathbb{R}^+, \beta : \binom{V}{3} \to \mathbb{R}^+$ and $B : \binom{V}{2} \times V \to \mathbb{R}^+$ be defined as*

$$
\begin{aligned}
L_\psi(u,v) &:= \psi(u,v)\overline{\tau(u,v)} + \overline{\psi(u,v)}\tau(u,v) \\
\beta(u,v;w) &:= \overline{\psi(w,u)}\;\overline{\psi(w,v)}\tau(u,v) + \psi(w,u)\overline{\psi(w,v)}\;\overline{\tau(u,v)} \\
&\quad\;\; + \overline{\psi(w,u)}\psi(w,v)\;\overline{\tau(u,v)} \\
B(u,v,w) &:= \frac{1}{3}\left[\beta(u,v;w) + \beta(v,w;u) + \beta(w,u;v)\right] .
\end{aligned}
$$

Then $E_{x\sim QC(h)}[f(\tau,x)] = \sum_{u<v} p_{uv} L_\psi(u,v) + \sum_{u<v<w} p_{uvw} B(u,v,w)$, where $x \in \mathcal{C}$ is a random clustering obtained as the output of QuickCluster.

Proof. Following the above discussion:

$$E_{x\sim QC(h)}[x(u,v)] = p_{uv}\psi(u,v) + \sum_{w\neq u,v} \frac{1}{3} p_{uvw}[\psi(w,u)\overline{\psi(w,v)} + \overline{\psi(w,u)}\psi(w,v)]$$

$$E_{x\sim QC(h)}[\overline{x(u,v)}] = p_{uv}\overline{\psi(u,v)} + \sum_{w\neq u,v} \frac{1}{3} p_{uvw}[\overline{\psi(w,u)}\ \overline{\psi(w,v)}] \ .$$

And so by linearity of expectation $E_{x\sim QC(h)}[\tau(u,v)\overline{x(u,v)} + \overline{\tau(u,v)}x(u,v)] = p_{uv}L_\psi(u,v) + \sum_{w\neq u,v} \frac{1}{3} p_{uvw}\beta(u,v;w)$.

$$\begin{aligned} E[f(\tau,x)] &= \sum_{u<v} p_{uv}L_\psi(u,v) + \sum_{u<v}\sum_{w\neq u,v} \frac{1}{3} p_{uvw}\beta(u,v;w) \\ &= \sum_{u<v} p_{uv}L_\psi(u,v) + \sum_{u<v<w} p_{uvw}\frac{1}{3}[\beta(u,w;v) + \beta(u,v;w) + \beta(v,w;u)] \\ &= \sum_{u<v} p_{uv}L_\psi(u,v) + \sum_{u<v<w} p_{uvw}B(u,v;w) \ , \end{aligned}$$

as required.

4.2 QuickCluster Decomposition

In order to compute $f(h,\tau)$ we introduce a general decomposition for the sum $\sum_{u<v} Z(u,v)$ for any function $Z : \binom{V}{2} \to \mathbb{R}$. Then, we apply our decomposition to $Z(u,v) = L_h(u,v) = h(u,v)\overline{\tau(u,v)} + \overline{h(u,v)}\tau(u,v)$.

Lemma 5. *Let Z be any function $Z : \binom{V}{2} \to \mathbb{R}$. Let $C(u,v;w) := \overline{\psi(w,u)}\ \overline{\psi(w,v)} + \psi(w,u)\overline{\psi(w,v)} + \overline{\psi(w,u)}\psi(w,v)$. Define the operator $A_Z : (\binom{V}{2} \to \mathbb{R}) \to (\binom{V}{3} \to \mathbb{R})$ on Z as:*

$$A_Z(u,v,w) := \frac{1}{3}\Big[C(u,v;w)Z(u,v) + C(v,w;u)Z(v,w) + C(w,u;v)Z(w,u)\Big] \ . \tag{3}$$

Then one has:

$$\sum_{u<v} Z(u,v) = \sum_{u<v} p_{uv}Z(u,v) + \sum_{u<v<w} p_{uvw}A_Z(u,v,w) \ .$$

Proof. The term $C(u,v;w)$ gives the probability that the node w determines the relation between u and v given that u, v and w are free and w is chosen as center. Since the relation between u and v is determined only once either indirectly (via w) or directly (either u or v are centers) we have:

$$p_{uv} + \sum_{w\neq u,v} \frac{1}{3} p_{uvw}C(u,v;w) = 1. \tag{4}$$

By (4), $Z(u,v) = 1 \cdot Z(u,v) = \left[p_{uv} + \sum_{w \neq u,v} \frac{1}{3} p_{uvw} C(u,v;w)\right] Z(u,v)$. Hence,

$$\begin{aligned}
\sum_{u<v} Z(u,v) &= \sum_{u<v} p_{uv} Z(u,v) + \sum_{u<v} \sum_{w \neq u,v} \frac{1}{3} p_{uvw} C(u,v;w) Z(u,v) \\
&= \sum_{u<v} p_{uv} Z(u,v) + \sum_{u<v<w} \frac{1}{3} p_{uvw} C(u,v;w) Z(u,v) \\
&\quad + \sum_{u<w<v} \frac{1}{3} p_{uvw} C(u,v;w) Z(u,v) + \sum_{w<u<v} \frac{1}{3} p_{uvw} C(u,v;w) Z(u,v) \\
&= \sum_{u<v} p_{uv} Z(u,v) + \sum_{u<w<v} p_{uvw} A_Z(u,v,w) .
\end{aligned}$$

Applying Lemma 5 to the cost function $f(h,\tau)$ we gain:

$$f(h,\tau) = \sum_{u<v} L_h(u,v) = \sum_{u<v} p_{uv} L_h(u,v) + \sum_{u<w<v} p_{uvw} A_{L_h}(u,v,w). \quad (5)$$

4.3 Bounded Ratio Argument

To bound the ratio $f(x,\tau)/f(h,\tau)$ using Equation (5) and Lemma 4 it suffices to bound $L_\psi(u,v)/L_h(u,v)$ for every pair $\{u,v\}$ and $B(u,v,w)/A_{L_h}(u,v,w)$ for every triplet $\{u,v,w\}$.

In the case where $h \in \Delta$ we have that $L_\psi(u,v)/L_h(u,v) \leq 6/5$ and that $B(u,v,w)/A_{L_h}(u,v,w) \leq 3/2$. Showing this entails breaking the polytope defining $(h(u,v), h(v,w), h(w,u))$ into 27 smaller polytopes in which each $h(\cdot,\cdot)$ is constrained to lie in $[0,1/6]$, $(1/6,5/6]$, or $(5/6,1]$. On each of these smaller polytopes and for each one of 5 possibilities for τ on u,v,w, the functions L_h, L_ψ are linear, and B and A_{L_h} are multinomials of total degree two and three respectively.[4] A computer aided proof was used to obtain the bound of 3/2 using standard polynomial maximization techniques on each one of the polytopes. We refer the reader to [17] for details. This proves Theorem 2.

When $h \in \mathcal{I}$, enumerating over all possible choices of h and τ gives that $L_\psi(u,v)/L_h(u,v) = 1$ and $B(u,v,w)/A_{L_\psi}(u,v,w) \leq 2$. This shows that performing QuickCluster directly on h without solving the LP gives a $C = 2$ approximation ratio. This proves Theorem 3.

5 Short Discussion

Our algorithm trivially also gives an expected factor of $2+1=3$ approximation to the traditional Correlation Clustering by triangle inequality of f. Note that the best known approximation factor for Correlation Clustering is 2.5 [2], raising the question of whether it is possible to obtain a 1.5 approximation for CorrelationClusteringX.

[4] The 5 possibilities for τ are: One single cluster, 3 singleton clusters, and the 3 ways to get a singleton and a pair.

Finding a specific instance h for which our algorithm achieves the 2 approximation bound for CorrelationClusteringX will show that our analysis is tight. The worst input known to the authors is h corresponding to the balanced complete bipartite graph ($h(u, v) = 0$ if $\{u, v\} \in e$) for which QuickCluster gives a 1.5 approximation factor (for τ which puts all of V into one cluster).

Acknowledgments. The authors would like to thanks Eyal Even-Dar, Mehryar Mohri, and Elad Hazan for sharing their insights and expertise.

References

1. Bansal, N., Blum, A., Chawla, S.: Correlation clustering. Machine Learning Journal (Special Issue on Theoretical Advances in Data Clustering) 56(1–3), 89–113 (2004); Extended abstract appeared in FOCS 2002, pp. 238–247
2. Ailon, N., Charikar, M., Newman, A.: Aggregating inconsistent information: ranking and clustering. In: Proceedings of the 37th Annual ACM Symposium on Theory of Computing (STOC), pp. 684–693 (2005)
3. Bonizzoni, P., Della Vedova, G., Dondi, R., Jiang, T.: On the approximation of correlation clustering and consensus clustering. Journal of Computer and System Sciences 74(5), 671–696 (2008)
4. Emanuel, D., Fiat, A.: Correlation clustering – minimizing disagreements on arbitrary weighted graphs. In: Di Battista, G., Zwick, U. (eds.) ESA 2003. LNCS, vol. 2832, pp. 208–220. Springer, Heidelberg (2003)
5. Charikar, M., Guruswami, V., Wirth, A.: Clustering with qualitative information. In: Proceedings of the 44th Annual IEEE Symposium on Foundations of Computer Science (FOCS), Boston, pp. 524–533 (2003)
6. Giotis, I., Guruswami, V.: Correlation clustering with a fixed number of clusters. In: SODA 2006: Proceedings of the seventeenth annual ACM-SIAM symposium on Discrete algorithm, pp. 1167–1176. ACM Press, New York (2006)
7. Ailon, N., Charikar, M.: Fitting tree metrics: Hierarchical clustering and phylogeny. In: Proceedings of the 46th Annual IEEE Symposium on Foundations of Computer Science, FOCS (2005)
8. Gionis, A., Mannila, H., Tsaparas, P.: Clustering aggregation. In: Proceedings of the 21st International Conference on Data Engineering (ICDE), Tokyo (to appear, 2005)
9. Filkov, V., Skiena, S.: Integrating microarray data by consensus clustering. In: Proceedings of International Conference on Tools with Artificial Intelligence (ICTAI), Sacramento, pp. 418–425 (2003)
10. Strehl, A.: Relationship-based clustering and cluster ensembles for high-dimensional data mining. PhD Dissertation, University of Texas at Austin (May 2002)
11. McSherry, F.: Spectral partitioning of random graphs. In: FOCS 2001: Proceedings of the 42nd IEEE symposium on Foundations of Computer Science, Washington, p. 529 (2001)
12. Aslam, J., Leblanc, A., Stein, C.: A new approach to clustering. In: 4th International Workshop on Algorithm Engineering (2000)
13. Ailon, N., Mohri, M.: Efficient reduction of ranking to classification. In: The 21st Annual Conference on Learning Theory (COLT), Helsinki, Finland (to appear, 2008)

14. Balcan, M.-F., Blum, A., Gupta, A.: Approximate clustering without the approximation. In: SODA 2009, New York (2009)
15. Ailon, N., Liberty, E.: Correlation clustering revisited: The "true" cost of error minimization problems. Yale University Tecnical Report 1214 (2008)
16. Ailon, N.: Aggregation of partial rankings, p-ratings and top-m lists. In: SODA (2007)
17. Ailon, N., Liberty, E.: Mathematica program (2008), http://www.cs.yale.edu/homes/el327/public/prove32/

Sorting and Selection with Imprecise Comparisons

Miklós Ajtai[1], Vitaly Feldman[1], Avinatan Hassidim[2], and Jelani Nelson[2]

[1] IBM Almaden Research Center, San Jose CA 95120, USA
[2] MIT, Cambridge MA 02139, USA

Abstract. In experimental psychology, the method of paired comparisons was proposed as a means for ranking preferences amongst n elements of a human subject. The method requires performing all $\binom{n}{2}$ comparisons then sorting elements according to the number of wins. The large number of comparisons is performed to counter the potentially faulty decision-making of the human subject, who acts as an imprecise comparator.

We consider a simple model of the imprecise comparisons: there exists some $\delta > 0$ such that when a subject is given two elements to compare, if the values of those elements (as perceived by the subject) differ by at least δ, then the comparison will be made correctly; when the two elements have values that are within δ, the outcome of the comparison is unpredictable. This δ corresponds to the *just noticeable difference unit (JND)* or *difference threshold* in the psychophysics literature, but does not require the statistical assumptions used to define this value.

In this model, the standard method of paired comparisons minimizes the errors introduced by the imprecise comparisons at the cost of $\binom{n}{2}$ comparisons. We show that the same optimal guarantees can be achieved using $4n^{3/2}$ comparisons, and we prove the optimality of our method. We then explore the general tradeoff between the guarantees on the error that can be made and number of comparisons for the problems of sorting, max-finding, and selection. Our results provide close-to-optimal solutions for each of these problems.

1 Introduction

Let $x_1, \ldots, x_n$ be n elements where each x_i has an unknown value $\mathrm{val}(x_i)$. We want to find the element with the maximum value using only pairwise comparisons. However, the outcomes of comparisons are imprecise in the following sense. For some fixed $\delta > 0$, if $|\mathrm{val}(x_i) - \mathrm{val}(x_j)| \leq \delta$, then the result of the comparison can be either "$\geq$" or "$\leq$". Otherwise, the result of the comparison is correct. It is easy to see that in such a setting it might be impossible to find the true maximum (for example when the values of all the elements are within δ). It might however be possible to identify an approximate maximum, that is an element x_{i^*} such that for all x_i, $\mathrm{val}(x_i) - \mathrm{val}(x_{i^*}) \leq k\delta$ for some, preferably small, value k. In addition, our goal is to minimize the number of comparisons performed to find x_{i^*}. We refer to the minimum value k such that an algorithm's

S. Albers et al. (Eds.): ICALP 2009, Part I, LNCS 5555, pp. 37–48, 2009.

output is always guaranteed to be $k\delta$-close to the maximum as the *error* of the algorithm in this setting. Similarly, to sort the above elements with error k we need to find a permutation π such that if $\pi(i) < \pi(j)$ then $\mathrm{val}(x_i) - \mathrm{val}(x_j) \leq k\delta$.

A key issue that our work addresses is that in any sorting (or max-finding) algorithm, errors resulting from imprecise comparisons might accumulate, causing the final output to have high error. Consider, for example, applying the classical bubble sort algorithm to a list of elements that are originally sorted in the reverse order and the difference between two adjacent elements is exactly δ. All the comparisons will be between elements within δ and therefore, in the worst case, the order will not be modified by the sorting, yielding error $(n-1)\delta$. Numerous other known algorithms that primarily optimize the number of comparisons can be easily shown to incur a relatively high error. As can be easily demonstrated (Theorem 1), performing all $\binom{n}{2}$ comparisons then sorting elements according to the number of wins, a "round-robin tournament", achieves error $k = 2$, which is lowest possible (Theorem 2). A natural question we ask here is whether $\binom{n}{2}$ comparisons are necessary to achieve the same error. We explore the same question for all values of k in the problems of sorting, max-finding, and general selection.

One motivation for studying this problem comes from social sciences. A common problem both in experimental psychology and sociology is to have a human subject rank preferences amongst many candidate options. It also occurs frequently in marketing research [20, Chapter 10], and in training information retrieval algorithms using human evaluators [1, Section 2.2]. The basic method to elicit preferences is to present the subject two alternatives at a time and ask which is the preferred one. The common approach to this problem today was presented by Thurstone as early as 1927, and is called the "method of paired comparisons" (see [8] for a thorough treatment). In this method, one asks the subject to give preferences for all pairwise comparisons amongst n elements. A ranked preference list is then determined by the number of "wins" each candidate element receives. A central concept in these studies introduced as far back as the 1800s by Weber and Fechner is that of *just noticeable difference (JND)* unit or *difference threshold* δ. If two physical stimuli with intensities $x < y$ have $|x - y| \leq \delta$, a human will not be able to reliably distinguish which intensity is greater[1]. The idea was later generalized by Thurstone to having humans not only compare physical stimuli, but also abstract concepts [21].

Most previous work on the method of paired comparisons has been through the lens of statistics. In such work the JND is modeled as a random variable and the statistical properties of Thurstone's method are studied [8]. Our problem corresponds to a simplified model of this problem which does not require any statistical assumptions, and is primarily from a combinatorial perspective.

Another context that captures the intuition of our model is that of designing a sporting tournament based on win/lose games. There, biases of a judge and unpredictable events can change the outcome of a game when the strengths of

---

[1] The JND is typically defined relative to x rather than as an absolute value. This is identical to absolute difference in the logarithmic scale and hence our discussion extends to this setting.

the players are close. Hence one cannot necessarily assume that the outcome is truly random in such a close call. It is clear that both restricting the influence of the faulty outcomes and reducing the total number of games required are important in this scenario, and hence exploring the tradeoff between the two is of interest. For convenience, in the rest of the paper we often use the terminology borrowed from this scenario.

1.1 Our Results

We first examine the simpler problem of finding only the maximum element. For this problem, we give a deterministic max-finding algorithm with error 2 using $2n^{3/2}$ comparisons. This contrasts with the method of paired comparisons, which makes $(n^2 - n)/2$ comparisons to achieve the same error. Using our algorithm recursively, we build deterministic algorithms with error k that require $O(n^{1+1/((3/4)\cdot 2^k - 1)})$ comparisons. We also give a lower bound of $\Omega(n^{1+1/(2^k-1)})$. The bounds are almost tight — the upper bound for our error-k algorithm is less than our lower bound for error-$(k-1)$ algorithms. We also give a linear-time randomized algorithm that achieves error 3 with probability at least $1 - 1/n^2$, showing that randomization greatly changes the complexity of the problem.

We then study the problem of sorting. For $k = 2$, we give an algorithm using $4 \cdot n^{3/2}$ comparisons. For general k, we show $O((n^{1+1/(3\cdot 2^{\lfloor k/2 \rfloor -1}-1)} + nk)\log n)$ comparisons is achievable, and we show a lower bound of $\Omega(n^{1+1/2^{k-1}})$ comparisons. When $k = O(1)$, or if only a single element of specific order needs to be selected, the $\log n$ factor disappears from our upper bound. Our lower bounds for selection depend on the order of the element that needs to be selected and interpolate between the lower bounds for max-finding and the lower bounds for sorting. For $k \geq 3$, our lower bound for finding the median (and also for sorting) is strictly larger than our upper bound for max-finding. For example, for $k = 3$ the lower bound for sorting is $\Omega(n^{5/4})$, whereas max-finding requires only $O(n^{6/5})$ comparisons.

Note that we achieve $\log\log n$ error for max-finding in $O(n)$ comparisons, and $2\log\log n$ error for sorting in $O(n \log n)$ comparisons. Standard methods using the same number of comparisons (e.g. a binary tournament tree, or Mergesort) can be shown to incur at least $\log n$ error. Also, all the algorithms we give are efficient in that their running times are of the same order as the number of comparisons they make.

The main idea in our deterministic upper bounds for both max-finding and selection is to develop efficient algorithms for a small value of k ($k = 2$), then for larger k show how to partition elements, recursively use algorithms for smaller k, then combine results. Achieving nearly tight results for max-finding requires in part relaxing the problem to that of finding a small *k-max-set*, or a set which is guaranteed to contain at least one element of value at least $x^* - k\delta$, where x^* is the maximum value of an element (we interchangeably use x^* to refer to an element of maximum value as well). It turns out we can find a k-max-set in a fewer number of comparisons than the lower bound for error-k max-finding algorithms. Exploiting this allows us to develop an efficient recursive max-finding algorithm. We note a similar approach of finding a small set of "good" elements was used by Borgstrom and Kosaraju [6] in the context of noisy binary search.

For our randomized max-finding algorithm, we use a type of tournament with random seeds at each level, in combination with random subsampling at each level of the tournament tree. By performing a round-robin tournament on the top few tournament players together with the subsampled elements, we obtain an element of value at least $x^* - 3\delta$ with high probability.

To obtain lower bounds we translate our problems into problems on directed graphs in which the goal is to ensure existence of short paths from a certain node to most other nodes. Using a comparison oracle that always prefers elements that had fewer wins in previous rounds, we obtain bounds on the minimum of edges that are required to create the paths of desired length. Such bounds are then translated back into bounds on the number of comparisons required to achieve specific error guarantees for the problems we consider. We are unaware of directly comparable techniques having been used before.

Some of the proofs are omitted from this extended abstract and appear in the full version of the paper.

1.2 Related Work

Handling noise in binary search procedures was first considered by Rényi [18] and by Ulam [22]. An algorithm for solving Ulam's game was proposed by Rivest et. al. in [19], where an adversarial comparator can err a bounded number of times. They gave an algorithm with query complexity $O(\log n)$ which succeeds if the number of adversarial errors is constant.

Yao and Yao [24] introduced the problem of sorting and of finding the maximal element in a sorting network when each comparison gate either returns the right answer or does not work at all. For finding the maximal element, they showed that it is necessary and sufficient to use $(e+1)(n-1)$ comparators when e comparators can be faulty. Ravikumar, Ganesan and Lakshmanan extended the model to arbitrary errors, showing that $O(en)$ comparisons are necessary and sufficient [17]. For sorting, Yao and Yao showed that $O(n \log n + en)$ gates are sufficient. In a different fault model, and with a different definition of a successful sort, Finocchi and Italiano [11] showed an $O(n \log n)$ time algorithm resilient to $(n \log n)^{1/3}$ faults. An improved algorithm handling $(n \log n)^{1/2}$ faults was later given by Finocchi, Grandoni and Italiano [10].

In the model where each comparison is incorrect with some probability p, Feige et al. [9] and Assaf and Upfal [2] give algorithms for several comparison problems, and [3,15] give algorithms for binary search. We refer the reader interested in the history of faulty comparison problems to a survey of Pelc [16].

We point out that some of the bounds we obtain appear similar to those known for max-finding, selection, and sorting in parallel in Valiant's model [23]. In particular, our bounds for max-finding are close to those obtained by Valiant for the parallel analogue of the problem (with the error used in place of parallel time) [23], and our lower bound of $\Omega(n^{1+1/(2^k-1)})$ for max-finding with error k is identical to a lower (and upper) bound given by Häggkvist and Hell [14] for merging two sorted arrays each of length n using a k-round parallel algorithm. Despite these similarities in bounds, our techniques are different, and we are not aware of any deep connections. For example, sorting in k parallel rounds $\Omega(n^{1+1/k})$ comparisons are required [5,13], whereas in our model, for constant k, we can sort with error k in $n^{1+1/2^{\Theta(k)}}$ comparisons. For a survey on parallel sorting algorithms, the reader is referred to [12].

2 Notation

Throughout this document we let x^* denote some x_i of the maximum value (if there are several such elements, we choose one arbitrarily). Furthermore, we use x_i interchangeably to refer to the both the ith element and its value, e.g. $x_i > x_j$ should be interpreted as $\mathrm{val}(x_i) > \mathrm{val}(x_j)$.

We assume $\delta = 1$ without loss of generality, since the problem with arbitrary $\delta > 0$ is equivalent to the problem with $\delta = 1$ and input values x_i/δ.

We say x *defeats* y when the comparator claims that x is larger than y (and we similarly use the phrase y *loses to* x). We say x is *k-greater* than y ($x \geq_k y$) if $x \geq y - k$. The term *k-smaller* is defined analogously. We say an element is a *k-max* of a set if it is k-greater than all other elements in the set, and a permutation $x_{\pi(1)}, \ldots, x_{\pi(n)}$ is *k-sorted* if $x_{\pi(i)} \geq_k x_{\pi(j)}$ for every $i > j$.

All logarithms throughout this document are base-2. For simplicity of presentation, we frequently omit floors and ceilings and ignore rounding errors when they have an insignificant effect on the bounds.

3 Max-finding

In this section we give deterministic and randomized algorithms for max-finding.

3.1 Deterministic Algorithms

We start by showing that the method of paired comparisons provides an optimal error guarantee, not just for max-finding, but also for sorting.

Theorem 1. *Sorting according to the number of wins in a round-robin tournament yields error* 2.

Proof. Let x, y be arbitrary elements with y strictly less than $x - 2$. For any z that y defeats, x also defeats z. Furthermore, x defeats y, and thus x has strictly more wins than y, implying y is placed lower in the sorted order.

Theorem 2. *No deterministic max-finding algorithm has error less than* 2.

Proof. Given three elements a, b, c, the comparator can claim $a > b > c > a$, making the elements indistinguishable. Without loss of generality, suppose A outputs a. Then the values could be $a = 0$, $b = 1$, $c = 2$, implying A has error 2.

In Figure 1 we give an error-2 algorithm for max-finding.

Algorithm A_2: `// Returns an element of value at least` x^*-2. `The value` $s>1$ `is a parameter which is by default` $\lceil\sqrt{n}\rceil$ `when not specified.`

1. Label all x_i as candidates.
2. **while** there are more than s candidate elements:
   (a) Pick an arbitrary set of s candidate elements and play them in a round-robin tournament. Let x have the most number of wins.
   (b) Compare x against all candidate elements and eliminate all elements that lose to x.
3. Play the final at most s candidate elements in a round-robin tournament and return the element with the most wins.

Fig. 1. The algorithm A_2 for finding a 2-max

Lemma 1. *The max-finding algorithm A_2 has error 2 and makes at most $ns+n^2/s$ comparisons. In particular, the number of comparisons is at most $2n^{3/2}$ for $s=\lceil\sqrt{n}\rceil$.*

Proof. We analyze the error in two cases. If x^* is never eliminated then x^* participates in Step 3. Theorem 1 then ensures that the final output is of value at least x^*-2. Otherwise, consider the iteration when x^* is eliminated. In this iteration, it must be the case that the x chosen in Step 2(b) has $x\geq x^*-1$, and thus any element with value less than x^*-2 was also eliminated in this iteration. In this case all future iterations only contain elements of value at least x^*-2, and so again the final output has value at least x^*-2. In each step at least $(s-1)/2$ elements are eliminated implying the given bound on the total number of comparisons.

The key recursion step of our general error max-finding is the algorithm 1-Cover of Lemma 3 which is based on A_2 and the following lemma.

Lemma 2. *There is a deterministic algorithm which makes $\binom{n}{2}$ comparisons and outputs a 1-max-set of size at most $\lceil\log n\rceil$.*

The algorithm performs a round-robin tournament and then iteratively greedily picks an element which defeats as many thus-far undefeated elements as possible. We now obtain 1-Cover by setting $s=\lceil\sqrt{n}\rceil/8$ in Figure 1, then returning the union of the x that were chosen in any iteration of Step 2(a), in addition to the output of Lemma 2 on the elements in the final tournament in Step 3.

Lemma 3. *There is an algorithm 1-Cover making $O(n^{3/2})$ comparisons which finds a 1-max-set of size at most $\sqrt{n}/4$ (for sufficiently large n).*

We are now ready to present our main algorithm for finding a k-max.

Theorem 3. *For every $3\leq k\leq\log\log n$, there exists an algorithm A_k finds a k-max element using $O(n^{1+1/((3/4)2^k-1)})$ comparisons.*

Corollary 1. *There is a max-finding algorithm using $O(n)$ comparisons with error $\log\log n$.*

Algorithm A_k: // **Returns a k-max for $k \geq 3$**

1. **return** $A_2(A'_{k-1}(x_1, x_2, \ldots, x_n))$

Algorithm A'_k: // **Returns a $(k-1)$-max set of size $O(n^{2^k/(3 \cdot 2^k - 4)})$ for $k \geq 2$**

1. **if** $k = 2$, return 1-COVER$(x_1, x_2, \ldots, x_n)$.
2. **else**
   (a) Equipartition the n elements into $t = b_k n^{2^{k-1}/(2^k - 4/3)}$ sets $S_1, \ldots, S_t$.
   (b) Recursively call A'_{k-1} on each set S_i to recover a $(k-2)$-max set T_i.
   (c) Return the output of 1-COVER with $\cup_{i=1}^t T_i$ as input.

Fig. 2. The algorithm A_k for finding a k-max based on a recursive algorithm A'_k for finding a $(k-1)$-max-set. The value b_k is $(1/2) \cdot (3/4)^{k-3}$ for $k \leq 10$ and $2^{-(3/4)^{k+5}(3 \cdot 2^{k-1} - 4)/4}$ otherwise.

3.2 Randomized Max-finding

We now show that randomization can significantly reduce the number of comparisons required to find an approximate maximum. We emphasize that although an adversary is not allowed to adaptively change the input values during the course of an algorithm's execution, the adversary can adaptively choose how to err when two elements are close. In particular, the classic randomized selection algorithm can take quadratic time since for an input with all equal values, the adversary can claim that the randomly chosen pivot is smaller than all other elements. Nevertheless, we show the following.

Theorem 4. *There exists a linear-time randomized algorithm which finds a 3-max with probability at least $1 - n^{-c}$ for any constant c and n large enough.*

Taking $c > 1$, and using the fact that the error of our algorithm can never be more than $n - 1$, this gives an algorithm which finds an element with expected value at least $x^* - 4$. The high-level idea of the algorithm is as follows. We randomly equipartition the elements into constant-sized sets. In each set we play a round-robin tournament and advance everyone who was not the absolute loser in their set. We also randomly subsample a set of players at each level of the tournament tree. We show that either (1) at some round of the tournament there is an abundance of elements with value at least $x^* - 1$, in which case at least one such element is subsampled with high probability, or (2) x^* makes it as one of the top few tournament players with high probability. We describe the properties of the tournament in Lemma 4. In Figure 3 we present the subroutine SAMPLEDTOURNAMENT for the tournament.

Lemma 4. SAMPLEDTOURNAMENT *outputs a W of size $O(n^{0.3} \log n)$ after $O(n)$ comparisons such that W is a 1-max-set with probability at least $1 - n^{-c}$.*

Theorem 4 follows immediately from Lemma 4: run the algorithm SAMPLEDTOURNAMENT, then return the winner of W in a round-robin tournament.

Algorithm SampledTournament: `// For constant` c `and` n `sufficiently large, returns a 1-max-set with probability at least` $1-n^{-c}$.

1. Initialize $N_0 \leftarrow \{x_1, \ldots, x_n\}$, $W \leftarrow \emptyset$, and $i \leftarrow 0$.
2. **if** $|N_i| \le n^{0.3}$, insert N_i into W and **return** W.
3. **else** randomly sample $n^{0.3}$ elements from N_i and insert them into W.
4. Randomly partition the elements in N_i into sets of size $80(c+2)$. In each set, perform a round-robin tournament to find the minimal element (the element with the fewest wins, with ties broken arbitrarily).
5. Let N_{i+1} contain all of N_i except for the minimal elements found in Step 4. That is, from each set of $80(c+2)$ elements of N_i, only one does not belong to N_{i+1}. Increment i and **goto** Step 2.

Fig. 3. The algorithm SampledTournament

4 Sorting and Selection

Definition 1. *Element x_j in the set $x_1, \ldots, x_n$ is of k-*order *i if there exists a partition S_1, S_2 of $[n]$ with $j \in S_1$, $|S_1| = i$, $x_\ell \le_k x_j$ for all $\ell \in S_1$, and $x_\ell \le_k x_{\ell'}$ for all $\ell \in S_1, \ell' \in S_2$. A k-*median *is an element of k-order $\lfloor n/2 \rfloor$.*

Our sorting and selection algorithms are based on the following lemma.

Lemma 5. *In a round-robin tournament on n elements, the element with the median number of wins has at least $(n-2)/4$ wins and at least $(n-2)/4$ losses.*

Algorithm C_k: `// Returns an element of` k`-order` i.

1. **if** $k \le 3$, sort the elements using B_2 then return the element with index i.
2. **else**
   (a) Set $k' = \lfloor k/2 \rfloor + 1$.
   (b) Equipartition the n elements into $t = b_{k'} n^{2^{k'-1}/(2^{k'}-4/3)}$ sets $S_1, \ldots, S_t$.
   (c) Recursively call C_{k-2} on each set S_i to obtain a $(k-2)$-median y_i.
   (d) Play the $y_1, \ldots, y_t$ in a round-robin tournament and let y be the element with the median number of wins.
   (e) Compare y with each of the other $n-1$ elements. If y defeats at least $(n-1)/2$ elements then let X_2 be a set of $d = (t-2)/4 \cdot ((n/t)-1)/2$ elements $(k-1)$-greater than y, and let X_1 be the set of at least $(n-1)/2 - d$ elements that y defeats which are not in X_2 (thus every $(x, x') \in (X_1 \cup \{y\}) \times X_2$ satisfies $x' \ge_k x$). If y defeats less than $(n-1)/2$ elements, X_1 and X_2 are defined symmetrically.
      i. **if** $|X_1| = i - 1$, return i.
      ii. **else if** $i \le |X_1|$, recursively find an element of k-order i in X_1.
      iii. **else** recursively find an element of k-order $(i - |X_1| - 1)$ in X_2.

Fig. 4. The algorithm C_k. The value $b_{k'}$ is chosen as in the algorithm $A'_{k'}$ (see Figure 2).

We can now obtain an error-2 sorting algorithm B_2 which needs only $4 \cdot n^{3/2}$ comparisons. The idea is to modify A_2 so that the x found in Step 2(a) of Figure 1 is a pivot in the sense of Lemma 5. We then compare this x against all elements and pivot into two sets, recursively sort each, then concatenate.

Lemma 6. *There is a deterministic sorting algorithm B_2 with error 2 that requires at most $4 \cdot n^{3/2}$ comparisons.*

At a high level our algorithm for k-order selection is similar to the classical selection algorithm of Blum et al. [4], in that in each step we try to find a pivot that allows us to recurse on a problem of geometrically decreasing size. In our scenario though, a good pivot must not only partition the input into nearly equal-sized chunks, but must itself be of $(k-2)$-order $c \cdot n$ for some constant $0 < c < 1$. The base case $k = 2$ can be solved by Lemma 6 since the element placed in position i of the sorted permutation is of 2-order i. Our algorithm is given in Figure 4.

Lemma 7. *For any $i \in [n]$ and $2 \leq k \leq 2 \log \log n$, the deterministic algorithm C_k finds an element of k-order i in $O(n^{1+1/(3 \cdot 2^{\lfloor k/2 \rfloor - 1} - 1)})$ comparisons.*

Theorem 5. *For any $2 \leq k \leq 2 \log \log n$, there is a deterministic sorting algorithm B_k with error k using $O((n^{1+1/(3 \cdot 2^{\lfloor k/2 \rfloor - 1} - 1)} + nk) \log n)$ comparisons. If $k = O(1)$, the number of comparisons reduces to $O(n^{1+1/(3 \cdot 2^{\lfloor k/2 \rfloor - 1} - 1)})$.*

Proof. We find a k-median using C_k, equipartition the elements into sets S_1, S_2 such that every element of S_2 is k-greater than every element of $S_1 \cup \{x\}$, recursively sort each partition, then concatenate the sorted results. The upper bound on the number of comparisons follows from the Master theorem (see [7, Theorem 4.1]), and correctness is immediate from the definition of a k-median.

5 Lower Bounds

Here we prove lower bounds against deterministic max-finding, sorting, and selection algorithms. In particular, we show that Theorem 3 and Theorem 5 achieve almost optimal trade-off between error and number of comparisons.

Lemma 8. *Suppose a deterministic algorithm A upon given n elements guarantees that after m comparisons it can list r elements, each of which is guaranteed to be k-greater than at least q elements. Then $m = \Omega(\max\{q^{1+1/(2^k-1)}, q \cdot r^{1/(2^{k-1})}\})$.*

Proof. We define a comparator that decides how to answer queries online in such a way that we can later choose values for the elements which are consistent with the given answers, while maximizing the error of the algorithm.

Let G_t be the comparison graph at time t. That is, G_t is a digraph whose vertices are the x_i and which contains the directed edge (x_i, x_j) if and only if before time t a comparison between x_i and x_j has been made, and the comparator has responded with "$x_i \geq x_j$". We denote the out-degree of x_i in G_t by $d_t(x_i)$. Assume that at time t the algorithm wants to compare some x_i and x_j. If $d_t(x_i) \geq d_t(x_j)$ then the comparator responds with "$x_j \geq x_i$", and it responds with "$x_i \geq x_j$" otherwise. (The response is arbitrary when $d_t(x_i) = d_t(x_j)$.) Let x be an element that is declared by A to be k-greater than at least q elements.

Let $y_i = \text{dist}(x, x_i)$, where dist gives the length of the shortest (directed) path in the final graph G_m. If no such path exists, we set $y_i = n$. After the algorithm is done, we define $\text{val}(x_i) = y_i$, We first claim that the values are consistent with the responses of the comparator. If for some pair of objects x_i, x_j the comparator has responded with "$x_i \geq x_j$", then G_m contains edge (x_i, x_j). This implies that for any x, $\text{dist}(x, x_j) \leq \text{dist}(x, x_i) + 1$, or $y_i \geq y_j - 1$. Therefore the answer "$x_i \geq x_j$" is consistent with the given values.

Consider the nodes x_i that x can reach via a path of length at most k. These are exactly the elements k-smaller than x, and thus there must be at least q of them. For $i \leq k$ let $S_i = \{x_j | y_j = i\}$ and $s_i = |S_i|$. We claim that for every $i \in [k]$, $m \geq s_i^2/(2s_{i-1}) - s_i/2$. For a node $u \in S_i$, let $\text{pred}(u)$ be a node in S_{i-1} such that the edge $(\text{pred}(u), u)$ is in the graph. For a node $v \in S_{i-1}$, let $S_{i,v} = \{u \in S_i \mid v = \text{pred}(u)\}$. Further, let $\text{d}_o(\text{pred}(u), u)$ be the out-degree of $\text{pred}(u)$ when the comparison between $\text{pred}(u)$ and u was made (as a result of which the edge was added to G_m). Note that for any distinct nodes $u, u' \in S_{i,v}$, $\text{d}_o(v, u) \neq \text{d}_o(v, u')$ since the out-degree of v grows each time an edge to a node in $S_{i,v}$ is added. This implies that

$$\sum_{u \in S_{i,v}} \text{d}_o(v, u) \geq \sum_{d \leq |S_{i,v}|-1} d = |S_{i,v}|(|S_{i,v}| - 1)/2 \ .$$

By the definition of our comparator, for every $u \in S_i$, $d_m(u) \geq \text{d}_o(\text{pred}(u), u)$. This implies that

$$m \geq \sum_{v \in S_{i-1}} \sum_{u \in S_{i,v}} d_m(u) \geq \sum_{v \in S_{i-1}} \frac{|S_{i,v}|(|S_{i,v}| - 1)}{2} = \frac{\sum_{v \in S_{i-1}} |S_{i,v}|^2 - |S_i|}{2} \ .$$

Using the inequality between the quadratic and arithmetic means,

$$\sum_{v \in S_{i-1}} |S_{i,v}|^2 \geq \left(\sum_{v \in S_{i-1}} |S_{i,v}| \right)^2 / |S_{i-1}| = s_i^2 / s_{i-1}.$$

This implies that $m \geq \frac{s_i^2}{2s_{i-1}} - \frac{s_i}{2}$.

We can therefore conclude that $s_i \leq \sqrt{(2m + s_i)s_{i-1}} \leq \sqrt{3ms_{i-1}}$ since $s_i \leq n \leq m$. By applying this inequality and using the fact that $s_0 = 1$ we obtain that $s_1^2/3 \leq m$ and $s_i \leq 3m \cdot (3m/s_1)^{2^{-(i-1)}}$ for $i > 1$. Since $\sum_{i \leq k} s_i \geq q + 1$, we thus find that $q \leq 12 \cdot m \cdot (3m/s_1)^{2^{-(k-1)}}$. This holds since either

1. $(3m/s_1)^{2^{-(k-1)}} > 1/2$ and then $12 \cdot m \cdot (3m/s_1)^{2^{-(k-1)}} \geq 6m > n$, or
2. $(3m/s_1)^{-2^{-(k-1)}} \leq 1/2$ and then $(3m/s_1)^{-2^{-i+1}}/(3m/s_1)^{-2^{-i}} = (3m/s_1)^{-2^{-i}} \leq (3m/s_1)^{-2^{-(k-1)}} \leq 1/2$ for $i \leq k-1$, where the penultimate inequality holds since $s_1 < 3m$. In this case

$$q - s_1 \leq \sum_{i=2}^{k} s_i \leq \sum_{i=2}^{k} (3m)(3m/s_1)^{-2^{-(i-1)}} \leq \sum_{i \leq k} 2^{i-k}(3m)(3m/s_1)^{-2^{-(k-1)}}$$
$$< 2(3m)^{1-2^{-(k-1)}} s_1^{2^{-(k-1)}}$$

If $s_1 \geq q/2$, then $m = \Omega(q^2)$ since $m \geq s_1^2/3$. Otherwise we have that $m \geq (q/4)^{1/(1-2^{-(k-1)})}/(3s_1^{1/(2^{(k-1)}-1)})$, implying

$$m = \Omega(\max\{s_1^2, q^{1/(1-2^{-(k-1)})}/s_1^{1/(2^{(k-1)}-1)}\}) = \Omega(q^{1+1/(2^k-1)})$$

where the final equality can be seen by making the two terms in the max equal.

Also, note that the choice of x amongst the r elements of the theorem statement was arbitrary, and that s_1 is just the out-degree of x. Let $s_{\min}$ be the minimum out-degree amongst the r elements. Then we trivially have $m \geq r \cdot s_{\min}$. Thus, if $s_{\min} \geq q/2$ then $m \geq qr/2$, and otherwise

$$m = \Omega(\max\{r \cdot s_{\min}, q^{1/(1-2^{-(k-1)})}/s_{\min}^{1/(2^{(k-1)}-1)}\}) = \Omega(q \cdot r^{1/(2^{k-1})})$$

where the final equality is again seen by making the two terms in the max equal.

From Lemma 8 we immediately obtain a lower bound for max-finding by setting $r = 1$, $q = n-1$, and for median-finding and sorting by setting $r = q = n/2$. In general, the sorting lower bound holds for k-order selection of the ith element for any $i = c \cdot n$ for constant $0 < c < 1$.

Theorem 6. *Every deterministic max-finding algorithm A with error k requires $\Omega(n^{1+1/(2^k-1)})$ comparisons.*

Theorem 7. *Every deterministic algorithm A which k-sorts n elements, or finds an element of k-order i for $i = c \cdot n$ with $0 < c < 1$ a constant, requires $\Omega(n^{1+1/2^{k-1}})$ comparisons.*

Theorem 6 implies the following, showing that Corollary 1 is tight.

Corollary 2. *Let A be a deterministic max-finding algorithm that makes $O(n)$ comparisons. Then A has error at least* $\log\log n - O(1)$.

References

1. Aggarwal, G., Ailon, N., Constantin, F., Even-Dar, E., Feldman, J., Frahling, G., Henzinger, M.R., Muthukrishnan, S., Nisan, N., Pál, M., Sandler, M., Sidiropoulos, A.: Theory research at Google. SIGACT News 39(2), 10–28 (2008)
2. Assaf, S., Upfal, E.: Fault tolerant sorting networks. SIAM J. Discrete Math 4(4), 472–480 (1991)

3. Ben-Or, M., Hassidim, A.: The bayesian learner is optimal for noisy binary search (and pretty good for quantum as well). In: FOCS, pp. 221–230 (2008)
4. Blum, M., Floyd, R.W., Pratt, V.R., Rivest, R.L., Tarjan, R.E.: Time bounds for selection. J. Comput. Syst. Sci. 7(4), 448–461 (1973)
5. Bollobás, B., Thomason, A.: Parallel sorting. Discrete Appl. Math. 6, 1–11 (1983)
6. Borgstrom, R.S., Kosaraju, S.R.: Comparison-based search in the presence of errors. In: STOC, pp. 130–136 (1993)
7. Cormen, T.H., Leiserson, C.E., Rivest, R.L., Stein, C.: Introduction to Algorithms, 2nd edn. MIT Press, Cambridge (2001)
8. David, H.A.: The Method of Paired Comparisons, 2nd edn. Charles Griffin & Company Limited (1988)
9. Feige, U., Raghavan, P., Peleg, D., Upfal, E.: Computing with noisy information. SIAM J. Comput. 23(5) (1994)
10. Finocchi, I., Grandoni, F., Italiano, G.F.: Optimal resilient sorting and searching in the presence of memory faults. In: Bugliesi, M., Preneel, B., Sassone, V., Wegener, I. (eds.) ICALP 2006. LNCS, vol. 4051, pp. 286–298. Springer, Heidelberg (2006)
11. Finocchi, I., Italiano, G.F.: Sorting and searching in the presence of memory faults (without redundancy). In: STOC, pp. 101–110 (2004)
12. Gasarch, W.I., Golub, E., Kruskal, C.P.: Constant time parallel sorting: an empirical view. J. Comput. Syst. Sci. 67(1), 63–91 (2003)
13. Häggkvist, R., Hell, P.: Parallel sorting with constant time for comparisons. SIAM J. Comput. 10(3), 465–472 (1981)
14. Häggkvist, R., Hell, P.: Sorting and merging in rounds. SIAM Journal on Algebraic and Discrete Methods 3(4), 465–473 (1982)
15. Karp, R.M., Kleinberg, R.: Noisy binary search and its applications. In: SODA, pp. 881–890 (2007)
16. Pelc, A.: Searching games with errors—fifty years of coping with liars. Theor. Comput. Sci. 270(1-2), 71–109 (2002)
17. Ravikumar, B., Ganesan, K., Lakshmanan, K.B.: On selecting the largest element in spite of erroneous information. In: Brandenburg, F.J., Wirsing, M., Vidal-Naquet, G. (eds.) STACS 1987. LNCS, vol. 247, pp. 88–99. Springer, Heidelberg (1987)
18. Rényi, A.: On a problem in information theory. Magyar Tud. Akad. Mat. Kutató Int. Közl 6, 505–516 (1962)
19. Rivest, R.L., Meyer, A.R., Kleitman, D.J., Winklmann, K., Spencer, J.: Coping with errors in binary search procedures. J. Comput. Sys. Sci. 20(3), 396–405 (1980)
20. Smith, S.M., Albaum, G.S.: Fundamentals of Marketing Research, 1st edn. Sage Publications, Thousand Oaks (2005)
21. Thurstone, L.L.: A law of comparative judgment. Psychological Review 34, 273–286 (1927)
22. Ulam, S.M.: Adventures of a Mathematician. Scribner's, New York (1976)
23. Valiant, L.G.: Parallelism in comparison problems. SIAM J. Comput. 4(3), 348–355 (1975)
24. Yao, A.C., Yao, F.F.: On fault-tolerant networks for sorting. SIAM J. Comput. 14(1), 120–128 (1985)

Fast FAST

Noga Alon[1,*], Daniel Lokshtanov[2], and Saket Saurabh[2,**]

[1] Tel Aviv University, Tel Aviv 69978, Israel and IAS, Princeton, NJ, 08540, USA
nogaa@tau.ac.il
[2] Department of Informatics, University of Bergen, N-5020 Bergen, Norway
{daniello,saket.saurabh}@ii.uib.no

Abstract. We present a randomized subexponential time, polynomial space parameterized algorithm for the k-Weighted Feedback Arc Set in Tournaments (k-FAST) problem. We also show that our algorithm can be derandomized by slightly increasing the running time. To derandomize our algorithm we construct a new kind of universal hash functions, that we coin *universal coloring families*. For integers m, k and r, a family $\mathcal{F}$ of functions from $[m]$ to $[r]$ is called a universal (m, k, r)-coloring family if for any graph G on the set of vertices $[m]$ with at most k edges, there exists an $f \in \mathcal{F}$ which is a proper vertex coloring of G. Our algorithm is the first non-trivial subexponential time parameterized algorithm outside the framework of bidimensionality.

1 Introduction

In a competition where everyone plays against everyone it is uncommon that the results are acyclic and hence one cannot rank the players by simply using a topological ordering. A natural ranking is one that minimizes the number of upsets, where an upset is a pair of players such that the lower ranked player beats the higher ranked one. The problem of finding such a ranking given the match outcomes is the Feedback Arc Set problem restricted to tournaments.

A *tournament* is a directed graph where every pair of vertices is connected by exactly one arc, and a *feedback arc set* is a set of arcs whose removal makes the graph acyclic. Feedback arc sets in tournaments are well studied, both from the combinatorial [16,17,20,21,22,28,31,32,35], statistical [29] and algorithmic [1,2,9,25,33,34] points of view. The problem has several applications - in psychology it occurs in relation to *ranking by paired comparisons*: here you wish to rank some items by an objective, but you don't have access to the objective function, only to pairwise comparisons of the objects in question. An example for this setting is measuring people's preferences for food. The weighted generalization of the problem, Weighted Feedback Arc Set in Tournaments

* Research supported in part by a USA Israeli BSF grant, by a grant from the Israel Science Foundation, by an ERC advanced grant, by NSF grant CCF 0832797 and by the Ambrose Monell Foundation.
** Research supported by the Research Council of Norway.

S. Albers et al. (Eds.): ICALP 2009, Part I, LNCS 5555, pp. 49–58, 2009.

is applied in *rank aggregation*: Here we are given several rankings of a set of objects, and we wish to produce a single ranking that on average is as consistent as possible with the given ones, according to some chosen measure of consistency. This problem has been studied in the context of voting [5,8], machine learning [7], and search engine ranking [14,15]. A natural consistency measure for rank aggregation is the number of pairs that occur in different order in the two rankings. This leads to *Kemeney-Young rank aggregation* [23,24], a special case of WEIGHTED FEEDBACK ARC SET IN TOURNAMENTS.

Unfortunately, the problem of finding a feedback arc set of minimum size in an unweighted tournament is NP-hard [2]. However, even the weighted version of the problem admits a polynomial time approximation scheme [25] and has been shown to be fixed parameter tractable [27]. One should note that the weighted generalization shown to admit a PTAS in [25] differs slightly from the one considered in this paper. We consider the following problem:

> k-WEIGHTED FEEDBACK ARC SET IN TOURNAMENTS (k-FAST)
> INSTANCE: A tournament $T = (V, A)$, a weight function $w : A \to \{x \in \mathbb{R} : x \geq 1\}$ and an integer k.
> QUESTION: Is there an arc set $S \subseteq A$ such that $\sum_{e \in S} w(e) \leq k$ and $T \setminus S$ is acyclic?

The fastest previously known parameterized algorithm for k-FAST by Raman and Saurabh [27] runs in time $O(2.415^k \cdot k^{4.752} + n^{O(1)})$, and it was an open problem of Guo et al. [19] whether k-FAST can be solved in time $2^k \cdot n^{O(1)}$. We give a randomized and a deterministic algorithm both running in time $2^{O(\sqrt{k}\log^2 k)} + n^{O(1)}$. Our algorithms run in subexponential time, a trait uncommon to parameterized algorithms. In fact, to the authors best knowledge the only parameterized problems for which non-trivial subexponential time algorithms are known are *bidimensional* problems in planar graphs or graphs excluding a certain fixed graph H as a minor [10,11,13].

Our randomized algorithm is based on a novel version of the color coding technique initiated in [4] combined with a divide and conquer algorithm and a k^2 kernel for the problem, due to Dom et al. [12]. In order to derandomize our algorithm we construct a new kind of universal hash functions, that we coin *universal coloring families*. For integers m, k and r, a family $\mathcal{F}$ of functions from $[m]$ to $[r]$ is called a universal (m, k, r)-coloring family if for any graph G on the set of vertices $[m]$ with at most k edges, there exists an $f \in \mathcal{F}$ which is a proper vertex coloring of G. In the last section of the paper we give an explicit construction of a $(10k^2, k, O(\sqrt{k}))$-coloring family $\mathcal{F}$ of size $|\mathcal{F}| \leq 2^{\tilde{O}(\sqrt{k})}$ and an explicit universal $(n, k, O(\sqrt{k}))$-coloring family $\mathcal{F}$ of size $|\mathcal{F}| \leq 2^{\tilde{O}(\sqrt{k})} \log n$. We believe that these constructions can turn out to be useful to solve other edge subset problems in dense graphs.

2 Preliminaries

For an arc weighted tournament we define the weight function $w^* : V \times V \to \mathbb{R}$ such that $w^*(u, v) = w(uv)$ if $uv \in A$ and 0 otherwise. Given a directed graph

1. Perform a data reduction to obtain a tournament T' of size $O(k^2)$.
2. Let $t = \sqrt{8k}$. Color the vertices of T' uniformly at random with colors from $\{1, \ldots, t\}$.
3. Let A_c be the set of arcs whose endpoints have different colors. Find a minimum weight feedback arc set contained in A_c, or conclude that no such feedback arc set exists.

Fig. 1. Outline of the algorithm for k-FAST

$D = (V, A)$ and a set F of arcs in A define $D\{F\}$ to be the directed graph obtained from D by reversing all arcs of F. In our arguments we will need the following characterization of minimal feedback arc sets in directed graphs.

Proposition 1. *Let $D = (V, A)$ be a directed graph and F be a subset of A. Then F is a minimal feedback arc set of D if and only if F is a minimal set of arcs such that $D\{F\}$ is a directed acyclic graph.*

Given a minimal feedback arc set F of a tournament T, the ordering σ corresponding to F is the unique topological ordering of $T\{F\}$. Conversely, given an ordering σ of the vertices of T, the feedback arc set F corresponding to σ is the set of arcs whose endpoint appears before their startpoint in σ. The cost of an arc set F is $\sum_{e \in F} w(e)$ and the cost of a vertex ordering σ is the cost of the feedback arc set corresponding to σ.

For a pair of integer row vectors $\hat{p} = [p_1, \ldots, p_t]$, $\hat{q} = [q_1, \ldots, q_t]$ we say that $\hat{p} \leq \hat{q}$ if $p_i \leq q_i$ for all i. The transpose of a row vector $\hat{p}$ is denoted by $\hat{p}^\dagger$. The t-sized vector $\hat{e}$ is $[1, 1, \ldots, 1]$, $\hat{0}$ is $[0, 0, \ldots, 0]$ and $\hat{e}_i$ is the t-sized vector with all entries 0 except for the i'th which is 1. Let $\tilde{O}(\sqrt{k})$ denote, as usual, any function which is $O(\sqrt{k}(\log k)^{O(1)})$. For any positive integer m put $[m] = \{1, 2, \ldots, m\}$.

3 Color and Conquer

Our algorithm consists of three steps. In the first step we reduce the instance to a problem kernel with at most $O(k^2)$ vertices, showing how to efficiently reduce the input tournament into one with $O(k^2)$ vertices, so that the original tournament has a feedback arc set of weight at most k, if and only if the new one has such a set. In the second step we randomly color the vertices of our graph with $t = \sqrt{8k}$ colors, and define the arc set A_c to be the set of arcs whose endpoints have different colors. In the last step the algorithm checks whether there is a weight k feedback arc set $S \subseteq A_c$. A summary of the algorithm is given in Figure 1.

3.1 Kernelization

For the first step of the algorithm we use the kernelization algorithm provided by Dom et al. [12]. They only show that the data reduction is feasible for the unweighted case, while in fact, it works for the weighted case as well. For completeness we provide a short proof of this. A triangle in what follows means a directed cyclic triangle.

Lemma 1. k-FAST *has a kernel with* $O(k^2)$ *vertices.*

Proof. We give two simple reduction rules.

1. If an arc e is contained in at least $k+1$ triangles reverse the arc and reduce k by $w(e)$.
2. If a vertex v is not contained in any triangle, delete v from T.

The first rule is safe because any feedback arc set that does not contain the arc e must contain at least one arc from each of the $k+1$ triangles containing e and thus must have weight at least $k+1$. The second rule is safe because the fact that v is not contained in any triangle implies that all arcs between $N^-(v)$ and $N^+(v)$ are oriented from $N^-(v)$ to $N^+(v)$. Hence for any feedback arc set S_1 of $T[N^-(v)]$ and feedback arc set S_2 of $T[N^+(v)]$, $S_1 \cup S_2$ is a feedback arc set of T.

Finally we show that any reduced yes instance T has at most $k(k+2)$ vertices. Let S be a feedback arc set of T with weight at most k. The set S contains at most k arcs, and for every arc $e \in S$, aside from the two endpoints of e, there are at most k vertices that are contained in a triangle containing e, because otherwise the first rule would have applied. Since every triangle in T contains an arc of S and every vertex of T is in a triangle, T has at most $k(k+2)$ vertices. □

3.2 Probability of a Good Coloring

We now proceed to analyze the second step of the algorithm. What we aim for, is to show that if T does have a feedback arc set S of weight at most k, then the probability that S is a subset of A_c is at least $2^{-c\sqrt{k}}$ for some fixed constant c. We show this by showing that if we randomly color the vertices of a k edge graph G with $t = \sqrt{8k}$ colors, then the probability that G has been properly colored is at least $2^{-c\sqrt{k}}$.

Lemma 2. *If a graph on q edges is colored randomly with $\sqrt{8q}$ colors then the probability that G is properly colored is at least* $(2e)^{-\sqrt{q/8}}$.

Proof. Arrange the vertices of the graph by repeatedly removing a vertex of lowest degree. Let $d_1, d_2, \ldots, d_s$ be the degrees of the vertices when they have been removed. Then for each i, $d_i(s-i+1) \leq 2q$, since when vertex i is removed each vertex had degree at least d_i. Furthermore, $d_i \leq s-i$ for all i, since the degree of the vertex removed can not exceed the number of remaining vertices at that point. Thus $d_i \leq \sqrt{2q}$ for all i. In the coloring, consider the colors of each vertex one by one starting from the last one, that is vertex number s. When vertex number i is colored, the probability that it will be colored by a color that differs from all those of its d_i neighbors following it is at least

$(1 - \frac{d_i}{\sqrt{8q}}) \geq (2e)^{-d_i/\sqrt{8q}}$ because $\sqrt{8q} \geq 2d_i$. Hence the probability that G is properly colored is at least

$$\prod_{i=1}^{s}(1-\frac{d_i}{\sqrt{8q}}) \geq \prod_{i=1}^{s}(2e)^{-d_i/\sqrt{8q}} = (2e)^{-\sqrt{q/8}}. \qquad \square$$

3.3 Solving a Colored Instance

Given a t-colored tournament T, we will say that an arc set F is colorful if no arc in F is monochromatic. An ordering σ of T is colorful if the feedback arc set corresponding to σ is colorful. An optimal colorful ordering of T is a colorful ordering of T with minimum cost among all colorful orderings. We now give an algorithm that takes a t-colored arc weighted tournament T as input and finds a colorful feedback arc set of minimum weight, or concludes that no such feedback arc set exists.

Observation 1. *Let $T = (V_1 \cup V_2 \cup \ldots \cup V_t, A)$ be a t-colored tournament. There exists a colorful feedback arc set of T if and only if $T[V_i]$ induces an acyclic tournament for every i.*

We say that a colored tournament T is *feasible* if $T[V_i]$ induces an acyclic tournament for every i. Let $n_i = |V_i|$ for every i and let $\hat{n}$ be the vector $[n_1, n_2 \ldots n_t]$. Let $\sigma = v_1 v_2 \ldots v_n$ be the ordering of V corresponding to a colorful feedback arc set F of T. For every color class V_i of T, let $v_i^1 v_i^2 \ldots v_i^{n_i}$ be the order in which the vertices of V_i appear according to σ. Observe that since F is colorful, $v_i^1 v_i^2 \ldots v_i^{n_i}$ must be the unique topological ordering of $T[V_i]$. We exploit this to give a dynamic programming algorithm for the problem.

Lemma 3. *Given a feasible t-colored tournament T, we can find a minimum weight colorful feedback arc set in $O(t \cdot n^{t+1})$ time and $O(n^t)$ space.*

Proof. For an integer $x \geq 1$, define $S_x^i = \{v_1^i, \ldots, v_x^i\}$ and $S_0^i = \emptyset$. Given an integer vector $\hat{p}$ of length t in which the i'th entry is between 0 and n_i, let $T(\hat{p})$ be $T[S_{p_1}^1 \cup S_{p_2}^2 \ldots \cup S_{p_t}^t]$. Observe that for any ordering $\sigma = v_1 v_2 \ldots v_n$ of V corresponding to a colorful feedback arc set F of T and any integer x there is a $\hat{p}$ such that $\{v_1, \ldots, v_x\} = S_{p_1}^1 \cup S_{p_2}^2 \ldots \cup S_{p_t}^t$.

For a feasible t-colored tournament T, let $\textsc{Fas}(T)$ be the weight of the minimum weight colorful feedback arc set of T. Notice that if a t-colored tournament T is feasible then so are all induced subtournaments of T, and hence the function $\textsc{Fas}$ is well defined on all induced subtournaments of T. We proceed to prove that the following recurrence holds for $\textsc{Fas}(T(\hat{p}))$.

$$\textsc{Fas}(T(\hat{p})) = \min_{i\,:\,\hat{p}_i>0} (\textsc{Fas}(T(\hat{p} - \hat{e}_i)) + \sum_{u \in V(T(\hat{p}))} w^*(v_{\hat{p}_i}^i, u)) \qquad (1)$$

The idea behind Recurrence 1 is to try all possible candidates for the last vertex v of an optimal ordering of $T(\hat{p})$. For every i the vertex $v_{p_i}^i$ is the only candidate for v with color i. First we prove that the left hand side is at most the right hand side. Let i be the integer that minimizes the right hand side. Taking the optimal ordering of $T(\hat{p}-\hat{e}_i)$ and appending it with $v^i_{\hat{p}_i}$ gives an ordering of $T(\hat{p})$ with cost at most $\text{FAS}(T(\hat{p}-\hat{e}_i)) + \sum_{u \in V(T(\hat{p}))} w^*(v^i_{\hat{p}_i}, u)$.

To prove that the right hand side is at most the left hand side, take an optimal colorful ordering σ of $T(\hat{p})$ and let v be the last vertex of this ordering. There is an i such that $v = v^i_{\hat{p}_i}$. Thus σ restricted to $V(T(\hat{p}-\hat{e}_i))$ is a colorful ordering of $T(\hat{p}-\hat{e}_i)$ and the total weight of the edges with startpoint in v and endpoint in $V(T(\hat{p}-\hat{e}_i))$ is exactly $\sum_{u \in V(T(\hat{p}))} w^*(v^i_{\hat{p}_i}, u)$. Thus the cost of σ is at least the value of the right hand side of the inequality, completing the proof.

Recurrence 1 naturally leads to a dynamic programming algorithm for the problem. We build a table containing $\text{FAS}(T(\hat{p}))$ for every $\hat{p}$. There are $O(n^t)$ table entries, for each entry it takes $O(nt)$ time to compute it giving the $O(t{\cdot}n^{t+1})$ time bound. □

In fact, the algorithm provided in Lemma 3 can be made to run slightly faster by pre-computing the value of $\sum_{u \in V(T(\hat{p}))} w^*(v^i_{\hat{p}_i}, u))$ for every $\hat{p}$ and i using dynamic programming, and storing it in a table. This would let us reduce the time to compute a table entry using Recurrence 1 from $O(nt)$ to $O(t)$ yielding an algorithm that runs in time and space $O(t \cdot n^t)$.

Lemma 4. k-FAST *(for a tournament of size $O(k^2)$) can be solved in expected time $2^{O(\sqrt{k}\log k)}$ and $2^{O(\sqrt{k}\log k)}$ space.*

Proof. Our algorithm proceeds as described in Figure 1. The correctness of the algorithm follows from Lemma 3. Combining Lemmata 1, 2, 3 yields an expected running time of $O((2e)^{\sqrt{k/8}}) \cdot O(\sqrt{8k} \cdot (k^2+2k)^{1+\sqrt{8k}}) \le 2^{O(\sqrt{k}\log k)}$ for finding a feedback arc set of weight at most k if one exists. The space required by the algorithm is $O((k^2+2k)^{1+\sqrt{8k}}) \le 2^{O(\sqrt{k}\log k)}$. □

The dynamic programming algorithm from Lemma 3 can be turned into a divide and conquer algorithm that runs in polynomial space, at a small cost in the running time.

Lemma 5. *Given a feasible t-colored tournament T, we can find a minimum weight colorful feedback arc set in time $O(n^{1+(t+2)\cdot\log n})$ in polynomial space.*

Proof. By expanding Recurrence (1) $\lfloor n/2 \rfloor$ times and simplifying the right hand side we obtain the following recurrence.

$$\text{FAS}(T(\hat{p})) = \min_{\substack{\hat{q} \ge \hat{0} \\ \hat{q}^\dagger \cdot \hat{e} = \lceil n/2 \rceil}} \{\text{FAS}(T(\hat{q})) + \text{FAS}(T \setminus V(T(\hat{q}))) + \sum_{\substack{u \in V(T(\hat{q})) \\ v \notin V(T(\hat{q}))}} w^*(v,u)\} \quad (2)$$

Recurrence 2 immediately yields a divide and conquer algorithm for the problem. Let $\mathcal{T}(n)$ be the running time of the algorithm restricted to a subtournament of T with n vertices. For a particular vector $\hat{q}$ it takes at most n^2 time to find the value of $\sum_{u \in V(T(\hat{q})), v \notin V(T(\hat{q}))} w^*(v,u)$. It follows that $\mathcal{T}(n) \le n^{t+2} \cdot 2 \cdot \mathcal{T}(n/2) \le 2^{\log n} \cdot n^{(t+2)\cdot\log n} = n^{1+(t+2)\cdot\log n}$. □

Theorem 1. k-FAST *(for a tournament of size* $O(k^2)$*) can be solved in expected time* $2^{O(\sqrt{k}\log^2 k)}$ *and polynomial space. Therefore,* k-FAST *for a tournament of size* n *can be solved in expected time* $2^{O(\sqrt{k}\log^2 k)} + n^{O(1)}$ *and polynomial space.*

4 Derandomization with Universal Coloring Families

For integers m, k and r, a family $\mathcal{F}$ of functions from $[m]$ to $[r]$ is called a universal (m, k, r)-coloring family if for any graph G on the set of vertices $[m]$ with at most k edges, there exists an $f \in \mathcal{F}$ which is a proper vertex coloring of G. An explicit construction of a $(10k^2, k, O(\sqrt{k}))$-coloring family can replace the randomized coloring step in the algorithm for k-FAST. In this section, we provide such a construction.

Theorem 2. *There exists an explicit universal* $(10k^2, k, O(\sqrt{k}))$*-coloring family* $\mathcal{F}$ *of size* $|\mathcal{F}| \le 2^{\tilde{O}(\sqrt{k})}$.

For simplicity we omit all floor and ceiling signs whenever these are not crucial. We make no attempt to optimize the absolute constants in the $\tilde{O}(\sqrt{k})$ or in the $O(\sqrt{k})$ notation. Whenever this is needed, we assume that k is sufficiently large.

Proof. Let $\mathcal{G}$ be an explicit family of functions g from $[10k^2]$ to $[\sqrt{k}]$ so that every coordinate of g is uniformly distributed in $[\sqrt{k}]$, and every two coordinates are pairwise independent. There are known constructions of such a family $\mathcal{G}$ with $|\mathcal{G}| \le k^{O(1)}$. Indeed, each function g represents the values of $10k^2$ pairwise independent random variables distributed uniformly in $[\sqrt{k}]$ in a point of a small sample space supporting such variables; a construction is given, for example, in [3]. The family $\mathcal{G}$ is obtained from the family of all linear polynomials over a finite field with some $k^{O(1)}$ elements, as described in [3].

We can now describe the required family $\mathcal{F}$. Each $f \in \mathcal{F}$ is described by a subset $T \subset [10k^2]$ of size $|T| = \sqrt{k}$ and by a function $g \in \mathcal{G}$. For each $i \in [10k^2]$, the value of $f(i)$ is determined as follows. Suppose $T = \{i_1, i_2, \ldots, i_{\sqrt{k}}\}$, with $i_1 < i_2 < \ldots < i_{\sqrt{k}}$. If $i = i_j \in T$, define $f(i) = \sqrt{k} + j$. Otherwise, $f(i) = g(i)$. Note that the range of f is of size $\sqrt{k} + \sqrt{k} = 2\sqrt{k}$, and the size of $\mathcal{F}$ is at most

$$\binom{10k^2}{\sqrt{k}}|\mathcal{G}| \le \binom{10k^2}{\sqrt{k}}k^{O(1)} \le 2^{O(\sqrt{k}\log k)} \le 2^{\tilde{O}(\sqrt{k})}.$$

To complete the proof we have to show that for every graph G on the set of vertices $[10k^2]$ with at most k edges, there is an $f \in \mathcal{F}$ which is a proper vertex coloring of G. Fix such a graph G.

The idea is to choose T and g in the definition of the function f that will provide the required coloring for G as follows. The function g is chosen at random in $\mathcal{G}$, and is used to properly color all but at most $\sqrt{k}$ edges. The set T is chosen to contain at least one endpoint of each of these edges, and the vertices in the set T will be re-colored by a unique color that is used only once by f. Using the properties of $\mathcal{G}$ we now observe that with positive probability the number of edges of G which are monochromatic is bounded by $\sqrt{k}$.

Claim. If the vertices of G are colored by a function g chosen at random from $\mathcal{G}$, then the expected number of monochromatic edges is $\sqrt{k}$.

Proof. Fix an edge e in the graph G and $j \in [\sqrt{k}]$. As g maps the vertices in a pairwise independent manner, the probability that both the end points of e get mapped to j is precisely $\frac{1}{(\sqrt{k})^2}$. There are $\sqrt{k}$ possibilities for j and hence the probability that e is monochromatic is given by $\frac{\sqrt{k}}{(\sqrt{k})^2} = \frac{1}{\sqrt{k}}$. Let X be the random variable denoting the number of monochromatic edges. By linearity of expectation, the expected value of X is $k \cdot \frac{1}{\sqrt{k}} = \sqrt{k}$. □

Returning to the proof of the theorem, observe that by the above claim, with positive probability, the number of monochromatic edges is upper bounded by $\sqrt{k}$. Fix a $g \in \mathcal{G}$ for which this holds and let $T = \{i_1, i_2, \ldots, i_{\sqrt{k}}\}$ be a set of $\sqrt{k}$ vertices containing at least one endpoint of each monochromatic edge. Consider the function f defined by this T and g. As mentioned above f colors each of the vertices in T by a unique color, which is used only once by f, and hence we only need to consider the coloring of $G \setminus T$. However all edges in $G \setminus T$ are properly colored by g and f coincides with g on $G \setminus T$. Hence f is a proper coloring of G, completing the proof of the theorem. □

Remarks

- Each universal $(n, k, O(\sqrt{k}))$-coloring family must also be an $(n, \sqrt{k}, O(\sqrt{k}))$-hashing family, as it must contain, for every set S of $\sqrt{k}$ vertices in $[n]$, a function that maps the elements of S in a one-to-one manner, since these vertices may form a clique that has to be properly colored by a function of the family. Therefore, by the known bounds for families of hash functions (see, e.g., [26]), each such family must be of size at least $2^{\tilde{\Omega}(\sqrt{k})} \log n$.

Although the next result is not required for our results on the feedback arc set problem, we present it here as it may be useful in similar applications.

Theorem 3. *For any $n > 10k^2$ there exists an explicit universal $(n, k, O(\sqrt{k}))$-coloring family $\mathcal{F}$ of size $|\mathcal{F}| \leq 2^{\tilde{O}(\sqrt{k})} \log n$.*

Proof. Let $\mathcal{F}_1$ be an explicit $(n, 2k, 10k^2)$-family of hash functions from $[n]$ to $10k^2$ of size $|\mathcal{F}_1| \leq k^{O(1)} \log n$. This means that for every set $S \subset [n]$ of size at most $2k$ there is an $f \in \mathcal{F}_1$ mapping S in a one-to-one fashion. The existence of such a family is well known, and follows, for example, from constructions of small spaces supporting n nearly pairwise independent random variables taking values in $[10k^2]$. Let $\mathcal{F}_2$ be an explicit universal $(10k^2, k, O(\sqrt{k}))$-coloring family, as described in Theorem 2. The required family $\mathcal{F}$ is simply the family of all compositions of a function from $\mathcal{F}_2$ followed by one from $\mathcal{F}_1$. It is easy to check that $\mathcal{F}$ satisfies the assertion of Theorem 3. □

Finally, combining the algorithm from Theorem 1 with the universal coloring family given by Theorem 2 yields a deterministic subexponential time polynomial space algorithm for k-FAST.

Theorem 4. k-FAST *can be solved in time* $2^{\tilde{O}(\sqrt{k})} + n^{O(1)}$ *and polynomial space.*

5 Concluding Remarks

In this article, we have shown that k-FAST can be solved in time $2^{\tilde{O}(\sqrt{k})} + n^{O(1)}$ and polynomial space. To achieve this we introduced a new variant of randomized color coding, and showed that this approach could be derandomized with an explicit construction of universal coloring families. We find it surprising that the problem admits a subexponential time parameterized algorithm, as even the existence of a $2^k \cdot n^{O(1)}$ time algorithm was an open problem until now.

At the end of the introduction of the paper in which it was proved that FEEDBACK ARC SET IN TOURNAMENTS admits a PTAS [25], Mathieu and Schudy write "*We can feel lucky that the FAS problem on tournaments turns out to be so easy as to have an approximation scheme: In contrast to Theorem* 1, *the related problem of feedback vertex set is hard to approximate even on tournaments.*" Interestingly, a similar remark can be made in our setting - a simple reduction from VERTEX COVER [30] shows that k-FEEDBACK VERTEX SET in tournaments can not be solved in subexponential time unless the Exponential Time Hypothesis [6,18] fails.

The results of Section 4 can be extended to universal coloring families of uniform hypergraphs. These families can also be useful in tackling several parameterized algorithmic problems. The details will appear in the full version of this paper.

References

1. Ailon, N., Charikar, M., Newman, A.: Aggregating inconsistent information: ranking and clustering. In: ACM Symposium on Theory of Computing (STOC), pp. 684–693 (2005)
2. Alon, N.: Ranking tournaments. SIAM J. Discrete Math. 20, 137–142 (2006)
3. Alon, N., Babai, L., Itai, A.: A fast and simple randomized parallel algorithm for the maximal independent set problem. Journal of Algorithms 7, 567–583 (1986)
4. Alon, N., Yuster, R., Zwick, U.: Color-coding. J. Assoc. Comput. Mach. 42, 844–856 (1995)
5. Borda, J.: Mémoire sur les élections au scrutin, Histoire de l'Académie Royale des Sciences (1781)
6. Cai, L., Juedes, D.W.: On the existence of subexponential parameterized algorithms. J. Comput. Syst. Sci. 67, 789–807 (2003)
7. Cohen, W.W., Schapire, R.E., Singer, Y.: Learning to order things. In: Advances in neural information processing systems (NIPS), pp. 451–457 (1997)
8. Condorcet, M.: Essai sur l'application de l'analyse à la probabilité des décisions rendues à la pluralité des voix (1785)
9. Coppersmith, D., Fleischer, L., Rudra, A.: Ordering by weighted number of wins gives a good ranking for weighted tournaments. In: ACM-SIAM Symposium on Discrete Algorithms (SODA), pp. 776–782 (2006)
10. Demaine, E.D., Fomin, F.V., Hajiaghayi, M.T., Thilikos, D.M.: Subexponential parameterized algorithms on bounded-genus graphs and -minor-free graphs. J. ACM 52, 866–893 (2005)

11. Demaine, E.D., Hajiaghayi, M.: The bidimensionality theory and its algorithmic applications. Comput. J. 51, 292–302 (2008)
12. Dom, M., Guo, J., Hüffner, F., Niedermeier, R., Truß, A.: Fixed-parameter tractability results for feedback set problems in tournaments. In: Calamoneri, T., Finocchi, I., Italiano, G.F. (eds.) CIAC 2006. LNCS, vol. 3998, pp. 320–331. Springer, Heidelberg (2006)
13. Dorn, F., Fomin, F.V., Thilikos, D.M.: Subexponential parameterized algorithms. Computer Science Review 2, 29–39 (2008)
14. Dwork, C., Kumar, R., Naor, M., Sivakumar, D.: Rank aggregation revisited (2001)
15. Dwork, C., Kumar, R., Naor, M., Sivakumar, D.: Rank aggregation methods for the web. In: WWW10 (2001)
16. Erdös, P., Moon, J.W.: On sets on consistent arcs in tournaments. Canadian Mathematical Bulletin 8, 269–271 (1965)
17. Fernandez de la Vega, W.: On the maximal cardinality of a consistent set of arcs in a random tournament. J. Combinatorial Theory, Ser. B 35, 328–332 (1983)
18. Flum, J., Grohe, M.: Parameterized Complexity Theory. Texts in Theoretical Computer Science. An EATCS Series. Springer, Secaucus (2006)
19. Guo, J., Moser, H., Niedermeier, R.: Iterative compression for exactly solving np-hard minimization problems. In: Algorithmics of Large and Complex Networks. LNCS, Springer, Heidelberg (to appear, 2008)
20. Jung, H.: On subgraphs without cycles in tournaments. Combinatorial Theory and its Applications II, pp. 675–677 (1970)
21. Reid, K.B.: On sets of arcs containing no cycles in tournaments. Canad. Math Bulletin 12, 261–264 (1969)
22. Reid, K.D., Parker, E.T.: Disproof of a conjecture of Erdös and Moseron tournaments. J. Combin. Theory 9, 225–238 (1970)
23. Kemeny, J.: Mathematics without numbers. Daedalus 88, 571–591 (1959)
24. Kemeny, J., Snell, J.: Mathematical models in the social sciences, Blaisdell (1962)
25. Kenyon-Mathieu, C., Schudy, W.: How to rank with few errors. In: STOC 2007, pp. 95–103. ACM Press, New York (2007)
26. Nilli, A.: Perfect hashing and probability. Combinatorics, Probability and Computing 3, 407–409 (1994)
27. Raman, V., Saurabh, S.: Parameterized algorithms for feedback set problems and their duals in tournaments. Theoretical Computer Science 351, 446–458 (2006)
28. Seshu, S., Reed, M.: Linear Graphs and Electrical Networks. Addison-Wesley, Reading (1961)
29. Slater, P.: Inconsistencies in a schedule of paired comparisons. Biometrika 48, 303–312 (1961)
30. Speckenmeyer, E.: On feedback problems in diagraphs. In: Nagl, M. (ed.) WG 1989. LNCS, vol. 411, pp. 218–231. Springer, Heidelberg (1990)
31. Spencer, J.: Optimal ranking of tournaments. Networks 1, 135–138 (1971)
32. Spencer, J.: Optimal ranking of unrankable tournaments. Period. Math. Hungar. 11, 131–144 (1980)
33. van Zuylen, A.: Deterministic approximation algorithms for ranking and clusterings, Tech. Report 1431, Cornell ORIE (2005)
34. van Zuylen, A., Hegde, R., Jain, K., Williamson, D.P.: Deterministic pivoting algorithms for constrained ranking and clustering problems. In: ACM-SIAM Symposium on Discrete Algorithms (SODA), pp. 405–414 (2007)
35. Younger, D.: Minimum feedback arc sets for a directed graph. IEEE Trans. Circuit Theory 10, 238–245 (1963)

Bounds on the Size of Small Depth Circuits for Approximating Majority

Kazuyuki Amano

Dept of Comput Sci, Gunma Univ, Tenjin 1-5-1, Kiryu, Gunma 376-8515, Japan
amano@cs.gunma-u.ac.jp

Abstract. In this paper, we show that for every constant $0 < \epsilon < 1/2$ and for every constant $d \geq 2$, the minimum size of a depth d Boolean circuit that ϵ-approximates Majority function on n variables is $\exp(\Theta(n^{1/(2d-2)}))$. The lower bound for every $d \geq 2$ and the upper bound for $d = 2$ have been previously shown by O'Donnell and Wimmer [ICALP'07], and the contribution of this paper is to give a matching upper bound for $d \geq 3$.

1 Introduction and Results

An investigation of the construction of small circuits for computing Majority function in various computational models has attracted many researchers for a long time. Interesting positive results (e.g., for comparator networks [3] or for monotone formulae [8]) as well as some negative results (e.g., for constant depth circuits [5]) have been obtained so far.

There also have been many researches on the construction of circuits to *approximate* the majority function. In this paper, we consider this problem in the model of constant depth circuits consisting of AND and OR gates with unbounded fan-in.

It seems that there are two major notions of "approximate-Majority" in this model. The first meaning of "approximate-Majority" is to compute a function that coincides with the majority function on every point including at least 2/3 fraction of 1's, or at most 1/3 fraction of 1's. The complexity of approximate-Majority of this notion is closely related to the complexity of probabilistic computations, and has been widely investigated (see e.g., [1,2,9].)

The second meaning of "approximate-Majority", which we focus on in this paper, is to compute a function that disagrees with the majority function on at most ϵ fraction of all points. We call such a function an ϵ-approximation of the majority function.

O'Donnell and Wimmer [7] first investigated this problem and obtained the following: (i) For every constant $0 < \epsilon < 1/2$ and every constant $d \geq 2$, any depth-d circuit computing an ϵ-approximation of the majority function on n variables has size $\exp(\Omega(n^{1/(2d-2)}))$, and (ii) When $d = 2$, this lower bound is optimal up to a constant factor in the exponent. The lower bound is proved by a combination of the argument based on the Håstad's switching lemma [5] (see

S. Albers et al. (Eds.): ICALP 2009, Part I, LNCS 5555, pp. 59–70, 2009.

also [4]) and the Kruskal-Katona Theorem developed in extremal set theory. The upper bound is proved by showing the existence of a DNF formula of size $\exp(O(\sqrt{n}))$ that ϵ-approximates the majority function for every constant $0 < \epsilon < 1/2$. Since the majority function has the largest *total influence* among all monotone Boolean functions, a good solution to this problem would help to a better understanding of the relationship between the total influence of monotone functions and the size of small depth circuits for approximating them, which has been widely investigated (see [7] and the references therein).

In this paper, we extend their results and show that their lower bound is in fact optimal (again, up to a constant factor in the exponent) *for every constant d*. Precisely, we give a probabilistic construction of depth d circuits of size $\exp(O(n^{1/(2d-2)}))$ that ϵ-approximates the majority function on n variables, for every constant $0 < \epsilon < 1/2$ and for every constant $d \geq 3$. This is a main (and essentially only) result of this paper. Note that the minimum size of a depth d circuit that *exactly* computes the majority function is known to be between $\exp(\Omega(n^{1/(d-1)}))$ and $\exp(O(n^{1/(d-1)}(\log n)^{1-1/(d-1)}))$ (see [5] or [10, Theorem 4.4, p.333] for the lower bound, and [6] for the upper bound).

The proof of our result is a simple generalization of the technique used in a beautiful construction of $O(n^{5.3})$ size monotone formulas for the majority function by Valiant [8]. It should be noted that our circuit is monotone (i.e., without negated literals) and is formula (i.e., every gate has fan-out one). In addition, our approach can also be used for constructing a small circuit for the alternate version of approximate-majority, which will be discussed in the last part of this paper.

The organization of the paper is as follows. In Section 2, we give some notations and definitions. In Section 3, we describe the framework of our construction. The proof of the main result is described in Section 4. Finally, in Section 5, we show that our approach can also yield a small circuit for approximating majority of the first kind, together with some open problems.

2 Notations and Definitions

For a binary string $x \in \{0,1\}^n$, $|x|$ denotes the number of 1's in x. The *majority function* on n variables, which is denoted by Maj_n, is a Boolean function defined by $\mathsf{Maj}_n(x) = 1$ iff $|x| \geq n/2$. For $0 < \epsilon < 1$, a Boolean function $f : \{0,1\}^n \to \{0,1\}$ is said an *ϵ-approximation* for Maj_n if f and Maj_n disagree on at most ϵ fraction of all inputs, i.e.,

$$\Pr_x[f(x) \neq \mathsf{Maj}_n(x)] \leq \epsilon,$$

where the probability is over the uniform distribution on $\{0,1\}^n$. For a set S, $\sharp S$ denotes the cardinality of S.

We consider single-output circuits that consists of unbounded fan-in AND and OR gates over the input literals, i.e., input variables and their negations. The *depth* of a circuit is the number of gates in a longest path from the output to an input. The *size* of a circuit is the number of AND and OR gates in it.

Throughout the paper, e denotes the base of the natural logarithm.

3 Random Circuits

Let $W = (w_1, \ldots, w_d)$ be a d-tuple of integers such that $w_i \geq 2$ for every $1 \leq i \leq d$. The values of w_i will be determined later. Define a sequence of random circuits $\mathbf{f}_0, \mathbf{f}_1, \ldots, \mathbf{f}_d$ on $X = \{x_1, \ldots, x_n\}$ recursively as follows:

1. $\mathbf{f}_0$ is a Boolean variable chosen uniformly from $X = \{x_1, \ldots, x_n\}$.
2. For odd k, $\mathbf{f}_k$ is an AND of w_k independent copies of $\mathbf{f}_{k-1}$. For even k, $\mathbf{f}_k$ is an OR of w_k independent copies of $\mathbf{f}_{k-1}$.

It is clear that $\mathbf{f}_d$ is a random circuit (in fact, formula) of depth d, where the bottom level consists of AND gates, and the fan-in of each gate at the k-th level is w_k. The number of gates in $\mathbf{f}_d$ is given by $1 + w_d + w_{d-1}w_d + \cdots + \prod_{k=2}^{d} w_k < 2\prod_{k=2}^{d} w_k$.

For $k = 0, \ldots, d$ and $i \in \{0, 1\}$, let $A_k^i(p) : [0,1] \to [0,1]$ be a function defined as follows:

$$\begin{aligned}
&A_0^1(p) = p, \quad \text{for every } p \in [0,1] \\
&A_k^1(p) = (A_{k-1}^1(p))^{w_k} \quad \text{for every odd } k, \text{and for every } p \in [0,1] \\
&A_k^0(p) = (A_{k-1}^0(p))^{w_k} \quad \text{for every even } k \text{ with } k \geq 2, \text{and for every } p \in [0,1] \\
&A_k^0(p) + A_k^1(p) = 1 \qquad \text{for every } k, \text{and for every } p \in [0,1].
\end{aligned}$$

When $\mathbf{f}_0$ gets one with probability p then $\mathbf{f}_k$ outputs $i \in \{0,1\}$ with probability $A_k^i(p)$. Note that $A_k^1(\cdot)$ ($A_k^0(\cdot)$, resp.) is monotonically increasing (decreasing, resp.).

The following simple lemma relates the value of $A_k^i(\cdot)$'s with the size of ϵ-approximator circuits for the majority function.

Lemma 1. *Suppose that, for a given $W = (w_1, \ldots, w_d)$, we have*

$$A_d^1\left(1/2 - \epsilon/\sqrt{n}\right) \leq \epsilon, \tag{1}$$

and

$$A_d^0\left(1/2 + \epsilon/\sqrt{n}\right) \leq \epsilon. \tag{2}$$

Then there is a depth d circuit of size less than $2\prod_{k=2}^{d} w_k$ that computes a 3ϵ-approximation for Maj_n.

Proof. For every $x \in \{0,1\}^n$ with $|x| \leq n/2 - \epsilon\sqrt{n}$, we have

$$\Pr_{\mathbf{f}_d}[\mathbf{f}_d(x) \neq \mathsf{Maj}_n(x)] \leq A_d^1\left(1/2 - \epsilon/\sqrt{n}\right) \leq \epsilon,$$

since Eq. (1) and $A_d^1(\cdot)$ is monotonically increasing. Similarly, for every $x \in \{0,1\}^n$ with $|x| \geq n/2 + \epsilon\sqrt{n}$, we have

$$\Pr_{\mathbf{f}_d}[\mathbf{f}_d(x) \neq \mathsf{Maj}_n(x)] \leq A_d^0\left(1/2 - \epsilon/\sqrt{n}\right) \leq \epsilon,$$

since Eq. (2) and $A_d^0(\cdot)$ is monotonically decreasing. These immediately implies that there is a depth d circuit of size less than $2\prod_{k=2}^{d} w_k$ whose output disagrees with the majority function on at most

$$\epsilon\frac{2^n}{2} + \epsilon\frac{2^n}{2} + \sharp\{x \in \{0,1\}^n \mid n/2 - \epsilon\sqrt{n} < |x| < n/2 + \epsilon\sqrt{n}\}$$

inputs. The last term is upper bounded by

$$2\epsilon\sqrt{n}\binom{n}{n/2} \leq 2\epsilon\sqrt{n}\cdot\frac{2^n}{\sqrt{n}} = 2\epsilon\cdot 2^n,$$

where the first inequality follows from the Stirling formula. This completes the proof of the lemma. □

Note that, in this notation, a famous construction of $O(n^{5.3})$ size monotone formulae by Valiant [8] can be written as: $A_d^0(\alpha(1/2+1/n)) < 2^{-n}$ and $A_d^1(\alpha(1/2-1/n)) < 2^{-n}$ for $W = (2, 2, \ldots, 2)$ with $d \sim 5.3\log_2 n$ and $\alpha = (\sqrt{5}-1)/2$.

4 Bounds for Approximating Majority

In this section, we show our main theorem:

Theorem 2. *For every constant $0 < \epsilon < 1/2$ and for every constant $d \geq 3$, the majority function on n variables can be ϵ-approximated by a depth d circuit of size $2^{O(n^{1/(2d-2)})}$. This is optimal up to a constant factor in the exponent.*

By Lemma 1, all we have to do is choose a suitable parameter $W = (w_1, \ldots, w_d)$ and verify that $A_d^0(1/2 + \epsilon/\sqrt{n}) \leq \epsilon$ and $A_d^1(1/2 - \epsilon/\sqrt{n}) \leq \epsilon$.

We first give a proof for the case $d = 3$ as an illustrative example in Section 4.1, and then give a proof for general cases in Section 4.2. The proof for general cases includes also the case $d = 3$, and so a reader can skip Section 4.1 and go directly to Section 4.2. The key ingredient of the proof is Lemmas 3 and 4 in Section 4.2.

4.1 Construction of Depth Three Circuits

We pick $W = (w_1, w_2, w_3)$ with $w_1 = \frac{1}{\epsilon}n^{1/4}$, $\tilde{w}_2 = 2^{w_1}w_1$, $w_2 = (\ln 2)\tilde{w}_2$, $\tilde{w}_3 = 2^{w_1}$ and $w_3 = (\ln 2)\tilde{w}_3$. The number of gates in a circuit that will be constructed is less than $2w_2w_3 = 2(\ln 2)^2(2^{w_1})^2 w_1 = 2^{O(n^{1/4}/\epsilon)}$.

Note that $A_1^1(1/2) = (1/2)^{w_1}$, $A_2^0(1/2) = (1-(1/2)^{w_1})^{w_2} \sim (1/2)^{w_1}$ and $A_3^1(1/2) = (1-(1/2)^{w_1})^{w_3} \sim 1/2$, which is a key property of our parameter. Put $p_h := 1/2 + \epsilon/\sqrt{n}$ and $p_\ell := 1/2 - \epsilon/\sqrt{n}$. Below we give a proof for $A_3^0(p_h) \leq \epsilon$ and $A_3^1(p_\ell) \leq \epsilon$, which is a bit long but uses only elementary calculations.

We first show that $A_3^0(p_h) \leq \epsilon$. By the definition, we have

$$A_2^0(p_h) = (1-p_h^{w_1})^{w_2} = \left\{(1-p_h^{w_1})^{(\ln 2/p_h^{w_1})}\right\}^{p_h^{w_1}\tilde{w}_2} \leq \left(\frac{1}{2}\right)^{p_h^{w_1}\tilde{w}_2}. \quad (3)$$

Here we use the inequality $(1-q)^{1/q} \leq 1/e$ for $q<1$. The exponent in the last term of Eq. (3) is

$$p_h^{w_1}\tilde{w_2} = \left(\frac{1}{2}+\frac{\epsilon}{\sqrt{n}}\right)^{w_1}\tilde{w_2} = \tilde{w_2}\left\{\left(\frac{1}{2}\right)^{w_1}\left(1+\frac{2\epsilon}{\sqrt{n}}\right)^{w_1}\right\}$$
$$\geq \tilde{w_2}\left(\frac{1}{2}\right)^{w_1}\left(1+\frac{2\epsilon}{\sqrt{n}}w_1\right) = w_1\left(1+\frac{2}{n^{1/4}}\right). \qquad (4)$$

Here we use the inequality $(1+q)^r \geq 1+qr$ for $q>0$ and $r \geq 1$.

We proceed to the estimation of $A_3^0(p_h)$. Since $(1-q)^r \geq 1-qr$ for $q<1$ and $r\geq 1$, we have

$$A_3^0(p_h) = 1-(1-A_2^0(p_h))^{w_3} \leq 1-(1-A_2^0(p_h)w_3) = A_2^0(p_h)w_3. \qquad (5)$$

By plugging Eqs. (3) and (4) into Eq. (5), we have

$$A_3^0(p_h) \leq A_2^0(p_h)w_3 \leq (\ln 2)\left(\frac{1}{2}\right)^{w_1}\left(\frac{1}{2}\right)^{w_1\frac{2}{n^{1/4}}}2^{w_1}$$
$$= (\ln 2)\left(\frac{1}{2}\right)^{\frac{2}{\epsilon}} < (\ln 2)\frac{\epsilon}{2} < \epsilon,$$

where the second last inequality follows from $(1/2)^{2/\epsilon} < \epsilon/2$ which is equivalent to $(1/2) < (\epsilon/2)^{\epsilon/2}$. This holds since the minimum value of the function q^q is $(1/e)^{1/e} \sim 0.6922 > (1/2)$.

We now turn to show $A_3^1(p_\ell) \leq \epsilon$, in which we should bound the value of A_2^0 *from below.*

$$A_2^0(p_\ell) = (1-p_\ell^{w_1})^{w_2} = \left\{(1-p_\ell^{w_1})^{(\ln 2/p_\ell^{w_1})}\right\}^{p_\ell^{w_1}\tilde{w_2}}$$
$$\geq \left\{(1-p_\ell^{w_1})\frac{1}{e}\right\}^{(\ln 2)\cdot p_\ell^{w_1}\tilde{w_2}} > \left\{(1-p_\ell^{w_1})\frac{1}{2}\right\}^{p_\ell^{w_1}\tilde{w_2}}. \qquad (6)$$

We use $(1-1/q)^q \geq (1-1/q)(1/e)$ for $q>1$ to derive the first inequality[1], and use $(1-q)^{\ln 2} > 1-q$ to the second. The exponent in the last term is

$$p_\ell^{w_1}\tilde{w_2} = \left(\frac{1}{2}-\frac{\epsilon}{\sqrt{n}}\right)^{w_1}\tilde{w_2} = \tilde{w_2}\left(\frac{1}{2}\right)^{w_1}\left(1-\frac{2\epsilon}{\sqrt{n}}\right)^{w_1}$$
$$\leq w_1\left(\frac{1}{2}\right)^{\frac{2\epsilon}{\sqrt{n}}\frac{1}{\ln 2}w_1} = w_1\left(\frac{1}{2}\right)^{\frac{2}{\ln 2}\frac{1}{n^{1/4}}} \leq w_1\left(1-\frac{1}{\ln 2}\frac{1}{n^{1/4}}\right). \qquad (7)$$

We use $(1-1/q)^q \leq 1/e$ for $q>1$ to derive the first inequality, and use $(1/2)^{2q} \leq (1-q)$ for $q \leq 1/2$, which is equivalent to $(1/4) \leq (1-q)^{1/q}$, to derive the

[1] Proof: $(1-1/q)^q = (1-1/q)(1-1/q)^{q-1} = (1-1/q)(1+1/(q-1))^{-(q-1)} \geq (1-1/q)(1/e)$.

last inequality. By plugging Eq. (7) into Eq. (6), we can show that, for every sufficiently large n,

$$A_2^0(p_\ell) \geq \left(\frac{1}{2}\right)^{w_1(1-1/n^{1/4})}. \tag{8}$$

The proof of the above inequality is described in Appendix (Section A.1).

We now proceed to the estimation of $A_3^1(p_\ell)$. Since $(1-q)^r \leq (1/e)^{qr}$ for $q<1$ and $r>0$, we have

$$A_3^1(p_\ell) = (1 - A_2^0(p_\ell))^{w_3} \leq \left(\frac{1}{2}\right)^{\tilde{w}_3 A_2^0(p_\ell)}.$$

In order to show $A_3^1(p_\ell) \leq \epsilon$, it is sufficient to show that $\tilde{w}_3 A_2^0(p_\ell) \geq \log_2(1/\epsilon)$. By Eq. (8), we have

$$\tilde{w}_3 A_2^0(p_\ell) \geq 2^{w_1}\left(\frac{1}{2}\right)^{w_1(1-1/n^{1/4})} = \left(\frac{1}{2}\right)^{-w_1/n^{1/4}} = 2^{1/\epsilon} > \log_2(1/\epsilon).$$

This completes the proof of Theorem 2 for $d=3$.

4.2 Construction for General Depths

We pick $W=(w_1,w_2,\ldots,w_d)$ such that

- $w_1 = (1/\epsilon)n^{1/(2d-2)}$,
- $\tilde{w}_k = 2^{w_1}w_1$ and $w_k = (\ln 2)\tilde{w}_k$ for $k=2,\ldots,d-1$,
- $\tilde{w}_d = 2^{w_1}$ and $w_d = (\ln 2)\tilde{w}_d$.

As for the case $d=3$, we choose parameters so that $A_1^1(1/2) = (1/2)^{w_1}$, $A_2^0(1/2) = (1-(1/2)^{w_1})^{w_2} \sim (1/2)^{w_1}$, $A_3^1(1/2) = (1-(1/2)^{w_1})^{w_3} \sim (1/2)^{w_1}$, and so on. It should be noted that the asymptotically optimal construction of depth two circuit of size $\exp(\Theta(\sqrt{n}))$ by O'Donnell and Wimmer [7] is a random DNF of width $w_1 = (1/\epsilon)\sqrt{n}$ and size $w_2 = (\ln 2)2^{w_1}$. Hence, for $d=2$, our construction completely matches their construction, and so this can also be viewed as a natural extension of their construction.

The following two lemmas are almost all that we need. The proof of these two lemmas is described in Appendix (Sections A.2 and A.3).

Lemma 3. *Let $w = (\ln 2)2^{w_1}w_1$. Suppose that n is sufficiently large. Suppose also that*

$$A \geq \left(\frac{1}{2}\right)^{w_1}\left(1 + cn^{\frac{\alpha-d}{2(d-1)}}\right)$$

for some $\alpha \in \{2,\ldots,d-1\}$ and some positive constant c. If $\alpha < d-1$, then

$$(1-A)^w \leq \left(\frac{1}{2}\right)^{w_1}\left(1 - \frac{c}{2\epsilon}n^{\frac{(\alpha+1)-d}{2(d-1)}}\right).$$

If $\alpha = d-1$, then

$$(1-A)^w \le \left(\frac{1}{2}\right)^{w_1}\left(\frac{1}{2}\right)^{\frac{c}{\epsilon}}.$$

Lemma 4. *Let $w = (\ln 2)2^{w_1}w_1$. Suppose that n is sufficiently large. Suppose also that*

$$A \le \left(\frac{1}{2}\right)^{w_1}\left(1 - cn^{\frac{\alpha-d}{2(d-1)}}\right),$$

for some $\alpha \in \{2,\ldots,d-1\}$ and some positive constant c. If $\alpha < d-1$, then

$$(1-A)^w \ge \left(\frac{1}{2}\right)^{w_1}\left(1 + \frac{c}{2\epsilon}n^{\frac{(\alpha+1)-d}{2(d-1)}}\right).$$

If $\alpha = d-1$, then

$$(1-A)^w \ge \left(\frac{1}{2}\right)^{w_1}\left(\frac{1}{2}\right)^{-\frac{c}{1.1\epsilon}}.$$

Proof of Theorem 2. Let $W = (w_1,\ldots,w_d)$ be as described at the beginning of this subsection. The size of a circuit that will be constructed is less than $2\prod_{k=2}^{d} w_k = 2(\ln 2)^{d-1}(2^{w_1})^{d-1}(w_1)^{d-2} = 2^{O(n^{1/(2d-2)}/\epsilon)}$. Put $p_h := 1/2+\epsilon/\sqrt{n}$ and $p_\ell := 1/2-\epsilon/\sqrt{n}$. Below, we will show that $A_d^0(p_h) \le \epsilon$ and $A_d^1(p_\ell) \le \epsilon$.

We first show that $A_d^0(p_h) \le \epsilon$. We start with

$$\begin{aligned} A_1^1(p_h) &= \left(\frac{1}{2}\right)^{w_1}\left(1+\frac{2\epsilon}{\sqrt{n}}\right)^{w_1} \\ &\ge \left(\frac{1}{2}\right)^{w_1}\left(1+2n^{\frac{2-d}{2(d-1)}}\right) > \left(\frac{1}{2}\right)^{w_1}\left(1+n^{\frac{2-d}{2(d-1)}}\right), \end{aligned} \tag{9}$$

where the first inequality follows from the inequality $(1+q)^r \ge 1+qr$ for $q \ge 0$ and $r \ge 1$. We use Lemma 3 to get

$$A_2^0(p_h) = (1-A_1^1(p_h))^{w_2} \le \left(\frac{1}{2}\right)^{w_1}\left(1-\frac{1}{2\epsilon}n^{\frac{3-d}{2(d-1)}}\right).$$

Then we use Lemma 4 to get

$$A_3^1(p_h) = (1-A_2^0(p_h))^{w_3} \ge \left(\frac{1}{2}\right)^{w_1}\left(1+\frac{1}{(2\epsilon)^2}n^{\frac{4-d}{2(d-1)}}\right).$$

By applying Lemmas 3 and 4 alternatively, we have

$$A_{d-2}^0(p_h) \le \left(\frac{1}{2}\right)^{w_1}\left(1-\frac{1}{(2\epsilon)^{d-3}}n^{\frac{-1}{2(d-1)}}\right) \tag{10}$$

when d is even, or we have

$$A_{d-2}^1(p_h) \ge \left(\frac{1}{2}\right)^{w_1}\left(1+\frac{1}{(2\epsilon)^{d-3}}n^{\frac{-1}{2(d-1)}}\right) \tag{11}$$

when d is odd. Note that when $d = 3$ we have already obtained Eq.(11) as Eq.(9). By applying Lemma 3 or 4 once again, we obtain

$$A^1_{d-1}(p_h) \geq \left(\frac{1}{2}\right)^{w_1} \left(\frac{1}{2}\right)^{-\frac{1}{(1.1\epsilon)(2\epsilon)^{d-3}}}$$
$$\geq \left(\frac{1}{2}\right)^{w_1} \left(\frac{1}{2}\right)^{-\frac{1}{2\epsilon}} = \left(\frac{1}{2}\right)^{w_1} 2^{\frac{1}{2\epsilon}} \geq \left(\frac{1}{2}\right)^{w_1} \cdot \log_2(1/\epsilon) \qquad (12)$$

when d is even, and

$$A^0_{d-1}(p_h) \leq \left(\frac{1}{2}\right)^{w_1} \left(\frac{1}{2}\right)^{\frac{1}{(\epsilon)(2\epsilon)^{d-3}}} \leq \left(\frac{1}{2}\right)^{w_1} \left(\frac{1}{2}\right)^{\frac{1}{\epsilon}} \qquad (13)$$

when d is odd. The case for even d is finished by using Eq. (12):

$$A^0_d(p_h) = (1 - A^1_d(p_h))^{w_d}$$
$$\leq \left\{1 - \left(\frac{1}{2}\right)^{w_1} \log_2(1/\epsilon)\right\}^{(\ln 2)2^{w_1}} \leq \left(\frac{1}{2}\right)^{\log_2(1/\epsilon)} = \epsilon,$$

where the first inequality follows from the inequality $(1-q)^r \leq (1/e)^{qr}$ for $q \leq 1$ and $r \geq 0$. The case for odd d is finished by using Eq. (13):

$$A^1_d(p_h) \geq \left\{1 - \left(\frac{1}{2}\right)^{w_1} \left(\frac{1}{2}\right)^{\frac{1}{\epsilon}}\right\}^{(\ln 2)2^{w_1}}$$
$$\geq 1 - (\ln 2) \left(\frac{1}{2}\right)^{\frac{1}{\epsilon}} > 1 - (\ln 2)\epsilon > 1 - \epsilon.$$

Here we use the inequality $(1 - q)^r \geq 1 - qr$ for $q \leq 1$ and $r \geq 1$ to derive the first inequality, and use $(1/2)^{(1/q)} < q$, which is equivalent to $(1/2) < q^q$, to the second. This holds since the minimum value of the function q^q is $(1/e)^{(1/e)} \sim 0.6922$.

We now turn to show $A^1_d(p_\ell) \leq \epsilon$. The proof is almost analogous to the proof for $A^0_d(p_h) \leq \epsilon$. The "base" is

$$A^1_1(p_\ell) = \left(\frac{1}{2}\right)^{w_1} \left(1 - \frac{2\epsilon}{\sqrt{n}}\right)^{w_1} \leq \left(\frac{1}{2}\right)^{w_1} \left(\frac{1}{2}\right)^{\frac{2}{\ln 2} n^{\frac{2-d}{2(d-1)}}}$$
$$\leq \left(\frac{1}{2}\right)^{w_1} \left(1 - \frac{1}{\ln 2} n^{\frac{2-d}{2(d-1)}}\right) < \left(\frac{1}{2}\right)^{w_1} \left(1 - n^{\frac{2-d}{2(d-1)}}\right), \qquad (14)$$

where the first inequality follows from the inequality $(1-q)^r \leq (1/e)^{qr}$ for $q \leq 1$ and $r \geq 0$, and the second inequality follows from the inequality $(1/2)^{2q} \leq 1 - q$ for $q \leq 1/2$, which is equivalent to $(1/4) \leq (1 - q)^{1/q}$. By applying Lemmas 3 and 4 alternatively, we have

$$A^1_{d-2}(p_\ell) \leq \left(\frac{1}{2}\right)^{w_1} \left(1 - \frac{1}{(2\epsilon)^{d-3}} n^{\frac{-1}{2(d-1)}}\right),$$

when d is odd (note again that when $d = 3$, we have already obtained this as Eq.(14)), or we have

$$A^0_{d-2}(p_\ell) \geq \left(\frac{1}{2}\right)^{w_1} \left(1 + \frac{1}{(2\epsilon)^{d-3}} n^{\frac{-1}{2(d-1)}}\right)$$

when d is even. These inequalities are identical to Eqs. (10) and (11) if we swap p_h and p_ℓ, "odd" and "even", and the role of 0 and 1. This immediately implies the desired bound, i.e., $A^1_d(p_\ell) \leq \epsilon$, since we have shown $A^0_d(p_h) \leq \epsilon$ from Eqs. (10) and (11). □

5 Bottom Fan-In and Depth-3 Circuit Size

Our approach can also handle the problem for constructing small circuits to compute an "Approximate-Majority" of the first kind described in Introduction, i.e., to compute a function $f : \{0,1\}^n \to \{0,1\}$ such that $f(x) = 1$ for every x with $|x| \geq (2/3)n$ and $f(x) = 0$ for every x with $|x| \leq (1/3)n$. If we restrict ourselves to depth $d = 3$, the conditions that should be satisfied are now

$$A^1_3(1/3) < \left\{ \sum_{i=0}^{(1/3)n} \binom{n}{i} \right\}^{-1} \sim 2^{-H(1/3)n}, \tag{15}$$

and

$$A^0_3(2/3) < \left\{ \sum_{i=(2/3)n}^{n} \binom{n}{i} \right\}^{-1} \sim 2^{-H(1/3)n}, \tag{16}$$

where $H(p) := -p\log_2 p - (1-p)\log_2(1-p)$ denotes the binary entropy function. An easy calculation shows that these are satisfied by the parameter $W = (w_1, w_2, w_3) := (\log_2 n, (\ln 2)(\log_2 n)n^{\log_2 3}, n^2)$, which implies that there is a depth-3 circuit with bottom fan-in $\log_2 n$ that approximates the majority (in the meaning of the first kind of approximation) whose size is $O(n^{2+\log_2 3+\epsilon})$, i.e., polynomial in n. The calculation for verifying this is described in Appendix (Section A.4).

It has been recently shown by Viola [9] that every depth-3 circuit with bottom fan-in at most $(\log_2 n)/2$ that approximates the majority on n variables has size at least $2^{n^{0.1}}$. This means that a sharp threshold phenomenon (i.e., the size of circuits becomes polynomial from exponential as the bottom fan-in increases) occurs at somewhere between $(\log_2 n)/2$ and $\log_2 n$. A careful inspection of the proof by Viola [9] can improve the lower limit to $\{1/(\log_2 3) - \epsilon\}n \sim 0.631\log_2 n$, but still has a gap. If we decrease the value of w_1 from $\log_2 n$ to $\alpha \log_2 n$ with $\alpha < 1$, then it will not be possible to satisfy Eqs. (15) and (16) by parameters w_2 and w_3 whose values are polynomial in n. Hence, the problem to determine the true threshold value, or to see what happen around the threshold would be interesting.

Acknowledgments

The author would like to thank anonymous reviewers for their helpful comments.

References

1. Ajtai, M.: Σ_1^1-formulae on Finite Structures. Annals of Pure and Applied Logic 24, 1–48 (1983)
2. Ajtai, M.: Approximate Counting with Uniform Constant-Depth Circuits. Advances in Computational Complexity Theory, pp. 1–20. Amer. Math. Soc., Providence (1993)
3. Ajtai, M., Komlós, J., Szemerédi, E.: Sorting in $c \log n$ Parallel Steps. Combinatorica 3, 1–19 (1983)
4. Boppana, R.: The Average Sensitivity of Bounded-Depth Circuits. Inf. Proc. Lett. 63(5), 257–261 (1997)
5. Håstad, J.: Almost Optimal Lower Bounds for Small Depth Circuits. In: Proc. of 18th ACM Symposium on Theory of Computing (STOC 1986), pp. 6–20 (1986)
6. Klawe, M., Paul, W.J., Pippenger, N., Yannakakis, M.: On Monotone Formulae with Restricted Depth. In: Proc. of 16th ACM Symposium on Theory of Computing (STOC 1984), pp. 480–487 (1984)
7. O'Donnell, R., Wimmer, K.: Approximation by DNF: Examples and Counterexamples. In: Arge, L., Cachin, C., Jurdziński, T., Tarlecki, A. (eds.) ICALP 2007. LNCS, vol. 4596, pp. 195–206. Springer, Heidelberg (2007)
8. Valiant, L.G.: Short Monotone Formulae for the Majority Function. J. Algorithms 5(3), 363–366 (1984)
9. Viola, E.: On Approximate Majority and Probabilistic Time. In: Proc. of 22nd IEEE Conference on Computational Complexity (CCC 2007), pp. 155–168 (2007)
10. Wegener, I.: The Complexity of Boolean Functions. Willey-Teubner (1987)

A Appendix

A.1 Proof of Eq. (8)

What we want to show is

$$\left\{(1-p_\ell^{w_1})\frac{1}{2}\right\}^{w_1(1-\frac{1}{\ln 2}\frac{1}{n^{1/4}})} \geq \left(\frac{1}{2}\right)^{w_1(1-\frac{1}{n^{1/4}})}.$$

This is equivalent to

$$1-p_\ell^{w_1} \geq \left(\frac{1}{2}\right)^{\frac{1-\ln 2}{(\ln 2)n^{1/4}-1}}. \tag{17}$$

Since $1-q \geq (1/2)^{2q}$ for $q \leq 1/2$, we have

$$1-\Theta\left(\frac{1}{n^{1/4}}\right) = 1-\frac{1}{2}\cdot\frac{1-\ln 2}{(\ln 2)n^{1/4}-1} \geq \text{RHS of Eq. (17)}.$$

Since $p_\ell^{w_1}$ is exponentially small in n, i.e., $p_\ell^{w_1} = O(1/2^{n^{1/4}}) = o(1/n^{1/4})$, Eq. (17) holds for sufficiently large n. □

A.2 Proof of Lemma 3

Since $(1-q)^r \le (1/e)^{qr}$ for $q \le 1$ and $r \ge 0$, we have

$$(1-A)^w = (1-A)^{(\ln 2)2^{w_1}w_1} \le \left(\frac{1}{2}\right)^{A\cdot 2^{w_1}w_1} \le \left(\frac{1}{2}\right)^{w_1(1+cn^{\frac{\alpha-d}{2(d-1)}})}$$
$$= \left(\frac{1}{2}\right)^{w_1}\left(\frac{1}{2}\right)^{c\cdot w_1 n^{\frac{\alpha-d}{2(d-1)}}} = \left(\frac{1}{2}\right)^{w_1}\left(\frac{1}{2}\right)^{\frac{c}{\epsilon}n^{\frac{(\alpha+1)-d}{2(d-1)}}}.$$

This completes the proof for $\alpha = d-1$. If $\alpha < d-1$, then the exponent of the last term converges to 0 as $n \to \infty$. Hence, we use the inequality $(1/2)^q \le (1-q/2)$ for $q \le 1$, which is equivalent to $(1/4) \le (1-q/2)^{2/q}$, to show

$$(1-A)^w \le \left(\frac{1}{2}\right)^{w_1}\left(1-\frac{c}{2\epsilon}n^{\frac{(\alpha+1)-d}{2(d-1)}}\right) \quad \text{(for sufficiently large } n\text{)},$$

which completes the proof of the lemma. □

A.3 Proof of Lemma 4

By using the inequality $(1-1/q)^q \ge (1-1/q)(1/e)$ for $q > 1$ (whose proof is in the footnote in Section 4.1), we have

$$(1-A)^w = (1-A)^{(\ln 2)2^{w_1}w_1} \ge \left\{(1-A)\left(\frac{1}{2}\right)\right\}^{A\cdot 2^{w_1}w_1}$$
$$\ge \left\{(1-A)\left(\frac{1}{2}\right)\right\}^{w_1(1-cn^{\frac{\alpha-d}{2(d-1)}})}$$
$$\ge \left(\frac{1}{2}\right)^{w_1(1-\frac{c}{1.1}n^{\frac{\alpha-d}{2(d-1)}})} \quad \text{(for sufficiently large } n\text{)}$$
$$= \left(\frac{1}{2}\right)^{w_1}\left(\frac{1}{2}\right)^{-\frac{c}{1.1\epsilon}n^{\frac{(\alpha+1)-d}{2(d-1)}}},$$

where the third inequality can be derived by a similar calculation as the proof of Eq. (8) in Section A.1. This completes the proof for the case $\alpha = d-1$. When $\alpha < d-1$, the exponent of the last term converges to 0 as $n \to \infty$. Hence, we can use the inequality $2^q \ge (1+(\ln 2)q)$ for $q < 1$, which is equivalent to $e \ge (1+q)^{1/q}$, to show

$$(1-A)^w \ge \left(\frac{1}{2}\right)^{w_1}\left\{1+\frac{(\ln 2)c}{1.1\epsilon}n^{\frac{(\alpha+1)-d}{2(d-1)}}\right\} \quad \text{(for sufficiently large } n\text{)}$$
$$> \left(\frac{1}{2}\right)^{w_1}\left\{1+\frac{c}{2\epsilon}n^{\frac{(\alpha+1)-d}{2(d-1)}}\right\}.$$

This completes the proof of the lemma. □

A.4 Depth-3 Circuits for Approximating Majority of the First Kind

This section describes the calculation for verifying Eqs. (15) and (16) when we set $W := (\log_2 n, (\ln 2)(\log_2 n)n^{\log_2 3}, n^2)$ (See Section 5). Eq. (15) is verified by the following series of calculations.

$$A_1^1(1/3) = \left(\frac{1}{3}\right)^{\log_2 n} = \frac{1}{n^{\log_2 3}},$$

$$A_2^0(1/3) = \left(1 - \frac{1}{n^{\log_2 3}}\right)^{(\ln 2)(\log_2 n)n^{\log_2 3}} \sim \left(\frac{1}{e}\right)^{(\ln 2)(\log_2 n)} = \frac{1}{n},$$

$$A_3^1(1/3) = \left(1 - \frac{1}{n}\right)^{n^2} \sim \left(\frac{1}{e}\right)^{n} < 2^{-n} < 2^{-H(1/3)n}.$$

Eq. (16) is verified by the following series of calculations.

$$A_1^1(2/3) = \left(\frac{2}{3}\right)^{\log_2 n} = \frac{1}{n^{\log_2 3-1}},$$

$$A_2^0(2/3) = \left(1 - \frac{1}{n^{\log_2 3-1}}\right)^{(\ln 2)(\log_2 n)n^{\log_2 3}} \sim \left(\frac{1}{e}\right)^{(\ln 2)(\log_2 n)n} = \frac{1}{n^n},$$

$$A_3^0(2/3) = 1 - \left(1 - \frac{1}{n^n}\right)^{n^2} < 1 - \left(1 - \frac{n^2}{n^n}\right) = \frac{n^2}{n^n} < 2^{-n} < 2^{-H(1/3)n}.$$

Strictly speaking, this is not a formal proof since we use an asymptotic estimation (i.e., $(1 - 1/n)^n \sim 1/e$) here. However, it will be obtained by some more careful estimations.

Counting Subgraphs via Homomorphisms

Omid Amini[1], Fedor V. Fomin[2], and Saket Saurabh[2]

[1] CNRS-DMA, École Normale Supérieure, Paris, France
omid.amini@m4x.org
[2] Department of Informatics, University of Bergen, Norway
{fedor.fomin,saket.saurabh}@ii.uib.no

Abstract. Counting homomorphisms between graphs has applications in variety of areas, including extremal graph theory, properties of graph products, partition functions in statistical physics and property testing of large graphs. In this work we show a new application of counting graph homomorphisms to the areas of exact and parameterized algorithms.

We introduce a generic approach for counting subgraphs in a graph. The main idea is to relate counting subgraphs to counting graph homomorphisms. This approach provides new algorithms and unifies several well known results in algorithms and combinatorics including the recent algorithm of Björklund, Husfeldt and Koivisto for computing the chromatic polynomial, the classical algorithm of Kohn, Gottlieb, Kohn, and Karp for counting Hamiltonian cycles, Ryser's formula for counting perfect matchings of a bipartite graph, and color coding based algorithms of Alon, Yuster, and Zwick.

1 Introduction

Given two undirected graphs F and G, a *homomorphism* from F to G is a mapping from the vertex set of F to that of G such that the image of every edge of F is an edge of G. Many combinatorial structures in F, for example independent sets and proper vertex colorings, may be viewed as graph homomorphisms to a particular graph G, see the book of Hell and Nešetřil [20] for a thorough introduction to the topic. Counting homomorphisms between graphs has applications in a variety of areas, including extremal graph theory, properties of graph products, partition functions in statistical physics and property testing of large graphs. We refer to the excellent survey of Borgs *et al.* [11] for more references on counting homomorphisms.

There is an extensive literature on the computational complexity of graph homomorphism and counting homomorphisms. Hell and Nešetřil showed that for any fixed simple graph G, the problem whether there exists a homomorphism from F to G is solvable in polynomial time if G is bipartite, and NP-complete if G is not bipartite [19]. Dyer and Greenhill [15] completely characterized the dichotomy between P and #P-complete for counting homomorphisms from F to G. It appears that polynomial-time solvable cases arise only when G is an isolated vertex, a complete graph with all loops present, a complete bipartite

S. Albers et al. (Eds.): ICALP 2009, Part I, LNCS 5555, pp. 71–82, 2009.

graph without loops, or a disjoint union of these graphs. Dalmau and Jonsson [13], extending a result of Grohe [18], proved that counting homomorphisms from a graph $F \in \mathcal{F}$ to an arbitrary graph G is in P if and only if all graphs in the family $\mathcal{F}$ have bounded treewidth (up to the assumption from parameterized complexity that FPT $\neq$ #W[1]).

In this paper we design *exact* and *parameterized* algorithms for counting the number of subgraphs isomorphic to a given graph F in a general graph. For any graph G with n vertices and m edges, all subgraphs of G isomorphic to a given graph F can be counted by trying all possible edge subsets of G and for each subset checking if the obtained graph is isomorphic to F. This algorithm runs in time $2^{m+o(n)}$ by making use of an algorithm due to Babai [2] to check isomorphism in time subexponential in n. Another approach is to try all permutations of G and F, and for each fixed permutation, to compare vertex neighborhoods. This will give us running time $\mathcal{O}(n!n^2) = 2^{\mathcal{O}(n \log n)}$. While it is an open question in the area whether subgraph isomorphism can be solved in time $2^{\mathcal{O}(n)}$, there are many special cases, depending on the structure of graph F, for which $2^{\mathcal{O}(n)}$ algorithms are known. Many natural problems like Hamiltonian Cycle, Perfect Matching, Graph Coloring, Bandwidth Minimization, Triangle Packing, and many others can be seen as subgraph isomorphism problems and for each of these problems there are $2^{\mathcal{O}(n)}$ time algorithms known in the literature. However, known algorithms for these problems are tailored to their specific properties.

The main idea behind our results is to reduce the problem of counting subgraphs of G isomorphic to a graph F to counting homomorphisms from F to G. Let $\text{sub}(F,G)$ denote the number of distinct copies of a graph F contained in a graph G. Let also $\text{hom}(F,G)$ and $\text{inj}(F,G)$ be the number of homomorphisms and injective homomorphisms from F to G respectively. The idea of relating $\text{hom}(F,G)$ and $\text{inj}(F,G)$ is not new in Graph Theory. Lovász [24,11] gave the following identities relating $\text{hom}(F,G)$ and $\text{inj}(F,G)$. For an equivalence relation Θ on $V(F)$ let F/Θ denote the graph obtained by identifying nodes that belong to the same class of Θ. Then $\text{inj}(F,G) = \sum_{\Theta} \mu(\Theta)\text{hom}(F/\Theta, G)$, where $\mu(\Theta) = \prod_{A\in\Theta}^{k} \left((-1)^{|A|-1}(|A|-1)!\right)$ with the product running over all equivalence classes of Θ and sum running over all equivalence relations. From algorithmic point of view the above formula is not efficient because the number of equivalence relations even for simple graphs like graph containing n isolated vertices is too large to be meaningful. We give an alternate formula which helps us in counting "simple structures" in time vertex exponential in the right hand graph G. Let us denote by $\text{aut}(F,F)$ the number of automorphisms from F to itself, that is bijective homomorphisms. Our result shows that if $|V(F)| = |V(G)|$, then

$$\text{sub}(F,G) = \frac{\text{inj}(F,G)}{\text{aut}(F,F)} = \frac{\sum_{W\subseteq V(G)} (-1)^{|W|}\, \text{hom}(F, G[V(G)\setminus W])}{\text{aut}(F,F)}. \quad (1)$$

This formula can be seen as a generalization of inclusion-exclusion based formulas which were used for many problems including counting the number of perfect matchings in a graph [7,28], counting Hamiltonian cycles [4,21,22], and computing chromatic polynomial of a graph [8,23]. The basic idea of inclusion-exclusion based approach is that we express the number of objects we want to count as a sum of other objects which are easier to count. The main advantage of using graph homomorphisms is that despite of their expressive power, graph homomorphisms from many structures can be counted efficiently.

Our results and related work. We start by proving (1), Theorem 1, which is our main tool in the design of exact algorithms. We observe that a number of well-known classical and recent results can be obtained as corollaries of Theorem 1. We demonstrate its power by reproving the following results. Let G be a graph on n vertices. Then the number of Hamiltonian cycles in G can be computed in time $2^n n^{\mathcal{O}(1)}$ and in polynomial space (this result was rediscovered several times [4,22,21]). The chromatic polynomial of G can be computed in time $2^{n+\mathcal{O}(\sqrt{n})}$ (this almost matches the running times of the celebrated result of Björklund, Husfeldt, and Koivisto [8,23]). The number of perfect matchings in a *bipartite* graph can be counted in time $2^{n/2} n^{\mathcal{O}(1)}$ (the classical result of Ryser [28], see also Björklund and Husfeldt [8]). Then we use Theorem 1 and its variants to obtain improvements on the following.

Number of optimal permutations for bandwidth. The BANDWIDTH problem is a famous combinatorial problem where given an undirected graph G on n vertices, we wish to embed its vertices onto an integer line such that the maximum stretch of any edge of G is minimized.

Feige and Kilian [16] provided an exact algorithm computing the optimal bandwidth in time $10^n n^{\mathcal{O}(1)}$. Recently an improved algorithm with running time $5^n n^{\mathcal{O}(1)}$ was given by Cygan and Pilipczuk [12]. None of the known algorithms can be adapted to count the number of optimal bandwidth assignments.

The BANDWIDTH problem can be seen as a subgraph isomorphism problem, and by combining Theorem 1 with the techniques of counting homomorphisms on graphs of bounded treewidth, we obtain the following. The number of optimal bandwidth permutations of a graph on n vertices of treewidth at most t can be counted in time $2^{t \log_2 n + n} n^{\mathcal{O}(1)}$ and space $2^{t \log_2 n} n^{\mathcal{O}(1)}$.

Number of perfect matchings. While a perfect matching in a graph can be found in polynomial time, the problem of counting the number of perfect matchings is #P-complete [29]. For bipartite graphs, the best known exact algorithm for counting perfect matchings is to apply the Ryser's formula for the permanent [28], which runs in time $\mathcal{O}(1.414^n)$. Björklund and Husfeldt [7] showed how to compute the number of perfect matchings of a graph in time $2^n n^{\mathcal{O}(1)}$ and polynomial space. They also showed how to count perfect matchings in time $\mathcal{O}(1.732^n)$ and exponential space.

We generalize the classical result of Ryser by showing that if G contains an independent set of size k then the number of perfect matchings in G can be found in time $\mathcal{O}(2^{n-k}n^3)$. Let us remark that the case of bipartite graphs is a special case as $k \geq \lfloor n/2 \rfloor$.

Counting maximum subgraphs with a given property. Combining algorithms for counting homomorphisms with ideas from data structures, we give algorithms running in time $2^{\mathcal{O}(n)}$ for various problems asking to count the number of subgraph with specific properties in an n-vertex graph. For example, it is possible to count in time $2^{\mathcal{O}(n)}$ maximum planar subgraphs, subgraphs of bounded genus, or, even more generally, subgraphs excluding a fixed graph M as a minor. The last result requires a new combinatorial bound on the number of non-isomorphic unlabeled M-minor-free graphs which implies as a corollary the main theorem of Norine, Seymour, Thomas, and Wollan from [27] on minor-closed families. This resolves an open problem of Bernardi, Noy and Welsh [6]. We also obtain a number of algorithms for counting spanning trees with different degree conditions. These structures can be seen as generalizations of Hamiltonian paths. Let us remark that prior to our work for many of the problems above the only known algorithm was the trivial edge subset enumerating algorithm running in time $2^{\mathcal{O}(n^2)}$.

Packing problems. We show how to solve in time $2^{\mathcal{O}(n)}$ even more difficult problems, like finding a maximum vertex disjoint packing from a class $\mathcal{H}$, where $\mathcal{H}$ is a graph class excluding a fixed graph M as a minor. In particular the Maximum Vertex Disjoint Cycles problem can be solved in time $2^{n+\mathcal{O}(\sqrt{n})}$. Again, for many of these problems no $2^{\mathcal{O}(n)}$ time algorithm were known.

Parameterized algorithms. By applying inclusion-exclusion it is possible to refine the celebrated Color Coding technique of Alon, Yuster, and Zwick [1]. The probabilistic algorithm of Alon et al. determines whether a given graph G contains a fixed graph F as a subgraph and works in two stages. First we color the vertices of G at random and then perform dynamic programming on the colored graph in order to find an isomorphic subgraph of F whose vertices have distinct colors. In [1] Alon et al. provide an algorithm for the case when F is a forest, and then mention that this algorithm can be generalized to an algorithm that finds a k-vertex graph F of treewidth t in an n-vertex graph G (if such a copy exists) in expected time $2^{\mathcal{O}(k)}n^{t+1}$. One of the significant disadvantages of using dynamic programming with color coding is that it requires exponential space. By combining ideas based on inclusion-exclusion and graph homomorphisms with color coding, we provide a polynomial space algorithm that in expected time $\mathcal{O}((2e)^k \cdot k \cdot t \cdot n^{t+1})$ finds a k-vertex graph F of treewidth t in an n-vertex graph G.

2 Preliminaries

Let G be a simple undirected graph without self loops and multiple edges. We denote the vertex set of G by $V(G)$ and the set of edges by $E(G)$. For a subset $W \subseteq V(G)$, by $G[W]$ we mean the subgraph of G induced by W. By $deg_W(v)$, $v \in V(G)$ and $W \subseteq V(G)$, we denote the number of vertices adjacent to v in

W. Given two graphs F and G, a graph *homomorphism* from F to G is a map f from $V(F)$ to $V(G)$, that is $f: V(F) \to V(G)$, such that if $uv \in E(F)$, then $f(u)f(v) \in E(G)$. Furthermore, when the map f is injective, f is called an *injective homomorphism.* Given two graphs F and G, the problem of SUBGRAPH ISOMORPHISM asks whether there exists an injective homomorphism from F to G. We also recall that $\hom(F, G)$, $\text{inj}(F, G)$ and $\text{sub}(F, G)$ denote the number of homomorphisms from F to G, the number of injective homomorphisms from F to G and the number of distinct copies of F in G, respectively. We need the following result relating $\text{sub}(F, G)$ and $\text{inj}(F, G)$ for our purpose and provide its proof in appendix.

Proposition 1. $\text{sub}(F, G) = \text{inj}(F, G)/\text{aut}(F, F)$.

This proposition allows us to focus on computing the value of $\text{inj}(F, G)$, as one can compute $\text{aut}(F, F)$ for a graph F on n_F vertices in time $2^{\mathcal{O}(\sqrt{n_F \log n_F})}$ [3][1], which is subexponential in n_F.

3 Relating Counting Subgraphs to Counting Homomorphism

We first give a formula expressing the number of injective homomorphisms from F to G in terms of the number of graphs homomorphisms from F to G, using the principle of inclusion-exclusion.

Theorem 1. *Let F and G be two graphs with $|V(G)| = |V(F)|$. Then*

$$\text{inj}(F, G) = \sum_{W \subseteq V(G)} (-1)^{|W|} \hom(F, G[V(G) \setminus W]).$$

Proof. To prove the theorem, we first show that if there is an injective homomorphism f from F to G then its contribution to the sum is exactly one. Notice that since $|V(G)| = |V(F)|$, an injective homomorphism only contributes when $W = \emptyset$. From this we conclude that injective homomorphisms are counted only once in the right hand side. Since we are counting homomorphisms, in the right hand side sum we also count maps which are not injective. Next we show that if a map h is not an injective homomorphism then its total contribution to the sum is zero, which will conclude the proof of the theorem. Observe that since h is not an injective homomorphism it misses some vertices of $V(G)$. Let $\widetilde{V} = im(h)$ be the image of h in $V(G)$. Since h is not an injective homomorphism, we infer that $X = V(G) \setminus \widetilde{V} \neq \emptyset$. We now observe that h is counted only when we are counting homomorphisms from $V(F)$ to $G[V(G) \setminus W]$ such that $W \subseteq X$. The total contribution of h in the sum, taking into account the signs, is $\sum_{i=0}^{|X|} \binom{|X|}{i} (-1)^i = (1-1)^{|X|} = 0$. Thus, we have shown that if h is not an

[1] In fact, here for a given graph F they solve a harder problem of computing its automorphism group and its generators. Please refer to Section 7 of [5] for further discussion.

injective homomorphism then its contribution to the sum is zero. This completes the proof of the theorem. □

For the rest of the section, we assume that we can count the number of graph homomorphisms from F to all the graphs $G[W]$ in time $t(n)$, where $|F| \leq |G| = n$ and $W \subseteq V(G)$. Then as a consequence of Theorem 1 we can compute the value of $\text{inj}(F, G)$ in time $\mathcal{O}(2^n \cdot t(n))$ when the size of $V(F)$ and $V(G)$ is n. A natural question then would be: what happens when the size of $V(F)$, say n_F is less than the size of $V(G)$, say n. An easy solution will be to enumerate all subsets V' of size n_F of $V(G)$ and then compute $\text{inj}(F, G[V'])$. But this will take time $\mathcal{O}(\binom{n}{n_F}2^{n_F}t(n))$, which in the worst case, could be equal to $\mathcal{O}(3^n \cdot t(n))$. In the remaining of this section let us observe how to reduce the time complexity to $\mathcal{O}(2^n \cdot t(n))$ by using the recently developed $\mathcal{O}(2^n n)$ algorithm for subset convolution [9].

Let f and g be functions that associate with every subset $S \subseteq V(G)$ elements $f(S)$ and $g(S)$ respectively of the ring $\mathbb{Z}$. Then the *convolution* of f and g is defined as follows: $\forall S \subseteq V(G) \quad (f * g)(S) = \sum_{T \subseteq S} f(T)g(S \setminus T)$. Let F and G be graphs of order n_F and n respectively. For a subset $S \subseteq V(G)$, we put $f(S) = (-1)^{|S|}$, and $g(S) = \text{hom}(F, G[S])$. Let $T \subseteq V(G)$ be a subset of size n_F. Then $\text{inj}(F, G[T]) = \sum_{Q \subseteq T}(-1)^{|Q|}\,\text{hom}(F, G[T \setminus Q]) = (f * g)(T)$. Now we compute the multi-set $X = \{\text{inj}(F, G[T]) \mid |F| = |T| = n_F \ \wedge \ T \subseteq V(G)\}$ in time $\mathcal{O}(2^n \cdot t(n))$ using the result from [9]. Moreover, $\text{inj}(F, G) = \sum_{x \in X} x$. The above discussion brings us to the following theorem.

Theorem 2. *Let F and G be two graphs with n_F and n vertices respectively, with $n_F \leq n$. Furthermore, we assume that the value* $\text{inj}(F, G[S])$ *is computable in time $t(n)$ for every $S \subseteq V(G)$. Then* $\text{inj}(F, G)$ *is computable in time $\mathcal{O}(2^n \cdot t(n))$.*

4 Classical Results

In this section we give alternative algorithms for a few classical algorithms through the method of counting homomorphisms.

Counting Hamiltonian Cycles – Kohn-Gottlieb-Kohn-Karp Algorithm. Let $\#\text{HAM}(G)$ denote the number of Hamiltonian cycles in a graph G and F be a cycle of length n then $\text{sub}(F, G) = \#\text{HAM}(G)$. It is also known that if F is a cycle C_n, then $\text{hom}(C_n, H) = \sum_{i=1}^{n} \lambda_i^n$, where $\lambda_1, \ldots, \lambda_n$ are the eigenvalues of the adjacency matrix of H and $\text{aut}(C_n, C_n) = 2n$ [11]. Using these results and Theorem 1, we can compute $\#\text{HAM}(G)$ in time $2^n n^{\mathcal{O}(1)}$ and polynomial space.

Chromatic Polynomial – Björklund-Husfeldt-Koivisto Algorithm. A *proper k-coloring* of a graph G is a function $f : V(G) \to \{1, \ldots, k\}$ such that for every edge $uv \in E(G)$, $f(u) \neq f(v)$. A well known polynomial associated with a graph G, is its CHROMATIC POLYNOMIAL. The *rank* of the graph G is $r(G) = |V(G)| - \eta(G)$, where $\eta(G)$ is the number of connected components of G. The CHROMATIC POLYNOMIAL of G is defined as $\chi(G; x) =$

$\sum_{E' \subseteq E(G)} (-1)^{|E'|} x^{|V(G)|-r(E')}$, where $r(E')$ is equal to the rank of the subgraph of G with vertex set $V(G)$ and the edge set E'. The polynomial derives its name due to the fact that for every fixed $k \geq 0$, $\chi(G;k)$ is the number of proper k-colorings of G. The *chromatic number* of G is the smallest integer $k > 0$ for which $\chi(G;k) > 0$. It is well known that for every $k > 0$, $\chi(G;k) = \hom(G, K_k)$. Here K_k is a clique of size k. However, to compute the chromatic polynomial of a graph, we have to look at homomorphisms from "the other side". A k-coloring of a graph G can also be viewed as a partition of the vertex set of the given graph into k independent sets, that is, a partition $(V_1, \ldots, V_k)$ of $V(G)$ such that for every $i \in \{1, \ldots, k\}$, $G[V_i]$ has no edges. For our purpose we reformulate the problem of coloring, as a problem of partitioning into k cliques in the complement graph. We model this as a problem of subgraph isomorphism as follows: we guess the sizes $t_1, t_2, \ldots, t_k$ of these cliques, where $\sum_i t_i = n$. By making use of this observation, it is possible to compute the chromatic polynomial of an n-vertex graph in time $2^{n+\mathcal{O}\sqrt{n}}$ and by making use of $2^n \times (n+1)$ space.

Number of Perfect Matchings in Bipartite Graphs – Ryser's Formula. Let G be a bipartite graph on an even number of vertices, say n, with $V(G)$ being partitioned into L and R of the same size. Then the RYSER'S FORMULA says that $\#\ \mathrm{PM}(G) = \sum_{X \subseteq R} (-1)^{|X|} \prod_{u \in L} \left(\sum_{v \notin X} 1_{[uv \in E(G)]} \right)$, where $\#\mathrm{PM}(G)$ is the number of perfect matchings in G. The sum $\sum_{v \notin X} 1_{[uv \in E(G)]}$ counts the number of u's neighbors not in X. Thus, we can count the number of perfect matchings in a bipartite graph in time $\mathcal{O}(2^{n/2} n^2)$. If we take F as $n/2$ disjoint copies of an edge then $\#\ \mathrm{PM}(G) = \mathrm{sub}(F, G)$. By using Theorem 1, it is easy to obtain an algorithm to compute the value of $\#\ \mathrm{PM}(G)$ in time $2^n n^{\mathcal{O}(1)}$. We will use the notion of saturating homomorphism in Section 5 to compute $\#\mathrm{PM}(G)$ in faster time, and in particular in time $\mathcal{O}(2^{n/2} n^2)$ in bipartite graphs.

5 New Applications

In this section we give new applications of Theorems 1 and 2 and show their wider applicability.

Set Saturating Homomorphisms and Ryser's Formula. In this subsection we give a faster poly-space algorithm for counting perfect matchings in graphs with large independent sets. To do so we first generalize the notion of graph homomorphism and prove a generalization of Theorem 1. Let S be a given subset of $V(G)$, then a homomorphism f from F to G is called *S-saturating* if (*a*) $S \subseteq f(V(F))$ and (*b*) for all $v \in S$, $|f^{-1}(v)| = 1$. By S-hom(F, G) we denote the number of S-saturating homomorphisms. Observe that for $S = \emptyset$ an S-saturating homomorphism is simply a homomorphism. Along the lines of Theorem 1 and using the method of inclusion-exclusion we can prove the following theorem.

Theorem 3. *Let F and G be two graphs with $|V(G)| = |V(F)|$, and $S \subseteq V(G)$. Then*

$$\mathrm{inj}(F, G) = \sum_{W \subseteq V(G) \setminus S} (-1)^{|W|} S\text{-hom}(F, G[V(G) \setminus W]).$$

Theorem 4. *Let G be an n-vertex graph and $S \subseteq V(G)$ be an independent set of G. There is an algorithm which counts the number of perfect matchings of G in time $2^{n-|S|} \cdot n^{\mathcal{O}(1)}$.*

It is well known that the chromatic number of a graph is always at most its average degree (or degeneracy) plus one. Also by Brooks theorem, the chromatic number of a graph is at most the maximum vertex degree, unless the graph is complete or an odd cycle. Then by Theorem 4, we obtain the following result.

Corollary 1. *Let G be an n-vertex graph and let δ and Δ be its average and maximum degrees. Then $\#\mathrm{PM}(G)$ is computable in time $\min\{2^{n-\frac{n}{\delta+1}}, 2^{n-\frac{n}{\Delta}}\}n^{\mathcal{O}(1)}$. In particular if G is a bipartite graph then we can find $\#\mathrm{PM}(G)$ in time $2^{n/2}n^{\mathcal{O}(1)}$.*

Subgraph Isomorphism when F has bounded Treewidth. Here, we give an algorithm for counting subgraphs isomorphic to F in G, when F is given together with a tree-decomposition of width t. We first mention an algorithm to compute $\hom(F, G)$, when F is a graph of bounded treewidth.

Proposition 2 ([14]). *Let F and G be two graphs on n_F and n vertices respectively, given together with a tree-decomposition of width t of F. Then $\hom(F, G)$ is computable in time $\mathcal{O}(n_F \cdot n^{t+1} \min\{t, n\})$ and space $\mathcal{O}(\log n_F \cdot n^{t+1})$.*

Theorem 5. *Let F and G be two graphs on n_F and n vertices respectively, given together with a tree-decomposition of width t of F. Then $\mathrm{sub}(F, G)$ is computable in time $\mathcal{O}(2^n \cdot n_F \cdot n^{t+1} \min\{t, n\})$. The space requirement is $\mathcal{O}(\log n_F \cdot n^{t+1})$ if $n_F = n$ else it is $\mathcal{O}(\log n_F \cdot n^{t+1} + 2^n)$.*

Bandwidth is one of the well studied graph layout problems. We begin the section by defining it formally. A *layout* of a graph G on n vertices is a map $f : V(G) \to \{1, \ldots, n\}$. In the Bandwidth problem, the objective is to find a layout function for a given graph G, such that $\max_{uv \in E(G)} |f(u) - f(v)|$ is minimized.

The following lemma formulates the Bandwidth problem as an instance of the Subgraph Isomorphism problem. By P_n we denote a path on n vertices. For a graph G, the rth power of the graph is denoted by G^r. This graph is on the same vertex set $V(G)$, but we add an edge between two distinct vertices u and v if there is a path of length at most r between them in G.

Lemma 1. *Let G be a graph on n vertices. Then G has a layout of bandwidth b if and only if there is an injective homomorphism from G to P_n^b.*

Using Lemma 1 together with Theorem 5 we obtain the following theorem.

Theorem 6. *Given a graph G on n vertices together with a tree decomposition of width t, it is possible to find the number of optimum bandwidth layouts in time $2^{n+t\log_2 n}n^{\mathcal{O}(1)}$ and space $\mathcal{O}(\log n \cdot n^{t+1})$. In particular if G is a tree then we can compute the number of optimum bandwidth layouts in time $\mathcal{O}(2^n \cdot n^3)$.*

Corollary 2. *Given a graph G on n vertices excluding a graph M on r vertices as a minor. It is possible to compute the number of optimum bandwidth layouts in time* $2^{n+r^{3/2}\sqrt{n}\log_2 n} \cdot n^{\mathcal{O}(1)}$.

Degree Constrained Spanning Tree Problem. HAMILTONIAN PATH is one of the earliest known problems for which an exact algorithm with time complexity $2^n n^{\mathcal{O}(1)}$ was known. This problem can also be seen as a special case of finding a spanning tree with certain degree constrains on the vertices. More precisely the DEGREE CONSTRAINED SPANNING TREE (DCST) problem is defined as follows: Given a connected undirected graph G and a vector of size n, $\hat{a} = (a_1, a_2, \cdots, a_n)$, find a spanning tree T of G (if one exists), such that there is a bijective mapping $g : V(G) \rightarrow \{a_1, a_2, \ldots, a_n\}$ with the property that $deg_T(v) = g(v)$. A variation of DCST called MODIFIED DEGREE CONSTRAINED SPANNING TREE (MDCST) is defined by replacing the condition of $deg_T(v) = g(v)$ with $deg_T(v) \leq g(v)$ in DCST.

Theorem 7. *Let G be a graph on n vertices and $\hat{a} = (a_1, \ldots, a_n)$ be a vector of length n. Then we can count the number of feasible solutions to DCST and MDCST in* $5.912^n \cdot n^{\mathcal{O}(1)}$ *time.*

We solve the MINIMUM DEGREE SPANNING TREE problem by finding the smallest $2 \leq i \leq n-1$ for which MDCST problem returns yes with $\hat{a} = (i, i, \cdots, i)$ resulting in the following.

Corollary 3. MINIMUM DEGREE SPANNING TREE *on a graph on n vertices can be solved in time* $5.912^n \cdot n^{\mathcal{O}(1)}$.

Counting Graphs Excluding a Fixed Minor. In this section we apply our results to count planar subgraphs of maximum size or more generally maximum sized subgraphs that do not contain some fixed graph M as a minor. More precisely, we consider MAXIMUM PLANAR SUBGRAPH and MAXIMUM M-MINOR FREE SUBGRAPH problems. Here given a graph G the objective is to find a subset $E' \subseteq E(G)$ of maximum size such that the graph $G_{E'}$ on the vertex set $V(G)$ and the edge set E' is planar and M-minor free respectively.

A naïve algorithm for the above problems is to enumerate all edge subsets of the given graph, for each subset test whether the subgraph induced by the edge set has the desired properties and output the feasible subgraph with the maximum number of edges. For a graph G on n vertices and m edges this algorithm will take $2^m \cdot n^{\mathcal{O}(1)}$ time. Let us remark that even for the decision version of these problems, no vertex exponential (i.e. $c^n n^{\mathcal{O}(1)}$) time algorithms were known. The basic ideas used here are similar to the ones used for trees, namely to prove that all unlabeled graphs on n vertices in the considered class can be enumerated in time $\mathcal{O}(c^n)$ for some constant c, and then for each element of the enumerated class, applying Theorem 5 to count the number of subgraphs of G isomorphic to it.

Let M be a fixed graph. Norine, Seymour, Thomas, and Wollan [27] proved that the number of labelled n-vertex graphs of size n in a family of graphs excluding M as a minor is at most $n!c^n$ for some constant c (depending on M).

We prove a more general result here, namely that the number of unlabelled M-minor-free n-vertex graphs is at most c^n for some constant c depending only on M. Let us remark that since the number of labelings is at most $n!$, our result immediately yields the main theorem from [27].

Theorem 8. *Let $\mathcal{G}$ be a family of unlabelled n-vertex graphs that do not contain some fixed graph M as a minor. Then there exists a constant c_M such that the number of non-isomorphic graphs in $\mathcal{G}$ is at most c_M^n. Moreover, the elements of $\mathcal{G}$ can be enumerated in time $\mathcal{O}(c_M^n)$.*

As far as we have Theorem 8, by making use of Theorem 5 and the fact, that the treewidth of an n-vertex graph excluding a fixed graph as a minor is $\mathcal{O}(\sqrt{n})$, we conclude with the main result of this section.

Theorem 9. *Given a graph G on n vertices the counting version of the* Maximum M-minor Free Subgraph *problem can be solved in time $\mathcal{O}(c^n) = 2^{\mathcal{O}(n)}$ for some constant $c = c_M$. All these algorithms use $n^{\mathcal{O}(\sqrt{n})}$ space.*

$\mathcal{H}$-Packing and some of its Variants. Let $\mathcal{H}$ and $\mathcal{G}$ be two graph classes. By *$\mathcal{H}$-subgraph* of G we mean any subgraph of G that belongs to $\mathcal{H}$. Given a graph $G \in \mathcal{G}$, the Covering problem asks for finding a subset W of $V(G)$ of minimum size which covers all the $\mathcal{H}$-subgraphs of G. On the other hand the Packing problem asks for finding a maximum number of vertex disjoint copies of $\mathcal{H}$-subgraphs in G. In other words, the packing number of G with respect to the class $\mathcal{H}$ is defined as

$$\begin{aligned}\mathbf{pack}_{\mathcal{H}}(G) &= \max \{k \mid \exists \text{ a partition} V_1, \cdots, V_k \text{ of } V(G) \text{ such that} \\ &\quad \forall i \in \{1, \cdots, k\},\ \exists_{H \in \mathcal{H}} H \subseteq G[V_i]\}.\end{aligned}$$

Let M be a fixed graph. In this section we show that if $\mathcal{H}$ is a graph class excluding M as a minor (that is no $H \in \mathcal{H}$ contains M as a minor) then there exists a constant c depending only on M such that it is possible to compute the value of $\mathbf{pack}_{\mathcal{H}}(G)$ in time $c^n n^{\mathcal{O}(1)}$ for any graph G on n vertices.

Theorem 10. *Let G be a graph on n vertices and $\mathcal{H}$ be a graph class excluding a fixed graph M as a minor then the value of $\mathbf{pack}_{\mathcal{H}}(G)$ can be computed in time $c^n n^{\mathcal{O}(1)} = 2^{\mathcal{O}(n)}$, where c is a constant depending only on M.*

Corollary 4. *Given a graph G on n vertices,* Maximum Vertex Disjoint Cycles *and* Maximum Odd Sized Vertex Disjoint Cycles *problems can be solved in time $2^{n+\mathcal{O}(\sqrt{n})}$, whereas* Maximum l-Cycle Packing *problem can be solved in time $\mathcal{O}(n^6 \cdot 2^n)$.*

6 Color Coding

Let $c\colon V(G) \to \{1, 2, \ldots, k\}$ be a coloring (not necessarily proper) of the vertex set of a graph G in k colors. Thus $V_i = c^{-1}(i)$ is not necessarily an independent set. For a graph F on k vertices, we say that an injective homomorphism f from

F to G is *colorful*, if each vertex of the image of F is colored by a distinct color. We denote the number of colorful injective homomorphisms from a graph F to a colored graph G by col-inj(F, G). Let us remark, that the number of colorful copies of F in G is equal to col-inj$(F, G)/\text{aut}(F, F)$. Let G^* be the graph obtained from G by deleting all the mono-chromatic edges, that is, by making each color class V_i into an independent set.

Theorem 11. *Let $c\colon V(G) \to \{1, 2, \ldots, k\}$ be a coloring of G and $V_i = c^{-1}(i)$. Then*

$$\text{col-inj}(F, G) = \text{col-inj}(F, G^*) = \sum_{I \subseteq \{1,2,\ldots,k\}} (-1)^{|I|}\text{hom}(F, G^*[V(G^*) \setminus \cup_{i \in I} V_i]).$$

Combined with Proposition 2, Theorem 11 yields the following result.

Theorem 12. *Let F be a k-vertex graph given with its tree decomposition of width t and let G be an n-vertex graph. A subgraph of G isomorphic to F (if one exists) can be found in either $\mathcal{O}((2e)^k \cdot k \cdot t \cdot n^{t+1})$ expected time and $\mathcal{O}(\log k \cdot n^{t+1})$ space or deterministically in time $\mathcal{O}((2e)^{k+o(k)} \cdot k \cdot t \cdot n^{t+1})$ and space $\mathcal{O}(\log k \cdot n^{t+1})$. Here, e is the base of natural logarithm.*

7 Conclusion and Discussions

In this paper we introduced an approach for counting subgraphs in a graph via counting graph homomorphisms in the realm of exact and parameterized algorithms. This approach yields various new algorithms for many basic problems like counting the number of perfect matchings, optimum bandwidth layouts, degree constrained spanning trees, maximum planar subgraphs beside others. On the other hand it also unified several well-known results in exact algorithms including counting coloring, Hamiltonian cycles and perfect matchings of bipartite graphs. These alternate algorithms generalize and unify algorithms for many well known problems. Let us remark that most of our results can be easily extended to weighted directed graphs. We believe that our method is generic and will find many more applications. One important question which remains unanswered is: Can sub(F, G) be computed in $2^{\mathcal{O}(n)}$ time? In particular, we do not know the answer to this question even for the very special case, when the maximum degree of F is 3.

References

1. Alon, N., Yuster, R., Zwick, U.: Color-coding. J. ACM 42, 844–856 (1995)
2. Babai, L.: Moderately exponential bound for graph isomorphism. In: Gecseg, F. (ed.) FCT 1981. LNCS, vol. 117, pp. 34–50. Springer, Heidelberg (1981)
3. Babai, L., Kantor, W.M., Luks, E.M.: Computational complexity and the classification of finite simple groups. In: FOCS, pp. 162–171 (1983)
4. Bax, E.T.: Algorithms to count paths and cycles. Inform. Process. Lett. 52, 249–252 (1994)
5. Beals, R., Chang, R., Gasarch, W.I., Torán, J.: On finding the number of graph automorphisms. Chicago J. Theor. Comput. Sci. (1999)
6. Bernardi, O., Noy, M., Welsh, D.: On the growth rate of minor-closed classes of graphs, `http://arxiv.org/abs/0710.2995`

7. Björklund, A., Husfeldt, T.: Exact algorithms for exact satisfiability and number of perfect matchings. Algorithmica 52(2), 226–249 (2008)
8. Björklund, A., Husfeldt, T.: Inclusion-exclusion algorithms for counting set partitions. In: FOCS, pp. 575–582 (2006)
9. Björklund, A., Husfeldt, T., Kaski, P., Koivisto, M.: Fourier meets möbius: fast subset convolution. In: STOC, pp. 67–74 (2007)
10. Blankenship, R., Oporowski, B.: Book embeddings of graphs and minor-closed classes. In: Proceedings of the 32nd Southeastern International Conference on Combinatorics, Graph Theory and Computing
11. Borgs, C., Chayes, J., Lovász, L., Sós, V.T., Vesztergombi, K.: Counting graph homomorphisms. In: Topics in discrete mathematics. Algorithms Combin., vol. 26, pp. 315–371. Springer, Berlin (2006)
12. Cygan, M., Pilipczuk, M.: Faster exact bandwidth. In: Broersma, H., Erlebach, T., Friedetzky, T., Paulusma, D. (eds.) WG 2008. LNCS, vol. 5344, pp. 101–109. Springer, Heidelberg (2008)
13. Dalmau, V., Jonsson, P.: The complexity of counting homomorphisms seen from the other side. Theoret. Comput. Sci. 329, 315–323 (2004)
14. Díaz, J., Serna, M.J., Thilikos, D.M.: Counting h-colorings of partial k-trees. Theor. Comput. Sci. 281, 291–309 (2002)
15. Dyer, M., Greenhill, C.: The complexity of counting graph homomorphisms. Random Structures Algorithms 17, 260–289 (2000)
16. Feige, U.: Coping with the NP-Hardness of the Graph Bandwidth Problem. In: Halldórsson, M.M. (ed.) SWAT 2000. LNCS, vol. 1851, pp. 10–19. Springer, Heidelberg (2000)
17. Feige, U., Hajiaghayi, M.T., Lee, J.R.: Improved approximation algorithms for minimum-weight vertex separators. In: STOC, pp. 563–572 (2005)
18. Grohe, M.: The complexity of homomorphism and constraint satisfaction problems seen from the other side. J. ACM 54, Art. 1, 24 (electronic) (2007)
19. Hell, P., Nešetřil, J.: On the complexity of H-coloring. J. Combin. Theory Ser. B 48, 92–110 (1990)
20. Hell, P., Nešetřil, J.: Graphs and homomorphisms. Oxford Lecture Series in Mathematics and its Applications, vol. 28. Oxford University Press, Oxford (2004)
21. Karp, R.M.: Dynamic programming meets the principle of inclusion and exclusion. Oper. Res. Lett. 1, 49–51 (1982)
22. Kohn, S., Gottlieb, A., Kohn, M.: A generating function approach to the traveling salesman problem. In: Proceedings of the annual ACM conference, pp. 294–300 (1977)
23. Koivisto, M.: An $O(2^n)$ algorithm for graph coloring and other partitioning problems via inclusion-exclusion. In: FOCS, pp. 583–590 (2006)
24. Lovász, L.: Operations with structures. Acta Math. Hung. 18, 321–328 (1967)
25. Malitz, S.M.: Genus g graphs have pagenumber $O(\sqrt{g})$. J. Algorithms 17, 85–109 (1994)
26. Naor, M., Schulman, L.J., Srinivasan, A.: Splitters and near-optimal derandomization. In: FOCS, pp. 182–191 (1995)
27. Norine, S., Seymour, P.D., Thomas, R., Wollan, P.: Proper minor-closed families are small. J. Comb. Theory, Ser. B 96, 754–757 (2006)
28. Ryser, H.J.: Combinatorial mathematics. The Carus Mathematical Monographs, vol. 14. The Mathematical Association of America (1963)
29. Valiant, L.G.: The complexity of computing the permanent. Theoret. Comput. Sci. 8, 189–201 (1979)

External Sampling*

Alexandr Andoni[1], Piotr Indyk[1], Krzysztof Onak[1], and Ronitt Rubinfeld[1,2]

[1] Massachusetts Institute of Technology, Cambridge, MA, USA
[2] Tel Aviv University, Tel Aviv, Israel
{andoni,indyk,konak}@mit.edu, ronitt@csail.mit.edu

Abstract. We initiate the study of sublinear-time algorithms in the external memory model [1]. In this model, the data is stored in blocks of a certain size B, and the algorithm is charged a unit cost for each block access. This model is well-studied, since it reflects the computational issues occurring when the (massive) input is stored on a disk. Since each block access operates on B data elements in parallel, many problems have external memory algorithms whose number of block accesses is only a small fraction (e.g. $1/B$) of their main memory complexity.

However, to the best of our knowledge, no such reduction in complexity is known for *any* sublinear-time algorithm. One plausible explanation is that the vast majority of sublinear-time algorithms use random sampling and thus exhibit no locality of reference. This state of affairs is quite unfortunate, since both sublinear-time algorithms and the external memory model are important approaches to dealing with massive data sets, and ideally they should be combined to achieve best performance.

In this paper we show that such combination is indeed possible. In particular, we consider three well-studied problems: testing of *distinctness*, *uniformity* and *identity* of an empirical distribution induced by data. For these problems we show random-sampling-based algorithms whose number of block accesses is up to a factor of $1/\sqrt{B}$ smaller than the main memory complexity of those problems. We also show that this improvement is optimal for those problems.

Since these problems are natural primitives for a number of sampling-based algorithms for other problems, our tools improve the external memory complexity of other problems as well.

1 Introduction

Random sampling is one of the most fundamental methods for reducing task complexity. For a wide variety of problems, it is possible to infer an approximate solution from a random sample containing only a small fraction of the data, yielding algorithms with sublinear running times. As a result, sampling is often the method of choice for processing massive data sets. Inferring properties of data from random sample has been a major subject of study in several areas, including statistics, databases [2,3], theoretical computer science [4,5,6,7], ...

* The research was supported in part by David and Lucille Packard Fellowship, by MADALGO (Center for Massive Data Algorithmics, funded by the Danish National Research Association), by Marie Curie IRG Grant 231077, by NSF grants 0514771, 0728645, and 0732334, and by a Symantec Research Fellowship.

S. Albers et al. (Eds.): ICALP 2009, Part I, LNCS 5555, pp. 83–94, 2009.

However, using random sampling for massive data sets encounters the following problem: typically, massive data sets are not stored in main memory, where each element can be accessed at a unit cost. Instead, the data is stored on external storage devices, such as a hard disk. There, the data is stored in blocks of certain size (say, B), and each disk access returns a block of data, as opposed to an individual element. In such models [1], it is often possible to solve problems using roughly T/B disk accesses, where T is the time needed to solve the problem in main memory. The $1/B$ factor is often crucial to the efficiency of the algorithms, given that (a) the block size B tends to be large, on the order of thousands and (b) each block access is many orders of magnitude slower than a main memory lookup. Unfortunately, implementations of sampling algorithms typically need to perform[1] one block access per each sampled element [2]. Effectively, this means that out of B data elements retrieved by each block access, $B-1$ elements are discarded by the algorithm. This makes sampling algorithms a much less attractive option for processing massive data sets.

Is it possible to improve the sampling algorithms by utilizing the *entire* information stored in each accessed block? At the first sight, it might not seem so. For example, consider the following basic sampling problem: the input data is a binary sequence such that the fraction of ones is either at most f or at least $2f$, and the goal is to detect which of these two cases occurs. A simple argument shows that any sampling algorithm for this problem requires $\Omega(1/f)$ samples to succeed with constant probability, since it may take that many trials to even retrieve one 1. It is also easy to observe that the same lower bound holds even if all elements within each block are equal (as long as the total number of blocks is $\Omega(1/f)$), in which case sampling blocks is equivalent to sampling elements. Thus, even for this simple problem, sampling blocks does not yield any reduction in the number of accesses.

Our Results. Contrary to the above impression, in this paper we show that there are natural problems for which it is possible to reduce the number of sampled blocks. Specifically, we consider the problem of testing properties of empirical distributions induced by the data sets. Consider a data set of size m with support size (i.e., the number of distinct elements) equal to n. Let p_i be the fraction of times an element i occurs in the data set. The vector p then defines a probability distribution over a set of distinct elements in the data set. We address the following three well-studied problems:

- Distinctness: are all data elements distinct (i.e., $n = m$), or are there at least ϵm duplicates?
- Uniformity: is p uniform over its support, or is it ϵ-far[2] from the uniform distribution?
- Identity: is p identical to an explicitly given distribution q, or is it ϵ-far from q?

[1] It is possible to retrieve more samples per block if the data happens to be stored in a random order. Unfortunately, this is typically not guaranteed.

[2] We measure the distance between distribution using the standard variational distance, which is the maximum probability with which a statistical test can distinguish the two distributions. Formally, a distribution p is ϵ-far from a distribution q, if $\|p-q\|_1 \geq \epsilon$, where p and q are interpreted as vectors.

Note that testing identity generalizes the first two problems. However, the algorithms for distinctness and uniformity are simpler and easier to describe.

It is known [8,9,10] that, if the elements are stored in main memory, then $\tilde{\Theta}(\sqrt{n})$ memory accesses are sufficient and necessary to solve both uniformity and identity testing. In this paper we give an external memory algorithm which uses only $\tilde{O}(\sqrt{m/B})$ block accesses. Thus, for m comparable to n, the number of accesses is reduced by a factor of $\sqrt{B}$. It also can be seen that this bound cannot be improved in general: if $B = m/n$, then each block could consist of equal elements, and thus the $\tilde{\Theta}(\sqrt{n}) = \tilde{\Theta}(\sqrt{m/B})$ main memory lower bound would apply.

From the technical perspective, our algorithms mimic the sampling algorithms of [10,11,9]. The key technical contribution is a careful analysis of those algorithms. In particular, we show that the additional information obtained from sampling blocks of data (as opposed to the individual elements) yields a substantial reduction of the variance of the estimators used by those algorithms.

Applications to Other Problems. The three problems from above are natural primitives for a number of other sampling-based problems. Thus, our algorithms improve the external memory complexity of other problems as well. Below we describe two examples of problems where our algorithms and techniques apply immediately to give improved guarantees in the external memory model.

The first such problem is testing graph isomorphism. In this problem, the tester is to decide, given two graphs G and H on n vertices, whether G are H are isomorphic or at least ϵn^2 edges of the graphs must be modified to achieve a pair of isomorphic graphs. Suppose one graph, G, is known to the tester (for instance, it is a fixed graph with an easily computable adjacency relation), and the other graph, H, is described by the adjacency matrix written in the row-major order on the disk. Then, our algorithm for identity testing improves the sample complexity of the Fischer and Matsliah algorithm [12] by essentially a factor of $\sqrt{B}$. Formally, in the main memory, the Fischer and Matsliah algorithm uses $O(\sqrt{n} \cdot \text{poly}(\log n, 1/\epsilon))$ queries to H. Combined with our external memory identity tester, algorithm will use only $O((\sqrt{n/B}+1) \cdot \text{poly}(\log n, 1/\epsilon))$ samples.

The second application is a set of questions on testing various properties of metric spaces, such as testing whether a metric is a tree-metric or ultra-metric. In [13], Onak considers several such properties, for which he gives algorithms whose sampling complexity in main memory is of the form $O(\alpha/\epsilon + n^{(\beta-1)/\beta}/\epsilon^{1/\beta})$, where $\alpha \geq 1$ and $\beta \geq 2$ are constant integers. The additive term $n^{(\beta-1)/\beta}/\epsilon^{1/\beta}$ corresponds to sampling for a specific β-tuple. Using our techniques for distinctness testing, it can easily be shown that whenever an algorithm from [13] requires $O(\alpha/\epsilon + n^{(\beta-1)/\beta}/\epsilon^{1/\beta})$ samples, the sample complexity in external memory can be improved to $O(\alpha/\epsilon + (n/B)^{(\beta-1)/\beta}/\epsilon^{1/\beta})$, provided a single disk block contains B points.

2 Distinctness Problem: Finding a Single Repetition

We start with the distinctness problem, which is the easiest problem where we can show how to harness the power of block queries. The distinctness problem consists of distinguishing inputs representing sets of m distinct elements, from those inputs representing multisets which have at least $\epsilon \cdot m$ repetitions. This problem was studied in [14] in the standard memory model and is known to have worst-case complexity $\Theta(\sqrt{m/\epsilon})$. In the main memory setting, the solution is a variant of the birthday paradox argument, an argument that will be needed (indirectly) for our analysis as well.

Fact 1 (The Birthday Paradox). *Let S be a set of size m. For $\alpha \in (0,1)$, let P be a set of αm disjoint pairs of elements in S. With probability $1/2$, a random subset of size $O(\sqrt{m/\alpha})$ contains two elements that belong to the same pair in P.*

The following theorem shows that it is enough to sample $O(\sqrt{m/\epsilon B})$ blocks to solve the distinctness problem.

Theorem 2. *Let m be the length of the sequence stored on disk in blocks of length B. If at least ϵm elements must be removed from the sequence to achieve a sequence with no repetitions, then $O(\sqrt{m/\epsilon B})$ block queries suffice to find with constant probability an element that appears at least twice in the sequence.*

Proof. For each element that appears at least twice in the sequence, we divide all its occurrences into pairs. This gives us at least $\Omega(\epsilon m)$ pairs of identical elements, and it suffices to detect any of them. Let us now consider possible layouts of these pairs into blocks. Let f_1 be the number of the pairs that have both elements in the same block, and let f_2 be the number of the pairs with elements in two different blocks. At least one of f_1 and f_2 must be $\Omega(\epsilon m)$.

If $f_1 = \Omega(\epsilon m)$, then the pairs must occupy at least an ϵ-fraction of all the m/B blocks, so $O(1/\epsilon)$ block-queries suffice to find at least one of the pairs.

Suppose now that $f_2 = \Omega(\epsilon m)$. For each block that contains an element of one of the pairs counted by f_2, there is a block that contains the other element of the pair. Consider the following procedure that creates a set P of disjoint pairs of blocks of size $\Omega(\epsilon m/B)$. Initially, let S be the set of all blocks. As long as there is a pair of blocks in S that contain two corresponding elements of a pair counted by f_2, we add the pair to P and remove the pair from S. Note that one such step erases at most $4B$ of the pairs counted by f_2. Thus, at termination, P contains at least $\Omega(\epsilon m/B)$ disjoint pairs of blocks such that each of the pairs exhibits a repetition. By Fact 1, it suffices to sample $O(\sqrt{m/\epsilon B})$ blocks to find such a pair, and hence, a repetition of elements. □

3 Testing Uniformity

In this section we show that we can test uniformity of a distribution with $O(\sqrt{\frac{m}{B}} \cdot \frac{1}{\epsilon} \cdot \log B)$ queries. Note that, for $m = \Theta(n)$, this improves over the usual (in main memory) testing by a factor of $\tilde{\Theta}(\sqrt{B})$.

Theorem 3. *Let m be the length of a sequence of elements in $[n]$, and assume $m \le nB$. The sequence is stored in blocks on disk, and each block contains B elements of the sequence. Let p be the empirical distribution of the sequence. There is an algorithm that samples $O\left(\frac{1}{\epsilon}\sqrt{\frac{m}{B}} \cdot \log B\right)$ blocks, and:*

– *accepts with probability* $\geq 2/3$ *if* p *is* $O\left(\frac{\epsilon}{\sqrt{Bm}\cdot\log B}\right)$*-close to uniformity on* $[n]$,
– *rejects with probability* $\geq 2/3$ *if* p *is* ϵ*-far from uniform on* $[n]$.

Our algorithm is given in Figure 1. Let $Q = C \cdot \frac{1}{\epsilon} \cdot \sqrt{\frac{m}{B}} \cdot \log B$ for a sufficiently big constant C.

UNIFORMITY-TEST(n, m, B, ϵ)

Let $Q = C \cdot \frac{1}{\epsilon} \cdot \sqrt{\frac{m}{B}} \cdot \log B$ for a sufficiently big constant C.
Pretest
1. Sample a set S of Q blocks (the set is sampled without replacement).
2. Fail if an element appears more than m/n times.

Test
3. Sample two sets S_1, S_2 of Q blocks each. The sets are sampled with replacement.
4. Count the number of collisions (of elements) between the two sets of samples. Let this number be W.
5. Fail if $W > (1 + \epsilon/2)\frac{(QB)^2}{n}$.

Fig. 1. Algorithm for testing uniformity in the block query model

3.1 Analysis: Proof of Theorem 3

We use the following notation. Set $s = m/n$. Let $P_{\alpha,i}$ be the number of occurrences of an element $i \in [n]$ in a block $\alpha \in [m/B]$. Note that $p_i = \frac{1}{m}\sum_\alpha P_{\alpha,i}$. Generally, $i, j \in [n]$ will denote elements (from the support of p), and $\alpha, \beta, \gamma, \delta \in [m/B]$ will denote indexes of blocks.

Proposition 4. *If* p *is uniform,* **Pretest** *stage passes (with probability 1).*

Lemma 5. *If* **Pretest** *stage passes with probability* $\geq 1/3$*, then:*

– *There are at most* $O(\frac{m/B}{Q}) = O(\frac{\epsilon}{\log B} \cdot \sqrt{\frac{m}{B}})$ *blocks that have more than* s *occurrences of some element.*
– p *is* $O(\epsilon\sqrt{B/m})$*-close to a distribution* q *such that* $\max_i q_i \leq O(\epsilon\sqrt{s/nB} \cdot \log B)$.

Proof. The first bullet is immediate. Let us call "bad" blocks the ones that have more than s occurrences of some element. "Good" ones are the rest of them.

To proceed with the second bullet, consider only the good blocks, since the bad ones contribute at most $O(B \cdot \epsilon\sqrt{m/B} \cdot \frac{1}{m}) = O(\epsilon\sqrt{B/m})$ fraction of the mass. We claim that for every i, and for every k such that $2^k \leq s$, there are at most $C_1 \cdot \frac{s}{2^k} \cdot \epsilon\sqrt{m/B}$ blocks that have at least 2^k occurrences of element i, for some sufficiently large C_1. Suppose for contradiction that for some i and k, there are more than $C_1 \cdot \frac{s}{2^k} \cdot \epsilon\sqrt{m/B}$ blocks that have $\geq 2^k$ occurrences of element i. The expected number of such blocks that the algorithm samples in the Pretest phase is $\geq C_2 s/2^k$, for some sufficiently large constant C_2. By the

Chernoff bound, the algorithm will sample more than $s/2^k$ such blocks with probability greater than 2/3. This means that the algorithm will sample more than s copies of i, and will reject the input with probability greater than 2/3, which contradicts the hypothesis.

We can now show that the number of occurrences of each element in good blocks is bounded. For each $k \in \{0, 1, \ldots, \lceil \log s \rceil\}$, there are at most $C_1 \frac{s}{2^k} \cdot \epsilon\sqrt{m/B}$ good blocks in which the number of occurrences of i is in the range $[2^k, 2^{k+1})$. Summing over all ranges, we see that the total number of occurrences of i is at most $O(s\epsilon\sqrt{m/B}\log s) = O(s\epsilon\sqrt{m/B}\log B)$, which implies that the distribution q defined by the good blocks is such that for each i,

$$q_i \le \frac{O(s\epsilon\sqrt{m/B} \cdot \log B)}{m(1 - O(\epsilon\sqrt{B/m}))} = O(\epsilon\sqrt{s/Bn} \cdot \log B). \qquad \square$$

Using the above lemma, we can assume for the rest of the proof that the input only contains blocks that have at most s occurrences of any element $i \in [n]$. If the input is ϵ-far from uniform, the number of blocks in S_1 and S_2 with more than s occurrences of an element is greater than a sufficiently high constant with a very small probability. Therefore, those blocks can decrease W by only a tiny amount, and have a negligible impact on the probability of rejecting an input that is ϵ-far from uniform. Moreover, this step changes the distribution by only at most $O(\epsilon\sqrt{B/m})$ probability mass. If the distribution is uniform, then the assumption holds *a priori.*

Let $w = W/B^2$ denote the sum of "weighted" collisions between blocks (i.e., if blocks α and β have z collisions, the pair (α, β) contributes z/B^2 to the "weighted" collision count w).

Proposition 6. *The expected number of weighted collisions is* $\mathbb{E}[w] = Q^2 \sum_i p_i^2$.

Let's call $t = \sum_i p_i^2$.

Lemma 7. *The variance of* w *is at most* $O(Q^2 ts/B + Q^3 t \max_i p_i)$.

Proof. Let $C_{\alpha,\beta}$ be the number of collisions between block α and β, divided by B^2. Thus $0 \le C_{\alpha,\beta} \le s/B$ (we would have $C_{\alpha,\beta} \le 1$ if the block could contain any number of occurrences of an element).

Note that $\mathbb{E}_{\alpha,\beta}[C_{\alpha,\beta}] = \sum_i p_i^2 = t$.

Let $\bar{C}_{\alpha,\beta} = C_{\alpha,\beta} - t$. We note that for any $\alpha, \beta, \gamma \in [m/B]$, we have that

$$\mathbb{E}\left[\bar{C}_{\alpha,\beta}\bar{C}_{\alpha,\gamma}\right] = \mathbb{E}\left[C_{\alpha,\beta}C_{\alpha,\gamma}\right] - t^2 \le \mathbb{E}\left[C_{\alpha,\beta}C_{\alpha,\gamma}\right]. \tag{1}$$

We bound the variance of w as follows.

$$\begin{aligned}
\mathbf{Var}[w] &= \mathbb{E}_{S_1,S_2}\left[\left(\sum_{\alpha,\beta \in S_1 \times S_2} \bar{C}_{\alpha,\beta}\right)^2\right] \\
&= Q^2 \mathbb{E}_{\alpha,\beta \in [m/B]}\left[(\bar{C}_{\alpha,\beta})^2\right] + \mathbb{E}\left[\sum_{\substack{(\alpha,\beta),(\delta,\gamma) \in S_1 \times S_2 \\ (\alpha,\beta) \ne (\delta,\gamma)}} \bar{C}_{\alpha,\beta} \cdot \bar{C}_{\delta,\gamma}\right] \\
&\le Q^2 \mathbb{E}_{\alpha,\beta \in [m/B]}\left[(\bar{C}_{\alpha,\beta})^2\right] + 2Q^3 \mathbb{E}_{\alpha,\beta,\gamma \in [m/B]}\left[\bar{C}_{\alpha,\beta} \cdot \bar{C}_{\alpha,\gamma}\right],
\end{aligned}$$

where we have used the fact that, if $\{\alpha,\beta\}\cap\{\gamma,\delta\}=\emptyset$, then $\mathbb{E}\left[\bar{C}_{\alpha,\beta}\cdot\bar{C}_{\delta,\gamma}\right]=0$. We upper bound each of the two terms of the variance separately. For the first term, we have

$$\begin{aligned}\mathbb{E}_{\alpha,\beta}\left[(\bar{C}_{\alpha,\beta})^2\right] \le \mathbb{E}_{\alpha,\beta}\left[(C_{\alpha,\beta})^2\right] &= \frac{B^2}{m^2}\sum_{\alpha,\beta}\left(\sum_i \frac{P_{\alpha,i}P_{\beta,i}}{B^2}\right)^2\\ &= \frac{1}{B^2m^2}\sum_i\sum_{\alpha,\beta}\sum_j P_{\alpha,i}P_{\beta,i}P_{\alpha,j}P_{\beta,j}\\ &\le \frac{1}{B^2m^2}\sum_i\sum_{\alpha,\beta} P_{\alpha,i}P_{\beta,i}\cdot Bs\\ &= \frac{1}{B^2}\sum_i p_i^2\cdot Bs = t\cdot\frac{s}{B},\end{aligned}$$

where for the last inequality we use the fact that $\sum_j P_{\alpha,j}P_{\beta,j}\le s\sum_j P_{\alpha,j}=sB$. To bound the second term of $\mathbf{Var}\,[w]$, we use Equation (1):

$$\begin{aligned}\mathbb{E}_{\alpha,\beta,\gamma\in[m/B]}\left[\bar{C}_{\alpha,\beta}\cdot\bar{C}_{\alpha,\gamma}\right] &\le \mathbb{E}_{\alpha,\beta,\gamma\in[m/B]}\left[C_{\alpha,\beta}\cdot C_{\alpha,\gamma}\right]\\ &= \frac{B^3}{m^3}\sum_{\alpha,\beta,\gamma}\sum_{i,j}\frac{P_{\alpha,i}P_{\beta,i}}{B^2}\cdot\frac{P_{\alpha,j}P_{\gamma,j}}{B^2}\\ &= \frac{1}{Bm^3}\sum_i\sum_{\alpha,\beta}P_{\alpha,i}P_{\beta,i}\sum_{j,\gamma}P_{\alpha,j}P_{\gamma,j}\\ &\le \frac{1}{m^2}\sum_i\sum_{\alpha,\beta}P_{\alpha,i}P_{\beta,i}\sum_j\frac{P_{\alpha,j}}{B}\cdot\max_j p_j\\ &\le \frac{\max_j p_j}{m^2}\sum_i\sum_{\alpha,\beta}P_{\alpha,i}P_{\beta,i} = t\max_j p_j.\end{aligned}$$ □

The following proposition gives bounds on t. Its proof is immediate.

Proposition 8. *If p is uniform, then $t=\sum p_i^2=\frac{1}{n}$. If p is ϵ-far from uniformity, then $t\ge(1+\epsilon)\frac{1}{n}$.*

Lemma 9. *If p is ϵ-far from uniformity, then the algorithm rejects with probability at least $2/3$.*

Proof. We have that $t\le\max_i p_i\le O(\epsilon\cdot\sqrt{s/nB}\cdot\log B)$, and thus

$$\begin{aligned}\Pr[w\le(1+\epsilon/2)\tfrac{1}{n}\cdot Q^2] &= \Pr[tQ^2-w\ge tQ^2-(1+\epsilon/2)Q^2\tfrac{1}{n}]\\ &= \Pr[tQ^2-w\ge Q^2(t-\tfrac{1}{n}-\tfrac{\epsilon/2}{n})]\\ &\le \frac{\mathbb{E}\left[(w-tQ^2)^2\right]}{(Q^2(t-\frac{1}{n}-\frac{\epsilon/2}{n}))^2}.\end{aligned}$$

If $t > \frac{2}{n}$, then

$$\begin{aligned}\Pr[w \leq (1+\epsilon/2)\tfrac{1}{n} \cdot Q^2] &\leq O(1) \cdot \frac{Q^2 st/B + Q^3 t \cdot \epsilon \cdot \sqrt{s/nB} \cdot \log B}{Q^4 t^2} \\ &\leq O(1) \cdot \left(\frac{s}{BQ^2 t} + \tfrac{\epsilon}{Q} \cdot \sqrt{s/nB} \cdot \log B\right) < 1/3.\end{aligned}$$

Otherwise, if $(1+\epsilon)\frac{1}{n} \leq t \leq \frac{2}{n}$, then

$$\begin{aligned}\Pr[w \leq (1+\epsilon/2)\tfrac{1}{n} \cdot Q^2] &\leq \frac{\mathbb{E}\left[(w - tQ^2)^2\right]}{(Q^2(t - \frac{1}{n} - \frac{\epsilon/2}{n}))^2} \\ &\leq O(1) \cdot \frac{Q^2 st/B + Q^3 t \cdot \epsilon \cdot \sqrt{s/nB} \cdot \log B}{Q^4 \epsilon^2 / n^2} \\ &\leq O(1) \cdot \frac{Q^2 s/B + Q^3 \epsilon \cdot \sqrt{s/nB} \cdot \log B}{Q^4 \epsilon^2 / n} \\ &= O(1) \cdot \left(\frac{m}{BQ^2\epsilon^2} + \frac{\sqrt{m}\log B}{Q\epsilon\sqrt{B}}\right) < 1/3. \qquad \square\end{aligned}$$

Lemma 10. *If p is uniform, then the algorithm passes with probability at least* 5/6.

Proof. Since $t = \frac{1}{n}$, we have

$$\begin{aligned}\Pr[w \geq (1+\epsilon/3)\tfrac{1}{n} \cdot Q^2] = \Pr[w - Q^2 t \geq \tfrac{\epsilon}{3} Q^2 t] &\leq \frac{\mathbf{Var}\,[w]}{(\frac{\epsilon}{3} Q^2 t)^2} \\ &= O(1) \cdot \frac{Q^2 st/B + Q^3 t/n}{(\frac{\epsilon}{3} Q^2 t)^2} \\ &= O(1) \cdot \left(\frac{m}{BQ^2\epsilon^2} + \frac{1}{\epsilon Q}\right) < 1/6.\end{aligned}$$

$\square$

Lemma 11. *If p is $O(\frac{\epsilon}{\sqrt{Bm}\log B})$-close to uniform, then the algorithm accepts with probability at least* 2/3.

Proof. The probability that the algorithm will see the difference between p and the uniform distribution is bounded by

$$3Q \cdot B \cdot O\left(\frac{\epsilon}{\sqrt{Bm}\log B}\right) = O(1).$$

Since the constant in $O(1)$ can be made arbitrarily small, we can assume that the probability of seeing a difference is at most $1/6$. The uniform distribution passes the test with probability at least $5/6$, so if p is as close to uniformity as specified above, it must be accepted with probability at least $5/6 - 1/6 = 2/3$. $\square$

This finishes the proof of Theorem 3.

4 Testing Identity

Now we show that we can test identity of a distribution with $O(\sqrt{\frac{m}{B}} \cdot \text{poly}(1/\epsilon) \cdot \text{polylog}(Bn))$ queries. As for uniformity testing, when $m = \Theta(n)$, this improves over the usual (in main memory) sampling complexity by $\tilde{\Theta}(\sqrt{B})$.

Theorem 12. *Let m be the length of a sequence of elements in $[n]$ and assume that $m \leq nB/2$ and $n^{-0.1} < \epsilon < 0.1$. The sequence is stored in blocks on disk, and each block contains B elements of the sequence. Let p be the empirical distribution of the sequence. There is an algorithm that given some explicit distribution q on $[n]$, samples $O\left(\frac{1}{\epsilon^3} \cdot \sqrt{\frac{m}{B}} \cdot \text{polylog}(Bn)\right)$ blocks, and:*

- *accepts with probability $\geq 2/3$ if $p = q$;*
- *rejects with probability $\geq 2/3$ if p is ϵ-far from q.*

Our algorithm is based on the identity-testing algorithm of [11] and is given in Figure 2.

The high-level idea of the algorithm is the following. We use the distribution q to partition the support into sets R_l, where R_l contains the elements with weight between $(1+\epsilon')^{-l}$ and $(1+\epsilon')^{-l+1}$, where ϵ' is proportional to ϵ, and l ranges from 1 to some $L = O(\frac{1}{\epsilon}\log n)$. Abusing notation, let R_l denote the size of the set R_l. Then, note that $R_l \leq (1+\epsilon')^l$. Let $p_{|R_l}$ be the restriction of p to the support R_l ($p_{|R_l}$ is not a distribution anymore). It is easy to see that: if $p = q$ then $p_{|R_l} = q_{|R_l}$ for all l, and if p and q are far, then there is some l such that $p_{|R_l}$ and $q_{|R_l}$ are roughly ϵ/L-far in the ℓ_1 norm. For all l's such that $R_l \leq m/B$, we use the standard identity testing (ignoring the power of the blocks). The standard identity testing takes only $O(\sqrt{m/B}(\epsilon^{-1}\log n)^{O(1)})$ samples because the support is bounded by m/B (instead of n).

To test identity for l's such that $R_l > m/B$, we harness the power of blocks. In fact, for each level R_l, we (roughly) test uniformity on $p_{|R_l}$ (as in the algorithm of [11]). Using our own uniformity testing with block queries from Theorem 3, we can test uniformity of $p_{|R_l}$ using only $O(\sqrt{m/B}(\epsilon^{-1}\log n)^{O(1)})$ samples. In the our algorithm, we do not use Theorem 3 directly because of the following technicality: when we consider the restriction $p_{|R_l}$, the blocks generally have less than B elements as the block also contains elements outside R_l. Still, it suffices to consider only $p_{|R_l}$ of "high" weight (roughly ϵ/L), and such $p_{|R_l}$ must populate many blocks with at least a fraction of ϵL of elements in each. This means that the same variance bound holds (modulo small additional factors).

We present the complete algorithm in Figure 2. Assume C is a big constant and c is a small constant.

4.1 Analysis: Proof of Theorem 12

We now proceed to the analysis of the algorithm. Suppose, by rescaling, that if $p \neq q$, they are 6ϵ-far (as opposed to ϵ-far). We can decompose the distribution p into two components by partitioning the support $[n]$: p' is the restriction of p on elements heavier than B/m (i.e., $i \in [n]$ such that $p_i \geq B/m$), and p'' on elements lighter than B/m. Clearly, running identity testing on both components is sufficient. Abusing notation, we refer to vectors $\tilde{p} \in (\mathbb{R}^+)^n$ as distributions

IDENTITY-TEST(n, m, B, ϵ, q)

1. Let $Q = C\frac{1}{\epsilon^3} \cdot \sqrt{\frac{m}{B}} \cdot \log^{O(1)} nB$ and $\epsilon' = c\epsilon$;
2. Let $L = 5\log_{1+\epsilon'} n$ and $L_0 = \log_{1+\epsilon'} m/B < L$;
3. Let $R_l = \{i \in [n] : (1+\epsilon')^{-l} < q_i \leq (1+\epsilon')^{-l+1}\}$, for $l \in \{1, 2, \ldots L-1\}$, and $R_L = \{i \in [n] : q_i \leq (1+\epsilon')^{-L}\}$;
4. If $\sum_{l \leq L_0} \|q_{|R_l}\|_1 \geq \epsilon/4$, then run the standard identity testing algorithm for the distribution defined on elements $\cup_{l \leq L_0} R_l$ (disregarding block structure), and, if it fails, FAIL;
5. Let $w_l = \sum_{i \in R_l} q_i$ for $l \in \{L_0+1, \ldots L\}$;
6. For each $L_0 < l < L$ such that $w_l \geq \epsilon/L$, do the following:
   (a) Query Q blocks, and restrict the elements to elements in R_l; call these blocks $V_1^l, \ldots V_Q^l$;
   (b) Let v_l be the total number of elements in $V_1^l, \ldots V_Q^l$ (counting repetitions of blocks);
   (c) If $|v_l - w_l QB| > (\epsilon/L)^2 w_l QB$, then FAIL;
   (d) For each element $i \in R_l$, compute how many times it appears in the blocks $V_1^l, \ldots V_Q^l$ (where we ignore repeats of a block);
   (e) If an element $i \in R_l$ appears more that $(1+\epsilon')^{-l+1} m$ times, then FAIL.
   (f) Let $N_2(l)$ be the number of collisions in $V_1^l, \ldots V_Q^l$ (with replacement);
   (g) If $N_2(l) > (1 + \frac{\epsilon}{2}) \cdot \frac{w_l^2}{R_l} \cdot \binom{QB}{2}$, then FAIL;

Fig. 2. Algorithm for identity testing in the block query model

as well, and, for $\tilde{p}, \tilde{q} \in (\mathbb{R}^+)^n$, we say the distributions $\tilde{p}$ and $\tilde{q}$ are ϵ-far if $\|\tilde{p} - \tilde{q}\|_1 \geq \epsilon \|\tilde{q}\|_1$.

Testing identity for distribution p' is handled immediately, by Step 4 (as long as $\|q'\|_1 \geq \epsilon/4$). Note that it requires only $\tilde{O}(\epsilon^{-2}\sqrt{m/B})$ samples because the support of the distribution is at most m/B. Let's call $p_{|R_l}$ the restriction of the probability distribution p to the elements from R_l. For all $i \in R_L$, we have $p_i \leq (1+\epsilon')^{-L} < n^{-5}$. We define $w_l = \|q_{|R_l}\|_1$, and then $w_L \leq n \cdot n^{-5} \leq n^{-4}$.

The main observation is the following:

- if $p = q$, then $p_{|R_l} = q_{|R_l}$ for all l;
- if p is 6ϵ-far from q, and $\|p' - q'\|_1 \leq \epsilon \|q'\|_1$ (or $\|q'\|_1 < \epsilon/4$), then there exist some l, with $L_0 < l < L$, such that $w_l \geq \epsilon/L$, and $p_{|R_l}$ and $q_{|R_l}$ are at least ϵ-far.

From now on, by "succeeds" we mean "succeeds with probability at least 9/10".

Proposition 13. *If $p = q$, then step 6c succeeds. If step 6c succeeds, then $|\sum_{i \in R_l} p_i - w_l| \leq (\epsilon/L)^2/2 \cdot w_l$ for all l, $L_0 < l < L$, such that $w_l \geq \epsilon/L$.*

The proof of the proposition is immediate by the Chernoff bound.

If step 6e passes, then we can assume that for each block, and each $i \in R_l$, $L_0 < l < L$, the element i appears at most $(1+\epsilon')^{-l+1}m$ times in that block—employing exactly the same argument as in Lemma 5. Furthermore, in this case, for each $i \in R_l$, $L_0 < l < L$, the value of p_i is $p_i \leq w_l \cdot$

$O(\frac{(1+\epsilon')^{-l+1}m\cdot m/B/Q}{m}\log^2 nB) = w_l\cdot O((1+\epsilon')^{-l}\sqrt{\frac{m}{B}}(\epsilon^{-1}\log nB)^{O(1)})$. Both conditions hold *a priori* if $p = q$.

As in the case of uniformity testing, we just need to test the ℓ_2 norm of $p_{|R_l}$ for at each level l.

Suppose $p_{|R_l}$ is ϵ-far from $q_{|R_l}$. Then $\|p_{|R_l}\|_2^2 \geq \|p_{|R_l}\|_1^2 \cdot \frac{1+\epsilon}{R_l} \cdot (1 - O(\epsilon'))$ by Proposition 8. Furthermore, by Proposition 13, we have $\|p_{|R_l}\|_2^2 \geq \frac{w_l^2}{R_l} \cdot (1+\epsilon)(1-O(\epsilon'))(1-(\epsilon/L)^2) > (1+\frac{2}{3}\epsilon)\cdot\frac{w_l^2}{R_l}$.

Now suppose $p_{|R_l} = q_{|R_l}$. Then $\|p_{|R_l}\|_2^2 \leq R_l \cdot (1+\epsilon')^{2(-l+1)} \leq \frac{(1+\epsilon')^2}{R_l}\cdot w_l^2 \leq (1+3c\epsilon)\cdot\frac{w_l^2}{R_l}$.

Finally, the step 6g verifies that the estimate $N_2(l)$ of $\|p_{|R_l}\|_2^2$ is close to its expected value when $p = q$. Indeed, if $N_2(l)$ were a faithful estimate — that is if $N_2(l) = \|p_{|R_l}\|_2^2 \cdot \binom{QB}{2}$ — then we would be done. However, as with uniformity testing, we have only $\mathbb{E}[N_2(l)] = \|p_{|R_l}\|_2^2 \cdot \binom{QB}{2}$. Still the bound is almost faithful as we can bound the standard deviation of $N_2(l)$. Exactly the same calculation as in Lemma 7 holds. Specifically, note that the variance is maximized when the elements of R_l appear in $v_l/B \geq \frac{w_l}{2}Q$ of the blocks $V_1^l, \ldots V_Q^l$, while the rest are devoid of elements from R_l. This means that the estimate $N_2(l)$ is effectively using only $\frac{w_l}{2}Q = \Omega(\epsilon^{-2}\sqrt{m/B}\log^{O(1)} n)$ blocks, which is enough for variance estimate of Lemma 7.

This finishes the proof of Theorem 12.

5 Lower Bounds

We show that for all three problems, distinctness, uniformity and identity testing, the $\sqrt{B}$ improvement is essentially optimal. Our lower bound is based on the following standard lower bound for testing uniformity.

Theorem 14 (Folklore). *For some $\epsilon > 0$, any algorithm for testing uniformity on $[n]$ needs $\Omega(\frac{1}{\epsilon}\sqrt{n})$ samples.*

From the above theorem we conclude the following lower bound for testing uniformity with block queries. Naturally, the bound also applies to identity testing, as uniformity testing is a particular case of identity testing. Similarly, the lower bound for the distinctness problem follows from the lower bound below for $m = n$.

Corollary 15. *For some $\epsilon > 0$, any algorithm for testing uniformity on $[n]$ in the block query model must use $\Omega(\sqrt{\frac{m}{B}})$ samples, for $n \geq m/B$.*

Proof. By Theorem 14, $\Omega(\sqrt{\frac{m}{B}})$ samples are required to test if a distribution on $[m/B]$ is uniform. We now show how a tester for uniformity in the block model can be used to test uniformity on $[m/B]$. We replace each occurrence of element $i \in [m/B]$ by a block with m/n copies of each of the elements $(i-1)\cdot\frac{nB}{m} + j$, for $j \in [nB/m]$. If the initial distribution was ϵ-far from uniformity on $[m/B]$, the new distribution is ϵ-far from uniformity on $[n]$. If the initial distribution was uniform on $[m/B]$, the new distribution is uniform on $[n]$. Hence, $\Omega(\frac{1}{\epsilon}\sqrt{\frac{m}{B}})$ samples are necessary to test uniformity and identity in the block query model for $m/B \leq n$. □

6 Open Problems

There is a vast literature on sublinear-time algorithms, and it is likely that other problems are amenable to approaches presented in this paper. From both the theoretical and practical perspective it would be very interesting to identify such problems.

References

1. Vitter, J.S.: External memory algorithms and data structures. ACM Comput. Surv. 33(2), 209–271 (2001)
2. Olken, F., Rotem, D.: Simple random sampling from relational databases. In: VLDB, pp. 160–169 (1986)
3. Olken, F.: Random Sampling from Databases. PhD thesis (1993)
4. Fischer, E.: The art of uninformed decisions: A primer to property testing. Bulletin of the European Association for Theoretical Computer Science 75, 97–126 (2001)
5. Ron, D.: Property testing (a tutorial). In: Rajasekaran, S., Pardalos, P.M., Reif, J.H., Rolim, J.D.P. (eds.) Handbook on Randomization, vol. II, pp. 597–649. Kluwer Academic Publishers, Dordrecht (2001)
6. Goldreich, O.: Combinatorial property testing—a survey. In: Randomization Methods in Algorithm Design, pp. 45–60 (1998)
7. Bar-Yossef, Z., Kumar, R., Sivakumar, D.: Sampling algorithms: lower bounds and applications. In: STOC, pp. 266–275 (2001)
8. Goldreich, O., Ron, D.: On testing expansion in bounded-degree graphs. Electronic Colloqium on Computational Complexity 7(20) (2000)
9. Batu, T.: Testing Properties of Distributions. PhD thesis, Cornell University (August 2001)
10. Batu, T., Fortnow, L., Rubinfeld, R., Smith, W.D., White, P.: Testing that distributions are close. In: FOCS, pp. 259–269 (2000)
11. Batu, T., Fortnow, L., Fischer, E., Kumar, R., Rubinfeld, R., White, P.: Testing random variables for independence and identity. In: FOCS, pp. 442–451 (2001)
12. Fischer, E., Matsliah, A.: Testing graph isomorphism. SIAM J. Comput. 38(1), 207–225 (2008)
13. Onak, K.: Testing properties of sets of points in metric spaces. In: Aceto, L., Damgård, I., Goldberg, L.A., Halldórsson, M.M., Ingólfsdóttir, A., Walukiewicz, I. (eds.) ICALP 2008, Part I. LNCS, vol. 5125, pp. 515–526. Springer, Heidelberg (2008)
14. Ergün, F., Kannan, S., Kumar, S.R., Rubinfeld, R., Viswanathan, M.: Spotcheckers. Journal of Computer and System Sciences 60(3), 717–751 (2000)

Functional Monitoring without Monotonicity*

Chrisil Arackaparambil, Joshua Brody, and Amit Chakrabarti

Department of Computer Science, Dartmouth College, Hanover, NH 03755, USA

Abstract. The notion of *distributed functional monitoring* was recently introduced by Cormode, Muthukrishnan and Yi [4] to initiate a formal study of the communication cost of certain fundamental problems arising in distributed systems, especially sensor networks. In this model, each of k sites reads a stream of tokens and is in communication with a central coordinator, who wishes to continuously monitor some function f of σ, the union of the k streams. The goal is to minimize the number of bits communicated by a protocol that correctly monitors $f(\sigma)$, to within some small error. As in previous work, we focus on a threshold version of the problem, where the coordinator's task is simply to maintain a single output bit, which is 0 whenever $f(\sigma) \leq \tau(1-\varepsilon)$ and 1 whenever $f(\sigma) \geq \tau$. Following Cormode et al., we term this the $(k, f, \tau, \varepsilon)$ functional monitoring problem.

In previous work, some upper and lower bounds were obtained for this problem, with f being a frequency moment function, e.g., F_0, F_1, F_2. Importantly, these functions are *monotone*. Here, we further advance the study of such problems, proving three new classes of results. First, we provide nontrivial monitoring protocols when f is either H, the empirical Shannon entropy of a stream, or any of a related class of entropy functions (Tsallis entropies). These are the first nontrivial algorithms for distributed monitoring of non-monotone functions. Second, we study the effect of non-monotonicity of f on our ability to give nontrivial monitoring protocols, by considering $f = F_p$ with deletions allowed, as well as $f = H$. Third, we prove new lower bounds on this problem when $f = F_p$, for several values of p.

Keywords: Communication complexity, distributed algorithms, data streams, sensor networks.

1 Introduction

Energy efficiency is a key issue in sensor network systems. Communication, which typically uses power-hungry radio, is a vital resource whose usage needs to be minimized [7]. Several other distributed systems have a similar need for minimizing communication. This is the primary motivation for our present work, which is a natural successor to the recent work of Cormode, Muthukrishnan

* Work supported in part by an NSF CAREER Award CCF-0448277 and NSF grant EIA-98-02068.

S. Albers et al. (Eds.): ICALP 2009, Part I, LNCS 5555, pp. 95–106, 2009.

and Yi [4], who introduced a clean formal model to study this issue. The formalization, known as *distributed functional monitoring*, involves a multi-party communication model consisting of k *sites* (the sensors, in a sensor network) and a single central *coordinator*. Each site asynchronously receives "readings" from its environment, formalized as a *data stream* consisting of *tokens* from a discrete universe. The union of these streams defines an overall input stream σ that the coordinator wishes to monitor continuously, using an appropriate protocol involving private two-way communication channels between the coordinator and each site. Specifically, the coordinator wants to continuously maintain approximate knowledge of some nonnegative real-valued function f of σ. (We assume that f is invariant under permutations of σ, which justifies our use of "union" above, rather than "concatenation.")

As is often the case in computer science, the essence of this problem is captured by a threshold version with Boolean outputs. Specifically, we have a threshold $\tau \in \mathbb{R}_+$ and an approximation parameter $\varepsilon \in \mathbb{R}_+$, and we require the coordinator to continuously maintain an output bit, which should be 0 whenever $f(\sigma) \leq \tau(1-\varepsilon)$ and 1 whenever $f(\sigma) \geq \tau$.[1] Following [4], we call this the $(k, f, \tau, \varepsilon)$ functional monitoring problem. This formulation of the problem combines aspects of streaming algorithms, sketching and communication complexity.

Motivation. Plenty of recent research (e.g., in network and database settings) has studied such continuous monitoring problems, for several special classes of functions f (see, e.g., [2,6,5,12]). However, *formal* worst-case bounds on communication cost were first addressed in detail in [4]. We continue to address this here. Philosophically, the study of such monitoring problems is a vast generalization of Slepian-Wolf-style distributed source coding [13] in much the same way that communication complexity is a vast generalization of basic source coding in information theory. Furthermore, while the problems and the model we consider here are strongly reminiscent of streaming algorithms, there are notable additional challenges: for instance, maintaining an approximate count of the total number of tokens received is a nontrivial problem in our setting, but is trivial in the streaming model. For a more detailed discussion of prior research, we refer the reader to [4] and the references therein.

Our Results and Comparison with Prior Work. Our work studies $(k, f, \tau, \varepsilon)$ functional monitoring for two natural classes of functions f: the empirical Shannon entropy H (and its generalization: Tsallis entropy) and the frequency moments F_p. For an input stream σ of tokens from the universe $[n] := \{1, 2, \ldots, n\}$, let f_i denote the number of appearances of i in σ, where $i \in [n]$. For $p \geq 0$, the pth frequency moment $F_p(\sigma)$ is defined to be $\sum_{i=1}^n f_i^p$. Note that p can be non-integral or zero: indeed, using the convention $0^0 = 0$ makes $F_0(\sigma)$ equal to the number of distinct tokens in σ. These functions F_p capture important statistical properties

[1] Clearly, a solution to the value monitoring problem solves this threshold version, and the value monitoring problem can be solved by running, in parallel, several copies of a solution to this threshold version with geometrically spaced thresholds.

of the stream and have been studied heavily in the streaming algorithms literature [1,9]. The stream σ also implicitly defines a probability distribution over $[n]$, given by $\Pr[i] = f_i/m$, where m is the length of σ. For various applications, especially ones related to anomaly detection in networks, the entropy of this distribution — also called the empirical entropy of the stream — is a measure of interest. Abusing notation somewhat, we denote this as $H(\sigma)$, when the underlying entropy measure is Shannon entropy: thus, $H(\sigma) = \sum_{i=1}^{n}(f_i/m)\log(m/f_i)$.[2] We also consider the family of functions $T_\alpha(\sigma) = (1 - \sum_{i=1}^{n}(f_i/m)^\alpha)/(\alpha - 1)$, which are collectively known as Tsallis entropy [14] and which generalize Shannon entropy, as shown by considering the limit as $\alpha \to 1$.

We study the effect of *non-monotonicity* of f on the $(k, f, \tau, \varepsilon)$ problem: the bounds of Cormode et al. [4] crucially exploited the fact that the functions being monitored were monotone nondecreasing. We obtain three new classes of results. First, we provide nontrivial monitoring protocols for H, and the related functions T_α. For this, we suitably extend recent sketching algorithms such as those due to Bhuvanagiri and Ganguly [3] and Harvey et al. [8]. These are the first nontrivial algorithms for monitoring non-monotone functions.[3] Our algorithms, which are simple and easily usable, can monitor continuously until the end of the stream, even as the $f(\sigma)$ crosses the threshold multiple times. This is the desired behavior when monitoring non-monotone functions.

Secondly, we prove lower bounds for monitoring $f = F_p$ with *deletions* allowed: i.e., the stream can contain "negative tokens" that effectively delete earlier tokens. In contrast with the good upper bounds in [4] for monitoring F_p *without* deletions (a monotone problem), we show that essentially no good upper bounds are possible. Using similar techniques, we also give a lower bound for monitoring H that is necessarily much milder, and in the same ballpark as our upper bound.

Thirdly, we prove new lower bounds for the monotone problems $f = F_p$, without deletions, for various values of p. These either improve or are incomparable with previous bounds [4]; see Table 1 for a side-by-side comparison.

Notation, etc. We now define some notation that we use at various points. We use $|\sigma|$ to denote the length of the stream σ and $\sigma_1 \circ \sigma_2$ to denote the concatenation: σ_1 followed by σ_2. We typically use $S_1, \ldots, S_k$ to denote the k sites, and C to denote the coordinator, in a $(k, f, \tau, \varepsilon)$ functional monitoring protocol. We tacitly assume that randomized protocols use a public coin and err with probability at most $1/3$ at each point of time. These assumptions do not lose generality, as shown by appropriate parallel repetition and the private-versus-public-coin theorem of Newman [11]. We use m to denote the overall input length (i.e., number of tokens) seen by the protocol under consideration. We state our communication bounds in terms of m, k and ε, and sometimes τ.

[2] Throughout this paper we use "log" to denote logarithm to the base 2 and "ln" to denote natural logarithm.

[3] Muthukrishnan [10] gives an upper bound for monitoring a non-monotone function, but with additive error.

Table 1. Summary of our results (somewhat simplified) and comparison with previous work [4]. Dependence on τ is not shown here, but is stated in the relevant theorems.

Problem	Previous Results	Our Results
H, deterministic	$\tilde{O}(m)$, trivially	$\Omega(k\varepsilon^{-1/2}\log m)$
H, randomized		$\tilde{O}(k\varepsilon^{-3}\log^4 m)$, $\Omega(\varepsilon^{-1/2}\log m)$
F_p, dels., determ.		$\Omega(m)$
F_p, dels., rand.		$\Omega(m/k)$
F_1, deterministic	$O(k\log(1/\varepsilon))$, $\Omega(k\log(1/(\varepsilon k)))$	$\Omega(k\log(1/\varepsilon))$
F_0, randomized	$\tilde{O}(k/\varepsilon^2)$, $\Omega(k)$	$\Omega(1/\varepsilon)$, $\Omega(1/\varepsilon^2)$ if round-based
$F_p, p > 1$, rand.	$\tilde{O}(k^2/\varepsilon + (\sqrt{k}/\varepsilon)^3)$, $\Omega(k)$, for $p = 2$	$\Omega(1/\varepsilon)$, $\Omega(1/\varepsilon^2)$ if round-based

2 An Algorithm for Monitoring Entropy

We discuss randomized algorithms for monitoring Shannon entropy, H, and Tsallis entropies, T_α. These provide the first nontrivial communication upper bounds for the monitoring of non-monotone functions. At a high level, our algorithms monitor changes (in the L_1 sense) in the empirical probability distribution defined by the input streams. For probability distributions μ, ν on the set $[n]$, we write $\|\mu - \nu\|_1 = \sum_{i=1}^{n} |\mu(i) - \nu(i)|$. We use the following three technical lemmas, whose proofs are left to the full version of the paper.

Lemma 1. *Let σ and σ' be streams of tokens from $[n]$, and μ and ν denote the empirical distributions induced by σ and $\sigma \circ \sigma'$ respectively. Let $m = |\sigma|$. If $|\sigma'| \le m$, then, $|H(\sigma \circ \sigma') - H(\sigma)| \le \|\nu - \mu\|_1 \log(2m)$.*

Lemma 2. *Let $\sigma, \sigma', \mu, \nu$ and m be defined as in Lemma 1. Then, for all $\alpha > 1$, $|T_\alpha(\sigma \circ \sigma') - T_\alpha(\sigma)| \le \|\nu - \mu\|_1 \cdot \min\{\log(2m), \alpha/(\alpha - 1)\}$.*

Lemma 3. *Let $\sigma, \sigma', \mu, \nu$ and m be defined as in Lemma 1. Then if $|\sigma'| < \ell$, then $\|\nu - \mu\|_1 < 2\ell/m$.*

We also need an *entropy sketching scheme*, such as the one provided by the following result, due to Harvey, Nelson and Onak [8].

Fact 1. *Let $\varepsilon > 0$. There is an algorithm that maintains a data structure (called a "sketch") $\mathcal{S}_H(\sigma)$, based on an input stream σ, such that (1) based on $\mathcal{S}_H(\sigma)$, we can compute an estimate $\hat{H}(\sigma) \in [H(\sigma) - \varepsilon, H(\sigma) + \varepsilon]$, (2) we can suitably combine $\mathcal{S}_H(\sigma_1)$ and $\mathcal{S}_H(\sigma_2)$ to obtain $\mathcal{S}_H(\sigma_1 \circ \sigma_2)$, and (3) $\mathcal{S}_H(\sigma)$ can be stored using $\tilde{O}(\varepsilon^{-2}\log m \log n \log(mn))$ bits.*[4] *Here, the $\tilde{O}$ notation hides factors polynomial in $\log\log m$ and $\log(1/\varepsilon)$.* □

[4] The $\tilde{O}(\varepsilon^{-2}\log m)$ bound in [8] is on the number of words of storage, each $O(\log(mn))$ bits long, and does not include $O(\log n)$ space for a pseudorandom generator.

The Algorithm. We proceed in multiple *rounds*. At the end of the ith round, let ρ_{ij} be the overall stream seen at site S_j, let $\sigma_i = \rho_{i1} \circ \cdots \circ \rho_{ik}$, and let $m_i = |\sigma_i|$. In round 0, sites directly forward input tokens to the coordinator C, who ends the round after seeing a $c_0 := 100$ items. Then, C uses $\mathcal{S}_H$ from Fact 1 to get an estimate $\hat{H}(\sigma_0)$ of $H(\sigma_0)$ with an additive error of $\hat{\varepsilon} := \varepsilon\tau/4$.

For rounds $i > 0$, C and $S_1, \ldots, S_k$ simulate a $(k, F_1, \tau_i, \frac{1}{2})$ monitoring algorithm, such as the one from [4], using error $\frac{1}{2}$ and threshold $\tau_i := \min\{m_{i-1}, m_{i-1}\lambda_i/(2\log(2m_{i-1}))\}$, where $\lambda_i = \tau(1-\frac{\varepsilon}{4}) - \hat{H}(\sigma_{i-1})$ if $\hat{H}(\sigma_{i-1}) < \tau(1-\frac{\varepsilon}{2})$, and $\lambda_i = \hat{H}(\sigma_{i-1}) - \tau(1-\frac{3\varepsilon}{4})$ otherwise. λ_i is the slack of the estimate $\hat{H}(\sigma_{i-1})$ from τ (or $\tau(1-\varepsilon)$ in the latter case), while allowing for error of the estimates. The choice of τ_i ensures that the simulated F_1 monitoring algorithm notifies the coordinator by outputting 1 when too many items (as determined from the technical lemmas) have been received in the round. When this happens, C signals each S_j that round i is ending, whereupon S_j sends it $\mathcal{S}_H(\rho_{ij})$. Then, C computes $\mathcal{S}_H(\sigma_i)$, updates its estimate $\hat{H}(\sigma_i)$, and outputs 1 iff $\hat{H}(\sigma_i) \geq \tau(1-\frac{\varepsilon}{2})$.

Theorem 1. *The above is a randomized algorithm for $(k, H, \tau, \varepsilon)$ functional monitoring that communicates $\widetilde{O}(k\varepsilon^{-3}\tau^{-3}\log^3 m \log n \log(mn))$ bits.*

Proof. We first analyze the correctness. In round 0, it is trivial for the coordinator to output the correct answer. Now, for round $i > 0$, suppose the coordinator outputs 0 at the end of round $i-1$. Then, we must have $\hat{H}(\sigma_{i-1}) \leq \tau(1-\frac{\varepsilon}{2})$, whence $H(\sigma_{i-1}) < \tau(1-\frac{\varepsilon}{4})$ by the bound on the sketching error. By the correctness of the F_1 monitoring algorithm, we receive at most τ_i items during round i. Therefore by Lemmas 1 and 3, when going from σ_{i-1} to σ_i, the total entropy will be less than τ throughout round i. Hence, the coordinator is free to output zero through the end of round i. If the coordinator instead outputs 1 at the end of round $i-1$, we are guaranteed to remain above $\tau(1-\varepsilon)$ similarly.

To bound the communication cost, we need to estimate both the number of rounds, and the number of bits exchanged in each round. It is easy to see that for each round i, $\lambda_i \geq \varepsilon\tau/4$. Suppose the stream ends during round $r+1$. Then,

$$\begin{aligned} m \;\geq\; m_r \;\geq\; m_{r-1} + \tau_r/2 \;&\geq\; m_{r-1}\left(1 + \min\{1/2, \tau\varepsilon/(16\log(2m_{r-1}))\}\right) \\ &\geq\; m_{r-1}\left(1 + \min\{1/2, \tau\varepsilon/(16\log(2m))\}\right) \;=\; m_{r-1}\beta \quad \text{(say)}, \end{aligned}$$

where the second inequality follows from the guarantee of the F_1 monitoring algorithm. Iterating the above recurrence for m_r, we get $m \geq c_0\beta^r$, whence $r \leq \log(m/c_0)/\log\beta = O(\max\{\log m, \log^2 m/(\tau\varepsilon)\})$, where the final bound uses $\ln(1+x) \geq x/(x+1)$ for all $x > 0$. In each round, we use $O(k\log m)$ bits to send τ_i to the sites and $O(k)$ bits for the F_1 algorithm. These terms are dominated by the sizes of the sketches that the sites send. Using the size bound from Fact 1 and the above bound on r, we can bound the total communication by $\widetilde{O}(k\varepsilon^{-3}\tau^{-3}\log^3 m \log n \log(mn))$, for m large enough (i.e., if $\log m \geq \tau\varepsilon$). □

Our algorithm for monitoring Tsallis entropy is similar. Lemma 2 bounds T_α just as Lemma 1 bounds H, and a suitable sketch $\mathcal{S}_{T_\alpha}$, analogous to $\mathcal{S}_H$, can be obtained from [8]. We postpone the details to the full paper.

Theorem 2. *There is a randomized algorithm for $(k, T_\alpha, \tau, \varepsilon)$ functional monitoring that communicates $\widetilde{O}(k\varepsilon^{-3}\tau^{-3}\log^3 m \log n \log(mn))$ bits.* □

3 Lower Bounds for Non-monotone Functions

We now give lower bounds for estimating entropy, and later, F_p. We give deterministic bounds first, and then randomized bounds. We abuse notation and let H denote both the empirical entropy of a stream and the binary entropy function $H : [0,1] \to [0,1]$ given by $H(x) = -x \log x - (1-x)\log(1-x)$.

Theorem 3. *For any $\varepsilon < 1/2$ and $m \geq k/\sqrt{\varepsilon}$, a deterministic algorithm solving $(k, H, \tau, \varepsilon)$ functional monitoring must communicate $\Omega(k\varepsilon^{-1/2}\log(\varepsilon m/k))$ bits.*

Proof. We use an adversarial argument that proceeds in rounds. Each round, the adversary will force the protocol to send at least one bit. The result will follow by showing a lower bound on the number of rounds r that the adversary can create, using no more than m tokens. Let $\tau = 1$, and let z be the unique positive real such that $H(\frac{z}{2z+1}) = 1 - \varepsilon$. Note that this implies $H(\frac{z}{2z+1}) > 1/2 > H(1/10)$, whence $\frac{z}{2z+1} > 1/10$, hence $z > 1/8$. An estimation of H using calculus shows that $z = \Theta(1/\sqrt{\varepsilon})$. Fix a monitoring protocol $\mathcal{P}$. The adversary only uses tokens from $\{0,1\}$, i.e., the stream will induce a two-point probability distribution.

The adversary starts with a "round 0" in which he sends nine 1s followed by a 0 to site S_1. Note that at the end of round 0, the entropy of the stream is $H(1/10) < 1/2$. For $i \in \{0, 1, \ldots, r\}$, let a_i denote the number of 0s and b_i the number of 1s in the stream at the end of round i. Then $a_0 = 1$ and $b_0 = 9$. For all $i > 0$, the adversary maintains the invariant that $b_i = \lceil a_i(z+1)/z \rceil$. This ensures that at the end of round i, the empirical entropy of the stream is

$$H\left(\frac{a_i}{a_i + b_i}\right) \leq H\left(\frac{a_i}{a_i(1 + (z+1)/z)}\right) = H\left(\frac{z}{2z+1}\right) = 1 - \varepsilon\,,$$

which requires the coordinator to output 0.

Consider the situation at the start of round i, where $i \geq 1$. If each player were to receive $\lceil (b_{i-1} - a_{i-1})/k \rceil$ 0-tokens in this round, then at some point the number of 0s in the stream would equal the number of 1s, which would make the empirical entropy equal to 1 and require the coordinator to change his output to 1. Therefore, there must exist a site S_{j_i}, $j_i \in [k]$, who would communicate upon receiving these many 0-tokens in round i. In actuality, the adversary does the following in round i: he sends these many 0s to S_{j_i}, followed by as many 1s as required to restore the invariant, i.e., to cause $b_i = \lceil a_i(z+1)/z \rceil$. Clearly, this strategy forces at least one bit of communication per round. It remains to bound r from below. Note that the adversary's invariant implies $b_i - a_i \leq a_i/z + 1$ and $a_i + b_i \leq a_i(2z+1)/z + 1 = a_i(2 + 1/z) + 1$. Therefore, we have

$$a_i = a_{i-1} + \left\lceil \frac{b_{i-1} - a_{i-1}}{k} \right\rceil \leq a_{i-1} + \left\lceil \frac{1 + a_{i-1}/z}{k} \right\rceil \leq a_{i-1}\left(1 + \frac{1}{zk}\right) + 2\,.$$

Setting $\alpha = (1 + 1/zk)$ and iterating gives $a_r \le a_0\alpha^r + 2(\alpha^r - 1)/(\alpha - 1) = a_0\alpha^r + 2zk(\alpha^r - 1) = \alpha^r(a_0 + 2zk) - 2zk$. Using our upper bound on $a_i + b_i$, the above inequality, and the facts that $a_0 = 1$ and that $z > 1/8$, we obtain

$$\begin{aligned} a_r + b_r &\le \alpha^r (1 + 2zk)(2 + 1/z) - 2zk(2 + 1/z) + 1 \\ &\le (2 + 1/z)(1 + 2zk)\alpha^r \le (2 + 1/z)(1 + 2zk)e^{r/zk} \le 60zke^{r/zk}. \end{aligned}$$

Therefore, we can have $a_r + b_r \le m$, provided $r \le zk\ln(m/(60zk))$. Recalling that $z = \Theta(1/\sqrt{\varepsilon})$, we get the claimed lower bound of $\Omega(k\varepsilon^{-1/2}\log(\varepsilon m/k))$. □

Our next lower bounds are for functional monitoring of frequency moments when we allow for deletions. Specifically, we now consider *update streams* that consist of tokens of the form (i, v), where $i \in [n]$ and $v \in \{-1, 1\}$, to be thought of as updates to a vector $(f_1, \ldots, f_n)$ of frequencies. The vector is initially zero and is updated using $f_i \leftarrow f_i + v$ upon receipt of the token (i, v): in English, each update either adds or deletes one copy of item i.

As usual, we let m denote the length of an update stream whose tokens are distributed amongst several sites. Our next results essentially show that no nontrivial savings in communication is possible for the problem of monitoring frequency moments in this setting. These bounds highlight the precise problem caused by the non-monotonicity of the function being monitored. They should be contrasted with the much smaller upper bounds achievable in the monotone case, when there are no deletions (see Table 1).

Our proofs are again adversarial and proceed in rounds. They use appropriate instantiations of the following generic lemma.

Definition 1. *An update stream is said to be* positive *if it consists entirely of tokens from $[n] \times \{1\}$, i.e., insertions only. The* inverse *of an update stream $\sigma = \langle(i_1, v_1), \ldots, (i_m, v_m)\rangle$ is defined to be $\sigma^{-1} := \langle(i_m, -v_m), \ldots, (i_1, -v_1)\rangle$. A function $G : \mathbb{Z}_+^n \to \mathbb{R}_+$ on frequency vectors is said to be* monotone *if G is nondecreasing in each parameter, separately. We extend such a G to a function on streams (or update streams) in the natural way, and write $G(\sigma)$ to denote $G(\boldsymbol{f})$, where $\boldsymbol{f}$ is the frequency vector determined by σ.*

Lemma 4. *Let $G : \mathbb{Z}_+^n \to \mathbb{R}_+$ be monotone and let $\mathcal{P}$ be a protocol for the $(k, G, \tau, \varepsilon)$ functional monitoring problem with deletions allowed. Let $\sigma_0, \sigma_1, \ldots, \sigma_k$ be a collection of positive update streams such that (1) $G(\sigma_0) \le \tau(1 - \varepsilon)$, and (2) $G(\sigma_0 \circ \sigma_1 \circ \ldots \circ \sigma_k) \ge \tau$. If $\mathcal{P}$ is a deterministic protocol, then the number of bits communicated is at least $\lfloor(m - |\sigma_0|) / (2 \cdot \max_{j\in[k]}\{|\sigma_j|\})\rfloor$. If $\mathcal{P}$ is a δ-error randomized protocol, then the expected number of bits communicated is at least $((1 - \delta)/k) \cdot \lfloor(m - |\sigma_0|) / (2 \cdot \max_{j\in[k]}\{|\sigma_j|\})\rfloor$.*

Proof. Let $S_1, \ldots, S_k$ be the k sites involved in $\mathcal{P}$. The adversary will send certain tokens to certain sites, maintaining the invariant that the coordinator is always required to output 0. In round 0, the adversary sends σ_0 to S_1; by condition (1), this maintains the invariant.

Let $s = \max_{j\in[k]}\{|\sigma_j|\}$ and $r = \lfloor(m - |\sigma_0|)/2s\rfloor$. The adversary uses r additional rounds maintaining the additional invariant that at the start of each such round the value of G is $G(\sigma_0)$. Consider round i, where $i \in [r]$. By condition (2), if the adversary were to send σ_j to S_j in this round, for each $j \in [k]$, the coordinator's output would have to change to 1.

Suppose $\mathcal{P}$ is a deterministic protocol. Then, since the coordinator's output would have to change to 1, there must exist a site S_{j_i}, with $j_i \in [k]$, that would have to communicate upon receiving σ_{j_i} in this round. In actuality, the adversary sends $\sigma_{j_i} \circ \sigma_{j_i}^{-1}$ to S_{j_i} and nothing to any other site in round i. Clearly, this maintains both invariants and causes at least one bit of communication. Also, this adds at most $2s$ tokens to the overall input stream. Thus, the adversary can cause r bits of communication using $|\sigma_0| + 2sr \le m$ tokens in all, which proves the claim for deterministic protocols.

The proof when $\mathcal{P}$ is a δ-error randomized protocol proceeds in a similar manner. The difference is that each round i has an associated collection of probabilities $(p_{i1}, \ldots, p_{ik})$, where $p_{ij} = \Pr[S_j$ communicates in round i upon receiving $\sigma_j]$. As before, condition (2) implies that were each S_j to receive σ_j in this round, correctness would require C's output to change to 1. Thus,

$$1 - \delta \;\le\; \Pr[\mathcal{P} \text{ is correct}] \;\le\; \Pr[C \text{ receives a bit in round } i] \;\le\; \textstyle\sum_{j=1}^{k} p_{ij}\,,$$

where the final inequality uses a union bound. Therefore, there exists a site S_{j_i}, with $j_i \in [k]$, having $p_{ij_i} \ge (1-\delta)/k$. Again, as in the deterministic case, the adversary actually sends $\sigma_{j_i} \circ \sigma_{j_i}^{-1}$ to S_{j_i} and nothing to any other site in round i. By linearity of expectation, the expected total communication with r rounds is at least $r(1-\delta)/k$, which proves the lemma. □

The theorems that follow are for randomized protocols with error $\delta = 1/3$. We prove only the first of these, leaving the (similar) proofs of the other two to the full paper.

Theorem 4. *The expected communication cost of a randomized $(k, F_0, \tau, \varepsilon)$ functional monitoring protocol that allows for deletions is $\Omega(\min\{m/k, m/\varepsilon\tau\})$.* □

Proof. Let $a := \max\{1, \lceil \frac{\tau\varepsilon}{k} \rceil\}$, and instantiate σ_0 as a stream of $\tau - ka$ distinct elements and $\sigma_1, \ldots, \sigma_k$ each as a stream of a distinct elements. Note that $ka \ge \tau\varepsilon$, so $F_0(\sigma_0) = \tau - ka \le \tau(1-\varepsilon)$. Furthermore, note that $F_0(\sigma_0 \circ \sigma_1 \circ \cdots \circ \sigma_k) = \tau$, hence the streams satisfy the conditions of Lemma 4 with $G = F_0$. Applying that lemma, and noting that $|\sigma_j| = a$ gives us a lower bound of $((1-\delta)/k) \cdot \lfloor (m - |\sigma_0|)/(2a) \rfloor = \Omega(\min\{m/k, m/\varepsilon\tau\})$ for m large enough. □

Theorem 5. *The expected communication cost of a randomized $(k, F_p, \tau, \varepsilon)$ monitoring protocol (with $p > 0$) that allows deletions is $\Omega(\min\{m/k, m/\tau^{1/p}\varepsilon\})$.* □

Theorem 6. *The expected communication cost of a randomized $(k, H, \tau, \varepsilon)$ functional monitoring protocol is $\Omega(\varepsilon^{-1/2} \log(\varepsilon m/k))$ bits.* □

We note that the deterministic part of Lemma 4 implies corresponding lower bounds that are k times larger. Of course, $\Omega(m)$ bounds were already known for F_p, even without deletions, by the techniques of [1] for deterministic streaming algorithms. We also note that Yi and Zhang [16] study problems similar to ours but in terms of competitive ratio. The bounds in this section rely on the construction of hard instances which might not be possible in their case.

4 Frequency Moments without Deletions: New Bounds

We finish with another set of lower bounds, this time for monitoring F_p (for various p) without deletions. Our bounds either improve or are incomparable with previous lower bounds: see Table 1.

Theorem 7. *A deterministic protocol that solves $(k, F_1, \tau, \varepsilon)$ functional monitoring must communicate at least $\Omega(k \log \frac{k+\tau}{k+\varepsilon\tau})$ bits. In particular, when $\tau \geq k/\varepsilon^{\Omega(1)}$, it must communicate $\Omega(k \log(1/\varepsilon))$ bits.*

Proof. Again we use an adversary, who proceeds in rounds: each round, he gives just enough tokens to a single site to force that site to communicate.

Let $a_0 = 0$ and, for $i \geq 1$, let a_i be the total number of tokens received by all sites (i.e., the value of F_1 for the input stream) at the end of round i. The adversary maintains the invariant that $a_i \leq \tau(1-\varepsilon)$, so that the coordinator must always output 0. For $j \in [k]$, let b_{ij} be the maximum number of tokens that site j can receive in round i without being required to communicate. The correctness of the protocol requires $a_{i-1} + \sum_{j=1}^{k} b_{ij} < \tau$, for otherwise the desired output can change from 0 to 1 without the coordinator having received any communication. Let $j^* = \text{argmin}_{j\in[k]}\{b_{ij}\}$. In round i, the adversary sends $b_{ij^*} + 1$ tokens to site j^*, forcing it to communicate. We have

$$a_i = a_{i-1} + b_{ij^*} + 1 \leq a_{i-1} + \frac{\tau - a_{i-1}}{k} + 1 = 1 + \frac{\tau}{k} + \left(1 - \frac{1}{k}\right) a_{i-1}.$$

Letting $\alpha = 1 - 1/k$ and iterating the above recurrence gives $a_i \leq (1+\tau/k)(1-\alpha^i)/(1-\alpha) = (k+\tau)(1-\alpha^i)$. Now note that $\alpha \geq e^{-2/k}$, so when $i \leq r := \frac{k}{2} \ln \frac{k+\tau}{k+\varepsilon\tau}$, we have $\alpha^i \geq \frac{k+\varepsilon\tau}{k+\tau}$, so that $a_i \leq (\tau+k)\cdot(k+\tau-k-\varepsilon\tau/(k+\tau) = \tau(1-\varepsilon)$.

This shows that the adversary can maintain the invariant for up to r rounds, forcing $\Omega(r)$ bits of communication, as claimed. □

Our next lower bounds use reductions from a fundamental problem in communication complexity: the "gap Hamming distance" problem, denoted GHD_c, where $c \in \mathbb{R}_+$ is a parameter. In this problem, Alice and Bob are given $x, y \in \{0,1\}^n$ respectively and want to output 1 if $\Delta(x,y) \geq \frac{n}{2} + c\sqrt{n}$ and 0 if $\Delta(x,y) \leq \frac{n}{2} - c\sqrt{n}$; they don't care what happens if the input satisfies neither of these conditions. We shall need the following lower bounds on the randomized communication complexity $\text{R}(\text{GHD}_c)$, as well as the one-way randomized communication complexity (where the only communication is from Alice to Bob) $\text{R}^{\rightarrow}(\text{GHD}_c)$. Proofs of these bounds, as well as further background on the problem, can be found in Woodruff [15].

Theorem 8. *Suppose $c > 0$ is a constant. Then $\text{R}(\text{GHD}_c) = \Omega(\sqrt{n})$ and $\text{R}^{\rightarrow}(\text{GHD}_c) = \Omega(n)$. Here, the Ω notation hides factors dependent upon c.*[5] □

[5] The bounds in [15] restrict the range of c, but this turns out not to be necessary.

It is conjectured that the general randomized bound is in fact as strong as the one-way version. This is not just a tantalizing conjecture about a basic communication problem. Settling it would have important consequences because, for instance, the gap Hamming distance problem is central to a number of results in streaming algorithms. As we shall soon see, it would also have consequences for our work here.

Conjecture 1. For sufficiently small constants c, we have $\mathrm{R}(\mathrm{GHD}_c) = \Omega(n)$.

Theorem 9. *For any $\varepsilon \leq 1/2$, a randomized protocol for $(k, F_0, \tau, \varepsilon)$ functional monitoring must communicate $\Omega(1/\varepsilon)$ bits.*

Proof. We give a reduction from GHD_1. Let $\mathcal{P}$ be a randomized protocol for $(k, F_0, \tau, \varepsilon)$ functional monitoring. Set $N := \lfloor 1/\varepsilon^2 \rfloor$ and $\tau = 3N/2 + \sqrt{N}$. We design a two-party public coin randomized communication protocol $\mathcal{Q}$ for GHD_1 on N-bit inputs that simulates a run of $\mathcal{P}$ involving the coordinator, C, and two sites, S_1 and S_2. Let $x \in \{0,1\}^N$ be Alice's input in $\mathcal{Q}$ and let $y \in \{0,1\}^N$ be Bob's input. Alice creates a stream $\sigma_a := \langle a_1, \ldots, a_N \rangle$ of tokens from $[N] \times \{0,1\}$ by letting $a_i := (i, x_i)$ and Bob similarly creates a stream $\sigma_b := \langle b_1, \ldots, b_N \rangle$, where $b_i := (i, y_i)$. They then simulate a run of $\mathcal{P}$ where S_1 first receives all of σ_a after which S_2 receives all of σ_b. They output whatever the coordinator would have output at the end of this run.

The simulation itself occurs as follows: Alice maintains the state of S_1, Bob maintains the state of S_2, and they *both* maintain the state of C. Clearly, this can be done by having Alice send to Bob all of S_1's messages to C plus C's messages to S_2 (and having Bob act similarly). The total communication in $\mathcal{Q}$ is at most that in $\mathcal{P}$.

We now show that $\mathcal{Q}$ is correct. By construction, the combined input stream $\sigma = \sigma_a \circ \sigma_b$ seen by $\mathcal{P}$ has $2\Delta(x,y)$ tokens with frequency 1 each and $N - \Delta(x,y)$ tokens with frequency 2 each. Therefore $F_0(\sigma) = N + \Delta(x,y)$. When $\Delta(x,y) \geq N/2 + \sqrt{N}$, we have $F_0(\sigma) \geq \tau$ and $\mathcal{Q}$, following $\mathcal{P}$, correctly outputs 1. On the other hand, when $\Delta(x,y) \leq N/2 - \sqrt{N}$, we have

$$F_0(\sigma) \leq \frac{3N}{2} - \sqrt{N} = \tau\left(1 - \frac{2\sqrt{N}}{3N/2 + \sqrt{N}}\right) \leq \tau\left(1 - \frac{1}{\sqrt{N}}\right) \leq \tau(1-\varepsilon).$$

Thus $\mathcal{Q}$ correctly outputs 0. Since $\mathcal{Q}$ is correct, by Theorem 8, it must communicate at least $\Omega(\sqrt{N}) = \Omega(1/\varepsilon)$ bits. Therefore, so must $\mathcal{P}$. □

Theorem 10. *For any $\varepsilon < 1/2$ and any constant $p > 1$, a randomized protocol for $(k, F_p, \tau, \varepsilon)$ functional monitoring must communicate $\Omega(1/\varepsilon)$ bits.*

Proof. For simplicity, we assume here that $p \geq 2$. As before, we reduce from GHD_1 on $N := \lfloor 1/\varepsilon^2 \rfloor$-bit inputs. For this reduction, we set $\tau := (N/2 + \sqrt{N})2^p + (N - 2\sqrt{N})$. Let $\mathcal{P}$ be a protocol for $(k, F_p, \tau, \varepsilon)$ functional monitoring. We design a protocol $\mathcal{Q}$ for GHD_1 on input (x,y) that simulates a run of $\mathcal{P}$ involving two sites, creating two streams $\langle (i, x_i) \rangle_{i \in [N]}$ and $\langle (i, y_i) \rangle_{i \in [N]}$, exactly as before; however, in this reduction, the output of $\mathcal{Q}$ is the *opposite* of the coordinator's output at the end of the run of $\mathcal{P}$.

We now show that $\mathcal{Q}$ is correct. The input stream σ seen by $\mathcal{P}$ has the same frequency distribution as before, which means that $F_p(\sigma) = 2\Delta(x,y) + (N - \Delta(x,y))\cdot 2^p = N\cdot 2^p - \Delta(x,y)(2^p - 2)$. When $\Delta(x,y) \le N/2 - \sqrt{N}$, we have

$$F_p(\sigma) \ \ge\ N\cdot 2^p - (N/2 - \sqrt{N})(2^p - 2) \ =\ (N/2 + \sqrt{N})2^p + (N - 2\sqrt{N}) \ =\ \tau\,.$$

Therefore $\mathcal{P}$ outputs 1, which means $\mathcal{Q}$ correctly outputs 0. On the other hand, when $\Delta(x,y) \ge N/2 + \sqrt{N}$, we have

$$\begin{aligned} F_p(\sigma) &\le\ N\cdot 2^p - (N/2 + \sqrt{N})(2^p - 2) \\ &=\ \tau\left(1 - \frac{2\sqrt{N}2^p - 4\sqrt{N}}{(N/2 + \sqrt{N})\cdot 2^p + (N - 2\sqrt{N})}\right) \ \le\ \tau(1 - 1/\sqrt{N}) \ \le\ \tau(1-\varepsilon)\,, \end{aligned}$$

where the penultimate inequality uses $p \ge 2$. Therefore $\mathcal{P}$ outputs 0, whence $\mathcal{Q}$ correctly outputs 1. Theorem 8 now implies that $\mathcal{Q}$, and hence $\mathcal{P}$, must communicate $\Omega(\sqrt{N}) = \Omega(1/\varepsilon)$ bits. □

We remark that if Conjecture 1 holds (for a favorable c), then the lower bounds in Theorems 9 and 10 would improve to $\Omega(1/\varepsilon^2)$. This further strengthens the motivation for settling the conjecture.

We also consider a restricted, yet natural, class of protocols that we call *round-based* protocols; the precise definition follows. Note that all nontrivial protocols in [4] are round-based, which illustrates the naturalness of this notion.

Definition 2. *A* round-based *protocol for $(k, f, \tau, \varepsilon)$ functional monitoring is one that proceeds in a series of rounds numbered $1,\ldots,r$. Each round has the following four stages. (1) Coordinator C sends messages to the sites S_i, based on the past communication history. (2) Each S_i read its tokens and sends messages to C from time to time, based on these tokens and the Stage* 1 *message from C to S_i. (3) At some point, based on the messages it receives, C decides to end the current round by sending a special, fixed, end-of-round message to each S_i. (4) Each S_i sends C a final message for the round, based on all its knowledge, and then resets itself, forgetting all previously read tokens and messages.*

It is possible to improve the lower bounds above by restricting to round-based protocols, as in Definition 2. The key is that if the functional monitoring protocol $\mathcal{P}$ in the proofs of Theorems 9 and 10 is round-based, then the corresponding communication protocol $\mathcal{Q}$ only requires messages from Alice to Bob. This is because Alice can now simulate the coordinator C and *both* sites S_1 and S_2, during $\mathcal{P}$'s processing of σ_a: she knows that S_2 receives no tokens at this time, so she has the information needed to compute any messages that S_2 might need to send. Consider the situation when Alice is done processing her tokens. At this time the Stage 4 message (see Definition 2) from S_1 to C in the current round has been determined, so Alice can send this message to Bob. From here on, Bob has all the information needed to continue simulating S_1, because he knows that S_1 receives no further tokens. Thus, Bob can simulate $\mathcal{P}$ to the end of the run.

Theorem 11. *Suppose p is either 0 or a constant greater than 1. For any $\varepsilon \leq 1/2$, a round-based randomized protocol for $(k, F_p, \tau, \varepsilon)$ functional monitoring must communicate $\Omega(1/\varepsilon^2)$ bits.*

Proof. We use the observations in the preceding paragraph, proceed as in the proofs of Theorems 9 and 10 above, and plug in the one-way communication lower bound from Theorem 8. □

References

1. Alon, N., Matias, Y., Szegedy, M.: The space complexity of approximating the frequency moments. J. Comput. Syst. Sci. 58(1), 137–147 (1999); Preliminary version In: Proc. 28th Annu. ACM Symp. Theory Comput., pp. 20–29 (1996)
2. Babcock, B., Olston, C.: Distributed top-k monitoring. In: Proc. Annual ACM SIGMOD Conference, pp. 28–39 (2003)
3. Bhuvanagiri, L., Ganguly, S.: Estimating entropy over data streams. In: Proc. 14th Annual European Symposium on Algorithms, pp. 148–159 (2006)
4. Cormode, G., Muthukrishnan, S., Yi, K.: Algorithms for distributed functional monitoring. In: Proc. 19th Annual ACM-SIAM Symposium on Discrete Algorithms, pp. 1076–1085 (2008)
5. Cormode, G., Muthukrishnan, S., Zhuang, W.: What's different: Distributed, continuous monitoring of duplicate-resilient aggregates on data streams. In: Proc. 22nd International Conference on Data Engineering, p. 57 (2006)
6. Das, A., Ganguly, S., Garofalakis, M.N., Rastogi, R.: Distributed set expression cardinality estimation. In: Proc. 30th International Conference on Very Large Data Bases, pp. 312–323 (2004)
7. Estrin, D., Govindan, R., Heidemann, J.S., Kumar, S.: Next century challenges: Scalable coordination in sensor networks. In: MOBICOM, pp. 263–270 (1999)
8. Harvey, N.J.A., Nelson, J., Onak, K.: Sketching and streaming entropy via approximation theory. In: Proc. 49th Annual IEEE Symposium on Foundations of Computer Science, pp. 489–498 (2008)
9. Muthukrishnan, S.: Data streams: Algorithms and applications. In: Proc. 14th Annual ACM-SIAM Symposium on Discrete Algorithms, p. 413 (2003)
10. Muthukrishnan, S.: Some algorithmic problems and results in compressed sensing. In: Proc. 44th Annual Allerton Conference (2006)
11. Newman, I.: Private vs. common random bits in communication complexity. Information Processing Letters 39(2), 67–71 (1991)
12. Sharfman, I., Schuster, A., Keren, D.: A geometric approach to monitoring threshold functions over distributed data streams. ACM Trans. Database Syst. 32(4) (2007)
13. Slepian, D., Wolf, J.K.: Noiseless coding of correlated information sources. IEEE Trans. Inf. Theory 19(4), 471–480 (1973)
14. Tsallis, C.: Possible generalization of Boltzmann-Gibbs statistics. J. Stat. Phys. 52, 479–487 (1988)
15. Woodruff, D.P.: Efficient and Private Distance Approximation in the Communication and Streaming Models. PhD thesis, MIT (2007)
16. Yi, K., Zhang, Q.: Multi-dimensional online tracking. In: Proc. 19th Annual ACM-SIAM Symposium on Discrete Algorithms, pp. 1098–1107 (2009)

De-amortized Cuckoo Hashing: Provable Worst-Case Performance and Experimental Results*

Yuriy Arbitman[1], Moni Naor[2,**], and Gil Segev[2,***]

[1] yuriy.arbitman@gmail.com

[2] Department of Computer Science and Applied Mathematics, Weizmann Institute of Science, Rehovot 76100, Israel
{moni.naor,gil.segev}@weizmann.ac.il

Abstract. Cuckoo hashing is a highly practical dynamic dictionary: it provides amortized constant insertion time, worst case constant deletion time and lookup time, and good memory utilization. However, with a noticeable probability during the insertion of n elements some insertion requires $\Omega(\log n)$ time. Whereas such an amortized guarantee may be suitable for some applications, in other applications (such as high-performance routing) this is highly undesirable.

Kirsch and Mitzenmacher (Allerton '07) proposed a de-amortization of cuckoo hashing using queueing techniques that preserve its attractive properties. They demonstrated a significant improvement to the worst case performance of cuckoo hashing via experimental results, but left open the problem of constructing a scheme with provable properties.

In this work we present a de-amortization of cuckoo hashing that *provably* guarantees constant worst case operations. Specifically, for any sequence of polynomially many operations, with overwhelming probability over the randomness of the initialization phase, each operation is performed in constant time. In addition, we present a general approach for proving that the performance guarantees are preserved when using hash functions with limited independence instead of truly random hash functions. Our approach relies on a recent result of Braverman (CCC '09) showing that poly-logarithmic independence fools AC^0 circuits, and may find additional applications in various similar settings. Our theoretical analysis and experimental results indicate that the scheme is highly efficient, and provides a practical alternative to the only other known approach for constructing dynamic dictionaries with such worst case guarantees, due to Dietzfelbinger and Meyer auf der Heide (ICALP '90).

---

* Due to space limitations we refer the reader to a longer version available at http://www.wisdom.weizmann.ac.il/~naor

** Incumbent of the Judith Kleeman Professorial Chair. Research supported in part by a grant from the Israel Science Foundation.

*** Research supported by the Adams Fellowship Program of the Israel Academy of Sciences and Humanities, and by a grant from the Israel Science Foundation.

S. Albers et al. (Eds.): ICALP 2009, Part I, LNCS 5555, pp. 107–118, 2009.

1 Introduction

A dynamic dictionary is a fundamental data structure used for maintaining a set of elements under insertions and deletions, while supporting membership queries. The performance of a dynamic dictionary is measured mainly by its update time, lookup time, and memory utilization. Extensive research has been devoted over the years for studying dynamic dictionaries both on the theoretical side by exploring upper and lower bounds on the performance guarantees, and on the practical side by designing efficient dynamic dictionaries that are suitable for real-world applications.

The most efficient dictionaries, in theory and in practice, are based on various forms of hashing techniques. Specifically, in this work we focus on cuckoo hashing, a hashing approach introduced by Pagh and Rodler [17]. Cuckoo hashing is an efficient dynamic dictionary with highly practical performance guarantees. It provides amortized constant insertion time, worst case constant deletion time and lookup time, and good memory utilization. Additional attractive features of cuckoo hashing are that no dynamic memory allocation is performed, and that the lookup procedure queries only two memory entries which are independent and can be queried in parallel.

Although the insertion time of cuckoo hashing is essentially constant, with a noticeable probability during the insertion of n elements into the hash table, some insertion requires $\Omega(\log n)$ time. Whereas such an amortized performance guarantee is suitable for a wide range of applications, in other applications this is highly undesirable. For these applications, the time per operation must be bounded in the worst case, or at least, the probability that some operation requires a significant amount of time must be negligible. For example, Kirsch and Mitzenmacher [9] considered the context of router hardware, where hash tables implementing dynamic dictionaries are used for a variety of operations, including various network measurements and monitoring tasks. In this setting, routers must keep up with line speeds and memory accesses are at a premium.

Clocked Adversaries. An additional motivation for the construction of dictionaries with worst case guarantees was first suggested by Lipton and Naughton [12]. One of the basic assumptions in the analysis of probabilistic data structures is that the elements that are inserted into the data structure are chosen independently of the randomness used by the data structure. This assumption is violated when the set of elements inserted might be influenced by the time it took the data structure to complete previous operations. Such timing information may reveal sensitive information on the randomness used by the data structure. For example, if the data structure is used for an operating system, then the time a process took to perform an operation affects which process is scheduled and that in turns affects the values of the inserted elements.

This motivates considering "clocked adversaries" – adversaries that can measure the exact time for each operation. Lipton and Naughton showed that several dynamic hashing schemes are susceptible to attacks by clocked adversaries, and demonstrated that clocked adversaries can identify elements whose insertion results in poor running time. The concern regarding timing information is even more acute in a cryptographic environment with an active adversary who might use timing information to compromise the system. The adversary might use the timing information to figure out sensitive information on the identity of the elements inserted, or as in the Lipton-Naughton case, to come up with a bad set of elements where even the amortized performance is bad. Note that timing attacks have been shown to be quite powerful and detrimental in cryptography (see, for example, [11,16]). To combat such attacks, at the very least we want the data structure to devote a fixed amount of time for each operation. There are further concerns such as cashing effects, but these are beyond the scope of this paper. Having a fixed upper bound on the time each operation (insert, delete, and lookup) takes, and an exact clock we can, in principle, make the response for each operation be independent of the input and the randomness used.

Dynamic Real-Time Hashing. Dietzfelbinger and Meyer auf der Heide [5] constructed the first dynamic dictionary with worst case time per operation and linear space (the construction is based on the dynamic dictionary of Dietzfelbinger et al. [4]). Specifically, for any constant $c > 0$ (determined prior to the initialization phase) and for any sequence of operations involving n elements, with probability at least $1 - n^{-c}$ each operation is performed in constant time (that depends on c). While this construction is a significant theoretical contribution, it may be unsuitable for highly demanding applications. Most notably, it suffers from hidden constant factors in its running time and memory utilization, and from an inherently hierarchal structure. We are not aware of any other dynamic dictionary with such provable performance guarantees which is not based on the approach of Dietzfelbinger and Meyer auf der Heide (but see Section 1.2 for a discussion on the hashing scheme of Dalal et al. [2]).

De-amortized Cuckoo Hashing. Motivated by the problem of constructing a *practical* dynamic dictionary with constant worst-case operations, Kirsch and Mitzenmacher [9] suggested an approach for de-amortizing the insertion time of cuckoo hashing, while essentially preserving the attractive features of the scheme. Specifically, Kirsch and Mitzenmacher suggested an approach for limiting the number of moves per insertion by using a small content-addressable memory as a queue for elements being moved. They demonstrated a significant improvement to the worst case performance of cuckoo hashing via experimental results, but left open the problem of constructing a scheme with provable properties.

1.1 Our Contributions

In this work we construct the first practical and efficient dynamic dictionary that provably supports constant worst case operations. We follow the approach of Kirsch and Mitzenmacher [9] for de-amortizing the insertion time of cuckoo hashing using a queue, while preserving many of the attractive features of the scheme. Specifically, for any polynomial $p(n)$ and constant $\epsilon > 0$ the parameters of our dictionary can be set such that the following properties hold[1]:

1. For any sequence of $p(n)$ insertions, deletions, and lookups, in which at any point in time at most n elements are stored in the data structure, with probability at least $1 - 1/p(n)$ each operation is performed in constant time, where the probability is over the randomness of the initialization phase[2].
2. The memory utilization is essentially 50%. Specifically, the dictionary utilizes $2(1+\epsilon)n + n^{\epsilon}$ words.

An additional attractive property is that we never perform rehashing. Rehashing is highly undesirable in practice for various reasons, and in particular, it significantly hurts the worst case performance. We avoid rehashing by following Kirsch, Mitzenmacher and Wieder [10] who suggested an augmentation to cuckoo hashing: exploiting a secondary data structure for "stashing" problematic elements that cannot be otherwise stored. We show that in our case, this can be achieved very efficiently by implicitly storing the stash inside the queue.

We provide a formal analysis of the worst-case performance of our dictionary, by generalizing known results in the theory of random graphs. In addition, our analysis involves an application of a recent result due to Braverman [1], to prove that polylog(n)-wise independent hash functions are sufficient for our dictionary. We note that this is a rather general technique, that may find additional applications in various similar settings. Our extensive experimental results clearly demonstrate that the scheme is highly practical. This seems to be the first dynamic dictionary that simultaneously enjoys all of these properties.

1.2 Related Work

Cuckoo Hashing. Several generalizations of cuckoo hashing circumvent the 50% memory utilization barrier: Fotakis et al. [8] suggested to use more than two hash functions; Panigrahy [18] and Dietzfelbinger and Weidling [6] suggested to store more than one element in each entry. These generalizations led to essentially optimal memory utilization, while preserving the update time and lookup time. Kirsch, Mitzenmacher and Wieder [10] provided an augmentation to cuckoo hashing in order to avoid rehashing. Their idea is to exploit a secondary data structure, referred to as a *stash*, for storing elements that cannot be stored without rehashing. Kirsch et al. proved that for cuckoo hashing with overwhelming probability the number of stashed elements is a very small constant. This augmentation was a crucial ingredient in the work of Naor, Segev, and Wieder [14], who constructed a history independent variant of cuckoo hashing.

[1] This is the same flavor of worst case guarantee as in the dynamic dictionary of Dietzfelbinger and Meyer auf der Heide [5].

[2] We note that the non-constant lower bound of Sundar for the membership problem in deterministic dictionaries implies that this type of guarantee is essentially the best possible (see the survey of Miltersen [13] who reports on Sundar's unpublished paper).

Dictionaries with Constant Worst-Case Guarantees. Dalal et al. [2] suggested an interesting alternative to the scheme of Dietzfelbinger and Meyer auf der Heide [5] by combining the two-choice paradigm with chaining. For each entry in the table there is a doubly-linked list and each element appears in one of two linked lists. In some sense the lists act as queues for each entry. Their scheme provides worst case constant insertion time, and with high probability lookup queries are performed in worst case constant time as well. However, their scheme is not fully dynamic since it does not support deletions, has memory utilization lower than 20%, allows only short sequences of insertions (no more than $O(n \log n)$, if one wants to preserve the performance guarantees), and requires dynamic memory allocation. Since lookup requires traversing two linked lists, it appears less practical than cuckoo hashing and its variants.

Demaine et al. [3] proposed a dynamic dictionary with memory consumption that asymptotically matches the information-theoretic lower bound (i.e., n elements from a universe of size u are stored using $O(n \log(u/n))$ bits instead of $O(n \log u)$ bits), where each operation is performed in constant time with high probability. Their construction extends the one of Dietzfelbinger and Meyer auf der Heide [5], and is the first dynamic dictionary that simultaneously provides asymptotically optimal memory consumption and constant time operations (in fact, when $u \geq n^{1+\alpha}$ for some constant $\alpha > 0$, the memory consumption of the dynamic dictionary of Dietzfelbinger and Meyer auf der Heide is already asymptotically optimal since in this case $O(n \log u) = O(n \log(u/n))$, and therefore Demaine et al. only had to address the case $u < n^{1+\alpha}$). Note, however, that asymptotically optimal memory consumption does not necessarily imply a practical memory utilization due to large hidden constants.

Paper Organization. The remainder of this paper is organized as follows. In Section 2 we provide a high-level overview of our construction. In Section 3 we present experimental results, and in Section 4 we discuss concluding remarks and open problems.

2 Overview of the Construction

In this section we provide an overview of our construction. We first provide a high-level description of cuckoo hashing, and of the approach of Kirsch and Mitzenmacher [9] for de-amortizing it. Then, we present our approach together with the main ideas underlying its analysis (due to space limitations we refer the reader to the full version of this paper for a complete description).

Cuckoo Hashing. Cuckoo hashing uses two tables T_0 and T_1, each consisting of $r = (1 + \epsilon)n$ entries for some constant $\epsilon > 0$, and two hash functions $h_0, h_1 : \mathcal{U} \to \{0, \ldots, r-1\}$. An element $x \in \mathcal{U}$ is stored either in entry $h_0(x)$ of table T_0 or in entry $h_1(x)$ of table T_1, but never in both. The lookup procedure is straightforward: when given an element $x \in \mathcal{U}$, query the two possible memory entries in which x may be stored. The deletion procedure deletes x from the entry in which it is stored. As for insertions, Pagh and Rodler [17] proved that the "cuckoo approach", kicking other elements away until every element has its own "nest", leads to a highly efficient insertion procedure. More specifically, in order to insert an element $x \in \mathcal{U}$ we first query entry $T_0[h_0(x)]$. If this entry is not occupied, we store x in that entry. Otherwise, we store x in that entry anyway, thus making the previous occupant "nestless". This element is then inserted to T_1 in the same manner, and so forth iteratively. We refer the reader to [17] for a more comprehensive description of cuckoo hashing.

De-amortization Using a Queue. Although the amortized insertion time of cuckoo hashing is constant, with a noticeable probability during the insertion of n elements into the hash table, some insertion requires moving $\Omega(\log n)$ elements before identifying an unoccupied entry. We follow the approach of Kirsch and Mitzenmacher [9] for de-amortizing cuckoo hashing by using a queue. The main idea underlying the construction of Kirsch and Mitzenmacher is as follows. A new element is always inserted to the queue. Then, an element x is chosen from the queue, according to some queueing policy, and is inserted into the tables. If this is the first insertion attempt for the element x (i.e., x was never stored in one of the tables), then we store it in entry $T_0[h_0(x)]$. If this entry is not occupied, we are done. Otherwise, the previous occupant y of that entry is inserted into the queue, together with an additional information bit specifying that the next insertion attempt for y should begin with table T_1. The queueing policy then determines the next element to be chosen from the queue, and so on. To fully specify a scheme in the family suggested by [9] one then needs to specify two issues: the queuing policy and the number of operations that are performed upon the insertion of a new element. In their experiments, Kirsch and Mitzenmacher loaded the queue with many insert operations, and let the system run. The number of operations that are performed upon the insertion of a new element depends on the success (small queue size) of the experiment.

Our Approach. In this work we propose a de-amortization of cuckoo hashing that provably guarantees worst case constant insertion time (with overwhelming probability over the randomness of the initialization phase). Our insertion procedure is parameterized by a constant L, and is defined as follows. Given a new element $x \in \mathcal{U}$, we place the pair $(x, 0)$ at the *back* of the queue (the additional bit 0 indicates that the element should be inserted to table T_0). Then, we carry out the following procedure as long as no more than L moves are performed in the cuckoo tables: we take the pair from the *head* of the queue, denoted (y, b), and place y in entry $T_b[h_b(y)]$. If this entry was unoccupied then we are done with the current element y, this is counted as one move and the next element is fetched from the *head* of the queue. However, if the entry $T_b[h_b(y)]$ was occupied, we place its previous occupant z in entry $T_{1-b}[h_{1-b}(z)]$ and so on, as in the above description of the standard cuckoo hashing. After L elements have been moved, we place the current "nestless" element at the *head* of the queue, together with a bit indicating the next table to which it should be inserted, and terminate the insertion procedure (note that it may take less than L moves, if the queue becomes empty).

The deletion and lookup procedures are naturally defined by the property that any element x is stored in one of $T_0[h_0(x)]$ and $T_1[h_1(x)]$, or in the queue. However, unlike the standard cuckoo hashing, here it is not clear that these procedures run in constant time. It may be the case that the insertion procedure causes the queue to contain many elements, and then the deletion and lookup procedures of the queue will require a significant amount of time.

The main property underlying our construction is that the constant L (i.e., the number of iterations of the insertion procedure) can be chosen such that with overwhelming probability the queue does not contain more than a logarithmic number of elements at any point in time. In this case we show that simple and efficient instantiations of the queue can indeed support insertions, deletions and lookups in worst case constant time. This is proved by considering the distribution of the *cuckoo graph*, formally defined as follows:

Definition 2.1. *Given a set $S \subseteq \mathcal{U}$ and two hash functions $h_0, h_1 : \mathcal{U} \to \{0, \ldots, r-1\}$, the* cuckoo graph *is the bipartite graph $G = (L, R, E)$, where $L = R = \{0, \ldots, r-1\}$ and $E = \{(h_0(x), h_1(x)) : x \in S\}$.*

The main idea of our analysis is to consider $\log n$ insertions each time, and to examine the total number of moves in the cuckoo graph that these $\log n$ insertions require. Our main technical contribution in this setting is proving that the sum of sizes of any $\log n$ connected components in the cuckoo graph is upper bounded by $O(\log n)$ with overwhelming probability. This is a generalization of a well-known bound in graph theory on the size of a single connected component. A corollary of this result is that in the standard cuckoo hashing the insertion of $\log n$ elements takes $O(\log n)$ time with high probability (ignoring the problem of rehashing, which is discussed below).

Avoiding Rehashing. It is rather easy to see that a set S can be successfully stored in the cuckoo graph using hash functions h_0 and h_1 if and only if no connected component in the graph has more edges then nodes. In other words, every component contains at most one cycle (unicyclic). It is known, however, that even if h_0 and h_1 are completely random functions, then with probability $\Theta(1/n)$ there will be a connected component with more than one cycle. In this case the given set cannot be stored using h_0 and h_1. The standard solution for this scenario is to choose new functions and rehash the entire data, but this significantly hurts the worst case performance.

To overcome this difficulty, we follow the approach of Kirsch et al. [10] who suggested an augmentation to cuckoo hashing in order to avoid rehashing: exploiting a secondary data structure, referred to as a *stash*, for storing elements that create cycles, starting from the second cycle of each component. That is, whenever an element is inserted into a unicyclic component and creates an additional cycle in this component, the element is stashed. Kirsch et al. showed that this approach performs remarkably well by proving that for any fixed set S of size n, the probability that at least k elements are stashed is $O(n^{-k})$. In our setting, however, where the data structure has to support delete operations in constant time, it is not straightforward to use a stash explicitly. Specifically, for the stash to remain of constant size, after every delete operation it may be required to move some element back from the stash to one of the two tables. Otherwise, the analysis of Kirsch et al. on the size of the stash no longer holds when considering long sequences of operations on the data structure.

We overcome this difficulty by storing the stashed elements in the queue. That is, whenever we identify an element that closes a second cycle in the cuckoo graph, this element is placed at the back of the queue. Informally, this guarantees that any stashed element is given a chance to be inserted back to the tables after essentially $\log n$ invocations of the insertion procedure. This implies that the number of stashed elements in the queue roughly corresponds to the number of elements that close a second cycle in the cuckoo graph (up to intervals of $\log n$ insertions). We can then use the result of Kirsch et al. [10] to argue that there is a very small number of such elements in the queue at any point.

For detecting cycles in the cuckoo graph we implement a simple *cycle detection mechanism* (CDM), as suggested by Kirsch et al. [10]. When inserting an element we insert to the CDM all the elements that are encountered in its connected component during the insertion process. Once we identify that a component has more than one cycle we stash the current nestless element (i.e., place it in the back of the queue) and reset the CDM. We note that in the classical cuckoo hashing cycles are detected by allowing the insertion procedure to run for $O(\log n)$ steps, and then announcing failure (which is followed by rehashing). In our case, however, it is crucial that a cycle is detected in time that is linear in the size of its connected component in the cuckoo graph.

A formal description of the lookup and update procedures is provided in Figure 1. In the full version of the paper we propose simple instantiations of the auxiliary data structures (i.e., the queue and the cycle-detection mechanism), and prove the following theorem:

Theorem 2.2. *For any polynomial $p(n)$ and constant $\epsilon > 0$, the parameters of the dictionary can be set such that the following properties hold:*

1. *For any sequence of at most $p(n)$ insertions, deletions, and lookups, in which at any point in time at most n elements are stored in the dictionary, with probability at least $1 - 1/p(n)$ each operation is performed in constant time, where the probability is over the randomness of the initialization phase.*
2. *The dictionary utilizes $2(1+\epsilon)n + n^{\epsilon}$ words.*

Using polylog(n)-wise Independent Hash Functions. When analyzing the performance of our scheme, we first assume the availability of truly random hash functions. Then, we apply a recent result of Braverman [1] and show that the same performance guarantees hold when instantiating our scheme with hash functions that are only polylog(n)-wise independent (see [7,15] for efficient constructions of such functions with succinct representations and constant evaluation time). Informally, Braverman proved that for any Boolean circuit C of depth d, size m, and unbounded fan-in, and for any k-wise distribution X with $k = (\log m)^{O(d^2)}$, it holds that $\mathbb{E}[C(U_n)] \approx \mathbb{E}[C(X)]$. That is, X "fools" the circuit C into behaving as if X is the uniform distribution U_n over $\{0,1\}^n$.

```
Lookup(x):
  if T0[h0(x)] = x or T1[h1(x)] = x then
      return true
  if LookupQueue(x) then
      return true
  return false

Delete(x):
  if T0[h0(x)] = x then
      T0[h0(x)] ← ⊥
      return
  if T1[h1(x)] = x then
      T1[h1(x)] ← ⊥
      return
  DeleteFromQueue(x)

Insert(x):
  InsertIntoBackOfQueue(x, 0)
  y ← ⊥ // y denotes the current element we work with
  for i = 1 to L do
      if y = ⊥ then // Fetching element y from the head of the queue
          if IsQueueEmpty() then
              return
          else
              (y, b) ← PopFromQueue()
      if Tb[hb(y)] = ⊥ then // Successful insert
          Tb[hb(y)] ← y
          ResetCDM()
          y ← ⊥
      else
          if LookupInCDM(y, b) then // Found the second cycle
              InsertIntoBackOfQueue(y, b)
              ResetCDM()
              y ← ⊥
          else // Evict existing element
              z ← Tb[hb(y)]
              Tb[hb(y)] ← y
              InsertIntoCDM(y, b)
              y ← z
              b ← 1 − b
  if y ≠ ⊥ then
      InsertIntoHeadOfQueue(y, b)
```

Fig. 1. The `LookUp`, `Delete` and `Insert` procedures

Specifically, in our analysis we define a "bad" event with respect to the hash values of h_0 and h_1, and prove that: (1) this event occurs with probability at most n^{-c} (for an arbitrarily large constant c) assuming truly random hash functions, and (2) as long as this event does not occur each operation is performed in constant time. We show that this event can be recognized by a Boolean circuit of constant depth, size $m = n^{O(\log n)}$, and unbounded fan-in. In turn, Braverman's result implies that it suffices to use k-wise independent hash functions for $k = \text{polylog}(n)$. We note that applying Braverman's result in such setting is quite a general technique and may be found useful in other similar scenarios. In

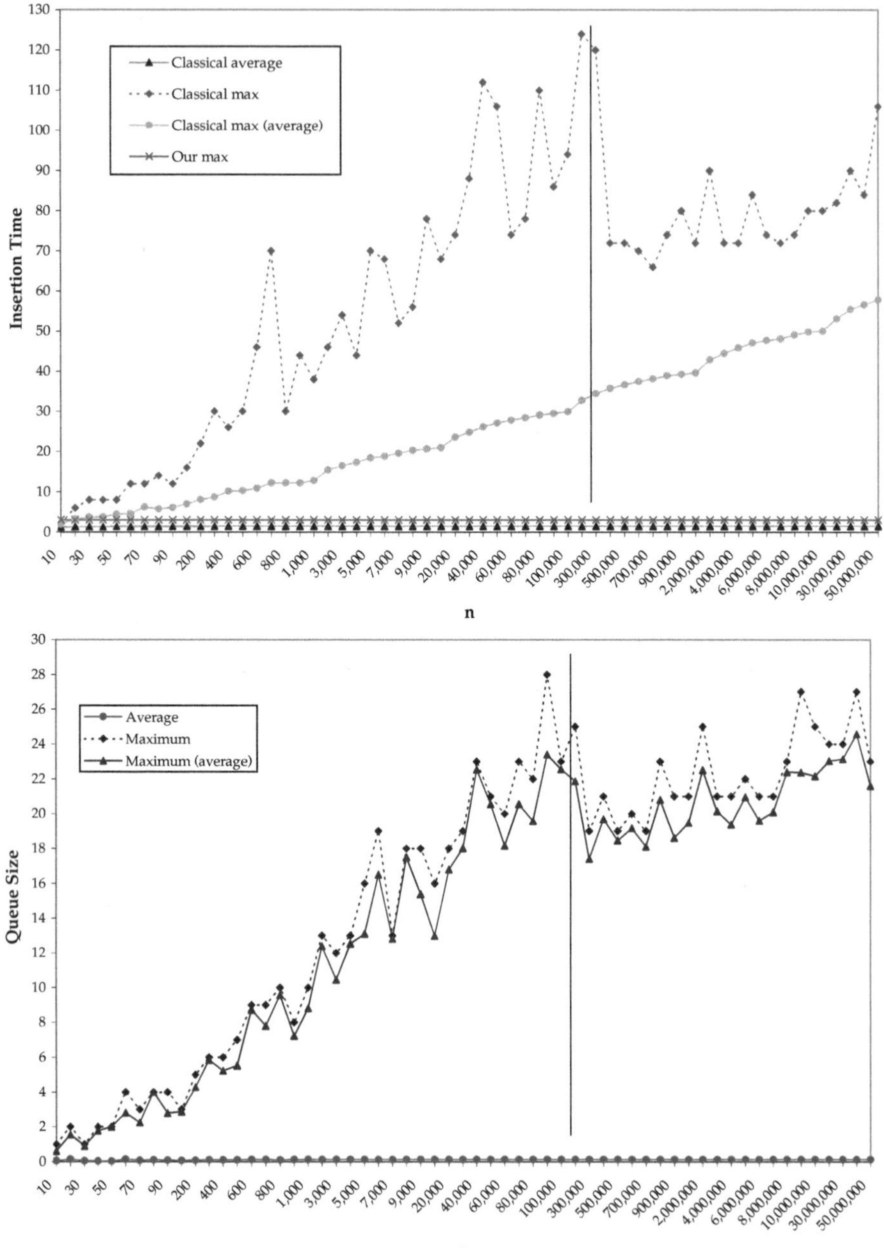

Fig. 2. The insertion time of classical cuckoo hashing vs. our dictionary (upper graph), and the size of the queue in our dictionary (lower graph)

particular, our argument implies that the same holds for the analysis of Kirsch et al. [10], who proved the above-mentioned bound on the number of stashed elements assuming that the underlying hash functions are truly random.

3 Experimental Results

We performed experiments to demonstrate the efficiency of the data structure in practice (we refer the reader to the full version for more details on the implementation and the results). The upper graph in Figure 2 presents a comparison between the insertion time of our scheme and the classical cuckoo hashing, and the lower graph shows the size of the queue in our dictionary (n denotes the number of inserted elements, we used two tables of size $1.2n$, and the vertical line represents a decrease in the number of experiments due to the large number of elements). The insertion time is measured as the length of the path that an element has traversed in the cuckoo graph. For our dictionary we plotted the maximal insertion time that was set to 3. Note that the scale in the graphs is logarithmic. Our experiments showed an excellent performance of our dictionary, noting that 3 moves are sufficient per insertion. This clearly demonstrates that adding an auxiliary memory of small (up to logarithmic) size reduces the worst case insertion time from logarithmic to a small constant.

4 Concluding Remarks

Clocked Adversaries. The worst case guarantees of our dictionary are important if one wishes to protect against "clocked adversaries", as discussed in Section 1. In the traditional RAM model, such guarantees are also sufficient for protecting against such attacks. However, for modern computer architectures the RAM model has limited applicability, and is nowadays replaced by more accurate hierarchical models, that capture the effect of several cache levels. Although our construction enables the "brute force" solution that measures the exact time every operation takes (see Section 1), a more elegant solution is desirable, which will make a better utilization of the cache hierarchy. We believe that our dictionary is an important step in this direction.

Memory Utilization. Our construction achieves memory utilization of essentially 50%. More efficient variants of cuckoo hashing [8,18,6] circumvent the 50% barrier and achieve better memory utilization by either using more than two hash functions, or storing more than one element in each entry. As demonstrated by Kirsch and Mitzenmacher [9], queue-based de-amortization performs very well in practice on these generalized variants, and it would be interesting to extend our analysis to these variants.

Optimal Memory Consumption. The memory consumption of our dictionary is $2(1 + \epsilon)n + n^{\epsilon}$ words, and each word is represented using $\log u$ bits where u is the size of the universe of elements (recall Theorem 2.2). As discussed in Section 1.2, when $u \geq n^{1+\alpha}$ for some constant $\alpha > 0$, this asymptotically matches the information-theoretic lower bound since in this case $O(n \log u) = O(n \log(u/n))$. An interesting open problem is to construct a dynamic dictionary with asymptotically optimal memory consumption also for $u < n^{1+\alpha}$ that will provide a practical alternative to the construction of Demaine et al. [3].

Acknowledgments. We thank Michael Mitzenmacher, Eran Tromer and Udi Wieder for very helpful discussions concerning queues, stashes and caches, and Pat Morin for pointing out the work of Dalal et al. [2]. We thank the anonymous reviewers for helpful remarks and suggestions.

References

1. Braverman, M.: Poly-logarithmic independence fools AC^0 circuits. In: 24th CCC 2009 (to appear)
2. Dalal, K., Devroye, L., Malalla, E., McLeis, E.: Two-way chaining with reassignment. SIAM J. Comput. 35(2), 327–340 (2005)
3. Demaine, E.D., Meyer auf der Heide, F., Pagh, R., Pătraşcu, M.: De dictionariis dynamicis pauco spatio utentibus. In: Correa, J.R., Hevia, A., Kiwi, M. (eds.) LATIN 2006. LNCS, vol. 3887, pp. 349–361. Springer, Heidelberg (2006)
4. Dietzfelbinger, M., Karlin, A.R., Mehlhorn, K., Meyer auf der Heide, F., Rohnert, H., Tarjan, R.E.: Dynamic perfect hashing: Upper and lower bounds. SIAM J. Comput. 23(4), 738–761 (1994)
5. Dietzfelbinger, M., Meyer auf der Heide, F.: A new universal class of hash functions and dynamic hashing in real time: In: Paterson, M. (ed.) ICALP 1990. LNCS, vol. 443, pp. 6–19. Springer, Heidelberg (1990)
6. Dietzfelbinger, M., Weidling, C.: Balanced allocation and dictionaries with tightly packed constant size bins. Theor. Comput. Sci. 380(1-2), 47–68 (2007)
7. Dietzfelbinger, M., Woelfel, P.: Almost random graphs with simple hash functions. In: 35th STOC, pp. 629–638 (2003)
8. Fotakis, D., Pagh, R., Sanders, P., Spirakis, P.G.: Space efficient hash tables with worst case constant access time. Theory Comput. Syst. 38(2), 229–248 (2005)
9. Kirsch, A., Mitzenmacher, M.: Using a queue to de-amortize cuckoo hashing in hardware. In: 45th Allerton, pp. 751–758 (2007)
10. Kirsch, A., Mitzenmacher, M., Wieder, U.: More robust hashing: Cuckoo hashing with a stash. In: Halperin, D., Mehlhorn, K. (eds.) Esa 2008. LNCS, vol. 5193, pp. 611–622. Springer, Heidelberg (2008)
11. Kocher, P.C.: Timing attacks on implementations of Diffie-Hellman, RSA, DSS, and other systems. In: Koblitz, N. (ed.) CRYPTO 1996. LNCS, vol. 1109, pp. 104–113. Springer, Heidelberg (1996)
12. Lipton, R.J., Naughton, J.F.: Clocked adversaries for hashing. Algorithmica 9(3), 239–252 (1993)
13. Miltersen, P.B.: Cell probe complexity - a survey. In: Pandu Rangan, C., Raman, V., Ramanujam, R. (eds.) FST TCS 1999. LNCS, vol. 1738, Springer, Heidelberg (1999)
14. Naor, M., Segev, G., Wieder, U.: History-independent cuckoo hashing. In: Aceto, L., Damgård, I., Goldberg, L.A., Halldórsson, M.M., Ingólfsdóttir, A., Walukiewicz, I. (eds.) ICALP 2008, Part II. LNCS, vol. 5126, pp. 631–642. Springer, Heidelberg (2008)
15. Ostlin, A., Pagh, R.: Uniform hashing in constant time and linear space. In: 35th STOC, pp. 622–628 (2003)
16. Osvik, D.A., Shamir, A., Tromer, E.: Cache attacks and countermeasures: The case of AES. In: Pointcheval, D. (ed.) CT-RSA 2006. LNCS, vol. 3860, pp. 1–20. Springer, Heidelberg (2006)
17. Pagh, R., Rodler, F.F.: Cuckoo hashing. J. Algorithms 51(2), 122–144 (2004)
18. Panigrahy, R.: Efficient hashing with lookups in two memory accesses. In: 16th SODA, pp. 830–839 (2005)

Towards a Study of Low-Complexity Graphs*

Sanjeev Arora[1], David Steurer[1], and Avi Wigderson[2]

[1] Computer Science Dept, Princeton University, Princeton NJ 08544, USA
{arora,dsteurer}@cs.princeton.edu
[2] Institute for Advanced Study, Princeton NJ
avi@ias.edu

Abstract. We propose the study of graphs that are defined by low-complexity distributed and deterministic agents. We suggest that this viewpoint may help introduce the element of *individual choice* in models of large scale social networks. This viewpoint may also provide interesting new classes of graphs for which to design algorithms.

We focus largely on the case where the "low complexity" computation is $\mathbf{AC}^0$. We show that this is already a rich class of graphs that includes examples of lossless expanders and power-law graphs. We give evidence that even such low complexity graphs present a formidable challenge to algorithms designers. On the positive side, we show that many algorithms from property testing and data sketching can be adapted to give meaningful results for low-complexity graphs.

1 Introduction

This paper tries to highlight some interesting families of graphs that we call *low complexity graphs*. These are graphs whose vertices are vectors of attributes (without loss of generality, vectors in $\{0,1\}^n$ for some n) and whose edge structure has low computational complexity, namely, determining whether (i,j) is an edge or not is a low-complexity computation involving i,j. (This definition assumes an *adjacency matrix* representation of the graph; we have an alternative definition for an *adjacency list* representation.)

There are many motivations for studying this concept. First, from a complexity-theoretic view such graphs are natural if we think of the graph as being defined by computationally bounded distributed agents, and they only use their personal information (namely, their names i,j) while computing their decision. Concretely, one could hope that many graph problems are easier to solve on low-complexity graphs than they are on general graphs (analogously to say, fixed treewidth graphs or minor-excluded graphs).

Second, such low-complexity graphs represent a plausible way to incorporate the concept of *individual choice* in models of formation of social network graphs (e.g., graphs that arise on the Web, Myspace, Facebook, etc.). Empirically these graphs are found to exhibit some strong properties such as *power law* distribution of degrees and top eigenvalues. A large body of work has been used to describe how such graphs can arise naturally as a result of distributed actions. The dominant model is some variant of a *preferential attachment* process, in

* Supported by NSF grants 0830673, 0832797, 528414.

S. Albers et al. (Eds.): ICALP 2009, Part I, LNCS 5555, pp. 119–131, 2009.

which nodes attach to the graph one by one, and when they try to (probabilistically) decide which other nodes to attach to, they prefer nodes with higher degree (i.e., "popular nodes"). By varying model parameters and assumptions (see the survey by Mitzenmacher [1]) a rich variety of graphs can be obtained.

While it is highly plausible that random choice coupled with a "follow the herd" behavior should be an important part of the story of how such networks come about, it is unclear if this is the complete story. For instance, it has been empirically observed that many graph problems are trivial to solve on the above random models but quite difficult on real-life graphs obtained from social networks [2].

One obvious missing element in the above models is that of *individual choice.* Clearly, people link to other people on Myspace or Facebook or the Web at least in part due to their interest profile, and they seek out interesting people or webpages using search tools (e.g., Google and other search engines) that allow them to explicitly or implicitly express (via boolean expressions) their choice function.

Our model of low-complexity graphs gives one way to incorporate the element of individual choice: the choice of a person with attribute vector i to link to a person with vector j is a low complexity computation involving i, j. For sake of clarity the model has been kept very bare: (i) The attributes are distributed uniformly in the population (i.e., the node set is $\{0,1\}^n$); (ii) The choice function for all nodes uses the same low-complexity computation —only their inputs are different, namely, their own attribute vectors. Note that this allows the adjacency list of each node i to be different because its attribute vector is unique.

(iii) The issue of random choice has been left out entirely, or to be more precise, is allowed only in very small doses as part of the choice function. The point in this paper is that even with these restrictions, great expressive power still remains —one can realize complicated graphs like power-law graphs and lossless expanders and extractors (see Section 2).

Now we formally define low complexity graphs. If $\mathcal{C}$ is a complexity class then an infinite family of graphs $\{G_n\}$ is said to be a *$\mathcal{C}$-graph family in the adjacency matrix representation* if for every $n \geq 1$, G_n has 2^n nodes and there is an algorithm in class $\mathcal{C}$ that, given indices $i, j \in \{0,1\}^n$, can compute whether or not (i, j) is an edge in the graph. A *$\mathcal{C}$-graph family in the adjacency list representation* is similarly defined, except we restrict it to be d-regular for some d (which could be a slowly growing function of n) and we require for each input size n a sequence of dn algorithms from $\mathcal{C}$ that given i compute the individual bits of the adjacency list of node i. (The output of each algorithm consists of one bit.) We can similarly define other low-complexity objects such as hypergraphs, circuits, etc.

Of course, similar definitions have been used in complexity theory before. For instance, the case $\mathcal{C} = \mathbf{DSPACE}(n)$ corresponds to *logspace uniform* circuits/graphs, and also corresponds to a common notion of *strongly explicit constructions* (used for example when we talk about explicit constructions of expanders). The case $\mathcal{C} = \mathbf{P}/poly$ coincides with *succinctly described* graphs [3].

It is known that many standard problems (e.g., connectivity) are intractable for succinctly represented graphs, when the input to the algorithm is the circuit that represents the graph. Classes such as **PPAD** [4] are also defined in this context. The difference in our paper is that we are interested in letting $\mathcal{C}$ be a very low complexity classes: say, $\mathbf{NC}^0, \mathbf{AC}^0$, or the set of graphs with $\log n$ decision tree complexity. $\mathbf{AC}^0$, the class of functions computable by bounded depth circuits of unlimited fan-in and polynomial size, will be a frequent choice in this paper. Furthermore, in our lower bound results we usually think of the entire graph as being presented to the algorithm (as opposed to just the circuit). The algorithms we present in Section 4 work even in the model where the algorithm is presented the circuit defining the input graph.

In fact, one motivation for studying low-complexity graphs is that this fits in the longer-term program of understanding low-complexity graphs (which are just truth tables of low-complexity functions) as a prelude to proving complexity lower bounds. Of course, the results of Razborov and Rudich [5] caution us against hoping for quick successes in this endeavor if the $\mathcal{C}$ is too complex. Accordingly, this paper generally restricts attention to classes no more powerful than $\mathbf{AC}^0$. Our results suggest that even these graphs can be quite complex and difficult. In Section 3 we show that solving any of the standard **NP** or **P** problems on $\mathbf{AC}^0$-graphs is probably no easier than solving them on general graphs, *even when the entire graph is given as input.*

In light of the hardness results one should relax one's expectations of what kind of algorithms can be possible. One could try for approximation algorithms. We observe using the current proof of the PCP Theorem [6] that approximation is intractable for a host of standard problems on $\mathbf{NC}^1$-graphs (Section 3.1). It is an open problem whether this result can be extended to $\mathbf{AC}^0$-graphs, and a positive resolution seems to require a new proof of the PCP Theorem. We can show however that $\mathbf{AC}^0$ graphs can prove to be difficult (ie have high integrality gaps) for current approximation algorithms that are based upon semidefinite programming (Theorem 9).

In light of the results about seeming difficulties of approximation, one can try to further relax what it means to solve the problem. An attractive approach is to try *sampling-based* algorithms such as the ones developed in property testing and streaming algorithms. We show that many known algorithms can be interpreted as follows: given a circuit representing a low-complexity graph, these algorithms return another low-complexity circuit that approximately represents a solution. (See Section 4.)

Though many known sampling-based algorithms are tight when the graph is given as a black box, it is conceivable that they can be improved upon when the graph is given as a small circuit. (Theorem 11 explains why the black box lower bounds don't apply.) Perhaps designing such algorithms can be a new motivation for revisiting lowerbound techniques for classes such as $\mathbf{AC}^0$, or even $\mathbf{NC}^0$. For instance we do not currently know if constant degree $\mathbf{NC}^0$-graphs can be expanders in the adjacency list representation. We observe in Section 2.1 that expanders of logarithmic degree can be realized as $\mathbf{NC}^0$ graphs.

This paper spans ideas from multiple areas, and consequently we have to assume familiarity with many standard terms and results and necessarily omit many details. We also see the results in this manuscript as representative rather than exhaustive. Even the basic model can be extended in many ways. For instance, if nodes are allowed a small amount of randomness and a small amount of computation then one obtains complexity-based extension of the theory of $G(n, p)$ graphs. We hope that many such extensions will be studied in future.

The recently studied model of random dot product graphs [7,8] shares some ideas with our low complexity graphs. There, the vertices are sampled from some distribution on the sphere, and the probability of an edge is proportional to the inner product of the endpoints. The inner product can be seen as a low-complexity function of the endpoints (albeit not $\mathbf{AC}^0$) that determines the edge probability.

2 Constructions of Interesting $\mathbf{AC}^0$ Graphs

In this section we show how some well-known graphs can be realized as $\mathbf{AC}^0$ graphs. Throughout, N denotes the number of nodes in the graph, and the nodes are assumed to be vectors of n bits (i.e., binary-valued attributes).

To set the context, we start with some simple examples of graphs that are *not* $\mathbf{AC}^0$, by the well-known results of Ajtai [9] and of Furst, Saxe, and Sipser [10].

Example 1 (Graphs that are not $\mathbf{AC}^0$). *Parity Graph*: Its edge set is $\{(x, y) : \oplus_i(x_i \oplus y_i) = 1\}$ where $x, y \in GF(2)^n$ and $\oplus$ addition mod 2.

Inner Product Graph: Its edge set is $\{(x, y) : x \odot y = 1\}$ where $x, y \in GF(2)^n$ and $\odot$ is inner product mod 2.

Threshold graph: Its edge set is $\{(x, y) :$ x and y agree on at least $2/3$rd of their bits.$\}$.

The fact that such simple operations are impossible in $\mathbf{AC}^0$ makes the task of designing $\mathbf{AC}^0$ graphs difficult. Next, we list simple graphs that *are* $\mathbf{AC}^0$. These will be building blocks in more complicated constructions later on.

Example 2 (Some $\mathbf{AC}^0$ graphs). *The N-cycle.* The attribute vector labeling the nodes can be interpreted as a binary number in $[0, N-1]$, and the set of edges is $\{x, x+1 \bmod N\}$. Since $x \to x+1$ is computable in $\mathbf{AC}^0$, we conclude that this is an $\mathbf{AC}^0$ graph.

The r-dimensional grid on N vertices. Reasoning similarly as above, this is also an $\mathbf{AC}^0$ graph for every fixed r. This will be useful in Theorem 5.

The following is also $\mathbf{AC}^0$ for every $k \le n$: the graph whose edge set is $\{(x, y) :\ x, y$ agree on the first k bits$\}$.

Next, since *approximate thresholds* can be computed probabilistically in $\mathbf{AC}^0$ (actually even in depth 3 [11]), we can construct an $\mathbf{AC}^0$ graph that is an approximate and noisy version of the Threshold graph of Example 1. (This could be useful in modeling social networks because "approximate threshold" of shared attributes could be a plausible strategy for setting up connections.) For any constant ϵ (to be thought of as $o(1)$), the following holds. For (x, y)'s such that x, y

agree in $\geq 2/3+\epsilon$ of the bits, $\{x,y\}$ is an edge. For (x,y)'s such that x,y agree in $< 2/3-\epsilon$ of the bits, (x,y) is a non-edge. All other pairs (x,y) may or may not be edges.

Finally, by definition, the set of $\mathbf{AC}^0$ graphs is closed under taking graph complements, union and intersection of edge sets, and taking AND product $G_1 \times G_2$, which is the graph whose vertex set is $V(G_1)\times V(G_2)$ and the edge set is $\{((u_1,v_1),(u_2,v_2)) : (u_1,u_2)\in E(G_1) \text{ AND } (v_1,v_2)\in E(G_2)\}$.

The following fact, a consequence of Håstad's [12] lower bound for $\mathbf{AC}^0$, suggests that every $\mathbf{AC}^0$ graph has some structure. More precisely, it shows that $\mathbf{AC}^0$-graphs are never good Ramsey graphs (in contrast to random graphs where the largest bipartite clique and the largest bipartite independent set are of size $O(\log N)$).

Proposition 1. *There is a polynomial-time algorithm that given an $\mathbf{AC}^0$ graph on N nodes (in the adjacency matrix representation) finds either a bipartite clique or independent set of size $2^{\Omega(\log N/\mathrm{poly}(\log\log N))}$ (in fact the algorithm can find many —say, superlogarithmic—number of these).*

2.1 Expanders and Lossless Expanders

In this subsection we show $\mathbf{AC}^0$-graphs can exhibit highly "pseudorandom" behavior. For a long time the only known explicit constructions of these graphs involved algebraic operations that are provably impossible in $\mathbf{AC}^0$. We use recent constructions involving the zig-zag product. For background and myriad applications of expanders please see the extensive survey of Hoory, Linial, Wigderson [13].

These graphs will be sparse and we choose to describe them in the adjacency list representation. Note that for constant degree graphs the adjacency matrix representation is at least as rich as the adjacency list representation. A d-regular graph $G=(V,E)$ is an *eigenvalue c-expander* if the second eigenvalue of the normalized Laplacian of G is at least c (sometimes called the "spectral gap"). In a *lossless expander*, vertex neighborhoods are essentially as large as possible. Graph G is an expander with *loss* ϵ if every set S of size at most $\alpha|V|$ has $|\Gamma(S)| \geq d(1-\epsilon)|S|$ (where $\alpha=\alpha(\epsilon,d)$ does not depend on the size of the graph). Lossless expanders have proved useful in a host of settings. See Capalbo et al. [14] for the first explicit construction and further references.

Theorem 3. *For every $\epsilon>0$ there is a $d=d(\epsilon)>0$ such that there is an $\mathbf{AC}^0$ family of d-regular expanders with loss ϵ.*

Proof. It will be convenient to think of the graph being represented by a circuit that given (v,i) with $v\in V$ and $i\in[d]$ (given in binary), and its output will be a vertex $u\in V$ which is the ith neighbor of v.

Let $G: N\times M\to N$ and $H: M\times D\to M$ be two circuits (note that the degree of G is the size of H). Then their zig-zag product $G \textcircled{z} H : (N\times M)\times(D^2)\to(N\times M)$ is defined by $(G \textcircled{z} H)((v,k),(i,j)) = (G(v,H(k,i)),H(H(k,i),j))$.

Note that the output of $G ⓩ H$ is simply a composition of the given circuits in serial, and its depth is the sum of depths of their depths (and size at most twice the sum of their sizes).

Below we rely upon two works. We use the simplified expander construction of Alon et al. [15] that gives a c-eigenvalue expander using only two applications of the zig-zag product. Then we do a zigzag product with a constant size graph as in Capalbo et al. [14] to end up with with a lossless expander. We assume familiarity with these constructions and only describe why they work for us.

As in [15] start with a Cayley expander G on $N = (F_2)^n$ of degree $M = n^2$, arising from construction of ϵ-biased sets. As addition in $GF(2)^n$ is carried out in $\mathbf{NC}^0$, this graph is an $\mathbf{AC}^0$-graph (we must move from $\mathbf{NC}^0$ to $\mathbf{AC}^0$ when using the edge index to obtain the actual representation of that vector from the generating set). This graph is composed once with a smaller copy of itself with n^2 vertices, and then again with another expander that is small enough to be trivially an $\mathbf{AC}^0$-graph. This gives an eigenvalue expander that is clearly an $\mathbf{AC}^0$-graph.

Now let us say a few words about the construction of lossless expanders on [14]. For these constructions it is useful to have the circuits output "extra information." Above, given (v, i) they produce only the ith neighbor of v, and now we'll ask them to output more information (whose main function is "keeping the entropy" in the input) beyond the neighbor. E.g. in [16] the circuit G will also include the index of the edge along which that neighboring vertex was reached. This modified circuit, $G : N \times M \to N \times M$ is called there a "rotation map". The objects which are multiplied using the extension of zig-zag in [14] are not expanders, but other pseudorandom objects related to extractors. In some of them as well the output has two parts, carrying a similar information as in the "rotation map".

With this in mind, the construction of lossless expanders in [14] has the exact same structure as the one above, and explicitly describes them as composition of functions (and as noted above, easily interpreted in circuit terms). The two components needed are an eigenvalue expander, which was constructed above in $\mathbf{AC}^0$ and a constant-sized graph, which is in $\mathbf{AC}^0$ trivially. □

Open questions for $\mathbf{NC}^0$ graphs. Can $\mathbf{NC}^0$-graphs be constant-degree expanders? We note that in the adjacency matrix representation the answer is "No", since any nontrivial $\mathbf{NC}^0$ circuit cannot even represent a constant-degree graph.

However, in the adjacency list representation the answer to the question is less clear (e.g., logarithmic degree $\mathbf{NC}^0$ expanders exist in this representation).

2.2 Constructions of Power-Law Graphs

Many real-life graphs have a degree distribution that satisfies the *power law: for some parameter $\alpha > 0$, the number of nodes of degree $> x$ is proportional to $x^{-\alpha}$.* As mentioned in the introduction, the standard explanation for these is some form of randomized attachment process where nodes favor other nodes with high degree ("popular nodes"). It has also been observed empirically that the largest few eigenvalues also satisfy a power law with exponent $\alpha/2$: *if the largest degrees* *are* $d_1, d_2, d_3, \ldots$, *the largest eigenvalues are close to* $\sqrt{d_1}, \sqrt{d_2}, \sqrt{d_3}, \ldots$. This has also been explained for random graph models by Mihail and Papadimitriou [17].

Theorem 4. *For every* $\alpha > 2$ *there is an* $\mathbf{AC}^0$ *family of graphs whose degrees satisfy the power-law with exponent* α*, and furthermore the top* $k = N^{1/2\alpha}$ *eigenvalues satisfy the eigenvalue power law.*

Proof. Many constructions are possible but since we only have to ensure power-laws for the degrees and some top eigenvalues we give a particularly easy one.

We note that since $\alpha > 2$ the number of edges is $\sum_d N/d^{\alpha-1} = O(N)$ and the max degree is less than $N^{1/\alpha}$. For simplicity we make all degrees powers of 2. We divide vertices into $O(\log n)$ classes $S_1, S_2, \ldots, S_k$, where $|S_i| = N/2^{i\alpha}$ and degree of vertices in S_i is close to 2^i (we say "close to" because the construction will allow degrees to deviate a little). Thus the maximum degree vertex is in S_k and has degree $N^{1/\alpha}$.

Vertices in S_i consist of vectors in $GF(2)^n$ whose first $i\alpha$ bits are 0 and the bit immediately after these is 1. Thus the S_i's are disjoint, and furthermore verifying whether a vector x lies in S_i is an $\mathbf{AC}^0$ computation.

For now we describe the construction as randomized and allow poly(n) random bits, which can obviously be appropriately chosen and hardwired so that certain bad events listed below do not happen. For each i pick a random shift $a_i \in GF(2)^n$ and connect $x \in S_i$ to $x + a_i + b$ where b is a vector whose last $n - i$ bits are zero. If $x + a_i + b$ is in S_j for $j \geq 5$ then leave out the edge. Note that the adjacency matrix of these connections is computable in $\mathbf{AC}^0$.

By construction, the degree of each node in S_i is almost 2^i, except that an occasional node x may have some missing edges because $x+a_i+b$ happened to lie in $\cup_{j\geq 5} S_j$. Since this set has size $\sum_{j\geq 5} N/2^{j\alpha} < N/2^{5\alpha-1}$ the chance that this happens (since shift a_i is random) is less than $1/2^{5\alpha-1}$. By Markov's inequality the fraction of nodes in S_i (for $i \geq 7$) that have more than $1/10$th of their edges missing is less than $1/4$.

Now we argue about the top eigenvalues along the lines of Mihail and Papadimitriou. One can show that the edges of k highest-degree vertices form a union of (mostly) disjoint stars with high probability. Since the largest eigenvalue of a d-star is $\sqrt{d-1}$, the graph formed by these edges has largest eigenvalues roughly $\sqrt{d_1}, \ldots, \sqrt{d_k}$. Furthermore, it is true that with high probability, for all $i \in [k]$, our graph contains much less than $\sqrt{d_i}$ edges between the leaves of the d_i-star. Hence, by eigenvalue interlacing, even including these edges does not spoil the eigenvalue power law. Finally, one can show that the maximum degree of the remaining edges is too small to influence the first k eigenvalues. □

3 Hardness Results

One reason to study specific graph families such as bounded tree-width graphs is that one can often solve certain problems more easily on them as compared to general graphs. In this section we explore whether such better algorithms exist for low-complexity graphs. The negative results in this section will guide the choice of what kind of algorithms to expect.

Note that N is the number of nodes in the graph, $n = \log_2 N$ is the number of labels in the vertices, and the algorithm is provided the entire graph as input.

The first result shows that there are unlikely to be efficient algorithms for the usual **NP**-complete problems on $\mathbf{AC}^0$-graphs. We say the **NP**-complete problem is *usual* if its **NP**-completeness can be proved using the classical Cook-style proof followed by a simple gadget-based reduction in the style of Karp. (This notion is obviously not precise. The precise version of the next theorem would involve replacing "usual **NP**-complete" by a list of a few thousand explicit problems.)

Theorem 5. *If* **NEXP** $\neq$ **EXP** *then none of the usual **NP**-complete problem can be solved on* $\mathbf{AC}^0$*-graphs in polynomial time.*

Proof (sketch). Suppose a polynomial-time algorithm were to solve a usual **NP**-complete problem, say CLIQUE, on $\mathbf{AC}^0$-graphs. We show how to use it to decide any language $L \in \mathbf{NTIME}(2^{n^c})$ deterministically in time $2^{O(n^c)}$, which contradicts the Theorem's hypothesis.

For any input $x \in \{0,1\}^n$ we can use the standard Cook-Karp style reduction to produce a graph with $2^{O(n^c)}$ nodes which has a clique of a certain size iff $x \in L$. Let us show that this instance is an $\mathbf{AC}^0$-graph, and therefore amenable to be solved by the algorithm.

Recall that the Cook-style reduction consists of writing constraints for the 2×3 "windows" in the tableau of M's computation on x. The window —and hence also each edge in the instance of CLIQUE—is defined by indices of the type $(i + b_1, j + b_2)$ where $b_1 \in \{0, 1, 2\}, b_2 \in \{0, 1\}$, and $i, j \leq 2^{n^c}$. As noted in Example 2 the function $i \to i + 1$ is computable in $\mathbf{AC}^0$ when i is given in binary, we can easily design an $\mathbf{AC}^0$ circuit of size poly(n^c) (with x hardwired in it) that represents the CLIQUE instance. This shows that the instance is an $\mathbf{AC}^0$-graph. □

At first glance the use of the conjecture **NEXP** $\neq$ **EXP** (which implies **P** $\neq$ **NP**) in the previous result may seem like overkill. But there is a (folklore) converse of sorts known.

Theorem 6. *If* **NEXP** $=$ **EXP** *then every **NP** problem on* **P**/poly*-graphs (thus, also on* $\mathbf{AC}^0$*-graphs) can be solved deterministically in* $n^{\text{poly}(\log n)}$ *time.*

Proof (sketch). Suppose A is an exponential time deterministic algorithm for a **NEXP**-complete problem. Let L be any **NP** language. To solve it on **P**/poly-graphs of size N, we note that the input can be represented by the circuit whose size is poly$(\log N)$. Though the circuit is not provided as part of the input, it can be recovered in exp(poly$(\log N)$) time by exhaustive search. Now we can think of the problem as really one in **NEXP** where the input is this circuit. Now we can use A to solve this problem in time exp(poly$(\log N)$). □

How about **P**-complete problems on $\mathbf{AC}^0$ graphs? Can we solve them more efficiently than on general graphs? The following theorem –proved completely analogously to the previous two theorems—suggests that we cannot.

Theorem 7. *If* **EXP** $\neq$ **PSPACE** *then no usual* ***P****-complete problem can be solved on* $\mathbf{AC}^0$*-graphs in polylog space. If* **EXP** $=$ **PSPACE** *then every problem in* ***P*** *can be solved on* ***P****/*poly*-graphs in polylog space.*

3.1 Results about Hardness of Approximation

We already saw that problems like CLIQUE cannot be optimally solved on $\mathbf{AC}^0$-graphs in polynomial time if **NEXP** $\neq$ **EXP**. What about approximation? We note that the current proof of the PCP Theorem (specifically, the generalization from **NP** to **NEXP**) and related results imply that current inapproximability results can be transfered to show hardness of the same problem on $\mathbf{NC}^1$-graphs. The following is a sample result.

Theorem 8. *If* **NEXP** $\neq$ **EXP** *then the MAX-CLIQUE problem on* $\mathbf{NC}^1$*-graphs cannot be approximated within any constant factor in polynomial time.*

Proof (sketch). The only known proof [18,19,20] of the result **NEXP** $=$ **PCP**(poly(n), 1) involves taking polynomial extension and then using procedures for polynomial testing/correcting and sum-check. These involve finite field arithmetic on n-bit vectors (where the graph size is 2^n), which is possible in $\mathbf{NC}^1$. The proof has a second step involving verifier composition which is trivially in $\mathbf{NC}^1$ because of the smaller inputs involved.

With these observations and the known reduction from **PCP**(poly(n), 1) to approximating MAX-CLIQUE [21] we can finish the proof as in Theorem 5. □

The stumbling block in proving a similar result for $\mathbf{AC}^0$ graphs is that current PCP constructions rely upon field operations that cannot be done in $\mathbf{AC}^0$. It is conceivable that this is inherent, and one way to show this would be to give a better CLIQUE approximation algorithm for $\mathbf{AC}^0$ graphs. This would probably involve an interesting new result about $\mathbf{AC}^0$.

At the same time, simple approximation algorithms such as basic SDP relaxations will probably not lead to better approximation in most cases. For example, for MAX-CUT the integrality gap of the standard SDP relaxation is achieved on instance of finite size, which are clearly $\mathbf{AC}^0$.

Theorem 9. *The worst-case integrality gaps of the standard SDP relaxation on* $\mathbf{AC}^0$ *graphs for MAX-CUT (see [22]) is* 0.878.. *(i.e., same as for general graphs).*

4 Algorithms for Low-Complexity Graphs

The hardness results in Section 3 greatly constrict our options in terms of what kinds of algorithms to shoot for on $\mathbf{AC}^0$ graphs.

Of course, the only hope in designing such algorithms is to exploit something about the structure of $\mathbf{AC}^0$ graphs (e.g., Proposition 1). Unfortunately, this hope is somewhat dashed in Section 3.

In light of these hardness results, it is reasonable to turn to sampling-based algorithms from areas such as property testing and sublinear algorithms, which result in a fairly weak approximation (necessarily so, since only a small portion of the graph is examined). Since the introduction of the property testing idea by Goldwasser, Goldreich, and Ron [23], a large body of literature has grown up in this area. Note that such algorithms are natural to consider for $\mathbf{AC}^0$ graphs, since we are given a circuit oracle for the edges.

We notice that many algorithms in this area implicitly give a stronger result than formally stated: in many cases if the graph is $\mathbf{AC}^0$, then the approximate *solution* to the problem (namely, a cut, assignment, etc.), which is an object of size $\exp(n)$, can also be represented as an $\mathbf{AC}^0$ circuit. As an illustrative example, we use MAX-CUT.

Theorem 10. *For any $\epsilon > 0$ there is a polynomial-time randomized algorithm that, given an $\mathbf{AC}^0$ circuit computing $f\colon \{0,1\}^n \times \{0,1\}^n \to \{0,1\}$ representing a graph in the adjacency matrix representation, produces an $\mathbf{AC}^0$ circuit computing some $g\colon \{0,1\}^n \to \{0,1\}$ circuit such that the cut $(g^{-1}(0), g^{-1}(1))$ in the graph has capacity at least $OPT - \epsilon n^2$ where OPT is the capacity of a MAX CUT in the graph.*

Proof. For any $\epsilon > 0$, the sampling algorithm of Alon et al. [24] samples $k = O(1/\epsilon^5)$ vertex pairs and examines whether or not the corresponding edges are present in the graph. This suffices for it to estimate the capacity of the maximum cut within additive error ϵN^2. In fact the proof of correctness shows something stronger. It gives a function $h\colon \{0,1\}^{\binom{k}{2}} \times \{0,1\}^k \to \{0,1\}$ such that the following is true for any graph $G = (V, E)$ of any size. For any subset of k vertices S let $E_S \in \{0,1\}^{\binom{k}{2}}$ be the characteristic vector showing which of the $\binom{k}{2}$ pairs of S are connected by an edge. If v is any other vertex, let v_S be a characteristic vector showing which of the k nodes of S are connected to v by an edge. For every S let $g_S\colon V \to \{0,1\}$ be defined as $g_S(v) = h(E_S, v_S)$. (Thus g_S implicitly defines a cut of the graph.) Then if S is chosen uniformly at random, then the probability is at least 0.99 that the cut represented by g_S has capacity at least $OPT - \epsilon n^2$

Thus the randomized algorithm is to sample S at random and output the function g_S. Note that if the graph was $\mathbf{AC}^0$ then so is g_S. □

In fact, using the subsequent work of Alon et al. [25] there is a similar algorithm (though again not mentioned explicitly) as in Theorem 10 for every testable graph property. For example, the property of being C-colorable is testable for any constant C. The analog of Theorem 10 for this problem shows that the final $\mathbf{AC}^0$ circuit will compute a C-coloring that is a proper coloring of some graph that differs in at most ϵn^2 edges from G. The sampling complexity is much worse than for MAX-CUT, though still constant for every constant $\epsilon > 0$.

4.1 Algorithms for the Adjacency List Representation

Property testing has also been studied in the adjacency list representation, and some algorithms transfer to the $\mathbf{AC}^0$ setting. For instance the work of [26] implies that it is possible to test in $\mathrm{poly}(n/\epsilon)$ time whether the $\mathbf{AC}^0$-graph is ϵ-close to a graph that is minor-free (ϵ-close means in this context that the symmetric difference of the edge sets of the two graphs is at most ϵNd).

However, it has been shown that testing many interesting graph properties require examining $\Omega(\sqrt{N})$ vertices in the adjacency list model. We note that this need not rule out existence of good algorithms for $\mathbf{AC}^0$ graphs in the sense of Theorem 10. The reason is that the lowerbounds assume an edge oracle that is *black-box*, whereas $\mathbf{AC}^0$ circuits of size n are learnable (albeit only with a very inefficient algorithm) with $\mathrm{poly}(n) = \mathrm{poly}(\log N)$ queries.

Theorem 11. *Every graph property is ϵ-testable (with 2-sided error) using* $\mathrm{poly}(\log N/\epsilon)$ *evaluations of the* $\mathbf{AC}^0$ *circuit representing the adjacency list of the d-regular graph.*

Proof. The adjacency list consists of $d \log N$ bits, and each is computed by a circuit of size $\mathrm{poly}(\log N)$ whose output is a single bit. Each of these $d \log N$ circuits can be learnt after evaluating it on $\mathrm{poly}(\log N/\epsilon)$ random points. Namely, if we simply pick the circuit that best fits those $\mathrm{poly}(\log N/\epsilon)$ values, then the standard Chernoff bound calculation shows that this circuit is with high probability (over the choice of the sample points) correct for all but $\epsilon/10d \log N$ fraction of the inputs. Doing this for all $d \log N$ circuits ensures that we get an $\mathbf{AC}^0$ representation of the graph that is $\epsilon/10$-close to the true graph. Having obtained such a representation, we can try all graphs that are $\epsilon/9$-close to our graph and check if any of them have the property. □

4.2 Adapting Sketching Algorithms to Low-Complexity Graphs

Sketching algorithms are given a data matrix M, and using random sampling they construct a *sketch* $S(M)$ of this data which can be used to approximate the value of some specific function f on M.

Here we note that if M is a low-complexity matrix —in other words, $M(i, j)$ can be produced by a low-complexity computation given i, j — many known sketching algorithms produce the sketch $S(M)$ that is also a low-complexity object. Usually this is trivial to see if by "low complexity" we mean $\mathbf{NC}^1$ but sometimes it is true for even $\mathbf{AC}^0$ thanks to the following fact about $\mathbf{AC}^0$.

Theorem 12 (Approximate counting). *Let* $f\colon \{0,1\}^n \times \{0,1\}^n \to \{0,1\}$ *be computable by an* $\mathbf{AC}^0$ *circuit and define* $g\colon \{0,1\}^n \to \{0,1\}$ *as* $g(x) = \sum_y f(x, y)$. *Then for every* $\epsilon, c > 0$ *there is a* $\mathbf{AC}^0$ *circuit that given x as input outputs* 1 *if* $g(x) > c$ *and* 0 *if* $g(x) < c - \epsilon$, *and an arbitrary value otherwise.*

Proof. We give a probabilistic construction. Pick $m = \mathrm{poly}(n)$ random vectors $y_1, y_2, \ldots, y_m$ and for any x try to compute $\sum_{i \le m} f(x, y_i)$. Using our observation in Example 2 there is an $\mathbf{AC}^0$ circuit (even explicit thanks to [11]) that outputs 1 if $\sum_{i \le m} f(x, y_i) > c$ and 0 if $\sum_{i \le m} f(x, y_i) < c - \epsilon/2$. Letting $m > n^2/\epsilon^2$ allows us to conclude via Chernoff bounds that $\sum_i f(x, y_i)$ correctly estimates $g(x) = \sum_y f(x, y)$ for *all* $x \in \{0,1\}^n$. □

As an example we describe how to compute a special sketch of a matrix M called *low rank decomposition.*

Theorem 13. *Let A be an $N \times N$ matrix that is $\mathbf{AC}^0$, and whose entries have absolute value at most $\log N$. Then there is a randomized $\mathrm{poly}(n)$ time algorithm that given $\epsilon > 0$ produces an $\mathbf{AC}^0$ matrix M with all entries of absolute values $O(\log N)$ such that $|A - M|_F \leq |A - M^*|_F + \epsilon N^2$, where $|\cdot|_F$ denotes Frobenius (i.e. sum of squares) norm and M^* is the rank k matrix that minimizes $|A - M^*|_F$.*

Proof. The theorem is implicit in the paper of Frieze, Kannan, Vempala [27] together with some simple observations. For simplicity we assume the matrix is 0/1. The FKV algorithm starts by sampling $s = \mathrm{poly}(k, \epsilon)$ columns according to the probability distribution where column i is picked with probability proportional to its squared ℓ_2 norm. Since we can estimate the squared ℓ_2 norm up to additive error ϵ by Theorem 12 the sampling of columns is easily implemented using standard rejection sampling. Next the FKV algorithm samples s rows using a similar sampling. Let S be the submatrix defined by the sampled columns and W be the final $s \times s$ matrix.

Then it computes the top k singular vectors $u_1, u_2, \ldots, u_k$ of the above $s \times s$ submatrix and scaling factors $c_1, c_2, \ldots, c_k$ (depending only on W) and lets $v_t = c_t S u_t$. The final approximation is $M = A \cdot (\sum_t v_t v_t^T)$.

Now we observe that if suffices to compute the u_i's up to precision $1/\mathrm{poly}(s)$ which is a constant. Thus the u_i's are computable in $\mathbf{AC}^0$ via brute force. Then $v_t = c_t S u_t$ is computable up to precision ϵ/k^2 in $\mathbf{AC}^0$ by Theorem 12, which implies that M is also computable up to precision ϵ in $\mathbf{AC}^0$. □

Acknowledgment. We thank Noga Alon, Avrim Blum, Russell Impagliazzo, Adam Klivans, Rocco Servedio and Commandur Seshadri for useful conversations.

References

1. Mitzenmacher, M.: A brief history of generative models for power law and lognormal distributions. Internet Mathematics 1(2) (2003)
2. Hopcroft, J.: Personal communication (2008)
3. Galperin, H., Wigderson, A.: Succinct representations of graphs. Inf. Control 56(3), 183–198 (1983)
4. Papadimitriou, C.H.: On the complexity of the parity argument and other inefficient proofs of existence. J. Comput. Syst. Sci. 48(3), 498–532 (1994)
5. Razborov, A.A., Rudich, S.: Natural proofs. J. Comput. Syst. Sci. 55(1), 24–35 (1997)
6. Arora, S., Barak, B.: Computational Complexity: A modern approach. Cambridge University Press, Cambridge (2009)
7. Nickel, C.L.M.: Random dot product graphs: A model for social networks. PhD thesis, Johns Hopkins University (2008)
8. Young, S.J., Scheinerman, E.R.: Random dot product graph models for social networks. In: Bonato, A., Chung, F.R.K. (eds.) WAW 2007. LNCS, vol. 4863, pp. 138–149. Springer, Heidelberg (2007)

9. Ajtai, M.: σ_1^1 formulae on finite structures. Annals of Pure Appl. Logic 24 (1983)
10. Furst, M.L., Saxe, J.B., Sipser, M.: Parity, circuits, and the polynomial-time hierarchy. Mathematical Systems Theory 17(1), 13–27 (1984)
11. Viola, E.: On approximate majority and probabilistic time. In: IEEE Conference on Computational Complexity, pp. 155–168 (2007)
12. Hastad, J.: Almost optimal lower bounds for small depth circuits. In: Randomness and Computation, pp. 6–20. JAI Press (1989)
13. Hoory, S., Linial, N., Wigderson, A.: Expander graphs and their applications. Bull. AMS (43), 439–561 (2006)
14. Capalbo, M.R., Reingold, O., Vadhan, S.P., Wigderson, A.: Randomness conductors and constant-degree lossless expanders. In: STOC, pp. 659–668 (2002)
15. Alon, N., Schwartz, O., Shapira, A.: An elementary construction of constant-degree expanders. In: SODA, pp. 454–458 (2007)
16. Reingold, O., Vadhan, S.P., Wigderson, A.: Entropy waves, the zig-zag graph product, and new constant-degree expanders and extractors. In: FOCS, pp. 3–13 (2000)
17. Mihail, M., Papadimitriou, C.H.: On the eigenvalue power law. In: Rolim, J.D.P., Vadhan, S.P. (eds.) RANDOM 2002. LNCS, vol. 2483, pp. 254–262. Springer, Heidelberg (2002)
18. Babai, L., Fortnow, L., Lund, L.: Non-deterministic exponential time has two-prover interactive protocols. Computational Complexity 1, 3–40 (1991); Prelim version FOCS 1990
19. Arora, S., Safra, S.: Probabilistic checking of proofs: A new characterization of NP. Journal of the ACM 45(1), 70–122 (1998); Prelim version FOCS 1992
20. Arora, S., Lund, C., Motwani, R., Sudan, M., Szegedy, M.: Proof verification and the hardness of approximation problems. Journal of the ACM 45(3), 501–555 (1998); Prelim version FOCS 1992
21. Feige, U., Goldwasser, S., Lovász, L., Safra, S., Szegedy, M.: Interactive proofs and the hardness of approximating cliques. Journal of the ACM 43(2), 268–292 (1996); Prelim version FOCS 1991
22. Goemans, M.X., Williamson, D.P.: Improved approximation algorithms for maximum cut and satisfiability problems using semidefinite programming. J. ACM 42(6), 1115–1145 (1995)
23. Goldreich, O., Goldwasser, S., Ron, D.: Property testing and its connection to learning and approximation. J. ACM 45(4), 653–750 (1998)
24. Alon, N., de la Vega, W.F., Kannan, R., Karpinski, M.: Random sampling and approximation of max-csps. J. Comput. Syst. Sci. 67(2), 212–243 (2003)
25. Alon, N., Fischer, E., Newman, I., Shapira, A.: A combinatorial characterization of the testable graph properties: it's all about regularity. In: STOC, pp. 251–260 (2006)
26. Benjamini, I., Schramm, O., Shapira, A.: Every minor-closed property of sparse graphs is testable. In: STOC, pp. 393–402 (2008)
27. Frieze, A.M., Kannan, R., Vempala, S.: Fast monte-carlo algorithms for finding low-rank approximations. In: FOCS, pp. 370–378 (1998)

Decidability of Conjugacy of Tree-Shifts of Finite Type

Nathalie Aubrun and Marie-Pierre Béal

Université Paris-Est
CNRS, Laboratoire d'informatique Gaspard-Monge

Abstract. A one-sided (resp. two-sided) shift of finite type of dimension one can be described as the set of infinite (resp. bi-infinite) sequences of consecutive edges in a finite-state automaton. While the conjugacy of shifts of finite type is decidable for one-sided shifts of finite type of dimension one, the result is unknown in the two-sided case.

In this paper, we study the shifts of finite type defined by infinite trees. Indeed, infinite trees have a natural structure of one-sided shifts, between the shifts of dimension one and two. We prove a decomposition theorem for these tree-shifts, *i.e.* we show that a conjugacy between two tree-shifts of finite type can be broken down into a finite sequence of elementary transformations called in-splittings and in-amalgamations. We prove that the conjugacy problem is decidable for tree-shifts of finite type. This result makes the class of tree-shifts closer to the class of one-sided shifts of dimension one than to the class of two-sided ones. Our proof uses the notion of bottom-up tree automata.

1 Introduction

Sofic shifts are bi-infinite sequences labeling paths in a finite automaton. Shifts of finite type are a particular important subclass of sofic shifts. Two-sided (resp. one-sided) shifts of finite type are bi-infinite (resp. right-infinite) sequences of consecutive edges in a finite-state automaton (see [8, 13.8], [5]). They are well understood in the one-sided case since the conjugacy is decidable for such shifts [14]. The proof uses the decomposition theorem (see for instance [5]). This theorem states that every conjugacy between two one-sided shifts of finite type can be decomposed into a finite sequence of splittings and amalgamations, which are elementary operations on automata presenting the two shifts.

In the two-sided case, the decidability of the conjugacy problem between two shifts of finite type is still an open question. In higher dimension, many questions become more difficult. The main reason is that there exists no good representation of multidimensional shifts comparable to finite automata in dimension one. Even if there exists a generalization of finite automata to dimension two, which are called textile systems (see [10], see also the automata for tiling systems in [2]), results are more complex than in dimension one. The decomposition theorem can be extended to two-sided multidimensional shifts of finite type, but an additional operation, called an inversion, is needed (see [4]).

S. Albers et al. (Eds.): ICALP 2009, Part I, LNCS 5555, pp. 132–143, 2009.

In this paper, we introduce the notion of shifts of finite type defined on infinite trees, that we call tree-shifts. Indeed, infinite trees have a natural structure of one-sided symbolic systems equipped with several shift transformations. The ith shift transformation applied to a tree gives the subtree rooted at the child number i of the tree. This defines a new class of shifts between the class of one-sided shifts of dimension one and the class of one-sided shifts of higher dimension. Tree-shifts can be described thanks to top-down or bottom-up tree automata which are used in automata theory for many purposes. Tree automata have applications to logic and game theory (see [13], [1], [11], and [12]). The tree automata that we consider here are bottom-up tree automata. They are simpler than Büchi or Muller tree automata since they have all their states final.

We define two elementary operations on tree automata: the in-splitting operation and the in-amalgamation operation. They are very close to those existing on finite automata. In particular two in-amalgamations commute. We prove a decomposition theorem for tree-shifts of finite type, *i.e.* we show that a conjugacy between two tree-shifts of finite type can be broken down into a finite sequence of in-splittings and in-amalgamations. We then prove that the conjugacy problem is decidable for this class of shifts. The heart of the proof is the commutation property of in-amalgamations. We prove that two tree-shifts of finite type are conjugate if and only if they have the same minimal in-am algamation. Furthermore, the minimal in-amalgamation of a tree automaton can be computed in a polynomial time in the number of states of the automaton.

The paper is organized as follows. In Section 2 we give basic definitions about tree-shifts and tree automata. The decomposition theorem is proved in Section 3. Our main result together with an example are given in Section 4. We end the paper with some concluding remarks.

2 Shifts, Automata and Infinite Trees

2.1 Tree-Shifts

We give here some basic definitions from symbolic dynamics which apply to infinite trees. We consider infinite trees whose nodes have a fixed number of children and are labeled in a finite alphabet.

Let $\Sigma = \{0, 1, .., d-1\}$ be a finite alphabet of cardinal d. An *infinite tree* t over a finite alphabet A is a complete function from Σ^* to A. Unless otherwise stated, a tree is an infinite tree. A node of a tree is a word of Σ^*. The empty word, that corresponds to the root of the tree, is denoted by ϵ. If x is a node, its children are xi with $i \in \Sigma$. Let t be a tree and let x be a node, we shall sometimes denote $t(x)$ by t_x.

When Σ is fixed, we denote by $\mathcal{T}(A)$ the set of all infinite trees on A, hence the set A^{Σ^*}. On this set we have a natural metric. If t, t' are two trees, we define the distance $d(t, t') = \frac{1}{n+1}$, where n is the length of the shortest word x in Σ^* such that $t(x) \neq t'(x)$ if such a word exists, and $d(t, t) = 0$. This metric induces a topology equivalent to the usual product topology, where the topology in A is the discrete one.

We define the shift transformations σ_i for $i \in \Sigma$ from $\mathcal{T}(A)$ to itself as follows. If t is a tree, $\sigma_i(t)$ is the tree rooted at the i-th child of t, *i.e.* $\sigma_i(t)_x = t_{ix}$ for any $x \in \Sigma^*$. The set $\mathcal{T}(A)$ equipped with the shift transformations σ_i is called the *full shift* of infinite trees over A.

A *pattern* is a function $p : L \to A$, where L is a finite subset of Σ^* containing the empty word. The set L is called the *support of the pattern.* A *block* of height n is a pattern with support $\Sigma^{\leq n}$, where n is some nonnegative integer, and $\Sigma^{\leq n}$ denotes the words of length at most n of letters of Σ.

We say that a pattern b of support L is a *block of a tree* t if there is a word $x \in \Sigma^*$ such that $t_{xy} = b_y$ for any $y \in \Sigma^*$. We say that b is a block of t rooted at the node x. If b is not a block of t, one says that t *avoids* p. If b is a block of some tree of tree-shift X, it is called an *allowed block* of X.

We define a *tree-subshift* (or *tree-shift*) X of $\mathcal{T}(A)$ as the set $\mathsf{X}_{\mathcal{F}}$ of all trees avoiding each pattern of a set of blocks $\mathcal{F}$. This tree-shift X is closed and for any shift transformation σ_i, $\sigma_i(X) \subseteq X$. A *tree-shift of finite type* X of $\mathcal{T}(A)$ is a set $\mathsf{X}_{\mathcal{F}}$ of all trees avoiding each block of a *finite* set of blocks $\mathcal{F}$. The set $\mathcal{F}$ is called *a set of forbidden blocks* of X.

We denote by $\mathcal{L}(X)$ the set of blocks of all trees of a tree-shift X, and by $\mathcal{L}_n(X)$ the set of all blocks of height n of X. If b is a block of height n with $n \geq 1$, we denote by $\sigma_i(b)$ the block of height $n-1$ such that $\sigma_i(b)_x = b_{ix}$ for $x \in \Sigma^{\leq n-1}$. The block b is written $b = (b_\varepsilon, \sigma_0(b), .., \sigma_{d-1}(b))$.

Example 1. In figure 1 is pictured an infinite tree of a tree-shift of finite type $\mathsf{X}_{\mathcal{F}}$ on the binary alphabet $\{0,1\}$ defined by a finite set $\mathcal{F}$ of forbidden blocks of height 1. The forbidden blocks are those whose labels have a sum equal to 1 modulus 2.

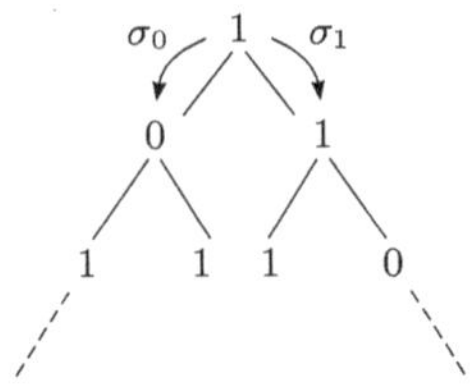

Fig. 1. A tree of the tree-shift of finite type $\mathsf{X}_{\mathcal{F}}$ on the alphabet $\{0,1\}$, where $\mathcal{F}$ is the set of blocks of height 1 whose sum of labels is 1 modulus 2

Let A, A' be two finite alphabets, X be a tree-shift of $\mathcal{T}(A)$ and m be a nonnegative integer. A map $\Phi : X \subseteq \mathcal{T}(A) \to \mathcal{T}(A')$ is called a $(m+1)$-*local map* (or a $(m+1)$-*block map*) if there exists a function $\phi : \mathcal{L}_{m+1}(X) \to A'$ such that, for any $x \in \Sigma^*$, $\Phi(t)_x = \phi(t_{|x\Sigma^{\leq m+1}})$, where $t_{|x\Sigma^{\leq m+1}}$ is the pattern q such that $q_y = t_{xy}$ for any $y \in \Sigma^{\leq m+1}$. The smallest integer m satisfying this property is called the *memory* of the block map. A *block map* is a map which is $(m+1)$-local for some nonnegative integer m.

It is known from the Curtis-Lyndon-Hedlund theorem (see [3]) that block maps are exactly the maps $\Phi : X \to Y$ which are continuous and commute with all tree-shifts transformations, *i.e.* such that $\sigma_i(\Phi(t)) = \Phi(\sigma_i(t))$ for any $t \in X$ and any $i \in \Sigma$. The image of X by a block map is also a tree-shift. A one-to-one and onto block map from a tree-shift X onto a tree-shift Y has an inverse which is also a block map. It is called a *conjugacy* from X onto Y. The tree-shifts X and Y are then *conjugate.*

Example 2. Let X the tree-shift of finite type defined in Example 1. Let Y be the tree-shift of finite type over the alphabet $\{a, b, c\}$, where the allowed blocks of height 2 are (a, a, a), (a, b, c), (a, c, b), (a, c, c), (b, b, a), (b, c, a), (c, a, b) and (c, a, c). The 1-block map $\Phi : X \to Y$, defined by $\phi(0, 0, 0) = a$, $\phi(0, 1, 1) = a$, $\phi(1, 1, 0) = b$, and $\phi(1, 0, 1) = c$, is pictured in Figure 2. The map Φ is a conjugacy. Its inverse is a 0-block map Ψ defined by $\psi(a) = 0$ and $\psi(b) = \psi(c) = 1$.

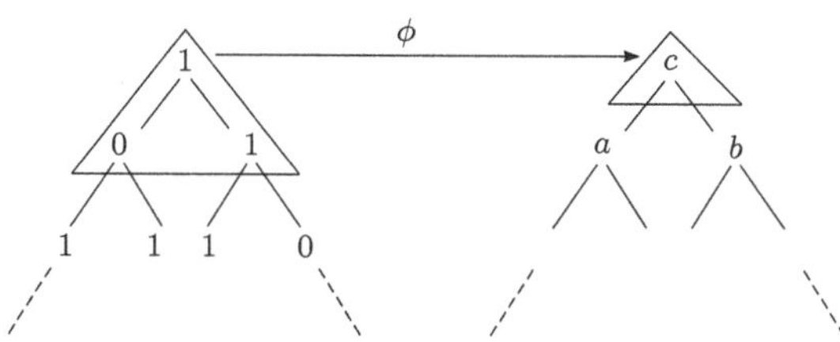

Fig. 2. A 1-block map $\Phi : X \to Y$, where X is the tree-shift of Figure 1 and Y a tree-shift of finite type over the alphabet $\{a, b, c\}$. The map Φ is a conjugacy.

Let X be a tree-shift on A. Let n be a positive integer. The *higher block presentation* of order n of X is the tree-shift $\hat{X}$ on the alphabet $\mathcal{L}_n(X)$ made of trees t such there is a tree $t' \in X$ such that, for any node x, t_x is the block of height n of t' rooted at x. A tree-shift is conjugate to any of its higher block presentations.

2.2 Tree Automata

In this section we consider bottom-up automata for infinite trees. Such an automata its computation from the infinite branches and moves upward. A *tree automaton* is here a structure $\mathcal{A} = (V, A, \Delta)$ where V is a finite set of states (or vertices), A is a finite set of input symbols, and Δ is a set of transitions of the form $(q_0, \dots, q_{d-1}), a \to q$, with $q, q_i \in V$, $a \in A$. A transition $(q_0, \dots, q_{d-1}), a \to q$ is called a transition *labeled by* a, *going out of* the d-tuple of states $(q_0, \dots, q_{d-1})$ and *coming in* the state q. Note that no initial nor final states are specified. This means that all states are both initial and final.

Such an automaton is *deterministic* if for all d-tuple of states $(q_0, \dots, q_{d-1})$ and for all $a \in A$, there is at most one transition $(q_0, \dots, q_{d-1}), a \to q$. Then the set of transitions defines a partial function δ from $V^d \times A$ to V.

A (bottom-up) *computation* of $\mathcal{A}$ on the infinite tree t is an infinite tree C on V such that, for each node x, there is a transition $(C_{x0}, \dots, C_{x(d-1)}, t_x) \to C_x \in \Delta$.

A tree t is *accepted* by $\mathcal{A}$ if there exists a computation of $\mathcal{A}$ on t. The set of infinite trees accepted by $\mathcal{A}$ is a tree-shift.

Let m be a nonnegative integer. An *m-deterministic local tree automaton* (or an *m-definite tree automaton*) is a tree automaton $\mathcal{A} = (V, A, \delta)$ such that whenever t and t' are two trees accepted by $\mathcal{A}$ with a same block b of height m rooted at the node x in t and rooted at the node x' in t', for any computation C of t and any computation C' of t', we have $C_x = C'_{x'}$. A tree automaton is *local* (or *definite*) if it is m-local for some nonnegative integer m.

Proposition 1. *Any tree-shift of finite type is accepted by a deterministic local tree automaton. Reciprocally any tree-shift accepted by a deterministic local tree automaton is of finite type.*

Let $\mathcal{A} = (V, A, \Delta)$ be a tree automaton such that whenever

$$(q_0, \ldots, q_{d-1}, a) \to q \in \Delta \text{ and } (q_0, \ldots, q_{d-1}, b) \to q \in \Delta, \text{ one has } a = b.$$

Note that the automaton built in the proof of Proposition 1 satisfies this condition. The set of computations in the tree automaton $\mathcal{A}$ defines a tree-shift called the *vertex tree-shift* (or *Markov tree-shift*) defined by $\mathcal{A}$. Equivalently, a vertex tree-shift is a tree-shift accepted by a 2-local automaton, *i.e.* the tree-shift of finite type $\mathsf{X}_{\mathcal{F}}$ where $\mathcal{F}$ is a set of forbidden blocks of height 2.

A vertex tree-shift is the set of computations of the unlabeled automaton $\mathcal{B} = (V, \Gamma)$ with transitions $(q_0, \ldots, q_{d-1}) \to q$.

Example 3. The tree-shift X of Example 1 is a vertex tree-shift accepted by the automaton $\mathcal{A} = (V, \Delta)$ with transitions $(0,0) \to 0$, $(1,1) \to 0$, $(1,0) \to 1$ and $(0,1) \to 1$. These transitions are given in the following table t where $(p, q) \to t[p, q]$ is a transition.

	0	1
0	0	1
1	1	0

Proposition 2. *Any tree-shift of finite type is conjugate to a vertex tree-shift.*

Proof. Let $X = \mathsf{X}_{\mathcal{F}}$ be a tree-shift of finite type defined by a finite set of forbidden blocks of height m for some nonnegative integer m. Let $\mathcal{A} = (V, A, \delta)$ be the deterministic m-local automaton such that $V = \mathcal{L}_m(X)$ and, for $p_i \in V, a \in A$, the block $q = (a, p_0, \ldots, p_{d-1})$ of height $m+1$ is an allowed block of X, then $\delta(p_0, \ldots, p_{d-1}), a) = \text{trunc}(q)$, where $\text{trunc}(q)$ is the block of height m such that $\text{trunc}(q)_x = q_x$ for $x \in \Sigma^{\leq m}$. The automaton $\mathcal{A}$ accepts X. Let Y be the vertex tree-shift made of the all computations on $\mathcal{A}$. Note that any tree t of X has a unique computation C in $\mathcal{A}$.

We define an $m+1$-block map Φ from X to Y via $\phi : \mathcal{L}_m(X) \to Y$ by setting $\phi(p) = p$. The map Φ associate to each tree of X its computation in $\mathcal{A}$. The one-block map Ψ from Y to X given by $\psi : Y \to A$ with $\psi(p) = p_\varepsilon$ is the inverse of ϕ. The tree-shifts X and Y are thus conjugate.

In the sequel, in order to simplify the notations, we restrict us to binary trees ($\Sigma = \{0, 1\}$) but all results extend trivially to the case of trees with d children for any $d \geq 1$.

3 A Decomposition Theorem for Trees

The decomposition theorem for shifts of infinite words states that any conjugacy between shifts of finite type can be decomposed into a finite sequence of splittings and amalgamations. We will prove an analogous theorem for infinite trees. The crucial lemma will show that the memory of a block map can be reduced using a notion of (input) splittings on tree automata defined below. We first consider the case of vertex tree-shifts.

Let X be a binary vertex tree-shift defined by a deterministic 2-local automaton $\mathcal{A} = (V, \Delta)$. Set $V = \{p_1, .., p_n\}$. We define an *in-splitting* of $\mathcal{A}$ as an automaton $\tilde{\mathcal{A}} = (\tilde{V}, \tilde{\Delta})$ obtained as follows. First, we in-split the automaton $\mathcal{A}$ by refining the natural partition of Δ; for each vertex $p \in V$, we partition the set Δ_p of transitions coming in p into subsets $\Delta_p^1, .., \Delta_p^{l(p)}$. We set

$$\tilde{V} = \{p_1^1, .., p_1^{l(1)}, p_2^1, .., p_2^{l(2)}, \dots, p_n^1, .., p_n^{l(n)}\} \tag{1}$$

and $(p^i, q^j) \to r^k \in \tilde{\Delta}$ if $(p, q) \to r \in \Delta_r^k$.

The tree-shift accepted by $\tilde{\mathcal{A}}$ is a vertex tree-shift denoted by $\tilde{X}$ and called the *in-splitting* of X defined by the above partitioning of the transitions.

The same notion of in-splitting is defined for tree-shifts of finite type as follows. Let X be a tree-shift of finite type accepted by a deterministic automaton $\mathcal{A} = (V, A, \Delta)$. An in-splitting of $\mathcal{A}$ is $\tilde{\mathcal{A}} = (\tilde{V}, \tilde{\Delta})$. For each vertex $p \in V$, we partition the set Δ_p of transitions coming in p into subsets $\Delta_p^1, .., \Delta_p^{l(p)}$ and set $\tilde{V}$ as in Equation 1. We set $(p^i, q^j), a \to r^k \in \tilde{\Delta}$ if $(p, q), a \to r \in \Delta_r^k$.

A tree *in-amalgamation* of an tree automaton $\mathcal{A}$ is an automaton $\mathcal{B}$ such that $\mathcal{B}$ is an in-splitting of $\mathcal{A}$. An *in-amalgamation* of an tree-shift X is a tree-shift Y such that Y is an in-splitting of X.

Lemma 1. *Let $\tilde{X}$ be an in-splitting of a vertex tree-shift X. Then $\tilde{X}$ and X are conjugate.*

Proof. Let $\tilde{X}$ be an in-splitting of X accepted by $\mathcal{A} = (V, \Delta)$. Define a 1-block map $\Phi : \tilde{X} \to X$ via $\phi(p^i) = p$ for each state $p \in V$, and a 2-block map $\Psi : X \to \tilde{X}$ via $\psi(r, p, q) = r^i$, where $(p, q) \to r \in \Delta_r^i$.

It is not difficult to check that $\Phi(\tilde{X}) \subseteq X$ and $\Psi(X) \subseteq \tilde{X}$. It is clear that $\Phi(\Psi(t))_u = t_u$ for all trees $t \in X$ and each word $u \in \Sigma^*$ since adding and removing superscripts has no effect. Thus we only need to check that $\Psi(\Phi(t))_u = t_u$ for all trees $t \in \tilde{X}$ and each word $u \in \Sigma^*$. Let $t \in \tilde{X}$ and $u \in \Sigma^* = \{0, 1\}^*$. We need to show that

$$\Psi(\Phi(t))_u = \psi((\phi(t_u), \phi(t_{u0}), \phi(t_{u1})) = t_u.$$

Indeed, suppose $t_{u0} = p^i$, $t_{uj} = q^j$ and $t_u = r^k$ where $p^i, q^j \in \tilde{V}$. Hence we have $(p, q) \to r \in \Delta_r^k$. We get $\psi((\phi(t_u), \phi(t_{u0}), \phi(t_{u1})) = \psi(r, p, q) = r^k = t_u$.

When the partition of the set of transitions consists of singleton sets, then the in-splitting $\tilde{X}$ of X is called the *complete in-splitting* of X.

In the remainder of this section, we give a proof of the decomposition theorem of vertex tree-shifts. In Lemma 3 we show that a higher block presentation of the tree-shift is the composition of a finite sequence of in-splittings and in-amalgamations. Then, by moving to a higher block presentation if necessary, we may assume that the conjugacy Φ between two tree-shifts is a 1-block map with an n-block map inverse. If $n = 1$, then this conjugacy is just a relabeling of the symbols of the states and as such a trivial splitting. So we need a way to reduce the memory of the inverse of Φ. We will reduce the memory in Lemmas 2 by using tree in-splittings and in-amalgamations.

Lemma 2. *Let X_k for $k = 1$ and 2, be two vertex tree-shift defined by the two automata $\mathcal{A}_1 = (V_1, \Delta)$ and $\mathcal{A}_2 = (V_2, \Lambda)$ respectively. Suppose $\Phi : X_1 \to X_2$ is a 1-block conjugacy with an n-block inverse. If $n \geq 1$, then there are vertex tree-shifts $\tilde{X}_k$ such that the following diagram commutes:*

$$\begin{array}{ccc} X_1 & \xrightarrow{\Phi} & X_2 \\ {\scriptstyle \Psi_1}\big\downarrow & & \big\downarrow{\scriptstyle \Psi_2} \\ \tilde{X}_1 & \xrightarrow{\tilde{\Phi}} & \tilde{X}_2 \end{array}$$

where Ψ_1 and Ψ_2 are in-splittings of X_1 and X_2 respectively, and where $\tilde{\Phi}$ is a 1-block conjugacy with an $(n-1)$-block inverse.

Proof. For $p \in V_1$, we partition Δ_p into $\bigcup_{(s,t) \in V_2 \times V_2} \{(q,r) \to p \mid \phi(q) = s \text{ and } \phi(r) = t\}$. Then we set

$$\tilde{V_1} = \{p^{s,t} : p \in V_1, s,t \in V_2 \text{ with } \phi(q) = s \text{ and } \phi(r) = t \text{ for some } (q,r) \to p \in \Delta_p\},$$

We set $(p^{s,t}, q^{u,v}) \to r^{p,q} \in \tilde{\Delta}$ if $(p,q) \to r \in \tilde{\Delta}$.

As shown in Lemma 1, $\Psi_1 : X_1 \to \tilde{X}_1$ is a conjugacy via $\Psi_1(t)_u = \psi_1(t_u, t_{u0}, t_{u1}) = t_u^{\phi(t_{u0}),\phi(t_{u1})}$, where $u \in \{0,1\}^*$.

Now let $\tilde{X}_2$ be the complete in-splitting of X_2. So

$$\tilde{V_2} = \{r^{p,q} : p,q,r \in V_2, \text{ with } (p,q) \to r \in \Lambda_r\},$$

We set $(r^{p,q}, s^{u,v}) \to t^{r,s} \in \tilde{\Lambda}$ if $(r,s) \to t \in \Lambda$. As shown in Lemma 1, $\Psi_2 : X_2 \to \tilde{X}_2$ is a conjugacy via $\Psi_2(t)_u = \psi_2(t_u, t_{u0}, t_{u1}) = t_u^{t_{u0},t_{u1}}$, where $u \in \{0,1\}^*$.

Now we define the 1-block map $\tilde{\Phi} : \tilde{X}_1 \to \tilde{X}_2$ via

$$\tilde{\Phi}(t)_u = \phi(t_u)^{\phi(t_{u0}),\phi(t_{u1})},$$

where $u \in \{0,1\}^*$. Clearly, the diagram commutes and thus Φ is one-to-one and onto. It remains to check that $\tilde{\Phi}^{-1} = \Psi_1 \circ \Phi^{-1} \circ \Psi_2$ is an $(n-1)$-block map. That is, we must show that for any tree $t \in \tilde{X}_2$, the coordinates in a block of height $n-1$ of t rooted at the node ε determines $\tilde{\Phi}^{-1}(t)_\varepsilon$. But this follows from the observation that the block of height $n-1$ of t rooted at ε determines all $\Psi_2^{-1}(t)_{v0}$ and all $\Psi_2^{-1}(t)_{v1}$ for $v \in \{0,1\}^{n-1}$, and therefore the block of height n at the root of t.

The proof of the above lemma is similar to the proof of the analogous result for shifts of $\Sigma^{\mathbb{Z}}$ (see [8, Lemma 7.3.1]) or shifts of $\Sigma^{\mathbb{Z}^2}$ (see [4]).

In general a conjugacy between vertex tree-shifts is an n-block map but the following lemma shows that moving to a higher block presentation we may assume it is a 1-block map.

Lemma 3. *Let X_k, for $k = 1$ and 2, be vertex tree-shifts. Let n be a positive integer. Suppose $\Phi : X_1 \to X_2$ is an n-block conjugacy and let $\hat{X}_1$ be the higher block presentation of X_1 of order n. There exists a map $\eta : X_1 \to \hat{X}_1$ which is a sequence of tree in-splittings, such that $\Phi \circ \eta^{-1}$ is a 1-block conjugacy.*

Proof. Clearly, $\Phi \circ \eta^{-1}$ is a 1-block conjugacy. We need to show that η is a sequence of tree in-splitting.

If X is the vertex tree-shift defined by $\mathcal{A} = (V, \Delta)$ and n is a nonnegative integer, a higher block presentation of X of order n is the vertex tree-shift $\hat{X}$ defined by $\hat{\mathcal{A}} = (\hat{V}, \hat{\Delta})$, where $\hat{V}$ is the set of allowed blocks of X of height n. There is a transition $(p, q) \to r$ in $\hat{\Delta}$, where $p, q, r \in \mathcal{L}_n(X)$ if and only if $r_{0u} = p_u$ and $r_{1u} = q_u$ for any $u \in \{0, 1\}^{n-1}$.

A complete in-splitting of the tree-shift X yields a higher block presentation of X of order 2 by $\Psi(t)_u = t_u^{t_{u0}, t_{u1}}$.

By iterating this construction on $\Psi(X)$, we can find a sequence of in-splittings η such that $\bar{\eta} = \eta \circ \Psi_1^{-1}$ is a 1-block conjugacy, or simply a relabeling. Then $\eta = \bar{\eta} \circ \Psi_1$ and we have the result.

Lemma 4. *Let X be tree-shift of finite type, there is a vertex tree-shift Y and a conjugacy from X to Y which is a sequence of in-splittings.*

Proof. Let $X = \mathsf{X}_{\mathcal{F}}$ be a tree-shift of finite type defined by a finite set of forbidden blocks of height m for some nonnegative integer m.

Let $\hat{X}$ be the higher block presentation of X of order m. The tree-shift $\hat{X}$ is the vertex tree-shift defined by $\hat{\mathcal{A}} = (\hat{V}, \hat{\Delta})$, where $\hat{V}$ is the set of allowed blocks of X of height m. There is a transition $(p, q) \to r$ in $\hat{\Delta}$, where $p, q, r \in \mathcal{L}_m(X)$ if and only if $r_{0u} = p_u$ and $r_{1u} = q_u$ for any $u \in \{0, 1\}^{m-1}$. We know from Lemma 3 that $\hat{X}$ is obtained from X with a sequence of in-splittings.

We are now ready to state the main result of this section.

Theorem 1. *Let X_1 and X_2 be two tree-shifts of finite type. Every conjugacy between X_1 and X_2 is the composition of a finite sequence of tree in-splittings and tree in-amalgamations.*

Proof. By Lemma 4, we can view any tree-shift of finite type as a vertex tree-shift. The theorem follows then from Lemma 1 and Lemma 3.

4 Commutation of In-Amalgamations

Proposition 3. *Suppose X_1 is a vertex tree-shift and X_2, X_3 are vertex tree-shifts obtained from X_1 by in-amalgamations. Then there is a vertex tree-shift X_4 that can be obtained from both X_3 and X_4 by in-amalgamations.*

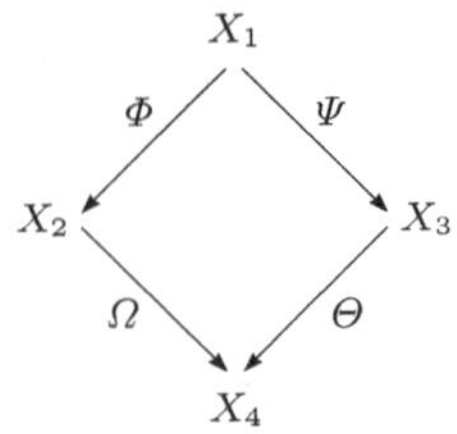

Fig. 3. The commutation of in-amalgamations. If X_2, X_3 are vertex tree-shifts which are in-amalgamations of X_1, the there is a vertex tree-shift X_4 which is a common amalgamation of X_2, X_3.

In Figure 3, the maps Φ and Ψ are in-amalgamations. As a consequence of Proposition 3 the maps Ω and Θ are also in-amalgamations.

Proof. Let us first assume that there is an in-amalgamation $\Phi : X_1 \to X_2$ and a in-amalgamation $\Psi : X_1 \to X_3$. Let us assume that states $p^1, .., p^{l(p)}$ of V_1 are amalgamated to a state p of V_2.

By definition of an in-amalgamation, this implies that if $(q, r) \to p^i \in \Delta_1$, then $(q, r) \to p^j \notin \Delta_1$ for any states $q, r \in V_1$ and any $1 \leq i \neq j \leq l(p)$. This implies also that $(p^i, q) \to r \in \Delta_1$ if an only if $(p^j, q) \to r \in \Delta_1$, and $(q, p^i) \to r \in \Delta_1$ if and only if $(q, p^j) \to r \in \Delta_1$ for any states $q, r \in V_1$ and any $1 \leq i, j \leq l(p)$.

Suppose also that states $q^1, .., q^{l(q)}$ of V_1 are amalgamated to a state q of V_3.

Let us first assume that the states $p^1, .., p^{l(p)}$ and $q^1, .., q^{l(q)}$ are all distinct. We define X_4 as the in-amalgamation of X_2 obtained by amalgamating the states $p, q^1, .., q^{l(q)}$ to a state q. It is also the in-amalgamation of X_3 obtained by amalgamating the states $q, p^1, .., p^{l(p)}$ to a state q.

Let us now assume that $p^1 = q^1, .., p^l = q^l$ for some integer $1 \leq l \leq \min(l(p), l(q))$. This implies that, for any $1 \leq i \leq l(p)$, $1 \leq j \leq l(q)$, one has $(p^i, q) \to r \in \Delta_n$ if and only if $(p^j, q) \to r \in \Delta_n$, and $(q, p^i) \to r \in \Delta_n$ if and only if $(q, p^j) \to r \in \Delta_n$ for $n = 1$ and $n = 2$.

We define X_4 as the in-amalgamation of X_2 obtained by amalgamating the states $p, q^{l+1}, .., q^{l(q)}$ to the state p. It is also the in-amalgamation of X_3 obtained by amalgamating the states $q, p^{l+1}, .., p^{l(p)}$ to a state p. Hence, if Φ and Ψ are in-amalgamations, then Ω and Θ also.

The previous theorem allows us to define the notion of *minimal in-amalgamation* of a edge tree-shift X. It is defined as the vertex tree-shift defined by an automaton $\mathcal{A} = (V, \Delta)$ with the smallest number of vertices which is obtained by in-amalgamations of X.

Corollary 1. *Any vertex tree-shift has a unique minimal in-amalgamation.*

Proof. Let us assume that X has two minimal amalgamations X_2 and X_3. By Proposition 3, X_2 and X_3 have a common in-amalgamation Y. By minimality, $Y = X_2 = X_3$.

We present the sketch of an algorithm that computes the minimal amalgamation of a vertex tree-shift. Let us assume that X is a vertex shift defined by an n-vertex automaton $\mathcal{A} = (V, \Delta)$. We call *out-degree* of $\mathcal{A}$ the maximal number of edges $(p, q) \to r$ or $(q, p) \to r$ for all pair of vertices (p, q). Let assume that the out-degree of the automata that we consider is bounded by K. This assumption is quite reasonable in our context.

We say that two vertices p, q are *pre-mergeable* if for any pair of vertices (s, t), $(s, t) \to p \in \Delta$ implies $(s, t) \to q \notin \Delta$, and $(s, t) \to q \in \Delta$ implies $(s, t) \to p \notin \Delta$. We call the *signature* of a vertex p the sequence $((0, q, r) \mid (p, q) \to r \in \Delta).((1, q, r) \mid (q, p) \to r \in \Delta)$ in lexicographic order. One can merge two states p and q if they are pre-mergeable and $\sigma(p) = \sigma(q)$. An algorithm based on this idea can compute the minimal amalgamation of a vertex tree-shift with a $O(Kn^3)$ overall time complexity of the procedure.

Theorem 2. *Let X_1 and X_2 be two tree-shifts of finite type. It is decidable whether X_1 and X_2 are conjugate.*

Proof. By Proposition 2, one may assume that X_1 and X_2 are vertex tree-shifts. By Theorem 1, there is a sequence of tree in-splittings and tree in-amalgamations from X_1 to X_2.

Let us consider first that this sequence is decomposed into a sequence of tree in-splittings from X_1 to X followed (up to a relabeling of X), by a sequence of tree in-amalgamations from X to X_2. This case is illustrated in Figure 4. By Proposition 3, there are vertex tree-shifts at the confluence of any two dashed edges of Figure 4. As a consequence, X_1 and X_2 have a common amalgamation and thus the same minimal amalgamation. Conversely, if X_1 and X_2 have the same minimal amalgamation, there is a sequence of tree in-splittings and tree in-amalgamations from X_1 to X_2.

We now consider the case where there is a sequence of tree in-splittings and tree in-amalgamations from X_1 to X_2. This sequence is decomposed into a sequence of the form described in the previous case and the same result holds by transitivity.

We can deduce from this proof an algorithm that computes, if it exists, a conjugacy between two vertex tree-shifts of finite type X_1 and X_2: by Proposition 3 , $\phi_1(\mathsf{X}_1) = \mathsf{Y}_1$ and $\phi_2(\mathsf{X}_2) = \mathsf{Y}_2$ where ϕ_i is a sequence of in-amalgamations and Y_i a minimal in-amalgamation. Then compare the Y_i: if they are different then X_1 and X_2 are not conjugated, if they are equal then $\mathsf{X}_1 = {\phi_1}^{-1} \circ \psi \circ \phi_2(\mathsf{X}_2)$ where ψ is a renimbering of the vertices. Let $\mathcal{A}_1 = (V_1, \Delta_1)$ and $\mathcal{A}_2 = (V_2, \Delta_2)$ be two tree automata defining respectively Y_1 and Y_2. Let us assume that $V = V' = \{0, 1, 2, .., n-1\}$. Checking whether $\mathcal{A}_1$ and $\mathcal{A}_2$ are equal up to a renumbering of their vertices consists in finding a permutation σ of $\{0, 1, 2, .., n-1\}$ compatible with the transitions: $(p, q) \to r \in \Delta$ if and only if $(\sigma(p), \sigma(q)) \to \sigma(r) \in \Delta'$. This can be done in an exponential time, so that the global time complexity of this procedure is exponential.

Example 4. Let X_1 and X_2 be two vertex tree-shifts over the alphabet $V = \{a, b, c\}$. The tree-shift X_1 is accepted by $\mathcal{A}_1 = (V, \Delta_1)$ and the tree-shift X_2

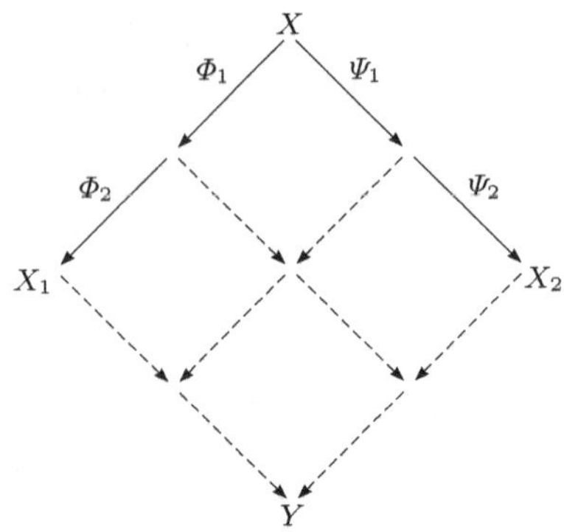

Fig. 4. A sequence of tree in-splittings from X_1 to X is followed (up to a relabeling of X), by a sequence of tree in-amalgamations from X to X_2. Any edge represents an in-amalgamation. The tree-shifts X_1 and X_2 have the same minimal amalgamation Y.

is accepted by $\mathcal{A}_2 = (V, \Delta_2)$ where Δ_1 and Δ_2 are given in the two following tables.

$$\Delta_1 = \begin{array}{c|c|c|c|} & a & b & c \\ \hline a & a & c & c \\ \hline b & b & a & a \\ \hline c & b & a & a \\ \hline \end{array} \qquad \Delta_2 = \begin{array}{c|c|c|c|} & a & b & c \\ \hline a & c & a & a \\ \hline b & a & b & b \\ \hline c & a & b & b \\ \hline \end{array} \qquad \Delta_3 = \begin{array}{c|c|c|} & a & b \\ \hline a & a & b \\ \hline b & b & a \\ \hline \end{array} \qquad \Delta_4 = \begin{array}{c|c|c|} & a & b \\ \hline a & b & a \\ \hline b & a & b \\ \hline \end{array}$$

Since the second and third row of Δ_1 and the second and third column of Δ_1 are equal, the vertices b and c can be amalgamated. There is an in-amalgamation from $\mathcal{A}_1$ to $\mathcal{A}_3 = (V_3, \Delta_3)$ where $V_3 = \{a, b\}$ and Δ_3 is given by the following tables.

No more in-amalgamation is possible from $\mathcal{A}_3$ and thus $\mathcal{A}_3$ is minimal. Similarly, the second and third row of Δ_2 and the second and third column of Δ_2 are equal, the vertices b and c can be amalgamated. There is an in-amalgamation from $\mathcal{A}_2$ to $\mathcal{A}_4 = (V_4, \Delta_4)$ where $V_4 = \{a, b\}$ and Δ_4 is given by the following tables.

Finally, relabeling the states of $\mathcal{A}_4$ by exchanging a and b gives $\mathcal{A}_3$. Hence, X_1 and X_2 have the same minimal amalgamation and are conjugate.

The 2-block map $\Phi : X_2 \to X_1$ of Figure 5 is a conjugacy. It is defined by $\phi(a, a, b) = b$, $\phi(a, b, a) = c$, $\phi(a, a, c) = b$, $\phi(a, c, a) = c$, $\phi(b, b, b) = a$, $\phi(b, b, c) = a$, $\phi(b, c, b) = a$, $\phi(b, c, c) = a$, $\phi(c, a, a) = a$.

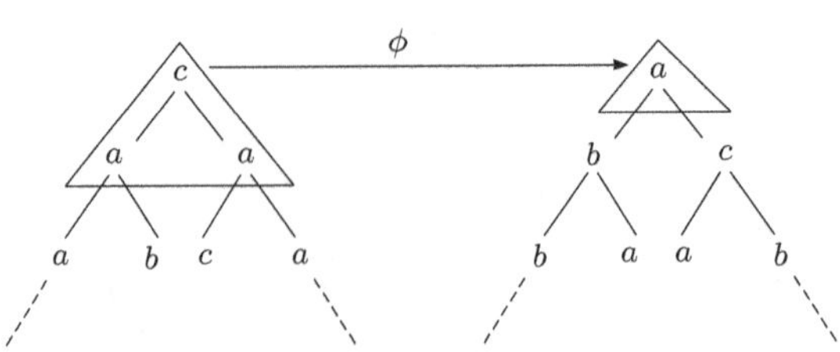

Fig. 5. A 2-block map $\Phi : X \to Y$, where X is the tree-shift of Figure 1 and Y a tree-shift of finite type over the alphabet $\{a, b, c\}$

5 Conclusion

We have shown that it decidable whether two tree-shifts of finite type are conjugate. Further work will include the case of sofic trees. We conjecture that the results that we have obtained for tree-shifts of finite type can be extended to sofic tree-shifts using techniques similar to the one used for shifts of sequences (see [9], [6], [7]). The decomposition theorem that we have proved for tree-shifts of finite type will also allow us to define a notion of strong tree-shift equivalence between tree-shifts and to deduce that two tree-shifts of finite type are equivalent if and only if their transition matrices are related by a sequence of simple algebraic matrix conditions.

References

1. Comon, H., Dauchet, M., Gilleron, R., Löding, C., Jacquemard, F., Lugiez, D., Tison, S., Tommasi, M.: Tree automata techniques and applications (2007), `http://www.grappa.univ-lille3.fr/tata` (release October 12, 2007)
2. Coven, E., Johnson, A., Jonoska, N., Madden, K.: The symbolic dynamics of multidimensional tiling systems. Ergodic Theory and Dynamical Systems 23(02), 447–460 (2003)
3. Hedlund, G.: Endomorphisms and automorphisms of the shift dynamical system. Theory of Computing Systems 3(4), 320–375 (1969)
4. Johnson, A., Madden, K.: The decomposition theorem for two-dimensional shifts of finite type. In: Proceedings-American Mathematical Society, vol. 127, pp. 1533–1544 (1999)
5. Kitchens, B.P.: Symbolic dynamics. Universitext, One-sided, two-sided and countable state Markov shifts. Springer, Berlin (1998)
6. Krieger, W.: On sofic systems. I. Israel J. Math. 48(4), 305–330 (1984)
7. Krieger, W.: On sofic systems. II. Israel J. Math. 60(2), 167–176 (1987)
8. Lind, D., Marcus, B.: An Introduction to Symbolic Dynamics and Coding. Cambridge University Press, Cambridge (1995)
9. Nasu, M.: Topological conjugacy for sofic systems and extensions of automorphisms of finite subsystems of topological markov shifts. In: Proceedings of Maryland special year in Dynamics 1986–1987. Lecture Notes in Mathematics, vol. 1342, pp. 564–607. Springer, Heidelberg (1988)
10. Nasu, M.: Textile Systems for Endomorphisms and Automorphisms of the Shift. American Mathematical Society (1995)
11. Nivat, M., Podelski, A. (eds.): Tree automata and languages. Studies in Computer Science and Artificial Intelligence, vol. 10. North-Holland Publishing Co., Amsterdam (1992); Papers from the workshop held in Le Touquet (June 1990)
12. Perrin, D., Pin, J.: Infinite words. Elsevier, Boston (2004)
13. Thomas, W.: Automata on infinite objects. In: Handbook of theoretical computer science, vol. B, pp. 133–191. Elsevier, Amsterdam (1990)
14. Williams, R.F.: Classification of subshifts of finite type. In: Recent advances in topological dynamics (Proc. Conf. Topological Dynamics, Yale Univ., New Haven, Conn., 1972; in honor of Gustav Arnold Hedlund). Lecture Notes in Math., vol. 318, pp. 281–285. Springer, Berlin (1973)

Improved Bounds for Speed Scaling in Devices Obeying the Cube-Root Rule

Nikhil Bansal[1], Ho-Leung Chan[2], Kirk Pruhs[3], and Dmitriy Katz[4,⋆]

[1] IBM T.J. Watson, Yorktown Heights, NY
nikhil@us.ibm.com
[2] Max-Planck-Institut für Informatik
hlchan@mpi-inf.mpg.de
[3] Computer Science Dept., Univ. of Pittsburgh
kirk@cs.pitt.edu
[4] IBM T.J. Watson, Yorktown Heights, NY
dkatzrog@us.ibm.com

Abstract. Speed scaling is a power management technique that involves dynamically changing the speed of a processor. This gives rise to dual-objective scheduling problems, where the operating system both wants to conserve energy and optimize some Quality of Service (QoS) measure of the resulting schedule. In the most investigated speed scaling problem in the literature, the QoS constraint is deadline feasibility, and the objective is to minimize the energy used. The standard assumption is that the power consumption is the speed to some constant power α. We give the first non-trivial lower bound, namely $e^{\alpha-1}/\alpha$, on the competitive ratio for this problem. This comes close to the best upper bound which is about $2e^{\alpha+1}$.

We analyze a natural class of algorithms called qOA, where at any time, the processor works at $q \geq 1$ times the minimum speed required to ensure feasibility assuming no new jobs arrive. For CMOS based processors, and many other types of devices, $\alpha = 3$, that is, they satisfy the cube-root rule. When $\alpha = 3$, we show that qOA is 6.7-competitive, improving upon the previous best guarantee of 27 achieved by the algorithm Optimal Available (OA). So when the cube-root rule holds, our results reduce the range for the optimal competitive ratio from $[1.2, 27]$ to $[2.4, 6.7]$. We also analyze qOA for general α and give almost matching upper and lower bounds.

1 Introduction

Current processors produced by Intel and AMD allow the speed of the processor to be changed dynamically. Intel's SpeedStep and AMD's PowerNOW technologies allow the Windows XP operating system to dynamically change the speed

⋆ The work of H.L.Chan was done when he was a postdoc in University of Pittsburgh. K. Pruhs was supported in part by NSF grants CNS-0325353, CCF-0514058, IIS-0534531, and CCF-0830558, and an IBM Faculty Award.

S. Albers et al. (Eds.): ICALP 2009, Part I, LNCS 5555, pp. 144–155, 2009.

of such a processor to conserve energy. In this setting, the operating system must not only have a *job selection policy* to determine which job to run, but also a *speed scaling policy* to determine the speed at which the job will be run. All theoretical studies we know of assume a speed to power function $P(s) = s^\alpha$, where s is the speed and $\alpha > 1$ is some constant. Energy consumption is power integrated over time. The operating system is faced with a dual objective optimization problem as it both wants to conserve energy, and optimize some Quality of Service (QoS) measure of the resulting schedule.

The first theoretical study of speed scaling algorithms was in the seminal paper [16] by Yao, Demers, and Shenker. In the problem introduced in [16] the QoS objective was deadline feasibility, and the objective was to minimize the energy used. To date, this is the most investigated speed scaling problem in the literature [16,2,9,6,5,14,17,12,1,11]. In this problem, each job i has a release time r_i when it arrives in the system, a work requirement w_i, and a deadline d_i by which the job must be finished. The deadlines might come from the application, or might arise from the system imposing a worst-case quality-of-service metric, such as maximum response time or maximum slow-down. It is clear that an optimal job selection policy is Earliest Deadline First (EDF). Thus the remaining issue is to find an online speed scaling policy to minimize energy.

1.1 The Story to Date

Yao, Demers and Shenker showed that the optimal offline schedule can be efficiently computed by a greedy algorithm [16]. [16] proposed two natural online speed scaling algorithms, Average Rate (AVR) and Optimal Available (OA). Conceptually, AVR is oblivious in that it runs each job in the way that would be optimal if there were no other jobs in the system. That is, AVR runs each job i (in parallel with other jobs) at the constant speed $w_i/(d_i - r_i)$ throughout interval $[r_i, d_i]$. The algorithm OA maintains the invariant that the speed at each time is optimal given the current state, and under the assumption that no more jobs will arrive in the future. In particular, let $w(x)$ denote the amount of unfinished work that has deadline within x time units from the current time. Then the current speed of OA is $\max_x w(x)/x$. Another online algorithm BKP is proposed in [5]. BKP runs at speed $e \cdot v(t)$ at time t, where $v(t) = \max_{t'>t} w(t, et - (e-1)t', t')/(e(t' - t))$ and $w(t, t_1, t_2)$ is the amount of work that has release time at least t_1, deadline at most t_2, and that has already arrived by time t. Clearly, if $w(t_1, t_2)$ is the total work of jobs that are released after t_1 and have deadline before t_2, then any algorithm must have an average speed of at least $w(t_1, t_2)/(t_2 - t_1)$ during $[t_1, t_2]$. Thus BKP can be viewed as computing a lower bound on the average speed in an online manner and running at e times that speed.

Table 1 summarizes the previous results. The competitive ratio of AVR is at most $2^{\alpha-1}\alpha^\alpha$. This was first shown in [16], and a simpler amortized local competitiveness analysis was given in [2]. The competitive ratio of AVR is least $(2-\delta)^{\alpha-1}\alpha^\alpha$, where δ is a function of α that approaches zero as α approaches infinity [2]. The competitive ratio of OA is exactly α^α [5], where the upper

Table 1. Results on the competitive ratio for energy minimization with deadline feasibility

Previous Results

Algorithm	General α		$\alpha = 2$		$\alpha = 3$	
	Upper	Lower	Upper	Lower	Upper	Lower
General		$\left(\frac{4}{3}\right)^{\alpha}/2$		1.1		1.2
AVR	$2^{\alpha-1}\alpha^{\alpha}$	$(2-\delta)^{\alpha-1}\alpha^{\alpha}$	8	4	108	48.2
OA	α^{α}	α^{α}	4	4	27	27
BKP	$2(\alpha/(\alpha-1))^{\alpha}e^{\alpha}$		59.1		135.6	

Our Contributions

Algorithm	General α		$\alpha = 2$		$\alpha = 3$	
	Upper	Lower	Upper	Lower	Upper	Lower
General		$e^{\alpha-1}/\alpha$		1.3		2.4
qOA	$4^{\alpha}/(2\sqrt{e\alpha})$	$\frac{1}{4\alpha}4^{\alpha}(1-\frac{2}{\alpha})^{\alpha/2}$	2.4		6.7	

bound is proved using an amortized local competitiveness argument. Thus the competitive ratio of AVR is strictly inferior to that of OA. The competitive ratio of BKP is at most $2(\alpha/(\alpha-1))^{\alpha}e^{\alpha}$ [5], which is about $2e^{\alpha+1}$ for large α . It is better than that of OA only for $\alpha \geq 5$. On the other hand, the lower bounds for general algorithms are rather weak. Somewhat surprisingly, the best known lower bound instance is the worst-possible instance consisting of two jobs. [4] shows a lower bound of $\left(\frac{4}{3}\right)^{\alpha}/2$ on the competitive ratio using a two job instance. If one tries to find the worst 3, 4, ... job instances, the calculations get messy quickly.

The most interesting value of α seems to be three. Most importantly, in current CMOS based processors, the speed satisfies the well-known cube-root-rule, that the speed is approximately the cube root of the power [8]. The power is also roughly proportional to the cube of the speed in many common devices/machines, such as vehicles/automobiles, and some types of motors. It seems likely that α would be in the range $[2, 3]$ for most conceivable devices. The best known guarantee for α in this range is α^{α} achieved by OA, which evaluates to 4 for $\alpha = 2$ and 27 for $\alpha = 3$. Our motivating goal is to focus on the case that $\alpha = 3$, and to a lesser extent on $\alpha = 2$, and to obtain better algorithms and lower bounds in these cases.

1.2 Our Contributions

We show, using an amortized local competitiveness analysis, that if q is set to $2 - \frac{1}{\alpha}$, then the competitive ratio of qOA is at most $4^{\alpha}/(2\sqrt{e\alpha})$. This bound is approximately 3.4 when $\alpha = 2$, and 11.2 when $\alpha = 3$. Using an analysis specialized to the specific cases that $\alpha = 2$ and $\alpha = 3$, we show that qOA is at worst 2.4-competitive when $\alpha = 2$, and at worst 6.7-competitive when $\alpha = 3$.

Our main technical idea is to introduce a new potential function which is quite different from the one used in the analysis of OA in [5] (and the potential function used to analyze AVR in [2]). This is necessary, since potential functions similar to those used earlier cannot yield guarantees of the form c^α where c is independent of α. The potential function we use is more similar to the one used in [7] to analyze a speed scaling algorithm for the (different) objective of minimizing flow time plus energy. However, here we will need a different analysis approach. The analysis in [7], and almost all of the amortized local competitiveness analyses in the speed scaling literature, rely critically on the Young's inequality. However, in the current setting, Young's inequality gives a bound that is too weak to be useful when analyzing qOA. The key insight that allows us to avoid the use of Young's inequality was to observe that certain expressions that arise in the analysis are convex, which allows us to reduce the analysis of the general case down to just two extreme cases. To the best of our knowledge, this convexity technique can replace all of the uses of Young's inequality in the speed scaling literature. In all cases, the resulting bound that one obtains using this convexity technique is at least as good as the bound that one obtains using Young's inequality, and the resulting proof is simpler and more intuitive. In some cases, this convexity technique gives a better bound. For example, if one applies this convexity technique to the analysis of the LAPS algorithm in [10], one obtains a bound on the competitive ratio of $O(\alpha^2/\log^2\alpha)$, whereas using Young's technique one can only get a bound of $O(\alpha^3)$.

In Section 4 we consider lower bounds. We give the first non-trivial lower bound on the competitive ratio for any algorithm. We show that every deterministic algorithm must have a competitive ratio of at least $e^{\alpha-1}/\alpha$. The base of the exponent, e, is the best possible since BKP achieves a ratio of about $2e^{\alpha+1}$. For $\alpha = 3$, this raises the best known lower bound a modest amount, from 1.2 to 2.4. The instance is identical to the one used in [5] to lower bound the competitive ratio with respect to the objective of minimizing the maximum speed. The innovation required to get a lower bound for energy is to categorize the variety of possible speed scaling policies in such a way that one can effectively reason about them.

Given the general lower bound of $e^{\alpha-1}/\alpha$, and that BKP achieves a ratio with base of exponent e, a natural question is whether there is some choice of the parameter q for which the competitive ratio of qOA varies with e as the base of the exponent. Somewhat surprisingly, we show that this is not the case and the base of the exponent cannot be improved beyond 4. In particular, we show that the competitive ratio of qOA is at least $\frac{1}{4\alpha}4^\alpha(1-\frac{2}{\alpha})^{\alpha/2}$. We note that this lower bound is quite close to our upper bound for qOA, especially as α increases.

Our results are summarized in the last two rows of table 1. In particular we give asymptotically matching upper and lower bounds for qOA and reduce the range for the optimal competitive ratio in the case that the cube-root rule holds from $[1.2, 27]$ to $[2.4, 6.7]$ and in the case that $\alpha = 2$ from $[1.1, 4]$ (obtained in [16]) to $[1.3, 2.4]$. Due to the limitation of space, some proofs are omitted and will be given in the full paper.

1.3 Other Related Results

There are now enough speed scaling papers in the literature that it is not practical to survey all such papers here. We limit ourselves to those papers most related to the results presented here.

A naive implementation of YDS runs in time $O(n^3)$. This can be improved to $O(n^2)$ if the intervals have a tree structure [12]. Li, Yao and Yao [13] gave an implementation that runs in $O(n^2 \log n)$ time for the general case. For hard real-time jobs with fixed priorities, Yun and Kim [17] showed that it is NP-hard to compute a minimum-energy schedule. They also gave a fully polynomial time approximation scheme for the problem. Kwon and Kim [11] gave a polynomial time algorithm for the case of a processor with discrete speeds. Li and Yao [14] gave an algorithm with running time $O(d \cdot n \log n)$ where d is the number of speeds. A simpler algorithm with this running time can be found in [13].

Albers, Müller, and Schmelzer [1] consider the problem of finding energy-efficient deadline-feasible schedules on multiprocessors. [1] showed that the offline problem is NP-hard, and gave $O(1)$-approximation algorithms. [1] also gave online algorithms that are $O(1)$-competitive when job deadlines occur in the same order as their release times. Chan et al. [9] considered the more general and realistic speed scaling setting where there is an upper bound on the maximum processor speed. They gave an $O(1)$-competitive algorithm based on OA. Recently, Bansal, Chan and Pruhs [3] investigated speed scaling for deadline feasibility in devices with a regenerative energy source such as a solar cell.

2 Formal Problem Statement

A problem instance consists of n jobs. Job i has a release time r_i, a deadline $d_i > r_i$, and work $w_i > 0$. In the online version of the problem, the scheduler learns about a job only at its release time; at this time, the scheduler also learns the exact work requirement and the deadline of the job. We assume that time is continuous. A schedule specifies for each time a job to be run and a speed at which to run the job. The speed is the amount of work performed on the job per unit time. A job with work w run at a constant speed s thus takes $\frac{w}{s}$ time to complete. More generally, the work done on a job during a time period is the integral over that time period of the speed at which the job is run. A schedule is *feasible* if for each job i, work at least w_i is done on job i during $[r_i, d_i]$. Note that the times at which work is performed on job i do not have to be contiguous. If the processor is run at speed s, then the power is $P(s) = s^\alpha$ for some constant $\alpha > 1$. The energy used during a time period is the integral of the power over that time period. Our objective is to minimize the total energy used by the schedule. An algorithm A is said to be c-competitive if for any job sequence, the energy usage of A is at most c times that of the optimal schedule.

3 Upper Bound Analysis of qOA

Our goal in this section is to show that qOA is about $4^\alpha/(2\sqrt{e\alpha})$-competitive when $q = 2 - (1/\alpha)$. We wish to point out that $q = 2 - 1/\alpha$ is not necessarily the optimum value of q. For general α it is not clear how to obtain the optimum choice of q since it involves solving a system of high degree algebraic inequalities. However, the lower bound for qOA will imply that the choice $q = 2 - 1/\alpha$ is close to optimum. For the case of $\alpha = 3$ and that of $\alpha = 2$, we can explicitly determine the optimum choice of q which gives better competitive ratios for these cases.

We use an amortized local competitiveness analysis, and use a potential function $\Phi(t)$ that is a function of time. In this setting, the value of $\Phi(t)$ will be energy, and thus, the derivative of $\Phi(t)$ with respect to time will be power. We need that Φ is initially and finally zero. Let s_a and s_o be the current speed of the online algorithm (qOA in our case) and the optimal algorithm OPT respectively. Then in order to establish that the online algorithm is c-competitive, it is sufficient to show that the following key equation holds at all times:

$$s_a^\alpha + \frac{d\Phi}{dt} \leq c \cdot s_o^\alpha \tag{1}$$

The fact that equation (1) establishes c-competitiveness follows by integrating this equation over time, and from the fact that Φ is initially and finally 0. For more information on amortized local competitiveness arguments see [15].

Before defining the potential function Φ that we use, we need to introduce some notation. We always denote the current time as t_0. For any $t_0 \leq t' \leq t''$, let $w_a(t', t'')$ denote the total amount of work remaining in qOA at t_0 with deadline in $(t', t'']$. Define $w_o(t', t'')$ similarly for OPT. Recall that qOA runs at speed $q \cdot \max_t w_a(t_0, t)/(t - t_0)$, which is q times the speed that OA would run. Let $d(t', t'') = \max\{0, w_a(t', t'') - w_o(t', t'')\}$, denote the amount of additional work left under the online algorithm that has deadline in $(t', t'']$. We define a sequence of time points $t_1 < t_2 < \ldots$ iteratively as follows: Let t_1 be the time such that $d(t_0, t_1)/(t_1 - t_0)$ is maximized. If there are several such points, we choose the furthest one. Given t_i, let $t_{i+1} > t_i$ be the furthest point that maximizes $d(t_i, t_{i+1})/(t_{i+1} - t_i)$. We use g_i to denote $d(t_i, t_{i+1})/(t_{i+1} - t_i)$. Note that g_i is a non-negative monotonically decreasing sequence.

We first bound the offline and online speed, which will be useful in our analysis:

Lemma 1. **(i)** *$s_o \geq \max_t w_o(t_0, t)/(t - t_0)$.* **(ii)** *$s_a \geq q g_0$ and $s_a \leq q g_0 + q s_o$.*

We are now ready to define the potential function Φ that we use in our analysis of qOA:

$$\Phi = \beta \sum_{i=0}^{\infty} ((t_{i+1} - t_i) \cdot g_i^\alpha) \ ,$$

where β is some constant (which will be set to $q^\alpha(1 + \alpha^{-1/(\alpha-1)})^{\alpha-1}$).

We now make some observations about the potential function Φ. Φ is obviously zero before any jobs are released, and after the last deadline. Job arrivals do not affect Φ since $d(t', t'')$ does not change upon a job arrival for any t' and t''. Similarly, job completions by either qOA or optimal do not change Φ since it is a continuous function of the unfinished work, and the unfinished work on a job continuously decreases to 0 as it completes. Finally, structural changes in the t_i

and g_i do not change the value of Φ. In particular, if g_0 decreases (for instance if online is working faster than offline on jobs with deadline in $[t_0, t_1]$), then at some point g_0 becomes equal to g_1, and the intervals $[t_0, t_1]$ and $[t_1, t_2]$ merge together. Upon this merge, the potential does not change as $g_0 = g_1$ at this point. Similarly, if offline works too fast, the interval $[t_k, t_{k+1}]$ (which contains the earliest deadline among the unfinished jobs under offline) might split into two critical intervals, $[t_k, t']$ and $[t', t_{k+1}]$, but again this change does not affect Φ since at the time of splitting, the value of g for the newly formed intervals is identical to the value of the interval $[t_k, t_{k+1}]$

Thus to complete our analysis, we are left to show the following lemma:

Lemma 2. *For general $\alpha > 1$, set $q = 2 - (1/\alpha)$, $\beta = c = (2 - (1/\alpha))^\alpha (1 + \alpha^{-1/(\alpha-1)})^{\alpha-1}$. Consider a time t where no jobs are released, no jobs are completed by qOA or optimal, and there are no structural changes to the t_i's nor g_i's. Then equation (1), $s_a^\alpha + d\Phi/dt - c \cdot s_o^\alpha \le 0$, holds at time t.*

Proof. Suppose first that $w_a(t_0, t_1) < w_o(t_0, t_1)$. In this case, $d(t_0, t_1) = 0$, $g_0 = 0$ and t_1 is basically infinity. Note that $d\Phi/dt = 0$ since Φ remains zero until $w_a(t_0, t_1) \ge w_o(t_0, t_1)$. Therefore, $s_a^\alpha + d\Phi/dt - c \cdot s_o^\alpha \le 0$ because $s_a \le q s_o$ and $c = q^\alpha (1 + \alpha^{-1/(\alpha-1)})^{\alpha-1} > q^\alpha$.

Hence we assume $w_a(t_0, t_1) \ge w_o(t_0, t_1)$ in the following. Without loss of generality, both OPT and qOA schedule jobs according to Earliest Deadline First, and hence qOA is working on a job with deadline at most t_1. Let t' be deadline of the job that OPT is working on, and let k be such that $t_k < t' \le t_{k+1}$.

First consider the case that $k > 0$. When both qOA and OPT work, g_0 decreases, the quantities $g_1, \ldots, g_{k-1}$, and $g_{k+1}, \ldots$ stay unchanged, and g_k increases. Note that $(t_1 - t_0)$ is decreasing, and the rate of decrease is the same as the rate that time passes. Therefore, the rate of change of $(t_1 - t_0) \cdot g_0^\alpha$ is

$$\begin{aligned} \frac{d}{dt_0}((t_1 - t_0) \cdot g_0^\alpha) &= (t_1 - t_0) \cdot \alpha g_0^{\alpha-1} \left(\frac{(t_1 - t_0)(-s_a) + d(t_0, t_1)}{(t_1 - t_0)^2} \right) - g_0^\alpha \\ &= \alpha g_0^{\alpha-1}(-s_a) + (\alpha - 1) g_0^\alpha \end{aligned}$$

For the rate of change of $(t_{k+1} - t_k) \cdot g_k^\alpha$, we note that $t_{k+1} - t_k$ stays unchanged. Also, the rate of change of $d(t_k, t_{k+1})$ may be $-s_o$ or 0, depending on whether $w_a(t_k, t_{k+1})$ is greater than $w_o(k_k, t_{k+1})$. Therefore,

$$\begin{aligned} \frac{d}{dt_0}((t_{k+1} - t_k) \cdot g_k^\alpha) &\le (t_{k+1} - t_k) \cdot \alpha g_k^{\alpha-1} \left(\frac{(t_{k+1} - t_k)(s_o)}{(t_{k+1} - t_k)^2} \right) \\ &= \alpha g_k^{\alpha-1}(s_o) \quad \le \quad \alpha g_0^{\alpha-1}(s_o) \end{aligned}$$

Thus to show $s_a^\alpha + d\Phi/dt - c \cdot s_o^\alpha \le 0$, it suffices to show that

$$s_a^\alpha + \beta(\alpha g_0^{\alpha-1}(-s_a + s_o) + (\alpha - 1) g_0^\alpha) - c \cdot s_o^\alpha \le 0. \qquad (2)$$

Now consider the case that $k = 0$. Note that for $i \ge 1$, neither g_i nor $t_{i+1} - t_i$ changes, so we need not consider these terms in the potential function. The rate of change of $(t_1 - t_0) \cdot g_o^\alpha$ is

$$\frac{d}{dt_0}((t_1 - t_0) \cdot g_0^\alpha) = (t_1 - t_0) \cdot \alpha g_0^{\alpha-1} \cdot \left(\frac{(t_1 - t_0)(-s_a + s_o) + d(t_0, t_1)}{(t_1 - t_0)^2} \right) - g_0^\alpha$$
$$= \alpha g_0^{\alpha-1}(-s_a + s_o) + (\alpha - 1) g_0^\alpha$$

which leads to the same inequality as equation (2).

Hence, we will focus on equation (2), and show that it is true for the stated values of q, c and β. We consider the left hand side of equation (2) as a function of s_a while g and s_o are fixed. Note that it is a convex function of s_a. Since $s_a \in [qg_0, qg_0 + qs_o]$, it suffices to show that equation (2) holds at the endpoints $s_a = qg_0$ and $s_a = qg_0 + qs_o$.

If $s_a = qg_0$, the left hand side of equation (2) becomes

$$q^\alpha g_0^\alpha - \beta q \alpha g_0^\alpha + \beta \alpha g_0^{\alpha-1} s_o + \beta(\alpha - 1) g_0^\alpha - c s_o^\alpha =$$
$$(q^\alpha - \beta \alpha q + \beta(\alpha - 1)) g_0^\alpha + \beta \alpha g_0^{\alpha-1} s_o - c s_o^\alpha$$

Taking derivative with respect to s_o, we get that this is maximized at s_o satisfying $cs_o^{\alpha-1} = \beta g_0^{\alpha-1}$, and hence $s_o = \left(\frac{\beta}{c}\right)^{1/(\alpha-1)} g_0$. Substituting this for s_o and canceling g_0^α, it follows that we need to satisfy the following equation:

$$(q^\alpha - \beta \alpha q + \beta(\alpha - 1)) + \beta(\alpha - 1)\left(\frac{\beta}{c}\right)^{1/(\alpha-1)} \leq 0. \tag{3}$$

If $s_a = qg_0 + qs_o$, the left hand side of equation (2) becomes

$$q^\alpha(g_0 + s_o)^\alpha - \beta q \alpha (g_0 + s_o) g_0^{\alpha-1} + \beta \alpha g_0^{\alpha-1} s_o + \beta(\alpha - 1) g_0^\alpha - c s_o^\alpha$$
$$= q^\alpha(g_0 + s_o)^\alpha - \beta(q\alpha - (\alpha - 1)) g_0^\alpha - \beta\alpha(q - 1) g_0^{\alpha-1} s_o - c s_o^\alpha$$

Setting $s_o = x \cdot g_0$ and canceling g_0^α, it follows that we need to satisfy

$$q^\alpha(1 + x)^\alpha - \beta(q\alpha - (\alpha - 1)) - \beta\alpha(q - 1)x - cx^\alpha \leq 0. \tag{4}$$

We set $q = 2 - (1/\alpha)$ and $\beta = c = q^\alpha \eta^{\alpha-1}$ where $\eta = 1 + \alpha^{-1/(\alpha-1)}$. With these choices of q, β and c, $\alpha q = 2\alpha - 1$, and to establish equation (3) it is sufficient to show that $q^\alpha - \beta \leq 0$, which is trivially true since $\eta > 1$. Similarly, equation (4) is equivalent to $(1 + x)^\alpha - \alpha\eta^{\alpha-1} - \eta^{\alpha-1}(\alpha - 1)x - \eta^{\alpha-1}x^\alpha \leq 0$ for all $x \geq 0$. Since $\alpha \geq 1$, it suffices to show that

$$(1 + x)^\alpha - \alpha\eta^{\alpha-1} - \eta^{\alpha-1}x^\alpha \leq 0. \tag{5}$$

To see this, note that if we take the derivative of the left side of equation (5), we obtain that the maximum is attained at x such that $(1 + x)^{\alpha-1} - \eta^{\alpha-1}x^{\alpha-1} = 0$ and hence $x = 1/(\eta - 1)$. For this value of x, the left side of equation (5) evaluates to 0 and hence the result follows. Hence equation (2) is satisfied and the lemma follows. □

Now our main theorem follows as a direct consequence.

Theorem 1. qOA *is* $(2 - \frac{1}{\alpha})^\alpha(1 + \alpha^{-1/(\alpha-1)})^{\alpha-1}$*-competitive for general* $\alpha > 1$.

Note that for large values of α, this bound on the competitive ratio of qOA is approximately $4^\alpha/(2\sqrt{e\alpha})$. For $\alpha = 3$ this bound on the competitive ratio of qOA evaluates to $(5/3)^3(1+1/\sqrt{3})^2 \approx 11.52$ (which is already better than the best known bound of 27). However, for the cases of $\alpha = 2$ and $\alpha = 3$, we can determine the optimum values of q and β to obtain Theorems 2 and 3.

Theorem 2. *If* $q = 1.54$, *then* qOA *is 6.73-competitive for* $\alpha = 3$.

Proof. We follow the same proof structure as that for Lemma 2 to obtain the inequalities (3) and (4). By putting $\alpha = 3$, it follows that we need to satisfy:

$$(q^3 - 3\beta q + 2\beta) + 2\beta\left(\frac{\beta}{c}\right)^{1/2} \leq 0$$
$$q^3(1+x)^3 - \beta(3q-2) - 3\beta(q-1)x - cx^3 \leq 0$$

We wrote a program to determine the values of q and β that minimize c. The best values we obtained are $q = 1.54, \beta = 7.78$ and $c = 6.73$. It is easy to check that the first inequality is satisfied. The left hand side of the second inequality becomes $-3.08x^3 + 10.96x^2 - 1.65x - 16.73$, which can be shown to be negative by differentiation. Hence (3) and (4) are satisfied and the theorem follows. □

Theorem 3. *If* $q = 1.46$ *and* $\beta = 2.7$, *then* qOA *is 2.391-competitive for* $\alpha = 2$.

4 Lower Bounds

In this section, we show that any algorithm is at least $\frac{1}{\alpha}e^{\alpha-1}$-competitive. Note that we assume α is fixed and is known to the algorithm. We first give an adversarial strategy for constructing a job instance such that any algorithm uses at least $\frac{1}{\alpha}e^{\alpha-1}$ times the energy of the optimal.

Adversarial Strategy: Let $\epsilon > 0$ be some small fixed constant. Work is arriving during $[0, \ell]$, where $0 < \ell \leq 1 - \epsilon$. The rate of work arriving at time $t \in [0, \ell]$ is

$$a(t) = \frac{1}{1-t}$$

So the work that arrives during any time interval $[u, v]$ is $\int_u^v a(t)dt$. All work has deadline 1. Let A be any online algorithm. The value of ℓ will be set by the adversary according to the action of A. Intuitively, if A spends too much energy initially, then ℓ will be set to be small. If A doesn't spend enough energy early on, then ℓ will be set to $1 - \epsilon$. In this case, A will have a lot of work left toward the end and will have to spend too much energy finishing this work off. To make this more formal, consider the function

$$E(t) = \int_0^t \left((1 + \frac{b}{\ln \epsilon})\frac{1}{1-x}\right)^\alpha dx \ ,$$

where b is a constant (set to $\frac{1}{(\alpha-1)^{1/\alpha}}$ later). This is the total energy usage up to time t if A runs at speed $s(t) = (1+\frac{b}{\ln\epsilon})\frac{1}{1-t}$. Of course, A may run at speed other than $s(t)$. If there is a first time $0 < h \le 1-\epsilon$ such that total energy usage of A up to h is at least $E(h)$, then the value of ℓ is set to h. If no such event occurs, then $\ell = 1-\epsilon$. ■

In Lemma 5 we show that if the adversary ends the arrival of work at some time $0 < h \le 1-\epsilon$ because the total energy usage of A is at least $E(h)$, then A must have used at least $\frac{1}{\alpha}e^{\alpha-1}$ times as much energy as optimal. Similarly, in Lemma 7, we show that if the adversary doesn't end the arrival of work until the time $1-\epsilon$, then the online algorithm uses at least $\frac{1}{\alpha}e^{\alpha-1}$ times as much energy as optimal. Then our main result, that any algorithm is at least $\frac{1}{\alpha}e^{\alpha-1}$-competitive, follows immediately. We start with two technical lemmas.

Lemma 3. *For any* $0 < h \le 1-\frac{1}{e}$, $(-\ln(1-h))^\alpha \le \frac{\alpha}{e^{\alpha-1}}(\frac{1}{(\alpha-1)(1-h)^{\alpha-1}} - \frac{1}{\alpha-1})$.

Lemma 4. *Let* $s_1(t)$ *and* $s_2(t)$ *be non-negative functions, and let* $\alpha > 1$ *and* $x > 0$ *be some real numbers. If* $s_2(t)$ *is continuous and monotonically increasing and if* $\int_0^y (s_1(t)^\alpha - s_2(t)^\alpha)dt < 0$ *for all* $0 < y \le x$, *then* $\int_0^x (s_1(t) - s_2(t))dt < 0$.

Lemma 5. *If there is a time* $0 < h \le 1-\epsilon$ *such that the total energy usage of* A *is at least* $E(h)$, *then* A *is at least* $\frac{1}{\alpha}e^{\alpha-1}$*-competitive.*

Proof. Let E_A be the total energy usage of A. Then,

$$E_A \ge E(h) = \int_0^h \left((1+\frac{b}{\ln\epsilon})\frac{1}{1-x}\right)^\alpha dx = (1+\frac{b}{\ln\epsilon})^\alpha(\frac{1}{(\alpha-1)(1-h)^{\alpha-1}} - \frac{1}{\alpha-1})$$

Let E_{opt} be the energy usage of the optimal algorithm. There are two cases for the value of E_{opt}: (i) $0 < h \le 1-\frac{1}{e}$ and (ii) $1-\frac{1}{e} < h \le 1-\epsilon$.

(i) If $0 < h \le 1-\frac{1}{e}$, the total amount of work released is $\int_0^h \frac{1}{1-x}dx = -\ln(1-h) \le 1$. Thus, the optimal algorithm can run at speed $-\ln(1-h)$ throughout $[0,1]$ to completes all work. Then

$$E_{opt} = (-\ln(1-h))^\alpha \le \frac{\alpha}{e^{\alpha-1}}(\frac{1}{(\alpha-1)(1-h)^{\alpha-1}} - \frac{1}{\alpha-1})$$

where the inequality comes from Lemma 3. The competitive ratio is $E_A/E_{opt} \ge (1+\frac{b}{\ln\epsilon})^\alpha \frac{1}{\alpha}e^{\alpha-1}$, which is again at least $\frac{1}{\alpha}e^{\alpha-1}$ when ϵ tends to 0.

(ii) If $1-\frac{1}{e} < h \le 1-\epsilon$, the optimal algorithm runs at speed $a(t)$ for $t \in [0, 1-e(1-h)]$ and run at speed $\frac{1}{e(1-h)}$ for $t \in [1-e(1-h), 1]$. It is easy to check that this schedule completes all work. Then,

$$E_{opt} = \int_0^{1-e(1-h)} (\frac{1}{1-x})^\alpha dx + (\frac{1}{e(1-h)})^\alpha \cdot e(1-h) \quad (6)$$

$$= \frac{\alpha}{e^{\alpha-1}}\frac{1}{(\alpha-1)(1-h)^{\alpha-1}} - \frac{1}{\alpha-1} \quad (7)$$

The competitive ratio is $\frac{E_A}{E_{opt}} \ge (1+\frac{b}{\ln\epsilon})^\alpha \frac{1}{\alpha}e^{\alpha-1}$, which is at least $\frac{1}{\alpha}e^{\alpha-1}$ when ϵ tends to 0. □

We now turn attention to the case that the energy usage of A is less than $E(t)$ for all $0 < t \leq 1-\epsilon$. We first show in Lemma 6 that A cannot complete too much work by time $1-\epsilon$.

Lemma 6. *Assume at any time $0 < t \leq 1-\epsilon$, the energy usage of A up to time t is less than $E(t)$. Then, the amount of work done by A up to time $1-\epsilon$ is less than $\int_0^{1-\epsilon}(1+\frac{b}{\ln\epsilon})\frac{1}{1-x}dx$.*

Proof. Let $s_1(y)$ be the speed of the algorithm A and consider the algorithm B that works at speed $s_2(t) = (1+\frac{b}{\ln\epsilon})\frac{1}{1-t}$. The energy consumed by B by time t is exactly $\int_0^t s_2(y)^\alpha dy = E(t)$. The result now follows by applying Lemma 4 with $x = 1-\epsilon$ and observing that $s_2(t)$ is monotonically increasing. □

We are now ready to show, in Lemma 7, that if the adversary doesn't end the arrival of work until time $1-\epsilon$ then the online algorithm uses at least $\frac{1}{\alpha}e^{\alpha-1}$ times as much energy as optimal.

Lemma 7. *If at any time $0 < t \leq 1-\epsilon$, the total energy usage of A is less than $E(t)$, then A is at least $\frac{1}{\alpha}e^{\alpha-1}$-competitive.*

Proof. Note that the adversary ends the arrival of work at time $1-\epsilon$ and the total amount of work arrived is $\int_0^{1-\epsilon}\frac{1}{1-x}dx = -\ln\epsilon$. By Lemma 6, the maximum amount of work completed by A up to time $1-\epsilon$ is

$$\int_0^{1-\epsilon}(1+\frac{b}{\ln\epsilon})\frac{1}{1-x}dx = (1+\frac{b}{\ln\epsilon})[-\ln(1-x)]_0^{1-\epsilon} = -\ln\epsilon - b$$

Hence, A has at least b units of work remaining at time $1-\epsilon$. To finish it, the total energy usage of A is at least $\frac{b^\alpha}{\epsilon^{\alpha-1}}$, which equals $\frac{1}{(\alpha-1)\epsilon^{\alpha-1}}$ by setting $b = \frac{1}{(\alpha-1)^{1/\alpha}}$. Using equation 7 we find that the energy usage of the optimal algorithm is at most $\frac{\alpha}{e^{\alpha-1}}\frac{1}{(\alpha-1)\epsilon^{\alpha-1}}$. Thus, the competitive ratio is at least $\frac{1}{\alpha}e^{\alpha-1}$. □

Theorem 4. *Any algorithm is at least $\frac{1}{\alpha}e^{\alpha-1}$-competitive.*

Lower bound for qOA. Finally, we give a job instance to show that qOA is at least $\frac{1}{4\alpha}4^\alpha(1-\frac{2}{\alpha})^{\alpha/2}$ competitive. The analysis is left to the full paper.

Job Instance: Let $1 > \epsilon > 0$ be some small fixed constant. Consider the input job sequence where work is arriving during $[0, 1-\epsilon]$ and the rate of arrival at time t is

$$a(t) = \frac{1}{(1-t)^x},$$

where $x > \frac{1}{\alpha}$ is a constant (which will be set to $\frac{2}{\alpha}$ later). All work has deadline 1. Finally, a job is released at time $1-\epsilon$ with work ϵ^{1-x} and deadline 1. ■

Theorem 5. *Let α be a known constant. For any choice of q,* qOA *is at least $\frac{1}{4\alpha}4^\alpha(1-\frac{2}{\alpha})^{\alpha/2}$ competitive.*

References

1. Albers, S., Müller, F., Schmelzer, S.: Speed scaling on parallel processors. In: Proc. ACM Symposium on Parallel Algorithms and Architectures (SPAA), pp. 289–298 (2007)
2. Bansal, N., Bunde, D., Chan, H.L., Pruhs, K.: Average rate speed scaling. In: Latin American Theoretical Informatics Symposium (2008)
3. Bansal, N., Chan, H., Pruhs, K.: Speed scaling with a solar cell. In: International Conference on Algorithmic Aspects in Information and Management (submitted, 2008)
4. Bansal, N., Kimbrel, T., Pruhs, K.: Dynamic speed scaling to manage energy and temperature. In: Proc. IEEE Symp. on Foundations of Computer Science, pp. 520–529 (2004)
5. Bansal, N., Kimbrel, T., Pruhs, K.: Speed scaling to manage energy and temperature. JACM 54(1) (2007)
6. Bansal, N., Pruhs, K.R.: Speed scaling to manage temperature. In: Diekert, V., Durand, B. (eds.) STACS 2005, vol. 3404, pp. 460–471. Springer, Heidelberg (2005)
7. Bansal, N., Pruhs, K., Stein, C.: Speed scaling for weighted flow time. In: SODA 2007: Proceedings of the eighteenth annual ACM-SIAM symposium on Discrete algorithms, pp. 805–813 (2007)
8. Brooks, D.M., Bose, P., Schuster, S.E., Jacobson, H., Kudva, P.N., Buyuktosunoglu, A., Wellman, J.-D., Zyuban, V., Gupta, M., Cook, P.W.: Power-aware microarchitecture: Design and modeling challenges for next-generation microprocessors. IEEE Micro 20(6), 26–44 (2000)
9. Chan, H.L., Chan, W.-T., Lam, T.-W., Lee, L.-K., Mak, K.-S., Wong, P.W.H.: Energy efficient online deadline scheduling. In: SODA 2007: Proceedings of the eighteenth annual ACM-SIAM symposium on Discrete algorithms, pp. 795–804 (2007)
10. Chan, H.-L., Edmonds, J., Lam, T.-W., Lee, L.-K., Marchetti-Spaccamela, A., Pruhs, K.: Nonclairvoyant speed scaling for flow and energy. In: STACS (2009)
11. Kwon, W.-C., Kim, T.: Optimal voltage allocation techniques for dynamically variable voltage processors. In: Proc. ACM-IEEE Design Automation Conf., pp. 125–130 (2003)
12. Li, M., Liu, B.J., Yao, F.F.: Min-energy voltage allocation for tree-structured tasks. Journal of Combinatorial Optimization 11(3), 305–319 (2006)
13. Li, M., Yao, A.C., Yao, F.F.: Discrete and continuous min-energy schedules for variable voltage processors. Proc. of the National Academy of Sciences USA 103, 3983–3987 (2006)
14. Li, M., Yao, F.F.: An efficient algorithm for computing optimal discrete voltage schedules. SIAM J. on Computing 35, 658–671 (2005)
15. Pruhs, K.: Competitive online scheduling for server systems. SIGMETRICS Performance Evaluation Review 34(4), 52–58 (2007)
16. Yao, F., Demers, A., Shenker, S.: A scheduling model for reduced CPU energy. In: Proc. IEEE Symp. Foundations of Computer Science, pp. 374–382 (1995)
17. Yun, H., Kim, J.: On energy-optimal voltage scheduling for fixed priority hard real-time systems. ACM Trans. on Embedded Computing Systems 2(3), 393–430 (2003)

Competitive Analysis of Aggregate Max in Windowed Streaming*

Luca Becchetti[1] and Elias Koutsoupias[2]

[1] "Sapienza" University of Rome
becchett@dis.uniroma1.it
[2] University of Athens
elias@di.uoa.gr

Abstract. We consider the problem of maintaining a fixed number k of items observed over a data stream, so as to optimize the maximum value over a fixed number n of recent observations. Unlike previous approaches, we use the competitive analysis framework and compare the performance of the online streaming algorithm against an optimal adversary that knows the entire sequence in advance. We consider the problem of maximizing the *aggregate max*, i.e., the sum of the values of the largest items in the algorithm's memory over the entire sequence. For this problem, we prove an asymptotically tight competitive ratio, achieved by a simple heuristic, called partition-greedy, that performs stream updates efficiently and has almost optimal performance. In contrast, we prove that the problem of maximizing, for every time t, the value maintained by the online algorithm in memory, is considerably harder: in particular, we show a tight competitive ratio that depends on the maximum value of the stream. We further prove negative results for the closely related problem of maintaining the aggregate minimum and for the generalized version of the aggregate max problem in which every item comes with an individual window.

1 Introduction

In streaming applications, a sequence or *stream* of items arrives online to be processed by an algorithm with memory much smaller than the length of the sequence or the cardinality of the universe of possible items. The objective of the algorithm is to maintain some statistical information about the stream. For many statistical properties, only some recent part—called window—of the stream is important. A typical assumption is that the window of interest has fixed size n. In many cases of practical interest the window size is much larger than the available memory size. This is the sliding window streaming model [12,23].

A typical application in the sliding window streaming model is the following: A sensor measures continuously the temperature and maintains the maximum temperature of the last hour. It should be clear that with limited memory, it is

* Partially supported by EU Projects AEOLUS and FRONTS and by MIUR FIRB project N. RBIN047MH9: "Tecnologia e Scienza per le reti di prossima generazione".

S. Albers et al. (Eds.): ICALP 2009, Part I, LNCS 5555, pp. 156–170, 2009.

not possible for the sensor to know at every time step the maximum temperature of the past hour. The question is "how well can this task be done with limited memory?" This is the main question that we address in this work. We naturally treat this as an online problem and we use competitive analysis [6].

Our main focus is on a simple online problem defined by two parameters: n, the window size, and k, the memory size of the algorithm. An online algorithm with memory of size k processes a sequence of positive items with values $a_1, a_2, \ldots$. At every time t, the algorithm maintains in each of its k memory slots an item from the recent window $\{a_{t-n+1}, \ldots, a_t\}$ of length n (for simplicity, we assume that the stream extends to negative times with $a_t = 0$ for $t \leq 0$). The objective at every time t is to maximize the maximum value of the items in memory. More precisely, if g_t denotes the maximum value of the items in memory at time t, the objective of the algorithm is to maximize $\sum_t g_t$. We call this the *online aggregate max problem* and study it using competitive analysis.

This kind of sliding window streaming problems have been addressed before in the literature on streaming algorithms. In fact, much more general questions have been answered with spectacular success. For example, the paper by Datar, Gionis, Indyk, and Motwani [12] showed how to extend the pioneering paper of Alon, Matias, and Szegedy [2] to estimate many statistics on sliding windows; these results were improved and extended in many directions (for example, in [7]). The difference between this body of work and our work lies mainly in the approach: we use competitive analysis to study the performance of an algorithm. To do this, we consider online algorithms with memory of fixed size k (in which each memory position can hold one item) and then ask how good its competitive ratio can be, whereas previous approaches impose a bound on the accuracy, described by a parameter ϵ, and then try to minimize the amount of memory necessary to guarantee (worst case or with high probability) the desired accuracy.

We summarize the main differences of our approach with the approaches in the existing literature: We consider mainly the aggregate value of the maximum value in memory (although we also give tight bounds for the worst case). This is a weaker objective than requiring the maximum value in memory to be *always* within a fraction ϵ from the optimal. The aggregate objective is more appropriate for economic applications in which we care about the total value, rather than the worst case. We measure memory in slots not bits; each slot can keep exactly one item. This assumption is more realistic in practical settings where the items carry large satellite information. We fix the memory size to some constant k and we prefer the competitive ratio to depend mainly on k, not the window length n. As it is usually the case with competitive analysis our focus is on the information-theoretic constraints rather than the computational complexity constraints; in particular, we don't pay much attention to the execution time of the online algorithms, although all our algorithms are very efficient. As a result, the techniques of competitive analysis seem to be more appropriate to address the kind of questions we are interested in than those usually adopted in streaming algorithms, which often rely on embeddings (e.g., [2]).

Our results are not directly comparable to the existing results on streaming algorithms. To make this point clear, consider the results in [10] where they study streaming algorithms that estimate the diameter of a set of points. Their result for the 1-dimensional case (which resembles a lot our question about max) is roughly as follows: There is an efficient ϵ-approximation algorithm with memory that stores of $O(\frac{1}{\epsilon}\log M)$ points, where M is the maximum diameter. Our result is that for every k, there is a competitive algorithm of memory size of k slots with approximation ratio $1+O(1/k)$. There is no direct translation of one result to the other, mainly because one is a worst-case result while the other is an aggregate result. The competitive ratio implied by the above result in [10] is $O(\frac{\log M}{k})$, which is very high compared to our result (and depends on the values of the stream).

Our contributions. We study the online aggregate max problem and give tight results on its competitive ratio: We show that the competitive ratio of online algorithms with memory size k is surprisingly low: $1 + \Theta(1/k)$. The constants inside our lower and upper bounds inside the Θ expression are quite different. In particular, for upper bound, we give an intuitive deterministic online algorithm (which we call partition-greedy), and show that it has competitive ratio $k/(k-1)$. The lower bound is technically more interesting: We show that every randomized online algorithm has competitive ratio $1 + \Omega(\frac{1}{66k})$. These bounds hold in the strongest possible sense: the upper bound holds even against offline algorithms with unlimited memory, whereas the lower bound holds for offline algorithms with memory k.

We also study from a competitive point of view the *anytime max problem* in the sense that the competitive ratio is the worst-case ratio over all times t of the optimal value at time t over the maximum value of the online algorithms memory at time t. Naturally, this tighter objective results in much higher competitive ratio which is not independent of the values of the stream. We show a tight bound on the competitive ratio which is $\sqrt[k+1]{M}$, where M is the maximum value in the stream.

We also explore natural extensions of the online aggregate max problem. An interesting extension comes from viewing the items of the stream as expiring after exactly n steps. We ask the question what happens when each item has its own expiration time (selected by an adversary and revealed to the online algorithm together with its value). We show that in this case the competitive ratio is unbounded. We also explore the competitive ratio for the aggregate min problem and show that it is completely different than the aggregate max objective because its competitive ratio is unbounded.

Related work. We summarize here some relevant publications.

In [12], the authors introduce the sliding window model for streaming. They also prove that maintaining the maximum/minimum over a size n sliding window requires at least $n\log(M/n)$, where M is the size of the universe, i.e., the set of all possible values for items in the stream.

In [16] the authors consider the problem of estimating the diameter of a set of points in the windowed streaming model. The authors address mainly the

2-dimensional case. This is a generalization of the problem of maintaining the maximum. The authors propose streaming algorithms that use polylogarithmic space with respect to the window size and the diameter.

The result above was improved in [10]. In particular, for the 1-dimensional case, the authors provide an algorithm that maintains the diameter of a set of points with ϵ-approximation storing in its memory $O(\frac{1}{\epsilon}\log M)$ points, M being the ratio between the diameter and the minimum distance between any non-coincident pair of points. Notice that the memory in the above expression is measured in number of points not number of bits, the same with this work.

In [7], the authors recast these results in a more general framework, proving among others that for a large class of "smooth" functions, including sum, distance, maximum, and minimum, it is possible to maintain polylogarithmic space sketches that allow to estimate their value within a sliding window with ϵ-approximation.

2 Model and Notation

Windowed streaming. We consider a streaming model in which the algorithm observes a sequence $\{a_1, a_2, \ldots\}$ of items over time, a_t being the item observed during the t-th *time step*. In practical settings, every item is a record consisting of several possible fields, including a distinguished *value* field. In the sequel we assume without loss of generality that the value field of a_j belongs to the (integer) interval $[1, M]$ for some $M > 1$ and we denote by a_j *both the j-th item and its value.* We assume that, at every time t, we are only interested in maintaining statistics over the (values of the) last n observations, i.e., over the window of items observed in the interval $\{t-n+1, \ldots, t\}$. For simplicity when $t < n$, we assume that the sequence extends to negative times with value 0. In the sequel, g_t denotes the item of maximum value maintained by the algorithm at time t and m_t denotes the maximum value in the window at time t.

Since we study the problems using competitive analysis, we are not only interested in the maximum value m_t of the window at time t, but also in the maximum value $\hat{m}_t$ which is stored in the memory of an offline algorithm; the offline algorithm has the same memory restrictions with the online algorithm, but it knows the future. In the definition of the competitive ratio, we should use $\hat{m}_t$. However, we study both ratios (against m_t and $\hat{m}_t$) and we show that the competitive ratio is essentially the same. We use m_t to denote both values.

In the literature of streaming algorithms, a streaming algorithm is always compared against the absolute optimum (such as m_t). It may be useful for other problems to adopt the competitive point of view and judge an algorithm against the value (such as $\hat{m}_t$) of an offline algorithm with the same limitations on its resources.

Memory space. We are interested in maintaining the items themselves, which can in general be complex data structures, not just their values. As a result, the required memory space is measured in units, each unit being the amount of memory necessary to exactly store an item. We assume that the algorithm

Table 1. Notation

Symbol	Meaning
a_j	The j-th item observed in the stream
n	Window size
k	Max. no. items kept by algorithm
M	Maximum item value
g_t	Max. value in algorithm's memory at t
m_t	Max. value in algorithm's memory at t
$\mathbf{r}(k,n)$	Competitive ratio

maintains, at any time t, at most k items among those observed in $\{t-n+1,\dots,t\}$ (where, typically, $k << n$). Table 1 summarizes the notation we use.

Objective functions. While approximating the maximum value over a sliding window can be done using polylogarithmic space [7], the basic task of maintaining the maximum value exactly is not feasible, unless one stores a number of elements in the order of n and this result also holds for randomized algorithms [12]. We consider the following objective functions, that measure how far we are from achieving this goal, both at every point in time and in the average over the entire sequence.

Aggregate max: Maximize $\sum_t g_t$, i.e., the average value of the largest item maintained by the algorithm.

Anytime max: For every t, maximize g_t, i.e., maximize the value of the largest item in the algorithm's memory at time t. As shown further, this function is harder to maintain for every t.

Competitive analysis. We compare the performance of the algorithm against the optimal algorithm that knows the entire sequence in advance [6]. Hence, in this case we define the competitive ratio $\mathbf{r}(k,n)$ as:

$$\mathbf{r}(k,n) = \max_{a \in \mathcal{S}} \frac{\sum_t g_t(a)}{\sum_t \hat{m}_t(a)},$$

where $\mathcal{S}$ is the set of possible input sequences and $g_t(a)$ (respectively, $\hat{m}_t(a)$) is the online algorithm's (respectively the offline algorithm's) maximum value at time t when input sequence a is observed. We also study the maximum value $m_t(a) = \max\{a_{t-n+1},\dots,a_t\}$ of the last n values instead of $\hat{m}_t$. In the sequel we drop the name of the sequence a to simplify the notation.

We note that our definition of the competitive ratio does not have an additive constant [6]. It is a simple observation that such an additive constant cannot play any role in this type of problems: the reason is that we can repeat a sequence many times (probably by appropriate scaling) to make the effects of an additive constant insignificant.

3 Competitive Analysis of the Online Aggregate Max Problem

In this section we study the online aggregate max problem. We first prove a $1 + \Omega(\frac{1}{k})$ lower bound on the competitive ratio of any randomized on line algorithm and then we prove an asymptotically tight deterministic upper bound for the partition-greedy algorithm.

3.1 Randomized Lower Bound

Theorem 1. *Every randomized online algorithm for the aggregate max problem with memory k has competitive ratio $1 + \Omega(1/k)$.*

Proof. The proof is based on Yao's Lemma [6]: We fix a probability distribution on inputs and compute the ratio of the gain of the optimal online algorithm (which knows the distribution but not the outcome of the random experiment) against the optimal algorithm (which knows the exact input).

We select a simple probability distribution of inputs: The input consists of two parts: the first part of n items has values $f(t)$, $t = 0, \ldots, n-1$, where f is a decreasing function (to be determined later); the second part of the input consists of x items of value 0, where x is random value uniformly distributed in $1, \ldots, n$. Thus the online algorithm knows everything, except of when the input sequence stops. For the above sequence, we compute the gain of the optimal online algorithm–which is the hard part—and we lower bound the gain of the optimal offline algorithm to get the desired ratio.

We can assume that the above sequence repeats arbitrarily many times (with independent values x each time) so that we can ignore additive constants in the definition of the competitive ratio.

Essentially, the online algorithm has only to decide which items to keep in memory from the first part of the sequence. To simplify the situation, we assume that the online gain during the first n steps is exactly $n \cdot f(0)$; this is an overestimate when the online algorithm drops the value $f(0)$ at some point, but we are fine since we care to bound its cost from above. With this assumption, an online algorithm is determined completely by the values $f(t_1)$,…,$f(t_k)$ which has in its memory at the end of the first part. To compute the expected online gain we define $t_0 = 0$, $t_{k+1} = n$, and

$$h(t) = \begin{cases} f(t_1) & t_0 \leq t \leq t_1 \\ \vdots & \\ f(t_{k+1}) & t_k \leq t \leq t_{k+1}. \end{cases}$$

To simplify the presentation, we treat f and h as real functions. Also, by scaling appropriately the time (by a factor $1/n$) and the values (by a factor $1/f(0)$), we assume that $n = 1$, $f(0) = 1$ and that the values t_i are real numbers in $[0, 1]$ with $t_0 = 0$ and $t_{k+1} = 1$ (see Figure 1). The effects of these assumptions can be safely ignored for large n and k. The function h approximates from below

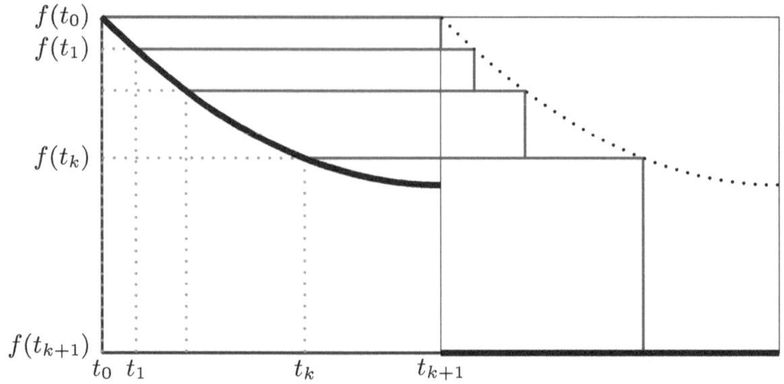

Fig. 1. The lower bound construction. The stream values follow the bold descending line. The descending staircase line shows the maximum value in the memory of the online algorithm. The area below the staircase is the online gain until some random point of the right part.

the function f in $k+2$ points; the situation resembles the Gaussian-Legendre quadrature for approximating integrals, but with a different objective.

When the input sequence ends at time $1+x$, the online gain is $1+\int_0^x h(t)\,dt$. Therefore the expected online gain is given by

$$E[g] = 1+\int_0^1 h(t)(1-t)\,dt = 1+\int_0^1 h(t)d(1-(1-t)^2) = 1+\int_0^1 h(1-\sqrt{1-r})dr.$$

The change in the variable suggests to consider $r_i = 1-(1-t_i)^2$ (and consequently $t_i = 1-\sqrt{1-r_i}$). Notice that $r_0 = 0$ and $r_{k+1} = 1$. The expected online gain then is

$$E[g] = 1+\sum_{i=0}^{k}\int_{r_i}^{r_{i+1}} f(t_{t+1})dr = 1+\sum_{i=0}^{k}(r_{i+1}-r_i)f(t_{i+1}).$$

We now want to select an appropriate f which simplifies the computations. In fact, it is very possible that many choices of the function f result in a similar lower bound—or even with a better coefficient of $1/k$ in the competitive ratio—and in particular linear functions such as $f(t) = 1-t/2$, but computing the optimal online gain appears to be very complicated. The difficulty lies in finding the optimum values for t_i. We however choose

$$f(t) = \frac{1}{2}+\frac{1}{2}(1-t)^2.$$

With this choice, we get that $f_i = 1-r_i/2$, and the expected online cost is

$$E[g] = 1+\sum_{i=0}^{k}(r_{i+1}-r_i)(1-\frac{r_{i+1}}{2}) = 1+\frac{1}{2}\left(\frac{3}{4}-\frac{1}{4}\sum_{i=0}^{k}(r_{i+1}-r_i)^2\right)$$
$$= \frac{11}{8}-\frac{1}{8}\sum_{i=0}^{k}(r_{i+1}-r_i)^2.$$

It is now very easy to select t_i's (or equivalently r_i's) to maximize the above expression: Since $\sum_{i=0}^{k}(r_{i+1}-r_i)=r_{k+1}-r_0=1$, the optimal online algorithm for this particular f should select $r_{i+1}-r_i=1/(k+1)$, or equivalently $r_i=i/(k+1)$. With this choice, the expected online cost is

$$E[g]=\frac{11}{8}-\frac{1}{8}\sum_{i=0}^{k}\frac{1}{(k+1)^2}=\frac{11}{8}-\frac{1}{8(k+1)}.$$

We now turn our attention to the offline algorithm. The advantage of this algorithm over the online algorithm is that it knows x; therefore it keeps in memory only items in $[0,x]$ (as compared to the online algorithm which keeps in memory items in $[0,1]$ but the items after x are useless). We do not need to compute the optimal offline algorithm, it suffices to compute some lower bound[1]. We consider the offline (suboptimal) algorithm which keeps in memory the equidistant values $t_i'=\frac{i}{k-1}\cdot x$. For a given x, the gain of this algorithm is

$$1+\sum_{i=1}^{k-1}(t_i'-t_{i-1}')f(t_i')=1+\sum_{i=1}^{k-1}\frac{x}{k-1}\left(\frac{1}{2}+\frac{1}{2}\left(1-\frac{i}{k-1}\cdot x\right)^2\right).$$

For uniformly distributed x in $[0,1]$, we get that the expected offline gain is

$$1+\sum_{i=1}^{k-1}\int_0^1\frac{x}{k-1}\left(\frac{1}{2}+\frac{1}{2}\left(1-\frac{i}{k-1}\cdot x\right)^2\right)dx=\frac{11}{8}-\frac{1}{48}\frac{5k-6}{(k-1)^2}.$$

Dividing it by the expected online cost $\frac{11}{8}-\frac{1}{8(k+1)}$, we get the ratio

$$1+\frac{1}{66k}+O(\frac{1}{k^2}),$$

which proves the theorem.

3.2 Deterministic Upper Bound

In this subsection we give (almost) tight bounds on the competitive ratio for large k. We will prove the result in the strongest possible sense: We will show that the deterministic competitive ratio is $\mathbf{r}(k,n)=1+O(1/k)$.

We analyze a very natural algorithm which we call partition-greedy: We partition the sequence into parts of size n/k. Part s starts at time $(s-1)n/k+1$ and ends at time sn/k. For every part s of the sequence, a particular slot of memory is active, the memory slot $i=1+(s \pmod k)$. For each part of the sequence, the active slot of the memory accepts the first item. In every other respect, the active slot in each part is updated greedily: the algorithm updates

[1] It is not hard to compute that even with arbitrarily large memory the expected optimal gain is $1+\int_0^1 f(t)(1-t)dt=11/8$ which also gives the same lower bound, although with an improved coefficient of $1/k$ in the competitive ratio.

```
PartitionGreedy(n, k)
t = 1
while Stream not finished {a is current item} do
  for j: 0...k - 1 do
    B[j].time = B[j].time + 1 {Window shifts 1 slot to the
    right}
  end for
  i = ⌈tk/n⌉ {Compute part to which t belongs}
  if B[i].time > n OR B[i].val ≤ a.val then
    B[i] = a
  end if
   t = t + 1
end while
```

Fig. 2. Partition greedy algorithm. For an item a, a.val and a.time repsectively denote the value of a and the time units elapsed since a was observed.

the slot value whenever an item of larger (or the same) value appears. Clearly, the partition-greedy algorithm can be implemented very efficiently (using two counters to keep track of the current active slot of memory and the number of items seen in each part).

Theorem 2. *The partition-greedy algorithm has competitive ratio $k/(k-1)$.*

Proof. Let m_t denote the maximum value in $\{a_{t-n+1}, \ldots, a_t$ and let g_t denote the maximum value in the online algorithm memory at time t. Let also P_t denote the values in the part of size n/k in which t belongs. If the value m_t appeared in the last $k-1$ parts (i.e., in parts $P_{t-(k-1)n/k}, \ldots, P_t$), it must be still in the online memory. Therefore, when $g_t < m_t$, it must be the case that the value m_t appeared in part P_{t-n} and it was dropped in the current part P_t. Roughly speaking, in a decreasing sequence the online algorithm maintains the maximum value in memory for $k-1$ parts, which gives the competitive ratio. However, we need to be more careful as the sequence of values may not be decreasing. To do this, we "charge" $k-1$ online values $g_t, g_{t-n/k}, \ldots, g_{t-n(k-1)/k}$ to the optimal value m_t, as follows: We first observe that the above implies

$$\begin{aligned} &\text{either } g_t = m_t, \\ &\quad \text{or } g_{t-n/k} \geq m_t \text{ and } \cdots \text{ and } g_{t-n(k-1)/k} \geq m_t \end{aligned}.$$

To simplify the presentation, we may use negative indices and we assume that both m_t and g_t are 0 for negative t. Since the above condition compares only values that differ by multiples of n/k (which we can assume to be an integer), we will use the notation $g'_t = g_{\lfloor tk/n \rfloor + t \pmod{n/k}}$, so that the above condition becomes

$$\begin{aligned} &\text{either } g'_t = m'_t, \\ &\quad \text{or } g'_{t-1} \geq m'_t \text{ and } \cdots \text{ and } g'_{t-(k-1)} \geq m'_t \end{aligned}.$$

To prove the theorem, it suffices to show that $k \sum_t g'_t \geq (k-1) \sum_t m'_t$, and this is what we will do.

We first show the following claim: For every t and every $r = 0, \ldots, k-1$, there is an index $\ell(t,r) \in \{t-r, \ldots, t\}$ such that

$$-g'_{\ell(t,r)} + \sum_{i=t-r}^{t} g'_{t-i} \geq \sum_{i=t-r+1}^{t} m'_{t-i}. \tag{1}$$

In words, for r consecutive optimal values there are equal values in $r+1$ consecutive online values (and $\ell(t,r)$ denotes the unmatched online value). The proof of the claim is by induction on r: For $r = 0$ the claim is trivial by taking $\ell(t,r) = t$. For the induction step,

- either $g'_t \geq m'_t$, in which case we take $\ell(t,r) = \ell(t-1, r-1)$; i.e., we match g'_t to m'_t and leave the same element $\ell(t-1, m-1)$ unmatched.
- or $g'_t < m'_t$, in which case we take $\ell(t,r) = t$; i.e., we match g'_t to the unmatched element $\ell(t-1, r-1)$ and leave t unmatched. For this, notice that $\ell(t-1, r-1) \geq t-(k-1)$ and therefore $g'_{\ell(t-1,r-1)} = m'_t$.

From (1), by dropping the term involving the missing element, we get

$$\sum_{i=t-r}^{t} g'_{t-i} \geq \sum_{i=t-r+1}^{t} m'_{t-i}, \tag{2}$$

for every $r = 0, \ldots, k-1$.

By summing up the above inequalities for every t and for $r = k-1$, we get that the left side is approximately $k \sum_{t=0}^{T} g'_t$ and the right hand side is approximately $(k-1) \sum_{t=1}^{T} m_t$, which would immediately prove the upper bound. There is a slight complication, in that the last values in the above sums appear fewer than k and $k-1$ times respectively; for example, g'_T and m'_T appears only once. To take care of this, we also add the inequalities (2) for $t = T$ and for every $r = 0, \ldots, k-2$. In detail, we sum up the following inequalities (2)

$$\sum_{t=0}^{T} \sum_{i=t-(k-1)}^{t} g'_{t-i} + \sum_{r=0}^{k-2} \sum_{i=T-r}^{t} g'_{T-i} \leq \sum_{t=0}^{T} \sum_{i=t-(k-2)}^{t} m'_{t-i} + \sum_{r=0}^{k-2} \sum_{i=T-r+1}^{t} m'_{T-i},$$

which simplifies to the desired inequality

$$k \sum_{t=0}^{T} g'_t \geq (k-1) \sum_{t=0}^{T} m'_t.$$

It is easy to give an example where the partition-greedy has competitive ratio $k/(k-1)$: The sequence has length $n(k+1)/k$ (equal to $k+1$ parts) and it consists entirely of 0's, except of the value at end of the first part which has value 1, i.e. $a_{n/k-1} = 1$. The online algorithm retains this value in memory for $k-1$ parts, while he optimal algorithm retains it for k parts.

4 Generalizations and Other Variants of the Problem

In this section we investigate generalizations and variants of the problem.

4.1 Items with Different Expiration Times

A natural generalization of the aggregate max problem is when each item has its own expiration time. In this generalization the input is a sequence $(a_1, n_1), \ldots, (a_n, n_n)$, where n_i is the time after which a_i expires. The online algorithm must decide for each item whether to keep it in memory or not. Again the restriction is that only k items can be kept in memory at any time and that an expired item has no value (or equivalently, that item i cannot be kept in memory for more than n_i steps). In the previous sections we have considered the special case $n_i = n$. We will show that no online algorithm can have bounded competitive ratio for this problem.

Theorem 3. *The deterministic aggregate max problem with variable expiration times has unbounded competitive ratio.*

Proof. The adversary gives $k+1$ different types of items: items of type i, $i = 1, \ldots, k+1$, have value v_i and duration T_i; the two sequences $v_1, v_2, \ldots, v_{k+1}$ and $v_{k+1}T_{k+1}, v_kT_k, \ldots, v_1T_1$ are steeply increasing. The items of every type are repeated periodically so that most of the time there is exactly one alive (not expired) item of each type. However, when the online algorithm rejects some item of type $j \in \{1, \ldots, k\}$ to accept an item of type $k+1$, the adversary stops giving more items of types $j+1, \ldots, k+1$ until the item of type j expires; the adversary resumes giving all types of tasks once the item of type j expires.

The online algorithm faces the following dilemma: To optimize the long-term gain, it should keep the items of the first k types in memory and reject the items of type $k+1$; this is so because the values of the items are such that v_iT_i is a steeply decreasing sequence. On the other hand, to optimize the short-term gain, the items of type $k+1$ must be kept because their value is much higher than the value of the rest. We show that there is no good way to resolve this dilemma. To do this, we show by induction on i that an online algorithm with bounded competitive ratio cannot reject an item of type $i \le k$. But then, the competitive ratio against an offline algorithm which keeps the items of type $k+1$ is at least $v_{k+1}/(v_1 + \cdots + v_k)$; this ratio can be also unbounded.

We first show only the basis of the induction for $i = 1$, i.e., that an online algorithm with bounded competitive ratio has to keep the items of type $i = 1$ in memory. Suppose that at some time t the online algorithm accepts the item of type $k+1$ by rejecting an item of type 1. Then no new items arrive until the expiration of the item of type 1. The online gain is at most $t(v_1 + \ldots + v_k) + (v_2T_2 + \cdots + v_{k+1}T_{k+1})$, while the optimal gain is at least $\max(v_1T_1, tv_{k+1}) \ge (v_1T_1 + tv_{k+1})/2$. If we select values with $v_1T_1 \ge \lambda(v_2T_2 + \cdots + v_{k+1}T_{k+1})$ and $v_{k+1} \ge \lambda(v_1 + \ldots + v_k)$ for some positive parameter λ, then the competitive ratio is at least λ, which can be arbitrarily high.

For the induction step, we focus only on online and offline algorithms that always keep all items of type $l < i$ in memory. Since the values $v_1, \ldots, v_{i-1}$ are much smaller than the values of higher-type items, they have small effect on the competitive ratio and we can essentially repeat the above argument of the basis case.

4.2 The Aggregate Min Problem

We show that when we change our objective from keeping the maximum value to keeping the minimum value, the problem is transformed completely. In particular, we show that the competitive ratio is unbounded for the aggregate min case.

Proposition 1. *The aggregate min problem has unbounded competitive ratio.*

Proof. Fix some online algorithm. The adversary gives a steeply increasing sequence $a_1, a_2, \ldots$. Let a_t (for some $t \leq k+1$) be the first item that is not kept by the online algorithm. The adversary ends the sequence at time $t+n-1$. The main point is that at time $t+n-1$ the online algorithm has in its memory only items that appeared after time $t+1$, whereas the offline algorithm can have the item a_t which has much smaller value. More precisely, the total online cost is at least $na_1 + a_2 + \ldots + a_{t-1} + a_{t+1}$, while the cost of an offline algorithm who keeps a_t (by replacing a_{t-1}, if it is needed) is $na_1 + a_2 + \ldots + a_{t-2} + 2a_t$[2]. By selecting a_{t+1} to be much higher than $na_1 + a_2 + \ldots + a_{t-2} + 2a_t$, we get an unbounded competitive ratio.

4.3 The Anytime Max Ratio

In this section we address the question of approximating at any time t the maximum value of the current window. In the aggregate max problem, the objective is to optimize the *sum* of the maximum value whereas here it is to optimize the *maximum*. That is, we are interested in minimizing

$$\max_a \max_t \frac{g_t}{m_t}.$$

We show that the competitive ratio in this case cannot be independent of the values. More precisely, we show that the competitive ratio is $O(\sqrt[k+1]{M})$, where M is the maximum value in the stream. A similar result has been shown also in [10] where they give an $1+\epsilon$-approximation algorithm with memory $\frac{1}{\epsilon}\log M$. Although the results are essentially the same, this seems to be a cleaner (and simpler) problem when we cast it in the framework of competitive analysis.

Proposition 2. *The competitive ratio* $\max_a \max_t \frac{g_t}{m_t}$ *is*

$$O(\sqrt[k+1]{M}),$$

where M is the maximum value in the stream.

[2] This expressions hold for $t > 1$. The case $t = 1$ is similar.

Proof. To show the lower bound, consider a geometrically decreasing sequence with $a_j = M\alpha^{j-1}$, where $\alpha = M^{1/n}$. Let $t+1$ be the first time that the online algorithm does not accept an item. In all previous times, the online algorithm accepts the current item by possibly replacing previously kept items. From then on, the adversary gives items with value 1. This means in particular that at time $t+n$ the online algorithm has only items with value 1 in its memory, while the offline algorithm could have kept $a_{t+1} = M/\alpha^t$. This gives a lower bound M/α^t for the online algorithm.

To obtain a different lower bound, observe that at time $t+1$ there are at most k items in the online memory. This means that there is a sequence of t/k consecutive items $a_i, \ldots, a_{i+t/k-1}$ none of which is in the online memory. Then at time $i+n-1$, the maximum item in the online memory is the item $a_{i+t/k}$, while an offline algorithm could have the item a_i. This gives a lower bound on the competitive ratio $\alpha^{t/k}$.

In summary, the competitive ratio is at least

$$\max\{M/\alpha^t, \alpha^{t/k}\},$$

which is at least $M^{1/(k+1)}$ when $\alpha = M^{1/n}$.

This result is tight: we show that a simple, deterministic, bucket-based algorithm can achieve the same approximation ratio. The algorithm arranges the interval $[1, \ldots, M]$ of possible values into k *buckets* as follows: for $i = 1, \ldots, k$, bucket i contains all values in the interval $[(\sqrt[k]{M})^{i-1}, (\sqrt[k]{M})^i)$. The algorithm maintains a list of *representatives*, one for every bucket. Clearly, the competitive ratio cannot be more than $M^{1/k}$. When the maximum value M is not known in advance, the online algorithm starts by assuming $M = k$; whenever a value higher than the current M appears, the online algorithm accepts it (in its top bucket) and updates M. The competitive ratio is not affected.

5 Conclusions and Open Problems

In this paper we considered windowed streaming problems from a competitive analysis perspective. We proved (positive and negative) results for several problems, revolving around or extending the central problem of maintaining the maximum value observed over the last n observations, while obeying stringent memory constraints, consistent with realistic streaming applications.

Many interesting open questions remain. First, there is the problem of tightening the results of Theorems 1 and 2, especially for the case of constant k and in particular for $k = 1$. Also, an interesting direction is to study other statistical questions using the competitive framework we considered in this work. One such question, which is directly relevant to this work, is to maintain some aggregate of the r largest values observed in the window, for some $r \leq k$. Another direction is to consider the order statistics of the items, rather than their actual value. This is useful in settings where we are interested in maintaining a subset of elements that are some of the largest ones in the current window. In this case, the quality of the solution depends on how far the ranks of the top items maintained by the algorithm are from the actual top items in each window of interest. Performance indices for this kind of questions are known in information retrieval and data mining. Analyzing them in a competitive scenario is a challenging task.

Finally an interesting framework to study this type of problems is to assume that the items appearing in the stream are chosen by an adversary, but then presented to the algorithm in a random permutation. This assumption is common in a related, important problem in decision theory, namely the *secretary problem* [17]. Variants of the secretary problem are to maximize the probability to select the top r items [20] or to minimize the expected sum of the ranks of the selected items [1]. The question is how to address these problems in a sliding window setting.

References

1. Ajtai, M., Megiddo, N., Waarts, O.: Improved algorithms and analysis for secretary problems and generalizations. SIAM Journal on Discrete Mathematics 14(1), 1–27 (2000)
2. Alon, N., Matias, Y., Szegedy, M.: The space complexity of approximating the frequency moments. In: Proc. of the ACM Symposium on the Theory of Computing, pp. 20–29 (1996)
3. Arasu, A., Manku, G.S.: Approximate counts and quantiles over sliding windows. In: Proceedings of the twenty-third ACM SIGMOD-SIGACT-SIGART Symposium on Principles of Database Systems (PODS 2004), pp. 286–296. ACM Press, New York (2004)
4. Babcock, B., Datar, M., Motwani, R., O'Callaghan, L.: Maintaining variance and k-medians over data stream windows. In: Proc. of the twenty-second ACM SIGMOD-SIGACT-SIGART symposium on Principles of database systems (PODS 2003), pp. 234–243. ACM Press, New York (2003)
5. Baeza-Yates, R., Ribeiro-Neto, B.: Modern Information Retrieval. Addison-Wesley, Reading (1999)
6. Borodin, A., El-Yaniv, R.: Online computation and competitive analysis. Cambridge University Press, New York (1998)
7. Braverman, V., Ostrovsky, R.: Smooth histograms for sliding windows. In: Proceedings of the IEEE Symposium on Foundations of Computer Science, pp. 283–293 (2007)
8. Braverman, V., Ostrovsky, R., Zaniolo, C.: Succinct sampling on streams. Computing Research Repository (CoRR), abs/cs/0702151 (2007)
9. Broder, A.Z., Kirsch, A., Kumar, R., Mitzenmacher, M., Upfal, E., Vassilvitskii, S.: The hiring problem and lake wobegon strategies. In: Proceedings of the nineteenth annual ACM-SIAM Symposium on Discrete Algorithms (SODA 2008), pp. 1184–1193. Society for Industrial and Applied Mathematics, Philadelphia (2008)
10. Chan, T.M., Sadjad, S.B.S.: Geometric optimization problems over sliding windows. In: Fleischer, R., Trippen, G. (eds.) ISAAC 2004. LNCS, vol. 3341, pp. 246–258. Springer, Heidelberg (2004)
11. Cormode, G., Muthukrishnan, S.: What's hot and what's not: tracking most frequent items dynamically. ACM Trans. Database Syst. 30(1), 249–278 (2005)
12. Datar, M., Gionis, A., Indyk, P., Motwani, R.: Maintaining stream statistics over sliding windows. SIAM Journal of Computing 31(6), 1794–1813 (2002)

13. Datar, M., Muthukrishnan, S.: Estimating rarity and similarity over data stream windows. In: Möhring, R.H., Raman, R. (eds.) ESA 2002. LNCS, vol. 2461, pp. 323–334. Springer, Heidelberg (2002)
14. El-yaniv, R., Fiat, A., Karp, R.M., Turpin, G.: Optimal search and one-way trading online algorithms. Algorithmica 30, 101–139 (2001)
15. Fagin, R., Kumar, R., Sivakumar, D.: Comparing top k lists. In: Proceedings of the fourteenth annual ACM-SIAM symposium on Discrete algorithms (SODA 2003), pp. 28–36. Society for Industrial and Applied Mathematics, Philadelphia (2003)
16. Feigenbaum, J., Kannan, S., Zhang, J.: Computing diameter in the streaming and sliding-window models. Algorithmica 41(1), 25–41 (2004)
17. Ferguson, T.S.: Who solved the secretary problem? Statistical Science 3(4), 282–296 (1988)
18. Golab, L., DeHaan, D., Demaine, E.D., Lopez-Ortiz, A., Munro, J.I.: Identifying frequent items in sliding windows over on-line packet streams. In: Proceedings of the 3rd ACM SIGCOMM conference on Internet Measurement (IMC 2003), pp. 173–178. ACM Press, New York (2003)
19. Guha, S., McGregor, A.: Approximate quantiles and the order of the stream. In: Proceedings of the twenty-fifth ACM SIGMOD-SIGACT-SIGART Symposium on Principles of Database Systems (PODS 2006), pp. 273–279. ACM, New York (2006)
20. Kleinberg, R.: A multiple-choice secretary algorithm with applications to online auctions. In: Proceedings of the sixteenth annual ACM-SIAM symposium on Discrete algorithms (SODA 2005), pp. 630–631. Society for Industrial and Applied Mathematics, Philadelphia (2005)
21. Lee, L.K., Ting, H.F.: A simpler and more efficient deterministic scheme for finding frequent items over sliding windows. In: Proceedings of the twenty-fifth ACM SIGMOD-SIGACT-SIGART Symposium on Principles of Database Systems (PODS 2006), pp. 290–297. ACM, New York (2006)
22. Manjhi, A., Shkapenyuk, V., Dhamdhere, K., Olston, C.: Finding (recently) frequent items in distributed data streams. In: Proceedings of the 21st International Conference on Data Engineering (ICDE 2005), pp. 767–778. IEEE Computer Society Press, Los Alamitos (2005)
23. Muthukrishnan, S.: Data streams: algorithms and applications. Now Publishers Inc. (2005)

Faster Regular Expression Matching

Philip Bille[1,*] and Mikkel Thorup[2]

[1] Technical University of Denmark, DK-2800 Kgs. Lyngby, Denmark
phbi@imm.dtu.dk
[2] AT&T Labs—Research, Shannon Laboratory,
180 Park Avenue, Florham Park, NJ 07932, USA
mthorup@research.att.com

Abstract. Regular expression matching is a key task (and often the computational bottleneck) in a variety of widely used software tools and applications, for instance, the **unix grep** and **sed** commands, scripting languages such as **awk** and **perl**, programs for analyzing massive data streams, etc. We show how to solve this ubiquitous task in linear space and $O(nm(\log\log n)/(\log n)^{3/2}+n+m)$ time where m is the length of the expression and n the length of the string. This is the first improvement for the dominant $O(nm/\log n)$ term in Myers' $O(nm/\log n+(n+m)\log n)$ bound [JACM 1992]. We also get improved bounds for external memory.

1 Introduction

Problem. Regular expression matching is performed commonly as a primitive by many of today's computer systems, and has been so for almost half a century. With `unix/linux`, or on a Mac, we match regular expressions in large file systems using the command line utility `grep`. With the stream editor `sed` we further specify a replacement of the occurrences of the regular expression. A more integrated use of regular expression matching is found in the general text processing language `perl` [22] which is commonly used to convert between input/output formats. One of the basic features in `perl` is to match regular expressions and substitute some of the subexpressions. The regular expression matching is often the hard part whereas the subsequent substitution is easy, so if we could improve regular expression matching, we would improve a lot of data processing.

Historically, regular expression matching goes back to Kleene in the 1950s [11]. It became popular for practical use in text editors in the 1960s [21]. Compilers use regular expression matching for separating tokens in the lexical analysis phase [1]. New and interesting applications continue to appear in diverse areas such as XML querying [12,15], protein searching [18], and Internet traffic analysis [9,24].

Computational model. We study the complexity of regular expression matching on the RAM model with standard word operations. This means that our

* Supported by the Danish Agency for Science, Technology, and Innovation. Work done while the author was at the IT University of Copenhagen.

S. Albers et al. (Eds.): ICALP 2009, Part I, LNCS 5555, pp. 171–182, 2009.

algorithms can be implemented directly in standard imperative programming languages such as C [10] or C++ [20]. For the last thirty years, these programming languages have been commonly used to write efficient and portable code. Even if we code the main program in some higher level programming language, it is normally possible to invoke subroutines written in C for parts that have to run efficiently, and computational efficiency is what we study here. Most open source implementations of algorithms for regular expression, e.g., `grep`, `sed` and `perl`, are written in C. In practice, it would be impressive to gain a factor 2 in speed for this well-studied problem, but as theoreticians we will focus on getting the best asymptotic worst-case running time in terms of the problem size $n \to \infty$.

Current Bounds. Let n and m be the lengths of the string Q and the regular expression R, respectively. Typically we assume $n > m$. The classical textbook solution to the problem by Thompson [21] from 1968 takes $O(nm)$ time. It uses a standard state-set simulation of a non-deterministic finite automaton (NFA) with $O(m)$ states produced from R. Using the NFA, we scan Q. Each of the n characters is processed in $O(m)$ time by progression of the NFA state set.

In 1985 Galil [7] asked if a faster algorithm could be derived. Myers [16] met this challenge in 1992 with an $O(nm/\log n + (n+m)\log n)$ time solution, thus improving the time complexity of Thompson algorithm by a $\log n$ factor for most values of n and m. Since then there has been several works mostly addressing issues of space and larger word length [3,4,19]. However, with no special assumptions on the word length the $O(nm/\log n)$ term from Myers' algorithms has not been touched. The fundamental reason is that all the previous approaches aim to speed-up the $O(m)$ time the NFA needs to process one character from the string. To do that we need at least $\Omega(m/w)$ time just to read or write the state set of the NFA. In fact, Bille [3] obtained this bound within a $\log w$ factor. However, we may have $w = O(\log n)$, so in terms of m and n, there is no hope of bypassing Myers' logarithmic improvement when working on one character at the time.

New Bounds. In this paper we present a linear space regular expression matching algorithm with a running time of

$$O\left(\frac{nm \log\log n}{\log^{3/2} n} + n + m\right).$$

If the alphabet is bounded or if $m \le n^{1-\varepsilon}$ for some positive constant ε, we can avoid the $\log\log n$ factor, getting a cleaner time bound of $O(nm/(\log n)^{3/2} + n + m)$. In any case, assuming $m, n \ge \log^{3/2} n$, this increases the improvement over Thompson's $O(nm)$ algorithm from the $\log n$ factor of Myers to almost a factor $(\log n)^{3/2}$.

We bypass the logarithmic limitation of previous approaches with a speed-up for the NFA processing of multiple characters at the time. This is, in itself, an obvious idea, but challenging to realize due to the complex interaction between the NFA and the input string (c.f. discussion at the end of Sec. 3).

Tabulation. Our improved time bound is achieved via a better tabulation technique. The idea of using tables to improve calculations is old. As a nice early example, in 1792, Gaspard de Prony started preparing nineteen volumes of trigonometric and logarithm tables for the revolutionary French government. With the assistance of a small group of mathematicians, Prony divided the computations into a series of additions and subtractions. He then hired about eighty (human) computers to do the arithmetic.

Tabulation is also used on today's computers for the fastest regular expression matching. In particular, the implementation of Myers' algorithm [16], the `agrep` tool [23] and the `nrgrep` tool [17] are all based on tabulations of NFAs, and they outperform other tools. The basic point is to use table look-ups to replace a complicated NFA simulation that would look up transitions from multiple states.

Tabulation is a speed-up technique that only makes sense after we have settled on the basic combinatorial algorithm, in this case Thompson's algorithm which has stood unbeaten since 1968. In fact, this might very well be the asymptotically fastest worst-case efficient algorithm for regular expression matching if we ignore polylogarithmic factors. The goal in tabulation is now to compactly represent as complex subproblems as possible. Our contribution is to show that we with x bits can represent subproblems requiring $\Theta(x^{3/2}/\log x)$ steps in Thompson's algorithm. The limit of all previous tabulation approaches to this problem was to represent x steps with $\Theta(x)$ bits. We are breaking the linear bound by moving into a higher dimension of encoding, representing both part of the NFA and part of the input string in the x bits. Higher dimensional encodings are known for several other problems [2, 5, 13] but combining the NFA and string dimensions has been a challenge for regular expression matching.

Our encoding allows us to construct fixed universal tables that can later be used to solve arbitrary regular expression problems. The tables are constructed once and for all in $O(2^x)$ time and space. Using them we can match an expression of size m in a text of size n in $O(nm(\log x)/x^{3/2} + n + m)$ time, and this works well even in a streaming context. For the previous mentioned bounds, we could set $x = (\log_2 n)/2$ thus using $O(\sqrt{n})$ time and space on the tables. However, the same tables can be used to solve many regular expression matching problems. Therefore we may chose a larger x to create some large and powerful tables once and for all. These could be used when running `perl` where we often need to match many different regular expressions in a text.

External memory. Our coding technique gives corresponding speed-ups for external memory with block size B and internal memory size $M \geq B$. Of course M and B could also represent smaller units in the memory hierarchy, e.g., cache of size M registers and cache line of length B. If $m = o(M)$, it is trivial to solve the regular expression matching problem with $O(n/B)$ I/O operations. Assuming $m = \Omega(M)$ we solve the problem with $O(nm/(\sqrt{M}B))$ I/O operations. This is a factor $\sqrt{M}$ better than what would be possible with previous algorithms regardless of the block size. Technically speaking, we use the internal memory to pack subproblems of complexity $M^{3/2}$ in Thompson's algorithm. Each subproblem is read with M/B I/O operations, so the saving is a factor $M^{3/2}/(M/B) = \sqrt{M}B$. Previous solutions could only pack problems of complexity linear in M, so they could only save a factor $M/(M/B) = B$. In particular, our algorithm is the first to gain substantially in I/O operations from a large internal memory even if the blocks are small.

Summing up. The theorem below formally states our results in terms of a resource parameter t capturing how much time and space we are willing to spend on the global tables.

Theorem 1. *On a unit-cost RAM with word length w and standard instruction set, for any parameter $t < 2^w$, we can do a general preprocessing for regular expression matching using $O(t)$ time and space. Subsequently, given any regular expression of length m and string of length n, we can perform the matching in*

$$O\left(\frac{nm \log \log t}{(\log t)^{3/2}} + n + m\right)$$

time using $O(t + m)$ space. Our matching only makes a single pass through the string, which may hence be presented as a stream.

Cast in terms of external memory with block size B and internal memory size M, $2B \leq M = O(m)$, we solve the regular expression matching problem with $O(nm/(\sqrt{M}B) + m/B \cdot \log_{M/B}(m/B))$ I/O operations using $O(m)$ space. Again we only perform a single pass over the string. After an $O(m/B \cdot \log_{M/B}(m/B))$ preprocessing of the pattern, we process string segments of length $\Theta(\sqrt{M})$ using $O(m/B)$ I/O operations.

As mentioned above, in many practical applications it makes sense once and for all to do the preprocessing with a fairly large t, fitting within the bounds of fast memory, and subsequently be able to solve lots of smaller regular expression matching problems quickly.

In connection with very large data, note how the unit-RAM result and the external memory result complement each other. In both cases, the string may be a stream passed only once, which is perfect. In the normal case where $m = o(M)$, we pick a large t such that the $O(t+m)$ space fits in internal memory, and use the unit-RAM algorithm to process the stream in $O(n/B)$ I/O operations. However, in the extreme event that $m = \Omega(M)$, we apply our external memory algorithm.

Overview. In this extended abstract, we only have room to present our algorithm for bounded size alphabets on the unit-cost word RAM. In Sec. 2 we define Thompson's standard automaton construction for regular expressions [21] and in Sec. 3 we describe how we decompose this automaton as done in the previous algorithms with logarithmic speedup. After that, we embark on our new attack on the problem. Our decomposition is different from that of Myers [16] because we want to tabulate the action of subautomatons, not just on a single input character, but on substrings of input characters. In Sec. 4 we describe the actual action of subautomatons on substrings and how to tabulate it.

2 Regular Expressions and Finite Automata

First we briefly review the classical concepts used in the paper. For more details see, e.g., Aho et al. [1]. The set of *regular expressions* over Σ are defined recursively as follows: A character $\alpha \in \Sigma$ is a regular expression, and if S and T are regular expressions then so is the *concatenation*, $(S) \cdot (T)$, the *union*, $(S)|(T)$, and the *star*, $(S)^*$. The *language* $L(R)$ generated by R is defined as follows: $L(\alpha) = \{\alpha\}$, $L(S \cdot T) = L(S) \cdot L(T)$, that is, any string formed by the concatenation of a string in $L(S)$ with a string in $L(T)$, $L(S)|L(T) = L(S) \cup L(T)$, and $L(S^*) = \bigcup_{i \geq 0} L(S)^i$, where $L(S)^0 = \{\epsilon\}$ and $L(S)^i = L(S)^{i-1} \cdot L(S)$, for $i > 0$. Here ϵ denotes the empty string. The *parse tree* $T(R)$ for R is the unique rooted binary tree representing the hierarchical structure of R. The leaves of $T(R)$ are labeled by a character from Σ and internal nodes are label by either $\cdot$, $|$, or $*$.

A *finite automaton* is a tuple $A = (V, E, \Sigma, \theta, \phi)$, where V is a set of nodes called *states*, E is a set of directed edges between states called *transitions* each labeled by a character from $\Sigma \cup \{\epsilon\}$, $\theta \in V$ is a *start state*, and $\phi \in V$ is an *accepting state*[1]. In short, A is an edge-labeled directed graph with a special start and accepting node. A is a *deterministic finite automaton* (DFA) if A does not contain any ϵ-transitions, and all outgoing transitions of any state have different labels. Otherwise, A is a *non-deterministic automaton* (NFA). If dealing with multiple automatons, we use a subscript $_A$ to indicate information associated with automaton A, e.g., θ_A is the start state of automaton A.

Given a string q and a path p in A we say that p and q *match* if the concatenation of the labels on the transitions in p is q. We also say that two paths p and p' match if they match the same string. The set of strings matching some path between states s and s' in A is denoted by $P_A(s, s')$. For state-sets S and S' we define $P_A(S, S') = \bigcup_{s \in S, s' \in S'} P_A(s, s')$. For a subset S of states in A and a string Q, define the *state-set transition*, $\delta_A(S, q)$, as the of states reachable from S through a path matching q. We say that A *accepts* the string q if $q \in P_A(\theta_A, \phi_A)$. Otherwise A *rejects* q. One may use a sequence of state-set transitions for a single character to test acceptance of a string Q of length n as follows. First set $S_0 := \{\theta_A\}$. For $i = 1, \ldots, n$ compute $S_i := \{\delta_A(S_{i-1}, Q[i])\}$. It follows inductively that q is accepted if and only if $\phi_A \in S_n$.

Given a regular expression R, an NFA A accepting precisely the strings in $L(R)$ can be obtained by several classic methods [14,8,21]. In particular, Thompson [21] gave the simple well-known construction in Fig. 1. We will call an automaton constructed with these rules a *Thompson NFA* (TNFA). Fig. 2 shows the TNFA for the regular expression $R = \texttt{a}\cdot(\texttt{a}^*)\cdot(\texttt{b}|\texttt{c})$, along with lots of other information to be discussed in later sections.

A TNFA $N(R)$ for R has at most $2m$ states, at most $4m$ transitions, and can be computed in $O(m)$ time. With a breadth-first search of A we can compute a state-set transition for a single character in $O(m)$ time. Hence, we can test acceptance of a string Q of length n in $O(nm)$ time. This is Thompson's algorithm [21].

[1] Sometimes NFAs are allowed a *set* of accepting states, but this is not necessary for our purposes.

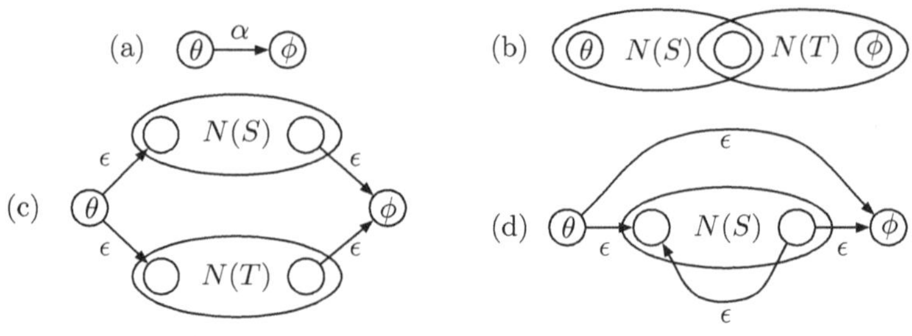

Fig. 1. Thompson's recursive NFA construction. The regular expression for a character $\alpha \in \Sigma$ corresponds to NFA (a). If S and T are regular expressions then $N(ST)$, $N(S|T)$, and $N(S^*)$ correspond to NFAs (b), (c), and (d), respectively. In each of these figures, the leftmost node θ and rightmost node ϕ are the start and the accept nodes, respectively. For the top recursive calls, these are the start and accept nodes of the overall automaton. In the recursions indicated, e.g., for $N(ST)$ in (b), we take the start node of the subautomaton $N(S)$ and identify with the state immediately to the left of $N(S)$ in (b). Similarly the accept node of $N(S)$ is identified with the state immediately to the right of $N(S)$ in (b).

3 1-Dimensional Speed-Up

Myers' algorithm [16] and later variants [3, 4] all aim to speed-up Thompson's algorithm by improving the $O(m)$ bound for a state-set transition for a single character. We describe this approach in some detail below as we are going to reuse much of it our own algorithm. The version we present is for bounded size alphabets, i.e., each character can be coded with a constant number of bits.

For a desired speed-up of $x < w$, we will need some global tables of size $2^{O(x)}$. First we *decompose* $N(R)$ into a tree AS of $O(\lceil m/x \rceil)$ micro TNFAs, each of size at most x. In each $A \in AS$, each child TNFA C is represented by a start and accepting state and a *pseudo-transition* labeled $\beta \notin \Sigma$ connecting these. For example, Fig. 2, shows the TNFAs for the regular expression $R = \mathtt{a}\cdot(\mathtt{a}^*)\cdot(\mathtt{b}|\mathtt{c})$ divided into 3 TNFAs $AS = \{A_1, A_2, A_3\}$, where $A_1 = N(\mathtt{a}^*)$, $A_2 = N(\mathtt{b}|\mathtt{c})$, and $A_3 = N((\mathtt{a} \cdot \beta \cdot \beta)^*)$. Here the βs represent A_1 and A_2, respectively. We can always construct a decomposition of $N(R)$ as described above since we can partition the parse tree into subtrees using standard techniques and build the decomposition from the TNFAs induced by the subtrees, see e.g., Myers [16] for details.

The point in such a micro TNFA A is that we can code it uniquely via its parse tree using $O(x)$ bits. We shall use "A" to denote the bit coding of A. If $x = o(\log m)$, there will be many micro TNFAs with the same bit code. Technically A is an index to all information associated with A including both the code "A" and the children and parents of A. Since A has only x states, we can also code its local state set S_A using x bits. Thus we can make a universal table T of size $2^{O(x)}$ that for every possible micro TNFA A of size $\leq x$, local state set S_A, and $\alpha \in \Sigma \cup \{\epsilon\}$, provides $T[\text{“}A\text{”}, S_A, \alpha] = \delta_A(S_A, \alpha)$ in constant time. Note that parents and children share some local states, and these states will have to be copies between their local state bitmaps.

Recall that our target is to compute a state-set transition $\delta_{N(R)}(S, \alpha)$ for a single character $\alpha \in \Sigma$. First we should traverse transitions labeled α from S and then traverse paths of ε-transitions. The challenge here is that paths of ε-transition may lead to distant TNFAs in the decomposition resulting in non-trivial dependencies between TNFAs. For example, suppose we start with state set $S = \{1\}$ in Fig. 2, and that we get the input character a. First we follow the only relevant a-transition in A_3 to state 2. Next we follow ε-transitions in A_1, leading us to states 3 and 5, and from there we follow ε-transitions in A_2 to states 6 and 8. We end up concluding that $\delta_{N(R)}(\{1\}, a) = \{2, 3, 5, 6, 8\}$.

Following the character transitions of α is in itself easy using our global table T. For every micro TNFA independently, we take the current local state set S_A and compute $S'_A = T[\text{"}A\text{"}, S_A, a] = \delta_A(S_A, \alpha)$. To handle the subsequent ε-transitions, Myers prove that any cycle-free path of ε-transition in a TNFA uses at most one of the back transitions we get from the star operators. This implies that we can compute the ε-closure in two depth-first traversals of the decomposition. Each of these traversals starts at the root, and are defined recursively for subtrees. When the traversal visits a micro TNFA A, it first sets $S'_A = T[A, S'_A, \varepsilon]$. Next, considering the children in order, one C at the time, it copies the accept state θ_C from S'_A to S'_C, perform an depth-first traversal down from C, and copy the accept state ϕ_C from S'_C to S_A. Finally, it sets $S'_A = T[A, S'_A, \varepsilon]$, and continue to the next child if any. If there were no back transitions, we would need only one such depth-first traversal. The total time we spend on a micro TNFA A is a constant plus a constant per child, adding up to $O(|AS|) = O(\lceil m/x \rceil)$ total time.

The above algorithm assumes that we have built the global table T in $2^{O(x)}$ time and space. It also requires that we for a new regular expression R first construct the decomposition AS, and for each micro TNFA $A \in AS$ find the bit code "A", but all this takes linear time. Thus, in total we can perform regular expression matching in time $O(nm/x + n + m)$.

To get a better time bound, we have to deal with an input segment q of super-constant length, yet we may only use constant time per micro TNFA. One basic problem is that we have to consider matching paths that leave a micro TNFA and returns as many times as there are characters in q, making a constant number of depth-first traversal sound elusive. Also, each state can correspond to multiple positions in q, but generally, we can only use a constant number of bits per state.

4 2-Dimensional Speed-Up

We will now show how to extend the 1-dimensional speed-up algorithm from the previous section to handle an input segment of length $y = \sqrt{x}$ within the same $O(\lceil m/x \rceil)$ time bound that we used before on a single character. We are going to use some different global tables, but their total size will still be $2^{O(x)}$.

We will need to impose the restriction on the decomposition that each micro TNFA has at most two children. We get this if we decompose the parse tree using the tree partition technique that Frederickson [6] used for topology trees.

(a) $R = \mathtt{a}\cdot(\mathtt{a}^*)\cdot(\mathtt{b}\,|\,\mathtt{c})$
$A_1 = N(\mathtt{a}^*)$
$A_2 = N(\mathtt{b}\,|\,\mathtt{c})$
$A_3 = N(\mathtt{a}\cdot\beta\cdot\beta)$

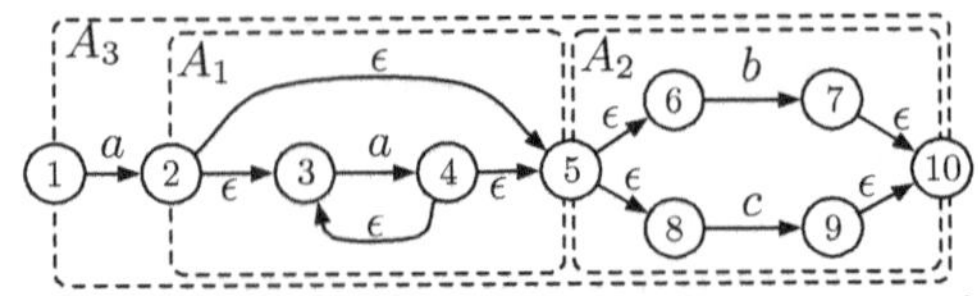

$q = \mathtt{aab}$
$S = \{1, 3\}$

(b) $\mathrm{SS}_{A_1} = \{(1,0),(1,1),(1,2),(2,2)\}$ $\mathrm{SS}_{A_2} = \{(3,3)\}$ $\mathrm{SS}_{A_3} = \{(1,3),(2,3)\}$
$A_1(q) = N(\epsilon|\mathtt{a}|\mathtt{aa})$ $A_2(q) = N(\mathtt{b})$ $A_3(q) = N(\mathtt{a}\cdot(\epsilon|\mathtt{a}|\mathtt{aa})\cdot\mathtt{b})$

(c) $\mathrm{PP}^{\phi}_{A_1} = \{1,2\}$ $\mathrm{PP}^{\phi}_{A_2} = \emptyset$ $\mathrm{PP}^{\phi}_{A_3} = \{3\}$
$\mathrm{PP}^{\theta}_{A_1} = \{1\}$ $\mathrm{PP}^{\theta}_{A_2} = \{1,2\}$ $\mathrm{PP}^{\theta}_{A_3} = \{0\}$
$S'_{A_1} = \emptyset$ $S'_{A_2} = \{7,10\}$ $S'_{A_3} = \{10\}$

Fig. 2. (a) TNFA $N(R)$ for regular expression $R = \mathtt{a}\cdot(\mathtt{a}^*)\cdot(\mathtt{b}\,|\,\mathtt{c})$. $N(R)$ is decomposed into 3 TNFAs $AS = \{A_1, A_2, A_3\}$, where $A_1 = N(\mathtt{a}^*)$, $A_2 = N(\mathtt{b}\,|\,\mathtt{c})$, and $A_3 = N((\mathtt{a}\cdot\beta\cdot\beta)^*)$. Here the βs represent A_1 and A_2, respectively. The root of AS is A_3 and A_1 and A_2 are children of A_3. (b) The accepted substrings of $q = \mathtt{aab}$ (the index $(1,0)$ indicates the empty string) and $A_1(q)$, $A_2(q)$, and $A_3(q)$. (c) The state-set transition computation for each phase on q on the state-set $S = \{1,3\}$.

As before, we get a decomposition tree AS of $O(\lceil m/x \rceil)$ micro TNFAs, each of size at most x, and now of with at most two children.

As useful new definitions, for any $A \in AS$, define $\overline{A}$ to be the TNFA induced by all states in A and descendants of A in the decomposition, and for any state-set S define $S|\overline{A}$ to be the restriction of S to states in $\overline{A}$.

Recall that one of our challenges we have is that a path p matching q may go in and out of the same micro TNFA many times. We are going to shortcut any downwards loop, that is, a segment s of p that leaves a micro TNFA A to go to a child C and later returns from C to A. When all downwards loops have been shortcut, we are left with a path p with a first part going up the decomposition tree, and a second part going down the decomposition tree. Either part may be empty. The basic point here is that we can follow all such shortcut paths if we do a bottom-up traversal followed by a top-down traversal.

The special thing about the downwards loop s going from A to a child C and later returning is that it matches an interval $q[i,j]$ of q that is accepted by $\overline{C}$. To shortcut the loop, we augment A with the information about all substrings accepted by $\overline{C}$, and do that for each of the at most 2 children of A. Even though $\overline{C}$ may be very large, there are only $\binom{|q|}{2} + |q| + 1 = \binom{|q|+1}{2} = O(y^2) = O(x)$ possible intervals of q.

In Sec. 4.1 we first show how to compute the augmented micro TNFAs and in Sec. 4.2 we show how to use this information to compute state-set transitions. We put the pieces together in Sec. 4.3 to get the full algorithm for regular expression matching.

4.1 Computing Accepted Substrings

In a single bottom-up traversal of the decomposition we construct for each $A \in AS$ the set SS_A of substrings of q accepted by $\overline{A}$. These are represented as pairs of indices (i,j), that is, $(i,j) \in \mathrm{SS}_A$ iff $\overline{A}$ accepts $q[i,j]$. To compute SS_A, we first construct a "local" representation $A(q)$ of $\overline{A}$ such that for any states s, s' in A, we have

$$P_{A(q)}(s, s') = P_{\overline{A}}(s, s').$$

We construct $A(q)$ from A and the set SS_C from each child C of A if any. More precisely, we replace the pseudo-edge for child C with an NFA accepting the set of substrings of q indexed by SS_C. Having constructed $A(q)$, we compute the set SS_A from $A(q)$ as

$$\mathrm{SS}_A := \{(i,j) \mid q[i,j] \text{ is accepted by } A(q)\}.$$

We use the pair $(1,0)$ as a unique representation of the empty substring. In our example in Fig. 2, A_1 accepts $q[1,0] = \epsilon$, $q[1,1] = \mathtt{a}$, $q[1,2] = \mathtt{aa}$, and $q[2,2] = \mathtt{a}$ and A_2 accepts $q[3,3] = \mathtt{b}$. Therefore $\mathrm{SS}_{A_1} = \{(1,0),(1,1),(1,2),(2,2)\}$ and $\mathrm{SS}_{A_2} = \{(3,3)\}$. Thus $A_1(q) = N(\epsilon|\mathtt{a}|\mathtt{aa})$ and $A_2(q) = N(\mathtt{b})$. Inserting $A_1(q)$ and $A_2(q)$ in A_3 we get $A_3(q) = N(\mathtt{a} \cdot (\epsilon|\mathtt{a}|\mathtt{aa}) \cdot \mathtt{b})$, accepting $q[1,3] = \mathtt{aab}$ and $q[2,3] = \mathtt{ab}$, hence $\mathrm{SS}_{A_3} = \{(1,3),(2,3)\}$.

Encoding and Tabulation. We encode q as a bit string "q" of length $O(y)$ and SS_A as a bit string "SS_A" with a bit for each possible interval of q, hence of length $\binom{|q|}{2} + |q| + 1 = \binom{|q|+1}{2} = O(y^2) = O(x)$. Our encoding "$A(q)$" of $A(q)$ consists of the bit-coding "A" of A from Sec. 3 followed by "q" and then the interval end-points in "SS_C" for each child C of A, that is,

$$\text{“}A(q)\text{”} = (\text{“}A\text{”}, \text{“}q\text{”}, \{\text{“}\mathrm{SS}_C\text{”} \mid C \text{ child of } A\}).$$

Thus "$A(q)$" is represented with $O(x + y + y^2) = O(x)$ bits. We precompute a global table SS providing

$$\text{“}\mathrm{SS}_A\text{”} = \mathrm{SS}[\text{“}A(q)\text{”}]$$

as a constant time look-up. The entries take $O(x)$ bits, so the table can be constructed in $2^{O(x)}$ time and space. Subsequently, for any substring q of length at most y, we can compute "SS_A" and "$A(q)$" for every $A \in AS$ in a bottom-up traversal in $O(\lceil m/x \rceil)$ total time.

4.2 Computing State-Set Transitions

We now show how to compute state-set transitions using the "SS_A" and "$A(q)$" computed above. We divide the algorithm into 3 phases. The first phase computes the set of prefixes of q that match a path from $S|\overline{A}$ to ϕ_A within $\overline{A}$, the second phase computes the set of prefixes of q that match a path from S to ϕ_A in $N(R)$, and finally, the third phase computes the local $S'_A = \delta_{N(R)}(S,q))|A$.

Phase 1: Computing path prefixes in $\overline{A}$ to ϕ_A. The first phase is a bottom-up traversal. For each $A \in AS$, it computes the set PP_A^{ϕ} of prefixes of q matched by a path in $\overline{A}$ from $S|\overline{A}$ to the accept state ϕ_A of A. The prefixes are represented via their end indices, so $j \in \mathrm{PP}^{\phi}_{A}$ iff $q[1,j] \in P_{\overline{A}}(S|\overline{A}, \phi_A)$. The set PP^{ϕ}_{A} is constructed from $A(q)$ and the sets PP^{ϕ}_{C} from the children C of A:

$$\mathrm{PP}^{\phi}_{A} := \{j \mid q[1,j] \in P_{A(q)}(S_A, \phi_A)\} \cup \bigcup_{C \text{ child of } A} \{j \mid i \in \mathrm{PP}^{\phi}_{C} \text{ and } q[i+1,j] \in P_{A(q)}(\phi_C, \phi_A)\}$$

In Fig. 2 we can reach ϕ_{A_1} from state 3 in $A_1(q)$ using $q[1,1]$ or $q[1,2]$ and therefore $\mathrm{PP}^{\phi}_{A_1} = \{1,2\}$. There are no states from S in A_2 so $\mathrm{PP}^{\phi}_{A_2} = \emptyset$. Finally, we can reach ϕ_{A_3} both from state 1 and 3 parsing $q[1,3]$ so $\mathrm{PP}^{\phi}_{A_3} = \{3\}$.

Phase 2: Computing path prefixes in $N(R)$ to θ_A. The second phase is a top-down traversal of the decomposition. For each $A \in AS$ we compute the set PP^{θ}_{A} of prefixes of q matched by a path in all of $N(R)$ from S to the start state θ_A of A. The prefixes are represented via their end indices, so $j \in \mathrm{PP}^{\phi}_{A}$ iff $q[1,j] \in P_N(S, \theta_A)$.

If A is the root automaton, we trivially have $\mathrm{PP}^{\theta}_{A} = \{0\}$ if $\theta_A \in S$ and otherwise $\mathrm{PP}^{\theta}_{A} = \emptyset$. Hence we may assume that A has a parent B, and we will use PP^{θ}_{B} to compute PP^{θ}_{A}. We also use $B(q)$ and PP^{ϕ}_{C} from phase 1 for each child C of B, including $C = A$. The computation is now done as

$$\begin{aligned}\mathrm{PP}^{\theta}_{A} :=& \{j \mid q[1,j] \in P_{B(q)}(S_B, \theta_A)\} \\ &\cup \{j \mid i \in \mathrm{PP}^{\theta}_{B} \text{ and } q[i+1,j] \in P_{B(q)}(\theta_B, \theta_A)\} \\ &\cup \bigcup_{C \text{ child of } B} \{j \mid i \in \mathrm{PP}^{\phi}_{C} \text{ and } q[i+1,j] \in P_{B(q)}(\phi_C, \theta_A)\}\end{aligned}$$

In Fig. 2 we have that $\mathrm{PP}^{\theta}_{A_3} = \{0\}$. We can reach θ_{A_1} with $q[1,1]$ from state 1 in $A_3(q)$ and hence $\mathrm{PP}^{\theta}_{A_1} = \{1\}$. We can reach θ_{A_2} with $q[1,1]$ and $q[1,2]$ in $A_3(q)$ from state 1 and hence $1,2 \in \mathrm{PP}^{\theta}_{A_2}$. From $\mathrm{PP}^{\phi}_{A_1} = \{1,2\}$ we also get that $1,2 \in \mathrm{PP}^{\theta}_{A_2}$. Hence, $\mathrm{PP}^{\theta}_{A_2} = \{1,2\}$.

Phase 3: Updating state-sets. The third and final phase traverses the decomposition in any order. For each $A \in AS$, we compute the $S'_A = \delta_{N(R)}(S,q)|A$. Then the desired state set transition $S' = \delta_{N(R)}(S,q)$ is just the union of these sets. Recall that $y = |q|$.

The set S' is now computed based on $A(q)$, PP^{ϕ}_{C} from phase 1 for each child C of A, and PP^{θ}_{A} from phase 2:

$$S'_A := \{s' \mid q \in P_{A(q)}(S_A, s')\} \cup \{s' \mid i \in \mathrm{PP}^{\theta}_{A} \text{ and } q[i+1,y] \in P_{A(q)}(\theta_A, s')\} \cup \bigcup_{C \text{ child of } A} \{s' \mid i \in \mathrm{PP}^{\phi}_{C} \text{ and } q[i+1,y] \in P_{A(q)}(\phi_C, s')\}$$

In Fig. 2 we have that $S'_{A_1} = \emptyset$, $S'_{A_2} = \{7,10\}$, and $S'_{A_3} = \{10\}$, so $S' = \delta_{N(R)}(S,q) = \{7,10\}$.

Encoding and Tabulation. We encode each of the sets PP_A^ϕ and PP_A^θ as bit strings "PP_A^ϕ" and "PP_A^θ" of length $y = \sqrt{x}$ where the jth bit is set iff the index j is in the set. The output set S'_A is represented as the input set S_A with a local bit string "S'_A" of length x.

For phase 1, we have a table PP^ϕ so that we can set

$$\text{“}\mathrm{PP}_A^\phi\text{”} := \mathrm{PP}^\phi[\text{“}A(q)\text{”}, \text{“}S_A\text{”}, \{\text{“}\mathrm{PP}_C^\phi\text{”} \mid C \text{ child of } A\}].$$

For phase 2, we have a table PP^θ so that for a child A of B, we can set

$$\text{“}\mathrm{PP}_A^\theta\text{”} := \mathrm{PP}^\theta[\text{“}B(q)\text{”}, \text{“}S_B\text{”}, \text{“}\mathrm{PP}_B^\theta\text{”}, \{\text{“}\mathrm{PP}_C^\phi\text{”} \mid C \text{ child of } B\}].$$

Finally, for phase 3, we have a table S', so that we can set

$$\text{“}S'_A\text{”} := S'[\text{“}A(q)\text{”}, \text{“}S_A\text{”}, \text{“}\mathrm{PP}_A^\theta\text{”}, \{\text{“}\mathrm{PP}_C^\phi\text{”} \mid C \text{ child of } B\}].$$

The entries in each table use $O(x)$ bits and hence we can construct the tables in $2^{O(x)}$ time and space. It follows that given any state-set S and input segment q of length at most y, we can compute $\delta_{N(R)}(S, q)$ in $O(\lceil m/x \rceil)$ total time.

Note that the construction only depends on the fixed alphabet Σ, and the parameters x and y, so the tables may be reused for any regular expression matching problem over the same alphabet.

4.3 The Algorithm

Let $t < 2^w$ be a bound on the space devoted to tables. Choosing $x = y^2 = \varepsilon \log t$ we construct and build all tables in $O(t)$ time and space. Given a regular expression R of length m and a string Q of length n, we solve the regular matching problem using $\lceil n/y \rceil$ state-set transitions computations as described above. To process a new input segment, we only need the global tables, the $O(m)$ space decomposition of R, and the state-set resulting from the preceeding input. Hence, the full algorithm uses space $O(t + m)$ and time $O(m + (m/x + y) \cdot n/y) = O(n + m + nm/\log^{3/2} t)$. Note that if Q is represented as a stream, we are simply processing substrings of length y, one at the time in a single pass, and hence our algorithm also works in this context.

References

1. Aho, A.V., Sethi, R., Ullman, J.D.: Compilers: principles, techniques, and tools. Addison-Wesley Longman Publishing Co., Inc., Boston (1986)
2. Arlazarov, V.L., Dinic, E.A., Kronrod, M.A., Faradzev, I.A.: On economic construction of the transitive closure of a directed graph. Dokl. Acad. Nauk. 194, 487–488 (1970)
3. Bille, P.: New algorithms for regular expression matching. In: Bugliesi, M., Preneel, B., Sassone, V., Wegener, I. (eds.) ICALP 2006. LNCS, vol. 4051, pp. 643–654. Springer, Heidelberg (2006)

4. Bille, P., Farach-Colton, M.: Fast and compact regular expression matching. Theoret. Comput. Sci. 409, 486–496 (2008)
5. Chan, T.M.: More algorithms for all-pairs shortest paths in weighted graphs. In: Proc. 39th STOC, pp. 590–598 (2007)
6. Frederickson, G.N.: Ambivalent data structures for dynamic 2-edge-connectivity and k smallest spanning trees. SIAM J. Comput. 26(2), 484–538 (1997); announced at FOCS 1991
7. Galil, Z.: Open problems in stringology. In: Apostolico, A., Galil, Z. (eds.) Combinatorial problems on words. NATO ASI Series, vol. F12, pp. 1–8 (1985)
8. Glushkov, V.M.: The abstract theory of automata. Russian Math. Surveys 16(5), 1–53 (1961)
9. Johnson, T., Muthukrishnan, S., Rozenbaum, I.: Monitoring regular expressions on out-of-order streams. In: Proc. 23rd ICDE, pp. 1315–1319 (2007)
10. Kernighan, B., Ritchie, D.: The C Programming Language, 2nd edn. Prentice-Hall, Englewood Cliffs (1988) (1st edn., 1978)
11. Kleene, S.C.: Representation of events in nerve nets and finite automata. Automata Studies, Ann. Math. Stud. 34, 3–41 (1956)
12. Li, Q., Moon, B.: Indexing and querying XML data for regular path expressions. In: Proc. 27th VLDB, pp. 361–370 (2001)
13. Masek, W., Paterson, M.: A faster algorithm for computing string edit distances. J. Comput. System Sci. 20, 18–31 (1980)
14. McNaughton, R., Yamada, H.: Regular expressions and state graphs for automata. IRE Trans. on Electronic Computers 9(1), 39–47 (1960)
15. Murata, M.: Extended path expressions of XML. In: Proc. 20th PODS, pp. 126–137 (2001)
16. Myers, E.W.: A four-russian algorithm for regular expression pattern matching. J. ACM 39(2), 430–448 (1992)
17. Navarro, G.: NR-grep: a fast and flexible pattern-matching tool. Softw. Pract. Exper. 31(13), 1265–1312 (2001)
18. Navarro, G., Raffinot, M.: Fast and simple character classes and bounded gaps pattern matching, with applications to protein searching. J. Comp. Biology 10(6), 903–923 (2003)
19. Navarro, G., Raffinot, M.: New techniques for regular expression searching. Algorithmica 41(2), 89–116 (2004)
20. Stroustrup, B.: The C++ Programming Language: Special Edition, 3rd edn. Addison-Wesley, Reading (2000) (1st. edn., 1985)
21. Thompson, K.: Regular expression search algorithm. Comm. ACM 11, 419–422 (1968)
22. Wall, L.: The Perl Programming Language. Prentice Hall Software Series (1994)
23. Wu, S., Manber, U.: Agrep – a fast approximate pattern-matching tool. In: Proc. USENIX, pp. 153–162 (1992)
24. Yu, F., Chen, Z., Diao, Y., Lakshman, T.V., Katz, R.H.: Fast and memory-efficient regular expression matching for deep packet inspection. In: Proc. ANCS, pp. 93–102 (2006)

A Fast and Simple Parallel Algorithm for the Monotone Duality Problem

Endre Boros[1] and Kazuhisa Makino[2]

[1] RUTCOR, Rutgers University, 640 Bartholomew Road, Piscataway NJ 08854-8003
boros@rutcor.rutgers.edu
[2] Department of Mathematical Informatics, University of Tokyo, Tokyo, 113-8656, Japan
makino@mist.i.u-tokyo.ac.jp

Abstract. We consider the monotone duality problem i.e., checking whether given monotone CNF φ and DNF ψ are equivalent, which is a prominent open problem in NP-completeness. We construct a fast and simple parallel algorithms for the problem, that run in polylogarithmic time by using quasi-polynomially many processors. The algorithm exhibits better parallel time complexity of the existing algorithms of Elbassioni [11]. By using a different threshold of the degree parameter ϵ of φ in the algorithm, we also present a stronger bound on the number of processors for polylogarithmic-time parallel computation and improves over the previously best known bound on the sequential time complexity of the problem in the case when the magnitudes of $|\varphi|$, $|\psi|$ and n are different, e.g., $|\psi| = |\varphi|^\alpha \gg n$ for $\alpha > 1$, where n denotes the number of variables. Furthermore, we show that, for several interesting well-known classes of monotone CNFs φ such as bounded degree, clause-size, and intersection-size, our parallel algorithm runs polylogarithmic time by using polynomially many processors.

1 Introduction

Let $V = \{1, 2, \ldots, n\}$ and let $f : \{0,1\}^V \to \{0,1\}$ be a Boolean function. A function is called *monotone* (also called *positive*) if for every pair of vectors $x, y \in \{0,1\}^V$, $x \leq y$ (i.e., $x_i \leq y_i$ for all i) always implies $f(x) \leq f(y)$. Any monotone function f has unique prime *conjunctive normal form* (*CNF*) and *disjunctive normal form* (*DNF*) expressions

$$\varphi(x) = \bigwedge_{H \in \mathcal{H}} \Big(\bigvee_{i \in H} x_i \Big) \tag{1}$$

$$\psi(x) = \bigvee_{G \in \mathcal{G}} \Big(\bigwedge_{i \in G} x_i \Big), \tag{2}$$

where $\mathcal{H}, \mathcal{G} \, (\subseteq 2^V)$ are both *Sperner* (i.e., $I \not\subseteq J$ and $I \not\supseteq J$ holds for any two distinct sets I and J in the family). We note that $\mathcal{H}$ and $\mathcal{G}$ respectively correspond to the sets of all *prime implicates* and *prime implicants* of f.

S. Albers et al. (Eds.): ICALP 2009, Part I, LNCS 5555, pp. 183–194, 2009.

The well-known *monotone duality problem* is to decide whether $\varphi \equiv \psi$ for given prime monotone CNF φ and DNF ψ. Equivalently, the problem is to check if for two given Sperner hypergraphs $\mathcal{H}$ and $\mathcal{G}$, $\mathcal{G}$ is the transversal hypergraph $\mathcal{H}^d$, consisting of all minimal transversals G of $\mathcal{H}$ (i.e., all subsets $G \subseteq V$ such that $G \cap H \neq \emptyset$ for all $H \in \mathcal{H}$). This problem has received considerable attention in the literature (see e.g., [1,7,9,27,29]), since it is known to be polynomially or quasi-polynomially equivalent with many problems in diverse areas, such as artificial intelligence (e.g., [7,20]), database theory (e.g., [28]), distributed systems (e.g., [14,17]), machine learning and data mining (e.g., [3,16,25]), and mathematical programming (e.g., [2]), matroid theory (e.g., [22,21]). However, it is open (for more than 25 years now, e.g., [1,8,18,26,27,29]) whether monotone duality problem is solvable in polynomial time or not. Any polynomial time algorithm for the monotone duality problem would significantly advance the state of the art of the problems in the application areas mentioned above. This is witnessed by the fact that these problems are cited in a rapidly growing body of literature and have been referenced in various survey papers and complexity theory retrospectives, e.g. [8,10,18,27,28,29].

In 1996, Fredman and Khachiyan [13] established a remarkable result that the monotone duality problem can be solved in quasi-polynomial time $O(nm) + m^{O(\log m/\log\log m)}$, where $m = |\varphi| + |\psi|$, thus putting the problem somewhere between polynomiality and NP-completeness. We note that the monotone duality problem is solvable in (quasi-)polynomial time if and only if the monotone dualization problem, i.e., to compute from a prime CNF φ of a monotone function f a prime DNF ψ of f, can be solved in output (quasi-)polynomial time [1], and thus providing an efficient algorithm for the decision problem is equivalent to providing the one for the computation problem. After Fredman-Khachiyan result [13], several quasi-polynomial algorithms were proposed [12,15,30], but it was open whether the monotone duality problem admits an efficient parallel computation, i.e., it is computable in polylogarithmic time by using quasi-polynomially many processors. In 2008, Elbassioni [11] proved this by constructing two algorithms: (i) $O(\log n + \chi(m,2)\log^2 m)$ time using $O(nm^{4\chi(m,2)+1})$ processors and (ii) $O(\log n + \chi(|\mathcal{G}|, 3(\ln n + 1))\log|\mathcal{H}|\log|\mathcal{G}|/\log n)$ time using $O(n|\mathcal{G}|^2|\mathcal{H}|^{\chi(\mathcal{G},3(\ln n+1))})$ processors,[1] which are regarded as generalizations of the sequential Fredman-Khachiyan algorithm. Here $\mathcal{H}$ and $\mathcal{G}$ are given in (1) and (2), and for two positive reals a, b, $\chi(a,b)$ denotes the unique positive root of the equation $\left(\frac{\chi(a,b)}{b}\right)^{\chi(a,b)} = a$.

We further note that, the monotone duality problem is known to be solvable in polynomial time, even NC solvable (i.e., solvable in polylogarithmic time using polynomially many processors) for several special classes of monotone CNFs, e.g., when every clause has bounded-size [4,7,19], when every variable has bounded degree [5,9,24], when clauses have bounded intersection-size [23], for read-once formulae, etc.

---

[1] We note that Theorem 1 in [11] only analyzes the number of nodes and depth of the parallel computational tree.

Our Results. We provide a fast and simple parallel algorithm for the monotone duality problem that runs in $O(\log n + \log|\mathcal{H}|\log|\mathcal{G}|)$ time by using $O(n|\mathcal{H}||\mathcal{G}|^{1+\lfloor\log|\mathcal{H}|\rfloor})$ many processors. This exhibits better parallel time complexity than the existing two algorithms of Elbassioni [11].

Our algorithm recursively decomposes the monotone duality problem into polynomially many smaller problems which together are equivalent with the original problem. The decomposition scheme makes use of *full covers* of φ based on the degree information of φ, where Elbassioni [11] also used a particular kind of full covers for the monotone duality problem. In order to have good full covers for the decompositions, we apply a classical bottleneck minmax theorem by Edmonds and Fulkerson [6].

We also show that, by using a different threshold of the degree parameter ϵ of φ, the monotone duality problem can be solved in $O\Big(\log n + \chi\big(|\mathcal{G}|, \frac{|\mathcal{H}|}{|\mathcal{H}|-1}(\ln n + 1)\big)\log|\mathcal{H}|\log|\mathcal{G}|/\log n\Big)$ time by using $O\Big(n|\mathcal{G}||\mathcal{H}|^{\chi\left(|\mathcal{G}|, \frac{|\mathcal{H}|}{|\mathcal{H}|-1}(\ln n+1)\right)+1}\Big)$ many processors. This yields a stronger bound on the number of processors for polylogarithmic-time parallel algorithm for the monotone duality problem, when the magnitudes of $|\mathcal{G}|$, $|\mathcal{H}|$ and n are different, e.g., $|\mathcal{G}| = |\mathcal{H}|^\alpha \gg n$ for $\alpha > 1$. We note that the sequential implementation of this algorithm improves over the previously best known bound on the sequential time complexity of the problem in the case when the magnitudes of $|\mathcal{G}|$, $|\mathcal{H}|$ and n are different.

Furthermore, we show that, for several interesting well-known classes of monotone CNF formulae such as bounded degree, clause-size, and intersection-size, our parallel algorithm solves the monotone duality problem in NC (i.e., in polylogarithmic time by using polynomially many processors).

Organization. The rest of the paper is organized as follows. In the next section, we fix our notations and show some basic properties on monotone duality. In Section 3, we define full covers and present our decomposition rule. In Section 4, we present our algorithm and analyze the complexity for a general input, and Section 5 discusses the time complexity of our algorithm when the input CNF has some natural property such as bounded clause-size, bounded degree, bounded clause-intersections, or read-once.

Due to space constraints, some proofs are omitted.

2 Notations and Basic Properties

We denote by V the set of variables (or their indices), and represent a monotone CNF by the hypergraph $\mathcal{A} \subseteq 2^V$ of its clauses. We let $n = |V|$. A hypergraph $\mathcal{A}$ is called *Sperner* if $A \not\subseteq A'$ and $A \not\supseteq A'$ hold for all $A, A' \in \mathcal{A}$ with $A \neq A'$. We call a subset $T \subseteq V$ a *transversal* of $\mathcal{A}$ if $T \cap A \neq \emptyset$ for all $A \in \mathcal{A}$, and denote by $\mathcal{A}^d$ the family of all minimal transversals of $\mathcal{A}$. It is easy to see that $\mathcal{A}^d$ is Sperner for any $\mathcal{A} \subseteq 2^V$, and $\mathcal{A}^d = \mathcal{B}$ if and only if $\mathcal{A} = \mathcal{B}^d$ for any Sperner hypergraphs $\mathcal{A}, \mathcal{B} \subseteq 2^V$. As is well-known, the monotone duality problem is the problem of deciding whether $\mathcal{B} = \mathcal{A}^d$ for two given Sperner hypergraphs $\mathcal{A}, \mathcal{B} \subseteq 2^V$.

For a subset $S \subseteq V$, we denote by $S^c = V \setminus S$ its complementary set, and for a family $\mathcal{A} \subseteq 2^V$, we denote by $\mathcal{A}^c$ the family of complementary sets, i.e., $\mathcal{A}^c = \{V \setminus A \mid A \in \mathcal{A}\}$. For $\mathcal{A} \subseteq 2^V$, we define the upward closure of $\mathcal{A}$ by $\mathcal{A}^+ = \{X \subseteq V \mid \exists\, A \in \mathcal{A} \textit{ s.t. } A \subseteq X\}$, and the downward closure of $\mathcal{A}$ by $\mathcal{A}^- = \{X \subseteq V \mid \exists\, A \in \mathcal{A} \textit{ s.t. } X \subseteq A\}$. Let us note that $(\mathcal{A}^d)^+ = \mathcal{A}^{d+}$ is the family of all transversals of $\mathcal{A}$, while $\mathcal{A}^{dc-}$ is the family of all *independent sets* of $\mathcal{A}$, i.e., all subsets $X \subseteq V$ for which $X \not\supseteq A$ for all $A \in \mathcal{A}$, and $\mathcal{A}^{dc}$ is the family of all *maximal independent sets* of $\mathcal{A}$.

Given a hypergraph $\mathcal{A} \subseteq 2^V$ and a subfamily $\mathcal{B}$ of its transversals, $\mathcal{B} \subseteq \mathcal{A}^{d+}$, we say that a subset $X \subseteq V$ is a *new transversal* of $\mathcal{A}$ (with respect to $\mathcal{B}$) if $X \in \mathcal{A}^{d+} \cap \mathcal{B}^{dc-}$, i.e.,

$$X \cap A \neq \emptyset \text{ for all } A \in \mathcal{A} \quad \text{and} \quad X \not\supseteq B \text{ for all } B \in \mathcal{B}. \tag{3}$$

Clearly, if $\mathcal{B} = \mathcal{A}^d$, then we cannot have any new transversal of $\mathcal{A}$ with respect to $\mathcal{B}$.

For a hypergraph $\mathcal{A} \subseteq 2^V$ and a subset $S \subseteq V$, let us define the contraction and deletion of S^c for $\mathcal{A}$ by

$$\mathcal{A}^S = \{A \cap S \mid A \in \mathcal{A}\} \quad \text{and} \quad \mathcal{A}_S = \{A \in \mathcal{A} \mid A \subseteq S\},$$

respectively. By definition, if $\mathcal{A}$ is Sperner, then so is $\mathcal{A}_S$, while $\mathcal{A}^S$ is not Sperner in general. Let us also note that for subsets $S, Q \subseteq V$ we have $(\mathcal{A}^S)^Q = \mathcal{A}^{S \cap Q}$ and $(\mathcal{A}_S)_Q = \mathcal{A}_{S \cap Q}$.

The following statement is a useful and important property of these subfamilies.

Lemma 1. *For all subsets $S \subseteq V$ we have $(\mathcal{A}^S)^d = (\mathcal{A}^d)_S$ $(\subseteq \mathcal{A}^d)$.*

3 Full Covers and Decomposition Rules

Given a hypergraph $\mathcal{A} \subseteq 2^V$, we say that a family $\mathcal{C} \subseteq 2^V$ forms a *full cover* of the dual $\mathcal{A}^d$ if

$$\mathcal{A}^d = \bigcup_{C \in \mathcal{C}} (\mathcal{A}^d)_C. \tag{4}$$

For example, $\{V\}$ and $\mathcal{A}^d$ are clearly full covers of $\mathcal{A}^d$. Here we describe an interesting characterization of full covers of $\mathcal{A}^d$ in terms of operations c, d and $+$.

Lemma 2. *Given $\mathcal{A} \subseteq 2^V$, a family $\mathcal{C} \subseteq 2^V$ is a full cover of $\mathcal{A}^d$ if and only if*

$$\mathcal{C}^{cd} \subseteq \mathcal{A}^{dcd+}.$$

The following lemma shows that full covers $\mathcal{C}$ can be used to decompose the monotone duality problem into $|\mathcal{C}|$ many ones.

Lemma 3. *Let $\mathcal{A} \subseteq 2^V$. Then $\mathcal{C} \subseteq 2^V$ is a full cover of $\mathcal{A}^d$ if and only if for any $\mathcal{B} \subseteq \mathcal{A}^d$, the following statement holds.*

$$\mathcal{B} = \mathcal{A}^d \qquad \textit{if and only if} \qquad \mathcal{B}_C = (\mathcal{A}^C)^d \quad \textit{for all} \quad C \in \mathcal{C}. \tag{5}$$

Proof. Let us assume that $\mathcal{C}$ is a full cover of $\mathcal{A}^d$. If $\mathcal{B} = \mathcal{A}^d$, then we must have $(\mathcal{A}^C)^d = \mathcal{B}_C$ for all $C \subseteq V$ by Lemma 1. Assume for the reverse direction that there exists a minimal transversal $T \in \mathcal{A}^d \setminus \mathcal{B}$. Since $\mathcal{C}$ is a full cover of $\mathcal{A}^d$, by (4) we have that there must exists a set $C \in \mathcal{C}$ such that $T \in (\mathcal{A}^C)^d \setminus \mathcal{B}_C$, implying that such a C satisfies $(\mathcal{A}^C)^d \neq \mathcal{B}_C$.

We next assume that $\mathcal{C}$ is not a full cover of $\mathcal{A}^d$, i.e., there exists a $T \in \mathcal{A}^d \setminus (\bigcup_{C \in \mathcal{C}} (\mathcal{A}^d)_C)$. Let $\mathcal{B} = \mathcal{A}^d \setminus \{T\}$. Then it is clear that $\mathcal{B} \neq \mathcal{A}^d$, and moreover we have $\mathcal{B}_C = (\mathcal{A}^C)^d$ for all $C \in \mathcal{C}$, since $\mathcal{B}_C = (\mathcal{A}^d)_C$ holds for all $C \in \mathcal{C}$. □

Let us now show two methods to construct full covers.

Lemma 4. *For a hypergraph $\mathcal{A} \subseteq 2^V$ and a hyperedge $A_0 \in \mathcal{A}$, the family*

$$\mathcal{C}(A_0) = \{V \setminus (A \setminus \{i\}) \mid A \in \mathcal{A} \setminus \mathcal{A}_{V \setminus A_0} \text{ and } i \in A \cap A_0\} \tag{6}$$

forms a full cover of $\mathcal{A}^d$.

Lemma 5. *For a hypergraph $\mathcal{A} \subseteq 2^V$ and a minimal transversal $B_0 \in \mathcal{A}^d$, the family*

$$\mathcal{C}(B_0) = \{V \setminus \{i\} \mid i \in B_0\} \cup \{B_0\} \tag{7}$$

forms a full cover of $\mathcal{A}^d$.

We remark that full covers of Lemma 5 are special types of full covers used in [11].

We shall below show that Lemmas 4 and 5 are enough to construct a good full cover for parallel computation of the monotone duality problem.

Lemma 6. *Given $\mathcal{A} \subseteq 2^V$, $\mathcal{B} \subseteq \mathcal{A}^{d+}$, and $I \subseteq V$, we have exactly one of the following three cases.*

(i) *there exists a $A_0 \in \mathcal{A}$ such that $A_0 \cap I = \emptyset$,*
(ii) *there exists a $B_0 \in \mathcal{B}$ such that $B_0 \subseteq I$,*
(iii) *I is a new transversal of $\mathcal{A}$ with respect to $\mathcal{B}$ and its complement $V \setminus I$ is a new transversal of $\mathcal{B}$ with respect to $\mathcal{A}$.*

Proof. Immediate from the definitions. □

Since (i), (ii), and (iii) in the above lemma are mutually exclusive, if $\mathcal{B} = \mathcal{A}^d$, then only (i) or (ii) is possible.

For a hypergraph $\mathcal{B} \subseteq 2^V$ and a real ϵ with $0 \leq \epsilon \leq 1$, let us define

$$I(\mathcal{B}, \epsilon) = \{i \in V \mid deg_{\mathcal{B}}(i) > \epsilon|\mathcal{B}|\}, \tag{8}$$

where $deg_{\mathcal{B}}(i)$ denotes the number of hyperedges in $\mathcal{B}$ that contains vertex i.

Lemma 7. *Given $\mathcal{A} \subseteq 2^V$, $\mathcal{B} \subseteq \mathcal{A}^{d+}$, and ϵ with $0 \leq \epsilon \leq 1$, let us assume that $A_0 \in \mathcal{A}$ satisfies condition* (i) *of Lemma* 6 *for $I = I(\mathcal{B}, \epsilon)$. Then we have*

$$|\mathcal{B}_C| \leq \epsilon|\mathcal{B}| \quad \text{for all} \quad C \in \mathcal{C}(A_0). \tag{9}$$

Lemma 8. *Given $\mathcal{A} \subseteq 2^V$, $\mathcal{B} \subseteq \mathcal{A}^{d+}$, and ϵ with $0 \leq \epsilon \leq 1 - \frac{1}{|\mathcal{B}|}$, let us assume that a minimal transversal $B_0 \in \mathcal{B}$ of $\mathcal{A}$ satisfies condition* (ii) *of Lemma* 6 *for $I = I(\mathcal{B}, \epsilon)$. Then we have*

$$|\mathcal{B}_C| \ \leq \ (1-\epsilon)|\mathcal{B}| \qquad \text{for all} \qquad C \in \mathcal{C}(B_0). \tag{10}$$

Lemma 9. *Let $\mathcal{A} \subseteq 2^V$ and $\mathcal{B} \subseteq \mathcal{A}^d$ with $|\mathcal{B}| \geq 2$. Then either a new transversal I of $\mathcal{A}$ with respect to $\mathcal{B}$ (i.e., a new transversal $V \setminus I$ of $\mathcal{B}$ with respect to $\mathcal{A}$) or a full cover $\mathcal{C}$ of $\mathcal{A}^d$ such that $|\mathcal{B}_C| \leq 1/2|\mathcal{B}|$ for all $C \in \mathcal{C}$ can be computed in $O(\log n + \log|\mathcal{A}| + \log|\mathcal{B}|))$ time with $O(n(|\mathcal{A}| + |\mathcal{B}|))$ processors.*

Before concluding this section, let us remark that Lemma 6 with $I = I(\mathcal{B}, \epsilon)$ can be regarded as a corollary of a classical bottleneck minmax theorem by Edmonds and Fulkerson [6]:

Theorem 1 (Edmonds and Fulkerson, 1970). *For Sperner families $\mathcal{A}, \mathcal{B} \subseteq 2^V$, we have $\mathcal{B} = \mathcal{A}^d$ if and only if*

$$\min_{A \in \mathcal{A}} \max_{i \in A} w(i) \ = \ \max_{B \in \mathcal{B}} \min_{i \in B} w(i),$$

holds for any $w \in \mathbb{R}^V$.

Namely, if $\mathcal{B} = \mathcal{A}^d$ and $\min_{A \in \mathcal{A}} \max_{i \in A} deg_{\mathcal{B}}(i) (= \max_{B \in \mathcal{B}} \min_{i \in B} deg_{\mathcal{B}}(i)) \leq \epsilon|\mathcal{B}|$, then Theorem 1 implies that Lemma 6 (i) holds, but (ii) does not. On the other hand if $\mathcal{B} = \mathcal{A}^d$ and $\min_{A \in \mathcal{A}} \max_{i \in A} deg_{\mathcal{B}}(i) (= \max_{B \in \mathcal{B}} \min_{i \in B} deg_{\mathcal{B}}(i)) > \epsilon|\mathcal{B}|$, then Theorem 1 implies that Lemma 6 (ii) holds, but (i) does not.

4 An Algorithm Based on Lemma 9

By Lemma 9, the monotone duality problem can be decomposed into polynomially many smaller problems which together are equivalent with the original problem (by Lemma 3). This immediately implies that the monotone duality problem can be solved in polylogarithmic time with quasi-polynomially many processors. In this section, we give a more careful analysis for our algorithm based on Lemma 9.

For given Sperner families $\mathcal{H}, \mathcal{G} \subseteq 2^V$, let $n_{\mathcal{H}} = |\bigcup_{H \in \mathcal{H}} H|$ and $n_{\mathcal{G}} = |\bigcup_{G \in \mathcal{G}} G|$. Then we first assume that

$$n_{\mathcal{H}} = n_{\mathcal{G}} \ \leq \ |\mathcal{H}||\mathcal{G}|, \tag{11}$$

where we assume that V contains no redundant element, i.e., $n = n_{\mathcal{H}} (= n_{\mathcal{G}})$. For, if $n_{\mathcal{H}} < n_{\mathcal{G}}$, some hyperedge $G \in \mathcal{G}$ contains an element $i \in V \setminus (\bigcup_{H \in \mathcal{H}} H)$. This means that G is not a minimal transversal of $\mathcal{H}$, since the transversality of G implies that of $G \setminus \{i\}$. Therefore, this G is a witness of nonduality of $\mathcal{H}$ and $\mathcal{G}$. Similarly, we can prove the case of $n_{\mathcal{H}} > n_{\mathcal{G}}$. Moreover, for any minimal transversal G of $\mathcal{H}$ and $i \in G$, there exists an $H \in \mathcal{H}$ such that $H \cap G = \{i\}$, which implies $|G| \leq |\mathcal{H}|$ and hence we have $n_{\mathcal{G}} \leq |\mathcal{H}||\mathcal{G}|$.

We also assume that

$$\mathcal{H} \subseteq \mathcal{G}^d \quad \text{and} \quad \mathcal{G} \subseteq \mathcal{H}^d. \tag{12}$$

By Assumption (12), we can interchange the roles of $\mathcal{H}$ and $\mathcal{G}$, when applying some of the above lemmas. We assume without loss of generality that $|\mathcal{H}| \leq |\mathcal{G}|$ and make use of the lemmas with $\mathcal{A} = \mathcal{G}$ and $\mathcal{B} = \mathcal{H}$. Note that Assumptions (11) and (12) can be checked (if the input does not satisfies at least one of the assumptions, we can find an witness of the nonduality) in $O(\log n + \log|\mathcal{H}| + \log|\mathcal{G}|)$ time by using $O(n|\mathcal{H}||\mathcal{G}|)$ many processors.

Our algorithm recursively builds decomposition tree for this input. Every node of this tree will correspond to the pair of hypergraphs $(\mathcal{G}^{S_\alpha}, \mathcal{H}_{S_\alpha})$, and we say that this node is labeled by $S_\alpha \subseteq V$, where for convenience, we index the subsets by integer sequences $\alpha \in \aleph = \mathbb{Z}^0 \cup \mathbb{Z} \cup \mathbb{Z}^2 \cup \cdots \cup \mathbb{Z}^{\lfloor \log|\mathcal{H}| \rfloor}$ (Here $\mathbb{Z}^0$ stands for the sequence $\emptyset$). We denote by $|\alpha|$ the length of α, and we say that the node labeled by S_α is at depth $|\alpha|$.

We start building our decomposition tree with a root, labeled by $S_\emptyset = V$, and we denote at any stage by $\Lambda \subseteq \aleph$ the subset of indices which correspond to the leaves of our decomposition tree. We maintain the following two properties throughout the procedure.

(I) $\mathcal{S} = \{S_\alpha \mid \alpha \in \Lambda\}$ is a full cover of $\mathcal{G}^d$, i.e., $\mathcal{G}^d = \bigcup_{\alpha \in \Lambda} (\mathcal{G}^d)_{S_\alpha}$.

(II) $|\mathcal{H}_{S_\alpha}| \leq \frac{1}{2^{|\alpha|}} |\mathcal{H}|$ for all $\alpha \in \Lambda$.

Clearly, these properties hold initially, when $\Lambda = \{\emptyset\}$ since $\mathcal{G}^d = (\mathcal{G}^d)_V$, and $|\emptyset| = 0$. In a general step of the algorithm for every $\alpha \in \Lambda$ for which $|\mathcal{H}_{S_\alpha}| \geq 2$, we apply Lemma 6 with $V = S_\alpha$, $\mathcal{A} = \mathcal{G}^{S_\alpha}$, $\mathcal{B} = \mathcal{H}_{S_\alpha}$ and $I = I(\mathcal{B}, \frac{1}{2})$. If case (iii) occurs, then we output "NO" and halt. If case (i) occurs, then we consider a full cover $\mathcal{C}$ of $\mathcal{H}_{S_\alpha}$ given by (6), while if case (ii) occurs, then we consider a full cover $\mathcal{C}$ of $\mathcal{H}_{S_\alpha}$ given by (7). In either case, we denote by $k(\alpha) = |\mathcal{C}|$ the size of this full cover, and index its elements as $\mathcal{C} = \{C_1, C_2, ..., C_{k(\alpha)}\}$. We then create and add to our decomposition tree $k(\alpha)$ new leaves, as children of the node labeled by S_α. We label these new leaves respectively by $S_{(\alpha,j)} = C_j$ for $j = 1, \ldots, k(\alpha)$. Furthermore, we update $\Lambda = (\Lambda \setminus \{\alpha\}) \cup \{(\alpha, j) \mid j = 1, \ldots, k(\alpha)\}$, and $\mathcal{S} = (\mathcal{S} \setminus \{S_\alpha\}) \cup \{S_{(\alpha,j)} \mid j = 1, \ldots, k(\alpha)\}$. Note that in case of (i) we have $k(\alpha) \leq n$, while in case of (ii) we have $k(\alpha) \leq |S_\alpha||\mathcal{G}^{S_\alpha}| \leq n|\mathcal{G}|$. Furthermore, by Lemmas 7 and 8 we have $|\mathcal{H}_{S_{\alpha,j}}| \leq \frac{1}{2}|\mathcal{H}_{S_\alpha}|$ for all $j = 1, \ldots, k(\alpha)$. These inequalities, and Lemmas 4 and 5 guarantee that properties (I) and (II) remain valid.

For indices $\alpha \in \Lambda$ with $|\mathcal{H}_{S_\alpha}| \leq 1$ we proceed as follows:

Case 1. If $|\mathcal{H}_{S_\alpha}| = 0$ and $\emptyset \notin \mathcal{G}^{S_\alpha}$, then we can output "NO" and halt, since $V \setminus S_\alpha$ is a new transversal of $\mathcal{H}$ with respect to $\mathcal{G}$.

Case 2. If $|\mathcal{H}_{S_\alpha}| = 0$ and $\emptyset \in \mathcal{G}^{S_\alpha}$, then we have $\mathcal{H}^d_{S_\alpha} = \mathcal{G}^{S_\alpha}$, we can mark this leaf as "DONE," and update $\Lambda := \Lambda \setminus \{\alpha\}$ and $\mathcal{S} := \mathcal{S} \setminus \{S_\alpha\}$.

Case 3. If $\mathcal{H}_{S_\alpha} = \{H\}$, by Assumption (12), we have $\emptyset \notin \mathcal{G}^{S_\alpha}$ and for every $i \in H$, there exists a subset $G \in \mathcal{G}^{S_\alpha}$ such that $G \cap H = \{i\}$. If $\{\{i\} \mid i \in H\} \subseteq \mathcal{G}^{S_\alpha}$, then we have $(\mathcal{H}_{S_\alpha})^d = \mathcal{G}^{S_\alpha}$, we can mark this leaf as "DONE," and update

$\Lambda := \Lambda \setminus \{\alpha\}$ and $\mathcal{S} := \mathcal{S} \setminus \{S_\alpha\}$. Otherwise, we can output "NO" and halt, since the set $S_\alpha \cup \{i\}$ is a new transversal of $\mathcal{H}$ with respect to $\mathcal{G}$ for every element $i \in H$ with $\{i\} \notin \mathcal{G}^{S_\alpha}$.

Finally, if we never have "NO" and if all leaves are labeled "DONE," then we have a proof, by Lemma 3 that $\mathcal{H}^d = \mathcal{G}$, and hence we output "YES" and halt.

To see the complexity of the above algorithm, let us first note that by Lemma 9 we have $|\alpha| \leq \lfloor \log |\mathcal{H}| \rfloor$ for all $\alpha \in \Lambda$ throughout the algorithm. Thus the decomposition tree we built has depth at most $\lfloor \log |\mathcal{H}| \rfloor$. Furthermore, each node has at most n children (by decomposition (7)) or $n|\mathcal{G}|$ children (by decomposition (6)), and thus the total number of nodes in the tree is $O\left((n|\mathcal{G}|)^{\lfloor \log |\mathcal{H}| \rfloor}\right)$.

Let us note next that we can slightly improve on this bound. Namely, when we decompose a node by a hyperedge $H_0 \in \mathcal{H}$, as in Lemma 5, then each hyperedge $H \in \mathcal{H}$ will belong to at most $|H_0 \setminus H| \leq |\mathcal{G}|$ children, since $H_0 \in \mathcal{G}^d$ implies $|H_0| \leq |\mathcal{G}|$. Furthermore, when we decompose a node labeled by S_α by a minimal transversal $G_0 \in \mathcal{G}$, as in Lemma 4, then each hyperedge $H \in \mathcal{H}$ will belong to at most $|\mathcal{G}^{S_\alpha}| \leq |\mathcal{G}|$ children of this node, because $H \cap G = \{i\}$, $i \in G_0$ can happen at most $|\mathcal{G}|$ times. Thus, the total number of copies of a hyperedge $H \in \mathcal{H}$ in the decomposition tree we built is limited by $O\left(|\mathcal{G}|^{\lfloor \log |\mathcal{H}| \rfloor}\right)$. Since every non-leaf of the complete decomposition tree contains at most one copy of each $H \in \mathcal{H}$, we get that the total number of nodes is $O\left(|\mathcal{H}|(|\mathcal{G}|)^{\lfloor \log |\mathcal{H}| \rfloor}\right)$.

Let us finally note that for each S_α, either applying Lemma 6 if $|\mathcal{H}_{S_\alpha}| \geq 2$, or processing it as a terminal leaf if $|\mathcal{H}_{S_\alpha}| \leq 1$, can be implemented in $O(\log |\mathcal{G}|)$ time by using $O(n|\mathcal{G}|)$ processors, since we have $|\mathcal{H}| \leq |\mathcal{G}|$ and $n \leq |\mathcal{H}||\mathcal{G}|$ by our assumption and initialization.

Therefore we get for the total time τ of our parallel algorithm that

$$\tau = O(\log n + \log |\mathcal{H}| \log |\mathcal{G}|),$$

where $O(\log n)$ is required by the initialization, and that the total number of processors we need is $O\left(n|\mathcal{H}||\mathcal{G}|^{\lfloor \log |\mathcal{H}| \rfloor + 1}\right)$. This proves the following theorem.

Theorem 2. *Given two Sperner families $\mathcal{H}, \mathcal{G} \subseteq 2^V$, we can check whether $\mathcal{G} = \mathcal{H}^d$ in $O(\log n + \log |\mathcal{H}| \log |\mathcal{G}|)$ time by using $O\left(n|\mathcal{H}||\mathcal{G}|^{\lfloor \log |\mathcal{H}| \rfloor + 1}\right)$ many processors.*

Let us recall the complexity of two existing parallel algorithms for the monotone dualization problem [11]: (i) $O(\log n + \chi(m, 2) \log^2 m)$ time by using $O(nm^{4\chi(m,2)+1})$ processors and (ii) $O(\log n + \chi(|\mathcal{G}|, 3(\ln n + 1)) \log |\mathcal{H}| \log |\mathcal{G}| / \log n)$ time by using $O(n|\mathcal{G}|^2 |\mathcal{H}|^{\chi(\mathcal{G}, 3(\ln n+1))})$ processors,[2] where $m = |\mathcal{H}| + |\mathcal{G}|$, and for two positive reals a, b, $\chi(a, b)$ denotes the unique positive root of the equation $\left(\frac{\chi(a,b)}{b}\right)^{\chi(a,b)} = a$. We note that our algorithm is faster than these algorithms, since $\chi(|\mathcal{G}|, 3(\ln n+1)) > 3(\ln n+1)$. But our algorithm uses $O\left(n|\mathcal{H}||\mathcal{G}|^{\lfloor \log |\mathcal{H}| \rfloor + 1}\right)$ many processors. To reduce the number of the processors in our algorithm, let us fix ϵ as

[2] We note that Theorem 1 in [11] only analyzes the number of nodes and depth of the parallel computational tree.

$$\epsilon = \frac{\ln n + 1}{\chi\left(|\mathcal{G}|, \frac{|\mathcal{H}|}{|\mathcal{H}|-1}(\ln n + 1)\right)}.$$

Then we have the following theorem.

Theorem 3. *Given two Sperner families $\mathcal{H}, \mathcal{G} \subseteq 2^V$, we can check whether $\mathcal{G} = \mathcal{H}^d$ in $O\Big(\log n + \chi(|\mathcal{G}|, \frac{|\mathcal{H}|}{|\mathcal{H}|-1}(\ln n + 1)) \log|\mathcal{H}| \log|\mathcal{G}|/\log n\Big)$ time by using $O\Big(n|\mathcal{G}||\mathcal{H}|^{\chi\left(|\mathcal{G}|, \frac{|\mathcal{H}|}{|\mathcal{H}|-1}(\ln n+1)\right)+1}\Big)$ many processors.*

Note that this algorithm is slower than the original one, but this yields a stronger bound on the number of processors for polylogarithmic-time parallel algorithms for the monotone duality problem, when the magnitudes of $|\mathcal{G}|$, $|\mathcal{H}|$ and n are different, e.g., $|\mathcal{G}| = |\mathcal{H}|^\alpha \gg n$ for $\alpha > 1$.

We also note that the sequential implementation of this algorithm improves over the previously best known bound on the sequential time complexity of the problem in the case when the magnitudes of $|\mathcal{G}|$, $|\mathcal{H}|$ and n are different.

Corollary 1. *Given two Sperner families $\mathcal{H}, \mathcal{G} \subseteq 2^V$, we can check whether $\mathcal{G} = \mathcal{H}^d$ in $O\Big(n|\mathcal{G}||\mathcal{H}|^{\chi\left(|\mathcal{G}|, \frac{|\mathcal{H}|}{|\mathcal{H}|-1}(\ln n+1)\right)+1}\Big)$ time.*

5 Polynomially Solvable Cases

In this section, we show that the algorithm described in the previous section is in fact efficient for several natural classes of hypergraphs.

5.1 Bounded Degree and Dimension

For a hypergraph $\mathcal{H} \subseteq 2^V$, let us denote by

$$\Delta(\mathcal{H}) = \max_{i \in V} deg_{\mathcal{H}}(i)$$

the maximum degree of $\mathcal{H}$. For a given k, we say that $\mathcal{H}$ has *bounded degree* if $\Delta(\mathcal{H}) \leq k$.

Theorem 4. *Given two Sperner families $\mathcal{H}, \mathcal{G} \subseteq 2^V$, if $\mathcal{H}$ has bounded degree, i.e., $\Delta(\mathcal{H}) \leq k$ for a fixed constant k, then Algorithm* DualityTest *can check if $\mathcal{G} = \mathcal{H}^d$ in $O(\log n + \log k \max\{\log|\mathcal{H}|, \log|\mathcal{G}|\})$ time by using $O\left(n \max\{|\mathcal{H}|, |\mathcal{G}|\}|\mathcal{H}||\mathcal{G}|^{\lfloor \log k \rfloor + 1}\right)$ many processors.*

Given $\mathcal{H} \subseteq 2^V$, we call

$$D_1(\mathcal{H}) = \max_{H \in \mathcal{H}} |H| \tag{13}$$

the *dimension* of $\mathcal{H}$, and we say that $\mathcal{H}$ is of *bounded dimension*, if $D_1(\mathcal{H}) \leq k$ for a given constant k.

Theorem 5. *Given two Sperner families $\mathcal{H}, \mathcal{G} \subseteq 2^V$, if $\mathcal{H}$ has bounded dimension, i.e., $D_1(\mathcal{H}) \leq k$ for a fixed constant k, then Algorithm* DualityTest *can check if $\mathcal{G} = \mathcal{H}^d$ in $O(\log n + \log|\mathcal{H}| \max\{\log|\mathcal{H}|, \log|\mathcal{G}|\})$ time by using $O(n \max\{|\mathcal{H}|, |\mathcal{G}|\}|\mathcal{H}|^{\log k+1}|\mathcal{G}|^k)$ many processors.*

5.2 Bounded (p, r)-Intersections

Given a hypergraph $\mathcal{H}$, let $D_2(\mathcal{H})$ denote the *intersection size* of $\mathcal{H}$, i.e.,

$$D_2(\mathcal{H}) = \max_{\substack{H,H'\in \mathcal{H}:\\ H\neq H'}} |H \cap H'|.$$

For a given constant k, we say that $\mathcal{H}$ has *bounded intersections* if $D_2(\mathcal{H}) \leq k$.

Note that Algorithm DUALITYTEST cannot show the NC solvability of the monotone duality problem for hypergraphs $\mathcal{H}$ with bounded intersections. In this section, we show that by using different thresholds ϵ, the problem for hypergraphs $\mathcal{H}$ with bounded intersections is NC solvable. More generally, we consider the following class of hypergraphs.

Let $p \geq 1$ and $r \geq 0$ be integers. We denote by $\mathbb{A}(p, r)$ the class of of hypergraphs with *(p, r)-bounded intersections* [23]: $\mathcal{H} \in \mathbb{A}(p, r)$ if for any p distinct hyperedges $H_{j_1}, \ldots, H_{j_p}$ in $\mathcal{H}$, we have $|\bigcap_{\ell=1}^{p} H_{j_\ell}| \leq r$. Note that $\Delta(\mathcal{H}) \leq k$ iff $\mathcal{H} \in \mathbb{A}(k+1, 0)$, $D_1(\mathcal{H}) \leq k$ iff $\mathcal{H} \in \mathbb{A}(1, k)$, and $D_2(\mathcal{H}) \leq k$ iff $\mathcal{H} \in \mathbb{A}(2, k)$, and hence the class $\mathbb{A}(p, r)$ contains hypergraphs of bounded degree, dimension, and intersections.

Let us modify Algorithm DUALITYTEST as follows.

We fix ϵ by $\epsilon := 1 - 1/|\mathcal{H}|$ if $|\alpha| \leq r$, and $\epsilon := 1/2$, otherwise.

Theorem 6. *Given two Sperner families $\mathcal{H}, \mathcal{G} \subseteq 2^V$, if $\mathcal{H}$ is of bounded (p, r)-intersections, then the modified version of Algorithm* DUALITYTEST *can check if $\mathcal{G} = \mathcal{H}^d$ in $O(\log n + \max\{\log|\mathcal{G}|, \log|\mathcal{H}|\}(r + \log p))$ time by using $O(n \max\{|\mathcal{H}|, |\mathcal{G}|\}|\mathcal{H}||\mathcal{G}|^{r+\log p+1})$ many processors.*

5.3 Read-Once Expressions

A Boolean formula φ is called *read-once* if it is an $\wedge - \vee$ formula in which every variable appears at most once. In this section, we consider those hypergraphs $\mathcal{H}$ such that the corresponding CNFs can be represented by read-once formulae. A well-known equivalent definition of $\mathcal{H}$ is that $\varphi = \bigwedge_{H\in\mathcal{H}}\left(\bigvee_{i\in H} x_i\right)$ can be represented by a read-once expression if and only if

$$|\mathcal{H} \cap \mathcal{G}| = 1 \quad \text{for every } H \in \mathcal{H} \text{ and } G \in \mathcal{H}^d. \tag{14}$$

Note that $\mathcal{H}$ is read-once if and only if is $\mathcal{H}^d$ is read-once.

If we know that $\mathcal{A}$ is read-once, then we can simplify Lemma 4 as follows.

Lemma 10. *For a hypergraph $\mathcal{A} \subseteq 2^V$ and a hyperedge $A_0 \in \mathcal{A}$, the family*

$$\mathcal{C}(A_0) \ = \ \{V \setminus (A_0 \setminus \{i\}) \mid i \in A_0\} \tag{15}$$

forms a full cover of $\mathcal{A}^d$, if $\mathcal{A}$ is read-once.

Note that this full cover constructs a partition of $\mathcal{B}$, i.e., $\mathcal{B} = \cup_{C\in\mathcal{C}(A_0)}\mathcal{B}_C$ and $\mathcal{B}_C \cap \mathcal{B}_{C'} = \emptyset$ for $C, C' \in \mathcal{C}(A_0)$ with $C \neq C'$. Therefore, if we fix ϵ by $\epsilon := 1 - 1/|\mathcal{H}|$, we have the following theorem.

Theorem 7. *Given two Sperner families $\mathcal{H}, \mathcal{G} \subseteq 2^V$, if $\mathcal{H}$ is read-once, then the monotone duality problem can be solved in $O(n|\mathcal{H}||\mathcal{G}|)$ time.*

Acknowledgement. The authors thank Y. Crama, V. Gurvich and K. Elbassioni for helpful discussions.

References

1. Bioch, J.C., Ibaraki, T.: Complexity of identification and dualization of positive Boolean functions. Information and Computation 123, 50–63 (1995)
2. Boros, E., Elbassioni, K., Gurvich, V., Khachiyan, L., Makino, K.: Dual-bounded generating problems: All minimal integer solutions for a monotone system of linear inequalities. SIAM J. Comput. 31(5), 1624–1643 (2002)
3. Boros, E., Gurvich, V., Khachiyan, L., Makino, K.: On maximal frequent and minimal infrequent sets in binary matrices. Ann. Math. Artif. Intell. 39, 211–221 (2003)
4. Dahlhaus, E., Karpinski, M.: A fast parallel algorithm for computing all maximal cliques in a graph and the related problems. In: Karlsson, R., Lingas, A. (eds.) SWAT 1988. LNCS, vol. 318, pp. 139–144. Springer, Heidelberg (1988)
5. Domingo, C., Mishra, N., Pitt, L.: Efficient read-restricted monotone CNF/DNF dualization by learning with membership queries. Machine learning 37, 89–110 (1999)
6. Edmonds, J., Fulkerson, D.R.: Bottleneck extrema. J. Combinatorial Theory 8, 299–306 (1970)
7. Eiter, T., Gottlob, G.: Identifying the minimal transversals of a hypergraph and related problems. SIAM J. Comput. 24, 1278–1304 (1995)
8. Eiter, T., Gottlob, G.: Hypergraph transversal computation and related problems in logic and AI. In: Flesca, S., Greco, S., Leone, N., Ianni, G. (eds.) JELIA 2002. LNCS, vol. 2424, pp. 549–564. Springer, Heidelberg (2002)
9. Eiter, T., Gottlob, G., Makino, K.: New results on monotone dualization and generating hypergraph transversals. SIAM J. Comput. 32, 514–537 (2003)
10. Eiter, T., Makino, K., Gottlob, G.: Computational aspects of monotone dualization: A brief survey. Discrete Applied Mathematics 156, 2035–2049 (2008)
11. Elbassioni, K.: On the complexity of monotone dualization and generating minimal hypergraph transversals. Discrete Applied Mathematics 156, 2109–2123 (2008)
12. Elbassioni, K.: On the complexity of the multiplication method for monotone CNF/DNF dualization. In: Azar, Y., Erlebach, T. (eds.) ESA 2006. LNCS, vol. 4168, pp. 340–351. Springer, Heidelberg (2006)
13. Fredman, M.L., Khachiyan, L.: On the complexity of dualization of monotone disjunctive normal forms. J. Algorithms 21, 618–628 (1996)
14. Garcia-Molina, H., Barbara, D.: How to assign votes in a distributed system. Journal of the ACM 32, 841–860 (1985)
15. Gaur, D., Krishnamurti, R.: Average case self-duality of monotone Boolean functions. In: Proceedings of the 17th Conference of the Canadian Society for Computational Studies of Intelligence on Advances in Artificial Intelligence, pp. 322–338 (2004)
16. Gunopulos, D., Khardon, R., Mannila, H., Toivonen, H.: Data mining, hypergraph transversals and machine learning. In: PODS 1997, pp. 12–15 (1997)

17. Ibaraki, T., Kameda, T.: A theory of coteries: Mutual exclusion in distributed systems. IEEE Transactions on Parallel and Distributed Systems 4, 779–794 (1993)
18. Johnson, D.S.: Open and closed problems in NP-completeness. Lecture given at the International School of Mathematics "G. Stampacchia": Summer School "NP-Completeness: The First 20 Years", Erice, Sicily, Italy, June 20-27 (1991)
19. Johnson, D.S., Yannakakis, M., Papadimitriou, C.H.: On generating all maximal independent sets. Info. Process. Lett. 27, 119–123 (1988)
20. Kavvadias, D.J., Papadimitriou, C., Sideri, M.: On Horn envelopes and hypergraph transversals. In: Ng, K.W., Balasubramanian, N.V., Raghavan, P., Chin, F.Y.L. (eds.) ISAAC 1993. LNCS, vol. 762, pp. 399–405. Springer, Heidelberg (1993)
21. Khachiyan, L., Boros, E., Borys, K., Elbassioni, K., Gurvich, V., Makino, K.: Enumerating spanning and connected subsets in graphs and matroids. In: Azar, Y., Erlebach, T. (eds.) ESA 2006. LNCS, vol. 4168, pp. 444–455. Springer, Heidelberg (2006)
22. Khachiyan, L., Boros, E., Elbassioni, K., Gurvich, V., Makino, K.: On the complexity of some enumeration problems for matroids. SIAM J. Discrete Math. 19, 966–984 (2005)
23. Khachiyan, L., Boros, E., Elbassioni, K., Gurvich, V.: On the dualization of hypergraphs with bounded edge-intersections and other related classes of hypergraphs. Theoretical Computer Science 382, 139–150 (2007)
24. Khachiyan, L., Boros, E., Elbassioni, K., Gurvich, V.: Computing many maximal independent sets for hypergraphs in parallel. Parallel Processing Letters 17, 141–152 (2007)
25. Khachiyan, L., Boros, E., Elbassioni, K., Gurvich, V., Makino, K.: Dual-bounded generating problems: Efficient and inefficient points for discrete probability distributions and sparse boxes for multidimensional data. Theor. Comput. Sci. 379, 361–376 (2007)
26. Lawler, E., Lenstra, J.K., Rinnooy Kan, A.H.G.: Generating all maximal independent sets: NP-hardness and polynomial-time algorithms. SIAM J. Comput. 9, 558–565 (1980)
27. Lovász, L.: Combinatorial optimization: some problems and trends, DIMACS Technical Report 92-53, Rutgers University (1992)
28. Mannila, H.: Global and local methods in data mining: basic techniques and open problems. In: Widmayer, P., Triguero, F., Morales, R., Hennessy, M., Eidenbenz, S., Conejo, R. (eds.) ICALP 2002. NCS, vol. 2380, pp. 57–68. Springer, Heidelberg (2002)
29. Papadimitriou, C.: NP-completeness: A retrospective. In: Degano, P., Gorrieri, R., Marchetti-Spaccamela, A. (eds.) ICALP 1997. LNCS, vol. 1256, pp. 2–6. Springer, Heidelberg (1997)
30. Tamaki, H.: Space-efficient enumeration of minimal transversals of a hypergraph. IPSJ-AL 75, 29–36 (2000)

Unconditional Lower Bounds against Advice

Harry Buhrman[1], Lance Fortnow[2], and Rahul Santhanam[3]

[1] CWI, Amsterdam
buhrman@cwi.nl
[2] Northwestern University
fortnow@eecs.northwestern.edu
[3] University of Edinburgh
rsanthan@inf.ed.ac.uk

Abstract. We show several unconditional lower bounds for exponential time classes against polynomial time classes with advice, including:

1. For any constant c, $\mathsf{NEXP} \not\subseteq \mathsf{P}^{\mathsf{NP}[n^c]}/n^c$
2. For any constant c, $\mathsf{MAEXP} \not\subseteq \mathsf{MA}/n^c$
3. $\mathsf{BPEXP} \not\subseteq \mathsf{BPP}/n^{o(1)}$

It was previously unknown even whether $\mathsf{NEXP} \subseteq \mathsf{NP}/n^{0.01}$. For the probabilistic classes, no lower bounds for uniform exponential time against advice were known before.

We also consider the question of whether these lower bounds can be made to work on almost all input lengths rather than on infinitely many. We give an oracle relative to which $\mathsf{NEXP} \subseteq \mathsf{i.o.NP}$, which provides evidence that this is not possible with current techniques.

1 Introduction

Lower bounds are the holy grail of complexity theory. Showing $\mathsf{P} \neq \mathsf{NP}$ or separating BPP from NEXP requires one to establish unconditional super polynomial lower bounds, which are currently beyond our abilities. Most techniques we currently have run into "obstacles" such as relativization [BGS75], natural proofs [RR97] and algebrization [AW08]. Consequently research has focused mainly on conditional results, such as most work in derandomization and PCPs.

In this paper, we show unconditional lower bounds for exponential-time classes such as NEXP, MAEXP, BPEXP and REXP against their polynomial-time versions with advice. We describe our results in more detail in the next subsection.

Lower bounds against advice are closely tied to derandomization. By showing strong enough lower bounds, we hope to get new and interesting derandomization results. Indeed, we illustrate this by showing that our results imply a new separation of a probabilistic class from NEXP.

Unconditional separations are valuable not just in and of themselves, but as components of more involved arguments. A number of major results in complexity theory in recent years have required a hierarchy theorem or unconditional separation to finish the proof [AG94, FLvMV05, IKW02, KI03]. By giving stronger

S. Albers et al. (Eds.): ICALP 2009, Part I, LNCS 5555, pp. 195–209, 2009.

versions of such separations, we hope to derive tighter results. We illustrate this in our paper by deriving an improved result of the form that derandomization implies circuit lower bounds using the machinery of Impagliazzo, Kabanets and Wigderson [IKW02].

Our results for probabilistic classes have an added significance. Probabilistic classes are instances of semantic classes–classes for which there is no recursive enumeration of machines defining languages in the classes and may not have complete sets. We don't understand semantic classes very well, indeed we do not have a strict time hierarchy theorem for any semantic class that is not known to be equal to a syntactic class. Much recent work [FS04, FST05, vMP06] has been focused on showing hierarchies for semantic classes with advice. However, none of the separation results in this line of work hold for uniform semantic classes. By showing separation results for uniform semantic classes here, we extend our understanding of them and are led to pose further questions which might lead to more progress in this area.

How do our techniques face up to the lower bound obstacles mentioned earlier? The natural proofs obstacle does not trouble us because our proofs ultimately rely on diagonalization, which does not naturalize. Diagonalization does relativize, however our proofs of the lower bounds for BPEXP and MAEXP use an indirect diagonalization technique which takes advantage of the non-relativizing PCP machinery [BFL91]. Thus our proofs of these results do not relativize. We suspect that all our proofs do algebrize in the sense of Aaronson and Wigderson [AW08]. This is not necessarily cause for pessimism since the Aaronson-Wigderson results do not rule out the possibility of using non-relativizing results in more than one place in a proof to derive a stronger lower bound, perhaps even one as strong as $\mathsf{NEXP} \not\subseteq \mathsf{SIZE}(poly)$.

1.1 Results

It is straightforward to show that $\mathsf{NEXP} \neq \mathsf{NP}$, either by using the nondeterministic hierarchy theorem or even more simply by a translation argument. This translation argument can be pushed to give $\mathsf{NEXP} \not\subseteq \mathsf{NP}/n^{o(1)}$. However, this is the limit of this translation argument, and it was unknown whether $\mathsf{NEXP} \not\subseteq \mathsf{NP}/n^{\alpha}$ for any fixed constant $\alpha > 0$. We settle this, and in fact prove a much stronger result.

Theorem. For any constant c, $\mathsf{NEXP} \not\subseteq \mathsf{P}^{\mathsf{NP}[n^c]}/n^c$.

Our proof relies on a combination of ideas, including careful use of the facts that NE has a complete set with respect to linear-time reductions and that SAT is NP-complete, together with translation, a census trick and diagonalization.

As a consequence of our lower bound for NEXP, we get that for each c, the class $\mathsf{BPTIME}(n^c)^{\mathsf{NP}}$ is different from NEXP–a previously unknown separation. We also get that even a mild derandomization of MA in NP with n^c bits of advice for some fixed c, or even MA in $\mathsf{P}^{\mathsf{NP}[n^c]}/n^c$, implies super-polynomial circuit lower bounds for NEXP.

For probabilistic classes, translation arguments with advice are not known to work. Furthermore, these classes are not known to have complete languages either. Thus we are forced to use very different techniques to prove analogous results for these classes.

We use indirect arguments which take advantage of advice elimination for carefully chosen complexity classes D. An advice elimination result for D with respect to a class C says that if D is solvable in C with a certain amount of advice, then D is in C uniformly. Among our contributions is a new advice elimination result with respect to MA. Our techniques allow us to derive the following separations.

Theorem. For any constant c, $\mathsf{MAEXP} \not\subseteq \mathsf{MA}/n^c$.

Theorem. $\mathsf{BPEXP} \not\subseteq \mathsf{BPP}/n^{o(1)}$.

The proofs of the two results above do not relativize. We also get a relativizing separation for REXP which is somewhat weaker in terms of the advice lower bound.

Finally, we consider the question of whether our lower bounds can be strengthened by making them hold on almost all input lengths rather than on infinitely many. We give some evidence that this is hard with current techniques by constructing an oracle with respect to which NEXP is infinitely often in NP, even without advice. Note that all known techniques for separating non-deterministic time classes relativize.

Theorem. There is an oracle respect to which $\mathsf{NEXP} \subseteq \mathsf{i.o.NP}$.

The rest of the paper is organized as follows. We give definitions and preliminaries in Section 2, our lower bounds for NEXP and their consequences in Section 3, our lower bounds for probabilistic classes in Section 4, and our oracle results in Section 5. We state a few open questions in Section 6.

2 Preliminaries

2.1 Complexity Classes, Promise Problems and Advice

We assume a basic familiarity with complexity classes such as P, RP, BPP, NP, MA, AM, and their exponential-time versions. The Complexity Zoo (which can be found at `http://qwiki.caltech.edu/wiki/ComplexityZoo`) is an excellent resource for basic definitions and statements of results.

The class $\mathsf{P}^{\mathsf{NP}[q(n)]}$ is the class of languages accepted by polynomial-time oracle machines making at most $q(n)$ queries to an NP oracle on any input of length n.

Given a complexity class C, coC is the class of languages L such that $\bar{L} \in \mathsf{C}$. Given a function $s : \mathbb{N} \to \mathbb{N}$, $\mathsf{SIZE}(s)$ is the class of Boolean functions $f = \{f_n\}$ such that for each n, f_n has Boolean circuits of size $O(s(n))$. Given a language L and an integer n, $L_n = L \cap \{0,1\}^n$. Given a class C, i.o.C is the class of languages

L for which there is a language $L' \in \mathsf{C}$ such that $L_n = L'_n$ for infinitely many length n.

In order to deal with promise classes in a general way, we take as fundamental the notion of a complexity measure. A complexity measure CTIME is a mapping which assigns to each pair (M, x), where M is a time-bounded machine (here a time function $t_M(x)$ is implicit) and x an input, one of three values "0" (accept), "1" (reject) and "?" (failure of CTIME promise). We distinguish between *syntactic* and *semantic* complexity measures. Syntactic measures have as their range $\{0, 1\}$ while semantic measures may map some machine-input pairs to "?". The complexity measures DTIME and NTIME are syntactic (each halting deterministic or non-deterministic machine either accepts or rejects on each input), while complexity measures such as BPTIME and MATIME are semantic (a probabilistic machine may accept on an input with probability 1/2, thus failing the bounded-error promise). For syntactic measures, any halting machine defines a language, while for semantic measures, only a subset of halting machines define languages.

A promise problem is a pair (Y, N), where $Y, N \subseteq \{0, 1\}^*$ and $Y \cap N = \varnothing$. We say that a promise problem (Y, N) belongs to a class $\mathsf{CTIME}(t)$ if there is a machine M halting in time t on all inputs of length n such that M fulfils the CTIME promise on inputs in $Y \cup N$, accepting on inputs in Y and rejecting on inputs in N.

A language L is in $\mathsf{CTIME}(t)/a$ if there is a machine M halting in time $t(\cdot)$ taking an auxiliary *advice* string of length $a(\cdot)$ such that for each n, there is some advice string $b_n, |b_n| = a(n)$ such that M fulfils the CTIME promise for each input x with advice string b_n and accepts x iff $x \in L$.

For syntactic classes, a lower bound with advice or for the promise version of the class translates to a lower bound for the class itself.

3 Lower Bounds for NEXP

It can be shown that $\mathsf{NEXP} \neq \mathsf{NP}$ by using the non-deterministic time hierarchy theorem [Coo72, SFM78, Ž83], or even by a simpler translation argument. In fact, a modification of the translation argument also gives a separation against $n^{o(1)}$ bits of advice.

Proposition 1. $\mathsf{NEXP} \not\subseteq \mathsf{NP}/n^{o(1)}$.

Until now, it's been open whether the lower bound in terms of advice can be pushed to linear or higher. The methods used in the translation argument or in the proof of the non-deterministic hierarchy theorem do not give this. Here, we prove the lower bound by an application of the fact that NE has complete sets with respect to linear-time reductions.

Our method yields somewhat more general results. We state the simpler result with proof first, and then show how to generalize it.

We first need the following lemma (a slightly stronger version of a result in [HM95]) about lower bounds for deterministic exponential time against advice. The proof we give is folklore.

Lemma 1. *For any constant d,* $\mathsf{EXP} \not\subseteq \mathsf{i.o.DTIME}(2^{n^d})/n^d$.

Proof. The proof is by diagonalization. We define a diagonalizing language L which is not in $\mathsf{i.o.DTIME}(2^{n^d})/n^d$ by defining a machine M which runs in exponential time and decides L.

M operates as follows on input x of length n. It enumerates advice taking machines $M_1, M_2 \ldots M_{\log(n)}$ each running in time at most 2^{n^d} and taking advice of length n^d. It then enumerates all $\log(n)2^{n^d}$ truth tables computed by these machines when every possible string of length n^d is given as advice. It then computes a truth table of an n-bit function f which is distinct from all the truth tables enumerated so far–this can be done in exponential time by a simple pruning strategy. Finally it outputs $f(x)$.

Now we are ready to state and prove our lower bound for NEXP.

Theorem 1. *For any constant c,* $\mathsf{NEXP} \not\subseteq \mathsf{NP}/n^c$.

Proof. We will show that either $\mathsf{NEXP} \not\subseteq \mathsf{NP}/poly$ or $\mathsf{NEXP} \not\subseteq \mathsf{NE}/n^c$. From this, the result follows.

Assume, to the contrary, that both these inclusions hold, i.e., $\mathsf{NEXP} \subseteq \mathsf{NP}/poly$ and $\mathsf{NEXP} \subseteq \mathsf{NE}/n^c$. We will derive a contradiction. Let L be a complete language for NE with respect to linear-time reductions. Since $\mathsf{NEXP} \subseteq \mathsf{NP}/poly$, we get that $L \in \mathsf{NTIME}(n^k)/n^k$ for some constant k. Since L is complete for NE with respect to linear-time reductions, we get that $\mathsf{NE} \subseteq \mathsf{NTIME}(n^k)/O(n^k)$.

By translation, we get that $\mathsf{NE}/n^c \subseteq \mathsf{NTIME}(n^{kc})/O(n^{kc})$. To see this, let L' be a language in NE/n^c, and let M' be an advice-taking NE machine accepting L' with advice length n^c. Define a language $L'' \in \mathsf{NE}$ as follows: a string $< x, a >$ is in L'' iff M' accepts x with advice a. Since M' is an NE machine, it follows that $L'' \in \mathsf{NE}$. Thus, by assumption $L'' \in \mathsf{NTIME}(m^k)/O(m^k)$, where m is the input length for L''. Let M'' be an advice-taking machine solving L'' using resources as stated. Now we can solve L' in $\mathsf{NTIME}(n^{kc})/O(n^{kc})$ as follows. The advice-taking machine M we construct for solving L' interprets its advice as consisting of two parts: the first part is an advice string a of length n^c, where n is the input size, and the second part is an advice string b of length $O((n + n^c)^k) = O(n^{kc})$. M simulates M'' on input $< x, a >$ with advice string b, where x is the input for L'. M accepts iff M'' accepts. M operates within time $O(n^{kc})$ (since it simulates an $O(n^k)$ time machine on an input of length $O(n^c)$), uses advice of length $O(n^{kc})$, and decides L' correctly, by definition of L'' and the assumption on M''.

Thus, we have $\mathsf{NEXP} \subseteq \mathsf{NE}/n^c$ and $\mathsf{NE}/n^c \subseteq \mathsf{NTIME}(n^{kc})/O(n^{kc})$, which together imply $\mathsf{NEXP} \subseteq \mathsf{NTIME}(n^{kc})/O(n^{kc})$. But since $\mathsf{EXP} \subseteq \mathsf{NEXP}$ and $\mathsf{NTIME}(n^{kc})/O(n^{kc}) \subseteq \mathsf{DTIME}(2^{n^{kc}})/O(n^{kc})$, we get that $\mathsf{EXP} \subseteq \mathsf{DTIME}(2^{n^{kc}})/O(n^{kc})$, which is a contradiction to Lemma 1.

Note that Theorem 1 is nearly optimal both with respect to the advice, and with respect to the class for which we show a separation, modulo our inability to prove superpolynomial circuit lower bounds for NEXP. If the advice allowed could be increased to an arbitrary polynomial, we would obtain non-trivial derandomizations of MA and AM, which is a long-standing open problem. In terms of proving a separation for a weaker class, if we could separate say NE from NP/n^c for any c, this would also imply a superpolynomial circuit lower bound for NEXP.

We generalize Theorem 1 in two ways. First we observe that an analogous result holds for any syntactic complexity measure CTIME which is stronger than deterministic time and can be simulated by deterministic exponential time.

Theorem 2. *Let* CTIME *be any syntactic complexity measure such that for any constructible* t, $\mathsf{DTIME}(t) \subseteq \mathsf{CTIME}(t) \subseteq \mathsf{DTIME}(2^{O(t)})$. *Then, for any constant* c, $\mathsf{CEXP} \not\subseteq \mathsf{CP}/n^c$.

Proof. The proof goes through exactly like the proof of Theorem 1, since CE has a linear-time complete set, and the same kind of translation argument can be applied as CTIME is syntactic.

Theorem 2 can also be extended to a separation for promise classes satisfying very general properties.

For the specific case of non-deterministic time, the lower bound holds not just against NP with advice, but against $P^{\mathsf{NP}[n^c]}$ with advice, where $\mathsf{P}^{\mathsf{NP}[n^c]}$ is the class of languages accepted by polynomial-time oracle Turing machines making at most n^c queries to an NP oracle. Here c is a fixed constant.

To derive this extension, we'll need to use the following lemma.

Lemma 2. *For any* $c \geqslant 1$, $\mathsf{E}^{SAT[O(n^c)]} \subseteq \mathsf{NTIME}(2^{O(n^c)})/O(n^c)$

Proof. Let L be any language in $\mathsf{E}^{SAT[O(n^c)]}$. Let M be an exponential linear-time oracle machine making $O(n^c)$ queries to SAT on inputs of length n and deciding L. We define an advice-taking non-deterministic machine M' which operates in time $O(2^{n^c})$ and decides L using at most $O(n^c)$ bits of advice.

On an input y of length n, M' first generates the query tree for M for every possible input x of length n. By a query tree for x, we mean the tree whose nodes are queries with the first query at the root, and for each query q, the children of that query are the queries asked depending on whether query q is answered 0 or 1. For each x of length n, the query tree for x has size at most $2^{O(n^c)}$. Since $c \geqslant 1$, the total number of queries asked in any query tree for an input of length n is $2^{O(n^c)}$. The advice for M' specifies how many of these queries are in SAT–this is a number describable in $O(n^c)$ bits. In a computation path of M', for each query q in a query tree for some x of length n in turn, M' guesses a witness for q and verifies that $q \in SAT$. If the number of correct verifications is not the number coded in the advice string, M' rejects. Otherwise, M' stores a list of all queries that have been verified to be in SAT on the current computation path, and then runs M on y. If M makes a query, M' takes the answer to be yes if the query is in its list, otherwise it takes the answer to be no. Finally, M' accepts iff M accepts.

Assuming the advice is correct, M' will have an accepting computation on y iff the oracle machine M accepts y. This accepting computation will correspond to forming the correct list of queries in SAT among all queries asked in query trees of length n inputs. It can be verified that M' runs in time $2^{O(n^c)}$.

We now use this fact to generalize Theorem 1. The following result also strengthens a result of Mocas [Moc96] separating NEXP from $\mathsf{P}^{\mathsf{NP}[n^c]}$ for fixed c.

Theorem 3. *For any c,* $\mathsf{NEXP} \not\subseteq \mathsf{P}^{\mathsf{NP}[n^c]}/n^c$.

Proof. On the contrary, assume $\mathsf{NEXP} \subseteq \mathsf{P}^{\mathsf{NP}[n^c]}/n^c$. We derive a contradiction. Since SAT is complete for NP under m-reductions, we get that $\mathsf{NEXP} \subseteq \mathsf{P}^{SAT[n^c]}/n^c$. Now let L be a language complete for NE under linear-time reductions. By assumption on NEXP, there is a constant k such that $L \in \mathsf{DTIME}(n^k)^{SAT[n^c]}/n^c$. Since L is complete for NE under linear-time reductions, it follows that $\mathsf{NE} \subseteq \mathsf{DTIME}(n^k)^{SAT[O(n^c)]}/O(n^c)$.

Now we use the assumption on NEXP again, in a different way. Since $\mathsf{NEXP} \subseteq \mathsf{P}^{\mathsf{NP}[n^c]}/n^c$, we get that $\mathsf{NEXP} \subseteq \mathsf{P}^{SAT[n^c]}/n^c$, by NP-completeness of SAT. Hence $\mathsf{NEXP} \subseteq \mathsf{E}^{SAT[n^c]}/n^c$, which implies $\mathsf{NEXP} \subseteq \mathsf{NTIME}(2^{O(n^{c^2})})/O(n^{c^2})$, using Lemma 2 and a translation argument exactly as in the proof of Theorem 1. From the conclusion in the previous paragraph, and using a translation argument again, we get that $\mathsf{NEXP} \subseteq \mathsf{DTIME}(n^{kc^2})^{SAT[O(n^{c^3})]}/O(n^{c^3})$. But the latter class is in $\mathsf{DTIME}(2^{n^{kc^2+1}})/O(n^{c^3})$, just by answering all SAT queries by exhaustive search, since the length of any SAT query asked by a machine running in time $O(n^{kc^2})$ is $O(n^{kc^2})$. Thus we get $\mathsf{EXP} \subseteq \mathsf{DTIME}(2^{n^{kc^2+1}})/O(n^{c^3})$, which is a contradiction to Lemma 1.

Theorems 1 and 3 have some interesting consequences for the connection between circuit lower bounds and derandomization. Impagliazzo, Kabanets and Wigderson [IKW02] showed that if MA is in NSUBEXP, then there is a language in NEXP that is not computable with polynomial-size circuits. However, their result does not say anything about a derandomization of MA where the simulating algorithm uses advice.

We will need to use the main result of [IKW02].

Theorem 4. *[IKW02] if* $\mathsf{NEXP} \subseteq \mathsf{SIZE}(poly)$, *then* $\mathsf{NEXP} = \mathsf{MA}$.

Theorem 5. *For any constant c, if* $\mathsf{MA} \subseteq \mathsf{NP}/n^c$, *then* $\mathsf{NEXP} \not\subseteq \mathsf{SIZE}(poly)$.

Proof. If $\mathsf{MA} \subseteq \mathsf{NP}/n^c$, then $\mathsf{NEXP} \not\subseteq \mathsf{MA}$, since otherwise we have a contradiction to Theorem 1. But this implies $\mathsf{NEXP} \not\subseteq \mathsf{SIZE}(poly)$ by Theorem 4, and thus the result follows.

In fact a circuit lower bound follows even from a simulation of MA in $\mathsf{P}^{\mathsf{NP}[n^c]}/n^c$, by using Theorem 3 rather than Theorem 1.

It is known that $\mathsf{MA} \subseteq \mathsf{NP}/poly$, since the randomness of a Merlin-Arthur machine can be simulated by a polynomial-size advice string. Theorem 5 shows that this simulation is essentially optimal with respect to advice–if we could simulate MA in NP with *fixed* polynomial advice, that would imply a long sought-after circuit lower bound for NEXP.

Theorem 3 gives a separation for NEXP and against fixed polynomial advice, but does it imply anything new for separating uniform classes? The answer is yes: consider the class $\mathsf{BPTIME}(n^c)^{\mathsf{NP}}$ of languages accepted by probabilistic oracle machines running in time n^c for some fixed c, and with access to an NP oracle. This class lies between NP and $\mathsf{BPP}^{\mathsf{NP}}$, and seems incomparable to BPP. It is contained in NEXP, but it's unclear if the containment is strict. This is because the NP oracle could have arbitrarily high non-deterministic polynomial time complexity.

Theorem 6. *For any* c, $\mathsf{NEXP} \not\subseteq \mathsf{BPTIME}(n^c)^{\mathsf{NP}}$.

Proof. $\mathsf{BPTIME}(n^c)^{\mathsf{NP}} \subseteq \mathsf{DTIME}(n^{c+1})^{\mathsf{NP}}/O(n^{c+1})$. This follows simply by amplifying the acceptance probability of the probabilistic oracle machine so that it is at most 2^{-n} or at least $1-2^{-n}$, and then using Adleman's trick to encode the randomness in the advice. Note that this transformation relativizes. By Theorem 3, we have that $\mathsf{NEXP} \not\subseteq \mathsf{DTIME}(n^{c+1})^{\mathsf{NP}}/O(n^{c+1})$, which yields the desired separation.

4 Lower Bounds for Probabilistic Exponential Time Classes

Here we prove results analogous to those in the previous section for probabilistic classes. Unlike in the case of non-deterministic time, it is not easy to show even that BPEXP is not in BPP/1. It can be shown using a translation argument that BPEXP is not in BPP [KV87], but this translation argument does not extend to showing a lower bound against advice. Since BPP and BPEXP are semantic classes, it is unknown whether for instance $\mathsf{BPEXP} \subseteq \mathsf{BPP}/1$ implies $\mathsf{BPEXP}/1 \subseteq \mathsf{BPP}/2$.

Recently, there's been a lot of work on hierarchies for semantic classes with advice, and one might ask whether similar techniques are applicable here. The generic methods used in [FST04, vMP06] fail to work here for two reasons. First, their upper bounds always require advice. Second, those methods are inherently incapable of accommodating more than $\log(n)$ bits of advice in the lower bound.

We circumvent both these difficulties by using specific properties of the semantic measures for which we prove bounds. Our general strategy is as follows. Let's say we're trying to show $\mathsf{CEXP} \not\subseteq \mathsf{CP}/a(n)$, where CEXP and CP are the exponential-time and polynomial-time versions respectively of a semantic measure CTIME, and $a(n) \geqslant \log(n)$ is an advice bound. We choose a syntactic class D such that $\mathsf{D} \subseteq \mathsf{CEXP}$. We then argue that if $\mathsf{D} \subseteq \mathsf{CP}/a(n)$, then the advice can be eliminated to place D uniformly in CTIME with some sub-exponential time bound. Now there are two cases: if $\mathsf{D} \not\subseteq \mathsf{CP}/a(n)$, we're done since $\mathsf{D} \subseteq \mathsf{CEXP}$. In case $\mathsf{D} \subseteq \mathsf{CP}/a(n)$, we can use the advice elimination together with a translation argument to diagonalize in CEXP against $\mathsf{CP}/a(n)$.

Our proofs vary depending on the class D, which needs to be chosen judiciously, and the specific form of the advice elimination, which also dictates how much advice we can accommodate in our lower bound.

We now show how this general strategy works for the classes BPEXP, MAEXP and REXP.

4.1 Lower Bound for BPEXP

Before we discuss the advice elimination strategy for BPEXP, we need a definition.

Definition 1. *A language L is said to have instance-checkers if there is a probabilistic polynomial-time oracle machine M outputting 1, 0 or ? such that:*

1. *If M is given L as oracle, then $M(x) = L(x)$ with probability 1*
2. *For any oracle A, when M is given A as oracle, for any input x, either $M(x) = L(x)$ or $M(x) = ?$ with probability at least $1 - 2^{-\Omega(n)}$.*

It follows from the work of Babai, Fortnow and Lund [BFL91] on multi-prover interactive protocols that EXP-complete languages have instance checkers.

Theorem 7. *[BFL91] All EXP-complete languages have instance-checkers.*

We will use Theorem 7 to eliminate the advice from a probabilistic polynomial-time machine accepting an EXP-complete language. We use Theorem 7 in the same way as Trevisan and Vadhan [TV02], but our choice of parameters is different.

Lemma 3. *If* $\mathsf{EXP} \subseteq \mathsf{BPP}/n^{o(1)}$, *then* $\mathsf{EXP} \subseteq \mathsf{BPSUBEXP}$.

Proof. Let L be an EXP-complete language. By Theorem 7, L has instance-checkers. Assume $L \in \mathsf{BPP}/n^{o(1)}$, and let M' be a probabilistic polynomial-time machine deciding L with $n^{o(1)}$ bits of advice and error bound $\leqslant 2^{-\omega(n)}$. Let M be an oracle machine witnessing the fact that L is instance-checkable. We define a probabilistic sub-exponential time machine N deciding L. On input x, N simulates the oracle machine M. Let n^k be a bound on the size of queries asked by M on inputs of length n. N uses each possible sequence of advice strings $a_1, a_2 \ldots a_{n^k}$ in turn as oracle for M', where for each i, $|a_i| \leqslant i^{o(1)}$. Namely, when M asks a query of size i to the oracle, N simulates M' with advice a_i to answer the query. When the simulation of the oracle machine M is finished for a certain oracle O, N returns an answer if the answer is either 0 or 1, otherwise it moves on to the next possible sequence of advice strings. If M still doesn't give a 0 or 1 answer after all possible sequences of advice strings have been used as oracles, N outputs an arbitrary value, say 0.

Since there are at most sub-exponentially many sequences of advice strings and each simulation of M with a sequence of advice strings takes polynomial time, N halts in sub-exponential time, i.e. time $2^{n^{o(1)}}$. We need to argue that N decides L with low error. For this, we simply use the properties of the instance-checker M. When the wrong sequence of advice strings is used as oracle, the probability that M outputs a 0 or 1 answer wrongly is $2^{-\Omega(n)}$, hence by a union bound, the probability that M outputs a 0 or 1 answer wrongly on *some* sequence of advice strings is also $2^{-\Omega(n)}$. When the right sequence of advice strings is used, M outputs the correct answer with probability at least $1-2^{-\Omega(n)}$, since the error bound of M' is at most $2^{-\Omega(n)}$. Thus the right answer is always output with probability at least $1-2^{-\Omega(n)}$.

Thus we get $L \in \mathsf{BPSUBEXP}$, and since L is complete for EXP under polynomial-time reductions, we also get $\mathsf{EXP} \subseteq \mathsf{BPSUBEXP}$.

Now we are ready to prove our separation.

Theorem 8. $\mathsf{BPEXP} \not\subseteq \mathsf{BPP}/n^{o(1)}$.

Proof. We consider two cases: either $\mathsf{EXP} \subseteq \mathsf{BPP}/n^{o(1)}$ or not. In the first case, by Lemma 3, we have that $\mathsf{EXP} \subseteq \mathsf{BPSUBEXP}$. By translation, this implies that there is a super-exponential time bound t such that $\mathsf{DTIME}(t) \subseteq \mathsf{BPEXP}$. But we can diagonalize in $\mathsf{DTIME}(t)$ against $\mathsf{BPP}/n^{o(1)}$, since $\mathsf{BPP}/n^{o(1)} \subseteq \mathsf{EXP}/n^{o(1)}$. Hence, in this case, we are done.

If $\mathsf{EXP} \not\subseteq \mathsf{BPP}/n^{o(1)}$, we are done immediately, since $\mathsf{EXP} \subseteq \mathsf{BPEXP}$.

4.2 Lower Bound for MAEXP

The key to our separation for MAEXP is the following advice elimination result, which involves strengthening work of Impagliazzo, Kabanets and Wigderson [IKW02].

Lemma 4. *For any constant $c > 0$, if $\mathsf{NEXP} \subseteq \mathsf{MA}/n^c$, then $\mathsf{NEXP} \subseteq \mathsf{MA}$.*

Proof. We show that $\mathsf{NEXP} \subseteq \mathsf{MA}/n^c$ implies $\mathsf{NEXP} \subseteq \mathsf{SIZE}(poly)$. $\mathsf{NEXP} \subseteq \mathsf{SIZE}(poly$ implies $\mathsf{NEXP} = \mathsf{MA}$, by Theorem 4.

Assume, to the contrary, that $\mathsf{NEXP} \subseteq \mathsf{MA}/n^c$ and $\mathsf{NEXP} \not\subseteq \mathsf{SIZE}(poly)$. We derive a contradiction. Since $\mathsf{NEXP} \subseteq \mathsf{MA}/n^c$, we have that $\mathsf{NEXP} \subseteq \mathsf{EXP}/n^c$. Thus we have $\mathsf{EXP}/n^c \not\subseteq \mathsf{SIZE}(poly)$. But this is equivalent to saying that $\mathsf{EXP} \not\subseteq \mathsf{SIZE}(poly)$. To see this, let L be a language in EXP/n^c such that $L \notin \mathsf{SIZE}(poly)$, and let M be a deterministic exponential-time machine taking n^c bits of advice and deciding L. We define a new language $L' \in \mathsf{EXP}$ such that $L' \notin \mathsf{SIZE}(poly)$. L' consists of all pairs $< x, a >$ such that M accepts x given advice a. $L' \in EXP$ since deciding L merely involves simulating the exponential-time machine M. Now suppose $L' \in \mathsf{SIZE}(poly)$. Then we could define polynomial-size circuits for L which simulate the polynomial-size circuits for L' on $< x, a >$, where a is the correct advice for L at length $|x|$ - a is of polynomial size in n and depends only on n.

If $\mathsf{EXP} \not\subseteq \mathsf{SIZE}(poly)$, we have that $\mathsf{MA}/n^c \subseteq \mathsf{i.o.NSUBEXP}/n^c \subseteq \mathsf{i.o.NE}/n^c$. This follows essentially from the connection between circuit lower bounds and derandomization [NW94, KvM99] because circuit lower bounds derandomize not just MA but also the promise version of MA. Now we use the assumption that $\mathsf{NEXP} \subseteq \mathsf{EXP}/n^c$ again, along with the fact that there is a language Q which is complete for NE with respect to linear-time reductions. Since $Q \in \mathsf{EXP}/n^c$, $Q \in \mathsf{DTIME}(2^{n^k})/n^c$ for some fixed k. By the fact that Q is complete for NE with respect to linear-time reductions, we have that $\mathsf{NE} \subseteq \mathsf{DTIME}(2^{n^k})/O(n^c)$. By using the same translation argument as in the proof of Theorem 1, we get that $\mathsf{NE}/n^c \subseteq \mathsf{DTIME}(2^{n^{kc}})/O(n^{c^2})$. Hence $i.o.\mathsf{NE}/n^c \subseteq i.o.\mathsf{DTIME}(2^{n^{kc}})/O(n^{c^2})$. By combining this with the derandomization result, we get that $\mathsf{MA}/n^c \subseteq i.o.\mathsf{DTIME}(2^{n^{kc}})/O(n^{c^2})$. By the assumption that $\mathsf{NEXP} \subseteq \mathsf{MA}/n^c$, we get that $\mathsf{NEXP} \subseteq i.o.\mathsf{DTIME}(2^{n^{kc}})/O(n^{c^2})$, which is a contradiction to Lemma 1.

Note that Theorem 4 is an unusually strong advice elimination result. Typically, for polynomial-time classes, advice elimination cannot handle more than $\Omega(\log(n))$ advice bits without blowing up the time to super-polynomial. Also, Theorem 4 requires the assumption to hold for *all* languages in NEXP, not just a complete language. Indeed, for a NEXP-complete language to be contained in MA/n^c for a fixed c is equivalent to $\mathsf{NEXP} \subseteq \mathsf{NP}/poly$.

Theorem 9. *For any constant c,* $\mathsf{MAEXP} \not\subseteq \mathsf{MA}/n^c$.

Proof. We consider two cases. The first case is that $\mathsf{NEXP} \not\subseteq \mathsf{MA}/n^c$. In this case, we have that $\mathsf{MAEXP} \not\subseteq \mathsf{MA}/n^c$, since $\mathsf{NEXP} \subseteq \mathsf{MAEXP}$.

The other case is that $\mathsf{NEXP} \subseteq \mathsf{MA}/n^c$. In this case, by Lemma 4, $\mathsf{NEXP} = \mathsf{MA}$. By translation, we get that $\mathsf{NEEXP} = \mathsf{MAEXP}$. In this case too, $\mathsf{MAEXP} \not\subseteq \mathsf{MA}/n^c$, since we can diagonalize against MA/n^c even in deterministic double-exponential time.

Theorems 8 and 9 can be strengthened so that the lower bound holds against certain superpolynomial time bounds, rather than just polynomial time. However, these time bounds are not easy to state, so we defer the statement of this extension to the full version of the paper.

4.3 Lower Bound for REXP

The result we obtain for REXP which is somewhat weaker with respect to the advice bound.

Theorem 10. $\mathsf{REXP} \not\subseteq \mathsf{RP}/O(\log(n))$.

Proof. We consider two cases: $\mathsf{NP} \subseteq \mathsf{BPP}/O(\log(n))$ or not. In the first case, we can use the downward self-reducibility of SAT to eliminate the advice and get $\mathsf{NP} \subseteq \mathsf{BPP}$, but this implies $\mathsf{NP} = \mathsf{RP}$, again using downward self-reducibility to eliminate wrong accepting paths. Thus we get $\mathsf{NEXP} = \mathsf{REXP}$, and in this case we even have $\mathsf{REXP} \not\subseteq \mathsf{RP}/n^c$ by using Theorem 1.

If $\mathsf{NP} \not\subseteq \mathsf{BPP}/O(\log(n))$, then we directly have $\mathsf{REXP} \not\subseteq \mathsf{RP}/O(\log(n))$, since $\mathsf{NP} \subseteq \mathsf{REXP}$ and $\mathsf{RP}/O(\log(n)) \subseteq \mathsf{BPP}/O(\log(n))$.

5 Relativizations

The results in Section 4 rely on tools from the theory of interactive proofs that don't relativize. But our results (and all other known results) on separating nondeterministic and deterministic time exponential and polynomial-time classes do relativize. By examining known and new oracle results we see that we've come close to the limit of what can be done using these relativizable techniques.

In Theorem 3 we showed that for any constant c, $\mathsf{NEXP} \not\subseteq \mathsf{P}^{\mathsf{NP}[n^c]}/n^c$. One cannot remove the n^c limit on queries even without the advice using non-relativizable techniques because of the following result due to Buhrman, Fenner, Fortnow and Torenvliet [BFFT01].

Theorem 11 (BFFT). *There exists a relativized world where* $\mathsf{NEXP} = \mathsf{P}^{\mathsf{NP}}$.

Eric Allender asked whether even Theorem 1 ($\mathsf{NEXP} \not\subseteq \mathsf{NP}/n^c$) can be strengthened to a lower bound that works on almost all input lengths, rather than on infinitely many. Direct diagonalizations tend to work on almost all input lengths– our separation is indirect, and technique does not give this stronger property. We give a new relativized world showing that relativizing techniques cannot get the stronger separation even without the advice.

Theorem 12. *There exists a relativized world such that* $\mathsf{NEXP} \subseteq \mathsf{i.o.NP}$.

Proof. Let M_i be a standard enumeration of non-deterministic relativized Turing machines that runs in time at most 2^{n^i}. Since these machines are paddable, for any A and any $L \in \mathsf{NEXP}^A$ there will some i such that $L = L(M_i^A)$. We will create A such that for every i there are an infinite number of n such that for all x of length n,

$$x \in L(M_i^A) \Leftrightarrow \text{ there exists a } y \text{ with } |y| = 2|x|^i \text{ and } (i, x, y) \in A$$

which immediately implies Theorem 12.

Start with $A = \varnothing$. We construct A in stages (i, j) chosen in any order that cover all possible (i, j).

Stage (i, j): Pick n such that n is larger than any frozen string as well as the n chosen in any previous stage.

Set all strings x of length n to be unmarked.

Repeat the following as long as there is an unmarked x of length n such that $M_i^A(x)$ accepts: Fix an accepting path of $M_i^A(x)$ and freeze every string queried along that path. Mark x. Pick a y, $|y| = 2|x|^i$ such that (i, x, y) is not frozen and let $A = A \cup \{(i, x, y)\}$.

We can always find such a y since we have 2^{2n^i} possible (i, x, y) and at this point since we have frozen at most 2^{n^i} strings for at most 2^n possible x's for a total of $2^{n^i}2^n < 2^{2n^i}$ frozen strings.

By adding every (i, x, y) that is non frozen in the proof above one can get an even stronger oracle.

Corollary 1. *There exists a relativized world such that* $\mathsf{NEXP} \subseteq \mathsf{i.o.RP}$.

Theorem 12 has independent interest. Consider the nondeterministic time hierarchy [Coo72, SFM78, Žŝ3].

Theorem 13 (Cook,Seiferas-Fischer-Meyer,Žák)
For any time-constructible functions t_1 and t_2 such that $t_1(n+1) = o(t_2(n))$,

$$\mathsf{DTIME}(t_1(n)) \subsetneq \mathsf{DTIME}(t_2(n)).$$

The proofs of Theorem 13 work by looking at collapsing several input lengths. While Rackoff and Seiferas [RS81] showed that the bounds are tight in relativized worlds, Theorem 12 directly shows that one needs to look at many input lengths for a relativizable diagonalization.

Theorem 12 should be contrasted with the fact that by direct diagonalization, it is easy to prove that $\mathsf{NEXP} \not\subseteq \mathsf{i.o.coNP}$ and even that $\mathsf{NEXP} \not\subseteq \mathsf{i.o.coNP}/(n - o(n))$. However, even the latter separation is nearly tight with respect to advice.

Corollary 2. *There exists a relativized world such that* $\mathsf{NEXP} \subseteq \mathsf{i.o.coNP}/n{+}1$.

This corollary immediately follows from Theorem 12 and the following folklore theorem.

Theorem 14. $\mathsf{NEXP} \subseteq \mathsf{coNEXP}/n+1$.

Proof. Let L be a language in NEXP and let the advice for length n be the number of strings in L of length n. Our NEXP algorithm with advice for $\overline{L}$ simply guesses and verifies all the strings in L of length n (since we know how many there are) and accepts x is x is not among them.

6 Conclusions and Open Questions

There are several open questions that remain. Are there explicit languages that witness the lower bounds we obtain? Is $\mathsf{BPEXP} \subseteq \mathsf{i.o.BPP}$, or $\mathsf{MAEXP} \subseteq \mathsf{i.o.MA}$? Is $\mathsf{NEXP} \subseteq \mathsf{NE}/n^c$, for any fixed c? Is there a generic method for obtaining separations for exponential time against polynomial time with advice that works for semantic as well as syntactic classes? Can the advice bounds against which separations are obtained for probabilistic time with two-sided error and one-sided error be improved?

We showed a new advice elimination result for MA in this paper, but there are several other semantic classes for which there are no advice elimination results known: ZPP, $\mathsf{NP} \cap \mathsf{coNP}$, AM etc.

Acknowledgments

We thank Eric Allender for helpful discussions and the question about infinitely-often separation that led to Theorem 12. We also thank many other participants of Dagstuhl Seminar 08381 on the Computational Complexity of Discrete Problems, discussions with whom helped shape this paper.

References

[AG94] Allender, E., Gore, V.: A uniform circuit lower bound for the permanent. SIAM Journal on Computing 23(5), 1026–1049 (1994)

[AW08] Aaronson, S., Wigderson, A.: Algebrization: A New Barrier in Complexity Theory. ACM Trans. Comput. Theory 1(1), 1–54 (2009), `http://doi.acm.org/10.1145/1490270.1490272`

[BFFT01] Buhrman, H., Fenner, S., Fortnow, L., Torenvliet, L.: Two oracles that force a big crunch. Computational Complexity 10(2), 93–116 (2001)

[BFL91] Babai, L., Fortnow, L., Lund, C.: Non-deterministic exponential time has two-prover interactive protocols. Computational Complexity 1, 3–40 (1991)

[BGS75] Baker, T., Gill, J., Solovay, R.: Relativizations of the P =? NP question. SIAM Journal on Computing 4(4), 431–442 (1975)

[Coo72] Cook, S.: A hierarchy for nondeterministic time complexity. In: Conference Record, Fourth Annual ACM Symposium on Theory of Computing, Denver, Colorado, May 1-3, 1972, pp. 187–192 (1972)

[FLvMV05] Fortnow, L., Lipton, R., van Melkebeek, D., Viglas, A.: Time-space lower bounds for satisfiability. Journal of the ACM 52(6), 833–865 (2005)

[FS04] Fortnow, L., Santhanam, R.: Hierarchy theorems for probabilistic polynomial time. In: Proceedings of the 45th IEEE Symposium on Foundations of Computer Science, pp. 316–324 (2004)

[FST04] Fortnow, L., Santhanam, R., Trevisan, L.: Promise hierarchies. Electronic Colloquium on Computational Complexity (ECCC) 11(98) (2004)

[FST05] Fortnow, L., Santhanam, R., Trevisan, L.: Hierarchies for semantic classes. In: Proceedings of the Thirty-Seventh Annual ACM Symposium on Theory of Computing (2005)

[HM95] Homer, S., Mocas, S.: Nonuniform lower bounds for exponential time classes. In: 20th International Symposium on Mathematical Foundations of Computer Science, pp. 159–168 (1995)

[IKW02] Impagliazzo, R., Kabanets, V., Wigderson, A.: In search of an easy witness: Exponential time vs. probabilistic polynomial time. Journal of Computer and System Sciences 65(4), 672–694 (2002)

[KI03] Kabanets, V., Impagliazzo, R.: Derandomizing polynomial identity tests means proving circuit lower bounds. In: Proceedings of the 35th Annual ACM Symposium on the Theory of Computing, pp. 355–364 (2003)

[KV87] Karpinski, M., Verbeek, R.: On the monte carlo space constructible functions and separation results for probabilistic complexity classes. Information and Computation 75 (1987)

[KvM99] Klivans, A.R., van Melkebeek, D.: Graph Nonisomorphism Has Subexponential Size Proofs Unless the Polynomial-Time Hierarchy Collapses. SIAM J. Comput. 31(5), 1501–1526 (2002), `http://dx.doi.org/10.1137/S0097539700389652`

[Moc96] Mocas, S.E.: Separating classes in the exponential-time hierarchy from classes in PH. Theoretical Computer Science 158(1-2), 221–231 (1996)

[NW94] Nisan, N., Wigderson, A.: Hardness vs randomness. Journal of Computer and System Sciences 49(2), 149–167 (1994)

[RR97] Razborov, A., Rudich, S.: Natural proofs. Journal of Computer and System Sciences 55(1), 24–35 (1997)

[RS81] Rackoff, C., Seiferas, J.: Limitations on separating nondeterministic complexity classes. SIAM Journal on Computing 10(4), 742–745 (1981)

[SFM78] Seiferas, J., Fischer, M., Meyer, A.: Separating nondeterministic time complexity classes. Journal of the ACM 25(1), 146–167 (1978)

[TV02] Trevisan, L., Vadhan, S.: Pseudorandomness and average-case complexity via uniform reductions. In: Proceedings of the 17th Annual IEEE Conference on Computational Complexity, vol. 17 (2002)

[vMP06] van Melkebeek, D., Pervyshev, K.: A generic time hierarchy for semantic models with one bit of advice. In: Proceedings of 21st Annual IEEE Conference on Computational Complexity, pp. 129–144 (2006)

[Ž83] Žák, S.: A Turing machine time hierarchy. Theoretical Computer Science 26(3), 327–333 (1983)

Approximating Decision Trees with Multiway Branches

Venkatesan T. Chakaravarthy, Vinayaka Pandit, Sambuddha Roy, and Yogish Sabharwal

IBM India Research Lab, New Delhi, India
{vechakra,pvinayak,sambuddha,ysabharwal}@in.ibm.com

Abstract. We consider the problem of constructing decision trees for entity identification from a given table. The input is a table containing information about a set of entities over a fixed set of attributes. The goal is to construct a decision tree that identifies each entity unambiguously by testing the attribute values such that the average number of tests is minimized. The previously best known approximation ratio for this problem was $O(\log^2 N)$. In this paper, we present a new greedy heuristic that yields an improved approximation ratio of $O(\log N)$.

1 Introduction

In many situations (such as machine fault detection, medical diagnosis, species identification), one is required to identify an unknown entity (among a possible set of known entities), by performing a sequence of tests. Typically, each test is restricted to checking the value of an attribute of the unknown entity. Naturally, one wishes to design a testing procedure that minimizes the expected number of tests needed to identify the unknown entity. As an example, consider a typical medical diagnosis application. A hospital maintains a table containing information about diseases. Each row represents a disease and each column is a medical test and the corresponding entry specifies the outcome of the test for a person suffering from the given disease. When the hospital receives a new patient whose disease has not been identified, it would like to determine the shortest sequence of tests which can unambiguously determine the disease of the patient. Motivated by such applications, the problem of constructing *decision trees for entity identification* from given data, has been well studied (See the surveys by Murthy [1] and Moret [2]). In this paper, we present an improved approximation algorithm for the above problem, when the tests are multi-valued.

Decision Trees for Entity Identification - Problem Statement. The input is a table $\mathcal{D}$ having N rows and m columns. Each row is called an *entity* and the columns are the *attributes* of these entities. A solution is a decision tree in which each internal node is labeled by an attribute and its branches are labeled by the values that the attribute can take. The entities are the leaves of the tree. The tree should identify each entity correctly. The goal is to construct a decision tree in which the average distance of a leaf from the root is minimized. Equivalently,

S. Albers et al. (Eds.): ICALP 2009, Part I, LNCS 5555, pp. 210–221, 2009.

	A	B	C
e_1	1	1	1
e_2	1	2	0
e_3	1	3	1
e_4	2	4	0
e_5	2	4	1
e_6	2	5	0

Input Table

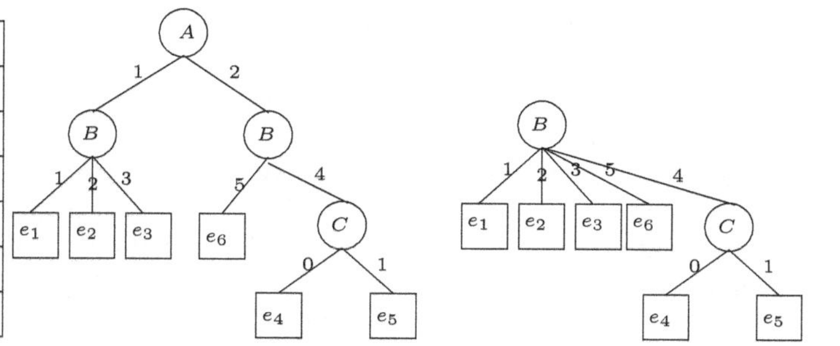

A Decision Tree of cost 14 Optimal Decision Tree of Cost 8

Fig. 1. Example decision trees

we want to minimize the *total external path length* defined as the sum of the distances of all the entities to the root. We call this the $\mathcal{DT}$ problem.

Example: Figure 1 shows an example table and two decision trees. The total external path length of the first tree is 14 and that of the second (optimal) tree is 8. An equivalent way of computing the cost is to sum over the internal nodes v, the number leaves under v. For instance, in the first tree, the cost can be computed as $6 + 3 + 3 + 2 = 14$ and for the second tree, it is $6 + 2 = 8$. □

A special case of the $\mathcal{DT}$ problem is when every attribute can take only two values (say $\{0, 1\}$); this restricted version is called the $2 - \mathcal{DT}$ problem. In general, for some $K \geq 2$, if every attribute is restricted to take at most K distinct values (say $\{0, 1, \ldots, K - 1\}$), we get the $K - \mathcal{DT}$ problem. The general $\mathcal{DT}$ problem has no such restriction on the possible values of attributes.

Hyafil and Rivest [3] showed that the $2 - \mathcal{DT}$ problem is NP-Hard. Garey [4] presented a dynamic programming based algorithm for the $2 - \mathcal{DT}$problem that finds the optimal solution, but the algorithm runs in exponential time in the worst case. Kosaraju et al. [5] studied the $2-\mathcal{DT}$ problem and showed that a popular greedy heuristic gives an $O(\log N)$ approximation. Adler and Heeringa [6] gave an alternative proof and improved the approximation ratio by a constant factor to $(1 + \ln N)$ (See also [7]).

Garey and Graham [8] showed that the natural greedy algorithm considered in [5,6] has approximation ratio $\Omega(\log N/ \log\log N)$. Chakaravarthy et al. [9] studied the $K - \mathcal{DT}$ problem and gave an $r_K(1 + \ln N)$-approximation algorithm, where r_K is a suitably defined Ramsey number with a known bound that $r_K \leq 2 + 0.64 \log K$; thus, their approximation ratio is $O(\log K \log N)$. On the hardness front, they also showed that the $2 - \mathcal{DT}$ problem and the general $\mathcal{DT}$ problem cannot be approximated within factors of 2 and 4 respectively unless NP=P. The weighted version of the $\mathcal{DT}$ problem has also been studied. Here, there is a probability distribution associated with the set of entities and the goal is to minimize the expected path length. (The unweighted version that we defined corresponds to the case where the distribution is uniform). The $O(\log N)$-approximation algorithm by Kosaraju et al. [5] can also handle the weighted $2-\mathcal{DT}$ problem. For the weighted $K-\mathcal{DT}$ problem, Chakaravarthy et al. [9] gave a $O(\log K \log N)$-approximation algorithm. On the hardness front, it is known that the weighted $2-\mathcal{DT}$ problem cannot be approximated within a factor of $\Omega(\log N)$, unless NP=P [9].

In the general $\mathcal{DT}$ problem, K can be very large (up to N), in which case the approximation ratio of [9] is $O(\log^2 N)$. So, a natural question is whether there exists an approximation algorithm for the general $\mathcal{DT}$ problem matching the approximation ratio of $O(\log N)$ for the $2-\mathcal{DT}$ problem. In this paper, we present an algorithm with an approximation ratio of $O(\log N)$ for the general (unweighted) $\mathcal{DT}$ problem based on a new greedy heuristic.

We now mention the main ingredients of our algorithm and its analysis. Essentially, each internal node of a decision tree narrows down the possibilities for the unknown entity. The greedy heuristic considered in [9], at each internal node, chooses the attribute that maximizes the number of pairs separated (two entities are said to be separated if they do not take the same branch). In contrast, the greedy heuristic that we propose chooses an attribute which minimizes the size of the heaviest branch. In the case of the $2-\mathcal{DT}$ problem, these two heuristics are equivalent. One of the central ideas used in the analysis of [5] is that of "centroidal path". Our main insight lies in drawing a connection between the notion of centroidal path and the objective function of the Minimum Sum Set Cover (MSSC) problem [10]. This insight allows us to utilize some ideas from the analysis of the greedy algorithm for the MSSC problem by Feige et al. [10]. It is interesting to note that the hardness results in [9] are obtained via reductions from the MSSC problem. In this paper, we utilize the MSSC problem in the other direction, namely we use ideas from the analysis of the greedy algorithm for the MSCC problem to obtained improved upper bounds on the approximation ratio.

Our analysis, in its current form, does not extend to the weighted $\mathcal{DT}$ problem. It would be interesting to obtain a $O(\log N)$ approximation algorithm for the weighted $\mathcal{DT}$ problem as that would give matching upper and lower bounds for the problem.

2 Preliminaries

In this section, we define the $\mathcal{DT}$ problem and develop some notations. Let $\mathcal{D}$ be a table having N rows and m columns. Each row is called an *entity* and each column is called an *attribute*. Let $\mathcal{E}$ denote the set of all N entities and $\mathcal{A}$ denote the set of m attributes. For $e \in \mathcal{E}$ and $a \in \mathcal{A}$, denote by $e.a$ the value taken by e on attribute a. We are guaranteed that no two entities are identical. For $a \in A$, let $\mathcal{V}_a$ denote the set of values taken by the attribute a. Let $K = \max_a |\mathcal{V}_a|$. We have $K \leq N$. For instance, if the table is binary, $K = 2$.

A *decision tree* T for the table $\mathcal{D}$ is a rooted tree satisfying the following properties. Each internal node u is labeled by an attribute a and has at most $|\mathcal{V}_a|$ children. Every branch (edge) out of u is labeled by a distinct value from the set $\mathcal{V}_a$. The entities are the leaves of the tree and thus the tree has exactly N leaves. The tree should identify every entity correctly. In other words, for any entity e, the following traversal process should correctly lead to e. The process starts at the root node. Let u be the current node and a be the attribute label of u. Take the branch out of u labeled by $e.a$ and move to the corresponding child of u. The requirement is that this traversal process should reach the entity e.

Observe that the values of the attributes are used only for taking the correct branch during the traversal. So, we can map each value of an attribute to a distinct number from 1 to K and assume that $\mathcal{V}_a$ is a subset of $\{1, 2, \ldots, K\}$. Therefore, we assume that for any $e \in \mathcal{E}$ and $a \in \mathcal{A}$, $e.a \in \{1, 2, \ldots, K\}$.

Let T be a decision tree for $\mathcal{D}$. For an entity $e \in \mathcal{E}$, *path length* of e is defined to be the number of internal nodes in the path from the root to e; it is denoted $\ell_T(e)$ (Note that this is the same as the number of edges on the path). The sum of all path lengths is called *total path length* and is denoted cost(T), i.e., $\text{cost}(T) = \Sigma_{e \in \mathcal{E}} \ell_T(e)$. We say that e pays a cost of $\ell_T(e)$ in T.

$\mathcal{DT}$ **Problem:** Given a table $\mathcal{D}$, construct a decision tree for $\mathcal{D}$ having the minimum cost.

Remark: In general, we can define a decision tree for any subset of entities of the table ($E \subseteq \mathcal{E}$) and its cost will be defined analogously.

3 Greedy Algorithm

In this section, we describe our greedy algorithm. Let $\mathcal{D}$ be the input table over N entities $\mathcal{E}$ and m attributes $\mathcal{A}$.

Consider a subset of entities $E \subseteq \mathcal{E}$ and an attribute $a \in \mathcal{A}$. The *coverage* of a on E is defined as below. The attribute a partitions the set E into K parts, given by $E_j = \{e \in E \ : \ e.a = j\}$, for $1 \leq j \leq K$. Let j^* be denote the part having the largest size, i.e., $j^* = \text{argmax}_j |E_j|$. We say that a *covers* all the entities in E, except those in E_{j^*} and define

$$\text{Cover}(E, a) = \bigcup_{i \in (\{1,2,\ldots,K\} - j^*)} E_i$$

The entities in E_{j^*} are said to be left *uncovered.* We note that some of the set E_j could be empty.

The greedy algorithm works as follows. For the root of the decision tree, we pick the attribute $\widehat{a}$ having the maximum coverage; equivalently, we minimize the size of the heaviest branch (containing the uncovered entities). The attribute $\widehat{a}$ partitions $\mathcal{E}$ into $E_1, E_2, \ldots E_K$, where $E_i = \{e \ : \ e.\widehat{a} = i\}$. We recursively apply the greedy procedure on each of these subsets and get the required decision tree (See Figure 2). Let $\widetilde{\mathcal{T}}$ denote the decision tree obtained by this procedure. Let $\mathcal{T}_{\text{opt}}$ denote the optimal decision tree for $\mathcal{D}$. We claim that the greedy algorithm has an approximation ratio of $4 \log N$. The theorem is proved in the next section.

Theorem 1. *The greedy algorithm has an approximation ratio of* $4 \log N$, *meaning* $\text{cost}(\widetilde{\mathcal{T}}) \leq (4 \log N)\text{cost}(\mathcal{T}_{\text{opt}})$.

Procedure Greedy(E)
Input: $E \subseteq \mathcal{E}$, a set of entities in $\mathcal{D}$; **Output:** A decision tree T for the set E
Begin
If E is empty, Return an empty tree.
If $|E| = 1$, Return a tree with $x \in E$ as a singleton node.
Let $\widehat{a}$ be the attribute that covers the maximum number of entities:
$\widehat{a} = \text{argmax}_{a \in \mathcal{A}} \text{Cover}(E, a)$
Create the root node r with $\widehat{a}$ as its attribute label.
For $1 \le j \le K$,
Let $E_j = \{e \in E | e.\widehat{a} = j\}$
$T_j = \text{Greedy}(E_j)$
Let r_j be the root of T_j. Add a branch from r to r_j with label j and join T_j to T.
Return T with r as the root.
End

Fig. 2. The Greedy Algorithm

4 Analysis of the Greedy Algorithm: Proof of Theorem 1

We extend the notion of cost to all subtrees of the greedy tree $\widetilde{T}$. Consider an internal node $v \in \widetilde{T}$ and let $\widetilde{T}_v$ be the subtree rooted at v. Let $L(v)$ denote the entities (or leaves) in $\widetilde{T}_v$; these are the entities that appear below v in $\widetilde{T}$. For instance, if r is the root of $\widetilde{T}$, then $\widetilde{T} = \widetilde{T}_r$ and $L(r) = \mathcal{E}$. Let $I(\widetilde{T}_v)$ denote the internal nodes of $\widetilde{T}_v$. Define

$$\text{cost}(\widetilde{T}_v) = \sum_{e \in L(v)} (\text{length of path from } v \text{ to } e).$$

Here, length refers to the number of internal nodes in the path from v to e.

Remark: Notice that $\widetilde{T}_v$ is the output of the greedy procedure, if $L(v)$ is the input set of entities. We say that v *handles* the set of entities given by $L(v)$.

Proposition 1. *For any internal node* $v \in \mathcal{T}$, $\text{cost}(\widetilde{T}_v) = \Sigma_{x \in I(\widetilde{T}_v)} |L(x)|$.

Proof. Consider an entity $e \in L(v)$. Let ℓ be the length of the path from v to e; this is the cost contributed by e in $\text{cost}(\mathcal{T}_v)$. Note that there are ℓ internal nodes on this path. We charge a unit cost to each of these internal nodes. Then, the charge accumulated at each internal node x is exactly $|L(x)|$. The proposition follows by summing up these accumulated charges over all the internal nodes of $\widetilde{T}_v$. □

The following proposition is useful in proving Theorem 1.

Proposition 2. *Let $E \subseteq \mathcal{E}$ be a set of entities. Let $E_1, E_2, \ldots, E_d$ be any disjoint subsets of E. Let $T_1^*, T_2^*, \ldots T_d^*$ be the optimal decision trees for $E_1, E_2, \ldots, E_d$, respectively and let T^* be the optimal decision tree for E. Then,*

$$\text{cost}(T_1^*) + \text{cost}(T_2^*) + \cdots + \text{cost}(T_d^*) \le \text{cost}(T^*).$$

Proof. Consider any E_i. We can construct a decision tree T_i for E_i from T^* as follows. Remove from T^* any entity (i.e, leaf) not appearing in E_i; iteratively remove any internal node having no children. Cleary, the resulting tree T_i is a decision tree for E_i. For any entity $e \in E_i$, the cost paid by e in T_i is the same as the cost paid by e in T^*. Since E_i's are disjoint, we have that

$$\text{cost}(T_1) + \text{cost}(T_2) + \cdots + \text{cost}(T_d) \leq \text{cost}(T^*).$$

The lemma now follows directly. □

We shall prove following lemma which generalizes Theorem 1. For a subset of entities $E \subseteq \mathcal{E}$, let T^*_E denote the optimal decision tree for E. Consider an internal node $u \in \widetilde{\mathcal{T}}$. $T^*_{L(u)}$ is the optimal decision tree for $L(u)$. Theorem 1 is obtained by applying the following lemma to the root node r of $\widetilde{\mathcal{T}}$.

Lemma 1. *For any internal node* $u \in \widetilde{\mathcal{T}}$, $\text{cost}(\widetilde{T}_u) \leq (4 \log n)\text{cost}(T^*_{L(u)})$ *where* $n = |L(\widetilde{u})|$.

Proof. The lemma is proved by induction on the number of internal nodes in the subtree under consideration. If $\widetilde{T}_u$ has only one internal node, then each entity pays a cost of 1; this is optimal. Now, let us prove the inductive step.

We state some important definitions first. An internal node x of $\widetilde{\mathcal{T}}$ is said to be *terminal*, if all its children are leaves. For a non-terminal internal node x of $\widetilde{\mathcal{T}}$, its *centroidal child* is defined to be the child of x having the maximum number of leaves (breaking ties arbitrarily). We now extend this to the notion of *centroidal path.* Consider the subtree $\widetilde{T}_v$ rooted at some internal node v of $\widetilde{\mathcal{T}}$. Starting at the root node of $\widetilde{T}_v$ (i.e., the node v), iteratively keep picking the centroidal child till we reach an internal node having only leaves as its children. Such a path is called the centroidal path of $\widetilde{T}_v$ and is denoted by $\text{Cent}(\widetilde{T}_v)$. Cost of the centroidal path is defined as

$$\text{cost}(\text{Cent}(\widetilde{T}_v)) = \sum_{x \in \text{Cent}(\widetilde{T}_v)} |L(x)|.$$

Remark: It is easier to visualize the centroidal path if we make the centroidal child of each node as the right most child. Then, the $\text{Cent}(\widetilde{T}_v)$ will be the right extreme path starting at v.

We now prove the inductive step in Lemma 1. Consider an internal node u. We wish to show that $\text{cost}(\widetilde{T}_u) \leq (4 \log n)\text{cost}(T^*_{L(u)})$, where $n = |L(u)|$

Cost of $\widetilde{T}_u$ can be decomposed into two parts by considering cost incurred along the centroidal path of $\widetilde{T}_u$ and the remaining cost. We say that an internal node v of $\widetilde{T}_u$ is *adjacent* to the centroidal path $\text{Cent}(\widetilde{T}_u)$, if v is not part of the centroidal path, but it is a child of some node in the centroidal path. Let Adj denote the set of all nodes adjacent to $\text{Cent}(\widetilde{T}_u)$. Then, the cost of $\widetilde{T}$ can expressed as below; the expression follows from Proposition 1.

$$\text{cost}(\widetilde{T}_u) = \text{cost}(\text{Cent}(\widetilde{T}_u)) + \sum_{v \in \text{Adj}} \text{cost}(\widetilde{T}_v). \quad (1)$$

Consider any $v \in \text{Adj}$. Since v is a non-centroidal node, we have $|L(v)| \leq n/2$. Moreover, the number of internal nodes in $\widetilde{T}_v$ is less than that of u. Thus, by the inductive hypothesis, we have that

$$\text{cost}(\widetilde{T}_v) \leq 4\log(n/2)T^*_{L(v)}. \quad (2)$$

In Lemma 2 (see Section 5), we show that

$$\text{cost}(\text{Cent}(\widetilde{T}_u)) \leq 4\text{cost}(T^*_{L(u)}) \quad (3)$$

Assuming the above bound, we can complete the proof of Lemma 1. So,

$$\begin{aligned}
\text{cost}(\widetilde{T}_u) &= \text{cost}(\text{Cent}(\widetilde{T}_{L(u)})) + \sum_{v\in\text{Adj}} \text{cost}(\widetilde{T}_v) && \text{(from Eqn. 1)}\\
&\leq 4\text{cost}(T^*_{L(u)}) + \sum_{v\in\text{Adj}} \text{cost}(\widetilde{T}_v) && \text{(from Eqn. 3)}\\
&\leq 4\text{cost}(T^*_{L(u)}) + \sum_{v\in\text{Adj}} 4\log(n/2)\text{cost}(T^*_{L(v)}) && \text{(from Eqn. 2)}\\
&\leq 4\text{cost}(T^*_{L(u)}) + 4\log(n/2)\text{cost}(T^*_{L(u)})\\
& && \text{(from Prop. 2, taking } E = L(u))\\
&\leq 4\log(n)\text{cost}(T^*_{L(u)})
\end{aligned}$$

We have completed the proof of Lemma 1 and hence, Theorem 1, assuming Lemma 2. □

5 Bound on the Centroidal Cost

In this section, we will prove the following lemma.

Lemma 2. *For any internal node* $u \in \widetilde{T}$, $\text{cost}(\text{Cent}(\widetilde{T}_u)) \leq 4\text{cost}(T^*_{L(u)})$.

Proof. Let us start by simplifying the notation. Consider any internal node $u \in \widetilde{T}$. Let $\widetilde{T} = \widetilde{T}_u$ be the greedy subtree rooted at u. and let $E = L(u)$ be the entities handled by $\widetilde{T}$. Notice that $\widetilde{T}$ is nothing but the tree constructed by the greedy algorithm, if E is the set of input entities. Let $T^* = T^*_{L(u)}$ be the optimal tree for E. Our goal is to argue that $\text{cost}(\text{Cent}(\widetilde{T}_u)) \leq 4\text{cost}(T^*)$.

Let r be the length of $\text{Cent}(\widetilde{T})$ and let the nodes on the path be $u_1, u_2, \ldots, u_r$. For $1 \leq i \leq r$, write $E_i = L(u_i)$. Let $a_1, a_2, \ldots a_r$ be the attributes associated with $u_1, u_2, \ldots, u_r$, respectively. Consider a node u_i, for some $1 \leq i < r$. Let $C_i = \text{Cover}(E_i, a_i)$; we say that u_i *covers* the entities in C_i. The node u_r is a boundary case, and is handled separately. The children of u_r are all leaves and these are given by E_r. Recall that $\text{Cover}(E_r, a_r)$ includes all the entities in E_r, except one. For the node u_r, we define $C_r = E_r$. Note that each entity in E is covered by exactly one node among $u_1, u_2, \ldots, u_r$. See Figure 3(a); here, $r = 4$.

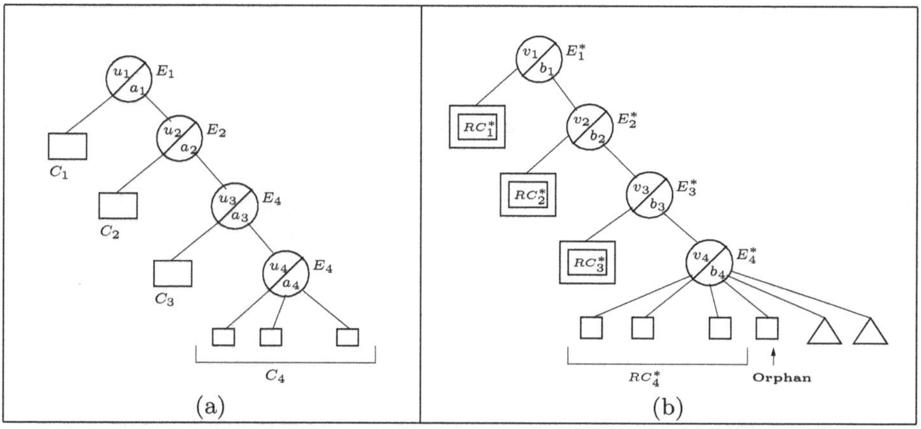

Fig. 3. Centroidal paths of the greedy and the optimal tree

Remark: Each node u_i handles the set of entities given by E_i. The node splits E_i into two parts: the set C_i covered by u_i and the set E_{i+1} handled by u_{i+1}. An alternative way to view $\text{cost}(\text{Cent}(\widetilde{T}))$ is as follows. Each entity in C_i pays a cost of i. Now, $\text{cost}(\text{Cent}(\widetilde{T}))$ is the sum of the cost paid by all the entities, i.e., $\text{cost}(\text{Cent}(\widetilde{T})) = \sum_{i=1}^{r} i|C_i|$. The above function is similar to the objective function in the MSSC problem. This connection allows us to utilize ideas from the analysis of the greedy algorithm for MSSC by Feige et al. [10].

Recall that the cost of the centroidal path $\text{cost}(\text{Cent}(\widetilde{T}))$ is given by $\Sigma_{i=1}^{r}|E_i|$. Consider a node u_i, for some $1 \leq i \leq r$. We imagine that the node u_i pays a cost of $|E_i|$ towards the centroidal cost. Distribute this cost equally among the entities covered by u_i: for each $e \in C_i$, define its price

$$p_e = \frac{|L(u_i)|}{|C_i|}$$

Notice that

$$\text{cost}(\text{Cent}(\widetilde{T})) = \sum_{e \in E} p_e.$$

Now, let us consider the optimal tree T^*. For each entity $e \in E$, let $\ell^*(e)$ denote length of the path from the root to e in T^*; this is viewed as the cost paid by e in T^*. Recall that $\text{cost}(T^*) = \Sigma_{e \in E}\ell^*(e)$.

We will now bound the price paid by any entity e in the greedy solution $\widetilde{T}$ by the cost paid by some entity e^* in the optimal solution T^* (within a factor of two). For this, we define a mapping $\pi : E \to E$ that maps entities in $\widetilde{T}$ to entities in T^* such that an entity e in $\widetilde{T}$ will charge its price p_e to the entity $\pi(e)$ in T^*. Our mapping π will satisfy the following two properties:

1. For any entity $e^* \in E$, there are at most two entities mapped to it (i.e., $|\{e : \pi(e) = e^*\}| \leq 2$).
2. For any entity $e \in E$, $p_e \leq 2\ell^*(\pi(e))$.

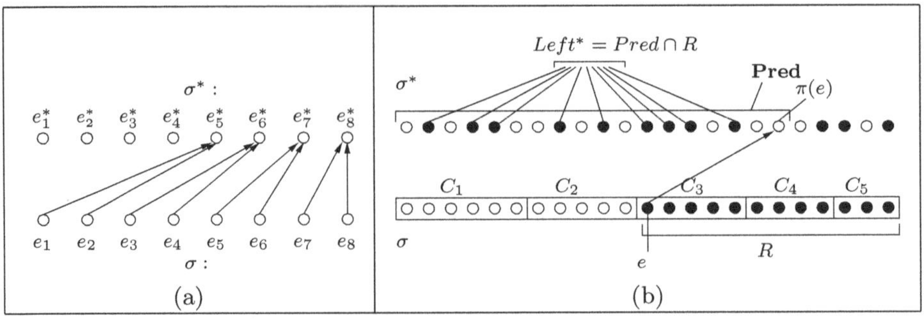

Fig. 4. Illustration of the mapping π and *Left**

We call such a mapping to be *nice*. We shall exhibit a nice mapping π in Section 6. Given such a mapping, we can now see that $\text{cost}(\text{Cent}(\widetilde{T})) \leq 4\text{cost}(T^*)$. This completes the proof of Lemma 2. □

6 A Nice Mapping

In this section, we exhibit a nice mapping π claimed in the proof of Lemma 2.

We arrange the nodes in E in the order in which they are covered by the nodes $u_1, u_2, \ldots, u_r$. Namely, first place all the nodes in C_1 (in some arbitrary order), followed by those in C_2 and so on, finally ending with those in C_r. Let this ordering be denoted $\sigma = e_1, e_2, \ldots e_n$ (where $n = |E|$). In this ordering, the entities come in blocks: first C_1, then C_2 and so on, ending with C_r.

Now, let us consider the optimal tree T^*. Arrange the entities in ascending order of cost $\ell^*(\cdot)$ they pay in T^*. Namely, all the entities that pay a cost of 1 (if any) are placed first (in some arbitrary order), followed by those paying a cost of 2 (if any) and so on. Let this ordering by denoted as $\sigma^* = e_1^*, e_2^*, \ldots, e_n^*$. In σ^*, as we scan from left to right, the cost paid by the entities increases monotonically.

We now define the mapping π as follows. Scan the lists σ and σ^* from right to left, picking two entities from σ and mapping them both to a single entity in σ^*. Formally, if e is an entity appearing in position i from the right in σ, it will be attached to the entity appearing in position $\lceil i/2 \rceil$ from the right in σ^*. See Figure 4(a). Clearly, every entity in σ^* has at most two entities from σ mapped to it. Thus, the mapping π satisfies the Part (1) of the niceness property. Now, we shall show that π also satisfies Part (2) of the niceness property.

6.1 Bounding Greedy Price

In this section, we will show that for any entity e in σ, $p(e) \leq 2\ell^*(\pi(e))$. We will use the following notations.

Consider any subset of entities $S \subseteq \mathcal{E}$. We extend the notion of coverage considering only the entities in S and define the concept of $\text{Cover}_S(\cdot,\cdot)$. Let $X \subseteq \mathcal{E}$ be a subset of entities and a be an attribute. The attribute a partitions the entities X into $X_1, X_2, \ldots, X_K$, where $X_j = \{e \in X : e.a = j\}$. Let $X_{j'}$ be set among these that has the maximum size, taking only entities in S into account; meaning, let $j' = \text{argmax}_i |X_i \cap S|$. Then, $\text{Cover}_S(X, a)$ is defined to include every set X_i, except $X_{j'}$:

$$\text{Cover}_S(X, a) = \bigcup_{i \in (\{1,2,\ldots,K\} - j')} (X_i \cap S)$$

Consider the optimal tree T^* and $S \subseteq \mathcal{E}$ be some subset of entities. In T^*, we define the notion of *centroidal path with respect to* S. For any internal node $y \in T^*$, let $L(y)$ be the leaves in the subtree rooted at y. By considering only those entities in S, we get $L_S(y)$; i.e., define $L_S(y) = L(y) \cap S$. Consider an internal node $v \in T^*$. Among the children of v, let y be node having the maximum value of $L_S(y)$ (breaking ties arbitrarily); such a y is called the *S-centroidal child of v*. The *S-centroidal path* of T^* is defined next: starting with the root node of T^*, we traverse the tree by picking up the S-centroidal child in each step, until we reach some node y in which the entities in $L_S(y)$ are all children of y.

We are now ready to prove the Part (2) of the niceness property.

Lemma 3. *For any entity e in σ, $\ell^*(\pi(e)) \geq p(e)/2$.*

Proof. Consider some entity e in σ. Let u_h be the node covering e, so that $e \in C_h$. Recall that u_h lies on the centroidal path of $\widetilde{T}$. It handles the entities $E_h = \cup_{i=h}^r C_i$. If $h < r$, u_h splits E_h into C_h (which it covers) and $E_{h+1} = \cup_{i=h+1}^r C_i$. Let $R = E_h$. Note that $p_e = |R|/|C_h|$.

Let s be the length of R-centroidal path of T^* and let $v_1, v_2, \ldots, v_s$ be nodes appearing on the path. For $1 \leq i \leq s$, let $E_i^* = L(v_i)$ be the entities handled by v_i. Let $b_1, b_2, \ldots, b_s$ be the attributes associated with these nodes. We next define the coverage obtained by each node along the R-centroidal path. For $1 \leq i \leq s$, define $RC_i^* = \text{Cover}_R(E_i^*, b_i)$. Consider $1 \leq t \leq s$. We say that the node v_t *covers* all the entities in RC_t^* and that each entity $\alpha \in RC_t^*$ is said to be covered in the time-step t. The last node v_s presents a boundary case. The entities in $L_R(v_s)$ are all children of v_s; among these all, except one, are included in RC_s^* and covered by v_s; we call the excluded entity as *orphan in* T^*. So, all entities in σ^*, except the orphan, are covered by node v_i. See Figure 3(b); here $s = 4$.

Let *Left** is the entities in R that appear before $\pi(e)$ in σ^*. Formally, let *Pred* be the set of all entities that appear before $\pi(e)$ in σ^* (including $\pi(e)$). Then, *Left** = *Pred* $\cap R$. See Figure 4(b).

Observation 1: Notice that for any entity $e' \in$ *Left**, $\ell^*(e') \leq \ell^*(\pi(e))$ (since, cost of entities in σ^* is monotonically increasing).

We now present an outline of the proof. Intuitively, we will show the following three claims: (i) $|$*Left*$^*| \geq |R|/2$; (ii) an entity covered in time-step t pays a cost of at least t in T^*; (iii) in each time-step t, the optimal tree T^* can cover at most $|C_h|$ entities from R. It would follow that it takes at least $|R|/(2|C_h|)$ time-steps to cover *Left**. Therefore, some entity in *Left** pays a cost of at least

$|R|/(2|C_h|)$. From Observation 1, it follows that $\ell^*(\pi(e)) \geq |R|/(2|C_h|)$. We now formally prove the claims.

Claim 1: $|Left^*| \geq \lceil R/2 \rceil$.

Proof. Since $e \in C_h$ and $R = \cup_{i=h}^{r} C_i$, we have that at most R entities (including e) appear to the right of e in σ. By the definition of the mapping π, we have that in σ^*, at most $\lceil R/2 \rceil$ entities appear to the right of $\pi(e)$ including $\pi(e)$. Thus, in σ^*, at most $\lceil R/2 \rceil - 1$ entities appear to the right of $\pi(e)$ excluding $\pi(e)$. It follows that $|Left^*| \geq \lfloor R/2 \rfloor + 1 \geq \lceil R/2 \rceil$. □

Claim 2: Consider any $1 \leq t \leq s$. Any entity α covered in the time-step t in T^* must satisfy $\ell^*(\alpha) \geq t$.

Proof. To reach to root of T^*, α must climb up along the first t nodes of the R-centroidal path. □

Claim 3: For $1 \leq t \leq s$, at most $|C_h|$ entities from R can be covered in time-step t, i.e, $|RC_t^*| \leq |C_h|$.

Proof. The claim is proved using the fact that the greedy procedure picked attributes that give maximum coverage. Refer to Figure 5 for an illustration. Recall that $L_R(v_t) \subseteq R$ is the entities from R handled by v_t and b_t is the attribute associated with v_t. The attribute b_t partitions $L_R(v_t)$ into $X_1^*, X_2^*, \ldots, X_K^*$, where $X_i^* = \{\alpha \in L_R(v_t) \ : \ \alpha.b_t = i\}$. Let $j^* = \text{argmax}_i |X_i^*|$. Then, note that

$$RC_t^* = \bigcup_{i \in \{1,2,\ldots,K\} - j^*} X_i^*.$$

By contradiction, suppose $|RC_t^*| > |C_h|$. It is easy to see that the union of any $K-1$ distinct sets from $X_1^*, X_2^*, \ldots, X_K^*$ contains at least $|C_h|$ entities. In the greedy tree, imagine what would happen if we picked b_t (instead a_h) as the attribute at the node u_h to split $R = E_h$. Let the partition obtained be $Y_1, Y_2, \ldots, Y_K$, where $Y_i = \{\alpha \in R \ : \ \alpha.b_t = i\}$. So, $\text{Cover}(R, b_t)$ is the union of some $K-1$ distinct subsets from $Y_1, Y_2, \ldots, Y_K$. Note that $X_i^* \subseteq Y_i$, for all $1 \leq i \leq K$. Therefore, $\text{Cover}(R, b_t)$ is a superset of the union of some $K-1$ subsets from $X_1^*, X_2^*, \ldots, X_K^*$. This implies that $|\text{Cover}(R, b_t)| > |C_h|$. This contradicts the fact that greedy's choice of a_h has the maximum coverage. □

The proof proceeds intuitively as follows. By Claim 3, at most C_h entities from R are covered in each time step. Since $Left^* \subseteq R$, it follows that at most C_h entities in $Left^*$ are covered in each time step. Therefore, it takes at least $|Left^*|/|C_h|$ time-steps to cover all the entities in $Left^*$. Since $|Left^*| \geq \lceil R/2 \rceil$ (by Claim 1), it takes at least $\frac{\lceil R/2 \rceil}{|C_h|}$ steps to cover all the entities in $Left^*$. Thus, by Claim 2, some entity in $Left^*$ pays a cost of at least $\frac{\lceil |R|/2 \rceil}{|C_h|}) \geq \frac{|R|}{2|C_h|}$. From Observation 1, it follows that $\ell^*(\pi(e)) \geq \frac{|R|}{2|C_h|}$.

The only issue in the above argument is that by definition, orphan is not said to be covered by any node. This is a problem if orphan belongs to $Left^*$.

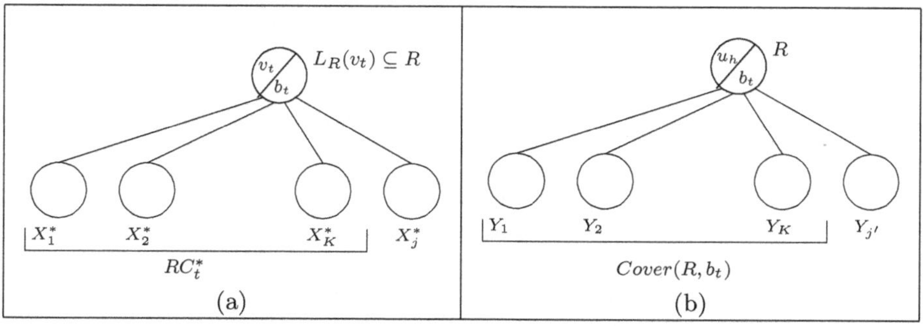

Fig. 5. Illustration for Claim 3 of Lemma 3

However, this needs only a minor argument. Notice that orphan pays a cost of at least s. Recall that any internal node of the R-centroidal path of T^* can cover at most $|C_h|$ entities from R. Therefore at most $s|C_h|$ entities of R can be covered in s time-steps. But, in the s time-steps all the entities in R, except the orphan get covered. Therefore, $s|C_h| \geq R - 1$ and hence, $s \geq (R-1)/|C_h|$. It follows that $s \geq R/(2|C_h|)$, for $R \geq 2$. It is assured that $R \geq 2$, since u_r is an internal node of $\widetilde{T}$ and therefore, has at least two children. Therefore, $s \geq \frac{R}{2|C_h|}$. Thus, orphan pays a cost of at least $\frac{R}{2|C_h|}$. Since orphan is in *Left**, by observation 1, we conclude that $\ell^*(\pi(e)) \geq \frac{|R|}{2|C_h|} = p_e/2$. □

References

1. Murthy, S.: Automatic construction of decision trees from data: A multi-disciplinary survey. Data Mining and Knowledge Discovery 2(4), 345–389 (1998)
2. Moret, B.: Decision trees and diagrams. ACM Computing Surveys 14(4), 593–623 (1982)
3. Hyafil, L., Rivest, R.: Constructing optimal binary decision trees is NP-complete. Information Processing Letters 5(1), 15–17 (1976)
4. Garey, M.: Optimal binary identification procedures. SIAM Journal on Applied Mathematics 23(2), 173–186 (1972)
5. Kosaraju, S., Przytycka, M., Borgstrom, R.: On an optimal split tree problem. In: Workshop on Algorithms and Data Structures (1999)
6. Adler, M., Heeringa, B.: Approximating optimal binary decision trees. In: Goel, A., Jansen, K., Rolim, J.D.P., Rubinfeld, R. (eds.) APPROX and RANDOM 2008. LNCS, vol. 5171, pp. 1–9. Springer, Heidelberg (2008)
7. Heeringa, B.: Improving Access to Organized Information. Ph.D. thesis, University of Massachusetts, Amherst (2006)
8. Garey, M., Graham, R.: Performance bounds on the splitting algorithm for binary testing. Acta Informatica 3, 347–355 (1974)
9. Chakaravarthy, V., Pandit, V., Roy, S., Awasthi, P., Mohania, M.: Decision trees for entity identification: approximation algorithms and hardness results. In: ACM Symposium on Principles of Database Systems (2007)
10. Feige, U., Lovász, L., Tetali, P.: Approximating min sum set cover. Algorithmica 40(4), 219–234 (2004)

Annotations in Data Streams

Amit Chakrabarti[1], Graham Cormode[2], and Andrew McGregor[3]

[1] Dartmouth College
ac@cs.dartmouth.edu
[2] AT&T Labs–Research
graham@research.att.com
[3] University of Massachusetts, Amherst
mcgregor@cs.umass.edu

Abstract. The central goal of data stream algorithms is to process massive streams of data using *sublinear* storage space. Motivated by work in the database community on outsourcing database and data stream processing, we ask whether the space usage of such algorithms be further reduced by enlisting a more powerful "helper" who can *annotate* the stream as it is read. We do not wish to blindly trust the helper, so we require that the algorithm be convinced of having computed a correct answer. We show upper bounds that achieve a non-trivial tradeoff between the amount of annotation used and the space required to verify it. We also prove lower bounds on such tradeoffs, often nearly matching the upper bounds, via notions related to Merlin-Arthur communication complexity. Our results cover the classic data stream problems of selection, frequency moments, and fundamental graph problems such as triangle-freeness and connectivity. Our work is also part of a growing trend — including recent studies of multi-pass streaming, read/write streams and randomly ordered streams — of asking more complexity-theoretic questions about data stream processing. It is a recognition that, in addition to practical relevance, the data stream model raises many interesting theoretical questions in its own right.

1 Introduction

The data stream model has become a popular abstraction when designing algorithms that process network traffic and massive data sets [4, 21]. The computational restrictions that define this model are severe: algorithms must use a relatively small amount of working memory and process input in whatever order it arrives. This captures constraints in high-throughput data processing settings. For example, network monitoring often requires (near) real-time response to anomalies and hence traffic must be processed as it arrives, rather than being stored and processed offline. For massive data sets stored in external memory, being able to process the data in any order avoids the I/O bottlenecks that arise with algorithms that assume random access. Unfortunately, while some problems admit efficient streaming algorithms, many others provably require a lot of working memory or multiple passes over the data, which is typically not feasible.

This paper considers the potential for off-loading stream computation to a more powerful "helper" so that single pass, small-space stream computation is possible even for such "hard" functions. The additional power of the helper can arise in a variety of situations, e.g., multiple processing units, special purpose hardware, or a third party who provide a commercial stream processing service. This last case has recently garnered

S. Albers et al. (Eds.): ICALP 2009, Part I, LNCS 5555, pp. 222–234, 2009.

attention in the context of outsourcing database processing [27, 29, 34]. A key issue is that we do not want to blindly trust the helper: hardware faults or outright deception by a third-party would lead to incorrect results. So our protocols must have sufficient information contained in the help to allow the "verifier" to be convinced that they have obtained the correct answer. We think of this help as annotations augmenting the original stream. Our goal is to design protocols so that the verifier finds the correct answer with an honest helper, and is likely not fooled by a dishonest helper. The primary metrics are the amount of annotations provided by the helper and the amount of working space used by the verifier.

Our approach is naturally related to Interactive Proofs and Merlin-Arthur communication protocols [1, 5, 25] but differs in two important regards. Firstly, the verifier must process both the original data and the advice provided by the helper under the usual restrictions of the data stream model. Secondly, we focus on annotations that can be provided *online*. Note that in Merlin-Arthur communication, it is assumed that the helper is omniscient and that the advice he provides can take into account data held by any of the players. In the stream model, this would correspond to *prescience* where the annotation in the stream at position t may depend on data that is yet to arrive. In contrast we are primarily interested in designing algorithms with online annotation, i.e., annotation that only depends on data that has arrived before the annotation is written. This corresponds to a helper who sees the data concurrently with the verifier.

Our Contributions: We first formally define the relevant models: traditional and online Merlin-Arthur communication, and streaming models with either prescient or online annotations. We then investigate the complexity of a range of problems in these models, including selection, frequency moments, and graph problems such as triangle-counting and connectivity. Estimating frequency moments in particular has become a canonical problem when exploring variants of the data stream model such as random order streams [10] and read/write streams [7]. Our results include:

- *Selection.* The problem of finding the median of m values in the range $[n]$ highlights the difference between prescient and online annotation. For any h, v such that $hv \geq m$ we present an $O(v \log m)$-space algorithm that uses $O(h \log m \log n)$ bits of online annotation. Furthermore, we show that this trade-off is optimal up to polylogarithmic factors. In contrast, a trivial $O(\log mn)$ space algorithm can verify $O(\log n)$ bits of prescient annotation.
- *Frequency Moments and Frequent Items.* We next consider properties of $\{f_i\}_{i \in [n]}$ where f_i is the frequency of the token "i". For any h, v such that $hv \geq n$, we present an $O(h \log m)$-space algorithm that uses $(\phi^{-1} v \log m)$ bits of online annotation and returns exactly the tokens whose frequency exceeds ϕm. We also show an $O(\log m)$ space algorithm that uses $O(\varepsilon^{-1} \log^2 m)$ bits of online annotation and returns a set of tokens containing $\{i : f_i \geq \phi m\}$ and no elements from $\{i : f_i \leq (\phi - \varepsilon) m\}$. This algorithm relies on a powerful way that annotation can be used in conjunction with sketch-based algorithms. For any h, v such that $hv \geq n$, we present an $O(kv \log m)$-space algorithm that uses $O(k^2 h \log m)$ bits of online annotation and computes $F_k = \sum_i f_i^k$ exactly ($k \in \mathbb{Z}_+$). The trade-off is optimal up to polylogarithmic factors even if the algorithm is allowed to use prescient annotation. To prove this we present the first Merlin-Arthur communication bounds for multi-party set-disjointness.

- *Graph Problems.* For graphs defined by streams of m edges on n nodes, we show that only $O(\log n)$ space is needed by the verifier to determine whether a graph is connected, contains a perfect matching, or is triangle-free, with annotation proportional to the input size. We show that our algorithms are optimal in many cases. For any h, v such that $hv \geq n^3$, we also present an $\tilde{O}(v)$ space algorithm for counting triangles that uses $\tilde{O}(h)$ bits of annotation where $\tilde{O}$ hides poly-logarithmic factors.

Related Work: When multiple passes over the input are allowed, it is natural to consider annotations that can be written to the "input tape" and are available to the stream algorithm in subsequent passes [3,14,15]. The read/write stream model, which provides both multiple passes and multiple working tapes, can be viewed as a natural extension of the multi-pass annotation model [7,8,20]. However, such annotations are of no use if only a single pass over the input is allowed.

Few examples of prior work have explicitly considered annotations that are provided by an (untrusted) third party. Gertner et al. [19] showed that the set of languages recognized by a verifier with logarithmic space given annotation polynomial in the input size is exactly NP. In contrast, our focus is on the case where the annotation is (sub)linear in the input size and can be provided online; the distinction between prescient and online annotation was not relevant in their results because with polynomial annotation, the entire input could be repeated. Feigenbaum et al. [17] observe that a logarithmic space verifier can check a linear space annotation for the disjointness problem. In communication complexity, the role of non-deterministic advice has been studied more extensively, see e.g., [5,26]. Recent works of Aaronson and Widgerson [1] and Klauck [25] are particularly relevant. They resolve the MA complexity of two-party set disjointness — we extend some of their techniques to our streaming model.

There has also been more applied work which implicitly defines annotation protocols. The notion of *stream punctuations* are, in our terminology, simple prescient annotations, indicating facts such as that there are no more tuples relevant to timestamp t in the remainder of the stream [33]. Work on stream outsourcing studies the problem of verifying that a claimed "grouping" corresponds to the input data [34]. They solve exact and approximate versions of the problem by using a linear amount of annotation. Lastly, work on *proof infused streams* answers various selection and aggregation queries over sliding windows [27] with logarithmic space and linear annotation. However, a critical difference is that this work requires that the helper and verifier agree on a one-way hash function, for which it is assumed the helper cannot find collisions. Our results are in a stronger model without this assumption.

2 Models and Definitions

2.1 Communication Models

Let $f : X_1 \times \cdots \times X_t \to \{0,1\}$ be a function, where each X_i is a finite set. This naturally gives a t-player number-in-hand communication problem, where Player i holds an input $x_i \in X_i$ and the players wish to output $f(x_1, \ldots, x_t)$ correctly, with high probability.

MA Communication: We first consider a variant of this communication model. A Merlin-Arthur protocol (henceforth, "MA protocol") for f is one that involves the usual t players, plus a "super-player," called Merlin, who knows the entire input $\mathbf{x} = (x_1, \ldots, x_t)$. The protocol works as follows: first Merlin deterministically writes a help message $\mathfrak{h}$ on the blackboard, and then Players 1 through t run a randomized protocol $\mathscr{P}$, using a public random string R, eventually outputting a bit $\mathrm{out}(\mathscr{P}; \mathbf{x}, R, \mathfrak{h})$. To clarify, R is not known to Merlin at the time he writes $\mathfrak{h}$. An MA protocol is δ-error if there exists a function $\mathfrak{h} : X_1 \times \ldots \times X_t \to \{0,1\}^*$, such that:

1. If $f(\mathbf{x}) = 1$ then $\Pr_R[\mathrm{out}(\mathscr{P}; \mathbf{x}, R, \mathfrak{h}(\mathbf{x})) = 0] \leq \delta$.
2. If $f(\mathbf{x}) = 0$ then $\forall \mathfrak{h}'$ $\Pr_R[\mathrm{out}(\mathscr{P}; \mathbf{x}, R, \mathfrak{h}') = 1] \leq \delta$.

We define $\mathrm{err}(\mathscr{P})$ to be the minimum δ such that the above conditions are satisfied. We also define the *help cost* $\mathrm{hcost}(\mathscr{P})$ to be the maximum length of $\mathfrak{h}$, over all $\mathbf{x}$, and the *verification cost* $\mathrm{vcost}(\mathscr{P})$ to be the maximum number of bits communicated by Players 1 through t over all $\mathbf{x}$ and R. Finally, we define the *cost* of $\mathscr{P}$ to be $\mathrm{cost}(\mathscr{P}) = \mathrm{hcost}(\mathscr{P}) + \mathrm{vcost}(\mathscr{P})$. We then define the δ-error MA-complexity of f as $\mathrm{MA}_\delta(f) = \min\{\mathrm{cost}(\mathscr{P}) : \mathscr{P}$ is an MA protocol for f with $\mathrm{err}(\mathscr{P}) \leq \delta\}$. Further, we define $\mathrm{MA}(f) = \mathrm{MA}_{1/3}(f)$.

Online-MA Communication: We also consider a variant of the above model, specific to *one-way protocols* (i.e., protocols where the players speak once each, in increasing order), where Merlin constructs t help messages $\mathfrak{h}_1, \ldots, \mathfrak{h}_t$ so that the ith message is only a function of the first i inputs. To make this precise we need to amend the definition of δ-error: An online-MA protocol is δ-error if there exists a family of functions $\mathfrak{h}_i : X_1 \times \ldots \times X_i \to \{0,1\}^*$, such that:

1. If $f(\mathbf{x}) = 1$ then $\Pr_R[\mathrm{out}(\mathscr{P}; \mathbf{x}, R, \mathfrak{h}_1(x_1), \mathfrak{h}_2(x_1, x_2), \ldots, \mathfrak{h}_t(x_1, \ldots, x_t)) = 0] \leq \delta$.
2. If $f(\mathbf{x}) = 0$ then $\forall \mathfrak{h}' = (\mathfrak{h}'_1, \mathfrak{h}'_2, \ldots, \mathfrak{h}'_t)$ $\Pr_R[\mathrm{out}(\mathscr{P}; \mathbf{x}, R, \mathfrak{h}') = 1] \leq \delta$.

The message $\mathfrak{h}_i$ is revealed privately to the ith player. We define the help cost, $\mathrm{hcost}(\mathscr{P})$, to be the maximum length of $\sum_{i \in [t]} |\mathfrak{h}_i|$. We define $\mathrm{err}(\mathscr{P}), \mathrm{vcost}(\mathscr{P})$, and $\mathrm{cost}(\mathscr{P})$ as for MA. Define $\mathrm{MA}^{\rightarrow}_\delta(f) = \min\{\mathrm{cost}(\mathscr{P}) : \mathscr{P}$ is an online MA protocol for f with $\mathrm{err}(\mathscr{P}) \leq \delta\}$ and write $\mathrm{MA}^{\rightarrow}(f) = \mathrm{MA}^{\rightarrow}_{1/3}(f)$.

2.2 Data Stream Models

The annotated data-stream models are most conveniently defined relative to the above communication models. Again we consider the computation of a function f on a t-tuple $\mathbf{x} \in \mathscr{U}^t$ for some universe $\mathscr{U}$, e.g., $\{0,1\}$ or $[n]$. The main difference from the communication model is that we further insist that the message sent by player i must be computed with limited memory and only sequential access to x_i and $\mathfrak{h}_i$. Without advice, this is equivalent to the usual definition of the single-pass data stream model. We will also consider non-Boolean functions f and a notion of approximation: we say f is computed correctly if the answer returned is in some pre-defined set $C(f(\mathbf{x}))$, e.g., $\{a : |a - f(\mathbf{x})| \leq \varepsilon f(\mathbf{x})\}$.

Stream Model with Prescient Annotations: In the context of the stream model we consider the help $\mathfrak{h}$ provided by Merlin to be decomposed into t (deterministic) functions that map the input to binary help strings: $\mathfrak{h}_1 : \mathscr{U}^t \to \{0,1\}^*, \ldots, \mathfrak{h}_t : \mathscr{U}^t \to \{0,1\}^*$.

Let $\mathfrak{h}(\mathbf{x}) := (\mathfrak{h}_1(\mathbf{x}), \ldots, \mathfrak{h}_t(\mathbf{x}))$. We then consider a randomized protocol, $\mathscr{A}$, with oracle access to a random string R, where Player i computes a message of size at most w given only w bits of working memory and only sequential access to the bit stream $\langle x_i, \mathfrak{h}_i(\mathbf{x})\rangle$. The output of this protocol is allowed to include the special symbol $\perp$ if the verifier is not convinced of the validity of the annotation. Such a protocol is said be δ-error if $\Pr_R[\text{out}(\mathscr{A};\mathbf{x},R,\mathfrak{h}) \notin C(f(\mathbf{x}))] \leq \delta$ and $\Pr_R[\text{out}(\mathscr{A};\mathbf{x},R,\mathfrak{h}') \neq \perp] \leq \delta$ for any $\mathfrak{h}' = (\mathfrak{h}'_1, \mathfrak{h}'_2, \ldots, \mathfrak{h}'_t) \neq \mathfrak{h}(\mathbf{x})$. We define $\text{err}(\mathscr{A})$ to be the minimum δ such that the above conditions are satisfied. We define the *help cost* $\text{hcost}(\mathscr{A})$ to be the maximum length of $\sum_i |\mathfrak{h}_i|$, over all $\mathbf{x}$, and the *verification cost* $\text{vcost}(\mathscr{A}) = w$. We say that $\mathscr{A}$ and $\mathfrak{h}$ forms an (h,v) *prescient scheme* if $\text{hcost}(\mathscr{A}) = O(h+1)$, $\text{vcost}(\mathscr{A}) = O(v+1)$ and $\text{err}(\mathscr{A}) < 1/3$.

Stream Model with Online Annotations: For online annotations we insist that the ith help function is only a function of $(x_1, \ldots, x_i)$. The other definitions are as above. We say that $\mathscr{A}$ and $\mathfrak{h}$ form an (h,v) *online scheme* as above if $\text{hcost}(\mathscr{A}) = O(h+1)$, $\text{vcost}(\mathscr{A}) = O(v+1)$ and $\text{err}(\mathscr{A}) < 1/3$.

2.3 Preliminary Lemmas

In multiple places we make use of basic fingerprinting techniques which enable a verifier to test whether two large streams represent the same object using small space. Let $\mathbb{Z}_+$ denote the set of non-negative integers, and let $\mathbb{F}_q$ denote the finite field with q elements (whenever it exists). Let $A = \langle a_1, \ldots, a_m\rangle$ denote a data stream, with each $a_i \in [n]$. Then A implicitly defines a frequency distribution $\mathbf{f}(A) := (f_1, \ldots, f_n)$, where $f_j = |\{i \in [m] : a_i = j\}|$. Fingerprints are formed by computations over $\mathbb{F}_q$, as $\text{BF}_q(r,\mathbf{f}) := \prod_{j=1}^n (r-j)^{f_j}$. To make fingerprints, we choose q based on an *a priori* bound m on $\|\mathbf{f}\|_1$.

Lemma 1. *Let $q \geq m$ be a prime, and choose r uniformly at random from $\mathbb{F}_q$. Given an input stream A of length m, the fingerprint $\text{BF}_q(r,\mathbf{f}(A))$ can be computed using $O(\log q)$ storage. Suppose $\mathbf{f}' \in \mathbb{Z}_+^n$ is a vector with $\mathbf{f}' \neq \mathbf{f}(A)$ and $\|\mathbf{f}'\|_1 \leq m$. Then the "collision probability" $\Pr_{r \in_R \mathbb{F}_q}[\text{BF}_q(r,\mathbf{f}') = \text{BF}_q(r,\mathbf{f}(A))] \leq m/q$.*

The proof of this fact, along with other proofs, is deferred to the full version. This fingerprint implies a prescient protocol for a multi-set inclusion problem:

Lemma 2. *Let $A \subset \mathscr{U}$ be a set of size n and let $B \subset \mathscr{U}$ be multi-set of size t. Let B' be the set formed by removing all duplicate elements from B. Then, given a stream which begins with the elements of A followed by the elements of B, there is a $(t\log t, \log t)$ prescient scheme that establishes whether $B' = A$.*

3 Warm-Up: Index and Selection

In this section, we present an online scheme for the SELECTION problem: Given desired rank $\rho \in [m]$, output an item a_k from the stream $A = \langle a_1, \ldots, a_m\rangle \in [n]^m$, such that $|\{i : a_i < a_k\}| < \rho$ and $|\{i : a_i > a_k\}| \leq m - \rho$. We assume $m = \Theta(n)$ to simplify the statement of bounds. An easy $(\log m, \log m)$ prescient scheme is for the helper to give an answer s as annotation at the start of the stream. The verifier need only count how many items in the stream are (a) smaller than s and (b) greater than s. The verifier returns s if the rank of s satisfies the necessary conditions. Next, we present (almost) matching upper and lower bounds when only online annotation is allowed.

To do this, we first consider the online MA complexity of the communication problem of INDEX: Alice holds a string $x \in \{0,1\}^N$, Bob holds an integer $i \in [N]$, and the goal is for Bob to output $\text{INDEX}(x,i) := x_i$. The lower bound for SELECTION will follow from the lower bound for INDEX and a key idea for the SELECTION upper bound follows from the communication protocol for INDEX.

Theorem 1 (Online MA complexity of INDEX). *Let h and v be integers such that $hv \geq N$. There is a online MA protocol $\mathscr{P}$ for* INDEX, *with* $\text{hcost}(\mathscr{P}) \leq h$ *and* $\text{vcost}(\mathscr{P}) = O(v \log h)$*; and any online MA protocol $\mathscr{Q}$ for* INDEX *must have* $\text{hcost}(\mathscr{Q})\,\text{vcost}(\mathscr{Q}) = \Omega(N)$. *So, in particular,* $\text{MA}^{\rightarrow}(\text{INDEX}) = \tilde{\Theta}(\sqrt{N})$.

Proof. For the lower bound, we use the given online MA protocol $\mathscr{Q}$ to build a randomized one-way INDEX protocol $\mathscr{Q}'$. Let $h = \text{hcost}(\mathscr{Q})$. Let $\mathscr{B}(n,p)$ denote the binomial distribution with parameters n and p, and let k be the smallest integer such that $X \sim \mathscr{B}(k,1/3) \Rightarrow \Pr[X > k/2] \leq 2^{-h}/3$. A standard tail estimate gives $k = \Theta(h)$. Let $a(x,R)$ denote the message that Alice sends in $\mathscr{Q}$ when her random string is R, and let $b(\mathfrak{a},i,\mathfrak{h})$ be the bit Bob outputs upon receiving message $\mathfrak{a}$ from Alice and $\mathfrak{h}$ from Merlin. In the protocol $\mathscr{Q}'$, Alice chooses k independent random strings $R_1,\ldots,R_k$ and sends Bob $a(x,R_1),\ldots,a(x,R_k)$. Bob then outputs 1 iff there exists a h-bit string $\mathfrak{h}$ such that $\text{MAJORITY}\,(b(a(x,R_1),i,\mathfrak{h}),\ldots,b(a(x,R_k),i,\mathfrak{h})) = 1$. Clearly, $\text{cost}(\mathscr{Q}') \leq k \cdot \text{vcost}(\mathscr{Q}) = O(\text{hcost}(\mathscr{Q})\,\text{vcost}(\mathscr{Q}))$. We claim that $\mathscr{Q}'$ is a $\frac{1}{3}$-error protocol for INDEX whence, by a standard lower bound (see, e.g., Ablayev [2]), $\text{cost}(\mathscr{Q}') = \Omega(N)$.

To prove the claim, consider the case when $x_i = 1$. By the correctness of $\mathscr{Q}$ there exists a suitable help message $\mathfrak{h}$ from Merlin that causes $\Pr[b(a(x,R),i,\mathfrak{h}) = 0] \leq 1/3$. Thus, by construction and our choice of k, the probability that Bob outputs 0 in $\mathscr{Q}'$ is at most $2^{-h}/3$. Now suppose $x_i = 0$. Then, *every* possible message $\mathfrak{h}$ from Merlin satisfies $\Pr[b(a(x,R),i,\mathfrak{h}) = 1] \leq 1/3$. Arguing as before, and using a union bound over all 2^h possible messages $\mathfrak{h}$, we see that Bob outputs 1 with probability at most $2^h \cdot 2^{-h}/3 = \frac{1}{3}$.

The upper bound follows as a special case of the two-party set-disjointness protocol in [1, Theorem. 7.4] since the protocol there is actually online. We give a more direct protocol which establishes intuition for our SELECTION result. Write Alice's input string x as $x = y^{(1)} \cdots y^{(v)}$, where each $y^{(j)}$ is a string of at most h bits, and fix a prime q with $3h < q < 6h$. Let $y^{(k)}$ be the substring that contains the desired bit x_i. Merlin sends Bob a string z of length at most h, claiming that it equals $y^{(k)}$. Alice picks a random $r \in \mathbb{F}_q$ and sends Bob r and the strings $\text{BF}_q(r,y^{(1)}),\ldots,\text{BF}_q(r,y^{(v)})$, thus communicating $O(v \log q) = O(v \log h)$ bits. Bob checks if $\text{BF}_q(r,z) = \text{BF}_q(r,y^{(k)})$, outputting 0 if not. If the check passes, Bob assumes that $z = y^{(k)}$, and outputs x_i from z under this assumption. By Lemma 1, the error probability is at most $h/q < 1/3$.

Remark 1. The above lower bound argument in fact shows that an online MA protocol $\mathscr{P}$ for an *arbitrary* two-party communication problem f satisfies $\text{hcost}(\mathscr{P})\,\text{vcost}(\mathscr{P}) = \Omega(\text{R}^{\rightarrow}(f))$. Thus, $\text{MA}^{\rightarrow}(f) = \Omega(\sqrt{\text{R}^{\rightarrow}(f)})$ where $\text{R}^{\rightarrow}(f)$ is the one-way, randomized communication complexity of f.

Theorem 2. *For any h, v s.t. $hv \geq m$ there is a $(h\log m, v\log m)$ online scheme for* SELECTION *and any (h,v) online scheme for* SELECTION *must have $hv = \Omega(m)$.*

Proof. Conceptually, the verifier builds a vector $\mathbf{r} = (r_1, \ldots, r_n) \in \mathbb{Z}_+^n$ where $r_k = |\{j \in [m] : a_j < k\}|$. This is done by inducing a new stream A' from the input stream A: each token a_j in A causes virtual tokens $a_j+1, a_j+2, \ldots, n$ to be inserted into A'. Then $\mathbf{r} = \mathbf{f}(A')$; note that $\|\mathbf{r}\|_1 = O(m^2)$. As in the INDEX protocol, the vector $\mathbf{r}$ is arranged into v subvectors of dimension h, and the verifier retains only fingerprints — based on a prime $q = O(m^2)$ — on each subvector. After the stream is seen, the helper claims that the answer is s, by providing the values of r_i for all i in the subvector containing s. The verifier fingerprints the provided block, and outputs s if it agrees with their stored fingerprint, otherwise it returns $\perp$. For the lower bound, we use a standard reduction from the INDEX problem and this is deferred until the full version.

4 Frequency Moments and Frequent Items

In this section we consider properties of $\mathbf{f} = \{f_i : i \in [n]\}$ where f_i is the frequency of the token "i" in the stream. In particular, the kth frequency moment is defined as $F_k = \sum_{i\in[n]} f_i^k$ and the frequent items are defined as the set $\{i : f_i > T\}$, for some threshold T. Both problems have a long history in the data streams literature. It is well known that in the traditional data stream model, exact computation of F_k ($k \neq 1$) requires $\Omega(n)$ space. Even constant factor approximation requires $\Omega(n^{1-2/k})$ space [11].

Frequent Items. We prove results on finding exact and approximate frequent items. The approximate result relies on a powerful way that annotation can be used in conjunction with sketch based algorithms (such as Count-Sketch [12] and Count-Min [13]) and we expect this will have other applications. The approximate case is more complicated than the exact case and further discussion is deferred to the full version.

A prescient helper can list the set of claimed frequent items, along with their frequencies, for the verifier to check against the stream. But we must also ensure that the helper is not able to omit any items that exceed the threshold. Our result shows a compact witness set for the exact case, which leads to online schemes for the exact and approximate versions of the problem.

Theorem 3. *There exists a $(\phi^{-1}\log^2 m, \phi^{-1}\log^2 m)$ prescient scheme and a $(\phi^{-1}n^{\alpha}\log m,\ n^{1-\alpha}\log m)$ online scheme ($\alpha \in [0,1]$) for finding $\{i : f_i > T := \phi m\}$. Any (h,v) online scheme for this must have $hv = \Omega(n)$.*

Proof. The lower bound follows from the hardness of INDEX and we omit the simple reduction from this presentation. For the upper bound consider a binary tree whose leaves are the elements of the universe $[n]$. Associate each node v with the set of elements at the leaves of the the subtree rooted at v. Call this set $S(v)$ where $S(u) = \{i\}$ if u is the ith leaf. Let $g(v) = \sum_{i\in S(v)} f_i$. Note that if u is a node and v is any ancestor of u, then $g(u) \leq g(v)$. Now observe that there is a witness set of size $O(\phi^{-1}\log n)$ to identify all leaves i with $f_i > T$: this consists of the set W of all such is in addition to pairs of nodes (u,v) such that u is the child of v, and $g(u) \leq T$ but $g(v) > T$. Here, each pair $(u,v) \in W$ is witness to the fact that no leaves $i \in S(u)$ can have $f_i > T$. The sets $S(u)$ for such u together with $\{i : f_i > T\}$ form a partition of $[n]$. Further, there can be at most ϕ^{-1} such nodes v at any level of the binary tree, as the sum of $g(v)$ is at most m. This bounds the size of this witness set to $|W| = O(\phi^{-1}\log n)$. This leads to two schemes for the problem. In the first,prescient scheme, the helper lists the members of W and their corresponding frequencies. The verifier remembers this information, and ensures that it agrees with the frequencies in the stream. Assuming $m = \Omega(n)$ gives hcost = vcost = $\phi^{-1}\log^2 m$. In the second, online scheme, the $2n-1$ nodes in the tree are divided into v groups of h such that $hv \geq 2n$. The verifier keeps a fingerprint of the frequency vector of each group. After the stream is seen, the helper provides the witness set W, sorted by the natural order on nodes, plus the frequency vector of all groups containing items named in W. This totals $\min\{O(|W|h), n\}$ items, yielding a $(\min\{n\log m, h\phi^{-1}\log m\}, v\log m)$ online scheme. A subtlety here is that the output size can exceed the verifier's memory, so the verifier may output a partial result before returning $\perp$.

Frequency Moments. We now show a family of algorithms that exhibit an optimal verification/annotation trade-off for the exact computation of F_k. Our algorithm is inspired by the "algebrization" results of Aaronson and Wigderson [1] but the key idea can be traced back to classic interactive proof protocols of Lund et al. [28] and Shamir [31].

Theorem 4. *Suppose h and v are positive integers with $hv \geq n$. Then, for integers $k \geq 1$, there exists a $(k^2 h\log m, kv\log m)$ online scheme for computing F_k exactly.*

Proof. Let A be the input stream. We map the length n vector $\mathbf{f}(A)$ into an $h \times v$ matrix $(f(x,y))_{x\in[h],y\in[v]}$, using any canonical bijection between $[n]$ and $[h]\times[v]$. Pick a prime $q \geq \max\{m^k, 3kh\}$; since $m \geq n$, this can be done while ensuring that $\log q = O(k\log m)$. We shall work in the field $\mathbb{F}_q$, which is safe because q exceeds the maximum possible value of $F_k(A)$. Let $\tilde{f}(X,Y) \in \mathbb{F}_q[X,Y]$ be the unique polynomial satisfying $\deg_X(\tilde{f}) = h-1$, $\deg_Y(\tilde{f}) = v-1$ and $\tilde{f}(x,y) = f(x,y)$ for all $(x,y) \in [h]\times[v]$. The verifier picks a random $r \in \mathbb{F}_q$. As the stream is read, the verifier maintains a sketch consisting of the v quantities $\tilde{f}(r,1),\ldots,\tilde{f}(r,v)$. Clearly, this sketch fits in $O(v\log q)$ bits of storage.

At the end of the stream, the annotator provides a polynomial $s'(X) \in \mathbb{F}_q[X]$ that is claimed to be equal to $s(X) := \sum_{y\in[v]}\tilde{f}(X,y)^k$, which has degree at most $k(h-1)$, thus using $O(kh\log q)$ bits of annotation. The verifier evaluates $s'(r)$ from the supplied annotation and computes $s(r) = \sum_{y\in[v]}\tilde{f}(r,y)^k$ from his sketch, checks that $s'(r) = s(r)$ and outputs $\perp$ if not. If the check passes, the verifier outputs $\sum_{x\in[h]} s'(x)$ as the final answer. Clearly, this answer is correct if the annotation was honest. Further, the verifier is fooled only if $s' \neq s$, but $s'(r) = s(r)$; the probability of this is at most $k(h-1)/q \leq \frac{1}{3}$, by choice of q.

It remains to show that the sketch can be computed incrementally in $O(v\log q)$ space. To maintain each $\tilde{f}(r,y)$ for $y \in [v]$, note that upon reading a new token $i \in [n]$ that maps to $(a,b) \in [h]\times[v]$, the necessary update is of the form $\tilde{f}(r,y) \leftarrow \tilde{f}(r,y) + p_{a,b}(r,y)$, where $p_{a,b}(X,Y) = \prod_{i\in[h]\setminus\{a\}}(X-i)(a-i)^{-1}\cdot\prod_{j\in[v]\setminus\{b\}}(Y-j)(b-j)^{-1}$. Since $p_{a,b}(r,y) = 0$ for any $y \in [v]\setminus\{b\}$, the verifier need only update the single value $\tilde{f}(r,b)$, by adding $p_{a,b}(r,b)$, upon reading this token. Note that using a table of $O(v)$ appropriate precomputed values, this update can be computed efficiently. For $h = v = \sqrt{n}$, this takes a constant number of arithmetic operations per update.

Numerous problems such as computing Hamming distance and Inner Product, and approximating F_2 and F_∞, can be solved using F_k as a primitive or using related techniques. We defer discussion to the full version. We next present lower bounds on the trade-off possible for computation of F_k.

Theorem 5. *Any (h,v) scheme that exactly computes F_k requires $hv = \Omega(n)$ and any (h,v) scheme that approximates F_k up to a constant factor requires $hv = \Omega(n^{1-5/k})$.*

These bounds are based on bounds we prove on the MA complexity of $\text{DISJ}_{n,t} : \{0,1\}^{nt} \to \{0,1\}$, the t-party communication problem defined as follows. The input is a $t \times n$ Boolean matrix, with Player i holding the ith row, for $i \in [t]$. The desired output is $\wedge_{i=1}^{t} \vee_{j=1}^{n} \neg x_{ij}$, i.e., 1 iff the subsets of $[n]$ represented by the rows are disjoint. We call an input $\mathbf{x} = (x_{ij})_{i\in[t],j\in[n]}$ *valid* if every column of $\mathbf{x}$ has weight either 0 or 1 or t, and at most one column has weight t. Note that $\text{DISJ}_{n,t}$ is naturally related to frequency moments: for any valid input $\mathbf{x}$, $F_k(S) \geq t^k$ if $\text{DISJ}_{n,t}(\mathbf{x}) = 0$ and $F_k(S) \leq n$ if $\text{DISJ}_{n,t}(\mathbf{x}) = 1$ where S is the multi-set $\{j : x_{ij} = 1\}$. The next theorem, a generalization of a result by Klauck [25], and reductions from $\text{DISJ}_{n,2}$ or $\text{DISJ}_{n,O(n^{1/k})}$ establish the first and second parts of Theorem 5 respectively in a straightforward manner. The next theorem also resolves a question of Feigenbaum et al. [17].

Theorem 6. *Let $\mathscr{P}$ be an ε-error MA protocol for $\text{DISJ}_{n,t}$, where $\varepsilon \leq 1/3$. Then* $\text{hcost}(\mathscr{P}) \cdot \text{vcost}(\mathscr{P}) = \Omega(n/t^4)$. *In particular,* $\text{MA}(\text{DISJ}_{n,t}) = \Omega(\sqrt{n}/t^2)$.

Proof. A rectangle is defined as a subset of inputs of the form $\mathscr{X}_1 \times \cdots \times \mathscr{X}_t$, where each $\mathscr{X}_i \subseteq \{0,1\}^n$ is a subset of all possible inputs for Player i. In deterministic communication protocols, the inverse image of any transcript of such a protocol must be a rectangle. Let $A = \text{DISJ}_{n,t}^{-1}(1)$ and $B = \text{DISJ}_{n,t}^{-1}(0)$.

Lemma 3 (Alon-Matias-Szegedy [4], generalizing Razborov [30]). *There exists distribution μ over valid inputs with 1) $\mu(A) = \mu(B) = 1/2$ and 2) $\mu(T \cap B) = (2e)^{-1}\mu(T \cap A) - t2^{-n/2t^4}$ for each rectangle T.* □

Assume $t = \omega(n^{1/4})$ since otherwise the bound is trivial. Put $h = \text{hcost}(\mathscr{P})$ and $v = \text{vcost}(\mathscr{P})$. An input $\mathbf{x} \in A$ is said to be *covered* by a message $\mathfrak{h}$ from Merlin if $\Pr_R[\text{out}(\mathscr{P};\mathbf{x},R,\mathfrak{h}) = 0] \leq \varepsilon$. By correctness, every such input must be covered, so there exists a help message $\mathfrak{h}^*$ that covers every input in a set $G \subseteq A$, with $\mu(G) \geq 2^{-h}\mu(A) = 2^{-h-1}$. Fix Merlin's message in $\mathscr{P}$ to $\mathfrak{h}^*$ and amplify the correctness of the resulting randomized Merlin-free protocol by repeating it $O(h)$ times and taking the majority of the outputs. This gives us a randomized protocol $\mathscr{P}'$ for $\text{DISJ}_{n,t}$ with communication cost $c = O(hv)$ whose error, on every input in $G \cup B$, is at most 2^{-2h}. Let μ' denote the distribution μ conditioned on $G \cup B$. Note that, by condition (1) of Lemma 3,

$$\forall \mathbf{x} \in \{0,1\}^{nt} : \quad \text{either } \mu'(\mathbf{x}) = 0 \text{ or } \mu(\mathbf{x}) \leq \mu'(\mathbf{x}) \leq 2\mu(\mathbf{x}) . \tag{1}$$

By fixing the random coins of $\mathscr{P}'$ we can obtain a deterministic protocol $\mathscr{Q}$, for $\text{DISJ}_{n,t}$, such that $\text{err}_{\mu'}(\mathscr{Q}) \leq 2^{-2h}$ and $\text{cost}(\mathscr{Q}) = c$. By the rectangle property, there exist disjoint rectangles $T_1, T_2, \ldots, T_{2^c}$ such that $\text{out}(\mathscr{Q};\mathbf{x}) = 1$ iff $\mathbf{x} \in \bigcup_{i=1}^{2^c} T_i$. Therefore

$$\sum_{i=1}^{2^c} \mu'(T_i \cap B) \le 2^{-2h} \quad (2) \quad \text{and} \quad \mu'\left(A \setminus \bigcup_{i=1}^{2^c} T_i\right) \le 2^{-2h} \quad (3)$$

By (1), $\mu'(A) = \mu'(G) \ge \mu(G) \ge 2^{-h-1}$. Using (1), and a rearrangement of (3):

$$\sum_{i=1}^{2^c} \mu(T_i \cap A) \ge \frac{1}{2}\sum_{i=1}^{2^c} \mu'(T_i \cap A) \ge \frac{1}{2}\left(\mu'(A) - 2^{-2h}\right) \ge 2^{-h-3}.$$

Suppose $c \le n/5t^4$ and n is large enough. Applying condition (2) of Lemma 3 we get $\sum_{i=1}^{2^c} \mu(T_i \cap B) \ge 2^{-h-3}/(2e) - 2^c t 2^{-n/2t^4} \ge 2^{-h-6}$. However, by (1) and (2), we have $\sum_{i=1}^{2^c} \mu(T_i \cap B) \le 2^{-2h}$, a contradiction. Hence $hv = \Omega(c) = \Omega(n/t^4)$.

5 Graph Problems

In this section we consider computing properties of graphs on n nodes, determined by a stream of m edges [16,21]. We present tight results for testing connectivity of sparse graphs, determining if a bipartite graph has a perfect matching, and counting triangles. Almost all proofs are deferred to the full version.

Triangles via Matrix Multiplication. Estimating the number of triangles in a graph has received significant attention because of its relevance to database query planning (knowing the degree of transitivity of a relation is useful when evaluating relational queries) and investigating structure properties of the web-graph [6,9,23]. In the absence of annotation, any single pass algorithm to determine if there is a non-zero number of triangles requires $\Omega(n^2)$ bits of space [6]. We show that the answer can be verified with $O(n^2)$ annotation in logarithmic space. The following theorem, proved using ideas from [6] coupled with Theorem 6, shows that this is best possible.

Theorem 7. *Any (h, v) scheme for counting triangles must have $hv = \Omega(n^2)$.*

We now outline an online scheme with vcost $= O(\log n)$ and hcost $= O(n^2)$. A major subroutine of our algorithm is the verification of matrix multiplication in our model. That is, given $n \times n$ matrices A, B and C, verify that $AB = C$. Our technique extends the classic result of Frievalds [18] by showing that if the helper presents the results in an appropriate order, the verifier needs only $O(\log n)$ bits to check the claim. Note that this much annotation is necessary if the helper is to provide C in his stream.

Theorem 8. *There exists a $(n^2, \log n)$ online scheme for matrix multiplication.*

With this primitive, arbitrary matrix products $A_\ell, A_{\ell-1} \ldots A_2 A_1$ are verified with $O(\ell n^2)$ annotation by verifying $A_{2,1} := A_2 A_1$, then $A_{3,2,1} := A_3 A_{2,1}$, etc. Matrix powers A^ℓ are verified with $O(n^2 \log \ell)$ annotation.

Theorem 9. *There is a $(n^2, \log n)$ online scheme for counting triangles.*

Proof. Denote the graph adjacency matrix by A, with $A_{i,i} := 0$. The helper lists $A_{v,w}$ and $A^2_{v,w}$ for all pairs (v, w) in some canonical order. The verifier computes $\sum_{v,w} A_{v,w} A^2_{v,w}$ as the number of triangles. The verifier uses fingerprints to check that A matches the original set of edges, and the protocol in Theorem 8 to ensure that A^2 is as claimed.

We also show that it is possible to trade-off the computation with the helper in a "smooth" manner. The approach is based on an observation of Bar-Yossef et al. [6]:

The frequency moments of a derived stream can be expressed in terms of the number of triples of nodes with exactly $\{0,1,2,3\}$ edges between them. In small space we can induce a length $m(n-2)$ stream by replacing each edge (u,v) by the set of triples $\{(u,v,w) : w \neq u,v\}$. It follows that the number of triangles can be expressed in terms of the frequency moments of this derived stream, as $(F_3 - 2F_2 + F_1)/12$. By using the protocol of Theorem 4, we obtain the following theorem.

Theorem 10. *There is a* $(n^{3\alpha}, n^{3-3\alpha})$ *online scheme for counting triangles* ($\alpha \in [0,1]$).

Bipartite Perfect Matchings. We now present an online scheme for testing whether a bipartite graph has a perfect matching. Graph matchings have been considered in the stream model [16,35] and it can be shown that any single pass algorithm for determining the exact size of the maximum matching requires $\Omega(n^2)$ space. We show that we can off-load this computation to the helper such that, with only $O(n^2)$ annotation, the answer can be verified in $O(\log n)$ space. This is shown to be best possible by combining a reduction from [16] coupled with Theorem 1.

Theorem 11. *There exists a* $(m, \log n)$ *online scheme for bipartite perfect matching and any* (h,v) *online scheme for bipartite perfect matching requires* $hv = \Omega(n^2)$.

Connectivity. The problem of determining if a graph is connected was considered in the standard stream model [16, 21] and the multi-pass W-stream model [15]. In both models, it can be shown that any constant pass algorithm without annotations needs $\Omega(n)$ bits of space. In our model, the helper can convince a verifier with $O(\log n)$ space whether a graph is connected with only $O(m)$ annotation. This is the best possible for sparse graphs where $m = O(n)$ by combining a reduction from [16] with Theorem 1.

Theorem 12. *There exists a* $(m, \log n)$ *online scheme for connectivity and any* (h,v) *online scheme for connectivity requires* $hv = \Omega(n)$ *even when* $m = O(n)$.

Acknowledgements. We thank Yael Gertner, Sampath Kannan, and Mahesh Viswanathan for sharing [19]. We also thank Sudipto Guha and T. S. Jayram for helpful discussions.

References

1. Aaronson, S., Wigderson, A.: Algebrization: a new barrier in complexity theory. In: ACM STOC (2008)
2. Ablayev, F.: Lower bounds for one-way probabilistic communication complexity and their application to space complexity. Theoretical Computer Science 175(2), 139–159 (1996)
3. Aggarwal, G., Datar, M., Rajagopalan, S., Ruhl, M.: On the streaming model augmented with a sorting primitive. In: IEEE FOCS (2004)
4. Alon, N., Matias, Y., Szegedy, M.: The space complexity of approximating the frequency moments. Journal of Computer and System Sciences 58(1), 137–147 (1999)
5. Babai, L., Frankl, P., Simon, J.: Complexity classes in communication complexity theory (preliminary version). In: IEEE FOCS (1986)
6. Bar-Yossef, Z., Kumar, R., Sivakumar, D.: Reductions in streaming algorithms, with an application to counting triangles in graphs. In: ACM-SIAM SODA (2002)

7. Beame, P., Huynh-Ngoc, D.-T.: On the value of multiple read/write streams for approximating frequency moments. In: IEEE FOCS (2008)
8. Beame, P., Jayram, T.S., Rudra, A.: Lower bounds for randomized read/write stream algorithms. In: ACM STOC (2007)
9. Buriol, L.S., Frahling, G., Leonardi, S., Marchetti-Spaccamela, A., Sohler, C.: Counting triangles in data streams. In: ACM PODS (2006)
10. Chakrabarti, A., Cormode, G., McGregor, A.: Robust lower bounds for communication and stream computation. In: ACM STOC (2008)
11. Chakrabarti, A., Khot, S., Sun, X.: Near-optimal lower bounds on the multi-party communication complexity of set disjointness. In: IEEE CCC (2003)
12. Charikar, M., Chen, K., Farach-Colton, M.: Finding frequent items in data streams. Theor. Comput. Sci. 312(1), 3–15 (2004)
13. Cormode, G., Muthukrishnan, S.: An improved data stream summary: the count-min sketch and its applications. J. Algorithms 55(1), 58–75 (2005)
14. Demetrescu, C., Escoffier, B., Moruz, G., Ribichini, A.: Adapting parallel algorithms to the W-stream model, with applications to graph problems. In: Kučera, L., Kučera, A. (eds.) MFCS 2007. LNCS, vol. 4708, pp. 194–205. Springer, Heidelberg (2007)
15. Demetrescu, C., Finocchi, I., Ribichini, A.: Trading off space for passes in graph streaming problems. In: ACM-SIAM SODA (2006)
16. Feigenbaum, J., Kannan, S., McGregor, A., Suri, S., Zhang, J.: On graph problems in a semi-streaming model. Theoretical Computer Science 348(2-3), 207–216 (2005)
17. Feigenbaum, J., Kannan, S., Zhang, J.: Annotation and computational geometry in the streaming model. Technical Report YALEU/DCS/TR-1249, Yale University (2003)
18. Freivalds, R.: Fast probabilistic algorithms. In: Becvar, J. (ed.) MFCS 1979. LNCS, vol. 74, Springer, Heidelberg (1979)
19. Gertner, Y., Kannan, S., Viswanathan, M.: NP and streaming verifiers (manuscript, 2002)
20. Grohe, M., Hernich, A., Schweikardt, N.: Randomized computations on large data sets: tight lower bounds. In: ACM PODS (2006)
21. Henzinger, M.R., Raghavan, P., Rajagopalan, S.: Computing on data streams. In: External memory algorithms (1999)
22. Johnson, W., Lindenstrauss, J.: Extensions of Lipshitz mapping into Hilbert space. Contemporary Mathematics 26, 189–206 (1984)
23. Jowhari, H., Ghodsi, M.: New streaming algorithms for counting triangles in graphs. In: Wang, L. (ed.) COCOON 2005. LNCS, vol. 3595, pp. 710–716. Springer, Heidelberg (2005)
24. Kimbrel, T., Sinha, R.K.: A probabilistic algorithm for verifying matrix products using $o(n^2)$ time and $\log_2 n + o(1)$ random bits. Inf. Process. Lett. 45(2), 107–110 (1993)
25. Klauck, H.: Rectangle size bounds and threshold covers in communication complexity. In: IEEE CCC (2003)
26. Kushilevitz, E., Nisan, N.: Communication Complexity. CUP (1997)
27. Li, F., Yi, K., Hadjieleftheriou, M., Kollios, G.: Proof-infused streams: Enabling authentication of sliding window queries on streams. In: VLDB (2007)
28. Lund, C., Fortnow, L., Karloff, H., Nisan, N.: Algebraic methods for interactive proof systems. J. ACM 39(4), 859–868 (1992)
29. Papadopoulos, S., Yang, Y., Papadias, D.: Cads: Continuous authentication on data streams. In: VLDB (2007)
30. Razborov, A.: On the distributional complexity of disjontness. In: Paterson, M. (ed.) ICALP 1990. LNCS, vol. 443, Springer, Heidelberg (1990)

31. Shamir, A.: IP = PSPACE. J. ACM 39(4), 869–877 (1992)
32. Thorup, M., Zhang, Y.: Tabulation based 4-universal hashing with applications to second moment estimation. In: ACM-SIAM SODA (2004)
33. Tucker, P.A., Maier, D., Delcambre, L.M.L., Sheard, T., Widom, J., Jones, M.P.: Punctuated data streams (2005)
34. Yi, K., Li, F., Hadjieleftheriou, M., Kollios, G., Srivastava, D.: Randomized synopses for query assurance on data streams. In: IEEE ICDE (2008)
35. Zelke, M.: Weighted matching in the semi-streaming model. In: STACS, pp. 669–680 (2008)

The Tile Complexity of Linear Assemblies

Harish Chandran, Nikhil Gopalkrishnan, and John Reif

Department of Computer Science, Duke University, Durham, NC 27707
{harish,nikhil,reif}@cs.duke.edu

Abstract. The conventional Tile Assembly Model (TAM) developed by Winfree using Wang tiles is a powerful, Turing-universal theoretical framework which models varied self-assembly processes. We describe a natural extension to TAM called the Probabilistic Tile Assembly Model (PTAM) to model the inherent probabilistic behavior in physically realized self-assembled systems. A particular challenge in DNA nanoscience is to form linear assemblies or *rulers* of a specified length using the smallest possible tile set. These rulers can then be used as components for construction of other complex structures. In TAM, a deterministic linear assembly of length N requires a tile set of cardinality at least N. In contrast, for any given N, we demonstrate linear assemblies of expected length N with a tile set of cardinality $\Theta(\log N)$ and prove a matching lower bound of $\Omega(\log N)$. We also propose a simple extension to PTAM called κ-pad systems in which we associate κ pads with each side of a tile, allowing abutting tiles to bind when at least one pair of corresponding pads match and prove analogous results. All our probabilistic constructions are free from co-operative tile binding errors and can be modified to produce assemblies whose probability distribution of lengths has arbitrarily small tail bounds dropping exponentially with a given multiplicative factor increase in number of tile types. Thus, for linear assembly systems, we have shown that randomization can be exploited to get large improvements in tile complexity at a small expense of precision in length.

1 Introduction

Biological systems show a remarkable range of form and function. How are these multitude of systems constructed? What are the principles that govern them? In particular, as computer scientists, we ask if there are simple rules whose repeated application can give rise to such complex systems. This leads us to the study of self-assembly.

1.1 Fundamental Nature of Self-assembly

Self-assembly is a fundamental pervasive natural phenomenon that gives rise to complex structures and functions. It describes processes in which a disordered system of pre-existing components form organized structures as a consequence of specific, local interactions among the components themselves, without any external direction. In its most complex form, self-assembly encompasses the processes involved in growth and reproduction of higher order life. A simpler example of self-assembly is the orderly growth of crystals. In the laboratory, self-assembly

S. Albers et al. (Eds.): ICALP 2009, Part I, LNCS 5555, pp. 235–253, 2009.

techniques have produced increasingly complex structures [1,2] and dynamical systems [3]. The roots of attempts to model and study self-assembly begin with the study of tilings.

A *Wang tile* [4] is an oriented unit square with a pad associated with each side. Any two tiles with the same pad on corresponding sides are said to be of the same *tile type*. Tile orientation is fixed, they cannot be rotated or reflected[1]. Given a finite set S of Wang tiles types, a valid arrangement of S on a planar unit square grid consists of copies of Wang tiles from the set S such that abutting pads of all pairs of neighboring tiles match. The *tiling* or *domino* problem for a set of Wang tiles is: can tiles from S (chosen with replacement) be arranged to cover the entire planar grid? Berger [5] proved the undecidability of the tiling problem by reducing the halting problem [6] to it. Robinson [7] gave an alternative proof involving a simulation of any single tape deterministic Turing Machine by some set of Wang tiles. Garey and Johnson [8] and Lewis and Papadimitrou [9] proved that the problem of tiling a finite rectangle is NP-complete. These results paved the way for Wang tiling systems to be used for computation. But, Wang tilings do not model coordinated growth and hence do not describe complex self-assembly processes. Winfree [10] extended Wang tilings to the Tile Assembly Model (TAM) with a view to model self-assembly processes, laying a theoretical foundation [11,12] for a form of DNA based computation, in particular, molecular computation via assembly of DNA lattices with tiles in the form of DNA motifs.

The *tile complexity* [13] of assembling a shape is defined as the minimum number of tile types for assembling that shape. Tile complexity, apart from capturing the information complexity of shapes, is also important as there exist fundamental limits on the number of tile types one can design using DNA sequences of fixed length. Various ingenious constructions for shapes like squares [14], rectangles and computations like counting [15], XOR [16] etc. exist in this model. Lower bounds on tile complexity have also been shown for various shapes. Stochastic processes play a major role in self-assembly and have been investigated theoretically by Winfree [17] and Adleman [11] and in the laboratory by Schulman et al. [18]. However, TAM is deterministic in the sense that it produces exactly one terminal assembly given a tile set. This is because at most one type of tile is allowed to attach at any position in a partially formed assembly. See Section 2 for more details. This work investigates the effects of relaxing these constraints and reduces the number of tile types required to form linear assemblies of given length. In contrast to earlier work in stochastic self-assembly, we make tile attachments irreversible (as in TAM) and allow multiple tile types to attach at any position.

1.2 Motivation

A particular challenge in DNA nanoscience is to form linear assemblies or *rulers* of a specified length from unit sized square tiles. These rulers can then be used

[1] This is a valid assumption when implementing Wang tiles in the laboratory using DNA due to the complimentary nature of DNA strand binding.

as a component for construction of other complex structures. One can use these structures as beams and struts within the nanoscale (See Fig.1a). Linear assemblies can also serve as boundaries [18] and nucleation sites for more complex nanostructures. (Note that due to the inherent flexible nature of linear nanostructures, most complex nanostructures will generally tolerate small deviations from the intended lengths of these substructures). Various tile based techniques for constructing linear assemblies have been successfully explored in the laboratory [19,18]. Hence, tile assembly models for linear assemblies are apt theoretical frameworks for exploring a fundamental and important challenge in DNA nanoscience. In TAM, rulers of length N can be trivially constructed by deterministic assembly of N distinct tile types. This is also the matching linear lower bound for size of tile sets in deterministic TAM, as shown in Section 4. Thus, it is impractical to form large linear structures using the deterministic techniques of TAM. Long thin rectangles (which are approximations of linear assemblies) can be formed using $\Theta(\frac{\log N}{\log \log N})$ tile types but they suffer errors due to co-operative tile binding. In contrast with linear assemblies, the number of tile types to form an $N \times N$ square is only $\Theta(\frac{\log N}{\log \log N})$ [14], which is exponentially better than the lower bound for linear assemblies. This bound for squares is asymptotically tight for almost all N as dictated by information theory[13] while the one for linear assemblies is not. This begs the question: why are we not able to reach information theoretic limit of $\Theta(\frac{\log N}{\log \log N})$ in linear structures using TAM? Is this lower bound tight? What is the longest (finite) linear assembly one can assemble with a set of n tile types in realistic tiling models? What changes to TAM will give us the power to specify the linear systems using a smaller tile set? While square assemblies have been extensively studied [13,14,20,21], many questions remain about linear assemblies, which are simpler constructs yet are fundamental building blocks at the nanoscale. We answer a number of these questions and show novel, interesting results using techniques that differ considerably from existing ones. While there have been numerous variations on TAM in recent years, their impact on laboratory techniques in DNA self-assembly are minimal. At the same time, design principles used in DNA self-assembly do not fully leverage the programmability and stochasticity inherent to self-assembly. Hence, our goal is to develop a simple model that directs design principles of experimental DNA self-assembly by taking advantage of inherent stochasticity of self-assembly. It is noteworthy that the techniques for designing and analyzing these simple constructs under our simple model are non-trivial and theoretically rich.

(a) (b)

Fig. 1. (a) Possible nanostructures using rulers as substructures. (b) Diagonal tiles: Colors indicate pad type. Red pads are implemented using complimentary DNA. Strands for other pads are omitted.

1.3 Related Work in Self-assembly Using Probabilistic and Randomized Models

Non-determinism was used in tiling by Lagoudakis et al. [22] for implementing an algorithm for SAT. Recently, Becker et al. [23] describe probabilistic tile systems that yield squares, rectangles and diamond in expectation using $O(1)$ tile types. This work was extended by Ming-Yang Kao et al. [21] to yield arbitrarily close approximations to squares with arbitrarily high probability using $O(1)$ tile types. Both these papers allow precise arbitrary relative concentrations of tile types with no cost incurred in tile complexity. In the laboratory, achieving precise arbitrary relative concentrations between tiles is infeasible. Also, the descriptional complexity of tile systems in such models include not just the descriptional complexity of the tile set, but also the descriptional complexity of the concentration function. Thus, size of tile set producing an assembly is not a true indicator of its complexity. In PTAM, the set of tiles is a multi-set that implicity defines relative concentrations and precludes arbitrary relative concentrations. Thus, size of the tile set producing an assembly is a true indicator of its complexity. In addition, all our constructions have equimolar tile concentration and hence are experimentally feasible. Reif [24], and later Demaine et al. [25] discuss *staged self-assembly*. Demaine et al. [25] show how to get various shapes using $O(1)$ pad types. Aggarwal et al. [20] introduce various extensions to TAM and study the impact of these extension on both running time and the number of tile types. Compared to the above, PTAM is a simple extension to TAM that requires no laboratory techniques beyond those used to implement TAM. In particular, we consider standard one pot reaction mixtures with no intermediate purification steps. The Kinetic Tile Assembly Model (kTAM) proposed by Winfree [17] models kinetics and thermodynamics of DNA hybridization reactions. Schulman et al. [18] used DX tiles consisting of DNA stands to create one dimensional boundaries within the nanoscale. Adleman [11] proposed a mathematical theory of self-assembly which is used to investigate linear assemblies. While many fundamental theoretical questions arise in these models, the question of tile complexity of linear assemblies is uninteresting due the existence of the trivial lower bound mentioned in Section 1.2. Thus, the questions about linear self-assemblies examined in this paper are original and the constructions presented are novel.

1.4 Main Results

We describe a natural extension to TAM in Section 3 to allow randomized assembly, called the Probabilistic Tile Assembly Model (PTAM). A restriction of the model to diagonal, haltable, uni-seeded, and east-growing systems (defined in Section 3), which we call the standard PTAM is considered in this paper. Prior work in DNA self-assembly strongly suggests that standard PTAM can be realized in the laboratory. We show various non-trivial probabilistic constructions in PTAM for forming linear assemblies with a small tile set in Section 4, using techniques that differ considerably from existing assembly techniques. In particular, for any given N, we demonstrate linear assemblies of expected length N with tile set of cardinality $\Theta(\log N)$ using one pad per side of each tile in Section 4.2. We derive a matching lower bound of $\Omega(\log N)$ on the tile complexity of linear assemblies of any given expected length N in standard PTAM systems using one pad per side of each tile in Section 4.3. This lower bound, which holds for all N, is tight and better than the information theoretic lower bound of $\Omega(\frac{\log N}{\log \log N})$ which holds only for almost all N. We also propose a simple extension to PTAM in Section 5 called κ-pad systems in which we associate κ pads with each side of a tile, allowing abutting tiles to bind when at least one pair of corresponding pads match. This gives linear assemblies of expected length N with 2-pad (two pads per side of each tile) tile set of cardinality $\Theta(\frac{\log N}{\log \log N})$ tile types for infinitely many N. We show that we cannot achieve smaller tile complexity by proving a lower bound of $\Omega(\frac{\log N}{\log \log N})$ for each N on the cardinality of the κ-pad (κ pads per side of each tile) tile set required to form linear assemblies of expected length N in standard κ-pad PTAM systems for any constant κ. The techniques used for deriving these lower bounds are notable as they are stronger and differ from traditional Kolmogorov complexity based information theoretic methods used for lower bounds on tile complexity. Kolmogorov complexity based lower bounds do not preclude the possibility of achieving assemblies of very small tile multiset cardinality for infinitely many N while our lower bounds do, as they hold for every N. All our probabilistic constructions can be modified to produce assemblies whose probability distribution of lengths has arbitrarily small tail bounds dropping exponentially with a given k at the cost of a multiplicative factor k increase in number of tile types, as proved in Section 6.

2 The Tile Assembly Model for Linear Assemblies

This section describes the Tile Assembly Model (TAM) by Winfree for linear (1D) assemblies (henceforth referred to as LTAM). For a complete and formal description of the model see [13]. LTAM describes deterministic linear assemblies. The next section extends the model by introducing randomization. This paper considers only one-dimensional grid of integers $\mathbb{Z}$ which simplifies the definitions of the model. The directions $\mathfrak{D} = \{\text{East}, \text{West}\}$ are functions from $\mathbb{Z}$ to $\mathbb{Z}$, with $\text{East}(x) = x + 1$ and $\text{West}(x) = x - 1$. We say that x and x' are neighbors if $x' \in \{\text{West}(x), \text{East}(x)\}$. Note that $\text{East}^{-1} = \text{West}$ and vice versa. $\mathbb{N}$ is the set of natural numbers.

A *Wang tile* over the finite set of distinct *pads* Σ is a unit square where two opposite sides have pads from the set Σ^2. Formally, a tile t is an ordered pair of pads $(W_t, E_t) \in \Sigma^2$ indicating pad types on the West and East sides respectively. Thus, a tile cannot be reflected. For each tile t, we define $\text{pad}_{\text{East}}(t) = E_t$ and $\text{pad}_{\text{West}}(t) = W_t$. Σ contains a special *null pad*, denoted by ϕ. The *empty* tile (ϕ, ϕ) represents the absence of any tile. Pads determine when two tiles attach.

[2] In general, for two dimensional assemblies, tiles have pads on all four sides. However, we do not use any pads on the North and South sides in this paper and hence omit them. Also, we allow for multiple pads on the sides of a tile in Section 5.

A function $g : \Sigma \times \Sigma \to \{0, 1\}$ is a binary *pad strength function* if it satisfies $\forall x, y \in \Sigma$, $g(x, y) = g(y, x)$ and $g(\phi, x) = 0$. Linear assemblies do not have co-operative tile binding, i.e, interactions of more than one pair of pads at a given step. Hence the temperature parameter used in TAM is redundant in linear assemblies where tiles have only one pad per side. Throughout this paper we assume only a binary pad strength function. In this model each tile has only a single pad on each of its sides (West and East) whereas in Section 5 we allow multiple pads per side for each tile.

A *linear tiling system*, $\mathbb{T}$, is a tuple $\langle T, S, g \rangle$ where T containing the empty tile is the finite set of tile types, $S \subset T$ is the set of *seed* tiles and g is the binary pad strength function. A *configuration* of T is a function $A : \mathbb{Z} \to T$ with $A(0) = s$ for some $s \in S$. For $D \in \mathfrak{D}$ we say the tiles at x and $D(x)$ *attach* if $g(\text{pad}_D(A(x)), \text{pad}_{D^{-1}}(A(D(x)))) = 1$. *Self-assembly* is defined by a relation between configurations, $A \to B$, if there exists a tile $t \in T$, a direction $D \in \mathfrak{D}$ and an empty position x such that t attaches to $A(D(x))$. We define $A \xrightarrow{*} B$ as the reflexive transitive closure of $\to$ and say B is *derived* from A. For all $s \in S$ a *start* configuration start_s is given by $\text{start}_s(0) = s$ and $\forall x \neq 0 : \text{start}(x) = \text{empty}$. A configuration B is *produced* if $\text{start}_s \xrightarrow{*} B$ for some $s \in S$. A configuration is *terminal* if it is produced from start_s for some $s \in S$ and no other configuration can be derived from it. Term($\mathbb{T}$) is the set of terminal configurations of $\mathbb{T}$. In TAM, a terminal configuration is thought of as the output of a tiling system given a seed tile $s \in S$. TAM requires that there be a unique terminal configuration for each seed. Note that it allows different attachment orders as long as they produce the same terminal configuration. This unique terminal configuration requirement means that given any non terminal configuration A, at most one $t \in T$ can attach at any given position. In this sense, TAM is deterministic. In the next section we will explore the effect of relaxing this condition of TAM.

DNA nanostructures can physically realize TAM as shown by Winfree et al. [10] with the DX tile and LaBean et al. [26] with the TX tile. Like the square tile in TAM, the DX and TX have *pads* that specify their interaction with other tiles. The pads are DNA sequences that attach via hybridization of complimentary nucleotides. Mao et al. [27] performed a laboratory demonstration of computation via tile assembly using TX tiles. Yan et al. [16] performed parallel XOR computation in the test-tube using Winfree's DX tile. Other simple computations have also been demonstrated. However, large and more complex computations are beset by errors and error correction remains a challenge towards general computing using DNA tiles.

3 The Probabilistic Tile Assembly Model

In TAM, the output of a tile system is said to be a shape of given fixed size (for example, square of side N, linear assemblies of length N) if the tile system *uniquely* produces it. In this paper, we consider some implications of relaxing this requirement. Instead of asking that a set of tiles produce a *unique* shape, we allow the set of terminal assemblies to contain *more than one shape* by designing tile systems which admit multiple tile type attachment at a given position in a configuration. Note that we do not allow pad mismatch errors. We also associate a probability of formation with each terminal assembly. These extensions and modifications to TAM are formalized for linear assemblies. Note that the definitions given below can be easily extended to assemblies in two-dimensions by introducing pads on North and South sides of tiles and including a temperature parameter τ as in [13] for co-operative binding effects.

3.1 The Probabilistic Tile Assembly Model (PTAM) for Linear Assemblies

A *probabilistic linear tiling system* $\mathbb{T}$ is given by the tuple $\langle T, S, g\rangle$, where T is a (finite) multiset of tile types, $S \subset T$ is the multiset of seed tiles and g is the binary pad strength function. The set of pad types Σ, tiles and configurations for $\mathbb{T}$ are defined as in Section 2. The *multiplicity* $\mathcal{M} : \Sigma \times \Sigma \to \mathbb{N}$ of a tile type is the number of times it occurs in T. T contains the empty tile type with $\mathcal{M}(empty) = 1$. Multiplicity models concentration. We assume a well-mixed, one pot reaction environment in which at each step some member of T is copied (chosen with replacement) from the pot with uniform probability. If the tile thus obtained can attach to the produced configuration, it does so, else a new member of T is copied with uniform probability in the next step. This continues till either a match is found or none exists, in which case the system halts. Note that this is a Gillespie simulation [28] with a seed serving as a nucleation site. A system with only one seed, $S = \{s\}$, is called *uni-seeded*. We consider only uni-seeded systems in this paper. The function *type(t)*, type : $T \to \Sigma \times \Sigma$, returns the tile type for any $t \in T$.

Self-assembly of a linear tiling system $\mathbb{T}$ is defined by a relation between set of positive probabilities and pair of configurations A and B as: $A \to^p_{\mathbb{T}} B$ (read as A gives B with probability p) if there exists a tile $t \in T$, a direction $D \in \mathcal{D}$ and an empty position x such that t attaches to $A(D(x))$ with positive probability p to give B where $p = \mathcal{M}(\text{type}(t))/\sum_{j\in\Delta} \mathcal{M}(\text{type}(j))$ where $\Delta = \{j|\text{ type}(j)$ attaches to $A(D(x))\}$. The closure of $\to^p_{\mathbb{T}}$, denoted by $\xrightarrow{*\,\hat{p}}_{\mathbb{T}}$ (read as 'derives'), is defined by the following transitive law: if $A \to^{p_1}_{\mathbb{T}} B$ and $B \to^{p_2}_{\mathbb{T}} C$ then $A \to^{p_1p_2}_{\mathbb{T}} C$. A configuration B is *produced* with positive probability p if $\text{start}_s \xrightarrow{*\,p}_{\mathbb{T}} B$. A configuration is *terminal* if it is produced from start_s and no other configuration can be derived from it with positive probability. Term($\mathbb{T}$) is the set of terminal configurations of $\mathbb{T}$. We associate a *probability of formation*, $P(A)$ to each produced configuration A recursively, as follows: $P(\text{start}_s) = 1$ and $P(B) = \sum_\Gamma p_k P(A_k)$ where $\Gamma = \{k|A_k \to^{p_k}_{\mathbb{T}} B\}$. *Length* of a produced configuration A, written as $|A|$, is the number of non-empty tiles in it.

A configuration A is called a *linear assembly of length N* if it is terminal and $|A| = N$. Following Rothemund and Winfree's terminology [13], a linear tiling system is defined to be *diagonal* iff $g(x, y) = 0$ for all x, y with $x \neq y$ and $g(x, x) = 1$ for all $x \neq \phi$. A tile t is *reachable* in $\mathbb{T}$ if it is part of some produced configuration. A tile $t \in T$ is a *capping tile* if t is reachable and there exists $D \in \mathfrak{D}$ such that $g(\text{pad}_D(t), \text{pad}_{D^{-1}}(t')) = 0$ for each $t' \in T$. For $D = $ East the tile is called *East capping* and for $D =$ West it is called *West capping.* A capping tile halts growth in either the East or West direction. Note that a tile other than the seed cannot be both East and West capping. A linear probabilistic tiling system $\mathbb{T}$ is *haltable* iff for each produced configuration A, there exists a terminal configuration B such that $A \xrightarrow{*\,p}_{\mathbb{T}} B$ with positive probability p. Each terminal configuration has a probability of formation associated with it. If $\mathbb{T}$ is haltable, some terminal configuration occurs with certainty as stated without proof in the following Lemma.

Lemma 1. *If* $\mathbb{T}$ *is a haltable probabilistic linear tiling system, then* $\sum_{A \in Term(\mathbb{T})} P(A) = 1$.

A linear tiling system is called *east-growing* if the West pad of the seed tile is ϕ. A *simulation* of a probabilistic tile system $\mathbb{T}$ by a probabilistic tile system $\mathbb{Q}$ is a bijection f between terminal configurations that preserves lengths and probabilities of formation of assemblies, i.e. $f : \text{Term}(\mathbb{T}) \to \text{Term}(\mathbb{Q})$ satisfying $|A| = |f(A)|$ and $P(A) = P(f(A))$ for each $A \in \text{Term}(\mathbb{T})$. Any probabilistic linear tiling system $\mathbb{T}$ can be simulated by an east-growing probabilistic linear tiling system $\mathbb{Q}$ using no more than twice the number of tile types of $\mathbb{T}$, in the following manner. For the seed $s = (W_s, E_s)$ of $\mathbb{T}$, let $s' = (\phi, E'_s)$ be the seed of $\mathbb{Q}$ and for each East-capping tile $c = (W_c, E_c)$ of $\mathbb{T}$ let $\mathbb{Q}$ contain tile $c' = (W'_c, W''_s)$. For all other tiles $t = (W_t, E_t)$ of $\mathbb{T}$, let $\mathbb{Q}$ contain tiles $t_r = (W'_t, E'_t)$ and $t_l = (E''_t, W''_t)$. The reader may verify that this is a simulation. Hence, we consider only east-growing tile systems in this paper. A probabilistic linear tiling system is *equimolar* if $\forall t \in T : \mathcal{M}(t) = 1$. Thus, for an equimolar tile system, the cardinality of T equals the number of tile types in it. A probabilistic linear tiling system is *two-way branching* if at most two tile types can attach at any given position for any given configuration. A probabilistic linear tiling system is *standard* if it is diagonal, haltable, uni-seeded and east-growing.

Diagonal tile systems were suggested by Winfree and Rothemund [13]. These systems are implementable using DNA tiles. Matching pads are implemented as perfect Watson-Crick complimentary DNA sequences (see Fig.1b). Non-diagonal tile systems are not implementable using this technique. For tile systems producing linear assemblies that are not haltable, the expected length of the assembly diverges. For linear assemblies, no advantage in tile complexity or tail bounds on length of assemblies results from using multiple seeds. *Thus, we consider only standard systems in this paper.* Achieving arbitrary concentration vectors is infeasible in laboratory implementations using molecules. In contrast, equimolar systems are frequently achieved by chemists for various reactions. We demonstrate a equimolar standard linear tiling system whose tile complexity matches the more general lower bound of $\Omega(\log N)$ applicable to all standard linear tiling systems.

3.2 Complexity Measures for Tile Systems

Recall that the tile complexity of a shape in TAM is defined as the number of different tile types in the smallest tile set that realizes the shape. The tile complexity in TAM is closely related to the size of the smallest Turing machine describing the shape [29]. While in TAM the shape is realized deterministically, in PTAM we drop the requirement that a shape be obtained uniquely and instead ask that it be approximated by our probabilistic tile systems. What should be the correct measure of descriptional complexity of shapes in such probabilistic systems? Consider a probabilistic linear tiling system with with three tile types (Seed, Growth and Halt) at a $1 : N : 1$ relative concentration such that it assembles into a linear assembly of expected length $N + 2$. Clearly, the number of distinct tile types does not completely describe the assembly process in the absence of information about relative concentrations. Thus concentrations must be taken into account in any measure that hopes to intrinsically capture descriptional complexity. There exist modifications of TAM [21,25,20] where the number of tile types does not correspond to the descriptional complexity of the shape. These systems encode the complexity elsewhere, like in the concentration, temperature, mechanism etc. In contrast, *the standard systems of* PTAM *encode all the description of the shape in the tile multiset* through multiplicity of a tile type which models its concentration. Thus, the (probabilistic) descriptional complexity of shapes corresponds to the cardinality of the tile multiset which we call tile complexity. Note that multiplicity of tiles in the multiset count distinctly towards tile complexity.

What is the effect of the probabilistic model on tile complexity? We demonstrate linear assemblies of fixed expected length N using a tile set of small cardinality. In general, we are asking if there is any benefit in sacrificing the exact description of a shape for a probabilistic description. For linear assemblies, the answer is yes, as we show in the next section.

4 Constructing Linear Assemblies of Expected Length N

In the standard TAM, the tile complexity for a linear assembly of length N is N. This is because if a tile type occurs at more than one position in the assembly, the sub-unit between these two positions can repeat infinitely often. This does not produce a linear assembly of length N. The PTAM does not suffer from this drawback. By making longer and longer chains less likely, we ensure that most chains are of length close to N. All our constructions can be shown to have exponentially decaying tail with a linear multiplicative increase in the number of tile types. So we focus on the expected lengths of linear assemblies in the following sections. All of our constructions for linear assemblies of expected length $N \in \mathbb{N}$ are standard, equimolar and two-way branching. The random variable L always denotes the length of the assembly. Specific tiles systems in the rest of this section are illustrated using *tile binding diagrams.* Each tile type is represented by a square, with labels distinguishing different tile types. All possible interactions among tiles are denoted via arrows that originate at the West side of some tile and terminate on the East side of some tile, indicating pad strengths of 1 between these tiles along these sides. Absence of arrows indicate that no possible attachment can occur, i.e. pad strength is 0. Thus, all our systems are temperature 1 assemblies which are more resilient to errors than assemblies at greater temperatures. The latter suffer errors due to co-operative tile binding [30,31]. Moreover, temperature 1 systems are easier to implement in the laboratory than higher temperature systems. Since we consider only equimolar systems for the rest of this section, the cardinality of our tile multisets equal the number of tile types. We use these terms interchangeably for equimolar systems.

4.1 Linear Assemblies of Expected Length N Using $O(\log^2 N)$ Tile Types

In this section we present a standard linear tiling system that achieves a linear assembly of expected length N for any given N using $O(\log^2 N)$ tile types. First, we give a construction for powers of two, i.e. for any given $N = 2^i$ for some $i \in \mathbb{N}$, we show how to construct N length linear assemblies using $\Theta(\log N)$ tile types. Then we extend this construction to all N by expressing N in binary and linking together the chains corresponding to 1s in the binary representation of N.

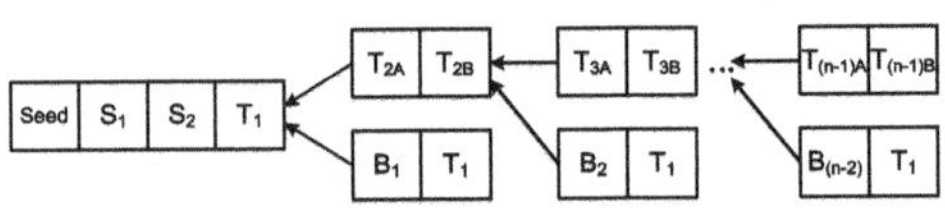

Fig. 2. Tile Binding Diagram for Powers of Two Construction

Powers of Two Construction: Fig.2 illustrates the tile set of size $3n + 2 = \Theta(n)$, used in a powers of two construction. The assembly halts only when the sequence $T_1, T_{2A}, T_{2B}, \ldots T_{(n-1)A}, T_{(n-1)B}$ of attachments is achieved. The bridge tiles B_i, $i = 1, 2 \ldots, n-2$, act as reset tiles at each stage of the assembly. Each probabilistic choice is between a reset in the form of B_i and progress towards completion in the form of $T_{(i+1)A}$. Attachment of T_1 to B_i and of T_{iB} to T_{iA} is deterministic.

Lemma 2. *Let L be the random variable equal to the length of the assembly. Then, $E[L] = 2^n$. Thus, an assembly of expected length 2^n can be constructed using $\Theta(n)$ tile types for any given $n \in \mathbb{N}$.*

Proof. We associate a sequence of independent Bernoulli trials, say coin flips, with the assembly process. Let the addition of the $\langle B_i, T_1 \rangle$ complex correspond to *Tails* and the addition of the $\langle T_{iA}, T_{iB} \rangle$ complex correspond to *Heads*. Halting of the assembly then corresponds to achieving a sequence of $n-2$ successive heads, corresponding to the sequence $\langle T_{2A}, T_{2B} \rangle, \ldots \langle T_{(n-1)A}, T_{(n-1)B} \rangle$ of attachments. The expected number of fair coin tosses for this to happen is $2(2^{(n-2)} - 1)$ [32]. Each coin toss adds two tiles to the linear assembly. Hence $E[L] = 4 + 4(2^{(n-2)} - 1) = 2^n$.

Extension to Arbitrary N: We extend the powers of two construction to all N by expressing N in binary, denoted by $B(N)$. For each ith bit of $B(N)$ $(i > 2)$ equal to 1, we have a power of two construction of expected length 2^i, using $3i-2$ tile types as in 4.1. We simply append these various constructions deterministically, and rely on linearity of expectation to achieve a linear assembly of length N in expectation.

Theorem 1. *An assembly of expected length N can be constructed using $O(\log^2 N)$ tile types for any given $N \in \mathbb{N}$.*

4.2 Linear Assemblies of Expected Length N Using $\Theta(\log N)$ Tile Types

In this section we present a standard linear tiling system that achieves linear assembly of length N in expectation for any given N using $\Theta(\log N)$ tile types. For powers of two, this construction reduces to one similar to that in Section 4.1. Our construction for general N is a more succinct than the one presented in Section 4.1. This new construction rests on the observation that the expected number of tiles of each type present in the powers of two construction decrease geometrically.

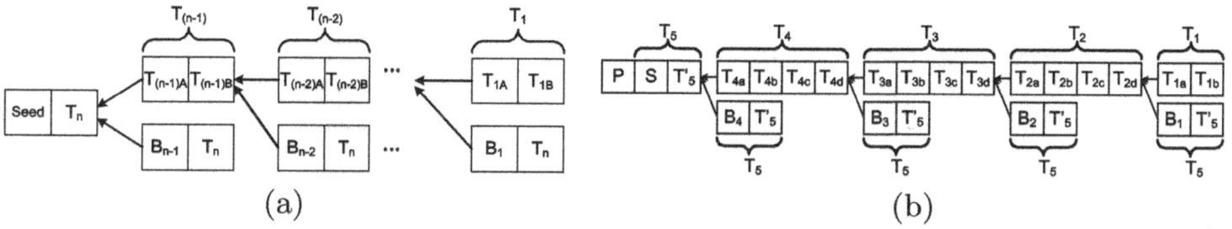

Fig. 3. Tile binding diagrams for $O(\log N)$ construction. (a) Tile Binding Diagram for Section 4.2. (b) Tile Binding Diagram: $N = 91$; $N'' = 90$; $N' = \frac{N''}{2} = 45 = (12221)_{\text{alt}2}$. P is the prefix tile.

Consider the linear tiling system depicted in Fig.3a. The size of the tile set is $3n-1 = \Theta(n)$. The expected length of the assembly is $N = 2^{n+1} - 2 = \Theta(2^n)$ [32]. We observe that the number of *bi-tiles* [3] of type T_i decrease geometrically as i decreases as stated below.

Lemma 3. *Let X_i be the random variable equal to the number of bi-tiles of type T_i in the final assembly. Then $E[X_{i-1}] = \frac{E[X_i]}{2}$ and hence $E[X_i] = 2^{i-1}$ for $i = 2, 3, \ldots, n$*

Proof. Every time a bi-tile of type T_i appears, a bi-tile of type T_{i-1} follows immediately in the resulting assembly with probability $1/2$ for $i = 2, 3 \ldots n$. So $E[X_{i-1}] = \frac{E[X_i]}{2}$. This property allows us to calculate the expected number of bi-tiles of each type. T_1 is the terminal bi-tile and appears exactly once. Hence its expectation is $1 = 2^0$. Repeated application of the above geometric decrease property proves the claim.

[3] A bi-tile T_i is a deterministic two tile complex T_{iA}, T_{iB}.

Next, we give an alternate binary encoding [33] for all non-zero natural numbers using $\{1, 2\}$ instead of the standard $\{0, 1\}$ encoding. This encoding will allow us to exploit the geometric decay property to build succinct constructions. The encoding of any non-zero natural number N is the Nth string in the lexicographic ordering of strings in $\{\mathbf{1}, \mathbf{2}\}^+$. An equivalent characterization s given below.

Lemma 4. *$\{1, 2\}$-Binary Encoding: For all non-zero natural numbers N, $\exists b_i \in \{1, 2\} : N = \sum_{i=0}^{n-1} b_i 2^i$ where $n \leq \lceil \log N \rceil$. Every N has a unique $\{\mathbf{1}, \mathbf{2}\}$-binary encoding.*

Now we show how to encode any N using $\Theta(\log N)$ tile types using the above two Lemmas. Fig.3b is an example illustrating the construction for $N = 91$. For any given N, let N'' be the greatest even number less than or equal to N. For $N' = \frac{N''}{2}$, let $B(N') = b_{n-1}b_{n-2} \dots b_0$ be its $\{\mathbf{1}, \mathbf{2}\}$-binary encoding of size n. For each bit b_i with $i \in \{0, 1, \dots, n-2\}$, our construction has a tile complex T_{i+1} of size $2b_i$ tiles that occurs X_{i+1} times with $E[X_{i+1}] = 2^i$. For the bit b_{n-1}, the tile complex T'_n of size $2b_{n-1} - 1$ tiles occurs X'_n times with $E[X'_n] = 2^{n-1}$. Each time T'_n is deterministically preceded by either the seed or one of the bridge tiles. Each such complex is called T_n. Thus T_n is of size of $2b_{n-1}$ tiles and occurs X_n times with $E[X_n] = 2^{n-1}$. For odd N, we deterministically prefix a single tile to the West of the seed tile.

Theorem 2. *The above construction has an expected length $E[L] = N$ tiles and uses $\Theta(\log N)$ tile types.*

Proof. The length of the assembly L is given by $L = X_1 + X_2 + \cdots + X_n + (N \mod 2)$ and hence by linearity of expectation, $E[L] = 2(\sum_{i=0}^{n-1} b_i 2^i) + (N \mod 2) = N$. The number of tile types is $\Theta(n) = \Theta(\log N)$.

4.3 Lower Bounds on the Tile Complexity of Linear Assemblies of Expected Length N

In this section we prove that for all N the cardinality of any tile multiset that forms linear assemblies of expected length N in standard PTAM systems is $\Omega(\log N)$. The techniques that we use for deriving these tile complexity lower bounds are notable as they differ from traditional information theoretic methods used for lower bounds on tile complexity and furthermore our low bound results hold for each N, rather than for almost all N.

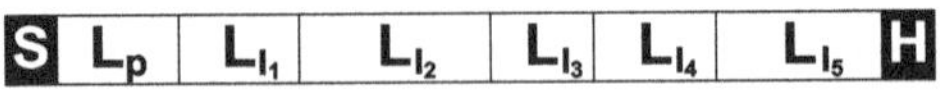

Fig. 4. $\mathbb{T}$ split into prefix and intermediates

Theorem 3. *For any N, the cardinality of any tile multiset that forms linear assemblies of expected length N in standard PTAM systems is $\Omega(\log N)$.*

Proof. We will show that any standard linear PTAM system with tile multiset cardinality n has expected length of assembly at most $O(2^n)$. This implies our result via the contrapositive. Recall that multiplicity of tiles in the multiset count distinctly towards tile complexity. Any standard PTAM linear tiling multiset with cardinality n that produces linear assemblies of greatest (finite) expected length is called *n-optimal.* Optimal linear tiling multisets must contain exactly one capping tile. If one had multiple capping tiles, say $term_1, \ldots, term_k$, replacing the East pads of $term_1, \ldots, term_{k-1}$ with the West pad of $term_k$ gives a modified tile multiset of same cardinality, which is still standard, and has a higher finite expected length, which is a contradiction. Define Ψ_n to be the expected length of the assembly produced by an n-optimal linear tiling multiset. We will prove $\Psi_n = O(2^n)$ by a recursive argument on n.

Let $\mathbb{T} = \langle T, \{s\}, g\rangle$ be any n-optimal linear tiling multiset. Let L be the random variable equal to the length of the linear assembly produced by $\mathbb{T}$ and so $E[L] = \Psi_n$. A *run* of a PTAM linear tiling system is a finite sequence of attachment of tile types resulting in a terminal assembly. A run might be alternatively thought of as a finite sequence of pad types where the number of pads in a run is one more than the number of tiles. For any run of $\mathbb{T}$, consider the pad type λ appearing on the West side of the capping tile. Let $\Lambda \subset T$ be the multiset of k_1 ($0 < k_1 < n - 1$[4]) tiles with λ as their West pad, not including the capping tile. Pad type λ might occur at many positions in this run. Define the *prefix* of the run as the subsequence from the West pad of the seed tile to the first occurrence of λ. Consider the subsequences that start and end in λ with no occurrence of λ within. Such a subsequence, excluding the first λ, is called an *intermediate* (See Fig.4). Define the following random variables: L_P equal to the length of the prefix, L_{I_i} equal to the length of the ith intermediate subsequence and r equal to number of intermediates. The L_{I_i} are independent identical random variables and let L_I be a representative random variable with the same distribution. Length of the assembly equals the sum of the lengths of the prefix and the intermediates. Thus, $L = L_P + \sum_{i=1}^{r}(L_{I_i})$. For every i, the random variables r and L_{I_i} are independent because of the memoryless property of linear tiling systems. Thus, by linearity of expectation we get, $\Psi_n = E[L] = E[L_P] + E[\sum_{i=1}^{r}(L_{I_i})] = E[L_P] + E[r]E[L_I]$

Since $\mathbb{T}$ is standard, each of the tiles in Λ and the capping tile can attach with equal probability $\frac{1}{k_1+1}$ to any tile with λ as its East pad. Thus, r is a geometric random variable, with parameter $\frac{1}{k_1+1}$, counting the number of times the capping tile fails to attach. Thus $E[r] = k_1$. We will show that $E[L_P]$ and $E[L_I]$ are at most Ψ_{n-k_1} by simulating the assemblies that produce these subsequences via linear tiling multisets of cardinality at most $n - k_1$. The prefix is simulated by the linear tiling system $\mathbb{T}_P$ obtained from $\mathbb{T}$ in the following manner. Drop the

---

[4] Note that Λ cannot be empty for an optimal linear tiling system. Suppose it were: let $\Lambda' = \{t_1, \ldots, t_k\}$, be the set of tile types with λ as their East pad. Replacing the East pads of $t_1, \ldots, t_{k-1}$ with the West pad of t_k gives a modified tile multiset of same cardinality, which is still standard, and has a higher finite expected length, which is a contradiction. The same arguments hold as we recurse. The seed s and the capping tile are never part of Λ.

tiles in Λ from $\mathbb{T}$. Observe that there is a run of $\mathbb{T}_P$ for every possible prefix and vice-versa, with the same probabilities of formation. Thus, the expected length of assembly produced by $\mathbb{T}_P$ is equal to $E[L_P]$. Also, the cardinality of tile multiset for $\mathbb{T}_P$ is $n - k_1$ and hence $E[L_P] \leq \Psi_{n-k_1}$ by definition. The intermediate sub-assemblies are simulated by a family of k_1 different tile systems. Each tile system has a tile multiset of cardinality $n - k_1$ obtained by (*i.*) dropping the tiles in Λ from $\mathbb{T}$ and (*ii.*) replacing the seed tile by some $t \in \Lambda$ and making $\text{pad}_{\text{West}}(t) = \phi$. Each intermediate sub-assembly is simulated by some tile system from this family. Thus $E[L_I] \leq \Psi_{n-k_1} - 1$. Thus, $\Psi_n = E[L_P] + E[r]E[L_I] \leq (k_1+1)\Psi_{n-k_1}$. In the next level of recursion, we drop $k_2 > 0$ tiles to get $\Psi_n \leq (k_1+1)\Psi_{n-k_1} \leq (k_1+1)(k_2+1)\Psi_{n-k_1-k_2}$. In general, we drop k_i tiles in the ith level of recursion to get $\Psi_n \leq \prod_{j=1}^{i}(k_j+1)\Psi_{n-\sum_{j=1}^{i} k_j}$. The base case is $\Psi_2 = 2$ since the best one can do with a single seed and capping tile is assembly of length 2. Also, let there be z levels of recursion. Thus $\Psi_n \leq \prod_{i=1}^{z}(k_i+1)$ with $\sum_{i=1}^{z} k_i = n-2$. The product $\prod_{i=1}^{z}(k_i+1)$ constrained by $\sum_{i=1}^{z} k_i = n-2$ has a maximum value of 2^{n-2}. Hence $\Psi_n \leq O(2^n)$.

5 κ-Pad Systems for Linear Assembly

In this section we will extend PTAM by modifying each tile to accommodate multiple pads on each side. Tiles bind when one pair of adjacent pads match (see Fig.5a). To ensure that tiles align fully and are not offset, each pad on a side of a tile are drawn from different sets of pad types. Using such multi-padded tiles, we will show it is possible to reduce the number of tile types to get linear assemblies of expected length N.

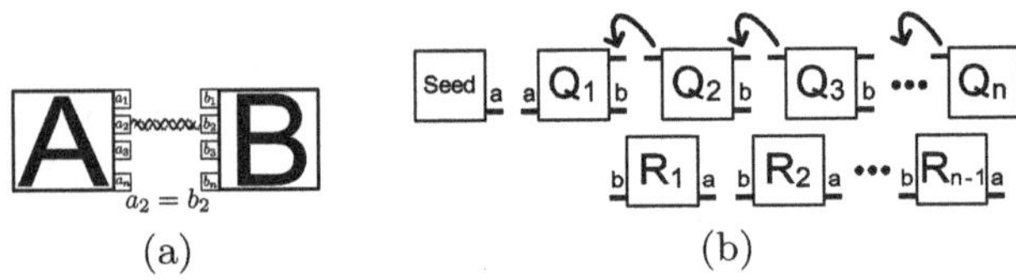

Fig. 5. κ-pad Systems. (a) κ-pad tiles A and B. (b) Pad binding diagram for linear tiling system using $\Theta_{i.o}(\frac{\log N}{\log\log N})$ 2-pad tile types. Small labeled rectangles on the sides of the tiles indicate various types of pads. Arrows indicate possible attachment. Absent pads are ϕ.

5.1 Definitions

A *κ-pad tile* t over the cartesian product $\Sigma = \Sigma_1 \times \Sigma_2 \times \cdots \times \Sigma_\kappa$ is a unit square whose two opposite sides each have a κ tuple of pads from Σ. Thus, tile $t \in T$ is an ordered pair[5] (W_t, E_t) where W_t and E_t are row vectors of size κ, where the ith component of each vector is from the set Σ_i. Thus, the East

[5] Again, for two dimensional assemblies, tiles have pads on all four sides and the model can be extended to include a temperature parameter τ for co-operative binding interactions with multiple tiles.

and West sides of each tile has κ pads. $\Sigma_1, \ldots, \Sigma_\kappa$ are finite, mutually disjoint set of distinct pad types. A *κ-pad* linear tiling system $\mathbb{T}$ is given by the tuple $\langle T, S, g \rangle$ where T is the finite multiset of κ-pad tile types, $S \subset T$ is the set of seed tiles and g is the binary pad strength function. Definitions from Section 3 hold with appropriate modifications to incorporate multiple pads on sides of each tile. For each tile t, we define $\text{pad}_{\text{East}}(t, i) = (E_t)_i$ and $\text{pad}_{\text{West}}(t, i) = (W_t)_i$ where $(E_t)_i$ and $(W_t)_i$ denote the ith component of the respective pad vectors. For $D \in \mathfrak{D}$ we say the tiles at x and $D(x)$ *attach* if there exists an i such that $g(\text{pad}_D(A(x), i), \text{pad}_{D^{-1}}(A(D(x)), i)) = 1$. (See Fig.5a).

With these modifications, *diagonal*, *uni-seeded* and *haltable* linear tiling systems and self-assembly of κ-pad tiles are defined as in Sections 2 and 3. In particular, *probabilities of attachment* of tiles is given by the same formula as in Section 3 and Lemma 1 holds for κ-pad systems. We restrict ourselves to studying diagonal, uni-seeded and haltable κ-pad linear tiling systems. Note that for assemblies in Section 5.3, adjacent tiles that bind have exactly one match among corresponding pads.

5.2 Implementing κ-Pad Systems Using DNA Self-assembly

κ-pad tiles can be feasibly realized using carefully designed self-assembled DNA motifs. Indeed, the DX motif [10], one of the early demonstrations of DNA motifs that self-assemble into two dimensional lattices, can serve as a 2-pad tile. Other similar motifs that also self-assemble into two dimensional lattices, like the TX [26], can serve as multipad systems. These motifs can be easily modified to self-assemble in one dimension, as a linear structure. On a much larger scale, Rothemund's origami technique [1] can be used to manufacture tiles with hundreds of pads. A drawback of such a system would be that the connection between adjacent tiles will be quite flexible, making a linear assembly behave more as a chain rather than a rigid ruler.

5.3 Linear Assemblies of Expected Length N Using $\Theta_{i.o}(\frac{\log N}{\log \log N})$ 2-Pad Tile Types

In this section we present an equimolar, standard κ-pad linear tiling system with $\kappa = 2$, i.e a 2-pad system, that achieves for any given $N' \in \mathbb{N}$, a linear assembly of expected length $N > N'$ using $\Theta(\frac{\log N}{\log \log N})$ 2-pad tile types, i.e., arbitrary long fixed length assemblies of expected length N using $\Theta(\frac{\log N}{\log \log N})$ 2-pad tile types. Fig.5b illustrates the tile set used in our construction. $Q_2, Q_3 \ldots Q_n$ are bi-tiles with deterministic internal pads and so for simplicity we will treat them as a single tile of length two. R is a tile type with multiplicity $n-1$, drawn as $R_1, \ldots, R_{n-1}$ in Fig.5b. Q_{i+1} can attach to Q_i's East side via the upper pad. For $j \in \{1, 2, \ldots, n-1\}$, $R_1, R_2, \ldots, R_{n-1}$ can attach to Q_j's East side via the lower pad and Q_1 attaches deterministically to R_j's East side via the lower pad. Q_1 attaches deterministically to the seed's East side while Q_n is the capping tile. The assembly halts iff the consecutive sequence $Q_1, Q_2, \ldots, Q_n$ occurs. At each stage, the assembly can restart by the attachment of Q_1 via any of the $n-1$ bridge tiles R_j. The number of tile types is $2n = \Theta(n)$.

Theorem 4. *Let X be the random variable that equals the length of the tile system illustrated in Fig.5b. Then $E[X] = N = \Theta(n^n)$ using $\Theta(\frac{\log N}{\log \log N}) = \Theta(n)$ 2-pad tile types.*

Proof. We can think of the process as a series of Bernoulli trials, say biased coin tosses. A *Head* corresponds to attachment of some Q_i $(i \neq 1)$ and a *Tail* to some R_j, Q_1 complex. The probability of a *Head* is $\frac{1}{n}$. The assembly halts iff $n-1$ successive heads occur. Each toss adds exactly 2 tiles to the assembly and the seed and Q_1 appear once before the first toss. So, from [32], the expected length of the assembly is given by $E[X] = N = \Theta(n^n)$. The number of tile types used is $\Theta(n) = \Theta(\frac{\log N}{\log \log N})$.

5.4 Lower Bounds for κ-Pad Systems

In this section we prove for each N that the cardinality of κ-pad tile multiset required to form linear assemblies of expected length N in standard PTAM systems is $\Omega(\frac{\log N}{\log \log N})$. Prior self-assembly lower bounds on numbers of tiles for assembly used information theoretic methods, whereas this proof is via a reduction to a problem in linear algebra, and furthermore holds for each N, rather than for almost all N.

Theorem 5. *For each N, the cardinality of the smallest κ-pad tile multiset required to form linear assemblies of expected length N in standard PTAM systems is $\Omega(\frac{\log N}{\log \log N})$.*

Proof. As in the Theorem 3, we will show that any κ-pad standard linear PTAM system with tile multiset of cardinality n has expected length of assembly at most $O(n^{2n})$ and this implies our result via the contrapositive. The proof uses a reduction to determining the expected time to first arrival at a vertex in a random walk over a graph, which is further reduced to a problem in linear algebra, namely determining magnitude bounds on the solution of a linear system, which is bounded by the magnitude of a ratio of two determinants.

Any n-optimal κ-pad system $\mathbb{T} = \langle T, \{s\}, g \rangle$ has exactly one seed and one capping tile, by an argument similar to the one in Section 4.3. Let L be the random variable equal to the length of linear assembly produced by $\mathbb{T}$. Consider the directed weighted graph $G = (V, E, w)$ constructed from $\mathbb{T}$ as follows: *i.* V is in one-to-one correspondence with T where vertices in V have distinct labels for repeated tile types in T, *ii.* directed edge $(u, v) \in E$ iff the East face of tile corresponding to u and West face of tile corresponding to v can attach and *iii.* for each $(u, v) \in E$, edge weights indicating transition probability are given by $w(u, v) = (\text{outdegree(u)})^{-1}$. Note that the sum of edge weights of all edges leaving a node is 1 and all edges leaving a vertex have equal transition probability. G has a *start* vertex corresponding to the seed s and a *destination* vertex corresponding to the capping tile. Self-assembly is a random walk on G

from the start to the destination, where paths in G from the start correspond to produced configurations in $\mathbb{T}$. Let *expected time to destination*, $\delta(u)$, be the expected length of the random walk from u to destination for some $u \in V$. The expected length of the assembly, $E[L] = \delta(start)$ and $\delta(destination) = 0$.

For any $u \in V$ (other than the destination), with edges to the k vertices $\{v_1, \ldots, v_k\}$, $\delta(u) = 1 + \sum_{i=1}^{k} \frac{1}{k}\delta(v_i)$. Writing such equations for each vertex in V, with $\delta(destination) = 0$ gives a system of n linear equations in n variables, say $\mathbf{A}\delta = \mathbf{b}$ where $\mathbf{A}$ is an $n \times n$ matrix of transition probabilities with values from the set $\{0, \frac{1}{n}, \frac{1}{n-1}, \ldots, 1\}$, $\mathbf{b} = [\mathbf{1}\ \mathbf{1}\ \ldots\ \mathbf{1}\ \mathbf{0}]^{\mathrm{T}}$ is a vector of size n and δ is the vector of expected times to destination. $\mathbf{A}$ is non-singular and therefore the system has a unique solution[6]. Using Cramer's rule $\delta(s) = \frac{|\mathbf{A_b}|}{|\mathbf{A}|}$ where $\mathbf{A_b}$ is the appropriate column of $\mathbf{A}$ substituted by $\mathbf{b}$. We upper and lower bound the two determinants using Leibniz's formula, $|C| = \sum_{\pi \in S_n} sgn(\pi) \prod_{i=1}^{n} C_{i,\pi(i)}$ where the sum is computed over all $n!$ permutations of S_n, where S_n is the permutations of the set $\{1, 2, \ldots, n\}$ and $sgn(\sigma)$ denotes the signature of the permutation σ: $+1$ if σ is an even permutation and -1 if it is odd. Note that the maximum value of the product $\prod_{i=1}^{n} C_{i,\pi(i)}$ is 1 since the values in each of the determinants are from the set $\{0, \frac{1}{n}, \frac{1}{n-1}, \ldots, 1\}$. Thus $|\mathbf{A_b}| \leq n!$ and similarly $|\mathbf{A}| \geq (1/n)^n$. Hence, $\delta(s) \leq O(n^{2n})$. Thus the expected length of an assembly of any κ-pad standard linear PTAM system with tile multiset of cardinality n is at most $O(n^{2n})$ which implies a lower bound of $\Omega(\frac{\log N}{\log\log N})$.

6 Improving Tail Bounds of Distribution of Lengths of Assembly

Linear tile systems that do not give assemblies with exponential tail bounds on length can be modified by concatenating k independent, distinct versions of the tile system into a new tile system with tail bounds that drop exponentially with k. Both the central limit theorem and Chernoff bounds are used for bounding the tail of this new distribution.

Given a tile multiset T (with single or κ-pads on each side of each tile) for a linear assembly, let $\hat{L}$ be the random variable equal to the length of the assembly with mean $\lfloor \frac{N}{k} \rfloor$ and variance $\frac{\sigma^2}{k}$, and let $f(\lfloor \frac{N}{k} \rfloor)$ be the cardinality of T. Consider k distinct versions of T, say $T_1, T_2, \ldots, T_k$, each mutually disjoint. We deterministically concatenate the assemblies produced by these tile multisets by introducing pads that allow the East side of each capping tile of T_i to attach to the West side of the seed tile of T_{i+1} for $i = 1, 2, \ldots, n-1$. We then add $N - k\lfloor \frac{N}{k} \rfloor \leq k$ distinct tiles that deterministically extend the assembly beyond the capping tile of T_k. Let L the random variable equal to the length of the assembly produced by this construction. This new multiset, T_{sh} of cardinality $f_{\mathrm{sh}}(N) \leq kf(\lfloor \frac{N}{k} \rfloor) + k$ gives linear assemblies of expected length $E[L] = N$ and variance σ^2. $k \in \{1, \ldots, N\}$ determines how sharp the overall probability distribution is.

[6] The solution is unique as the expected number of transitions from any vertex to the capping vertex is well defined.

The central limit theorem gives: $\forall\delta \geq 0 : P(|L - N| \leq \delta\sigma) \rightarrow \Phi(\delta)$ as $k \rightarrow \infty$, where Φ and ψ are the probability density function and cumulative distribution function respectively of the standard normal distribution. Thus, $P(|L - N| \geq \delta\sigma) \rightarrow 2(1 - \Phi(\delta)) \leq 2\psi(\delta)/\delta \leq \sqrt{2/\pi}(e^{-\delta^2/2}/\delta)$ as $k \rightarrow \infty$. Thus, we achieve an exponentially decaying tail bound with a linear multiplicative increase in tile complexity for large k. Since T_{sh} is the concatenation of independent assemblies T_i, Chernoff bounds for sums of independent random variables gives $\forall\delta, t > 0$: $P(L > (1+\delta)N) \leq (M(t)/e^{(1+\delta)\lfloor\frac{N}{k}\rfloor t})^k$ and $\forall\delta > 0, t < 0 : P(L < (1-\delta)N) \leq (M(t)/e^{(1-\delta)\lfloor\frac{N}{k}\rfloor t})^k$ where $M(t)$ is the moment generating function of the random variable $\hat{L}$. If $M(t)/e^{(1+\delta)\lfloor\frac{N}{k}\rfloor t} < 1$ for some $t > 0$ and $M(t)/e^{(1-\delta)\lfloor\frac{N}{k}\rfloor t} < 1$ for some $t < 0$, we get tail bounds dropping exponentially with k.

Acknowledgements

We wish to thank our anonymous reviewers for helpful comments, especially for pointing out the error resilience of temperature 1 assemblies. We also thank Josh Letchford and Thom LaBean for helpful discussions and Sudheer Sahu, Manoj Gopalkrishnan, Jeff Phillips, Shashidhara Ganjugunte and Urmi Majumder for their comments on an earlier draft of this paper. This work was supported by NSF EMT Grant CCF-0829797, CCF-0829798 and AFSOR Contract FA9550-08-1-0188.

References

1. Rothemund, P.: Folding DNA to Create Nanoscale Shapes and Patterns. Nature 440, 297–302 (2006)
2. Yan, H., Yin, P., Park, S.H., Li, H., Feng, L., Guan, X., Liu, D., Reif, J., LaBean, T.: Self-assembled DNA Structures for Nanoconstruction. American Institute of Physics Conference Series, vol. 725, pp. 43–52 (2004)
3. Zhang, D.Y., Turberfield, A., Yurke, B., Winfree, E.: Engineering Entropy-Driven Reactions and Networks Catalyzed by DNA. Science 318, 1121–1125 (2007)
4. Wang, H.: Proving Theorems by Pattern Recognition II (1961)
5. Berger, R.: The Undecidability of the Domino Problem, vol. 66, pp. 1–72 (1966)
6. Papadimitriou, C.: Computational Complexity. Addison-Wesley, Reading (1993)
7. Robinson, R.: Undecidability and Nonperiodicity for Tilings of the Plane. Inventiones Mathematicae 12, 177–209 (1971)
8. Garey, M., Johnson, D.: Computers and Intractability: A Guide to the Theory of NP-Completeness. W.H. Freeman, New York (1981)
9. Lewis, H., Papadimitriou, C.: Elements of the Theory of Computation. Prentice-Hall, Englewood Cliffs (1981)
10. Winfree, E., Liu, F., Wenzler, L., Seeman, N.: Design and Self-Assembly of Two-Dimensional DNA Crystals. Nature 394, 539–544 (1999)
11. Adleman, L.: Towards a mathematical theory of self-assembly. Technical report, University of Southern California (2000)
12. Winfree, E.: DNA Computing by Self-Assembly. In: NAE's The Bridge, vol. 33, pp. 31–38 (2003)
13. Rothemund, P., Winfree, E.: The Program-Size Complexity of Self-Assembled Squares. In: STOC, pp. 459–468 (2000)

14. Adleman, L., Cheng, Q., Goel, A., Huang, M.D.: Running Time and Program Size for Self-Assembled Squares. In: STOC, pp. 740–748 (2001)
15. Barish, R., Rothemund, P., Winfree, E.: Two Computational Primitives for Algorithmic Self-Assembly: Copying and Counting. Nano Letters 5(12), 2586–2592 (2005)
16. Yan, H., Feng, L., LaBean, T., Reif, J.: Parallel Molecular Computation of Pair-Wise XOR using DNA String Tile. Journal of the American Chemical Society (125) (2003)
17. Winfree, E.: Simulations of Computing by Self-Assembly. Technical report, Caltech CS Tech Report (1998)
18. Schulman, R., Lee, S., Papadakis, N., Winfree, E.: One Dimensional Boundaries for DNA Tile Self-Assembly. In: Chen, J., Reif, J.H. (eds.) DNA 2003. LNCS, vol. 2943, pp. 108–126. Springer, Heidelberg (2004)
19. Park, S.H., Yin, P., Liu, Y., Reif, J., LaBean, T., Yan, H.: Programmable DNA Self-assemblies for Nanoscale Organization of Ligands and Proteins. Nano Letters 5, 729–733 (2005)
20. Aggarwal, G., Goldwasser, M., Kao, M.Y., Schweller, R.: Complexities for Generalized Models of Self-Assembly. In: SODA, pp. 880–889 (2004)
21. Kao, M.Y., Schweller, R.: Randomized Self-Assembly for Approximate Shapes. In: Aceto, L., Damgård, I., Goldberg, L.A., Halldórsson, M.M., Ingólfsdóttir, A., Walukiewicz, I. (eds.) ICALP 2008, Part I. LNCS, vol. 5125, pp. 370–384. Springer, Heidelberg (2008)
22. Lagoudakis, M., LaBean, T.: 2D DNA Self-Assembly for Satisfiability. In: DIMACS Workshop on DNA Based Computers (1999)
23. Becker, F., Rapaport, I., Rémila, É.: Self-assemblying classes of shapes with a minimum number of tiles, and in optimal time. In: Arun-Kumar, S., Garg, N. (eds.) FSTTCS 2006. LNCS, vol. 4337, pp. 45–56. Springer, Heidelberg (2006)
24. Reif, J.: Local parallel biomolecular computation. In: NA-Based Computers, III, pp. 217–254. American Mathematical Society, Providence, RI (1997)
25. Demaine, E.D., Demaine, M.L., Fekete, S.P., Ishaque, M., Rafalin, E., Schweller, R.T., Souvaine, D.L.: Staged Self-assembly: Nanomanufacture of Arbitrary Shapes with $O(1)$ Glues. In: Garzon, M.H., Yan, H. (eds.) DNA 2007. LNCS, vol. 4848, pp. 1–14. Springer, Heidelberg (2008)
26. LaBean, T., Yan, H., Kopatsch, J., Liu, F., Winfree, E., Reif, J., Seeman, N.: Construction, Analysis, Ligation, and Self-Assembly of DNA Triple Crossover Complexes. Journal of the American Chemical Society 122(9), 1848–1860 (2000)
27. Mao, C., LaBean, T., Reif, J., Seeman, N.: Logical Computation Using Algorithmic Self-Assembly of DNA Triple-Crossover Molecules. Nature 407, 493–496 (2000)
28. Gillespie, D.: Exact Stochastic Simulation of Coupled Chemical Reactions. The Journal of Physical Chemistry 81, 2340–2361 (1977)
29. Soloveichik, D., Winfree, E.: Complexity of Self-Assembled Shapes. SIAM Journal of Computing 36(6), 1544–1569 (2007)
30. Winfree, E., Bekbolatov, R.: Proofreading Tile Sets: Error Correction for Algorithmic Self-Assembly. In: Chen, J., Reif, J.H. (eds.) DNA 2003. LNCS, vol. 2943, pp. 126–144. Springer, Heidelberg (2004)
31. Chen, H.-L., Goel, A.: Error free self-assembly using error prone tiles. In: Ferretti, C., Mauri, G., Zandron, C. (eds.) DNA 2004. LNCS, vol. 3384, pp. 62–75. Springer, Heidelberg (2005)
32. Gordan, H.: Discrete Probability. Springer, Heidelberg (1997)
33. Li, M., Vitanyi, P.: An Introduction to Kolmogorov Complexity and Its Applications, 2nd edn. Springer, Heidelberg (1997)

A Graph Reduction Step Preserving Element-Connectivity and Applications

Chandra Chekuri* and Nitish Korula**

Dept. of Computer Science, University of Illinois, Urbana, IL 61801
{chekuri,nkorula2}@illinois.edu

Abstract. Given an undirected graph $G = (V, E)$ and subset of terminals $T \subseteq V$, the *element-connectivity* $\kappa'_G(u, v)$ of two terminals $u, v \in T$ is the maximum number of u-v paths that are pairwise disjoint in both edges and non-terminals $V \setminus T$ (the paths need not be disjoint in terminals). Element-connectivity is more general than edge-connectivity and less general than vertex-connectivity. Hind and Oellermann [18] gave a graph reduction step that preserves the *global* element-connectivity of the graph. We show that this step also preserves *local* connectivity, that is, all the pairwise element-connectivities of the terminals.

We give two applications of the step to connectivity and network design problems: First, we show a polylogarithmic approximation for the problem of packing element-disjoint Steiner forests in general graphs, and an $O(1)$-approximation in planar graphs. Second, we find a very short and intuitive proof of a spider-decomposition theorem of Chuzhoy and Khanna [10] in the context of the single-sink *k-vertex-connectivity* problem. Our results highlight the effectiveness of the element-connectivity reduction step; we believe it will find more applications in the future.

1 Introduction

In this paper we consider several connectivity and network design problems. Given an undirected graph G and two nodes u, v we let $\lambda_G(u, v)$ and $\kappa_G(u, v)$ denote the edge and vertex connectivities between u and v in G. It is well-known that edge-connectivity problems are "easier" than their vertex-connectivity counterparts. Vertex-connectivity exhibits less structure than edge-connectivity and this often translates into significant differences in the algorithmic and computational difficulty of the corresponding problems. As an example, consider the well-known survivable network design problem (SNDP): the input consists of an undirected edge-weighted graph G and connectivity requirements $r : V \times V \rightarrow Z^+$ between each pair of vertices. The goal is to find a min-cost subgraph H of G such that each pair u, v has $r(u, v)$ disjoint paths between them in H. If the paths are required to be edge-disjoint ($\lambda_H(u, v) \geq r(u, v)$) then the problem is referred to as EC-SNDP and if the paths are required to be vertex-disjoint the problem is

* Partially supported by NSF grants CCF 07-28782 and CNS-0721899.
** Partially supported by NSF grant CCF 07-28782.

S. Albers et al. (Eds.): ICALP 2009, Part I, LNCS 5555, pp. 254–265, 2009.

referred to as VC-SNDP. Jain [19] gave a 2-approximation for EC-SNDP based on the powerful iterated rounding technique. On the other hand, VC-SNDP is known to be hard to within polynomial factors [24,3]. To address this gap, Jain *et al.* [21] introduced a connectivity measure intermediate to edge and vertex connectivities known as *element-connectivity*. The vertices are partitioned into terminals $T \subseteq V$ and non-terminals $V \setminus T$. The element-connectivity between two terminals u, v, denoted by $\kappa'_G(u,v)$ is defined to be the maximum number of paths between u and v that are pairwise disjoint in edges and non-terminals (the paths can share terminals). In some respects, element-connectivity resembles edge-connectivity: For example, $\kappa'(u,w) \geq \min(\kappa'(u,v), \kappa'(v,w))$ for any three terminals u, v, w; this triangle inequality holds for edge-connectivity but does not for vertex-connectivity. In element-connectivity SNDP (ELC-SNDP) the requirements are only between terminals and the goal is to find a min-cost subgraph H such that $\kappa'_H(u,v) \geq r(u,v)$ for each $u, v \in T$. Fleischer, Jain and Williamson [13] (see also [9]) generalized the iterated rounding technique of Jain for EC-SNDP to give a 2-approximation for ELC-SNDP. In other respects, element-connectivity is related to vertex connectivity. One class of problems motivating this paper is on generalizing the classical theorem of Menger on s-t vertex-connectivity; we discuss this below.

In studying element-connectivity, we often assume without loss of generality that there are no edges between terminals (by subdividing each such edge) and hence $\kappa'(u,v)$ is the maximum number of non-terminal disjoint u-v paths. Menger's theorem shows that the maximum number of internally vertex-disjoint s-t paths is equal to $\kappa(s,t)$. Hind and Oellermann [18] considered a natural generalization to multiple terminals. Given a terminal set $T \subseteq V$, what is the maximum number of trees that each contain T and are disjoint in $V \setminus T$? The natural upper bound here is the element connectivity of T in G, in other words, $k = \min_{u,v \in T} \kappa'(u,v)$. In [18] a graph reduction step was introduced to answer this question. Cheriyan and Salavatiour [8] called this the problem of packing element-disjoint Steiner trees; crucially using the graph reduction step, they showed that there always exist $\Omega(k/\log|T|)$ element-disjoint Steiner trees and moreover, this bound is tight (up to constant factors) in the worst case. In contrast, if we seek edge-disjoint Steiner trees then Lau [27] has shown that if T is $26k$ edge-connected in G, there are k edge-disjoint trees each of which spans T.

Finally, we remark that in some recent work Chuzhoy and Khanna [10] gave an $O(k \log |T|)$ approximation for the special case of VC-SNDP in which a terminal set T needs to be k-vertex-connected (this is equivalent to the single-sink problem). Their algorithm and analysis are based on a structural characterization of feasible solutions — they use element-connectivity (they call it weak connectivity) as a key stepping stone. Subsequent to this paper, Chuzhoy and Khanna [11] gave a simple and elegant reduction from the the *general* VC-SNDP problem to ELC-SNDP, obtaining an $O(k^3 \log n)$-approximation and reinforcing the connection between element- and vertex-connectivity.

The discussion above suggests that it is fruitful to study element-connectivity as a way to generalize edge-connectivity and attack problems on vertex-connectivity. In this paper we consider the graph reduction step for element-connectivity introduced by Hind and Oellermann [18] (and rediscovered by Cheriyan and Salavatipour [8]). We generalize the applicability of the step and demonstrate applications to several problems.

A Graph Reduction Step Preserving Element Connectivity. The well-known *splitting-off* operation introduced by Lovász [29] is a standard tool in the study of (primarily) edge-connectivity problems. Given an undirected multi-graph G and two edges su and sv incident to s, the splitting-off operation replaces su and sv by the single edge uv. Lovász proved the following theorem on splitting-off to preserve *global* edge-connectivity.

Theorem 1 (Lovász). *Let $G = (V \cup \{s\}, E)$ be an undirected multi-graph in which V is k-edge-connected for some $k \geq 2$ and degree of s is even. Then for every edge su there is another edge sv such that V is k-edge-connected after splitting-off su and sv.*

Mader strengthened Theorem 1 to show the existence of a pair of edges incident to s that when split-off preserve the *local* edge-connectivity of the graph.

Theorem 2 (Mader [30]). *Let $G = (V \cup \{s\}, E)$ be an undirected multi-graph, where $deg(s) \neq 3$ and s is not incident to a cut edge of G. Then s has two neighbours u and v such that the graph G' obtained from G by replacing su and sv by uv satisfies $\lambda_{G'}(x, y) = \lambda_G(x, y)$ for all $x, y \in V \setminus \{s\}$.*

Generalization to directed graphs are also known [30,14,22]. The splitting-off theorems have numerous applications in graph theory and combinatorial optimization. See [29,15,26,20,5,27,28,23] for various pointers and applications. Although splitting-off techniques can be sometimes be used in the study of vertex-connectivity, their use is limited and no generally applicable theorem akin to Theorem 2 is known. On the other hand, Hind and Oellermann [18] proved an elegant theorem on preserving global element connectivity. In the sequel we use $\kappa'_G(S)$ to denote $\min_{u,v \in S} \kappa'_G(u, v)$ and G/pq to denote the graph obtained from G by contracting vertices p, q.

Theorem 3 (Hind & Oellermann [18]). *Let $G = (V, E)$ be an undirected graph and $T \subseteq V$ be a terminal-set such that $\kappa'_G(T) \geq k$. Let (p, q) be any edge where $p, q \in V \setminus T$. Then $\kappa'_{G_1}(T) \geq k$ or $\kappa'_{G_2}(T) \geq k$ where $G_1 = G - pq$ and $G_2 = G/pq$.*

This theorem has been used in two applications on element-connectivity [8,23]. We strengthen it to handle local connectivity, increasing its applicability.

Reduction Lemma. *Let $G = (V, E)$ be an undirected graph and $T \subseteq V$ be a terminal-set. Let (p, q) be any edge where $p, q \in V \setminus T$ and let $G_1 = G - pq$ and $G_2 = G/pq$. Then one of the following holds: (i) $\forall u, v \in T$, $\kappa'_{G_1}(u, v) = \kappa'_G(u, v)$ (ii) $\forall u, v \in T$, $\kappa'_{G_2}(u, v) = \kappa'_G(u, v)$.*

The Reduction Lemma, applied repeatedly, transforms a graph into another graph in which the non-terminals form a stable set. Moreover, the reduced graph is a minor of the original graph. Below, we discuss two problems, briefly alluded to earlier, to which we apply the Reduction Lemma.

Packing Element-Disjoint Steiner Trees and Forests. There has been much interest in the recent past on algorithms for (integer) packing of disjoint Steiner trees in both the edge and element-connectivity settings [26,20,27,28,7,8,5]. (A *Steiner tree* is simply a tree containing the entire terminal set T.) See [17] for applications of Steiner tree packing to VLSI design. An outstanding open problem is Kriesell's conjecture which states that if the terminal set T is $2k$-edge-connected then there are k-edge-disjoint Steiner trees each of which spans T; this would generalize a classical theorem of Nash-Williams and Tutte on edge-disjoint spanning trees. Lau made substantial progress [27] and proved that $26k$-connectivity suffices for k edge-disjoint Steiner trees; he extended his result for packing Steiner forests [28]. We remark that Mader's splitting-off theorem plays an important role in Lau's work. The element-disjoint Steiner tree packing problem was first considered by Hind and Oellermann. As we mentioned, Cheriyan and Salavatipour [8] gave a nearly tight bound for this problem, since the problem of packing element-disjoint Steiner trees is hard to approximate to within an $\Omega(\log n)$ factor [7]. Their result relies crucially on Theorem 3 followed by a simple randomized coloring algorithm whose analysis extends that of a similar algorithm for computing the domatic number of a graph [12]. Here, we consider the more general problem posed by [8] of packing Steiner forests. The input consists of a graph $G = (V, E)$ and disjoint terminal sets $T_1, T_2, \ldots, T_m$, such that $\kappa'_G(T_i) \geq k$ for $1 \leq i \leq k$. What is the maximum number of element disjoint forests such that in each forest T_i is connected for $1 \leq i \leq k$? Our Reduction Lemma is primarily motivated by this question. For general graphs we prove that there exist $\Omega(k/(\log |T| \log m))$ element disjoint forests, where $T = \bigcup_i T_i$. This can also be viewed as an $O(\log |T| \log m)$ approximation for the problem.

We also study the packing problem in planar graphs and prove substantially stronger results: We show that there exist $\lceil k/5 \rceil - 1$ disjoint forests. Our method also extends to give an $\Omega(k)$ bound for graphs of fixed genus and an $\Omega(k)$ bound for packing Steiner trees in graphs of fixed treewidth; we conjecture that one can find $\Omega(k)$ disjoint forests in graphs excluding a fixed minor. These are the first non-trivial bounds for packing element-disjoint Steiner forests in general graphs or planar graphs; the bounds also imply corresponding approximation algorithms for maximizing the number of disjoint forests. Since element-connectivity generalizes edge-connectivity, our bounds in planar graphs are considerably stronger than those of Lau [27,28] for *edge*-connectivity. Our proof is simple, and the simplicity comes from thinking about element-connectivity (using the Reduction Lemma) instead of edge-connectivity! Our proof also gives the strong property that the non-terminals in the forests all have degree 2.

Single-Sink k-vertex-connectivity. Polynomial factor inapproximability results for VC-SNDP [24,3] have focused attention on restricted, yet useful, special cases of the problem. A recent focus of attention has been the single-sink k-vertex-connectivity problem for small k; the goal is to k-vertex-connect a set of terminals T to a given root r. The recent $O(k \log |T|)$-approximation of [10] relied on a beautiful decomposition result for k-connectivity which is independently interesting from a graph theoretic view point. The proof of this theorem in [10] is long and complicated although it is based on only elementary operations. Using the Reduction Lemma, we give a very short proof in Section 4 of the main technical result of [10]. Due to space constraints, we omit further discussion of this application in this paper.

Related Work. We have already mentioned most of the closely related papers. Our work on packing Steiner forests in planar graphs was inspired by a question of Joseph Cheriyan [6]. Independent of our work, Aazami, Cheriyan and Jampani [1] proved that if a terminal set T is k-element-connected in a planar graph, then there exist $k/2 - 1$ element-disjoint Steiner trees. Moreover this is tight; they also prove that it is NP-hard to obtain a $(1/2+\varepsilon)$ approximation. Our bound for packing Steiner Trees in planar graphs is slightly weaker than theirs; however, our algorithms and proofs are simple and intuitive, and generalize to packing Steiner forests. Their algorithm uses Theorem 3, followed by a reduction to a theorem of Frank *et al.* [16] that uses Edmonds' matroid partition theorem. One could attempt to pack Steiner forests using their approach (with the stronger Reduction Lemma in place of Theorem 3), but the theorem of [16] does not have a natural generalization for Steiner forests. The techniques of both [1] and this paper extend to graphs of small genus or treewidth; we discuss this further in Section 3.2. We refer the reader to [3,10,4,31] for discussion of recent work on single-sink vertex connectivity, including hardness results [3] and extensions to related problems such as the node-weighted case [10] and buy-at-bulk network design [4]. See also the survey on network design by Kortsarz and Nutov [25].

Several proofs have been omitted from this extended abstract; a longer version can be found at `http://arxiv.org/abs/0902.2795` or on the authors' websites.

2 The Reduction Lemma

Let $G(V, E)$ be a graph, with a given set $T \subseteq V(G)$ of terminals. For ease of notation, we subsequently refer to terminals as *black* vertices, and non-terminals (also called Steiner vertices) as *white*. The elements of G are white vertices and edges; two paths are *element-disjoint* if they have no white vertices or edges in common. Recall that the element-connectivity of two black vertices u and v, denoted by $\kappa'_G(u, v)$, is the maximum number of element-disjoint (that is, disjoint in edges and white vertices) paths between u and v in G.

For this section, to simplify the proof, we will assume that G has no edges between black vertices; any such edge can be subdivided, with a white vertex inserted between the two black vertices. It is easy to see that two paths are element-disjoint in the original graph iff they are element-disjoint in the modified graph. Thus, we can say that paths are element disjoint if they share no white vertices, or that u and v are k-element-connected if the smallest set of white vertices whose deletion separates u from v has size k.

Recall that our lemma strengthens Theorem 3 on preserving global connectivity. We remark that our proof is based on a cutset argument unlike the path-based proofs in [18,8] for the global case.

Proof of the Reduction Lemma: Consider an arbitrary edge pq. Deleting or contracting an edge can reduce the element-connectivity of a pair by at most 1. Suppose the lemma were not true; there must be pairs s,t and x,y of black vertices such that $\kappa'_{G_1}(s,t) = \kappa'_G(s,t) - 1$ and $\kappa'_{G_2}(x,y) = \kappa'_G(x,y) - 1$. The pairs have to be distinct since it cannot be the case that $\kappa'_{G_1}(u,v) = \kappa'_{G_2}(u,v) = \kappa'_G(u,v) - 1$ for any pair u,v. (To see this, if one of the $\kappa'_G(u,v)$ u-v paths uses pq, contracting the edge will not affect that path, and will leave the other paths untouched. Otherwise, no path uses pq, and so it can be deleted.). Note that one of s,t could be the same vertex as one of x,y; for simplicity we will assume that $\{s,t\} \cap \{x,y\} = \emptyset$, but this does not change our proof in any detail. We show that our assumption on the existence of s,t and x,y with the properties above leads to a contradiction. Let $\kappa'_G(s,t) = k_1$ and $\kappa'_G(x,y) = k_2$. We use the following facts several times:

1. Any cutset of size less than k_1 that separates s and t in G_1 cannot include p or q. (If it did, it would also separate s and t in G.)
2. $\kappa'_{G_1}(x,y) = k_2$ since $\kappa'_{G_2}(x,y) = k_2 - 1$.

We define a vertex tri-partition of a graph G as follows: (A,B,C) is a vertex tri-partition of G if A, B, and C partition $V(G)$, B contains only white vertices, and there are no edges between A and C. (That is, removing the white vertices in B disconnects A and C.)

Since $\kappa'_{G_1}(s,t) = k_1 - 1$, there is a vertex-tri-partition (S,M,T) such that $|M| = k_1 - 1$ and $s \in S$ and $t \in T$. From Fact 1 above, M cannot contain p or q. For the same reason, it is also easy to see that p and q cannot both be in S (or both in T); otherwise M would be a cutset of size $k_1 - 1$ in G. Therefore, assume w.l.o.g. that $p \in S, q \in T$.

Similarly, since $\kappa'_{G_2}(x,y) = k_2 - 1$, there is a vertex-tri-partition (X,N',Y) in G_2 with $|N'| = k_2 - 1$ and $x \in X$ and $y \in Y$. We claim that N' contains the contracted vertex pq for otherwise N' would be a cutset of size $k_2 - 1$ in G. Therefore, it follows that (X,N,Y) where $N = N' \cup \{p,q\} - \{pq\}$ is a vertex-tri-partition in G that separates x from y. Note that $|N| = k_2$ and N includes *both* p and q. For the latter reason we note that (X,N,Y) is a vertex-tri-partition also in G_1.

Subsequently, we work with the two vertex tri-partitions (S,M,T) and (X,N,Y) in G_1 (we stress that we work in G_1 and not in G or G_2). Recall that $s,p \in S$, and $t,q \in T$, and that M has size $k_1 - 1$; also, N separates x from y, and $p,q \in N$. Fig. 1 (a) above shows these vertex tri-partitions. Since M and N contain only white vertices, all terminals are in S or T, and in X or Y. We say that $S \cap X$ is *diagonally opposite* from $T \cap Y$, and $S \cap Y$ is diagonally opposite from $T \cap X$. Let A,B,C,D

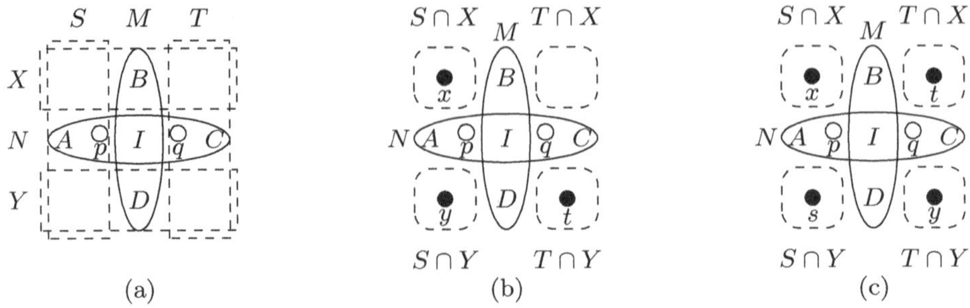

Fig. 1. Part (a) illustrates the vertex tri-partitions (S, M, T) and (X, N, Y). In parts (b) and (c), we consider possible locations of the terminals s, t, x, y.

denote $S \cap N, X \cap M, T \cap N$ and $Y \cap M$ respectively, with I denoting $N \cap M$; note that A, B, C, D, I partition $M \cup N$.

We assume w.l.o.g. that $x \in S$. If we also have $y \in S$, then $x \in S \cap X$ and $y \in S \cap Y$; therefore, one of x, y is diagonally opposite from t, suppose this is x. Fig. 1 (b) illustrates this case. Observe that $A \cup I \cup B$ separates x from y; since x and y are k_2-connected and $|N = A \cup I \cup C| = k_2$, it follows that $|B| \geq |C|$. Similarly, $C \cup I \cup D$ separates t from s, and since C contains q, Fact 1 implies that $|C \cup I \cup D| \geq k_1 > |B \cup I \cup D = M| = k_1 - 1$. Therefore, $|C| > |B|$, and we have a contradiction.

Hence, it must be that $y \notin S$; so $y \in T \cap Y$. The argument above shows that x and t cannot be diagonally opposite, so $t \in T \cap X$. Exactly as before, s and y cannot be diagonally opposite, so $s \in S \cap Y$. Fig. 1 (c) shows the required positions of the vertices. Now, N separates s from t and contains p, q; so from fact 1, $|N| \geq k_1 > |M|$. But M separates x from y, and fact 2 implies that x, y are k_2-connected in G_1; therefore, $|M| \geq k_2 = |N|$, giving a contradiction. □

3 Packing Element-Disjoint Steiner Trees and Forests

Consider a graph $G(V, E)$, with its vertex set V partitioned into $T_1, T_2, \ldots T_m, W$. We refer to each T_i as a group of *terminals*, and W as the set of Steiner or white vertices; we use $T = \bigcup_i T_i$ to denote the set of all terminals. A Steiner Forest for this graph is a forest that is a subgraph of G, such that each T_i is entirely contained in a single tree of this forest. (Note that T_i and T_j can be in the same tree.) For any group T_i of terminals, we define $\kappa'(T_i)$, the element-connectivity of T_i, as the largest k such that for every $u, v \in T_i$, the element-connectivity of u and v in the graph G is at least k.

Two Steiner Forests for G are element-disjoint if they share no edges or Steiner vertices. (Every Steiner Forest must contain all the terminals.) The Steiner Forest packing problem is to find as many element-disjoint Steiner Forests for G as possible. By inserting a Steiner vertex between any pair of adjacent terminals, we can assume that there are no edges between terminals, so the problem of finding element-disjoint Steiner forests is simply that of finding Steiner forests that do not share any Steiner vertices. In the special case when $m = 1$, we seek a maximum number of element-disjoint Steiner trees. It is easy to see that if $k = \min_i \kappa'_G(T_i)$, there are at most k element-disjoint Steiner Forests in G.

Cheriyan and Salavatipour [8] proved that if there is a single group T of terminals, with $\kappa'(T) = k$, then there always exist $\Omega(k/\log|T|)$ Steiner trees. Their algorithm uses Theorem 3, the global element-connectivity reduction of [18], to delete and contract edges between Steiner vertices, while preserving $\kappa'(T) = k$. After obtaining a bipartite graph G' with terminals on one side and Steiner vertices on the other side, randomly color the Steiner vertices using $k/6\log|T|$ colors; w.h.p., each color class connects the terminal set T. The bipartite case can be cast as a special case of packing bases of a polymatroid and a variant of the random coloring idea is applicable in this more general setting [2]; a derandomization is also provided in [2], yielding a deterministic polynomial-time algorithm to find $\Omega(k/\log|T|)$ element-disjoint Steiner trees.

In this section, we give algorithms for packing element-disjoint Steiner Forests, where we are given m groups of terminals $T_1, T_2, \ldots T_m$. The approach of [8] encounters two difficulties. First, we cannot reduce to a bipartite instance, using only the global-connectivity version of the Reduction Lemma. In fact, our strengthening of the Reduction Lemma to preserve local connectivity was motivated by this; using it allows us once again assume that we have a bipartite graph $G'(T \cup W, E)$. Second, we cannot apply the random coloring algorithm on the bipartite graph G' directly. One reason for this is that, unlike the Steiner tree case, it is no longer a problem of packing bases of a submodular function. To overcome this second difficulty we use a decomposition technique followed by the random coloring algorithm to prove that there always exist $\Omega(k/(\log|T|\log m))$ element-disjoint forests. We believe that the bound can be improved to $\Omega(k/\log|T|)$.

We also consider the packing problem in restricted classes of graphs, in particular planar graphs. We obtain a much stronger bound, showing the existence of $\lceil k/5 \rceil - 1$ Steiner forests. The (simple) technique extends to graphs of fixed genus to prove the existence of $\Omega(k)$ Steiner forests where the constant depends mildly on the genus. We believe that there exist $\Omega(k)$ Steiner forests in any H-minor-free graph where H is fixed. Our technique for planar graphs does not extend directly, but generalizing this technique allows us to make partial progress; we can prove that in graphs of any fixed treewidth, there exist $\Omega(k)$ element-disjoint Steiner Trees if the terminal set is k-element-connected.

3.1 An $O(\log|T|\log m)$-Approximation for General Graphs

In order to pack element-disjoint Steiner forests we borrow the basic idea from [5] in the *edge-connectivity* setting for Eulerian graphs; this idea was later used by Lau [28] in the much more difficult non-Eulerian case. The idea at a high level is as follows: If all the terminals are nearly k-connected then we can treat the terminals as forming one group and reduce the problem to that of packing Steiner trees. Otherwise, we can find a cut $(S, V \setminus S)$ that separates some groups from others. By choosing the cut appropriately, *all* terminals on one side, say S, are $\Omega(k/2\log m)$-element-connected, and we can pack Steiner *trees* in them *without* using the edges crossing the cut. Then we can shrink S and find Steiner forests in the reduced graph; unshrinking of S is possible since we have many trees on S. In [5,28] this scheme works to give $\Omega(k)$ edge-disjoint Steiner forests. However, the approach relies strongly on properties of edge-connectivity as well as the properties of the packing algorithm for Steiner trees. These do not generalize easily for element-connectivity. Nevertheless, we show that the basic idea can be applied in a slightly weaker way (resulting in the loss of an $O(\log m)$ factor over the Steiner tree packing factor); the reduction to a bipartite instance using the Reduction Lemma plays a critical role.

Theorem 4. *Given a graph $G(V, E)$, with terminal sets $T_1, T_2, \ldots T_m$, such that for all i, $\kappa'(T_i) \geq k$, there is a polynomial-time algorithm to pack $\Omega(k/\log|T|\log m)$ element-disjoint Steiner Forests in G.*

3.2 Packing Steiner Trees and Forests in Planar Graphs

We now prove much improved results for restricted classes of graphs, in particular planar graphs. If G is planar, we show the existence of $\lceil k/5 \rceil - 1$ element-disjoint Steiner Forests. The intuition and algorithm are easier to describe for the Steiner tree packing problem and we do this first. We achieve the improved bound by observing that planarity restricts the use of many white vertices as "branch points" (that is, vertices of degree ≥ 3) in forests. Intuitively, even in the case of packing trees, if there are terminals $t_1, t_2, t_3, \ldots$ that must be in every tree, and white vertices $w_1, w_2, w_3 \ldots$ that all have degree 3, it is difficult to avoid a $K_{3,3}$ minor. Note, however, that degree 2 white vertices behave like edges and do not form an obstruction. We capture this intuition more precisely by showing that there must be a pair of terminals t_1, t_2 that are connected by $\Omega(k)$ degree-2 white vertices; we can contract these "parallel edges", and recurse.

We now describe an algorithm for packing Steiner Trees: Given an instance of the Steiner Tree packing problem in planar graphs, we construct a *reduced instance* as follows: Use the Reduction Lemma to delete and contract edges between white vertices to obtain a planar graph with vertex set $T \cup W$, such that W is a stable set. Now, for each vertex $w \in W$ of degree 2, connect the two terminals that are its endpoints directly with an edge, and delete w. (All edges have unit capacity.) We now have a planar *multigraph*, though the only parallel edges are between terminals, as these were the only edges added while deleting degree-2 vertices in W. Note that this reduction preserves the element-connectivity of each pair of terminals; further, any set of element-disjoint trees in this reduced instance corresponds to a set of element-disjoint trees in the original instance. We can now prove the following lemma:

Lemma 1. *In a reduced instance of the Planar Steiner Tree Packing problem, if $\kappa'(T) \geq k$, there are two terminals t_1, t_2 with at least $\lceil k/5 \rceil - 1$ parallel edges between them.*

It is now easy to prove by induction that we can pack $\lceil k/5 \rceil - 1$ disjoint trees; simply contract the terminals t_1, t_2 with many edges between them, and recurse.

Theorem 5. *Given an instance of the Steiner Tree packing problem on a planar graph G with terminal set T, if $\kappa'(T) \geq k$, there is a polynomial-time algorithm to find at least $\lceil k/5 \rceil - 1$ element-disjoint Steiner trees in G. Moreover, in each tree, the white (non-terminal) vertices all have degree 2.*

Extensions. Our result for planar graphs can be generalized to graphs of fixed genus; these results also hold for packing Steiner forests. These techniques can also be used to pack Steiner trees in graphs of bounded treewidth; see the full version for details. Aazami *et al.* [1] also give algorithms for packing Steiner trees in these graph classes, and graphs excluding a fixed minor. We thus make the following natural conjecture:

Conjecture 1. Let $G = (V, E)$ be a H-minor-free graph, with terminal sets $T_1, T_2, \ldots T_m$, such that for all i, $\kappa'(T_i) \geq k$. There exist $\Omega(k/c)$ element-disjoint Steiner forests in G, where c depends only on the size of H.

4 A Spider Decomposition Theorem

In this section, we sketch a very short proof of the beautiful Spider Decomposition Theorem of [10]. A *spider* is a tree with at most one vertex of degree greater than 2. If such a vertex exists, it is referred to as the *head* of the spider, and each leaf is referred to as a *foot*. Thus, a spider may be viewed as a collection of disjoint paths (called *legs*) from its feet to its head. If the spider has no vertex of degree 3 or more, any vertex of the spider may be considered its head. Vertices that are not the head or feet are called intermediate vertices of the spider.

Theorem 6 ([10]). *Let $G(V, E)$ be a graph with a set $B \subseteq V$ of black vertices such that every pair of black vertices is k-element connected. There is a subgraph H of G whose edges can be partitioned into spiders such that: (i) For each spider, its feet are distinct black vertices, and all intermediate vertices are white. (ii) Each black vertex is a foot of exactly k spiders, and each white vertex appears in at most one spider. (iii) If a white vertex is the head of a spider, the spider has at least two feet.*

Proof Sketch: Use induction on the number of edges between white vertices in G: As the base case, we have a graph G with no edges between white vertices; therefore, G is bipartite. (Recall that there are no edges between black vertices.) Each pair of black vertices is k-element connected, and hence every black vertex has at least k white neighbors. Let every $b \in B$ mark k of its (white) neighbors arbitrarily. Every white vertex w that is marked at least twice becomes the head of a spider, the feet of which are the black vertices that marked w. For each white vertex w marked only once, let b be its neighbor that marked it, and b' be another neighbor. We let $b - w - b'$ be a spider with foot b and head b'. It is easy to see that the spiders are disjoint and satisfy all the other desired conditions.

For the inductive step, consider a graph G with an edge pq between white vertices. If all black vertices are k-element connected in $G_1 = G - pq$, then we can apply induction, and find the desired subgraph of G_1 and hence of G.

Otherwise, by the Reduction Lemma , we can find the desired set of spiders in $G_2 = G/pq$. If the new vertex $v = pq$ is not in any spider, this set of spiders exists in G, and we are done. Otherwise, let S be the spider containing v. If v is not the head of S, let x, y be its neighbors in S. We can replace the path $x - v - y$ in S with one of $x - p - y$, $x - q - y$, or $x - p - q - y$. If v is the head of S, it is easy to see that we can either split S into two disjoint spiders with heads at p and q, or create a single new spider S' with head at p or q. □

5 Conclusions

Having strengthened the reduction step of [18] to handle local element connectivity, we demonstrated applications of this stronger Reduction Lemma to connectivity and network design problems. We believe that the Reduction Lemma will find other applications in the future. We close with several open questions:

- We believe that our bound on the number of element-disjoint Steiner forests in a general graph can be improved from $\Omega(k/(\log |T| \log m))$ to $\Omega(k/\log |T|)$.
- Prove or disprove Conjecture 1.
- In a natural generalization of the Steiner Forest packing problem, each non-terminal/white vertex has a *capacity*, and the goal is to pack forests subject to these capacity constraints. In general graphs, it is easy to reduce this problem to the uncapacitated/unit-capacity version, but this is not necessarily the case for restricted classes of graphs. In particular, it would be interesting to pack $\Omega(k)$ forests for the capacitated planar Steiner Forest problem.

Acknowledgements. We thank Anupam Gupta for discussions on single-sink k-vertex-connectivity and connectivity-preserving reductions. We thank Sanjeev Khanna and Julia Chuzhoy for sharing a preliminary version of [10]. We thank Joseph Cheriyan for asking about planar packing of Steiner Trees which inspired our work on that problem. We also thank Oleg Borodin, Dan Cranston, Alexandr Kostochka and Doug West for pointers to structural results on planar graphs.

References

1. Aazami, A., Cheriyan, J., Jampani, K.R.: Approximation Algorithms and Hardness Results for Packing Element-Disjoint Steiner Trees. Algorithmica (manuscript submitted, 2008)
2. Calinescu, G., Chekuri, C., Vondrak, J.: Disjoint Bases in a Polymatroid. Random Structures & Algorithms (to appear)
3. Chakraborty, T., Chuzhoy, J., Khanna, S.: Network Design for Vertex Connectivity. In: Proc. of ACM STOC, pp. 167–176 (2008)
4. Chekuri, C., Korula, N.: Single-Sink Network Design with Vertex-Connectivity Requirements. In: Proc. of FST&TCS (2008)
5. Chekuri, C., Shepherd, B.: Approximate Integer Decompositions for Undirected Network Design Problems. SIAM J. on Disc. Math. 23(1), 163–177 (2008)
6. Cheriyan, J.: Personal Communication (October 2008)
7. Cheriyan, J., Salavatipour, M.: Hardness and Approximation Results for Packing Steiner Trees. Algorithmica 45(1), 21–43 (2006)

8. Cheriyan, J., Salavatipour, M.: Packing Element-disjoint Steiner Trees. ACM Trans. on Algorithms 3(4) (2007)
9. Cheriyan, J., Vempala, S., Vetta, A.: Network design via iterative rounding of setpair relaxations. Combinatorica 26(3), 255–275 (2006)
10. Chuzhoy, J., Khanna, S.: Algorithms for Single-Source Vertex-Connectivity. In: Proc. of IEEE FOCS, pp. 105–114 (October 2008)
11. Chuzhoy, J., Khanna, S.: An O(k^3 logn)-Approximation Algorithm for Vertex-Connectivity Survivable Network Design, arXiv.org preprint. arXiv:0812.4442v1
12. Feige, U., Halldórsson, M., Kortsarz, G., Srinivasan, A.: Approximating the Domatic Number. SIAM J. Comput. 32(1), 172–195 (2002)
13. Fleischer, L., Jain, K., Williamson, D.P.: Iterative Rounding 2-approximation Algorithms for Minimum-cost Vertex Connectivity Problems. J. of Computer and System Sciences 72(5), 838–867 (2006)
14. Frank, A.: On Connectivity Properties of Eulerian Digraphs. Annals of Discrete Mathematics 41, 179–194 (1989)
15. Frank, A.: Augmenting graphs to meet edge-connectivity requirements. SIAM J. on Discrete Math. 5(1), 22–53 (1992)
16. Frank, A., Király, T., Kriesell, M.: On decomposing a hypergraph into k connected sub-hypergraphs. Discr. Applied Math. 131(2), 373–383 (2003)
17. Grötschel, M., Martin, A., Weismantel, R.: The Steiner tree packing problem in VLSI-design. Math. Programming 78, 265–281 (1997)
18. Hind, H.R., Oellermann, O.: Menger-type results for three or more vertices. Congressus Numerantium 113, 179–204 (1996)
19. Jain, K.: A factor 2 approximation algorithm for the generalized Steiner network problem. Combinatorica 21(1), 39–60 (2001)
20. Jain, K., Mahdian, M., Salavatipour, M.: Packing Steiner Trees. In: Proc. of ACM-SIAM SODA, pp. 266–274 (2003)
21. Jain, K., Mandoiu, I.I., Vazirani, V.V., Williamson, D.P.: A Primal-Dual Schema Based Approximation Algorithm for the Element Connectivity Problem. In: Proc. of ACM-SIAM SODA, pp. 484–489 (1999)
22. Jackson, B.: Some Remarks on Arc-connectivity, Vertex splitting, and Orientation in Digraphs. Journal of Graph Theory 12(3), 429–436 (1998)
23. Király, T., Lau, L.C.: Approximate Min-Max Theorems of Steiner Rooted-Orientations of Hypergraphs. J. of Combin. Theory B 98(6), 1233–1252 (2008)
24. Kortsarz, G., Krauthgamer, R., Lee, J.R.: Hardness of Approximation for Vertex-Connectivity Network Design Problems. SIAM J. Comput. 33(3), 704–720 (2004)
25. Kortsarz, G., Nutov, Z.: Approximating min-cost connectivity problems. In: Handbook on Approx. Algorithms and Metaheuristics, CRC Press, Boca Raton (2006)
26. Kriesell, M.: Edge-disjoint trees containing some given vertices in a graph. J. of Combin. Theory B 88, 53–63 (2003)
27. Lau, L.C.: An approximate max-Steiner-tree-packing min-Steiner-cut theorem. Combinatorica 27(1), 71–90 (2007)
28. Lau, L.C.: Packing steiner forests. In: Jünger, M., Kaibel, V. (eds.) IPCO 2005. LNCS, vol. 3509, pp. 362–376. Springer, Heidelberg (2005)
29. Lovász, L.: On some connectivity properties of Eulerian graphs. Acta Math. Hung. 28, 129–138 (1976)
30. Mader, W.: A reduction method for edge connectivity in graphs. Ann. Discrete Math. 3, 145–164 (1978)
31. Nutov, Z.: An almost $O(\log k)$-approximation for k-connected subgraphs. In: Proc. of ACM-SIAM SODA, pp. 912–921 (2009)

Approximating Matches Made in Heaven

Ning Chen[1], Nicole Immorlica[2], Anna R. Karlin[3], Mohammad Mahdian[4], and Atri Rudra[5]

[1] Nanyang Technological University, Singapore
[2] Northwestern University, Chicago, IL, USA
[3] University of Washington, Seattle, WA, USA
[4] Yahoo! Research, Santa Clara, CA, USA
[5] University at Buffalo, State University of New York, Buffalo, NY, USA

Abstract. Motivated by applications in online dating and kidney exchange, we study a stochastic matching problem in which we have a random graph G given by a node set V and probabilities $p(i,j)$ on all pairs $i,j \in V$ representing the probability that edge (i,j) exists. Additionally, each node has an integer weight $t(i)$ called its patience parameter. Nodes represent agents in a matching market with dichotomous preferences, i.e., each agent finds every other agent either *acceptable* or *unacceptable* and is indifferent between all acceptable agents. The goal is to maximize the welfare, or produce a matching between acceptable agents of maximum size. Preferences must be solicited based on probabilistic information represented by $p(i,j)$, and agent i can be asked at most $t(i)$ questions regarding his or her preferences.

A stochastic matching algorithm iteratively probes pairs of nodes i and j with positive patience parameters. With probability $p(i,j)$, an edge exists and the nodes are irrevocably matched. With probability $1 - p(i,j)$, the edge does not exist and the patience parameters of the nodes are decremented. We give a simple greedy strategy for selecting probes which produces a matching whose cardinality is, in expectation, at least a quarter of the size of this optimal algorithm's matching. We additionally show that variants of our algorithm (and our analysis) can handle more complicated constraints, such as a limit on the maximum number of rounds, or the number of pairs probed in each round.

1 Introduction

Matching is a fundamental primitive of many markets including job markets, commercial markets, and even dating markets [3,4,5,14,15,16]. While matching is a well understood graph-theoretic concept, its stochastic variants are considerably less well-developed. Yet stochastic variants are precisely the relevant framework for most markets which incorporate a degree of uncertainty regarding the preferences of the agents. In this paper we study a stochastic variant of matching motivated by applications in the kidney exchange and online dating markets, or more generally, for matching markets with dichotomous preferences in which each agent finds every other agent either *acceptable* or *unacceptable*

S. Albers et al. (Eds.): ICALP 2009, Part I, LNCS 5555, pp. 266–278, 2009.

and is indifferent between acceptable agents (see, e.g., [6]). The basic stochastic matching problem, which is the main focus of this paper, can be stated as follows:

> Let G be a random undirected graph given by a node set V (representing agents in the matching market) and a probability $p(i, j)$ on any pair i, j of nodes, representing the probability that an edge exists between that pair of nodes (i.e., the probability that the corresponding agents find each other acceptable). Whether or not there is an edge between a pair of nodes is not revealed to us unless we *probe* this pair (solicit the preference information from the relevant agents). Upon probing a pair, if there is an edge between them, they are matched and removed from the graph. In other words, when a pair (i, j) is probed, a coin is flipped with probability $p(i, j)$. Upon heads, the pair is matched and leaves the system. In addition, for every node i, we are given a number $t(i)$ called the *patience parameter* of i, which specifies the maximum number of failed probes i is willing to participate in.
>
> The goal is to maximize the welfare, i.e., design a probing strategy to maximize the expected number of matches.

The above formulation of the problem is similar in nature to the formulation of other stochastic optimization problems such as stochastic shortest path [10,7] and stochastic knapsack [8]. The stochastic matching problem is an exponential-sized Markov Decision Process (MDP) and hence has an optimal dynamic program, also exponential. Our goal is to approximate the expected value of this dynamic program in polynomial time. We show that a simple non-adaptive greedy algorithm that runs in near-linear time is a 4-approximation (Section 3). The algorithm simply probes edges in order of decreasing probability. Our algorithm is practical, intuitive, and near-optimal. Interestingly, the algorithm need not even know the patience parameters, but just which edges are more probable.

It is easy to see that the above greedy algorithm is a good approximation when the patience parameters are all one or all infinite: when the patience parameters are all one, the optimal algorithm clearly selects a maximum matching and so the maximal matching selected by the greedy algorithm is a 2-approximation; when the patience parameters are all infinite, for any instantiation of the coin flips, the greedy algorithm finds a maximal matching and hence is a 2-approximation to the (ex-post) maximum matching. To prove that the greedy algorithm is a constant approximation in general, we can no longer compare our performance to the expected size of the maximum matching. Actually, the gap between the expected size of the maximum matching and the expected value of the optimum algorithm may be larger than any constant. Instead, we compare the decision tree of the greedy algorithm to the decision tree of the optimum algorithm. Using induction on the graph as well as a careful charging scheme, we are able to show that the greedy algorithm is a 4-approximation for general patience parameters. Unfortunately, we do not know if computing the optimal solution is even NP-hard. Further, we do not know whether if the analysis of the greedy algorithm is tight. We leave these as open questions and conjecture that (i) computing the optimal strategy is indeed NP-hard and (ii) the greedy algorithm is indeed a 2-approximation.

We also show that our algorithm and analysis can be adapted to handle more complicated constraints (Section 4). In particular, if probes must be performed in a limited number of rounds, each round consisting of probing a matching, a natural generalization of the greedy algorithm gives a 6-approximation in the uniform probability case. For this generalization, the problem does turn out to be NP-hard. We can also generalize the algorithm to a case where we only probe a limited number of edges in each round (Section 4).

1.1 Motivation

In addition to being an innately appealing and natural problem, the stochastic matching problem has important applications. We outline here two applications to kidney exchange and online dating.

Kidney Exchange. Currently, there are 98,167 people in need of an organ in the United States. Of these, 74,047 patients are waiting for a kidney.[1] Every healthy person has two kidneys, and only needs one kidney to survive. Hence it is possible for a living friend or family of the patient to donate a kidney to the patient. Unfortunately, not all patients have compatible donors. At the recommendation of the medical community [12,13], in year 2000 the United Network for Organ Sharing (UNOS) began performing *kidney exchanges* in which two incompatible patient/donor pairs are identified such that each donor is compatible with the other pair's patient. Four simultaneous operations are then performed, exchanging the kidneys between the pairs in order to have two successful transplants.

To maximize the total number of kidney transplants in the kidney exchange program, it is important to match the maximum number of pairs. This problem can be phrased as that of maximum matching on graphs in which the nodes represent incompatible pairs and the edges represent possible transplants based on medical tests [15,16]. There are three main tests which indicate the likelihood of successful transplants. The first two tests, the blood-type test and the antibody screen, compare the blood of the recipient and donor. The third test, called *crossmatching*, combines the recipient's blood serum with some of the donor's red blood cells and checks to see if the antibodies in the serum kill the cells. If this happens (the crossmatch is *positive*), the transplant can not be performed. Otherwise (the crossmatch is *negative*), the transplant may be performed.[2]

Of course, the feasibility of a transplant can only be determined after the final crossmatch test. As this test is time-consuming and must be performed close to the surgery date [2,1], it is infeasible to perform crossmatch tests on all nodes

[1] Data retrieved on November 19th, 2007 from United Network for Organ Sharing (UNOS) — The Organ Procurement and Transplantation Network (OPTN), `http://www.optn.org/data`

[2] Recent advances in medicine actually allow positive crossmatch transplants as well, but these are significantly more risky.

in the graph. Furthermore, due to incentives facing doctors, it is important to perform a transplant as soon as a pair with negative crossmatch tests is identified. Thus the edges are really stochastic; they only reflect the *probability*, based on the initial two tests and related factors, that an exchange is possible. Based on this information alone, edges must be selected and, upon a negative crossmatch test, the surgery performed. Hence the matching problem is actually a stochastic matching problem. The patience parameters in the stochastic matching problem can be used to model the unfortunate fact that patients will eventually die without a successful match.

Online Dating. Another relevant marketplace for stochastic matching is the online dating scene, the second-largest paid-content industry on the web, expected to gross around \$600 million in 2008 [9]. In many online dating sites, most notably eHarmony and Just Lunch, users submit profiles to a central server. The server then estimates the compatibility of a couple and sends plausibly compatible couples on blind dates (and even virtual blind dates). The purported goal of these sites is to create as many happily married couples as possible.

Again, this problem may be modeled as a stochastic matching problem. Here, the people participating in the online match-making program are the nodes in the graph. From the personal characteristics of these individuals, the system deduces for each pair a probability that they are a good match. Whether or not a pair is actually successful can only be known if they are sent on a date. In this case, if the pair is a match, they will immediately leave the program. Also, each person is willing to participate in at most a given number of unsuccessful dates before he/she runs out of patience and leaves the match-making program. The online dating problem is to design a schedule for dates to maximize the expected number of matched couples.

2 Preliminaries

The stochastic matching problem can be represented by a random graph $G = (V, E)$, where for each pair (α, β) of vertices, there is an undirected edge between α and β with a probability $p(\alpha, \beta) \in [0, 1]$.[3] For the rest of the paper, without loss of generality we assume that E contains exactly the pairs that have positive probability. These probabilities are all independent. Additionally, for each vertex $\gamma \in V$ a number $t(\gamma)$ called the *patience parameter* of γ is given. The existence of an edge between a pair of vertices of the graph is only revealed to us after we *probe* this pair. When a pair (α, β) is probed, a coin is flipped with probability $p(\alpha, \beta)$. Upon heads, the pair is matched and is removed from the graph. Upon tails, the patience parameter of both α and β are decremented by one. If the patience parameter of a node reaches 0, this node is removed from the graph. This guarantees that each vertex γ can be probed at most $t(\gamma)$ times. The problem

[3] Note that here we do not impose any constraint that the graph G should be bipartite. In settings such as heterosexual dating where such a constraint is natural, it can be imposed by setting the probabilities between vertices on the same side to zero.

is to design (possibly adaptive) strategies to probe pairs of vertices in the graph such that the expected number of matched pairs is maximized.

An instance of our problem is thus a tuple (G, t). For a given algorithm ALG, let $\mathbf{E}_{\text{ALG}}(G, t)$ (or $\mathbf{E}_{\text{ALG}}(G)$ for simplicity, when t is clear from the context) be the expected number of pairs matched by ALG, where the expectation is over the realizations of probes and (possible) coin tosses of the algorithm itself.

Decision Tree Representation. For any deterministic algorithm ALG and any instance (G, t) of the problem, the entire operation of ALG on (G, t) can be represented as an (exponential-sized) *decision tree* T_{ALG}. The root of T_{ALG}, r, represents the first pair $e = (\alpha, \beta) \in E$ probed by ALG. The *left* and the *right* subtrees of r represent *success* and *failure* for the probe to (α, β), respectively. In general, each node of this tree corresponds to a probe and the left and the right subtrees correspond to the respective success or failure.

For each node $v \in T_{\text{ALG}}$, a corresponding sub-instance (G_v, t_v) of the problem can be defined recursively as follows: The root r corresponds to the initial instance (G, t). If a node v that represents a probe to a pair (α, β) corresponds to (G_v, t_v),

- the left child of v corresponds to $(G_v \setminus \{\alpha, \beta\}, t_v)$, and
- the right child of v corresponds to $(G_v \setminus \{(\alpha, \beta)\}, t'_v)$, where $G_v \setminus \{(\alpha, \beta)\}$ denotes the instance obtained from G_v by setting the probability of the edge (α, β) to zero, and $t'_v(\alpha) = t_v(\alpha) - 1$, $t'_v(\beta) = t_v(\beta) - 1$ and $t'_v(\gamma) = t_v(\gamma)$ for any other vertex γ.

For each node $v \in T_{\text{ALG}}$, let T_v be the subtree rooted at v. Let $T_{L(v)}$ and $T_{R(v)}$ be the left and right subtree of v, respectively. Observe that T_v essentially defines an algorithm ALG′ on the sub-instance (G_v, t_v) corresponding to v. Define $\mathbf{E}_{\text{ALG}}(T_v)$ to be the expected value generated by the algorithm corresponding to ALG′, i.e. $\mathbf{E}_{\text{ALG}}(T_v) = \mathbf{E}_{\text{ALG}'}(G_v, t_v)$.

The stochastic matching problem can be viewed as the problem of computing the optimal policy in an exponential-sized Markov Decision Process (for more details on MDPs, see the textbook by Puterman [11]). The states of this MDP correspond to subgraphs of G that are already probed, and the outcome of these probes. The actions that can be taken at a given state correspond to the choice of the next pair to be probed. Given an action, the state transitions probabilistically to one of two possible states, one corresponding to a success, and the other corresponding to a failure in the probe. We denote by OPT the optimal algorithm, i.e., the solution of this MDP. Note that we can assume without loss of generality that OPT is deterministic, and therefore, a decision tree T_{OPT} representing OPT can be defined as described above. Observe that by definition, for any node v of this tree, if the probability of reaching v from the root is non-zero, the algorithm defined by T_v must be the optimal for the instance (G_v, t_v) corresponding to v. To simplify our arguments, we assume without loss of generality that the algorithm defined by T_v is optimal for (G_v, t_v) for *every* $v \in T_{\text{OPT}}$, even for nodes v that have probability zero of being reached. Note that such nodes can exist in $T_{\texttt{OPT}}$, since OPT can probe edges of probability 1, in which case the corresponding right subtree is never reached.

It is not even known if the optimal strategy OPT can be described in polynomial space. Therefore, one might hope to use other benchmarks such as the optimal offline solution (i.e., the expected size of maximum matching in G) as an upper bound on OPT. However, in the full version of the paper, we show that the gap between OPT and the optimal offline solution can be unbounded.

3 Greedy Algorithm

We consider the following greedy algorithm.

```
GREEDY
1. Sort all edges in E by probabilities, say, p(e_1) ≥ p(e_2) ≥ ··· ≥
   p(e_m) (ties are broken arbitrarily)
2. For i = 1,...,m, if the two endpoints of e_i are available, probe e_i
```

Our main result is as follows.

Theorem 1. *For any instance graph* (G, t), GREEDY *is a 4-approximation to the optimal algorithm, i.e.* $\mathbf{E}_{\texttt{OPT}}(G, t) \leq 4 \cdot \mathbf{E}_{\texttt{GREEDY}}(G, t)$.

In the rest of this section, we will prove Theorem 1. The proof is inductive and based on carefully charging the value obtained at different nodes of $T_{\texttt{OPT}}$ to $T_{\texttt{GREEDY}}$. We begin by stating two lemmas that will be useful for the proof.

Lemma 1. *For any node* $v \in T_{\texttt{OPT}}$, $\mathbf{E}_{\texttt{OPT}}\left(T_{L(v)}\right) \leq \mathbf{E}_{\texttt{OPT}}\left(T_{R(v)}\right) \leq 1 + \mathbf{E}_{\texttt{OPT}}\left(T_{L(v)}\right)$.

Lemma 2. *For any node* $v \in T_{\texttt{OPT}}$, *assume* v *represents the edge* $e = (\alpha, \beta) \in E$, *and let* $p = p(\alpha, \beta)$ *be the probability of* e. *If we increase the probability of* v *to* $p' > p$ *in* $T_{\texttt{OPT}}$, *then* $\mathbf{E}_{\texttt{OPT}}(T_{\texttt{OPT}})$ *will not decrease.*

Note that Lemma 2 does not mean we increase the probability of edge e in graph G. It only says for a particular probe of e in $T_{\texttt{OPT}}$, which corresponds to node v in the claim, if the probability of e is increased, the expected value of OPT will not decrease. (The proofs of the lemmas are deferred to the full version of the paper.)

These two lemmas provide the key ingredients of our proof. To get an idea of the proof, imagine that the first probe of the greedy algorithm is to edge (α, β) represented by node r at the root of $T_{\texttt{GREEDY}}$ as in Figure 1 and suppose that $T_{\texttt{OPT}}$ is as in Figure 2. Let p_r be the probability of success of probe (α, β).

Note that the algorithm $\texttt{ALG}_1$ defined by subtree A in $T_{\texttt{OPT}}$ is a valid algorithm for the left subtree L of greedy (since the optimum algorithm has already matched nodes α and β upon reaching subtree A, all probes in subtree A are valid probes for the left-subtree L of $T_{\texttt{GREEDY}}$). Furthermore, $\texttt{ALG}_1$ achieves the same value, in expectation, as the optimum algorithm on subtree A. Similarly the algorithm $\texttt{ALG}_2$ defined by subtree D in $T_{\texttt{OPT}}$ is a valid algorithm for the right

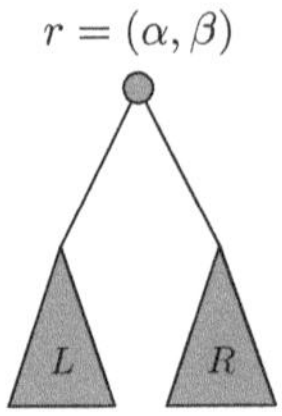

Fig. 1. Greedy tree $T_{\texttt{GREEDY}}$

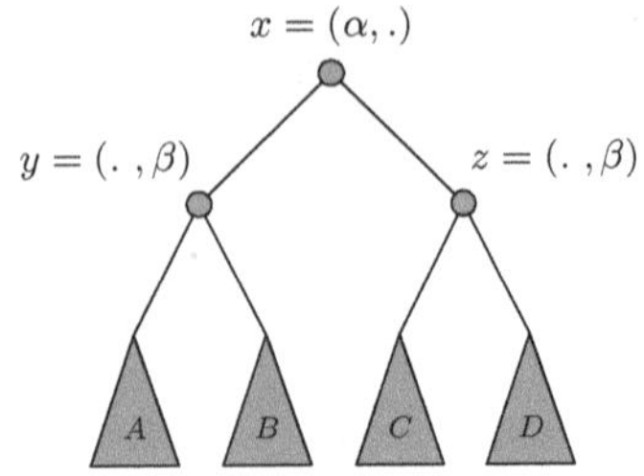

Fig. 2. Optimum tree $T_{\texttt{OPT}}$

subtree R of greedy *except* $\texttt{ALG}_2$ may perform a probe to (α, β). Thus we define a secondary (randomized) algorithm $\texttt{ALG}'_2$ which follows $\texttt{ALG}_2$ but upon reaching a probe to (α, β) simply flips a coin with probability p_r to decide which subtree to follow and does not probe the edge. Hence $\texttt{ALG}'_2$ is a valid algorithm for the right subtree R of greedy, and gets the same value as the optimum algorithm on subtree D minus a penalty of p_r for the missed probe to (α, β). The value of $\texttt{ALG}_1$ and $\texttt{ALG}'_2$ on the left and right subtree L and R of $T_{\texttt{GREEDY}}$ respectively is at most the value of the optimum algorithm on those subtrees and so, by the inductive hypothesis, at most four times the value of the greedy algorithm on those subtrees. By Lemma 2, we can assume that the probes at nodes x, y, and z in $T_{\texttt{OPT}}$ have probability p_r of success. Furthermore, we can use Lemma 1 to bound the value of the optimum algorithm in terms of the left-most subtree A and the right-most subtree D. With a slight abuse of notation, we use A to denote the expected value of the optimum algorithm on subtree A (and similarly, B, C, and D). Summarizing the above observations, we then get:

$$\begin{aligned}
\mathbf{E}_{\texttt{OPT}}(G,t) &\le p_r^2(A+2) + p_r(1-p_r)(B+1) + p_r(1-p_r)(C+1) + (1-p_r)^2 D \\
&= 2p_r + p_r^2 A + p_r(1-p_r)B + p_r(1-p_r)C + (1-p_r)^2 D \\
&\le 2p_r + p_r^2 A + p_r(1-p_r)(A+1) + p_r(1-p_r)D + (1-p_r)^2 D \\
&= 3p_r - p_r^2 + p_r A + (1-p_r)D \\
&\le 4p_r + p_r A + (1-p_r)(D - p_r) \\
&= 4 \cdot \left(p_r(1 + \mathbf{E}_{\texttt{ALG}_1}) + (1-p_r)\mathbf{E}_{\texttt{ALG}'_2}\right) \\
&\le 4\mathbf{E}_{\texttt{GREEDY}}(G,t)
\end{aligned}$$

where the first inequality is by Lemma 2, the second inequality is by Lemma 1, and the last inequality is by the inductive hypothesis.

The above sketch represents the crux of the proof. To formalize the argument, we must account for all possibilities of $T_{\texttt{OPT}}$. We do this by considering "frontiers" in $T_{\texttt{OPT}}$ representing initial probes to α and β, and then follow the general accounting scheme suggested above via slightly more complicated algebraic manipulations.

Proof of Theorem 1. The proof is by induction on the set of edges in the graph G and the patience parameters. In particular, (G', t') is a sub-instance of (G, t)

if G' is an edge subgraph of G and for every vertex $v \in V(G')$, $t'(v) \leq t(v)$. In the base case where the graph has only one edge, the claim is obviously true. Assume that for any sub-instance (G', t') of instance (G, t),

$$\mathbf{E}_{\texttt{OPT}}(G', t') \leq 4 \cdot \mathbf{E}_{\texttt{GREEDY}}(G', t')$$

Given the induction hypothesis, we will show that $\mathbf{E}_{\texttt{OPT}}(G, t) \leq 4 \cdot \mathbf{E}_{\texttt{GREEDY}}(G, t)$.

Let r be the root of $T_{\texttt{GREEDY}}$, which represents probing the edge $(\alpha, \beta) \in E$, and p_r be the probability of edge (α, β). Let (G_L, t_L) and (G_R, t_R) be the sub-instances corresponding to the left and right child of r, respectively. Note that

$$\mathbf{E}_{\texttt{GREEDY}}(G, t) = p_r + p_r \cdot \mathbf{E}_{\texttt{GREEDY}}(G_L, t_L) + (1 - p_r) \cdot \mathbf{E}_{\texttt{GREEDY}}(G_R, t_R) \quad (1)$$

We consider two cases based on whether $p_r = 1$ or $p_r < 1$. If $p_r = 1$, then it is easy to see that the inductive hypothesis holds. Namely, let (G', t') be the sub-instance of (G, t) obtained by removing edge (α, β). Then,

$$\mathbf{E}_{\texttt{OPT}}(G, t) \leq \mathbf{E}_{\texttt{OPT}}(G', t') + 1 \leq 4 \cdot \mathbf{E}_{\texttt{GREEDY}}(G', t') + 1 \leq 4 \cdot \mathbf{E}_{\texttt{GREEDY}}(G, t)$$

where the second inequality follows from the inductive hypothesis.

If $p_r < 1$, then for every node $v \in T_{\texttt{OPT}}$, the probability p_v of the edge corresponding to v satisfies $0 < p_v \leq p_r < 1$ by the definition of the greedy algorithm.

We define q_v to be the probability that OPT reaches node v in $T_{\texttt{OPT}}$. That is, q_v is the product of probabilities of all edges on the path from the root of $T_{\texttt{OPT}}$ to v. The following equality follows from the definition of q_v:

$$\mathbf{E}_{\texttt{OPT}}(G, t) = \sum_{v \in T_{\texttt{OPT}}} p_v \cdot q_v \quad (2)$$

Define $X \subseteq T_{\texttt{OPT}}$ to be the set of nodes that correspond to the first time where OPT probes an edge incident to α or β. In other words, X is the set of nodes $v \in T_{\texttt{OPT}}$ such that OPT probes an edge incident to α or β (or both) at v and at none of the vertices on the path from the root to v. Observe that no node in X lies in the subtree rooted at another node in X. Thus, X essentially defines a "frontier" in $T_{\texttt{OPT}}$.

Take a node $v \in X$. If v represents probing an edge incident to α, consider the set of all nodes in $T_{L(v)}$ that correspond to the *first* time an edge incident to β is probed; otherwise, consider all nodes in $T_{L(v)}$ that correspond to the first time an edge incident to α is probed. Let Y_1 be the union of all these sets, taken over all $v \in X$. Define $Y_2 \subseteq \bigcup_{v \in X} T_{R(v)}$ similarly, with $L(v)$ replaced by $R(v)$.

For any subset of nodes $S \subseteq T_{\texttt{OPT}}$, define $T(S) = \bigcup_{v \in S} T_v$. We show in the full version of the paper that

$$\begin{aligned}\mathbf{E}_{\texttt{OPT}}(G, t) \leq 3p_r &+ \sum_{v \in Y_1} q_v \cdot \mathbf{E}_{\texttt{OPT}}\left(T_{L(v)}\right) + \sum_{u \in X} \sum_{v \in T_{L(u)} \setminus T(Y_1)} p_v \cdot q_v \\ &+ \sum_{v \in Y_2} q_v \cdot \mathbf{E}_{\texttt{OPT}}\left(T_{R(v)}\right) + \sum_{u \in X} \sum_{v \in T_{R(u)} \setminus T(Y_2)} p_v \cdot q_v + \sum_{v \in T_{\texttt{OPT}} \setminus T(X)} p_v \cdot q_v \quad (3)\end{aligned}$$

Define an algorithm $\texttt{ALG}_1$ that works as follows: $\texttt{ALG}_1$ follows the decision tree of $\texttt{OPT}$ except that when the algorithm reaches a node $v \in X \cup Y_1$, it will not probe the edge corresponding to v and go to the left subtree $T_{L(v)}$ directly. Since in $\texttt{ALG}_1$, every path from the root to a node in $T(Y_1)$ (and $\bigcup_{u \in X} T_{L(u)} \setminus T(Y_1)$) has two (and one respectively) less successful probes in $X \cup Y_1$ than $\texttt{OPT}$, it follows that

$$\begin{aligned}
\mathbf{E}_{\texttt{ALG}_1} &= \sum_{u \in X} \sum_{v \in Y_1 \cap T_{L(u)}} \sum_{w \in T_{L(v)}} p_w \cdot \frac{q_w}{p_u p_v} + \sum_{u \in X} \sum_{w \in T_{L(u)} \setminus T(Y_1)} p_w \cdot \frac{q_w}{p_u} \\
&+ \sum_{w \in T_{\texttt{OPT}} \setminus T(X)} p_w \cdot q_w = \sum_{u \in X} \sum_{v \in Y_1 \cap T_{L(u)}} \frac{q_v}{p_u} \cdot \mathbf{E}_{\texttt{OPT}}\left(T_{L(v)}\right) \\
&+ \sum_{u \in X} \sum_{w \in T_{L(u)} \setminus T(Y_1)} p_w \cdot \frac{q_w}{p_u} + \sum_{w \in T_{\texttt{OPT}} \setminus T(X)} p_w \cdot q_w
\end{aligned} \tag{4}$$

(recall $p_u > 0$, and hence the division is valid). In the second equality above, we have used the following fact: Fix $v \in Y_1$. For every $w \in T_{L(v)}$, let $q_w = q_v \cdot p_v \cdot q'_w$. Then $\mathbf{E}_{\texttt{OPT}}\left(T_{L(v)}\right) = \sum_{w \in T_{L(u)}} q'_w \cdot p_w$.

On the other hand, by the definition of X and Y_1, $\texttt{ALG}_1$ will not probe any edge incident to α and β. Thus it is a valid algorithm for the instance (G_L, t_L). By the induction hypothesis, we have

$$\mathbf{E}_{\texttt{ALG}_1} \leq \mathbf{E}_{\texttt{OPT}}(G_L, t_L) \leq 4 \cdot \mathbf{E}_{\texttt{GREEDY}}(G_L, t_L) \tag{5}$$

Define an algorithm $\texttt{ALG}_2$ that works as follows: $\texttt{ALG}_2$ follows the decision tree of $\texttt{OPT}$ except that when the algorithm reaches a node $v \in X \cup Y_2$, it will not probe the edge corresponding to v and proceed to the right subtree $T_{R(v)}$ directly. Using an argument similar to the one used for $\mathbf{E}_{\texttt{ALG}_1}$, we get

$$\begin{aligned}
\mathbf{E}_{\texttt{ALG}_2} &= \sum_{u \in X} \sum_{v \in Y_2 \cap T_{R(u)}} \sum_{w \in T_{R(v)}} p_w \cdot \frac{q_w}{(1-p_u)(1-p_v)} \\
&+ \sum_{u \in X} \sum_{w \in T_{R(u)} \setminus T(Y_2)} p_w \cdot \frac{q_w}{1-p_u} + \sum_{w \in T_{\texttt{OPT}} \setminus T(X)} p_w \cdot q_w \\
&= \sum_{u \in X} \sum_{v \in Y_2 \cap T_{R(u)}} \frac{q_v}{1-p_u} \cdot \mathbf{E}_{\texttt{OPT}}\left(T_{R(v)}\right) \\
&+ \sum_{u \in X} \sum_{w \in T_{R(u)} \setminus T(Y_2)} p_w \cdot \frac{q_w}{1-p_u} + \sum_{w \in T_{\texttt{OPT}} \setminus T(X)} p_w \cdot q_w
\end{aligned} \tag{6}$$

(recall $p_u < 1$, and hence the division is valid).

We define a variant $\texttt{ALG}'_2$ from $\texttt{ALG}_2$ where whenever $\texttt{ALG}_2$ reaches a node corresponding to edge (α, β), $\texttt{ALG}'_2$ will only make a coin toss with the same distribution to decide which subtree to go, but not probe the edge (α, β). That is, the contribution of edge (α, β) is not included in $\texttt{ALG}'_2$. It is easy to see that

$$\mathbf{E}_{\texttt{ALG}_2} \leq \mathbf{E}_{\texttt{ALG}'_2} + p_r \tag{7}$$

By the definition of X and Y_2, $\texttt{ALG}'_2$ is a valid algorithm for the instance (G_R, t_R). By the induction hypothesis, we have

$$\mathbf{E}_{\texttt{ALG}'_2} \leq \mathbf{E}_{\texttt{OPT}}(G_R, t_R) \leq 4 \cdot \mathbf{E}_{\texttt{GREEDY}}(G_R, t_R) \tag{8}$$

Now consider nodes $u \in X$ and imagine increasing the probability of success, p_u, to p_r for each such node. By Lemma 2, this can only increase the value of $\mathbf{E}_{\mathtt{OPT}}(G,t)$ but it clearly does not change the value of $\mathtt{ALG}_1$ or $\mathtt{ALG}_2$. Let $\mathbf{E}_{\mathtt{OPT}}(G,t)'$ be the value of the algorithm $T_{\mathtt{OPT}}$ on this new instance. From (3), we have

$$\begin{aligned}
\mathbf{E}_{\mathtt{OPT}}(G,t)' &\le 3p_r + \sum_{u\in X}\sum_{v\in Y_1\cap T_{L(u)}} p_r\frac{q_v}{p_u}\cdot \mathbf{E}_{\mathtt{OPT}}\left(T_{L(v)}\right) + \sum_{u\in X}\sum_{v\in T_{L(u)}\setminus T(Y_1)} p_v\cdot p_r\frac{q_v}{p_u}\\
&\quad + \sum_{u\in X}\sum_{v\in Y_2\cap T_{R(u)}} (1-p_r)\frac{q_v}{1-p_u}\cdot \mathbf{E}_{\mathtt{OPT}}\left(T_{R(v)}\right)\\
&\quad + \sum_{u\in X}\sum_{v\in T_{R(u)}\setminus T(Y_2)} p_v\cdot(1-p_r)\frac{q_v}{1-p_u} + \sum_{v\in T_{\mathtt{OPT}}\setminus T(X)} p_v\cdot q_v\\
&= 3p_r + p_r\cdot \mathbf{E}_{\mathtt{ALG}_1} + (1-p_r)\cdot \mathbf{E}_{\mathtt{ALG}_2} \qquad (9)
\end{aligned}$$

where the last line follows from (4) and (6). Therefore, we have

$$\begin{aligned}
\mathbf{E}_{\mathtt{OPT}}(G,t) &\le \mathbf{E}_{\mathtt{OPT}}(G,t)' \le 3p_r + p_r\cdot \mathbf{E}_{\mathtt{ALG}_1} + (1-p_r)\cdot \mathbf{E}_{\mathtt{ALG}_2}\\
&\le 4p_r + p_r\cdot \mathbf{E}_{\mathtt{ALG}_1} + (1-p_r)\cdot \mathbf{E}_{\mathtt{ALG}_2'} \qquad (10)\\
&\le 4p_r + 4p_r\cdot \mathbf{E}_{\mathtt{GREEDY}}(G_L,t_L) + 4(1-p_r)\cdot \mathbf{E}_{\mathtt{GREEDY}}(G_R,t_R) \qquad (11)\\
&= 4\cdot \mathbf{E}_{\mathtt{GREEDY}}(G,t) \qquad (12)
\end{aligned}$$

where (10) follows from (7), (11) follows from (5) and (8), and (12) follows from (1). This completes the proof.

4 Multiple Rounds Matching

In this section, we consider a generalization of the stochastic matching problem defined in Section 2. In this generalization, the algorithm proceeds in rounds, and is allowed to probe a set of edges (which have to be a matching) in each round. The additional constraint is a bound, k, on the maximum number of rounds. In the full version of the paper, we show that finding the optimal strategy in this new model is NP-hard. Note that when k is large enough, the problem is equivalent to the model discussed in previous sections.

In the rest of this section, we will study approximation algorithms for the problem. By looking at the probabilities as the weights on edges, we have the following natural generalization of the greedy algorithm.

```
GREEDY_k

1. For each round i = 1,...,k
   (a) compute the maximum weighted matching in the current graph
   (b) probe all edges in the matching
```

Let $\mathtt{OPT}_k$ be the optimal algorithm under this setting. We would like to compare $\mathbf{E}_{\mathtt{GREEDY}_k}$ against $\mathbf{E}_{\mathtt{OPT}_k}$. Unfortunately, with no restriction on the instance, $\mathtt{GREEDY}_k$ can be arbitrarily bad.

However, we can still prove that $\texttt{GREEDY}_k$ is a constant-factor approximation algorithm in two important special cases: when all nodes have infinite patience, and when nodes have arbitrary patience but all non-zero probability edges of G have bounded probability (which contains the equal probability case). Furthermore, we observe that the latter result can be used to give a logarithmic approximation for the general case of the problem.

Special Cases. When the patience of all vertices are infinity, we can show that $\texttt{GREEDY}_k$ is a 4-approximation.

Theorem 2. *For any graph* $G = (V, E)$, $\mathbf{E}_{\text{OPT}_k}(G) \leq 4 \cdot \mathbf{E}_{\text{GREEDY}_k}(G)$, *when the patience of all vertices are infinity.*

Next, we study the approximability of $\texttt{GREEDY}_k$ on instances where nodes have arbitrary patience, but all edges of G have probabilities in a bounded range.

Theorem 3. *Let* (G, t) *be an instance such that for all pairs* α, β *of vertices,* $p(\alpha, \beta)$ *is either 0 or in* $[p_{\min}, p_{\max}]$, *for* $0 < p_{\min} \leq p_{\max} \leq 1$. *Then* $\mathbf{E}_{\text{OPT}_k}(G) \leq (4 + 2p_{\max}/p_{\min}) \cdot \mathbf{E}_{\text{GREEDY}_k}(G)$.

Note that this implies that $\texttt{GREEDY}_k$ is a 6-approximation in the uniform probability case, i.e. $p_{\min} = p_{\max}$.

The General Case. Theorem 3 can be used to obtain a (randomized) approximation algorithm for the general case of the multi-round stochastic matching problem with an approximation factor of $O(\log n)$. This follows from the observations that one can delete all edges with probability less than $p_{\max}/n^2$ and the fact that Theorem 3 gives a constant factor approximation on subgraphs of G with edge weight in the range $(p_{max}/2^i, p_{\max}/2^{i+1}]$, for some integer $i \geq 0$.

A Further Extension. We also consider the following extension of the multi-round model. In each round, an algorithm is only allowed to probe a matching of size at most C, where $1 \leq C \leq \lfloor |V|/2 \rfloor$ is another parameter (V is the set of vertices in the graph). Note that till now we have only considered the cases $C = 1$ and $C = \lfloor |V|/2 \rfloor$. Theorems 2 and 3 for the natural extension of the $\texttt{GREEDY}_k$ algorithm also hold in this model. Further, for the arbitrary patience and probability case, $\texttt{GREEDY}_k$ is a $\Theta(\min(k, C))$-approximation algorithm. (The details are deferred to the full version of the paper.)

5 Conclusions

We studied natural greedy algorithms for the stochastic matching problem with patience parameters and proved that these algorithms are constant factor approximations. A natural question to ask is if designing the optimal strategy is computationally hard (this is even unknown for infinite patience parameters). Actually, we can show the following two variants are NP-hard: (i) The algorithm can probe a matching in at most k rounds (the model we studied in

Section 4) and (ii) the order in which the edges need to be probed are fixed (and the algorithm just needs to decide whether to probe an edge or not). In terms of positive results, it is well known that the greedy algorithm in Section 3 for the special cases of (i) all probabilities being 1 and (ii) all patience parameters being infinity is a 2-approximation. However, we proved that the greedy algorithm is a factor of 4-approximation. We conjecture that the greedy algorithm is in fact a 2-approximation even for the general stochastic matching problem.

Another interesting variant of the problem is when edges also have weights associated with them and the objective is to maximize the (expected) total weight of the matched edges. In the full version of the paper, we exhibit an example that shows that the natural greedy algorithm has an unbounded approximation ratio. In addition, the greedy algorithm considered in Section 3 is *non-adaptive*, that is, the order of edges to probe are decided before the first probe. A natural question to ask is what is the "gap" between the non-adaptive and *adaptive* optimal values (e.g. [8])? In the full version of the paper, we present an example to show that the adaptive optimal is strictly larger than the non-adaptive optimal.

References

1. `http://www.aakp.org/aakp-library/transplant-compatibility/index.cfm`
2. `http://www.online-family-doctor.com/medical-tests/crossmatching.html`
3. Abdulkadiroglu, A., Pathak, P.A., Roth, A.E.: The new york city high school match. American Economic Review 95(2), 364–367 (2005)
4. Abdulkadiroglu, A., Pathak, P.A., Roth, A.E., Sonmez, T.: The boston public school match. American Economic Review 95(2), 368–371 (2005)
5. Abdulkadiroglu, A., Sonmez, T.: School choice: A mechanism design approach. American Economic Review 93(3), 729–747 (2003)
6. Bogomolnaia, A., Moulin, H.: Random matching under dichotomous preferences. Econometrica 72(1), 257–279 (2004)
7. Nikolova, E., Kelner, J.A., Brand, M., Mitzenmacher, M.: Stochastic shortest paths via quasi-convex maximization. In: Azar, Y., Erlebach, T. (eds.) ESA 2006. LNCS, vol. 4168, pp. 552–563. Springer, Heidelberg (2006)
8. Dean, B., Goemans, M., Vondrak, J.: Approximating the stochastic knapsack: the benefit of adaptivity. In: Proceedings of the 45th Annual IEEE Symposium on Foundations of Computer Science (FOCS), pp. 208–217 (2004)
9. Epstein, R.: The truth about online dating. Scientific American (February 2007)
10. Papadimitriou, C.H., Yannakakis, M.: Shortest paths without a map. Theoretical Computer Science 84(1), 127–150 (1991)
11. Puterman, M.: Markov Decision Processes: Discrete Stochastic Dynamic Programming. Wiley Series in Probability and Statistics (2005)
12. Rapaport, F.T.: The case for a living emotionally related international kidney donor exchange registry. Transplantation Proceedings 18, 5–9 (1986)
13. Ross, L.F., Rubin, D.T., Siegler, M., Josephson, M.A., Thistlethwaite, J.R., Woodle, E.S.: Ethics of a paired-kidney-exchange program. The New England Journal of Medicine 336(24), 1752–1755 (1997)

14. Roth, A.E.: The evolution of the labor market for medical interns and residents: A case study in game theory. Journal of Economic Theory 92, 991–1016 (1984)
15. Roth, A.E., Sonmez, T., Unver, U.: Kidney exchange. Quarterly Journal of Economics 119(2), 457–488 (2004)
16. Roth, A.E., Sonmez, T., Unver, U.: Pairwise kidney exchange. Journal of Economic Theory 125(2), 151–188 (2005)

Strong and Pareto Price of Anarchy in Congestion Games

Steve Chien[1] and Alistair Sinclair[2,*]

[1] Microsoft Research Silicon Valley Campus
schien@microsoft.com
[2] Computer Science Division, University of California at Berkeley
sinclair@cs.berkeley.edu

Abstract. Strong Nash equilibria and Pareto-optimal Nash equilibria are natural and important strengthenings of the Nash equilibrium concept. We study these stronger notions of equilibrium in congestion games, focusing on the relationships between the price of anarchy for these equilibria and that for standard Nash equilibria (which is well understood). For *symmetric* congestion games with polynomial or exponential latency functions, we show that the price of anarchy for strong and Pareto-optimal equilibria is much smaller than the standard price of anarchy. On the other hand, for asymmetric congestion games with polynomial latencies the strong and Pareto prices of anarchy are essentially as large as the standard price of anarchy; while for asymmetric games with exponential latencies the Pareto and standard prices of anarchy are the same but the strong price of anarchy is substantially smaller. Finally, in the special case of linear latencies, we show that in asymmetric games the strong and Pareto prices of anarchy coincide exactly with the known value $\frac{5}{2}$ for standard Nash, but are strictly smaller for symmetric games.

1 Introduction

1.1 Background

In algorithmic game theory, the *price of anarchy* [14] is defined as the ratio of the social cost of a worst Nash equilibrium to that of a social optimum (i.e., an assignment of strategies to players achieving optimal social cost). This highly successful and influential concept is frequently thought of as the standard measure of the potential efficiency loss due to individual selfishness, when players are concerned only with their own utility and not with the overall social welfare. However, because a Nash equilibrium guarantees only that no single player (as opposed to no coalition) can improve his utility by moving to a new strategy, the price of anarchy arguably conflates the effects of selfishness and lack of coordination. Indeed, for several natural classes of games, the worst-case price of anarchy is achieved at a Nash equilibrium in which a *group* of selfish players can all improve their individual utilities by moving simultaneously to new strategies; in some cases, the worst Nash equilibrium may not even be *Pareto-optimal*—i.e., it may be possible that a group of players can move to new strategies so that *every* player is better off (or no worse off) than before.

* Supported in part by NSF grant 0635153. Work done in part while this author was visiting Microsoft Research Silicon Valley Campus.

S. Albers et al. (Eds.): ICALP 2009, Part I, LNCS 5555, pp. 279–291, 2009.

In this context, two stronger equilibrium concepts perhaps better isolate the efficiency loss due only to selfishness. A *strong Nash equilibrium* [5] is defined as a state in which no subset of the players may simultaneously change their strategies so as to improve all of their costs. The *strong price of anarchy* (e.g., [3]) is the ratio between the cost of the worst strong equilibrium and the optimum cost. A weaker concept that is very widely studied in the economics literature (see, e.g., [15]) is that of a *Pareto-optimal Nash equilibrium*, which is defined as a Nash equilibrium for which there is no other state in which every player is better off. (Equivalently, one may think of a Pareto-optimal equilibrium as being stable under moves by single players or the coalition of all players, but not necessarily arbitrary coalitions.) One can argue that Pareto-optimality should be a minimum requirement for any equilibrium concept intended to capture the notion of selfishness, in that it should not be in every player's self-interest to move to another state. The *Pareto price of anarchy* is then defined in the obvious way.[1]

A natural question to ask is whether the strong and/or Pareto prices of anarchy are significantly less than the standard price of anarchy. In other words, does the requirement that the equilibrium be stable against coalitions lead to greater efficiency? We note that this question has been addressed recently for several specific families of games in the case of the strong (though not Pareto) price of anarchy [2, 3, 10]; see the related work section below. In this paper, we investigate the question for the large and well-studied class of *congestion games* with linear, polynomial or exponential latency functions.

A *congestion game* is an n-player game in which each player's strategy consists of a set of resources, and the cost of the strategy depends only on the number of players using each resource, i.e., the cost takes the form $\sum_r \ell_r(f(r))$, where $f(r)$ is the number of players using resource r, and ℓ_r is a non-negative increasing function. A standard example is a *network congestion game* on a directed graph, in which each player selects a path from some source to some destination, and each edge has an associated cost function, or "latency", ℓ_r that increases with the number of players using it. (Throughout, we shall use the term "latency" even though we will always be discussing general (non-network) congestion games.) Frequently the latencies are assumed to have a simple form, such as linear, polynomial, or exponential.

Congestion games were introduced in Economics by Rosenthal [18], and further studied in an influential paper by Monderer and Shapley [16]. They have since featured prominently in algorithmic game theory, partly because they capture a large class of routing and resource allocation scenarios, and partly because they are known to possess *pure* Nash equilibria [18]. The price of anarchy for congestion games is by now quite well understood, starting with Koutsoupias and Papadimitriou [14] who considered a (weighted) congestion game on a set of parallel edges. The seminal work of Roughgarden and Tardos [20] established the value $\frac{4}{3}$ as the price of anarchy of network congestion games with linear latencies in the *nonatomic* or Wardrop case [7] (where there are infinitely many players, each of whom controls an infinitesimal amount of traffic); this was extended to polynomial latencies in [21]. The more delicate n-player case was solved independently by Awerbuch, Azar and Epstein [6] and by Christodoulou and Koutsoupias [8], who obtained the tight value $\frac{5}{2}$ for the price of anarchy in the linear

[1] Note that throughout we are assuming that cost (or utility) is *non-transferable*, i.e., players in a coalition cannot share their costs with each other. If costs can be shared, the situation is very different; see, e.g., [12] for a discussion of this alternative scenario.

case, and the asymptotically tight value $k^{k(1-o(1))}$ for the case of polynomial latencies. Subsequently Aland et al. [1] gave an exact value for the polynomial case.

Much less is known about strong or Pareto-optimal Nash equilibria in congestion games. Note that such equilibria need not exist. Holzman and Law-Yone [13] give a sufficient condition for the existence of a strong equilibrium based on the absence of a certain structural feature in the game, and also discuss the uniqueness and Pareto-optimality of Nash equilibria under the same condition. For the strong or Pareto price of anarchy, however, there appear to be no results for general congestion games.

1.2 Results

We investigate the strong and Pareto price of anarchy for congestion games with linear, polynomial and exponential latencies. Roughly speaking, we find that in symmetric[2] games the resulting price of anarchy can be much less than the standard (Nash) price of anarchy, while in asymmetric games the behavior is more complicated: for linear and polynomial latencies all three prices of anarchy are essentially the same, but for exponential latencies the standard and Pareto prices of anarchy are equal, while the strong price of anarchy is substantially smaller. (We note that this gap between symmetric and asymmetric games does not appear for standard Nash equilibria. Understanding the reason for this difference may be worthy of further study.)

More specifically, we show that the strong and Pareto prices of anarchy for symmetric congestion games with polynomial latencies of degree k are at most 2^{k+1} (and that this is tight up to a constant factor); this is in sharp contrast to the Nash price of anarchy of $k^{k(1-o(1))}$ obtained in [1, 6, 8]. In the special case of linear latency, we show that the strong and Pareto prices of anarchy are strictly less than the exact value $\frac{5}{2}$ for standard Nash obtained in [6, 8]. For symmetric games with exponential latency α^t, we show that the strong and Pareto prices of anarchy are at most $\max\{\alpha, n\}$, while the standard Nash price of anarchy is at least β^n, where $\beta > 1$ is a constant that depends on α.

On the other hand, for *asymmetric* games with polynomial latency of degree k, we show that the strong (and therefore also the Pareto) price of anarchy is $k^{k(1-o(1))}$, matching the asymptotic value for standard Nash derived in [6, 8]. Moreover, in the linear case all three prices of anarchy coincide exactly. For exponential latencies, we show that the Pareto price of anarchy is the same as for standard Nash (which is exponentially large), and also that the strong price of anarchy is much smaller; thus we exhibit a separation between strong and Pareto prices of anarchy for a natural class of games.

Since strong and Pareto-optimal equilibria do not always exist, we should clarify the meaning of the above statements. An upper bound on the strong (respectively, Pareto) price of anarchy for a certain class of games bounds the price of anarchy *whenever a strong (respectively, Pareto-optimal) equilibrium exists.* A lower bound means that there is a specific game in the class that has a strong (respectively, Pareto-optimal) equilibrium achieving the stated price of anarchy.

We now briefly highlight a few of our proof techniques. To obtain upper bounds on the Pareto (and hence also strong) price of anarchy in symmetric games, we show that this price of anarchy can always be bounded above by the maximum ratio of the costs of

[2] A game is *symmetric* if all players have the same sets of allowable strategies.

individual players at equilibrium and the same ratio at the social optimum. This allows us to study the equilibrium and the optimum separately, greatly simplifying the analysis. We note that this fact holds for arbitrary symmetric games, not only congestion games, and thus may be of wider interest. Our upper bound on the Pareto price of anarchy for linear latencies requires a much more intricate analysis, and makes use of a matrix $M = (m_{ij})$, where m_{ij} is the relative cost increase to player i's cost at optimum when a new player moves to player j's strategy. This turns out to be a stochastic matrix with several useful properties. Finally, our lower bound arguments are based on constructions used in [1, 6, 8], suitably modified so as to handle the stronger requirements of strong and Pareto equilibria. (These constructions typically have the property that the social optimum is a strong Nash equilibrium, so they are not applicable in our setting.)

1.3 Related Work

We briefly mention here some other related results not discussed above. A fair amount is known about the standard price of anarchy for variations of congestion games. The previously mentioned [6] and [1] both extend their results to congestion games with weighted players, while [8] handles the case where social cost is defined as the *maximum*, rather than total player cost. This latter case is also addressed by the papers [9, 19] for the Wardrop model.

For the strong equilibrium concept, several authors have considered the strong price of anarchy and the existence of strong Nash equilibria in various specific classes of games, often deriving significant gaps between the strong and standard price of anarchy. For example, Andelman et al. [3] study job scheduling and network creation games, Epstein et al. [10] cost-sharing connection games, and Albers [2] network design games.

Other measures stronger than the standard Nash price of anarchy have also been studied recently. Anshelevich et al. [4] consider the *price of stability*, or the ratio of the cost of a *best* Nash equilibrium to the social optimum, for network design games. And Hayrepeyan et al. [12] define and study the "price of collusion" in analogous fashion to the strong price of anarchy, with the crucial difference that coalitions aim to minimize not the cost of each of their members but the combined cost of all members.

On the issue of existence of strong Nash equilibria in congestion games, Rozenfeld and Tennenholtz [17] follow on from the above-mentioned [13] and consider the case where the "latencies" are monotonically decreasing.

2 Preliminaries

A *game* consists of a finite set of players $P = \{1, \ldots, n\}$, each of which is assigned a finite set of *strategies* S_i and a cost function $c_i : S_1 \times \cdots \times S_n \to \mathbb{N}$ that he wishes to minimize. A game is called *symmetric* if all of the S_i are identical. A *state* $s = (s_1, \ldots, s_n) \in S_1 \times \cdots \times S_n$ is any combination of strategies for the players. A state s is a *pure Nash equilibrium* if for all players i, $c_i(s_1, \ldots, s_i, \ldots, s_n) \leq c_i(s_1, \ldots, s'_i, \ldots, s_n)$ for all $s'_i \in S_i$; thus at a Nash equilibrium, no player can improve his cost by unilaterally changing his strategy. It is well known that while every (finite) game has a *mixed* Nash equilibrium[3], not every game has a pure Nash equilibrium. A

[3] In a mixed Nash equilbrium, a player's strategy is any probability distribution over available strategies, and no player can improve his expected cost by choosing another distribution.

state $s = (s_1, \ldots, s_n)$ is a *Pareto-optimal Nash equilibrium* if it is a pure Nash equilibrium and there is no other state in which *every* player has lower[4] cost than at s; in other words, for all $s' = (s'_1, \ldots, s'_n) \in S_1 \times \cdots \times S_n$, there exists some player $j \in P$ such that $c_j(s') \geq c_j(s)$. A state $s = (s_1, \ldots, s_n)$ is a *strong Nash equilibrium* if there does not exist any *coalition* of players that can move in such a way that *every member of the coalition* pays lower cost than at equilibrium; i.e. for any other state $s' \neq s$, there exists a player j such that $s_j \neq s'_j$ and $c_j(s') \geq c_j(s)$. Finally, for any given state s, we define the *social cost* $c(s)$ to be the sum of the players' costs in s, i.e., $c(s) = \sum_{i \in P} c_i(s)$. A state minimizing the social cost in a game is called a *social optimum*.

We will focus on the class of games known as *congestion games*, which are known to always possess a pure Nash equilibrium [18]. In a congestion game, players' costs are based on the shared usage of a common set of *resources* $R = \{r_1, \ldots, r_m\}$. A player's strategy set $S_i \subseteq 2^R$ is a collection of subsets of R; his strategy $s_i \in S_i$ will thus be a subset of R. Each resource $r \in R$ has an associated non-decreasing cost or "latency" function $\ell_r : \{1, \ldots, n\} \to \mathbb{N}$; if t players are using resource r, they each incur a cost of $\ell_r(t)$. Thus in a state $s = (s_1, \ldots, s_n)$, the cost of player i is $c_i(s) = \sum_{r \in s_i} \ell_r(f_s(r))$, where $f_s(r) = |\{j : r \in s_j\}|$ is the number of players using r under s.

Of particular interest are congestion games where the latency functions are linear ($\ell_r(t) = \alpha_r t + \beta_r$), polynomial ($\ell_r(t)$ is a degree-k polynomial in t with non-negative coefficients), or exponential ($\ell_r(t) = \alpha_r^t$ for $1 \leq \alpha_r \leq \alpha$.) For simplicity of notation, we shall assume that $\ell_r(t) = t$ for all r in the linear case, $\ell_r(t) = t^k$ for all r in the polynomial case, and $\ell_r(t) = \alpha^t$ for all r in the exponential case. This will not affect our lower bounds, which are based on explicit constructions of this restricted form, and it is not hard to check that the upper bounds go through as well. We omit the details, which are technical but standard. We note that in our congestion games, Pareto-optimal and strong equilibria may not exist, and games may have the first without the second (see the full version for such an example).

As is standard, we measure the relative efficiency loss for a specific type of equilibrium for a given family of games $\mathcal{G}$ as the maximum possible ratio, over all games in the family, of the social cost of an equilibrium state e in that game to the cost of a social optimum o of the same game, or $\sup_{\mathcal{G},e} \frac{c(e)}{c(o)}$. This measure is known as the *price of anarchy* (or *coordination ratio*) in the case of Nash equilibria, and the *strong price of anarchy* for strong Nash equilibria. In addition to these, we will also consider Pareto-optimal Nash equilibria, in which case we call the above ratio the *Pareto price of anarchy*. Clearly the strong price of anarchy is no larger than the Pareto price of anarchy, which in turn is no larger than the standard (Nash) price of anarchy.

3 Symmetric Games

In this section we prove upper bounds on the strong and Pareto price of anarchy for symmetric congestion games with polynomial and exponential latencies. We shall see

[4] Some definitions of Pareto-optimality require there to be no other state in which no player has higher cost than at s and at least one player has lower cost. It is easy to check that our results carry over to this alternative definition with minor modifications to the proofs.

that these are much smaller than the known values for the standard Nash price of anarchy. Thus for symmetric games, increased coordination, when possible, leads to greater efficiency.

3.1 The Basic Framework

The main vehicle for these proofs is a simple framework that allows us to bound the price of anarchy in terms of the maximum ratio of the player costs at equilibrium and the maximum ratio of the player costs at a social optimum. This is the content of the following theorem, which applies to *all* symmetric games, not only congestion games.

Theorem 3.1. *Given a particular symmetric game with n players, let the state e be a Pareto-optimal Nash equilibrium and s be any other state. Let ρ_e be defined as* $\max_{i,j} c_i(e)/c_j(e)$ *over all players i, j, and ρ_s be defined as* $\max_{i,j} c_i(s)/c_j(s)$. *Then* $\frac{c(e)}{c(s)} \leq \max\{\rho_e, \rho_s\}$.

Proof. By symmetry, we can assume w.l.o.g. that the players are ordered by cost in both e and s: that is, $c_1(e) \leq \cdots \leq c_n(e)$ and $c_1(s) \leq \cdots \leq c_n(s)$. Thus $c_n(e)/c_1(e) = \rho_e$, and $c_n(s)/c_1(s) = \rho_s$. We start from e, and consider the hypothetical move in which every player i moves from e_i to his corresponding strategy s_i in s. Since e is Pareto-optimal, there must exist some player j for whom $c_j(s) \geq c_j(e)$.

We now upper bound the social cost of equilibrium, $c(e) = \sum_i c_i(e)$, and lower bound the social cost $c(s) = \sum_i c_i(s)$ of state s. Consider first the $c_i(e)$ values. We have $c_1(e) \leq \cdots \leq c_j(e) \leq \cdots \leq c_n(e) = \rho_e c_1(e)$. The sum $\sum_i c_i(e)$ is therefore maximized when $c_1(e) = c_2(e) = \cdots = c_j(e)$ and $c_{j+1}(e) = \cdots = \rho_e c_1(e)$, giving an upper bound of $jc_j(e)+(n-j)\rho_e c_j(e)$. Similarly, for the $c_i(s)$ values, we have $c_1(s) \leq \cdots \leq c_j(s) \leq \cdots \leq \rho_s c_n(s)$. The sum $\sum_i c_i(s)$ is minimized when $c_1(s) = \cdots = c_{j-1}(s) = c_j(s)/\rho_s$ and $c_j(s) = \cdots = c_n(s)$, and is therefore at least $\frac{(j-1)c_j(s)}{\rho_s} + (n-j+1)c_j(s)$. Recalling that $c_j(s) \geq c_j(e)$ and combining the two bounds, we obtain

$$\frac{\sum_i c_i(e)}{\sum_i c_i(s)} \leq \frac{jc_j(s) + (n-j)\rho_e c_j(s)}{\frac{(j-1)c_j(s)}{\rho_s} + (n-j+1)c_j(s)} \leq \frac{j + (n-j)\rho_e}{\frac{(j-1)}{\rho_s} + (n-j+1)}. \tag{1}$$

Differentiating with respect to j, we find that this expression is maximized at $j = 1$ or $j = n$. In the former case the quotient is at most $\frac{1+(n-1)\rho_e}{n} \leq \rho_e$, while in the latter case it is at most $\frac{n}{(n-1)/\rho_s+1} \leq \rho_s$. □

Thus, for a family of symmetric games, if we can find ρ_e and ρ_o such that $c_i(e)/c_j(e) \leq \rho_e$ for all Pareto-optimal equilibria e, and $c_i(o)/c_j(o) \leq \rho_o$ for all social optima o, we can bound the Pareto (and hence the strong) price of anarchy by $\max\{\rho_e, \rho_o\}$. We now proceed to do this for polynomial and exponential symmetric congestion games.

3.2 Polynomial Latencies

For the case of polynomial latencies, where each resource r has latency function $\ell_r(t) = t^k$, we show the following:

Theorem 3.2. *For symmetric congestion games with polynomial latencies of degree k, the Pareto price of anarchy (and hence also the strong price of anarchy) is at most 2^{k+1}.*

Remarks. (i) Note that the Pareto price of anarchy is much smaller than the known value of $k^{k(1-o(1))}$ for the standard Nash price of anarchy [6, 8, 1]. **(ii)** The upper bound in Theorem 3.2 is tight up to a constant factor; see the full version for an example.

Proof of Theorem 3.2. Following the framework of Theorem 3.1, it suffices to derive upper bounds on the ratios of player costs both at equilibrium and at a social optimum. This we do in the following two claims.

Claim 3.3. *In the situation of Theorem 3.2, we have* $\max_{i,j \in P} \frac{c_i(e)}{c_j(e)} \leq 2^k$.

Proof. Consider any two players i and j at an equilibrium e, and the hypothetical move in which player i switches from his current strategy e_i to j's strategy e_j, resulting in the new state e'. We can now bound $c_i(e')$ in terms of $c_j(e)$. Note that

$$c_i(e') = \sum_{r \in e_j} f_{e'}(r)^k = \sum_{r \in e_j \setminus e_i} (f_e(r)+1)^k + \sum_{r \in e_j \cap e_i} f_e(r)^k \leq \sum_{r \in e_j} (f_e(r)+1)^k.$$

(This captures the intuition that in switching to e_j, player i pays at most what player j would pay if there were one more player using each resource.) From this, it follows that

$$\frac{c_i(e')}{c_j(e)} \leq \frac{\sum_{r \in e_j} (f_e(r)+1)^k}{\sum_{r \in e_j} f_e(r)^k} \leq \max_{r \in e_j} \frac{(f_e(r)+1)^k}{f_e(r)^k} \leq 2^k.$$

Since e is a Nash equilibrium, we have $c_i(e) \leq c_i(e')$, and thus $c_i(e) \leq 2^k c_j(e)$. □

Claim 3.4. *In the situation of Theorem 3.2, we have* $\max_{i,j \in P} \frac{c_i(o)}{c_j(o)} \leq 2^{k+1}$.

Proof. As in the proof of Claim 3.3, consider any two players i and j at a social optimum o. Assume that $c_i(o) \geq c_j(o)$ as the claim is immediately true otherwise. Again, consider the move in which i moves from his current strategy o_i to j's strategy o_j, resulting in the new state o'.

Since o is a social optimum, the social cost of o' must be at least that of o; i.e., $\sum_l c_l(o') - \sum_l c_l(o) \geq 0$. Also, as $\sum_l c_l(s) = \sum_r f_s(r)^{k+1}$ for any state s, we have

$$\begin{aligned} 0 \leq \sum_r f_{o'}(r)^{k+1} - \sum_r f_o(r)^{k+1} &= \sum_{r \in o_i \oplus o_j} f_{o'}(r)^{k+1} - \sum_{r \in o_i \oplus o_j} f_o(r)^{k+1} \\ &= \sum_{r \in o_j \setminus o_i} ((f_o(r)+1)^{k+1} - f_o(r)^{k+1}) - \sum_{r \in o_i \setminus o_j} (f_o(r)^{k+1} - (f_o(r)-1)^{k+1}), \end{aligned}$$

where the second equality in the first line follows since $f_{o'}(r) = f_o(r)$ for $r \notin o_i \oplus o_j$.

Now observe that $c_i(o) = \sum_{r \in o_i} f_o(r)^k = \sum_{r \in o_i \setminus o_j} f_o(r)^k + \sum_{r \in o_i \cap o_j} f_o(r)^k$, and add this to both sides of the above to get

$$\begin{aligned} c_i(o) \leq &\sum_{r \in o_j \setminus o_i} ((f_o(r)+1)^{k+1} - f_o(r)^{k+1}) + \sum_{r \in o_i \cap o_j} f_o(r)^k \\ &- \sum_{r \in o_i \setminus o_j} (f_o(r)^{k+1} - (f_o(r)-1)^{k+1} - f_o(r)^k). \end{aligned}$$

It is not hard to verify that the last summation here is always non-negative.

Since $c_j(o) = \sum_{r \in o_j \setminus o_i} f_o(r)^k + \sum_{r \in o_i \cap o_j} f_o(r)^k$, we have

$$\begin{aligned}\frac{c_i(o)}{c_j(o)} &\leq \frac{\sum_{r \in o_j \setminus o_i}((f_o(r)+1)^{k+1} - f_o(r)^{k+1}) + \sum_{r \in o_i \cap o_j} f_o(r)^k}{\sum_{r \in o_j \setminus o_i} f_o(r)^k + \sum_{r \in o_i \cap o_j} f_o(r)^k} \\ &\leq \frac{\sum_{r \in o_j \setminus o_i}((f_o(r)+1)^{k+1} - f_o(r)^{k+1})}{\sum_{r \in o_j \setminus o_i} f_o(r)^k} \\ &\leq \max_{r \in o_j \setminus o_i} \frac{(f_o(r)+1)^{k+1} - f_o(r)^{k+1}}{f_o(r)^k},\end{aligned}$$

where the second line follows because the ratio of the first sums in the numerator and denominator is greater than 1. This last quantity is at most 2^{k+1}, proving the claim. □

Combining these claims with Theorem 3.1 finishes the proof of Theorem 3.2. □

3.3 Exponential Latencies

For the case of exponential latencies $\ell_r(t) = \alpha^t$, we show the following upper bound on the Pareto and strong prices of anarchy. The proof follows the same structure as that of Theorem 3.2 and is left to the full version.

Theorem 3.5. *For symmetric congestion games with n players and exponential latencies α^t, the Pareto (and hence also the strong) price of anarchy is at most* $\max\{\alpha, n\}$.

In contrast to the above upper bound, we now show that the standard Nash price of anarchy in exponential congestion games is much larger—indeed, exponential in n.

Proposition 3.6. *For symmetric n-player congestion games with exponential latencies α^t, the (standard Nash) price of anarchy is at least $(\frac{\alpha}{2})\alpha^{(\frac{\alpha/2-1}{\alpha-1})n}$.*

Proof. Our construction is based on that of [8] for the case of linear latencies. Our game contains m groups of resources, and $n = mt$ players divided evenly into m equivalence classes, labeled $\{1, \ldots, m\}$, with t players per class. Each of the m groups of resources consists of $\binom{m}{k}$ resources, each labeled with a different k-tuple of equivalence classes. The available strategies for all players are to take either (1) all resources in a single group of resources, or (2) for any i in $\{1, \ldots, m\}$, all resources that are labeled with i.

Given the value of α, we choose m and k so that $m \leq 1 + (\frac{m}{k} - 1)\alpha$; for example, $k = \frac{\alpha}{2}$ and $m = \alpha - 1$ for integer $\alpha \geq 4$. It can be verified that, with these settings, the state in which each player takes all resources labeled with his equivalence class number is a Nash equilibrium, while the state in which each player takes the group of resources corresponding to his equivalence class is a social optimum. A straightforward calculation then shows that the ratio of a player's Nash cost to his cost at social optimum is $k\alpha^{t(k-1)}$, which is $(\frac{\alpha}{2})\alpha^{(\frac{\alpha/2-1}{\alpha-1})n}$ for the above values of k and m. □

For completeness, we show that this same price of anarchy for standard Nash is upper bounded by α^n:

Proposition 3.7. *For asymmetric (and hence also symmetric) congestion games with exponential latencies, the (standard Nash) price of anarchy is at most α^n.*

Proof (sketch). Following [6, 8, 1], in a game with latencies $\ell(t)$ we can prove an upper bound on the price of anarchy by finding $c_1, c_2 \geq 0$ such that the inequality $y\ell(x+1) \leq c_1 x\ell(x) + c_2 y\ell(y)$ holds for all $0 \leq x \leq n, 1 \leq y \leq n$; this implies a price of anarchy of at most $\frac{c_2}{1-c_1}$. For $\ell(t) = \alpha^t$, this clearly holds with $c_1 = 0$ and $c_2 = \alpha^x \leq \alpha^n$. □

Remark. With some extra work, this bound can be improved to $O(\alpha^{(1-\frac{1}{\alpha})n})$.

4 Asymmetric Games

In this section we extend the investigation of the previous section to asymmetric games, and find that the situation is quite different. First we will see that, for asymmetric congestion games with polynomial latencies, the strong (and therefore also the Pareto) price of anarchy is essentially the same as the standard Nash price of anarchy. We will then go on to consider exponential latencies, where we find that the Pareto price of anarchy is the same as standard Nash, but the strong price of anarchy is significantly smaller.

Theorem 4.1. *For asymmetric congestion games with polynomial latencies t^k, the strong price of anarchy is at least $\lfloor \Phi_k \rfloor^k$, where Φ_k is the positive solution of $(x+1)^k = x^{k+1}$.*

Remark. Φ_k is a generalization of the golden ratio (which is just Φ_1); its value is $\frac{k}{\log k}(1+o(1))$. Hence the lower bound of Theorem 4.1 is of the form $k^{k(1-o(1))}$, which is asymptotically the same value for the Nash price of anarchy obtained in [6, 8], and very close to the exact value obtained by Aland et al. [1].

Proof. Our lower bound construction is based on that of Aland et al., extended so as to handle the stricter requirement of a strong equilibrium. Consider an (asymmetric) game with n players (n assumed sufficiently large). Each player i has exactly two possible strategies, e_i and o_i. There are $n+m$ resources labeled $\{r_1, \ldots, r_{n+m}\}$, where m is a constant to be chosen later. For each player i, strategy o_i consists of the single resource r_i. (We shall modify this slightly for some of the players shortly.) Strategy e_i consists of the resources $\{r_{i+1}, \ldots, r_{i+m}\}$.

We claim that the state $e = (e_1, \ldots, e_n)$ is a strong Nash equilibrium. To see this, note that under e the cost for player i is $c_i(e) = \sum_{j=i+1}^{i+m} \min\{j-1, m\}^k$. If now player i moves to his alternative strategy o_i, resulting in a new state $e^{(i)}$, his cost becomes $c_i(e^{(i)}) = \min\{i, m+1\}^k$. To show that e is a Nash equilibrium, we need to show that $c_i(e^{(i)}) \geq c_i(e)$ for all i.

Now note that, for all players $i \geq m+1$, we have $c_i(e^{(i)}) - c_i(e) = (m+1)^k - m^{k+1}$. Thus if we choose $m = \lfloor \Phi_k \rfloor$ to be the smallest integer such that $(m+1)^k \geq m^{k+1}$, we ensure that $c_i(e^{(i)}) \geq c_i(e)$ for all $i \geq m+1$. To obtain the same condition for players $1 \leq i \leq m$, we append to the strategy o_i the minimum number a_i of additional resources (unique to i) so that $c_i(e^{(i)}) = i^k + a_i \geq c_i(e)$. (Note that all the a_i are less than m^{k+1}.) This ensures that e is a Nash equilibrium.

To see that it is a strong equilibrium, consider a move by an arbitrary coalition of players to their alternative strategies o_i. We claim that the lowest numbered player in the coalition does not see an improvement in cost. This follows because the resource r_i, which i occupies under o_i, is still occupied by the same players as under e, so by the Nash property i's cost does not decrease.

Thus the strong price of anarchy is bounded below by $\frac{c(e)}{c(o)}$. But $c(e) \geq (n-m)m^k$, and $c(o) \leq mm^{k+1} + (n-m)$. Thus $\frac{c(e)}{c(o)} \geq \frac{(n-m)m^k}{m^{k+2}+n-m} \rightarrow \lfloor \Phi_k \rfloor^k$ as $n \rightarrow \infty$. □

We now turn to exponential latencies. Our next result shows that the Pareto price of anarchy is bounded below by, and hence equal to, the standard Nash price of anarchy (which we showed to be exponential in n in Proposition 3.6).

Theorem 4.2. *For asymmetric congestion games with exponential latencies α^t, the Pareto price of anarchy is equal to the standard Nash price of anarchy.*

Proof. Consider any n-player congestion game with exponential latencies α^t with Nash equilibrium e. We create a modified game in which the Pareto price of anarchy is only a $(1 - O(\frac{1}{n}))$-factor smaller than the Nash price of anarchy of the original game.

To do this, we first replace each resource in the original game with a set of n resources in the modified game; strategies in the modified game correspond to those in the original game, except that the former include all n copies of the resources of the latter. This has the effect of multiplying player costs by a factor of n, but does not change the set of Nash and Pareto-optimal Nash equilibria. We then add one more player, $n+1$, to the modified game; this player has a single strategy s_{n+1} consisting of new resources $\{\hat{r}_i : i = 1, \ldots, n\}$. Also, for players $1, \ldots, n$, we append resource $\hat{r}_i$ to every strategy of player i except for the equilibrium strategy e_i. Note that this makes the modified game asymmetric even if the original one is not. There is an obvious bijection between states of the original game and those of the modified game, and it can be verified that the original equilibrium is now a Pareto-optimal equilibrium (as any move increases the cost of player $n+1$). A routine calculation now shows that the ratio of the cost of the modified equilibrium to the cost of the modified social optimum is at least $\frac{1}{1+\frac{2\alpha}{n}} \frac{c(e)}{c(o)}$, which approaches the Nash price of anarchy as n increases. □

Finally, we exhibit a separation between the Pareto and strong price of anarchy by showing that the latter (while still exponential) is significantly smaller than the value we obtained for the standard Nash price of anarchy in Proposition 3.6. We make the reasonable assumption that the number of resources is polynomially bounded in the number of players, as is the case in our lower bound construction in Proposition 3.6.

Theorem 4.3. *For asymmetric congestion games with n players and exponential latencies α^t, in which every strategy contains at most $p(n)$ resources for some fixed polynomial p, the strong price of anarchy is at most $\alpha^{(\frac{1}{3}+o(1))n}$.*

Proof. Let e be a strong equilibrium state and o be a social optimum state. As before, we will consider moves in which subsets of players move from their equilibrium strategies e_i to their strategies at optimum o_i. Let $c_*(o)$ denote the maximum cost of any player in state o. We need the following lemma, whose proof we leave to the full version:

Lemma 4.4. *Let S be a subset of players each of whose strategies at e contains at least one resource that is shared by at least u_e players at e (i.e., for all $i \in S$, $\exists r \in e_i$ such that $f_e(r) \geq u_e$). Then at least one of the following must be true: (1) $\alpha^{u_e} \leq c_*(o)$; or (2) there exist at least $u'_e = u_e - \log_\alpha p(n) - \log_\alpha c_*(o)$ players outside of S, each of whose strategies at e contains a resource that is shared by at least u'_e players not in S.*

(The intuition is as follows: let the players holding a resource r form a coalition C and deviate by taking their strategies at o. Since e is a strong equilibrium, one of these players must have higher cost as a result. This might happen if this player shares a resource at o with many members of C, in which case (1) holds; or if one of this player's resources at o is shared by enough non-coalition players at e, in which case (2) holds.)

Suppose now that there exists a strong equilibrium e in a game fitting the description of the theorem such that $\frac{c_e}{c_*(o)} \geq \alpha^{\delta n}$ for a social optimum o. Then there must exist a player j for whom $c_j(e) \geq \frac{\alpha^{\delta n}}{n} c_*(o)$. Thus e_j must contain a resource r for which $f_r(e) \geq \log_\alpha \left(\frac{\alpha^{\delta n}}{np(n)} c_*(o) \right) = \delta n + \log_\alpha c_*(o) - \log_\alpha(np(n))$.

Let S_1 denote the players holding this resource. Consider the move in which we try to move all players in S_1 to their strategies at o. Applying Lemma 4.4 to these players, we find that either (1) $\alpha^{\delta n} \leq np(n)$, in which case the theorem is proven; or (2) there exist $\delta n - \log_\alpha(np^2(n))$ additional players, each of whose equilibrium strategies contains a resource shared by at least that many players not in S_1. Let these additional players form the set S_2. Since the game has n players, this implies that

$$n \geq |S_1 \cup S_2| \geq 2\delta n + \log_\alpha c_*(o) - \log_\alpha(n^2 p(n)^3). \quad (2)$$

We can then apply Lemma 4.4 again to $S_1 \cup S_2$, which again yields two possible outcomes. In case (1), we have that $\alpha^{\delta n} \leq c_*(o)np(n)^2$, or $\delta n \leq \log_\alpha c_*(o) + \log_\alpha(np(n)^2)$. Combining this with inequality (2) gives $3\delta n \leq n + \log_\alpha(n^3 p(n)^5)$, or $\delta \leq \frac{1}{3} + o(1)$, as claimed. In case (2), we are guaranteed the existence of $\delta n - \log_\alpha c_*(o) - \log_\alpha(np(n)^3)$ players not in $S_1 \cup S_2$. Combining this with the lower bound on $|S_1 \cup S_2|$ from Eqn (2), we must have at least $3\delta n - \log_\alpha(n^3 p(n)^6)$ players. Since this cannot exceed n, we have $\delta \leq \frac{1}{3} + o(1)$, again as claimed. □

5 Linear Latencies

This section presents more detailed results for the special case of linear latencies.

Exact price of anarchy for asymmetric games. We first show that the strong (and thus also the Pareto) price of anarchy for asymmetric games with linear latencies coincides exactly with the standard Nash price of anarchy, which is known to be $\frac{5}{2}$ [6, 8]. To do this, we exhibit a lower bound of $\frac{5}{2}$ on the strong price of anarchy, using a refinement of the construction in the proof of Theorem 4.1; the proof is left to the full version.

Theorem 5.1. *For asymmetric linear congestion games, the strong price of anarchy is at least* $\frac{5}{2}$.

Upper bound for symmetric games. We now show that, for *symmetric* linear congestion games, the Pareto (and hence also strong) price of anarchy is less than the known value $\frac{5}{2}$ for standard Nash equilibria in both symmetric and asymmetric games. For linear latencies, the framework of Theorem 3.1 only gives an upper bound of 3, so we must resort to a more involved analysis. We prove the following, stressing that our goal is not to find the best possible upper bound but to show that it is strictly less than $\frac{5}{2}$.

Theorem 5.2. *For symmetric congestion games with linear latencies, the Pareto price of anarchy (and hence also the strong price of anarchy) is strictly less than* $\frac{5}{2}$.

The proof of this is quite involved and is left to the full version; we roughly outline the approach here. As in the proof of Theorem 3.1 we first sort the strategies at equilibrium e and optimum o by cost. A key idea in that proof is the hypothetical move from e in which every player moves from his current strategy e_i to his strategy o_i at o, and the realization that there exists at least one player j for whom $\frac{c_j(e)}{c_j(o)} \leq 1$. Here we extend this to a more complicated sequence of player moves in which, after the players move to o, we attempt to reassign a subset of them, including all those for whom the ratio $\frac{c_i(e)}{c_i(o)}$ is at most 1, to the strategies of other players at o for whom the corresponding ratios are larger than $\frac{5}{2}$. In a technical analysis that again exploits the fact that at least one player must pay higher cost at the end of this sequence than at e, we are able to use the probabilistic method to bound the number of large-ratio players to the number of small-ratio players, and thus show that the Pareto price of anarchy is strictly below $\frac{5}{2}$.

Acknowledgement

We thank Tim Roughgarden for helpful input.

References

[1] Aland, S., Dumrauf, D., Gairing, M., Monien, B., Schoppmann, F.: Exact price of anarchy for polynomial congestion games. In: Durand, B., Thomas, W. (eds.) STACS 2006. LNCS, vol. 3884, pp. 218–229. Springer, Heidelberg (2006)
[2] Albers, S.: On the value of coordination in network design. In: Proceedings of the 19th Annual ACM-SIAM Symposium on Discrete Algorithms (SODA), pp. 294–303 (2008)
[3] Andelman, N., Feldman, M., Mansour, Y.: Strong price of anarchy. In: Proceedings of the 18th Annual ACM-SIAM Symposium on Discrete Algorithms (SODA), pp. 189–198 (2007)
[4] Anshelevich, E., Dasgupta, A., Kleinberg, J., Tardos, E., Wexler, T., Roughgarden, T.: The price of stability for network design with fair cost allocation. In: FOCS, pp. 59–73 (2004)
[5] Aumann, R.J.: Acceptable points in general cooperative n-person games. In: Contribution to the Theory of Games. Annals of Mathematics Studies, vol. IV, 40, pp. 287–324 (1959)
[6] Awerbuch, B., Azar, Y., Epstein, A.: The price of routing unsplittable flow. In: Proc. 37th ACM Symposium on Theory of Computing, pp. 57–66 (2005)
[7] Beckmann, M., McGuire, C.B., Winsten, C.B.: Studies in the Economics of Transportation. Yale University Press, New Haven (1956)
[8] Christodoulou, G., Koutsoupias, E.: The price of anarchy of finite congestion games. In: Proc. 37th ACM Symposium on Theory of Computing, pp. 67–73 (2005)
[9] Correa, J.R., Schulz, A.S., Stier Moses, N.E.: Computational complexity, fairness, and the price of anarchy of the maximum latency problem. In: Bienstock, D., Nemhauser, G.L. (eds.) IPCO 2004. LNCS, vol. 3064, pp. 59–73. Springer, Heidelberg (2004)
[10] Epstein, A., Feldman, M., Mansour, Y.: Strong equilibrium in cost sharing connection games. In: Proceedings of the 8th ACM Symposium on Electronic Commerce, pp. 84–92 (2007)
[11] Fabrikant, A., Papadimitriou, C.H., Talwar, K.: The complexity of pure Nash equilibria. In: Proc. 36th ACM Symposium on Theory of Computing, pp. 604–612 (2004)
[12] Hayrapetyan, A., Tardos, E., Wexler, T.: The effect of collusion in congestion games. In: Proceedings of the 38th STOC, pp. 89–98 (2006)

[13] Holzman, R., Law-Yone, N.: Strong equilibrium in congestion games. Games and Economic Behaviour 21, 85–101 (1997)
[14] Koutsoupias, E., Papadimitriou, C.: Worst-case equilibria. In: Meinel, C., Tison, S. (eds.) STACS 1999. LNCS, vol. 1563, pp. 404–413. Springer, Heidelberg (1999)
[15] Mas-Colell, A., Whinston, M.D., Green, J.R.: Microeconomic Theory. Oxford University Press, Oxford (1995)
[16] Monderer, D., Shapley, L.S.: Potential games. Games and Economic Behavior 14, 124–143 (1996)
[17] Rozenfeld, O., Tennenholtz, M.: Strong and correlated strong equilibria in monotone congestion games. In: Spirakis, P.G., Mavronicolas, M., Kontogiannis, S.C. (eds.) WINE 2006. LNCS, vol. 4286, pp. 74–86. Springer, Heidelberg (2006)
[18] Rosenthal, R.W.: A class of games possessing pure-strategy Nash equilibria. International Journal of Game Theory 2, 65–67 (1973)
[19] Roughgarden, T.: The maximum latency of selfish routing. In: Proceedings of the 15th Annual ACM-SIAM Symposium on Discrete Algorithms (SODA), pp. 980–981 (2004)
[20] Roughgarden, T., Tardos, E.: How bad is selfish routing? Journal of the ACM 49, 236–259 (2002)
[21] Roughgarden, T., Tardos, E.: Bounding the inefficiency of equilibria in nonatomic congestion games. Games and Economic Behavior 47, 389–403 (2004)

A Better Algorithm for Random k-SAT

Amin Coja-Oghlan*

University of Edinburgh, School of Informatics, Edinburgh EH8 9AB, UK
acoghlan@inf.ed.ac.uk

Abstract. Let $\boldsymbol{\Phi}$ be a uniformly distributed random k-SAT formula with n variables and m clauses. We present a polynomial time algorithm that finds a satisfying assignment of $\boldsymbol{\Phi}$ with high probability for constraint densities $m/n < (1 - \varepsilon_k)2^k \ln(k)/k$, where $\varepsilon_k \to 0$. Previously no efficient algorithm was known to find solutions with non-vanishing probability beyond $m/n = 1.817 \cdot 2^k/k$ [Frieze and Suen, Journal of Algorithms 1996].

1 Introduction

The k-SAT problem is well known to be NP-hard for $k \geq 3$. But this merely indicates that no algorithm can solve *all* possible inputs efficiently. Therefore, a significant amount of research has been conducted on *heuristics* for k-SAT, i.e., algorithms that solve 'most' inputs efficiently (where the meaning of 'most' depends on the scope of the respective paper). While some heuristics for k-SAT are very sophisticated, virtually all of them are based on at least one of the following basic paradigms.

Pure literal rule. If a variable x occurs only positively (resp. negatively) in the formula, set it to true (resp. false). Simplify the formula by substituting the newly assigned value for x and repeat.

Unit clause propagation. If the formula contains a clause that consists of only a single literal ('unit clause'), then set the underlying variable so as to satisfy this clause. Simplify and repeat.

Walksat. Initially pick a random assignment. Then repeat the following. While there is an unsatisfied clause, pick one at random, pick a variable occurring in the chosen clause randomly, and flip its value.

Backtracking. Assign a variable x, simplify the formula, and recurse. If the recursion fails to find a satisfying assignment, assign x the opposite value and recurse.

Heuristics based on these paradigms can be surprisingly successful (given that k-SAT is NP-hard) on certain types of inputs. However, it remains remarkably simple to generate formulas that elude all known algorithms/heuristics. Indeed, the simplest conceivable type of *random* instances does the trick: let $\boldsymbol{\Phi}$ denote a

* Supported by EPSRC grant EP/G039070/1.

S. Albers et al. (Eds.): ICALP 2009, Part I, LNCS 5555, pp. 292–303, 2009.

k-SAT formula over the variable set $V = \{x_1, \dots, x_n\}$ that is obtained by choosing m clauses uniformly at random and independently from the set of all $(2n)^k$ possible clauses. Then for a large regime of densities m/n satisfying assignments are known to exist due to non-constructive arguments, but no efficient algorithm is known to find one.

To be precise, keeping k fixed and letting $m = \lceil rn \rceil$ for a fixed $r > 0$, we say that $\boldsymbol{\Phi}$ has some property *with high probability* ('w.h.p.') if the probability of the property tends to one as $n \to \infty$. Via the (non-algorithmic) second moment method [3,4] it can be shown that $\boldsymbol{\Phi}$ has a satisfying assignment w.h.p. if $m/n < (1 - \varepsilon_k)2^k \ln 2$. Here ε_k tends to 0 for large k. On the other hand, a simple first moment argument shows that no satisfying assignment exists w.h.p. if $m/n > 2^k \ln 2$. In summary, the threshold for $\boldsymbol{\Phi}$ being satisfiable is asymptotically $2^k \ln 2$.

Yet for densities m/n beyond $c \cdot 2^k/k$, where c is a constant (independent of k), no algorithm has been known to find a satisfying assignment in polynomial time with a probability that does not tend to zero. Using merely the Unit Clause rule yields a linear time algorithm that succeeds up to $m/n = c \cdot 2^k/k$ with $c \sim \mathrm{e}/2 \approx 1.36$. The best previous rigorous result, based on a somewhat more involved algorithm, achieved $c \sim 1.817$ (cf. Section 2). Conversely, many algorithms, including Pure Literal, Unit Clause, and DPLL, are known to fail or exhibit an exponential running time beyond $c \cdot 2^k/k$. There is experimental evidence that the same is true of Walksat. In effect, devising an algorithm to solve random formulas w.h.p. for densities m/n up to $2^k \omega(k)/k$ for *any* (howsoever slowly growing) $\omega(k) \to \infty$ has been a prominent open problem [3,4,8,15].

Theorem 1. *There are a sequence $\varepsilon_k \to 0$ and a polynomial time algorithm* `Fix` *such that* `Fix` *applied to a random formula $\boldsymbol{\Phi}$ with $m/n \leq (1 - \varepsilon_k)2^k \ln(k)/k$ outputs a satisfying assignment w.h.p.*

`Fix` is a deterministic local search algorithm and runs in time $O(n + m)^{3/2}$. The recent paper [2] provides evidence that the density $m/n = 2^k \ln(k)/k$ may be a barrier for (at least) a large class of algorithms to find satisfying assignments in polynomial time. Hence, Theorem 1 may mark, at least up to the precise second order term hidden in the ε_ks, the end of the algorithmic road for random k-SAT. To explain this, we need to discuss a concept that originates from statistical physics.

A digression: replica symmetry breaking. For the last decade random k-SAT has been studied by statistical physicists via sophisticated, insightful, but mathematically non-rigorous techniques from the theory of spin glasses. Their results suggest that below the threshold density $2^k \ln 2$ for the existence of satisfying assignments various other phase transitions take place that affect the performance of algorithms.

To us the most important one is the *dynamic replica symmetry breaking* (dRSB) transition. Let $S(\boldsymbol{\Phi}) \subset \{0,1\}^V$ be the set of all satisfying assignments of the random formula $\boldsymbol{\Phi}$. Very roughly speaking, according to the dRSB hypothesis there is a density r_{RSB} such that for $m/n < r_{RSB}$ the correlations that shape the set $S(\boldsymbol{\Phi})$ are purely local, whereas for densities $m/n > r_{RSB}$ long range correlations occur. Furthermore, $r_{RSB} \sim 2^k \ln(k)/k$.

Confirming and elaborating on this hypothesis, we recently established a good part of the dRSB phenomenon rigorously [2]. In particular, we proved that there is a sequence $\varepsilon_k \to 0$ such that for $m/n > (1+\varepsilon_k)2^k \ln(k)/k$ the values that the solutions $\sigma \in S(\boldsymbol{\Phi})$ assign to the variables are mutually heavily correlated in the following sense. Let us call a variable x *frozen* in a satisfying assignment σ if any satisfying assignment τ such that $\sigma(x) \neq \tau(x)$ is at Hamming distance at least $\Omega(n)$ from σ. Then for $m/n > (1+\varepsilon_k)2^k \ln(k)/k$ in all but a $o(1)$-fraction of all solutions $\sigma \in S(\boldsymbol{\Phi})$ all but an ε_k-fraction of the variables are frozen w.h.p., where $\varepsilon_k \to 0$.

This suggests that on random formulas with density $m/n > (1+\varepsilon_k)2^k \ln(k)/k$ local search algorithms (such as Pure Literal, Unit Clause, or Walksat) are unlikely to succeed. For think of the *factor graph*, whose vertices are the variables and the clauses, and where a variable is adjacent to all clauses in which it occurs. Then a local search algorithm assigns a value to a variable x on the basis of the values of variables that have distance $O(1)$ from x in the factor graph. But in the random formula $\boldsymbol{\Phi}$ with $m/n > (1+\varepsilon_k)2^k \ln(k)/k$ assigning one variable x is likely to impose constraints on the values that can be assigned to variables at distance $\Omega(\ln n)$ from x in the factor graph (due to the occurrence of frozen variables).

The above discussion applies to 'large' values of k (say, $k \geq 8$). In fact, non-rigorous arguments as well as experimental evidence [5] suggest that the picture is quite different and rather more complicated for 'small' k (say, $k = 3, 4, 5$). In this case the various phenomena that occur at (or very near) the point $2^k \ln(k)/k$ for $k \geq 8$ appear to happen at vastly different points in the satisfiable regime, and in a different order. To keep matters as simple as possible we focus on 'large' k in this paper.

Notation. We let $V = V_n = \{x_1, \ldots, x_n\}$ be a set of propositional variables. For a set $Z \subset V$ let $\bar{Z} = \{\bar{x} : x \in Z\}$ contain the corresponding set of negative literals. If l is a literal, then $|l|$ signifies the underlying variable. Let $[\mu] = \{1, 2, \ldots, \mu\}$ for integers μ.

Let $\Omega_k(n,m)$ be the set of all k-SAT formulas over V. Throughout the paper we denote a random element of $\Omega_k(n,m)$ by $\boldsymbol{\Phi}$; unless otherwise specified, $\boldsymbol{\Phi}$ is uniformly distributed. We use the letter Φ to denote specific (i.e., non-random) elements of $\Omega_k(n,m)$. Further, Φ_i denotes the ith clause of Φ, and Φ_{ij} is the jth literal of Φ_i.

2 Related Work

Quite a few papers deal with efficient algorithms for random k-SAT, contributing either rigorous results, non-rigorous evidence based on physics arguments, or experimental evidence. Table 1 summarizes the part of this work that is most relevant to us. The best rigorous result prior to this work is due to Frieze and Suen [11], who proved that 'SCB' succeeds for densities $\eta_k 2^k/k$, where η_k increases to 1.817 as $k \to \infty$. SCB combines the shortest clause rule, which is a generalization of Unit Clause, with (very limited) backtracking.

Table 1. Algorithms for random k-SAT

Algorithm	Density $m/n < \cdots$	Success probability	Ref., year
Pure Literal	$o(1)$ as $k \to \infty$	w.h.p.	[14], 2006
Walksat, rigorous	$\frac{1}{6} \cdot 2^k/k^2$	w.h.p.	[9], 2009
Walksat, non-rigorous	$2^k/k$	w.h.p.	[16], 2003
Unit Clause	$\frac{1}{2}\left(\frac{k-1}{k-2}\right)^{k-2} \cdot \frac{2^k}{k}$	$\Omega(1)$	[7], 1990
Shortest Clause	$\frac{1}{8}\left(\frac{k-1}{k-3}\right)^{k-3} \frac{k-1}{k-2} \cdot \frac{2^k}{k}$	w.h.p.	[8], 1992
SCB	$\sim 1.817 \cdot \frac{2^k}{k}$	w.h.p.	[11], 1996
BP+decimation (non-rigorous)	$\mathrm{e} \cdot 2^k/k$	w.h.p.	[15], 2007

Montanari, Ricci-Tersenghi, and Semerjian [15] provide evidence that Belief Propagation guided decimation may succeed up to density $\mathrm{e} \cdot 2^k/k$. This algorithm is based on a very different paradigm than the others mentioned in Table 1. The basic idea is to run a message passing algorithm ('Belief Propagation') to compute for each variable the marginal probability that this variable takes the value true/false in a uniformly random satisfying assignment. Then, the decimation step selects a variable, assigns it the value true/false with the corresponding marginal probability, and simplifies the formula. Ideally, repeating this procedure will lead to a satisfying assignment, provided that Belief Propagation keeps yielding the correct marginals. Proving (or disproving) this remains a major open problem.

Survey Propagation is a modification of Belief Propagation that aims to approximate the marginal probabilities induced by a particular (non-uniform) probability distribution on the set of satisfying assignments [6]. It can be combined with a decimation procedure as well to obtain a heuristic for finding a satisfying assignment. Analyzing Survey Propagation guided decimation is a further outstanding open problem.

The discussion so far concerns the case of general k. In addition, a large number of papers deal with the case $k = 3$. Flaxman [10] provides a survey. Currently the best rigorously analyzed algorithm for random 3-SAT is known to succeed up to $m/n = 3.52$ [12,13]. This is also the best known lower bound on the 3-SAT threshold. Non-rigorous arguments suggest the threshold to be ≈ 4.267 [6]. As mentioned earlier, there is non-rigorous evidence that the structure of the set of all satisfying assignment evolves differently in random 3-SAT than in random k-SAT for 'large' k. This may be why experiments suggest that Survey Propagation guided decimation for 3-SAT succeeds for densities m/n up to 4.2 [6].

3 The Algorithm `Fix`

In this section we present the algorithm `Fix`. To establish Theorem 1 we will prove the following: for any $0 < \varepsilon < 0.1$ there is $k_0 = k_0(\varepsilon) > 3$ such that for all $k \geq k_0$ the algorithm `Fix` outputs a satisfying assignment w.h.p. when applied to Φ with $m = \lfloor n\cdot(1-\varepsilon)2^k k^{-1}\ln k\rfloor$. Thus, we assume that k exceeds some large enough number k_0 depending on ε only. In addition, we assume throughout that $n > n_0$ for some large $n_0 = n_0(\varepsilon, k)$. We set $\omega = (1-\varepsilon)\ln k$ and $k_1 = \lceil k/2 \rceil$.

When applied to a k-SAT instance Φ the algorithm basically tries to 'fix' the all-true assignment by setting 'a few' variables $Z \subset V$ to false so as to satisfy all clauses. Obviously, the set Z will have to contain one variable from each clause consisting of negative literals only. The key issue is to pick 'the right' variables. To this end, the algorithm goes over the all-negative clauses in the natural order. If the present all-negative clause Φ_i does not contain a variable from Z yet, `Fix` (tries to) identify a 'safe' variable in Φ_i, which it then adds to Z. Here 'safe' means that setting the variable to false does not create new unsatisfied clauses. More precisely, we say that a clause Φ_i is *Z-unique* if Φ_i contains exactly one positive literal from $V \setminus Z$ and no negative literal whose underlying variable is in Z. Moreover, $x \in V \setminus Z$ is *Z-unsafe* if it occurs positively in a Z-unique clause, and *Z-safe* if this is not the case. Then in order to fix an all-negative clause Φ_i we prefer Z-safe variables.

To implement this idea, `Fix` proceeds in three phases. Phase 1 performs the operation described in the previous paragraph: try to identify a Z-safe variable in each all-negative clause. Of course, not every all-negative clause will contain one. In this case `Fix` just picks the variable in position k_1. This entails that the assignment constructed in Phase 1 will not satisfy *all* clauses. However, we will prove that the number of unsatisfied clauses is very small, and the purpose of Phases 2 and 3 is to deal with them. Before we come to this, let us describe Phase 1 precisely.

Algorithm 2. `Fix`(Φ)
Input: A k-SAT formula Φ. *Output:* Either a satisfying assignment or 'fail'.

1a. Let $Z = \emptyset$.
1b. For $i = 1, \ldots, m$ do
1c. If Φ_i is all-negative and contains no variable from Z
1d. If there is $1 \leq j < k_1$ such that $|\Phi_{ij}|$ is Z-safe, then pick the least such j and add $|\Phi_{ij}|$ to Z.
1e. Otherwise add $|\Phi_{i\,k_1}|$ to Z.

Let σ_Z be the assignment that sets all variables in $V \setminus Z$ to true and all variables in Z to false.

Proposition 3. *At the end of the first phase of* `Fix`(Φ) *the following statements are true w.h.p.*

1. *We have* $|Z| \leq 4nk^{-1}\ln\omega$.
2. *At most* $(1+\varepsilon/3)\omega n$ *clauses are Z-unique.*
3. *At most* $\exp(-k^{\varepsilon/8})n$ *clauses are unsatisfied under* σ_Z.

Since the probability that a random clause is all-negative is 2^{-k}, under the all-true assignment $(1+o(1))2^{-k}m \sim \omega n/k$ clauses are unsatisfied w.h.p. Hence, the outcome σ_Z of Phase 1 is already a lot better than the all-true assignment w.h.p.

Phase 2 deals with the clauses that are unsatisfied under σ_Z. The general plan is similar to Phase 1: we (try to) identify a set Z' of 'safe' variables that can be used to satisfy the σ_Z-unsatisfied clauses without 'endangering' further clauses. More precisely, we say that a clause Φ_i is (Z, Z')-*endangered* if there is no $1 \leq j \leq k$ such that the literal Φ_{ij} is true under σ_Z and $|\Phi_{ij}| \in V \setminus Z'$. In words, Φ_i is (Z, Z')-endangered if it is unsatisfied under σ_Z or it relies on one of the variables in Z' to be satisfied. Call Φ_i (Z, Z')-*secure* if it is not (Z, Z')-endangered. Phase 2 will construct a set Z' such that for all $1 \leq i \leq m$ either Φ_i is (Z, Z')-secure, or there are at least three indices $1 \leq j \leq k$ such that $|\Phi_{ij}| \in Z'$. To achieve this, we say that a variable x is (Z, Z')-*unsafe* if $x \in Z \cup Z'$ or there are indices $(i, l) \in [m] \times [k]$ such that the following two conditions hold:

a. For all $j \neq l$ we have $\Phi_{ij} \in Z \cup Z' \cup \overline{V \setminus Z}$.
b. $\Phi_{il} = x$.

(In words, x occurs positively in Φ_i, and all other literals of Φ_i are either positive but in $Z \cup Z'$ or negative but not in Z.) Otherwise we call x (Z, Z')-*safe*. `Fix` greedily tries to add as few (Z, Z')-unsafe variables to Z' as possible.

2a. Let Q consist of all $i \in [m]$ such that Φ_i is unsatisfied under σ_Z. Let $Z' = \emptyset$.
2b. While $Q \neq \emptyset$
2c. Let $i = \min Q$.
2d. If there are indices $k_1 < j_1 < j_2 < j_3 \leq k - 5$ such that $|\Phi_{ij_l}|$ is (Z, Z')-safe for $l = 1, 2, 3$,
pick the lexicographically first such sequence and add the variables $|\Phi_{ij_1}|, |\Phi_{ij_2}|, |\Phi_{ij_3}|$ to Z'.
2e. else
let $k - 5 < j_1 < j_2 < j_3 \leq k$ be the lexicographically first sequence such that $|\Phi_{ij_l}| \notin Z'$ and add $|\Phi_{ij_l}|$ to Z' $(l = 1, 2, 3)$.
2f. Let Q be the set of all (Z, Z')-endangered clauses that contain less than 3 variables from Z'.

Note that the While-loop gets executed at most $n/3$ times, because Z' gains three new elements in each iteration. Actually the final set Z' is fairly small w.h.p.:

Proposition 4. *The set Z' obtained in Phase 2 of* **Fix**(Φ) *has size* $|Z'| \leq nk^{-12}$ *w.h.p.*

After completing Phase 2, `Fix` is going to set the variables in $V \setminus (Z \cup Z')$ to true and the variables in $Z \setminus Z'$ to false. This will satisfy all (Z, Z')-secure clauses. In order to satisfy the (Z, Z')-endangered clauses as well, `Fix` needs to set the variables in Z' appropriately. Since each (Z, Z')-endangered clause contains three variables from Z', this is essentially equivalent to solving a 3-SAT problem, in which Z' is the set of variables. As we shall see, w.h.p. the resulting formula is sufficiently sparse for the following 'matching heuristic' to succeed: set up a bipartite graph $G(\Phi, Z, Z')$ whose vertex set consists of the (Z, Z')-endangered clauses and the set Z'. Each (Z, Z')-endangered clause is adjacent to the variables from Z' that occur in it. If M is a matching in $G(\Phi, Z, Z')$ that covers all (Z, Z')-endangered clauses, we construct an assignment $\sigma_{Z,Z',M}$ as follows: for each variable $x \in V$ we define

$$\sigma_{Z,Z',M}(x) = \begin{cases} \text{false} & \text{if } x \in Z \setminus Z' \\ \text{false} & \text{if } \{\Phi_i, x\} \in M \text{ for some } i \text{ and } x \text{ occurs negatively in } \Phi_i, \\ \text{true} & \text{otherwise.} \end{cases}$$

To be precise, Phase 3 proceeds as follows.

3. If $G(\Phi, Z, Z')$ has a matching that covers all (Z, Z')-endangered clauses, then compute an (arbitrary) such matching M and output $\sigma_{Z,Z',M}$. If not, output 'fail'.

Proposition 5. *W.h.p. $G(\boldsymbol{\Phi}, Z, Z')$ has a matching that covers all (Z, Z')-endangered clauses.*

Proof of Theorem 1. `Fix` is clearly a deterministic algorithm with running time $O(n + m)^{3/2}$ (if we use the Hopcroft-Karp algorithm to compute the matching in Phase 3). It remains to show that `Fix`($\boldsymbol{\Phi}$) outputs a satisfying assignment w.h.p. By Proposition 5 Phase 3 will find a matching M that covers all (Z, Z')-endangered clauses w.h.p., and thus the output will be the assignment $\sigma = \sigma_{Z,Z',M}$ w.h.p. Assume that this is the case. Then σ sets all variables in $Z \setminus Z'$ to false and all variables in $V \setminus (Z \cup Z')$ to true, thereby satisfying all (Z, Z')-secure clauses. Furthermore, for each (Z, Z')-endangered clause $\boldsymbol{\Phi}_i$ there is an edge $\{\boldsymbol{\Phi}_i, |\boldsymbol{\Phi}_{ij}|\}$ in M. If $\boldsymbol{\Phi}_{ij}$ is negative, then $\sigma(|\boldsymbol{\Phi}_{ij}|) =$ false, and if if $\boldsymbol{\Phi}_{ij}$ is positive, then $\sigma(\boldsymbol{\Phi}_{ij}) =$ true. In either case σ satisfies $\boldsymbol{\Phi}_i$. □

In the next section we sketch the analysis of Phase 1, i.e., the proof of Proposition 3. The analysis of Phase 2 (Proposition 4) is based on very similar ideas (details omitted). Furthermore, the proof of Proposition 5 combines ideas from the analysis of Phase 1 with a first moment argument.

4 Analyzing Phase 1

In this section we let $0 < \varepsilon < 0.1$ and assume that $k \geq k_0$ for a sufficiently large $k_0 = k_0(\varepsilon)$. Moreover, we assume that $m = \lfloor (1 - \varepsilon) 2^k k^{-1} \ln k \rfloor$ and that $n > n_0$ for some large enough $n_0 = n_0(\varepsilon, k)$. Let $\omega = (1 - \varepsilon) \ln k$ and $k_1 = \lceil k/2 \rceil$.

It is worthwhile giving a brief intuitive explanation as to why Phase 1 'works'. Namely, let us just consider the *first* all-negative clause $\boldsymbol{\Phi}_i$ of the random input formula. Assume that $i = 1$. If we condition on $\boldsymbol{\Phi}_1$ being all-negative, the k-tuple of variables $(|\boldsymbol{\Phi}_{1j}|)_{j \in [k]}$ is uniformly distributed. Furthermore, at this point $Z = \emptyset$. Hence, a variable x is Z-safe unless it occurs as the unique positive literal in some clause. For any x the expected number of such clauses is $k 2^{-k} m/n \sim \omega$ (for in each clause there are k slots where to put x, the probability that x occurs in any slot is $1/n$, and the probability that x occurs positively and all other literals are negative is 2^{-k}). In fact, for each variable the number of such clauses is asymptotically Poisson. Consequently, the probability that x is Z-safe is $\exp(-\omega)$. Returning to the clause $\boldsymbol{\Phi}_1$, we conclude that the *expected* number of indices $1 \leq j \leq k_1$ such that $|\boldsymbol{\Phi}_{1j}|$ is Z-safe is $k_1 \exp(-\omega)$. Since $\omega = (1-\varepsilon)\ln k$, we have $k_1 \exp(-\omega) \geq k^{\varepsilon}/3$. Indeed, the number of indices $1 \leq j \leq k_1$ so that $|\boldsymbol{\Phi}_{1j}|$ is Z-safe is binomially distributed, and hence the probability that there is no Z-safe $|\boldsymbol{\Phi}_{1j}|$ is at most $\exp(-k^{\varepsilon}/3)$. Thinking of k 'large' (in terms of ε), we see that there is a good chance that $\boldsymbol{\Phi}_1$ can be satisfied by setting some variable to false without creating any new unsatisfied clauses. Of course, this argument only applies to the first all-negative clause, and the challenge lies in dealing with the stochastic dependencies that arise in the course of the execution of the algorithm.

To this end, we need to investigate how the set Z computed in Phase 1 evolves over time. Thus, we will analyze the execution of Phase 1 as a stochastic process, in which Z corresponds to a sequence $(Z_t)_{t \geq 0}$ of sets. The time parameter t is the number of all-negative clauses for which either Step 1d or 1e has been executed. We will represent the execution of Phase 1 on input $\boldsymbol{\Phi}$ by a sequence of (random) maps

$$\pi_t : [m] \times [k] \to \{-1, 1\} \cup V \cup \bar{V}.$$

The map π_t is meant to capture the information that has determined the first t steps of the process. If $\pi_t(i,j) = 1$ (resp. $\pi_t(i,j) = -1$), then `Fix` has only taken into account that $\boldsymbol{\Phi}_{ij}$ is a positive (negative) literal, but not what the underlying variable is. If $\pi_t(i,j) \in V \cup \bar{V}$, then `Fix` has revealed the actual literal $\boldsymbol{\Phi}_{ij}$.

Let us define the sequence $\pi_t(i,j)$ precisely. Let $Z_0 = \emptyset$. Moreover, let U_0 be the set of all i such that there is exactly one j such that $\boldsymbol{\Phi}_{ij}$ is positive. Further, define $\pi_0(i,j)$ for $(i,j) \in [m] \times [k]$ as follows. If $i \in U_0$ and $\boldsymbol{\Phi}_{ij}$ is positive, then let $\pi_0(i,j) = \boldsymbol{\Phi}_{ij}$. Otherwise, let $\pi_0(i,j)$ be 1 if $\boldsymbol{\Phi}_{ij}$ is a positive literal and -1 if $\boldsymbol{\Phi}_{ij}$ is a negative literal. In addition, for $x \in V$ let $U_0(x)$ be the number of $i \in U_0$ such that x occurs positively in $\boldsymbol{\Phi}_i$. For $t \geq 1$ we define π_t as follows.

PI1. If there is no index $i \in [m]$ such that $\boldsymbol{\Phi}_i$ is all-negative but contains no variable from Z_{t-1}, the process stops. Otherwise let ϕ_t be the smallest such index.

PI2. If there is $1 \leq j < k_1$ such that $U_{t-1}(|\boldsymbol{\Phi}_{\phi_t j}|) = 0$, then choose the smallest such index; otherwise let $j = k_1$. Let $z_t = \boldsymbol{\Phi}_{\phi_t j}$ and $Z_t = Z_{t-1} \cup \{z_t\}$.

PI3. Let U_t be the set of all $i \in [m]$ such that $\boldsymbol{\Phi}_i$ is Z_t-unique. For $x \in V$ let $U_t(x)$ be the number of indices $i \in U_t$ such that x occurs positively in $\boldsymbol{\Phi}_i$.

PI4. For any $(i,j) \in [m] \times [k]$ let

$$\pi_t(i,j) = \begin{cases} \boldsymbol{\Phi}_{ij} & \text{if } (i = \phi_t \wedge j \leq k_1) \vee |\boldsymbol{\Phi}_{ij}| \in Z_t \\ & \qquad \vee (i \in U_t \wedge \pi_0(i,j) = 1), \\ \pi_{t-1}(i,j) & \text{otherwise.} \end{cases}$$

Let T be the total number of iterations before the process stops and define $\pi_t = \pi_T$, $Z_t = Z_T$, $U_t = U_T$, $U_t(x) = U_T(x)$, $\phi_t = z_t = 0$ for all $t > T$.

The process mirrors Phase 1 of `Fix` as follows. Step **PI1** selects the least index ϕ_t such that clause $\boldsymbol{\Phi}_{\phi_t}$ is all-negative but contains none of the variables Z_{t-1} that have been selected to be set to false so far. In terms of `Fix`, this corresponds to fast-forwarding to the next execution of Steps 1d–e. Since $U_{t-1}(x)$

is the number of Z_{t-1}-unique clauses in which variable x occurs positively, **PI2** applies the same rule as steps 1d–e of `Fix` to select the new element z_t to be included in the set Z_t. Step **PI3** then 'updates' the numbers $U_t(x)$. Finally, step **PI4** sets up the map π_t to represent the information that has guided the process so far: we reveal the first k_1 literals of the current clause $\boldsymbol{\Phi}_{\phi_t}$, all occurrences of the variable z_t, and all positive literals of Z_t-unique clauses.

The process **PI1**–**PI4** can be applied to any concrete k-SAT formula Φ (rather than the random $\boldsymbol{\Phi}$). It then yields a sequence $\pi_t\,[\Phi]$ of maps, variables $z_t\,[\Phi]$, etc. For each integer $t \geq 0$ we define an equivalence relation $\equiv_t$ on the set $\Omega_k(n,m)$ of k-SAT formulas by letting $\Phi \equiv_t \Psi$ iff $\pi_s\,[\Phi] = \pi_s\,[\Psi]$ for all $0 \leq s \leq t$. Let $\mathcal{F}_t$ be the σ-algebra generated by the equivalence classes of $\equiv_t$. Then (loosely speaking) a random variable $X(\boldsymbol{\Phi})$ is $\mathcal{F}_t$-measurable if its value is determined by time t.

Fact 6. For any $t \geq 0$ the random map π_t, the random variables ϕ_{t+1}, z_t, the random sets U_t and Z_t, and the random variables $U_t(x)$ for $x \in V$ are $\mathcal{F}_t$-measurable.

The first t steps of the process **PI1**–**PI4** are only driven by the information encoded in the map π_t, Hence, for (i,j) such that $\pi_t(i,j) = \pm 1$ the process has only taken into accout the *sign* of the literal $\boldsymbol{\Phi}_{ij}$ and the fact that $|\boldsymbol{\Phi}_{ij}| \notin Z_t$. But the process has been oblivious to the actual underlying variable $|\boldsymbol{\Phi}_{ij}|$. This implies the following.

Proposition 7. *Let $\mathcal{E}_t$ be the set of all pairs (i,j) such that $\pi_t(i,j) \in \{-1,1\}$. The conditional joint distribution of the variables $(|\boldsymbol{\Phi}_{ij}|)_{(i,j)\in\mathcal{E}_t}$ given $\mathcal{F}_t$ is uniform over $(V \setminus Z_t)^{\mathcal{E}_t}$. That is, for any map $f : \mathcal{E}_t \to V \setminus Z_t$ we have*

$$\mathrm{P}\left[\forall (i,j) \in \mathcal{E}_t : |\boldsymbol{\Phi}_{ij}| = f(i,j) | \mathcal{F}_t\right] = |V \setminus Z_t|^{-|\mathcal{E}_t|}.$$

In each step of the process **PI1**–**PI4** one variable z_t is added to Z_t. There is a chance that this variable occurs in several other all-negative clauses. Hence, the stopping time T should be smaller than the total number of all-negative clauses. To prove this, we need the following lemma.

Lemma 8. *W.h.p. the following is true for all $1 \leq t \leq \min\{T,n\}$: the number of indices $i \in [m]$ such that $\pi_t(i,j) = -1$ for all $1 \leq j \leq k$ is at most $2n\omega \exp(-kt/n)/k$.*

Proof. The proof illustrates the use of Proposition 7. Let $\mathcal{N}_{tij} = 1$ if $\pi_t(i,j) = -1$ and $t \leq T$, and let $\mathcal{N}_{tij} = 0$ otherwise. Let $t \leq n$, $\mu = \lceil \ln^2 n \rceil$, and let $\mathcal{I} \subset [m]$ be a set of size μ. Let $Y_i = 1$ if $t \leq T$ and $\pi_t(i,j) = -1$ for all $j \in [k]$, and let $Y_i = 0$ otherwise. Set $\mathcal{J} = [t] \times \mathcal{I} \times [k]$. If $Y_i = 1$ for all $i \in \mathcal{I}$, then $\mathcal{N}_{0ij} = 1$ for all $(i,j) \in \mathcal{I} \times [k]$ and $\mathcal{N}_{sij} = 1$ for all $(s,i,j) \in \mathcal{J}$. We will prove below that

$$\mathrm{E}\left[\prod_{(i,j)\in\mathcal{I}\times[k]} \mathcal{N}_{0ij} \cdot \prod_{(t,i,j)\in\mathcal{J}} \mathcal{N}_{tij}\right] \leq 2^{-k|\mathcal{I}|}(1-1/n)^{|\mathcal{J}|}, \text{ whence} \tag{1}$$

$$\mathrm{E}\left[\prod_{i\in\mathcal{I}} Y_i\right] \leq \lambda^{\mu}, \text{ where } \lambda = 2^{-k}\exp(-kt/n). \tag{2}$$

Let $Y = \sum_{i \in [m]} Y_i$. Then (2) entails that $\mathrm{E}[Y^\mu] \le (1+o(1))(\lambda m)^\mu$. Therefore, Markov's inequality yields $\mathrm{P}[Y > 2n\omega \exp(-kt/n) \ge 1.9\lambda m] \le 1.9^{-\mu}$, and thus the assertion follows from the union bound.

To complete the proof, we need to establish (1). Let

$$\mathcal{N}_0 = \prod_{(i,j) \in \mathcal{I} \times [k]} \mathcal{N}_{0ij}, \ \mathcal{J}_s = \{(i,j) : (s,i,j) \in \mathcal{J}\}, \text{ and } \mathcal{N}_s = \prod_{(i,j) \in \mathcal{J}_s} \mathcal{N}_{sij}.$$

Since the signs of the literals $\boldsymbol{\Phi}_{ij}$ are mutually independent, we have $\mathrm{E}[\mathcal{N}_0] = 2^{-k|\mathcal{I}|}$. Furthermore, we will prove below that $\mathrm{E}[\mathcal{N}_s | \mathcal{F}_{s-1}] \le (1-1/n)^{|\mathcal{J}_s|}$. Since $\mathcal{N}_s$ is $\mathcal{F}_s$-measurable for any s, we obtain

$$\mathrm{E}\left[\prod_{s=0}^{t} \mathcal{N}_s\right] = \mathrm{E}\left[\mathrm{E}[\mathcal{N}_t | \mathcal{F}_{t-1}] \cdot \prod_{s=0}^{t-1} \mathcal{N}_s\right] \le (1-1/n)^{|\mathcal{J}_t|} \cdot \mathrm{E}\left[\prod_{s=0}^{t-1} \mathcal{N}_s\right].$$

Proceeding inductively, we obtain (1).

Finally, we bound $\mathrm{E}[\mathcal{N}_s | \mathcal{F}_{s-1}]$ for $s \ge 1$. If $T < s$ or $\pi_{s-1}(i,j) \ne -1$ for some $(i,j) \in \mathcal{J}_s$, then $\mathcal{N}_s = \mathcal{N}_{sij} = 0$. Hence, suppose that $T \ge s$ and $\pi_{s-1}(i,j) = -1$ for all $(i,j) \in \mathcal{J}_s$. Then at time s **PI2** selects some variable $z_s \in V \setminus Z_{s-1}$, and $\mathcal{N}_{sij} = 1$ only if $|\boldsymbol{\Phi}_{ij}| \ne z_s$. As $\pi_{t-1}(i,j) = -1$ for all $(i,j) \in \mathcal{J}_s$, given $\mathcal{F}_{s-1}$ the variables $(|\boldsymbol{\Phi}_{ij}|)_{(i,j) \in \mathcal{J}_s}$ are independently uniformly distributed over $V \setminus Z_{s-1}$ by Proposition 7. Therefore, for each $(i,j) \in \mathcal{J}_s$ we have $|\boldsymbol{\Phi}_{ij}| = z_s$ with probability at least $1/n$. Hence, $\mathrm{E}[\mathcal{N}_s | \mathcal{F}_{s-1}] \le (1-1/n)^{|\mathcal{J}_s|}$. □

Corollary 9. *W.h.p. we have* $T < 4nk^{-1} \ln \omega$.

Proof. Let $t_0 = 2nk^{-1} \ln \omega$ and let I_t be the number of indices i such that $\pi_t(i,j) = -1$ for all $1 \le j \le k$. By **PI2** $I_t \le I_{t-1} - 1$ for all $t \le T$. Consequently, if $T \ge 2t_0$, then $0 \le I_T \le I_{t_0} - t_0$, and thus $I_{t_0} \ge t_0$. But Lemma 8 entails that $I_{t_0} < t_0$ w.h.p. □

Let $\theta = \lfloor 4nk^{-1} \ln \omega \rfloor$. The next goal is to estimate the number of Z_t-unique clauses, i.e., the size of the set U_t. Using a similar (if slightly more involved) argument as in the proof of Lemma 8, we can infer the following.

Lemma 10. *W.h.p.* $\max_{0 \le t \le T} |U_t| \le (1+\varepsilon/3)\omega n$.

Let us think of the variables $x \in V \setminus Z_t$ as bins and of the clauses $\boldsymbol{\Phi}_i$ with $i \in U_t$ as balls. If we place each ball i into the (unique) bin x such that x occurs positively in $\boldsymbol{\Phi}_i$, then by Lemma 10 and Corollary 9 for $t \le T$ the average number of balls in a bin is $\le (1+\varepsilon/3)\omega n / |V \setminus Z_t| \le (1-0.6\varepsilon)\ln k$ w.h.p. Hence, if the balls were thrown uniformly at random into the bins, we would expect

$$|V \setminus Z_t| \exp(-|U_t|/|V \setminus Z_t|) \ge (n-t)k^{0.6\varepsilon - 1} \ge nk^{\varepsilon/2-1}$$

bins to be empty (i.e., $U_t(x) = 0$). The next corollary shows that this is accurate.

Corollary 11. *Let* $\mathcal{Q}_t = |\{x \in V \setminus Z_t : U_t(x) = 0\}|$. *W.h.p. we have*

$$\min_{t \le T} \mathcal{Q}_t \ge nk^{\varepsilon/2-1}.$$

Now that we know that w.h.p. there are 'a lot' of variables $x \in V \setminus Z_{t-1}$ such that $U_t(x) = 0$, we expect that it is quite likely for clause $\boldsymbol{\Phi}_{\phi_t}$ to contain one. More precisely, we have the following.

Corollary 12. *Let $\mathcal{B}_t = 1$ if $\min_{j<k_1} U_{t-1}(|\boldsymbol{\Phi}_{\phi_t j}|) > 0$, $\mathcal{Q}_{t-1} \geq nk^{\varepsilon/2-1}$, $|U_t| \leq (1+\varepsilon/3)\omega n$, and $T \geq t$. Let $\mathcal{B}_t = 0$ otherwise. Then $\mathrm{E}[\mathcal{B}_t|\mathcal{F}_{t-1}] \leq \exp(-k^{\varepsilon/6})$ for all $1 \leq t \leq \theta$.*

Proof of Proposition 3. The definition of the process **PI1**–**PI4** mirrors the execution of the algorithm, i.e., the set Z obtained after Steps 1a–1d of `Fix` equals the set Z_T. Therefore, the first assertion is a consequence of Corollary 9 and the fact that $|Z_t| = t$ for all $t \leq T$. Furthermore, the second assertion follows directly from Lemma 10.

To prove the third claim, we need to bound the number of clauses that are unsatisfied under σ_{Z_T}. It is not difficult to see that the construction **PI1**–**PI4** ensures that for any $i \in [m]$ such that $\boldsymbol{\Phi}_i$ is unsatisfied under σ_{Z_T} one of the following is true.

a. There is $t \leq T$ such that $i \in U_{t-1}$ and z_t occurs positively in $\boldsymbol{\Phi}_i$.
b. There are $1 \leq j_1 < j_2 \leq k$ such that $\boldsymbol{\Phi}_{ij_1} = \boldsymbol{\Phi}_{ij_2}$.

Let $\mathcal{X}$ be the number of indices $i \in [m]$ such that a. occurs. We will show that

$$\mathcal{X} \leq n\exp(-k^{\varepsilon/7}) \qquad \text{w.h.p.} \tag{3}$$

Since the number of 'degenerate' $i \in [m]$ for which b. occurs is $O(\ln n)$ w.h.p. (by a simple first moment argument), (3) implies the third assertion.

To establish (3), let $\mathcal{B}_t$ be as in Corollary 12 and set $\mathcal{D}_t = \mathcal{B}_t \cdot U_{t-1}(z_t)$. Invoking Corollary 11 and Lemma 10, it is easy to show that $\mathcal{X} \leq \sum_{1\leq t\leq\theta} \mathcal{D}_t$ w.h.p. Further, the random variable $\mathcal{D}_t$ is $\mathcal{F}_t$-measurable and $\mathcal{D}_t = 0$ for all $t > \theta$. Let

$$\bar{\mathcal{D}}_t = \mathrm{E}[\mathcal{D}_t|\mathcal{F}_{t-1}] \text{ and } \mathcal{M}_t = \sum_{s=1}^{t} \mathcal{D}_s - \bar{\mathcal{D}}_s.$$

Then $\mathcal{M}_1, \ldots, \mathcal{M}_\theta$ is a martingale with $\mathrm{E}[M_\theta] = 0$. Azuma's inequality entails that $\mathcal{M}_\theta = o(n)$ w.h.p. Hence, w.h.p. $\sum_{1\leq t\leq\theta} \mathcal{D}_t = o(n) + \sum_{1\leq t\leq\theta} \bar{\mathcal{D}}_t$.

We claim that $\bar{\mathcal{D}}_t \leq 2\omega\exp(-k^{\varepsilon/6})$ for all $1 \leq t \leq \theta$. For by Corollary 12 we have $\mathrm{E}[\mathcal{B}_t|\mathcal{F}_{t-1}] \leq \exp(-k^{\varepsilon/6})$. Moreover, given $\mathcal{F}_{t-1}$ we have $\pi_{t-1}(\phi_t, k_1) = -1$, whence z_t is uniformly distributed over $V \setminus Z_{t-1}$ (by Proposition 7). Since $\mathcal{B}_t = 1$ implies $|U_{t-1}| \leq (1+\varepsilon/3)\omega n$, the conditional expectation of $U_{t-1}(z_t)$ is

$$\leq |U_{t-1}|/|V \setminus Z_{t-1}| \leq (1+\varepsilon/3)\omega n/(n-t) \leq 2\omega.$$

Combining these estimates, we obtain that w.h.p.

$$\sum_{1\leq t\leq\theta} \mathcal{D}_t \leq 2\omega\exp(-k^{\varepsilon/2}/3)\theta + o(n) \leq n\exp(-k^{\varepsilon/7}).$$

Thus, (3) follows from the fact that $\mathcal{X} \leq \sum_{1\leq t\leq\theta} \mathcal{D}_t$ w.h.p. □

References

1. Achlioptas, D., Beame, P., Molloy, M.: Exponential bounds for DPLL below the satisfiability threshold. In: Proc. 15th SODA, pp. 139–140 (2004)
2. Achlioptas, D., Coja-Oghlan, A.: Algorithmic barriers from phase transitions. In: Proc. 49th FOCS, pp. 793–802 (2008)
3. Achlioptas, D., Moore, C.: Random k-SAT: two moments suffice to cross a sharp threshold. SIAM Journal on Computing 36, 740–762 (2006)
4. Achlioptas, D., Peres, Y.: The threshold for random k-SAT is $2^k \ln 2 - O(k)$. Journal of the AMS 17, 947–973 (2004)
5. Ardelius, J., Zdeborova, L.: Exhaustive enumeration unveils clustering and freezing in random 3-SAT. Phys. Rev. E 78, 040101(R) (2008)
6. Braunstein, A., Mézard, M., Zecchina, R.: Survey propagation: an algorithm for satisfiability. Random Structures and Algorithms 27, 201–226 (2005)
7. Chao, M.-T., Franco, J.: Probabilistic analysis of a generalization of the unit-clause literal selection heuristic for the k-satisfiability problem. Inform. Sci. 51, 289–314 (1990)
8. Chvátal, V., Reed, B.: Mick gets some (the odds are on his side). In: Proc. 33th FOCS, pp. 620–627 (1992)
9. Coja-Oghlan, A., Feige, U., Frieze, A., Krivelevich, M., Vilenchik, D.: On smoothed k-CNF formulas and the Walksat algorithm. In: Proc. 20th SODA, pp. 451–460 (2009)
10. Flaxman, A.: Algorithms for random 3-SAT. Encyclopedia of Algorithms (2008)
11. Frieze, A., Suen, S.: Analysis of two simple heuristics on a random instance of k-SAT. Journal of Algorithms 20, 312–355 (1996)
12. Hajiaghayi, M., Sorkin, G.: The satisfiability threshold of random 3-SAT is at least 3.52. IBM Research Report RC22942 (2003)
13. Kaporis, A., Kirousis, L., Lalas, E.: The probabilistic analysis of a greedy satisfiability algorithm. Random Structures and Algorithms 28, 444–480 (2006)
14. Kim, J.H.: Poisson cloning model for random graph (preprint, 2006)
15. Montanari, A., Ricci-Tersenghi, F., Semerjian, G.: Solving constraint satisfaction problems through Belief Propagation-guided decimation. In: Proc. 45th Allerton (2007)
16. Semerjian, G., Monasson, R.: A study of pure random walk on random satisfiability problems with "Physical" methods. In: Giunchiglia, E., Tacchella, A. (eds.) SAT 2003. LNCS, vol. 2919, pp. 120–134. Springer, Heidelberg (2004)

Exact and Approximate Bandwidth

Marek Cygan and Marcin Pilipczuk

Dept. of Mathematics, Computer Science and Mechanics, University of Warsaw, Poland
{cygan,malcin}@mimuw.edu.pl

Abstract. In this paper we gather several improvements in the field of exact and approximate exponential-time algorithms for the BANDWIDTH problem. For graphs with treewidth t we present a $O(n^{O(t)}2^n)$ exact algorithm. Moreover for the same class of graphs we introduce a subexponential constant-approximation scheme – for any $\alpha > 0$ there exists a $(1+\alpha)$-approximation algorithm running in $O(\exp(c(t + \sqrt{n/\alpha}) \log n))$ time where c is a universal constant. These results seem interesting since Unger has proved that BANDWIDTH does not belong to APX even when the input graph is a tree (assuming $\mathbf{P} \neq \mathbf{NP}$). So somewhat surprisingly, despite Unger's result it turns out that not only a subexponential constant approximation is possible but also a subexponential approximation scheme exists. Furthermore, for any positive integer r, we present a $(4r-1)$-approximation algorithm that solves BANDWIDTH for an arbitrary input graph in $O^*(2^{\frac{n}{r}})$ time and polynomial space.[1] Finally we improve the currently best known exact algorithm for arbitrary graphs with a $O(4.473^n)$ time and space algorithm.

In the algorithms for the small treewidth we develop a technique based on the Fast Fourier Transform, parallel to the Fast Subset Convolution techniques introduced by Björklund et al. This technique can be also used as a simple method of finding a chromatic number of all subgraphs of a given graph in $O^*(2^n)$ time and space, what matches the best known results.

1 Introduction

Notation. In this paper we focus on exponential-time exact and approximation algorithms for the BANDWIDTH problem. Let $G = (V, E)$ be an undirected graph with $n = |V|$. For a given one-to-one function $\pi : V \to \{1, 2 \ldots, n\}$ (called an *ordering*) its *bandwidth* is the maximum difference between positions of adjacent vertices, i.e. $\max_{uv \in E} |\pi(u) - \pi(v)|$. The *bandwidth* of the graph, denoted by $\mathrm{bw}(G)$, is the minimum bandwidth over all orderings. The BANDWIDTH problem asks to find an ordering with bandwidth $\mathrm{bw}(G)$. W.l.o.g. we can assume that G is connected, otherwise we can consider all connected components of G independently.

Motivation. The BANDWIDTH problem seems to be hard from many perspectives. It is NP-hard even on some subfamilies of trees [10,12], does not belong to APX even when G is a caterpillar [15] and is hard for any fixed level of the W hierarchy [3]. The best known polynomial-time approximation, due to Feige [7], has a $O(\log^3 n\sqrt{\log n \log\log n})$ approximation guarantee. At WG'08 we presented an exact algorithm for arbitrary graphs

[1] By O^* we denote standard big O notation but omitting polynomial factors.

S. Albers et al. (Eds.): ICALP 2009, Part I, LNCS 5555, pp. 304–315, 2009.

that runs in $O^*(5^n)$ time and $O^*(2^n)$ space [5]; we were able to enhance this algorithm to run in $O(4.83^n)$ time at the cost of $O^*(4^n)$ space complexity [4]. However, the fastest exact algorithm that runs in a polynomial space needs $O^*(10^n)$ time [8]. We are not aware of any *published* faster algorithms for trees or graphs with bounded treewidth, but we were informed [9] that very recently, Amini, Fomin and Saurabh independently developed a $O(n^{O(t)}2^n)$ algorithm for graphs with treewidth t, using completely different approach than ours.

Since a polynomial time constant approximation for the BANDWIDTH problem is probably not possible, it seems reasonable to search for approximation algorithms that are faster than the exact ones, but still give us a constant approximation guarantee at the cost of a superpolynomial time complexity. This area was considered recently by Bourgeois et al. [13].

Our results. We present several results in the field of exponential solutions to the BANDWIDTH problem, both approximate and exact.

First, we define a DISJOINT SET SUM problem and show how to solve it using the Fast Fourier Transform in a similar time complexity as the Fast Subset Convolution [2]. This problem appears to be a crucial part in some NP-hard problems like CHROMATIC NUMBER. We solve these problems in time matching the best known results.

Then we use DISJOINT SET SUM to develop a simple $O^*(2^n)$ time and space exact algorithm for the BANDWIDTH problem for trees. We enhance this algorithm to work in $O(n^{O(t)}2^n)$ time and space for graphs with treewidth t (for more about the treewidth see [6]). Later, we exploit the idea from the exact algorithm to obtain a subexponential approximation scheme for trees and graphs with bounded treewidth. Precisely, for a parameter $\alpha > 0$, we get a $(1 + \alpha)$-approximation algorithm that works in $O(\exp(c(t + \sqrt{n/\alpha}) \log n))$ time and space. As far as we know this is the first example of a problem that is not approximable with a constant factor in a polynomial time and admits a subexponential approximation scheme.

Further we switch to arbitrary graphs and develop, for a fixed integer $r \geq 1$, an exponential approximation algorithm with a $(4r - 1)$–approximation guarantee that works in $O^*(2^{n/r})$ time and polynomial space. Finally, we present a $O^*(20^{n/2}) = O(4.473^n)$ time and space exact algorithm for arbitrary graphs, improving the previous $O(4.83^n)$ bound from [4].

Organization. In Section 2 we present a definition and solution using the Fast Fourier Transform for DISJOINT SET SUM and discuss usage of the Fast Subset Convolution. In Section 3 we use DISJOINT SET SUM to develop an exact algorithm for graphs with bounded treewidth and in Section 4 we describe an approximation scheme for the same class of graphs. Sections 5 and 6 contain approximate and exact algorithms for arbitrary graphs respectively.

2 DISJOINT SET SUM and Its Solutions

In this section we focus on the DISJOINT SET SUM problem, defined as follows. Consider a set $N = \{1, \ldots, n\}$ and two sets $\mathcal{A}, \mathcal{B}$ containing subsets of N as its elements ($\mathcal{A}, \mathcal{B} \subseteq 2^N$). Our goal is to compute $\mathcal{A} \oplus \mathcal{B} = \{A \cup B : A \in \mathcal{A}, B \in \mathcal{B}, A \cap B = \emptyset\}$ —

explaining in words we would like to find all subsets of N which can be represented as a disjoint sum of two sets, where one of them is contained in $\mathcal{A}$ and one in $\mathcal{B}$. Moreover for each set in $\mathcal{A} \oplus \mathcal{B}$ we would like to know in how many ways one can obtain it as such a disjoint sum.

Let us first solve the DISJOINT SET SUM naively. Taking each pair of sets from $\mathcal{A}$ and $\mathcal{B}$ would take us $O^*(4^n)$ time, but we easily improve this result by considering all subsets of N and for each such set $X \subseteq N$ iterate through all its subsets $A \subseteq X$ and check whether $A \in \mathcal{A}$ and $X \setminus A \in \mathcal{B}$ (which can be decided in polynomial of n time using any balanced search tree). This improvement leads to $O^*(3^n)$ time, since $\sum_{i=0}^{n} \binom{n}{i} 2^i = (2+1)^n = 3^n$. However, this can be done much better.

Theorem 1. *Let $\mathcal{A}, \mathcal{B} \subseteq 2^N$, we can compute $\mathcal{A} \oplus \mathcal{B}$ in $O^*(2^n)$ time and space. Moreover for each element in $\mathcal{A} \oplus \mathcal{B}$ we obtain the number of ways it can be composed as a disjoint sum.*

Note that Theorem 1 is a quite straightforward corollary from the Fast Subset Convolution algorithm by Björklund et al. [2]. The SUBSET CONVOLUTION problem can be defined as follows: given functions f and g defined on the lattice of subsets of an n-element set N, compute their subset convolution $f * g$ defined for all $S \subseteq N$ by $(f * g)(S) = \sum_{T \subseteq S} f(T) g(S \setminus T)$. Björklund et al. [2] developed a very clever algorithm that computes SUBSET CONVOLUTION in $O^*(2^n)$ time. By taking $f = 1_{\mathcal{A}}$ and $g = 1_{\mathcal{B}}$ this algorithm solves DISJOINT SET SUM in $O^*(2^n)$ time.

Here we present a different approach to the DISJOINT SET SUM problem. In Section 2.1 we show how to solve it using the Fast Fourier Transform. The FFT solution has a few advantages over FSC. As to our best knowledge it is new, gives the same time complexity as FSC (up to a polynomial factor), FFT is probably more widely known in the computer science community than FSC and, most important, although the methods are somehow similar, FFT allows some extensions that seems hard for FSC approach; since we use in Section 4 the extension defined in Section 2.2, we prefer the FFT solution. On the other hand, note that the FSC solution allows a smaller polynomial factor hidden in the $O^*()$ notation.

2.1 Solution Using the Fast Fourier Transform

The Fast Fourier Transform is an efficient algorithm which computes the discrete Fourier transform and its inverse. Due to its efficiency it has several applications, including digital signal processing and big integer or polynomial multiplication. However surprisingly enough, FFT happened to be a handy tool in DISJOINT SET SUM.

We can group subsets of N belonging to $\mathcal{A}$ and $\mathcal{B}$ according to their cardinality, thus it is sufficient to solve the reduced problem, where sets in $\mathcal{A}$ and $\mathcal{B}$ have fixed sizes, denote them by $k_{\mathcal{A}}$ and $k_{\mathcal{B}}$. This grouping gives us only n^2 overhead which is omitted in the O^* notation. To use FFT we need to look at our problem from a different angle, we treat subset $A \subset N$ as a monomial $x^{bin(A)}$, where $bin(A) \in [0, 2^n - 1]$ is the binary representation of a subset A. Grouping sets according to their size allows us to use the following observation:

Lemma 2. *Let A, B be subsets of $\{1, \ldots, n\}$. Then $A \cap B = \emptyset$ iff the number of ones in the binary representation of their sum $bin(A) + bin(B)$ is equal to $|A| + |B|$.*

Now we represent $\mathcal{A}$ and $\mathcal{B}$ as polynomials (where each element from $\mathcal{A}$ and $\mathcal{B}$ is a monomial) and multiply those polynomials using FFT in $O^*(2^n)$ time. We iterate over all subsets $I \subseteq N$ having exactly $k_{\mathcal{A}} + k_{\mathcal{B}}$ elements and check what coefficient stands in front of $x^{bin(I)}$ in the resulting polynomial. This coefficient is clearly the number of ways in which I can be composed as a disjoint sum.

We perform n^2 polynomial multiplications using FFT. Each multiplication needs $O(2^n \log(2^n))$ arithmetic operations, which consume a polynomial time, thus we get a $O^*(2^n)$ time and space algorithm.

2.2 FFT and DISJOINT SET SUM Extension

We can extend the FFT solution to DISJOINT SET SUM to the following problem. Fix an integer $r \geq 1$. Instead of having $\mathcal{A}, \mathcal{B} \subseteq 2^N$ we can have $\mathcal{A}, \mathcal{B} \subseteq \{0, 1, \ldots, r\}^N$, i.e., $\mathcal{A}$ and $\mathcal{B}$ are sets of n-tuples of integers from the set $\{0, 1, \ldots, r\}$. In this setting, for every n-tuple $X \in \{0, 1, \ldots, r\}^N$ we ask for the number of partitions of X into a sum $A + B$, where $A \in \mathcal{A}$, $B \in \mathcal{B}$ and the addition is defined pointwise, i.e., $X(i) = A(i) + B(i)$ for $i \in N$. In other words, $\mathcal{A} \oplus \mathcal{B} = \{(A_1 + B_1, \ldots, A_n + B_n) : (A_i) \in \mathcal{A} \wedge (B_i) \in \mathcal{B}\}$. Note that classical DISJOINT SET SUM is the extended problem with $r = 1$.

Using FFT, we can solve the extended DISJOINT SET SUM in $O^*((r+1)^n)$, assuming $r = O(n^\gamma)$ for some constant γ (in our applications $r \leq n$). We simply apply the previous solution, but treat tuple $X \in \{0, 1, \ldots, r\}^n$ as a number written in the base $r + 1$. As before, we split summands in respect of their sum of digits.

2.3 Applications

We can use the disjoint sets merging in all kinds of partition problems which fit into the following framework. Consider an n-element set U and a family $\mathcal{S} \subseteq 2^U$ of its subsets.

Definition 3. *For a positive integer $k \leq n$ and subset $X \subseteq U$ let $p_k(X)$ be the number of ordered k-partitions of X into subsets being elements of $\mathcal{S}$, i.e., $p_k(X)$ is the number of ways to choose $S_1, \ldots, S_k \in \mathcal{S}$, which are pairwise disjoint and $\bigcup S_i = X$.*

Using tools introduced in the previous subsection we calculate a series of sets $\mathcal{P}_i$, where $\mathcal{P}_1 = \mathcal{S}$ and $\mathcal{P}_{i+1} = \mathcal{P}_i \oplus \mathcal{S}$ (for $i = 2, \ldots, n$). With each element of $\mathcal{P}_i$ we store the number of ways it can be represented as a disjoint sum of elements from $\mathcal{S}$ (which are extracted easily from the presented polynomials). Clearly $p_k(X)$ is the coefficient stored in $\mathcal{P}_k$ related to X (or $p_k(X) = 0$ if $X \notin \mathcal{P}_k$).

Theorem 4. *For each k and X we can compute all values $p_k(X)$ at once in $O^*(2^n)$ time and space.*

Despite simplicity, our tool can be used to solve the CHROMATIC NUMBER problem in $O^*(2^n)$ time, which was a major open problem in the exact algorithms field, solved in 2006 independently by Björklund and Husfeldt in [1] and by Koivisto in [11] via the inclusion-exclusion principle. To achieve it we simply take all independent sets of the given graph as a family $\mathcal{S}$. Another problems that fit into the partition problems category are DOMATIC NUMBER, BOUNDED COMPONENT SPANNING FOREST, PARTITION INTO HAMILTONIAN SUBGRAPHS and BIN PACKING (all of those were mentioned in [1]). For all those problems we get $O^*(2^n)$ time and space algorithms, which are the same as the best known results. Moreover this framework also counts the number of solutions for those problems and calculates the number of partitions not only for the whole set, but also for all its subsets.

3 Exact Algorithm for Trees

In this section let us assume that the input graph for the BANDWIDTH problem is a tree T. We are given an integer $1 \leq b < n$ and we are to decide whether $b \geq \mathrm{bw}(G)$. For convenience we root this tree at some arbitrary vertex r. By $parent(v)$ we denote a vertex which is a parent of the vertex v in the rooted tree, additionally we set $parent(r) = r$. For a vertex v by $T(v)$ we denote a set of vertices of the subtree rooted at v.

Algorithms that operate on trees usually perform some computation on nodes recursively and join results from children to obtain results for the currently processed node. This also is the case here, hence we firstly define what do we want to calculate for each subtree.

Definition 5. *Let v be a vertex of the given tree and pos_v be a positive integer ($1 \leq pos_v \leq n$). By $assignments(v, pos_v)$ we denote a set of all such subsets $mask \subseteq \{1, \ldots, n\}$ for which there exists a surjective function $f : T(v) \to mask$ satisfying:*

- $\forall_{w \in T(v) \setminus \{v\}} |f(w) - f(parent(w))| \leq b.$
- $f(v) = pos_v.$

Less formally $assignments(v, pos_v)$ is corresponding to the set of all legal (with bandwidth not greater than b) assignments of vertices $T(v)$ with the vertex v having a fixed position pos_v, where from each assignment we remember only the set of used positions (since relative order of vertices from $T(v)$ does not matter outside this subtree).

To find $assignments(v, pos_v)$ for the vertex v and every pos_v firstly we set $assignments(v, pos_v)$ to contain only the set $\{pos_v\}$, which is the position used by v. Now we have to join $assignments(v, pos_v)$ with assignments for each child, one by one. We can join assignments of v with assignments for a child u iff u and v are not too far away from each other (which is easy to control using positions of those vertices) and sets of used positions are disjoint, and this is the place where we can use disjoint sets merging via FFT or FSC (see Pseudocode 1 for details).

To check whether there exists an ordering of T with bandwidth not greater than b we run the GENERATEASSIGNMENTS(r) procedure and check whether there exists i such that $assignments(r, i)$ is not empty, since the only element that can be in this set is the set of all positions $\{1, \ldots, n\}$.

Theorem 6. *Algorithm 1 solves the Bandwidth problem for trees in $O^*(2^n)$ time and space. For each edge of our tree we solve $O(n^2)$* DISJOINT SET SUM *instances.*

Algorithm 1. Exact algorithm for trees

```
procedure GenerateAssignments(v)
    for pos_v ← 1 to n do
        assignments(v, pos_v) ← {{pos_v}}
    for each child u of v do
        GenerateAssignments(u)
        for pos_v ← 1 to n do
            temp_assignments ← ∅
            for pos_u ← max(1, pos_v − b) to min(n, pos_v + b) do
                temp_assignments ← temp_assignments∪
                (assignments(v, pos_v) ⊕ assignments(u, pos_u))
            assignments(v, pos_v) ← temp_assignments
procedure Bandwidth(T,b)
    GenerateAssignments(r)                ▷ We start from the root.
    for i ← 1 to n do
        if assignments(r, i) ≠ ∅ then
            return true
    return false
```

In case of a positive answer, we could also be interested in finding an ordering with bandwidth not greater than b. This is not hard, but due to the space limitations we omit details here.

At the end, let us note that the aforementioned algorithm, at the cost of a polynomial factor in the complexity, can count the number of orderings with bandwidth at most b. As mentioned in Section 2, in the DISJOINT SET SUM problem we can count the number of possible partitions, thus in the end obtaining the number of orderings. Note that in this approach the coefficients in FFT (FSC) are of size $O(n!)$, thus they have $O(n \log n)$ bits and still all arithmetic operations can be done in polynomial time.

It is not hard to guess that our algorithm can be modified to handle not only trees, but also graphs with a fixed treewidth. If a graph has treewidth at most t this leads to a $O(n^{O(t)}2^n)$ algorithm. Since this modification does not introduce more understanding for the bandwidth problem and can be easily performed by anyone familiar with the treewidth we omit details here.

4 Approximation Scheme for Trees

Now we modify the algorithm presented in the previous section to construct an approximation scheme, still assuming that the input graph is a tree. The bottleneck of the exact algorithm for trees is the number of elements in *assignments* set, which is 2^n, because we assign vertices to specific positions. We can, however, loose this constraint and instead of an ordering construct a coarse ordering.

Definition 7. *A* coarse ordering *is a function $A : V \to 2^{\mathbb{Z}}$, assigning to every vertex $v \in V$ an interval $A(v) = [a_v, b_v] \subset \mathbb{Z}$. An ordering π is* consistent *with the coarse ordering A if for all $v \in V$ we have $\pi(v) \in A(v)$.*

Assume we want to have a $(1 + \alpha)$-approximation algorithm for a constant $\alpha > 0$.

Let us recall that the exact algorithm introduced by Saxe (see [14]) performs well for a small bandwidth of the given graph. Saxe's algorithm checks whether graph has bandwidth at most b, and if so constructs an ordering, in $O^*((4n)^b)$ time and space.

Our algorithm, given an integer $1 \leq b < n$, produces an ordering π satisfying $\text{bw}(\pi) \leq (1+\alpha)b$ or states that $b < \text{bw}(G)$. Using a binary search over b, we find the smallest b for which the algorithm finds an ordering. This ordering has bandwidth at most $(1+\alpha)\text{bw}(G)$.

First, let $k := \lfloor 0.5b\alpha \rfloor$. If $b \leq \frac{n}{k}$ we simply use the Saxe's algorithm to determine if $b \geq \text{bw}(G)$ in time $O(\exp(cb \log n))$ for universal constant c. Note that for $b \leq \frac{n}{k}$ we have $b = O(\sqrt{n/\alpha})$ therefore Saxe's algorithm works in $O(\exp(c\sqrt{n/\alpha}\log n))$ time.

Now we deal with the case $b > \frac{n}{k}$.

We use coarse orderings where values of A can be intervals of length k (except possibly the last interval which has size $n \mod k$). The i-th interval contains positions $\{(i-1)k+1, \ldots, \min((i+1)k, n)\}$. Like in the previous section $assignments(v, pos_v)$ corresponds to the set of all legal orderings of $T(v)$ (trying to have bandwidth not greater than b), but this time from each assignment we remember only to which interval vertex v was assigned and the number of vertices assigned to each interval.

The following algorithm, given an integer $\frac{n}{k} < b < n$, produces an ordering π satisfying $\text{bw}(\pi) \leq b + 2k$ or states that $b < \text{bw}(G)$. Note that this means that the produced ordering satisfies $\text{bw}(\pi) \leq (1+\alpha)b$ by the choice of k.

Definition 8. *Lev v be a vertex of the given tree and pos_v be a positive integer ($1 \leq pos_v \leq \lceil \frac{n}{k} \rceil$). By $assignments(v, pos_v)$ we denote a set of all tuples $(a_1, \ldots, a_{\lceil \frac{n}{k} \rceil})$, where $0 \leq a_i \leq k$, such that there exists a function $f : T(v) \to \{1, ..., \lceil \frac{n}{k} \rceil\}$ satisfying:*

- $\forall_{w \in T(v) \setminus \{v\}} (|f(w) - f(parent(w))| - 1) \cdot k < b.$
- $f(v) = pos_v,$
- $\forall_{1 \leq i \leq \lceil \frac{n}{k} \rceil} |f^{-1}(i)| = a_i,$

One may notice that we allow assigning more vertices (up to k) to the last segment than the number of positions it contains ($n \mod k$), we do it for simplicity and this issue is fixed later.

Now let us recall the operation $\oplus$ defined on tuples in Section 2.2. It is easy to see that this $\oplus$ definition is exactly what we need when merging assignments, taking $r = k$ and using $\lceil n/k \rceil$-tuples. In the sum we take each pair of tuples, such that for each segment the number of vertices assigned in both tuples in sum is not greater than k, which means that we can merge those two assignments without exceeding the capacity of intervals. Therefore this merge can be done in $O^*((k+1)^{\lceil n/k \rceil})$ time.

To check whether a coarse ordering exists we run our algorithm and check whether there exists a tuple in $assignments(r, i)$ for any i, such that the number of vertices assigned to the last interval is equal to $n \mod k$. The whole algorithm works in $O^*((k+1)^{\lceil \frac{n}{k} \rceil})$.

Similarly as in the previous section the algorithm can be modified to return not only the binary answer, but also a coarse ordering (in case of a positive answer) A. Let $f_A : V \to \{1, 2, \ldots, \lceil \frac{n}{k} \rceil\}$ be such that $A(v) = \{(f_A(v)-1)k+1, \ldots, f_A(v)k\}$. Then, by Definition 8, we have for every edge uv in the tree $(|f_A(u) - f_A(v)| - 1) \cdot k < b$, thus

$$\max A(v) - \min A(u) = (f_A(v) - f_A(u)) \cdot k + k - 1 \leq 2k + b - 1.$$

Since segments — values of the coarse ordering — are disjoint, one can easily find any ordering consistent with the given coarse ordering. This ordering has bandwidth at most $2k + b - 1$.

Note that if $b \geq \text{bw}(G)$, the algorithm does not fail to find a coarse ordering. Indeed, if π is an ordering with $\text{bw}(\pi) \leq b$, the algorithm may generate an assignment $f_A(v) = \lceil \frac{\pi(v)}{k} \rceil$.

Note that $\frac{n}{k} < b$ means that $b = \Omega(\sqrt{n/\alpha})$ and $k = \lfloor 0.5\alpha b \rfloor \geq c\sqrt{n\alpha}$ for some universal constant c. Thus the algorithm works in $O(\exp(c\sqrt{n/\alpha}\log n))$ time.

Theorem 9. *For arbitrary b at least one algorithm, Saxe's or ours, runs in $O(\exp(c\sqrt{n/\alpha}\log n))$ time.*

Note that, similarly as the algorithm described in Section 3, this algorithm extends to the class of graphs with treewidth not greater than t at the cost of a $O(n^{O(t)})$ time and space overhead. In this case the presented algorithm works in $O(n^{O(t)}(k+1)^{\lceil \frac{n}{k} \rceil})$ time and space, whereas the Saxe's algorithm does not use any treewidth assumption, thus the approximation scheme is still subexponential.

5 Approximation Algorithm for General Graphs

In this section we describe an approximation algorithm that, for a fixed integer $r \geq 1$, computes, given an undirected connected graph $G = (V, E)$, an ordering π satisfying $\text{bw}(G) \leq \text{bw}(\pi) \leq (4r-1)\text{bw}(G)$. The algorithm works in $O^*(2^{n/r})$ time and polynomial space and uses *coarse orderings* (see Definition 7).

Note that, given a coarse ordering A, checking if there exists any ordering consistent with A can be done in $O(n \log n)$ time. We use a simple greedy algorithm that assigns vertices to successive positions $1, 2, \ldots, n$. For any position i, it chooses still unassigned vertex v such that $i \in A(v)$ and for any other unassigned vertex w with $i \in A(w)$, $\max A(v) \leq \max A(w)$.

Lemma 10. *Let A be a coarse ordering for an input graph $G = (V, E)$. Let s be the size of the largest interval in A, i.e., $s = \max_{v \in V} |A(v)|$. If there exists an ordering π^* of bandwidth at most b consistent with A, then one can find in a polynomial time an ordering π that is consistent with A and has bandwidth at most $s + b$.*

Proof. For every edge uv replace $A(u)$ by $A(u) \cap [\min A(v) - b, \max A(v) + b]$ and similarly for $A(v)$. This operation maintains the invariant that π^* is consistent with A. Then find any ordering consistent with the new coarse ordering.

Let us now describe our algorithm. For a fixed positive integer r we obtain a $(4r-1)$-approximation algorithm. Let T be any fixed spanning tree of G and take any vertex of G as a root of T. For any vertex v different from the root, by *parent* of v we denote the parent of v in the rooted tree T.

The algorithm, given an integer $1 \leq b < n$, produces an ordering π satisfying $\text{bw}(\pi) \leq (4r-1)b$ or states that $b < \text{bw}(G)$. By a binary search over b we find the smallest b for which the algorithm produces an ordering. For this b we have $b \leq \mathrm{bw}(G)$ and $\mathrm{bw}(\pi) \leq (4r-1)b$.

By $I_{j,2i}$ we denote the interval of length $2ib$ starting at $jb+1$, for $j \in \mathbb{Z}$ and $r \leq i \leq 2r-1$. The algorithm is sketched in Pseudocode 2.

Algorithm 2. Generating assignments in the $(4r-1)$-approximation algorithm

procedure GENERATEASSIGNMENTS(A)
    if all nodes in T are assigned **then**
        Cut all intervals in A to make them contained in $\{1,\ldots,n\}$.
        Use Lemma 10 to find ordering consistent with A.
    else
        $v \leftarrow$ a node in T such that v's parent w is assigned; let $I_{j,2i} = A(w)$.
        if $i+1 \leq 2r-1$ **then**
            GENERATEASSIGNMENTS($A \cup \{(v, I_{j-1,2(i+1)})\}$)
        else
            GENERATEASSIGNMENTS($A \cup \{(v, I_{j-1,2r})\}$)
            GENERATEASSIGNMENTS($A \cup \{(v, I_{j-1+2r,2r})\}$)

procedure MAIN(i_0)
    for $j \leftarrow 0$ **to** $\lceil n/b \rceil - 1$ **do**
        GENERATEASSIGNMENTS ($\{(r, I_{j,2i_0})\}$) ▷ Generate all assignments with root in $I_{j,2i_0}$

Let $i_0 \in \{r,\ldots,2r-1\}$ be a parameter that we determine later. The algorithm assigns the root of T to all possible intervals of size $2i_0b$ that overlap with $\{1,\ldots,n\}$ and extends each of these partial assignments recursively.

To extend a given assignment, the algorithm chooses a node v of T such that the parent w of v has been already assigned an interval, say $I_{j,2i}$. Consider the interval $I_{j-1,2(i+1)}$ which is obtained from $I_{j,2i}$ by "extending" it by b positions both at the left and right side. Note that in any ordering consistent with the current assignment, v is put in a position from $I_{j-1,2(i+1)}$. Hence, if $I_{j-1,2(i+1)}$ is not too big, i.e. $i+1 \leq 2r-1$, the algorithm simply assigns $I_{j-1,2(i+1)}$ to v and proceeds with no branching (just one recursive call). Otherwise, if $i+1 = 2r$, the interval $I_{j-1,2(i+1)}$ is split into two intervals of size $2r$, namely $I_{j-1,2r}$ and $I_{j-1+2r,2r}$ and two recursive calls follow: with v assigned to $I_{j-1,2r}$ and $I_{j-1+2r,2r}$ respectively.

For every generated assignment (after cutting the intervals to make them contained in $\{1,\ldots,n\}$) the algorithm applies Lemma 10 to verify whether it is consistent with an ordering of bandwidth $[2(2r-1)+1]b = (4r-1)b$.

It is clear that the produced ordering has bandwidth at most $(4r-1)b$. The following lemma finishes the proof of correctness of the Algorithm 2.

Lemma 11. *If there exists an ordering π with $\mathrm{bw}(\pi) \leq b$, then Algorithm 2 produces at least one coarse ordering consistent with π.*

Proof. To produce such a coarse ordering, just simulate Algorithm 2 and at each step, whenever the algorithm has a choice for a vertex v between $I_{j-1,2r}$ and $I_{j-1+2r,2r}$, choose the interval containing $\pi(v)$. Since $\mathrm{bw}(\pi) \leq b$, the produced coarse ordering is consistent with π.

We conclude with the time complexity analysis. Observe that the nodes at tree distance d from the root are assigned intervals of size $2[(i_0 + d) \bmod r + r]$. It follows that branching appears only when $i_0 + d \equiv 0 \pmod{r}$. Let $\hat{n}(i_0)$ denote the number of nodes at tree distance d satisfying this condition. It is clear that the above algorithm works in time $O^*(2^{\hat{n}(i_0)})$. Since $\sum_{i \in \{r,\ldots,2r-1\}} \hat{n}(i) = n$, for some $i \in \{r, \ldots, 2r-1\}$, $\hat{n}(i) \leq n/r$. By choosing this value as i_0, we get the $O^*(2^{n/r})$ time bound.

Theorem 12. *For any positive integer r, there is a $(4r-1)$-approximation algorithm for* BANDWIDTH *running in $O^*(2^{n/r})$ time and polynomial space.* □

6 Exact Algorithm for General Graphs

In this section we focus on the exact algorithm for the BANDWIDTH problem on general graphs. This algorithm works in $O^*(20^{n/2}) \leq O(4.473^n)$ time, improving previous results [4] [5]. As an input, the algorithm takes a connected undirected graph $G = (V, E)$, where $|V| = n$, and a number b, $1 \leq b < n$ and checks if there exists an ordering π with $\text{bw}(\pi) \leq b$.

This algorithm uses major ideas from algorithms $O^*(5^n)$ [5] and $O(4.83^n)$ [4], i.e., placing vertices in the *color order* of positions (all important definitions are recalled in Section 6.1). However, the major difference is that we perform both phases (segment placing and depth-first search over states) of the previous algorithms at once, thus reducing the total number of possible states of the algorithm, at the cost of space complexity — we need to store all used states in memory.

6.1 Segments and Colors

First, let us recall some important observations made in [5]. An ordering π is called a b-ordering if $\text{bw}(\pi) \leq b$. Let $\mathbf{Pos} = \{1, 2, \ldots, n\}$ be the set of possible positions and for every position $i \in \mathbf{Pos}$ we define the *segment* it belongs to by $\texttt{segment}(i) = \lceil \frac{i}{b+1} \rceil$ and the *color* of it by $\texttt{color}(i) = (i-1) \mod (b+1) + 1$. By $\mathbf{Seg} = \{1, 2, \ldots, \lceil \frac{n}{b+1} \rceil\}$ we denote the set of possible segments, and by $\mathbf{Col} = \{1, 2, \ldots, b+1\}$ the set of possible colors. The pair $(\texttt{color}(i), \texttt{segment}(i))$ defines the position i uniquely. We order positions lexicographically by pairs $(\texttt{color}(i), \texttt{segment}(i))$, i.e., the color has higher order that the segment number, and call this order the *color order* of positions. By $\mathbf{Pos}_i$ we denote the set of the first i positions in the color order. Given some (maybe partial) ordering π, and $v \in V$ for which $\pi(v)$ is defined, by $\texttt{color}(v)$ and $\texttt{segment}(v)$ we understand $\texttt{color}(\pi(v))$ and $\texttt{segment}(\pi(v))$ respectively.

Let us recall the crucial observation made in [5].

Lemma 13 ([5], Lemma 8). *Let π be an ordering. It is a b-ordering iff, for every $uv \in E$, $|\texttt{segment}(u) - \texttt{segment}(v)| \leq 1$ and if $\texttt{segment}(u) + 1 = \texttt{segment}(v)$ then $\texttt{color}(u) > \texttt{color}(v)$ (equivalently, $\pi(u)$ is later in color order than $\pi(v)$).*

6.2 The Algorithm

In the algorithm, the *state* and *extension* are defined as follows.

Definition 14. *A pair (A, f) where $A \subset V$ and $f : A \to \mathbf{Seg}$ is called a state iff the following condition holds:*

1. *The multiset $\{f(v) : v \in A\}$ is equal to the multiset $\{\mathtt{segment}(i) : i \in \mathbf{Pos}_{|A|}\}$.*
2. *If $uv \in E$ and $u, v \in A$, then $|f(u) - f(v)| \leq 1$.*
3. *There exists $\bar{f} : V \to \mathbf{Seg}$ with $f \subset \bar{f}$ such that for every edge $uv \in E$, we have $|\bar{f}(u) - \bar{f}(v)| \leq 1$ and if $u \in A$, $v \notin A$ then $\bar{f}(u) \geq \bar{f}(v)$.*

Definition 15. *A state $(A \cup \{v\}, f')$ is an extension of a state (A, f) with a vertex $v \notin A$ iff $f = f'|_A$ and for all $uv \in E$ with $u \in A$ we have $f(u) \geq f'(v)$.*

The following equivalence holds (compare to [5], Lemma 11).

Lemma 16. *Let π be a b-ordering. For $0 \leq k \leq n$ let $A_k = \{v \in V : \pi(v) \in \mathbf{Pos}_k\}$ and $f_k = \mathtt{segment}|_{A_k}$. Then every (A_k, f_k) is a state and for every $0 \leq k < n$ state (A_{k+1}, f_{k+1}) is an extension of state (A_k, f_k).*

Lemma 17. *Assume we have states (A_k, f_k) for $0 \leq k \leq n$ and for all $0 \leq k < n$ state (A_{k+1}, f_{k+1}) is an extension of state (A_k, f_k) by vertex v_{k+1}. Let π be an ordering assigning v_k to the k-th position in the color order. Then π is a b-ordering.*

Theorem 18. *There are at most $O^*(20^{n/2})$ states.*

Due to the space limit, we omit details of the proof of Theorem 18, but we quickly sketch it. First, note that if we remove some edges of G, the number of states does not decrease, so we can assume that G is a tree and we root it at some vertex v_r. Let B be the set of vertices that have at least two children in G. We define *prestate* (A, f) as a *state*, but with the function f defined on $A \cup B$ instead of A. The number of prestates is not smaller that the number of states and the number of prestates can be bounded inductively for every subtree of G.

Note that checking if a pair (A, f) is a state can be done in a polynomial time. Points 1 and 2 are trivial to check. For Point 3 we can iteratively calculate for every $v \in V \setminus A$ function $p(v) \subset \mathbf{Seg}$, intuitively the set of possible values for $\bar{f}(v)$, by the following algorithm.

Algorithm 3. Check if (A, f) satisfies Point 3 of the state definition

Set $p(v) := \mathbf{Seg}$ for all $v \in V \setminus A$.
repeat
  for all $v \in V \setminus A$ **do**
    $p(v) := p(v) \cap \bigcap_{u \in N(v) \cap A} \{f(u) - 1, f(u)\} \cap \bigcap_{u \in N(v) \setminus A} \bigcup_{i \in p(u)} \{i - 1, i, i + 1\}$
until some $p(v)$ is empty or we do not made any change of any $p(v)$ in the inner loop
return True iff all $p(v)$ remain nonempty.

This algorithm performs at most $|\mathbf{Seg}|n \leq n^2$ runs of the outer loop, since reevaluated $p(v)$ can only be a subset of the previous value. The proof of correctness of Algorithm 3 is quite straightforward, but due to the space limit we omit it here. It is easy to see that checking if (A', f') is an extension of (A, f) and generating all extensions of the given state (A, f) can be done in polynomial time, too.

The algorithm is quite simple now. We do the depth-first search over states, starting from the state $(\emptyset, \emptyset)$. From a state (A, f) we can move to any of its extension and we seek for any state (A, f) with $A = V$. We remember (in some balanced search tree) all visited states and do not go into into the same state twice. By Lemmas 16 and 17, we reach such state iff there exists a b-ordering in G. Therefore Theorem 18 implies that:

Theorem 19. *The Bandwidth problem can be solved in* $O^*(20^{n/2})$ *time and space.*

Acknowledgments. We would like to deeply thank Lukasz Kowalik for supervising our work in the BANDWIDTH problem area and giving us a lot of valuable comments and remarks.

References

1. Björklund, A., Husfeldt, T.: Inclusion–exclusion algorithms for counting set partitions. In: Proc. FOCS 2006, pp. 575–582 (2006)
2. Björklund, A., Husfeldt, T., Kaski, P., Koivisto, M.: Fourier meets möbius: fast subset convolution. In: STOC 2007: Proceedings of the thirty-ninth annual ACM symposium on Theory of computing, pp. 67–74. ACM Press, New York (2007)
3. Bodlaender, H.L., Fellows, M.R., Hallett, M.T.: Beyond NP-completeness for problems of bounded width: Hardness for the W hierarchy (extended abstract). In: ACM Symposium on Theory of Computing, pp. 449–458 (1994)
4. Cygan, M., Pilipczuk, M.: Even faster exact bandwidth, arxiv:0902.1661v1 (unpublished, 2008), `http://arxiv.org/abs/0902.1661v1`
5. Cygan, M., Pilipczuk, M.: Faster exact bandwidth. In: Broersma, H., Erlebach, T., Friedetzky, T., Paulusma, D. (eds.) WG 2008. LNCS, vol. 5344, pp. 101–109. Springer, Heidelberg (2008)
6. Diestel, R.: Graph Theory. Springer, Heidelberg (2005)
7. Feige, U.: Approximating the bandwidth via volume respecting embeddings. J. Comput. Syst. Sci. 60(3), 510–539 (2000)
8. Feige, U.: Coping with the NP-hardness of the graph bandwidth problem. In: Halldórsson, M.M. (ed.) SWAT 2000. LNCS, vol. 1851, pp. 10–19. Springer, Heidelberg (2000)
9. Fomin, F.: Private communication (2009)
10. Garey, M., Graham, R., Johnson, D., Knuth, D.: Complexity results for bandwidth minimization. SIAM J. Appl. Math. 34, 477–495 (1978)
11. Koivisto, M.: An $O^*(2^n)$ algorithm for graph coloring and other partitioning problems via inclusion–exclusion. In: Proc. FOCS 2006, pp. 583–590 (2006)
12. Monien, B.: The bandwidth-minimization problem for caterpillars with hairlength 3 is np-complete. SIAM J. Algebraic Discrete Methods 7, 505–512 (1986)
13. Escoffier, B., Bourgeois, N., Paschos, V.T.: Efficient approximation by "low-complexity" exponential algorithms. Cahier du LAMSADE 271, LAMSADE, Universite Paris-Dauphine (2007)
14. Saxe, J.: Dynamic programming algorithms for recognizing small-bandwidth graphs in polynomial time. SIAM Journal on Algebraic Methods 1, 363–369 (1980)
15. Unger, W.: The complexity of the approximation of the bandwidth problem. In: Proc. FOCS 1998, pp. 82–91 (1998)

Approximation Algorithms via Structural Results for Apex-Minor-Free Graphs

Erik D. Demaine[1], MohammadTaghi Hajiaghayi[2,⋆], and Ken-ichi Kawarabayashi[3]

[1] MIT Computer Science and Artificial Intelligence Laboratory, 32 Vassar St., Cambridge, MA 02139, USA
edemaine@mit.edu

[2] AT&T Labs — Research, 180 Park Ave., Florham Park, NJ 07932, USA
hajiagha@research.att.com

[3] National Institute for Informatics, 2-1-2 Hitotsubashi, Chiyoda-ku, Tokyo 101-8430, Japan
k_keniti@dais.is.tohoku.ac.jp

Abstract. We develop new structural results for apex-minor-free graphs and show their power by developing two new approximation algorithms. The first is an additive approximation for coloring within 2 of the optimal chromatic number, which is essentially best possible, and generalizes the seminal result by Thomassen [32] for bounded-genus graphs. This result also improves our understanding from an algorithmic point of view of the venerable Hadwiger conjecture about coloring H-minor-free graphs. The second approximation result is a PTAS for unweighted TSP in apex-minor-free graphs, which generalizes PTASs for TSP in planar graphs and bounded-genus graphs [20,2,24,15].

We strengthen the structural results from the seminal Graph Minor Theory of Robertson and Seymour in the case of apex-minor-free graphs, showing that apices can be made adjacent only to vortices if we generalize the notion of vortices to "quasivortices" of bounded treewidth, proving a conjecture from [10]. We show that this structure theorem is a powerful tool for developing algorithms on apex-minor-free graphs, including for the classic problems of coloring and TSP. In particular, we use this theorem to partition the edges of a graph into k pieces, for any k, such that contracting any piece results in a bounded-treewidth graph, generalizing previous similar results for planar graphs [24] and bounded-genus graphs [15]. We also highlight the difficulties in extending our results to general H-minor-free graphs.

1 Introduction

Structural graph theory provides powerful tools for designing efficient algorithms in large families of graphs. The seminal work about the structure of graphs is

⋆ Work done while at MIT.

S. Albers et al. (Eds.): ICALP 2009, Part I, LNCS 5555, pp. 316–327, 2009.

Robertson and Seymour's Graph Minors series of over twenty papers over the past twenty years. From this work, particularly the decomposition theorem for graphs excluding any fixed minor H [28], has been made increasingly algorithmic and has led to increasingly general approximation and fixed-parameter algorithms; see, e.g., [14,21,8,7,12,1]. In general, it is interesting to explore how the combinatorial structure of the graph family influences the approximability of classic computational problems.

One such structural graph family that has played an important role in theoretical computer science (e.g., in [17,9,10,7]) is *apex-minor-free graphs*: graphs excluding a fixed apex graph H, where the removal of some vertex of H results in a planar graph. Apex-minor-free graphs include all bounded-genus graphs and many, many more graph families, almost to the extent of general H-minor-free graphs. For example, K_5 is an apex graph, and the class of K_5-minor-free graphs includes $K_{3,n}$ for all n, but the genus of $K_{3,n}$ goes to infinity as n grows.[1] Another example is $K_{3,k}$-minor-free graphs, which according to personal communication were Robertson and Seymour's first step toward their core decomposition result for H-minor-free graphs, because $K_{3,k}$-minor-free graphs can have arbitrarily large genus. More generally, apex-minor-free graphs have all the structural elements of H-minor-free graphs: clique-sums, bounded-genus graphs, apices, and vortices. Thus apex-minor-free graphs serve as an important testbed for algorithmic graph minor theory.

Eppstein [17] showed that apex-minor-free graphs are the largest minor-closed family of graphs that have a property called *bounded local treewidth*. This property has important algorithmic implications: such graphs admit a general family of PTASs following a generalization of Baker's approach for planar graphs [3]. Since this work, apex-minor-free graphs have been studied extensively, in particular in the bidimensionality theory (see [13]), with many algorithmic applications including more general PTASs and subexponential fixed-parameter algorithms [10,7,11,12,9,13]. On the structural side, it has been shown that apex-minor-free graphs have *linear local treewidth*, i.e., every radius-r neighborhood of every vertex has treewidth $O(r)$ [10]. This bound is best possible and substantially improves the bounds on the running of PTASs based on the generalized Baker's approach, from the previous bound of $2^{2^{2^{O(1/\varepsilon)}}} n^{O(1)}$ to the likely best possible bound of $2^{O(1/\varepsilon)} n^{O(1)}$.

To advance our understanding of how structural graph theory impacts approximation algorithms, we develop new such tools for apex-minor-free graphs. In particular, we develop two new decomposition results, strengthening previous results from Graph Minors [28] and from [15], and proving a conjecture from [10]. We use these decompositions to obtain an additive 2-approximation for graph coloring, improving previous results from [14], and to obtain a PTAS for unweighted TSP, generalizing results from [20] and from [15].

[1] Also, because $K_{3,k}$ has arbitrarily large genus, and $K_{3,k}$ is itself an apex graph, for any genus g there is a k such that $K_{3,k}$ has genus more than g and thus $K_{3,k}$-minor-free graphs include all genus-g graphs.

Graph coloring. Graph coloring is one of the hardest problems to approximate. In general graphs, the chromatic number is inapproximable within $n^{1-\varepsilon}$ for any $\varepsilon > 0$, unless ZPP = NP [18]. Even for 3-colorable graphs, the best approximation algorithm achieves a factor of $O(n^{3/14} \lg^{O(1)} n)$ [5]. In planar graphs, the problem is 4/3-approximable in the multiplicative sense, but more interestingly can be approximated within an additive 1, essentially because all planar graphs are 4-colorable; these approximations are the best possible unless P = NP. In contrast, graphs excluding a fixed minor H (or even graphs embeddable on a bounded-genus surface) are not $O(1)$-colorable for a constant independent of H (or the genus); the worst-case chromatic number is between $\Omega(|V(H)|)$ and $O(|V(H)|\sqrt{\lg |V(H)|})$. Therefore we need different approaches to approximate coloring of such graphs within small factors (independent of H or genus).

In a seminal paper, Thomassen [32] gives an additive approximation algorithm for coloring graphs embeddable on bounded-genus surfaces that is within 2 of optimal. More precisely, for any $k \geq 5$, he gives a polynomial-time algorithm to test k-colorability of graphs embeddable on a bounded-genus surface. Thus, for bounded-genus graphs, we do not know how to efficiently distinguish between 3, 4, and 5 colorability, but we can otherwise compute the chromatic number, and in all cases we can color within an additive 2 of the chromatic number. This result is essentially best possible: distinguishing between 3 and 4 colorability is NP-complete on any fixed surface, and distinguishing between 4 and 5 colorability would require a significant generalization of the Four Color Theorem characterizing 4-colorability in fixed surfaces.

More recently, a 2-approximation to graph coloring has been obtained for the more general family of graphs excluding any fixed minor H [14]. However, additive approximations have remained elusive for this general situation.

The challenge of an improved approximation for H-minor-free graphs is particularly interesting given its relation to Hadwiger's conjecture, one of the major unsolved problems in graph theory. This conjecture states that every H-minor-free graph has a vertex coloring with $|V(H)| - 1$ colors. Hadwiger [22] posed this problem in 1943, and proved the conjecture for $|V(H)| \leq 4$. As mentioned above, the best general upper bound on the chromatic number is $O(|V(H)|\sqrt{\lg |V(H)|})$ [26,29]. Thus, Hadwiger's conjecture is not resolved even up to constant factors, and the conjecture itself is only a worst-case bound, while an approximation algorithm must do better when possible.

Here we develop an additive approximation for graph coloring in apex-minor-free graphs, which are between bounded-genus graphs and H-minor-free graphs. We obtain the same additive error of 2 that Thomassen does for bounded-genus graphs:

Theorem 1. *For any apex graph H, there is an additive approximation algorithm that colors any given H-minor-free graph using at most* 2 *more colors than the optimal chromatic number.*

As mentioned above, the additive constant of 2 is essentially best possible. Also, Thomassen's proof method for bounded-genus graphs [32] does not work for apex-minor-free graphs. More precisely, Thomassen's main result in [32] says that, for any $k \geq 6$, there are only finitely many "k-color-critical graphs" embeddable in a fixed surface. Here a graph G is *k-color-critical* if G is not $(k-1)$-colorable, but removing any edge from G makes it $(k-1)$-colorable. However, for any k, there is any apex graph H such that there are infinitely many k-color-critical H-minor-free graphs.[2]

TSP and related problems. The Traveling Salesman Problem is a classic problem that has served as a testbed for almost every new algorithmic idea over the past 50 years. It has been considered extensively in planar graphs and its generalizations, starting with a PTAS for unweighted planar graphs [20], then a PTAS for weighted planar graphs [2], recently improved to linear time [24], then a quasi-polynomial-time approximation scheme (QPTAS) for weighted bounded-genus graphs [19], recently improved to a PTAS [15]. Grohe [21] posed as an open problem whether TSP has a PTAS in general H-minor-free graphs.

We advance the state-of-the-art in TSP approximation by obtaining a PTAS for unweighted apex-minor-free graphs. Furthermore, we obtain a PTAS for minimum c-edge-connected submultigraph[3] in unweighted apex-minor-free graphs, for any constant $c \geq 2$, which generalizes and improves previous algorithms for $c = 2$ on planar graphs [4,6] and for general c on bounded-genus graphs [15].

Theorem 2. *For any fixed apex graph H, any constant $c \geq 2$, and any $0 < \varepsilon \leq 1$, there is a polynomial-time $(1+\varepsilon)$-approximation algorithm for TSP, and for minimum c-edge-connected submultigraph, in unweighted H-minor-free graphs.*

TSP and minimum c-edge-connected submultigraph are examples of a general class of problems called *contraction-closed problems*, where the optimal solution only improves when contracting an edge. Many other classic problems are contraction-closed but not minor-closed, for example, dominating set (and its many variations) and minimum chordal completion. For this reason, contraction-closed problems have been highlighted as particularly interesting in the bidimensionality theory (see [13]).

To obtain the PTASs for TSP and minimum c-edge-connected submultigraph, as well as general family of approximation algorithms for contraction-closed problems, we study a structural decomposition problem introduced in [15]: partition the edges of a graph into k pieces such that contracting any one of the pieces

[2] Start with any k-color-critical graph G, e.g., K_k. Pick an apex graph H so that G is H-minor-free, which is possible because there is an apex graph of arbitrarily large genus. Now we can combine two vertex-disjoint copies G_1 and G_2 of G into an H-minor-free k-color-critical graph G' using the following *Hajós construction*. Start from the union graph $G_1 \cup G_2$. Pick any edge $\{v_1, w_1\}$ in G_1 and any edge $\{v_2, w_2\}$ in G_2. Remove both of these edges, identify v_1 and v_2, and join w_1 and w_2 by a new edge. The resulting graph G' is H-minor-free and k-color-critical provided G is. By repeating this construction starting from G', etc., we obtain infinitely many k-color-critical H-minor-free graphs.

[3] This problem allows using multiple copies of an edge in the input graph—hence sub*multi*graph—but the solution must pay for every copy.

results in a bounded-treewidth graph (where the bound depends on k). Such a result has been obtained for bounded-genus graphs [15] and for planar graphs with a variation of contraction called compression (deletion in the dual graph) [24,25]. These results parallel similar decomposition results for edge deletions in planar graphs [3], apex-minor-free graphs [17], and H-minor-free graphs [16,14]. However, contraction decomposition results are not known for graphs beyond bounded genus.

In this paper, we prove such a contraction decomposition result for apex-minor-free graphs:

Theorem 3. *For any fixed apex graph H, any integer $k \geq 2$, and every H-minor-free graph G, the edges of G can be partitioned into k sets such that contracting any one of the sets results in a graph of treewidth at most $f(k, H)$. Furthermore, such a partition can be found in polynomial time.*

In [15], it is shown how Theorem 3 leads to a general family of PTASs for any contraction-closed problem satisfying a few simple criteria, including TSP and minimum c-edge-connected submultigraph, thus proving Theorem 2.

Structural results. In order to prove both Theorem 1 about coloring approximation and Theorem 3 about contraction decomposition, we need to strengthen the structural results from the seminal Graph Minor Theory in the case of apex-minor-free graphs. Roughly speaking, Robertson and Seymour [28] prove that every H-minor-free graph is a clique sum of graphs "almost embeddable" into bounded-genus surfaces, with the exception of a bounded number of "apex" vertices and a bounded number of "local areas of non-planarity", called "vortices", which have bounded pathwidth. See [14] for the relevant definitions. More recently, this result was made algorithmic [14]. We prove that, when H is an apex graph, the apex vertices can be constrained to have edges only to vertices of vortices, but we have to generalize vortices to what we call "quasivortices", which have bounded treewidth instead of pathwidth. Our result is also algorithmic:

Theorem 4. *For any fixed apex graph H, there is a constant h such that any H-minor-free graph can be written as a clique-sum of h-almost embeddable graphs such that the apex vertices in each piece are only adjacent to quasivortices. Moreover, apices in each piece are not involved in the surface part of other pieces. Furthermore, there is a polynomial-time algorithm to construct this clique-sum decomposition for a given H-minor-free graph.*

Let us observe that it is obviously necessary for the running time of the coloring algorithm to depend exponentially on the excluded apex graph H.

This theorem is a powerful tool for the design of approximation algorithms in apex-minor-free graphs, as we show here for graph coloring and TSP. By analogy, the structural result for H-minor-free graphs has already proved critical throughout graph algorithms—see, e.g., [21,8,12,14,1]—and the additional structure we establish for apex-minor-free graphs seems essential. Indeed, this theorem was conjectured in the context of proving that apex-minor-free graphs have linear local treewidth [10], where it was suggested that this theorem would make it substantially easier to prove linear local treewidth and thereby improve the running time of many PTASs from $2^{2^{2^{O(1/\varepsilon)}}} n^{O(1)}$ to $2^{O(1/\varepsilon)} n^{O(1)}$. In fact, the conjecture of [10] used the standard notion of vortices, and one of our insights is to introduce quasivortices, which are just as good for algorithmic purposes; there is evidence that the use of quasivortices in Theorem 4 is necessary.

In Section 3, we describe the significant difficulties in generalizing our results to general H-minor-free graphs, as we crucially rely on our structure theorems for apex-minor-free graphs.

2 Overview

In this section, we give overviews of the proofs and algorithms behind our three main results. We start in Section 2.1 with our structural theorem about apex-minor-free graphs, which strengthens the Robertson-Seymour clique-sum decomposition. The structure determined by this theorem can be computed in polynomial time, and forms the foundation for our other results. The most direct use is our additive 2-approximation for coloring apex-minor-free graphs, described in Sections 2.2–2.3, which shows how to combine Thomassen's bounded-genus techniques over the new structure of apex-minor-free graphs. The PTASs for unweighted TSP and related problems are based on the contraction decomposition result, as described in Section 2.4, which in turn is based on the new structure of apex-minor-free graphs.

2.1 Overview of Structure Theorem for Apex-Minor-Free Graphs

Our main structure theorem, Theorem 4, builds upon the seminal Graph Minor decomposition theorem together with a new technique that is also developed in the Graph Minors series. The main challenge for our structure theorem is to control the neighborhood of apex vertices. How do we do it? Consider an apex v in the apex set. If v is adjacent to a vertex set W in the surface part in such a way that each vertex in W is far apart from other vertices in W, then we should be able to find a desired apex graph minor, because one of lemmas in Graph Minors tells us that, if a given vertex set is located far apart from each other on a planar graph or a graph on a fixed surface with sufficiently large representativity, then we can find any desired "rooted" planar graph minor. By choosing v to be the apex vertex of H, if there is a rooted $H - v$ minor in the surface part of some piece, we find that the graph actually has an H minor.

Otherwise, the neighbors of each apex vertex in the surface part of each piece are covered by a bounded number of bounded-radius disks. Because the number of apex vertices in each piece is bounded, this implies that there are bounded number of bounded-radius disks in the surface part such that these disks take care of all the neighbors of apices in the surface part of each piece. Now these bounded-radius disks become quasivortices after some modification. More precisely, these disks can decompose into a linear decomposition such that the intersection of two consecutive bags has bounded size. Then we can decompose each bag to extend our structure in such a way that each apex vertex is adjacent only to quasivortices.

This result is quite powerful. For example, it follows that apices in each piece are not involved in the surface part of other pieces. This is because there are no ≤ 3-separations in the surface part (excluding vortices) that have a neighbor in the apex vertex set. Otherwise, we could find a "neighbor" of some apex vertex v by finding a path from v to the surface part through the component. So the surface part could involve a clique-sum, but this is really a ≤ 3-separation. This property helps a lot in our algorithm.

2.2 Overview of Coloring Algorithm

Next we turn to our coloring algorithm. As pointed out before, Thomassen's proof method in [32] does not work for H-minor-free graphs for a fixed apex graph H, because there are infinitely many k-color-critical graphs without H minors for a fixed apex graph H. Nonetheless, some of the results proved in [32] are useful for us. Let us point out that Thomassen's result [32] depends on many other results, mostly by Thomassen himself; see [31,30,33,27]. The result in [32] is considered by many to be one of the deepest results in chromatic graph theory. This is because the series of results obtained by Thomassen opens up how topological graph theory can be used in chromatic graph theory.

At a high level, by our structure theorem, we have a clique-sum decomposition such that each torso (intersection of two pieces) in the surface part involves at most three vertices (in the surface), and no other vertices at all. Because there are at most three vertices in the intersection of pieces in the surface part, we can focus on each bag individually, and the coloring of each bag can be matched nicely by putting cliques in the intersections of two pieces. Let us observe that the clique size here is at most three.

Hence we can really focus on one piece, which has h-almost embeddable structure without any 3-separations in the surface part. Call this graph G. Roughly, what we will do is to decompose G into two parts V_1 and V_2 such that V_1 has bounded treewidth, and V_2 is a union of bounded-genus graphs. We can find such a decomposition such that the boundary of V_1 in V_2 is, roughly, the vertices on the cuffs to which quasivortices are attached. We shall add these boundary vertices of V_2 to V_1, and let V_1' be the resulting graph obtained from V_1. It turns out to be possible to prove that V_1' also has bounded treewidth. To find such a partition, we use the properties of our structure theorem for apex-minor-free graphs.

Next, we color $V_1 \cup V_2$. It is well-known that we can color graphs of bounded treewidth in polynomial time. So we can color V_1' using at most $\chi(G)$ colors. The main challenge is how to extend the precoloring of the vertices in $V_1 \cap V_2$ to the rest of vertices in V_2. To achieve this, we shall need some deep results by Thomassen [32], as described in the next subsection.

2.3 Coloring Extensions in Bounded-Genus Graphs

As we pointed out above, we shall decompose a given graph G into two parts V_1 and V_2 (possibly with $V_1 \cap V_2 \neq \emptyset$) such that V_1 has bounded treewidth, V_2 is union of bounded-genus graphs, and $V_1 \cap V_2$ is, roughly, the vertices on the cuffs to which quasivortices are attached. So we can color the vertices in V_1 using the bounded-treewidth method. Then the vertices in $V_1 \cap V_2$ are precolored.

Our challenge is the following. Suppose the vertices in the bounded number of cuffs are precolored. Can we extend this precoloring to the whole surface part? To answer this question, we use the following tool developed by Thomassen [32]. In fact, our statement below is different from the original by Thomassen. The list-coloring version of the following theorem is proved in [23]. So this immediately implies Theorem 5. But if we only need graph-coloring version of the result, let us point out that it follows from the same proof as in [32] by combining with the metric of Robertson and Seymour.

Theorem 5. *For any two nonnegative integer g, q, d, there exists a natural number $r(g, q, d)$ such that the following holds: Suppose that G is embedded on a fixed surface S of genus g and of the representativity at least $r(g, q, d)$ and there are d disjoint cuffs $S_1, S_2, \ldots, S_d$ such that the distance (in a sense of Robertson and Seymour's metric) of any two cuffs of $S_1, S_2, \ldots, S_d$ is at least q. Suppose furthermore that all the vertices in $S_1, S_2, \ldots, S_d$ are precolored with at most five colors such that*

1. *all the faces except for $S_1, S_2, \ldots, S_d$ are triangles, and*
2. *no vertex v of $G - (S_1 \cup S_2 \cup \cdots \cup S_d)$ is joined to more than two colors unless v has degree 4 or v has degree 5 and v is joined to two vertices of the same color.*

Then the precoloring of $S_1, S_2, \ldots, S_d$ can be extended to a 5-coloring of G. Also, there is a polynomial time algorithm to 5-color G.

For the reader's convenience, let us make some remarks for the proof of Theorem 5. The above statement follows from Theorem 8.1 (by putting $p = 0$) of [32]. The large distance of the metric of Robertson and Seymour implies the existence of large number of canonical cycles in the statement of Theorem 8.1. Then we can extend the result of Theorem 8.1 to the above statement. But as we pointed out above, there is now a stronger theorem, which is the list-coloring version of Theorem 5, see [23].

By a small modification to the theorem above, we obtain the following theorem which we will use:

Theorem 6. *For any two nonnegative integer g, q, d, there exists a natural number $r(g, q, d)$ such that the following holds: Suppose that G is embedded on a fixed surface S of genus g and of the representativity at least $r(g, q, d)$ and there are d disjoint cuffs $S_1, S_2, \ldots, S_d$ such that the distance (in a sense of Robertson and Seymour's metric) of any two cuffs of $S_1, S_2, \ldots, S_d$ is at least q. Suppose furthermore that all the vertices in $S_1, S_2, \ldots, S_d$ are precolored such that*

1. *all the faces except for $S_1, S_2, \ldots, S_d$ are triangles, and*
2. *no vertex v of $G - (S_1 \cup S_2 \cup \ldots \cup S_d)$ is joined to more than $\chi(G) - 1$ colors unless v has degree 4 or v has degree 5 and v is joined to two vertices of the same color.*

Then the precoloring of $S_1, S_2, \ldots, S_d$ using at most $\chi(G)$ colors can be extended to a $(\chi(G)+2)$-coloring of G. In fact, in the surface part, we only need five colors for most of vertices. (Precoloring may use more than five colors and vertices that are adjacent to precolored vertices may need some other color, though.) Also, there is a polynomial time algorithm for such a coloring G.

The list-coloring version of Theorem 6 is also proved in [23]. So this immediately implies Theorem 6, but for the reader's convenience, let us make some remarks. By the assumption of Theorem 6, each vertex not on the cuffs $S_1, S_2, \ldots, S_d$ has a list with at least 5 colors available, and each vertex that has a neighbor in the cuffs $S_1, S_2, \ldots, S_d$ has a list with at least 3 colors available. If we delete all the cuffs $S_1, S_2, \ldots, S_d$ in Theorem 5, then each vertex that has a neighbor in the cuffs $S_1, S_2, \ldots, S_d$ has a list with at least 3 colors available. So it follows that the conditions in Theorem 6 are equivalent to that in Theorem 5. Hence Theorem 6 follows from Theorem 5.

Let us observe that the coloring of Theorem 6 may use more than five colors because the precoloring vertices may use $\chi(G)$ colors. On the other hand, most of the vertices in the surface part use only five colors. The exceptional vertices are the precolored vertices on the cuffs, and vertices that are adjacent to precolored vertices.

2.4 Contraction Decomposition Result and PTASs

Finally, we sketch our proof of the contraction decomposition result and its applications to PTASs.

Our proof of Theorem 3 heavily depends on our decomposition theorem. Let us focus on one piece that has an h-almost embeddable structure without any 3-separations in the surface part. Call this graph G. As we did in our coloring algorithm, we decompose the graph G into two parts V_1 and V_2 such that V_1 has bounded treewidth, and V_2 is union of bounded-genus graphs. We can find such a decomposition such that the boundary of V_1 in V_2 is, roughly, the vertices on the cuffs to which quasivortices are attached. We shall add these boundary vertices of V_2 to V_1, and let V_1' be the resulting graph obtained from V_1. Again we can prove that V_1' also has bounded treewidth and V_2' is union of bounded-genus graphs.

Now our goal is to label the edges of the graph such that contracting any label set results in a bounded-treewidth graph. One challenge is the following. Suppose that the edges in the bounded number of cuffs are prelabeled. Can we extend this prelabeling to the whole surface part? Fortunately, this extension can be done by pushing the proof of the result in [15] just a little.

The main challenge is in handling the clique-sums. As we mentioned in the context of the coloring algorithm, we have a clique-sum decomposition such that each torso (intersection of two pieces) in the surface part involves at most three vertices, and nothing else. Because there are at most three vertices in the intersection of two pieces in the surface part, so for our coloring algorithm, we could really focus on each piece separately, and combine the colorings of each piece nicely by putting cliques in the intersection of two pieces. But for the contraction decomposition, this is a serious issue. Fortunately, the clique-sum in the surface part involves at most three vertices. We now put cliques in the intersection of two pieces (i.e., we add all the possible edges in the intersection of two pieces) and precolor these edges. What we need are edge-disjoint paths connecting these at most three vertices such that each path has the same label as the prelabeling of the corresponding edge. This is possible because we only need to control at most three edge-disjoint paths. All we need is to modify the argument in [15] to prove this, and the proof is identical to that in [15]. The details will be provided in the appendix.

As described in [15], the contraction decomposition result of Theorem 3 is strong enough to obtain a PTAS for TSP, minimum c-edge-connected submultigraph, and a variety of other contraction-closed problems, for an unweighted apex-minor-free graph, in particular proving Theorem 2. But it seems very hard to extend this result to H-minor-free graphs, because we do not know how many vertices are involved in the clique-sum—the number is bounded, but it seems to us that the precise number of vertices involved in the clique-sum is important—so we may not be able to control the neighbors of each apex vertex.

3 Difficulties with H-Minor-Free Graphs

One natural question is whether it is possible to extend our approach in this paper to H-minor-free graphs. More specifically, Robin Thomas (private communication) asked whether there is an additive c-approximation for chromatic number in H-minor-free graphs, where c is independent of H. One obvious way to attack this question is to prove the following conjecture:

Conjecture 1. Every H-minor-free graph has a partition of vertices into two vertex sets V_1 and V_2 such that V_1 has bounded treewidth and V_2 has chromatic number at most c for some absolute constant c.

We could add some moderate connectivity condition on the conjecture.

Our approach clearly breaks down for general H-minor-free graphs. Let us highlight some technical difficulties.

1. We need to consider separations of huge order, dependent on H instead of an absolute constant like 3.
2. We can no longer control the neighbors of apices: they can be all over the surface part of the piece.
3. Clique-sums become problematic. In particular, clique-sums involving vertices in the surface part are difficult to handle because of so-called *virtual edges*: edges present in the pieces but not in the clique-sum (the actual graph). Many pieces may be clique-summed to a common piece, making all of the surface edges in that piece virtual, effectively nonexistent.

So far we have been unable to surmount any of these difficulties, which is why apex-minor-free graphs seems like a natural limiting point.

Acknowledgments. We thank Paul Seymour for helpful suggestions and intuition about the structure of apex-minor-free graphs.

References

1. Abraham, I., Gavoille, C.: Object location using path separators. In: Proceedings of the 25th Annual ACM Symposium on Principles of Distributed Computing, pp. 188–197 (2006)
2. Arora, S., Grigni, M., Karger, D., Klein, P., Woloszyn, A.: A polynomial-time approximation scheme for weighted planar graph TSP. In: Proceedings of the 9th Annual ACM-SIAM Symposium on Discrete Algorithms, pp. 33–41 (1998)
3. Baker, B.S.: Approximation algorithms for NP-complete problems on planar graphs. Journal of the ACM 41, 153–180 (1994)
4. Berger, A., Czumaj, A., Grigni, M., Zhao, H.: Approximation schemes for minimum 2-connected spanning subgraphs in weighted planar graphs. In: Brodal, G.S., Leonardi, S. (eds.) ESA 2005. LNCS, vol. 3669, pp. 472–483. Springer, Heidelberg (2005)
5. Blum, A., Karger, D.: An $\tilde{O}(n^{3/14})$-coloring algorithm for 3-colorable graphs. Information Processing Letters 61, 49–53 (1997)
6. Czumaj, A., Grigni, M., Sissokho, P., Zhao, H.: Approximation schemes for minimum 2-edge-connected and biconnected subgraphs in planar graphs. In: Proceedings of the 15th Annual ACM-SIAM Symposium on Discrete Algorithms, pp. 496–505. Society for Industrial and Applied Mathematics, Philadelphia (2004)
7. Demaine, E.D., Fomin, F.V., Hajiaghayi, M., Thilikos, D.M.: Bidimensional parameters and local treewidth. SIAM Journal on Discrete Mathematics 18, 501–511 (2004)
8. Demaine, E.D., Fomin, F.V., Hajiaghayi, M., Thilikos, D.M.: Subexponential parameterized algorithms on graphs of bounded genus and H-minor-free graphs. Journal of the ACM 52, 866–893 (2005)
9. Demaine, E.D., Hajiaghayi, M.: Diameter and treewidth in minor-closed graph families, revisited. Algorithmica 40, 211–215 (2004)
10. Demaine, E.D., Hajiaghayi, M.: Equivalence of local treewidth and linear local treewidth and its algorithmic applications. In: Proceedings of the 15th ACM-SIAM Symposium on Discrete Algorithms (SODA 2004), January 2004, pp. 833–842 (2004)
11. Demaine, E.D., Hajiaghayi, M.: Bidimensionality: New connections between FPT algorithms and PTASs. In: Proceedings of the 16th Annual ACM-SIAM Symposium on Discrete Algorithms (SODA 2005), Vancouver, January 2005, pp. 590–601 (2005)
12. Demaine, E.D., Hajiaghayi, M.: Graphs excluding a fixed minor have grids as large as treewidth, with combinatorial and algorithmic applications through bidimensionality. In: Proceedings of the 16th Annual ACM-SIAM Symposium on Discrete Algorithms (SODA 2005), Vancouver, January 2005, pp. 682–689 (2005)
13. Demaine, E.D., Hajiaghayi, M.: The bidimensionality theory and its algorithmic applications. The Computer Journal 51, 292–302 (2008)
14. Demaine, E.D., Hajiaghayi, M., Kawarabayashi, K.: Algorithmic graph minor theory: Decomposition, approximation, and coloring. In: Proceedings of the 46th Annual IEEE Symposium on Foundations of Computer Science, Pittsburgh, October 2005, pp. 637–646 (2005)

15. Demaine, E.D., Hajiaghayi, M., Mohar, B.: Approximation algorithms via contraction decomposition. In: Proceedings of the 18th Annual ACM-SIAM Symposium on Discrete Algorithms, New Orleans, Louisiana, pp. 278–287 (2007)
16. DeVos, M., Ding, G., Oporowski, B., Sanders, D.P., Reed, B., Seymour, P., Vertigan, D.: Excluding any graph as a minor allows a low tree-width 2-coloring. Journal of Combinatorial Theory, Series B 91, 25–41 (2004)
17. Eppstein, D.: Diameter and treewidth in minor-closed graph families. Algorithmica 27, 275–291 (2000)
18. Feige, U., Kilian, J.: Zero knowledge and the chromatic number. Journal of Computer and System Sciences 57, 187–199 (1998)
19. Grigni, M.: Approximate TSP in Graphs with Forbidden Minors. In: Welzl, E., Montanari, U., Rolim, J.D.P. (eds.) ICALP 2000. LNCS, vol. 1853, pp. 869–877. Springer, Heidelberg (2000)
20. Grigni, M., Koutsoupias, E., Papadimitriou, C.: An approximation scheme for planar graph TSP. In: Proceedings of the 36th Annual Symposium on Foundations of Computer Science, pp. 640–645 (1995)
21. Grohe, M.: Local tree-width, excluded minors, and approximation algorithms. Combinatorica 23, 613–632 (2003)
22. Hadwiger, H.: Über eine Klassifikation der Streckenkomplexe. Vierteljschr. Naturforsch. Ges. Zürich 88, 133–142 (1943)
23. Kawarabayashi, K., Mohar, B.: List-color-critical graphs on a surface (preprint, 2008)
24. Klein, P.N.: A linear-time approximation scheme for TSP for planar weighted graphs. In: Proceedings of the 46th IEEE Symposium on Foundations of Computer Science, pp. 146–155 (2005)
25. Klein, P.N.: A subset spanner for planar graphs, with application to subset TSP. In: Proceedings of the 38th ACM Symposium on Theory of Computing (STOC 2006), pp. 749–756 (2006)
26. Kostochka, A.V.: Lower bound of the Hadwiger number of graphs by their average degree. Combinatorica 4, 307–316 (1984)
27. Mohar, B., Thomassen, C.: Graphs on surfaces. Johns Hopkins Studies in the Mathematical Sciences. Johns Hopkins University Press, Baltimore (2001)
28. Robertson, N., Seymour, P.D.: Graph minors. XVI. Excluding a non-planar graph. Journal of Combinatorial Theory, Series B 89, 43–76 (2003)
29. Thomason, A.: The extremal function for complete minors. Journal of Combinatorial Theory, Series B 81, 318–338 (2001)
30. Thomassen, C.: Five-coloring maps on surfaces. Journal of Combinatorial Theory. Series B 59, 89–105 (1993)
31. Thomassen, C.: Every planar graph is 5-choosable. Journal of Combinatorial Theory. Series B 62, 180–181 (1994)
32. Thomassen, C.: Color-critical graphs on a fixed surface. J. Combin. Theory Ser. B 70, 67–100 (1997)
33. Thomassen, C.: The chromatic number of a graph of girth 5 on a fixed surface. Journal of Combinatorial Theory. Series B, dedicated to Crispin St. J. A. Nash-Williams 87, 38–71 (2003)

Node-Weighted Steiner Tree and Group Steiner Tree in Planar Graphs

Erik D. Demaine[1], MohammadTaghi Hajiaghayi[2], and Philip N. Klein[3]

[1] MIT Computer Science and Artificial Intelligence Laboratory,
32 Vassar St., Cambridge, MA 02139, USA
edemaine@mit.edu

[2] AT&T Labs — Research,
180 Park Ave., Florham Park, NJ 07932, USA
hajiagha@research.att.com

[3] Department of Computer Science, Brown University,
Providence, RI 02912, USA
klein@cs.brown.edu

Abstract. We improve the approximation ratios for two optimization problems in planar graphs. For node-weighted Steiner tree, a classical network-optimization problem, the best achievable approximation ratio in general graphs is $\Theta(\log n)$, and nothing better was previously known for planar graphs. We give a constant-factor approximation for planar graphs. Our algorithm generalizes to allow as input any nontrivial minor-closed graph family, and also generalizes to address other optimization problems such as Steiner forest, prize-collecting Steiner tree, and network-formation games.

The second problem we address is group Steiner tree: given a graph with edge weights and a collection of groups (subsets of nodes), find a minimum-weight connected subgraph that includes at least one node from each group. The best approximation ratio known in general graphs is $O(\log^3 n)$, or $O(\log^2 n)$ when the host graph is a tree. We obtain an $O(\log n \operatorname{polyloglog} n)$ approximation algorithm for the special case where the graph is planar embedded and each group is the set of nodes on a face. We obtain the same approximation ratio for the minimum-weight tour that must visit each group.

1 Introduction

One of the most fundamental problems in combinatorial optimization is the *network Steiner tree problem.* This was one of the first problems shown NP-hard by Karp [19]. In the traditional formulation, we are given an undirected graph with edge costs and a subset of nodes called *terminals.* The goal is to find a minimum-cost subgraph of G that connects the terminals. A long sequence of papers give polynomial-time constant-factor approximation algorithms for this problem; the current best approximation ratio is 1.55 [3].

The generalization[1] of network Steiner tree in which the nodes are also assigned costs is of both practical and theoretical significance. On the practical

[1] The case of both edge costs and node costs can be reduced to the case of node costs.

S. Albers et al. (Eds.): ICALP 2009, Part I, LNCS 5555, pp. 328–340, 2009.

side, in telecommunications for example, expensive equipment such as routers and switches are at the nodes of the underlying network and it is natural to model such problems as node-weighted. On the theoretical side, node-weighted versions of many classic edge-weighted problems have been considered by many authors so far; see, e.g., [8,10,15,17,22] for some recent work.

Unfortunately, set cover can be reduced to node-weighted Steiner tree in general graphs, so an approximation ratio better than $\ln n$ is not achievable in polynomial time unless P = NP [9,26]. Klein and Ravi [20] gave a polynomial-time approximation algorithm with a performance ratio of $O(\log n)$, so this result is within a constant factor of optimal. To achieve a better bound, we must restrict the class of inputs.

A natural restriction is planarity. In practical scenarios of physical networking, with cable or fiber embedded in the ground, crossings are rare or nonexistent. Somewhat surprisingly, no one has yet addressed this classic problem of node-weighted Steiner tree in this natural class of graphs. In this paper, we achieve a much better approximation ratio for this problem:

Theorem 1. *There is a polynomial-time 6-approximation algorithm for node-weighted Steiner tree in planar graphs.*

Our algorithm is a simple and natural extension of primal-dual techniques to node weights, which to our knowledge has never been explored. We suspect that the factor 6 can be improved.

In fact, our result is more general in two senses: we show that a constant approximation ratio is achievable for a much broader family of graphs, and we show that a much more general optimization problem can be approximated.

1.1 Broader Graph Classes

A *minor* of a graph G is a graph obtainable from G by deleting and contracting edges.

It is well-known that planar graphs are the graphs with no minor isomorphic to $K_{3,3}$ or K_5. More generally, for any graph H, we can consider the family of graphs excluding H as a minor. We obtain the following generalization of Theorem 1.

Theorem 2. *For any graph H, there is a constant c_H and a polynomial-time c_H-approximation algorithm for node-weighted Steiner tree on H-minor-free graphs.*

1.2 More Network-Design Problems

Our algorithm can solve a broader range of network-design problems. The node-weighted *Steiner forest problem* takes as input an undirected graph with node costs and a set of unordered pairs $\{s_i, t_i\}$ of nodes. The goal is to find the minimum-cost network that includes a path between each given pair of nodes. For the edge-weighted case, there is a polynomial-time 2-approximation algorithm [2]. For the node-weighted case, an $O(\log n)$-approximation can be achieved [20]. In this paper, we achieve a constant-factor approximation for planar graphs, and more generally, graphs excluding a fixed minor.

Theorem 3. *There is a polynomial-time* 6*-approximation algorithm for node-weighted Steiner forest in planar graphs. For any graph* H*, there is a constant* c_H *and a polynomial-time* c_H*-approximation algorithm for this problem on* H*-minor-free graphs.*

More generally, in Section 2, we show how our framework applies to a node-weighted variation of *proper 0-1 functions* by Goemans and Williamson [12]. These functions model node-weighted versions of several other problems, e.g., T-joins and nonfixed point-to-point connections. We thus obtain constant-factor approximation algorithms for all of these problems when the input is restricted to be a planar graph or a graph excluding a fixed minor.

We also consider the *prize-collecting* version of Steiner tree, where some terminal pairs can remain disconnected, but we pay a specific penalty for each such pair. The best approximation algorithm for this problem achieves an $O(\log n)$ approximation ratio [25]. (See also the related work on the dual quota version of the problem [15,25].) Similar to [7,16], we can prove the following:

Theorem 4. *The prize-collecting Steiner forest problem has a constant-factor approximation algorithm in graphs excluding a fixed minor.*

1.3 Other Applications

Node-weighted Steiner tree is a network design problem with many practical applications and theoretical implications. Enumerating such applications is beyond the scope of this paper. We point out, however, that it has an application even in network formation games.

Anshelevich et al. [4] consider a network formation game in which k terminals (players) buy edges in a directed graph, equally sharing the unit cost of an edge bought by multiple players, to form a Steiner tree. They prove that the price of stability in this game is at most H_k (the kth harmonic number, which is within an additive 1 of $\ln k$) by defining a dynamics that converges to an equilibrium within an H_k factor of the social optimum. However, their dynamics [4, Theorem 2.2] starts by computing an optimal Steiner tree, which cannot even be efficiently approximated. With our results, we can obtain a polynomially computable Nash equilibrium within an H_k factor of the social optimum for their game in node-weighted undirected planar graphs. Furthermore, this bound is tight: there is a node-weighted graph whose only Nash equilibrium is a factor H_k worse than the social optimum.[2]

[2] The graph has terminals $t_1, t_2, \ldots, t_k$, additional nodes $u_1, u_2, \ldots, u_k$, where u_i has weight $1/i$, and another node v of weight $1+\varepsilon$. Each terminal t_i has two candidate paths to the root r, one through u_i and the other through v. (This construction is similar to [4, Figure 1].) The social optimum buys v at cost $1+\varepsilon$, but the Nash equilibrium buys $u_1, u_2, \ldots, u_k$ at cost H_k.

1.4 Planar Group Steiner Tree

In the wire-routing phase of VLSI design, a *net* is a set of pins on the boundaries of various components that must be connected. A minimum-length Steiner tree is a natural choice for routing the net. (The routing must avoid previously routed nets and other obstacles.) Reich and Widmayer [27] observed that, for each component, there is flexibility as to the location of the pin used, and that the routing of the net should exploit that flexibility.

With that as motivation, they introduced the *group Steiner tree problem*: we are given a graph G with edge weights, and a collection $g_1, g_2, \ldots, g_k$ of node sets called *groups*. The goal is to find a minimum-weight connected subgraph of G that contains at least one node from each group.

Much research has gone into finding good approximation algorithms for this problem. For general graphs, the best approximation ratio known to be achievable in polynomial time [11] is $O(\log^3 n)$, and for trees, the best known is $O(\log^2 n)$.

Even when the host graph is a tree and hence planar, the problem cannot be approximated better than $\Omega(\log^{2-\varepsilon} n)$ unless NP admits quasipolynomial-time Las Vegas algorithms [18]. It would thus appear that restricting the input to planar graphs cannot lead to a substantially improved approximation.

Returning to the origin of the problem provides some inspiration. In a VLSI instance, the elements of a single group are all located on the boundary of a component, which occupies a region on the plane. Motivated by this real-world restriction, we define an instance of the *planar group Steiner tree problem* to be a planar embedded graph G with edge weights, and a collection of groups $g_1, g_2, \ldots, g_k$ and corresponding distinct faces $f_1, f_2, \ldots, f_k$ of G, such that the nodes belonging to each group g_i lie on the boundary of the corresponding face f_i.

We can without loss of generality require that each group g_i consists of exactly the nodes on the boundary of f_i. (The more general problem can be reduced to this one by the introduction of high-weight edges.) Therefore, an equivalent and more concise definition of an instance of *planar group Steiner tree* is a planar embedded graph G and a set of faces $f_1, f_2, \ldots, f_k$, which implicitly define a group for each f_i consisting of the nodes on the boundary of f_i.

Theorem 5. *Planar group Steiner tree has a polynomial-time $O(\log n \operatorname{polyloglog} n)$-approximation algorithm.*

Our proof of this theorem uses probabilistic embedding into spanning trees with expected distortion $O(\log n \log\log n (\log\log\log n)^3)$ [1]. (We cannot use the original result of Bartal [6] which does not preserve the planar structure of the problem.) On trees, we can solve the problem via dynamic programming. Alternatively, because paths in trees cannot properly cross, we can use the following rounding result of independent interest:

Theorem 6. *Any solution f to the noncrossing-flow relaxation of directed Steiner tree can be converted in polynomial time into an integral solution f'' of weight $c(f'') \leq 6\, c(f)$.*

Recall the *directed Steiner tree problem*: we are given a directed graph G with edge costs, a sink node s, and a set T of *terminal* nodes. The goal is to find a minimum-cost subgraph of G that, for each terminal t_i, contains a directed path from t_i to s. The noncrossing-flow variation is a natural relaxation of directed Steiner tree in planar graphs, a kind of minimum-cost flow where flow paths cannot cross; see Appendix A. Using a novel approach for rounding such *planar* flows, we show that a (fractional) solution to this noncrossing relaxation can be converted into an (integral) directed Steiner tree whose cost is at most 6 times the value of the solution to the relaxation.

Unfortunately, we do not know a polynomial-time algorithm for finding an optimal solution to the noncrossing-flow formulation for an arbitrary planar instance of directed Steiner tree. Such a result would yield a constant-factor approximation algorithm for planar directed Steiner tree.

Related Work. Motivated in part by the VLSI application, Mata and Mitchell [23] consider the following problem: given a set of n polygonal regions in the plane, find a tour that visits at least one point from each region. They describe this problem as a special case of the problem *TSP with neighborhoods* (also called *group TSP*). They give a polynomial-time $O(\log n)$-approximation algorithm. Because the tour contains a spanning tree, and doubling each edge of a tree yields a tour, it is also an approximation algorithm for group Steiner tree where the groups are the polygonal regions. Gudmundsson and Levcopoulos [14] gave a faster algorithm for the same problem. No known polynomial-time algorithm achieves an approximation ratio better than $\Theta(\log n)$ for this problem. On the lower-bound side, unless P = NP, no constant-factor approximation is possible for disjoint disconnected regions, and no $(2 - \varepsilon)$-approximation is possible for (nondisjoint) connected regions [28].

Arkin and Hassin [5] gave constant-factor approximation algorithms for the special cases of parallel unit-length line segments, translated copies of a polygonal region, and circles. Mitchell [24] recently gave a PTAS for group TSP when the groups are disjoint and "fat".

Our bound for planar group Steiner tree nearly matches the bound of Mata and Mitchell. Our approach has the advantage that planar graphs can capture metrics that are not captured by the Euclidean metric, useful e.g. in the VLSI problem where the routing of a net must avoid obstacles and previously routed nets.

2 Node-Weighted Network Design in Planar and Minor-Excluding Graphs

For a graph G and a set S of nodes, we use $G[S]$ to denote the subgraph of G induced by the nodes of S.

To formulate the node-weighted network-design problems we address, we adapt an approach due to Goemans and Williamson [12].

Proper Function. Let V be the set of nodes in an undirected graph G. A function $f : 2^V \to \{0, 1\}$ is *proper* if $f(\emptyset) = 0$ and the following two properties hold:

1. (Symmetry) $f(S) = f(V - S)$.
2. (Disjointness) If S_1 and S_2 are disjoint, then $f(S_1) = f(S_2) = 0$ implies $f(S_1 \cup S_2) = 0$.

In [12], proper functions are used in a formulation of *edge-weighted* network-design problems as *cut-covering* problems. A proper function specifies a family of cuts, and the goal is to minimize the cost of a set of edges that covers all cuts in this family.

Following Klein and Ravi [20], we adapt this formulation for *node-weighted* problems. We use $\Gamma(S)$ to denote the set of nodes that are not in S but have neighbors in S. The problems we address can be formulated by the following integer linear program with a variable $x(v)$ for each node $v \in V$:

$$\begin{aligned} &\text{minimize } \textstyle\sum_{v \in V} w(v)\, x(v) \\ &\text{subject to } \textstyle\sum_{v \in \Gamma(S)} x(v) \geq f(S) \text{ for all } S \subseteq V, \\ &\qquad x(v) \in \{0,1\} \qquad\qquad \text{for all } v \in V. \end{aligned} \tag{1}$$

where f is a proper 0-1 function. The minimum solution x to this integer program assigns 1's to a subset X of nodes, and X is then considered the solution to the network-design problem. Conversely, a subset X of nodes is considered a *feasible solution* if the corresponding $\{0,1\}$-assignment to nodes of G satisfies the inequalities.

For example, consider the node-weighted Steiner forest problem. The input is an undirected graph G with node weights $w(v)$, and a set of pairs $\{s_i, t_i\}$ of nodes. For a set S of nodes, define $f(S)$ to be 1 if, for some pair $\{s_i, t_i\}$, S contains one element of the pair but not the other. Otherwise, define $f(S)$ to be 0. It is easy to verify that this function is proper. To see that the solution to the integer linear program is a solution to the Steiner forest instance, assume for a contradiction that some pair s_i, t_i are not connected via nodes assigned 1's by x. Let S be the set of nodes connected to s_i via such nodes. By our assumption, $f(S) \geq 1$ but by the choice of S, every node $v \in \Gamma(S)$ is assigned 0 by x, contradicting the linear constraint.

We assume in our algorithm that $f(\cdot)$ can be queried for a specific set S in polynomial time. For the analysis, we assume that each node v with $f(\{v\}) = 1$ has zero cost. For the Steiner forest problem, for example, each such node belongs to some pair, so must belong to the solution, so we can make this assumption without loss of generality.

For a subset X of nodes in a graph G, let $\mathrm{CC}(X)$ denote the node sets of connected components of the subgraph of G induced by X.

Lemma 1. *Let X be a subset of nodes of G that contains every node v such that $f(\{v\}) = 1$. Suppose that $f(C) = 0$ for every $C \in \mathrm{CC}(X)$. Then X is a feasible solution to the integer program.*

Proof. Let x be the function that assigns 1 to nodes in X and 0 to nodes not in X. Let S be any subset of nodes such that $\sum_{v \in \Gamma(S)} x(v) = 0$. We shall show that $f(S) = 0$.

For each $C \in \mathrm{CC}(X)$, we claim that $S \cap C$ is either $\emptyset$ or C. (Otherwise, there would be a pair u, v of adjacent nodes in X such that $u \in S$ and $v \notin S$, so $v \in \Gamma(S)$. Since $x(v) = 1$, this would contradict the choice of S.)

The claim implies that S is the disjoint union of some of the connected components $C \in \mathrm{CC}(X)$ together with some singletons $\{v\}$ with $v \notin X$. We assumed that X contains all nodes v with $f(\{v\}) = 1$, so $f(\{v\}) = 0$ for $v \notin X$. We also assumed that each connected component C of X has $f(C) = 0$. Combining these facts using the disjointness property of f, we infer that $f(S) = 0$. □

Dual. The linear relaxation of the above integer program is obtained by replacing the constraint $x(v) \in \{0, 1\}$ with the constraint $x(v) \geq 0$. The dual of the resulting linear program is as follows:

$$\begin{aligned} &\text{maximize } \textstyle\sum_{S \subseteq V} f(S)\, y(S) \\ &\text{subject to } \textstyle\sum_{S \subseteq V : v \in \Gamma(S)} y(S) \leq w(v) \quad \text{for all } v \in V, \\ &\qquad\qquad y(S) \geq 0 \qquad\qquad\qquad\quad \text{for all } S \subseteq V. \end{aligned}$$

There is a dual variable $y(S)$ for each subset S of V. However, the only such variables that affect the objective function (and therefore the only variables we need to consider) are those variables $y(S)$ where $f(S) = 1$. Intuitively, the goal of the dual linear program is to find a maximum-size family of node sets S with $f(S) = 1$ subject to the constraint that each node v is the neighbor of at most $w(v)$ sets in the family.

Primal-Dual Algorithm. Goemans and Williamson [13] gave a generic version of the primal-dual approximation algorithm. In this section, we give an algorithm that is a specialization (and slight modification) of their generic algorithm. We start with some terminology.

Let G be the input graph. A node set S is a *violated connected component with respect to* X if $S \in \mathrm{CC}(X)$ and $f(S) = 1$. Define a *partial solution* to be a set X of nodes containing $\{v : f(v) = 1\}$ such that there is some violated connected component with respect to X.

Goemans and Williamson's generic algorithm is defined in terms of an oracle. We will use an oracle $\mathrm{Viol}(\cdot)$ that takes a partial solution X as input, and that outputs the violated connected components with respect to X. This oracle can be implemented in polynomial time using a connected-components subroutine and queries to the function $f(\cdot)$.

Now we give our specialization and modification of the generic algorithm. The modification is as follows. Their algorithm maintains a solution X, initially empty, and adds to it in a series of iterations; finally, the algorithm removes some elements from it. In our modified version, X initially consists of all nodes v such that $f(\{v\}) = 1$, and these elements are never removed from X. However, these nodes are required to have weight zero, so their presence in X does not affect the approximation performance.

The algorithm also maintains a dual feasible solution $\mathbf{y}$. Recall that the dual linear program has a constraint $\sum_{S \subseteq V : v \in \Gamma(S)} y(S) \leq w(v)$ for each node $v \in V$.

1. $\mathbf{y} \leftarrow \mathbf{0}$
2. $X \leftarrow \{v : f(\{v\}) = 1\}$
3. While there is a violated connected component with respect to X:
   (a) Increase $y(S)$ uniformly for all sets $S \in \text{Viol}(X)$ until the dual linear-program inequality for some v becomes tight: $\sum_{S \subseteq V : v \in \Gamma(S)} y(S) = w(v)$.
   (b) Add v to X, i.e., $x(v) \leftarrow 1$.
4. For each v in X in the reverse of the order in which they were added during the while-loop:
   (a) If $f(S) = 0$ for every connected component $S \in \text{CC}(X - \{v\})$, then remove v from X.
5. Return X.

The algorithm above is almost an instantiation of an algorithm of Goemans and Williamson [13]. The difference is that, in our algorithm, X is required at all times to contain every node v such that $f(\{v\}) = 1$. Because these nodes are assumed to have zero cost, the proof of Theorem 4.2 of [13] can be adapted to show the theorem below (see the full paper). For a set X of nodes of G, a set F of nodes is a *feasible augmentation of* X if $F \supseteq X$ and F is a feasible solution. If in addition no proper subset of F is a feasible augmentation of X then F is a *minimal* feasible augmentation of X.

Theorem 7. *Suppose γ is a number such that, for any partial solution $X \subseteq V$ and any minimal feasible augmentation F of X, we have*

$$\sum\{|F \cap \Gamma(S)| : S \in \text{Viol}(X)\} \leq \gamma\, |\text{Viol}(X)|. \tag{2}$$

Then the algorithm described above returns a feasible solution of weight at most $\gamma \sum_{S \subseteq V} y(S) \leq \gamma$ LP-OPT where LP-OPT denotes the weight of an optimal solution to the linear program (1).

In order to apply Theorem 7, we need to prove (2) for some γ.

Theorem 8. *Let X be a partial solution, and let F be any minimal feasible augmentation of X.*

If G is planar, then $\sum\{|F \cap \Gamma(S)| : S \in \text{Viol}(X)\} \leq 6\, |\text{Viol}(X)|$.
If G is H-minor-free, then $\sum\{|F \cap \Gamma(S)| : S \in \text{Viol}(X)\} \leq O(|V(H)|\sqrt{\log |V(H)|})\, |\text{Viol}(X)|$.

The proof of Theorem 8 will be given in this section. By using the bounds proved in Theorem 8 in Theorem 7, we obtain

Theorem 9. *The primal-dual algorithm above is a 6-approximation on planar graphs and, more generally, an $O(1)$-approximation in H-minor-free graphs for any fixed graph H.*

Now we give the proof of Theorem 8.

The sum $\sum\{|F \cap \Gamma(S)| \ : \ S \in \text{Viol}(X)\}$ counts the number of adjacencies between F and violated connected components of X, counting multiply if one violated connected component of X is adjacent to several nodes in F, but counting only once if multiple nodes in a common violated connected component of X are adjacent to one node of F.

Let $\hat{F} = F - X$. For any $S \in \text{Viol}(X)$, since S is the node set of a connected component of $G[X]$, no neighbor of S belongs to X. This shows that $|\hat{F} \cap \Gamma(S)| = |F \cap \Gamma(S)|$. To prove Theorem 8, it therefore suffices to bound

$$\sum\{|\hat{F} \cap \Gamma(S)| \ : \ S \in \text{Viol}(X)\}. \tag{3}$$

Let $\hat{G}$ be the graph obtained from G by contracting each violated connected component of X to a single node, which we call a *terminal*. Let R be the set of terminals. Because of the contractions, no two nodes of R are adjacent in $\hat{G}$. We discard multiple copies of edges in $\hat{G}$ so that the sum (3) in G becomes the number of edges in $\hat{G}$ between nodes in $\hat{F}$ and terminals. Our goal is to bound the number of such edges in terms of $|R|$. We do this separately for each connected component of $\hat{G}[\hat{F} \cup R]$. Let $G' = \hat{G}[F' \cup R']$ be one such connected component, where $F' \subseteq \hat{F}$ and $R' \subseteq R$.

By minimality of F, R' is nonempty. Assign distinct integers as IDs to the nodes of G'. Let r be a node in R'. For each node v in $F' \cup R'$, define its *level* $\ell(v)$ to be its breadth-first-search distance from r in G'. We next define the *parent* $p(v)$ of each node $v \neq r$. For $v \in R' - \{r\}$, define $p(v)$ to be a neighbor of v in G' having level $\ell(v) - 1$, namely that neighbor having minimum ID. For each node $v \in F'$, select $p(v)$ as follows. If v has a neighbor w in R' such that $p(w) \neq v$, then $p(v)$ is any such neighbor. Suppose that v has no neighbor w in R' such that $p(w) \neq v$. By the properties of breadth-first-search distances, v has some neighbor w' such that $\ell(w') = \ell(v) - 1$. Let $p(v)$ be this node w'. We show that in this case that $w' \in F'$.

Assume for a contradiction that $w' \in R'$. Its parent would have level $\ell(w') - 1 = \ell(v) - 2$, so its parent could not be v. This implies that v has a neighbor in R' whose parent is not v, a contradiction.

Lemma 2. *The parent pointers do not form a cycle.*

Proof. Suppose for contradiction that $C = x_0 x_1 x_2 \ldots x_{k-1} x_0$ were a minimal (simple) cycle with $p(x_i) = x_{i+1}$ for all i (where the indices are taken modulo k). The root r cannot be in the cycle C because it has no parent. Because $x_i x_{i+1}$ is an edge of the graph, $\ell(x_{i+1}) \leq \ell(x_i) + 1$ for every i. By construction, $\ell(x_{i+1}) = \ell(x_i) - 1$ for each i where $x_i \in R'$ and for each i where $x_i \in F'$ and $x_{i+1} \in F'$, So the only case in which $\ell(x_{i+1})$ could be $\ell(x_i) + 1$ is when $x_i \in F'$ and $x_{i+1} \in R'$. But then by construction $\ell(x_{i+2}) = \ell(x_{i+1}) - 1 = \ell(x_i)$. Thus every increase in level is immediately followed in the cycle C by a strict decrease in level.

Suppose there were two consecutive strict decreases: $\ell(x_{i+1}) = \ell(x_i) - 1$ and $\ell(x_{i+2}) = \ell(x_{i+1}) - 1$. Then all nodes after x_{i+2} would have level at most $\ell(x_i) - 1$, contradicting the fact that C is a cycle. It follows that the x_i's must alternate between R' and F', and that, for some positive integer d, the levels of the nodes in $C \cap R'$ are d, and the levels of the nodes in $C \cap F'$ are $d-1$. Now consider the minimum-ID node x_i in $C \cap F'$. By construction, x_{i-1} and x_{i+1} must have x_i as their parent. But this contradicts $p(x_{i+1}) = x_{i+2}$ given that the cycle is simple. □

Lemma 2 shows that the parent pointers define a rooted spanning tree of $\hat{G}[F' \cup R']$. Let F'' be the subset of nodes in F' with a neighbor in R'. We need to compute the number of edges in $\hat{G}[F'' \cup R']$.

Lemma 3. $|F''| \leq 2\,|R'|$.

Proof. As we show in Figure 1, to each node u of R', we charge at most two nodes of F'': u's parent and the nearest ancestor of u whose parent is in R'. We claim that every node of F'' gets counted by this charging.

By minimality of F, every node of F' is on a path in the tree T from some terminal in R' to the root r. Let v be any node of F''. Let P be the shortest terminal-to-terminal path in T that includes v.

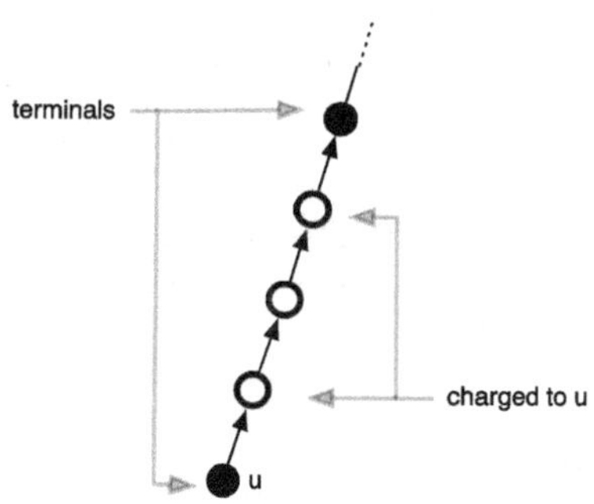

Fig. 1. Charging nodes of F'' to terminals

Since $v \in F''$, v has a neighbor w in R. If v is the second node of P then v is charged to the first node of P. Otherwise, by minimality of P, v is not the parent of a terminal, so $p(w) \neq v$, so v's parent is a terminal (not necessarily w), so again v is charged to the first node of P. □

Now we can complete the proof of Theorem 8. Since G is H-minor free, and $\hat{G}[F'' \cup R']$ is a minor of G, $\hat{G}[F'' \cup R']$ is also H-minor free. Hence the number of edges in $G[F' \cup R]$ is $O((|F'| + |R|)\,|V(H)|\sqrt{\log |V(H)|})$ [21,29], which is $O(|R|\,|V(H)|\sqrt{\log |V(H)|})$ as desired. If G is planar then the number of edges in $\hat{G}[F'' \cup R']$ is at most $2\,(|F''| + |R'|) \leq 6\,|R'|$ because $\hat{G}[F'' \cup R']$ is a simple planar bipartite graph [30].

Corollary 1. *Node-weighted Steiner tree, Steiner forest, T-join, point-to-point communication, exact tree/cycle/path partitioning problems, lower capacitated partitioning, and location-design and location-routing problems [12] have polynomial-time $O(1)$-approximation algorithms for any family of graphs excluding a fixed minor.*

References

1. Abraham, I., Bartal, Y., Neiman, O.: Nearly tight low stretch spanning trees. In: Proceedings of the 49th Annual Symposium on Foundations of Computer Science, pp. 781–790 (2008)

2. Agrawal, A., Klein, P., Ravi, R.: When trees collide: an approximation algorithm for the generalized Steiner problem on networks. In: STOC, pp. 134–144 (1991)
3. Andrews, M.: Hardness of buy-at-bulk network design. In: Proceedings of the 45th Symposium on Foundations of Computer Science, pp. 115–124 (2004)
4. Anshelevich, E., Dasgupta, A., Kleinberg, J., Tardos, E., Wexler, T., Roughgarden, T.: The price of stability for network design with fair cost allocation. In: Proceedings of the 45th Annual IEEE Symposium on Foundations of Computer Science, pp. 295–304 (2004)
5. Arkin, E.M., Hassin, R.: Approximation algorithms for the geometric covering salesman problem. Discrete Applied Mathematics 55(3), 197–218 (1994)
6. Bartal, Y.: On approximating arbitrary metrices by tree metrics. In: Proceedings of the 30th Annual ACM Symposium on Theory of Computing, pp. 161–168 (1998)
7. Bienstock, D., Goemans, M.X., Simchi-Levi, D., Williamson, D.: A note on the prize collecting traveling salesman problem. Math. Programming 59(3, ser. A), 413–420 (1993)
8. Chekuri, C., Khanna, S., Shepherd, F.B.: Multicommodity flow, well-linked terminals, and routing problems. In: Proceedings of the 37th Annual ACM Symposium on Theory of Computing, pp. 183–192 (2005)
9. Feige, U.: A threshold of $\ln n$ for approximating set cover. Journal of the ACM 45(4), 634–652 (1998)
10. Feige, U., Hajiaghayi, M., Lee, J.R.: Improved approximation algorithms for minimum-weight vertex separators. In: Proceedings of the 37th Annual ACM Symposium on Theory of Computing, pp. 563–572 (2005)
11. Garg, N., Konjevod, G., Ravi, R.: A polylogarithmic approximation algorithm for the group steiner tree problem. Journal of Algorithms 37(1), 66–84 (2000)
12. Goemans, M.X., Williamson, D.P.: A general approximation technique for constrained forest problems. SIAM J. Comput. 24(2), 296–317 (1995)
13. Goemans, M.X., Williamson, D.P.: The primal-dual method for approximation algorithms and its application to network design problems. In: Hochbaum, D.S. (ed.) Approximation Algorithms for NP-hard Problems, ch. 4, pp. 144–191. PWS, Boston (1997)
14. Gudmundsson, J., Levcopoulos, C.: A fast approximation algorithm for TSP with neighborhoods. Nordic Journal of Computing 6(4), 469–488 (Winter 1999)
15. Guha, S., Moss, A., Naor, J., Schieber, B.: Efficient recovery from power outage. In: STOC, pp. 574–582 (1999)
16. Hajiaghayi, M.T., Jain, K.: The prize-collecting generalized steiner tree problem via a new approach of primal-dual schema. In: Proceedings of the 17th Annual ACM-SIAM Symposium on Discrete Algorithm, pp. 631–640 (2006)
17. Hajiaghayi, M.T., Kleinberg, R.D., Leighton, T., Raecke, H.: Oblivious routing on node-capacitated and directed graphs. In: Proceedings of the 16th Annual ACM-SIAM Symposium on Discrete Algorithms, Philadelphia, pp. 782–790 (2005)
18. Halperin, E., Krauthgamer, R.: Polylogarithmic inapproximability. In: Proceedings of the the 35th Annual ACM Symposium on Theory of Computing, pp. 585–594 (2003)
19. Karp, R.M.: Reducibility among combinatorial problems. In: Complexity of Computer Computations (Proc. Sympos., IBM Thomas J. Watson Res. Center, Yorktown Heights, NY, 1972), pp. 85–103. Plenum, New York (1972)
20. Klein, P., Ravi, R.: A nearly best-possible approximation algorithm for node-weighted Steiner trees. Journal of Algorithms 19(1), 104–115 (1995)
21. Kostochka, A.V.: Lower bound of the Hadwiger number of graphs by their average degree. Combinatorica 4(4), 307–316 (1984)

22. Marathe, M.V., Ravi, R., Sundaram, R., Ravi, S.S., Rosenkrantz, D.J., Hunt III, H.B.: Bicriteria network design problems. J. Algorithms 28(1), 142–171 (1998)
23. Mata, C.S., Mitchell, J.S.B.: Approximation algorithms for geometric tour and network design problems (extended abstract). In: Proceedings of the 11th Annual Symposium on Computational Geometry, Vancouver, Canada, pp. 360–369 (1995)
24. Mitchell, J.S.B.: A PTAS for TSP with neighborhoods among fat regions in the plane. In: Proceedings of the 18th Annual ACM-SIAM Symposium on Discrete Algorithms, pp. 11–18 (2007)
25. Moss, A., Rabani, Y.: Approximation algorithms for constrained node weighted steiner tree problems. SIAM Journal on Computing 37(2), 460–481 (2007)
26. Raz, R., Safra, S.: A sub-constant error-probability low-degree test, and a sub-constant error-probability PCP characterization of NP. In: Proceedings of the 29th Annual ACM Symposium on Theory of Computing, pp. 475–484 (1997)
27. Reich, G., Widmayer, P.: Beyond steiner's problem: a VLSI oriented generalization. In: Proceedings of the 15th International Workshop on Graph-Theoretic Concepts in Computer Science, pp. 196–210 (1990)
28. Safra, S., Schwartz, O.: On the complexity of approximating TSP with neighborhoods and related problems. Computational Complexity 14(4), 281–307 (2006)
29. Thomason, A.: The extremal function for complete minors. Journal of Combinatorial Theory, Series B 81(2), 318–338 (2001)
30. West, D.B.: Introduction to Graph Theory. Prentice Hall Inc., Upper Saddle River (1996)

A Noncrossing-Flow Directed Steiner Tree

Our relaxation is related to a standard linear-program relaxation of directed Steiner tree, one based on min-cost flows. For each terminal $t \in T$ and each arc e, there is a variable $f_t(e)$. For each arc e, there is a variable $c(e)$. The linear program is as follows:

$$\begin{aligned} &\text{minimize} \sum_{e \in E} c(e) \max_t f_t(e) \\ &\text{subject to} \sum_{w:vw \in E} f_t(vw) - \sum_{u:uv \in E} f_t(uv) = \begin{cases} 1 \text{ if } v = t \\ -1 \text{ if } v = s \\ 0 \text{ otherwise} \end{cases} && \text{for all } t \in T, v \in V \\ &f_t(e) \geq 0 && \text{for all } t \in T, e \in E, \end{aligned} \tag{4}$$

(The inner max in the objective function can be removed using auxiliary variables, one for each arc.) We denote an assignment to all the variables $f_t(e)$ by f, and we denote by $c(f)$ the corresponding value of the objective function.

Consider the case where G is a planar embedded graph. We say that two paths P and Q in G *cross* if P enters Q on the left, shares zero or more edges with Q, and then exits Q on the right, or vice versa. For a terminal t, a *flow path for* t is a path consisting of arcs e such that $f_t(e) > 0$. We say that an assignment to the variables $f_t(e)$ of the linear program is *noncrossing* if, for every pair t, t' of distinct terminals, every flow path p for t, and every flow path q for t', p and

q do not cross. The *noncrossing-flow formulation* of directed Steiner tree refers to the linear program augmented with the (nonlinear) constraint that the flow assignment is noncrossing.

Any minimal solution to directed Steiner tree is a directed tree, so flow paths in the solution do not cross. It follows that the noncrossing-flow formulation is a relaxation for directed Steiner tree in a planar graph. In particular, the optimum of that noncrossing-flow formulation is a lower bound on the minimum cost of a directed Steiner problem. Theorem 6 provides a converse.

On Cartesian Trees and Range Minimum Queries

Erik D. Demaine[1,*], Gad M. Landau[2,3,**], and Oren Weimann[1,*]

[1] MIT Computer Science and Artificial Intelligence Laboratory, Cambridge, MA, USA
{edemaine,oweimann}@mit.edu
[2] Department of Computer Science, University of Haifa, Haifa, Israeal
[3] Department of Computer and Information Science, Polytechnic Institute of NYU
landau@cs.haifa.ac.il

Abstract. We present new results on Cartesian trees with applications in range minimum queries and bottleneck edge queries. We introduce a cache-oblivious Cartesian tree for solving the range minimum query problem, a Cartesian tree of a tree for the bottleneck edge query problem on trees and undirected graphs, and a proof that no Cartesian tree exists for the two-dimensional version of the range minimum query problem.

1 Introduction

In the *Range Minimum Query* (RMQ) problem, we wish to preprocess an array A of n numbers for subsequent queries asking for $\min\{A[i], \ldots, A[j]\}$. In the two-dimensional version of RMQ, an $n \times n$ matrix is preprocessed and the queries ask for the minimum element in a given rectangle. The *Bottleneck Edge Query* (BEQ) problem further generalizes RMQ to graphs. In this problem, we preprocess a graph for subsequent queries asking for the maximum amount of flow that can be routed between some vertices u and v along any single path. The capacity of a path is captured by its edge with minimal capacity (weight). Thus, RMQ can be seen as a special case of BEQ on line-like graphs.

In all solutions to the RMQ problem the *Cartesian tree* plays a central role. Given an array A of n numbers its Cartesian tree is defined as follows: The root of the Cartesian tree is $A[i] = \min\{A[1], \ldots, A[n]\}$, its left subtree is computed recursively on $A[1], \ldots, A[i-1]$ and its right subtree on $A[i+1], \ldots, A[n]$. In this paper we present new results on Cartesian trees with applications in RMQ and BEQ. We introduce a cache-oblivious version of the Cartesian tree that leads to an optimal cache-oblivious RMQ solution. We then give a natural generalization of the Cartesian tree from arrays to trees and show how it can be used for solving BEQ on undirected graphs and on trees. Finally, we show that there is no two-dimensional generalization of a Cartesian tree.

* Supported in part by the Center for Massive Data Algorithmics (MADALGO), a center of the Danish National Research Foundation.

** Supported in part by the Israel Science Foundation grant 35/05, the Israel-Korea Scientific Research Cooperation and Yahoo.

S. Albers et al. (Eds.): ICALP 2009, Part I, LNCS 5555, pp. 341–353, 2009.

Range Minimum Queries and Lowest Common Ancestors. The traditional RMQ solutions rely on the tight connection between RMQ and the *Lowest Common Ancestor* (LCA) problem. In LCA, we wish to preprocess a rooted tree T for subsequent queries asking for the common ancestor of two nodes that is located farthest from the root. RMQ and LCA were shown by Gabow*et al.* [18] to be equivalent in the sense that either one can be reduced in linear time to the other. An LCA instance can be obtained from an RMQ instance on an array A by letting T be the Cartesian tree of A that can be constructed in linear time [18]. It is easy to see that RMQ(i, j) in A translates to LCA$(A[i], A[j])$ in A's Cartesian tree. In the other direction, an RMQ instance A can be obtained from an LCA instance on a tree T by writing down the *depths* of the nodes visited during an *Euler tour* of T. That is, A is obtained by listing all first and last node-visitations in a DFS traversal of T that starts from the root. The LCA of two nodes translates to a range minimum query between the first occurrences of these nodes in A. An important property of the array A is that the difference between any two adjacent cells is ± 1.

As the classical RMQ and LCA problems are equally hard, it is tempting to think that this is also the case in the cache-oblivious model, where the complexity is measured in terms of the number of memory-blocks transferred between cache and disk (see Section 2 for a description of the cache-oblivious model). Indeed, to solve RMQ on array A, one could solve LCA on the Cartesian tree of A that can be constructed optimally using $O(\frac{n}{B})$ memory transfers (where B is the block-transfer size). An optimal $O(\frac{n}{B})$ cache-oblivious LCA solution follows directly from [9] after obtaining the Euler tour of the Cartesian tree. However, the best cache-oblivious algorithm for obtaining the Euler tour requires $\Theta(\frac{n}{B} \log_{M/B} \frac{n}{B})$ memory transfers [6] where M is the cache-size. The first result of our paper is an optimal cache-oblivious RMQ data structure that requires $O(\frac{n}{B})$ memory transfers and gives constant-time queries. This makes RMQ cache-obliviously easier than LCA (unless the LCA instance is given in Euler tour form).

Bottleneck Edge Queries on Graphs and Trees. Given an edge weighted graph, the *bottleneck edge* e between a pair of vertices s, t is defined as follows: If P is the set of all simple paths from s to t then e's weight is given by $\max_{p \in P}$(lightest edge in p). In the BEQ problem, we wish to preprocess a graph for subsequent bottleneck edge queries. Hu [20] proved that in undirected graphs bottleneck edges can be obtained by considering only the unique paths in a maximum spanning tree. This means that BEQ on undirected graphs can be solved by BEQ on trees. Another reason for the importance of BEQ on trees is the equivalent *online minimum spanning tree verification* problem. Given a spanning tree T of some edge weighted graph G, the problem is to preprocess T for queries verifying if an edge $e \in G - T$ can replace some edge in T and decrease T's weight. It is easy to see that this is equivalent to a bottleneck edge query between e's endpoints.

The second result in our paper is a natural generalization of the Cartesian tree from arrays to trees. We show that a Cartesian tree of a tree can be constructed in linear time plus the time required to sort the edges-weights. It can then be used to answer bottleneck edge queries in constant time for trees and undirected graphs. In the full version of this paper, we show how to maintain this Cartesian tree and constant-time bottleneck edge queries while leaf insertions and deletions are performed on the input tree. Insertions and deletions require $O(\log n)$ amortized time and $O(\log\log u)$ amortized time for integral edge weights bounded by u.

Two-dimensional RMQ. In the two-dimensional version of RMQ, we wish to preprocess an $n \times n$ matrix for subsequent queries asking for the minimum element in a given rectangle. Amir, Fischer, and Lewenstein [5] conjectured that it should be possible to show that in two dimensions there is no such nice relation as the one between RMQs and Cartesian Trees in the one-dimensional case. Our third result proves this conjecture to be true by proving that the number of *different* RMQ matrices is roughly $(n^2)!$, where two RMQ matrices are different if their range minimum is in different locations for some rectangular range.

1.1 Relation to Previous Work

Practically all known solutions to the range minimum query (RMQ) problem, its two-dimensional version (2D-RMQ), and its bottleneck edge version (BEQ) on trees share the same high-level description. They all partition the problem into some n/s smaller subproblems of size s each. From each of the small subproblems, one representative (the minimal element of the subproblem) is chosen and the problem is solved recursively on the n/s representatives. A similar recursion is applied on each one of the small subproblems. Besides different choices of s, the main difference between these solutions is the recursion's halting conditions that can take one of the following forms.

(I) *keep recursing* - the recursive procedure is applied until the subproblem is of constant size.
(II) *handle at query* - for small enough s do nothing and handle during query-time.
(III) *sort* - for small enough s use a linear-space $O(s \log s)$-time solution with constant query time.
(IV) *table-lookup* - for roughly logarithmic size s construct a lookup-table for all possible subproblems.
(V) *table-lookup & handle at query* - for roughly logarithmic size s construct a lookup-table. A query to this table will return a fixed number of candidates to be compared during query-time.

Table 1 describes how our solutions (in bold) and the existing solutions relate to the above four options. We next describe the existing solutions in detail.

RMQ and LCA. As we mentioned before, RMQ can be solved by solving LCA. Harel and Tarjan [19] were the first to show that the LCA problem can be optimally solved with linear-time preprocessing and constant-time queries by relying on word-level parallelism. Their data structure was later simplified by Schieber and Vishkin [28] but remained rather complicated and impractical. Berkman and Vishkin [10], and then Bender *et al.* [9] presented further simplifications and removed the need for word-level parallelism by actually reducing the LCA problem back to an RMQ problem. This time, the RMQ array has the property

Table 1. Our results (in bold) and their relation to the existing solutions. We denote a solution by ⟨*preprocessing time, space, query time*⟩ and by ⟨*memory transfers in preprocessing, memory transfers in query*⟩ for a cache-oblivious solution.

	RMQ	BEQ on trees	2D-RMQ
keep recursing		$\langle n\alpha_k(n), n\alpha_k(n), k\rangle$ [3]	$\langle n^2\alpha_k(n)^2, n^2\alpha_k(n)^2, k\rangle$ [13]
handle at query		$\langle n, n, \alpha(n)\rangle$ [3]	$\langle n^2, n^2, \alpha^2(n)\rangle$ [13]
sort		$\boldsymbol{\langle n\log^{[k]} n, n, k\rangle}$	$\langle n^2\log^{[k]} n, n^2, k\rangle$ [5]
table-lookup	$\langle n, n, 1\rangle$ [19] $\boldsymbol{\langle n/B, 1\rangle}$	impossible [24]	**impossible**
table-lookup & handle at query		impossible [24]	$\langle n^2, n^2, 1\rangle$ [7]

that any two adjacent cells differ by ± 1. This property was used in [9,10,16] to enable table lookups as we now explain.

It is not hard to see that two arrays that admit the same ± 1 vector have the same location of minimum element for every possible range. Therefore, it is possible to compute a lookup-table P storing the answers to all range minimum queries of all possible ± 1 vectors of length $s = \frac{1}{2}\log n$. Since there are $O(s^2)$ possible queries, the size of P is $O(s^2 \cdot 2^s) = o(n)$. Fischer and Heun [16] recently presented the first optimal RMQ solution that makes no use of LCA algorithms or the ± 1 property. Their solution uses the Cartesian tree but in a different manner. It uses the fact that the number of *different*[1] RMQ arrays is equal to the number of possible Cartesian trees and thus to the Catalan number which is $\frac{1}{s+1}\binom{2s}{s} = O(\frac{4^s}{s^{1.5}})$. This means that if we pick $s = \frac{1}{4}\log n$ then again the lookup-table P requires only $O(s^2 \cdot \frac{4^s}{s^{1.5}}) = O(n)$ space.

BEQ. On directed edge weighted graphs, the BEQ problem has been studied in its offline version, where we need to determine the bottleneck edge for every pair of vertices. Pollack [27] introduced the problem and showed how to solve it in $O(n^3)$ time. Vassilevska, Williams, and Yuster [31] gave an $O(n^{2+\omega/3})$-time algorithm, where ω is the exponent of matrix multiplication over a ring. This was recently improved by Duan and Pettie [14] to $O(n^{(3+\omega)/2})$. For the case of vertex weighted graphs, Shapira *et al.* [30] gave an $O(n^{2.575})$-time algorithm.

On trees, the BEQ problem was studied when the tree T is a spanning tree of some graph and a bottleneck edge query verifies if an edge $e \notin T$ can replace some edge in T and decrease T's weight. A celebrated result of Komlós [23] is a linear-time algorithm that verifies all edges in G. The idea of progressively improving an approximately minimum solution T is the basis of all recent minimum spanning tree algorithms [12,21,25,26]. Alon and Schieber [3] show that after an almost linear $O(n \cdot \alpha_k(n))$ preprocessing time and space bottleneck edge queries can be done in constant time for any fixed k. Here, $\alpha_k(n)$ is the inverse of the kth

[1] Two RMQ arrays are different if their range minimum is in different locations for some range.

row of Ackerman's function[2]: $\alpha_k(n) = 1 + \alpha_k(\alpha_{k-1}(n))$ so that $\alpha_1(n) = n/2$, $\alpha_2(n) = \log n$, $\alpha_3(n) = \log^* n$, $\alpha_4(n) = \log^{**} n$ and so on[3]. Pettie [24] gave a tight lower bound of $\Omega(n \cdot \alpha_k(n))$ preprocessing time required for $O(k)$ query time. Alon and Schieber further prove that if optimal $O(n)$ preprocessing space is required, it can be done with $\alpha(n)$ query time.

If the edge-weights are already sorted, our solution is better than the one of [3]. For arbitrary unsorted edge-weights, we can improve [3] in terms of space complexity. Namely, we give a linear space and constant query time solution that requires $O(n \log^{[k]} n)$ preprocessing time for any fixed k, where $\log^{[k]} n$ denotes the iterated application of k logarithms.

2D-RMQ. For the two-dimensional version of RMQ on an $n \times n$ matrix, Gabow, Bentley and Tarjan [18] suggested an $O(n^2 \log n)$ preprocessing time and space and $O(\log n)$ query time solution. This was improved by Chazelle and Rosenberg [13] to $O(n^2 \cdot \alpha_k(n)^2)$ preprocessing time and space and $O(1)$ query time for any fixed k. Chazelle and Rosenberg further prove that if optimal $O(n^2)$ preprocessing space is required, it can be done with $\alpha^2(n)$ query time. Amir, Fischer, and Lewenstein [5] showed that $O(n^2)$ space and constant query time can be obtained by allowing $O(n^2 \log^{[k]} n)$ preprocessing time for any fixed k.

Amir, Fischer, and Lewenstein also conjectured that in two dimensions there is no such nice relation as the one between the number of different RMQs and the number of different Cartesian Trees in the one-dimensional case. We prove this conjecture to be true thereby showing that $O(n^2)$ preprocessing time and constant query time can not be achieved using the existing methods for one-dimensional RMQ. Indeed, shortly after we proved this, Atallah and Yuan (in a yet unpublished result [7]) discovered a new optimal RMQ solution that does not use Cartesian trees and extends to two dimensions.

2 A Cache-Oblivious Cartesian Tree

While modern memory systems consist of several levels of cache, main memory, and disk, the traditional RAM model of computation assumes a flat memory with uniform access time. The I/O-model, developed by Aggarwal and Vitter [2], is a two-level memory model designed to account for the large difference in the access times of cache and disks. In this model, the disk is partitioned into blocks of B elements each, and accessing one element on disk copies its entire block to cache. The cache can store up to M/B blocks, for a total size of M. The efficiency of an algorithm is captured by the number of block transfers it makes between the disk and cache.

The cache-oblivious model, introduced by Frigo *et al.* [17], extends the I/O-model to a multi-level memory model by a simple measure: the algorithm is not

---

[2] We follow Seidel [29]. The function $\alpha(\cdot)$ is usually defined slightly differently, but all variants are equivalent up to an additive constant.

[3] $\log^{**} n$ is the number of times $\log^*$ function is applied to n to produce a constant, $\alpha_k(n) = \log^{\overbrace{**\cdots*}^{k \text{ times}}} n$ and the inverse Ackerman function $\alpha(n)$ is the smallest k such that $\alpha_k(n)$ is a constant.

allowed to know the value of B and M. More precisely, a cache-oblivious algorithm is an algorithm formulated in the standard RAM model, but analyzed in the I/O-model, with an analysis valid for any value of B and M and between any two adjacent memory-levels. When the cache is full, a cache-oblivious algorithm can assume that the *ideal* block in cache is selected for replacement based on the future characteristics of the algorithm, that is, an *optimal* offline paging strategy is assumed. This assumption is fair as most memory systems (such as LRU and FIFO) approximate the omniscient strategy within a constant factor. See [17] for full details on the cache-oblivious model.

It is easy to see that the number of memory transfers needed to read or write n contiguous elements from disk is $scan(n) = \Theta(\frac{n}{B})$, even if B is unknown. A *stack* that is implemented using a doubling array can support n push\pop operations in $scan(n)$ memory transfers. This is because the optimal paging strategy can always keep the last block of the array (accessed by both push and pop) in cache. Other optimal cache-oblivious data structures have been recently proposed, including priority queues [6], B-trees [8], string dictionaries [11], kd-trees and Range trees [1]. An important result of Frigo *et al.* shows that the number of memory transfers needed to sort n elements is $sort(n) = \Theta(\frac{n}{B} \log_{M/B} \frac{n}{B})$.

For the RMQ problem on array A, the Cartesian tree of A can be constructed optimally using $scan(n)$ memory transfers by implementing the following construction of [18]. Let C_i be the Cartesian tree of $A[1, \ldots, i]$. To build C_{i+1}, we notice that node $A[i+1]$ will belong to the rightmost path of C_{i+1}, so we climb up the rightmost path of C_i until we find the position where $A[i+1]$ belongs. It is easy to see that every "climbed" node will be removed from the rightmost path and so the total time complexity is $O(n)$. A cache-oblivious stack can therefore maintain the current rightmost path and the construction outputs the nodes of the Cartesian tree in postorder. However, in order to use an LCA data structure on the Cartesian tree we need an Euler tour order and not a postorder. The most efficient way to obtain an Euler tour [6] requires $sort(n)$ memory transfers. Therefore $sort(n)$ was until now the upper bound for cache-oblivious RMQ. In this section we prove the following result.

Theorem 1. *An optimal RMQ data structure with constant query-time can be constructed using $scan(n)$ memory transfers.*

We start by showing a simple constant-time RMQ data structure [9] that can be constructed using $scan(n) \log n$ memory transfers. The idea is to precompute the answers to all range minimum queries whose length is a power of two. Then, to answer $RMQ(i, j)$ we can find (in constant time) two such overlapping ranges that exactly cover the interval $[i, j]$, and return the minimum between them. We therefore wish to construct arrays $M_0, M_1, \ldots, M_{\log n}$ where $M_j[i] = \min\{A[i], \ldots, A[i + 2^j - 1]\}$ for every $i = 1, 2, \ldots, n$. M_0 is simply A. For $j > 0$, we construct M_j by doing two parallel scans of M_{j-1} using $scan(n)$ memory transfers (we assume $M \geq 2B$). The first scan starts at $M_{j-1}[1]$ and the second at $M_{j-1}[1 + 2^{j-1}]$. During the parallel scan we set $M_j[i] = \min\{M_{j-1}[i], M_{j-1}[i + 2^{j-1}]\}$ for every $i = 1, 2, \ldots, n$.

After describing this $scan(n) \log n$ solution, we can now describe the $scan(n)$ solution. Consider the partition of A into disjoint intervals (blocks) of $s = \frac{1}{4} \log n$ consecutive elements. The *representative* of every block is the minimal element in this block. Clearly, using $scan(n)$ memory transfers we can compute an array of the n/s representatives. We use the RMQ data structure above on the representatives array. This data structure is constructed with $scan(n/s)\log(n/s) = scan(n)$ memory transfers and is used to handle queries whose range spans more than one block. The additional in-block prefix and suffix of such queries can be accounted for by pre-computing $RMQ(i,j)$ of every block prefix or suffix. This again can easily be done using $scan(n)$ memory transfers.

We are therefore left only with the problem of answering queries whose range is entirely inside one block. Recall that two RMQ arrays are *different* if their range minima are in different locations for some range. Fischer and Heun [16] observed that the number of different blocks is equal to the number of possible Cartesian trees of s elements and thus to the s'th Catalan number which is $o(4^s)$. For each such unique block type, the number of possible in-block ranges $[i,j]$ is $O(s^2)$. We can therefore construct an $s^2 \times 4^s$ lookup-table P of size $O(s^2 \cdot 4^s) = o(n)$ that stores the locations of all range minimum queries for all possible blocks.

It remains to show how to index table P (i.e. how to identify the type of every block in the partition of A) and how to construct P using $scan(n)$ memory transfers. We begin with the former. The most naive way to calculate the block types would be to actually construct the Cartesian tree of each block in A, and then use an inverse enumeration of binary trees [22] to compute its type. This approach however can not be implemented via scans. Instead, consider the Cartesian tree *signature* of a block as the sequence $\ell_1\ell_2\cdots\ell_s$ where $0 \leq \ell_i < s$ is the number of nodes removed from the rightmost path of the block's Cartesian tree when inserting the i'th element. For example, the block "3421" has signature "0021".

Notice that for every signature $\ell_1\ell_2\cdots\ell_s$ we have $\sum_{k=1}^{i}\ell_k < i$ for every $1 \leq i \leq s$. This is because one cannot remove more elements from the rightmost path than one has inserted before. Fischer and Heun used this property to identify each signature by a special sum of the so-called *Ballot Numbers* [22]. We suggest a simpler and cache-oblivious way of computing a unique number $f(\ell_1\ell_2\cdots\ell_s) \in \{0,1,\ldots,4^s-1\}$ for every signature $\ell_1\ell_2\cdots\ell_s$. The binary representation of this number is simply

$$\overbrace{11\cdots1}^{\ell_1}0\overbrace{11\cdots1}^{\ell_2}0\cdots0\overbrace{11\cdots1}^{\ell_s}0.$$

Clearly, each signature is assigned a different number and since $\sum_{i=1}^{s}\ell_i < s$ this number is between 0 and $2^{2s}-1$ as its binary representation is of length at most $2s$. Notice that some binary strings of length at most $2s$ (for example, strings starting with 1 or with 011) are not really an $f(\ell_1\ell_2\cdots\ell_s)$ of a valid signature $\ell_1\ell_2\cdots\ell_s$ (i.e. f is not surjective). Using a stack, in one scan of A we can compute the signatures of all blocks in the partition of A in the order they appear. We refer to the sequence of signatures as $S(A)$, this sequence has n/s signatures each of length s. In a single scan of $S(A)$ we can compute $f(\ell_1\ell_2\cdots\ell_s)$ for all signatures in $S(A)$ thus solving our problem of indexing P.

We are left only with showing how to construct P using $scan(n)$ memory transfers. In the non cache-oblivious world (that Fischer and Heun consider) this is easy. We really only need to compute the entries in P that correspond to blocks that actually appear in the partition of A. During a scan of $S(A)$, for each signature $\ell_1\ell_2\cdots\ell_s$ we check if column $f(\ell_1\ell_2\cdots\ell_s)$ in P was already computed (this requires n/s such checks). If not, we can compute all $O(s^2)$ range minima of the block trivially in $O(s)$ time per range. In the cache-oblivious model however, each of the n/s checks might bring a new block to cache. If $s < B$ then this incurs more than $scan(n)$ memory transfers.

Therefore, for an optimal cache-oblivious performance, we must construct the entire table P and not only the columns that correspond to signatures in $S(A)$. Instead of computing P's entries for all possible RMQ arrays of length s we compute P's entries for all possible binary strings of length $2s$. Consider an RMQ block A' of length s with signature $\ell_1\ell_2\cdots\ell_s$. In Fig. 1 we give a simple procedure that computes $RMQ(i,j)$ in A' for any $1 \le i \le j \le s$ by a single scan of $f' = f(\ell_1\ell_2\cdots\ell_s)$. Again, we note that for binary strings f' that are not really an $f(\ell_1\ell_2\cdots\ell_s)$ of a valid signature $\ell_1\ell_2\cdots\ell_s$ this procedure computes "garbage" that will never be queried.

```
initialize min ← i and x ← 0
scan f' until the ith 0
for j' = i + 1, ..., j
    continue scanning f' until the next 0
    set x ← x + 1− the number of 1's read between the last two 0's
    if x ≤ 0 set min ← j' and x ← 0
return min as the location of RMQ(i, j)
```

Fig. 1. Pseudocode for computing $RMQ(i,j)$ for some $1 \le i \le j \le s$ using one scan of the binary string $f' = f(\ell_1\ell_2\cdots\ell_s)$ of some signature $\ell_1\ell_2\cdots\ell_s$

In order to use the procedure of Fig. 1 on all possible signatures, we construct a sequence S of all binary strings of length $2s$ in lexicographic order. S is the concatenation of 4^s substrings each of length $2s$, and S can be written using $scan(2s \cdot 4^s) = o(scan(n))$ memory transfers. For correctness of the above procedure, a parallel scan of S can be used to apply the procedure only on those substrings that have exactly s 0's. We can thus compute each row of P using $scan(2s \cdot 4^s)$ memory transfers and the entire table P using $s^2 \cdot scan(2s \cdot 4^s) = o(scan(n))$ memory transfers[4]. This concludes the description of our cache-oblivious RMQ data structure that can be constructed using a constant number of scans.

3 A Cartesian Tree of a Tree

In this section we address bottleneck edge queries on trees. We introduce the Cartesian tree of a tree and show how to construct it in $O(n)$ time plus the

[4] Since the (unknown) B can be greater than the length of S we don't really scan S for s^2 times. Instead, we scan once a sequence of s copies of S.

time required to sort the edge weights. Recall that an LCA data structure on the standard Cartesian tree can be constructed in linear time to answer range minimum queries in constant time. Similarly, an LCA data structure on our Cartesian tree can be constructed in linear time to answer bottleneck edge queries in constant time for trees.

Given an edge weighted input tree T, we define its Cartesian tree C as follows. The root r of C represents the edge $e = (u, v)$ of T with minimum weight (ties are resolved arbitrarily). The two children of r correspond to the two connected component of $T - e$: the left child is the recursively constructed Cartesian tree of the connected component containing u, and the right child is the recursive construction for v. Notice that C's internal nodes correspond to T's edges and C's leaves correspond to T's vertices.

Theorem 2. *The Cartesian tree of a weighted input tree with n edges can be constructed in $O(n)$ time plus the time required to sort the weights.*

Decremental connectivity in trees. The proof of Theorem 2 uses a data structure by Alstrup and Spork [4] for decremental connectivity in trees. This data structure maintains a forest subject to two operations: deleting an edge in $O(1)$ amortized time, and testing whether two vertices u and v are in the same connected component in $O(1)$ worst-case time.

The data structure is based on a *micro-macro decomposition* of the input tree T. The set of nodes of T is partitioned into disjoint subsets where each subset induces a connected component of T called the *micro tree*. The division is constructed such that each micro tree is of size $\Theta(\lg n)$ and at most two nodes in a micro tree (the *boundary nodes*) are incident with nodes in other micro trees. The nodes of the *macro tree* are exactly the boundary nodes and it contains an edge between two nodes iff T has a path between the two nodes which does not contain any other boundary nodes.

It is easy to see that deletions and connectivity queries can be performed by a constant number of deletions and connectivity queries on the micro and macro trees. The macro tree can afford to use a standard $O(\lg n)$ amortized solution [15], which explicitly relabels all nodes in the smaller of the two components resulting from a deletion. The micro trees use simple word-level parallelism to manipulate the logarithmic-size subtrees.

Although not explicitly stated in [4], the data structure can in fact maintain a canonical name (record) for each connected component, and can support finding the canonical name of the connected component containing a given vertex in $O(1)$ worst-case time. The macro structure explicitly maintains such names, and the existing tools in the micro structure can find the highest node (common ancestor) of the connected component of a vertex, which serves as a name. This slight modification enables us to store a constant amount of additional information with each connected component, and find that information given just a vertex in the component.

Proof of Theorem 2. We next describe the algorithm for constructing the Cartesian tree C of an input tree T. The algorithm essentially bounces around T, considering the edges in increasing weight order, and uses the decremental connectivity data structure on T to pick up where it left off in each component. Precisely:

1. initialize the decremental connectivity structure on T. Each connected component has two fields: "parent" and "side".
2. set the "parent" of the single connected component to null.
3. sort the edge weights.
4. for each edge $e = (u, v)$ in increasing order by weight:
   (a) make a vertex w in C corresponding to e, whose parent is the "parent" of e's connected component in the forest, and who is the left or right child of that parent according to the "side" of that component.
   (b) delete edge e from the forest.
   (c) find the connected component containing u and set its "parent" to w and its "side" to "left".
   (d) find the connected component containing v and set its "parent" to w and its "side" to "right".

After sorting the edge weights, this algorithm does $O(n)$ work plus the work spent for $O(n)$ operations in the decremental connectivity data structure, for a total of $O(n)$ time.

Optimality. It is not hard to see that sorting the edge weights is unavoidable when computing a Cartesian tree of a tree. Consider a tree T with a root and n children, where the ith child edge has weight $A[i]$. Then the Cartesian tree consists of a path, with weights equal to the array A in increasing order. Thus we obtain a linear-time reduction from sorting to computing a Cartesian tree. If you prefer to compute Cartesian trees only of bounded-degree trees, you can expand the root vertex into a path of n vertices, and put on every edge on the path a weight larger than $\max\{A[1], \ldots, A[n]\}$.

BEQ on Trees. For the BEQ problem on a tree T, if the edge weights are integers or are already sorted, our solution is optimal. We now show that for arbitrary unsorted edge-weights we can use our Cartesian tree to get an $O(n)$-space $O(1)$-query BEQ solution for trees that requires $O(n \lg^{[k]} n)$ preprocessing time for any fixed k (recall $\lg^{[k]} n$ denotes the iterated application of k logarithms). We present an $O(n \lg\lg n)$ preprocessing time algorithm, $O(n \lg^{[k]} n)$ is achieved by recursively applying our solution for k times.

Consider the micro-macro decomposition of T described above. Recall that each micro tree is of size $\Theta(\lg n)$. We can therefore sort the edges in each micro tree in $\Theta(\lg n \lg\lg n)$ time and construct the $\Theta(\frac{n}{\lg n})$ Cartesian trees of all micro trees in a total of $\Theta(n \lg\lg n)$ time. This allows us to solve BEQ within a micro tree in constant time. To handle BEQ between vertices in different micro trees, we construct the Cartesian tree of the macro tree. Recall that the macro tree contains $\Theta(\frac{n}{\lg n})$ nodes (all boundary nodes). The edges of the macro tree are of two types: edges between boundary nodes of different micro trees, and edges between boundary nodes of the same micro tree. For the former, we set their weights according to their weights in T. For the latter, we set the weight of an edge between two boundary nodes u, v of the same micro tree to be equal to

BEQ(u, v) in this micro tree. BEQ(u, v) is computed in constant time from the Cartesian tree of the appropriate micro tree. Thus, we can compute all edge weights of the macro tree and then sort them in $O(\frac{n}{\lg n} \lg \frac{n}{\lg n}) = O(n)$ time and construct the Cartesian tree of the macro tree.

To conclude, we notice that any bottleneck edge query on T can be solved by a constant number of bottleneck edge queries on the Cartesian tree of the macro tree and on two more Cartesian trees of micro trees.

In the full version of this paper, we show how to maintain the Cartesian tree along with its LCA data structure on a dynamic input tree.

Theorem 3. *We can maintain constant-time bottleneck edge queries on a tree while leaf insertions/deletions are performed in $O(\lg n)$ amortized time and in $O(\lg \lg u)$ amortized time when the edge-weights are integers bounded by u.*

4 Two-Dimensional RMQ

Apart from the new result of [7], the known solutions [9,10,16,19,28] to the standard one dimensional RMQ problem make use of the Cartesian tree. Whether as a tool for reducing the problem to LCA or in order to enable table-lookups for all different Cartesian trees. Amir, Fischer, and Lewenstein [5] conjectured that no Cartesian tree equivalent exists in the two-dimensional version of RMQ denoted 2D-RMQ. In the full version of this paper we prove this conjecture to be true.

Theorem 4. *The number of different[5] 2D-RMQ $n \times n$ matrices is $\Omega\left(\left(\frac{n}{4}!\right)^{n/4}\right)$.*

References

1. Agarwal, P.K., Arge, L., Danner, A., Holland-Minkley, B.: Cache-oblivious data structures for orthogonal range searching. In: Proceedings of the 19th annual ACM Symposium on Computational Geometry (SCG), pp. 237–245 (2003)
2. Aggarwal, A., Vitter, J.S.: The input/output complexity of sorting and related problems. Communications of the ACM 31(9), 1116–1127 (1988)
3. Alon, N., Schieber, B.: Optimal preprocessing for answering on-line product queries. Technical report, TR-71/87, Institute of Computer Science, Tel Aviv University (1987)
4. Alstrup, S., Spork, M.: Optimal on-line decremental connectivity in trees. Information Processing Letters 64(4), 161–164 (1997)
5. Amir, A., Fischer, J., Lewenstein, M.: Two-dimensional range minimum queries. In: Ma, B., Zhang, K. (eds.) CPM 2007. LNCS, vol. 4580, pp. 286–294. Springer, Heidelberg (2007)
6. Arge, L., Bender, M.A., Demaine, E.D., Holland-Minkley, B., Munro, J.I.: An optimal cache-oblivious priority queue and its application to graph algorithms. SIAM Journal on Computing 36(6), 1672–1695 (2007)

[5] Two matrices are different if their range minima are in different locations for some rectangular range.

7. Atallah, M.J., Yuan, H.: Data structures for range minimum queries in multidimensional arrays (manuscript, 2009)
8. Bender, M.A., Demaine, E.D., Farach-colton, M.: Cache-oblivious B-trees. SIAM Journal on Computing, 399–409 (2000)
9. Bender, M.A., Farach-Colton, M., Pemmasani, G., Skiena, S., Sumazin, P.: Lowest common ancestors in trees and directed acyclic graphs. Journal of Algorithms 57(2), 75–94 (2005)
10. Berkman, O., Vishkin, U.: Recursive star-tree parallel data structure. SIAM Journal on Computing 22(2), 221–242 (1993)
11. Brodal, G.S., Fagerberg, R.: Cache-oblivious string dictionaries. In: Proceedings of the 17th annual Symp. On Discrete Algorithms (SODA), pp. 581–590 (2006)
12. Chazelle, B.: A minimum spanning tree algorithm with inverse-ackermann type complexity. Journal of the ACM 47(6), 1028–1047 (2000)
13. Chazelle, B., Rosenberg, B.: Computing partial sums in multidimensional arrays. In: Proceedings of the 5th annual ACM Symposium on Computational Geometry (SCG), pp. 131–139 (1989)
14. Duan, R., Pettie, S.: Fast algorithms for (max,min)-matrix multiplication and bottleneck shortest paths. In: Proceedings of the 20th annual Symposium On Discrete Algorithms, SODA (2009)
15. Even, S., Shiloach, Y.: An on-line edge deletion problem. Journal of the ACM 28, 1–4 (1981)
16. Fischer, J., Heun, V.: Theoretical and practical improvements on the RMQ-problem, with applications to LCA and LCE. In: Lewenstein, M., Valiente, G. (eds.) CPM 2006. LNCS, vol. 4009, pp. 36–48. Springer, Heidelberg (2006)
17. Frigo, M., Leiserson, C.E., Prokop, H., Ramachandran, S.: Cache-oblivious algorithms. In: Proceedings of the 40th symposium on Foundations Of Computer Science (FOCS), pp. 285–298 (1999)
18. Gabow, H., Bentley, J.L., Tarjan, R.E.: Scaling and related techniques for geometry problems. In: Proceedings of the 16th annual ACM Symposium on Theory Of Computing (STOC), pp. 135–143 (1984)
19. Harel, D., Tarjan, R.E.: Fast algorithms for finding nearest common ancestors. SIAM Journal on Computing 13(2), 338–355 (1984)
20. Hu, T.C.: The maximum capacity route problem. Operations Research 9(6), 898–900 (1961)
21. Karger, D.R., Klein, P.N., Tarjan, R.E.: A randomized linear-time algorithm for finding minimum spanning trees. Journal of the ACM 42, 321–329 (1995)
22. Knuth, D.E.: The Art of Computer Programming Volume 4 Fascicle 4: Generating All Trees; History of Combinatorial Generation. Addison-Wesley, Reading (2006)
23. Komlós, J.: Linear verification for spanning trees. Combinatorica 5(1), 57–65 (1985)
24. Pettie, S.: An inverse-ackermann style lower bound for the online minimum spanning tree. In: Proceedings of the 43rd symposium on Foundations Of Computer Science (FOCS), pp. 155–163 (2002)
25. Pettie, S., Ramachandran, V.: Minimizing randomness in minimum spanning tree, parallel connectivity and set maxima algorithms. In: Proceedings of the 13th annual Symposium On Discrete Algorithms (SODA), pp. 713–722 (2002)
26. Pettie, S., Ramachandran, V.: An optimal minimum spanning tree algorithm. Journal of the ACM 49(1), 16–34 (2002)
27. Pollack, M.: The maximum capacity through a network. Operations Research 8(5), 733–736 (1960)

28. Schieber, B., Vishkin, U.: On finding lowest common ancestors: Simplification and parallelization. SIAM Journal on Computing 17, 1253–1262 (1988)
29. Seidel, R.: Understanding the inverse ackermann function. PDF presenttion, http://cgi.di.uoa.gr/~ewcg06/invited/Seidel.pdf
30. Shapira, A., Yuster, R., Zwick, U.: All-pairs bottleneck paths in vertex weighted graphs. In: Proceedings of the 18th annual Symposium On Discrete Algorithms (SODA), pp. 978–985 (2007)
31. Vassilevska, V., Williams, R., Yuster, R.: All-pairs bottleneck paths for general graphs in truly sub-cubic time. In: Proceedings of the 39th annual ACM Symposium on Theory Of Computing (STOC), pp. 585–589 (2007)

Applications of a Splitting Trick*

Martin Dietzfelbinger and Michael Rink

Technische Universität Ilmenau, 98693 Ilmenau, Germany
{martin.dietzfelbinger,michael.rink}@tu-ilmenau.de

Abstract. We study applications of a simple method for circumventing the "full randomness assumption" when building a hashing-based data structure for a set S of keys. The general approach is to "split" S into "pieces" S_i, by a splitting hash function. On a piece S_i, a method or data structure for generating full randomness is used that uses more space than $|S_i|$. Under certain circumstances, this data structure can be "shared" among the constructions for the pieces S_i, which leads to a tighter overall space bound. The method was introduced in the context of cuckoo hashing and its variants, but it seems to have wider applicability. To demonstrate its power and some subtleties, we study three new applications, improving previous constructions: (i) Space-efficient simulation of full randomness on S (following work by Pagh and Pagh (2003/08) and Dietzfelbinger and Woelfel (2003)); (ii) Construction of highly independent functions in the style of Siegel (1989/2004); (iii) One-probe schemes as in work by Buhrman, Miltersen, Radhakrishnan, and Venkatesh (2000/02) and Pagh and Pagh (2002).

1 Introduction

A hash function h, in the data structures setting, maps keys from a *universe* U to some range R. Normally, one would like such a hash function to satisfy certain randomness properties. (Often, but not always, it is sufficient to assume these properties on all keys from a set $S \subseteq U$, the keys that occur in an application.) In the data structures literature, there are two fundamentally different approaches to this situation.

"Full randomness assumption": *Assume* that $h(x)$, $x \in U$ (or $h(x), x \in S$) is fully random (uniform in R, independent), uses no (or marginal) space and has constant evaluation time.

"Universal hashing": Consider a set $\mathcal{H}$ of hash functions and choose h from $\mathcal{H}$ at random. Depending on this choice, at a certain cost in terms of time or space consumption, partial or full randomness is generated [1].

Let us take *cuckoo hashing* [2] as an example. This elegant implementation of a dynamic dictionary for a set $S \subseteq U$ of n keys using only slightly more than $2n$ memory cells offers constant lookup time and expected constant insertion time. The analysis requires that on each set of $c \log n$ keys in S the two hash functions used behave fully randomly, and that these hash functions can be evaluated in

* Research supported by DFG, Grant Di 412/10-1.

S. Albers et al. (Eds.): ICALP 2009, Part I, LNCS 5555, pp. 354–365, 2009.

constant time. One can now just assume this kind of randomness (and justify this by experience) or use universal hashing methods — and justify $(c \log n)$-wise independence by a sophisticated construction like in [3]. Generalizations of cuckoo hashing [4,5] explicitly require that the hash functions used are fully random on S and that new hash functions can be supplied if needed. A combination of the two approaches is suggested by a recent paper by Mitzenmacher and Vadhan [6], who showed that under certain circumstances the combination of relatively weak universal hash classes with key sets that have a certain partial randomness will result in a behavior close to full randomness.

A slightly different approach is the "split-and-share" trick originally observed by the first author, mentioned in [4] and sketched in [5], but present in principle in other works before, e. g., [7]. One uses a hash function h to split the set S of n "interesting" keys into subsets S_i, $0 \leq i < t$, for $t = n^{1-\delta}$, say. If h is chosen from a suitably strong universal class, which can be supplied easily, the pieces S_i can be made to have size close to n^{δ}, whp.[1] Then there is a simple randomized data structure D that (whp) provides hash functions that behave fully randomly on a single S_i, at the price of using space n^{α} for $\delta < \alpha < 1$. One runs cuckoo hashing (original [2] or generalized [4,5]) for each piece S_i separately, using these fully random hash functions. (It is very simple to provide even a whole sequence of independent hash functions in this manner.) The crux is that the data structure D can be "shared" among all S_i, and hence the extra space paid for getting full randomness on each S_i is $o(n)$ and hence negligible. Instead of the cuckoo hashing structure, we may use other data structures or combinatorial structures like expander graphs that are based on having fully random functions available. In [8] it was described how to use this idea for building space-efficient *static* hash tables, where the data structures can be checked against S. In the present paper we give three more involved applications, which improve existing data structures; the first two constructions are carried out completely *without knowledge of S* and still succeed whp.

1.1 Simulating Full Randomness

Only in 2003 [9,10], solutions to the following problem were provided. Construct a randomized data structure D with the following behavior: For each $S \subseteq U$, $|S| = n$, there is a $1/n^c$ probability of failure, but if failure does not occur, then D calculates a hash function $h\colon U \to R$ that is fully random on S. (R is some abelian group.) The space for D is $O(n \log |R|) + o(n) + O(\log \log |U|)$ bits.[2] In [11], the space bound was even reduced to $(1 + \sigma)n \log |R|$ for an arbitrary constant $\sigma > 0$, with evaluation time $O(1/\sigma^2)$. (Note that, as far as not stated otherwise, we measure time complexity in word operations and space complexity in bits.) For a thorough discussion of the relevance of simulating full randomness on a set S we refer to [11]. Using "split-and-share", we provide an alternative construction with the same functionality (an additional "cache friendly" variant will be given in the full version), following a totally new approach which, if one

[1] whp: with high probability, i.e., probability $1 - n^{-c}$ for some $c > 0$.

[2] In the following, we will ignore the summand $\log \log |U|$, which is relevant only for extremely long keys.

wishes to see it that way, leads to functions of a simpler structure than in [11] (especially, the use of Siegel's functions is avoided) and an exponentially faster evaluation time of $O(\log(1/\sigma))$.

Theorem 1. *For arbitrary* $\sigma \in (0,1)$ *and* $c > 0$ *there is a class* $\mathcal{H} \subseteq \{h \mid U \to R\}$ *of hash functions and a set* $\mathcal{B} \subseteq \mathcal{H}$ *such that for arbitrary* $S \subseteq U$*, with* $|S| = n$*, and* $h \in \mathcal{H}$ *chosen at random the following holds*: (i) $\Pr(h \in \mathcal{B}) = O(1/n^c)$; (ii) *under the condition* $h \notin \mathcal{B}$ *the function* h *behaves fully randomly on* S; (iii) *the space used to store* h *is asymptotically* $(1+\sigma)n \log |R|$ *bits*; (iv) *the evaluation time of* h *is* $O(\max\{c, \log(1/\sigma)\})$.

1.2 Siegel's Hash Functions without Graph Powers

Siegel's [12,3] important construction of 1989/2004 provides universal hash classes with a high (n^η) degree of independence on $U, |U| = n^r$, with functions that can be evaluated in constant time, and take space n^γ for some $\gamma \in (\eta, 1)$. The core of Siegel's construction $\mathcal{SI}$ is a bipartite graph $G_{\mathcal{SI}}$ (left side U, right side $[n^\gamma]$, left degree d, a constant) with certain expansion properties. If one insists on constant evaluation time, up to now no explicit constructions of such graphs are known; rather, the graph is supplied by the probabilistic method, and must be stored as part of the data structure. In order to avoid using space n^r for the neighborhood lists of the nodes in U, Siegel uses graph powers to "blow up" a small expander graph of description size $o(n)$ to obtain large ones. Unfortunately, this approach increases the left degree and hence the evaluation time drastically (although it remains constant as long as $|U|$ is polynomial in n), and requires that $\eta < 1/r$. Utilizing the "split-and-share" trick, to provide and share the randomness needed for the probabilistic expander graph construction, we show how n^η-wise independence on an arbitrary (unknown) set $S \subseteq U$ of size n can be achieved whp, for arbitrary $\eta < 1$, avoiding the graph power construction.

Theorem 2. *For arbitrary* $\eta \in (0,1)$ *and* $c > 0$ *there is a class* $\mathcal{H} \subseteq \{h \mid U \to R\}$ *of hash functions and a set* $\mathcal{B} \subseteq \mathcal{H}$ *such that for arbitrary* $S \subseteq U$*, with* $|S| = n$*, and* $h \in \mathcal{H}$ *chosen at random the following holds*: (i) $\Pr(h \in \mathcal{B}) = O(1/n^c)$; (ii) *under the condition* $h \notin \mathcal{B}$ *the function* h *is* n^η*-wise independent on* S; (iii) *the space used to store* h *is* $o(n \log |R|)$; (iv) *the evaluation time of* h *is constant. (No graph powering is used.)*

1.3 One-Probe Storage Schemes

Buhrman, Miltersen, Radhakrishnan, and Venkatesh [13] showed that there exist storage schemes that can solve the membership problem for a set $S \subseteq U$ by probing just one bit of a data structure in a lookup. The lookup procedure is randomized, and there is a two-sided error probability of ε for each lookup and each $x \in U$. The space needed is $s = O(\frac{n \log |U|}{\varepsilon^2})$ bits, which is close to the optimal lower bound of $\Omega(\frac{n \log |U|}{\varepsilon \log(1/\varepsilon)})$ shown in [13]. The core of the construction is a bipartite graph G_{BMRV}, with left side U and right side V, $|V| = s$, and left degree

$d = \Theta((\log |U|)/\varepsilon)$, that has certain expansion properties on subsets of U of size up to $2n$. A drawback of the BMRV construction is that it is nonuniform in that G_{BMRV} has to be assumed to be given for free and not be counted towards the space consumption. Uniform ("explicit") constructions with constant evaluation time need much more space. Ta-Shma [14] gave an extractor/condenser-based construction for such graphs, which makes it superfluous to store the graph, at the price of nonconstant evaluation time. The construction time in [13] is $O(|U| + \mathrm{poly}(s))$, in [14] it is $\mathrm{poly}(s)$. Östlin and Pagh [15] extended the BMRV construction (using essentially the same graph) to the problem of storing a function $f\colon S \to R$ and retrieving values by probing just one word of size $\log |R|$ in the range, i. e., a dictionary is realized. We use "split-and-share" to construct a data structure with a functionality close to the structures from [13,15], which makes it unnecessary to store the expander graph G_{BMRV} (at the cost of $o(n)$ extra random bits), and has constant evaluation time. The construction time is $\mathrm{poly}(s)$.

Theorem 3. *Let $|U| \leq n^r$ for a constant $r > 0$ and let n be sufficiently large. Then for every $\varepsilon \in (0, 1)$ there exists an* explicit *one-probe membership tester [dictionary] with size $s = O((n \log |U|)/\varepsilon^2)$ bits [words], construction time* $\mathrm{poly}(s)$ *and evaluation time $O(1)$, that returns the correct result with probability at least $1 - \varepsilon$, with probability at most ε a wrong result ["don't know"].*

The result can be extended to universes larger than $\mathrm{poly}(n)$.

2 Preliminaries

We introduce some notation and definitions and recall the high performance hash classes from [3,16] as well as the static membership tester from [13] and the static dictionary from [15]. Define $[i] = \{0, 1, 2, \ldots, i - 1\}$. We write log for the logarithm to the base 2 and ln for the logarithm to the base e. We omit $\lfloor\ \rfloor$ and $\lceil\ \rceil$ where they are not important for our argumentation. We will assume that the universe U has size n^r for some constant r. If this is not the case we can resort to a standard technique known as "collapsing the universe" [3, Appendix 1] for reducing the size of U, at the cost of extra but negligible components in space usage and error probability.

2.1 Random Hash Functions, κ-Wise Independence, Expanders

A hash function $h : U \to R$ determined by some random experiment is called *fully random* if the values $h(x), x \in U$, are independent and uniformly distributed in R. To circumvent the assumption of fully random hash functions one often utilizes the concept of *universal hashing* [1].

Definition 1. *A set $\mathcal{H}_r^\kappa \subseteq \{h \mid h : U \to [r]\}$ is called a* κ-wise independent *hash class if for each sequence $x_0, x_1, \ldots, x_{\kappa-1}$ of distinct elements from U and all $y_0, y_1, \ldots, y_{\kappa-1}$ from $[r]$ and h chosen uniformly at random from $\mathcal{H}_r^\kappa$ we have:*

$$\Pr\left(h(x_0) = y_0 \wedge \ldots \wedge h(x_{\kappa-1}) = y_{\kappa-1}\right) = r^{-\kappa} \ .$$

A hash function randomly chosen from a κ-wise independent hash class behaves fully randomly on each subset S of U of size at most κ. A simple (almost) κ-wise independent class is the set of polynomials over a field $\mathbb{F}_p$ of degree up to $\kappa - 1$, prime $p > |U|$, projected into $[r]$. The space needed to store such a function $h \in \mathcal{H}_r^\kappa$ is $O(\kappa \log |U|)$; the evaluation time is $O(\kappa)$.

Definition 2. *For integers $m, s, d, t \geq 1$ and real $\rho \geq 1$, a bipartite graph $G = (U, V, E)$ with node sets U, V and edge set E is an (m, s, d, t, ρ)-expander if:*

(i) $|U| = m$, $|V| = s$;
(ii) *every node $x \in U$ has exactly d distinct neighbors*;
(iii) $\forall T \subseteq U, |T| \leq t : |\Gamma(T)| \geq \rho \cdot |T|$, *for* $\Gamma(T) = \{y \in V \mid \exists x \in T : (x, y) \in E\}$.

The standard way to obtain an (m, s, d, t, ρ)-expander is the probabilistic method, where one assumes that for each node $x \in U$ a set of d neighbors is chosen uniformly at random. In known explicit constructions it is necessary to store the edges of the graph [3] or accept non-constant time for finding a neighbor [14]. In our constructions, the edges of the expander graphs are given by hash functions, rather than being stored explicitly. Using d hash functions $h_0, h_1, \ldots, h_{d-1}$ chosen uniformly at random from a t-wise independent hash class, with $h_k : U \to [s/d], k \in [d]$, one can define the edge set as $E = \left\{\left(x, k \cdot \frac{s}{d} + h_k(x)\right) \mid k \in [d]\right\}$ (thus avoiding collisions of the hash values of one node $x \in U$).

2.2 High Performance Hash Classes

In this paper we utilize two "high performance" hash classes, *viz.* class $\mathcal{R}$ from [16] and $\mathcal{SI}$ from [3], roughly described in the following.

Hash Class $\mathcal{R}$. A hash function h from $\mathcal{R}$ for range $[n^{1-\delta}], \delta \in (0, 1)$, is built as follows:

$$h(x) = (g(x) + V[f(x)]) \bmod n^{1-\delta}, \text{for } x \in U \ , \tag{1}$$

where g and f are chosen uniformly at random from $(\kappa + 1)$-wise independent classes $\mathcal{H}_{n^{1-\delta}}^{\kappa+1}$ and $\mathcal{H}_{n^\alpha}^{\kappa+1}$, resp., for $\alpha \in (\delta, 1)$, and V is a vector of size n^α filled with fully random elements from $[n^{1-\delta}]$. We will use $\mathcal{R}$ to *sharply split* a given key set $S, |S| = n$, into $n^{1-\delta}$ subsets $S_i, i \in [n^{1-\delta}]$, of size about n^δ via:

$$S_i = \{x \in S \mid h(x) = i\}, h \in \mathcal{R} \ . \tag{2}$$

The proof of the following lemma can be found in the full version of this paper:

Lemma 1 ([16, Theorem 4.6]). *Let $\lambda > 0$ be arbitrary. If h is chosen uniformly at random from $\mathcal{R}$, then:*

$$\Pr\left(\exists i : |S_i| \geq (1 + \lambda) n^\delta\right) = O(n^{1-(\alpha-\delta)\kappa} + n^{1-\delta} c^{n^\delta}), \text{ for } c = c(\kappa, \lambda) < 1 \ . \quad \square$$

If f and g are polynomials of degree κ and $|U| = n^r$, for constants κ and r, the description size for h is $O(n^\alpha \log n)$ and the evaluation time is $O(\kappa) = O(1)$.

Hash Class $\mathcal{SI}$. We outline Siegel's construction [3] of a κ-wise independent hash class $\mathcal{SI}$ for non-constant $\kappa = \kappa(n)$, but constant evaluation time. The central building block of $\mathcal{SI} \subseteq \{h \mid h : U \to R\}$ is a d-left-regular, bipartite graph $G_{\mathcal{SI}} = (U, [n^\gamma], E), \gamma \in (0,1)$. Each hash function h from $\mathcal{SI}$ uses $G_{\mathcal{SI}}$ and a vector V of size n^γ, which is filled with elements from a group $(R, \oplus)$; to choose h, one chooses $V \in R^{n^\gamma}$ at random. The hash function $h \in \mathcal{SI}$ is then defined as:

$$h(x) = \bigoplus_{y:(x,y)\in E} V[y], \text{ for } x \in U . \tag{3}$$

Siegel shows that to achieve κ-wise independence it is sufficient that for each $T \subseteq U$ with $|T| \leq \kappa$ the κ rows of the adjacency matrix of G corresponding to T contain a square submatrix that up to row/column permutations has triangular form. Siegel referred to this as *κ-locally peelable* and showed that if $G_{\mathcal{SI}}$ is an $(|U|, n^\gamma, d, \kappa, d/2 + \nu)$-expander, for some $\nu(n) > 0$, then it has this property [3, Lemma 3]. Since up to now no deterministic construction of such expanders (with constant degree) is known, the graph $G_{\mathcal{SI}}$ has to be chosen at random and then stored. In order to avoid storing a graph $G_{\mathcal{SI}}$ with left side as large as U, a smaller expander graph G_0 is chosen at random and stored instead. This graph makes it possible to compute the neighbors for all elements from U using a certain *graph product* [3, Definition 6]. This graph product operation $\otimes$ preserves the property of being κ-locally peelable [3, Lemma 5]; thus we get: If G_0 is an $(n^\delta, n^{\gamma \cdot \delta / r}, d, \kappa, d/2 + \nu)$-expander then the graph power $G_{\mathcal{SI}} = \bigotimes_{i=1}^{r/\delta} G_0 = ([n^r], [n^\gamma], E)$ is κ-locally peelable. If we use $G_{\mathcal{SI}}$ in (3), we obtain κ-wise independence and evaluation time $\Theta(d^{r/\delta})$.

2.3 One-Probe Schemes

The static one-probe schemes of Buhrman et al. [13] and Östlin and Pagh [15] rely on bipartite graphs with near optimal expansion. The one-probe scheme from [13] stores a data structure for a subset S, $|S| = n$, of a universe U, such that membership in S can be tested by reading one bit of this data structure. The lookup procedure is randomized and has some error probability ε. Consider the bipartite graph $G = (U, V, E)$ and think of the left nodes as elements from U and the right nodes as elements of a binary vector V. On *lookup*(x) one of the positions $V[y], y \in \Gamma(\{x\})$, is chosen randomly and answer "yes" is returned if $V[y] = 1$, otherwise "no". To get a two-sided error of at most ε, at least a $(1 - \varepsilon)$ fraction of the neighbors of each $x \in U$ must carry the correct bit. Buhrman et al. showed that a legal $\{0,1\}$-assignment of V exists if $G = (U, V, E)$ is an $(|U|, |V|, d, 2n, (1 - \varepsilon/2)d)$-expander. It is assumed that G is found (e.g. by a probabilistic construction) and is hardwired into the system. — The static dictionary from [15] has basically the same structure but stores key-value pairs, where the value for each $x \in S$ is individual and for each $x \in U - S$ the same ("element $\notin S$"). It is assumed that each key-value pair fits into one word of memory. The randomized lookup procedure never returns the wrong value but there is a small probability ε that the answer is "don't know". Again, as Östlin and Pagh showed, for a scheme as required to exist it is sufficient that G is an $(|U|, |V|, d, 2n, (1 - \varepsilon/2)d)$-expander.

3 Uniform Hashing in Close to Optimal Space

In this section we use "split-and-share" to prove Theorem 1. Our construction is similar to the construction of the "basic retrieval data structure" in [17, Section 2.2]. In fact we use exactly the same tools from linear algebra. The big difference is that in [17] the set S is known while here the construction must work whp for an arbitrary, unknown set S.

3.1 Splitting the Key Set Evenly

We need to partition the key set S into $t = n^{1-\delta}$ disjoint subsets $S_0, S_1, \ldots, S_{t-1}$ such that for all $i \in [t]$ it holds $|S_i| \leq (1+\lambda)n^\delta$ for an arbitrary $\lambda > 0$. For this we choose a function h' from class $\mathcal{R} \subseteq \{h \mid h : U \to [t]\}$ (Section 2.2) at random, and define S_i as in (2). Let δ, α and κ be constant, $0 < \delta < \alpha < 1$. According to Lemma 1 we have:

(i) $\varepsilon_1 = \Pr\left(\exists i : |S_i| \geq (1+\lambda)n^\delta\right) = O(n^{1-(\alpha-\delta)\kappa} + n^{1-\delta}c^{n^\delta}), c = c(\kappa, \lambda)$.
(ii) The space usage for h' is $O(n^\alpha \log n)$.
(iii) The evaluation time for h' is $O(1)$.

From now on we assume all subsets S_i have size at most $(1+\lambda)n^\delta$.

3.2 Fully Random Functions on S_i

First we focus on one subset S_i. It is easy to construct hash functions that are fully random on S_i if one allows spending many more random words than the size of S_i. We sketch one such (folklore) construction. Let κ be a suitable constant. Choose $h'' : U \to [n^\alpha]$, from a $(\kappa+1)$-wise independent hash class $\mathcal{H}_{n^\alpha}^{\kappa+1}$. This function splits S_i into n^α buckets $B_j = \{x \in S_i \mid j = h''(x)\}$. With probability $1 - O(n^{\delta-(\alpha-\delta)\kappa})$ the function h'' is κ-perfect for S_i, that is $\forall j \in [n^\alpha] : |B_j| \leq \kappa$. Under this assumption we obtain a hash function $\mathfrak{h}$ that is fully random on S_i via:

$$\mathfrak{h}(x) = h^j(x),\ j = h''(x)\ , \tag{4}$$

by using (and storing) $t = n^\alpha$ hash functions $h^0, h^1, \ldots, h^{t-1}$ drawn uniformly at random from a κ-wise independent hash class. Overall we get for such a function $\mathfrak{h} : U \to [s], s \leq |U|$:

(i) The probability ε_2 that the construction fails is $O(n^{\delta-(\alpha-\delta)\kappa})$.
(ii) The space usage for $\mathfrak{h}$ is $O(n^\alpha \log n)$.
(iii) The evaluation time for $\mathfrak{h}$ is $O(1)$.

If we need $b > 1$ many hash functions $\mathfrak{h}_k : U \to [s], k \in [b]$, that are fully random on S_i we simply use one splitting function h'' and b lists of t hash functions $h_k^0, h_k^1, \ldots, h_k^{t-1}$, maintaining the same error probability ε_2.

3.3 Linearly Independent Vectors

Still focussing on one S_i, we now show how the fully random hash functions from Section 3.2 can be used to assign a random vector $v_x \in \{0,1\}^s$ to each element $x \in S_i$ such that the family of the vectors v_x is linearly independent whp.

Random Weight-b Binary Vectors. One can use b random hash functions $\mathfrak{h}_0, \mathfrak{h}_1, \ldots, \mathfrak{h}_{b-1}$, where $\mathfrak{h}_k : U \to [s-k], k \in [b]$, to obtain a mapping $x \mapsto A_x \subseteq [s]$ with $|A_x| = b$ [8, Section 4.1]. Let v_x be the characteristic vector of A_x and write:

$$A_x = \phi(h_0^j(x), h_1^j(x), \ldots, h_{b-1}^j(x)),\ j = h''(x)\ . \tag{5}$$

Linear Independence over $\mathbb{F}_2$. We consider a family of random vectors as above and want to bound the probability that this family is linearly independent. Such bounds can be obtained as functions of dimension and number of non-zero entries of the vectors. According to [18, Theorem 1.2] it holds that if we have $|S_i|$ weight-b binary vectors chosen uniformly at random from $\mathbb{F}_2^s$ then they are linearly independent whp as long as the dimension/length s is large enough, for which it is sufficient that:

$$s \geq (1+\mu)\cdot(1+\lambda)n^\delta \geq (1+\mu)|S_i|\ , \tag{6}$$

for some suitable $\mu \in (0,1)$. More precisely it holds that for each $b \geq 3$ there is some $\eta_b > 0$ and a threshold β_b such that if $\frac{1}{1+\mu} < \beta_b$ then the probability that the family of random binary vectors of weight b is linearly independent is $1 - O(s^{-\eta_b})$. Suitable parameters, which can be derived from the proof of Theorem 1.2 in [18, Lemma 4.1] are $\eta_3 = \frac{2}{7}, \eta_4 = \frac{5}{7}, \eta_{\geq 5} = 1$ and $\beta_3 < 0.91, \beta_4 < 0.97, \beta_5 < 0.99$. Asymptotically, $\beta_b - (1 - e^{-b}/\ln 2) \to 0$ as $b \to \infty$, exponentially fast in b. Therefore the number b of non-zero entries needed can be bounded by:

$$b = c\cdot\log(\tfrac{1}{\mu}) \geq 3,\ \text{for some constant } c\ . \tag{7}$$

We obtain:

(i) $\varepsilon_3 = \Pr(\text{the family } (v_x)_{x\in S_i} \text{ is linearly dependent}) = O(s^{-\eta_b})$.
(ii) The space usage for the function ϕ is $O(b\cdot n^\alpha \log n)$ bits.
(iii) The evaluation time for ϕ is $O(\log(1/\mu))$.

From here on we assume that the keys from S_i are mapped to linearly independent weight-b vectors from $\mathbb{F}_2^s$.

3.4 Stochastic Independence

With the help of the family of linearly independent vectors $v_x, x \in S_i$, we can now construct a function $h_i : U \to R$ that is fully random on S_i. For that purpose we choose $R = \{0,1\}^q, q \geq 1$, and a vector $V \in R^s$ at random. We define $h_i(x)$ as the bitwise XOR ($\oplus$, addition in $(\mathbb{F}_2)^q$) of the components of V at places $a_k \in A_x$, $k \in [b]$, that is:

$$h_i(x) = V[a_0] \oplus V[a_1] \oplus \ldots \oplus V[a_{b-1}],\ \text{where } A_x = \{a_0, \ldots, a_{b-1}\}\ . \tag{8}$$

Proposition 1. *h_i is fully random on S_i.*

Proof. Assume first $R = \{0,1\}$, that is, V is a random binary vector. Then $h_i(x) = <v_x, V>$. Let $M = (v_x^\top)_{x\in S_i} \in \{0,1\}^{|S_i|\times s}$. Since the matrix M has full row rank, the linear mapping $\{0,1\}^s \ni V \mapsto M\cdot V$ is onto $\{0,1\}^{|S_i|}$. All $2^{|S_i|}$ cosets of the kernel of this mapping have the same cardinality. Hence the vector $M \cdot V$ is uniformly distributed. If $R = \{0,1\}^q$, we also get independence, since the operations in the q bit positions are carried out independently. □

3.5 Sharing the Hash Functions

We replicate the construction described so far for all subsets $S_i, i \in [n^{1-\delta}]$. The crucial observation is that since a κ-wise independent hash function with domain U is fully random not only on each subset of at most κ elements from S_i but on each subset of at most κ elements from U we can *share* h'' and the $b \cdot n^\alpha$ extra hash functions h_k^j among all subsets S_i. The vector V cannot be shared, and we need to choose a separate random vector V_i from R^s for each subset S_i. Hence the desired uniform hash function $h : U \to R$ is calculated as follows:

$$A_x = \phi(h_0^j(x), h_1^j(x), \ldots, h_{b-1}^j(x)), \quad j = h''(x) \in [n^\alpha]$$
$$h(x) = V_i[a_0] \oplus V_i[a_1] \oplus \ldots \oplus V_i[a_{b-1}], \; i = h'(x) \in [n^{1-\delta}]$$

with $A_x = \{a_0, \ldots, a_{b-1}\}$ and $h' \in \mathcal{R}, h'' \in \mathcal{H}_{n^\alpha}^{\kappa+1}$ and $h_k^j \in \mathcal{H}_s^\kappa, k \in [b]$. We have:

(i) The probability ε that the hash function $h : U \to R$ fails to provide the desired full randomness can be bounded by: $\varepsilon \le \varepsilon_1 + n^{1-\delta}(\varepsilon_2 + \varepsilon_3) = O(n^{1-2\delta})$, for $\frac{1}{2} < \delta < \alpha < 1$, sufficiently large κ and $\eta_b = 1$, i.e. $b \ge 5$.

(ii) The overall space usage for the hash functions h', h'', h_k^j is $o(n)$, the space usage for the vectors V_i is $(1+\mu)(1+\lambda)n \log |R| \le (1+\sigma)n \log |R|$, for $\sigma > \mu + \lambda$. By Lemma 1 we can achieve $\lambda \le \mu$.

(iii) The evaluation time of $h(x)$ is $O(\log(1/\mu)) = O(\log(1/\sigma))$ (by (7)).

Remark 1. A finer analysis of [18, Lemma 4.1] shows that ε_3 can be bounded by $n^{-\Omega(b)}$, which leads to an overall error ε of $O(1/n^c), c > 0$, and an evaluation time of $O(\max\{c, \log(1/\sigma)\})$. Details can be found in the full version.

This finishes the proof of Theorem 1. □

4 Siegel's Functions without Graph Powering

In this section we prove Theorem 2. The difference to Siegel's construction is that we get n^η-wise independence on an arbitrary key set S (whp), and not on the whole universe $U, |U| = n^r, r > 1$, but we can choose η from $(0,1)$ instead from $(0, 1/r)$. The graph powering construction is avoided.

4.1 Virtual Expander Graph

As explained in Section 2.2, the critical building block of an n^η-wise independent hash class $\mathcal{SI}$ is a bipartite left-regular graph $G_{\mathcal{SI}} = (U, [n^\gamma], E), \gamma \in (\eta, 1)$, with constant left degree d and expansion larger than $d/2 \cdot |T|$ for each $T \subseteq U$ with $|T| \le n^\eta$. Thus, for our hash class $\mathcal{SI}'$ we need a graph $G_{\mathcal{SI}'}$ that expands each $T \subseteq S$ with $|T| \le n^\eta$. We split the set S into subsets $S_i, i \in [n^{1-\delta}]$, of size close to $n^\delta, \delta \in (\gamma, 1)$, via a hash function h'. As described in Section 3.2, we get d hash functions $\mathfrak{h}_k : U \to [n^\gamma/d], k \in [d]$, that are fully random on an arbitrary but fixed S_i. With these functions we can build an expander graph $G = (U, [n^\gamma], E)$ for S_i, with $E = \{(x, k \cdot \frac{n^\gamma}{d} + \mathfrak{h}_k(x)) \mid k \in [d]\}$, utilizing the following Lemma.

Lemma 2. *Let η, γ, δ be fixed with $0 < \eta < \gamma < \delta < 1$ and let $\delta < (\gamma - \eta)d/2$ for an integer constant d. Let n be sufficiently large and $G = ([n^\delta], [n^\gamma], E)$ be a d-left regular bipartite graph with $E = \{(x, k \cdot \frac{n^\gamma}{d} + h_k(x)) \mid k \in [d]\}$ for n^η-wise independent hash functions $h_k : [n^\delta] \to [n^\gamma/d]$. Then G is an $(n^\delta, n^\gamma, d, n^\eta, d/2+\nu)$-expander with probability $1 - O(n^{2\delta-(\gamma-\eta)d})$.*

Proof. This is proved in essence as in [3, Lemma 4], using n^η-wise independence.

So for appropriate constants η, γ, δ the graph G has expansion larger than $d/2 \cdot |T|$ for all $T \subseteq S_i$ with $|T| \leq n^\eta$, whp.

4.2 Sharing the Expander

Using the same hash functions $\mathfrak{h}_k, k \in [d]$, we obtain an expander graph G_i for each subset S_i whp. We define $G_{\mathcal{SI}'} = (U, [n^{\gamma'}], E)$ as a certain kind of union of these G_i (left hand sides identical, right hand sides put next to each other), via:

$$E = \left\{ \left(x, i \cdot n^\gamma + k \cdot \tfrac{n^\gamma}{d} + \mathfrak{h}_k(x)\right) \mid i = h'(x) \in [n^{1-\delta}], k \in [d] \right\}, \tag{9}$$

for $\gamma' = \gamma + (1 - \delta) < 1$. We observe that for all $T = \bigcup_i T_i$, with $T_i \subseteq S_i$ and $|T| \leq n^\eta$ it holds: $|\Gamma(T)| = \sum_i |\Gamma(T_i)| > \frac{d}{2} \cdot |T|$. Besides the graph $G_{\mathcal{SI}'}$, the hash class $\mathcal{SI}'$ uses a vector of size $n^{1-\delta+\gamma}$ filled with random words (see Section 2.2). Hence for $1 > \alpha > \delta > \gamma > \eta > 0$ and integer constants d and κ we obtain the following properties for a randomly chosen $h \in \mathcal{SI}'$:

(i) h is n^η-wise independent on S with probability $1 - O(n^{1-(\alpha-\delta)\kappa} + n^{2-(\gamma-\eta)d})$.
(ii) The space usage for h is $O(n^\alpha \cdot \log n + n^{1-\delta+\gamma} \cdot \log |R|)$.
(iii) The evaluation time for h is $O(d \cdot \kappa) = O(1)$.

This finishes the proof of Theorem 2. □

5 Constant Time Explicit One-Probe Schemes

In this section we prove Theorem 3. The main building block of the one-probe schemes from [13] and [15] is an expander graph. As in Section 4 we construct this graph with hash functions and the help of "split-and-share". Since we need expansion not only on one set but on all sets of size $O(n^\delta)$ we must use Siegel's high performance hash class $\mathcal{SI}$ in the original version with the graph product operation. For using Siegel's class we have to assume that the universe U has size polynomial in the size of the key set S. So we restrict ourselves to the special case $|U| = n^r$ for a constant $r > 0$. At the end of this section we consider larger universes.

5.1 Virtual Expander Graph via Siegel's Hash Class

Both one-probe schemes rely on an $(n^r, s, d, 2n, (1 - \varepsilon/2)d)$-expander with $d = \Theta(\frac{\log |U|}{\varepsilon})$ and $s = O(n \cdot \frac{d}{\varepsilon})$. Once again we start by splitting the key set S (as in Section 3.1) into $n^{1-\delta}$ disjoint subsets S_i via a hash function $h' \in \mathcal{R}$. A minor difference is that now S is given and therefore we can keep choosing new split functions h' until $|S_i| \leq (1+\lambda)n^\delta$ for each $i \in [n^{1-\delta}]$. So for a universe with at most $(1+\lambda)n^\delta$ key elements we can consider the reduced problem of constructing an $(n^r, s', d, c \cdot n^\delta, (1-\varepsilon/2)d)$-expander with $s' = O(n^\delta \cdot \frac{d}{\varepsilon})$ and $c = 2(1+\lambda)$.

To get such an expander we use d hash functions $h_0, h_1, \ldots, h_{d-1}$ with disjoint ranges, $h_k : U \to [s'/d], k \in [d]$, from the $c \cdot n^\delta$-wise independent hash class $\mathcal{SI}$, where $\mathcal{SI}$ is based on a small random graph G_0 (as described in Section 2.2), which is stored explicitly in space $o(n)$.

Lemma 3. *Let* $|U| = n^r$, $d = \frac{\log|U|}{\varepsilon}$, $V = [s']$ *with* $s' = c' \cdot \frac{c \cdot n^\delta \cdot d}{\varepsilon}$, $c' \geq e2^{2l+1}$, $l \geq 2$, *and* $h_k : U \to [s'/d], k \in [d]$, *chosen uniformly at random from a* $c \cdot n^\delta$*-wise independent hash class. Then* $G = (U, V, E)$ *with* $E = \{(x, k \cdot \frac{s'}{d} + h_k(x)) \mid k \in [d]\}$, *is an* $(n^r, s', d, c \cdot n^\delta, (1-\varepsilon/2)d)$*-expander with probability* $1 - O(n^{2r(1-l)})$.

Proof. Exactly as in [13, Lemma 3.10], using $c \cdot n^\delta$-wise independence. □

5.2 Sharing the Expander

We replicate the construction just described for each S_i and thus obtain a storing scheme for the whole key set S by juxtaposing the distinct storing schemes for all sets S_i. However, we can share one expander G among these storing schemes, since the expansion in G holds *for all* $T \subseteq U$ with $|T| \leq c \cdot n^\delta$. The complete storing scheme consists of a vector V of size $s = n^{1-\delta} \cdot s'$, subdivided into $n^{1-\delta}$ parts, one for each subset S_i, and the $(n^r, s', d, c \cdot n^\delta, (1-\varepsilon/2)d)$-expander graph G. Using an offset of $i \cdot s'$ for the i-th partial storing scheme, the access to V during the randomized lookup for an element $x \in U$ is realized as follows:

$$\mathit{lookup}(x) = V[i \cdot s' + k \cdot \tfrac{s'}{d} + h_k(x)], \text{ for a random } k \in [d], i = h'(x) . \quad (10)$$

Our construction has the following properties:

(i) The probability that the construction fails (given $\mathcal{SI}$) is $O(n^{-2r})$.
(ii) The space usage is $o(n)$ for all hash functions (i.e the implicit expander) and $s = n^{1-\delta} \cdot s' = O((n \log n)/\varepsilon^2)$ bits or words of memory for V.
(iii) Since $h' \in \mathcal{R}$ and $h_k \in \mathcal{SI}$ have constant evaluation time, the evaluation time of the one-probe scheme is constant.

With the methods from [13,15] one can show that time $O(|U|) = \text{poly}(s)$ is sufficient to construct the entries of V. This finishes the proof of Theorem 3. □

Remark 2. If $|U|$ cannot be bounded by n^r for a constant r, we may use a collapse function $h : U \to [n^r]$ [3,11]. The additional space is negligible, but the change comes at the cost of an extra component in the error probability since keys in S and in $U - S$ may collide under h. In the case of the dictionary we can avoid errors by making sure that h is one-to-one on S and store for each $x \in S$ the original key x, not only $h(x)$. Details are provided in the full version. The following corollary summarizes the new error probabilities for both one-probe schemes.

Corollary 1. *Let* $|U|$ *be superpolynomial in* n. *Then there exists an explicit one-probe membership tester [dictionary] with properties as described in Theorem 3, except that for an arbitrary key* $x \in U$ *a query returns with probability:*

(i) *at least* $(1 - \varepsilon_{rev}) \cdot (1 - \varepsilon)$ *the correct result;*
(ii) *at most* $\varepsilon + \varepsilon_{rev}$ *a wrong result ["don't know"] ;*

where $\varepsilon_{rev} = O(n^{l+1-r})$, *if the number of queries during the lifetime of the one-probe scheme is at most* n^l *for a constant* $r > l + 1$.

References

1. Carter, L., Wegman, M.N.: Universal classes of hash functions. J. Comput. Syst. Sci. 18(2), 143–154 (1979)
2. Pagh, R., Rodler, F.F.: Cuckoo hashing. J. Algorithms 51(2), 122–144 (2004)
3. Siegel, A.: On universal classes of extremely random constant-time hash functions. SIAM J. Comput. 33(3), 505–543 (2004)
4. Fotakis, D., Pagh, R., Sanders, P., Spirakis, P.G.: Space efficient hash tables with worst case constant access time. Theory Comput. Syst. 38(2), 229–248 (2005)
5. Dietzfelbinger, M., Weidling, C.: Balanced allocation and dictionaries with tightly packed constant size bins. Theor. Comput. Sci. 380(1-2), 47–68 (2007)
6. Mitzenmacher, M., Vadhan, S.: Why simple hash functions work: exploiting the entropy in a data stream. In: Proc. 19th SODA, pp. 746–755. SIAM, Philadelphia (2008)
7. Hagerup, T., Tholey, T.: Efficient minimal perfect hashing in nearly minimal space. In: Ferreira, A., Reichel, H. (eds.) STACS 2001. LNCS, vol. 2010, pp. 317–326. Springer, Heidelberg (2001)
8. Dietzfelbinger, M.: Design strategies for minimal perfect hash functions. In: Hromkovič, J., Královič, R., Nunkesser, M., Widmayer, P. (eds.) SAGA 2007. LNCS, vol. 4665, pp. 2–17. Springer, Heidelberg (2007)
9. Östlin, A., Pagh, R.: Uniform hashing in constant time and linear space. In: Proc. 35th STOC, pp. 622–628. ACM Press, New York (2003)
10. Dietzfelbinger, M., Woelfel, P.: Almost random graphs with simple hash functions. In: Proc. 35th STOC, pp. 629–638. ACM Press, New York (2003)
11. Pagh, A., Pagh, R.: Uniform hashing in constant time and optimal space. SIAM J. Comput. 38(1), 85–96 (2008)
12. Siegel, A.: On universal classes of fast high performance hash functions, their time-space tradeoff, and their applications. In: Proc. 30th FOCS, pp. 20–25 (1989)
13. Buhrman, H., Miltersen, P.B., Radhakrishnan, J., Venkatesh, S.: Are bitvectors optimal? SIAM J. Comput. 31(6), 1723–1744 (2002)
14. Ta-Shma, A.: Storing information with extractors. Inf. Process. Lett. 83(5), 267–274 (2002)
15. Östlin, A., Pagh, R.: One-probe search. In: Widmayer, P., Triguero, F., Morales, R., Hennessy, M., Eidenbenz, S., Conejo, R. (eds.) ICALP 2002. LNCS, vol. 2380, pp. 439–450. Springer, Heidelberg (2002)
16. Dietzfelbinger, M., Meyer auf der Heide, F.: A new universal class of hash functions and dynamic hashing in real time. In: Paterson, M. (ed.) ICALP 1990. LNCS, vol. 443, pp. 6–19. Springer, Heidelberg (1990)
17. Dietzfelbinger, M., Pagh, R.: Succinct data structures for retrieval and approximate membership (Extended abstract). In: Aceto, L., Damgård, I., Goldberg, L.A., Halldórsson, M.M., Ingólfsdóttir, A., Walukiewicz, I. (eds.) ICALP 2008, Part I. LNCS, vol. 5125, pp. 385–396. Springer, Heidelberg (2008)
18. Calkin, N.J.: Dependent sets of constant weight binary vectors. Combinatorics, Probability & Computing 6(3), 263–271 (1997)

Quasirandom Rumor Spreading: Expanders, Push vs. Pull, and Robustness*

Benjamin Doerr[1], Tobias Friedrich[1,2], and Thomas Sauerwald[2]

[1] Max-Planck-Institut für Informatik, Saarbrücken, Germany
[2] International Computer Science Institute, Berkeley, CA, USA

Abstract. Randomized rumor spreading is an efficient protocol to distribute information in networks. Recently, a quasirandom version has been proposed and proven to work equally well on many graphs and better for sparse random graphs. In this work we show three main results for the quasirandom rumor spreading model.

We exhibit a natural expansion property for networks which suffices to make quasirandom rumor spreading inform all nodes of the network in logarithmic time with high probability. This expansion property is satisfied, among others, by many expander graphs, random regular graphs, and Erdős-Rényi random graphs.

For all network topologies, we show that if one of the push or pull model works well, so does the other. We also show that quasirandom rumor spreading is robust against transmission failures. If each message sent out gets lost with probability f, then the runtime increases only by a factor of $\mathcal{O}(1/(1-f))$.

1 Introduction

Randomized rumor spreading or *random phone call protocols* are simple randomized epidemic algorithms designed to distribute a piece of information in a network. They build on the simple approach that informed nodes call random neighbors and make them informed (*push model*), or that uninformed nodes call random neighbors and become informed if the neighbor was (*pull model*). In spite of the simple concept, these algorithms succeed in distributing information extremely fast. In contrast to many natural deterministic approaches, they are also highly robust against transmission failures.

Such algorithms have been applied successfully both in the context where a single news has to be distributed from one processor to all others (cf. [11]), and in the one where news may be injected at various nodes at different times. The latter problem occurs when maintaining data integrity in a distributed databases, e.g., name servers in large corporate networks [3, 15]. For a more extensive, but still concise discussion of various central aspects of this area, we refer the reader to the paper by Karp, Schindelhauer, Shenker, and Vöcking [14].

1.1 Quasirandom Rumor Spreading

Rumor spreading protocols often assume that all nodes have access to a central clock. The protocols then proceed in rounds, in each of which each node independent from the

* Tobias Friedrich and Thomas Sauerwald were partially supported by postdoctoral fellowships from the German Academic Exchange Service (DAAD).

S. Albers et al. (Eds.): ICALP 2009, Part I, LNCS 5555, pp. 366–377, 2009.

others can perform certain actions. In the classical randomized rumor spreading protocols, in each round each node contacts a neighbor chosen independently and uniformly at random. In the push model, the latter node becomes informed if the first was, and vice versa in the pull model.

In [4], we proposed a quasirandom version of randomized rumor spreading. It assumes that each node has a (cyclic) list of its neighbors. Except that each node starts at a random position in the list, this list describes the order in which the node contacts its neighbors. We make no particular assumption on the structure of the list. This allows to use any list that is already present to technically organize the communication. In such, this protocol is simpler than the classical, fully-random model.

Surprisingly, even though the amount of independent randomness is greatly reduced, similar or even better results could be shown. For the push model, with high probability $\log n$ rounds suffice to inform all nodes of an n–vertex hypercube or random graph $G \in \mathcal{G}(n,p)$, if $p \geqslant (\ln(n) + \omega(1))/n$. The same results are known for the classical model [8], except that for random graphs this only holds for $p \geqslant (1+\varepsilon)\ln(n)/n$, $\varepsilon > 0$ constant. For smaller p, $\Theta((\log n)^2)$ rounds are necessary.

These theoretical results are complemented by an experimental investigation [5], which observes that the quasirandom model typically needs less time than the fully-random one, e.g., by more than 10% for the 12-dimensional hypercube.

1.2 Our Results

In this paper, we greatly expand the first results of [4]. We *(a)* exhibit a natural expansion property that guarantees that quasirandom rumor spreading succeeds in $\mathcal{O}(\log n)$ iterations, *(b)* prove the surprising result that for each graph, the quasirandom push and pull model need the same time to inform all nodes with probability $1 - n^{-\Theta(1)}$, and *(c)* demonstrate that robustness is no problem for the quasirandom model in spite its greatly reduced use of independent randomness.

Our expansion properties (see Definition 3.1) are fulfilled by expander graphs (defined via the second largest eigenvalue, see Definition 3.4). In consequence, random regular graphs fulfill these properties with probability $1 - o(1)$. However, regularity is not necessary. These expansion properties are also satisfied by random graphs $\mathcal{G}(n,p)$ with probability $1 - o(1)$, where p can be as small as $(\ln(n) + \omega(1))/n$. Such graphs typically have vertices of constant degree and of logarithmic degree. Hence our result also subsumes (and improves in terms of the failure probability) the result on random graphs in [4], which gave a runtime of $\mathcal{O}(\log n)$ with probability $1 - o(1)$.

For all these graphs, we show that with probability $1 - n^{-\gamma}$, where γ can be an arbitrary constant, the quasirandom rumor spreading model succeeds in informing all vertices from a single initially informed one in $\mathcal{O}(\log n)$ rounds. This result holds independent of how the cyclic lists look like. To the best of our knowledge, this is first attempt to analyze rumor spreading on several diverse graph classes (some of which are even far from being regular) altogether.

We then show two results that hold for all graphs. The first concerns the pull model, where vertices call others to retrieve information. This model is traditionally regarded less frequently in the literature, though some beautiful results exist. In particular, it is known that combining both push and pull model can lead to a drastic reduction of the number of messages needed (some restrictions apply to the underlying model, though). The first result of this type is Karp et al. [14].

The pull model has quite different characteristics from the push model. For example, in the push model, the number of informed vertices can at most double each round. In the pull model, an increase by a factor of $\Delta(G)$ is possible.

In spite of these differences, we can show the following surprising result. If for some graph, one of the two quasirandom broadcasting models has the property that for all lists and all starting vertices all other vertices become informed in time T with probability $1 - n^{-\Theta(1)}$, then the other variant has this property as well. In consequence, our result that the expansion property implies efficient broadcasting holds as well for the pull model.

We finally analyze the robustness of quasirandom rumor spreading. By robustness we mean that we want the protocol still to work well, even if some transmissions get lost. Since quasirandom rumor spreading uses much fewer independent random bits, some colleagues after publication of [4] expressed the concern that robustness might be a problem here. However, we are now able to show that such problems do not occur. We prove that if each transmission independently fails with probability $f < 1$, the time needed to inform all vertices with high probability increases only by a factor of $\mathcal{O}(1/(1-f))$. This again holds for all graphs.

Due to lack of space, several of our proofs are abbreviated or deferred to the full version of the paper.

2 Precise Model and Preliminaries

In the quasirandom model, each vertex $v \in V$ is equipped with a cyclic permutation $\pi_v \colon \Gamma(v) \to \Gamma(v)$ of its neighbors $\Gamma(v)$. This can also be seen as a list of its neighbors.

At the start of the protocol each vertex chooses a first neighbor i_v uniformly at random from $\Gamma(v)$. This is the neighbor it contacts at time $t = 1$. In each following time step $t = 2, 3, \ldots$, the vertex v contacts a vertex $\pi_v^{t-1}(i_v)$. For the quasirandom push[1] model the result of one vertex contacting another one is as follows. If v was informed at time $t-1$, then $\pi_v^t(i_v)$ becomes informed at time t.

This model slightly deviates from the description in the introduction, where each vertex chooses the starting point on its list only when it gets informed. However, the two variants are clearly equivalent and in the following it will be advantageous to assume that all vertices start contacting their neighbors already when they are uninformed.

We shall analyze how long it takes until a rumor known to a single vertex is broadcasted to all other vertices. We adapt a worst-case view in that we aim at bounds that are independent of the starting vertex and of all the lists present in the model. For the quasirandom model the probability space consists of the initial positions of the fixed neighborhood lists of all vertices.

In the analysis it will occasionally be convenient to assume that a vertex after receiving the rumor does not transfer it on for a certain number of time steps. We call this a *delayed model*. Clearly, delaying only results in other vertices receiving the rumor

[1] Here we focus on the push model. In the *pull model* the result of a node u contacting a vertex v is opposite, that is, if v is informed, u gets informed. The differences are discussed in Section 5.

later. Consequently, the random variable describing the broadcast time of this model strictly dominates the original one. This, of course, also holds if several vertices delay the propagation of the rumor.

We will also need chains of contacting vertices. So, a vertex $u_1 \in V$ contacts another vertex $u_m \in V$ within the time-interval $[a, b]$, if there is a path $(u_1, u_2, \ldots, u_m)$ in G and $t_1 < t_2 < \cdots < t_{m-1} \in [a, b]$ such that for all $j \in [1, m-1]$, $\pi_{u_j}^{t_j}(i_{u_j}) = u_{j+1}$.

Throughout the paper, we use the following graph-theoretical notation. Let $n = |V|$ denote the number of vertices. For a vertex v of a graph $G = (V, E)$, let $\Gamma(v) := \{u \in V \mid \{u, v\} \in E\}$ the set of its *neighbors* and by $\deg(v) := |\Gamma(v)|$ its *degree*. For any $S \subseteq V$, let $\deg_S(v) := |\Gamma(v) \cap S|$. Let $\delta := \min_{v \in V} \deg(v)$ be the *minimum degree*, $d := 2|E|/n$ be the *average degree*, and $\Delta := \max_{v \in V} \deg(v)$ be the *maximum degree*. The *distance* $\mathrm{dist}(x, y)$ between vertices x and y is the length of the shortest path from x to y. The *diameter* $\mathrm{diam}(G)$ of a connected graph G is the largest distance between two vertices. We will also use $\Gamma^k(u) := \{v \in V \mid \mathrm{dist}(u, v) = k\}$ and $\Gamma^{\leqslant k}(u) := \{v \in V \mid \mathrm{dist}(u, v) \leqslant k\}$. For sets S we define $\Gamma(S) := \{v \in V \mid \exists u \in S, (u, v) \in E\}$ as the set of neighbors of S.

All logarithms $\log n$ are natural logarithms to the base e. As we are only interested in the asymptotic behavior, we will sometimes assume that n is sufficiently large.

3 Expanding Graphs

Instead of analyzing specific graphs, we distill three simple properties. For these properties we can prove that the quasirandom rumor spreading model succeeds in a logarithmic runtime to inform all vertices. This is independent of which vertex is initially informed and independent of the order of the lists. The properties are as follows.

Definition 3.1 (expanding graphs). *We call a connected graph expanding if the following properties hold:*

(P1) *For all constants C_α with $0 < C_\alpha \leqslant d/2$ there is a constant $C_\beta \in (0, 1)$ such that for any connected $S \subseteq V$ with $3 \leqslant |S| \leqslant C_\alpha\, n/d$, we have $|\Gamma(S) \setminus S| \geqslant C_\beta\, d\, |S|$.*

(P2) *There are constants $C_\delta \in (0, 1)$ and $C_\omega > 0$ such that for any $S \subseteq V$, the number of vertices in S^c which have at least $C_\delta d(|S|/n)$ neighbors in S is at least $|S|^c - \frac{C_\omega n^2}{d|S|}$.*

(P3) $d = \Omega(\Delta)$. *If $d = \omega(\log n)$ then $d = O(\delta)$.*

We will now describe the properties in detail and argue why each of them is intrinsic for the analysis. **(P1)** describes a vertex expansion which means that connected sets have a neighborhood which is roughly in the order of the average degree larger than the set itself. Without this property, the broadcasting process could end up in a set with a tiny neighborhood and slow down thereby too much. Note that in **(P1)**, C_β depends on C_α. As C_α has to be a constant, the upper limit on C_α only applies for constant d.

(P2) is a certain edge-expansion property implying that a large portion of uninformed vertices have a sufficiently number of informed neighbors. This avoids that the broadcasting process stumbles upon a point when it has informed many vertices but most of the remaining uninformed vertices have very few informed neighbors and therefore only a small chance to get informed. Note that **(P2)** is only useful for $|S| = \Omega(n/d)$ and $|S| \leqslant n/2$.

The last property **(P3)** demands a certain regularity of the graph. It is trivially fulfilled for regular graphs, which most definitions of expanders require. The condition $d = \Omega(\Delta)$ for the case $d = \mathcal{O}(\log n)$ does not limit any of our graph classes below. If the average degree is at most logarithmic, **(P3)** applies no further restrictions. Otherwise, we require δ, d and Δ to be of the same order of magnitude. Without this condition, there could be an uninformed vertex with δ informed neighbors of degree $\omega(\delta)$ which does not get informed in logarithmic time with a good probability. With an additional factor of Δ/δ this could be resolved, but as we aim at a logarithmic bound, we require $\delta = \Theta(\Delta)$ for $d = \omega(\log n)$. Note that we do not require $d = \omega(1)$, but the proof techniques for constant and non-constant average degrees will differ in Section 4.

We describe three important graph classes which are expanding, i.e., satisfy all three properties of Definition 3.1, with high probability.

Random Graphs $\mathcal{G}(n,p), p \geqslant (\log n + \omega(1))/n$. Here, we show that sparse and dense random graphs are expanding with probability $1 - o(1)$. We use the popular random graph model $\mathcal{G}(n,p)$ introduced by Erdős and Rényi [7] where each edge of an n-vertex graph is picked independently with probability p. We distinguish two kinds of random graphs with slightly different properties:

Definition 3.2 (sparse and dense random graph). *We call a random graph $\mathcal{G}(n,p)$* sparse *if $p = (\log n + f_n)/n$ with $f_n = \omega(1)$ and $f_n = \mathcal{O}(\log n)$, and* dense *if $p = \omega(\log(n)/n)$.*

Note that our definition of sparse random graph coincides with the one of Cooper and Frieze [2] who set $p = c_n \log(n)/n$ with $(c_n - 1)\log n \to \infty$ and $c_n = \mathcal{O}(1)$.

Theorem 3.3. *Sparse and dense random graphs are expanding with probability $1 - o(1)$.*

By setting $p = 1$ this also shows that complete graphs are expanding.

Expander Graphs. In order to define a (regular) expander graph formally (see Hoory, Linial, and Wigderson [12] for a survey on expander graphs), we have to introduce a bit of notation. For a d-regular graph, its adjacency matrix A of G is symmetric and has real eigenvalues $d = \lambda_1 \geqslant \lambda_2 \geqslant \cdots \geqslant \lambda_n$. Define $\lambda := \max\{|\lambda_2|, |\lambda_n|\}$.

Definition 3.4 (expander). *We call a d-regular graph $G = (V,E)$ expander if $\lambda(G) \leqslant \min\{d/C, C'\sqrt{d}\}$, where $C > 1, C' > 0$ are arbitrary constants.*

Hence for the case when $d = \mathcal{O}(1)$, Definition 3.4 requires that $\lambda(G) \leqslant d/C$, $C > 1$, which is the classic (algebraic) definition of expander graphs. For larger d, we require the bound $\lambda(G) \leqslant C'\sqrt{d}$. We point out that graphs that satisfy the even stronger condition $\lambda \leqslant 2\sqrt{d-1}$ are called *Ramanujan graphs* and the construction of them has received a lot of attention in mathematics and computer science (cf. Hoory et al. [12]).

Theorem 3.5. *Let G be a d-regular expander. Then G is expanding.*

Random Regular Graphs. Random regular graphs are a natural extension of the classic Erdős-Rényi-Graph model, satisfying the additional property of being regular.

Definition 3.6 (random regular graph). *For any even d, a random d-regular graph $G \in \mathcal{G}(n, d)$ is chosen uniformly among all labeled d-regular graphs with n vertices.*

Using a result of [9], we can prove that a random d-regular graph is also an expander in the sense of Definition 3.4. Hence, applying Theorem 3.5 gives the following.

Theorem 3.7. *A random d-regular graph $G \in \mathcal{G}(n, d)$ is expanding w. p. $1 - o(1)$.*

3.1 Previous Results on Expanding Graphs

We summarize what is known for the runtime of the fully-random and the quasirandom model for expanding graphs as defined in Definition 3.1.

The complete graph is the simplest expanding graph. Frieze and Grimmett [10] and later Pittel [16] analyzed the fully random model on this graph. It was shown that with probability $1 - o(1)$, $\log_2 n + \ln n + \omega(1)$ rounds suffice. For the quasirandom model, [4] proved a bound of $\mathcal{O}(\log n)$ rounds with probability $1 - o(1)$.

Feige et al. [8] showed that on random graphs $G(n, p)$, $p \geq (1+\varepsilon) \log n/n$, the fully random model satisfies a runtime bound of $\mathcal{O}(\log n)$ with probability $1-n^{-1}$. They also showed that this failure probability can be achieved for $p = (\log n + \mathcal{O}(\log\log n))/n$ only in $\Omega(\log^2 n)$ rounds. For the quasirandom model, [4] showed that a runtime of $O(\log n)$ holds with probability $1 - o(1)$, already if $p \geq (\log n + \omega(1))/n$.

For expanders with $\Delta/\delta = \mathcal{O}(1)$, it was shown in [17] that the fully random model completes its broadcast campaign in $\mathcal{O}(\log n)$ rounds with probability $1 - 1/n$. For the quasirandom model, no such results have been known so far.

The situation is the same for random regular graphs. Berenbrink, Elsässer, and Friedetzky [1] investigated the fully-random model on random regular graphs and proved, amongst other results, an upper bound of $\mathcal{O}(\log n)$ with probability $1 - n^{-1}$. However, the runtime of the quasirandom model was not considered therein.

As a unified answer to these open questions, this work shows for all aforementioned graphs a runtime of the quasirandom model of $\mathcal{O}(\log n)$ with probability $1 - n^{-\gamma}$, where $\gamma \geqslant 1$ is an arbitrary constant.

4 Analysis of the Quasirandom Push Model

In this section, we prove the following theorem, which is one of the three main results.

Theorem 4.1. *Let $\gamma \geqslant 1$ be a constant. The probability that the quasirandom push model started at an arbitrary vertex of an expanding graph informs all other vertices within $\mathcal{O}(\log n)$ rounds is $1 - \mathcal{O}(n^{-\gamma})$.*

To analyze the propagation process, we decompose it in a forward and a backward part. In the forward part we show that one informed vertex informs $n - \mathcal{O}(n/d)$ vertices in $\mathcal{O}(\log n)$ steps (cf. Theorem 4.2). In the backward part we show that if a vertex is uninformed, $\mathcal{O}(\log n)$ steps earlier, at least $\omega(n/d)$ vertices must be uninformed as well (cf. Theorem 4.7). Combining both yields Theorem 4.1.

We will show that all this holds with probability $1 - n^{-\gamma}$ for any $\gamma \geqslant 1$. As Theorem 4.1 is easy to show for $d = \mathcal{O}(1)$, we handle this case separately in Section 4.3 and now consider the case $d = \omega(1)$. This makes the proofs of the lemmas of this section shorter. Therefore in Sections 4.1 and 4.2 we may use the following adjusted property:

(P3') $d = \omega(1)$ and $d = \Omega(\Delta)$. If $d = \omega(\log n)$ then $d = \mathcal{O}(\delta)$.

As the precise constants will be crucial in parts of the following proofs, we use the following notation. Constants with a lower case Greek letter index (e.g., C_α and C_β) stem from Definition 3.1. Constants without an index or with a numbered index (e.g., C and C_1) are local constants in lemmas. K is used to denote a number of time steps.

4.1 Forward Analysis

Theorem 4.2. *Let $\gamma \geqslant 1$ be a constant. The probability that the quasirandom push model started in a fixed vertex u does not inform $n - \mathcal{O}(n/d)$ vertices within $\mathcal{O}(\log n)$ rounds, is at most $n^{-\gamma}$.*

In our analysis we will use the following two notations for sets of informed vertices. Let I_t be the set of vertices that know the rumor after the t-th step. Let $N_t \subseteq I_t$ be the set of "newly informed" vertices that know the rumor after the t-th step, but have not spread this information yet. The latter set will be especially important as these are the vertices which have preserved their independent random choice.

Each of the following Lemmas 4.3–4.6 examines one phase consisting of several steps. Within each phase, we will only consider information spread from vertices N_{t-1} that became informed in the previous phase.

Let u be (newly) informed at time step 0. To get a sufficiently large set of (newly) informed vertices to start with, we first show how to obtain a set N_t of size $\Theta(\log n)$ within $t = \mathcal{O}(\log n)$ steps. If $d = \omega(\log n)$, it suffices to inform enough vertices in $\Gamma(u)$.Otherwise, we use that **(P1)** implies that the neighborhoods $\Gamma^k(u)$ grow exponentially with k. Since within Δ steps, $\Gamma^k(u)$ can be informed if $\Gamma^{k-1}(u)$ was informed beforehand, the claim follows in this case. More precisely:

Lemma 4.3. *Let $C > 0$ be an arbitrary constant. Then with probability 1 there is a time step $t = \mathcal{O}(\log n)$ such that*

- $|N_t| \geqslant C \log n$,
- $|I_t \setminus N_t| = o(|N_t|)$.

We can now assume that we have a set N_t of size $\Omega(\log n)$. The next step aims at informing $\Omega(n/d)$ vertices. For the very dense case of $d = \Omega(n/\log n)$ it can obviously be skipped. Note that in the following we can always assume that we have not informed *too many* vertices as the number of informed vertices will always at most double in each time step. The following lemma shows that given a set of informed vertices matching the conditions of **(P1)** within a constant number of steps the set of informed vertices increases by a factor strictly larger than one.

Lemma 4.4. *For all constants $\gamma \geqslant 1$ and $C_\alpha > 0$ there are constants $K > 1$, $C_1 > 1$, $C_2 > 1$, and $C_3 \in (3/4, 1)$ such that for all time steps t, if*

- $C_1 \log n \leqslant |I_t| \leqslant C_\alpha\,(n/d)$,
- $|N_t| \geqslant C_3\,|I_t|$,

then with probability $1 - n^{-\gamma}$,

- $|I_{t+K}| \geqslant C_2\,|I_t|$,
- $|N_{t+K}| \geqslant C_3\,|I_{t+K}|$.

To avoid the process dying out, it is important that a large fraction of the vertices is newly informed in each phase. With every application of Lemma 4.4, the number of informed vertices increases by a factor of $C_2 > 1$ which depends on C_β which in turn depends on C_α. As the precondition of the next Lemma 4.5 is $|I_t| = 16\,C_\omega(n/d)$, we choose $C_\alpha = 16\,C_\omega$ in every application of Lemma 4.4. This implies a constant $C_2 > 0$ in every phase and therefore at most $\log_{C_2}\big(16\,C_\omega\,(n/d)\big) = \mathcal{O}(\log n)$ applications of Lemma 4.4 suffice to reach $16\,C_\omega(n/d)$ informed vertices with a constant fraction of them newly informed.

The next aim is informing a linear number of vertices. Note that as long as that is not achieved, **(P2)** says that there is always a large set of uninformed vertices which have many neighbors in N_t. Lemma 4.5 below shows that under these conditions, a phase of a constant number of steps suffices to triple the number of informed vertices.

Lemma 4.5. *For all constants* $\gamma \geqslant 1$ *there are constants* $K > 1$, $C > 1$, *and* $C_\omega > 0$ *such that for all time steps t, if*

- $C \log n \leqslant |I_t| \leqslant n/16$,
- $|I_t| \geqslant 16\,C_\omega(n/d)$,
- $|N_t| \geqslant 3/4\,|I_t|$,

then with probability $1 - n^{-\gamma}$,

- $|I_{t+K}| \geqslant 3\,|I_t|$,
- $|N_{t+K}| \geqslant 3/4\,|I_{t+K}|$.

Applying Lemma 4.5 at most $\mathcal{O}(\log n)$ times, a linear fraction of the vertices gets informed. In a final phase of $\mathcal{O}(\log n)$ steps, one can then inform all but $\mathcal{O}(n/d)$ vertices as shown in the following Lemma 4.6.

Lemma 4.6. *Let* $\gamma \geqslant 1$ *be a constant and t be a time step such that* $|N_t| = \Theta(n)$. *Then with probability* $1 - n^{-\gamma}$, $|I_{t+\mathcal{O}(\log n)}| = n - \mathcal{O}(n/d)$.

Combining all above phases, a union bound gives that $|I_{\mathcal{O}(\log n)}| = n - \mathcal{O}(n/d)$ with probability $1 - \mathcal{O}(\log(n)\,n^{-\gamma})$, and Theorem 4.2 follows.

4.2 Backward Analysis

The forward analysis has shown that within $\mathcal{O}(\log n)$ steps, at most $\mathcal{O}(n/d)$ vertices stay uninformed. We now analyze the reverse. The question here is, how many vertices have to be uninformed at time $t - \mathcal{O}(\log n)$ if there is an uninformed vertex at time t? We will show that this is at least $\omega(n/d)$ which finishes the overall proof. For this, we introduce a further piece of notation needed in the backward analysis. We denote by $U_{[t_1,t_2]}(w)$ the set of nodes that contact the vertex w within the time-interval $[t_1, t_2]$.

Theorem 4.7. *Let $\gamma \geqslant 1$ be a constant. If the quasirandom rumor spreading process does not inform a fixed vertex w at some time t, then there are $\omega(n/d)$ uninformed vertices at time $t - \mathcal{O}(\log n)$ with probability at least $1 - n^{-\gamma}$.*

To prove Theorem 4.7, we fix an arbitrary vertex w and a time t. Ignoring some technicalities, our aim is now to lower bound the number of vertices which have to be uniformed at times $< t$ to keep w uniformed at time t. As before in Lemma 4.3, we first show the set of uninformed vertices is at least of logarithmic size.

For $d = \mathcal{O}(\log n)$ this follows from **(P1)** as all vertices of $\Gamma^{\mathcal{O}(\log\log n/\log d)}(w) = \Omega(\log n)$ contact w within $\mathcal{O}(\log n)$ steps. For $d = \omega(\log n)$, simple Chernoff bounds show that enough vertices of $\Gamma(w)$ contact w within $\mathcal{O}(\log n)$ steps.

Lemma 4.8. *Let $\gamma \geqslant 1$ and $C \geqslant 1$ be constants, w a vertex, and $t_2 = \Omega(\log n)$ a time step. Then with probability $1 - 2\,n^{-\gamma}$ there is a time step $t_1 = t_2 - \mathcal{O}(\log n)$ such that*

$$|U_{[t_1,t_2]}(w)| \geqslant C\,\log n.$$

We now know that within a logarithmic number of time steps, there are at least $\log n$ vertices which have contacted w. Very similar to Lemmas 4.4 and 4.5 in the forward analysis, we can increase the set of vertices that contact w by a multiplicative factor within a constant number of time steps. The following lemma again mainly draws on **(P1)**. For the very dense case of $d = \Omega(n/\log n)$ it can again be ignored.

Lemma 4.9. *For all constants $\gamma \geqslant 1$ there is a time step K such that for all vertices w and time steps t_1, t_2, if*

$$\log n \leqslant |U_{[t_1,t_2]}(w)| = \mathcal{O}(n/d),$$

then with probability $1 - n^{-\gamma}$,

$$|U_{[t_1-K,t_2]}(w)| \geqslant 4\,|U_{[t_1,t_2]}(w)|.$$

Using Lemma 4.9 at most $\mathcal{O}(\log n)$ times reaches a set of vertices that contact w of size $\Omega(n/d)$. If we have already reached $\omega(n/d)$, we are done. Otherwise, the following Lemma 4.10 shows that a phase consisting of $\mathcal{O}(\log n)$ suffices to reach it. This is the only lemma which substantially draws on **(P3')**.

Lemma 4.10. *Let $\gamma \geqslant 1$ be a constant, w a vertex, and t_1, t_2 time steps such that*

$$|U_{[t_1,t_2]}(w)| = \Theta(n/d).$$

Then with probability $1 - n^{-\gamma}$,

$$|U_{[t_1-\mathcal{O}(\log n),t_2]}(w)| = \omega(n/d).$$

This finishes the backward analysis and shows that $\omega(n/d)$ vertices have to uninformed to keep a single vertex uninformed within $\mathcal{O}(\log n)$ steps. Together with the forward analysis which proved that only $\mathcal{O}(n/d)$ vertices remain uninformed after $\mathcal{O}(\log n)$ steps, this finishes the overall proof of Theorem 4.1 for $d = \omega(1)$.

4.3 Analysis for Graphs with Constant Degree

It remains to show that the quasirandom push model also succeeds on expanding graphs with constant degree $d = \mathcal{O}(1)$. For this, it is easy to see (cf. [4, Theorem 2]) that for any graph the quasirandom push model succeeds in time $\leqslant \Delta \cdot \mathrm{diam}(G)$ with probability 1.[2] Naturally, the diameter of expanding graphs can be bounded easily.

Lemma 4.11. *For any expanding graph G with* $d = \mathcal{O}(1)$, $\mathrm{diam}(G) = \mathcal{O}(\log n)$.

Plugging Lemma 4.11 into the bound $\Delta \cdot \mathrm{diam}(G)$ yields Theorem 4.1 for $d = \mathcal{O}(1)$.

5 Quasirandom Pull Model

In this section, we analyze a second broadcasting model called *pull model*. Its only difference to the push model defined in Section 2 is that here non-informed vertices call other vertices to gain the information. More precisely, let G be a graph equipped with a list of neighbors for each vertex as defined in Section 2. Again, each vertex chooses a random initial neighbor and contacts its neighbors in the order of the list.

The only difference is the result of such a contact. Assume that u and v are vertices in G and that u contacts v. Then if v is informed, u becomes informed (and not vice versa). To avoid misunderstanding, we say that v *asks* u instead of *contacts*, to stress the fact that v is the vertex possibly becoming informed.

Besides being equally natural as the push model, there is a second reason to analyze both models. For the fully-random broadcasting scenario, it is well known that typically the push model works more efficiently when still many vertices are uninformed, whereas the pull model is more efficient in informing the remaining few vertices after the majority is already informed. Combining the two models in a non-trivial manner allows to develop protocols that still work in time $\mathcal{O}(\log n)$, but need fewer messages sent than $\mathcal{O}(n \log n)$. See Karp et al. [14] for more details.

While we shall not go that far and discuss optimizing the number of messages sent, we shall prove bounds for the pull model along. In fact, we shall prove the surprising result that both models are equally efficient. This comes unexpected in the light, e.g., of the following simple example.

Assume that G is a star, that is, a tree with one central vertex c which is neighbor to all other vertices. If c is initially informed, then the quasirandom push model needs exactly $n - 1$ rounds with probability one. For the pull model, things are very different as in this case the quasirandom pull model[3] needs only a single round. At first sight this might suggest that the push and pull models are not related at all, but as the initial position is chosen worst-case we can in fact show the following result.

Theorem 5.1. *Let G be a graph such that for all lists and all initially informed vertices, the quasirandom push model needs T rounds to inform all other vertices with probability* $1 - n^{-\gamma}$, *where* $\gamma \geqslant 1$. *Then the quasirandom pull model, again for all lists and*

[2] The corresponding bound for the fully-random model is $\mathcal{O}(\Delta \cdot (\mathrm{diam}(G) + \log n))$ with probability $1 - n^{-1}$ [8, Theorem 2.2].

[3] The corresponding bound for the fully-random push model is $\Theta(n \log n)$ with probability $1 - n^{-1}$ while the fully-random pull model also succeeds within a single round.

starting vertices, within T rounds informs all other nodes with probability $1 - n^{-\gamma+1}$. The same holds with the roles of push and pull reversed.

6 Robustness

To demonstrate that quasirandom rumor spreading is robust against failures, we consider a natural model similar to the one considered in [6, 13]. Here, each sent message reaches its destination only with a certain probability. These failures are assumed to be stochastically independent.

We assume that each vertex sends an acknowledgment to each neighbor from which it has received a correctly sent message. However, also the acknowledged message may not be sent correctly. Only after a vertex has received the acknowledged message, it continues to broadcast the message to the next neighbor on its list, otherwise it tries to send the message to the same neighbor in the next round again. We assume that a node sends a message and receives an acknowledgment with probability f, $0 < f < 1$ (again independently of all other messages). Note that we do not require f to be a constant.

We believe that the assumption of acknowledged messages is inline with practical considerations since the required communication only increases by a constant factor.

Theorem 6.1. *Let G be any graph and let T such that the quasirandom model started at an arbitrary vertex needs T rounds to inform all vertices with probability $1-n^{-\gamma}$, where $\gamma \geqslant 1$. Then the quasirandom model started at an arbitrary vertex needs $4\gamma/(1-f)\,T$ rounds to inform all vertices with probability $1 - 2n^{-\gamma}$.*

7 Conclusion and Outlook

In this work, we made significant progress to understand quasirandom rumor spreading. In particular, we answered the question if it is robust affirmatively, and showed that both push and pull model achieve logarithmic broadcast times on the large class of expanding graphs. From the broader view-point of randomized algorithmics, this work again shows that a reduced amount of randomness can yield superior algorithms, but still can be analyzed with theoretical means.

One open problem is to analyze to what extent push and pull model can be combined to reduce the number of messages needed. We should note, though, that the quasirandom push model naturally never needs more than $2m = nd$ messages. Hence for really sparse networks, e.g., constant-degree expander graphs, the standard quasirandom model is both time and message efficient.

References

[1] Berenbrink, P., Elsässer, R., Friedetzky, T.: Efficient randomized broadcasting in random regular networks with applications in peer-to-peer systems. In: 27th ACM Symposium on Principles of Distributed Computing (PODC), pp. 155–164 (2008)

[2] Cooper, C., Frieze, A.M.: The cover time of sparse random graphs. Random Struct. Algorithms 30, 1–16 (2007)

[3] Demers, A.J., Greene, D.H., Hauser, C., Irish, W., Larson, J., Shenker, S., Sturgis, H.E., Swinehart, D.C., Terry, D.B.: Epidemic algorithms for replicated database maintenance. Operating Systems Review 22, 8–32 (1988)

[4] Doerr, B., Friedrich, T., Sauerwald, T.: Quasirandom rumor spreading. In: Proceedings of the 19th Annual ACM-SIAM Symposium on Discrete Algorithms (SODA), pp. 773–781 (2008)

[5] Doerr, B., Friedrich, T., Künnemann, M., Sauerwald, T.: Quasirandom rumor spreading: An experimental analysis. In: Proceedings of the 10th Workshop on Algorithm Engineering and Experiments (ALENEX), pp. 145–153 (2009)

[6] Elsässer, R., Sauerwald, T.: On the Runtime and Robustness of Randomized Broadcasting. In: Asano, T. (ed.) ISAAC 2006. LNCS, vol. 4288, pp. 349–358. Springer, Heidelberg (2006)

[7] Erdős, P., Rényi, A.: On random graphs. Publ. Math. Debrecen 6, 290–297 (1959)

[8] Feige, U., Peleg, D., Raghavan, P., Upfal, E.: Randomized broadcast in networks. Random Structures and Algorithms 1, 447–460 (1990)

[9] Friedman, J.: On the second eigenvalue and random walks in random d-regular graphs. Combinatorica 11, 331–362 (1991)

[10] Frieze, A., Grimmett, G.: The shortest-path problem for graphs with random arc-lengths. Discrete Applied Mathematics 10, 57–77 (1985)

[11] Hedetniemi, S.M., Hedetniemi, S.T., Liestman, A.L.: A survey of gossiping and broadcasting in communication networks. Networks 18, 319–349 (1988)

[12] Hoory, S., Linial, N., Wigderson, A.: Expander graphs and their applications. Bulletin of the American Mathematical Society 43, 439–561 (2006)

[13] Hromkovič, J., Klasing, R., Pelc, A., Ružička, P., Unger, W.: Dissemination of Information in Communication Networks (2005)

[14] Karp, R., Schindelhauer, C., Shenker, S., Vöcking, B.: Randomized Rumor Spreading. In: 41st IEEE Symposium on Foundations of Computer Science (FOCS), pp. 565–574 (2000)

[15] Kempe, D., Dobra, A., Gehrke, J.: Gossip-based computation of aggregate information. In: 44th IEEE Symposium on Foundations of Computer Science (FOCS), pp. 482–491 (2003)

[16] Pittel, B.: On spreading a rumor. SIAM Journal on Applied Mathematics 47, 213–223 (1987)

[17] Sauerwald, T.: On mixing and edge expansion properties in randomized broadcasting. In: Tokuyama, T. (ed.) ISAAC 2007. LNCS, vol. 4835, pp. 196–207. Springer, Heidelberg (2007)

Incompressibility through Colors and IDs

Michael Dom[1], Daniel Lokshtanov[2], and Saket Saurabh[2]

[1] Institut für Informatik, Friedrich-Schiller-Universität Jena, Ernst-Abbe-Platz 2, 07743 Jena, Germany
michael.dom@uni-jena.de

[2] Department of Informatics, University of Bergen, 5020 Bergen, Norway
{daniello,saket.saurabh}@ii.uib.no

Abstract. In parameterized complexity each problem instance comes with a parameter k, and a parameterized problem is said to admit a *polynomial kernel* if there are polynomial time preprocessing rules that reduce the input instance to an instance with size polynomial in k. Many problems have been shown to admit polynomial kernels, but it is only recently that a framework for showing the non-existence of polynomial kernels has been developed by Bodlaender et al. [4] and Fortnow and Santhanam [9]. In this paper we show how to combine these results with combinatorial reductions which use colors and IDs in order to prove kernelization lower bounds for a variety of basic problems:

- We show that the Steiner Tree problem parameterized by the number of terminals and solution size k, and the Connected Vertex Cover and Capacitated Vertex Cover problems do not admit a polynomial kernel. The two latter results are surprising because the closely related Vertex Cover problem admits a kernel of size $2k$.
- Alon and Gutner obtain a $k^{poly(h)}$ kernel for Dominating Set in H-Minor Free Graphs parameterized by $h = |H|$ and solution size k and ask whether kernels of smaller size exist [2]. We partially resolve this question by showing that Dominating Set in H-Minor Free Graphs does not admit a kernel with size polynomial in $k + h$.
- Harnik and Naor obtain a "compression algorithm" for the Sparse Subset Sum problem [13]. We show that their algorithm is essentially optimal since the instances cannot be compressed further.
- Hitting Set and Set Cover admit a kernel of size $k^{O(d)}$ when parameterized by solution size k and maximum set size d. We show that neither of them, along with the Unique Coverage and Bounded Rank Disjoint Sets problems, admits a polynomial kernel.

All results are under the assumption that the polynomial hierarchy does not collapse to the third level. The existence of polynomial kernels for several of the problems mentioned above were open problems explicitly stated in the literature [2,3,11,12,14]. Many of our results also rule out the existence of compression algorithms, a notion similar to kernelization defined by Harnik and Naor [13], for the problems in question.

1 Introduction

Polynomial time preprocessing to reduce instance size is one of the most widely used approaches to tackle computationally hard problems. A natural question

S. Albers et al. (Eds.): ICALP 2009, Part I, LNCS 5555, pp. 378–389, 2009.

in this regard is how to measure the quality of preprocessing rules. Parameterized complexity provides a natural mathematical framework to give performance guarantees of preprocessing rules: A parameterized problem is said to admit a *polynomial kernel* if there is a polynomial time algorithm, called a *kernelization*, that reduces the input instance to an instance with size bounded by a polynomial $p(k)$ in the parameter k, while preserving the answer. This reduced instance is called a $p(k)$ *kernel* for the problem. (See [6,8,15] for further introductions.) While positive kernelization results have appeared regularly over the last two decades, the first results establishing infeasibility of polynomial kernels for specific problems have appeared only recently. In particular, Bodlaender et al. [4] and Fortnow and Santhanam [9] have developed a framework for showing that a problem does not admit a polynomial kernel unless the polynomial hierarchy collapses to the third level ($PH = \Sigma_p^3$), which is deemed unlikely.

Bodlaender et al. [4] observed that their framework can be directly applied to show kernelization lower bounds for many parameterized problems, including Longest Path and Longest Cycle. To the authors' best knowledge, the only non-trivial applications of this framework are in a recent result of Fernau et al. [7] showing that the Directed Max Leaf Out-Branching problem does not have a polynomial kernel, and a result by Bodlaender et al. [5] showing that the Disjoint Paths and Disjoint Cycles problems do not admit a polynomial kernel unless $PH = \Sigma_p^3$.

Our Results & Techniques. At present, there are two ways of showing that a particular problem does not admit a polynomial kernel unless $PH = \Sigma_p^3$. One is to give a "composition algorithm" for the problem in question. The other is to reduce from a problem for which a kernelization lower bound is known to the problem in question, such that a polynomial kernel for the considered problem would transfer to a polynomial kernel for the problem we reduced from. Such a reduction is called a *polynomial parameter transformation* and was introduced by Bodlaender et al. [5]. In order to show our results, we apply both methods. First, we present in Section 3 a "cookbook" approach for showing kernelization lower bounds by using composition algorithms together with polynomial parameter transformations. In the subsequent sections, we apply our approach to show that Unique Coverage parameterized by solution size k and Hitting Set and Set Cover parameterized by solution size k and universe size $|U|$ do not admit polynomial kernels unless $PH = \Sigma_p^3$. These problems turn out to be useful starting points for polynomial parameter transformations, showing that a variety of basic problems do not have a polynomial kernel. All our results summarized below are under the assumption that $PH \neq \Sigma_p^3$ and unless explicitly stated otherwise, all the problems considered are parameterized by the solution size.

Connectivity and Covering Problems. In Section 4, we show that the Set Cover problem parameterized by solution size k and the size $|U|$ of the universe does not have a polynomial kernel. Using this result, we prove that Steiner Tree parameterized by the number of terminals and solution size k does not have a polynomial kernel, resolving an open problem stated in [3]. We proceed to show that the Connected Vertex Cover and Capacitated Vertex Cover problems do not admit a polynomial kernel for the parameter k. The existence of polynomial kernels for these problems was an open problem explicitly stated in the literature [11,12], and the negative answer is surprising because the closely related VERTEX COVER problem admits a kernel of size $2k$. Finally, BOUNDED RANK DISJOINT SETS and UNIQUE COVERAGE do not admit a polynomial kernel. The latter result resolves an open problem of Moser et al. [14].

Domination and Transversals. In Section 5, we show that the HITTING SET problem parameterized by solution size k and the size $|U|$ of the universe does not have a polynomial kernel. This implies that the DOMINATING SET problem parameterized by solution size k and the size of a minimum vertex cover of the input graph does not admit a polynomial kernel. The latter result in turn implies that DOMINATING SET IN H-MINOR FREE GRAPHS parameterized by $h = |H|$ and k does not admit kernel with size polynomial in $k + h$, partially resolving an open problem by Alon and Gutner [2], who obtain a $k^{poly(h)}$ kernel for DOMINATING SET IN H-MINOR FREE GRAPHS and ask whether kernels of smaller size exist. Another implication of the results in Sections 4 and 5 is that the HITTING SET and SET COVER problems parameterized by solution size k and maximum set size d do not have a kernel polynomial in k, d. Both HITTING SET and SET COVER admit a $k^{O(d)}$ kernel [1].

Numeric Problems. Harnik and Naor obtain a *compression algorithm* for the SPARSE SUBSET SUM problem [13]. Essentially, Harnik and Naor show that if the input instance to SUBSET SUM is a relatively small set of huge numbers, the instance can be reduced. In Section 6, we show in contrast that if the input instance is a huge set of relatively small numbers, the instance cannot be reduced.

Harnik and Naor [13] define *compression*, a notion with applications in cryptography and similar to kernelization in spirit. It is implicit from the discussion in [9] that for a large class of problems the notions of kernelization and compression are equivalent. Due to this, our kernelization lower bounds imply that several of the problems we considered do not admit compression to a language in NP. These problems are CONNECTED VERTEX COVER, CAPACITATED VERTEX COVER, STEINER TREE, UNIQUE COVERAGE, and SMALL SUBSET SUM.

2 Preliminaries

A parameterized problem L is a subset of $\Sigma^* \times \mathbb{N}$ for some finite alphabet Σ. An instance of a parameterized problem consists of (x, k), where k is called the parameter. A central notion in parameterized complexity is *fixed parameter tractability (FPT)*, which means solvability in time $f(k) \cdot p(|x|)$ for any instance (x, k), where f is an arbitrary function of k and p is a polynomial.

Definition 1. *A* kernelization algorithm, *or in short, a* kernel *for a parameterized problem $Q \subseteq \Sigma^* \times \mathbb{N}$ is an algorithm that, given $(x, k) \in \Sigma^* \times \mathbb{N}$, outputs in time polynomial in $|x| + k$ a pair $(x', k') \in \Sigma^* \times \mathbb{N}$ such that (a) $(x, k) \in Q$ if and only if $(x', k') \in Q$ and (b) $|x'| + k' \leq g(k)$, where g is an arbitrary computable function. The function g is referred to as the size of the kernel. If g is a polynomial function then we say that Q admits a polynomial kernel.*

Definition 2. [Composition [4]] *A composition algorithm for a parameterized problem* $L \subseteq \Sigma^* \times \mathbb{N}$ *is an algorithm that receives as input a sequence* $((x_1, k), \ldots, (x_t, k))$, *with* $(x_i, k) \in \Sigma^* \times \mathbb{N}^+$ *for each* $1 \leq i \leq t$, *uses time polynomial in* $\sum_{i=1}^{t} |x_i| + k$, *and outputs* $(y, k') \in \Sigma^* \times \mathbb{N}^+$ *with (a)* $(y, k') \in L \iff (x_i, k) \in L$ *for some* $1 \leq i \leq t$ *and (b)* k' *is polynomial in* k. *A parameterized problem is compositional if there is a composition algorithm for it.*

We utilize a recent result of Bodlaender et al. [4] and Fortnow and Santhanam [9] concerning the non-existence of polynomial kernels. To this end, we define the *unparameterized version* $\tilde{L}$ of a parameterized problem L as the language $\tilde{L} = \{x\#1^k \mid (x, k) \in L\}$, that is, the mapping of parameterized problems to unparameterized problems is done by mapping an instance (x, k) to the string $x\#1^k$, where 1 is an arbitrary fixed letter in Σ and $\# \notin \Sigma$.

Theorem 1 **([4,9]).** *Let* L *be a compositional parameterized problem whose unparameterized version* $\tilde{L}$ *is* NP*-complete. Then, unless* PH=Σ_p^3, *there is no polynomial kernel for* L.

Finally we define the notion of *polynomial parameter transformations.*

Definition 3 **([5]).** *Let* P *and* Q *be parameterized problems. We say that* P *is polynomial parameter reducible to* Q, *written* $P \leq_{Ptp} Q$, *if there exists a polynomial time computable function* $f : \Sigma^* \times \mathbb{N} \to \Sigma^* \times \mathbb{N}$ *and a polynomial* p, *such that for all* $(x, k) \in \Sigma^* \times \mathbb{N}$ *(a)* $(x, k) \in P$ *if and only* $(x', k') = f(x, k) \in Q$ *and (b)* $k' \leq p(k)$. *The function* f *is called* polynomial parameter transformation.

Proposition 1 **([5]).** *Let* P *and* Q *be the parameterized problems and* $\tilde{P}$ *and* $\tilde{Q}$ *be the unparameterized versions of* P *and* Q *respectively. Suppose that* $\tilde{P}$ *is* NP*-hard and* $\tilde{Q}$ *is in* NP. *Furthermore if there is a polynomial parameter transformation from* P *to* Q, *then if* Q *has a polynomial kernel then* P *also has a polynomial kernel.*

A notion similar to polynomial parameter transformation was independently used by Fernau et al. [7] albeit without being explicitly defined.

We close with some definitions from graph theory. For a vertex v in a graph G, we write $N_G(v)$ to denote the set of v's neighbors in G, and we write $\deg_G(v)$ to denote the *degree* of v. The subgraph of G induced by a vertex set V' is denoted with $G[V']$. A vertex v *dominates* a vertex u if $u \in N_G(v)$.

3 A Systematic Approach to Prove Kernelization Lower Bounds

In this section we describe a "cookbook" for showing kernelization and compressibility lower bounds. To show that a problem does not admit a polynomial size kernel we go through the following steps.

1. Define a suitable colored version of the problem. This is in order to get more control over how solutions to problem instances can look.

2. Show that the unparameterized version of the considered problem is in NP and that the unparameterized version of the colored version of the problem is NP-hard.
3. Give a polynomial parameter transformation from the colored to the uncolored version. This will imply that kernelization lower bounds for the colored version directly transfer to the original problem.
4. Show that the colored version parameterized by k is solvable in time $2^{k^c} \cdot n^{O(1)}$ for a fixed constant c.
5. Finally, show that the colored version is compositional and thus has no polynomial kernel. To do so, proceed as follows.

(a) If the number of instances in the input to the composition algorithm is at least 2^{k^c} then running the parameterized algorithm on each instance takes time polynomial in input size. This automatically yields a composition algorithm [5].
(b) If the number of instances is less than 2^{k^c}, every instance receives a unique identifier. Notice that in order to uniquely code the identifiers (ID) of all instances, k^c bits per instance is sufficient. The IDs are coded either as an integer, or as a subset of a *poly*(k) sized set.
(c) Use the coding power provided by colors and IDs to complete the composition algorithm.

4 Connectivity and Covering Problems

Set Cover, Steiner Tree, and Variants of Vertex Cover. The problems STEINER TREE, CONNECTED VERTEX COVER (CONVC), CAPACITATED VERTEX COVER (CAPVC), and SMALL UNIVERSE SET COVER are defined as follows. In STEINER TREE we are given a graph $G = (T \cup N, E)$ and an integer k and asked for a vertex set $N' \subseteq N$ of size at most k such that $G[T \cup N']$ is connected. In CONVC we are given a graph $G = (V, E)$ and an integer k and asked for a vertex cover of size at most k that induces a connected subgraph in G. A *vertex cover* is a set $C \subseteq V$ such that each edge in E has at least one endpoint in C. The problem CAPVC takes as input a graph $G = (V, E)$, a capacity function $cap : V \to \mathbb{N}^+$ and an integer k, and the task is to find a vertex cover C and a mapping from E to C in such a way that at most $cap(v)$ edges are mapped to every vertex $v \in C$. Finally, an instance of SMALL UNIVERSE SET COVER consists of a set family $\mathcal{F}$ over a universe U with $|U| \leq d$ and a positive integer k. The task is to find a subfamily $\mathcal{F}' \subseteq \mathcal{F}$ of size at most k such that $\cup_{S \in \mathcal{F}'} S = U$. All four problems are known to be NP-complete (e.g., see [10] and the proof of Theorem 2); in this section, we show that the problems do not admit polynomial kernels for the parameter $(|T|, k)$ (in the case of STEINER TREE), k (in the case of CONVC and CAPVC), and (d, k) (in the case of SMALL UNIVERSE SET COVER), respectively. To this end, we first use the framework presented in Section 3 to prove that another problem, which is called RBDS, does not have a polynomial kernel. Then, by giving polynomial parameter transformations from RBDS to the above problems, we show the non-existence of polynomial kernels for these problems.

In Red-Blue Dominating Set (RBDS) we are given a bipartite graph $G = (T \cup N, E)$ and an integer k and asked whether there exists a vertex set $N' \subseteq N$ of size at most k such that every vertex in T has at least one neighbor in N'. We show that RBDS parameterized by $(|T|, k)$ does not have a polynomial kernel. In the literature, the sets T and N are called "blue vertices" and "red vertices", respectively. In this paper we will call the vertices "terminals" and "nonterminals" in order to avoid confusion with the colored version of the problem that we are going to introduce. RBDS is equivalent to Set Cover and Hitting Set and, therefore, NP-complete [10]. In the colored version of RBDS, denoted by Colored Red-Blue Dominating Set (Col-RBDS), the vertices of N are colored with colors chosen from $\{1, \dots, k\}$, that is, we are additionally given a function $col\colon N \to \{1, \dots, k\}$, and N' is required to contain exactly one vertex of each color. We will now follow the framework from Section 3.

Lemma 1. [$\star$][1] (1) *The unparameterized version of* RBDS *is in* NP, *and the unparameterized version of* Col-RBDS *is* NP-*hard.* (2) *There is a polynomial parameter transformation from* Col-RBDS *to* RBDS. (3) Col-RBDS *is solvable in* $2^{|T|+k} \cdot |T \cup N|^{O(1)}$ *time.*

Lemma 2. Col-RBDS *parameterized by* $(|T|, k)$ *is compositional.*

Proof. For a sequence $(G_1 = (T_1 \cup N_1, E_1), k, col_1), \dots, (G_t = (T_t \cup N_t, E_t), k, col_t)$ of Col-RBDS instances with $|T_1| = |T_2| = \dots = |T_t| = p$, we show how to construct a Col-RBDS instance $(G = (T \cup N, E), k, col)$ as described in Definition 2.

For $i \in \{1, \dots, t\}$, let $T_i := \{u_1^i, \dots, u_p^i\}$ and $N_i := \{v_1^i, \dots, v_{q_i}^i\}$. We start with adding p vertices $u_1, \dots, u_p$ to the set T of terminals to be constructed. (We will add more vertices to T later.) Next, we add to the set N of nonterminals all vertices from the vertex sets $N_1, \dots, N_t$, preserving the colors of the vertices. That is, we set $N = \bigcup_{i \in \{1,\dots,t\}} N_i$, and $col(v_j^i) = col_i(v_j^i)$. Now, we add the edge set $\bigcup_{i \in \{1,\dots,t\}} \{\{u_{j_1}, v_{j_2}^i\} \mid \{u_{j_1}^i, v_{j_2}^i\} \in E_i\}$ to G (see Figure 1). The graph G and the coloring col constructed so far have the following property: If at least one of $(G_1, k, col_1), \dots, (G_t, k, col_t)$ is a *yes*-instance, then (G, k, col) is also a *yes*-instance. However, (G, k, col) may even be a *yes*-instance in the case where all instances $(G_1, k, col_1), \dots, (G_t, k, col_t)$ are *no*-instances, because in G one can select vertices into the solution that originate from different instances of the input sequence.

To ensure the correctness of the composition, we add more vertices and edges to G. We define for every graph G_i of the input sequence a unique identifier $\mathrm{ID}(G_i)$, which consists of a size-$(p + k)$ subset of $\{1, \dots, 2(p + k)\}$. Since $\binom{2(p+k)}{p+k} \geq 2^{p+k}$ and since we can assume that the input sequence does not contain more than 2^{p+k} instances, it is always possible to assign unique identifiers to all instances of the input sequence. For each color pair $(a, b) \in \{1, \dots, k\} \times \{1, \dots, k\}$ with $a \neq b$, we add a vertex set $W_{(a,b)} = \{w_1^{(a,b)}, \dots, w_{2(p+k)}^{(a,b)}\}$ to T, (see Figure 1), and we add to E the edge set

[1] Proofs of results labelled with [$\star$] have been omitted, whole or in part.

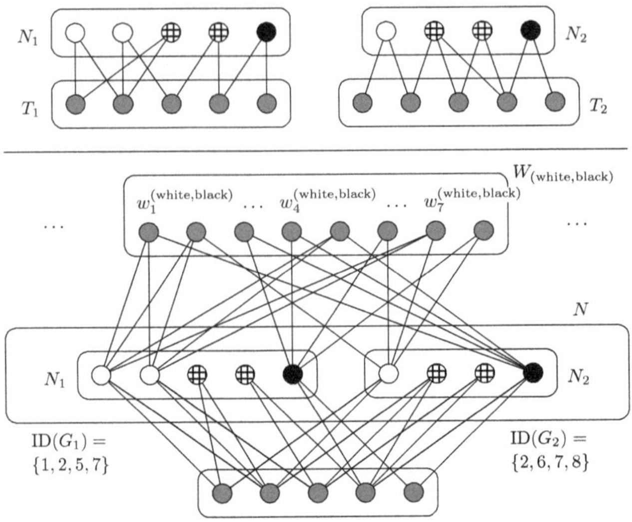

Fig. 1. Example for the composition algorithm for Col-RBDS. The upper part of the figure shows an input sequence consisting of two instances with $k = 3$ (there are three colors: white, checkered, and black). The lower part of the figure shows the output of the composition algorithm. For the sake of clarity, only the vertex set $W_{(\text{white,black})}$ is displayed, whereas five other vertex sets $W_{(a,b)}$ with $a, b \in \{\text{white, checkered, black}\}$ are omitted. Since $k = 3$ and $p = 5$, each ID should consist of eight numbers, and $W_{(\text{white,black})}$ should contain 16 vertices. For the sake of clarity, the displayed IDs consist of only four numbers each, and $W_{(\text{white,black})}$ contains only eight vertices.

$$\bigcup_{i \in \{1,\ldots,t\}, j_1 \in \{1,\ldots,q_i\}} \Big\{\{v^i_{j_1}, w^{(a,b)}_{j_2}\} \mid a = col(v^i_{j_1}) \wedge b \neq a \wedge j_2 \in \text{ID}(G_i)\Big\} \cup$$
$$\bigcup_{i \in \{1,\ldots,t\}, j_1 \in \{1,\ldots,q_i\}} \Big\{\{v^i_{j_1}, w^{(a,b)}_{j_2}\} \mid b = col(v^i_{j_1}) \wedge a \neq b \wedge j_2 \notin \text{ID}(G_i)\Big\}.$$

Note that the construction conforms to the definition of a composition algorithm; in particular, k remains unchanged and the size of T is polynomial in p, k because $|T| = p + k(k-1) \cdot 2(p+k)$. To prove the correctness of the construction, we show that (G, k, col) has a solution $N' \subseteq N$ if and only if at least one instance (G_i, k, col_i) from the input sequence has a solution $N'_i \subseteq N_i$.

In one direction, if $N'_i \subseteq N_i$ is a solution for (G_i, k, col_i), then the same vertex set chosen from N forms a solution for (G, k, col). To see this, note that for every color pair $(a, b) \in \{1, \ldots, k\} \times \{1, \ldots, k\}$ with $a \neq b$, each vertex from $W_{(a,b)}$ is either connected to all vertices v from N_i with $col(v) = a$ or to all vertices v from N_i with $col(v) = b$.

In the other direction, to show that any solution $N' \subseteq N$ for (G, k, col) is a solution for at least one instance (G_i, k, col_i), we prove that N' cannot contain vertices originating from different instances of the input sequence. To this end, note that each two vertices in N' must have different colors: Assume, for the sake of a contradiction, that N' contains a vertex v_{j1}^{i1} with $col(v_{j1}^{i1}) = a$ originating from the instance (G_{i_1}, k, col_{i_1}) and a vertex v_{j2}^{i2} with $col(v_{j2}^{i2}) = b$ originating from a different instance (G_{i_2}, k, col_{i_2}). Due to the construction of the IDs, we have $\mathrm{ID}(G_{i_1}) \setminus \mathrm{ID}(G_{i_2}) \neq \emptyset$ and $\mathrm{ID}(G_{i_2}) \setminus \mathrm{ID}(G_{i_1}) \neq \emptyset$. No vertex $w_j^{(a,b)}$ with $j \in \mathrm{ID}(G_{i_2}) \setminus \mathrm{ID}(G_{i_1})$ and no vertex $w_j^{(b,a)}$ with $j \in \mathrm{ID}(G_{i_1}) \setminus \mathrm{ID}(G_{i_2})$ is adjacent to one of v_{j1}^{i1} and v_{j2}^{i2}. Therefore, N' does not dominate all vertices from T—a contradiction. □

Theorem 2. *The problems* Red-Blue Dominating Set *and* Steiner Tree, *both parameterized by* $(|T|, k)$, *the problems* Connected Vertex Cover *and* Capacitated Vertex Cover, *both parameterized by* k, *the problem* Small Universe Set Cover *parameterized by* (k, d), *and the problem* Set Cover *parameterized by solution size* k *and the maximum size of any set in* $\mathcal{F}$ *do not admit polynomial kernels unless* $PH = \Sigma_p^3$.

Proof. For RBDS the statement of the theorem follows directly by Theorem 1 together with Lemmata 1 and 2.

To show that the statement is true for the other four problems, we give polynomial parameter transformations from RBDS to each of them—due to Proposition 1, this suffices to prove the statement. Let $(G = (T \cup N, E), k)$ be an instance of RBDS. To transform it into an instance $(G' = (T' \cup N, E'), k)$ of Steiner Tree, define $T' = T \cup \{\tilde{u}\}$ where $\tilde{u}$ is a new vertex and let $E' = E \cup \{\{\tilde{u}, v_i\} \mid v_i \in N\}$. It is easy to see that every solution for Steiner Tree on (G', k) one-to-one corresponds to a solution for RBDS on (G, k).

To transform (G, k) into an instance $(G'' = (V'', E''), k'')$ of ConVC, first construct the graph $G' = (T' \cup N, E')$ as described above. The graph G'' is then obtained from G' by attaching a leaf to every vertex in T'. Now, G'' has a connected vertex cover of size $k'' = |T'| + k = |T| + 1 + k$ iff G' has a Steiner tree containing k vertices from N iff all vertices from T can be dominated in G by k vertices from N.

Next, we describe how to transform (G, k) into an instance $(G''' = (V''', E'''), cap, k''')$ of CapVC. First, for each vertex $u_i \in T$, add a clique to G''' that contains four vertices $u_i^0, u_i^1, u_i^2, u_i^3$. Second, for each vertex $v_i \in N$, add a vertex v_i''' to G'''. Finally, for each edge $\{u_i, v_j\} \in E$ with $u_i \in T$ and $v_j \in N$, add the edge $\{u_i^0, v_j'''\}$ to G'''. The capacities of the vertices are defined as follows: For each vertex $u_i \in T$, the vertices $u_i^1, u_i^2, u_i^3 \in V'''$ have capacity 1 and the vertex $u_i^0 \in V'''$ has capacity $\deg_{G'''}(u_i^0) - 1$. Each vertex v_i''' has capacity $\deg_{G'''}(v_i''')$. Clearly, in order to cover the edges of the size-4 cliques inserted for the vertices of T, every capacitated vertex cover for G''' must contain all vertices $u_i^0, u_i^1, u_i^2, u_i^3$. Moreover, since the capacity of each vertex u_i^0 is too small to cover all edges incident to u_i^0, at least one neighbor v_j''' of u_i^0 must be selected into every capacitated vertex cover for G'''. Therefore, it is not hard to see that G''' has a capacitated vertex cover of size $k''' = 4 \cdot |T| + k$ iff all vertices from T can be dominated in G by k vertices from N.

The results for Small Universe Set Cover and Set Cover follow from the equivalence of Set Cover and RBDS. □

Unique Coverage. In the UNIQUE COVERAGE problem we are given a universe U, a family of sets $\mathcal{F}$ over U and an integer k. The problem is to find a subfamily $\mathcal{F}'$ of $\mathcal{F}$ and a set S of elements in U such that $|S| \geq k$ and every element of S appears in exactly one set in $\mathcal{F}'$, that is, the number of elements uniquely covered by $\mathcal{F}'$ is at least k.

In order to obtain our negative results we have to utilize positive kernelization results for the problem. In some sense, we have to compress our instances as much as possible in order to show that what remains is incompressible even though it is big. We utilize the following well-known and simple reduction rules for the problem: (a) If any set $S \in \mathcal{F}$ contains at least k elements, return yes; (b) If any element e is not contained in any set in $\mathcal{F}$, remove e from U; and (c) If none of the above rules can be applied and $|U| \geq k(k-1)$ return yes.

We show that the UNIQUE COVERAGE problem does not have a polynomial kernel unless PH=Σ_p^3. Notice that while the above reduction rules will compress the instance to an instance with at most $O(k^2)$ elements, this is not a polynomial kernel because there is no polynomial bound on the size of $\mathcal{F}$. We start by defining the colorful reduced version COLORED REDUCED UNIQUE COVERAGE (COL-RED-UC) of the UNIQUE COVERAGE problem. In this version the sets of $\mathcal{F}$ are colored with colors from $\{1,\ldots,k\}$ and $\mathcal{F}'$ is required to contain exactly one set of each color. Furthermore, in COL-RED-UC every set S in $\mathcal{F}$ has size at most $k-1$ and $|U| \leq k^2$.

Lemma 3. [$\star$] (1) *The unparameterized version of* UNIQUE COVERAGE *is in* NP, *and the unparameterized version of* COL-RED-UC *is* NP*-hard.* (2) *There is a polynomial parameter transformation from* COL-RED-UC *to* UNIQUE COVERAGE. (3) COL-RED-UC *parameterized by k is solvable in time $O(k^{2k^2})$.*

Lemma 4. [$\star$] *The* COL-RED-UC *problem is compositional.*

Proof. Given a sequence of COL-RED-UC instances $\mathcal{I}_1 = (U, \mathcal{F}_1, k), \ldots, \mathcal{I}_t = (U, \mathcal{F}_t, k)$, we construct a COL-RED-UC instance $\mathcal{I} = (U', \mathcal{F}, k')$. If the number of instances t is at least $2^{2k^2 \log k}$ then running the algorithm from Lemma 3 on all instances takes time polynomial in the input size yielding a trivial composition algorithm. Thus we assume that t is at most $2^{2k^2 \log k}$. We now construct ID's for for every instance, this is done in two steps. In the first step every instance i gets a unique small id $\mathrm{ID}'(\mathcal{I}_i)$ which is a subset of size $k^3/2$ of the set $\{1,\ldots,k^3\}$. The identifier of instance i is the set $\mathrm{ID}(\mathcal{I}_i)$ which is defined to be $\mathrm{ID}(\mathcal{I}_i) = \{x \in \mathbb{N} : \lfloor x/k^3 \rfloor \in \mathrm{ID}'(\mathcal{I}_i)\}$. In other words, $\mathrm{ID}(\mathcal{I}_i) = \{k^3 \cdot j + j' \mid j \in \mathrm{ID}'(\mathcal{I}_i) \wedge j' \in \{0,\ldots,k^3-1\}\}$. Notice that the identifier of every instance is now a subset of size $k^6/2$ of the set $\{1,\ldots,k^6\}$ and that the IDs of two different instances differ in at least k^3 places.

We start building the instance $\mathcal{I}$ by letting $U' = U$ and $\mathcal{F} = \mathcal{F}_1 \cup \mathcal{F}_2 \ldots \cup \mathcal{F}_t$. The sets have the same color as in their respective instance. For every distinct ordered pair of colors $i,j \leq k$ we add the set $U_{i,j} = \{u_{i,j}^1, \ldots, u_{i,j}^{k^6}\}$ to U'. For every instance $\mathcal{I}_p$ we consider the sets colored i and j respectively in $\mathcal{F}_p$. To every set S with color i in $\mathcal{F}_p$ we add the set $\{u_{i,j}^x : x \in \mathrm{ID}(\mathcal{I}_p)\}$. Also, to every set S with color j in $\mathcal{F}_p$ we add the set $\{u_{i,j}^x : x \notin \mathrm{ID}(\mathcal{I}_p)\}$. Finally we set $k' = k(k-1)k^6 + k$. This concludes the construction. The correctness proof is omitted. □

Theorem 3. *The* Unique Coverage *problem parameterized by k does not admit a polynomial kernel unless $PH = \Sigma_p^3$.*

Bounded Rank Disjoint Sets. In the Bounded Rank Disjoint Sets problem we are given a family $\mathcal{F}$ over a universe U with every set $S \in \mathcal{F}$ having size at most d together with a positive integer k. The question is whether there exists a subfamily $\mathcal{F}'$ of $\mathcal{F}$ with $|\mathcal{F}'| \geq k$ such that for every pair of sets $S_1, S_2 \in \mathcal{F}'$ we have that $S_1 \cap S_2 = \emptyset$. The problem can be solved in time $2^{O(dk)}n^{O(1)}$ using color-coding and an application of dk-perfect hash families. To show that this problem does not admit a poly(k, d) kernel, we define a variation of the Perfect Code problem on graphs: In Bipartite Regular Perfect Code we are given a bipartite graph $G = (T \cup N, E)$, where every vertex in N has the same degree, and an integer k and asked whether there exists a vertex set $N' \subseteq N$ of size at most k such that every vertex in T has *exactly one neighbor* in N'.

Theorem 4. [$\star$] Bipartite Regular Perfect Code *parameterized by* $(|T|, k)$ *and* Bounded Rank Disjoint Sets *parameterized by* (d, k) *do not have a polynomial kernel unless $PH = \Sigma_p^3$.*

5 Domination and Transversals

In the Small Universe Hitting Set problem we are given a set family $\mathcal{F}$ over a universe U with $|U| \leq d$ together with a positive integer k. The question is whether there exists a subset S in U of size at most k such that every set in $\mathcal{F}$ has a non-empty intersection with S. We show that the Small Universe Hitting Set problem parameterized by the solution size k and the size $d = |U|$ of the universe does not have a kernel of size polynomial in (k, d) unless $PH = \Sigma_p^3$. We define the colored version of Small Universe Hitting Set, called Col-SUHS as follows. We are given a set family $\mathcal{F}$ over a universe U with $|U| \leq d$, and a positive integer k. The elements of U are colored with colors from the set $\{1, \ldots, k\}$ and the question is whether there exists a subset $S \subseteq U$ containing exactly one element of each color such that every set in $\mathcal{F}$ has a non-empty intersection with S.

Lemma 5. [$\star$] (1) *The unparameterized version of* Small Universe Hitting Set *is in* NP, *and the unparameterized version of* Col-SUHS *is* NP*-hard.* (2) *There is a polynomial parameter transformation from* Col-SUHS *to* Small Universe Hitting Set. (3) Col-SUHS *parameterized by d, k is solvable in time $O(2^d \cdot n^{O(1)})$.*

Lemma 6. *The problem* Col-SUHS *is compositional.*

Proof. Given a sequence $(\mathcal{F}_1, U, d, k), \ldots, (\mathcal{F}_t, U, d, k)$ of Col-SUHS instances where $|U| \leq d$, we construct an instance $(\mathcal{F}, U', d', k')$ of Col-SUHS as described in Definition 2. Due to the time-$O(2^d \cdot n^{O(1)})$ algorithm from Lemma 5, we can assume that $t \leq 2^d$. Furthermore, we need the number of instances to be a power of 2. To make this true we add an appropriate number of no-instances, such that the total number of instances is 2^l. Since $t \leq 2^d$ we have that $l \leq d$. Now, let every instance be identified by a unique number from 0 to $t - 1$.

We let $k' = k+l$ and start building $(\mathcal{F}, U', d', k')$ from $(\mathcal{F}_1, U, d, k), \ldots, (\mathcal{F}_t, U, d, k)$ by letting $U' = U$ and letting elements keep their color. For every $i \leq t$ we add the family $\mathcal{F}_i$ to $\mathcal{F}$. We now add $2l$ new elements $C = \{a_1, b_1, \ldots, a_l, b_l\}$ to U' and for every $i \leq l$, $\{a_i, b_i\}$ comprise a new color class. We conclude the construction by modifying the sets in $\mathcal{F}$ that came from the input instances to the composition algorithm. For every $j \leq t$ we consider all sets in $\mathcal{F}_j$. For every such set S we proceed as follows. Let ID(j) be the identification number of instance number j. For every $i \leq l$ we look at the i'th bit in the binary representation of ID(j). If this bit is set to 1 we add a_i to S and if the bit is set to 0 we add b_i to S. This concludes the construction.

Now, if there is a colored hitting set S for $\mathcal{F}_j$ with $|S| \leq k$, one can construct a colored hitting set S' for $\mathcal{F}$ of size $k + l$ as follows. First, add S to S' and then consider the identification number ID(j) of instance j. For every i between 1 and l consider the i'th bit of ID(j). If this bit is set to 1 add b_j to S' else add a_j to S'. Clearly S' is a hitting set for $\mathcal{F}_i$, has size $k + l$ and contains one vertex of each color. Moreover, one can easily prove that S' hits all other sets of $\mathcal{F}$.

In the other direction, suppose there is a colored hitting set S' of size $l + k$ of $\mathcal{F}$. For every $i \leq l$, exactly one out of the vertices a_i and b_i is in S'. Let p be the number between 0 and $2^l - 1$ such that for every i the i'th bit of p is 1 if and only if $b_i \in S'$. Observe that the sets in $\mathcal{F}$ originating from the family $\mathcal{F}_j$ such that ID$(j) = p$ do not contain any of the elements of $S' \cap C$. Thus $S'' = S' \cap U$ is a colored hitting set for $\mathcal{F}_j$ containing at most one element from each color class. S'' can thus be extended to a colored hitting set S of $\mathcal{F}_j$ with $|S| = k$. □

Theorem 5. [$\star$] SMALL UNIVERSE HITTING SET *parameterized by solution size k and universe size $|U| = d$ does not have a polynomial kernel unless $PH = \Sigma_p^3$. The* DOMINATING SET *problem parameterized by the solution size k and the size c of a minimum vertex cover of the input graph does not have a polynomial kernel.*

Theorem 5 has some interesting consequences. For instance, the second part of Theorem 5 implies that the DOMINATING SET problem in graphs excluding a fixed graph H as a minor parameterized by $(k, |H|)$ does not have a kernel of size $poly(k, |H|)$ unless $PH = \Sigma_p^3$.

Theorem 6. *Unless $PH = \Sigma_p^3$ the problems* HITTING SET *parameterized by solution size k and the maximum size d of any set in $\mathcal{F}$,* DOMINATING SET IN H-MINOR FREE GRAPHS *parameterized by $(k, |H|)$, and* DOMINATING SET *parameterized by solution size k and degeneracy d of the input graph do not have a polynomial kernel.*

6 Numeric Problem: Small Subset Sum

In the SUBSET SUM problem we are given a set S of n integers and a target t and asked whether there is a subset S' of S that adds up to exactly t. In the most common parameterization of the problem one is also given an integer k, and S'

may contain not more than k numbers. This parameterization is $W[1]$-hard. We consider a stronger parameterization where in addition to k a parameter d is provided and the integers in S must have size at most 2^d. This version, SMALL SUBSET SUM, is trivially fixed parameter tractable by dynamic programming.

Theorem 7. [$\star$] SMALL SUBSET SUM *parameterized by* (d, k) *does not admit a kernel polynomial in* (d, k) *unless* $PH = \Sigma_p^3$.

References

1. Abu-Khzam, F.N.: Kernelization algorithms for d-hitting set problems. In: Dehne, F., Sack, J.-R., Zeh, N. (eds.) WADS 2007. LNCS, vol. 4619, pp. 434–445. Springer, Heidelberg (2007)
2. Alon, N., Gutner, S.: Kernels for the dominating set problem on graphs with an excluded minor. Technical Report TR08-066, Electronic Colloquium on Computational Complexity (ECCC) (2008)
3. Betzler, N.: Steiner tree problems in the analysis of biological networks. Diploma thesis, Wilhelm-Schickard-Institut für Informatik, Universität Tübingen, Germany (2006)
4. Bodlaender, H.L., Downey, R.G., Fellows, M.R., Hermelin, D.: On problems without polynomial kernels (Extended abstract). In: Aceto, L., Damgård, I., Goldberg, L.A., Halldórsson, M.M., Ingólfsdóttir, A., Walukiewicz, I. (eds.) ICALP 2008, Part I. LNCS, vol. 5125, pp. 563–574. Springer, Heidelberg (2008)
5. Bodlaender, H.L., Thomassé, S., Yeo, A.: Analysis of data reduction: Transformations give evidence for non-existence of polynomial kernels. Technical Report UU-CS-2008-030, Institute of Information and Computing Sciences, Utrecht University, Netherlands (2008)
6. Downey, R.G., Fellows, M.R.: Parameterized Complexity. Springer, Heidelberg (1999)
7. Fernau, H., Fomin, F.V., Lokshtanov, D., Raible, D., Saurabh, S., Villanger, Y.: Kernel(s) for problems with no kernel: On out-trees with many leaves. In: Proc. 26th STACS (to appear)
8. Flum, J., Grohe, M.: Parameterized Complexity Theory. Springer, Heidelberg (2006)
9. Fortnow, L., Santhanam, R.: Infeasibility of instance compression and succinct PCPs for NP. In: Proc. 40th STOC, pp. 133–142. ACM Press, New York (2008)
10. Garey, M.R., Johnson, D.S.: Computers and Intractability: A Guide to the Theory of NP-Completeness. Freeman, New York (1979)
11. Guo, J., Niedermeier, R.: Invitation to data reduction and problem kernelization. SIGACT News 38(1), 31–45 (2007)
12. Guo, J., Niedermeier, R., Wernicke, S.: Parameterized complexity of Vertex Cover variants. Theory Comput. Syst. 41(3), 501–520 (2007)
13. Harnik, D., Naor, M.: On the compressibility of NP instances and cryptographic applications. In: Proc. 47th FOCS, pp. 719–728. IEEE, Los Alamitos (2007)
14. Moser, H., Raman, V., Sikdar, S.: The parameterized complexity of the unique coverage problem. In: Tokuyama, T. (ed.) ISAAC 2007. LNCS, vol. 4835, pp. 621–631. Springer, Heidelberg (2007)
15. Niedermeier, R.: Invitation to Fixed-Parameter Algorithms. Oxford University Press, Oxford (2006)

Partition Arguments in Multiparty Communication Complexity

Jan Draisma[1,⋆], Eyal Kushilevitz[2,⋆⋆], and Enav Weinreb[2,⋆⋆]

[1] Department of Mathematics and Computer Science, Technische Universiteit Eindhoven, and CWI Amsterdam
j.draisma@tue.nl

[2] Computer Science Department, Technion Israel Institute of Technology, Haifa
eyalk@cs.technion.ac.il, weinreb@cs.technion.ac.il

Abstract. Consider the "Number in Hand" multiparty communication complexity model, where k players $P_1, \ldots, P_k$ holding inputs $x_1, \ldots, x_k \in \{0,1\}^n$ (respectively) communicate in order to compute the value $f(x_1, \ldots, x_k)$. The main lower bound technique for the communication complexity of such problems is that of *partition arguments*: partition the k players into two disjoint sets of players and find a lower bound for the induced two-party communication complexity problem. In this paper, we study the power of the partition arguments method. Our two main results are very different in nature:

(i) For *randomized* communication complexity we show that partition arguments may be exponentially far from the true communication complexity. Specifically, we prove that there exists a 3-argument function f whose communication complexity is $\Omega(n)$ but partition arguments can only yield an $\Omega(\log n)$ lower bound. The same holds for *nondeterministic* communication complexity.
(ii) For *deterministic* communication complexity, we prove that finding significant gaps, between the true communication complexity and the best lower bound that can be obtained via partition arguments, would imply progress on (a generalized version of) the "log rank conjecture" of communication complexity.

1 Introduction

Yao's two-party communication complexity [23] is a well studied model (see [16] for background). Several extensions of this model for multiparty settings were considered in the literature. In this paper, we consider the following extension that is arguably the simplest one (alternative mutltiparty models are discussed below): there are $k > 2$ players $P_1, \ldots, P_k$, where each player P_i holds an input $x_i \in \{0,1\}^n$. The players communicate by using a broadcast channel (sometimes referred to as a "blackboard" in the communication complexity literature) and their goal is to compute some function f

⋆ Supported by NWO mathematics cluster DIAMANT.
⋆⋆ Research supported by grant 1310/06 from the Israel Science Foundation (ISF).

S. Albers et al. (Eds.): ICALP 2009, Part I, LNCS 5555, pp. 390–402, 2009.

evaluated at their inputs, i.e., the value $f(x_1, \ldots, x_k)$, while minimizing the number of bits communicated.[1]

As in the two-party case, the most interesting question for such a model is proving lower bounds, with an emphasis on "generic" methods. The main lower bound method known for the above multiparty model is the so-called *partition argument* method. Namely, the k players are partitioned into two disjoint sets of players A and B and we look at the induced two-argument function $f^{A,B}$ defined by $f^{A,B}(\{x_i\}_{i \in A}, \{x_i\}_{i \in B}) \stackrel{\text{def}}{=} f(x_1, \ldots, x_k)$. Then, by applying any of the various lower-bound methods known for the two-party case, we obtain some lower bound $\ell_{A,B}$ on the (two-party) communication complexity of $f^{A,B}$. This value is obviously a lower bound also for the (multiparty) communication complexity of f. Finally, the partition arguments bound ℓ_{PAR} is the best lower bound that can be obtained in this way; namely, $\ell_{\text{PAR}} = \max_{A,B} \{\ell_{A,B}\}$, where the maximum is taken over all possible partitions A, B as above.

The fundamental question studied in this work is whether the partition argument method suffices for determining the multiparty communication complexity of every k-argument function f; or, how close is the partition argument bound to the true communication complexity of f. More specifically,

Question. Is there a constant $c \geq 1$ such that, for every k-argument function f, the k-party communication complexity of f is between ℓ_{PAR} and $(\ell_{\text{PAR}})^c$?

As usual, this question can be studied with respect to various communication complexity measures (deterministic, non-deterministic, randomized etc.).

Our Results. On one hand, for the deterministic case, we explain the current state of affairs where partition arguments seem to yield essentially the best known lower bounds. We do this by relating the above question, in the deterministic setting, to one of the central open problems in the study of communication complexity, the so-called "log-rank conjecture" (see [21, 20] and the references therein), stating that the deterministic communication complexity of every two-argument boolean function g is polynomially-related to the log of the algebraic rank (over the reals) of the matrix M_g corresponding to the function. Specifically, we show that if (a natural extension of) the log-rank conjecture holds then the answer to the above question is positive; namely, in this case, the partition arguments bound is polynomially close to the true multiparty communication complexity. In other words, a negative answer to the above question implies refuting the (generalized) log-rank conjecture. Furthermore, we characterize the collections of partitions one has to consider in order to decide if the rank lower bound is applicable for a given k-argument function. Specifically, these are the collections of partitions such that for every two players P_i and P_j there is a partition A, B such that $i \in A$ and $j \in B$. That is, if all induced functions in such a collection are easy, then, assuming the generalized "log rank" conjecture, the original function is easy as well.

On the other hand, we show that both in the case of randomized communication complexity and in the case of non-deterministic communication complexity, the answer to the above question is negative (in a very strong manner). Namely, there exists a

[1] If broadcast is not available, but rather the players are connected via point-to-point channels, this influences the communication complexity by a factor of at most k; we will mostly view k as a constant (e.g., $k = 3$) and hence the difference is minor.

3-argument function f, for which the partition-argument bound cannot yield a lower bound better than $\Omega(\log n)$ (or, in other words, each of the induced two-party functions has an upper bound of $O(\log n)$); however, the true 3-party communication complexity of f is exponentially larger, i.e. $\Omega(n)$ (of course, our lower bound should be proved using a different method; specifically, we pick the function f from a carefully designed family of functions, where the induced two-argument functions for all of them have low complexity, and show that with positive probability we will get a function with large multiparty communication complexity).[2] In the full version of the paper, we show that there exist search problems (relations) that require $\Omega(n)$ bits of communication while all their induced relations can be solved without communication at all. The full version also contains a comparison between the lower bounds achieved by applying "fooling set" technique to a function f and the bounds we get by applying the same lower bound technique to the induced functions of f.

Related work. Multiparty communication complexity was studied in other models as well. Dolev and Feder [9, 8] (see also [10, 11]) studied a k-party model where the communication is managed via an additional party referred to as the "coordinator". Their main result is a proof that the gap between the deterministic and the non-deterministic communication complexity of every function is quadratic even in this multiparty setting. Their motivation was bridging between the two-party communication complexity model and the model of decision trees, where both have such quadratic gaps. Our model differs from theirs in terms of the communication among players and in that we concentrate on the case of a small number of players.

Another popular model in the study of multiparty communication complexity is the so-called "Number On the Forehead" (NOF) model [6, 3], where each party P_i gets all the inputs $x_1, \ldots, x_k$ *except* for x_i. This model is less natural in distributed systems settings but it has a wide variety of other applications. It is interesting to note that in the NOF model, partition arguments are useless (because every two players when put together know the entire input to f).

Our results concern the "Number in Hand" k-party model. Lower bound techniques different from partition arguments were presented by Chakrabarti et al. [5], following [2, 4]. These lower bounds are for the "disjointness with unique intersection" *promise problem*. In this problem, the k inputs are subsets of a universe of size n, together with the promise that the k sets are either pairwise disjoint or uniquely intersecting, i.e. they have one element in common but are otherwise disjoint. Note that partition arguments are useless for this promise problem: any two inputs determine the output. Chakrabarti et al. prove a near optimal lower bound of $\Omega(n/k \log k)$ for this function, using information theoretical tools from [4]. Their result is improved to the optimal lower bound of $\Omega(n/k)$ in [13]. This problem has applications to the space complexity of approximating frequency moments in the data stream model (see [1, 2]). It should be noted, however, that there are several contexts where the communication complexity of relations and, in particular, of promise problems, seems to behave differently than that of *functions* (e.g, the context of the "direct-sum" problem [12]). Indeed, for functions, no lower bound technique different than partition arguments is known.

[2] Note that, in order to give a negative answer to the above question, it is enough to discuss the case $k = 3$. This immediately yields a gap also for larger values of k.

2 Preliminaries

Notation. For a positive integer m, we denote by $[m]$ the set $\{1, 2, \ldots, m\}$. All the logarithms in this paper are to the base 2. For two strings $x, y \in \{0,1\}^*$, we use $x \circ y$ to denote their concatenation. We refer by $poly(n)$ to the set of functions that are asymptotically bounded by a polynomial in n.

Two-Party Communication Complexity. For a Boolean function $g : \{0,1\}^n \times \{0,1\}^n \to \{0,1\}$, denote by $D(g)$ the deterministic communication complexity of g, i.e., the number of bits Alice, holding $x \in \{0,1\}^n$, and Bob, holding $y \in \{0,1\}^n$, need to exchange in order to jointly compute $g(x,y)$. Denote by $M_g \in \{0,1\}^{2^n \times 2^n}$ the matrix representing g, i.e., $M_g[x,y] = g(x,y)$ for every $(x,y) \in \{0,1\}^n \times \{0,1\}^n$.

k-Party Communication Complexity. Let $f : (\{0,1\}^n)^k \to \{0,1\}$ be a Boolean function. A set of k players $P_1, \ldots, P_k$ hold inputs $x_1, \ldots, x_k$ respectively, and wish to compute $f(x_1, \ldots, x_k)$. The means of communication is broadcast. Again, we denote by $D(f)$ the complexity of the best deterministic protocol for computing f in this model, where the complexity of a protocol is the number of bits sent on the worst-case input. Generalizing the two-argument case, we represent f using a k-dimensional tensor M_f.

Non-Deterministic Communication Complexity. For $b \in \{0,1\}$, a b-monochromatic (combinatorial) rectangle of a function $g : \{0,1\}^n \times \{0,1\}^n \to \{0,1\}$ is a set of pairs $R = X \times Y$, where $X, Y \subseteq \{0,1\}^n$, such that for every $x \in X$ and $y \in Y$ it holds that $g(x,y) = b$. A b-cover of g of size t is a set of (possibly overlapping) rectangles $\mathcal{R} = \{R_1, \ldots, R_t\}$ such that, for every pair $(x,y) \in \{0,1\}^n \times \{0,1\}^n$, if $g(x,y) = b$ then there exists an index $i \in [t]$ such that $(x,y) \in R_i$. Denote by $C^b(g)$ the size of the smallest b-cover of g. The non-deterministic communication complexity of g is denoted by $N^1(g) = \log C^1(g)$. Similarly, the co-non-deterministic communication complexity of g is denoted by $N^0(g) = \log C^0(g)$. Finally, denote $C(g) = C^0(g) + C^1(g)$ and $N(g) = \log C(g) \leq \max(N^0(g), N^1(g)) + 1$. (An alternative to this combinatorial definition asks for the number of bits that the parties need to exchange so as to verify that $f(x,y) = b$.) All these definitions generalize naturally to k-argument functions, where we consider combinatorial k-boxes $B = X_1 \times \cdots \times X_k$, rather than combinatorial rectangles.

Randomized Communication Complexity. For a function $g : \{0,1\}^n \times \{0,1\}^n \to \{0,1\}$ and a positive number $0 \leq \epsilon < \frac{1}{2}$, denote by $R_\epsilon(g)$ the communication complexity of the best randomized protocol for g that errs on every input with probability at most ϵ, and denote $R(g) = R_{\frac{1}{3}}(g)$. Newman [19] proved that the *public-coin* model, where the players share a public random string, is equivalent, up to an additive factor of $O(\log n)$ communication, to the *private-coin* model, where each party uses a private independent random string. Moreover, he proved that w.l.o.g, the number of random strings used by the players in the public-coin model is polynomial in n. All these results can be easily extended to k-argument functions.

Lemma 2.1 ([19]). *There exist constants $c > 0, c' \geq 1$ such that for every Boolean function $g : \{0,1\}^n \times \{0,1\}^n \to \{0,1\}$, if $R(g) = r(n)$ then there exists a protocol for*

g in the public-coin model with communication complexity $c' \cdot r(n)$ that uses random strings taken from a set of size $O(n^c)$.

3 The Deterministic Case

In this section we study the power of partition-argument lower bounds in the deterministic case.

Question 1. *Let $k \geq 3$ be a constant integer and f be a k-argument function. What is the maximal gap between $D(f)$ and the maximum $\max_{A,B} D(f^{A,B})$ over all partitions of $[k]$ into (disjoint) subsets A and B?*

In Section 3.2, we use properties of tensor algebra to show that under a generalized version of the well known "log rank" conjecture, partition arguments are universal for multi-party communication complexity. We also characterize the set of partitions one needs to study in order to analyze the communication complexity of a k-argument function. Before that, we give in Section 3.1 a simpler proof for the case $k = 3$. This proof avoids the use of tensor algebra that is needed in the general case.

Let $g : \{0,1\}^n \times \{0,1\}^n \to \{0,1\}$ be a Boolean two-argument function and $M_g \in \{0,1\}^{2^n \times 2^n}$ be the matrix representing it. It is well known that $\log \text{rank}(M_g)$ serves as a lower bound on the (two-party) deterministic communication complexity of g.

Theorem 3.1 ([18]). *For any function $g : \{0,1\}^n \times \{0,1\}^n \to \{0,1\}$, we have $D(g) \geq \log \text{rank}(M_g)$.*

An important open problem in communication complexity is whether the converse is true. This problem is known as the "log rank conjecture." Formally,

Conjecture 3.2 (Log Rank Conjecture). There exists a constant $c > 0$ such that every function $g : \{0,1\}^n \times \{0,1\}^n \to \{0,1\}$ satisfies $D(g) = O((\log^c \text{rank}(M_g))$.

It is known that if such a constant c exists, then $c > 1/0.58 = 1.724$ [20]. As in the two-party case, in k-party communication complexity still $\log \text{rank}(M_f) \leq D(f)$ (formal definition of $\text{rank}(M_f)$ appears in Subsection 3.2). This is true for exactly the same reason as in the two-party case: any deterministic protocol whose complexity is c induces a partition of the tensor M_f into 2^c monochromatic k-boxes. Such boxes are, in particular, rank-1 tensors whose sum is M_f. This, in turn, leads to the following natural generalization of the above conjecture.

Conjecture 3.3 (Log Rank Conjecture for k-Party Computation). Let k be a constant. There exists a constant $c' = c'(k) > 0$, such that for every function $f : (\{0,1\}^n)^k \to \{0,1\}$ it holds that $D(f) = O(\log^{c'} \text{rank}(M_f))$.

Computationally, even tensor rank in three dimensions is very different than rank in two dimensions. While the former is NP-Complete (see [14]), the latter can be computed very efficiently using, e.g., Gaussian elimination. However, in the (combinatorial) context of communication complexity, much of the properties are the same in two and three dimensions. We will show below that, assuming Conjecture 3.3 is correct, the answer to Question 1 is that the partition argument technique always produces a bound that is within a polynomial factor of the true bound. We start with the case $k = 3$ whose proof is similar in nature to the general case but is somewhat simpler and avoids the tensor notation.

3.1 The Three-Party Case

We start with the definition of a rank of three dimensinal matrices, known as *tensor rank*.

Definition 3.4 (Rank of A Three Dimensional Matrix). *A three dimensional matrix $M \in \mathbb{F}^{m\times m\times m}$ is of rank 1 if there exist three non-zero vectors $v, u, w \in \mathbb{F}^m$ such that, for every $x, y, z \in [m]$, it holds that $M[x, y, z] = v[x]u[y]w[z]$. In this case*[3]*, we write $M = v \otimes u \otimes w$. A matrix $M \in \mathbb{F}^{m\times m\times m}$ is of rank r if it can be represented as a sum of r rank 1 matrices (i.e., for some rank-1 three-dimensional matrices $M_1, \ldots, M_r \in \mathbb{F}^{m\times m\times m}$ we have $M = M_1 + \ldots + M_r$), but cannot be represented as the sum of $r - 1$ rank 1 matrices.*

The next theorem states that, assuming the log rank conjecture for 3-party protocols, partition arguments are universal. Furthermore, it is enough to study the communication complexity of *any two* of the three induced functions, in order to understand the communication complexity of the original function. We will use the notation $f^1 := f^{\{1\},\{2,3\}}$, $f^2 := f^{\{2\},\{1,3\}}$, and $f^3 := f^{\{3\},\{1,2\}}$.

Theorem 3.5. *Let $f : \{0,1\}^n \times \{0,1\}^n \times \{0,1\}^n \to \{0,1\}$ be a Boolean function. Consider any two induced functions of f, say f^1, f^2, and assume that Conjecture 3.3 holds with a constant $c' > 0$. Then $D(f) = O((D(f^1) + D(f^2))^{c'})$.*

Towards proving Theorem 3.5, we analyze the connection between the rank of a three-dimensional matrix $M \in \mathbb{F}^{m\times m\times m}$ and some related two-dimensional matrices. More specifically, given M, consider the following two-dimensional matrices $M_1, M_2, M_3 \in \mathbb{F}^{m\times m^2}$, which we call the *induced matrices* of M:

$$M_1[x, \langle y, z\rangle] = M[x, y, z], \quad M_2[y, \langle x, z\rangle] = M[x, y, z], \quad M_3[z, \langle x, y\rangle] = M[x, y, z]$$

We show that if M has "large" rank, then at least two of its induced matrices have large rank, as well [4].

Lemma 3.6. *Let $r_1 = \mathrm{rank}(M_1)$ and $r_2 = \mathrm{rank}(M_2)$. Then $\mathrm{rank}(M) \le r_1 r_2$.*

Proof. Let $v_1, \ldots, v_{r_1} \in \mathbb{F}^n$ be a basis to the columns of M_1. Let $u_1, \ldots, u_{r_2} \in \mathbb{F}^n$ be a basis to the columns of M_2. We claim that there are $r_1 r_2$ vectors $w_{1,1}, \ldots, w_{r_1,r_2}$

[3] Note that the definition is symmetric with respect to v, u, and w.

[4] It is possible to have *one* induced matrix with small rank. For example, if M is defined so that $M[x, y, z] = 1$ if $y = z$ and $M[x, y, z] = 0$ otherwise, then M has rank m while its induced matrix M_1 is of rank 1.

such that $M = \sum_{i=1}^{r_1} \sum_{j=1}^{r_2} v_i \otimes u_j \otimes w_{i,j}$. This implies that $\mathrm{rank}(M) \leq r_1 r_2$, as required.

Fix $z \in [m]$, and consider the matrix $A_z \in \mathbb{F}^{m \times m}$ defined by $A_z[x,y] = M[x,y,z]$. Observe that the columns of the matrix A_z belong to the set of columns of the matrix M_1 (note that along each column of A_z only the x coordinate changes, exactly as is the case along the columns of the matrix M_1). Therefore, the columns of A_z are spanned by the vectors $v_1, \ldots v_{r_1}$. Similarly, the rows of the matrix A_z belong to the set of columns of the matrix M_2 (in each row of A_z, the value x is fixed and y is changed as is the case along the columns of the matrix M_2) and are thus spanned by the vectors $u_1, \ldots u_{r_2}$

Let $V \in \mathbb{F}^{m \times r_1}$ be the matrix whose columns are the vectors $v_1, \ldots v_{r_1}$. Similarly, let $U \in \mathbb{F}^{r_2 \times m}$ be the matrix whose rows are $u_1, \ldots, u_{r_2}$. The above arguments show that there exists a matrix $Q'_z \in \mathbb{F}^{m \times r_2}$ such that $A_z = Q'_z U$ and $\mathrm{rank}(Q'_z) = \mathrm{rank}(A_z)$. This is since the row space of A_z is spanned by the rows of U, and since the rows of U are independent. Hence, the column space of the matrix Q'_z is identical to the column space of the matrix A_z, and so it is also spanned by the columns of the matrix V. Therefore, there exists a matrix $Q_z \in \mathbb{F}^{r_1 \times r_2}$ such that $Q'_z = VQ_z$. Altogether, we get that $A_z = VQ_zU$. Simple linear algebraic manipulations show that this means that $A_z = \sum_{i=1}^{r_1} \sum_{j=1}^{r_2} Q_z[i,j] v_i \otimes u_j$.

Now, for every $i \in [r_1]$ and $j \in [r_2]$, define $w_{i,j} \in \mathbb{F}^n$ such that, for every $z \in [m]$, it holds that $w_{i,j}[z] = Q_z[i,j]$. Then $M = \sum_{i=1}^{r_1} \sum_{j=1}^{r_2} v_i \otimes u_j \otimes w_{i,j}$. □

Proof. (of Theorem 3.5) By the rank lower bound, $\log \mathrm{rank}(M_{f^1}) \leq D(f^1)$ and $\log \mathrm{rank}(M_{f^2}) \leq D(f^2)$. By Lemma 3.6, $\mathrm{rank}(M_f) \leq \mathrm{rank}(M_{f^1}) \mathrm{rank}(M_{f^2})$. Therefore,

$$\log \mathrm{rank}(M_f) \leq \log \mathrm{rank}(M_{f^1}) + \log \mathrm{rank}(M_{f^2}) \leq D(f^1) + D(f^2).$$

Finally, assuming Conjecture 3.3, we get $D(f) = O(\log^{c'} \mathrm{rank}(M_f)) = O((D(f^1) + D(f^2))^{c'})$. □

3.2 The k-Party Case

We start with some mathematical background.

Tensors, Flattening, Pairing, and Rank. Let $V_1, \ldots, V_k$ be vector spaces over the same field $\mathcal{F}$; all tensor products are understood to be over that field. For any subset I of $[k]$ write $V_I := \bigotimes_{i \in I} V_i$. An element T of $V_{[k]}$ is called a *k-tensor*, and can be written as a sum of *pure* tensors $v_1 \otimes \cdots \otimes v_k$ where $v_i \in V_i$. The minimal number of pure tensors in such an expression for T is called the *rank* of T. Hence, pure tensors have rank 1.

If each V_i is some $\mathcal{F}^{n_i}$, then an element of the tensor product can be thought of as a k-dimensional array of numbers from $\mathcal{F}$, of size $n_1 \times \cdots \times n_k$. A rank-1 tensor is an array whose $(j_1, \ldots, j_k)$-entry is the product $a_{1,j_1} \cdots a_{k,j_k}$ where $(a_{i,j})_j$ is an element of $\mathcal{F}^{n_i}$.

For any partition $\{I_1, \ldots, I_m\}$ of $[k]$, we can view T as an element of $\bigotimes_{l \in [m]} (V_{I_l})$; this is called the *flattening* $\flat_{I_1,\ldots,I_m} T$ of T or just an m-flattening of T. It is the same tensor—or more precisely, its image under a canonical isomorphism—but the notion of rank changes: the rank of this m-flattening is the rank of T considered as an m-tensor in the space $\bigotimes_{l\in[m]} U_l$, where U_l happens to be the space V_{I_l}.

If one views a k-tensor as a k-dimensional array of numbers, then an m-flattening is an m-dimensional array. For instance, if $k = 3$ and $n_1 = 2, n_2 = 3, n_3 = 5$, then the partition $\{\{1,2\},\{3\}\}$ gives rise to the flattening where the $2 \times 3 \times 5$-array T is turned into a 6×5-matrix.

Another operation that we will use is *pairing*. For a vector space U, denote by U^* the dual space of functions $\phi : U \to \mathcal{F}$ that are $\mathcal{F}$-linear, i.e., that satisfy $\phi(u+v) = \phi(u) + \phi(v)$ and $\phi(cu) = c\phi(u)$ for all $u, v \in U$ and $c \in \mathcal{F}$. Let I be a subset of $[k]$, let $\xi = \otimes_{i\in I}\xi_i \in \bigotimes_{i\in I}(V_i^*)$ be a pure tensor, and let $T = \otimes_{i\in[k]} v_i \in V_{[k]}$ be a pure tensor. Then, the *pairing* $\langle T, \xi\rangle \in V_{[k]\setminus I}$ is defined as $\langle T, \xi\rangle = c \cdot \otimes_{i\in[k]\setminus I} v_i$; where $c \in \mathcal{F}$ is defined as $c := \left(\prod_{i\in I}\langle v_i, \xi_i\rangle\right)$. The pairing is extended bilinearly in ξ and T to general tensors. Note that ξ induces a natural linear map $V_{[k]} \to V_{[k]\setminus I}$, sending T to the pairing $\langle T, \xi\rangle$.

If one views a k-tensor as a k-dimensional array of numbers, then pairing also reduces the dimension of the array. For instance, pairing a tensor $T \in \mathcal{F}^2 \otimes \mathcal{F}^3 \otimes \mathcal{F}^5$ with a vector in the dual of the last factor $\mathcal{F}^5$ gives a linear combination of the five 2×3-matrices of which T consists. Pairing with pure tensors corresponds to a repeated pairing with dual vectors in individual factors.

Here are some elementary facts about tensors, rank, flattening, and pairing:

Submultiplicativity. If T is a k-tensor in $\bigotimes_{i\in[k]} V_i$ and S is an l-tensor in $\bigotimes_{j\in[l]} W_j$, then the rank of the $(k+l)$-tensor $T \otimes S$ is at most the product of the ranks of T and S.

Subadditivity. If T_1, T_2 are k-tensors in $\bigotimes_{i\in[k]} V_i$, then the rank of the k-tensor $T_1 + T_2$ is at most the sum of the ranks of T_1 and T_2.

Pairing with pure tensors does not increase rank. If $T \in V_{[k]}$ and $\xi = \otimes_{i\in I}\xi_i$ then the rank of $\langle T, \xi\rangle$ is at most that of T.

Linear independence for 2-tensors. If a 2-tensor T in $V_1 \otimes V_2$ has rank d, then in any expression $\sum_{p=1}^{d} R_p \otimes S_p = T$ with $R_p \in V_1$ and $S_p \in V_2$ the set $\{S_1, \ldots, S_p\}$ (and also the set $\{R_1, \ldots, R_p\}$) is linearly independent.

To state our theorem, we need the following definition.

Definition 3.7 (Separating Collection of Partitions). *Let k be a positive integer. Let C be a collection of partitions $\{I, J\}$ of $[k] = \{1, \ldots, k\}$ into two non-empty parts. We say that C is* separating *if, for every $i, j \in [k]$ such that $i \neq j$, there exists a partition $\{I, J\} \in C$ with $i \in I$ and $j \in J$.*

Theorem 3.8. *Let $f : (\{0,1\}^n)^k \to \{0,1\}$ be a Boolean function. Let C be a separating collection of partitions of $[k]$ and assume that Conjecture 3.3 holds with a constant $c' > 0$. Then*

$$D(f) = O(\ (2(k-1) \max_{\{I,J\}\in C} D(f^{I,J})\ )^{c'}\ ).$$

For a special separating collection of partitions we can give the following better bound.

Theorem 3.9. *Let* $f : (\{0,1\}^n)^k \to \{0,1\}$ *be a Boolean function. Set* $d_i := D(f^{\{i\},[k]\setminus\{i\}})$ *and assume that Conjecture 3.3 holds with a constant* $c' > 0$. *Then* $D(f) = O((\sum_{i=1}^{k-1} d_i)^{c'})$.

These results will follow from upper bounds on the rank of k-tensors, given upper bounds on the ranks of the 2-flattenings corresponding to $\mathcal{C}$.

Theorem 3.10. *Let* $V_1, \ldots, V_k$ *be finite-dimensional vector spaces and let* T *be a tensor in their tensor product* $\bigotimes_{i\in[k]} V_i$. *Let* $\mathcal{C}$ *be a separating collection of partitions of* $[k]$, *and let* $d_{\max}$ *be the maximal rank of any 2-flattening* $\flat_{I,J}T$ *with* $\{I,J\} \in \mathcal{C}$. *Then* $\text{rank}T \leq d_{\max}^{2(k-1)}$.

Proof. We prove the statement by induction on k. For $k = 1$ the statement is that $\text{rank}T \leq 1$, which is true. Now suppose that $k > 1$ and that the result is true for all l-tensors with $l < k$ and all separating collections of partitions of $[l]$. Pick $\{I,J\} \in \mathcal{C}$ and write $T = \sum_{p=1}^{d} R_p \otimes S_p$, where $R_p \in V_I$, $S_p \in V_J$, $d \leq d_{\max}$, and the sets $R_1, \ldots, R_d$ and $S_1, \ldots, S_d$ are both linearly independent. This is possible by the condition that the 2-tensor (or matrix) $\flat_{I,J}T$ has rank at most $d_{\max}$. As the S_p are linearly independent, we can find *pure* tensors $\zeta_1, \ldots, \zeta_d \in V_J^*$ such that the matrix $(\langle S_p, \zeta_q\rangle)_{p,q}$ is invertible.

For each $q = 1, \ldots, d$ set $T_q := \langle T, \zeta_q\rangle \in V_I$. By invertibility of the matrix $(\langle S_p, \zeta_q\rangle)_{p,q}$ every R_p is a linear combination of the T_q, so we can write T as $T = \sum_{p=1}^{d} T_p \otimes S'_p$, where $S'_1, \ldots, S'_d$ are the linear combinations of the S_i that satisfy $\langle S'_p, \zeta_q\rangle = \delta_{p,q}$. Now we may apply the induction hypothesis to each $T_q \in V_I$. Indeed, for every $\{I',J'\} \in \mathcal{C}$ such that $I \cap I', I \cap J' \neq \emptyset$ we have $\flat_{I\cap I', I\cap J'} T_q = \langle \flat_{I',J'} T, \zeta_q\rangle$, and since ζ_q is a pure tensor, the rank of the right-hand side is at most that of $\flat_{I',J'}T$, hence at most $d_{\max}$ by assumption. Moreover, the collection

$$\{\{I\cap I', I\cap J'\} \mid \{I',J'\} \in \mathcal{C} \text{ with } I\cap I', I\cap J' \neq \emptyset\}$$

is a separating collection of partitions of I. Hence each T_q satisfies the induction hypothesis and we conclude that $\text{rank}T_q \leq d_{\max}^{2(|I|-1)}$. A similar, albeit slightly asymmetric, argument shows that $\text{rank}S'_q \leq d_{\max}^{2(|J|-1)+1}$: there exist pure tensors $\xi_1, \ldots, \xi_d \in V_I^*$ such that the matrix $(\langle T_p, \xi_r\rangle)_{p,r}$ is invertible. This means that each S'_q is a linear combination of the d tensors $T'_r := \langle T, \xi_r\rangle \in V_J$. The induction hypothesis applies to each of these, and hence $\text{rank}S'_q \leq d_{\max} \cdot d_{\max}^{2(|J|-1)}$ by subadditivity. Finally, using submultiplicativity and subadditivity we find

$$\text{rank}T \leq d_{\max} \cdot d_{\max}^{2|I|-1} \cdot d_{\max} \cdot d_{\max}^{2|J|-1} = d_{\max}^{2(k-1)},$$

□

Proof. (of Theorem 3.8) By Conjecture 3.3, we have $D(f) = O(\log^{c'} \text{rank}(M_f))$. Theorem 3.10 yields $\log \text{rank}(M_f) \leq 2(k-1)\max_{\{I,J\}\in\mathcal{C}} \log \text{rank}(M_{f^{I,J}})$, which by the rank lower bound is at most $2(k-1)\max_{\{I,J\}\in\mathcal{C}} D(f^{I,J})$. This proves the theorem. □

Just like Theorem 3.8 follows from Theorem 3.10, Theorem 3.9 follows immediately from the following proposition.

Proposition 3.11. *Let* $V_1, \ldots, V_k$ *be finite-dimensional vector spaces and let* T *be a tensor in their tensor product* $\bigotimes_{i\in[k]} V_i$*. Denote the rank of the* 2*-flattening* $\flat_{\{i\},[k]\setminus\{i\}}T$ *by* d_i*. Then* $\text{rank}T \leq d_1 \cdots d_{k-1}$.

Proof. Denote by U_i the subspace of V_i consisting of all pairings $\langle T, \xi\rangle$ as ξ runs over $V^*_{[k]\setminus\{i\}}$. Then $\dim U_i = d_i$ and the k-tensor T already lies in $(\bigotimes_{i\in[k-1]} U_i)\otimes U_k$. After choosing a basis of $\bigotimes_{i\in[k-1]} U_i$ consisting of pure tensors T_l $(l = 1, \ldots, d_1 \cdots d_{k-1})$, T can be written (in a unique manner) as $\sum_l T_l \otimes u_l$ for some vectors $u_l \in U_k$. Hence, T has rank at most $d_1 \cdots d_{k-1}$. □

4 Other Models of Communication Complexity

4.1 The Nondeterministic Model

In this section, we show that in the nondeterministic model there are functions with large communication complexity, for which all induced functions have low communication complexity. As in the deterministic case, the non-deterministic communication complexity of the induced functions of f give a lower bound on the non-deterministic communication complexity of f. It is natural to ask the analogue of Question 1 for non-deterministic communication complexity. We will show however that there can be an exponential gap between the non-deterministic communication complexity of a function and its induced functions. Note that, for proving the existence of a gap, it is enough to present such a gap in the 3-party setting. Not being able to find an explicit function f for which partition arguments result in lower bounds that are exponentially weaker than the true non-deterministic communication complexity of f, we turn to proving that such functions *exist.* Towards this goal, we use a well known combinatorial object – Latin squares.

Definition 4.1 (Latin square). *Let* m *be an integer. A matrix* $L \in [m]^{m\times m}$ *is a* Latin square *of dimension* m *if every row and every column of L is a permutation of* $[m]$.

The following lemma gives a lower bound on the number of Latin squares of dimension m (see, for example, [22, Chapter 17]).

Lemma 4.2. *The number of Latin squares of dimension* m *is at least* $\prod_{j=0}^{m} j!$*. In particular, this is larger than* $2^{m^2/4}$.

Let n be an integer and set $m = 2^n$. Let L be a Latin square of dimension m. Define the function $f_L : [m] \times [m] \times [m] \to \{0,1\}$ such that $f_L(x, y, z) = 1$ if and only if $L[x, y] \neq z$. The non-deterministic communication complexity of f_L^1, f_L^2 and f_L^3 is at most $\log n = \log\log m$ since each of the induced functions locally reduces to the function $\text{NE}_n : \{0,1\}^n \times \{0,1\}^n \to \{0,1\}$, defined by $\text{NE}_n(a,b) = 1$ iff $a \neq b$ for which it is known that $N^1(\text{NE}_n) = \log n + 1$ (e.g., for f_L^1, the player holding (y, z) locally computes the unique value x_0 such that $L[x_0, y] = z$; then, the players verify that $x_0 \neq x$). It is left to prove that there exists a Latin square L such that the non-deterministic communication complexity of f_L is $\Omega(n)$.

Lemma 4.3. *The number of different covers of size t of the $[m] \times [m] \times [m]$ cube is at most 2^{3mt}. (The proof is by simple counting and is omitted for lack of space.)*

Theorem 4.4. *There exists a Latin square L of dimension $m = 2^n$ such that the non-deterministic communication complexity of f_L is $n - O(1)$.*

Proof. For two different Latin squares $L_0 \neq L_1$ of dimension m, it holds that $f_{L_0} \neq f_{L_1}$. In addition, no 1-cover $\mathcal{R}$ corresponds to two distinct functions $f_0 \neq f_1$. Hence, the number of covers needed to cover all the functions f_L, where L is a Latin square of dimension m, is at least $2^{m^2/4}$. Let t be the size of the largest 1-cover among this set of covers. Then we get that $2^{3mt} \geq 2^{m^2/4}$. Hence, $3mt \geq m^2/4$, which implies $t \geq m/12$. Therefore, $\log t \geq \log m - \log 12 = n - \log 12$. □

4.2 The Randomized Model

Next, we show that partition arguments are also not sufficient for proving tight lower bounds on the randomized communication complexity. Let $f : \{0,1\}^n \times \{0,1\}^n \rightarrow \{0,1\}$ be a Boolean function. Recall that $R(f)$ denotes the communication complexity of a best randomized protocol for f that errs with probability at most $1/3$. It is well known that $R(\text{NE}_n) = O(\log n)$. Again, we use the functions defined by Latin squares of dimension $m = 2^n$. Our argument follows the, somewhat simpler, non-deterministic case. On one hand, as before, the three induced functions are easily reduced to NE and hence their randomized communication complexity is $O(\log n)$. To prove that some of the functions f_L are hard (i.e., an analog of Theorem 4.4), we need to count the number of distinct randomized protocols of communication complexity $\log t$.

Lemma 4.5. *The number of different randomized protocols over inputs from $[m] \times [m] \times [m]$ of communication complexity r is $2^{m2^{O(r)}\text{poly}(\log m)}$.*

Proof. By Lemma 2.1, any randomized protocol $\mathcal{P}$, with communication complexity r can be transformed into another protocol $\mathcal{P}'$ with communication complexity $O(r)$ that uses just $O(\log n)$ random bits (or, alternatively, $\text{poly}(n) = \text{poly}(\log m)$ possible random tapes). Hence, we can view any randomized protocol of complexity r as a set of $\text{poly}(\log m)$ disjoint covers of the cube $[m] \times [m] \times [m]$, each consisting of at most $2^{O(r)}$ boxes. The number of ways for choosing each such box is 2^{3m} and so the total number of such protocols is $2^{m2^{O(r)}\text{poly}(\log m)}$. □

Theorem 4.6. *There exists a Latin square L of dimension $m = 2^n$ such that the randomized communication complexity of f_L is $\Omega(n)$.*

Proof. By Lemma 4.2, the number of randomized protocols needed to solve all the functions f_L where L is a Latin square of dimension m must be at least $2^{m^2/4}$ (again, each randomized protocol corresponds to at most one function, according to the majority value for each input). Let r be the maximum randomized complexity of a function f_L over the set of all Latin squares L. Then we get that $2^{m2^{O(r)}\text{poly}(\log m)} \geq 2^{m^2/4}$. Hence, $m2^{O(r)}\text{poly}(\log m) \geq m^2/4$, which implies $2^{O(r)} \geq m/\text{poly}(\log m)$. Therefore, $r = \Omega(\log m - \log\log m) = \Omega(n)$. □

Acknowledgments. We wish to thank Ronald de Wolf for useful discussions and, in particular, for referring us to [5].

References

[1] Ajtai, M., Jayram, T.S., Kumar, R., Sivakumar, D.: Approximate counting of inversions in a data stream. In: STOC, pp. 370–379 (2002)
[2] Alon, N., Matias, Y., Szegedy, M.: The space complexity of approximating the frequency moments. J. Comput. Syst. Sci. 58(1), 137–147 (1999)
[3] Babai, L., Nisan, N., Szegedy, M.: Multiparty protocols and logspace-hard pseudorandom sequences. In: Proc. of the 21st ACM Symp. on the Theory of Computing, pp. 1–11 (1989)
[4] Bar-Yossef, Z., Jayram, T.S., Kumar, R., Sivakumar, D.: An information statistics approach to data stream and communication complexity. J. Comput. Syst. Sci. 68(4), 702–732 (2004)
[5] Chakrabarti, A., Khot, S., Sun, X.: Near-optimal lower bounds on the multi-party communication complexity of set disjointness. In: IEEE Conference on Computational Complexity, pp. 107–117 (2003)
[6] Chandra, A., Furst, M., Lipton, R.: Multiparty protocols. In: Proc. of the 15th ACM Symp. on the Theory of Computing, pp. 94–99 (1983)
[7] Dietzfelbinger, M., Hromkovic, J., Schnitger, G.: A comparison of two lower-bound methods for communication complexity. Theor. Comput. Sci. 168(1), 39–51 (1996)
[8] Dolev, D., Feder, T.: Multiparty communication complexity. In: Proc. of the 30th IEEE Symp. on Foundations of Computer Science, pp. 428–433 (1989)
[9] Dolev, D., Feder, T.: Determinism vs. nondeterminism in multiparty communication complexity. SIAM J. Comput. 21(5), 889–895 (1992)
[10] Duris, P.: Multiparty communication complexity and very hard functions. Inf. Comput. 192(1), 1–14 (2004)
[11] Duris, P., Rolim, J.D.P.: Lower bounds on the multiparty communication complexity. J. Comput. Syst. Sci. 56(1), 90–95 (1998)
[12] Feder, T., Kushilevitz, E., Naor, M., Nisan, N.: Amortized communication complexity. SIAM J. Comput. 24(4), 736–750 (1995)
[13] Gronemeier, A.: Asymptotically optimal lower bounds on the nih-multi-party information complexity of the and-function and disjointness. In: STACS 2009, pp. 505–516 (2009)
[14] Håstad, J.: Tensor rank is NP-complete. J. Algorithms 11(4), 644–654 (1990)
[15] Karchmer, M., Wigderson, A.: Monotone circuits for connectivity require super-logarithmic depth. In: Proc. of the 20th ACM Symp. on the Theory of Computing, pp. 539–550 (1988)
[16] Kushilevitz, E., Nisan, N.: Communication Complexity. Cambridge University Press, Cambridge (1997)
[17] Lipton, R.J., Sedgewick, R.: Lower bounds for VLSI. In: Proc. of the 13rd ACM Symp. on the Theory of Computing, pp. 300–307 (1981)
[18] Mehlhorn, K., Schmidt, E.M.: Las vegas is better than determinism in VLSI and distributed computing. In: Proc. of the 14th ACM Symp. on the Theory of Computing, pp. 330–337 (1982)
[19] Newman, I.: Private vs. common random bits in communication complexity. Inf. Process. Lett. 39(2), 67–71 (1991)
[20] Nisan, N., Wigderson, A.: On rank vs. communication complexity. Combinatorica 15(4), 557–565 (1995)

[21] Raz, R., Spieker, B.: On the "log rank"-conjecture in communication complexity. Combinatorica 15(4), 567–588 (1995)
[22] van Lint, J.H., Wilson, R.M.: A Course in Combinatorics. Cambridge University Press, Cambridge (1992)
[23] Yao, A.C.: Some complexity questions related to distributed computing. In: Proc. of the 11th ACM Symp. on the Theory of Computing, pp. 209–213 (1979)

High Complexity Tilings with Sparse Errors*

Bruno Durand[1], Andrei Romashchenko[2], and Alexander Shen[3]

[1] LIF Marseille, CNRS & Univ. Aix–Marseille
Bruno.Durand@lif.univ-mrs.fr
[2] LIF Marseille, CNRS & Univ. Aix–Marseille, on leave from IITP RAS
andrei.romashchenko@gmail.com
[3] LIF Marseille, CNRS & Univ. Aix–Marseille, on leave from IITP RAS
alexander.shen@lif.univ-mrs.fr

Abstract. Tile sets and tilings of the plane appear in many topics ranging from logic (the Entscheidungsproblem) to physics (quasicrystals). The idea is to enforce some global properties (of the entire tiling) by means of local rules (for neighbor tiles). A fundamental question: Can local rules enforce a complex (highly irregular) structure of a tiling?

The minimal (and weak) notion of irregularity is *aperiodicity*. R. Berger constructed a tile set such that every tiling is aperiodic. Though Berger's tilings are not periodic, they are very regular in an intuitive sense.

In [3] a stronger result was proven: There exists a tile set such that all $n \times n$ squares in all tilings have Kolmogorov complexity $\Omega(n)$, i.e., contain $\Omega(n)$ bits of information. Such a tiling cannot be periodic or even computable.

In the present paper we apply the fixed-point argument from [5] to give a new construction of a tile set that enforces high Kolmogorov complexity tilings (thus providing an alternative proof of the results of [3]). The new construction is quite flexible, and we use it to prove a much stronger result: there exists a tile set such that all tilings have high Kolmogorov complexity even if (sparse enough) tiling errors are allowed.

1 Introduction

Tiles are unit squares with colored sides. We may place translated copies of the same tile into different cells of a cell paper (rotations are not allowed). Tiles in the neighbor cells should match (common side must have the same color in both).

More formally, we consider a finite set C of *colors*. A *tile* is a quadruple of colors (left, right, top and bottom ones), i.e., an element of C^4. A *tile set* is a subset $\tau \subset C^4$. A *tiling* of the plane with tiles from τ (τ-*tiling*) is a mapping $U \colon \mathbb{Z}^2 \to \tau$ that respects the color matching condition.

It is well known that local rules can enforce some kind of irregularity in tilings. This first and simplest example of *irregularity property* is aperiodicity. The following classical result was proven by Berger [1]:

* Supported by NAFIT ANR-08-EMER-008-01 and RFBR 09-01-00709-a grants.

S. Albers et al. (Eds.): ICALP 2009, Part I, LNCS 5555, pp. 403–414, 2009.

Theorem 1. *There exists a tile set τ such that τ-tilings exist and all of them are not periodic.* [1]

Berger's tilings are aperiodic, but their structure is very regular and rather simple to describe. A tile set which enforces irregularity in a much stronger sense was constructed in [3]: it is a tile set that accepts only tilings of high Kolmogorov complexity. As a tiling is an infinite object, we look at finite patterns in a tiling and measure their Kolmogorov complexity:

Theorem 2. *There exists a tile set τ such that τ-tilings exist and all $n \times n$ squares in all τ-tilings have Kolmogorov complexity $\Omega(n)$.*

More precisely the result can be reformulated as follows: there exists a tile set τ and constants c_1 and c_2 such that τ-tilings exist and in every τ-tiling every $n \times n$-square has Kolmogorov complexity at least $c_1 n - c_2$. The lower bound $\Omega(n)$ in this theorem is tight: if τ-tilings exist, in one of them every $n \times n$ square has complexity $O(n)$ (see the discussion in [3]).

We stress that it is more reasonable to investigate the minimal complexity of a τ-tiling (for a given tile set τ), not its maximal complexity. Indeed, the maximal complexity can be very large (of order $O(n^2)$) for a trivial tile set. E.g., for the set τ of *all* tiles with black and white edges, there is a τ-tiling that contains every $n \times n$ pattern, and maximal complexity is $\Omega(n^2)$.

We refer to [3] for a more detailed discussion and philosophical motivation of Theorem 2. In this paper we give a proof of Theorem 2 and generalize it for tilings *with sparse random errors*.

The precise statement of our result about tilings with random errors requires some technical definitions, see Section 4. Here we explain the intuitive idea. We consider 'faulty' tilings of the plane, where the local tiling rules are not true for the entire plane (as we required in the usual definition of tiling) but can be violated on a sparse randomly chosen set of cells. (This generalization of the standard definition of tiling looks rather natural: in physical crystal grids we usually expect that local rules can be violated at sparse points.) Then we construct such a tile set that even 'faulty' tilings (for *almost all* sets of faults) must have high Kolmogorov complexity.

The paper is organized as follows. In Section 2 we remind the fixed-point construction of an aperiodic tile set [5] that is used as a starting point.

In Section 3 we provide a new proof of Theorem 2 using fixed point construction with variable zoom factors. The new proof is simpler in some respects then the original one from [3], and can be generalized to the case when sparse errors are allowed.

Finally, in Section 4 we briefly explain our most difficult result: a "robust" tile set such that all tilings, even with a sparsely placed errors, have linear complexity of fragments. To achieve this result, we combine several ideas: (1) a fixed-point tile set with variable zoom factors; (2) calculation and propagation of Reed–Solomon's checksums, and (3) covering of a random sparse set by a hierarchical family of isolated islands (technically, we need to update the construction from [5,14] and use bi-islands instead of simple islands). For the lack of space, the technical details are omitted in the conference version of the paper.

2 Fixed-Point Aperiodic Tile Set

In this section we remind the fixed point construction of aperiodic tile sets from [5] (the reader familiar with the arguments from [5] can safely skip this part).

2.1 Macro-tiles

Fix a tile set τ and an integer $N > 1$ (*zoom factor*). A *macro-tile* is an $N \times N$ square tiled by matching τ-tiles. Every side of a macro-tile carries a sequence of N colors called a *macro-color.*

Let ρ be a set of τ-macro-tiles. We say that τ *simulates* ρ if (a) τ-tilings exist, and (b) for every τ-tiling there exists a unique grid of vertical and horizontal lines that cuts this tiling into $N \times N$ macro-tiles from ρ.

Example 1. Assume that we have only one ('white') color and τ consists of a single tile with 4 white sides. Fix some N. There exists a single macro-tile of size $N \times N$. Let ρ be a singleton that contains this macro-tile. Then every τ-tiling can be cut into macro-tiles from ρ. However, τ does not simulate ρ since the placement of cutting lines is not unique.

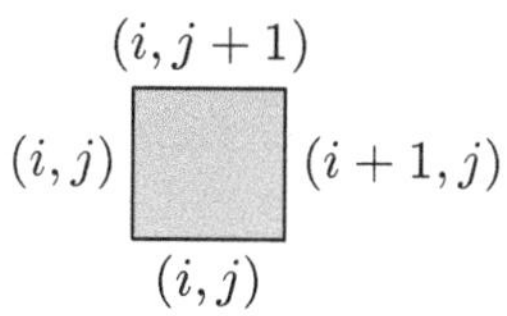

Fig. 1. Tiles of the set τ for Example 2

Example 2. In this example a set ρ that consists of a single macro-tile (that matches itself horizontally and vertically) is simulated by some tile set τ. The tile set τ consists of N^2 tiles indexed by pairs (i, j) of integers modulo N. A tile from τ has colors on its sides as shown (Fig. 1). The macro-tile in ρ has colors $(0, 0), \ldots, (0, N-1)$ and $(0, 0), \ldots, (N-1, 0)$ on its borders. (Fig. 2).

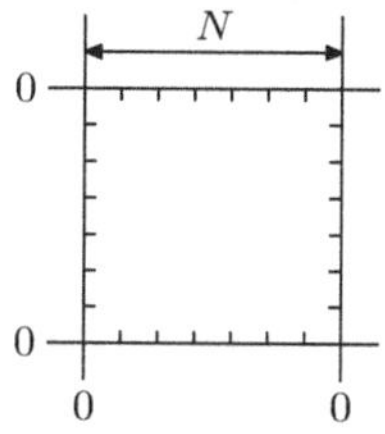

Fig. 2. Macro-tile of size $N \times N$ for Example 2

If a tile set τ simulates some set ρ of τ-macro-tiles with zoom factor $N > 1$ and ρ is isomorphic to τ, the set τ is called *self-similar* (an *isomorphism* between

τ and ρ is a bijection that respects the relations "one tile can be placed on the right of another one" and "one tile can be placed on the top of another one").

The idea of self-similarity is used (more or less explicitly) in most constructions of aperiodic tile sets ([8,2] are exceptions). The usage of self-similarity is based on the following remark:

Proposition 1 (folklore). *All self-similar tile sets τ have only aperiodic tilings.*

(A simple proof of this statement can be found, e.g., in [4] or [5].)

So to prove the existence of aperiodic tile sets it is enough to construct a self-similar tile set.

Theorem 3. *There exists a self-similar tile set τ.*

In the rest of this section we explain some technique (similar to the classical proof of Kleene's fixed-point theorem) that can be used to construct self-similar tile sets. In particular, we get a proof of Theorem 3. In the sequel we generalize this argument and use it in more complicated situations.

First of all, we explain some technique used in our construction: how to simulate a given tile set by embedding computations.

2.2 Simulating a Tile Set

For brevity we say that a tile set τ simulates a tile set ρ when τ simulates some set of macro-tiles $\tilde{\rho}$ isomorphic to ρ (e.g., we say that a self-similar tile set simulates itself).

Let us start with some informal discussion. Assume that we have a tile set ρ whose colors are k-bit strings ($C = \{0,1\}^k$) and the set of tiles $\rho \subset C^4$ is presented as a predicate $R(c_1, c_2, c_3, c_4)$ of four k-bit arguments. Assume that we have some Turing machine $\mathcal{R}$ that computes R. Let us show how to simulate ρ using some other tile set τ.

This construction extends Example 2, but simulates a tile set ρ that contains not a single tile but many tiles. We keep the coordinate system modulo N embedded into tiles of τ; these coordinates guarantee that all τ-tilings can be uniquely cut into blocks of size $N \times N$ and every tile "knows" its position in the block (as in Example 2). In addition to the coordinate system, now each tile in τ carries supplementary colors (from a finite set specified below) on its sides. These colors form a new "layer" superimposed with the old one, i.e., the set of colors is now a Cartesian product of the old one and the set of colors used in this layer.

On the border of a macro-tile (i.e., when one of the coordinates is zero) only two supplementary colors (say, 0 and 1) are allowed. So the macro-color encodes a string of N bits (where N is the size of macro-tiles). We assume that $N \geq k$ and let k bits in the middle of macro-tile sides represent colors from C. All other bits on the sides are zeros (this is a restriction on tiles: each tile "knows" its coordinates so it also knows whether non-zero supplementary colors are allowed).

Now we need additional restrictions on tiles in τ that guarantee that macro-colors on the sides of each macro-tile satisfy the relation R. To achieve this, we ensure that bits from the macro-tile sides are transferred to the central part of the tile where the checking computation of $\mathcal{R}$ is simulated.

For that we need to fix which tiles in a macro-tile form "wires" (this can be done in any reasonable way; let us assume that wires do not cross each other) and then require that each of these tiles carries equal bits on two sides (so some bit propagates along the entire wire); again it is easy to arrange since each tile knows its coordinates.

Then we check R by a local rule that guarantees that the central part of a macro-tile represents a time-space diagram of $\mathcal{R}$'s computation (the tape is horizontal, time goes up). This is done in a standard way. We require that computation terminates in an accepting state: if not, the tiling cannot be formed.

To make this construction work, the size of macro-tile (N) should be large enough: we need enough space for k bits to propagate and enough time and space (=height and width) for all accepting computations of $\mathcal{R}$ to terminate.

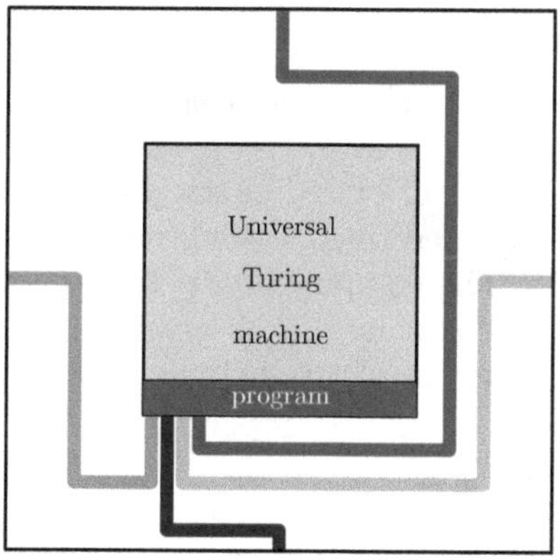

Fig. 3. Checking tiles with a universal TM

In this construction the number of supplementary colors depends on the machine $\mathcal{R}$ (the more states it has, the more colors are needed in the computation zone). To avoid this dependency, we replace $\mathcal{R}$ by a fixed universal Turing machine $\mathcal{U}$ that runs a *program* simulating $\mathcal{R}$. Let us agree that the tape of the universal Turing machine has an additional read-only layer. Each cell carries a bit that is not changed during the computation; these bits are used as a program for the universal machine $\mathcal{U}$ (Fig. 3). In terms of our simulation, the columns of the computation zone carry unchanged bits (considered as a program for U), and the tile set restrictions guarantee that the central zone represents the protocol of an accepting computation of $\mathcal{U}$ (with this program). In this way we get a tile set τ that simulates ρ with zoom factor N using $O(N^2)$ tiles. (Again we need N to be large enough, but the constant in $O(N^2)$ does not depend on N.)

2.3 Simulating Itself

We know how to simulate a given tile set ρ (represented as a program for the universal TM) by another tile set τ with a large enough zoom factor N. Now we want τ to be isomorphic to ρ (then Proposition 1 guarantees aperiodicity). For this we use a construction that follows Kleene's recursion (fixed-point) theorem [9].

Note that most rules of τ do not depend on the program for $\mathcal{R}$, dealing with information transfer along the wires, the vertical propagation of unchanged program bits, and the space-time diagram for the universal TM in the computation zone. Making these rules a part of ρ's definition (we let $k = 2\log N + O(1)$ and encode $O(N^2)$ colors by $2\log N + O(1)$ bits), we get a program that checks that macro-tiles behave like τ-tiles in this respect.

The only remaining part of the rules for τ is the hardwired program. We need to ensure that macro-tiles carry the same program as τ-tiles do. For that our program (for the universal TM) needs to access the bits of its own text. (This self-referential action is in fact quite legal: the program is written on the tape, and the machine can read it.) The program checks that if a macro-tile belongs to the first line of the computation zone, this macro-tile carries the correct bit of the program.

How should we choose N (hardwired in the program)? We need it to be large enough so the computation described (which deals with $O(\log N)$ bits) can fit in the computation zone. The computation is rather simple (polynomial in the input size, i.e., $O(\log N)$), so for large N it easily fits in $\Omega(N)$ available time.

This finishes the construction of a self-similar aperiodic tile set.

2.4 Variable Zoom Factor

This construction is flexible enough and can be used in other contexts. For example, the "zoom factor" N could depend on the level k (i.e., be different for macro-tiles, macro-macro-tiles etc.) For this each macro-tile should have k encoded at its sides; this labeling should be consistent when switching to the next level. Using the anthropomorphic terminology, we say that each macro-tile "knows" its level, i.e., the sequence of bits that form a binary representation of the level is transferred from the sides to the tape and the computation checks that all these numbers (level bits for all four sides) are the same. This is, so to say, a "conscious" information processed by a computation in the central region of the macro-tile. One may say also that a macro-tile of any level contains "subconscious" information ("existing in mind but not immediately available to consciousness", as the dictionary says): this is the information that is conscious for the sub-tiles that form a macro-tile, and their sub-tiles (all the way down to the ground level).

Using this terminology, we can say that each macro-tile knows its coordinates in the macro-tile of the next level: for a tile of level k these coordinates are integers modulo N_{k+1}, so in total $\log k+O(\log N_{k+1})$ bits are required for keeping both the level and these coordinates. Note that N_k steps should be enough to perform increment operation modulo N_{k+1}; we assume that both $\log k$ and $\log N_{k+1}$ are much less than N_k. This means that N_k should not increase too fast or too slow (say, $N_k = \log k$ is too slow and $N_{k+1} = 2^{N_k}$ is too fast). Also we need to compute N_{k+1} when k is known, so we assume that not only the size of N_{k+1} (i.e., $\log N_{k+1}$) but also the time needed to compute it given k are small compared to N_k. These restrictions still allow many possibilities, say, $N_k = \sqrt{k}$, $N_k = k$, $N_k = 2^k$, $N_k = 2^{(2^k)}$, $N_k = k!$ etc.

There is one more important point that needs to be covered. How do we guarantee that the bits representing the level k (on the tape of a macro-tile) are correct? In other terms, we need to ensure that the levels known to a macro-tile and to one of its tiles differ by one. (In psychoanalytic terms we need to check that conscious and subconscious information in a tile match each other.) This is done as follows. The tile knows its level and also knows its position in the macro-tile it belongs (its father). So it knows whether it is in the place where father should keep level bits, and can check whether indeed the level bit that father keeps in this place is consistent with the level information the tile has.

3 Tile Set That Has Only Complex Tilings

In this section we provide a new proof of Theorem 2.

3.1 A Bi-infinite Bit Sequence

Proof. We start the proof in the same way as in [3]: we assume that each tile keeps a bit that propagates (unchanged) in the vertical direction. Then any tiling contains a bi-infinite sequence of bits ω_i (where $i \in \mathbb{Z}$). Any $N \times N$ square contains a N-bit substring of this string, so if (for large enough N) any N-bit substring of ω has complexity at least $c_1 N$ for some fixed c_1, we are done.

Such a bi-infinite sequence indeed exists (see [3]; another proof can be obtained by using Lovasz local lemma, see [15]). So our goal is to formulate tilings rules in such a way that a correct tiling "ensures" that the bi-infinite sequence embedded in it indeed has this property.

The set of all "forbidden" binary strings, i.e., strings x such that $K(x) < c_1|x| - c_2$ (here $K(x)$ stands for Kolmogorov complexity of x and $|x|$ stands for the length of x) is enumerable: there is a program that generates all forbidden strings. It would be nice to embed into the tiling a computation that runs this program and compares its output strings with the substrings of ω; such a computation may blow up (create a tiling error) if a forbidden substring is found.

However, this is not easy. There are several difficulties.

- First of all, our self-similar tiling contains only finite computations; the duration depends on the zoom factor and may increase as the level increases (bigger macro-tiles keep longer computations), but at any level the computations are finite.
- The computation at some level deals with bits encoded in the cells of that level, i.e., with macro-tile states. So the computation cannot access the bits

of the sequence directly (they are "deep in the subconscious"), and some mechanism to dig them out is needed.

Let us explain how to overcome these difficulties.

3.2 Bits Delegation

Macro-tile of level k is a square whose side is $L_k = N_0 \cdot N_1 \cdot \ldots \cdot N_{k-1}$, so there are L_k vertical lines (carrying the bits of the sequence) that intersect this macro-tile. Let us delegate each of these bits to one of the macro-tiles of level k that cover the corresponding line. Note that a macro-tile of the next $(k+1)$-th level is made of $N_k \times N_k$ macro-tiles of level k. We assume that N_k is much bigger than L_k (more about choice of N_k later); this guarantees that there is enough macro-tiles of level k (in the next level macro-tile) to serve all bits that intersect them. Let us decide that ith macro-tile of level k (from bottom to top) in a $(k+1)$-level macro-tile serves (consciously knows, so to say) $(i \bmod L_k)$-th bit (from the left) in its zone. (In this way we have several macro-tiles of level k in each macro-tile of level $k+1$ that are responsible for the same bit, but this does not create any problems.)

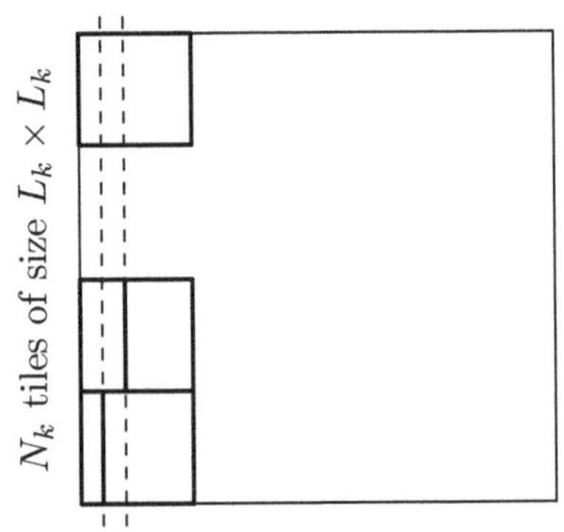

Fig. 4. Bit delegation

So each bit (each vertical line) has a representative on every level — a macro-tile that consciously knows this bit. However, we need some mechanisms that guarantee that this information is indeed true (consistent on different levels). On the bottom level it is easy, since the bits are available directly.

To guarantee the consistency we use the same trick as in Section 2.4: at each level we keep the information not only for this level but also for the next (father) level, and made necessary consistency checks. Namely, each macro-tile knows (has on its computation tape):

- the bit delegated to this macro-tile;
- the coordinates of this macro-tile in its father macro-tile (that are already used in the fixed-point construction); the y-coordinate determines the position of the bit delegated to this macro-tile (relative to the left boundary of the macro-tile).

- the bit delegated to the father of this macro-tile;
- the coordinates of the father macro-tile in the grandfather macro-tile.

This information is subject to consistency checks:

- the information about the father macro-tile should coincide with the same information in neighbor tiles (unless they have a different father, i.e., one of the coordinates is zero).
- if it happens that the bit delegated to the father macro-tile is the same bit as delegated for the tile, these bits should match;
- it can happen that the macro-tile occupies a place in its father macro-tile where some bits of its coordinates (inside grandfather macro-tile) or the bit delegated to the father are kept; then this partial information on the father level should be checked against the information about father coordinates and bit.

These tests guarantee that the information about father is the same in all brothers, and some of these brothers (that are located on the father tape) can check it against actual father information; at the same time some other brother (that has the same delegated bit as the father) checks the consistency of the delegated bits information.

Note that this scheme requires that both $\log N_k$ and $\log N_{k+1}$ are much smaller than N_{k-1}. This is the case, for example, if $N_k = 2^{4^k}$; note that $L_k = N_0 \cdot N_1 \cdot \ldots \cdot N_{k-1}$ is then less than $\sqrt{N_k}$ (which is even better than the requirement $L_k \leq N_k$ mentioned earlier).

In Section 4 we set $N_k = Q^{2.5^k}$ for some large enough Q. (In fact, any constant between 2 and 3 can be used instead of 2.5.)

3.3 Bit Blocks Checked

We explained how macro-tile of any level can get a true information about one bit (delegated to it). However, we need to check not bits, but substrings (and create a tiling error if a forbidden string appears). Note that it is OK to test only very short substrings compared to the macro-tile size (N_k): if this test is done on all levels, this short substring becomes long enough to detect any violation. (Also note the short forbidden substrings can appear very late in the generation process, so we need computation in arbitrary high levels for this reason, too.)

So we need to provide more information to tiles. It can be done in the following way. Let us assume that a tile contains not one bit but a group of bits that starts at the delegated bit and has length depending on the level k (and growing very slowly with k, say, $\log\log\log k$ is slow enough). If this group goes out of the region occupied by a tile, we truncate it.

Similarly, a macro-tile should have this information for the father macro-tile (even if the bits are outside its own region), this information should be the same for brothers and needs to be checked against the delegated bits on the macro-tile level and pieces of information on the father level.

Then the computation in the computation zone can start the generating process checking the forbidden strings that appear against all the substrings of the group of bits available to this computation. This process is time- and space-bounded, but this does not matter since every string if considered on a high enough level.

3.4 Last Correction

The argument explained above still needs some correction. We claim that every forbidden string will be detected at some level where it is short enough compared to the level parameters. However, there could be strings that never become a part of one macro-tile. Imagine that there is some vertical line that is a boundary between macro-tiles of all levels (so we have bigger and bigger tiles on both sides, and this line still separates them). Then a substring that crosses this line will be never checked and therefore we cannot guarantee that it is not forbidden.

There are several ways to get around this problem. One can decide that each macro-tile contains information not only about blocks inside its father macro-tile but in a wider regions (say, three times wider including uncle macro-tiles); this information should be checked for consistency between cousins, too.

But there is a simpler solution. Note that even if a string that crosses the boundary is never checked, its parts (on both sides of the boundary) are, so their complexity is proportional to their length. And one of the parts has length at least half of the original length, so we still have a complexity bound, just the constant is twice smaller.

This finishes the proof of Theorem 2. □

4 Robust Tile Set That Enforces Complex Tilings

We want to construct a "robustified" tile set such that any tiling with "sparse enough" errors or holes can be patched (by changing a small fraction of tiles). It does not matter whether we speak about errors (places where two neighbor tiles do not match) or holes (places without tiles). Indeed, we can convert a tiling error into a hole by deleting one of the two non-matching tiles and convert a hole into a small number of errors by placing an arbitrary tile there.

Let E be a subset of $\mathbb{Z}^2$ and let τ be a tile set.

Definition 1. *A (τ, E)-tiling is a mapping*

$$T\colon\ (\mathbb{Z}^2 \setminus E) \to \tau$$

such that for every two neighbor cells $x, y \in \mathbb{Z}^2 \setminus E$, the tiles $T(x)$ and $T(y)$ match.

In other terms, T is a τ-tiling of the complement of E.

We assume that a set of errors E is chosen at random, according to the Bernoulli distribution B_ε: every point in $\mathbb{Z}^2$ belongs to E with some probability $\varepsilon > 0$; the random choices at different points are done independently.

Theorem 4. *There exist a tile set τ and constants $c_1, c_2 > 0$ with the following properties:*

(1) *a τ-tiling of $\mathbb{Z}^2$ exists;*

(2) *for every τ-tiling T of the plane, every $N \times N$-square in T has Kolmogorov complexity at least $c_1 N - C_2$;*

(3) *for all sufficiently small ε for almost every (with respect to the Bernoulli distribution B_ε) subset $E \subset \mathbb{Z}^2$ every (τ, E)-tiling is at most $1/10$ Besicovitch-apart from some τ-tiling of the entire plane $\mathbb{Z}^2$;*

(4) *for all sufficiently small ε for almost every (with respect to B_ε) subset $E \subset \mathbb{Z}^2$ and every (τ, E)-tiling T, Kolmogorov complexity of centered frames of T of size $n \times n$ is $\Omega(n)$.*

Note that in (4) we speak about complexity of squares with "holes" understood as the minimal complexity for all possible ways to fill the holes. Note also that we cannot claim that every $n \times n$ square has high complexity since this square can be completely isolated from the rest of the tiling by elements of E and therefore can be simple: lexicographically first tiling of the $n \times n$ square has complexity $O(\log n)$.

This is the main result of our paper. Its proof is based on a generalization of the construction from Section 3. Here we outline the plan of the proof:

- **bi-islands (probabilistic part):** we prove that with probability 1 random errors can be split into isolated 'doubled islands' of different rank (an n-level doubled island, or a *bi-island*, consists of two sets of diameter $O(Q^{2.5^n})$ and is isolated from other bi-islands of the same rank). This construction slightly improves the technique used in [5] (the general idea of splitting a random sparse set in ranked 'islands' goes back to [7]).
- **robustification (combinatorial part):** we embed into the primary structure of a self-similar tiling (with variable zoom factors) some redundancy, so that every single isolated island (and even a bi-island) of errors can be patched. The patching procedure involves correction of the tiling only in a small neighborhood of the island (bi-island). So we can sequentially 'patch' all errors, starting from bi-islands of low rank.
- **checksums (error-correction trick):** high level macro-tiles calculate some checksums for the bits in their 'subconscious' and communicate them to their neighbors. This guarantees that most macro-tiles on all levels have coherent bits in their subconscious, even if there are sparse errors.
- **patching errors (join everything together):** we check that with probability 1, for a randomly chosen set of errors E every tiling of $\mathbb{Z}^2 \setminus E$ can be converted into a (close enough) tiling of the entire plane. Composing this fact with the proof from Section 3 we get the theorem.

The full proof of this theorem is rather technical, and exceeds the conference paper limits.

References

1. Berger, R.: The Undecidability of the Domino Problem. Mem. Amer. Math. Soc. 66 (1966)
2. Culik, K.: An Aperiodic Set of 13 Wang Tiles. Discrete Math. 160, 245–251 (1996)
3. Durand, B., Levin, L., Shen, A.: Complex Tilings. J. Symbolic Logic 73(2), 593–613 (2008); see also Proc. 33rd Ann. ACM Symp. Theory Computing, pp. 732–739 (2001), www.arxiv.org/cs.CC/0107008
4. Durand, B., Levin, L., Shen, A.: Local Rules and Global Order, or Aperiodic Tilings. Math. Intelligencer 27(1), 64–68 (2004)
5. Durand, B., Romashchenko, A., Shen, A.: Fixed point and aperiodic tilings. In: Ito, M., Toyama, M. (eds.) DLT 2008. LNCS, vol. 5257, pp. 276–288. Springer, Heidelberg (2008), http://arxiv.org/abs/0802.2432
6. Gács, P.: Reliable Cellular Automata with Self-Organization. In: Proc. 38th Ann. Symp. Found. Comput. Sci., pp. 90–97 (1997)
7. Gács, P.: Reliable Cellular Automata with Self-Organization. J. Stat. Phys. 103(1/2), 45–267 (2001)
8. Kari, J.: A Small Aperiodic Set of Wang tiles. Discrete Math. 160, 259–264 (1996)
9. Rogers, H.: The Theory of Recursive Functions and Effective Computability. MIT Press, Cambridge (1987)
10. Lafitte, G., Weiss, M.: Computability of Tilings. In: IFIP TCS 2008, pp. 187–201 (2008)
11. Levin, L.: Aperiodic Tilings: Breaking Translational Symmetry. Computer J. 48(6), 642–645 (2005); http://www.arxiv.org/cs.DM/0409024
12. von Neumann, J.: Theory of Self-reproducing Automata. In: Burks, A. (ed.), University of Illinois Press (1966)
13. Robinson, R.: Undecidability and Nonperiodicity for Tilings of the Plane. Inventiones Mathematicae 12, 177–209 (1971)
14. Bienvenu, L., Romashchenko, A., Shen, A.: Sparse sets. Journées Automates Cellulaires 2008 (Uzès), 18–28, MCCME Publishers (2008); http://hal.archives-ouvertes.fr/docs/00/27/40/10/PDF/18-28.pdf
15. Rumyantsev, A.Y., Ushakov, M.A.: Forbidden substrings, kolmogorov complexity and almost periodic sequences. In: Durand, B., Thomas, W. (eds.) STACS 2006. LNCS, vol. 3884, pp. 396–407. Springer, Heidelberg (2006)

Tight Bounds for the Cover Time of Multiple Random Walks*

Robert Elsässer[1] and Thomas Sauerwald[2]

[1] Department of Computer Science, University of Paderborn, Germany
[2] International Computer Science Institute, Berkeley, CA, USA

Abstract. We study the cover time of multiple random walks. Given a graph G of n vertices, assume that k independent random walks start from the same vertex. The parameter of interest is the *speed-up* defined as the ratio between the cover time of one and the cover time of k random walks. Recently Alon et al. developed several bounds that are based on the quotient between the cover time and maximum hitting times. Their technique gives a speed-up of $\Omega(k)$ on many graphs, however, for many graph classes, k has to be bounded by $\mathcal{O}(\log n)$. They also conjectured that, for any $1 \leqslant k \leqslant n$, the speed-up is at most $\mathcal{O}(k)$ on any graph. As our main results, we prove the following:

- We present a new lower bound on the speed-up that depends on the mixing-time. It gives a speed-up of $\Omega(k)$ on many graphs, even if k is as large as n.
- We prove that the speed-up is $\mathcal{O}(k \log n)$ on any graph. Under rather mild conditions, we can also improve this bound to $\mathcal{O}(k)$, matching exactly the conjecture of Alon et al.
- We find the correct order of the speed-up for any value of $1 \leqslant k \leqslant n$ on hypercubes, random graphs and expanders. For d-dimensional torus graphs ($d > 2$), our bounds are tight up to a factor of $\mathcal{O}(\log n)$.
- Our findings also reveal a surprisingly sharp dichotomy on several graphs (including d-dim. torus and hypercubes): up to a certain threshold the speed-up is k, while there is no additional speed-up above the threshold.

1 Introduction

Random walks come up and are studied in many sciences like mathematics, physics, computer science etc. While mathematicians have studied random walks on *infinite* graphs for a long time, computer scientists have spurred an interest on random walks on *finite* graphs during the last two decades. Roughly speaking,

* The first author was partially supported by the German Research Foundation under contract EL 399/2-1, and by the Integrated Project IST 15964 "Algorithmic Principles for Building Efficient Overlay Networks" of the EU. The second author was partially supported by a postdoctoral fellowship from the German Academic Exchange Service (DAAD).

S. Albers et al. (Eds.): ICALP 2009, Part I, LNCS 5555, pp. 415–426, 2009.

there have been two main lines of research. One is concerned with the development of rapidly mixing random walks, resulting in approximation schemes of #P hard problems (cf. [17] for more details and a survey on random walks). The second line of research deals with the time to explore a graph, formally known as cover time.

Random walks are an attractive tool for graph exploration due to their inherent simplicity, locality and robustness to dynamical changes. For example, Avin, Koucky, and Lotker [4] recently proved that a (slighly modified) random walk can still explore all vertices of a graph efficiently, even if the graph is dynamically changing during the covering procedure. Other algorithmic applications where random walks have been used are searching [13], routing [18], gossiping [16] and self-stabilization [12] etc.

Probably the first theoretical applications of the cover time traces back to Aleliunas, Karp, Lipton, Lovász, and Rackoff [2]. It was shown that by taking a random walk, it is possible to explore every undirected graph in polynomial time and logarithmic space. In response to their question about time-space tradeoffs, Broder, Karlin, Raghavan, and Upfal [7] studied the cover time of many, independent random walks, each of which starts from the stationary distribution.

Certainly, the situation becomes more challenging if all random walks start from the same vertex. Will they stick together and cover more or less the same set of vertices, or will they quickly disperse in different regions to ensure a fast covering? Alon, Avin, Koucky, Kozma, Lotker, and Tuttle [3] posed this question and studied the speed-up defined as the ratio between the cover time of a single random walk and the cover time of k random walks, where $1 \leqslant k \leqslant n$. As it turns out, the answer depends very much on the underlying graph: on complete graphs, a speed-up of k is always possible, while on the cycle the speed-up is only $\mathcal{O}(\log k)$. On certain graphs, there are even starting positions of the k walks such that the speed-up is $\Omega(2^k)$ (for small k).

Another reason why the cover time of random walks has been investigated is its intimate relation to other graph-theoretical parameters. For example, Broder and Karlin [6] gave a comprehensive collection of bounds relating the cover time to spectral properties of G. Chandra, Raghavan, Ruzzo, Smolensky, and Tiwari [9] established a tight connection between random walks and electrical networks and related the cover time to other properties such as the vertex-expansion.

1.1 Related Work

One slight drawback of the cover time of a single random walk is that it takes at least $\Omega(n \log n)$ steps on every graph, and may even increase to $\Omega(n^2)$ on regular and $\Omega(n^3)$ steps on non-regular graphs. This has led to several modified covering schemes. Adler, Halperin, Karp, and Vazirani [1] introduced a covering process where in each round one first chooses a vertex uniformly at random, and then chooses an uncovered neighbor of this vertex (if there is one). Later Dimitrov and Plaxton [11] proved that this process achieves a cover time of $\mathcal{O}(n + (n \log n)/d)$ on any d-regular graph. Note that in this scheme, one has to sample uniformly among *all* vertices (not just among visited neighbors) which is not completely inline with the scenario of a decentralized exploration process.

Another approach taken by Ikeda, Kubo, Okumoto, and Yamashita [14] and Avin and Krishnamachari [5] is to change the transition probabilities of the random walk. For example, Ikeda et al. [14] devised a way of locally computable transition probabilities which results in a cover time of $\mathcal{O}(n^2 \log n)$ on any graph. However, one limitation of all these approaches is that they can only explore a graph within $\Omega(n)$ steps.

Multiple random walks can break this barrier of $\Omega(n)$ and have been used by Broder et al. [7] to obtain tradeoffs between space and time for the s-t-connectivity problem. As mentioned before, they assumed that each random walk starts from an independent sample of the stationary distribution. While this indeed significantly speeds up the covering process, one has to sample again among *all* vertices. This could be one reason why researchers have recently studied multiple random walks which start all from the same vertex ([3, 10]). Alon et al. [3] derived several (asymptotic) lower and upper bounds on the speed-up on several graph classes, while Cooper et al. [10] focused on the class of random regular graphs and derived nearly exact bounds on the speed-up. Finally, multiple random walks starting from the same vertex are also a fundamental tool for property testing, cf. [15] for a recent analysis of a property tester of expanders. The basic idea is to count the collisions of random walks that start from the same vertex to estimate the expansion properties of a graph.

1.2 Our Contribution

Before describing our main results, we have to introduce a little bit of notation. Let G be any undirected, connected graph with n vertices. For any $1 \leqslant k \leqslant n$, let $\mathbf{E}\left[\,\mathsf{COV}_u^k\,\right]$ be the expected time for k random walks that start from u to cover all vertices. Let $\mathbf{E}\left[\,\mathsf{COV}_{\max}^k\,\right] = \max_{u \in V} \mathbf{E}\left[\,\mathsf{COV}_u^k\,\right]$ (we also use $\mathbf{E}\left[\,\mathsf{COV}_{\max}(G)\,\right] = \mathbf{E}\left[\,\mathsf{COV}_{\max}^1(G)\,\right]$ to stick to the common notation). For any undirected, connected graph G, we define the *speed-up* $S^k := \mathbf{E}\left[\,\mathsf{COV}_{\max}(G)\,\right] / \mathbf{E}\left[\,\mathsf{COV}_{\max}^k(G)\,\right]$. By $\mathsf{H}(u, v)$ we denote the expected time for the random walk to get from u to v; it is a well-known fact that $\max_{u,v} \mathsf{H}(u, v)$ approximates $\mathbf{E}\left[\,\mathsf{COV}_{\max}(G)\,\right]$ up to a factor of $\mathcal{O}(\log n)$ (see Theorem 2.2). The mixing time $\mathsf{MIX}_{1/2}(G)$ is the time required for the random walk to approach its stationary distribution (exact definition in Section 2).

We first present a general lower bound on the speed-up. It is based on the following upper bound on $\mathbf{E}\left[\,\mathsf{COV}_{\max}^k(G)\,\right]$:

Theorem 3.4 (from page 421). *For any graph G and any k with $1 \leqslant k \leqslant n$,*

$$\mathbf{E}\left[\,\mathsf{COV}_{\max}^k(G)\,\right] = \mathcal{O}\left(\frac{\log n \cdot (\max_{u,v} \mathsf{H}(u, v) + \mathsf{MIX}_{1/2}(G))}{k} + \mathsf{MIX}_{1/2}(G) \right).$$

Graph	$\mathsf{COV}(G)$	$\mathsf{H}_{\max}$	$\mathsf{MIX}_{1/2}(G)$	$k \in$	Speed up $S^k(G)$ bounds
cycle	n^2	n^2	n^2	$[1, n]$	$= \log k$ [3, Thm. 6]
2-dim. torus	$n \log^2 n$	$n \log n$	n	$[1, \log n]$ $[1, n]$	$\geqslant k$ [3, Thm. 4] $\leqslant \log^2 n \log k$ [3, Cor. 25]
d-dim. torus, $d > 2$	$n \log n$	n	$n^{2/d}$	$[1, \log n]$ $[1, n^{1-2/d} \log n]$ $[1, n]$ $[1, n]$	$\geqslant k$ [3, Thm. 4] $\geqslant k$ [★, Cor. 1] $\leqslant n^{1-2/d} \log n \log k$ [3, Thm. 24] $\leqslant k$ [★, Cor. 4.6]
Hypercube	$n \log n$	n	$\log n \log \log n$	$[1, \log n]$ $[1, \frac{n}{\log \log n}]$ $[\frac{n}{\log \log n}, n]$	$\geqslant k$ [3, Thm. 4] $= k$ [★, Thm. 3.4 & Cor. 4.6] $= \frac{n}{\log \log n}$ [★, Thm. 3.4 & 5.3]
Complete	$n \log n$	n	1	$[1, n]$	$= k$ [3, Lem. 12]
Expander	$n \log n$	n	$\log n$	$[1, n]$ $[1, n]$	$\geqslant k$ [3, Thm. 18] $= k$ [★, Cor. 5.1]
Random	$n \log n$	n	$\log n$	$[1, \log n]$ $[1, n]$	$\geqslant k$ [3, Thm. 4] $= k$ [★, Cor. 5.1]

Fig. 1. Summary of the new and old results for the graphs mentioned by [3], where constant factors are neglected in all columns. $\mathsf{H}_{\max}$ stands for $\max_{u,v} \mathsf{H}(u,v)$. Our new results are marked with ★. For torus graphs, the bounds are tight up to a logarithmic factor and for all other graphs, the bounds are tight (for each $1 \leqslant k \leqslant n$).

This shows that $\mathbf{E}\left[\mathsf{COV}^k_{\max}(G)\right]$ is upper bounded by $\log n \cdot \max_{u,v} \mathsf{H}(u,v)/k$, as long as $\max_{u,v} \mathsf{H}(u,v)/k$ is not smaller than the mixing time (see Corollary 3.5 for a simpler, but slightly weaker statement than Theorem 3.4).

We point out that most previous general upper bounds on $\mathbf{E}\left[\mathsf{COV}^k_{\max}(G)\right]$ in Alon et al. [3] are at least $\Omega(n)$ [3, Thm. 4,5,13,14], and therefore only useful on most graphs when $k = \mathcal{O}(\log n)$. A similar bound to Theorem 3.4 from [3] is:

Theorem 1.1 ([3, Proof of Theorem 9]). *For any graph G and $1 \leqslant k \leqslant n$,*

$$\mathbf{E}\left[\mathsf{COV}^k_{\max}(G)\right] = \mathcal{O}\left(\frac{\mathsf{MIX}_{n^{-1}}(G)\, n\, (\log n)^2}{k}\right).$$

Note that the bound of Alon et al. [3] includes the mixing time as a factor, while in our bound (Theorem 3.4 above), for any k, the mixing time does not come into play at all, as long as $\max_{u,v} \mathsf{H}(u,v)/k$ is larger than $\mathsf{MIX}_{1/2}(G)$. Since for most graph classes (cmp. Figure 1) $\max_{u,v} \mathsf{H}(u,v) = \Theta(n)$ and $\mathsf{MIX}_{1/2}(G) = o(n)$, our theorem gives a lower bound on the speed-up of k for a wide range of k (cf. Figure 1, or Section 5 for more details).

The main idea to establish Theorem 3.4 is based on a coupling argument between one random walk and k random walks (see Theorem 3.3 for details). We believe that this technique might be very useful for deriving further bounds on the cover time of one or many random walks.

We continue to prove a general upper bound for any graph, namely that $S^k = \mathcal{O}(k \log n)$ for any $1 \leqslant k \leqslant n$. This already matches the conjecture of Alon et al. [3] up to a logarithmic factor. Under a rather mild condition on the mixing-time and cover time of one random walk, we improve this upper bound to $S^k = \mathcal{O}(k)$, establishing the conjecture of [3] for a large class of graphs

(Corollary 4.7). Finally, we also present an upper bound based on the diameter of the graph (Theorem 4.8).

Applications of our lower and upper bounds to concrete graphs are summarized in Figure 1, completing Table 1 of [3]. As an example, consider the hypercube with n vertices. We prove that $S^k = \Theta(k)$ as long as $k = \mathcal{O}(n/\log\log n)$. However, for $k = \Omega(n/\log\log n)$, $S^k = \Theta(n/\log\log n)$. The same dichotomy is established for d-dimensional torus graphs ($d > 2$), where also $n/\mathsf{MIX}_{1/2}(G)$ represents as a "sharp threshold" on the speed-up.

1.3 Road Map

In Section 2 we introduce our notation and some preliminary results. Section 3 contains the proof of our upper bound on S^k. This is followed by Section 4 consisting of several lower bounds on S^k. In Section 5 we show how to apply our general results to obtain tight bounds on S^k for concrete graph classes. We close in Section 6 with the conclusions. Several proofs are omitted due to space limitations.

2 Notations, Definitions and Preliminaries

Random Walk. A *random walk* (cf. [17] for a survey) on an undirected, connected graph $G = (V, E)$ starts at some specified vertex $u \in V$ and moves in each step along some adjacent edge chosen uniformly at random. To ensure convergence also on non-bipartite graphs, a common way is to add loop probabilities: at each step the random walk stays with probability 1/2 at the current vertex and otherwise it moves to a randomly chosen neighbor. It is a well-known fact that the loops only increase the cover time by a factor of 2.

There are two ways to represent the walk. The first and concrete one is to view the walk as an infinite sequence of vertices $X_0, X_1, \ldots$, where $X_0 = u$ is the starting vertex and X_t is the vertex visited at step t.

A more abstract way is to only consider the distribution of the walk. To this end, let $\mathbf{P}$ be the transition matrix of the walk, i.e., $p_{u,v} = \frac{1}{2\deg(u)}$ if $\{u, v\} \in E$, $p_{u,u} = \frac{1}{2}$ and $p_{u,v} = 0$ otherwise. Note that $\mathbf{P}$ is symmetric if and only if G is regular. Now define for each pair of vertices u, v, $p_{u,v}^t$ as the probability that a random walk starting at u visits the vertex v at step t. Hence the vector $p_u^t = (p_{u,v}^t)_{v\in V}$ represents the distribution of X_t, i.e., the visited vertex at step t. It is a well-known fact that under our assumptions on G, $\mathbf{p}_u(t)$ converges for $t \to \infty$ towards the *stationary distribution* π given by $\pi(v) = \deg(v)/(2|E|)$.

Mixing Time. To quantify the convergence speed, we define the *relative pointwise distance* ([20, p. 45]) as

$$\Delta(t) := \max_{u,v\in V} \frac{|p_{u,v}^t - \pi(v)|}{\pi(v)}.$$

Definition 1. *The* mixing time *of a random walk on* G *with transition matrix* $\mathbf{P}$ *is defined for any* $0 < \varepsilon < 1$ *by*

$$\mathsf{MIX}_{\varepsilon}^{\mathbf{P}}(G) := \min\{t \in \mathbb{N} \colon \Delta(t) \leqslant \varepsilon\}.$$

If the reference to $\mathbf{P}$ is obvious, we shall also just write $\mathsf{MIX}_{\varepsilon}(G)$. Our definition of mixing time should be compared with the one based on the variation distance used by Alon et al. [3], $\overline{\mathsf{MIX}}_{\varepsilon}(G) := \max_{u \in V} \min\left\{t \in \mathbb{N} \colon \|p_u^t - \pi\|_1 \leqslant \varepsilon\right\}$. The next lemma shows that $\mathsf{MIX}_{n^{-1}}(G)$ is not larger than $\overline{\mathsf{MIX}}_{n^{-1}}(G)$.

Lemma 2.1. *For any graph* $G = (V, E)$, $\mathsf{MIX}_{n^{-1}}(G) = \mathcal{O}(\overline{\mathsf{MIX}}_{n^{-1}}(G))$.

Hitting Time and Cover Time. For two vertices $u, v \in V(G)$, we define the *hitting time* from u to v as $\mathsf{H}(u,v) := \mathbf{E}\left[\min\{t \in \mathbb{N}\backslash\{0\} : X_t = v, X_0 = u\}\right]$, i. e., the expected number of steps to reach v from u. Denote by $\mathsf{COV}_s(G)$ the first time when a (single) random walk starting from s has visited all n vertices of G. Then the cover time is defined as $\mathbf{E}\left[\mathsf{COV}_{\max}(G)\right] := \max_{u \in V} \mathbf{E}\left[\mathsf{COV}_u(G)\right]$. (We point out that in several previous work the cover time is written without $\mathbf{E}\left[\cdot\right]$, however, in this work we also have to deal with the random variable $\mathsf{COV}_u(G)$). The following well-known result relates the maximum hitting time to the cover time.

Theorem 2.2 ([9, 18]). *For any graph* $G = (V, E)$ *we have* $\max_{u,v \in V} \mathsf{H}(u,v) \leqslant \mathbf{E}\left[\mathsf{COV}_{\max}(G)\right] \leqslant 2e^3 \cdot \max_{u,v \in V} \mathsf{H}(u,v) \ln n + n$

We shall consider the cover time when k random walks start at the same vertex, where $1 \leqslant k \leqslant n$. To this end, we study $\mathbf{E}\left[\mathsf{COV}_u^k(G)\right]$, defined as the expected time for k random walks starting from u to cover all n vertices of G. Set $\mathbf{E}\left[\mathsf{COV}_{\max}^k(G)\right] = \max_{u \in V} \mathbf{E}\left[\mathsf{COV}_u^k(G)\right]$. Clearly, $\mathbf{E}\left[\mathsf{COV}_{\max}^k(G)\right]$ decreases in k. Hence several of our lower bounds stated for $\mathbf{E}\left[\mathsf{COV}_{\max}^n(G)\right]$ directly imply the same bound on $\mathbf{E}\left[\mathsf{COV}_{\max}^k(G)\right]$ with $k \leqslant n$. Sometimes, we will also consider $\mathbf{E}\left[\mathsf{COV}_{\pi}^k(G)\right]$. In this case, each starting point of the k random walks is chosen independently from the stationary distribution π. We recall:

Theorem 2.3 ([7, Theorem 1]). *Let* G *be any graph with* m *edges. Then we have for any* $1 \leqslant k \leqslant n$, $\mathbf{E}\left[\mathsf{COV}_{\pi}^k(G)\right] = \mathcal{O}\left(\frac{m^2}{k^2} \cdot \log^3 n\right)$.

We continue with an auxiliary lemma.

Lemma 2.4. *Let* X_1 *and* X_2 *be two random variables taking values in a finite set* S. *Assume that there is a number* $0 < C < 1$ *such that for every* $s \in S$, $\mathbf{Pr}\left[X_1 = s\right] \geqslant C\,\mathbf{Pr}\left[X_2 = s\right]$. *Then there exists a coupling* $\widehat{X} = (\widehat{X}_1, \widehat{X}_2)$ *of* X_1 *and* X_2 *such that* $\mathbf{Pr}\left[\widehat{X}_1 = \widehat{X}_2\right] \geqslant C$.

3 Lower Bounds on the Speed-Up

A natural relation that has been also used by Alon et al. [3] is the following.

Lemma 3.1. *For any* $1 \leqslant k \leqslant n$, $\mathbf{E}\left[\mathsf{COV}^k_{\max}(G)\right] \leqslant \mathbf{E}\left[\mathsf{COV}^k_{\pi}(G)\right] + \mathsf{MIX}_{n^{-3}}(G)$.

We prove an extension where the threshold for the mixing time is much smaller. This apparently small difference will be crucial to obtain tight bounds for hypercubes (Section 5).

Lemma 3.2. *For any* $1 \leqslant k \leqslant n$, $\mathbf{E}\left[\mathsf{COV}^k_{\max}(G)\right] \leqslant 16\,\mathbf{E}\left[\mathsf{COV}^{k/2}_{\pi}(G)\right] + 4\,\mathsf{MIX}_{1/2}(G)$.

Proof Sketch. The basic idea is as follows. Let $X^1, X^2, \ldots, X^k$ be k random walks starting from the same vertex u. Moreover, let $Y^1, Y^2, \ldots, Y^k$ be k random walks, all starting from independent samples of π. Our goal is to relate the set of covered vertices by $X^1, \ldots, X^k$ to the covered ones by $Y^1, \ldots, Y^{k/2}$ at the cost of an additional $\mathsf{MIX}_{1/2}(G)$-term. In order to do so, we will prove that at least half of the random walks among $X^1, \ldots, X^k$ are located on a uniformly chosen vertex after $\mathsf{MIX}_{1/2}(G)$ steps.

Theorem 3.3. *For every graph G and k with* $1 \leqslant k \leqslant n$,

$$\mathbf{E}\left[\mathsf{COV}^k_{\pi}(G)\right] = \mathcal{O}\left(\frac{\log n \cdot (\max_{u,v} \mathsf{H}(u,v) + \mathsf{MIX}_{1/2}(G))}{k} + \mathsf{MIX}_{1/2}(G)\right).$$

Before we outline the proof of Theorem 3.3, let us point out that the result also improves over Theorem 2.3 for a wide range of k, provided that $\mathsf{MIX}_{1/2}(G)$ and $\max_{u,v} \mathsf{H}(u,v)$ are not too large.

Proof Sketch. We devise a coupling of a single random walk X to k random walks, each of which starts according to π. We shall divide the single random walk X into consecutive sections of length $\mathsf{MIX}_{1/2}(G)$. We then argue that a random walk starting from the stationary distribution has (almost) the same chance of visiting a vertex within $\mathsf{MIX}_{1/2}(G)$ steps as the single random walk has in one fixed section. This implies that the probability that the k random walks visit this vertex is (nearly) the probability that X visits the same vertex in one of the even sections. Here it is crucial to consider only the even (or odd) sections, so that the random walk X is located on a vertex according to π each time a new section begins.

Combining this result with Lemma 3.2 we get immediately:

Theorem 3.4. *For any graph G and any k with* $1 \leqslant k \leqslant n$,

$$\mathbf{E}\left[\mathsf{COV}^k_{\max}(G)\right] = \mathcal{O}\left(\frac{\log n \cdot (\max_{u,v} \mathsf{H}(u,v) + \mathsf{MIX}_{1/2}(G))}{k} + \mathsf{MIX}_{1/2}(G)\right).$$

Let us state a simpler, slightly weaker bound on the speed-up that follows directly from Theorem 3.4:

Corollary 3.5. *Let G be a graph that satisfies $\mathsf{MIX}_{1/2}(G) = \mathcal{O}(\max_{u,v} \mathsf{H}(u,v))$ and $\mathbf{E}[\mathsf{COV}_{\max}(G)] = \Theta(\max_{u,v} \mathsf{H}(u,v) \log n)$. Then for any $1 \leqslant k \leqslant n$,*

$$S^k(G) = \Omega\left(\frac{k}{1 + \frac{\mathsf{MIX}_{1/2}(G)}{\mathbf{E}[\mathsf{COV}_{\max}(G)]} \cdot k}\right).$$

Hence as long as $k = \mathcal{O}\left(\frac{\mathbf{E}[\mathsf{COV}_{\max}(G)]}{\mathsf{MIX}_{1/2}(G)}\right)$, Corollary 3.5 yields a speed-up of $\Omega(k)$. Note that all graphs (except cycles and 2-dim. torus) in Figure 1 satisfy the conditions of Corollary 3.5.

4 Upper Bounds on the Speed-Up

Alon et al. [3] gave a graph G and vertex u such that $\frac{\mathbf{E}[\mathsf{COV}_u(G)]}{\mathbf{E}[\mathsf{COV}_u^k(G)]} = \Omega(2^k)$ for $k = \Theta(\log n)$, so the speed-up is exponential in k. However, their example does not work when u is replaced by a worst-case starting vertex. This lead to their conjecture that the speed-up is always polynomial in k, if the starting vertex is worst-case. More precisely, they conjectured that for any graph and any $1 \leqslant k \leqslant n$, $S^k = \mathcal{O}(k)$.

We shall prove that $S^k = \mathcal{O}(k \log n)$ for any graph and k, matching the conjecture up to a factor of $\mathcal{O}(\log n)$. This also shows that while for an arbitary starting vertex an *exponential* speed-up is possible, the speed-up is always *polynomial*, if the starting vertex is worst-case.

Proposition 4.1. *For any graph G and any $1 \leqslant k \leqslant n$, $S^k = \mathcal{O}(k \log n)$.*

Proof. Fix a vertex w. Choose a vertex u such that

$$\mathbf{Pr}\left[\text{walk of length } \mathbf{E}\left[\mathsf{COV}_u^k(G)\right] \text{ starting at } u \text{ visits } v\right]$$

is minimized. We claim by way of contradiction that

$$\mathbf{Pr}\left[\text{walk of length } 2\,\mathbf{E}\left[\mathsf{COV}_u^k(G)\right] \text{ starting at } u \text{ visits } v\right] \geqslant \frac{1}{4k}.$$

Assuming the converse, the probability that all k random walks starting at u do not cover w would be at least $\prod_{i=1}^{k}\left(1 - \frac{1}{4k}\right) \geqslant 1 - \sum_{i=1}^{k} \frac{1}{4k} = \frac{3}{4}$, which in turn would imply $\mathbf{E}\left[\mathsf{COV}_u^k(G)\right] \geqslant \frac{3}{2}\,\mathbf{E}\left[\mathsf{COV}_u^k(G)\right]$, a contradiction.

Consider now a single random walk of length $16\,\mathbf{E}\left[\mathsf{COV}_u^k(G)\right] k \ln n$. Then,

$$\mathbf{Pr}\left[\text{walk of length } 16\,\mathbf{E}\left[\mathsf{COV}_u^k(G)\right] k \ln n \text{ starting at } u \text{ visits } v\right]$$
$$\geqslant 1 - \left(1 - \frac{1}{4k}\right)^{8k \ln n} \geqslant 1 - \frac{1}{n^2}.$$

Taking the union bound over all n vertices yields the claim.

4.1 Special Upper Bounds

Additionally, we shall derive three more specific lower bounds on $\mathbf{E}\left[\mathsf{COV}^k_{\max}(G)\right]$. As a consequence, they are most useful to upper bound the speed-up when the graph satisfies $\mathbf{E}\left[\mathsf{COV}_{\max}(G)\right] = \mathcal{O}(n \log n)$ (which is the case for most interesting graphs (cmp. Figure 1)).

We start by deriving a lower bound of $\Omega((n/k)\log n)$ for not too small k by using a relatively simple coupon-collecting argument. After that we present a lower bound of $\Omega((n/k)\log n)$ for not too large k, requiring that the mixing time is sublinear. Combining these bounds, we obtain that $\mathbf{E}\left[\mathsf{COV}^k_{\max}(G)\right] = \Omega((n/k)\log n)$ for any $1 \leqslant k \leqslant n$ (if the mixing time is sublinear).

We start with a bound based on a coupon-collecting argument. We view each random walk as an independent string of n letters (corresponding to n vertices). Then we bound the probability that all letters occur in a sample of k random strings.

Theorem 4.2. *Let k be an arbitrary integer satisfying $k \geqslant n^{\varepsilon}$ for an arbitrary constant $0 < \varepsilon < 1$. Then,* $\mathbf{E}\left[\mathsf{COV}^k_{\max}(G)\right] = \Omega\left(\frac{n}{k}\log n\right)$.

For $k < n^{\varepsilon}$, we devise a lower bound on $\mathbf{E}\left[\mathsf{COV}^k_{\max}(G)\right]$ that requires a sublinear mixing time. We use the following result from Broder and Karlin [6].

Lemma 4.3 ([6, Lemma 12]). *Consider a single random walk $X_1, X_2, \ldots$ with a symmetric transition matrix $\mathbf{P}$. Let T_s be the first time when s different vertices are covered. Then for any $m \in \mathbb{N}$,*

$$\mathbf{E}\left[T_{\lfloor (m+1)n/(m+2)\rfloor} - T_{\lfloor (m)n/(m+1)\rfloor}\right] \geqslant \frac{1}{2}\frac{n}{m+2} - \mathcal{O}(\mathsf{MIX}_{n^{-1}}(G)\cdot m).$$

Using the lemma above, we can show the following corollary:

Corollary 4.4. *Let $X = (X_1, X_2, \ldots)$ be a random walk on regular graph. Let T_s be the first time that s different vertices are covered. Let $1 \leqslant m \leqslant n$ be any positive integer. Define $\gamma_m := \frac{1}{2}\frac{n}{m+2} - \mathcal{O}(\mathsf{MIX}_{n^{-1}}(G)\cdot m)$. Then,*

$$\mathbf{Pr}\left[T_{\lfloor (m+1)n/(m+2)\rfloor} - T_{\lfloor (m)n/(m+1)\rfloor} \geqslant \frac{1}{4}\gamma_m\right] \geqslant \frac{1}{16}.$$

Theorem 4.5. *Assume that $\mathsf{MIX}_{n^{-1}}(G) = \mathcal{O}(n^{1-\varepsilon})$ for a constant $\varepsilon > 0$. Then for any regular graph G and $k \leqslant \sqrt[4]{n/\mathsf{MIX}_{n^{-1}}(G)}$,* $\mathbf{E}\left[\mathsf{COV}^k_{\max}\right] = \Omega\left(\frac{n}{k}\log n\right)$.

Proof Sketch. As in [6] our goal is to divide the random walks viewing one after another into a certain number of epochs, where a new epoch starts if a certain number of new vertices has been covered. Then we can bound the remaining time in each epoch by Corollary 4.4. The technical difficulty arises when the lower bound by Corollary 4.4 is larger than the remaining time of the walk. In this case we assume (quite pessimistically) that the random walk has finished one epoch, but this suffices, since k is rather small.

Combining Theorem 4.2 and Theorem 4.5 we obtain immediately:

Corollary 4.6. *For any regular graph G with $\mathsf{MIX}_{n^{-1}}(G) = \mathcal{O}(n^{1-\varepsilon})$ for a constant $\varepsilon > 0$ and any $1 \leqslant k \leqslant n$, we have* $\mathbf{E}\left[\mathsf{COV}^k_{\max}(G)\right] = \Omega\left(\frac{n}{k}\log n\right)$.

Turning back to the original question on upper bounding S^k we get:

Corollary 4.7. *For any regular graph G that satisfies $\mathsf{MIX}_{n^{-1}}(G) = \mathcal{O}(n^{1-\varepsilon})$ and $\mathbf{E}[\mathsf{COV}_{\max}(G)] = \Theta(n\log n)$, we have for any $1 \leqslant k \leqslant n$, $S^k = \mathcal{O}(k)$.*

This establishes the conjecture of Alon et al. [3] for a large class of graphs including most graphs of Figure 1.

Obviously, $\mathrm{diam}(G)$ is a lower bound on $\mathbf{E}\left[\mathsf{COV}^k_{\max}(G)\right]$ for each k. Using a result of [8], we can prove the following improvement (if $\mathrm{diam}(G) \geqslant \log n$):

Theorem 4.8. *For any graph G with diameter* $\mathrm{diam}(G)$, $\mathbf{E}[\mathsf{COV}^n_{\max}(G)] = \Omega\left(\frac{\mathrm{diam}(G)^2}{\log n}\right)$.

5 Applications to Concrete Graphs

Expanders and Random Graphs. There are several (mostly equivalent) definitions of expanders. Here, we call a regular graph an expander if $\mathsf{MIX}_{n^{-1}}(G) = \mathcal{O}(\log n)$ (this is a more general definition than [3], where additionally the degree has to be constant). It is also a well-known fact that any expander graph satisfies $\max_{u,v}\mathsf{H}(u,v) = \mathcal{O}(n)$ (cf. [6, 17]). Hence Corollary 3.5 implies a speed-up of $\Omega(k)$ for any $1 \leqslant k \leqslant n$. Moreover, Corollary 4.6 establishes tightness.

For any given $(1+\varepsilon)\log(n)/n < p < 1, \varepsilon > 0$, an Erdős-Rényi random graph is constructed by placing an edge between each pair of vertices independently with probability p. Similar to regular expanders, we can prove the same result for random graphs leading to the following corollary.

Corollary 5.1. *For any regular expander graph and almost all Erdős-Rényi random graphs with $p \geqslant (1+\varepsilon)\log(n)/n$, we have for any $1 \leqslant k \leqslant n$ that $S^k = \Theta(k)$.*

Hypercubes. Let us consider the speed-up on the $\log n$-dimensional hypercube H_n with n vertices. It is known that $\max_{u,v\in V}\mathsf{H}(u,v) = \mathcal{O}(n)$ (cf. [17]) which readily implies that $\mathbf{E}[\mathsf{COV}_{\max}(H_n)] = \Theta(n\log n)$.

Lemma 5.2. *For the hypercube H_n, $\mathsf{MIX}_{1/2}(G) = \mathcal{O}(\log n\,\log\log n)$.*

We remark that $\overline{\mathsf{MIX}}_{n^{-1}}(H_n) = \Omega(\log^2 n)$, so it is crucial to use $\mathsf{MIX}_{1/2}(H_n)$. Hence, as long as $k \leqslant C_1 n/(\log n\log\log n)$ for a sufficiently small constant C_1, Corollary 3.5 and Corollary 4.7 imply that the speed up is $\Theta(k)$. (We point out that using the techniques of [10], a more precise bound on the speed could be obtained). Let us now consider the case when k is large.

Theorem 5.3. *For the hypercube H_n, $\mathbf{E}[\mathsf{COV}^n_{\max}(H_n)] = \Omega(\log n\log\log n)$.*

Hence the speed-up on hypercubes undergoes a surprisingly sharp transition: it equals k if $k = \mathcal{O}(n/\log\log n)$, but as soon as $k = \Theta(n/\log\log n)$ the speed-up does not increase further.

Cayley Graphs with Small Degree (including Torus Graphs). Let us now consider torus graphs. For cycles, Alon et al. [3, Theorem 6] prove that $S^k = \Theta(\log k)$ for any $1 \leqslant k \leqslant n$. For the two-dimensional torus graph, they proved [3, Theorem 4 & 8, Corollary 25] that $S^k(G) = \Omega(k)$ for $k \leqslant \log n$, but $S^k(G) = \mathcal{O}(\log^2 n \log k)$ for any $1 \leqslant k \leqslant n$. Therefore, we only have to consider the d-dimensional torus with $d \geqslant 3$ in the following. In fact, we shall look at Cayley graphs more generally. Recall that an undirected Cayley graph is a graph whose vertex set is equal to the elements of a finite group and the edge set is given by a set of self-inverse group generators (cf. [19]). We recall the following lemma.

Lemma 5.4 ([19]). *For any Cayley graph* G, $\mathsf{MIX}_{1/2}(G) = \mathcal{O}(\Delta \operatorname{diam}(G)^2 \log n)$.

Now applying Corollary 3.5 and Theorem 4.8 we obtain the following.

Theorem 5.5. *Let* G *be a* Δ*-regular Cayley graph such that* $\mathbf{E}[\,\mathsf{COV}_{\max}(G)\,] = \Theta(\max_{u,v} \mathsf{H}(u,v) \log n)$. *Then, for any* $k \leqslant \frac{\mathbf{E}[\,\mathsf{COV}_{\max}(G)\,]}{\Delta \operatorname{diam}(G)^2 \log n}$, $S^k(G) = \Omega(k)$. *Moreover for any* $1 \leqslant k \leqslant n$, $S^k(G) = \mathcal{O}\left(\frac{\mathbf{E}[\,\mathsf{COV}_{\max}(G)\,]}{\operatorname{diam}(G)^2} \log n\right)$.

Hence for any Cayley graph with small degree Δ, there is a sharp threshold point near $\frac{\mathbf{E}[\,\mathsf{COV}_{\max}(G)\,]}{\operatorname{diam}(G)^2}$. For d-dimensional torus with $d > 2$ we can prove a slightly stronger result, since it is known that $\max_{u,v} \mathsf{H}(u,v) = \Theta(n)$ and $\mathsf{MIX}_{1/2}(G) = \Theta(\operatorname{diam}(G)^2) = \Theta(n^{2/d})$ (cf. [3, 17]). Applying Corollary 3.5 for the lower bound, and, Theorem 4.8 and Corollary 4.7 for the upper bound gives:

Corollary 1. *Let* G *be a* d*-dimensional torus with* $d > 2$. *Then for any* $1 \leqslant k \leqslant n^{1-2/d} \log n$, $S^k(G) = \Omega(k)$. *Moreover for any* $1 \leqslant k \leqslant n$, $S^k(G) = \mathcal{O}\left(\min\{k, n^{1-2/d} \log^2 n\}\right)$.

6 Conclusions

We presented several lower and upper bounds on the speed-up defined as the ratio between the cover time of one and the cover time of k random walks. On a concrete level, our results fill several gaps left open in the previous work of Alon et al. [3] (cmp. Figure 1). From a higher perspective, our findings also provide an answer to the question raised by [3] about a good characterization of a best-possible speed-up. For a large class of graphs, a speed-up of $\Omega(k)$ is possible up to a certain threshold (roughly $n \log n$ divided by the mixing time), while above the threshold the speed-up does not increase further.

Bibliography

[1] Adler, M., Halperin, E., Karp, R., Vazirani, V.: A Stochastic Process on the Hypercube with Applications to Peer-to-Peer Networks. In: 35th Annual ACM Symposium on Theory of Computing (STOC 2003), pp. 575–584 (2003)

[2] Aleliunas, R., Karp, R., Lipton, R., Lovász, L., Rackoff, C.: Random Walks, Universal Traversal Sequences, and the Complexity of Maze Problems. In: 20th Annual IEEE Symposium on Foundations of Computer Science (FOCS 1979), pp. 218–223 (1979)

[3] Alon, N., Avin, C., Koucky, M., Kozma, G., Lotker, Z., Tuttle, M.: Many Random Walks are faster than one. In: 20th Annual ACM Symposium on Parallel Algorithms and Architectures (SPAA 2008), pp. 119–128 (2008)

[4] Avin, C., Koucky, M., Lotker, Z.: How to Explore a Fast-Changing World (Cover Time of a Simple Random Walk on Evolving Graphs). In: Aceto, L., Damgård, I., Goldberg, L.A., Halldórsson, M.M., Ingólfsdóttir, A., Walukiewicz, I. (eds.) ICALP 2008, Part I. LNCS, vol. 5125, pp. 121–132. Springer, Heidelberg (2008)

[5] Avin, C., Krishnamachari, B.: The power of choice in random walks: an empirical study. In: 9th International Symposium on Modeling Analysis and Simulation of Wireless and Mobile Systems, pp. 219–228 (2006)

[6] Broder, A., Karlin, A.: Bounds on the cover time. In: 29th Annual IEEE Symposium on Foundations of Computer Science (FOCS 1988), pp. 479–487 (1988)

[7] Broder, A., Karlin, A., Raghavan, P., Upfal, E.: Trading space for time in undirected s–t-connectivity. SIAM Journal on Computing 23(2), 324–334 (1994)

[8] Carne, T.: A Transmutation Formula for Markov Chains. Bulletin des Sciences Mathematiques 109, 399–405 (1985)

[9] Chandra, A., Raghavan, P., Ruzzo, W., Smolensky, R., Tiwari, P.: The Electrical Resistance of a Graph Captures its Commute and Cover Times. Computational Complexity 6(4), 312–340 (1997)

[10] Cooper, C., Frieze, A., Radzik, T.: Multiple random walks and interacting particle systems. In: 36th International Colloquium on Automata, Languages, and Programming (ICALP 2009) (2009) (to appear)

[11] Dimitrov, N.B., Plaxton, C.G.: Optimal Cover Time for a Graph-Based Coupon Collector Process. In: Caires, L., Italiano, G.F., Monteiro, L., Palamidessi, C., Yung, M. (eds.) ICALP 2005, vol. 3580, pp. 702–716. Springer, Heidelberg (2005)

[12] Dolev, S., Schiller, E., Welch, J.L.: Random walk for self-stabilizing group communication in ad hoc networks. IEEE Transactions on Mobile Computing 5(7), 893–905 (2006)

[13] Gkantsidis, C., Mihail, M., Saberi, A.: Random walks in peer-to-peer networks. Performance Evaluation 63(3), 241–263 (2006)

[14] Ikeda, S., Kubo, I., Okumoto, N., Yamashita, M.: Impact of Local Topological Information on Random Walks on Finite Graphs. In: Baeten, J.C.M., Lenstra, J.K., Parrow, J., Woeginger, G.J. (eds.) ICALP 2003. LNCS, vol. 2719, pp. 1054–1067. Springer, Heidelberg (2003)

[15] Kale, S., Seshadhri, C.: An expansion tester for bounded degree graphs. In: Aceto, L., Damgård, I., Goldberg, L.A., Halldórsson, M.M., Ingólfsdóttir, A., Walukiewicz, I. (eds.) ICALP 2008, Part I. LNCS, vol. 5125, pp. 527–538. Springer, Heidelberg (2008)

[16] Kempe, D., Dobra, A., Gehrke, J.: Gossip-based computation of aggregate information. In: 44th Annual IEEE Symposium on Foundations of Computer Science (FOCS 2003), pp. 482–491 (2003)

[17] Lovász, L.: Random walks on graphs: A survey. Combinatorics, Paul Erdös is Eighty 2, 1–46 (1993)

[18] Motwani, R., Raghavan, P.: Randomized Algorithms, 7th edn. Cambridge University Press, Cambridge (1995)

[19] Pak, I.: Mixing time and long paths in graphs. In: 13th Annual ACM-SIAM Symposium on Discrete Algorithms (SODA 2002), pp. 321–328 (2002)

[20] Sinclair, A.: Algorithms for Random Generation and Counting. Birkhäuser, Basel (1993)

Online Computation with Advice

Yuval Emek[1,*], Pierre Fraigniaud[2,**], Amos Korman[2,***], and Adi Rosén[3,†]

[1] Tel Aviv University, Tel Aviv, 69978 Israel
[2] CNRS and University Paris Diderot, France
[3] CNRS and University of Paris 11, France

Abstract. We consider a model for online computation in which the online algorithm receives, together with each request, some information regarding the future, referred to as *advice*. The advice provided to the online algorithm may allow an improvement in its performance, compared to the classical model of complete lack of information regarding the future. We are interested in the impact of such advice on the competitive ratio, and in particular, in the relation between the size b of the advice, measured in terms of bits of information per request, and the (improved) competitive ratio. Since $b = 0$ corresponds to the classical online model, and $b = \lceil \log |\mathcal{A}| \rceil$, where $\mathcal{A}$ is the algorithm's action space, corresponds to the optimal (offline) one, our model spans a spectrum of settings ranging from classical online algorithms to offline ones.

In this paper we propose the above model and illustrate its applicability by considering two of the most extensively studied online problems, namely, *metrical task systems (MTS)* and the *k-server* problem. For MTS we establish tight (up to constant factors) upper and lower bounds on the competitive ratio of deterministic and randomized online algorithms with advice for any choice of $1 \leq b \leq \Theta(\log n)$, where n is the number of states in the system: we prove that any randomized online algorithm for MTS has competitive ratio $\Omega(\log(n)/b)$ and we present a deterministic online algorithm for MTS with competitive ratio $O(\log(n)/b)$. For the k-server problem we construct a deterministic online algorithm for general metric spaces with competitive ratio $k^{O(1/b)}$ for any choice of $\Theta(1) \leq b \leq \log k$.

1 Introduction

Online algorithms are algorithms that receive their input piece by piece and have to act upon the receipt of each piece of input (a.k.a. request). Yet, their

* This work was partially done during this author's visit at LIAFA, CNRS and University Paris Diderot, supported by Action COST 295 DYNAMO. The author is also partially supported by The Israel Science Foundation, grants 664/05 and 221/07.
** Additional support from the ANR project ALADDIN, by the INRIA project GANG, and by COST Action 295 DYNAMO.
*** Additional support from the ANR project ALADDIN, by the INRIA project GANG, and by COST Action 295 DYNAMO.
† Research partially supported by ANR projects AlgoQP and ALADDIN.

S. Albers et al. (Eds.): ICALP 2009, Part I, LNCS 5555, pp. 427–438, 2009.

goal is usually to guarantee a performance which is as close as possible to the optimal performance achievable if the entire input is known in advance. How close do they get to this optimal performance is usually analyzed by means of competitive analysis (cf. [4]).

From a theoretical standpoint, the *complete* lack of knowledge about the future makes it many times impossible to achieve "reasonable" competitive ratios. From a practical standpoint, complete lack of knowledge about the future does not always accurately model realistic situations. Consequently, several attempts have been made in the literature to somewhat relax the "absolutely no knowledge" setting, and achieve better competitive ratios in such relaxed settings. Most notable are the setting where a limited number of steps into the future is known at any time (lookahead) (e.g., [1,7,19]), and the "access graph" setting for paging (e.g., [5,13]). These settings and their analyses are usually specific to the problem they address.

In this paper we study a new, general framework whose purpose is to model online algorithms which have access to *some* information about the future. This framework is intended to analyze the impact of such information on the achievable competitive ratio. One important feature of our framework is that it takes a *quantitative* approach for measuring the amount of information about the future available to an online algorithm. Roughly speaking, we define a finite *advice space* $\mathcal{U}$, and augment the power of the online algorithm `Alg` (and thus reduce the power of the adversary) by means of a series of *queries* u_t, $t = 1, 2, \ldots$, where u_t maps the whole request sequence σ (including the future requests) to an *advice* $u_t(\sigma) \in \mathcal{U}$ provided to `Alg` in conjunction with the tth request of σ. This advice can then be used by the online algorithm to improve its performance. At the risk of a small loss of generality, we assume that the advice space is of size 2^b for some integer $b \geq 0$ and consider the advice to be a string of b bits.

Example 1. For the paging problem, it is relatively easy to verify that the following is a 1-competitive algorithm which uses 1 bit of advice per request (i.e., $|\mathcal{U}| = 2$) [10]. The bit of advice indicates whether the optimal offline algorithm keeps in memory the requested page until the next request to that same page. The online algorithm tries to imitate the behavior of the optimal algorithm: if the optimal algorithm indeed keeps in memory the requested page until the next request to that same page, then so does the online algorithm. Whenever a page must be swapped out from memory, the online algorithm picks an arbitrary page among all pages that are not supposed to remain in memory until they are requested again.

Clearly, since for a "usual" online problem the set of all possible request sequences is often infinite and in any case would typically be larger than the advice space $\mathcal{U}$, our framework just imposes some "commitment" of the adversary regarding the future. This reduces the power of the adversary, and gives to the online algorithm some information about the future. Since (typically) an online algorithm has at any time a finite set of possible actions, our setting additionally provides a smooth spectrum of computation models whose extremes are (classical) online computation with no advice (advice space of size 1) and optimal, offline computation, where the advice is simply the optimal action (the advice space corresponds to the set of all possible actions).

The main motivation for studying online algorithms that receive a small advice with each request is purely theoretical. Nevertheless, this framework may be motivated by settings such as the following, which may be dubbed "spy behind enemy lines": an entity which is aware of the plans of the adversary collaborates with the online algorithm, however the communication between this entity and the online algorithm is limited in terms of its capacity.

In this work we concentrate on two classical and extensively studied online problems, *metrical task systems (MTS)* and the *k-server* problem. We establish several (upper and lower) bounds on the achievable competitive ratios for these problems by online algorithms with advice, thus demonstrating the applicability of our approach for online problems, and giving a more refined analysis for online algorithms having some information about the future. Specifically, for MTS we establish asymptotically tight upper and lower bounds by proving Theorems 1 and 2.

Theorem 1. *Any randomized online algorithm for uniform n-node MTS with $1 \leq b \leq \Theta(\log n)$ bits of advice per request cannot be ρ-competitive against an oblivious adversary unless $\rho = \Omega(\log(n)/b)$.*

Theorem 2. *For any choice of $1 \leq b \leq \log n$, there exists a deterministic online algorithm for general n-node metrical task systems that receives b bits of advice per request and whose competitive ratio is $O(\log(n)/b)$.*

For the k-server problem we first prove Theorem 3 and then generalize it to establish Theorem 4.

Theorem 3. *There exists an $O(\sqrt{k})$-competitive deterministic algorithm for the k-server problem that receives $O(1)$ bits of advice per request.*

Theorem 4. *For any choice of $\Theta(1) \leq b \leq \log k$, there exists a deterministic online algorithm for the k-server problem that receives b bits of advice per request and whose competitive ratio is $k^{O(1/b)}$.*

Related work. Online algorithms operating against restricted adversaries have been considered in the literature on many occasions, and under different settings. For example, online algorithms that operate against an adversary that has to provide some lookahead into the future have been considered, e.g., for the list accessing problem [1], the bin-packing problem [19], and the paging problem [7]. Another example is the model of "access graph" for the paging problem [5,13].

The notion of advice is central in computer science (actually, checking membership in NP-languages can be seen as computing with advice). In particular, the concept of advice and the analysis of its size and its impact on various computations has recently found various applications in distributed computing. It is for instance present in frameworks such as informative labeling for graphs [27], distance oracles [28], and proof labeling [22,23]. A formalism of computing with advice based on a pair of collaborative entities, usually referred to as an oracle and an algorithm, has been defined in [17] for the purpose of differentiating the broadcast problem from the wake-up problem. This framework has been recently used in [16] for the design of distributed algorithms for computing minimum spanning trees (MST), in [15] for tackling the distributed coloring problem, and in [26] for analyzing the graph searching problem (a.k.a. the cops-and-robbers problem). Other applications can be found in [8,18,20]. In the framework of computing with advice, the work probably most closely related to the present one is the work of Dobrev, Královič, and Pardubská [10] who essentially prove that there is a 1-competitive online algorithm for the paging problem, with 1 bit of advice[1] (see Example 1).

Online algorithms (without advice) for metrical task systems have been extensively studied. For deterministic algorithms it is known that the competitive ratio is exactly $2n-1$, where n is the number of states in the system [6]. For randomized algorithms, the known upper bound for general metrical task systems is $O(\log^2 n \log\log n)$ [12,14] and the known lower bound is $\Omega(\log n/\log\log n)$ [2,3]. For uniform metric spaces the randomized competitive ratio is known to be $\Theta(\log n)$ [6,21].

For the k-server problem the best competitive ratio for deterministic algorithms on general metric spaces is $2k-1$ [24], and the lower bound is k [25]. Randomized algorithms for the k-server problem (against oblivious adversaries) are not well understood: it is known that in general metric spaces no algorithm has competitive ratio better than $\Omega(\log k/\log\log k)$ [2,3], but no upper bound better than the one of [24] (that holds for deterministic algorithms) is known.

Organization. The rest of the paper is organized as follows. In Section 2 we give the necessary preliminaries. The lower bound for metrical task systems is presented in Section 3; the matching upper bound is established in Section 4. In Section 5 we prove Theorem 3 regarding the k-Server problem. Due to space limitations we only give outlines of these three results, while the proof of Theorem 4 is omitted entirely. We conclude in Section 6 with some further discussion and open problems.

2 Preliminaries

An online algorithm is an algorithm that receives its input piece by piece. Each such piece is an element in some set $\mathcal{S}$ and we refer to it as a *request*. Let σ be a finite request sequence. The tth request is denoted by $\sigma[t] \in \mathcal{S}$. The online algorithm has to perform an action upon the receipt of each request, that is, at

[1] The model and interests of [10] actually differ from ours in two aspects. First, they are interested in the amount of information required in order to obtain online algorithms with optimal performance, rather than improved competitive ratios. Second, they allow the advice to be of variable size, including size zero, and concentrate their work on the question of how much below 1 can the average size of the advice be. This is done by means of encoding methods such as encoding the 3-letter alphabet $\{\emptyset, 0, 1\}$ using one bit only.

round t, $1 \leq t \leq |\sigma|$, this action has to be performed when the online algorithm only knows the requests $\sigma[1], \ldots, \sigma[t]$. For this action it incurs some *cost* (in the case of minimization problems).

To formally define a deterministic online algorithm we use the formulation of [4] (cf. Chapter 7). A deterministic online algorithm is a sequence of functions $g_t : \mathcal{S}^t \to \mathcal{A}_t$, $t \geq 1$, where $\mathcal{A}_t$ is the set of possible actions for request t (in many cases all $\mathcal{A}_t$ are identical and we denote them by $\mathcal{A}$.) In this work we strengthen the online algorithm (and thus weaken the adversary) in the following manner. For some finite set $\mathcal{U}$, referred to as the *advice space*, the online algorithm is augmented by means of a sequence of *queries* $u_t : \mathcal{S}^* \to \mathcal{U}$, $t \geq 1$. The value of $u_t(\sigma)$, referred to as *advice*, is provided to the online algorithm in each round $1 \leq t \leq |\sigma|$. The complexity of the advice is defined to be $\log |\mathcal{U}|$. For simplicity of presentation, and at the risk of an inaccuracy in our results by a factor of at most 2, we only consider advice spaces of size 2^b for some integer $b \geq 0$, and view the advice as a string of b bits.

Formally, a deterministic online algorithm with advice is a sequence of pairs (g_t, u_t), $t \geq 1$, where $g_t : \mathcal{S}^t \times \mathcal{U}^t \to \mathcal{A}_t$, and $u_t : \mathcal{S}^* \to \mathcal{U}$. Given a finite sequence of requests $\sigma = (\sigma[1], \ldots, \sigma[\ell])$, the action taken by the online algorithm in round t is $g_t(\sigma[1], \ldots, \sigma[t], u_1(\sigma), \ldots, u_t(\sigma))$.

A *randomized* online algorithm with advice is allowed to make random choices (i.e., "toss coins") to determine its actions (the functions g_t) and the advice scheme (the queries u_t). Formally, then, a randomized online algorithm with advice is a probability distribution over deterministic online algorithms with advice.

A deterministic online algorithm $\texttt{Alg}$ (with or without advice) is said to be *c-competitive* if for all finite request sequences σ, we have $\texttt{Alg}(\sigma) \leq c \cdot \texttt{Opt}(\sigma) + \beta$, where $\texttt{Alg}(\sigma)$ is the cost incurred by $\texttt{Alg}$ on σ, $\texttt{Opt}(\sigma)$ is the cost incurred by an optimal (offline) algorithm on σ, and β is a constant which does not depend on σ. If the above holds with $\beta = 0$, then $\texttt{Alg}$ is said to be *strictly c-competitive*. For a randomized online algorithm (with or without advice) we consider the expectation (over the random choices of the algorithm) of the cost incurred by $\texttt{Alg}$ on σ. Therefore a randomized online algorithm $\texttt{Alg}$ (with or without advice) is said to be *c-competitive* (against an oblivious adversary) if for all finite request sequences σ, we have $\mathbb{E}[\texttt{Alg}(\sigma)] \leq c \cdot \texttt{Opt}(\sigma) + \beta$.

As commonly done for the analysis of online algorithms, one may view the setting as a game between the online algorithm and an adversary that issues the request sequence round by round. In this framework, the values of the queries u_t can be thought of as commitments made by the adversary to issue a request sequence which is consistent with the advice seen so far. For an online algorithm $\texttt{Alg}$, augmented with advices in $\mathcal{U}$, we are interested in the competitive ratio of $\texttt{Alg}$, the advice complexity $\log |\mathcal{U}|$, and the interplay between these values.

Metrical Task Systems. A *metrical task system (MTS)* is a pair $(\mathcal{M}, \mathcal{R})$, where $\mathcal{M} = (V, \delta)$ is an n-point metric space[2], and $\mathcal{R} \subseteq (\mathbb{R}_{\geq 0} \cup \{\infty\})^n$ is a set

[2] Throughout the paper, we use the standard definition of a metric space consisting of a set V of points and a distance function δ.

of allowable *tasks*. The points in V are usually referred to as *states*. We assume without loss of generality that $\mathcal{M}$ is scaled so that the minimum distance between two distinct states is 1.

An instance I of $(\mathcal{M}, \mathcal{R})$ consists of an initial state s_0 and a finite task sequence $r_1, \ldots, r_m$, where $r_t \in \mathcal{R}$ for all $1 \leq t \leq m$. Consider some algorithm Alg for $(\mathcal{M}, \mathcal{R})$ and suppose that Alg is in state s at the beginning of round t (the algorithm is in state s_0 at the beginning of round 1). In round t Alg first moves to some state s' (possibly equal to s), incurring a *transition cost* of $\delta(s, s')$, and then processes the task r_t in state s', incurring a *processing cost* of $r_t(s')$. The *cost* incurred by Alg on I is the sum of the total transition cost in all rounds and the total processing cost in all rounds.

The k-server problem. Let $\mathcal{M} = (V, \delta)$ be a metric space. We consider instances of the k-server problem on $\mathcal{M}$, and when clear from the context, omit the mention of the metric space. At any given time, each server resides in some node $v \in V$. A subset $X \subseteq V$, $|X| = k$, where the servers reside is called a *configuration*. The *distance* between two configurations X and Y, denoted by $D(X, Y)$, is defined as the weight of a minimum weight matching between X and Y.

An instance I of the k-server problem on $\mathcal{M}$ consists of an initial configuration X_0 and a finite request sequence $r_1, \ldots, r_m$, where $r_t \in V$ for all $1 \leq t \leq m$. Consider some algorithm Alg for the k-server problem on $\mathcal{M}$ and suppose that Alg is in configuration X at the beginning of round t (the algorithm is in configuration X_0 at the beginning of round 1). The request r_t must be *processed* by one of the k servers in round t, which means that Alg moves to some configuration Y such that $r_t \in Y$ (Y may be equal to X if $r \in X$), incurring a cost of $D(X, Y)$. The *cost* incurred by Alg on I is the total cost in all rounds.

3 A Lower Bound for MTS

In this section we sketch the proof of Theorem 1, that is, we show that if a randomized online algorithm for uniform n-node MTS with $1 \leq b \leq \Theta(\log n)$ bits of advice per request is ρ-competitive, then $\rho = 1 + \Omega(\log(n)/b)$. For the sake of the analysis, we consider a stronger model for the online algorithms, where the whole advice is provided at the beginning of the execution rather than round by round.

Before dwelling into the details of the lower bound, we describe a key ingredient in the proof that relates to a basic two-player zero-sum game, referred to as *generalized matching pennies (GMP)*. In GMP both the min-player and the max-player have the same discrete action space $\{1, \ldots, k\}$. The cost incurred by the min-player is 0 if both players play the same action $1 \leq i \leq k$; otherwise, it is 1. Let S be the random variable that takes on the action of the max-player (S reflects the mixed strategy of the max-player). Clearly, if $\mathrm{H}(S) = \log k$, namely, if the entropy of S is maximal (which means that the max-player chooses its action uniformly at random), then the expected cost incurred by the min-player is $1 - 1/k$. The following lemma, whose proof is omitted from this extended abstract, provides a lower bound on the expected cost incurred by the min-player when the entropy in S is not maximal.

Lemma 1. *For every mixed strategy of the max-player, if* $\mathrm{H}(S) \geq \delta \log k$*, where* $0 < \delta < 1$*, then the expected cost incurred by the min-player is greater than* $\delta - 1/\log k$*.*

Given some integer $1 \leq b \leq \Theta(\log n)$ (the size of the advice per round), fix $\phi = 2^{\Theta(b)}$ and $\tau = \lfloor \log_{\phi}(n/2) \rfloor$. The hidden constants in the Theta notations are chosen to guarantee the following properties: (P1) $4 \leq \phi < \sqrt{n/2}$; and (P2) $1 < \tau = \Omega(\log(n)/b)$.

Let L be a sufficiently large integer. By repeating the GMP game with $k = \phi$ for $L\tau$ rounds, we obtain the following *online GMP problem.* Each round $1 \leq t \leq L\tau$ of the online GMP problem is characterized by some *letter* ℓ_t in the alphabet $[\phi]$. This letter (chosen by an oblivious adversary) is revealed immediately after round t. The alphabet $[\phi]$ also serves as the action space of the algorithms for the online GMP problem, that is, the action of an algorithm in round t is characterized by some letter $\ell'_t \in [\phi]$. The cost incurred by the algorithm in round t is 0 if $\ell'_t = \ell_t$ and 1 if $\ell'_t \neq \ell_t$. The online GMP problem is defined such that any algorithm incurs additional L units of *dummy cost* regardless of its choices of letters ℓ'_t. Clearly, an optimal (offline) algorithm for the online GMP problem does not incur any cost other than the dummy cost as its action in round t is $\ell'_t = \ell_t$ for every $1 \leq t \leq L\tau$.

The remainder of the proof consists of two parts. First, we employ an information theoretic argument to show that if the request sequence $\sigma \in [\phi]^{L\tau}$ of the online GMP problem is chosen uniformly at random among all $L\tau$-letter words over the alphabet $[\phi]$, then the expected cost incurred on σ by any deterministic online algorithm that receives $bL(\tau + 1)$ bits of advice in advance is $L(1 + \Omega(\tau))$. Therefore by Yao's principle, it follows that for every randomized online algorithm `Alg` that receives $bL(\tau + 1)$ bits of advice in advance, there exists a request sequence $\sigma \in [\phi]^{L\tau}$ such that $\mathbb{E}[\mathtt{Alg}(\sigma)] = L(1 + \Omega(\tau))$. Second, we reduce the online GMP problem to uniform n-node MTS showing that a request sequence of length $L\tau$ for the former problem can be implemented as a request sequence of length $L(\tau + 1)$ for the latter. By combining these two parts, and since $\mathtt{Opt}(\sigma) = L$ for every $\sigma \in [\phi]^{L\tau}$, we conclude that the competitive ratio of any randomized algorithm for uniform n-node MTS with advice of b bits per round is $1 + \Omega(\tau) = 1 + \Omega(\log(n)/b)$ (even if the whole advice is provided to the algorithm at the beginning of the execution).

The information theoretic argument. Let $\sigma = (\sigma_1, \ldots, \sigma_{L\tau})$ be a sequence of $L\tau$ letters of the alphabet ϕ chosen independently and uniformly at random. Consider some deterministic online algorithm `Alg` for the online GMP problem that receives $bL(\tau + 1) = \Theta(L\tau \log \phi)$ bits of advice, denoted by U, at the beginning of the execution.

The entropy in σ is $\mathrm{H}(\sigma) = L\tau \log \phi$. By definition, the advice U is a random variable which is fully determined by σ, thus $\mathrm{H}(\sigma \mid U) = \mathrm{H}(\sigma, U) - \mathrm{H}(U) =$

$H(\sigma) - H(U) = \Omega(L\tau \log \phi)$. A straightforward variant of the chain rule of conditional entropy implies that $H(\sigma \mid U) = H(\sigma_1 \mid U) + H(\sigma_2 \mid \sigma_1, U) + \cdots + H(\sigma_{L\tau} \mid \sigma_1, \ldots, \sigma_{L\tau-1}, U)$. Since $H(\sigma_t) = \log \phi$ for all t, it follows by an averaging argument that in a constant fraction of the rounds t the entropy that remains in σ_t after `Alg` saw the advice U and the outcomes $\sigma_1, \ldots, \sigma_{t-1}$ of the previous rounds is $\Omega(\log \phi)$. Lemma 1 can now be used to deduce that the expected cost incurred by `Alg` in each such round t is $\Omega(1)$. Therefore the expected cost (including the dummy cost) incurred by `Alg` throughout the execution is $L(1 + \Omega(\tau))$.

The reduction. We now turn to reduce the online GMP problem to the MTS problem. The online GMP problem is implemented as an n-node MTS $(\mathcal{M}, \mathcal{R})$, where $\mathcal{R} = \{0, \infty\}^n$. That is, each task $r \in \mathcal{R}$ has 0 processing cost for some states and infinite processing cost for the rest. (It is assumed that in every task, at least one state has 0 processing cost.) Clearly, a competitive algorithm for $(\mathcal{M}, \mathcal{R})$ must have 0 total processing cost.

Given a request sequence of length $L\tau$ over the alphabet $[\phi]$ (for the online GMP problem), fix $n' = \phi^\tau$. The corresponding MTS request sequence is divided into L *cycles*, where each cycle consists of $\tau + 1$ rounds, so the total length of the request sequence is $L(\tau + 1)$. (Consequently, the advice provided to the online algorithm at the beginning of the execution contains $bL(\tau + 1)$ bits.) A request $r = \langle r(1), \ldots, r(n) \rangle$ in odd (respectively, even) cycles satisfies $r(i) = \infty$ for every $n'+1 \leq i \leq 2n'$ (resp., for every $1 \leq i \leq n'$). For states $2n'+1 \leq i \leq n$, we always have $r(i) = \infty$. Therefore, throughout an odd (respectively, even) cycle, every algorithm must be in state i for some $1 \leq i \leq n'$ (resp., for some $n'+1 \leq i \leq 2n'$). This means that between cycle c and cycle $c+1$ every algorithm must move from some state in $\{1, \ldots, n'\}$ to some state in $\{n' + 1, \ldots, 2n'\}$ or vice versa, which sums up to L units of cost referred to as the *dummy cost.*

In what follows we describe the structure of cycle c for some odd c. The structure of the even cycles is analogous. States $1, \ldots, n'$ are organized in contiguous *ranges.* For every round $1 \leq t \leq \tau + 1$ of cycle c, there exists some range $R_t \subseteq \{1, \ldots, n'\}$ such that the processing cost of state i is 0 if $i \in R_t$; and ∞ if $i \notin R_t$. Clearly, every competitive algorithm must process the request of round t in some state of R_t. In round 1 all states have zero processing cost, i.e., $R_1 = \{1, \ldots, n'\}$. For $t = 1, \ldots, \tau$, we have $R_{t+1} \subseteq R_t$. Specifically, the range R_t is partitioned into ϕ equally-sized contiguous *subranges*; range R_{t+1} will be one of these subranges (determined by the letter in $[\phi]$ that characterizes the corresponding round in the online GMP request sequence). Eventually, in the last round of the cycle, the range $R_{\tau+1}$ consists of a single state. This is consistent with our choice of parameters since $\tau = \log_\phi n'$.

Note that the only unknown in each round $1 \leq t \leq \tau$ of cycle c is which of the ϕ subranges of R_t corresponds to the range r_{t+1}. If at the end of round t the algorithm is located in some state of R_{t+1}, then no cost is incurred in round $t+1$; otherwise, a unit cost is incurred. So, we have $L\tau$ rounds characterized by some letter in $[\phi]$ and L additional rounds in which every algorithm incurs a unit cost. Therefore a ρ-competitive online algorithm for $(\mathcal{M}, \mathcal{R})$ implies a ρ-competitive online GMP algorithm.

4 An Upper Bound for MTS

In this section we establish Theorem 2 by presenting a deterministic online algorithm for general MTS that gets b, $1 \le b \le \log n$, bits of advice per request and achieves a competitive ratio of $\lceil \frac{\lceil \log n \rceil}{b} \rceil = O(\log(n)/b)$. We call this algorithm `Follow`.

Let $(\mathcal{M}, \mathcal{R})$ be a metrical task system. The request sequence is divided into *cycles*, each consisting of $\alpha = \lceil \frac{\lceil \log n \rceil}{b} \rceil$ requests, with the last cycle possibly shorter. The first cycle, cycle 0, consists of the first α requests, and so on. During cycle $i \ge 0$, `Follow` receives advice of $\lceil \log n \rceil$ bits, which indicate the state in which the optimal algorithm serves (will serve) the first request of cycle $i+1$.

For cycle i, let s_i be the state in which the optimal algorithm serves the first request of the cycle, and let OPT_i be the cost of the optimal algorithm during cycle i. Let OPT be the cost of the optimal algorithm on the whole request sequence.

Definition of `Follow`. Before starting to serve cycle i, $i \ge 0$, `Follow` places itself at state s_i. This is possible for cycle 0 because both the optimal algorithm and `Follow` start at the same initial state s_0. This is possible for any cycle $i > 0$ by moving, at the end of phase $i-1$, to state s_i, known to `Follow` by the advice given in cycle $i-1$.

To describe how `Follow` serves the requests in a cycle we give the following definition. Let $B_i(j)$, $j \ge 0$, be the set of states in the metrical task system that are at distance less than 2^j from s_i. I.e., $B_i(j) = \{s : d(s, s_i) < 2^j\}$. We now partition the (at most) α requests of cycle i, into *phases*. When the cycle starts, phase 0 starts. During phase j, `Follow` serves the requests by moving to the state, among the states in $B_i(j)$, which has the least processing cost for the given task, and serving the request there. A request no longer belongs to phase j, and phase $j+1$ starts, if serving the request according to the above rule will bring the total *processing cost* since the cycle started to be at least 2^j. Note that if a given request belongs to some phase j, the next request may belong to phase $j' > j+1$. That is, there may be phases with no request.

We first give a lower bound on the cost of the optimal algorithm in each cycle (proof omitted), and then give the main theorem of this section.

Lemma 2. *If the last request of cycle i belongs to phase k, $k \ge 1$, then $OPT_i \ge 2^{k-2}$.*

Theorem 5. *`Follow` is $O(\alpha)$-competitive.*

Proof sketch: We consider the cost incurred by `Follow` cycle by cycle. We denote by C_i the cost of `Follow` during cycle i. $C_i = C_i^s + C_i^t + C_i^*$, where C_i^s is the processing cost during cycle i, C_i^t is the transition cost during cycle i, and C_i^* is the cost incurred by `Follow`, at the end of cycle i, to move to state s_{i+1} (we do not count this cost in C_i^t). First note that by the triangle inequality $C_i^* \le C_i^t + d(s_i, s_{i+1})$. Then, for each cycle we consider two cases.

Case 1: The last request of cycle i is in phase $k = 0$. In this case $C_i^s + C_i^t \leq OPT_i$ because `Follow` does not incur any transition cost, and the optimal algorithm either moved away from s_i, incurring a cost of at least 1, or serves the whole cycle in s_i, incurring processing cost equal to that of `Follow`.
Case 2: The last request of cycle i is in phase $k > 0$. In this case we have $C_i^s + C_i^t \leq 8\alpha \cdot OPT_i$. This is because (1) by Lemma 2 we have that $OPT_i \geq 2^{k-2}$; and (2) $C_i^s < 2^k$ by the definition of the phases, and $C_i^t \leq (\alpha - 1)2^{k+1}$ since `Follow` does not leave $B_i(k)$ during the phase.

Since $\sum_i d(s_i, s_{i+1}) \leq OPT$, summing over all cycles gives the result. □

5 An Upper Bound for the k-Server Problem

In this section we present a deterministic algorithm for the k-server problem that receives 4 bits of advice per request and admits a competitive ratio of $O(\sqrt{k})$, thus establishing Theorem 3.

The algorithm, denoted `Partition`, works in *iterations*, where each iteration consists of k requests. Fix some optimal (offline) algorithm `Opt` and let A_i denote the configuration of `Opt` at the beginning of iteration i. `Partition` uses *two* bits of advice per round of iteration i to identify A_{i+1} as follows. The first bit of advice, received in round $1 \leq j \leq k$ of iteration i, determines whether the node corresponding to the current request belongs to A_{i+1}. The second bit of advice, received in round $1 \leq j \leq k$ of iteration i, determines whether the jth (according to some predefined order) node of A_i is still occupied in A_{i+1}. Based on these two bits of advice, and based on the knowledge of A_i, `Partition` computes the configuration A_{i+1} and moves to it at the end of iteration i. The cost incurred by this move is bounded from above by the sum of the total cost incurred by `Partition` in iteration i (tracing the steps of `Partition` back to configuration A_i) and the total cost incurred by `Opt` in iteration i (tracing the steps of `Opt` from A_i to A_{i+1}), thus increasing the overall competitive ratio of `Partition` only by a constant factor.

In order to serve the requests of iteration i, the k servers are partitioned into *heavy* servers and *light* servers according to the role they play in `Opt` during iteration i: those serving (in `Opt`) at least $\sqrt{k}$ requests are classified as heavy servers; the rest are classified as light servers. The third bit of advice, received in round $1 \leq j \leq k$ of iteration $i-1$, determines whether the server that occupies the jth node of configuration A_i is heavy or light. Consequently, at the beginning of iteration i `Partition` knows the heavy/light classification of all the servers. (The details related to the first iteration are omitted from this extended abstract.) The fourth bit of advice, received in round $1 \leq j \leq k$ of iteration i, determines whether `Opt` serves the current request with a heavy server or with a light one. Note that this partitions the requests of iteration i into *heavy* requests (served in `Opt` by a heavy server) and *light* requests (served in `Opt` by a light server).

To actually serve the requests of iteration i, `Partition` invokes the Work Function Algorithm (`WFA`) [9], with the heavy servers, on the subsequence consisting of the heavy requests of iteration i. `WFA` has a competitive ratio of $2k' - 1$, where k' is the number of servers in the problem [24]. Moreover, it is shown in

[11] that WFA is in fact strictly $O(k')$-competitive. In our case $k' \leq \sqrt{k}$ as there cannot be more than $\sqrt{k}$ heavy servers. Thus the cost incurred by Partition on the heavy requests of iteration i is at most $O(\sqrt{k})$ times larger than that of Opt.

The subsequence consisting of the light requests of iteration i is served by the light servers according to the following greedy strategy: each light request is served by the closest light server, which then immediately returns to its initial position in A_i. This strategy is strictly $O(\ell)$-competitive, where ℓ is the maximum number of requests served by any (light) server in Opt. By the definition of the light servers, ℓ in our case is at most $\sqrt{k}$, which implies that the cost incurred by Partition on the light requests of iteration i is at most $O(\sqrt{k})$ times larger than that of Opt. By summing over all iterations, we conclude that the competitive ratio of Partition is $O(\sqrt{k})$, thus establishing Theorem 3.

6 Conclusions

We define a model for online computation with advice. The advice provides the online algorithm with some (limited) information regarding future requests. Our model quantifies the amount of this information in terms of the size b of the advice measured in bits per request. This model does not depend on the specific online problem.

The applicability and usefulness of our model is demonstrated by studying, within its framework, two of the most extensively studied online problems: metrical task systems (MTS) and the k-server problem. For general metrical task systems we present a deterministic algorithm whose competitive ratio is $O(\log(n)/b)$. We further show that any online algorithm, even randomized, for MTS has competitive ratio $\Omega(\log(n)/b)$ if it receives b bits of advice per request. This lower bound is proved on uniform metric spaces. For the k-server problem we present a deterministic online algorithm whose competitive ratio is $k^{O(1/b)}$. Whether this is best possible is left as an open problem.

We believe that employing our model of online computation with advice may lead to other results, thus enhancing our understanding of the exact impact of the amount of knowledge an online algorithm has regarding the future on its competitive ratio.

References

1. Albers, S.: A competitive analysis of the list update problem with lookahead. Theor. Comput. Sci. 197(1–2), 95–109 (1998)
2. Bartal, Y., Bollobás, B., Mendel, M.: Ramsey-type theorems for metric spaces with applications to online problems. J. Comput. Syst. Sci. 72, 890–921 (2006)
3. Bartal, Y., Mendel, M., Linial, N., Naor, A.: On metric Ramsey-type phenomena. Annals of Mathematics 162, 643–710 (2005)
4. Borodin, A., El-Yaniv, R.: Online Computation and Competitive Analysis. Cambridge University Press, Cambridge (1998)
5. Borodin, A., Irani, S., Raghavan, P., Schieber, B.: Competitive paging with locality of reference. In: STOC, pp. 249–259 (1991)
6. Borodin, A., Linial, N., Saks, M.E.: An optimal on-line algorithm for metrical task system. J. ACM 39, 745–763 (1992)

7. Breslauer, D.: On competitive on-line paging with lookahead. Theor. Comput. Sci. 209(1-2), 365–375 (1998)
8. Cohen, R., Fraigniaud, P., Ilcinkas, D., Korman, A., Peleg, D.: Label-guided graph exploration by a finite automaton. In: Caires, L., Italiano, G.F., Monteiro, L., Palamidessi, C., Yung, M. (eds.) ICALP 2005. LNCS, vol. 3580, pp. 335–346. Springer, Heidelberg (2005)
9. Chrobak, M., Larmore, L.L.: The server problem and on-line games. In: On-line algorithms: Proc. of a DIMACS Workshop. DIMACS Series in Discrete Mathematics and Theoretical Computer Science, vol. 7, pp. 11–64 (1991)
10. Dobrev, S., Královič, R., Pardubská, D.: How much information about the future is needed? In: Geffert, V., Karhumäki, J., Bertoni, A., Preneel, B., Návrat, P., Bieliková, M. (eds.) SOFSEM 2008. LNCS, vol. 4910, pp. 247–258. Springer, Heidelberg (2008)
11. Emek, Y., Fraigniaud, P., Korman, A., Rosén, A.: On the additive constant of the k-server work function algorithm. A manuscript, `http://arxiv.org/abs/0902.1378v1`
12. Fakcharoenphol, J., Rao, S., Talwar, K.: A tight bound on approximating arbitrary metrics by tree metrics. J. Comput. Syst. Sci. 69, 485–497 (2004)
13. Fiat, A., Karlin, A.: Randomized and multipointer paging with locality of reference. In: STOC, pp. 626–634 (1995)
14. Fiat, A., Mendel, M.: Better algorithms for unfair metrical task systems and applications. SIAM J. Comput. 32, 1403–1422 (2003)
15. Fraigniaud, P., Gavoille, C., Ilcinkas, D., Pelc, A.: Distributed computing with advice: information sensitivity of graph coloring. In: Arge, L., Cachin, C., Jurdziński, T., Tarlecki, A. (eds.) ICALP 2007. LNCS, vol. 4596, pp. 231–242. Springer, Heidelberg (2007)
16. Fraigniaud, P., Korman, A., Lebhar, E.: Local MST computation with short advice. In: SPAA, pp. 154–160 (2007)
17. Fraigniaud, P., Ilcinkas, D., Pelc, A.: Oracle size: a new measure of difficulty for communication tasks. In: PODC, pp. 179–187 (2006)
18. Fraigniaud, P., Ilcinkas, D., Pelc, A.: Tree exploration with an oracle. In: Královič, R., Urzyczyn, P. (eds.) MFCS 2006. LNCS, vol. 4162, pp. 24–37. Springer, Heidelberg (2006)
19. Grove, E.F.: Online bin packing with lookahead. In: SODA, pp. 430–436 (1995)
20. Fusco, E.G., Pelc, A.: Trade-offs between the size of advice and broadcasting time in trees. In: SPAA (2008)
21. Irani, S., Seiden, S.S.: Randomized algorithms for metrical task systems. Theor. Comput. Sci. 194, 163–182 (1998)
22. Korman, A., Kutten, S.: Distributed verification of minimum spanning trees. In: PODC, pp. 26–34 (2006)
23. Korman, A., Kutten, S., Peleg, D.: Proof labeling schemes. In: PODC, pp. 9–18 (2005)
24. Koutsoupias, E., Papadimitriou, C.H.: On the k-server conjecture. J. ACM 42(5), 971–983 (1995)
25. Manasse, M.S., McGeoch, L.A., Sleator, D.D.: Competitive algorithms for server problems. Journal of Algorithms 11, 208–230 (1990)
26. Nisse, N., Soguet, D.: Graph searching with advice. In: Prencipe, G., Zaks, S. (eds.) SIROCCO 2007. LNCS, vol. 4474, pp. 51–65. Springer, Heidelberg (2007)
27. Peleg, D.: Informative labeling schemes for graphs. Theor. Comput. Sci. 340(3), 577–593 (2005)
28. Thorup, M., Zwick, U.: Approximate distance oracles. In: STOC, pp. 183–192 (2001)

Dynamic Succinct Ordered Trees

Arash Farzan and J. Ian Munro

Cheriton School of Computer Science,
University of Waterloo
{afarzan,imunro}@cs.uwaterloo.ca

Abstract. We study the problem of maintaining a dynamic tree succinctly, in $2n + o(n)$ bits, under updates of the following form: insertion or deletion of a leaf, insertion of a node on an edge (edge subdivision) or deletion of a node with only one child (the child becomes a child of the grandparent). We allow satellite data of a fixed (but not necessarily constant) size to be associated to the nodes of the tree.

We support update operations in constant amortized time and support access to satellite data and basic navigation operations in worst-case constant time; the basic navigation operations includes **`parent`**, **`first/last-child`**, **`previous/next-child`**. We demonstrate that to allow fast support for more extended operations such as determining the i-th child of a node, rank of a child among its siblings, or subtree size, we require a restrictive update strategy for which we propose the finger-update model where updates are performed by a finger which is only allowed to crawl on the tree (between a child and a parent or between consecutive siblings). Under this model, we describe how the named operations can be performed in worst-case constant time.

Previous work on dynamic succinct ordered trees [1,2] is mainly restricted to binary trees and achieves poly-logarithmic [1] or "poly-log-log" [2] update time under a more restricted model. Best previous result on ordinal trees achieves only sublinear amortized update time and "poly-log-log" query time [3].

1 Introduction

A succinct representation of a combinatorial object is an encoding which supports a reasonable set of operations on the object in constant time and has a storage requirement matching the information theoretic lower bound, to within lower order terms. Succinct data structures perform under the uniform-time word-RAM model with $\Theta(\lg n)$ word size.

Ordered trees are trees in which the order of children of nodes is significant and preserved. Ordered trees, as a fundamental data structure, have attracted a great deal of research on their succinct representation [4,5,6,7,8,9,10]. All such approaches achieves the information-theory lower bound of $2n$ bits (to within lower order terms) to encode an ordered tree with n nodes, and their difference is in what queries they support.

S. Albers et al. (Eds.): ICALP 2009, Part I, LNCS 5555, pp. 439–450, 2009.

The main drawback with most of the succinct tree representations is their lack of flexibility to allow dynamic updates, which necessitates an entire structure rebuild on modifications. Storm [11] in his thesis gives some main obstacles in dynamising the previously existing approaches and introduces a new succinct representation for binary trees allowing updates in poly-logarithmic time [1]. Raman and Rao [2] improve updates to poly-log-log time for binary trees. Gupta *et al.* [3] propose a framework for dynamising various succinct structures which, when applied to ordered trees, results in a dynamic succinct representation that performs updates in $O(n^{\epsilon})$ time for any constant $\epsilon > 0$ and performs basic navigation queries in $O(\log\log n)$ time. Arroyuelo [12] proposes a dynamic succinct structure for cardinal trees.

A recent decomposition-based approach exists [10] to represent static ordered trees which greatly simplifies the encoding of trees and implementation of many operations on such trees. This paper demonstrates that the approach is of use in attaining a dynamic succinct representation of ordered trees which allows fast updates and operations.

1.1 Overview of the Results

We use the general representation approach of [10] and show how to maintain the representation under insertions and deletions. The type of insertions and deletions supported is the same as in [1,2] and are as follows: insertion or deletion of a leaf, insertion of a node with one child on an edge or deletion of a node with only one child (by connecting its parent to its child directly).

Basic navigation operations which we support permit crawls on trees and include `parent`, `first-child`, `last-child`, `previous-child`, and `next-child` operations. These are supported in worst-case constant time. We note that these operations subsume the `parent`, `left-child`, and `right-child` operations in binary trees supported in previous work [1,2].

Associated with each node v is a piece of data of length b bits ($b = O(\log n)$). Given a node v, we support operations `access-data(v)` and `change-data(v)` in worst-case constant time.

We show that, to support more advanced operations, one needs to restrict the pattern of updates. The operations, we consider, are `child(v, i)` which returns the i-th child of a given node v, `child-rank(v)` which returns the rank of a given node v among its siblings, and `subtree-size(v)` which returns the number of descendants of a given node v. `child(v,i)`, `child-rank(v)` are two natural operations to consider as we move from binary trees to ordered trees where nodes can have arbitrary large degrees. The `subtree-size(v)` operation was considered and supported by previous work on dynamic binary trees [1,2] using the traversal pattern for update. Under the traversal pattern updates are performed during the course of a traversal which starts at the root and ends at the root and subtree-size queries can be answered only on completion of the traversal.

We relax the update pattern by introducing the finger-update model. In the finger-update model, there is a finger maintained on a node of the tree which is initially on the root and it can move from a parent to the leftmost or the rightmost child or can move from a node to its immediate next or previous siblings. Updates are only allowed on the node currently under the finger. Any number of updates can be performed as the finger crawls on the tree and queries (such as `subtree-size`) can be asked at any time at any node. Updates are supported in amortized constant time over the number of movements of the finger and the number of updates performed.

In the course of showing support for these operations, we design a partial-sum data structure which works optimally in the finger-update model. This structure is very useful and is of independent interest.

2 Succinct Representation

It is shown in [10] how an ordered tree with n nodes can be decomposed into n/L subtrees of size roughly L for any value of parameter L. Strictly speaking:

Theorem 1. *[10] A tree with n nodes can be decomposed into $\Theta(n/L)$ subtrees of size at most $2L$ which are pairwise disjoint aside from the subtree roots. Furthermore, aside from edges stemming from the subtree root nodes, there is at most one edge leaving a node of a component to its child in another component.*

The representation that we use is analogous to the static representation of [10]. It is based on a two-level recursive decomposition of a given tree. In the first level of recursion, the tree with n nodes is decomposed into subtrees using value $L = \lceil \lg^2 n \rceil$, and subsequently these subtrees are, in turn, decomposed into yet smaller subtrees using value $L = \lceil (\lg n)/16 \rceil$ to obtain the subtrees on the second level of recursion. Using the standard terminology of [8], we refer to the subtrees on the first level by *mini-trees* and the second level by *micro-trees*.

Micro-trees of size less than $\lceil (\lg n)/8 \rceil$ are small enough to be represented by a look-up table. The representation of a micro-tree with k nodes consists of two fields: the first field simply is the size of the micro-tree ($O(\log k) = O(\lg \lg n)$ bits) and the second field is an index to the look-up table ($2k$ bits). These indices sum up to $2n$ bits over all micro trees and are the dominant term in our representation; other auxiliary data amounts to $o(n)$ bits.

The table stores encodings of all trees with sizes up to $\lceil (\lg n)/8 \rceil$ along with answers to variety of queries types for each such tree. For all possible updates, each tree contains a pointer to the destination tree, *i.e.* the tree resulting from the application of that update. The size of the auxiliary data for each tree will be poly-logarithmic in n and thus the size of the entire table is $o(n)$.

Mini-trees consist of micro-trees and links between them. Links between different micro-trees can be in either of the three following forms: 1) a common root node, 2) edges from a single-node micro-tree to its children micro-tree or 3) an edge from a non-root node from a micro-tree to the root of another micro-tree. We represent the latter edges by introducing a *dummy node* on them. Due to the manner in which the tree is decomposed, there is at most one such edge leaving a non-root node in a micro-tree and hence, at most one dummy node in each micro-tree. We keep explicit pointers (of size $O(\log\log n)$) to represent dummy edges, which are edges from dummy nodes to their children.

To represent a common root among micro-trees, or edges emanating from a single-vertex micro-tree, we use explicit $O(\log\log n)$-bit pointers to form a doubly-linked list on sibling micro-trees such that two consecutive micro-trees on the chain contain two immediate sibling children of the root. This is in contrast to the use of dictionary structures in [10], as there is no dynamic dictionary structure with the desired functionalities. The linked list does not facilitate random accesses to children of a node, and we can overcome this issue by introducing a substitute structure that supports fast updates under the finger-update model in section 5.3.

The tree consists of mini-trees and links between them. The tree over mini-trees is represented analogously to the mini-tree representation over micro-trees: *i.e.* explicit pointers for edges coming out of mini-trees from non-root nodes and explicit $O(\log n)$-bit pointers to reference left and right sibling mini-trees.

3 Performing Updates

In this section, we show how the succinct structure of section 2 is maintained under insertions and deletions of nodes. To allow for dynamic updates, we relax the maximum size of a micro-tree to be $\lceil(\lg n)/4\rceil$ which is twice the maximum size we allowed in the static representation. Analogously, we allow the maximum size of a mini-tree to be $4\lceil\lg^2 n\rceil$, twice the maximum mini-tree size in the static representation. When a node is inserted or deleted, the topology of the corresponding micro-tree changes. The look-up table which lists all trees of size up to $\lceil(\lg n)/4\rceil$, also contains, for each tree and all possible insertions and deletions on the tree, the reference to the resulting tree under the update. Therefore, the update can be performed in constant time, provided the size of the corresponding micro-tree or mini-tree does not exceed the limits.

When a micro-tree exceeds the maximum size, it is decomposed into (at most 4) subtrees using theorem 1 with value $L = \lceil(\lg n)/16\rceil$. The decomposition is performed in constant time using the look-up table: each entry of maximum size in the look-up table contains its decomposition information. Some of the newly created micro-trees resulted from the micro-tree decomposition could be undersized. However, we can charge them to the $\Theta(\log n)$ insertions that made the original micro-tree exceed over limits. Since we rebuild the containing mini-tree after $O(\log^2 n)$ such insertions, there are $O(\log n)$ undersized micro-trees at any point and the representation remains valid.

Analogously, when a mini-tree exceeds the maximum mini-tree size, it is decomposed into (at most 4) subtrees using theorem 1 with value $L = \lceil\lg^2 n\rceil$. The decomposition is performed in $\Theta(\log^2 n)$ time which is amortized over the $\Theta(\log^2 n)$ insertions which brought the mini-tree over the limits. Some of the newly created mini-trees could be undersized. We charge these to the $\Theta(\log^2 n)$ insertions which caused the mini-tree to overflow. As we rebuild the entire tree after n insertions, there are $O(n/\log^2 n)$ such undersized mini-trees at any point and hence the representation remains valid. Together with creation of a mini-tree, micro-trees inside it are recomputed and represented.

Handling of deletions from micro-trees and mini-trees is easier. No particular action is required to the point that the micro-tree or the mini-tree being deleted from becomes empty. In this event, the corresponding entries are removed from the representation. The entry removal costs $\Theta(\log n)$ time for a micro-tree and $\Theta(\log^2 n)$ for a mini-tree. These costs are amortized over the deletions that emptied the micro-tree or mini-tree. This is justified if the micro-tree or the mini-tree are not undersized and have sizes $\Theta(\log n)$ and $\Theta(\log^2 n)$ respectively. If the original mini-tree or micro-tree are undersized, they are either created as a result of a split or they were initially undersized before any updates. In the former case, we showed there are $\Theta(\log n)$ and $\Theta(\log^2 n)$ insertions which push the micro-tree or the mini-tree over limits, we charge these insertions instead. In the latter case, a one-time linear fund of $\Theta(\log n)$ for each micro-tree and $\Theta(\log^2 n)$ for each mini-tree is enough to deal with deletions of these components.

One crucial part of dealing with updates is memory management. Micro-trees and mini-trees grow and shrink as a result of insertions and deletions, and new ones are created as a result of splits, and they disappear when they become empty. We show in section 3.1 how these structures are handled in memory.

3.1 Memory Management

Following the path of previous work [2], we assume the memory model in which the working space of the algorithm is from word 0 to the highest word h the algorithm is using at the time, and the space usage at any given time is the highest word being used.

In the work of Munro *et al.* [1], they insist on having blocks and subblocks sit on a continuous segment of memory. We completely revise the memory management part of the structure to obtain constant amortized update time. We relax the requirement that mini/micro trees be stored in a continuous segment of memory; each may be spread over many segments.

Mini trees are stored in *blocks* of memory and micro trees are stored in *subblocks* of memory. Each block consists of many continuous memory chunks called *blocklets* and similarly, each subblock consists of many continuous memory chunks called *subblocklets.*

Subblock memory management. Within a mini tree (of size $\Theta(\log^2 n)$), there are $\Theta(\log n)$ micro trees (of size $\Theta(\log n)$). Therefore, a block of size $O(\log^2 n)$ consists of $O(\log n)$ subblocks of size $O(\log n)$.

As depicted in figure 1, a block is divided into two areas: a clean area containing the main portion of subblocks and an overflow area of $\lg^{3/2} n$ bits which contains the overflow portion of subblocks that have overflown. Occasionally, once enough updates have been performed, the entire structure is rebuilt: everything is moved to the clean area and the overflow area is emptied.

A subblock can be spread over two subblocklets: one in the main area and the other one in the overflow area. Each subblock has a header which contains a table

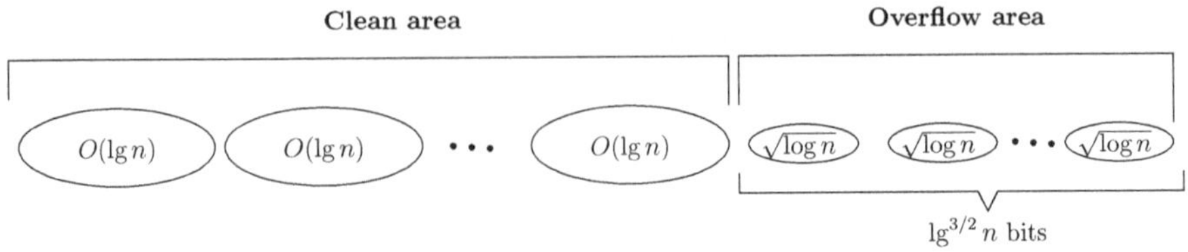

Fig. 1. Subblock layout within a block

of (two) references to the subblocklets and their sizes. This *subblock reference table* is used to resolve a memory address within a subblock. There are only two $O(\log\log n)$-bit pointers to store and the size of each blocklets can be stored in $O(\log\log n)$ bits and the space overhead is negligible.

We first explain how we handle insertions. We keep the count of the number of insertions and rebuild the entire structure at every $\frac{1}{16}\lg n$ insertions. The overflow area has $\lg^{3/2} n$ bits which are allocated in chunks of $\sqrt{\lg n}$ bits. If a subblocklet in the clean area overflows, a $\sqrt{\lg n}$-bit chunk is allocated as the second subblocklet and the subblock reference table is updated accordingly. If the subblocklet in the overflow area becomes full, we must allocate another chunk to the subblock. As we require this chunk to be the immediate following chunk, we may have to move several following chunks if they belong to the same subblocklet to the end of the allocated chunks in the overflow area. Moreover, the references to these chunks must be updated accordingly in the subblock's reference table. Moving takes $O\left(\sqrt{\log n}\right)$ time in the worst case if the entire overflow area must be moved. This cost can be amortized over the $O\left(\sqrt{\log n}\right)$ insertions in the overflowing subblocklet. Therefore, the amortized time is constant.

When the overflow area is full, we rebuild the entire block and the overload area is emptied. This takes $O(\log n)$ time which can be amortized over $O(\log n)$ insertions that filled up the overflow area.

Thus far, we have dealt only with insertions which cause subblocks to overflow. However, nodes can be deleted and as a result subblocks shrink. Deletions are easier to handle. We keep a count of the number of deletions in a block and if they go over $\frac{1}{48}\lg n$, we rebuild the entire block[1]. The key observation is that since there has been $O(\log n)$ deletions since last rebuilt, the structure remains dense enough.

Block memory management. In the previous section, we explained how subblocks are organized within an individual block. In this section, we explain how blocks are fragmented and laid out in memory. An outline of the memory layout is depicted in figure 2.

At the block level, memory is allocated in chunks of $\lg^{3/2} n$ bits; when a block outgrows its allocated space a chunk of $\lg^{3/2} n$ bits at the end of the currently used space is allocated for it. Since the block takes $\Theta\left(\log^2 n\right)$ bits and chunks

[1] Constants such as $\frac{1}{48}$ are chosen to make things work either by keeping table sizes sublinear or to allow for growth and contraction of substructures.

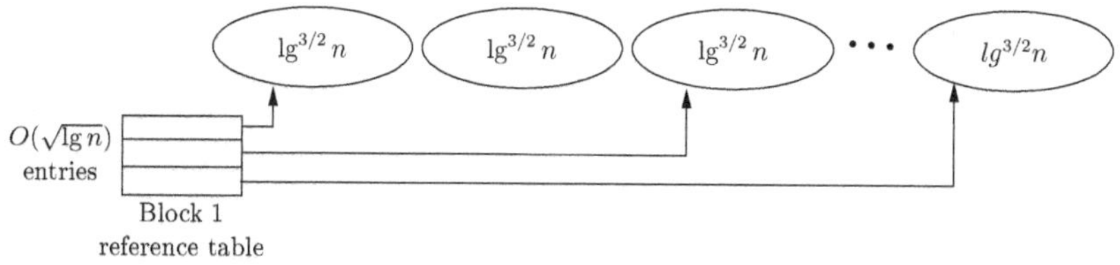

Fig. 2. Block memory layout: memory blocklets relative to a same block are not necessarily in a consecutive zone

are $\lg^{3/2} n$ bits, a block consists of $\Theta\left(\sqrt{\log n}\right)$ chunks (*i.e.* blocklets). Thus, a table can be stored for each block which keeps tracks of pointers to blocklets of a block. We refer to this table as *block reference table*. Since blocklets have the same size, addressing inside a block using the block reference table is easy.

Deletions are easier to deal with; We only keep a count of the number of deletions and if it exceeds $n/\lg n$, we compactify and rebuild the entire structure. Since, there can only be $n/\lg n$ deleted nodes (and thus bits) in the structure, it remains dense enough.

Furthermore, we can have keys and satellite data associated with each nodes and manage them efficiently in memory upon insertions and deletions. Due to lack of space, we defer the detailed discussion of this subject to a full version of this paper.

4 Navigation Queries

In this section, we explain how the basic navigation operations are supported. The basic navigational operations are parent (`parent`), leftmost/rightmost child (`first/last-child`) and right/left siblings (`previous/next-child`). More enhanced operations such as i-th child of a node are to be discussed in section 5. We note that previous work [1,2] functions only on binary trees and our basic navigation operations (leftmost child, right sibling) supports all of their supported navigational operations.

We recall that the tree decomposition of [10] guarantees that each component subtree C has at most one edge going out of the component with the exception of the root of the component which can be shared by many components or can have many edges emanating if it is a single-node component. We refer to component D which contains the other end of the edge from a non-root node as the child of component C (C is the component parent of D). If a component root has children in different components, these components are referred to as the sibling components. Two immediate sibling components are each other's left and right sibling components.

As mentioned in section 2, each mini-tree keeps (long) pointers to its parent, leftmost/rightmost children and left/right sibling mini-trees. Similarly, each micro-tree keeps (short) pointers to its parent, child, and left/right sibling micro-trees. These pointers can be trivially maintained under updates. Given these pointers, the basic navigation operations named are easy to implement. For instance, given a node to find its right sibling, we first consult the look-up table and if there is no right sibling there and the node is a micro-tree root, we locate the right sibling micro-tree and consult the look-up table for that micro-tree. If there is no right sibling micro-tree and the node is a mini-tree root, we find the right sibling mini-tree and find the leftmost child therein. The edge going out of a component from a non-root node v is dealt with as an exception and data is explicitly stored to accommodate for it. Thus, we can support basic navigation queries on the succinct tree:

Theorem 2. *A tree with n nodes and satellite data of length b bits associated with all nodes can be represented in $(2+b)n + o(bn)$ bits and maintained in constant amortized time per insertions and deletions such that it facilitates navigation by supporting* `parent`, `first/last-child`, `previous/next-child` *in worst-case constant time. Furthermore, given any node, its data can be accessed and modified in worst-case constant time.* □

5 Support for Enhanced Queries

We showed in section 4 how basic navigation on tree can be performed in constant time. Nevertheless, more enhanced operations that allow us to gallop in the tree (such as going directly to the i-th child of a node) or obtaining useful information on nodes of the tree (such as subtree size of a node) are of interest.

As we will discuss in section 5.1, it is infeasible to support such enhanced queries in constant time when insertion and deletions are performed in arbitrary locations. In previous work [1,2] on binary trees, the difficulty is felt and restrictions are put in place to support queries such as subtree size of a node in constant time. The restriction is updates are performed during a course of a traversal which starts and ends at the root, and furthermore, queries can be answered only after the traversal is completed.

We strengthen the model by separating the issue of updates from queries while supporting updates in amortized constant time and queries in worst-case constant time. We allow queries to be performed anywhere in the tree while restrict updates to a crawl on the tree.

5.1 Lower Bounds

In this section, we show the infeasibility of supporting enhanced operations in constant time, if dynamic updates are performed in arbitrary locations and not in form of a crawl. Our lower bounds derive from the lower bounds on the *List representation* and *subset rank* problems proved by Fredman and Saks [13]. Chan *et al.* [14] also derive lower bounds for dynamic succinct texts using the same set of results. The proofs are ommitted due to lack of space.

Theorem 3. *Maintaining an ordered tree under insertions and deletions of leaves to support navigation queries* `parent` *and* `child(v, k)` *and access satellite data requires $\Omega(\log n / \log\log n)$ amortized time per operation.* □

Theorem 4. *Maintaining an ordered tree under insertions and deletions in arbitrary locations to support sub-tree size queries requires $\Omega(\log n / \log \log n)$ amortized time per operation.* □

5.2 The Dynamic Model

As mentioned previously, we allow queries to be asked anywhere in the tree, however we require the updates not to jump from a node to a distant node and therefore only crawl on the tree. One can visualize this as if there is a finger which is initially at the root of the tree. The finger can travel from a node to its leftmost or rightmost child. Inversely, the finger can go from a child to its parent. The finger can move from a node to its immediate left/right siblings. We note that we charge for each movement of the finger and thus the update time is amortized over the movement of finger as well as the updates performed.

Queries other than updates, on the other hand, can be asked anywhere on the tree and a query can be asked on a node independent of the history of the locations where previous queries have been asked.

5.3 Succinct Dynamic Prefix Sum Data Structure

In support for enhanced queries, we use a data structure which indexes prefix sums in a dynamic array. The dynamic model is the *finger-update model* where there is a finger and only the position which is under the finger can be updated at any time and the finger moves from an element one step to the left or right at a time. The desired queries to support are `ps_rank` and `ps_select` queries. Query `ps_rank(t)` simply reports $\sum_{i=1}^{t} A[i]$ and `ps_select(s)` returns index k such that $\sum_{i=1}^{k} A[i] \leq s < \sum_{i=1}^{k+1} A[i]$. Hon *et al.* [15] studied the same problem, although in the fully-dynamic model where updates perform in arbitrary independent locations, and thus operations are supported in non-constant time as there are similar lower bounds as discussed in section 5.1.

We will show in the proof of theorem 5 that the finger dynamic array problem can be reduced to the case where the array is only updated from one end. This case is captured in the following lemma whose proof we omit due to lack of space.

Lemma 1. *Let $A[1..t]$ be a non-negative array of integers such that $\sum_{i=1}^{t} A[i] \leq n$ for some previously known n and $\forall i; A[i] < \lg^c n$ for some absolute constant $c > 0$. There exists a data structure which requires $O(t \log n) + o(n)$ bits and maintains the array under* insertions at the end, *and* deletions from the end *in worst-case constant time per update and it can answer* `ps_rank` *and* `ps_select` *queries in worst-case constant time.* □

Theorem 5. *Let $A[1..t]$ be a non-negative array of integers such that $\sum_{i=1}^{t} A[i] \leq n$ for some previously known n and $\forall i; A[i] < \lg^c n$ for some absolute constant $c > 0$. There exists a data structure which requires $O(t \log n) + o(n)$ bits and maintains the array under insertions/deletions in the finger-update model in worst-case constant time per update. This structure can answer* `ps_rank` *and* `ps_select` *queries in worst-case constant time.*

Proof. The idea is to break the array from where the finger lies into two arrays `Left` and `Right`. These two arrays grow/shrink from the end to simulate the finger-update array. Movement of the finger to left and right corresponds to insertion of an element from one piece and its deletion from the other.

A `ps_rank(i)` query where i is to the left of the update finger can be answered by the same query in `Left` array. In case i lies in the `Right array` the answer to `ps_rank(n-i)` in the `Right` array is deducted from the total sum of all elements and returned. Similarly, `ps_select(i)` can be derived from the answer to such queries in `Left` or `Right` array depending on whether i is less than or larger than the total sum of elements in array `Left`. □

As we delete and insert elements the sum of elements changes. In lemma 1 and theorem 5 we assumed prior knowledge of an upper bound on the sum. However, such knowledge does not always exist in advance and the structure must be adaptive to the total sum as it changes. We allow updates which are increments and decrements by one and insertion and deletions of zeros. Under this model, we can afford to rebuild the structure as the value of the total sum doubles or halves and always stay within the desired space bound.

Corollary 1. *Let $A[1..t]$ be an array of non-negative integers such that $\forall i; A[i] < \lg^c n$ for some absolute constant $c > 0$ where $n = \sum_{i=1}^{t} A[i]$. There exists a data structure which requires $O(t \log n) + o(n)$ bits and maintains the array under decrement and increment by one and insertion and deletion of zeros in the finger-update model in constant amortized time per update. This structure can answer* `ps_rank` *and* `ps_select` *queries in worst-case constant time.* □

5.4 Implementing `Child(v,i)` and `child_rank(v)` Operations

On a node which is not a micro-tree or a mini-tree root, the queries are answered from the look-up table. A micro/mini-tree root is dealt with using the prefix sum structure (corollary 1). For each mini/micro-tree root, we maintain an array which lists the number of children the roots has in different mini-trees (or micro-trees). As children are inserted or deleted one by one and in order, the increment and decrement model of the corollary applies.

Operation `child(v,i)` is implemented by a `ps_select` to find the right mini-tree followed by a `ps_select` to find the right micro-tree in which the a table look-up gives the answer. Similarly, operation `child_rank` is performed by `ps_rank` queries on the mini-tree and micro-tree and finally a table-lookup to find the rank within a micro-tree.

Using corollary 1, a structure on a node v which is a mini-tree root requires $O(\text{mini-deg(v)} * \log n) + o(\text{deg(v)})$ bits where mini-deg(v) is the number of mini-trees in which v has children and deg(v) is the actual degree of v. These terms sum to $o(n)$ over the entire tree. Similarly, a structure on a node u which is micro-tree root within a mini-tree requires $O(\text{micro-deg(v)} * \log\log n) + o(\log^2 n)$ bits where micro-deg(v) is the number of micro-trees within the mini-tre in which v has children. Over all mini-trees these terms sum to $o(n)$ bits.

5.5 Implementation of Subtree_size Operation

The implementation of **subtree_size** in a static tree is that each mini-tree explicitly stores its subtree size and, within a mini-tree, each micro-tree stores its subtree size within the mini-tree [10]. To determine **subtree_size(v)**, we determine the subtree size within the containing micro-tree using the look-up table, the descendent micro-tree (if any) has the subtree size within the mini-tree stored explicitly. Finally, the descendent mini-tree (if any) has the subtree size in the entire tree explicitly stored.

A dynamic implementation of **subtree_size** resembles the static tree representation: the same information is retained at the mini-tree and micro-tree roots. However, the information is not necessarily valid at all nodes at all times. Particularly, the subtree sizes stored at mini-tree and micro-tree roots which are ancestors of the update finger node are not up-to-date. We maintain the invariant that roots which are not an ancestor of the current node, with the update finger, possess up-to-date information. This is easily maintainable by updating the sub-tree size of a root when the finger is moved up from one of its children.

Thus, nodes other than those on the path from the root to the finger node can determine their subtree size as in the static case. We denote by $m_1, m_2, \ldots m_i$ the mini-trees that the root-to-finger path crosses. For each such mini-tree m_k, we maintain a value a_i which is the net additions to the mini-tree (*i.e.* number of insertions minus the number of deletions). We store values a_i in an array which is maintained as the **ps_rank** structure in lemma 1 which allows constant-time prefix sum and thus quick suffix sum computation (we note that as we only need the **ps_rank** functionality of the structure, numbers do not have to satisfy the non-negativity criteria).Within each mini-tree we repeat the same structure, for the micro-trees that the path crosses. Now given a node on the path, we determine the subtree size within the micro-tree using a table-lookup. We read the subtree size within the mini-tree from the descending micro-tree. This values is adjusted using the array structure on the micro-trees. Finally, the subtree size in the entire subtree is read from the descending mini-tree. This value is adjusted using the array structure on mini-trees.

6 Conclusion

We have studied the problem of maintaining an ordered tree under insertions and deletions of nodes where nodes have associated satellite data. We showed how insertions and deletions can be performed in constant amortized time to allow support for basic navigational queries and access to satellite data in worst-case constant time. This is an improvement over the poly-logarithmic update time [1] and a poly-log-log update time [2] presented for a more restricted trees (*i.e.* binary trees). It also improves a sublinear update time with poly-log-log query time on ordered trees [3].

We proved a lower bound to allow more enhanced operations such as determining the i-th child of a node or obtaining the subtree-size of a node. To allow constant time support for these operations, we introduced the finger-update model where updates are performed where the update finger lies, and the update finger crawls on the tree. Nevertheless, queries other than updates can be asked on any node of the tree independent of where the update finger is.

Extending permitted updates on trees to a more enhanced set, such as a set that also includes cutting and attaching entire subtrees or left/right rotations, remains open as a possible future work.

References

1. Munro, J.I., Raman, V., Storm, A.J.: Representing dynamic binary trees succinctly. In: SODA 2001: Proceedings of the twelfth annual ACM-SIAM symposium on Discrete algorithms, pp. 529–536 (2001)
2. Raman, R., Rao, S.S.: Succinct dynamic dictionaries and trees. In: Baeten, J.C.M., Lenstra, J.K., Parrow, J., Woeginger, G.J. (eds.) ICALP 2003. LNCS, vol. 2719, pp. 357–368. Springer, Heidelberg (2003)
3. Gupta, A., Hon, W.K., Shah, R., Vitter, J.S.: A framework for dynamizing succinct data structures. In: Arge, L., Cachin, C., Jurdziński, T., Tarlecki, A. (eds.) ICALP 2007. LNCS, vol. 4596, pp. 521–532. Springer, Heidelberg (2007)
4. Jacobson, G.: Space-efficient static trees and graphs. In: 30th Annual Symposium on Foundations of Computer Science (FOCS), pp. 549–554 (1989)
5. Clark, D.R., Munro, J.I.: Efficient suffix trees on secondary storage (extended abstract). In: Proceedings of the annual ACM-SIAM symposium on Discrete algorithms, pp. 383–391 (1996)
6. Munro, J.I., Raman, V.: Succinct representation of balanced parentheses, static trees and planar graphs. In: IEEE Symposium on Foundations of Computer Science, pp. 118–126 (1997)
7. Benoit, D., Demaine, E.D., Munro, J.I., Raman, R., Raman, V., Rao, S.S.: Representing trees of higher degree. Algorithmica 43(4), 275–292 (2005)
8. Geary, R.F., Raman, R., Raman, V.: Succinct ordinal trees with level-ancestor queries. ACM Transactions on Algorithms 2(4), 510–534 (2006)
9. He, M., Munro, J.I., Rao, S.S.: Succinct ordinal trees based on tree covering. In: Arge, L., Cachin, C., Jurdziński, T., Tarlecki, A. (eds.) ICALP 2007. LNCS, vol. 4596, pp. 509–520. Springer, Heidelberg (2007)
10. Farzan, A., Munro, J.I.: A uniform approach towards succinct representation of trees. In: Gudmundsson, J. (ed.) SWAT 2008. LNCS, vol. 5124, pp. 173–184. Springer, Heidelberg (2008)
11. Storm, A.J.: Representing dynamic binary trees succinctly. Master's thesis, School of Computer Science, University of Waterloo, Waterloo, Ontario, Canada (2000)
12. Arroyuelo, D.: An improved succinct representation for dynamic k-ary trees. In: Ferragina, P., Landau, G.M. (eds.) CPM 2008. LNCS, vol. 5029, pp. 277–289. Springer, Heidelberg (2008)
13. Fredman, M., Saks, M.: The cell probe complexity of dynamic data structures. In: STOC 1989: Proceedings of annual ACM symposium on Theory of computing, pp. 345–354. ACM, New York (1989)
14. Chan, H.L., Hon, W.K., Lam, T.W., Sadakane, K.: Compressed indexes for dynamic text collections. ACM Trans. Algorithms 3(2), 21 (2007)
15. Hon, W.K., Sadakane, K., Sung, W.K.: Succinct data structures for searchable partial sums. In: Ibaraki, T., Katoh, N., Ono, H. (eds.) ISAAC 2003. LNCS, vol. 2906, pp. 505–516. Springer, Heidelberg (2003)

Universal Succinct Representations of Trees?

Arash Farzan[1], Rajeev Raman[2], and S. Srinivasa Rao[3,⋆]

[1] David R. Cheriton School of Computer Science, University of Waterloo, Canada
[2] Department of Computer Science, University of Leicester, UK
[3] School of Computer Science and Engineering, Seoul National University, S. Korea

Abstract. We consider the succinct representation of *ordinal* and *cardinal* trees on the RAM with logarithmic word size. Given a tree T, our representations support the following operations in $O(1)$ time: (i) **BP-substring**(i, b), which reports the substring of length b bits (b is at most the wordsize) beginning at position i of the balanced parenthesis representation of T, (ii) **DFUDS-substring**(i, b), which does the same for the *depth first unary degree sequence* representation, and (iii) a similar operation for tree-partition based representations of T. We give:

- an asymptotically space-optimal $2n + o(n)$ bit representation of n-node ordinal trees that supports all the above operations with $b = \Theta(\log n)$, answering an open question from [He et al., ICALP'07].
- an asymptotically space-optimal $C(n, k) + o(n)$-bit representation of k-ary cardinal trees, that supports (with $b = \Theta(\sqrt{\log n})$) the operations (ii) and (iii) above, on the ordinal tree obtained by removing labels from the cardinal tree, as well as the usual label-based operations. As a result, we obtain a fully-functional cardinal tree representation with the above space complexity. This answers an open question from [Raman et al, SODA'02].

Our new representations are able to simultaneously *emulate* the BP, DFUDS and partitioned representations using a single instance of the data structure, and thus aim towards *universality*. They not only support the union of all the ordinal tree operations supported by these representations, but will also automatically inherit any new operations supported by these representations in the future.

1 Introduction

Succinct, or highly space-efficient, representations of trees have found an increasing number of applications in indexing massive collections of textual and semi-structured data [15], and have consequently been intensively studied in recent years. Using succinct representations, one can, for example, represent an n-node binary tree in $2n + o(n)$ bits and support standard navigational and other operations in $O(1)$ time on the RAM model with word size w bits, where $w = O(\lg n)$ [10,14,18]; by contrast, the standard representation takes $\Theta(n)$

⋆ Work supported by MADALGO (Center for Massive Data Algorithmics, a center of the Danish National Research Foundation), Aarhus University, Denmark.

S. Albers et al. (Eds.): ICALP 2009, Part I, LNCS 5555, pp. 451–462, 2009.

words, or $\Theta(n \lg n)$ bits, of memory. Minimizing the constant factor in the leading term of the space usage of a succinct data structure helps to show the asymptotic optimality of the space usage (and is important in practice); e.g., as there are $C_n = \frac{1}{n+1}\binom{2n}{n}$ binary trees on n nodes, there is an information-theoretic lower bound of $\lg C_n = 2n - O(\lg n)$ bits on any binary tree representation.

Succinct tree representations often store the *structure* of the tree as a bit-string of length $2n + o(n)$ bits — even fewer, if the tree is "compressible" [11] — in one of many ways, together with an *index* of $o(n)$ bits (the index depends upon the choice of structure bit-string and the operations to be supported). Operations are supported in $O(1)$ time on the RAM model with logarithmic word size by reading $O(1)$ words from the structure bit-string and/or the index. This approach gives rise to the following undesirable properties of succinct representations:

- The numbering of nodes is based upon the position of the representation of the nodes in the structure bit-string; different representations number nodes differently. This is problematic because one often uses the node-number to associate information with a node, and different numberings are convenient for associating different kinds of information with the nodes of the same tree (e.g. element labels [5], and text data [3] in XML documents).
- Certain operations can be implemented efficiently in one representation, but are hard or impossible to implement in another: to create a representation that supports the union of the sets of operations of two representations, one would need to represent the given tree as two separate copies, each using the respective structure bit-strings and index data structures, thereby doubling the space usage and losing optimality.

This situation is clearly unsatisfactory. For instance, in the case of succinct *ordinal* trees, there are at least three kinds of representations: the *balanced parenthesis (BP)*, the *depth-first unary degree sequence (DFUDS)* and various 'partitioned' representations (see below for definitions); a sequence of papers has been written that attempts to update each representation with the latest additional functionality supported by the others [6,8,11,12,13,16]. At the very least, this 'arms race' is confusing for someone wishing to use these results, or for someone seeking to understand the relative power of these representations.

In this paper, we take a step towards an *universal* encoding of succinct trees by giving encodings that can be used to *emulate* other encodings. Specifically, we provide optimal-space succinct encodings that can return b consecutive bits from the structure bit-strings of other encodings, where b is close to the word-size w, in $O(1)$ time. Since we can emulate access to the structure bit-strings of other encodings, by adding the appropriate index of $o(n)$ bits, one can directly support any operations supported by those encodings, with only a constant factor slowdown and negligible space cost. We consider representing *ordinal* and *cardinal* trees succinctly, and now summarize previous work and our results.

Ordinal Trees. An *ordinal tree* is an arbitrary rooted tree where the children of each node are ordered. As there are $\frac{1}{n}\binom{2n-2}{n-1}$ ordinal trees on n nodes, storing an ordinal tree requires $2n - O(\lg n)$ bits. A plethora of operations has been defined

	Structure	Reference	Functionality
(1)	BP		navigation, subtree size, leaf operations
(2)	DFUDS	[1]	(1) plus i-th child
(3)	BP	[2]	(1) plus degree
(4)	Tree covering	[6]	(2), (3) plus level-ancestor
(5)	BP	[16]	(2), (3) plus level-ancestor, level-successor/predecessor
(6)	BP	[12]	(4) plus i-th child, depth, height, distance
(7)	DFUDS	[11]	(4) plus depth, height, distance, leaf operations
(8)	Tree covering	[8]	(6) plus level-successor/predecessor, level-first/last
(9)	Tree covering	[4]	(8) plus level-descendant

Fig. 1. Functionality of $2n + o(n)$-bit ordinal tree representations

on ordinal trees (see Fig. 1). Equally, there are many $2n+O(1)$-bit alternatives for the structure bit-string of ordinal trees. The BP structure bit-string is obtained by traversing the tree in pre-order, outputting '(' when a node is visited for the first time, and ')' when leaving it. In BP, nodes may be numbered in either pre-order or post-order. The DFUDS [1] structure bit-string is obtained by visiting nodes in depth-first order, and outputting i '('s and one ')' if the current node has i children. The resulting parenthesis string is prefixed with '(' to balance it. In DFUDS numbering, nodes are visited in depth-first order, but upon visiting a node, all its children are numbered consecutively. Further structure bit-strings are obtained from the *level-order unary degree sequence* [10], or by viewing an ordinal tree as a binary tree and using the binary tree representation of [10], but these have limited functionality and are not considered here.

Ordinal trees can also be represented using a *tree covering* approach. These approaches are based on a two-level decomposition of trees into *mini*-trees and *micro*-trees [6,8,14] (an encoding of all microtrees is basically the structure bit-string). We select the *uniform* approach of Farzan and Munro [4] as the representative from this class of tree representations; as this approach satisfies relevant properties of earlier tree covering approaches, any operation supported in $O(1)$ time in previous tree covering representations can naturally be supported in $O(1)$ time in the uniform representation. Hence, in the rest of the paper, we refer to this uniform approach [4] as the tree covering (TC) approach. Fig. 1 summarizes the development of the functionality of ordinal tree representations.

In this paper, we consider $2n + o(n)$-bit representations of an ordinal tree T that support the following operations in $O(1)$ time (w is the word-size):

- `BP-substring`(i, b) - this returns the substring of b consecutive bits beginning at position i in the BP structure bit-string of T, for some $b \leq w$.
- `DFUDS-substring`(i, b) - as above, but for the DFUDS structure bit-string.
- `TC-microtree`(μ, m) - returns the μ-th microtree of the minitree numbered m in the TC representation.

Define `BP-word`(k) as `BP-substring`$((k-1)w + 1, w)$, i.e., `BP-word`(k) returns the k-th word in the BP structure bitstring (`DFUDS-word` is defined analogously).

The tree-covering representation of Geary et al. [6] was able to output the position of the i-th opening or closing parenthesis in the BP structure bit-string.

However, this functionality is too weak to replace the kind of access to the BP structure bit-string required by 'native' BP representations. He et al. [8] extended the functionality of this representation, and showed the following: for any given $f \leq w = \Theta(\lg n)$, one can support `BP-substring`(i, f) in $O(1)$ time using $O(nf/\lg n)$ additional bits. To support `BP-word` in $O(1)$ time, this requires $O(n)$ additional bits giving an overall space bound of $O(n)$ bits, rather than $2n + o(n)$ bits. Indeed, supporting `BP-word` while keeping the overall space at $2n + o(n)$ bits was stated as an open problem in [8].

We give a $2n+o(n)$-bit representation that not only supports `BP-word` in $O(1)$ time, but also `DFUDS-word` and `TC-microtree`.

Cardinal Trees. A *cardinal tree* (or trie) of degree k is a tree in which each node has k positions for an edge to a child. Each node has up to k children, each labeled by a unique integer from the set $\{1, 2, \ldots, k\}$ (a binary tree is a cardinal tree of degree 2). We assume $k \leq n$, but k is taken to be a nondecreasing function of n. Since there are $\frac{1}{n}\binom{kn}{n-1}$ cardinal trees of degree k [7], $\mathcal{C}(n, k) = \lg \binom{kn}{n-1} - \lg n$ bits is a lower bound on the space required to store an arbitrary k-ary cardinal tree. In addition to the above-mentioned set of ordinal tree operations, a cardinal tree representation must also support the operation of returning the child labelled i (if there is one). As all representations below support the latter operation in $O(1)$ time, we do not explicitly mention it.

Jacobson [9] gave a cardinal tree representation that uses $kn + o(kn)$ bits, but this is optimal only for $k = 2$. The representation of Benoit et al. [1] uses $\mathcal{C}(n, k) + O(n)$ bits and supports the full set of DFUDS operations. Raman et al. [17] improved the space bound to $\mathcal{C}(n, k) + o(n)$ bits, but their data structure only supports basic navigational operations. Obtaining space $\mathcal{C}(n, k) + o(n)$ bits while supporting a full range of operations was stated as an open problem in [17]. The representation of Farzan and Munro [4] supported the full range of known ordinal tree operations, but using $\mathcal{C}(n, k) + o(n \lg k)$ bits. In this paper, we give a cardinal tree representation that uses $\mathcal{C}(n, k) + o(n)$ bits for all values of k and supports `DFUDS-substring`(i, b) in $O(1)$ time[1], for some $b = \Theta(\sqrt{\lg n})$ (`TC-microtree` is also supported, if microtrees are of size $\Theta(\sqrt{\lg n})$). By storing indexes of size $o(n)$ bits for the DFUDS/TC representations, we can support all the DFUDS/TC operations in $O(1)$ time, thus solving the open problem in [17].

Preliminaries. Given a subset S from a universe U, we define a *fully indexable dictionary* (FID, from now on) on S to be any data structure that supports the following operations on S in constant time, for any $x \in U$, and $1 \leq i \leq |U|$:

- *rank*(x): return the number of elements in S that are less than x,
- *select*(i, S): return the i-th smallest element in S, and
- *select*$(i, \bar{S})$: return the i-th smallest element in $U \setminus S$.

Lemma 1. *[17] Given a subset of size n from the universe [m], there is a FID that uses* $\lg \binom{m}{n} + O(m \lg\lg m/\lg m)$ *bits.*

[1] More precisely, `BP/DFUDS-substring` for cardinal trees returns a substring of the DFUDS structure bit-string of the ordinal tree obtained by removing vertex labels from the cardinal tree.

Given a bitvector B, we define its FID to be the FID for the set S where B is the characteristic vector of set S.

2 Unifying Different Representations of Ordinal Trees

In this section, we present the new representation that unifies the three previous ones: BP, DFUDS, and TC. As noted already, those representations consist of two parts: the structure bit-string (occupying $2n + o(n)$ bits) and the index (occupying $o(n)$ bits). In the new unified representation, we replicate the indices of all the approaches as they only contribute to lower order terms. The challenge is to provide access to the structure bit-strings of all three representations while still using only $2n + o(n)$ bits. We show that the new unified representation supports `BP-word(`k`)`, `DFUDS-word(`k`)` and `TC-microtree(`m`,`μ`)` in $O(1)$ time, and thereby prove that the new representation can be used as a black box to emulate BP, DFUDS, and TC representations at the same time.

We decompose the tree into $\Theta(n/\lg^2 n)$ mini-trees each of size $O(\lg^2 n)$, using the decomposition algorithm in [4]. The representation of these mini-trees is described in Section 2.1.

We first demonstrate that the problem of supporting previous representations (BP, DFUDS, and TC) can be confined to within mini-trees. In other words, that if any BP or DFUDS word or any micro-tree encoding corresponding to a mini-tree can be generated in constant time, then any BP or DFUDS word or any micro-tree encoding corresponding to the entire tree can be generated in constant time. The claim is obvious for the micro-tree encodings of the TC representation as micro-tree encodings within a mini-tree are the same as for the entire tree. We prove the claim for the BP and DFUDS representations:

Theorem 1. *If there is a structure which represents a mini-tree with m nodes ($m = O(\lg^2 n)$) in $2m + o(m)$ bits and supports operations* `BP-word()` *and* `DFUDS-word()` *on the mini-tree in constant time, then the entire tree with n nodes can be represented in $2n+o(n)$ bits and operations* `BP-word()` *and* `DFUDS-word()` *on the entire tree can be supported in constant time.*

Proof. In both the BP and DFUDS representations of a given tree, each node corresponds to two bits (one open and one closing parenthesis); in the BP representation, these are the '(' output when the node is first discovered in the pre-order traversal, and the ')' output when the subtree of the node is fully discovered in the pre-order traversal. In the DFUDS representation, a node is represented by the '(' in the unary representation of its parent's degree, and by the ')' which concludes the unary representation of its own degree. (The root of the entire tree is an exception in that it misses the first bit; the exception is handled by adding an extra opening parenthesis at the beginning of the representation.) Thus given any subset of the nodes, one can talk about the set of bits corresponding to that subset of nodes.

We note that as mini-trees may share their roots, their corresponding set of bits may share bits which represent their roots. The key observation is Lemma 2, from which Theorem 1 follows.

Lemma 2. *In the BP or the DFUDS sequence of a tree, the bits corresponding to a mini-tree (micro-tree) form a set of constant number of substrings. Furthermore, these substrings concatenated together in order, form the BP or the DFUDS sequence of the mini-tree (micro-tree).* □

2.1 Supporting Representations within a Mini-Tree

In this section, we confine our attention to mini-trees. We give our new representation and argue how the representation supports the operations of `BP-word()`, `DFUDS-word()` and `TC-microtree()` within an individual mini-tree.

Support for the BP and DFUDS Encodings. Given a mini tree consisting of $k = O(\lg^2 n)$ nodes, we now describe how to represent it using $2k + o(k)$ bits to support `BP-word()` and `DFUDS-word()` in constant time.

Definition 1. *We call a node in the mini tree significant if its subtree size (with respect to the mini-tree) is larger than* $\frac{\lg n}{16}$. *All the significant nodes form a connected subtree as each of their ancestors is also significant. We call this subtree the skeleton of the mini tree. Since each leaf in the skeleton has at least* $\frac{\lg n}{16}$ *nodes in its subtree (which are not part of the skeleton), the number of leaves in the skeleton is* $O(\lg n)$. *We call a mini-tree* skinny *if its skeleton is only a path.*

Skinny Trees: We first start with the case of skinny trees and then focus on the general case.

Let u be the leaf of the skeleton (which is a path) and let v be the last (rightmost) child of u in the tree. Let S be the set of all immediate children of the nodes of the skeleton. We denote the set of all nodes of S whose preorder numbers are at most the preorder number of v (including v) by S_D, and the set of all nodes of S which are after v in preorder by S_U. The new representation consists of the following four components (see Fig. 2):

- *Path down,* P_D: Consists of unary representations of the number of children of each node of the skeleton in the set S_D, in order from the root down to leaf u.
- *Path up,* P_U: Consists of the unary representations of the number of children of each node of the skeleton in the set S_U, from leaf u up to the root.
- *Trees on path down,* T_D: Let $D_1, D_2, D_3, \ldots$ be all the subtrees attached to the nodes of S_D, ordered by the preorder numbers of their roots. The bit sequence T_D is obtained by concatenating the BP representations of each of the trees T_i with the first bit (opening parenthesis) removed.
- *Trees on path up,* T_U: Let $U_1, U_2, U_3, \ldots$ be all the subtrees attached to the nodes of S_U, ordered by the preorder numbers of their roots. The bit sequence T_U is obtained by concatenating the BP representations of each of the trees T_i with the first bit (opening parenthesis) removed.

We first show how to reconstruct the original tree from these four components. This can be done by first reconstructing the skeleton and all its immediate

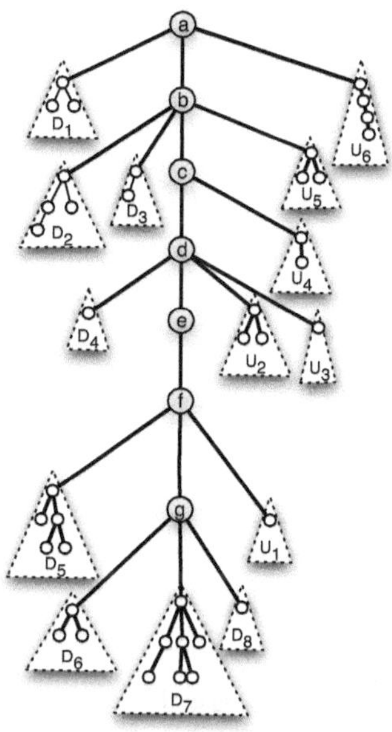

P_D = 101100100101110.

	a	b	c	d	e	f	g
P_D:	10	110	0	10	0	10	1110

P_U = 0100110101010.

	g	f	e	d	c	b	a
P_U:	0	10	0	110	10	10	10

T_D =101001100100100010110100010100110011010010000.

	D_1	D_2	D_3	D_4	D_5	D_6	D_7	D_8
T_D:	10100	1100100	100	0	101101000	10100	1100110100100	0

T_U = 01010001001010011100000.

	U_1	U_2	U_3	U_4	U_5	U_6
T_U:	0	10100	0	100	10100	1110000

Fig. 2. Four components of a skinny tree representation (P_D, P_U, T_D, and T_U) are given for a skinny tree

children using P_D and P_U. Then we attach the subtrees to the immediate children of the skeleton using T_D and T_U. The important fact to observe is that the representations of subtrees in both T_D and T_U are self-delimiting, as these are the BP representations of a tree with the first bit (open parenthesis) removed.

The sum of the sizes of the four components is exactly twice the number of nodes in the mini tree: each node of the skeleton is represented using one bit in P_D and one bit in P_U; each of the immediate children of the skeleton are represented using one bit in either P_D or P_U, and one bit in either T_D or T_U; and each of the other nodes is represented by two bits in either T_D or T_U.

We now show how to produce a word of the BP/DFUDS sequence from the four-component representation of the skinny tree. The proof starts by showing that each consecutive $\lceil (\lg n)/8 \rceil$ bits of the BP sequence can be generated in constant time. Analogously, we show that each consecutive block of $\lceil (\lg n)/24 \rceil$ bits of the DFUDS sequence can be produced in constant time. Thus we have

Theorem 2. *The new unified representation supports operations* `DFUDS-word()` *and* `BP-word()` *in $O(1)$ time in a skinny mini-tree using $o(\lg^2 n)$ extra space.*

General trees: In this case where the skeleton is an arbitrary tree, we decompose the skeleton into $O(\lg n)$ paths using the following recursive procedure.

If the given subtree of the skeleton is a path, then return it as the only path of that subtree. Otherwise, find the maximal leftmost path of the skeleton subtree from the root to the leftmost skeleton leaf, and remove it. The remaining nodes of the subtree form a set of disjoint subtrees. Among these subtrees, we first identify all the "rightmost" subtrees, i.e. subtrees whose roots are the rightmost

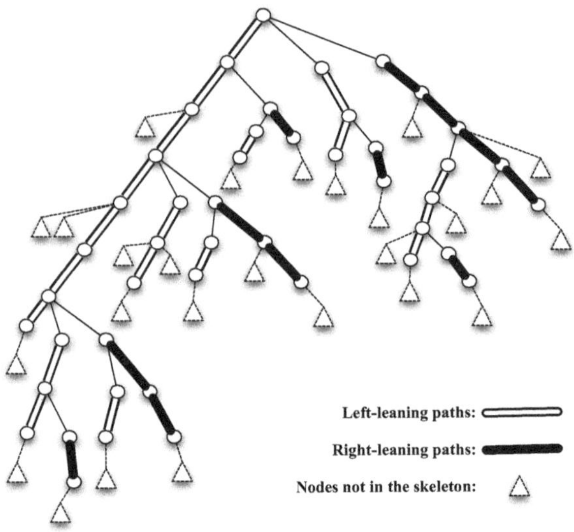

Fig. 3. The skeleton of a tree is decomposed into left-leaning and right-leaning paths to partition the tree into skinny trees

children of their parents, and remove the rightmost paths of each. For each of the disjoint subtrees thus obtained, we apply the decomposition algorithm recursively. Figure 3 shows the partitioning of a tree into these left-leaning and right-leaning paths. This recursive decomposition produces $O(\lg n)$ paths since each leaf in the skeleton is associated with exactly one path, and the number of leaves in the skeleton is $O(\lg n)$ (See Definition 1).

We associate each of the nodes of the mini-tree that is not part of the skeleton with its lowest ancestor that is in the skeleton. Thus the above procedure decomposes the mini-tree into $O(\lg n)$ skinny trees. We use the previously-described skinny tree representation for each of these skinny trees.

We now show how each word of $\lg n$ bits long from the BP/DFUDS sequence can be produced in constant time in a general tree. As the tree is partitioned into skinny trees, the BP/DFUDS sequences are split into parts each of which is obtained from a skinny tree.

Definition 2. *We split the BP/DFUDS sequences into maximal substrings such that bits of each substring can be extracted from the representation of the same skinny-tree. We refer to these maximal substrings as* skinny chunks.

The main feature with our way of decomposing a general tree into skinny-trees is that any $\frac{\lg n}{16}$-bit substring of the BP/DFUDS sequences consists of at most four skinny chunks.

Lemma 3. *Each $\lg n$-bit substring of the BP or the DFUDS sequences spans over $O(1)$ skinny chunks.* □

As a consequence of Lemma 3, it follows that the number of skinny chunks in the BP/DFUDS sequences of a mini-tree is $O(\lg n)$. We are now ready to present the main result of this section that any $\lg n$-bit substring of the BP/DFUDS sequences can be reported in constant time:

Theorem 3. *Any substring of length* $\lg n$ *from the BP/DFUDS sequences of a mini-tree can be reported in* $O(1)$ *time using* $o(\lg^2 n)$ *bits of extra storage.*

Proof. We build an FID structure over the universe of $2n$ bits of the BP and the DFUDS sequences which indicates the starting points of the skinny chunks. For each skinny chunk, we store a pointer to the representation of the corresponding skinny tree and the offset within the BP/DFUDS sequences of the skinny tree where the chunk starts. This dictionary requires $o(\lg^2 n)$ bits as the number of skinny chunks is $O(\lg n)$ and each pointer/offset requires only $O(\lg \lg n)$ bits.

Using the FID, for any skinny chunk c, we can produce the bits of the intersection of the BP or the DFUDS word and the chunk within the skinny tree in constant time by Theorem 2. Lemma 3 states that at most $O(1)$ skinny chunks can intersect a word of length $\lg n$ bits, so by repeating the procedure for each chunk that intersects the word, we discover the entire word in $O(1)$ time. □

Support for the Tree Covering Representation. We now show how the new unified representation can generate micro-trees of the tree-covering representation; specifically, that the new representation can produce the BP/DFUDS bit sequence of any micro-tree in constant time.

Lemma 4. *Within a mini-tree, the BP/DFUDS bit sequence of any micro-tree in the tree-covering representation can be produced in constant time using the new unified representation with an additional space of* $o(\lg^2 n)$ *bits.* □

Since we can generate the BP (or the DFUDS) sequence corresponding to any mico-tree in constant time, using a translation table we can get the actual micro-tree representation.

Theorem 4. *The encoding of any micro-tree in the tree-covering representation can be determined in the new unified representation in constant time by using an additional* $o(\lg^2 n)$ *bits for each mini-tree (of size* $\Theta(\lg^2 n)$*).* □

Theorems 1, 3 and 4 together imply the desired final result:

Theorem 5. *Given an ordinal tree on* n *nodes, the unified representation uses* $2n + o(n)$ *bits and supports the BP, DFUDS, and TC representations by supporting operations* `BP-word(`k`)`, `DFUDS-word(`k`)`, *and* `TC-microtree(`m,μ`)` *in constant time.* □

3 Cardinal Tree Representation

The BP and DFUDS representations of ordinal trees can be easily modified to obtain the following result:

Lemma 5. *Given an ordinal tree on n nodes, suppose there exists a structure that supports* `BP-substring`$(i, f(n))$ *(*`DFUDS-substring`$(i, f(n))$*) in $O(1)$ time, then one can augment the structure with an additional $O(n(\lg f(n))/f(n))$ bits to support all the navigational operations supported by the BP (DFUDS) representation, where $f(n) < \lg n$ is any increasing function of n.*

Proof (Sketch). The BP/DFUDS representations support queries by reading $O(1)$ words (of $\lg n$ bits each) from the BP/DFUDS structure bit-string and the indices. The indices can be easily modified so that the query algorithm reads smaller words (of $f(n)$ bits each) from the structure bit-string, by increasing the size of the index slightly. One can go through all the indices for supporting various operations on the BP/DFUDS representations and verify that the above statement is true for each of the operations. □

We use the following simple extension of Lemma 1.

Lemma 6. *Given a bitvector B of length m with n ones in it, there exists an FID for B that uses $\lg \binom{m}{n} + O(m \lg\lg m / \lg m)$ bits which also supports retrieving any $\lg m$-bit substring of B in constant time.*

We now prove the main result of this section.

Theorem 6. *A k-ary tree on n nodes can be represented using $\mathcal{C}(k, n) + o(n) + O(\lg\lg k)$ bits to support all the ordinal operations on the underlying tree structure that are supported by DFUDS representation, and also the cardinal operation of finding the child of a node with a given label, all in constant time.*

Proof. The k-ary tree representation is similar to the representation of Raman et al. [17, Lemma 6.2] with the difference that instead of numbering the nodes in level-order, we number them in depth-first order. More specifically, we number the n nodes of the given cardinal tree with the numbers from the set $[n]$ in depth-first order. Let S_x be the set of labels of the edges to the children of the vertex numbered x. Then the sets $S_0, S_1, \ldots S_{n-1}$ form a sequence of n sets of total cardinality $n-1$, each being a subset of $[k]$. We represent these subsets using the multiple dictionary structure of Raman et al. [17, Theorem 6.1], which occupies $\lg \binom{nk}{n-1} + o(n) + O(\lg\lg k) = \mathcal{C}(n, k) + o(n) + O(\lg\lg k)$ bits, and supports (partial) `rank` and `select` on each S_x in $O(1)$ time. These are enough to support the operation of finding the child of a node with a given label, in constant time. We now show how to support the operation `DFUDS-substring`$(i, \sqrt{\lg n})$ in constant time (over the DFUDS sequence of the underlying tree structure). The result then follows from Lemma 5.

The above multiple dictionary structure represents the set $S = \{(x, a) | x \in [n], a \in S_x\}$ (obtained by adding the pair (x, a) for each edge labeled a out of the node numbered x) as an indexable dictionary. This indexable dictionary structure in turn considers the following two cases depending on whether the universe size, kn, is large or small relative to the set size, $n-1$:

Dense case, $k \le \sqrt{\lg n}$: In this case, we represent S using the structure of Lemma 6. This structure enables us to extract any $\lg n$-bit substring of the characteristic vector B of S (which is a bit vector of length nk with $n-1$ ones in it) in constant time. Note that the $(ik+j)$th bit in B is a 1 if node $i+1$ in preorder has a child labeled j, and 0 otherwise. Thus from the sequence of bits $ik+1$ to $(i+1)k$ in the bit vector B, we can obtain the (unary) degree of the node $(i+1)$. And in fact from any $\lg n$-bit substring of B, we can obtain the unary degrees of $\Theta((\lg n)/k)$ consecutive nodes in preorder, in constant time using precomputed tables. Since $k \leq \sqrt{\lg n}$, we can extract any $\sqrt{\lg n}$-bit subsequence of the DFUDS sequence in constant time (recall that the DFUDS sequence is obtained by concatenating the unary degrees of the nodes in preorder). To find out which $\lg n$-bit subsequence of B to read to extract the required DFUDS subsequence, we store the following. Let p_i be the position in B which corresponds to the $(i\sqrt{\lg n})$-th bit in the DFUDS sequence. We store an FID for the set of all p_i's, $1 \leq i \leq 2n/\sqrt{\lg n}$. This FID, which stores a set of size $n/\sqrt{\lg n}$ from the universe $[nk]$, uses $O((n/\sqrt{\lg n}) \lg(k\sqrt{\lg n})) = o(n)$ bits, and enables us to find the position in B which corresponds to a given position in the DFUDS sequence, in constant time.

Sparse case, $k > \sqrt{\lg n}$: In this case, we divide universe $[nk]$ into $n\sqrt{\lg n}$ equal-sized buckets, and distribute the elements of S into the corresponding buckets. Let B_{top} be the bit vector representing the bucket cardinalities in unary (a number j is represented in unary by j ones followed by a zero). The bit vector B_{top} is stored using the structure of Lemma 6. From this FID structure one can extract any $\lg n$-bit substring of B_{top} in constant time. Note that B_{top} is a bit vector of length $n\sqrt{\lg n}+n-1$ containing $n-1$ ones. Also the degree of the ith node in preorder is stored between the $(i\sqrt{\lg n})$-th and the $((i+1)\sqrt{\lg n}-1)$-th zeroes in B_{top}, and the unary degree of this node can in fact be obtained by removing all but the last zero. In general, every one in B_{top} corresponds to an '(' in the DFUDS string, and every $\sqrt{\lg n}$-th zero corresponds to a ')'. Thus, using the FID for B_{top} we can obtain $\Omega(\sqrt{\lg n})$ bits of the DFUDS bit-string (using table lookup) and thereby support `DFUDS-substring`$(i, \sqrt{\lg n})$ in $O(1)$ time. As in the previous case, we also store an additional $o(n)$ bit structure to efficiently find the correspondence between B_{top} and the DFUDS sequence. □

From the proof of Lemma 4, if we can support `DFUDS-substring`(i, b) in $O(1)$ time, then using $o(n)$ additional bits we can also support `TC-microtree` in $O(1)$ time, where the micro-trees are of size $\Theta(b)$. Thus the structure of Theorem 6 can also support `TC-microtree` in $O(1)$ time, if the micro-trees are of size $\Theta(\sqrt{\lg n})$.

References

1. Benoit, D., Demaine, E.D., Munro, J.I., Raman, R., Raman, V., Rao, S.S.: Representing trees of higher degree. Algorithmica 43(4), 275–292 (2005)
2. Chuang, R.C., Garg, A., He, X., Kao, M., Lu, H.: Compact encodings of planar graphs via canonical orderings and multiple parentheses. In: Larsen, K.G., Skyum, S., Winskel, G. (eds.) ICALP 1998. LNCS, vol. 1443, pp. 118–129. Springer, Heidelberg (1998)

3. Delpratt, O., Raman, R., Rahman, N.: Engineering succinct DOM. In: EDBT. ACM Intl. Conference Proceeding Series, vol. 261, pp. 49–60. ACM Press, New York (2008)
4. Farzan, A., Munro, J.I.: A uniform approach towards succinct representation of trees. In: Gudmundsson, J. (ed.) SWAT 2008. LNCS, vol. 5124, pp. 173–184. Springer, Heidelberg (2008)
5. Ferragina, P., Luccio, F., Manzini, G., Muthukrishnan, S.: Structuring labeled trees for optimal succinctness, and beyond. In: FOCS, pp. 184–196. IEEE Computer Society Press, Los Alamitos (2005)
6. Geary, R.F., Raman, R., Raman, V.: Succinct ordinal trees with level-ancestor queries. ACM Transactions on Algorithms 2(4), 510–534 (2006)
7. Graham, R.L., Knuth, D.E., Patashnik, O.: Concrete Mathematics: A Foundation for Computer Science. Addison-Wesley Longman Publishing Co., Inc., Boston (1994)
8. He, M., Munro, J.I., Rao, S.S.: Succinct ordinal trees based on tree covering. In: Arge, L., Cachin, C., Jurdziński, T., Tarlecki, A. (eds.) ICALP 2007. LNCS, vol. 4596, pp. 509–520. Springer, Heidelberg (2007)
9. Jacobson, G.J.: Succinct static data structures. PhD thesis, Carnegie Mellon University, Pittsburgh, PA, USA (1988)
10. Jacobson, G.J.: Space-efficient static trees and graphs. In: IEEE Symposium on Foundations of Computer Science, 1989, pp. 549–554 (1989)
11. Jansson, J., Sadakane, K., Sung, W.: Ultra-succinct representation of ordered trees. In: SODA, pp. 575–584. SIAM, Philadelphia (2007)
12. Lu, H., Yeh, C.: Balanced parentheses strike back. ACM Trans. Algorithms 4(3), 1–13 (2008)
13. Munro, J.I., Raman, V.: Succinct representation of balanced parentheses, static trees and planar graphs. In: IEEE Symposium on Foundations of Computer Science, pp. 118–126 (1997)
14. Munro, J.I., Raman, V., Storm, A.J.: Representing dynamic binary trees succinctly. In: SODA, pp. 529–536. SIAM, Philadelphia (2001)
15. Munro, J.I., Rao, S.S.: Succinct Representation of Data Structures. In: Handbook of Data Structures and Applications, ch. 37, Chapman & Hall/CRC (2004)
16. Munro, J.I., Rao, S.S.: Succinct representations of functions. In: Díaz, J., Karhumäki, J., Lepistö, A., Sannella, D. (eds.) ICALP 2004. LNCS, vol. 3142, pp. 1006–1015. Springer, Heidelberg (2004)
17. Raman, R., Raman, V., Rao, S.S.: Succinct indexable dictionaries with applications to encoding k-ary trees, prefix sums and multisets. ACM Transactions on Algorithms 3(4), 43 (2002); Preliminary version in SODA 2002
18. Raman, R., Rao, S.S.: Succinct dynamic dictionaries and trees. In: Baeten, J.C.M., Lenstra, J.K., Parrow, J., Woeginger, G.J. (eds.) ICALP 2003. LNCS, vol. 2719, pp. 357–368. Springer, Heidelberg (2003)

Distortion Is Fixed Parameter Tractable

Michael R. Fellows[1], Fedor V. Fomin[2], Daniel Lokshtanov[2], Elena Losievskaja[3], Frances A. Rosamond[1], and Saket Saurabh[2]

[1] University of Newcastle, Newcastle, Australia
{michael.fellows,frances.rosamond}@newcastle.edu.au
[2] Department of Informatics, University of Bergen, N-5020 Bergen, Norway
{fedor.fomin,daniello,saket.saurabh}@ii.uib.no
[3] Department of Computer Science, University of Iceland, Iceland
elenal@hi.is

Abstract. We study low-distortion embedding of metric spaces into the line, and more generally, into the shortest path metric of trees, from the parameterized complexity perspective. Let $M = M(G)$ be the shortest path metric of an edge weighted graph G, with the vertex set $V(G)$ and the edge set $E(G)$, on n vertices. We give the first fixed parameter tractable algorithm that for an *unweighted* graph metric M and integer d either constructs an embedding of M into the line with distortion at most d, or concludes that no such embedding exists. Our algorithm requires $O(nd^4(2d+1)^{2d})$ time which is a significant improvement over the best previous algorithm of Bădoiu *et al.* that runs in time $O(n^{4d+2}d^{O(1)})$. We find it surprising that this problem turns out to be fixed parameter tractable, because of its apparent similarity to the notoriously hard BANDWIDTH MINIMIZATION problem.

We extend our results on embedding unweighted graph metric into the line in two ways. First, we give an algorithm to construct small distortion embeddings of *weighted* graph metrics. The running time of our algorithm is $O(n(dW)^4(2d+1)^{2dW})$ where W is the largest edge weight of the input graph. To complement this result, we show that the exponential dependence on the maximum edge weight is unavoidable. In particular, we show that deciding whether a weighted graph metric $M(G)$ with maximum weight $W < |V(G)|$ can be embedded into the line with distortion at most d is NP-Complete for every fixed rational $d \geq 2$. This rules out any possibility of an algorithm with running time $O((nW)^{h(d)})$ where h is a function of d alone. Secondly, we consider more general host metrics for which analogous results hold. In particular, we prove that for any tree T with maximum degree Δ, embedding M into a shortest path metric of T is fixed parameter tractable, parameterized by (Δ, d).

1 Introduction

Given an undirected graph G with the vertex set $V(G)$ and the edge set $E(G)$ together with a weight function w that assigns a positive weight $w(uv)$ to every edge $uv \in E(G)$, a natural metric associated with G is $M(G) = (V(G), D_G)$

S. Albers et al. (Eds.): ICALP 2009, Part I, LNCS 5555, pp. 463–474, 2009.

where the distance function D_G[1] is the weighted shortest path distance between u and v for each pair of vertices $u, v \in V(G)$. We call $M(G)$ as the (weighted) *graph metric* of G. If $w(uv) = 1$ for every edge $uv \in E(G)$, we say that $M(G) = (V(G), D_G)$ is an *unweighted graph metric*. For a subset S of $V(G)$, we say that $M[S] = (S, D'')$ (where D'' is D restricted to S^2) is the submetric of $M(G)$ induced by S. Given a graph metric M and another metric space M' with distance functions D and D', a mapping $f : M \to M'$ is called an *embedding* of M into M'. The mapping f has *contraction* c_f and *expansion* e_f if for every pair of points p, q in M, $D(p,q) \leq D'(f(p), f(q)) \cdot c_f$ and $D(p,q) \cdot e_f \geq D'(f(p), f(q))$ respectively. We say that f is *non-contracting* if c_f is at most 1. A non-contracting mapping f has *distortion* d if e_f is at most d.

Embedding a graph metric into a simple metric space like the real line has proved to be a useful tool in designing algorithms in various fields. A long list of applications given in [8] includes approximation algorithms for graph and network problems, such as sparsest cut, minimum bandwidth, low-diameter decomposition and optimal group steiner trees, and online algorithms for metrical task systems and file migration problems. These applications often require algorithms for finding low distortion embeddings, and the study of the algorithmic issues of metric embeddings has recently begun to develop [1,2,3,11]. For example, Bădoiu *et al.* [1,3] describe approximation algorithms and hardness results for embedding general metrics into the line and tree metrics respectively. In particular they show that the minimum distortion for a line embedding is hard to approximate up to a factor polynomial in n even for weighted trees with polynomial spread (the ratio of maximum/minimum weights). Hall and Papadimitriou [9] studied the hardness of approximation for bijective embeddings. Independently from the algorithmic viewpoint, the problem of finding a low-distortion embedding between metric spaces is a fundamental mathematical problem [10,12] that has been studied intensively.

In many applications one needs the distortion of the required embedding to be relatively small. Hence it is natural to study the algorithmic issues related to small distortion embeddings within the framework of parameterized complexity [6,7,13]. This paradigm associates a natural secondary measurement to the problem and studies the algorithmic behavior of the problem in terms of the associated measurement, called the *parameter*. In this paper we consider a natural parameter, the *distortion* d, and consider the feasibility of having an algorithm of time complexity $g(d) \cdot n^{O(1)}$ for the problem of embedding weighted graph metrics into the line with distortion at most d.

What would one expect about the complexity of embedding an *unweighted* graph metric into the line, parameterized by the distortion d? At a glance, the problem seems to closely resemble the BANDWIDTH MINIMIZATION problem. In the BANDWIDTH MINIMIZATION problem one is given a graph G and asked to find a bijective mapping $f : V(G) \to \{1, \ldots, n\}$, for which the bandwidth, i.e. $b = \max_{(u,v) \in E(G)} |f(u) - f(v)|$, is minimized. This problem is known to be

[1] We also denote the distance function D_G by D if the graph in consideration is clear from the context.

$W[t]$-hard for all $t \geq 1$ [4,5], when parameterized by b. Unless an unlikely collapse of parameterized complexity classes occurs, this rules out any possibility of having an algorithm with running time $g(b) \cdot n^{O(1)}$ for BANDWIDTH MINIMIZATION and thus the algorithm of Saxe [14] running in time $O(4^b n^{b+1})$ is essentially the best possible. Previous to this paper, the best algorithm (by Bădoiu *et al.* [2]) to decide whether an unweighted graph metric can be embedded into the line with distortion at most d has a running time where d appears in the exponent of n, that is $O(n^{4d+2} \cdot d^{O(1)})$. Because of the apparent similarity to the notoriously hard bandwidth problem, it is very surprising that, in fact, this fundamental problem of embedding unweighted graph metrics into the line turns out to be fixed parameter tractable (FPT).

Theorem 1. *Given an unweighted graph G on n vertices we can decide whether $M(G)$ can be embedded into the real line with distortion at most d in time $O(nd^4(2d+1)^{2d})$.*

The running time of the algorithm is linear for every fixed d and clearly improves the running time of the previously known algorithm. In fact, one can apply Theorem 1 in order to check whether the unweighted graph metric can be embedded into the line with distortion at most $\lg n / \lg \lg n$ in time polynomial in n.

Having coped with the unweighted case, we return to the study of low distortion embeddings of weighted graph metrics into the line. We show that if the maximum weight of any edge is bounded by W, then we can modify the algorithm presented in Theorem 1 to give an algorithm to decide whether $M(G)$ can be embedded into the line with distortion at most d in time $O(n(dW)^4(2d+1)^{2dW})$. However the weights in a graph metric do not need to be small, and hence this algorithm is not sufficient to give a $g(d) \cdot n^{O(1)}$ time algorithm for the problem of embedding weighted graph metrics into the line. Can such an algorithm exist? Unfortunately, it turns out that our $O(n(dW)^4(2d+1)^{2dW})$ algorithm essentially is the best one can hope for. In fact, our next result rules out not only any possibility of having an algorithm with running time of the form $g(d) \cdot n^{O(1)}$, but also any algorithm with running time $(nW)^{h(d)}$, where h only depends on d.

Theorem 2. *Deciding whether a weighted graph metric $M(G)$ with maximum weight $W < |V(G)|$ can be embedded into the line with distortion at most d is NP-Complete for every fixed rational $d \geq 2$.*

Another direction for generalizing Theorem 1 is to look for other simple topologies or host metrics for which an analogous result to Theorem 1 holds. Kenyon *et al.* [11] provided FPT algorithms for the *bijective* embedding of unweighted graph metrics into the metric of a tree with bounded maximum degree Δ. The running time of their algorithm is $n^2 \cdot 2^{\Delta^{\alpha^3}}$ where α is the maximum of c_f and e_f. An important point, observed in [2], is that constraining the embedding to be bijective (not just injective, as in our case) is crucial for the correctness of the algorithms from [11]. We complement the FPT result of Kenyon, Rabani and Sinclair [11] by extending our results to give an algorithm for the problem of embedding unweighted graph metrics into a metric generated by a tree with maximum degree bounded by Δ, parameterized by distortion d and Δ.

Theorem 3. *Given a graph G, a tree T with maximum degree Δ and an integer d we can decide whether G can be embedded into T with distortion at most d in time $n^2 \cdot |V(T)| \cdot 2^{O((5d)^{\Delta^{d+1}} \cdot d)}$.*

2 Algorithms for Embedding Graph Metrics into the Line

2.1 Unweighted Graph Metrics into the Line

In this section we give an algorithm for embedding unweighted graph metrics into the line. We slightly abuse the terminology here by saying *embedding of a graph G* instead of embedding of the unweighted graph metric $M(G)$ of G. Before we proceed to the details of the algorithm we need a few observations that allow us to only consider a specific kind of embeddings. For a non-contracting embedding f of a graph G into the line, we say that vertex u *pushes* vertex v if $D(u, v) = |f(u) - f(v)|$.

Observation 1. [$\star$] [2] *If $f(u) < f(v) < f(w)$ and u pushes w, then u pushes v and v pushes w.*

For an embedding f, let $v_1, v_2, \ldots, v_n$ be an ordering of the vertices such that $f(v_1) < f(v_2) < \ldots < f(v_n)$. We say that f is *pushing* if v_i pushes v_{i+1}, for each $1 \leq i \leq n-1$.

Observation 2. [$\star$] *If G can be embedded into the line with distortion d, then there is a pushing embedding of G into the line with distortion d. Furthermore, every pushing embedding of G into the line is non-contracting.*

Observation 3. *Let f be a pushing embedding of a connected graph G into the line with distortion at most d. Then $D(v_{i-1}, v_i) \leq d$ for every $1 \leq i \leq n$.*

By Observation 2, it is sufficient to work only with pushing embeddings. Our algorithm is based on dynamic programming over small intervals of the line. The intuition behind the algorithm is as follows. Let us consider a distortion d embedding of G into the line and an interval of length $2d + 1$ of the line. First, observe that no edge can have one end-point to the left of this interval and one end-point to the right. This means that if there is a vertex u embedded to the left of this interval and another vertex v that has been embedded to the right, then the set of vertices embedded into the interval form an u, v-separator. Moreover, for each edge, its end-points can be mapped at most d apart, and hence there is no edge with one end-point to the left of this interval and the other end-point in the rightmost part of this interval. Thus just by looking at the vertices mapped into an interval of length $2d + 1$, we deduce which of the remaining vertices of G were mapped to the left and which were mapped to the right of this interval. This is a natural division of the problem into independent subproblems and the solutions to these subproblems can be used to find an embedding of G. Next we

[2] Proofs of results labelled with [$\star$] will appear in the full version of the paper.

formalize this intuition by defining *partial embeddings* and showing how they are glued onto each other to form a distortion d embedding of the input graph.

It is well known (and it follows from Observation 2) that there always exists an optimal embedding with all the vertices embedded into *integer coordinates* of the line. Without loss of generality, in the rest of this section we only consider pushing embeddings of this type. We also assume that our input graph G is *connected.*

Definition 1. *For a graph G and a subset $S \subseteq V(G)$, a* partial embedding of S *is a function $f : S \rightarrow \{-(d+1), \ldots, d+1\}$. We define $S_f^{[a,b]}$, $-(d+1) \leq a \leq b \leq d+1$, to be the set of all vertices of S which are mapped into $\{a, \ldots, b\}$ by f (let us remark that this can be $\emptyset$). We also define $S_f^L = S_f^{[-(d+1),-1]}$ and $S_f^R = S_f^{[1,d+1]}$. For an integer x, $-(d+1) \leq x \leq d+1$, we put $S_f^x = S_f^{[x,x]}$. Finally, we put $L(f)$ ($R(f)$) to denote the union of the vertex sets of all connected components of $G \setminus S$ that have neighbors in S_f^L (S_f^R).*

Definition 2. *A partial embedding f of a subset $S \subseteq V(G)$ is called* feasible *if (1) f is a non-contracting distortion d embedding of S; (2) $L(f) \cap R(f) = \emptyset$; (3) Every neighbor of S_f^0 is in S; (4) if $R(f) = \emptyset$, then S_f^{d+1} is nonempty; (5) if $L(f) = \emptyset$, then $S_f^{-(d+1)}$ is nonempty; (6) if $f(u)+1 < f(v)$ and $S_f^{[f(u)+1,f(v)-1]} = \emptyset$, then $f(v) - f(u) = D(u,v)$. (Basically, u pushes v.)*

The properties 1, 2, and 3 of this definition will be used to show that every distortion d embedding of G into the line can be described as a sequence of feasible partial embeddings that have been glued onto each other. Properties 4, 5 and 6 are helpful to bound the number of feasible partial embeddings.

Definition 3. *Let f and g be feasible partial embeddings of a graph G, with domains S_f and S_g, respectively. We say that g* succeeds *f if (1) $S_f^{[-d,d+1]} = S_g^{[-(d+1),d]} = S_f \cap S_g$; (2) for every $u \in S_f \cap S_g$, $f(u) = g(u)+1$; (3) $S_g^{d+1} \subseteq R(f)$; (4) $S_f^{-(d+1)} \subseteq L(g)$.*

The properties 1 and 2 describe how one can glue a partial embedding g that has been shifted one to the right onto another partial embedding f. Properties 3 and 4 are employed to enforce "intuitive" behavior of the sets $L(f)$, $R(f)$, $L(g)$ and $R(g)$. That is, since g is glued on the right side of f, everything to the right of g should appear in the right side of f. Similarly, everything to the left of f should be to the left of g.

Lemma 1. [$\star$] *For every pair of feasible partial embeddings f and g of subsets S_f and S_g of $V(G)$ such that g succeeds f, we have $R(f) = R(g) \cup S_g^{d+1}$ and $L(g) = L(f) \cup S_f^{-(d+1)}$.*

Lemma 2. [$\star$] *For every integer d, a graph G has an embedding of distortion at most d if and only if there exists a sequence of feasible partial embeddings $f_0, f_1, f_2, \ldots, f_t$ such that for each $0 \leq i \leq t-1$, f_{i+1} succeeds f_i, and $L(f_0) = R(f_t) = \emptyset$.*

For a vertex v of a graph G and integer $r \geq 0$ we denote the ball of radius r centered in v, which is the set of vertices at distance at most r in G, by $B(v, r)$. The *local density* of a graph G is $\delta = \max_{v \in V(G), r>0} \frac{|B(v,r)-1|}{2r}$. We will apply the following well known lower bound on distortion.

Lemma 3 ([1]). [Local Density] *Let G be a graph that can be embedded into the line with distortion d. Then d is at least the local density δ of G.*

Applying Lemma 3 we can bound the number of possible feasible partial embeddings. Observe that each feasible partial embedding f can be represented as a number $1 \leq t \leq d$ and a sequence of vertices $v_0 v_1 \dots v_q$ such that $t + \sum_{i=1}^{q} D(v_{i-1}, v_i) \leq 2d + 1$ and $D(v_{i-1}, v_i) \leq d$ for every $i \geq 1$. This is done by simply saying that the domain S of f is the set $\{v_0, v_1, \dots, v_q\}$ and that $f(v_a) = -(d+1) + t + \sum_{i=1}^{a} D(v_{i-1}, v_i)$. Let $\mathcal{N}(x)$ be the maximum number of sequences $v_0 v_1 \dots v_q$ such that $\sum_{i=1}^{q} D(v_{i-1}, v_i) = x$, where maximum is taken over all $v_0 \in V(G)$. For any negative number x, $\mathcal{N}(x) = 0$.

Lemma 4. [$\star$] *For $x \in \mathbb{Z}$, $\mathcal{N}(x) \leq (2d+1)^x$.*

Corollary 1. [$\star$] *For a graph G with local density at most d the number of possible feasible partial embeddings of subsets of $V(G)$ is at most $O(n(2d+1)^{2d})$.*

Now we are in the position to prove Theorem 1.

Proof [**of Theorem 1**]. The algorithm proceeds as follows. First, check whether G has local density δ bounded by d. Checking the local density of G can be done in time linear in n because if $|E(G)| \geq nd$ we can immediately answer "no". If $\delta > d$, answer "no". Otherwise, we can test whether the conditions of Lemma 2 apply. That is, we construct a directed graph $\mathcal{D}$ where the vertices are feasible partial embeddings and there is an edge from a partial embedding f_x to a partial embedding f_y if f_y succeeds f_x. Checking the conditions of Lemma 2, reduces to checking for the existence of a directed path starting in a feasible partial embedding f_0 with $L(f_0) = \emptyset$ and ending in a feasible partial embedding f_t with $R(f_t) = \emptyset$. This can be done in linear time in the size of $\mathcal{D}$ by running a depth first search in $\mathcal{D}$. The number of vertices in $\mathcal{D}$ is at most $O(n(2d+1)^{2d})$. Every vertex of $\mathcal{D}$ has at most $O(d^2)$ edges going out of it, as a feasible partial embedding f_y succeeding another feasible partial embedding f_x is completely determined by f_x together with the vertex that f_y maps to $d+1$ (or the fact that f_y does not map anything there). Using prefix-tree-like data structures one can test whether a given partial embedding f_x succeeds another in $O(d^2)$ time. The total running time is then bounded by $O(nd^4(2d+1)^{2d})$. □

2.2 Weighted Graph Metrics into the Line Parameterized by d and W

In the previous section, we gave an FPT algorithm for embedding unweighted graph metrics into the line. Here, we generalize this result to handle metrics generated by weighted graphs. More precisely, let G be a graph with weight function $w: E(G) \rightarrow \mathbb{Z}^+ \setminus \{0\}$ and $M = (V(G), D)$ be the weighted shortest path distance metric of G. Now we give an outline of an algorithm for embedding M into the line, parameterized by the distortion d and the maximum edge weight W, that is, $W = \max_{e \in E(G)}\{w(e)\}$. The definition of a pushing embedding and Observations 1 and 2 work out even when G is a weighted graph. Once we define the notion of partial embeddings, other notions like feasibility and succession are adapted in an obvious way. Given a graph G and a subset $S \subseteq V(G)$, a *partial embedding* of S is a function $f : S \rightarrow \{-(dW+1), \ldots, (dW+1)\}$. We can prove results analogous to Lemma 1 and Theorem 2 with the new definitions of *partial embeddings, feasibility* and *succession*. Thus, we can give an algorithm for this problem similar to the algorithm presented in Theorem 1. The runtime of this algorithm is dominated by the number of different feasible partial embeddings. Let $B_w(v, r)$ denote the set of vertices at *weighted* distance at most r from v and δ_w be the analogous notion of *weighted local density* of a graph G. It is easy to see that if M can be embedded into the line with distortion at most d then $d \geq \delta_w$. This result immediately upper bounds the number of feasible partial embeddings by $n \cdot (dW)^{O(dW)}$. In what follows next we show that the number of feasible partial embeddings actually is bounded by $n \cdot (2d+1)^{2dW}$. Let $\mathcal{N}(x)$ be as in Lemma 4. For each fixed first vertex v_0 in the partial embedding, and each value of $1 \leq t \leq (2dW + 1)$, there are at most $\mathcal{N}(2dW + 1 - t)$ feasible partial embeddings that map v_0 to $-(dW + 1) + t$. Thus the number of feasible partial embeddings is at most $\sum_{t=1}^{dW} n \cdot \mathcal{N}(2dW + 1 - t)$. By Lemma 4, this is at most $n \cdot \sum_{t=1}^{dW}(2d + 1)^{2dW+1-t} \leq \frac{3}{2}n(2d + 1)^{2dW}$.

Theorem 4. *Given a weighted graph G with maximum edge weight W we can decide whether $M(G)$ can be embedded into the real line with distortion at most d in time $O(n(dW)^4(2d + 1)^{2dW})$.*

3 Graph Metrics into the Line Is Hard for Fixed Rational $d \geq 2$

We complement Theorem 4 by proving that deciding whether a given weighted graph metric can be embedded into the line with distortion at most d is NP-complete for every fixed rational $d \geq 2$. Our reduction is from 3-COLORING, one of the classical NP-complete problems. On input G to 3-COLORING we show how to construct an edge weighted graph G'. For an edge $uv \in E(G')$, $w(uv)$ will be the weight if the edge uv. The weighted shortest path metric $M(G')$ will then be the input to our embedding problem. Let $n = |V(G)|$, $m = |E(G)|$ and $d = \frac{a}{b} \geq 2$ for some integers a and b. Let $e_1, e_2, \ldots, e_m$ be an ordering of the edges of G, and choose the integers $g = 5a - 1$, $r = 10b$, $q = m(2n + 1)$, $L = 10qb$ and $t = abL + 1$. We start constructing G' by making two cliques C_1 and C_2 of size t. Let $C_1 = \{c_1, c_2, \ldots, c_t\}$ and $C_2 = \{c'_1, c'_2, \ldots, c'_t\}$. Let $w(c_i c_j) = w(c'_i c'_j) = \lceil |i - j|/d \rceil$. Now, we make $q - 1$ *separator vertices* and label them $s_1, \ldots, s_{q-1}$. We make q gadgets $T_1, \ldots, T_q$ encoding the edges of

G. For every edge $e_i = uv$ there are $2n+1$ gadgets, namely T_{i+mp} for every $0 \leq p < 2n+1$. Each such gadget, say T_{i+mp}, consists of three vertices, one vertex corresponding to u, one vertex corresponding to v and one vertex corresponding to e_i. These three vertices form a triangle with edges of weight 1. For every j between 1 and q we connect all vertices of T_j to s_{j-1} and s_j with edges of weight g. Whenever this implies that we need to connect something to the non-existing vertices s_0 and s_q we connect to c_t and c'_1 respectively. Now, for every pair of vertices $x \in T_i$ and $y \in T_j$ that correspond to the same vertex or edge of G we add an edge of weight $r|i-j|$ between x and y. Finally, we add an edge with weight L between c_t and c'_1. This concludes the construction of G'. The next lemma essentially shows that if there is an edge $uv \in E(G')$ then that is the shortest weight path between u and v in G'.

Lemma 5. [$\star$] *For every edge uv in $E(G')$, $D_{G'}(u,v) = w(uv)$.*

Lemma 6. [$\star$] *If G is 3-colorable then there is an embedding f of $M(G')$ into the line with distortion at most d.*

Lemma 7. [$\star$] *If there is an embedding f of $M(G')$ into the line with distortion at most d then G is 3-colorable.*

Together with the construction of G' from G, Lemmas 6 and 7 imply Theorem 2.

4 Embedding Graphs into Trees of Bounded Degree

Given a graph G with shortest path metric D_G and a tree T with maximum degree Δ, having shortest path metric D_T, we give an algorithm that decides whether G can be embedded into T with distortion at most d in time $n^2 \cdot |V(T)| \cdot 2^{O((5d)^{\Delta^{d+1}} \cdot d)}$. We assume that the tree T is rooted, and we will refer to the root of T as $r(T)$. For a vertex v in the tree, T_v is the subtree of T rooted at v, and $C(v)$ is the set of v's children. Finally, for an edge uv of T, let $T_u(uv)$ and $T_v(uv)$ be the tree of $T \setminus uv$ that contains u and v respectively. Notice that if u is the parent of v in the tree, then $T_v(uv) = T_v$ and $T_u(uv) = T \setminus V(T_v)$. As in the previous section, we need to define feasible partial embeddings together with the notion of succession. For a vertex $u \in V(T)$ and a subset S of $V(G)$, a *u-partial embedding* is a function $f_u : S \to B(u, d+1)$.

Definition 4. *For a u-partial embedding f_u of a subset $S \subseteq V(G)$ and a vertex $v \in N(u)$ we define $S[v, f_u] = \{x \in S : f_u(x) \in V(T_v(uv))\}$. Given two integers i and j, $0 \leq i \leq j \leq k$, let $S^{[i,j]}[f_u] = \{x \in S : i \leq D(f_u(x), u) \leq j\}$. Finally, let $S^{[i,j]}[v, f_u] = S^{[i,j]}[f_u] \cap S[v, f_u]$, $S^k[v, f_u] = S^{[k,k]}[v, f_u]$ for $k \geq 1$ and $S^0[f_u] = S^{[0,0]}[f_u]$.*

Definition 5. *For a u-partial embedding f_u of a subset $S \subseteq V(G)$ and a vertex $v \in N(u)$ we define $M[v, f_u]$ to be the union of the vertex sets of all connected components of $G \setminus S$ that have neighbors in $S[v, f_u]$.*

Definition 6. *A u-partial embedding f_u of a subset S of $V(G)$ is called feasible if (1) f_u is a non-contracting distortion d embedding of S into $B(u, d+1)$; (2) for any distinct pair $v, w \in N(u)$, $M[v, f_u] \cap M[w, f_u] = \emptyset$; (3) $N(S^0[f_u]) \subseteq S$.*

Definition 7. *For a feasible u-partial embedding f_u of a subset S_u of $V(G)$ and a feasible v-partial embedding f_v of a subset S_v of $V(G)$ with $v \in C(u)$ we say that f_v* succeeds *f_u if (1) $S_u \cap S_v = (S_u^{[0,d]}[f_u] \cup S_u^{d+1}[v, f_u]) = (S_v^{[0,d]}[f_v] \cup S_v^{d+1}[u, f_v])$; (2) for every $x \in S_u \cap S_v$, $f_u(x) = f_v(x)$; (3) $M[v, f_u] = \bigcup_{x \in N(v) \setminus u} (M[x, f_v] \uplus S_u^{d+1}[x, f_v])$; and (4) $M[u, f_v] = \bigcup_{x \in N(u) \setminus v} (M[x, f_u] \uplus S_v^{d+1}[x, f_u])$.*

Suppose we have picked out a subtree T_v for a vertex $v \in V(T)$ and found a non-contracting embedding f' with distortion at most d of a subset Z of $V(G)$ into $T' = T[\bigcup_{u \in V(T_v)} B(u, d+1)]$. We wish to find a non-contracting distortion d embedding of G into T such that for every vertex u with $f(u) \in V(T')$, we have that $u \in Z$ and such that if $u \in Z$ then $f(u) = f'(u)$. At this point, a natural question arises. Can we impose constraints on the restriction of f to $V(T) \setminus V(T_v)$ such that f restricted to $V(T) \setminus V(T_v)$ satisfies these conditions if and only if f is a non-contracting distortion d embedding of G into T? One necessary condition is that f restricted to $V(T) \setminus V(T_v)$ must be a non-contracting distortion d embedding of $\{u \in V(G) : f(u) \in V(T) \setminus V(T_v)\}$. We can obtain another condition by applying the definition of feasible u-partial embeddings. For each vertex u, we can use arguments similar to the ones in Section 2 in order to determine which connected components of $T \setminus V(T_v)$ f must map u to in order to be a non-contracting distortion d embedding of G into T.

For the line, these two conditions are both necessary and sufficient. Unfortunately, for the case of bounded degree trees, this is not the case. The reason the conditions are sufficient when we restrict ourselves to the line is that every embedding of a graph metric into the line that is *locally* non-contracting and *locally* expanding by a factor at most d, also is *globally* non-contracting and expanding by a factor at most d. When we embed into trees of bounded degree, every embedding that is locally expanding by a factor at most d, also has this property globally. However, every locally non-contracting embedding need not be *globally* non-contracting. To cope with this issue, we introduce the concept of vertex types. Intuitively, vertices of the same type in T_v are indistinguishable when viewed from $T \setminus V(T_v)$. We show that the set of possible vertex types can be bounded by a function of d and Δ. Then, to complete f from f' we only need to know the restriction of f' to $B(v, d+1)$ and which vertex types appear in T_v. Then the amount of information we need to pass on from f' to f is bounded by $n \cdot h(d, \Delta)$. We exploit this fact to give an algorithm for the problem. In the rest of this section, we formalize this intuition.

For a vertex $u \in V(T)$, a neighbor v of u and a feasible u-partial embedding f_u of a subset S of $V(G)$ we define a *$[v, f_u]$-type* to be a function $t : S[v, f_u] \rightarrow \{\infty, 3d+2, d, \ldots, -d, -(d+1)\}$ and a *$[v, f_u]$-typelist* to be a set of *$[v, f_u]$-types.* For an integer k let $\beta(k) = k$ if $k \leq 3d+2$ and $\beta(k) = \infty$ otherwise.

Definition 8. *For a vertex $u \in V(T)$ with two neighbors v and w, and a feasible u-partial embedding f_u of a subset S of $V(G)$ together with a $[v, f_u]$-typelist $\mathcal{L}_1$*

and a $[w, f_u]$*-typelist* $\mathcal{L}_2$ *we say that* $\mathcal{L}_1$ *and* $\mathcal{L}_2$ agree *if for every type* $t_1 \in \mathcal{L}_1$ *and* $t_2 \in \mathcal{L}_2$ *there is a vertex* $x \in S[v, f_u]$ *and a vertex* $y \in S[w, f_u]$ *such that* $t_1(x) + t_2(y) \geq D_G(x, y)$.

Definition 9. *For a vertex* $u \in V(T)$*, a neighbor* v *of* u*, a feasible* u*-partial embedding* f_u *of a subset* S *of* $V(G)$ *and a* $[v, f_u]$*-typelist* $\mathcal{L}$ *we say that* $\mathcal{L}$ *is* compatible *with* $S[v, f_u]$ *if for every vertex* x *in* $S[v, f_u]$ *there is a type* $t \in \mathcal{L}$ *such that for every* $y \in S[v, f_u]$, $D_T(f_u(x), u) - D_G(x, y) = t(y)$.

Definition 10. *A* feasible u-state *is a feasible partial embedding* f_u *of a subset* S *of* $V(G)$ *together with a* $[v, f_u]$*-typelist* $\mathcal{L}[v, f_u]$ *for every* $v \in N(u)$ *such that the following conditions are satisfied: (1)* $\mathcal{L}[v, f_u]$ *is compatible with* $S[v, f_u]$ *for every* $v \in N(u)$*; and (2) For every pair of distinct vertices* x *and* y *in* $N(u)$, $\mathcal{L}[x, f_u]$ *agrees with* $\mathcal{L}[y, f_u]$.

Definition 11. *Let* $u \in V(T)$, $v \in C(u)$*. Let* $\mathcal{X}_u$ *be a feasible* u*-state and* $\mathcal{X}_v$ *be a feasible* v*-state. We say that* $\mathcal{X}_v$ succeeds $\mathcal{X}_u$ *if*

1. f_v *succeeds* f_u;
2. *For every* $w \in (N(v) \setminus u)$ *and a type* $t_1 \in \mathcal{L}[w, f_v]$ *there is a type* $t_2 \in \mathcal{L}[v, f_u]$ *such that*
   (a) *For every node* $x \in S[v, f_u] \cap S[w, f_v]$, $t_2(x) = \beta(t_1(x) + 1)$;
   (b) *For every node* $x \in (S[v, f_u] \setminus S[w, f_v])$, $t_2(x) = \beta(\max_{y \in S[w, f_v]}(t_1(y) + 1 - D_G(x, y)))$.
3. *For every* $w \in (N(u) \setminus v)$ *and a type* $t_1 \in \mathcal{L}[w, f_u]$ *there is a type* $t_2 \in \mathcal{L}[u, f_v]$ *such that*
   (a) *For every node* $x \in S[u, f_v] \cap S[w, f_u]$, $t_2(x) = \beta(t_1(x) + 1)$;
   (b) *For every node* $x \in (S[u, f_v] \setminus S[w, f_u])$, $t_2(x) = \beta(\max_{y \in S[w, f_u]}(t_1(y) + 1 - D_G(x, y)))$.

The main result of this section relies on the next two lemmas.

Lemma 8. [$\star$] *If there is a distortion* d *embedding* F *of* G *into* T *then, for every vertex* u *of* $V(T)$ *there is a feasible* u*-state* $\mathcal{X}_u$ *such that for every vertex* $v \in V(T), w \in C(v)$, $\mathcal{X}_w$ *succeeds* $\mathcal{X}_v$.

Lemma 9. [$\star$] *If there is a feasible* u*-state* $\mathcal{X}_u$ *for every vertex* u *of* $V(T)$ *such that for every vertex* $v \in V(T), w \in C(v)$, $\mathcal{X}_w$ *succeeds* $\mathcal{X}_v$ *then there is a distortion* d *embedding* F *of* G *into* T.

Next we give the proof of Theorem 3.

Proof [**of Theorem 3**]. The algorithm proceeds as follows. First, check that $\Delta(G) \leq \Delta^d$ (follows from local density argument). Now, we do bottom up dynamic programming on the tree T. For each vertex u of the tree we make a boolean table with an entry for each possible feasible u-state. For every leaf of the tree all the entries are set to true. For an inner node u and a feasible u-state $\mathcal{X}_u$ we set $\mathcal{X}_u$'s entry to true if for each child v of u there is a feasible v-state $\mathcal{X}_v$ that succeeds $\mathcal{X}_u$ and so that $\mathcal{X}_v$'s entry is set to true. The algorithm returns "yes" if, at the termination of this procedure, there is a feasible $r(T)$-state $\mathcal{X}_{r(T)}$ with its table entry set to true. The algorithm clearly terminates, and correctness of this algorithm follows from Lemmas 9 and 8.

We now proceed to the running time analysis. In our bottom up sweep of T, we consider every edge and every vertex of T exactly once, which yields a factor of $n_t = |V(T)|$. For each vertex u we consider each feasible u-state $\mathcal{X}_u$ once, and for each such state and every child v of u of the state we need to enumerate all feasible v-states that succeed $\mathcal{X}_u$. In fact, we enumerate a larger set of candidate feasible v-states and for each such state $\mathcal{X}_v$ we check whether $\mathcal{X}_v$ succeeds $\mathcal{X}_u$.

First we show that the number of feasible u-partial embeddings is at most $n \cdot \Delta^{O(d^2 \cdot \Delta^{d+1})}$. This follows from the fact that for any vertex u of the tree $|B(u, d+1)| \leq \Delta^{d+1}$ and that the domain of a any feasible u-partial embedding f_u is contained in a ball of radius at most $2d + 2$ in G. Because the degree of G is bounded, a ball of radius $2d + 2$ in G can contain at most $\Delta^{O(d^2)}$ vertices.

One can easily prove that if the feasible partial embedding f_u is given, the number of types and typelists that can appear in a feasible u-state together with f_u is bounded by $(5d)^{\Delta^{d+1}}$ and $2^{O((5d)^{\Delta^{d+1}})}$ respectively. Thus, the number of feasible u-states is bounded by $2^{O((5d)^{\Delta^{d+1}} \cdot d)}$. If the domain S_v of a feasible partial embedding f_v for a child v of u is non-empty then we can use the fact that S_v must have a non-empty intersection with the domain of f_u to bound the number of potential successors of a u-state by $2^{O((5d)^{\Delta^{d+1}} \cdot d)} \cdot \Delta^d \leq 2^{O((5d)^{\Delta^{d+1}} \cdot d)}$. Since we can check whether a particular u-feasible state succeeds another in time $n \cdot 2^{O((5d)^{\Delta^{d+1}} \cdot d)}$ the overall running time of the algorithm is bounded by $n^2 n_t \cdot 2^{O((5d)^{\Delta^{d+1}} \cdot d)}$. □

5 Concluding Remarks and Open Problems

In this paper we described FPT algorithms for embedding unweighted graph metrics into a tree metric for a tree of maximum degree Δ, parameterized by (Δ, d) where d is the distortion. For the case when the host metric is the line, we generalized our result and showed that embedding weighted graph metrics into the line is FPT parameterized by distortion d and maximum edge weight W. A similar generalization can also be obtained for embedding weighted graph metrics into weighted bounded degree tree metrics, parameterized by d, Δ and W where W is the maximum edge weight in the input graph. We postpone the details for the full version of the paper. Our hardness result that embedding a weighted metric into the line is NP-hard for every fixed distortion $d \geq 2$ showed that our algorithms qualitatively are the best possible.

We believe that our results will lead to further investigation of the combinatorially challenging field of low distortion embeddings within the framework of multivariate algorithmics. We conclude with two concrete interesting open problems:

- What is the parameterized complexity of embedding unweighted graph metrics into unbounded degree trees, parameterized by distortion d?

- What is the parameterized complexity of embedding unweighted graph metrics into target metrics that are minimum distance metrics of cycles, parameterized by d?

References

1. Bădoiu, M., Chuzhoy, J., Indyk, P., Sidiropoulos, A.: Low-distortion embeddings of general metrics into the line. In: Proceedings of the STOC, pp. 225–233. ACM Press, New York (2005)
2. Bădoiu, M., Dhamdhere, K., Gupta, A., Rabinovich, Y., Räcke, H., Ravi, R., Sidiropoulos, A.: Approximation algorithms for low-distortion embeddings into low-dimensional spaces. In: Proceedings of the SODA, pp. 119–128. SIAM, Philadelphia (2005)
3. Badoiu, M., Indyk, P., Sidiropoulos, A.: Approximation algorithms for embedding general metrics into trees. In: Proceedings of the SODA, pp. 512–521. SIAM, Philadelphia (2007)
4. Bodlaender, H.L., Fellows, M.R., Hallett, M.T.: Beyond NP-completeness for problems of bounded width: Hardness for the w hierarchy (extended abstract). In: Proceedings of the STOC, pp. 449–458. ACM Press, New York (1994)
5. Bodlaender, H.L., Fellows, M.R., Hallett, M.T., Wareham, T., Warnow, T.: The hardness of perfect phylogeny, feasible register assignment and other problems on thin colored graphs. Theor. Comput. Sci. 244(1-2), 167–188 (2000)
6. Downey, R.G., Fellows, M.R.: Parameterized complexity. Springer, New York (1999)
7. Flum, J., Grohe, M.: Parameterized Complexity Theory. Texts in Theoretical Computer Science. An EATCS Series. Springer, Berlin (2006)
8. Gupta, A., Newman, I., Rabinovich, Y., Sinclair, A.: Cuts, trees and l_1-embeddings of graphs. Combinatorica 24(2), 233–269 (2004)
9. Hall, A., Papadimitriou, C.H.: Approximating the distortion. In: Chekuri, C., Jansen, K., Rolim, J.D.P., Trevisan, L. (eds.) APPROX 2005 and RANDOM 2005. LNCS, vol. 3624, pp. 111–122. Springer, Heidelberg (2005)
10. Indyk, P.: Algorithmic applications of low-distortion geometric embeddings. In: Proceedings of the FOCS, pp. 10–33. IEEE Computer Society Press, Los Alamitos (2001)
11. Kenyon, C., Rabani, Y., Sinclair, A.: Low distortion maps between point sets. In: Proceedings of the STOC, pp. 272–280. ACM Press, New York (2004)
12. Linial, N.: Finite metric-spaces—combinatorics, geometry and algorithms. In: Proceedings of the International Congress of Mathematicians, vol. III, pp. 573–586. Higher Ed. Press, Beijing (2002)
13. Niedermeier, R.: Invitation to fixed-parameter algorithms. Oxford Lecture Series in Mathematics and its Applications, vol. 31. Oxford University Press, Oxford (2006)
14. Saxe, J.B.: Dynamic programming algorithms for recognizing small bandwidth graphs in polynomial time. SIAM J. on Algebraic and Discrete Methods 1(4), 363–369 (1980)

Towards Optimal Range Medians

Beat Gfeller[1,*] and Peter Sanders[2,**]

[1] ETH Zürich, Switzerland
gfeller@inf.ethz.ch
[2] Universität Karlsruhe, Germany
sanders@ira.uka.de

Abstract. We consider the following problem: given an unsorted array of n elements, and a sequence of intervals in the array, compute the median in each of the subarrays defined by the intervals. We describe a simple algorithm which uses $O(n)$ space and needs $O(n \log k + k \log n)$ time to answer k such median queries. This improves previous algorithms by a logarithmic factor and matches a lower bound for $k = O(n)$. Since, in contrast to previous approaches, the algorithm decomposes the range of element values rather than the array, it has natural generalizations to higher-dimensional problems – it reduces a range median query to a logarithmic number of range counting queries.

1 Introduction and Related Work

The classic problem of finding the *median* is to find the element of rank $\lfloor n/2 \rfloor$ in an unsorted array of n elements.[1] Clearly, the median can be found in $O(n \log n)$ time by sorting the elements. However, a classic algorithm finds the median in $O(n)$ time [BFP+72], which is asymptotically optimal.

More recently, the following generalization, called the *Range Median Problem (RMP)*, has been considered [KMS05, HPM08]:

Input: An unsorted array A with n elements, each having a *value*. Furthermore, a sequence of k *queries* $Q_1, \ldots, Q_k$, each defined as an interval $Q_i = [L_i, R_i]$. In general, the sequence is given in an online fashion, we want an answer after every query, and k is not known in advance.

Output: A sequence $x_1, \ldots, x_k$ of values, where x_i is the median of the elements in $A[L_i, R_i]$. Here, $A[L, R]$ denotes the set of all elements whose index in A is at least L and at most R.

The RMP naturally fits into a larger group of problems, in which an unsorted array is given, and in a query one wants to compute a certain function of all the

* The author gratefully acknowledges the support of the Swiss SBF under contract no. C05.0047 within COST-295 (DYNAMO) of the European Union.

** Partially supported by DFG grant SA 933/3-1.

[1] An element has rank i if it is the i-th element in some sorted order. Actually, any specified rank might be of interest. We restrict ourselves to the median to simplify notation but a generalization to arbitrary ranks is straightforward for all our results except the ones in Section 6.

S. Albers et al. (Eds.): ICALP 2009, Part I, LNCS 5555, pp. 475–486, 2009.

elements in a given interval. Instead of the median, natural candidates for such a function are:

- Sum: this problem can be trivially solved with $O(n)$ preprocessing time and $O(1)$ query time by computing prefix sums.
- Semigroup operator: this problem is significantly more difficult than the sum. However, there exists a very efficient solution: for any constant c, preprocessing in $O(nc)$ time and space allows to answer queries in $O(\alpha_c(n))$ time, where α_c is the inverse of a certain function at the $(c/2)$th level of the primitive recursion hierarchy. In particular, using $O(n)$ processing time and space, each query can be answered in $O(\alpha(n))$ time, where $\alpha(n)$ is the inverse Ackerman function [Yao82]. A matching lower bound is known [Yao85].
- Maximum, Minimum: This is a special case of a semigroup operator, for which the problem can be solved slightly more efficiently: $O(n)$ preprocessing time and space is sufficient to allow $O(1)$ time queries (see e.g. [BFC04]).
- Mode: the problem of finding the most frequent element within a given array range is still rather open. Using $O(n^2 \log\log n/\log^2 n)$ space (in words), constant query time is possible [PG09], and with $O(n^{2-2\epsilon})$ space, $0 < \epsilon \leq 1/2$, $O(n^\epsilon)$ query time can be achieved [Pet08]. Some earlier space-time tradeoffs were given in [KMS05].

In addition to being a natural extension of the median problem, the RMP has applications in practice, namely obtaining a "typical" element in a given time series out of a given time interval [HPM08].

Natural special cases of the RMP are an *offline* variant, where all queries are given in a batch, and a variant where we want to do all *preprocessing up front* and are then interested in good worst case bounds for answering a single query.

The authors of [HPM08] give a solution of the online RMP which requires $O(n \log k + k \log n \log k)$ time and $O(n \log k)$ space. In addition, they give a lower bound of $\Omega(n \log k)$ time for comparison-based algorithms. They basically use a one-dimensional range tree over the input array, where each inner node corresponds to a subarray defined by an interval. Each such subarray is sorted, and stored with the node. A range median query then corresponds to selecting the median from $O(\log k)$ sorted subarrays (whose union is the queried subarray) of total length $O(n)$, which requires $O(\log n \log k)$ time. The main difficulty of their approach is to show that the subarrays need not be fully sorted, but only presorted in a particular way, which reduces the construction time of the tree from $O(n \log n)$ to $O(n \log k)$.

Concerning the preprocessing variant of the RMP, [KMS05] give a data structure to answer queries in $O(\log n)$ time, which uses $O(n \log^2 n/\log\log n)$ space. They do not analyze the required preprocessing time, but it is clearly at least as large as the required space in machine words. Moreover, they give a structure which uses only $O(n)$ space, but query time $O(n^\epsilon)$ for arbitrary $\epsilon > 0$. To obtain $O(1)$ query time, the best-known data structure [PG09] requires $O(n^2(\log\log n/\log n)^2)$ space (in words), which improves upon [KMS05] and [Pet08].[2]

[2] Note that the data structures in [KMS05, Pet08, PG09] work only for a specific quantile (e.g. the median), which must be the same for all queries.

Our Results. First, in Section 2 we give an algorithm for the pointer-machine model which solves the RMP for an online sequence of k queries in $O(n \log k + k \log n)$ time and $O(n \log k)$ space. This improves the running time of $O(n \log k + k \log n \log k)$ reported in [HPM08] for $k \in \omega(n/\log n)$. Our algorithm is also considerably simpler. The idea is to reduce a range median query to a logarithmic number of related range counting queries. Similar to quicksort, we descend a tree that stems from recursively partitioning the values in array A. The final time bound is achieved using the technique of Fractional Cascading. In Section 2.1, we explain why our algorithm is optimal for $k \in O(n)$ and at most $\Omega(\log n)$ from optimal for $k \in \omega(n)$.

In Section 3 we achieve linear space in the RAM model using techniques from succinct data structures – the range counting problems are reduced to rank computations in bit arrays. To achieve the desired bound, we compress the recursive subproblems in such a way that the bit arrays remain dense at all times. The latter algorithm can be easily modified to obtain a linear space data structure using $O(n \log n)$ preprocessing time that allows arbitrary subsequent range median queries to be answered in time $O(\log n)$. Note that the previously best linear-space data structure required $O(n^{\epsilon})$ query time [KMS05].

After a few remarks on generalizations for higher dimensional inputs in Section 4, we discuss dynamic variants of the RMP problem in Section 5. In Section 6, we consider random input arrays and give a construction with constant expected query time using $O(n^{3/2})$ space in expectation. Section 7 concludes with a summary and some open problems.

We would like to note that a preliminary version of our paper (which excludes Section 6) appeared earlier as a technical report [GS09]. Furthermore, during the preparation of the final version of this paper, we learned that Gerth Stølting Brodal and Allan Grønlund Jørgensen have independently obtained results similar to ours, which have not been published yet.

2 A Pointer Machine Algorithm

Our algorithm is based on the following key observation: Suppose we partition the elements in array A of length n into two smaller arrays: $A.low$ which contains all elements with the $n/2$ smallest[3] values in A, and $A.high$ which contains all elements with the $n/2$ largest values. The elements in $A.low$ and $A.high$ are sorted by their index in A, and each element e in $A.low$ and $A.high$ is associated with its index $e.i$ in the original input array, and its value $e.v$. Now, if we want to find the element of rank p in the subarray $A[L, R]$, we can do the following: We count the number m of elements in $A.low$ which are contained in $A[L, R]$. To obtain m, we do a binary search for both L and R in $A.low$ (using the $e.i$ fields). If $p \leq m$, then the element of rank p in $A[L, R]$ is the element of rank p

[3] To simplify notation we ignore some trivial rounding issues and also sometimes assume that all elements have unique values. This is without loss of generality because we could artificially expand the size of A to the next power of two and because we can use the index of an element in A to break ties in element comparisons.

in $A.low[L, R]$. Otherwise, the element of rank p is the element of rank $p - m$ in $A.high[L, R]$.

Hence, using the partition of A into $A.low$ and $A.high$, we can reduce the problem of finding an element of a given rank in array $A[L, R]$ to the same problem, but on a smaller array (either $A.low[L, R]$ or $A.high[L, R]$). Our algorithm applies this reduction recursively.

Algorithm overview. The basic idea is therefore to subdivide the n elements in the array into two parts of (almost) equal size by computing the median of their values and using it to split the list into a list of the $n/2$ elements with smaller values and a list of the $n/2$ elements with larger values. The two parts are recursively subdivided further, but only when required by a query. To answer a range median query, we determine in which of the two parts the element of the desired rank lies (initially, this rank corresponds to the median, but this may change during the search). Once this is known, the search continues recursively in the appropriate part until a trivial problem of constant size is encountered.

We will show that the total work involved in splitting the subarrays is $O(n \log k)$ and that the search required for any query can be completed in $O(\log n)$ time using Fractional Cascading [CG86]. Hence, the total running time is $O(n \log k + k \log n)$.

Algorithm 1. Query(A, L, R, p)

Input: range select data structure A, query range $[L, R]$, desired rank p
if $|A| = 1$ **then return** $A[1]$
if A.low *is undefined* **then**
    Compute median x value of the pairs in A
    A.low $:= \langle e \in A : e.v \leq x \rangle$
    A.high $:= \langle e \in A : e.v > x \rangle$
    { $\langle e \in A : \mathcal{Q} \rangle$ is an array containing all elements e of A satisfying the given condition $\mathcal{Q}$, ordered as in A }
{ Find(A, q) returns $\max\{j : A[j].i \leq q\}$ (with Find$(A, 0) = 0$) }
$l :=$ Find$(A.\text{low}, L - 1)$; `// # of low elements left of` L
$r :=$ Find$(A.\text{low}, R)$; `// # of low elements up to` R
$m := r - l$; `// # of low elements between` L `and` R
if $p \leq m$ **then return** Query$(A.\text{low}, L, R, p)$
else return Query$(A.\text{high}, L, R, p - m)$

Detailed description and analysis. Algorithm 1 gives pseudocode for the query, which performs preprocessing (i.e., splitting the array into two smaller arrays) only where needed. Note that we have to keep three things separate here: values that are relevant for median computation and partitioning the input, positions in the input sequence that are relevant for finding the elements within the range $[L, R]$, and positions in the subdivided arrays that are important for counting elements.

Let us first analyze the time required for processing a query not counting the 'preprocessing' time within lines 4–6: The query descends $\log_2 n$ levels of recursion. On each level, Find-operations for L and R are performed on the lower half of the current subproblem. If we used binary search, we would get a total execution time of up to $\sum_{i=1}^{\log_2 n} O(\log \frac{n}{2^i}) = \Theta(\log^2 n)$. However, the fact that in all these searches, we search for the same key (L or R) allows us to use a standard technique called Fractional Cascading [CG86] that reduces the search time to a constant, once the resulting position of the first search is known. Indeed, we only need a rather basic variant of Fractional Cascading, which applies when each successor list is a sublist of the previous one [dBvKOS00]. Here, it suffices to augment an element e of a list with a pointer to the position of some element e' in each subsequent list (we have two successors – $A.low$ and $A.high$). In our case, we need to point to the largest element in the successor that is no larger than e. We get a total search time of $O(\log n)$.

Now we turn to the preprocessing code in lines 4–6 of Algorithm 1. Let $s(i)$ denote the level of recursion at which query i encountered an undefined array $A.low$ for the first time. Then the preprocessing time invested during query i is $O(n/2^{s(i)})$ if a linear time algorithm is used for median selection [BFP+72] (note that we have a linear recursion with geometrically decreasing execution times). This preprocessing time also includes the cost of finding the pointers for Fractional Cascading while splitting the list in lines 4–6. Since the preprocessing time during query i decreases with $s(i)$, the total preprocessing time is maximized if small levels $s(i)$ appear as often as possible. However, level j can appear no more than 2^j times in the sequence $s(1), s(2), \ldots, s(k)$.[4] Hence, we get an upper bound for the preprocessing time when the smallest $\lfloor \log k \rfloor$ levels are used as often as possible ('filled') and the remaining levels are $\lceil \log k \rceil$. The preprocessing time at every used level is $O(n)$ giving a total time of $O(n \log k)$. The same bound applies to the space consumption since we never allocate memory that is not used later. We summarize the main result of this section in a theorem:

Theorem 1. *The online range median problem (RMP) on an array with n elements and k range queries can be solved in time $O(n \log k + k \log n)$ and space $O(n \log k)$.*

Another variant of the above algorithm invests $O(n \log n)$ time and space into complete preprocessing up front. Subsequently, any range median query can be answered in $O(\log n)$ time. This improves the preprocessing space of the corresponding result in [KMS05] by a factor $\log n / \log \log n$ and the preprocessing time by at least this factor.

2.1 Lower Bounds

We briefly discuss how far our algorithm is from optimality. In [HPM08], a comparison-based lower bound of $\Omega(n \log k)$ is shown for the range median

[4] Indeed, for $j > 0$ the maximal number is 2^{j-1} since the other half of the available subintervals have already been covered by the preprocessing happening in the layer above.

problem[5]. As our algorithm shows, this bound is (asymptotically) tight if $k \in O(n)$. For larger k, the above lower bound is no longer valid, as the construction requires $k < n$. Yet, a lower bound of $\Omega(n \log n)$ is immediate for $k \geq n$, considering only the first $n-1$ queries. Furthermore, $\Omega(k)$ is a trivial lower bound. Note that in our algorithm, the number of levels of the recursion is actually bounded by $O(\min\{\log k, \log n\})$, and thus for any $k \geq n$ our algorithm has running time $O(n \log n + k \log n)$, which is up to $\Omega(\log n)$ from the trivial linear bound.

In a very restricted model (sometimes called "Pointer Machine"), where a memory location can be reached only by following pointers, and not by direct addressing, our algorithm is indeed optimal also for $k \geq n$: it takes $\Omega(\log n)$ time to even access an arbitrary element of the input (which is initially given as a linked list). Since every element of the input is the answer to at least one range query (e.g. the query whose range contains only this element), the lower bound follows. For a matching upper bound, the array based algorithm described in this section can be transformed into an algorithm for the strict pointer machine model by replacing the arrays *A.low* and *A.high* by balanced binary search trees. An interesting question is whether a lower bound $\Omega(k \log n)$ could be shown in more realistic models. However, note that any comparison-based lower bound (as the one in [HPM08]) cannot be higher than $\Omega(n \log n)$: With $O(n \log n)$ comparisons, an algorithm can determine the permutation of the array elements, which suffices to answer any query without further element comparisons. Therefore, one would need to consider more realistic models (e.g. the "cell-probe" model), in which proving lower bounds is significantly more difficult.

3 A Linear Space RAM Implementation

Our starting point for a more space efficient implementation of Algorithm 1 is the observation that we do not actually need all the information available in the arrays stored at the interior nodes of our data structure. All we need is support for the operation *Find*(x) that counts the number of elements e in *A.low* that have index $e.i \leq x$. This information can already be obtained from a bit-vector where a 1-bit indicates whether an element of the original array is in *A.low*. For this bit-vector, the operation corresponding to *Find* is called *rank*. In the RAM model, there are data structures that need space $n + o(n)$ bits, can be constructed in linear time and support *rank* in constant time (e.g., [Cla88, OS06][6]). Unfortunately, this idea alone is not enough since we would need to store 2^j bit

[5] The authors derive a lower bound of $l := \frac{n!}{k!((n/k-1)!)^k}$, where n is a multiple of $k < n$. Unfortunately, the analysis of the asymptotics of l given in [HPM08] is erroneous; however, a corrected analysis shows that the claimed $\Omega(n \log k)$ bound holds.

[6] Indeed, since we only need the rank operation, there are very simple and efficient implementations: store a table with ranks for indices that are a multiple of $w = \Theta(\log n)$. General ranks are then the sum of the next smaller table entry and the number of 1-bits in the bit array between this rounded position and the query position. Some processors have a POPCNT instruction for this purpose. Otherwise we can use lookup tables.

Algorithm 2. Query(A, L, R, p)

```
Input: range select data structure A, query range [L, R] and desired rank p
if |A| = 1 then return A[1]
if A.low is undefined then
    Compute median x value of the values in A
    A.lowbits := BitVector(|A|, {i ∈ [1, |A|] : A[i] ≤ x})
    A.low  := ⟨A[i] : i ∈ [1, |A|], A[i] ≤ x⟩
    A.high := ⟨A[i] : i ∈ [1, |A|], A[i] > x⟩
    deallocate the value array of A itself
l := A.lowbits.rank(L − 1)
r := A.lowbits.rank(R)
m := r − l
if p ≤ m then   return Query(A.low, l + 1, r, p)
else            return Query(A.high, L − l, R − r, p − m)
```

arrays consisting of n positions each on every level j. Summed over all levels, this would still need $\Omega(n \log^2 n)$ bits of space even if optimally compressed data structures were used. This problem is solved using an additional idea: for a node of our data structure with value array A, we do not store a bit array with n possible positions but only with $|A|$ possible positions, i.e., bits represent positions in A rather than in the original input array. This way, we have n positions on every *level* leading to a total space consumption of $O(n \log n)$ bits. For this idea to work, we need to transform the query range in the recursive call in such a way that *rank* operations in the contracted bit arrays are meaningful. Fortunately, this is easy because the rank information we compute also defines the query range in the contracted arrays. Algorithm 2 gives pseudocode specifying the details. Note that the algorithm is largely analogous to Algorithm 1. In some sense, the algorithm becomes simpler because the distinction between query positions and array positions for counting disappears (If we still want to report the positions of the median values in the input, we can store this information at the leaves of the data structure using linear space). Using an analysis analogous to the analysis of Algorithm 1, we obtain the following theorem:

Theorem 2. *The online range median problem (RMP) on an array with n elements and k range queries can be solved in time $O(n \log k + k \log n)$ and space $O(n)$ words in the RAM model.*

By doing all the preprocessing up front, we obtain an algorithm with preprocessing time $O(n \log n)$ using $O(n)$ space and query time $O(\log n)$. This improves the space consumption compared to [KMS05] by a factor $\log^2 n / \log \log n$.

4 Higher Dimensions

Since our algorithm decomposes the values rather than the positions of elements, it can be naturally generalized to higher dimensional point sets. We obtain an algorithm that needs $O(n \log k)$ preprocessing time plus the time for supporting range counting queries on each level. The amortized query time is the time for $O(\log n)$ range counting queries. Note that query ranges can be specified in any way we wish: (hyper)-rectangles, circles, etc., without affecting the way we handle values. For example, using the data structure for 2D range counting from [JMS04] we obtain a data structure for the 2D rectangular range median problem that needs $O(n \log n \log k)$ preprocessing time, $O(n \log k / \log\log n)$ space, and $O(\log^2 n / \log\log n)$ query time. This not only applies to 2D arrays consisting of n input points but to arbitrary two-dimensional point sets with n elements.

Unfortunately, further improvements, e.g. to logarithmic query time seem difficult. Although the query range is the same at all levels of recursion, Fractional Cascading becomes less effective when the result of a rectangular range counting query is defined by more than a constant number of positions within the data structure because we would have to follow many forwarding pointers. Also, the array contraction trick that allowed us to use dense bit arrays in Section 3 does not work anymore because an array with half the number of bits need not contain any empty rows or columns.

Another indication that logarithmic query time in two dimensions might be difficult to achieve is that there has been intensive work on the more specialized median-filtering problem in image processing where we ask for *all* range medians with query ranges that are squares of size $(2r+1) \times (2r+1)$ in an image with n pixels. The best previous algorithms known here need time $\Theta(n \log^2 r)$ [GW93] unless the range of values is very small [PH07, CWE07]. Our result above improves this by a factor $\log\log r$ (by applying the general algorithm to input pieces of size $3r \times 3r$) but this seems to be of theoretical interest only.

5 Dynamic Range Medians

In this section, we consider a dynamic variant of the RMP, where we have a linked list instead of an array, and elements can be deleted or inserted arbitrarily. In this setting, we still want to answer median queries, whose range is given by two pointers to the first and the last element in the query range.

In the following, we sketch a solution which allows inserts and deletes in $O(\log^2 n)$ amortized time each, and range median queries in $O(\log^2 n)$ worst case time. The basic idea is to use a BB(α) tree [NR72] as a primary structure, in which all elements are ordered by their *value*. With each inner node, we associate a secondary structure, which contains all the elements of the node's subtree, ordered by their position in the input list. More precisely, we store these elements in a balanced binary search tree, where nodes are augmented with a field indicating the size of their subtree, see e.g. [Rou01]. This data structure permits to answer a range query by a simple adaptation of Algorithm 1: starting at the root, we determine the number of elements within the query range which are in the left subtree, and depending on the result continue the search for the median in the left or in the right subtree. The required counting in each secondary structure takes $O(\log n)$ time, and we need to perform at most $O(\log n)$ such searches for any query. When an element is inserted or deleted, we follow the search path in the BB(α) tree according to its value, and update all the $O(\log n)$ secondary structures of the visited nodes. The main difficulty arises when a rotation in the BB(α) tree is required: in this case, the secondary structures are rebuilt from scratch, which costs $O(p \log p)$ time if the subtree which is rotated contains p nodes. However, as shown in [Meh84, WL85], such rotations are required so rarely that the amortized time of such an event is only $O(\log p \log n) = O(\log^2 n)$.

We note that using this dynamic data structure for the one-dimensional RMP, we can implement a two-dimensional median filter, by scanning over the image, maintaining all the pixels in a strip of width r. In this way, we obtain a running time of $O(\log^2 r)$ per pixel, which matches the state-of-the-art solution for this problem [GW93]. This indicates that obtaining a solution with $O(\log n)$ time for all operations could be difficult.

6 Towards Constant Query Time

In this section we restrict ourselves to computing range medians rather than general range selection. Using $O(n^2)$ space, we can trivially precompute all medians so that the query time becomes constant. This space requirement is reduced by somewhat less than a logarithmic factor in [KMS05, Pet08]. An interesting question is whether we can save more than a polylogarithmic factor. We now outline an algorithm that needs space $O(n^{3/2})$ (machine words), preprocessing time $O(n^{3/2} \log n)$ and achieves constant query time on the average, i.e., the expected query time is constant for random inputs. [7]

We first consider median queries for a range $[L, R]$ where $L \leq a+1$ and $R \geq n-a$ with $a \in \Theta(\sqrt{n})$, i.e., the range contains a large middle part $C = A[a+1, n-a]$ of the input array. Furthermore let $B = A[1, a]$ and $D = A[n-a+1, n]$.

$$A = \overbrace{\boxed{1 \;\cdots\; L \;\cdots\; a}}^{B}\overbrace{\boxed{a+1 \qquad \cdots \qquad n-a}}^{C \supseteq C'}\overbrace{\boxed{n-a+1 \cdots R \cdots n}}^{D}$$

If the median value v of $A[L, R]$ comes from C, then its rank within C must be in $[\lfloor \frac{n}{2} \rfloor - 2a, \lfloor \frac{n}{2} \rfloor]$ because the median of the elements in C has rank $\lfloor \frac{n}{2} \rfloor - a$, and adding at most $2a$ elements outside C can increase or decrease the rank of the median by at most a. The basic idea is to precompute a sorted array $C'[1, 2a+1]$ of these *central elements*. The result of a query then only depends on $A[L, a]$, C', and $A[n-a+1, R]$. For a start, let us assume that all elements in B and D are either smaller than $C'[1]$ or larger than $C'[2a+1]$. Suppose $A[L, a]$ and $A[n-a+1, R]$ contain s_l and s_r values smaller than $C'[1]$, respectively. Then the median of $A[L, R]$ is $C'[a + 1 + \lfloor \frac{b-s}{2} \rfloor]$, where $s := s_l + s_r$ and $b := R - L + 1 + 2a - n - s$ (b is the number of values larger than $C'[2a+1]$ in $A[L, a]$ and $A[n-a+1, R]$). Note that s_l can be precomputed for all possible values of L using using time and space $O(\sqrt{n})$ (and the same is true for s_r

[7] The analysis is for inputs with distinct elements where every permutation of ranks is equally likely. The queries can be arbitrary.

and R). For general contents of B and D, elements in B and D with value between $C'[1]$ and $C'[2a+1]$ are stored explicitly. Since, for a random input, each element from B and D has probability $\Theta(1/a)$ to lie within this range, only $O(1)$ elements have to be stored on the average. During a query, these extra elements are scanned[8] and those with position within $[L, R]$ are moved to a sorted extra array X. The median of $A[L, R]$ is then the element with rank $a+1+\lfloor\frac{b-s}{2}\rfloor+\frac{|X|}{2}$ in $C' \cup X$. Equivalently, we can take the element with rank $|X|/2$ in $C'[a+1+\lfloor\frac{b-s}{2}\rfloor, a+1+\lfloor\frac{b-s}{2}\rfloor+|X|] \cup X$. We are facing a selection problem from two sorted arrays of size $|X|$ which is possible in time $O(\log|X|)$ (see e.g. [VSIR91]), i.e., $O(1)$ on the average.

To generalize for arbitrary ranges, we can cover the input array A with subarrays obeying the above rules in such a way that every possible query can be performed in one subarray. Here is one possible covering scheme: In category $i \in [1, \lfloor\sqrt{n}\rfloor]$ we want to cover all queries with ranges $R-L+1$ in $[i^2, i^2+2i]$. Note that these ranges cover all of $[1, n]$ since $i^2+2i+1 = (i+1)^2$, i.e. subsequent range intervals $[i^2, i^2+2i]$ and $[(i+1)^2, (i+1)^2+2(i+1)]$ are contiguous. In category i, we use subarrays of size $2i+(i^2-i)+2i$ such that the central part C of the j-th subarray starts at position $ij+1$ for $j \in [0, n/i-i]$.[9] A query $[L, R]$ is now handled by category $i = \lfloor\sqrt{R-L+1}\rfloor$.[10] It remains to determine the number j of the subarray within category i. If $R-L+1 \in [i^2, i^2+1]$ we use $j = \lfloor L/i \rfloor$, otherwise $j = \lfloor L/i \rfloor + 1$. In both cases, L is within part B of the j-th subarray and R is within part D of the j-th subarray.

A subarray of category i needs space $O(i)$ and there are $O(n/i)$ arrays from category i. Hence in total, the arrays of category i need space $O(n)$. Since $O(\sqrt{n})$ categories suffice to cover the entire array, the overall space consumption is $O(n^{3/2})$. Precomputing the arrays of category i can be done in time $O(n \log i)$ by keeping the elements of the current subarray in a search tree sorted by element values. Finding the central elements for the next subarray then amounts to i deletions, i insertions and one range reporting query in this search tree. This takes time $O(i \log i)$. The remaining precomputations can be performed in time $O(i)$. Summing over all categories yields preprocessing time $O(n^{3/2} \log n)$.

The above bounds for space and preprocessing time are deterministic worst case bounds. The average space consumption can be reduced by a factor $\log n$: First, instead of precomputing the counts for s_l and s_r they can be computed using a bit array with fast rank operation (see also Section 3). Second, instead of blindly storing the worst case number of $2a+1$ central elements, we may only store the elements actually needed by any query. This number is upper bounded by the number of elements needed for queries of the form $[a+1, R]$ plus the

[8] An algorithm that is more robust for nonrandom inputs could avoid scanning by using a selection algorithm working on one sorted array and a data structure that supports fast range-rank queries (see also Section 3). This way, we would get an algorithm running in time logarithmic in the number of extra elements.

[9] We pad the input array A on both sides with random values in order to avoid special cases.

[10] Note that we can precompute all required square roots if desired.

number of elements needed for queries of the form $[L, n - a]$. Let us consider the position of the median in queries of the form $[a + 1, R]$ as a function of R. This position (ignoring rounding) performs a random walk on the line with step-width 1/2. In $a = O(\sqrt{n})$ steps, the expected maximum distance from its starting point reached by such a random walk is $O(\sqrt{\sqrt{n}}) = O(n^{1/4})$. Queries of the form $[L, n - a]$ behave analogously.

7 Conclusion

We have presented improved upper bounds for the range median problem.

The query time of our solution is asymptotically optimal for $k \in O(n)$, or when all preprocessing has to be done up front. For larger values of k, our solution is at most a factor $\log n$ from optimal. In a very restricted model where no arrays are allowed, our solution is optimal for all k. Moreover, in the RAM model, our data structure requires only $O(n)$ space, which is clearly optimal. It is open whether the term $O(k \log n)$ in the query time could be reduced to $O(k)$ in the RAM model when k is sufficiently large. The interesting range here is when k lies between $\Theta(n)$ and $\Theta(n^2)$. Making the data structure dynamic adds a factor $\log n$ to the query time.

Given the simplicity of our data structure, a practical implementation would be easily possible. To avoid the large constants involved when computing medians for recursively splitting the array, one could use a pivot chosen uniformly at random. This should work well in expectation.

It would be interesting to find faster solutions for the dynamic RMP or the two-dimensional (static) RMP: Either would lead to a faster median filter for images, which is a basic tool in image processing.

References

[BFC04] Bender, M.A., Farach-Colton, M.: The Level Ancestor Problem simplified. Theor. Comput. Sci. 321(1), 5–12 (2004)

[BFP+72] Blum, M., Floyd, R.W., Pratt, V.R., Rivest, R.L., Tarjan, R.E.: Linear Time Bounds for Median Computations. In: 4th Annual Symp. on Theory of Computing (STOC), pp. 119–124 (1972)

[CG86] Chazelle, B., Guibas, L.J.: Fractional cascading: I. A data structuring technique. Algorithmica 1, 133–162 (1986)

[Cla88] Clark, D.R.: Compact Pat Trees. PhD thesis, University of Waterloo (1988)

[CWE07] Cline, D., White, K.B., Egbert, P.K.: Fast 8-bit median filtering based on separability. In: IEEE International Conference on Image Processing (ICIP), vol. 5, pp. 281–284 (2007)

[dBvKOS00] de Berg, M., van Kreveld, M., Overmars, M., Schwarzkopf, O.: Computational Geometry Algorithms and Applications, 2nd edn. Springer, Heidelberg (2000)

[GS09] Gfeller, B., Sanders, P.: Towards optimal range medians. CoRR, abs/0901.1761 (2009)

[GW93] Gil, J., Werman, M.: Computing 2-d min, median, and max filters. IEEE Transactions on Pattern Analysis and Machine Intelligence 15(5), 504–507 (1993)

[HPM08] Har-Peled, S., Muthukrishnan, S.M.: Range Medians. In: Halperin, D., Mehlhorn, K. (eds.) Esa 2008. LNCS, vol. 5193, pp. 503–514. Springer, Heidelberg (2008)

[JMS04] JáJá, J., Mortensen, C.W., Shi, Q.: Space-efficient and fast algorithms for multidimensional dominance reporting and counting. In: Fleischer, R., Trippen, G. (eds.) ISAAC 2004. LNCS, vol. 3341, pp. 558–568. Springer, Heidelberg (2004)

[KMS05] Krizanc, D., Morin, P., Smid, M.H.M.: Range Mode and Range Median Queries on Lists and Trees. Nord. J. Comput. 12(1), 1–17 (2005)

[Meh84] Mehlhorn, K.: Data structures and algorithms. Sorting and Searching, vol. 1. Springer, Berlin (1984)

[NR72] Nievergelt, J., Reingold, E.M.: Binary Search Trees of Bounded Balance. In: 4th Annual Symp. on Theory of Computing (STOC), pp. 137–142. ACM Press, New York (1972)

[OS06] Okanohara, D., Sadakane, K.: Practical entropy-compressed rank/select dictionary. The Computing Research Repository (CoRR), abs/cs/0610001 (2006)

[Pet08] Petersen, H.: Improved Bounds for Range Mode and Range Median Queries. In: Geffert, V., Karhumäki, J., Bertoni, A., Preneel, B., Návrat, P., Bieliková, M. (eds.) SOFSEM 2008. LNCS, vol. 4910, pp. 418–423. Springer, Heidelberg (2008)

[PG09] Petersen, H., Grabowski, S.: Range mode and range median queries in constant time and sub-quadratic space. Information Processing Letters 109(4), 225–228 (2009)

[PH07] Perreault, S., Hébert, P.: Median filtering in constant time. IEEE Transactions on Image Processing 16(9), 2389–2394 (2007)

[Rou01] Roura, S.: A New Method for Balancing Binary Search Trees. In: Orejas, F., Spirakis, P.G., van Leeuwen, J. (eds.) ICALP 2001. LNCS, vol. 2076, pp. 469–480. Springer, Heidelberg (2001)

[VSIR91] Varman, P.J., Scheufler, S.D., Iyer, B.R., Ricard, G.R.: Merging multiple lists on hierarchical-memory multiprocessors. J. Parallel Distrib. Comput. 12(2), 171–177 (1991)

[WL85] Willard, D.E., Lueker, G.S.: Adding range restriction capability to dynamic data structures. J. ACM 32(3), 597–617 (1985)

[Yao82] Yao, A.C.-C.: Space-time tradeoff for answering range queries (extended abstract). In: 14th Annual Symp. on Theory of Computing (STOC), pp. 128–136 (1982)

[Yao85] Yao, A.C.-C.: On the Complexity of Maintaining Partial Sums. SIAM J. Comput. 14(2), 277–288 (1985)

B-Treaps: A Uniquely Represented Alternative to B-Trees

Daniel Golovin*

Computer Science Department, Carnegie Mellon University, Pittsburgh, PA, USA
dgolovin@cs.cmu.edu

Abstract. We present the first uniquely represented data structure for an external memory model of computation, a B-tree analogue called a *B-treap*. Uniquely represented data structures represent each logical state with a unique machine state. Such data structures are *strongly history-independent*; they reveal no information about the historical sequence of operations that led to the current logical state. For example, a uniquely represented file-system would support the deletion of a file in a way that, in a strong information-theoretic sense, provably removes all evidence that the file ever existed. Like the B-tree, the B-treap has depth $O(\log_B n)$, uses linear space with high probability, where B is the block transfer size of the external memory, and supports efficient one-dimensional range queries.

1 Introduction

Most computer applications store a significant amount of information that is hidden from the application interface—sometimes intentionally but more often not. This information might consist of data left behind in memory or disk, but can also consist of much more subtle variations in the state of a structure due to previous actions or the ordering of the actions. To address the concern of releasing historical and potentially private information various notions of *history independence* have been derived along with data structures that support these notions [10,12,9,6,1]. Roughly, a data structure is history independent if someone with complete access to the memory layout of the data structure (henceforth called the "observer") can learn no more information than a legitimate user accessing the data structure via its standard interface (e.g., what is visible on screen). The most stringent form of history independence, *strong history independence*, requires that the behavior of the data structure under its standard interface along with a sequence of randomly generated bits, which are revealed to the observer, uniquely determine its memory representation. We say that such structures are *uniquely represented.*

Unique representation had been studied even earlier [15,16,2], in the context of theoretical investigations into the need for redundancy in efficient data structures. The results were mostly negative, however they were for various comparison-based and pointer machine-based models of computation, and did not hold for the RAM model.

There has been much recent progress on efficient uniquely represented data structures in the RAM model. Blelloch and Golovin [4] described a uniquely represented

* Computer Science Department, Carnegie Mellon University, Pittsburgh PA 15213. This work was supported in part by Microsoft Corporation through the Center for Computational Thinking at Carnegie Mellon and by NSF ITR grant CCR-0122581 (The Aladdin Center).

S. Albers et al. (Eds.): ICALP 2009, Part I, LNCS 5555, pp. 487–499, 2009.

hash table for the RAM model supporting insertion, deletion and queries in expected constant time, using linear space and only $O(\log n)$ random bits. They also provided a perfect hashing scheme that allows for $O(1)$ worst-case queries, and efficient uniquely represented data structures for ordered dictionaries and the order maintenance problem. Naor *et al.* [11] developed a second uniquely represented dynamic perfect hash table supporting deletions, based on cuckoo hashing. Finally, Blelloch *et al.* [5] developed efficient uniquely represented data structures for some common data structures in computational geometry. Several other results, as well as a more comprehensive discussion of uniquely represented data structures, may be found in the author's doctoral thesis [8].

Recent progress on uniquely represented data structures for the RAM opens up the possibility of full-fledged uniquely represented (and thus strongly history independent) systems. Consider a filesystem that supports a delete operation that provably leaves no trace that a file ever existed on the system, or a database that reveals nothing about the previous updates or queries to the system. Existing uniquely represented data structures allow us to build efficient versions of such systems on a RAM, however these systems are best modeled not in the RAM model of computation but in *external memory* (EM) models of computation [17] which account for the fact that modern computers have a memory hierarchy in which external memory (e.g., a disk drive) is several orders of magnitude slower than internal memory (e.g., DRAM).

To the best of my knowledge, there is no previous work on uniquely represented data structures in an EM model of computation. For background on the extensive work on conventional data structures in EM models, we refer the interested reader to the excellent book by Vitter [17]. Within this body of work, *extendible hash tables* and the *B-tree* and its variants (e.g., the B^+-tree and the B^*-tree) play a prominent role.

It is worth noting here that the extendible hashing construction of Fagin *et al.* [7] is almost uniquely represented, and in fact can be made uniquely represented with some minor modifications, the most significant of which is to use a uniquely represented hash table [4,11] to layout blocks on disk. However in this paper we focus on uniquely represented B-tree analogs, which can support efficient one-dimensional range queries.

The B-tree was invented by Bayer and McCreight [3], to organize information on magnetic disk drives so as to minimize disk I/O operations. The salient features of the B-tree are that it stores $\Theta(B)$ keys in each node and each node (other than the root) has degree $\Theta(B)$, where B is a parameter of the tree. Thus the B-tree has height roughly $\log_B(n)$ and $\Theta(n/B)$ nodes. We will construct a uniquely represented tree that is analogous to a B-tree, based on the *treap* data structure [14]. Recall a treap is a binary search tree where each key also has an associated priority. In addition to the standard search tree constraint on keys, the keys must also be in heap order with respect to their priorities; each key must have priority less than its parent. We call the resulting data structure a *B-treap*, for "bushy-treap." It supports the following operations.

- $\mathrm{insert}(x)$: insert key x into the B-treap.
- $\mathrm{delete}(x)$: delete key x from the B-treap.
- $\mathrm{lookup}(x)$: determine if x is present in the B-treap, and if so, return a pointer to it.
- $\mathrm{range\text{-}query}(x, y)$: return all keys between x and y in the B-treap.

It is easy to associate auxiliary data with the keys, though for simplicity of exposition we will assume there is no auxiliary data being stored. We will prove the following result.

Theorem 1. *There exists a uniquely represented B-treap that stores elements of a fixed size, such that if the B-treap contains n keys and $B = \Omega\left((\ln(n))^{1/(1-\epsilon)}\right)$ for some $\epsilon > 0$, then* lookup, insert, *and* delete *each touch at most $O(\frac{1}{\epsilon}\log_B(n))$ B-treap nodes in expectation, and* range-query *touches at most $O(\frac{1}{\epsilon}\log_B(n) + k/B)$ B-treap nodes in expectation where k is the size of the output. Furthermore, if $B = O(n^{\frac{1}{2}-\delta})$ for some $\delta > 0$, then with high probability the B-treap has depth $O(\frac{1}{\epsilon}\log_B(n))$ and requires only linear space to store.*

The External Memory Model. We use a variant of the parallel disk model of Vitter [17] with one processor and one disk, which measures performance in terms of disk I/Os. Internal memory is modeled as a 1-D array of data items, as in a RAM. External memory is modeled as a large 1-D array of *blocks* of data items. A block is a sequence of B data items, where B is a parameter called the *block transfer size*. The external memory can read (or write) a single block of data items to (or from) internal memory during a single I/O. Other parameters include the problem size, n, and the internal memory size m, both measured in units of data items. We will assume $m = \omega(B)$.

Uniquely Represented Memory Allocation. Uniquely represented hash tables [4,11] can be used as the basis for a uniquely represented memory allocator. Intuitively, if the nodes of a pointer structure can be labeled with distinct hashable labels in a uniquely represented (i.e., strongly history independent) manner, and the pointer structure itself is uniquely represented in a pointer based model of computation, then these hash tables provide a way of mapping the pointer structure into a one dimensional memory array while preserving unique representation. Pointers are replaced by labels, and pointer dereferencing is replaced by hash table lookups. We will assume that the data items have distinct hashable labels. Similarly, we assume that the B-treap nodes are assigned distinct hashable labels in a uniquely represented manner. This can be achieved in any number of ways. For example, if the B-treap is the only object on disk, we may use the label of the minimum data item stored in the B-treap node.

Given these labels, we will hash B-treap nodes (one per block) into external memory. If we use distinct random bits for the hash table and everything else, this inflates number of the expected I/Os by a constant factor. (For B-treaps, uniquely represented dynamic perfect hash tables [4,11] may be an attractive option, since in expectation most of the I/Os will involve reads, even in the case of insertions and deletions.) We thus focus on the problem of building a uniquely represented pointer structure with the desired properties.

Notation. For $n \in \mathbb{Z}$, let $[n]$ denote $\{1, 2, \ldots, n\}$. For a tree T, let $|T|$ be the number of nodes in T, and for a node $v \in T$, let T_v denote the subtree rooted at v. Finally, let $\mathrm{node}(k)$ denote the tree node with key k.

2 The Treap Partitioning Technique

The *treap partitioning* technique was introduced in [4], and is a crucial element in the design of B-treaps. It is a uniquely represented scheme that partitions a dynamic ordered set U into small contiguous subsets. It works as follows. Given a treap T and an element

$x \in T$, let its *weight* $|T_x|$ be the number of its descendants, including itself. For a parameter s, let

$$\mathcal{L}_s[T] := \{x \in T : |T_x| \geq s\} \cup \{\text{root}(T)\}$$

be the *weight s partition leaders* of T[1]. We will often refer to these nodes simply as *leaders*. For every $x \in T$ let $\ell(x, T)$ be the least (deepest) ancestor of x in T that is a partition leader. Here, each node is considered an ancestor of itself. We will call a node x a *follower* of its leader $\ell(x, T)$. The weight s partition leaders partition the treap into the sets $\{\{y \in T : \ell(y, T) = x\} : x \in \mathcal{L}_s[T]\}$, each of which is a contiguous block of keys from T consisting of the followers of some leader. It is not hard to see that each set in the partition has at most $2s - 1$ elements.

We implement treap partitioning by storing the set U in a treap, where each node v has a key field storing an element of U that induces an ordering on treap nodes. Also, the treap priority for a node u is generated by hashing u.key, and the treap nodes are hashed into memory using their keys. Each node v additionally has a leader field which stores the representative of the set it is in. With some additional subtree size information, we can support finger insertions, finger deletions, and leader queries in expected constant time. For simplicity, however, we will describe a variant that supports finger insertions and deletions in expected $O(\log s)$ time. Each node v will have a size field, and we will maintain the following invariant on the contents of these fields: *For all v with $|T_v| < s$, v.size $= |T_v|$. Otherwise v.size $= \infty$.*

The treap partitioning scheme has several useful properties. We will make use of the following lemmata, which are proved in Section 5.3.4 of [8].

Lemma 1. *Fix any treap T. Inserting or deleting a node u can alter the assignment of nodes to their weight s leaders in T or the size fields for at most $2s$ other nodes in T, namely those within distance s of u in key-order.*

Lemma 2. *Let T be a treap of size n with priorities generated by an 11-wise independent hash function h from keys to $[r]$ for some $r \geq n^3$. Then* $\mathbf{Pr}[|T_v| = k] = O(1/k^2)$ *for any $1 \leq k < n$,* $\mathbf{Pr}[|T_v| = n] = O(1/n)$, *and for any $k \geq 1$,* $\mathbf{Pr}[|T_v| \geq k] = O(1/k)$, *so for any $s \geq 1$ each node is a weight s partition leader with probability $O(1/s)$.*

Lemma 3. *Let T be a random treap with relative priorities determined by a random permutation selected uniformly at random. Let n be the number of nodes in T and fix $s \leq n$. Then $|\mathcal{L}_s[T]| = O(n/s)$ with probability* $1 - O\left(\exp\{-\frac{2n}{s(s+1)}\}\right)$.

3 B-Treap Organization

Fix a parameter α, such that $2 < \alpha = \Theta(B)$. We will say the B-treap described below is of *order* α. For convenience, we will analyze the B-treap in terms of α rather than B. Our construction will store at most $2\alpha - 1$ keys in any given node. See Figure 1 for a depiction of a B-treap. Suppose we wish to store a set of keys U. We first describe how the B-treap is organized, and then discuss how to implement the operations.

[1] For technical reasons we include root(T) in $\mathcal{L}_s[T]$ ensuring that $\mathcal{L}_s[T]$ is nonempty.

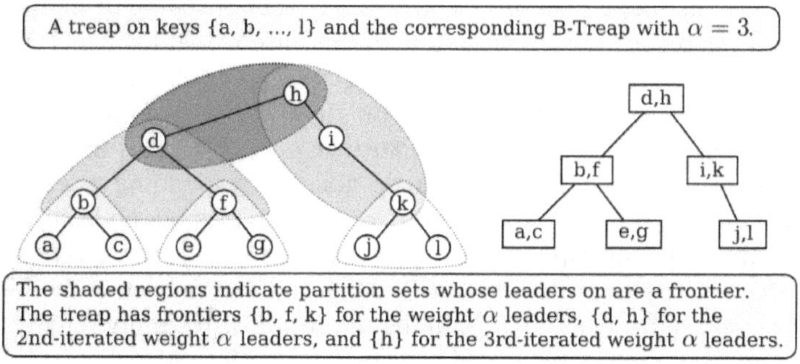

Fig. 1. A depiction of a B-treap

First consider a uniquely represented treap T storing U [4]. We first describe the organization of the B-treap informally, and then give a formal description. We make use of a refinement of the treap partitioning scheme. Let $\mathcal{L}_\alpha[T]$ be the weight α leaders of T, and let $\ell(u)$ be the leader of u in $\mathcal{L}_\alpha[T]$. We will make use of the following definition.

Definition 1 (Frontier). *For a set of nodes $S \subseteq U$, let the* frontier of S, *denoted $\mathcal{F}[S]$, be the nodes in S that have at least one child not in S.*

We will store the followers in the partition sets whose leaders are in the frontier, one set of followers per node of the B-treap. That is, for each $v \in \mathcal{F}[\mathcal{L}_\alpha[T]]$, we create a node for the B-treap and store $\{u : \ell(u) = v\} \setminus \{v\}$ in it. These will be the leaves of the B-treap. We then remove all the followers (i.e., elements of $U \setminus \mathcal{L}_\alpha[T]$), and perform the same procedure on the remaining portion of T. A B-treap leaf storing key set $\{u : \ell(u) = v\} \setminus \{v\}$ has as its parent the node containing key v. We repeat the process until T has less than α nodes left, in which case all the remaining nodes are assigned to the root node of the B-treap.

To formally describe the B-treap's organization, we will need the following definitions. Fix the random bits of the RAM. Given a set of keys $S \subseteq U$, let $\text{treap}(S)$ be the uniquely represented treap on key set S.

Definition 2 (Iterated Leaders of T). *The ith–iterated weight α leaders of T, denoted $\mathcal{L}^i_\alpha[T]$, are defined inductively as follows.*

- *If $i = 0$, $\mathcal{L}^i_\alpha[T]$ is the set of all keys in T.*
- *If $i \geq 1$ and $|\mathcal{L}^{i-1}_\alpha[T]| > 1$, then $\mathcal{L}^i_\alpha[T] = \mathcal{L}_\alpha[\text{treap}(\mathcal{L}^{i-1}_\alpha[T])]$.*
- *If $i \geq 1$ and $|\mathcal{L}^{i-1}_\alpha[T]| \leq 1$ then $\mathcal{L}^i_\alpha[T] = \emptyset$.*

Furthermore, let $\ell^i(u)$ be the deepest ancestor of u in T that is an ith–iterated weight α leader of T.

Definition 3 (Rank). *The* rank *of a node v in T, denoted* $\text{rank}(v)$, *is the maximum integer k such that $v \in \mathcal{L}^k_\alpha[T]$. The rank of a tree is the rank of its root.*

Let $k = \text{rank}(T)$ be the rank of the root of T. We store the keys in $\mathcal{L}^{k-1}_\alpha[T]$ and the root of T at the root of the B-treap $\bar{T}$. For $i = k - 1$ to 1 in decreasing order, for each node

$v \in \mathcal{F}[\mathcal{L}^i_\alpha[T]]$, construct a node $\bar{v}$ in $\bar{T}$ with a key set consisting of the followers of v in the ith level treap partition, excluding v. In this case we will say that $\bar{v}$ *corresponds* to v. Formally, the key set of $\bar{v}$ is $\{u : u \neq v, \ell^i(u) = v, \text{ and } u \in \mathcal{L}^{i-1}_\alpha[T]\}$. Finally, make $\bar{v}$ a child of the node in $\bar{T}$ corresponding to $\ell^{i+1}(v)$. Note that it is possible for a node v to be in two different frontiers, so that $v \in \mathcal{F}[\mathcal{L}^i_\alpha[T]]$ and $v \in \mathcal{F}[\mathcal{L}^j_\alpha[T]]$ for $i < j$. In this case, we create a B-treap node corresponding to each instance in which v is in some frontier. In other words, we create a B-treap node corresponding to (v, i) and another one corresponding to (v, j). In this case, the key set of the former is $\{u : u \neq v, \ell^i(u) = v, \text{ and } u \in \mathcal{L}^{i-1}_\alpha[T]\}$ and the key set of the latter is $\{u : u \neq v, \ell^j(u) = v, \text{ and } u \in \mathcal{L}^{j-1}_\alpha[T]\}$. In the future, we will simply refer to the B-treap node corresponding to v, since it can always be inferred from the context which copy is meant.

We have described above how to assign keys to B-treap nodes. In fact, the B-treap will not store merely a set of keys in each node, rather it will store the corresponding treap nodes, with left and right fields for the left and right child of the current node. Finally, if a treap node v stored in B-treap node $\bar{v}$ has a child u that is stored in a different B-treap node $\bar{u}$, we store the label of $\bar{u}$ with v.

By storing small regions of the treap in each B-treap nodes, and storing abstract pointers (in the form of labels) corresponding to treap edges that cross from one region to another, we can search the B-treap for a key by using the underlying treap that it stores. However, to dynamically maintain the B-treap's organization we also must implement several layers of treap partitioning. Specifically, we must dynamically maintain $\text{rank}(T)$ treap partitioning instances simultaneously on the same treap T, where the ith instance stores the weight α partition of $\text{treap}(\mathcal{L}^{i-1}_\alpha[T])$. Consider the treap partitioning scheme of Section 2, particularly the size field invariant. For the iterated treap partitioning scheme, we can modify the size fields to store a pair of integers and modify the invariant as follows. (Below, for a treap T and set of nodes S, we define $T \cap S$ as though T were the set of the nodes it contains.)

The size field invariant (iterated version). For each treap node v,

$$v.\mathsf{size} = (\text{rank}(v),\ |T_v \cap \mathcal{L}^{\text{rank}(v)}_\alpha[T]|).$$

The invariant describing what the leader fields should be set to must also be modified. In particular, each node v with $v.\mathsf{size} \in \{i\} \times \mathbb{N}$ should have its leader field set to $\ell^{i+1}(v)$, its deepest ancestor in $\mathcal{L}^{i+1}_\alpha[T]$. Then v will be stored in the B-treap node corresponding to $\text{node}(v.\mathsf{leader})$.

4 Implementing B-Treaps

Let $\bar{T}$ be a B-treap storing a treap T. We implement the operations as follows.

Lookup: Given an input key u, start at the root $\bar{r}$ of $\bar{T}$, find the root r of the treap T (by finding the highest priority node stored in $\bar{r}$) and proceed as in a regular treap lookup, jumping from one B-treap node to the next as necessary.

Insertion: To insert a key, create a new node u with that key and search for the leaf position $\text{leaf}(u)$ that u would occupy in the treap T, if it had the lowest priority of any node. Rotate u up to its proper position in T, updating the size fields appropriately during the rotations, so as to maintain the iterated size field invariant. This can be done in $O(1)$ time per rotation, and ensures the size fields of all descendants of u are correct.

Suppose u is a leader in some partition – that is, $\text{rank}(u) > 0$. Find the predecessor a and successor b of u in T. Proceed up the a-to-u and b-to-u paths looking for $\ell^i(a)$ and $\ell^i(b)$ for all $i \in \{1, 2, \ldots, \text{rank}(u)\}$. (Recall $\ell^i(v)$ is the deepest ancestor of v in $\mathcal{L}^i_\alpha[T]$.) These are easy to find given the size fields. If a node v was in $\mathcal{F}[\mathcal{L}^i_\alpha[T]]$ and no longer is, then the corresponding B-treap node must be destroyed. Similarly, if a node v is now in $\mathcal{F}[\mathcal{L}^i_\alpha[T]]$ and was not previously, then a corresponding B-treap node must be created. Furthermore, whenever a node has its leader field changed, we must move it to the B-treap node corresponding to its new leader.

Let a'_i and b'_i be the children of $\ell^i(a)$ and $\ell^i(b)$ that are ancestors of a and b, respectively. For each $i \in 1, 2, \ldots \text{rank}(u)$, update the leader fields of all of the descendants of a'_i in $\mathcal{L}^{i-1}_\alpha[T]$ to $\ell^i(a)$.key and move them to the B-treap node corresponding to $\ell^i(a)$. Do likewise for the the descendants of b'_i in $\mathcal{L}^{i-1}_\alpha[T]$, with $\ell^i(b)$ in place of $\ell^i(a)$. Make sure to destroy all B-treap nodes that become empty of treap nodes during this process.

That addresses the descendants of u with rank less than $\text{rank}(u)$. The descendants of u with rank equal to $\text{rank}(u)$ will have their fields set correctly unless inserting u causes some node to be "promoted" (i.e., its rank increases). We will deal with that possibility later, and will now focus on setting u's leader field correctly. To do so, we find $\ell^{\text{rank}(u)+1}(u)$. As in the case with treap partitioning, inserting u may cause a promotion of some ancestor of u into $\mathcal{L}^{\text{rank}(u)+1}_\alpha[T]$. We can determine this as in treap partitioning. Specifically, set the leader field of u equal to the leader field of its parent node. Find the child of $\text{node}(u.\text{leader})$ that is an ancestor of u, which we denote by u'. Then there is a promotion if and only if $u'.\text{size} \in \mathbb{N} \times \{\alpha - 1\}$ and $u' \neq u$. If there is no promotion, then merely increment the second coordinate of w.size for each w on the path from u to u' (excluding u and including u'), and insert u into the B-treap node corresponding to $\text{node}(u.\text{leader})$.

If u' is promoted, create a new B-treap node $\bar{u}'$ corresponding to it. Traverse $T_{u'} \cap \mathcal{L}^{\text{rank}(u)}_\alpha[T]$, move all nodes therein to $\bar{u}'$, and change their leader fields to u'.key. Note that this correctly updates the leader fields and placements of the descendants of u with rank equal to $\text{rank}(u)$. Make sure to destroy any B-treap node that is emptied of treap nodes in the process. Next, increment the second coordinate of w.size for each w on the path from u to u' (excluding u and u'), and update u'.size appropriately by incrementing the first coordinate and setting the second coordinate to one. Additionally, increment of second coordinate of the size field for each node on the path from u''s parent to its parent's leader (i.e., $\ell^{\text{rank}(u)+2}(u')$), excluding $\ell^{\text{rank}(u)+2}(u')$.

Finally, we must consider the possibility that the promotion of u' into $\mathcal{L}^{\text{rank}(u)+1}_\alpha[T]$ may cause cascading promotions, potentially all the way up to the root. However, the promotion of a node w into $\mathcal{L}^k_\alpha[T]$ is like an insertion of w into the weight α partition on $\mathcal{L}^k_\alpha[T]$ with the additional fact that w is a leaf of $\text{treap}(\mathcal{L}^k_\alpha[T])$. Hence we can handle it as discussed above.

Deletion: Let u be the node to be deleted. First, we handle potentially cascading "demotions" (i.e., decreases in rank). We repeat the following steps until the demotion or deletion of the current node w does not cause a demotion of its leader. Initialize w to u.

(1) Proceed up the path P from w to $w_l :=$ node(w.leader), decrementing the second coordinate of w'.size for each $w' \in P \setminus \{w, w_l\}$. Use the size fields of the children of w_l to check if $|T_{w_l} \cap \mathcal{L}_\alpha^{\text{rank}(w_l)-1}[T]| = \alpha$ before the deletion or demotion of w.
(2) If $|T_{w_l} \cap \mathcal{L}_\alpha^{\text{rank}(w_l)-1}[T]| = \alpha$ then w_l will be demoted, in which case we create a new B-treap node corresponding to the parent of w_l (if it does not already exist), traverse $T_{w_l} \cap \mathcal{L}_\alpha^{\text{rank}(w)}[T]$, move all of the nodes therein to the newly created B-treap node, and set their leader fields to the key of the parent of w_l. Delete any B-treap node that becomes empty. Also, decrement the first coordinate of w_l.size (i.e., its rank) and set its second coordinate to $\alpha - 1$.
(3) Set $w = w_l$.

Next, rotate u down to a leaf position, updating the subtree size and rank information appropriately on each rotation (this can be done in constant time per rotation). Make sure to account for the fact that u will ultimately be deleted from the treap when updating these size fields.

If initially rank(u) $= 0$, then just delete u. If initially rank(u) ≥ 1, maintain a list L of nodes x such that we rotated on edge $\{x, u\}$ when rotating u down to a leaf position. Delete u from the treap, but retain a temporary copy of its fields. Let rank(u) denote the initial rank of u. For $i \in \{1, 2, \ldots, \text{rank}(u)\}$ in increasing order, find the deepest element v_i of $\mathcal{L}_\alpha^i[T]$ in L using the recently updated size fields. For each i, if there is no B-treap node corresponding to v_i then create one. For each node $x \in L$, in order of increasing priority, if $x \in \mathcal{L}_\alpha^{i-1}[T]$ is a descendant of v_i and has a child $x' \in \mathcal{L}_\alpha^{i-1}[T]$ with x'.leader $\neq v_i$.key, update the leader field of each node in $T_{x'} \cap \mathcal{L}_\alpha^{i-1}[T]$ to v_i.key. Move all such nodes to the B-treap node corresponding to v_i. Also set x.leader $= v_i$.key and move it to the B-treap node corresponding to v_i. Throughout the whole operation, make sure to destroy all B-treap nodes that are emptied of treap nodes. If the deletion of u did not cause any demotions, then for each ancestor x of $v_{\text{rank}(u)}$ in L set x.leader $= u$.leader. Also, if x has a child y of with rank(y) $<$ rank(x) then for each such child y test if y.leader $\neq x$.key. If so, do a traversal of $T_y \cap \mathcal{L}_\alpha^{\text{rank}(y)}[T]$, update the leader field of each node in that subtree to x.key, and move each such node to the B-treap node corresponding to x.

Range Query: To return all keys between x and y, simply lookup x, then proceed to simulate an in-order traversal of the underlying treap until reaching y.

5 The Analysis of B-Treaps

For simplicity of exposition we assume the nodes have relative priorities determined by a random permutation selected uniformly at random. To remove this assumption and prove Theorem 1, it suffices to use the hash family of Östlin and Pagh [13] to generate priorities by hashing keys to a sufficiently large set of integers (e.g., $[n^3]$).

Unique Representation. The data structure is uniquely represented assuming the operations maintain the size and leader field invariants and the proper B-treap organization. In this case, the B-treap's representation in external memory is a deterministic function of the treap it stores and the hash table used to map it onto external memory. Thus it is independent of the historical sequence of operations that led to the current logical state, and must be the same for all such sequences of operations.

Bounding the Space Usage. If the treap priorities are generated via a random permutation and $\alpha = O(n^{\frac{1}{2}-\epsilon})$, then the space usage is linear with overwhelming probability (roughly $1 - \exp\left(n/\alpha^2\right)$). While it is unfortunate that the analysis depends on the branching factor, the assumption that $\alpha = O(n^{\frac{1}{2}-\epsilon})$ is not unreasonable, and I conjecture that this dependence can be removed. In any event, Proposition 1 bounds the space needed for any given B-treap node at $O(\alpha)$ data items, and Lemmas 4 and 5 together bound the number of B-treap nodes at $O(n/\alpha)$. If $B = \Theta(\alpha)$, this implies that a B-treap with n keys uses only $O(n/B)$ blocks of space with high probability, thus proving the space bounds of Theorem 1. The following proposition is straightforward to prove.

Proposition 1. *Each B-treap node $\bar{v}$ has at most $2\alpha - 1$ treap nodes stored in it.*

Lemma 4. *A B-treap $\bar{T}$ on $n \geq \alpha$ keys obtained from the iterated weight α partition of a random treap T has $O(n/\alpha + l)$ nodes, where l is the number of leaves in the B-treap.*

Proof. A *chain* C of T is a connected set of degree two nodes in T such that for all $u, v \in C$, u is an ancestor of v or v is an ancestor of u. Thus, each chain C can be written as $\text{ancestors}(u) \cap \text{descendants}(v)$ for some nodes $u, v \in C$, called the *endpoints* of C. It is not hard to see that any chain C of T has all of its nodes stored in at most $\left\lfloor \frac{|C|}{\alpha} \right\rfloor + 2$ nodes in the B-treap $\bar{T}$. If, for example, $|C| \geq \alpha/2$, then we may amortize the storage required for these $\left\lfloor \frac{|C|}{\alpha} \right\rfloor + 2$ nodes (each of which takes $O(\alpha)$ data items worth of space) against the $|C|$ treap nodes. We thus *mark* all B-treap nodes that store at least one treap node from any chain of T of length at least $\alpha/2$. As per our previous remarks, there are at most $O(n/\alpha)$ marked nodes.

Next, consider the unmarked nodes. We claim that the number of unmarked nodes is at most $2l$, where l is the number of leaves in the B-treap (either marked or unmarked). To prove this, we first prove that in the B-treap, each unmarked internal node $\bar{v}$ other than the root has at least two children. Since $\bar{v}$ is an internal node, each treap node u stored in it has $\text{rank}(u) \geq 1$, so that $|T_u| \geq \alpha$. Suppose the treap nodes in $\bar{v}$ have rank k. Then each $u \in \mathcal{F}[\mathcal{L}_\alpha^k[T]]$ stored in $\bar{v}$ corresponds to a child of $\bar{v}$ in $\bar{T}$. However, if there were only one such node, then $\bar{v}$ must store a chain of length at least $\alpha - 1$, contradicting the fact that $\bar{v}$ is an unmarked, internal, non-root node. This allows us to bound the number of unmarked nodes by $2l$ via a standard argument using the well-known facts that in any undirected graph $G = (V, E)$, $\sum_{v \in V} \deg(v) = 2|E|$ and in a tree $|E| = |V| - 1$.

Lemma 5. *Let $\bar{T}$ be a B-treap on $n \geq \alpha$ keys obtained from the iterated weight α partition of a random treap T with relative priorities determined by a random permutation selected uniformly at random. Let l be the number of leaves of $\bar{T}$. Then $l = O(n/\alpha)$ with probability $1 - O\left(\exp\{-\frac{2n}{\alpha(\alpha+1)}\}\right)$.*

Proof. The number of leaves in $\bar{T}$ equals $|\mathcal{F}[\mathcal{L}_\alpha[T]]|$, which is bounded by $|\mathcal{L}_\alpha[T]|$. The result thus follows from Lemma 3.

Bounding the Depth. The main purpose of the B-tree is to reduce the depth of the search tree from, e.g., $2\log_2(n)$ (for red-black trees) or $\sim 1.44\log_2(n)$ (for AVL trees), to $\log_\alpha n$ for a suitable parameter α. Indeed, α is often n^ϵ for some constant ϵ, so that the tree height is roughly $1/\epsilon$. So it is reasonable to require any proper B-tree alternative to also have height $O(\log_\alpha(n))$. The B-treap does indeed have this property with high probability if α is sufficiently large. Proving this fact involves some inductive probabilistic conditioning. Due to space limitations, we only sketch the proof below, and omit the proof of Lemma 6. For the full proof, see Section 5.5.4 of [8].

Theorem 2. *Fix a random treap T on n nodes with relative priorities determined by a random permutation selected uniformly at random. Let $\bar{T}$ be the corresponding B-treap generated from the iterated weight α partitioning of T. If $\alpha = \Omega\left(\ln(n)^{1/(1-\epsilon)}\right)$ for some positive constant ϵ, then $\text{rank}(T) = O\left(\frac{1}{\epsilon}\log_\alpha(n)\right)$ with high probability. Furthermore, since the B-treap depth is bounded by the rank of its root node, these bounds apply to the B-treap depth as well.*

Proof. Let $f(m,k) := \mathbf{Pr}[\text{A random treap } T \text{ on } m \text{ nodes has } \text{rank}(T) > k]$. In the definition of f, we assume the treap T has priorities determined by a random permutation on the keys. We claim that for any m and k, $f(m,r) \leq f(m+1,r)$, and then proceed by induction on the treap size. Specifically, we use the following induction hypothesis, for suitable constants c_1, c_2 and c_3.

$$f\left(\left(\frac{\alpha}{c_2\ln(n)}\right)^k, c_1 k\right) \leq \frac{(3(\alpha+1))^k}{n^{c_3}} \tag{5.1}$$

The basis $k = 1$ is straightforward. For the induction step, consider a treap T on $m = \left(\frac{\alpha}{c_2\ln(n)}\right)^k$ nodes, with keys $X := \{1, 2, \ldots, m\}$. Let Y be the α nodes of T with the highest priorities. Since the priorities are random, Y is will be distributed uniformly at random on $\{V : V \subset X, |V| = \alpha\}$. Let $Y = \{y_1, y_2, \ldots, y_\alpha\}$, with $y_1 < y_2 < \cdots < y_\alpha$, and let $y_0 := 0, y_{\alpha+1} := m+1$. Define $Z_i := \{z : y_i < z < y_{i+1}\}$ for all $0 \leq i \leq \alpha$. We claim that

$$\text{rank}(T) \leq \max_{0\leq i\leq\alpha}(\text{rank}(\text{treap}(Z_i))) + 2 \tag{5.2}$$

Let $\rho := \max_{0\leq i\leq\alpha}(\text{rank}(\text{treap}(Z_i)))$. Note that the only nodes in T with rank $\rho+1$ must be in Y itself, since all the nodes in any Z_i have rank at most ρ. Thus T contains at most α nodes of rank $\rho+1$, and hence the root of T can have rank at most $\rho+2$.

Next we obtain a high probability bound on $\max_{0\leq i\leq\alpha}(\text{rank}(\text{treap}(Z_i)))$ using Lemma 6 and the induction hypothesis. Let A_i be the event that $|Z_i| > \left(\frac{\alpha}{c_2\ln(n)}\right)^{k-1}$, and let $A := \cup_i A_i$. Use Lemma 6 below to bound $\mathbf{Pr}[A]$. Note that if we condition on all the Z_i's being sufficiently small (i.e., the event $\bar{A}$), then the priorities on the nodes of each Z_i are still random. Let B be the event that there exists an i such that $\text{rank}(\text{treap}(Z_i)) > c_1(k-1)$. Apply the induction hypothesis to bound $\mathbf{Pr}\left[B|\bar{A}\right]$. Equation (5.2) then implies that $f\left(\left(\frac{\alpha}{c_2\ln(n)}\right)^k, c_1(k-1)+2\right) \leq \mathbf{Pr}[A\cup B] \leq \mathbf{Pr}[A] + \mathbf{Pr}\left[B|\bar{A}\right]$. If $c_1 \geq 2$, then $c_1(k-1)+2 \leq c_1 k$, and thus we have a bound for $f\left(\left(\frac{\alpha}{c_2\ln(n)}\right)^k, c_1\cdot k\right)$. This completes the induction. Next, let $d(n) := \frac{\ln(n)}{\ln\left(\frac{3(\alpha+1)}{c_2\ln(n)}\right)}$

and use the fact that $f(m,r) \le f(m+1,r)$ to show $f(n,\ c_1 \lceil d(n) \rceil) \le \frac{(3(\alpha+1))^{\lceil d(n) \rceil}}{n^{c_3}}$. The rest of the proof follows from suitable assumptions on the size of α, c_1, c_2 and c_3.

Lemma 6. *Fix $\alpha, m \in \mathbb{N}$ with $\alpha < m$. Let $X = \{1, 2, \ldots, m\}$. Select a subset Y of X uniformly at random from $\{V : V \subset X, |V| = \alpha\}$. Let $Y = \{y_1, y_2, \ldots, y_\alpha\}$, with $y_1 < y_2 < \cdots < y_\alpha$, let $y_0 := 0, y_{\alpha+1} := m+1$, and let $\Delta = \max_i\{y_{i+1} - y_i - 1\}$, where i ranges from zero to α. Then for all c and n,* $\mathbf{Pr}\big[\Delta > c\frac{m+1}{\alpha}\ln(n)\big] \le \frac{m}{n^c}$.

Bounding the Running Time. We measure the running time of the B-treap operations in terms of the number t of B-treap nodes they inspect or alter. This is in line with the EM model, since the number of I/Os is $O(t)$ in expectation.

By this measure, lookups clearly inspect only $\mathrm{depth}(\bar{T})$ B-treap nodes. Inserting u may require inspecting up to $\mathrm{depth}(\bar{T})$ nodes to find $\mathrm{leaf}(u)$. After that, if P is the rotation path of u from $\mathrm{leaf}(u)$ to its proper location in treap T, then the operation might modify every B-treap node containing treap nodes in P, as well as B-treap nodes corresponding to elements of $F := P \cap \left(\cup_{i \ge 1} \mathcal{F}[\mathcal{L}^i_\alpha[T]]\right)$. Moreover, it is not hard to see that these are the only B-treap nodes that need to be updated. The number of B-treap nodes storing treap nodes in P is bounded by $\mathrm{depth}(\bar{T})$. Bounding $|F|$ is trickier. Note that $\mathbf{E}[|F|] = O(1)$, since $|F| \le |P|$ is bounded by the number of rotations in P, and $\mathbf{E}[|P|] = O(1)$ in a random treap [14]. The same argument holds true for deletions.

For range queries between x and y, it is not hard to see that the inspected B-treap nodes consist of the set Q of nodes containing treap nodes in the root-to-x and root-to-y paths, as well as the set of B-treap nodes R containing keys strictly between x and y. Using the analysis for the size of a B-treap on n treap nodes, it is not too difficult to show that $|R| = O(k/\alpha + \mathrm{depth}(\bar{T}))$ with high probability, where k is the number of treap nodes in the output. Clearly $|Q| \le 2\,\mathrm{depth}(\bar{T})$, so this implies only $O(k/\alpha + \mathrm{depth}(\bar{T}))$ B-treap nodes are inspected during the range query. If the internal memory has size $m \ge B \cdot \mathrm{depth}(\bar{T})$, and we store the $\mathrm{depth}(\bar{T})$ previously accessed B-treap nodes, then we need lookup each node in external memory at most once. Otherwise, if $m = O(B)$, we can bound the number of B-treap node lookups by the number of edges in the subtree of the B-treap containing $Q \cap R$, which is also $O(k/\alpha + \mathrm{depth}(\bar{T}))$ in expectation. We have now proven the following result.

Lemma 7. *In a B-treap $\bar{T}$, the lookup operation inspects* $\mathrm{depth}(\bar{T})$ *nodes, updates inspect or modify* $\mathrm{depth}(\bar{T}) + O(1)$ *nodes in expectation, and range queries inspect* $O(k/\alpha + \mathrm{depth}(\bar{T}))$ *nodes in expectation, where k is the size of the output.*

Combining Lemma 7 with Theorem 2 and noting that $\alpha = \Theta(B)$ then proves the running time bounds of Theorem 1.

Additional Supported Operations. There are several additional properties of treaps that the B-treap can exploit. Based on the reasoning above, it follows that finger insertions and deletions touch $O(1)$ B-treap nodes in expectation. Similarly, predecessor and successor queries can be answered by inspecting $O(1)$ B-treap nodes in expectation. Finally, given a sorted list of elements, a B-treap on them can be constructed in expected linear time.

6 Empirical Observations

Simulations in which the treap priorities were provided by a psuedorandom number generator suggest the B-treap has small depth (e.g., empirically bounded by $1.5 \log_\alpha(n)$ for $\alpha = 100$ and $n \leq 10^6$) and space utilization of roughly $1/3^{\text{rd}}$. Refer to Section 5.5.5 of [8] for more detail. If the $1/3^{\text{rd}}$ space utilization is judged unacceptably low, there are various ways to improve it, at the cost of increasing the average number of I/Os per operation. For example, for any fixed $k \in \mathbb{N}$ we may divide each block into k equally sized *block-parts*, and store a B-treap node containing q treap nodes in $\lceil kq/(2\alpha - 1) \rceil$ block-parts dynamically allocated to it. This ensures there is at most one block-part of unused space per B-treap node, rather than a whole block.

References

1. Acar, U.A., Blelloch, G.E., Harper, R., Vittes, J.L., Woo, S.L.M.: Dynamizing static algorithms, with applications to dynamic trees and history independence. In: 15th annual ACM-SIAM Symposium on Discrete Algorithms, pp. 531–540 (2004)
2. Andersson, A., Ottmann, T.: New tight bounds on uniquely represented dictionaries. SIAM Journal of Computing 24(5), 1091–1103 (1995)
3. Bayer, R., McCreight, E.M.: Organization and maintenance of large ordered indices. Acta Informatica 1, 173–189 (1972)
4. Blelloch, G.E., Golovin, D.: Strongly history-independent hashing with applications. In: 48th Annual IEEE Symposium on Foundations of Computer Science, pp. 272–282 (2007)
5. Blelloch, G.E., Golovin, D., Vassilevska, V.: Uniquely represented data structures for computational geometry. In: Gudmundsson, J. (ed.) SWAT 2008. LNCS, vol. 5124, pp. 17–28. Springer, Heidelberg (2008)
6. Buchbinder, N., Petrank, E.: Lower and upper bounds on obtaining history independence. Information and Computation 204(2), 291–337 (2006)
7. Fagin, R., Nievergelt, J., Pippenger, N., Strong, H.R.: Extendible hashing—a fast access method for dynamic files. ACM Trans. Database Syst. 4(3), 315–344 (1979)
8. Golovin, D.: Uniquely Represented Data Structures with Applications to Privacy. PhD thesis, Carnegie Mellon University, Pittsburgh, PA, Tech. Report CMU-CS-08-135 (2008)
9. Hartline, J.D., Hong, E.S., Mohr, A.E., Pentney, W.R., Rocke, E.: Characterizing history independent data structures. Algorithmica 42(1), 57–74 (2005)
10. Micciancio, D.: Oblivious data structures: applications to cryptography. In: STOC 1997: Proceedings of the twenty-ninth annual ACM Symposium on Theory of Computing, pp. 456–464. ACM Press, New York (1997)
11. Naor, M., Segev, G., Wieder, U.: History-independent cuckoo hashing. In: Aceto, L., Damgård, I., Goldberg, L.A., Halldórsson, M.M., Ingólfsdóttir, A., Walukiewicz, I. (eds.) ICALP 2008, Part II. LNCS, vol. 5126, pp. 631–642. Springer, Heidelberg (2008)
12. Naor, M., Teague, V.: Anti-persistence: history independent data structures. In: STOC 2001: Proceedings of the thirty-third annual ACM Symposium on Theory of Computing, pp. 492–501. ACM Press, New York (2001)
13. Östlin, A., Pagh, R.: Uniform hashing in constant time and linear space. In: STOC 2003: Proceedings of the thirty-fifth annual ACM Symposium on Theory of Computing, pp. 622–628. ACM Press, New York (2003)
14. Seidel, R., Aragon, C.R.: Randomized search trees. Algorithmica 16(4/5), 464–497 (1996)

15. Snyder, L.: On uniquely representable data structures. In: FOCS 1977: IEEE Symposium on Foundations of Computer Science, pp. 142–146. IEEE Computer Society Press, Los Alamitos (1977)
16. Sundar, R., Tarjan, R.E.: Unique binary search tree representations and equality-testing of sets and sequences. In: STOC 1990: Proceedings of the twenty-second annual ACM Symposium on Theory of Computing, pp. 18–25. ACM Press, New York (1990)
17. Vitter, J.S.: Algorithms and data structures for external memory. Foundations and Trends in Theoretical Computer Science 2(4), 305–474 (2006)

Testing Fourier Dimensionality and Sparsity

Parikshit Gopalan[1], Ryan O'Donnell[2], Rocco A. Servedio[4], Amir Shpilka[3], and Karl Wimmer[2]

[1] Microsoft Research, Silicon Valley
[2] Carnegie Mellon University
[3] Technion
[4] Columbia University

parik@microsoft.com, odonnell@cs.cmu.edu, rocco@cs.columbia.edu, shpilka@cs.technion.ac.il, kwimmer@andrew.cmu.edu

Abstract. We present a range of new results for testing properties of Boolean functions that are defined in terms of the Fourier spectrum. Broadly speaking, our results show that the property of a Boolean function having a concise Fourier representation is locally testable.

We first give an efficient algorithm for testing whether the Fourier spectrum of a Boolean function is supported in a low-dimensional subspace of $\mathbb{F}_2^n$ (equivalently, for testing whether f is a junta over a small number of parities). We next give an efficient algorithm for testing whether a Boolean function has a sparse Fourier spectrum (small number of nonzero coefficients). In both cases we also prove lower bounds showing that any testing algorithm — even an adaptive one — must have query complexity within a polynomial factor of our algorithms, which are nonadaptive. Finally, we give an "implicit learning" algorithm that lets us test *any* sub-property of Fourier concision.

Our technical contributions include new structural results about sparse Boolean functions and new analysis of the pairwise independent hashing of Fourier coefficients from [12].

1 Introduction

Recent years have witnessed broad research interest in the local testability of mathematical objects such as graphs, error-correcting codes, and Boolean functions. One of the goals of this study is to understand the minimal conditions required to make a property locally testable. For graphs and codes, works such as [1,5,3,4] and [16,17] have given fairly general characterizations of when a property is testable. For Boolean functions, however, testability is less well understood. On one hand, there are a fair number of testing algorithms for specific classes of functions such as $\mathbb{F}_2$-linear functions [10,6], dictators [7,21], low-degree $\mathbb{F}_2$-polynomials [2,22], juntas [14,9], and halfspaces [20]. But there is not much by way of general characterizations of what makes a property of Boolean functions testable. Perhaps the only example is the work of [11], showing that any class of functions sufficiently well-approximated by juntas is locally testable.

It is natural to think that general characterizations of testability for Boolean functions might come from analyzing the Fourier spectrum (see e.g. [13, Section 9.1]). For one thing, many of the known tests — for linearity, dictators, juntas, and halfspaces — involve a careful analysis of the Fourier spectrum. Further intuition comes from learning

S. Albers et al. (Eds.): ICALP 2009, Part I, LNCS 5555, pp. 500–512, 2009.

theory, where the class of functions that are learnable using many of the well-known algorithms [19,18,15] can be characterized in terms of the Fourier spectrum.

In this paper we make some progress toward this goal, by giving efficient algorithms for testing Boolean functions that have *low-dimensional* or *sparse* Fourier representations. These are two natural ways to formalize what it means for a Boolean function to have a "concise" Fourier representation; thus, roughly speaking our results show that the property of having a concise Fourier representation is efficiently testable. Further, as we explain below, Boolean functions with low-dimensional or sparse Fourier representations are closely related to linear functions, juntas, and low-degree polynomials whose testability has been intensively studied, and thus the testability of these classes is a natural question in its own right. Building on our testing algorithms, we are able to give an "implicit learner" (in the sense of [11]), which determines the "truth table" of a sparse Fourier spectrum without actually knowing the identities of the underlying Fourier characters. This lets us test *any* sub-property of having a concise Fourier representation. We view this as a step toward the goal of a more unified understanding of the testability of Boolean functions.

Our algorithms rely on new structural results on Boolean functions with sparse and close-to-sparse Fourier spectrums, which may find applications elsewhere. As one such application, we show that the well-known Kushilevitz-Mansour algorithm is in fact an exact proper learning algorithm for Boolean functions with sparse Fourier representations. As another application, we give polynomial-time unique-decoding algorithms for sparse functions and k-dimensional functions; due to space limitations these results will only appear in the full version of the paper.

1.1 The Fourier Spectrum, Dimensionality, and Sparsity

We are concerned with testing various properties defined in terms of the *Fourier representation* of Boolean functions $f : \mathbb{F}_2^n \to \{-1, 1\}$. Input bits will be treated as $0, 1 \in \mathbb{F}_2$, the field with two elements; output bits will be treated as $-1, 1 \in \mathbb{R}$. Every Boolean function $f : \mathbb{F}_2^n \to \mathbb{R}$ has a unique representation as

$$f(x) = \sum_{\alpha \in \mathbb{F}_2^n} \hat{f}(\alpha)\chi_\alpha(x) \text{ where } \chi_\alpha(x) \stackrel{\text{def}}{=} (-1)^{\langle \alpha, x\rangle} = (-1)^{\sum_{i=1}^n \alpha_i x_i}. \qquad (1)$$

The coefficients $\hat{f}(\alpha)$ are the *Fourier coefficients* of f, and the functions $\chi_\alpha(\cdot)$ are sometimes referred to as *linear functions* or *characters*. In addition to treating input strings x as lying in $\mathbb{F}_2^n$, we also index the characters by vectors $\alpha \in \mathbb{F}_2^n$. This is to emphasize the fact that we are concerned with the linear-algebraic structure. We write $\text{Spec}(f)$ for the Fourier spectrum of f, i.e. the set $\{\alpha \in \mathbb{F}_2^n : \hat{f}(\alpha) \neq 0\}$.

Dimensionality and sparsity (and degree). A function $f : \mathbb{F}_2^n \to \{-1, 1\}$ is said to be *k-dimensional* if $\text{Spec}(f)$ lies in a k-dimensional subspace of $\mathbb{F}_2^n$. An equivalent definition is that f is k-dimensional if it is a function of k characters $\chi_{\alpha_1}, \ldots, \chi_{\alpha_k}$, i.e. f is a junta over k parity functions (this is easily seen by picking $\{\alpha_i\}$ to be a basis for $\text{Spec}(f)$). We write $\dim(f)$ to denote the smallest k for which f is k-dimensional. A function f is said to be *s-sparse* if $|\text{Spec}(f)| \leq s$. We write $\text{sp}(f)$ to denote $|\text{Spec}(f)|$, i.e. the smallest s for which f is s-sparse.

We recall the notion of the $\mathbb{F}_2$-*degree* of a Boolean function, $\deg_2(f)$, which is the degree of the unique multilinear $\mathbb{F}_2$-polynomial representation for f when viewed as a function $\mathbb{F}_2^n \to \mathbb{F}_2$. (This should not be confused with the real-degree/Fourier-degree. For example, $\deg_2(\chi_\alpha) = 1$ for all $\alpha \neq 0$.) Let us note some relations between $\dim(f)$ and $\mathrm{sp}(f)$. For any Boolean function f, we have

$$\deg_2(f) \leq \log \mathrm{sp}(f) \leq \dim(f), \tag{2}$$

except that the first inequality fails when $\deg_2(f) = 1$. (Throughout this paper, log always means $\log_2$.) The first inequality above is not difficult (see e.g. [8, Lemma 3]) and the second one is essentially immediate. Either of the above inequalities can be quite loose; for the first inequality, the inner product function on n variables has $\deg_2(f) = 2$ but $\log \mathrm{sp}(f) = n$. For the second inequality, the addressing function with $\frac{1}{2}\log s$ addressing variables and $s^{1/2}$ addressee variables can be shown to be s-sparse but has $\dim(f) \geq s^{1/2}$. (It is trivially true that $\dim(f) \leq s$ for any s-sparse function.)

We may rephrase these bounds as containments between classes of functions:

$$\{k\text{-dimensional}\} \subseteq \{2^k\text{-sparse}\} \subseteq \{\mathbb{F}_2 - \text{degree-}k\}\} \tag{3}$$

where the right containment is proper for $k > 1$ and the left is proper for k larger than some small constant such as 6. Alon et al. [2] gave essentially matching upper and lower bounds for testing the class of $\mathbb{F}_2$-degree-k functions, showing that $2^{\Theta(k)}$ nonadaptive queries are necessary and sufficient. We show that $2^{\Theta(k)}$ queries are also necessary and sufficient for testing each of the first two classes as well; in fact, by our implicit learning result, we can test *any* sub-class of k-dimensional functions using $2^{O(k)}$ queries.[1]

1.2 Our Results and Techniques

Testing Low-Dimensionality. We give nearly matching upper and lower bounds for testing whether a function is k-dimensional:

Theorem 1 [Testing k-dimensionality – informal]. *There is a nonadaptive $O(k2^{2k}/\epsilon)$-query algorithm for ϵ-testing whether f is k-dimensional. Moreover, any algorithm (adaptive, even) for 0.49-testing this property must make $\Omega(2^{k/2})$ queries.*

We outline the basic idea behind our dimensionality test. Given $h \in \mathbb{F}_2^n$, we say that $f : \mathbb{F}_2^n \to \mathbb{R}$ is h-*invariant* if it satisfies $f(x + h) = f(x)$ for all $x \in \mathbb{F}_2^n$. We define the subspace $\mathrm{Inv}(f) = \{h : f \text{ is } h\text{-invariant}\}$. If f is truly k-dimensional, then $\mathrm{Inv}(f)$ has codimension k; we use this as the characterization of k-dimensional functions. We estimate the size of $\mathrm{Inv}(f)$ by randomly sampling vectors h and testing if they belong to $\mathrm{Inv}(f)$. We reject if the fraction of such h is much smaller than 2^{-k}. The crux of our soundness analysis is to show that if a function passes the test with good probability, most of its Fourier spectrum is concentrated on a k-dimensional subspace. From this we conclude that it must in fact be close to a k-dimensional function. Because of space constraints, this algorithm is omitted from this version of the paper.

Testing Sparsity. We next give an algorithm for testing whether a function is s-sparse. Its query complexity is $\mathrm{poly}(s)$, which is optimal up to the degree of the polynomial:

---

[1] We remind the reader that efficient testability does not translate downward: if C_1 is a class of functions that is efficiently testable and $C_2 \subsetneq C_1$, the class C_2 need not be efficiently testable.

Theorem 2 [Testing s-sparsity – informal]. *There is a nonadaptive* $\text{poly}(s, 1/\epsilon)$*-query algorithm for ϵ-testing whether f is s-sparse. Moreover, any algorithm (adaptive, even) for* 0.49*-testing this property must make $\Omega(\sqrt{s})$ queries.*

The high-level idea behind our tester is that of "hashing" the Fourier coefficients, following [12]. We choose a random subspace H of $\mathbb{F}_2^n$ with codimension $O(s^2)$. This partitions all the Fourier coefficients into the cosets (affine subspaces) defined by H. If f is s-sparse, then each vector in $\text{Spec}(f)$ is likely to land in a distinct coset. We define the "projection" of f to a coset $r + H$ to be the real-valued function given by zeroing out all Fourier coefficients not in $r + H$. Given query access to f, one can obtain approximate query access to a projection of f by a certain averaging. Now if each vector in $\text{Spec}(f)$ is hashed to a different coset, then each projection function will have sparsity either 1 or 0, so we can try to test that at most s of the projection functions have sparsity 1, and the rest have sparsity 0.

A similar argument to the one used for k-dimensionality shows that if f passes this test, most of its Fourier mass lies on a few coefficients. However, unlike in the low-dimensionality test, this is not *a priori* enough to conclude that f is close to a sparse Boolean function. The obvious way to get a Boolean function close to f would be to truncate the Fourier spectrum to its s largest coefficients and then take the sign, but taking the sign could destroy the sparsity and give a function which is not at all sparse.

We circumvent this obstacle by using some new structural theorems about sparse Boolean functions. We show that if most of the Fourier mass of a function f lies on its largest s coefficients, then these coefficients are close to being "$\lceil \log s \rceil$–granular," i.e. close to integer multiples of $1/2^{\lceil \log s \rceil}$. We then prove that truncating the Fourier expansion to these coefficients and rounding them to nearby granular values gives a sparse *Boolean*-valued function (Theorem 6). Thus our sparsity test and its analysis depart significantly from the tests for juntas [14] and from our test for low-dimensionality.

Testing subclasses of k-dimensional functions. Finally, we show that a broad range of subclasses of k-dimensional functions are also testable with $2^{O(k)}$ queries. Recall that k-dimensional functions are all functions $f(x) = g(\chi_{\alpha_1}(x), \ldots, \chi_{\alpha_k}(x))$ where g is any k-variable Boolean function. We say that a class $\mathcal{C}$ is an *induced subclass of k-dimensional functions* if there is some collection $\mathcal{C}'$ of k-variable Boolean functions such that $\mathcal{C}$ is the class of all functions $f = g(\chi_{\alpha_1}, \ldots, \chi_{\alpha_k})$ where g is any function in $\mathcal{C}'$ and $\chi_{\alpha_1}, \ldots, \chi_{\alpha_k}$ are any linear functions from $\mathbb{F}_2^n$ to $\mathbb{F}_2$ as before. For example, let $\mathcal{C}$ be the class of all k-sparse polynomial threshold functions over $\{-1, 1\}^n$; i.e., each function in $\mathcal{C}$ is the sign of a *real* polynomial with at most k nonzero terms. This is an induced subclass of k-dimensional functions, corresponding to the collection $\mathcal{C}' = \{$ all linear threshold functions over k Boolean variables$\}$.

We show that any induced subclass of k-dimensional functions can be tested:

Theorem 3 [Testing induced subclasses of k-dimensional functions – informal]. *Let $\mathcal{C}$ be any induced subclass of k-dimensional functions. There is a nonadaptive* $\text{poly}(2^k, 1/\epsilon)$*-query algorithm for ϵ-testing $\mathcal{C}$.*

We note that the upper bound of Theorem 3 is essentially best possible in general, by the $2^{\Omega(k)}$ lower bound for testing the whole class of k-dimensional functions.

Our algorithm for Theorem 3 extends the approach of Theorem 2 with ideas from the "testing by implicit learning" work of [11]. Briefly, by hashing the Fourier coefficients we are able to construct a matrix of size $2^k \times 2^k$ whose entries are the values taken by the characters χ_α in the spectrum of f. This matrix, together with a vector of the corresponding values of f, serves as a data set for "implicit learning" (we say the learning is "implicit" since we do not actually know the names of the relevant characters). Our test inspects sub-matrices of this matrix and tries to find one which, together with the vector of f-values, matches the truth table of some k-variable function $g \in \mathcal{C}'$.

Organization of the paper. We give standard preliminaries and an explanation of our techniques for hashing the Fourier spectrum in Section 2. Section 3 gives our new structural theorems about sparse Boolean functions, and Section 4 uses these theorems to give our test for s-sparse functions. Because of space constraints, our results for testing k-dimensional functions, for unique-decoding, for testing induced subclasses of k-dimensional functions, and our lower bounds are given in the full version.

2 Preliminaries

Throughout the paper we view Boolean functions as mappings from $\mathbb{F}_2^n$ to $\{-1,1\}$. We will also consider functions which map from $\mathbb{F}_2^n$ to $\mathbb{R}$. Such functions have a unique Fourier expansion as in Equation (1). For $\mathcal{A}$ a collection of vectors $\alpha \in \mathbb{F}_2^n$, we write $\mathrm{wt}(\mathcal{A})$ to denote the "Fourier weight" $\mathrm{wt}(\mathcal{A}) = \sum_{\alpha \in \mathcal{A}} \hat{f}(\alpha)^2$ on the elements of $\mathcal{A}$. This notation suppresses the dependence on f, but it will always be clear from context. We frequently use Parseval's identity: $\mathrm{wt}(\mathbb{F}_2^n) = \sum_{\alpha \in \mathbb{F}_2^n} \hat{f}(\alpha)^2 = \|f\|_2^2 \stackrel{\text{def}}{=} \mathbf{E}_{x \in \mathbb{F}_2^n}[f(x)^2]$. Here and elsewhere, an expectation or probability over "$x \in X$" refers to the uniform distribution on X.

As defined in the previous section, the sparsity of f is $\mathrm{sp}(f) = |\mathrm{Spec}(f)|$. We may concisely restate the definition of dimension as $\dim(f) = \dim(\mathrm{span}(\mathrm{Spec}(f)))$.

Given two Boolean functions f and g, we say that f and g are *ϵ-close* if $\mathbf{Pr}_{x \in \mathbb{F}_2^n}[f(x) \neq g(x)] \leq \epsilon$ and say they are *ϵ-far* if $\mathbf{Pr}_{x \in \mathbb{F}_2^n}[f(x) \neq g(x)] \geq \epsilon$. We use the standard definition of property testing:

Definition 1. *Let $\mathcal{C}$ be a class of functions mapping $\mathbb{F}_2^n$ to $\{-1,1\}$. A* property tester *for $\mathcal{C}$ is an oracle algorithm $\mathcal{A}$ which is given a distance parameter $\epsilon > 0$ and oracle access to a function $f : \mathbb{F}_2^n \to \{-1,1\}$ and satisfies the following conditions:*

1. *if $f \in \mathcal{C}$ then $\mathcal{A}$ outputs "accept" with probability at least $2/3$;*
2. *if f is ϵ-far from every $g \in \mathcal{C}$ then $\mathcal{A}$ outputs "accept" with probability at most $1/3$.*

We also say that $\mathcal{A}$ ϵ-tests $\mathcal{C}$. The main interest is in the number of queries *the testing algorithm makes.*

All of our testing upper and lower bounds allow "two-sided error" as described above. Our lower bounds are for adaptive query algorithms and our upper bounds are via non-adaptive query algorithms.

2.1 Projections of the Fourier Spectrum

The idea of "isolating" or "hashing" Fourier coefficients by projection, as done in [12] in a learning-theoretic context, plays an important role in our tests.

Definition 2. *Given a subspace $H \leq \mathbb{F}_2^n$ and a coset $r + H$, define the* projection operator P_{r+H} *on functions $f : \mathbb{F}_2^n \to \mathbb{R}$ as follows:*

$$\widehat{\mathrm{P}_{r+H} f}(\alpha) \stackrel{def}{=} \begin{cases} \widehat{f}(\alpha) & \text{if } \alpha \in r + H, \\ 0 & \text{otherwise.} \end{cases}$$

In other words, we have $\mathrm{P}_{r+H} f = A_{r+H} * f$, where $A_{r+H} \stackrel{def}{=} \sum_{\alpha \in r+H} \chi_\alpha$ and $*$ is the convolution operator: $f * g(x) = \mathbf{E}_y[f(x+y) \cdot g(y)]$.

Clearly $A_{r+H} = \chi_r \cdot \sum_{h \in H} \chi_h$, and it is a simple and well-known fact that $\sum_{h \in H} \chi_h = |H| \cdot \mathbf{1}_{H^\perp}$. Thus we conclude the following (see also Lemma 1 of [12]):

Fact 4. $\mathrm{P}_{r+H} f(x) = \mathbf{E}_{y \in H^\perp}[\chi_r(y) f(x+y)]$.

We now show that for any coset $r+H$, we can approximately determine both $\mathrm{P}_{r+H} f(x)$ and $\|\mathrm{P}_{r+H} f\|_2^2$.

Proposition 1. *For any $x \in \mathbb{F}_2^n$, the value $\mathrm{P}_{r+H} f(x)$ can be estimated to within $\pm\tau$ with confidence $1 - \delta$ using $O(\log(1/\delta)/\tau^2)$ queries to f.*

Proof. Empirically estimate the right-hand side in Fact 4. Since the quantity inside the expectation is bounded in $[-1, 1]$, the result follows from a Chernoff bound. □

Recall that $\mathrm{wt}(r + H) = \sum_{\alpha \in r+H} \hat{f}(\alpha)^2 = \|\mathrm{P}_{r+H} f\|_2^2$. We have:

Fact 5. $\mathrm{wt}(r + H) = \mathbf{E}_{x \in \mathbb{F}_2^n, z \in H^\perp}[\chi_r(z) f(x) f(x + z)]$.

Proof. Using Parseval and Fact 4, we have

$$\mathrm{wt}(r + H) = \mathop{\mathbf{E}}_{w \in \mathbb{F}_2^n}[(\mathrm{P}_{r+H} f(w))^2] = \mathop{\mathbf{E}}_{w \in \mathbb{F}_2^n, y_1, y_2 \in H^\perp}[\chi_r(y_1) f(w + y_1) \chi_r(y_2) f(w + y_2)],$$

which reduces to the desired equality upon writing $x = w + y_1$, $z = y_1 + y_2$. □

Proposition 2. *The value $\mathrm{wt}(r + H)$ can be estimated to within $\pm\tau$ with confidence $1 - \delta$ using $O(\log(1/\delta)/\tau^2)$ queries to f.*

Proof. Empirically estimate the right-hand side in Fact 5. Since the quantity inside the expectation is bounded in $[-1, 1]$, the result follows from a Chernoff bound. □

2.2 Hashing to a Random Coset Structure

In this section we present our technique for pairwise independently hashing the Fourier characters.

Definition 3. *For $t \in \mathbb{N}$, we define a* random t-dimensional coset structure $(H, \mathcal{C})$ *as follows: We choose vectors $\beta_1, \dots, \beta_t \in \mathbb{F}_2^n$ independently and uniformly at random and set $H = \mathrm{span}\{\beta_1, \dots, \beta_t\}^\perp$. For each $b \in \mathbb{F}_2^t$ we define the "bucket"*

$$C(b) \stackrel{def}{=} \{\alpha \in \mathbb{F}_2^n : \langle \alpha, \beta_i \rangle = b_i \text{ for all } i\}.$$

We take $\mathcal{C}$ to be the set of $C(b)$'s, which has cardinality 2^t.

Remark 1. Given such a random coset structure, if the β_i's are linearly independent then the buckets $C(b)$ are precisely the cosets in $\mathbb{F}_2^n/H$, and the coset-projection function $\mathrm{P}_{C(b)}f$ is defined according to Definition 2. In the (usually unlikely) case that the β_i's are linearly *dependent*, some of the $C(b)$'s will be cosets in $\mathbb{F}_2^n/H$ and some of them will be empty. For the empty buckets $C(b)$ we define $\mathrm{P}_{C(b)}f$ to be identically 0. It is algorithmically easy to distinguish empty buckets from genuine coset buckets.

We now derive some simple but important facts about this random hashing process:

Proposition 3. *Let $(H, \mathcal{C})$ be a random t-dimensional coset structure. Define the indicator random variable $I_{\alpha \to b}$ for the event that $\alpha \in C(b)$.*

1. *For each $\alpha \in \mathbb{F}_2^n \setminus \{0\}$ and each b we have $\mathbf{Pr}[\alpha \in C(b)] = \mathbf{E}[I_{\alpha \to b}] = 2^{-t}$.*
2. *Let $\alpha, \alpha' \in \mathbb{F}_2^n$ be distinct. Then $\mathbf{Pr}[\alpha, \alpha'$ belong to the same bucket$] = 2^{-t}$.*
3. *Fix any set $S \subseteq \mathbb{F}_2^n$ with $|S| \leq s + 1$. If $t \geq 2 \log s + \log(1/\delta)$ then except with probability at most δ, all vectors in S fall into different buckets.*
4. *For each b, the collection of random variables $(I_{\alpha \to b})_{\alpha \in \mathbb{F}_2^n}$ is pairwise independent.*

Proof. Part 1 is because for any $\alpha \neq 0$, each $\langle \alpha, \beta_i \rangle$ is an independent uniformly random bit. Part 2 is because each $\langle \alpha - \alpha', \beta_i \rangle$ is an independent uniformly random bit, and hence the probability that $\langle \alpha, \beta_i \rangle = \langle \alpha', \beta_i \rangle$ for all i is 2^{-t}. Part 3 follows from Part 2 and taking a union bound over the at most $\binom{s+1}{2} \leq s^2$ distinct pairs in S. For Part 4, assume first that $\alpha \neq \alpha'$ are both nonzero. Then from the fact that α and α' are linearly independent, it follows that $\mathbf{Pr}[\alpha, \alpha' \in C(b)] = 2^{-2t}$ as required. On the other hand, if one of $\alpha \neq \alpha'$ is zero, then $\mathbf{Pr}[\alpha, \alpha' \in C(b)] = \mathbf{Pr}[\alpha \in C(b)]\mathbf{Pr}[\alpha' \in C(b)]$ follows immediately by checking the two cases $b = 0, b \neq 0$. □

With Proposition 3 in mind, we give the following simple deviation bound for the sum of pairwise independent random variables:

Proposition 4. *Let $X = \sum_{i=1}^n X_i$, where the X_i's are pairwise independent random variables satisfying $0 \leq X_i \leq \tau$. Assume $\mu = \mathbf{E}[X] > 0$. Then for any $\epsilon > 0$, we have $\mathbf{Pr}[X \leq (1 - \epsilon)\mu] \leq \frac{\tau}{\epsilon^2 \mu}$.*

Proof. By pairwise independence, we have $\mathbf{Var}[X] = \sum \mathbf{Var}[X_i] \leq \sum \mathbf{E}[X_i^2] \leq \sum \tau \mathbf{E}[X_i] = \tau\mu$. The result now follows from Chebyshev's inequality. □

Finally, it is slightly annoying that Part 1 of Proposition 3 fails for $\alpha = 0$ (because 0 is always hashed to $C(0)$). However we can easily handle this issue by renaming the buckets with a simple random permutation.

Definition 4. *In a random* permuted *t-dimensional coset structure, we additionally choose a random $z \in \mathbb{F}_2^t$ and rename $C(b)$ by $C(b+z)$.*

Proposition 5. *For a random permuted t-dimensional coset structure, Proposition 3 continues to hold, with Part 1 even holding for $\alpha = 0$.*

Proof. Use Proposition 3 and the fact that adding a random z permutes the buckets. □

3 Structural Theorems about s-Sparse Functions

In this section we prove structural theorems about close-to-sparse Boolean functions. These theorems are crucial to the analysis of our test for s-sparsity; we also present a learning application in the full version.

Definition 5. *Let $B = \{\alpha_1, \cdots, \alpha_s\}$ denote the (subsets of $[n]$ with the) s largest Fourier coefficients of f, and let $S = \bar{B}$ be its complement. We say that f is* μ-close to s-sparse in ℓ_2 *if $\sum_{\alpha \in S} \hat{f}(\alpha)^2 \leq \mu^2$.*

Definition 6. *We say a rational number* has granularity $k \in \mathbb{N}$, *or is* k-granular, *if it is of the form (integer)$/2^k$. We say a function $f : \mathbb{F}_2^n \to \mathbb{R}$ is* k-granular *if $\widehat{f}(\alpha)$ is k-granular for every α. We say that a number v is* μ-close to k-granular *if $|v - j/2^k| \leq \mu$ for some integer j.*

The following structural result is the key theorem for the completeness of our sparsity test; it says that in any function that is close to being sparse in ℓ_2, all the large Fourier coefficients are close to being granular.

Theorem 1 *[Completeness Theorem]. If f is μ-close to s-sparse in ℓ_2, then each $\hat{f}(\alpha)$ for $\alpha \in B$ is $\frac{\mu}{\sqrt{s}}$-close to $\lceil \log s \rceil$-granular.*

Proof. Pick a set of $k = \lceil \log s \rceil + 1$ equations $A\alpha = b$ at random (i.e. pick a $k \times n$ random matrix A and a random vector $b \in \mathbb{F}_2^k$). Let $A^\perp \subset \mathbb{F}_2^n$ be the set of solutions to $A\alpha = 0$. Define H to be the coset of $A^\perp$ of solutions to $A\alpha = b$. We have

$$\mathrm{P}_H f(x) = \sum_{\alpha \in H} \hat{f}(\alpha)\chi_\alpha(x).$$

Fix $\alpha_i \in B$. We will show that with non-zero probability the following two events happen together: the vector α_i is the unique coefficient in $B \cap H$, and the ℓ_2 Fourier mass of the set $S \cap H$ is bounded by $\frac{\mu^2}{s}$. Clearly, $\mathbf{Pr}_{A,b}[A\alpha_i = b] = 2^{-k}$. Let us condition on this event. By pairwise independence, for any $j \neq i$, $\mathbf{Pr}_{A,b}[A\alpha_j = b | A\alpha_i = b] = 2^{-k} \leq \frac{1}{2s}$. Thus $\mathbf{E}_{A,b}\left[|\{j \neq i \text{ such that } A\alpha_j = b\}| \,\middle|\, A\alpha_i = b\right] = \frac{(s-1)}{2^k} < \frac{1}{2}$. Hence by Markov's inequality

$$\mathbf{Pr}_{A,b}[\exists j \neq i \text{ such that } A\alpha_j = b \,|\, A\alpha_i = b] < \frac{1}{2}. \quad (4)$$

Now consider the coefficients from S. We have

$$\mathbf{E}_{A,b}\left[\sum_{\beta\in S\cap H}\hat{f}(\beta)^2|A\alpha_i=b\right]=\sum_{\beta\in S}\mathbf{Pr}[\beta\in H|A\alpha_i=b]\hat{f}(\beta)^2\leq 2^{-k}\mu^2\leq\frac{\mu^2}{2s}.$$

Hence by Markov's inequality,

$$\mathbf{Pr}_{A,b}\left[\sum_{\beta\in S\cap H}\hat{f}(\beta)^2\geq\frac{\mu^2}{s}|A\alpha_i=b\right]\leq\frac{1}{2}. \quad (5)$$

Thus by applying the union bound to Equations 4 and 5, we have both the desired events (α_i being the unique solution from B, and small ℓ_2 mass from S) happening with non-zero probability over the choice of A, b. Fixing this choice, we have

$$\mathrm{P}_H f(x)=\hat{f}(\alpha_i)\chi_{\alpha_i}(x)+\sum_{\beta\in S\cap H}\hat{f}(\beta)\chi_\beta(x)\ \text{ where }\sum_{\beta\in S\cap H}\hat{f}(\beta)^2\leq\frac{\mu^2}{s}.$$

But by Fact 4 we also have $\mathrm{P}_H f(x)=\mathbf{E}_{y\in A}[\chi_b(y)f(x+y)]$ (here we abuse notations and think of A as both the matrix A and the space spanned by the rows of A. In particular, $A=(A^\perp)^\perp$). Thus the function $\mathrm{P}_H f(x)$ is the average of a Boolean function over 2^k points, hence it is $(k-1)$-granular.

We now consider the function $g(x)=\sum_{\beta\in S\cap H}\hat{f}(\beta)\chi_\beta(x)$. Since $\mathbf{E}_x[g(x)^2]\leq\frac{\mu^2}{s}$, for some $x_0\in\mathbb{F}_2^n$ we have $g(x_0)^2\leq\frac{\mu^2}{s}$, hence $|g(x_0)|\leq\frac{\mu}{\sqrt{s}}$. Fixing this x_0, we have $\mathrm{P}_H f(x_0)=\hat{f}(\alpha_i)\chi_{\alpha_i}(x_0)+g(x_0)$, and hence $|\hat{f}(\alpha_i)|=|\mathrm{P}_H f(x_0)-g(x_0)|$. Since $\mathrm{P}_H f(x_0)$ is $(k-1)$-granular and $|g(x_0)|\leq\frac{\mu}{\sqrt{s}}$, the claim follows. □

Thus, if f has its Fourier mass concentrated on s coefficients, then it is close in ℓ_2 to an s-sparse, $\lceil\log s\rceil$ granular real-valued function. We next show that this real-valued function must in fact be Boolean.

Theorem 6 *[Soundness Theorem]. Let $f:\mathbb{F}_2^n\rightarrow\{-1,1\}$ be $\mu\leq\frac{1}{20s^2}$ close to s-sparse in ℓ_2. Then there is an s-sparse Boolean function $F:\mathbb{F}_2^n\rightarrow\{-1,1\}$ within Hamming distance $\frac{\mu^2}{2}$ from f.*

Proof. Let $B=\{\alpha_1,\cdots,\alpha_s\}$ be the s largest Fourier coefficients of f and let $k=\lceil\log s\rceil$. By Theorem 1, each $\hat{f}(\alpha_i)$ is $\frac{\mu}{\sqrt{s}}$ close to k-granular. So we can write

$$\hat{f}(\alpha_i)=\hat{F}(\alpha_i)+\hat{G}(\alpha_i)$$

where $\hat{F}(\alpha_i)$ is k-granular and $|\hat{G}(\alpha_i)|\leq\frac{\mu}{\sqrt{s}}$. Set $\hat{F}(\beta)=0$ and $\hat{G}(\beta)=\hat{f}(\beta)$ for $\beta\in S=\bar{B}$. Thus we have $f(x)=F(x)+G(x)$, further F is s-sparse and k-granular, while

$$\mathbf{E}[G(x)^2]\leq s\frac{\mu^2}{s}+\mu^2\leq 2\mu^2.$$

It suffices to show that F's range is $\{-1,1\}$. In this case, G's range must be $\{-2,0,2\}$, the value $G(x)^2$ is exactly 4 whenever f and F differ, and therefore f and F satisfy

$$\mathbf{Pr}_x[f(x) \neq F(x)] = \mathbf{Pr}[|G(x)| = 2] = \frac{1}{4}\mathbf{E}_x[G(x)^2] \leq \frac{\mu^2}{2}.$$

As f is a Boolean function on $\mathbb{F}_2^n$ we have

$$1 = f^2 = F^2 + 2FG + G^2 = F^2 + G(2f - G). \tag{6}$$

Writing $H = G(2f - G)$, from Fact 7 below we have that for all α,

$$|\widehat{H}(\alpha)| \leq \|G\|_2\|2f - G\|_2 \leq \|G\|_2(\|2f\|_2 + \|G\|_2) \leq 2\sqrt{2}\mu + 2\mu^2 < 4\mu \leq \frac{1}{5s^2}.$$

On the other hand, since F has granularity k it is easy to see that F^2 has granularity $2k$; in particular, $|\widehat{F^2}(\alpha)|$ is either an integer or at least $2^{-2k} \geq \frac{1}{4s^2}$-far from being an integer. But for (6) to hold as a functional identity, we must have $\widehat{F^2}(0) + \widehat{H}(0) = 1$ and $\widehat{F^2}(\alpha) + \widehat{H}(\alpha) = 0$ for all $\alpha \neq 0$. It follows then that we must have $\widehat{F^2}(0) = 1$ and $\widehat{F^2}(\alpha) = 0$ for all $\alpha \neq 0$; i.e., $F^2 = 1$ and hence F has range $\{-1,1\}$, as claimed. □

Fact 7. *Let $f, g : \mathbb{F}_2^n \to \mathbb{R}$. Then $|\widehat{fg}(\alpha)| \leq \|f\|_2\|g\|_2$ for every α.*

Proof. Using Cauchy-Schwarz and Parseval,

$$|\widehat{fg}(\alpha)| = |\sum_{\beta} \widehat{f}(\beta)\widehat{g}(\alpha+\beta)| \leq \sqrt{\sum_{\beta} \widehat{f}(\beta)^2}\sqrt{\sum_{\beta} \widehat{g}(\alpha+\beta)^2} = \|f\|_2\|g\|_2.$$ □

4 Testing s-Sparsity

The following is our algorithm for testing whether $f : \mathbb{F}_2^n \to \{-1,1\}$ is s-sparse:

Algorithm 1. Testing s-sparsity

Inputs: s, ϵ

Parameters: $\mu = \min(\sqrt{2\epsilon}, \frac{1}{20s^2})$, $t = \lceil 2\log s + \log 100 \rceil$, $\tau = \frac{\mu^2}{100 \cdot 2^t}$.

1. Choose a random permuted t-dimensional coset structure $(H, \mathcal{C})$.
2. For each bucket $C \in \mathcal{C}$, estimate $\mathrm{wt}(C) = \sum_{\alpha \in C} \hat{f}(\alpha)^2$ to accuracy $\pm\tau$ with confidence $1 - (1/100)2^{-t}$, using Proposition 2.
3. Let $\mathcal{L}$ be the set of buckets where the estimate is at least 2τ. If $|\mathcal{L}| \geq s+1$, reject.

Roughly speaking, Step 1 pairwise independently hashes the Fourier coefficients of f into $\Theta(s^2)$ buckets. If f is s-sparse then at most s buckets have nonzero weight and the test accepts. On the other hand, if f passes the test with high probability then we show that almost all the Fourier mass of f is concentrated on at most s nonzero coefficients (one for each bucket in $\mathcal{L}$). Theorem 6 now shows that f is close to a sparse function. Our theorem about the test is the following:

Theorem 8. *Algorithm 1 ϵ-tests whether $f : \mathbb{F}_2^n \to \{-1,1\}$ is s-sparse (with confidence $3/4$), making $O\left(\frac{s^6 \log s}{\epsilon^2} + s^{14} \log s\right)$ nonadaptive queries.*

The query complexity of Theorem 8 follows immediately from Proposition 2 and the fact that there are $2^t = O(s^2)$ buckets. In the remainder of this section we present the completeness (Lemma 1) and the soundness (Lemma 4) of the test. We begin with the completeness, which is straightforward.

Lemma 1. *If f is s-sparse then the test accepts with probability at least* 0.9.

Proof. Write $f = \sum_{i=1}^{s'} \hat{f}(\alpha_i)\chi_{\alpha_i}$, where each $\hat{f}(\alpha_i) \neq 0$ and $s' \leq s$. Since there are 2^t buckets, all of the estimates in Step 2 are indeed τ-accurate, except with probability at most $1/100$. If the estimates are indeed accurate, the only buckets with weight at least τ are those that contain a nonzero Fourier coefficient, which are at most s in number. So f passes the test with probability at least 0.9. □

We now analyze the soundness. We partition the Fourier coefficients of f into two sets: B of big coefficients and S of small coefficients. Formally, let

$$B \stackrel{\text{def}}{=} \{\alpha : \hat{f}(\alpha)^2 \geq 3\tau\}, \qquad S \stackrel{\text{def}}{=} \{\alpha : \hat{f}(\alpha)^2 < 3\tau\}.$$

We observe that if there are too many big coefficients the test will probably reject:

Lemma 2. *If $|B| \geq s+1$ then the test rejects with probability at least* $3/4$.

Proof. Proposition 5(3) implies that after Step 1, except with probability at most $1/100$ there are at least $s+1$ buckets C containing an element of B. In Step 2, except with probability at most $1/100$, we get an estimate of at least $3\tau - \tau \geq 2\tau$ for each such bucket. Then $|\mathcal{L}|$ will be at least $s+1$ in Step 3. Hence the overall rejection probability is at least $1 - 2/100$. □

Next we show that if the weight on small coefficients, $\mathrm{wt}(S) = \sum_{\alpha \in S} \hat{f}(\alpha)^2$, is too large then the test will probably reject:

Lemma 3. *If $\mathrm{wt}(S) \geq \mu^2$ then the test rejects with probability at least* $3/4$.

Proof. Suppose that indeed $\mathrm{wt}(S) \geq \mu^2$. Fix a bucket index b and define the random variable $M_b := \mathrm{wt}(C(b) \cap S) = \sum_{\alpha \in C(b) \cap S} \hat{f}(\alpha)^2 = \sum_{\alpha \in S} \hat{f}(\alpha)^2 \cdot I_{\alpha \to b}$. Here the randomness is from the choice of $(H, \mathcal{C})$, and we have used the pairwise independent indicator random variables defined in Proposition 5(3). Let us say that the bucket $C(b)$ is *good* if $M_b \geq \frac{1}{2}\mathbf{E}[M_b]$. We have $\mathbf{E}[M_b] = 2^{-t}\,\mathrm{wt}(S) \geq 100\tau > 0$, and by Proposition 4 we deduce $\mathbf{Pr}[M_b \leq \frac{1}{2}\mathbf{E}[M_b]] \leq \frac{3\tau}{(1/2)^2\mathbf{E}[M_b]} \leq 3/25$. Thus the expected fraction of bad buckets is at most $3/25$, so by Markov's inequality there are at most $(3/5)2^t$ bad buckets except with probability at most $1/5$. But if there are at least $(2/5)2^t$ good buckets, we have at least $(2/5)(100s^2) \geq s+1$ buckets b with $\mathrm{wt}(C(b) \cap S) \geq \frac{1}{2}\mathbf{E}[M_b] \geq 50\tau$. Assuming all estimates in Step 2 of the test are accurate to within $\pm\tau$ (which fails with probability at most $1/100$), Step 3 of the test will reject. Thus we reject except with probability at most $1/5 + 1/100 < 1/4$. □

Now we put together the pieces to establish soundness of the test:

Lemma 4. *Suppose the test accepts f with probability exceeding $1/4$. Then f is ϵ-close to an s-sparse Boolean function.*

Proof. Assuming the test accepts f with probability exceeding $1/4$, by Lemma 2 we have $|B| \leq s$, by Lemma 3 we have $\mathrm{wt}(S) \leq \mu^2$. Thus f is $\mu \leq \frac{1}{20s^2}$ close in ℓ_2 to being s-sparse. We now apply the soundness theorem, Theorem 6 to conclude that f must be $\frac{\mu^2}{2} \leq \epsilon$-close in Hamming distance to an s-sparse Boolean function. □

References

1. Alon, N., Fischer, E., Newman, I., Shapira, A.: A combinatorial characterization of the testable graph properties: It's all about regularity. In: Proc. STOC (2006)
2. Alon, N., Kaufman, T., Krivelevich, M., Litsyn, S., Ron, D.: Testing low-degree polynomials over GF(2). In: Arora, S., Jansen, K., Rolim, J.D.P., Sahai, A. (eds.) RANDOM 2003. LNCS, vol. 2764, pp. 188–199. Springer, Heidelberg (2003)
3. Alon, N., Shapira, A.: A characterization of the (natural) graph properties testable with one-sided error. In: Proc. FOCS, pp. 429–438 (2005)
4. Alon, N., Shapira, A.: Every monotone graph property is testable. In: Proc. STOC, pp. 128–137 (2005)
5. Austin, T., Tao, T.: On the testability and repair of hereditary hypergraph properties. Random Structures and Algorithms (submitted, 2008)
6. Bellare, M., Coppersmith, D., Hastad, J., Kiwi, M., Sudan, M.: Linearity testing in characteristic two. IEEE Trans. on Information Theory 42(6), 1781–1795 (1996)
7. Bellare, M., Goldreich, O., Sudan, M.: Free bits, pcps and non-approximability-towards tight results. SIAM J. Comput. 27(3), 804–915 (1998)
8. Bernasconi, A., Codenotti, B.: Spectral analysis of boolean functions as a graph eigenvalue problem. IEEE Trans. Computers 48(3), 345–351 (1999)
9. Blais, E.: Improved bounds for testing juntas. In: Goel, A., Jansen, K., Rolim, J.D.P., Rubinfeld, R. (eds.) APPROX and RANDOM 2008. LNCS, vol. 5171, pp. 317–330. Springer, Heidelberg (2008)
10. Blum, M., Luby, M., Rubinfeld, R.: Self-testing/correcting with applications to numerical problems. J. Comp. Sys. Sci. 47, 549–595 (1993); Earlier version in STOC 1990
11. Diakonikolas, I., Lee, H., Matulef, K., Onak, K., Rubinfeld, R., Servedio, R., Wan, A.: Testing for concise representations. In: Proc. FOCS, pp. 549–558 (2007)
12. Feldman, V., Gopalan, P., Khot, S., Ponnuswami, A.: New results for learning noisy parities and halfspaces. In: Proc. FOCS, pp. 563–576 (2006)
13. Fischer, E.: The art of uninformed decisions: A primer to property testing. Bulletin of the European Association for Theoretical Computer Science 75, 97–126 (2001)
14. Fischer, E., Kindler, G., Ron, D., Safra, S., Samorodnitsky, A.: Testing juntas. J. Computer & System Sciences 68(4), 753–787 (2004)

15. Jackson, J.: An efficient membership-query algorithm for learning DNF with respect to the uniform distribution. Journal of Computer and System Sciences 55, 414–440 (1997)
16. Kaufman, T., Sudan, M.: Sparse random linear codes are locally decodable and testable. In: Proc. FOCS, pp. 590–600 (2007)
17. Kaufman, T., Sudan, M.: Algebraic property testing: the role of invariance. In: Proc. 40th Annual ACM Symposium on Theory of Computing (STOC), pp. 403–412 (2008)
18. Kushilevitz, E., Mansour, Y.: Learning decision trees using the fourier spectrum. SIAM Journal on Computing 22(6), 1331–1348 (1993)
19. Linial, N., Mansour, Y., Nisan, N.: Constant depth circuits, Fourier transform and learnability. Journal of the ACM 40(3), 607–620 (1993)
20. Matulef, K., O'Donnell, R., Rubinfeld, R., Servedio, R.: Testing halfspaces. In: Proc. SODA, pp. 256–264 (2009)
21. Parnas, M., Ron, D., Samorodnitsky, A.: Testing basic boolean formulae. SIAM J. Disc. Math. 16, 20–46 (2002)
22. Samorodnitsky, A.: Low-degree tests at large distances. In: Proc. 39th ACM Symposium on the Theory of Computing (STOC 2007), pp. 506–515 (2007)

Revisiting the Direct Sum Theorem and Space Lower Bounds in Random Order Streams

Sudipto Guha and Zhiyi Huang*

University of Pennsylvania, Philadelphia PA 19104, USA
{sudipto,hzhiyi}@cis.upenn.edu

Abstract. Estimating frequency moments and L_p distances are well studied problems in the adversarial data stream model and tight space bounds are known for these two problems. There has been growing interest in revisiting these problems in the framework of random-order streams. The best space lower bound known for computing the kth frequency moment in random-order streams is $\Omega(n^{1-2.5/k})$ by Andoni et al., and it is conjectured that the real lower bound shall be $\Omega(n^{1-2/k})$. In this paper, we resolve this conjecture. In our approach, we revisit the direct sum theorem developed by Bar-Yossef et al. in a random-partition private messages model and provide a tight $\Omega(n^{1-2/k}/\ell)$ space lower bound for any ℓ-pass algorithm that approximates the frequency moment in random-order stream model to a constant factor. Finally, we also introduce the notion of space-entropy tradeoffs in random order streams, as a means of studying intermediate models between adversarial and fully random order streams. We show an almost tight space-entropy tradeoff for L_∞ distance and a non-trivial tradeoff for L_p distances.

1 Introduction

The data stream model is a very useful computational model for designing efficient algorithms for massive data sets. In the data stream model, the algorithm can only access the data in a given order and for a limited number of times (passes). Designing sub-linear space algorithms and proving space lower bound for numerous problems have received a lot of attention.

The problem of estimating the Frequency Moments is one of the most studied problems in data stream model. Given an alphabet $\Sigma = \{\sigma_1, \sigma_2, \cdots, \sigma_m\}$ of size m and a sequence of n numbers $x_1, x_2, \cdots, x_n$ in Σ, y_i is the number of occurrence of σ_i in the sequence, and the kth frequency moment f_k is defined as $f_k = \sum_{i=1}^{m} y_i^k$.

Usually, it is assumed that the order is given by an adversary and the model is known as adversarially ordered streaming. In this model, there are approximation algorithms for computing the kth frequency moment using only $\tilde{O}(n^{1-2/k})$ space [4,13]. Alon et al. [1] proved the first lower bound of $\Omega(n^{1-5/k})$ for the space required to estimate the kth frequency moment to a constant factor. Bar-Yossef et al. [3] gave an improved lower bound of $\Omega(n^{1-3/k})$ via their direct

* This research was supported in part by NSF award CCF-0644119.

S. Albers et al. (Eds.): ICALP 2009, Part I, LNCS 5555, pp. 513–524, 2009.

sum theorem. And Chakrabarti et al. [7] showed that any single-pass algorithm required $\Omega(n^{1-2/k})$ space in order to approximate the kth frequency moment, while for algorithms with a constant number of passes require $\Omega(n^{1-2/k}/\log n)$ space. Very recently, Gronemeier [9] improved the lower bound for constant-pass algorithms to $\Omega(n^{1-2/k})$.

A related and almost equally well studied problem in the data stream model is the approximation of L_∞ and L_p distances. Given $x = (x_1, x_2, \cdots, x_n) \in [0, \ell]^n$ and $y = (y_1, y_2, \cdots, y_n) \in [0, \ell]^n$, the L_p distance between x and y is defined as $L_p(x, y) = (\sum_{i=1}^{n} (x_i - y_i)^p)^{1/p}$. The L_∞ distance between x and y is $\max_i |x_i - y_i|$. Saks and Sun [15] proved that any two-party one-way protocol that distinguishes $L_\infty(x, y) = 1$ from $L_\infty(x, y) = \ell$ with probability at least 2/3 uses at least $\Omega(n/\ell^2)$ communication. Later, Bar-Yossef et al. [3] use their direct sum theorem to prove the same space lower bound for general two-party protocols. Matching protocols for this problem are also known. Using a reduction from L_∞ to L_p proposed by Saks and Sun, a space lower bound of $\Omega(n^{1-2/p}/\ell^2)$ holds for L_p, $p > 2$.

In many scenarios, however, an adversarially ordered data stream is not the best model, and recently, random-order data streams has received a lot of attention [12,5,6]. The work which is closest to this paper, by Chakrabarti et al. [5] show that the space complexity of estimating the kth frequency moment is $\Omega(n^{1-3/k})$ and $\Omega(n^{1-3/k}/\log n)$ for single-pass and constant-passes algorithms respectively for the random order stream model. Andoni et al. [2] improve these lower bounds to $\Omega(n^{1-2.5/k}/\log n)$ and conjecture that the lower bound for adversarially ordered streams holds for random-order streams.

Communication complexity [14,16] plays a central role in proofs of most results on space lower bound results. There are two models of communication complexity which are useful in this context. The blackboard model refers to the communication games in which players can broadcast their message to all other players. In the private messages model, only one-to-one communication is allowed. In the literature to date, most lower bound results are based on reductions from various communication complexity problems in the blackboard model. And a key technique is the direct sum theorem developed by Bar-Yossef et al. [3]. In contrast, the private messages model has received less attention so far. The private messages model is more restrictive than the broadcast model, may lead to better space lower bounds; and further, to prove lower bounds in the streaming model, the private message model is more relevant (in fact the order in which he players speak is also preordained). To the best of our knowledge, the only effort on proving space lower bound from communication complexity in private messages model is the work on the longest increasing sequence problem by Gal and Gopalan [8]. We note that direct lower bounds for streaming problems that bypass communication games as in [11] also use ideas which are similar in spirit to the private messages model.

Our Contributions. In this paper, we revisit the notion of information cost and information complexity in the framework of private messages model. We prove that the private information cost of a decomposable function is at least as large as the sum of the private information costs of the primitive functions. Using this direct sum theorem, we prove a tight $\Omega(nm/t^2)$ lower bound for the communication complexity of random-partition multiparty set disjointness. Here n is the number of different items, m is the number of players. The players try to distinguish the case that all items are distinct and the case that there are t identical items. As a corollary of this result, we show that any ℓ-pass algorithm which gives constant factor approximation of the kth frequency moment in random-order stream model requires $\Omega(n^{1-2/k}/\ell)$ space. This result resolves the conjecture by Andoni et al. [2]. It also provides an alternate approach for space lower bound for constant-pass algorithms in adversarially ordered streams.

We then study protocols for L_∞ and the tradeoff of the entropy of the input order and the communication complexity used by the protocol. We show that if the protocol can distinguish $L_\infty = \ell$ and $L_\infty \leq 1$, and $2n \log n - E = \alpha n \log n$, then the $2n$-party communication complexity is at least $\Omega(n^{2-\alpha(1+\epsilon)}/\ell^2)$ for any constant $\epsilon > 0$. As a corollary, we have $\Omega(n^{1-\alpha(1+\epsilon)}/\ell^2)$ and $\Omega(n^{1-2/p-\alpha(1+\epsilon)}/\ell^2)$ space lower bounds for data stream algorithms which approximates L_∞ and L_p for $p > 2$ respectively. We also prove this tradeoff is essentially tight for L_∞ and give algorithm matching the lower bound.

2 Preliminaries

2.1 Definitions and Notations

Definition 1. *Suppose Σ is a finite set. A function $f : \Sigma^T \mapsto \{0,1\}$ is defined to be* decomposable *if there exists t, n and functions $g : \{0,1\}^n \mapsto \{0,1\}$ and $h : \Sigma^t \mapsto \{0,1\}$ such that $T = tn$ and the function f is of the form $f(x_1, x_2, \cdots, x_T) = g\left(h(x_1, x_2, \cdots, x_t), \cdots, h(x_{(n-1)t+1}, \cdots, x_T)\right)$. We call function h the* primitive function.

We shall consider the following two special cases of decomposable functions in this paper. If h is the $\textsc{And}_t$ function with t input bits, and g is function $\textsc{Or}_n$ with n input bits, the decomposable function f is denoted as the SET DISJOINTNESS function:

$$\textsc{SetDisj}_{n,t} = \textsc{Or}_n\left(\textsc{And}_t(x_1), \cdots, \textsc{And}_t(x_n)\right) .$$

If h is the bivariate gap function $\textsc{BiGap}_\ell$ such that $\textsc{BiGap}_\ell(x,y) = 1$ when $|x-y| = \ell$ and $\textsc{BiGap}_\ell(x,y) = 0$ when $|x-y| = 0, 1$, and g is function $\textsc{Or}$ with n input bits, then the decomposable function f is denoted as the GAP DISTANCE function:

$$\textsc{GapDist}_{n,\ell} = \textsc{Or}_n\left(\textsc{BiGap}_\ell(x_1, x_2), \cdots, \textsc{BiGap}_\ell(x_{2n-1}, x_{2n})\right) .$$

We use capital letters X, Y, and Z to denote random variables. We use boldface letters $\boldsymbol{X}$ and $\boldsymbol{Y}$ to denote vectors. Moreover, we shall let $\boldsymbol{X}_1, \boldsymbol{X}_2, \cdots, \boldsymbol{X}_n$ denote the input vectors of primitive functions and let $\boldsymbol{X} = \boldsymbol{X}_1 \times \boldsymbol{X}_2 \times \cdots \times \boldsymbol{X}_n$ denote the input vector of the decomposable function f. We let ν denote the input distribution of the primitive function and let $\boldsymbol{\mu}$ denote the input distribution of the decomposable function. Usually we shall have $\boldsymbol{\mu} = \nu^n$.

We use $[d]$ to denote the set $\{1, 2, \cdots, d\}$. We say a distribution $\boldsymbol{\mu}$ is *symmetric* if and only if for any permutation π of $[T]$, $\boldsymbol{X} \sim \pi(\boldsymbol{X}) \sim \boldsymbol{\mu}$.

Definition 2. *A distribution $\boldsymbol{\mu}$ is defined as a* collapsing distribution *if for any input $\boldsymbol{x}$ drawn from the distribution $\boldsymbol{\mu}$ and any $\boldsymbol{X}_i \in \Sigma^t$, we always have that $f(\boldsymbol{x}_1, \cdots, \boldsymbol{x}_{i-1}, \boldsymbol{X}_i, \boldsymbol{x}_{i+1}, \cdots, \boldsymbol{x}_n) = h(\boldsymbol{X}_i)$.*

We shall use η to denote the distribution of random variable Y_i and let $\boldsymbol{\zeta}$ to denote the distribution of random vector $\boldsymbol{Y}$. We shall have $\boldsymbol{\zeta} = \eta^n$. We will consider the distribution of random vector $\boldsymbol{X}$ conditioned on $\boldsymbol{Y}$.

Definition 3. *$\boldsymbol{Y}$ is defined to* partition *$\boldsymbol{X}$ if the distribution of $\boldsymbol{X}$ given $\boldsymbol{Y}$ is a product distribution.*

2.2 Communication Games and Various Models

We let $\mathcal{P}$ denote a communication protocol. We shall always use δ to denote the error rate of a protocol. Let Γ denote the set of all protocols and let Γ_δ denote the set of all protocols whose error rate is at most δ. Similarly we shall use Φ and Φ_δ to denote the set of all deterministic protocols and the set of all deterministic protocols with error rate at most δ.

The term ϵ be denote the relevant approximation parameter (we shall consider either $(1+\epsilon)$-approximation or n^ϵ-approximation depending on the problem we study). We use ρ to denote other small values.

Private Messages Model: We shall focus on the communication complexity of various (decomposable) functions in private messages model (with public coins) in this paper. A multiparty communication game in private messages model with m players is as follows. In step 1, the first player sends a message M_1^1 to the second player based merely on her own input. In general, in step $im+j$ such that $i \geq 0$ and $1 \leq j \leq m$ the jth player sends a message M_{i+1}^j to the $(j+1)^{th}$ player based on her own input and all messages she received from the $(j-1)^{th}$ player. Note that in private messages model, each message is known only by the sender and recipient. This is a major difference from the blackboard model. We shall use $\mathsf{CC}_\delta^P(f)$ to denote the multiparty communication complexity of computing a decomposable function f in private messages model with error rate at most δ.

The transcript of the ℓth player is the union of all messages sent by player ℓ and is denoted by $\Pi_\ell(\boldsymbol{X})$. The transcript $\Pi(\boldsymbol{X})$ is the union of $\Pi_\ell(\boldsymbol{X})$ for $1 \leq \ell \leq m$. We sometimes abbreviate these notations with Π_ℓ and Π. The communication complexity $\mathsf{CC}_\delta^P = \min_{\mathcal{P}\in\Gamma_\delta} \max_{\boldsymbol{x}\in\{0,1\}^T} |\Pi(\boldsymbol{x})|$.

Random Partitioned Communication Games: An allocation is a function $\sigma : [T] \mapsto [m]$. Let $[m]^T$ denote the set of all allocations. In a random partitioned communication game with respect to function f and a distribution Σ on $[m]^T$, an allocation σ is drawn from distribution Σ, and each input bit x_i is given to the $\sigma(i)^{th}$ player. The players then play a communication game in private messages model to compute the function value of f for the given input. Let $\mathcal{U}_{T,m}$ denote the uniformly random distribution over $[m]^T$. The special case when $\Sigma = \mathcal{U}_{T,m}$ is of particular interest in proving robust communication complexity and space lower bounds for various functions.

We shall use $\Gamma_{\delta,\Sigma}$ to denote the set of protocols whose error rate is at most δ in the random partitioned communication game with respect to function f and distribution Σ. And the communication complexity in a random partitioned communication game is $\mathsf{CC}^P_{\delta,\Sigma} = \min_{\mathcal{P}\in\Gamma_{\delta,\Sigma}} \max_{x\in\{0,1\}^T} |\Pi|$.

3 Revisiting the Direct Sum Theorem

Now we revisit the definition of information cost and information complexity in the literature of private messages model. A major difference between private messages model and blackboard model is that a player may need to forward information in the messages she received to the other players, while in blackboard model that information is already known by every player.

Therefore, any optimal protocol in blackboard model shall satisfies that $I(\Pi_i;\Pi_j) = 0$ for any $1 \le i \ne j \le m$. Thus we shall have that $I(\boldsymbol{X};\Pi) = \sum_{i=1}^m I(\boldsymbol{X};\Pi_i)$. However, similar statement is not true in private messages model. Based on this observation, we consider the following definition of information cost and information complexity in private messages model.

Definition 4. *Suppose $\mathcal{P}$ is a communication protocol and Π is its transcript. The* information cost *of $\mathcal{P}$ with respect to the input distribution $\boldsymbol{X} \sim \boldsymbol{\mu}$ is* $\mathrm{ICost}_{\boldsymbol{\mu}}(\boldsymbol{X};\Pi) = \sum_{i=1}^m I_{\boldsymbol{\mu}}(\boldsymbol{X};\Pi_i)$. *The δ-error* information complexity *with respect to function f and input distribution $\boldsymbol{X} \sim \boldsymbol{\mu}$ is the minimal information cost among all δ-error protocols, that is,* $\mathrm{IC}_{\boldsymbol{\mu},\delta}(f) = \min_{\mathcal{P}\in\Gamma_\delta} \mathrm{ICost}_{\boldsymbol{\mu}}(\boldsymbol{X};\Pi)$.

Similar to the results in blackboard model, we sometimes need to consider the conditional information cost and conditional information complexity, which are defined as follows.

Definition 5. *The* conditional information cost *of a protocol $\mathcal{P}$ with respect to distribution $\boldsymbol{X} \sim \boldsymbol{\mu}$ and $\boldsymbol{Y} \sim \boldsymbol{\zeta}$ is* $\mathrm{ICost}_{\boldsymbol{\mu},\boldsymbol{\zeta}}(\boldsymbol{X};\Pi|\boldsymbol{Y}) = \sum_{i=1}^m I_{\boldsymbol{\mu},\boldsymbol{\zeta}}(\boldsymbol{X};\Pi_i|\boldsymbol{Y})$. *The δ-error* conditional information complexity *with respect to function f and distribution $\boldsymbol{X} \sim \boldsymbol{\mu}$ and $\boldsymbol{Y} \sim \boldsymbol{\zeta}$ is* $\mathrm{IC}_{\boldsymbol{\mu},\boldsymbol{\zeta},\delta}(f|\boldsymbol{Y}) = \min_{\mathcal{P}\in\Gamma_\delta} \mathrm{ICost}_{\boldsymbol{\mu}}(\boldsymbol{X};\Pi|\boldsymbol{Y})$.

Given the modified definition of information cost and information complexity, we now rephrase the direct sum theorem in the context of private messages model as follows.

Theorem 1 (Direct Sum Theorem). *Recall that $f : \{0,1\}^T \mapsto \{0,1\}$ is a decomposable function with primitive function $h : \{0,1\}^t \mapsto \{0,1\}$. Suppose the input distribution $\boldsymbol{X} \sim \boldsymbol{\mu} = \nu^n$ is a collapsing distribution and random variable $\boldsymbol{Y} \sim \boldsymbol{\zeta} = \eta^n$ partitions $\boldsymbol{X}$. Consider a random partitioned communication game with respect to function f and distribution Σ, then* $\mathrm{IC}_{\boldsymbol{\mu},\boldsymbol{\zeta},\delta,\Sigma}(f|\boldsymbol{Y}) \ge \sum_{i=1}^n \mathrm{IC}_{\nu,\eta,\delta,\Sigma}(h|Y_i)$.

The proof of Theorem 1 is an analogue of the proof by Bar-Yossef et. al. [3], and can be found in the full version [10].

4 Near Optimal Lower Bound for Frequency Moments

In this section, we will prove the following asymptotically optimal space lower bound for computing the kth frequency moments for $k > 2$.

Theorem 2. *Suppose ϵ and δ are small constants. If an algorithm correctly gives a $(1+\epsilon)$-approximation of the kth frequency moment of n numbers with probability at least $1-\delta$ in a random order stream within ℓ passes, then the space it needs is at least $\Omega\left(n^{1-2/k}/\ell\right)$.*

We consider the decomposable function $\text{SETDISJ}_{n,t}$. The intuition is the following. Suppose $m = td$ and we shall assume that m is large enough such that if the allocation $\sigma \sim \mathcal{U}_{t,m}$, then with probability $1 - o(1)$ we have $\sigma(i) \neq \sigma(j)$ for all $1 \leq i \neq j \leq t$. Consider a collapsing and symmetric distribution $\boldsymbol{X} \sim \nu$ partitioned by $Y \sim \eta$, where η is a uniform distribution over $[t]$ and conditioned on $Y_i = j$ we have $\boldsymbol{X}_i = e^j$ with probability $1/2$ and $\boldsymbol{X}_i = 0$ with probability $1/2$. From Theorem 1, it suffices to prove lower bound for the primitive function. Recall that the information complexity for AND_t is at least $\text{IC}^B = \Omega(1/t)$ in a blackboard fixed-partition t-player communication game with respect to this input distribution [7,9]. Conditioned on a particular allocation σ, suppose the indexes of the players who get the t bits of the input $\boldsymbol{X}_i$ are $i_1 < i_2 < \cdots < i_t$. We can imagine that these t players play a communication game to compute the function value of AND_t and only the messages these t players receive contribute to the information cost. So the effective information cost is

$$I(\boldsymbol{X}_i; \Pi_{i_1-1}|Y_i) + I(\boldsymbol{X}_i; \Pi_{i_2-1}|Y_i) + \cdots + I(\boldsymbol{X}_i; \Pi_{i_t-1}|Y_i) \ .$$

Now we use the simple fact that the information cost in private messages model is at least as large as the information cost in blackboard model. We get that the above information cost is at least IC^B. Note that for each $1 \leq \ell \leq t$, player $i_\ell + 1, i_\ell + 2, \cdots, i_{\ell+1} - 1$ do not have any bit of the input $\boldsymbol{X}_i$, we have

$$I(\boldsymbol{X}_i; \Pi_{i_\ell}|Y_i) \geq I(\boldsymbol{X}_i; \Pi_{i_\ell+1}|Y_i) \geq \cdots \geq I(\boldsymbol{X}_i; \Pi_{i_{\ell+1}-1}|Y_i) \ .$$

Since the expected distance between i_j and i_{j+1} is d, the next lemma is intuitive.

Lemma 1. *Suppose $\boldsymbol{X}_i \sim \boldsymbol{\nu}$ is a collapsing symmetric distribution partitioned by $Y_i \sim \eta$, then the information cost of computing the function value of AND_t with small constant error rate δ is at least $\text{IC}(\text{AND}_t|Y_i) = \Omega(d/t)$.*

Now we formally prove this key lemma. Given an allocation $\sigma : [t] \to [m]$, $m = td$, let $\sigma(\ell)$ be the image of ℓ, and $\pi(\ell)$ be the smallest $\sigma(\ell')$ such that $\ell' \in [t] \setminus \{\ell\}$ and $\sigma(\ell') \geq \sigma(\ell)$ (if $\sigma(\ell) = \max_{\ell' \in [t]} \sigma(\ell')$ then $\pi(\ell) = \min_{\ell' \in [t]} \sigma(\ell') + m$). Let p_j denote the probability that $\pi(\ell) - \sigma(\ell) = j$ when $\sigma \sim \mathcal{U}_{t,m}$. We have $p_j = (t/m)\left(1 - j/m\right)^{t-1} = \left(1 - j/m\right)^{t-1}/d$. We first prove the following lemmas.

Lemma 2. *For any $0 \leq i < j \leq m-1$,*

$$p_j(p_i + p_{i+1} + \cdots + p_{m-1}) \geq p_i(p_j + p_{j+1} + \cdots + p_{m-1}) \ .$$

Proof. Consider the function $p(x) = (1-x)^{t-1}/d$. It is easy to verify that this function is log-concave. Note that $i < j$ and $i \le i+k < j+k$ for $k \ge 0$, we get that $p_j p_{i+k} \ge p_i p_{j+k}$ and thus $p_{i+k}/p_i \ge p_{j+k}/p_j$. So

$$\frac{p_i + p_{i+1} + \cdots + p_{m-1}}{p_i} \ge \frac{p_i + p_{i+1} + \cdots + p_{i+m-j-1}}{p_i} \ge \frac{p_j + p_{j+1} + \cdots + p_{m-1}}{p_j} . \quad \square$$

Lemma 3. *If $c_1 \ge c_2 \ge \cdots \ge c_{m-1} \ge 0$, then*

$$\sum_{i=1}^{m-1}\sum_{j=i}^{m-1} p_j c_i \ge \sum_{i=1}^{m-1} i p_i \sum_{j=1}^{m-1} p_j c_j .$$

Proof. Note $\sum_{j=1}^{m-1} p_j = 1$. Now,

$$\sum_{i=1}^{m-1}\sum_{j=i}^{m-1} p_j c_i - \sum_{i=1}^{m-1} i p_i \sum_{j=1}^{m-1} p_j c_j = \sum_{j=1}^{m-1}\sum_{i=1}^{j} p_j c_i - \sum_{i=1}^{m-1} i p_i \sum_{j=1}^{m-1} p_j c_j$$

$$= \sum_{i=1}^{m-1}\sum_{j=1}^{m-1}\sum_{\ell=1}^{i} p_i p_j c_\ell - \sum_{i=1}^{m-1}\sum_{j=1}^{m-1}\sum_{\ell=1}^{i} p_i p_j c_j$$

$$= \sum_{i=1}^{m-1}\sum_{j=1}^{m-1}\sum_{\ell=1}^{i} p_i p_j (c_\ell - c_j) = \sum_{j=1}^{m-1}\sum_{\ell=1}^{m-1}\sum_{i=\ell}^{m-1} p_i p_j (c_\ell - c_j)$$

$$= \sum_{\ell<j} [p_j(p_\ell + \cdots + p_{m-1}) - p_\ell(p_j + \cdots + p_{m-1})](c_\ell - c_j) \ge 0 .$$

The last step follows from Lemma 2. $\square$

Proof (of Lemma 1). Suppose $\mathcal{P}$ is a δ-error protocol and Π is its transcript. Let $1 \le \ell \le t$. Let c_j denote the expected communication cost contributed by player $\sigma(\ell)+j-1$ if $\pi(\ell)-\sigma(\ell) \ge j$, that is, $c_j = I(\boldsymbol{X}_i; \Pi_{\sigma(\ell)+j-1} | Y_i, \pi(\ell)-\sigma(\ell) \ge j)$. Since we consider private messages model we shall have that $c_1 \ge c_2 \ge \cdots \ge c_{m-1} \ge 0$. By Lemma 3 we get that

$$\sum_{i'=1}^{m-1} c_{i'} \sum_{j=i'}^{m-1} p_j = \sum_{i'=1}^{m-1}\sum_{j=i'}^{m-1} p_j c_{i'} \ge \sum_{i'=1}^{m-1} i' p_{i'} \sum_{j=1}^{m-1} p_j c_j . \tag{1}$$

Note that $\sum_{j=i'}^{m-1} p_j$ is the probability that $\pi(\ell)-\sigma(\ell) \ge i'$. The left-hand side of Equation 1 is the communication cost contributed by players $\sigma(\ell), \sigma(\ell)+1, \cdots, \pi(\ell)-1$. The first term on the right-hand side $\sum_{i'=1}^{m-1} i' p_{i'}$ is the expected distance between $\sigma(\ell)$ and $\pi(\ell)$, which equals d. The second term on the right-hand side $\sum_{j=1}^{m-1} p_j c_j$ is the information cost contributed by player $\pi(\ell)-1$. So we have $\sum_{j=\sigma(\ell)}^{\pi(\ell)-1} I(\boldsymbol{X}_i; \Pi_j | Y_i) \ge d \cdot I(\boldsymbol{X}_i; \Pi_{\pi(\ell)-1} | Y_i)$. Recall that $\sum_{\ell=1}^{t} I(\boldsymbol{X}_i; \Pi_{\pi(\ell)-1} | Y_i) \ge \mathrm{IC}^B = \Omega(1/t)$. We have

$$\mathrm{ICost}(\boldsymbol{X}_i; \Pi | Y_i) = \sum_{\ell=1}^{t}\sum_{j=\sigma(\ell)}^{\pi(\ell)-1} I(\boldsymbol{X}_i; \Pi_j | Y_i) \ge \sum_{\ell=1}^{t} d \cdot I(\boldsymbol{X}_i; \Pi_{\pi(\ell)-1} | Y_i) = \Omega\left(\frac{d}{t}\right) .$$

Since the above result is true for any δ-error protocol, we prove Lemma 1. $\square$

Remark 1. We realize the reduction technique we introduce in Lemma 1, 2, and 3 works for other decomposable functions if we can prove information complexity lower bound for some symmetric collapsing input distribution.

Using the direct sum theorem we have the following corollaries.

Corollary 1. *If a protocol $\mathcal{P}$ correctly computes the value of* $\text{SETDISJ}_{n,t}$ *with probability at least* $1 - \delta$ *in a random partition communication game, then the total communication complexity is at least*

$$\mathsf{CC}(\text{SETDISJ}_{n,t}) \geq \sum_{i=1}^{n} \text{IC}(\text{AND}_t | Y_i) = \Omega\left(\frac{nd}{t}\right) = \Omega\left(\frac{nm}{t^2}\right) .$$

Now we can prove Theorem 2 via the a reduction as follows.

Proof (of Theorem 2). Suppose an algorithm gives $(1 + \epsilon)$-approximation of the kth frequency moment using s bits of space and within ℓ passes. Consider the following ℓ-round protocol which compute the function value of $\text{SETDISJ}_{n,t}$ when $t = (5\epsilon \cdot n)^{1/k}$. Set m to be large[1], $m = \Omega(t^2)$, which rules out collisions with constant probability. Each player shall receive some bits of the input. For each bit of value 1, that indicates some value v in one of the set. And the player take that as probing a number v in the data stream. The first player runs the algorithm on the inputs she receives, then sends the s bits of memory and another $O(\log n)$ bits that indicates the number of 1's she receives to the second player. The second player continues the algorithm on her own inputs, then sends the memory bits and the number of 1's the first two players receive to the third player. And so on and so forth.

Now assume the number of 1's in the input is n', we get that $n' < n + t < (1 + \epsilon)n$. If the function value of $\text{SETDISJ}_{n,t}$ is 1, then one of the value appears t times in the data stream. So the frequency moment is at least $(n' - t) + t^k = n' - t + 5\epsilon \cdot n \geq n' + 4\epsilon \cdot n > n'(1 + \epsilon)^2$. On the other hand, if the function value of $\text{SETDISJ}_{n,t}$ is 0, then the frequency moment is n'. Therefore, if the last player claims the function value is 1 if the the frequency moment given by the algorithm is at least $(1 + \epsilon)n'$ and claims the function value is 0 otherwise, she will be correct with probability at least $1 - \delta$.

The total communication complexity of this protocol is $O\left(\ell m(s + \log n)\right)$. Recall that this value is at least $\Omega(nm/t^2)$, we get that

$$s = \Omega\left(\frac{n}{t^2 \ell}\right) = \Omega\left(\frac{n^{1-2/k}}{\ell}\right) .$$

□

[1] Note that the private communication model allows a large number of players, say even one corresponding to each input, which is one of the reasons for getting the improved space lower bounds for streaming algorithms compared to the blackboard model.

5 Entropy–Space Tradeoff for L_∞ and L_p Distances

In this section, we consider the entropy–space tradeoff of finding an n^ϵ-approximation of the L_∞ distance.

We consider the following communication game. The two vectors correspond to $\langle x_1, x_3, \ldots, x_{2n-1}\rangle$ and $\langle x_2, x_4, \ldots, x_{2n}\rangle$ (we can use any fixed permutation). There are $2n$ players. The input allocation $\sigma : [2n] \mapsto [2n]$ is drawn from a distribution over all permutations of $[2n]$. The entire input x_i is allocated to player $\sigma(i)$. The players then communicate in the private messages model in order to compute the function value of $\text{GAPDIST}_{n,\ell}$.

We shall show the following theorems.

Theorem 3. *Let $\delta > 0$ be a small constant. Let Σ be a distribution of input order with entropy E. Any δ-error n^ϵ-approximation algorithm for L_∞ distances with respect to input order distribution Σ requires space at least*

$$\Omega\left(\frac{n^{1-4\epsilon}}{2^{(2n\log n - E)/(1-2\delta)n}}\right) .$$

Theorem 4. *Theorem 3 is tight, given E there exists an order distribution Σ' with entropy at least E, and a δ-error n^ϵ-approximation algorithm of L_∞ distance with respect to Σ', using $O\left(\frac{n^{1-4\epsilon}}{2^{(2n\log n - E)/n}}\right)$ space.*

Proof (of Theorem 3). We consider the function $\text{GAPDIST}_{n,\ell}$. Recall that the function BIGAP_ℓ is defined as: $\text{BIGAP}_\ell(x,y) = 1$ when $|x-y| = \ell$ and $\text{BIGAP}_\ell(x,y) = 0$ when $|x-y| = 0, 1$. The decomposable function $\text{GAPDIST}_{n,\ell}$ is defined as $\text{GAPDIST}_{n,\ell} = \text{OR}_n\left(\text{BIGAP}_\ell(x_1,x_2), \cdots, \text{BIGAP}_\ell(x_{2n-1},x_{2n})\right)$. If an algorithm can correctly compute the L_∞ distance of two n dimensional vectors up to a n^ϵ factor, then it shall be able to distinguish whether the L_∞ distance is at most 1 or the L_∞ distance is at least $n^{2\epsilon}$. Therefore, the space needed by such an algorithm is as large as the space needed to compute the function value of $\text{GAPDIST}_{n,n^{2\epsilon}}$ with probability at least $1-\delta$. Hence to prove a space lower bound for computing the L_∞ distances, it suffices to show strong lower bound for the communication complexity of $\text{GAPDIST}_{n,\ell}$.

We shall consider the following input distribution of $\text{GAPDIST}_{n,\ell}$. For each $1 \le i \le n$, $Y_i \sim \eta$ is randomly drawn from $[2\ell]$. Conditioned on $Y_i = 2j+1$, $0 \le j < \ell$, $X_{2i-1} = j$ and X_{2i} is uniformly distributed in $\{j, j+1\}$. Conditioned on $Y_i = 2j$, $1 \le j \le \ell$, X_{2i-1} is uniformly distributed in $\{j, j-1\}$ and $X_{2i} = j$. It is clear that $\boldsymbol{X} \sim \mu = \nu^n$ is a collapsing distribution since we always have the value of each primitive function is $\text{BIGAP}_\ell = 0$. Bar-Yossef et. al. [3] shows the following lower bound for the primitive function BIGAP_ℓ in the literature of blackboard model:

Lemma 4 (Lemma 8.2 in [3]). *Suppose $0 < \delta < 1/4$ is a constant, the two-party communication complexity of computing the function value of* BIGAP_ℓ *with probability $1-\delta$ is $\text{IC}^B = \Omega\left(1/\ell^2\right)$.*

Now we consider the information complexity lower bound for the ith primitive function BIGAP_ℓ in the private messages model. Suppose player u and player v receive the input X_{2i-1} and X_{2i}. Effectively these two players play a communcation game to compute the primitive function and Π_{u-1} and Π_{v-1} are the effective transcripts. So from Lemma 4 we get that $I(X_{2i-1}, X_{2i}; \Pi_{u-1}) + I(X_{2i-1}, X_{2i}; \Pi_{v-1}) = \Omega(1/\ell^2)$. Moreover, we shall have $I(X_{2i-1}, X_{2i}; \Pi_u) \geq I(X_{2i-1}, X_{2i}; \Pi_{u+1}) \geq \cdots \geq I(X_{2i-1}, X_{2i}; \Pi_{v-1})$ as well as $I(X_{2i-1}, X_{2i}; \Pi_v) \geq I(X_{2i-1}, X_{2i}; \Pi_{v+1}) \geq \cdots \geq I(X_{2i-1}, X_{2i}; \Pi_{u-1})$. So the information cost in private messages model is at least $\Omega(\min\{|u-v|, n-|u-v|\}/\ell^2)$. If we can prove with some constant probability the value of $\min\{|u-v|, n-|u-v|\}$ is large and the protocol correctly gets the function value of BIGAP_ℓ, then we shall have a lower bound for the primitive function.

Suppose E_i is the entropy the allocation distribution for the ith primitive function. We let d' denote the value $n/2^{(2n\log 2n - E_1)/(1-2\delta)}$ for the sake of convenience. We shall prove by contradiction that with probability at least 2δ, $\min\{|u-v|, n-|u-v|\} \geq d'$.

Suppose not. Note that the total number of different allocations for a primitive function $\sigma_i : [2] \mapsto [2n]$ is $2n(2n-1)$, and the number of different allocations such that $\min\{|u-v|, n-|u-v|\} \geq d'$ is $2n(2n-2d'+1)$.

Hence if the probability of getting an allocation $\sigma_i \sim \Sigma_i$ satisfying $\min\{|u-v|, n-|u-v|\} \geq d'$ is at most 2δ, then the entropy of distribution is

$$\begin{aligned} E_i &< 2\delta \log 2n(2n-2d'+1) + (1-2\delta)\log\left(2n(2n-1) - 2n(2n-2d'+1)\right) \\ &< 2\log 2n + (1-2\delta)\log\left(\frac{d'}{n}\right) . \end{aligned}$$

Thus we have $d > n/2^{(2\log 2n - E_i)/(1-2\delta)}$, a contradiction. Therefore, the information cost for the ith primitive function is at least

$$\Omega(d'/\ell^2) = \Omega\left(\frac{n}{\ell^2 2^{(2\log 2n - E_i)/(1-2\delta)}}\right) .$$

Note that we shall have $\sum_{i=1}^{n} E_i \geq E$ and the function 2^x is convex. Using Theorem 1 and Jensen's inequality we get that

$$\text{IC}(\text{GAPDIST}_{n,\ell}|\boldsymbol{Y}) = \sum_{i=1}^{n} \text{IC}(\text{BIGAP}_\ell|Y_i) \geq \Omega\left(\frac{n^2}{\ell^2 2^{(2n\log n - E)/(1-2\delta)/n}}\right) .$$

Therefore, to compute the function value of $\text{GAPDIST}_{n,n^{2\epsilon}}$ or to compute the L_∞ distance of two n-dimensional vectors up to a n^ϵ factor, we shall need the memory space to be

$$\Omega\left(\frac{n^{1-4\epsilon}}{2^{(2n\log n - E)/(1-2\delta)n}}\right) . \qquad \square$$

Proof (of Theorem 4). We let d denote the value $c \cdot n/2^{(2n\log n - E)/n}$ for the sake of convenience, where c is a large constant, then we shall have $\log d =$

$\log c + E/n - n\log n$. Consider the distribution of allocations σ generated by the following algorithm:

```
Pick a random permutation π of [n].
Let σ(2j − 1) = 2π(j) − 1 for 1 ≤ j ≤ n.
for all 1 ≤ i ≤ n/d do
    Pick a random permutation π_i of [d]
    Let σ(2d · i + 2j) = 2π(d · i + π_i(j)) for 1 ≤ j ≤ d.
end for
```

This allocation distribution is a uniform distribution over $n!(d!)^{n/d}$ different allocations. So the entropy is $n\log n + (n/d)\cdot d\log d + O(n) > E$ for large c. Here we use the following simple corollary of Stirling's approximation for factorials.

Lemma 5. *Suppose $n > 0$ is a positive integer, then*

$$\log(n!) = n\log n + O(n) \ .$$

For each allocation in this distribution, the first $2d$ numbers are the inputs of d dimensions, and the next $2d$ numbers are the inputs of another d dimensions, and so on and so forth. Therefore, we can divide the original problem into n/d subproblems of computing the L_∞ distance for d dimensional vectors. And the space can be reused for each subproblem. Saks and Sun [15] showed these subproblems can be resolve using only $O(d/n^{4\epsilon})$ space. So we can n^ϵ-approximate the L_∞ distance using $O(d/n^{4\epsilon}) = O(n^{1-4\epsilon}/2^{(2n\log n - E)/n})$ of space. □

Using a reduction proposed by Saks and Sun [15] we get the following entropy space tradeoff for approximating L_p distances.

Theorem 5. *Let $\delta > 0$ be a small constant and $p > 2$. Let Σ be a distribution of input order with entropy E. Any δ-error n^ϵ-approximation algorithm for L_p distances with respect to input order distribution Σ requires space*

$$\Omega\left(\frac{n^{1-2/p-4\epsilon}}{2^{(2n\log n - E)/(1-2\delta)n}}\right) \ .$$

References

1. Alon, N., Matias, Y., Szegedy, M.: The Space Complexity of Approximating the Frequency Moments. Journal of Computer and System Sciences 58(1), 137–147 (1999)
2. Andoni, A., McGregor, A., Onak, K., Panigrahy, R.: Better Bounds for Frequency Moments in Random-Order Streams. Arxiv preprint arXiv:0808.2222 (2008)
3. Bar-Yossef, Z., Jayram, T.S., Kumar, R., Sivakumar, D.: An information statistics approach to data stream and communication complexity. Journal of Computer and System Sciences 68(4), 702–732 (2004)
4. Bhuvanagiri, L., Ganguly, S., Kesh, D., Saha, C.: Simpler algorithm for estimating frequency moments of data streams. In: Proceedings of the seventeenth annual ACM-SIAM symposium on Discrete algorithm, pp. 708–713. ACM, New York (2006)

5. Chakrabarti, A., Cormode, G., McGregor, A.: Robust lower bounds for communication and stream computation. In: Proceedings of the fourtieth annual ACM symposium on Theory of computing, pp. 641–650. ACM Press, New York (2008)
6. Chakrabarti, A., Jayram, T.S., Patraşcu, M.: Tight lower bounds for selection in randomly ordered streams. In: Proceedings of the nineteenth annual ACM-SIAM Symposium on Discrete Algorithms, pp. 720–729. Society for Industrial and Applied Mathematics, Philadelphia (2008)
7. Chakrabarti, A., Khot, S., Sun, X.: Near-optimal lower bounds on the multi-party communication complexity of set disjointness. In: IEEE Conference on Computational Complexity, pp. 107–117 (2003)
8. Gal, A., Gopalan, P.: Lower Bounds on Streaming Algorithms for Approximating the Length of the Longest Increasing Subsequence. In: 48th Annual IEEE Symposium on Foundations of Computer Science, 2007. FOCS 2007, pp. 294–304 (2007)
9. Gronemeier, A.: Asymptotically optimal lower bounds on the nih-multi-party information. In: 26th International Symposium on Theoretical Aspects of Computer, p. 505 (2009)
10. Guha, S., Huang, Z.: Revisiting the direct sum theorem and space lower bounds for random order streams. Technical Report (2009), `http://repository.upenn.edu/cis_papers/`
11. Guha, S., McGregor, A.: Tight lower bounds for multi-pass stream computation via pass elimination. In: Aceto, L., Damgård, I., Goldberg, L.A., Halldórsson, M.M., Ingólfsdóttir, A., Walukiewicz, I. (eds.) ICALP 2008, Part I. LNCS, vol. 5125, pp. 760–772. Springer, Heidelberg (2008)
12. Guha, S., McGregor, A.: Stream-Order and Order-Statistics: Quantile Estimation in Random-Order Streams. SIAM Journal of Computing 38(5), 2044–2059 (2009)
13. Indyk, P., Woodruff, D.: Optimal approximations of the frequency moments of data streams. In: Proceedings of the thirty-seventh annual ACM symposium on Theory of computing, pp. 202–208. ACM, New York (2005)
14. Kushilevitz, E., Nisan, N.: Communication Complexity. Cambridge University Press, Cambridge (1996)
15. Saks, M., Sun, X.: Space lower bounds for distance approximation in the data stream model. In: Proceedings of the thiry-fourth annual ACM symposium on Theory of computing, pp. 360–369. ACM, New York (2002)
16. Yao, A.C.: Some complexity questions related to distributed computing. In: Proceedings of the 11th Annual ACM Symposium on Theory of Computing, pp. 209–213 (1979)

Wireless Communication Is in APX

Magnús M. Halldórsson[1,⋆] and Roger Wattenhofer[2]

[1] School of Computer Science, Reykjavik University, 103 Reykjavik, Iceland
mmh@ru.is

[2] Computer Engineering and Networks Laboratory, ETH Zurich, Switzerland
wattenhofer@tik.ee.ethz.ch

Abstract. In this paper we address a common question in wireless communication: How long does it take to satisfy an arbitrary set of wireless communication requests? This problem is known as the wireless scheduling problem. Our main result proves that wireless scheduling is in APX. In addition we present a robustness result, showing that constant parameter and model changes will modify the result only by a constant.

1 Introduction

Despite the omnipresence of wireless networks, surprisingly little is known about their algorithmic complexity and efficiency: Designing and tuning a wireless network is a matter of experience, regardless whether it is a Wireless LAN in an office building, a GSM phone network, or a sensor network on a volcano.

We are interested in the fundamental communication limits of wireless networks. In particular, we would like to know what communication throughput can possibly be achieved. This question essentially boils down to spatial reuse, i.e., which devices can transmit concurrently, without interfering. More precisely, formulated as an optimization problem: Given a set of communication requests, how much time does it take to schedule them?

Evidently the answer to this question depends on the wireless transmission model. In the past, algorithmic research has focused on graph-based models, also known as protocol models. Unfortunately, graph-based models are too simplistic. Consider for instance a case of three wireless transmissions, every two of which can be scheduled concurrently without a conflict. In a graph-based model one will conclude that all three transmissions may be scheduled concurrently as well, while in reality this might not be the case since wireless signals sum up. Instead, it may be that two transmissions *together* generate too much interference, hindering the third receiver from correctly receiving the signal of its sender. This *many-to-many* relationship makes understanding wireless transmissions difficult – a model where interference sums up seems paramount to truly comprehending wireless communication. Similarly, a graph-based model oversimplifies wireless attenuation. In graph-based models the signal is "binary", as if there was an invisible wall at which the signal immediately drops. Not surprisingly, in reality the signal decreases gracefully with distance.

⋆ Work done while visiting Research Institute for Mathematical Sciences (RIMS) at Kyoto University.

S. Albers et al. (Eds.): ICALP 2009, Part I, LNCS 5555, pp. 525–536, 2009.

In contrast to the algorithmic ("CS") community which focuses on graph-based models, researchers in information, communication, or network theory ("EE") are working with wireless models that sum up interference and respect attenuation. The standard model is the signal-to-interference-plus-noise (SINR) model – we will formally introduce it in Section 3. The SINR model is reflecting the physical reality more precisely, it is therefore often simply called the physical model. On the other hand, "EE researchers" are not really looking for algorithmic results. Instead, they usually propose heuristics that are evaluated by simulation. Analytical work is only done for special cases, e.g. the network has a grid structure, or traffic is random. However, these special cases do neither give insights into the complexity of the problem, nor do they give algorithmic results that may ultimately lead to new protocols. Since the SINR model is somewhere between graph-theory and geometry, we believe that it will be interesting for the algorithms community, as a new set of tools will be necessary.

The specific question we are addressing in this paper is a classic question in wireless communication: How long does it take to satisfy an arbitrary set of wireless communication requests? This problem is known as the wireless scheduling problem. It is at the heart of wireless communication. Our solution is hopefully pleasing to the EE community as it is using their models, and it is hopefully pleasing to the CS community because we make no restrictions on the input. Our main result proves that wireless scheduling is in APX.

2 Related Work

Most work in wireless scheduling in the physical (SINR) model is of heuristic nature, e.g. [3,7]. Only after the work of Gupta and Kumar [14], analytical results became *en vogue*. The analytical results however are restricted to networks with a well-behaving topology and traffic pattern. On the one hand this restriction keeps the math involved tractable, on the other hand, it allows for presenting the results in a concise form, i.e., "the throughput capacity of a wireless network with X and Y is Z", where X and Y are some parameters defining the network, and Z is a function of the network size. This area of research has been exceptionally popular, with a multi-dimensional parameter space (e.g. node distribution, traffic pattern, transport layer, mobility), and consequently literally thousands of publications. The intrinsic problem with this line of research is that real networks often do not resemble the models studied here, so one cannot learn much about the capacity of a *real* network. Moreover, one cannot devise protocols since the results are not algorithmic.

In contrast there is a body of algorithmic work, however, mostly on graph-based models. Studying wireless communication in graph-based models commonly implies studying some variants of independent set, matching, or coloring [18,26]. Although these algorithms present extensive theoretical analysis, they are constrained to the limitations of a model that ultimately abstracts away the nature of wireless communication. The inefficiency of graph-based protocols in the SINR model is well documented and has been shown theoretically as well as experimentally [13,19,23].

Algorithmic work in the SINR model is fairly new; to the best of our knowledge it was started just three years ago [22]. In this paper Moscibroda et al. present an algorithm that successfully schedules a set of links (carefully chosen to strongly connect an arbitrary set of nodes) in polylogarithmic time, even in arbitrary worst-case networks. In contrast to our work the links themselves are *not* arbitrary (but do have structure that will simplify the problem). This work has been extended and applied to topology control [8,24], sensor networks [20], combined scheduling and routing [5], or ultra-wideband [15], or analog network coding [12]. Recently a moderately exponential-time algorithm has been proposed [16]. Apart from these papers, algorithmic SINR results also started popping up here and there, for instance in a game theoretic context or a distributed algorithms context, e.g., [1,2,4,10,17,25]

So far there are only a few papers that tackle the general problem of scheduling arbitrary wireless links. Goussevskaia et al. give a simple proof that the problem is NP-complete [11], and [21] test popular heuristics. Both papers also present approximation algorithms, however, in both cases the approximation ratio may grow linearly with the network size.

The work most relevant for this paper is by Goussevskaia et al. [9]. Among other things, [9] presents the first wireless scheduling algorithm with approximation guarantee independent of the topology of the network. The paper accomplishes a constant approximation guarantee for the problem of maximizing the number of links scheduled in one single time-slot. Furthermore, by applying that single-slot subroutine repeatedly the paper realizes a $O(\log n)$ approximation for the problem of minimizing the number of time slots needed to schedule a given set of arbitrary requests.

Our present paper removes the logarithmic approximation overhead of [9]. Hence the problem of wireless scheduling is in APX. Moreover, our algorithm is simpler than [9], and will be easier to build on. In addition we are able to present a quite general robustness result, showing that constant parameter and model changes will modify the result only by a constant.

3 Notation and Model

Given is a set of links $\ell_1, \ell_2, \ldots, \ell_n$, where each link ℓ_v represents a communication request from a sender s_v to a receiver r_v. We assume the senders and receivers are points in the Euclidean plane; this can be extended to other metrics. The Euclidean distance between two points p and q is denoted $d(p,q)$. The asymmetric distance from link v to link w is the distance from v's sender to w's receiver, denoted $d_{vw} = d(s_v, r_w)$. The length of link ℓ_v is denoted $d_{vv} = d(s_v, r_v)$. We shall assume for simplicity of exposition that all links are of different length; this does not affect the results. We assume that each link has a unit-traffic demand, and model the case of non-unit traffic demands by replicating the links. We also assume that all nodes transmit with the same power level P. We show later how to extend the results to variable power levels, with a slight increase in the performance ratio.

We assume the *path loss radio propagation* model for the reception of signals, where the received signal from w at receiver v is $P_{wv} = P/d_{wv}^\alpha$ and $\alpha > 2$ denotes the path-loss exponent. When $w \neq v$, we write $I_{wv} = P_{wv}$. We adopt the *physical interference model*, in which a node r_v successfully receives a message from a sender s_v if and only if the following condition holds:

$$\frac{P_{vv}}{\sum_{\ell_w \in S \setminus \{\ell_v\}} I_{wv} + N} \geq \beta, \tag{1}$$

where N is the ambient noise, β denotes the minimum SINR (signal-to-interference-plus-noise-ratio) required for a message to be successfully received, and S is the set of concurrently scheduled links in the same channel or *slot*. We say that S is *SINR-feasible* if (1) is satisfied for each link in S.

The problems we treat are the following. In all cases are we given a set of links of arbitrary lengths. In the Scheduling problem, we want to partition the set of input links into minimum number of SINR-feasible sets, each referred to as a *slot*. In the Single-Shot Scheduling (SSS) problem, we seek the maximum cardinality subset of links that is SINR-feasible. And, in the k-Thruput problem, for a positive integer k, we seek a collection of k disjoint SINR-feasible sets with maximum combined cardinality. Let χ denote the minimum number of slots in an SINR-feasible schedule.

We make crucial use of the following new definitions.

Definition 1. *The* relative interference *(RI) of link ℓ_w on link ℓ_v is the increase caused by ℓ_w in the inverse of the SINR at ℓ_v, namely $RI_w(v) = I_{wv}/P_{vv}$. For convenience, define $RI_v(v) = 0$. Let $c_v = \frac{\beta}{1-\beta N/P_{vv}} = \frac{1}{\frac{1}{\beta} - \frac{N}{P_{vv}}}$ be a constant that indicates the extent to which the ambient noise approaches the required signal at receiver r_v. The* affectance[1] *of link ℓ_v, caused by a set S of links, is the sum of the relative interferences of the links in S on ℓ_v, scaled by c_v, or*

$$a_S(\ell_v) = c_v \cdot \sum_{\ell_w \in S} RI_w(v).$$

For a single link ℓ_w, we use the shorthand $a_w(\ell_v) = a_{\{\ell_w\}}(\ell_v)$. We define a p-signal *set or schedule to be one where the affectance of any link is at most $1/p$.*

Observation 1. *The affectance function satisfies the following properties for a set S of links:*

1. (Range) *S is SINR-feasible iff, for all $\ell_v \in S$, $a_S(\ell_v) \leq 1$.*
2. (Additivity) *$a_S = a_{S_1} + a_{S_2}$, whenever (S_1, S_2) is a partition of S.*
3. (Distance bound) *$a_w(\ell_v) = c_v \cdot \left(\frac{d_{vv}}{d_{wv}}\right)^\alpha$, for any pair ℓ_w, ℓ_v in S.*

[1] Affectance is closely related to *affectedness*, defined in [9], but treats the effect of noise more accurately.

4 Robustness of SINR

We present here properties of schedules in the SINR model, which double as tools for the algorithm designer. The results of this section apply equally to scheduling links of different powers, including involving topology control. In the next subsection, we examine the desirable property of link dispersion, and how any schedule can be dispersed at a limited cost.

We now explore how signal requirements (in the value of β), or equivalently interference tolerance, affects schedule length. It is not *a priori* obvious that minor discrepancies cause only minor changes in schedule length, but by showing that it is so, we can give our algorithms the advantage of being compared with a stricter optimal schedule. This also has implications regarding the robustness of SINR models with respect to perturbations in signal transmissions.

The pure geometric version of SINR given in (1) is an idealization of true physical characteristics. It assumes, e.g., perfectly isotropic radios, no obstructions, and a constant ambient noise level. That begs the question, why move algorithm analysis from analytically amenable graph-based models to a more realistic model if the latter isn't all that realistic? Fortunately, the fact that schedule lengths are relatively invariant to signal requirements shows that these concerns are largely unnecessary.

The following result on signal requirement applies also to throughput optimization.

Theorem 1. *There is a polynomial-time algorithm that takes a p-signal schedule and refines into a p'-signal schedule, for $p' > p$, increasing the number of slots by a factor of at most $\lceil 2p'/p \rceil^2$.*

Proof. Consider a p-signal schedule $\mathcal{S}$ and a slot S in $\mathcal{S}$. We partition S into a sequence $S_1, S_2, \ldots$ of sets. Order the links in S in some order, e.g., decreasing order. For each link ℓ_v, assign ℓ_v to the first set S_j for which $a_{S_j}(\ell_v) \leq 1/2p'$, i.e. the accumulated affectance on ℓ_v among the previous, longer links in S_j is at most $1/2p'$. Since each link ℓ_v originally had affectance at most $1/p$, then by the additivity of affectance, the number of sets used is at most $\lceil \frac{1/p}{1/2p'} \rceil = \lceil \frac{2p'}{p} \rceil$.

We then repeat the same approach on each of the sets S_i, processing the links this time in increasing order. The number of sets is again $\lceil \frac{2p'}{p} \rceil$ for each S_i, or $\lceil \frac{2p'}{p} \rceil^2$ in total. In each final slot (set), the affectance on a link by shorter links in the same slot is at most $1/2p'$. In total, then, the affectance on each link is at most $2 \cdot 1/2p' = 1/p'$.

This result applies in particular to optimal solutions. Let OPT_p be an optimal p-signal schedule and let χ_p be the number of slots in OPT_p. It is not *a priori* clear that a smooth relationship exists between χ_p and χ, for $p > 1$.

Corollary 1. $\chi_p \leq \lceil 2p \rceil^2 \chi$.

This has significant implications. One regards the validity of studying the pure SINR model. As asked in [9], "what if the signal is attenuated by a certain factor in one direction but by another factor in another direction?" A *generalized physical model* was introduced in [24] to allow for such a deviation.

Theorem 1 implies that scheduling is relatively robust under discrepancies in the SINR model. This validates the analytic study of the pure SINR model, in spite of its simplifying assumptions.

Corollary 2. *If a scheduling algorithm gives a ρ-approximation in the SINR model, it provides a $O(\theta^2 \rho)$-approximation in variations in the SINR model with a discrepancy of up to a factor of θ in signal attenuation or ambient noise levels.*

This result can be contrasted with the strong $n^{1-\epsilon}$-approximation hardness of scheduling in an abstract (non-geometric) SINR model that allows for arbitrary distances between nodes [11]. Alternatively, Theorem 1 allows us to analyze algorithms under more relaxed situations than the optimal solutions that we compare to.

4.1 Dispersion Properties

One desirable property of schedules is that links in the same slot be spatially well separated. This blurs the difference in position between sender and receiver of a link, since it affects distances only by a small constant. Intuitively, we want to measure nearness as a fraction of the lengths of the respective links. Given the affectance measure, it proves to be useful to define it somewhat less restrictively.

Definition 2. *Link ℓ_w is said to be* q-near *link ℓ_v, if $d_{wv} < q \cdot c_v^{1/\alpha} \cdot d_{vv}$. A set of links is* q-dispersed *if no (ordered) pairs of links in the set are q-near. A schedule is q-dispersed if all the slots are formed by q-dispersed sets.*

Observation 1, item 3, states that link w is q-near a link ℓ_v iff $a_w(\ell_v) > q^{-\alpha}$. This immediately gives the following strengthening of Lemma 4.2 in [9].

Lemma 2. *Fewer than q^α senders in an SINR-feasible set S are q-near to any given link $\ell_v \in S$.*

At a cost of a constant factor, any schedule can be made dispersed.

Lemma 3. *There is a polynomial-time algorithm that takes a SINR-feasible schedule and refines it into a q-dispersed schedule, increasing the number of slots by a factor of at most $(q+2)^\alpha$.*

Proof. Let S be a slot in the schedule. We show how to partition S into sets $S_1, S_2, \ldots, S_t$ that are q-dispersed, where $t \le (q+2)^\alpha + 1$.

Process the links of S in increasing order of length, assigning each link ℓ_v "first-fit" to the first set S_j in which the receiver r_v is at least $\left(qc_v^{1/\alpha} + 2\right) \cdot d_{vv}$ away from any other link. Let ℓ_w be a link previously in S_j, and note that ℓ_w

is shorter than ℓ_v. By the selection rule, $d_{wv} \geq \left(qc_v^{1/\alpha} + 2\right) \cdot d_{vv} \geq qc_v^{1/\alpha} \cdot d_{vv}$. Also,

$$d_{vw} \geq d_{wv} - d_{ww} - d_{vv} \geq \left(qc_v^{1/\alpha} + 1\right) d_{vv} - d_{ww} \geq qc_v^{1/\alpha} d_{ww}.$$

Since this holds for every pair in the same set, the schedule is q-dispersed. Suppose S_t is the last set used by the algorithm, and let ℓ_v be a link in it. Then, each S_i, for $i = 1, 2, \ldots, t-1$, contains a link whose sender is closer than $(qc_v^{1/\alpha} + 2) \cdot d_{vv} \leq (q+2)c_v^{1/\alpha} d_{vv}$ to r_v, i.e., is $(q+2)$-near to ℓ_v. By Lemma 2, $t - 1 < (q+2)^\alpha$.

Let χ^q denote the minimum number of slots in a q-dispersed schedule.

Corollary 3. $\chi^q \leq (q+2)^\alpha \cdot \chi$.

Intuitively, there is a correlation between low affectance and high dispersion in schedules. The following result makes this connection clearer. The converse is, however, not true, since interference can be caused by far-away links.

Lemma 4. *A p-signal schedule is also $p^{1/\alpha}$-dispersed.*

Proof. Let ℓ_v and ℓ_w be an ordered pair of links in a slot S in a p-signal schedule. By definition, $a_w(\ell_v) \leq a_S(\ell_v) \leq 1/p$. By Observation 1, item 3, $d_{wv} \geq p^{1/\alpha} c_v^{1/\alpha} \cdot d_{vv}$. Hence, the lemma.

5 Scheduling Approximation

The algorithm we analyze is a slightly simplified version of the algorithm of [9]. It involves repeated application of the following algorithm for the Single-Shot Scheduling problem.

Let $c = 1/\tau^\alpha$, where $\tau = 2 + \max\left(2, \left(2^6 3\beta\frac{\alpha-1}{\alpha-2}\right)^{\frac{1}{\alpha}}\right)$.

```
A(c)
  sort the links ℓ1, ℓ2, ..., ℓn by non-decreasing order of length
  S ← ∅
  for v ← 1 to n do
    if (a_S(ℓv) ≤ c)
      add ℓv to S
  output S
```

We shortly show that this algorithm also gives a $O(1)$-approximation to the Single-Shot Scheduling problem. It is rather surprising that a $O(1)$-approximation algorithm can be obtained in a single sweep. This should help in applying the ideas further, e.g., in distributed implementations. Simulation results in [9] also indicate very good practical performance, in relation to previous algorithms, and the simplification given here is likely to perform at least as well.

Instead of applying algorithm **A** repeatedly, we equivalently implement it as the following algorithm **B**:

```
B(c)
  sort the links ℓ_1, ℓ_2, ..., ℓ_n by non-decreasing order of length
  S_i ← ∅, for i = 1, 2, ...
  for v ← 1 to n do
    assign ℓ_v to the first set S_i for which a_{S_i}(ℓ_v) ≤ c
  output 𝒮 = (S_1, S_2, ...)
```

It is not immediate that algorithm **A** (or, equivalently, **B**) produces a feasible solution.

Lemma 5. *Algorithms **A** and **B** produce a $\tau - 2$-dispersed solution.*

Proof. Let ℓ_w be a link in the set S output by algorithm **A**. Let N^- (N^+) be the set of links in S that are shorter (longer) than ℓ_w. Consider first a link $\ell_u \in N^-$. Since ℓ_w was added by the algorithm after adding ℓ_u, $a_u(\ell_w) \leq c = 1/\tau^\alpha$, which implies by Observation 1, item 3, that $d_{uw} \geq \tau c_w^{1/\alpha} d_{ww} > (\tau - 2) c_w^{1/\alpha} d_{ww}$. Consider next a link $\ell_v \in N^+$. Since ℓ_v was added after ℓ_w, it holds that $a_w(\ell_v) \leq c = 1/\tau^\alpha$. So by Observation 1, $d_{wv} \geq \tau \cdot c_v^{1/\alpha} d_{vv}$. Note that $c_v \geq c_w$ whenever $d_{vv} \geq d_{ww}$. Then, using the triangular inequality,

$$d_{vw} = d(s_v, r_w) \geq d_{wv} - d_{vv} - d_{ww} \geq \left(\tau c_v^{1/\alpha} - 2\right) d_{vv} \geq (\tau - 2) c_w^{1/\alpha} d_{ww}.$$

Since this holds for every ordered pair in S, we have that S is $(\tau - 2)$-dispersed.

The following appeared as part of Lemma 4.1 in [9], and has also been applied in similar forms directly or indirectly elsewhere (e.g. [6]).

Lemma 6. *Let ℓ_v be a link in an SINR-feasible set S. Let N_z^+ be the set of links in S that are at least as long as ℓ_v and whose senders are of distance greater than $z \cdot d_{vv}$ from r_v. Then,*

$$a_{N_z^+}(\ell_v) < \left(\frac{\alpha - 1}{\alpha - 2} 2^5 3\right) z^{-\alpha} c_v.$$

Theorem 2. *Algorithms **A** and **B** produce an SINR-feasible solution.*

Proof. Let ℓ_w be a link in the set S output by algorithm **A**. Let N^- (N^+) be the set of links in S that are shorter (longer) than ℓ_w. The links in N^- were processed before ℓ_w, so by the if-condition in the algorithm, $a_{N^-}(\ell_v) \leq c$. By Lemma 5, S is $\tau - 2$-dispersed, so by Lemma 6 and the definitions of τ and dispersion,

$$a_{N^+}(\ell_w) < \left(\frac{\alpha - 1}{\alpha - 2} 2^5 3\right) \frac{c_v}{(\tau - 2)^\alpha c_v} c_v \leq \frac{1}{2}.$$

Hence, the affectance of each link in S is at most $c + 1/2 < 1$.

5.1 Performance Analysis

We need an extension of a geometric result from [9]. Let $\mathcal{R}$ and $\mathcal{B}$ be two disjoint sets of points in a metric space, called the *red* and the *blue* points. A blue point g *guards* a red point w, with respect to a point b, if $d(g, w) \leq d(b, w)$ and the angle $\angle gwb$ is at most 30°. That is, g is contained in the 60° sector emanating from w whose centerline goes through b. See Fig. 1. We say that a point b in $\mathcal{B}$ is *blue-shadowed* if each red point has a private guard in $\mathcal{B}$ with respect to b; i.e., there is an injective function $f : \mathcal{R} \rightarrow \mathcal{B} \setminus \{b\}$ such that $f(w)$ guards w from b for any $w \in \mathcal{R}$.

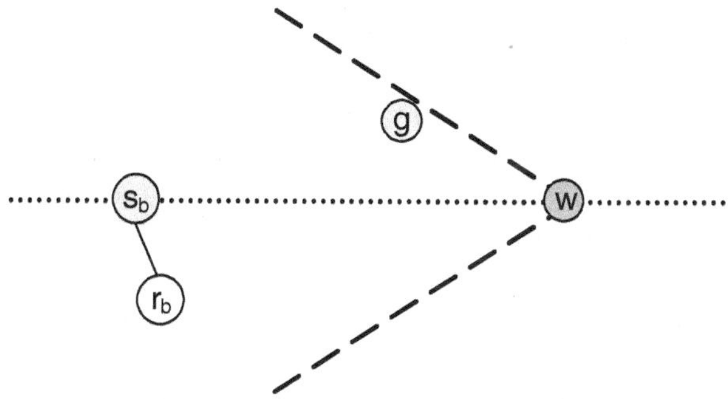

Fig. 1. Blue point g guards red point w from blue point s_b. If the blue points are sufficiently dispersed, then the receiver r_b will also be closer to g than to w.

The following result is a variation on Lemma 4.4 ("Blue-dominant centers lemma") of [9].

Lemma 7 (Blue-shadowed lemma). *Let $\mathcal{R}$ and $\mathcal{B}$ be two disjoint sets of red and blue points in 2-dimensional Euclidean space. If $|\mathcal{B}| > 12 \cdot |\mathcal{R}|$, then there is a blue-shadowed point in $\mathcal{B}$.*

Proof. Process the points in $\mathcal{R}$ in an arbitrary order, we work with a subset $\mathcal{B}'$ of $\mathcal{B}$ initially set at $\mathcal{B}' = \mathcal{B}$. We shall assign each $r \in \mathcal{R}$ a set $\{g_1^r, g_2^r, \ldots, g_{12}^r\}$ of guards.

For each point $r \in \mathcal{R}$ in order, let g_i^r be the blue point closest to r among the points in $\mathcal{B}'$ that are contained in the 30°-sector sec_i at angle in the range $[(i-1) \cdot 30°, i \cdot 30°)$ emanating from r. If a sector i contains no blue point in $\mathcal{B}'$, then no point is assigned as g_i^r. We then remove these points g_i^r from $\mathcal{B}'$ and continue with the next point in $\mathcal{R}$.

After going through all the points in $\mathcal{R}$, the set $\mathcal{B}'$ is still nonempty by the assumption on the relative sizes of $\mathcal{R}$ and $\mathcal{B}$. We claim that every point in $\mathcal{B}'$ is now blue-shadowed. Let b be such a point and consider a point $r \in \mathcal{R}$. Consider the 60°-sector emanating from r whose centerline goes through b. This sector properly contains one of the 30°-sectors sec_i, and thus contains one of r's guards. Since b was not selected as a guard, there was a guard selected for that sector and it is closer to r than b is. Since this holds for any point r, b is blue-shadowed.

The following lemma builds on Lemma 4.5 of [9]. Note that the straightforward modification of that lemma appears insufficient. Instead, we need something like our Theorem 1, allowing us to compare the algorithm's solution with the stricter optimal solution OPT_c. We also utilize the dispersion property to simplify the proof argument.

Lemma 8. *Let $\rho = 12$. Let S_k be the set of links scheduled by algorithm **B** in slot k, and let X_k be the set of links scheduled in slot k of OPT_c. Further, let $\mathcal{S}_k = \cup_{i=1}^k S_i$ and $\mathcal{X}_k = \cup_{i=1}^k X_i$. Then, for any positive integer k, $|\mathcal{S}_{\rho k}| \geq |\mathcal{X}_k|$.*

Proof. Suppose the claim is false for some integer k. Then, $|\mathcal{S}_{\rho k}| < |\mathcal{X}_k|$ or, equivalently, $|\mathcal{S}_{\rho k} \setminus \mathcal{X}_k| < |\mathcal{X}_k \setminus \mathcal{S}_{\rho k}|$. Thus, there are slots i_0, $1 \leq i_0 \leq \rho k$, and j_0, $1 \leq j_0 \leq k$, for which $|S_{i_0} \setminus \mathcal{X}_k| < |X_{j_0} \setminus \mathcal{S}_{\rho k}|/\rho$. Let $S = S_{i_0}$, $S' = S \setminus \mathcal{X}_k$, $X = X_{j_0}$, and $X' = X \setminus \mathcal{S}_{\rho k}$.

Since OPT_c is a $1/c$-signal schedule and $1/c = \tau^\alpha$, X' is also a τ^α-signal set. By Lemma 4, X' is then τ-dispersed. In particular, it is 3-dispersed.

Let $\mathcal{B} = \{s_v | \ell_v \in X'\}$ and $\mathcal{R} = \{s_w | \ell_w \in S'\}$ be the sets of senders in X' and S'; we call them blue and red points, respectively. By Lemma 7, there is a blue-shadowed point (sender) s_b in $\mathcal{B}$. We shall argue that the link $\ell_b = (s_b, r_b)$ would have been picked up by our algorithm for the slot i_0.

Consider any red point (sender) $w \in \mathcal{R}$, and let $g = f(w)$ be the guard for w guaranteed by the blue-shadowed lemma. Since g guards w, $d(s_b, w) \geq d(s_b, g)$. By the dispersion property, $d(g, r_b) \geq \tau \cdot d_{bb}$. Thus,

$$d(s_b, w) \geq d(s_b, g) \geq d(r_b, g) - d_{bb} \geq (\tau - 1) \cdot d_{bb} = (\tau - 1) d(s_b, r_b).$$

Then, the angle $\angle r_b w s_b$ is at most $\arcsin 1/(\tau - 1) \leq 30°$, since $\tau - 1 \geq 2$. That implies that r_b is contained in the 60°-sector emanating from w with centerline going through s_b, just like the guard g. See Fig. 1 for the relative positions of the points. Then, r_b is closer to g than to w. Thus, $a_g(\ell_b) > a_w(\ell_b)$. Summing up over all links w in $\mathcal{R}$ and their guards $f(w)$, we get

$$a_{S'}(\ell_b) = \sum_{w \in \mathcal{R}} a_w(\ell_b) < \sum_{w \in \mathcal{R}} a_{f(w)}(\ell_b) \leq \sum_{v \in \mathcal{B}} a_v(\ell_b) = a_{X'}(\ell_b).$$

Thus, since the affectance threshold of X is c,

$$a_S(\ell_b) = a_{S'}(\ell_b) + a_{S \cap X}(\ell_b) < a_{X'}(\ell_b) + a_{S \cap X}(\ell_b) = a_X(\ell_b) \leq c,$$

which contradicts the fact that ℓ_b was not selected into S.

The following result is largely immediate from Lemma 8.

Theorem 3. *Algorithm **B** outputs a schedule that approximates both the Scheduling and k-Thruput problems, for every $k \geq 1$, within a constant factor.*

Proof. By Lemma 8 and Theorem 1, the number ALG of slots used by algorithm **B** is bounded by

$$ALG \leq \rho \chi_c \leq \rho \left\lceil \frac{2}{c} \right\rceil^2 \chi.$$

Also, by Lemma 8, the number of links scheduled by **B** in the first $12k$ slots is at least the number of links in an optimal c-signal k-Thruput solution. Again by Theorem 1, we obtain a constant factor approximation to k-Thruput.

5.2 Handling Different Transmission Powers

We can treat the case when links transmit with different powers in two different ways. Let P_{max} (P_{min}) be the maximum (minimum) power used by a link, respectively. By introducing a factor of P_{min}/P_{max} into the affectance threshold c, the algorithm **B** still produces a feasible schedule, that is longer by a factor of at most P_{max}/P_{min}.

Alternatively, we can partition the instance into "power regimes", where each regime consists of links whose powers are equal up to a factor of 2. We schedule each power regime separately, obtaining an approximation factor of at most $\log P_{max}/P_{min}$, or at most the number of different power values.

6 Conclusions

This paper shows that wireless scheduling is in APX. Having a constant approximation algorithm for wireless scheduling implies that we can derive the single-hop throughput capacity of an arbitrary wireless network, up to a constant factor. As such this paper basically solves the scheduling complexity introduced by Moscibroda et al. [22]. However, various parameter combinations are still open, and deserve more research, e.g. power control, multi-hop traffic, scheduling and routing, analog network coding, models beyond SINR such as log-normal shadowing, to name just a few of the obvious ones.

References

1. Auletta, V., Moscardelli, L., Penna, P., Persiano, G.: Interference games in wireless networks. In: Papadimitriou, C., Zhang, S. (eds.) WINE 2008. LNCS, vol. 5385, pp. 278–285. Springer, Heidelberg (2008)
2. Avin, C., Emek, Y., Kantor, E., Lotker, Z., Peleg, D., Roditty, L.: SNR diagrams: Towards algorithmically usable SINR models of wireless networks. arXiv:0811.3284v2 (2008)
3. Brar, G., Blough, D., Santi, P.: Computationally Efficient Scheduling with the Physical Interference Model for Throughput Improvement in Wireless Mesh Networks. In: Mobicom (2006)
4. Brar, G.S., Blough, D.M., Santi, P.: The scream approach for efficient distributed scheduling with physical interference in wireless mesh networks. In: ICDCS, pp. 214–224 (2008)
5. Chafekar, D., Kumar, V., Marathe, M., Parthasarathy, S., Srinivasan, A.: Cross-layer Latency Minimization for Wireless Networks using SINR Constraints. In: Mobihoc (2007)
6. Chafekar, D., Kumar, V., Marathe, M., Parthasarathy, S., Srinivasan, A.: Approximation Algorithms for Computing Capacity of Wireless Networks with SINR Constraints. In: Infocom (2008)

7. ElBatt, T.A., Ephremides, A.: Joint scheduling and power control for wireless ad hoc networks. IEEE Transactions on Wireless Communications 3(1), 74–85 (2004)
8. Gao, Y., Hou, J.C., Nguyen, H.: Topology control for maintaining network connectivity and maximizing network capacity under the physical model. In: Infocom (2008)
9. Goussevskaia, O., Halldórsson, M., Wattenhofer, R., Welzl, E.: Capacity of Arbitrary Wireless Networks. In: 28th Annual IEEE Conference on Computer Communications (INFOCOM), Rio de Janeiro, Brazil (April 2009)
10. Goussevskaia, O., Moscibroda, T., Wattenhofer, R.: Local Broadcasting in the Physical Interference Model. In: ACM SIGACT-SIGOPT International Workshop on Foundations of Mobile Computing (DialM-POMC), Toronto, Canada (August 2008)
11. Goussevskaia, O., Oswald, Y.A., Wattenhofer, R.: Complexity in Geometric SINR. In: Mobihoc (2007)
12. Goussevskaia, O., Wattenhofer, R.: Complexity of scheduling with analog network coding. In: FOWANC (2008)
13. Gronkvist, J., Hansson, A.: Comparison between graph-based and interference-based STDMA scheduling. In: Mobihoc (2001)
14. Gupta, P., Kumar, P.R.: The Capacity of Wireless Networks. IEEE Trans. Information Theory (2000)
15. Hua, Q.-S., Lau, F.C.M.: The scheduling and energy complexity of strong connectivity in ultra-wideband networks. In: MSWiM, pp. 282–290 (2006)
16. Hua, Q.-S., Lau, F.C.M.: Exact and approximate link scheduling algorithms under the physical interference model. In: DIALM-POMC, pp. 45–54 (2008)
17. Katz, B., Völker, M., Wagner, D.: Link scheduling in local interference models. In: Fekete, S.P. (ed.) ALGOSENSORS 2008. LNCS, vol. 5389, pp. 57–71. Springer, Heidelberg (2008)
18. Kumar, V., Marathe, M., Parthasarathy, S., Srinivasan, A.: Algorithmic Aspects of Capacity in Wireless Networks. In: Sigmetrics (2005)
19. Maheshwari, R., Jain, S., Das, S.R.: A measurement study of interference modeling and scheduling in low-power wireless networks. In: SenSys, pp. 141–154 (2008)
20. Moscibroda, T.: The worst-case capacity of wireless sensor networks. In: IPSN (2007)
21. Moscibroda, T., Oswald, Y.A., Wattenhofer, R.: How optimal are wireless scheduling protocols? In: INFOCOM, pp. 1433–1441 (2007)
22. Moscibroda, T., Wattenhofer, R.: The Complexity of Connectivity in Wireless Networks. In: Infocom (2006)
23. Moscibroda, T., Wattenhofer, R., Weber, Y.: Protocol Design Beyond Graph-Based Models. In: Hotnets (November 2006)
24. Moscibroda, T., Wattenhofer, R., Zollinger, A.: Topology Control meets SINR: The Scheduling Complexity of Arbitrary Topologies. In: Mobihoc (2006)
25. Scheideler, C., Richa, A.W., Santi, P.: An o(log n) dominating set protocol for wireless ad-hoc networks under the physical interference model. In: MobiHoc, pp. 91–100 (2008)
26. Zussman, G., Brzezinski, A., Modiano, E.: Multihop Local Pooling for Distributed Throughput Maximization in Wireless Networks. In: Infocom (2008)

The Ehrenfeucht-Silberger Problem

Štěpán Holub[1,⋆] and Dirk Nowotka[2]

[1] Department of Algebra, Charles University of Prague, Czech Republic
holub@karlin.mff.cuni.cz
[2] Institute for Formal Methods in Computer Science, Universität Stuttgart, Germany
nowotka@fmi.uni-stuttgart.de

Abstract. We consider repetitions in words and solve a longstanding open problem about the relation between the period and the length of its longest unbordered factor. A word u is called bordered if there exists a proper prefix that is also a suffix of u, otherwise it is called unbordered. In 1979 Ehrenfeucht and Silberger raised the following problem: What is the maximum length of a word w, w.r.t. the length τ of its longest unbordered factor, still allowing that τ is shorter than the period π of w. We show that if w is longer than $7(\tau - 1)/3$ then $\tau = \pi$ which gives the optimal asymtotic bound.

Introduction

Combinatorial problems about repetitions lie at the core of algorithmic questions regarding strings (called words here), being it search, compression, or coding algorithms. Despite a long tradition of research many questions about the combinatorial properties of data structures as simple as words remain open. The focus of this paper is on the solution of such a question namely the problem by Ehrenfeucht and Silberger which had been open for about three decades.

When repetitions in words of symbols are considered then two notions are central: the *period*, which gives the least amount by which a word has to be shifted in order to overlap with itself, and the shortest *border*, which denotes the least (nontrivial) overlap of a word with itself. Both notions are related in several ways, for example, the length of the shortest border of a word w is not larger than the period of w, and hence, the period of an unbordered word is its length. Moreover, a shortest border itself is always unbordered. Deeper dependencies between the period of a word and its unbordered factors have been investigated and exploited in applications for decades; see also the references to related work below.

Let us recall the problem by Ehrenfeucht and Silberger. Let w be a (finite) word of length $|w|$, let $\tau(w)$ denote the length of the largest unbordered factor of w, and let $\pi(w)$ denote the period of w. Certainly, $\tau(w) \leq \pi(w)$ since the period of a factor of w cannot be larger than the period of w itself. Moreover,

⋆ The work on this article has been supported by the research project MSM 0021620839.

S. Albers et al. (Eds.): ICALP 2009, Part I, LNCS 5555, pp. 537–548, 2009.

it is well-known that $\tau(w) = \pi(w)$ when $|w| \geq 2\pi(w)$. So, the interesting cases are those where $|w| < 2\pi(w)$. Actually, the interesting cases are also the most common ones since by far most words have a period that is longer than one half of their length. When such words are considered, a bound on $|w|$, enforcing $\tau(w) = \pi(w)$, that depends on $\tau(w)$ becomes more interesting than one depending on $\pi(w)$.

The problem by Ehrenfeucht and Silberger asks about a bound of $|w|$ depending on $\tau(w)$ such that $\tau(w) = \pi(w)$ is enforced. In this paper we establish the following fact for all finite words w:

$$\text{If} \quad |w| > \frac{7}{3}(\tau(w) - 1) \quad \text{then} \quad \tau(w) = \pi(w) \ .$$

This bound on the length of w is asymtotically tight (see the following example).

Previous Work. Ehrenfeucht and Silberger raised the problem described above in [7]. They conjectured that $|w| \geq 2\,\tau(w)$ implies $\tau(w) = \pi(w)$. That conjecture was falsified shortly thereafter by Assous and Pouzet [1] by the following example:

$$w = a^n b a^{n+1} b a^n b a^{n+2} b a^n b a^{n+1} b a^n$$

where $n \geq 1$ and $\tau(w) = 3n + 6$ (note that $ba^{n+1}ba^nba^{n+2}$ and $a^{n+2}ba^nba^{n+1}b$ are the two longest unbordered factors of w) and $\pi(w) = 4n+7$ and $|w| = 7n+10$, that is, $\tau(w) < \pi(w)$ and $|w| = 7/3\,\tau(w) - 4 > 2\tau(w)$. Assous and Pouzet in turn conjectured that $3\tau(w)$ is the bound on the length of w for establishing $\tau(w) = \pi(w)$. Duval [5] did the next step towards answering the conjecture. He established that $|w| \geq 4\,\tau(w) - 6$ implies $\tau(w) = \pi(w)$ and conjectured that, if w possesses an unbordered prefix of length $\tau(w)$, then $|w| \geq 2\,\tau(w)$ implies $\tau(w) = \pi(w)$. Note that a positive answer to Duval's conjecture yields the bound $3\,\tau(w)$ for the general question. Despite some partial results [12,6,8] towards a solution Duval's conjecture was only solved in 2004 [9,10] with a new proof given in [11]. The proof of (the extended version of) Duval's conjecture lowered the bound for Ehrenfeucht and Silberger's problem to $3\,\tau(w) - 2$ as conjectured by Assous and Pouzet [1]. However, there remained a gap of $\tau(w)/3$ between that bound and the largest known example which is given above. The bound of $7\tau(w)/3$ has been conjectured in [9,10]. This conjecture is proved in this paper, and the problem by Ehrenfeucht and Silberger is finally solved.

Other Related Work. The result related most closely to the problem by Ehrenfeucht and Silberger is the so called critical factorization theorem (CFT).

What is the CFT? Let $w = uv$ be a factorization of a word w into u and v. The local period of w at the point $|u|$ is the length q of the shortest square centered at $|u|$. More formally, let x be the shortest word such that x is a prefix of vy and a suffix of zu for some y and z, then $q = |x|$. It is straightforward to see that q is not larger than the period of w. The factorization uv is called critical if q equals the period of w. The CFT states that a critical factorization exists for every nonempty word w, and moreover, a critical factorization uv can always be found such that $|u|$ is shorter than the period of w. The CFT was conjectured first by Schützenberger [13], proved by Césari and Vincent [2], and brought into its current form by Duval [4]. Crochemore and Perrin [3] found a new elegant proof of the CFT using lexicographic orders, and realized a direct application of the theorem in a new string-matching algorithm.

How does the CFT relate to the problem by Ehrenfeucht and Silberger? Observe that the shortest square x^2 centered at some point in w is always such that x is unbordered. If x results from a critical factorization and occurs in w, then w contains an unbordered factor of the length of its period. Therefore, it follows from the CFT that $|w| > 2\pi(w) - 2$ implies $\tau(w) = \pi(w)$. This bound is asymptotically optimal. In this paper, we establish the asymptotically optimal bound on $|w|$ enforcing the equality $\tau(w) = \pi(w)$ in terms of $\tau(w)$ instead of $\pi(w)$. This rounds off the long lasting research effort on the mutual relationship between the two basic properties of a word w, that is $\tau(w)$ and $\pi(w)$.

1 Notation and Basic Facts

Let us fix a finite set A of letters, called alphabet, for the rest of this paper. Let A^* denote the monoid of all finite words over A including the *empty word* denoted by ε. Let $w = uv \in A^*$. Then $u^{-1}w = v$ and $wv^{-1} = u$. In general, we denote variables over A by a, b, c, and d and variables over A^* are usually denoted by f, g, h, r through z, and α through δ, and ξ including their subscripted and primed versions. The letters i through q are to range over the set of nonnegative integers.

Let $w = a_1a_2\cdots a_n$. The word $a_na_{n-1}\cdots a_1$ is called the *reversal* of w denoted by $\overline{w}$. We denote the length n of w by $|w|$, in particular $|\varepsilon| = 0$. Let $0 \leq i \leq n$. Then $u = a_1a_2\cdots a_i$ is called a *prefix* of w, denoted by $u \leq_{\mathrm{p}} w$, and $v = a_{i+1}a_{i+2}\cdots a_n$ is called a *suffix* of w, denoted by $v \leq_{\mathrm{s}} w$. A prefix or suffix is called *proper* when $0 < i < n$. The *longest common prefix* w of two words u and v is denoted by $u \wedge_{\mathrm{p}} v$ and is defined by $w = u$, if $u \leq_{\mathrm{p}} v$, or $w = v$, if $v \leq_{\mathrm{p}} u$, or $wa \leq_{\mathrm{p}} u$ and $wb \leq_{\mathrm{p}} v$ for some different letters a and b. The *longest common suffix* of u and v, denoted $u \wedge_{\mathrm{s}} v$, is defined similarly, as one would expect. An integer $1 \leq p \leq n$ is a *period* of w if $a_i = a_{i+p}$ for all $1 \leq i \leq n - p$. The smallest period of w is called *the period* of w, denoted by $\pi(w)$. A nonempty word u is called a *border* of a word w, if $w = uy = zu$ for some nonempty words y and z. We call w *bordered*, if it has a border, otherwise w is called *unbordered.* Let $\tau(w)$ denote the maximum length of unbordered factors of w, and $\tau_2(w)$ denote the maximum length of unbordered factors occurring at least twice in w. We have that

$$\tau(w) \leq \pi(w) \ .$$

Indeed, let $u = b_1b_2\cdots b_{\tau(w)}$ be an unbordered factor of w. If $\tau(w) > \pi(w)$ then $b_i = b_{i+\pi(w)}$ for all $1 \leq i \leq \tau(w) - \pi(w)$ and $b_1b_2\cdots b_{\tau(w)-\pi(w)}$ is a border of u; a contradiction.

Let $\lhd$ be a total order on A. Then $\lhd$ extends to a *lexicographic order*, also denoted by $\lhd$, on A^* with $u \lhd v$ if either $u \leq_{\mathrm{p}} v$ or $xa \leq_{\mathrm{p}} u$ and $xb \leq_{\mathrm{p}} v$ and

$a \lhd b$. Let $\overline{\lhd}$ denote a lexicographic order on the reversals, that is, $u \mathrel{\overline{\lhd}} v$ if $\overline{u} \lhd \overline{v}$. Let $\lhd^a$ and $\lhd_b$ denote lexicographic orders where the maximum letter a or the minimum letter b is fixed in the respective orders on A. A $\lhd$-*maximum prefix* (*suffix*) α of a word w is defined as a prefix (suffix) of w such that $v \mathrel{\overline{\lhd}} \alpha$ ($v \lhd \alpha$) for all $v \leq_{\mathrm{p}} w$ ($v \leq_{\mathrm{s}} w$).

The notions of maximum pre- and suffix are symmetric. It is general practice that facts involving the maximum ends of words are mostly formulated for maximum suffixes. The analogue version involving maximum prefixes is tacitly assumed.

The following remarks state some facts about maximum suffixes which are folklore. They are included in this paper to make it self-contained.

Remark 1. Let w be a bordered word. The shortest border u of w is unbordered, and $w = uzu$. The longest border of w has length equal to $|w| - \pi(w)$.

Indeed, if u is a border of w, then each border of u is also a border of w. Therefore u is unbordered, and it does not overlap with itself. If v is a border of w then $|w| - |v|$ is a period of w. Conversely, the prefix of w of length $|w| - \pi(w)$ is a border of w.

Remark 2. Any maximum suffix of a word w occurs only once in w and is longer than $|w| - \pi(w)$.

Indeed, let α be the $\lhd$-maximum suffix of w for some order $\lhd$. Then $w = x\alpha y$ and $\alpha \lhd \alpha y$ implies $y = \varepsilon$ by the maximality of α. If $w = uv\alpha$ with $|v| = \pi(w)$, then $u\alpha \leq_{\mathrm{p}} w$ gives a contradiction again.

Remark 3. Let α be its own maximum suffix w.r.t. some order $\lhd$, and let x be a prefix of α of length $\pi(\alpha)$. Then x is unbordered.

Indeed, suppose on the contrary that x is bordered, that is, $x = ghg$ for some nonempty g. Let $\alpha = xy$. We have $gy \lhd \alpha$, by assumption, which implies $y \lhd hgy$. Note that gy is not a prefix of α otherwise $|gh| < |x|$ is a period of α contradicting the choice of x. Hence, $za \leq_{\mathrm{p}} y$ and $zb \leq_{\mathrm{p}} hgy$ for some different letters a and b with $a \lhd b$. But, $y \leq_{\mathrm{p}} \alpha$, since $|x| = \pi(w)$, implies $za \leq_{\mathrm{p}} \alpha$ which contradicts the maximality of α (since $za \leq_{\mathrm{p}} \alpha \lhd zb \leq_{\mathrm{p}} hgy$).

Let an integer q with $0 \leq q < |w|$ be called *point* in w. A nonempty word x is called a *repetition word* at point q if $w = uv$ with $|u| = q$ and there exist words y and z such that $x \leq_{\mathrm{s}} yu$ and $x \leq_{\mathrm{p}} vz$. Let $\pi(w, q)$ denote the length of the shortest repetition word at point q in w. We call $\pi(w, q)$ the *local period* at point q in w. Note that the repetition word of length $\pi(w, q)$ at point q is necessarily unbordered and $\pi(w, q) \leq \pi(w)$. A factorization $w = uv$, with $u, v \neq \varepsilon$ and $|u| = q$, is called *critical*, if $\pi(w, q) = \pi(w)$, and if this holds, then q is called a *critical point*. Let $\lhd$ be an order on A and $\blacktriangleleft$ be its inverse. Then the shorter of the $\lhd$-maximum suffix and the $\blacktriangleleft$-maximum suffix of some word w is called a *critical suffix* of w. Similarly, we define a *critical prefix* of w by the shorter of the two maximum prefixes resulting from some order and its inverse. This notation is justified by the following formulation of the so called critical factorization theorem (CFT) [3] which relates maximum suffixes and critical points.

Theorem 1 (CFT). *Let w be a nonempty word and γ be a critical suffix of w. Then $|w| - |\gamma|$ is a critical point.*

Remark 4. Let rs be an unbordered word where $|r|$ is a critical point. Then s and r do not overlap and sr is unbordered with $|s|$ as a critical point.

Let us highlight the following definitions. They are not standard but will be central in the proof of Theorem 2.

Definition 1. *Let words g and w be given. The longest prefix of g shorter than g that is also a suffix of w will be called the g-*suffix *of w.*

*The number $|ws^{-1}|$, where s is the g-suffix of w, is called the g-*period *of w, denoted by $\pi_g(w)$.*

*The shortest prefix w' of w satisfying $\pi_g(w') = \pi_g(w)$ is called the g-*critical prefix *of w.*

Remark 5. Note that zd, where d is a letter, is the g-critical prefix of w if and only if zd is the longest prefix of w satisfying $\pi_g(z) < \pi_g(zd)$.

Example 1. Consider $w = ababbaababab$ of length 12 and $g = ababb$. The g-suffix of w is $abab$, whence $\pi_g(w) = 8$. The g-critical prefix of w is $ababbaababa$ of length 11, since

$$\pi_g(ababbaababa) = 8\,, \quad \text{and} \quad \pi_g(ababbaabab) = 6\,.$$

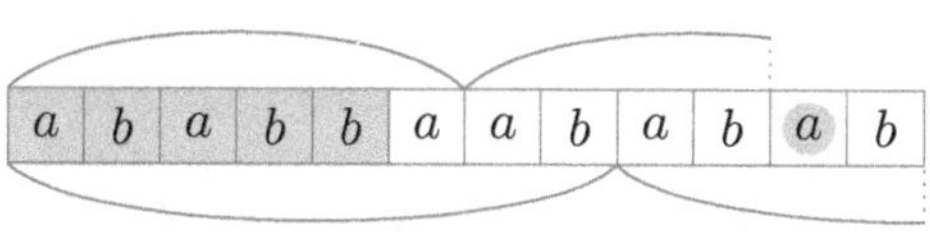

Note that, by definition, the g-suffix of w can be empty, but it cannot be equal to g. For example, the abb-suffix of $aabb$ is empty. Therefore, the abb-critical prefix of $aabb$ is $aabb$ itself.

2 Solution of the Ehrenfeucht-Silberger Problem

This entire section is devoted to the proof of the main result of this paper: the solution of the Ehrenfeucht-Silberger problem.

Theorem 2. *Let $w \in A^*$. If $|w| > \frac{7}{3}(\tau(w) - 1)$ then $\tau(w) = \pi(w)$.*

We identify two particular unbordered factors of w and show that the assumption of the theorem, namely that these factors are strictly smaller than $\frac{3}{7}|w|+1$, leads either to a contradiction or to $\tau(w) = \pi(w)$.

Note that the claim holds trivially if every letter in w occurs only once because $\tau(w) = \pi(w) = |w|$ holds in that case. Let

$$w = v'uzuv$$

such that $|u| = \tau_2(w)$ and z is of maximum length (recall that $\tau_2(w)$ denotes the maximum length of unbordered factors occurring at least *twice* in w). It is clear that such a factorization exists whenever a letter occurs more than once in w. Based on such a factorization of w we fix some more notation for the rest of this proof. Let

$$t = v \wedge_{\mathrm{p}} zu \quad \text{and} \quad t' = v' \wedge_{\mathrm{s}} uz \ .$$

If $t \neq v$, then let

- ta be a prefix of zu and tb be a prefix of v with $a \neq b$,
- δa be the $\lhd^a$-maximum suffix of $t'uta$ for some fixed order $\lhd^a$ such that a is the maximum in A,
- α be the $\lhd^a$-maximum suffix of $t'u$, and
- β be the $\blacktriangleleft_a$-maximum suffix of $t'u$ where $\blacktriangleleft_a$ is the inverse order of $\lhd^a$.

The notation introduced so far is exemplified by the following figure where we assume that $t \neq v$ and $t' \neq v'$ and $|t'| < |z| < |t| < |\delta| < |\alpha t|$ and $|\alpha| < |\beta| < |u|$.

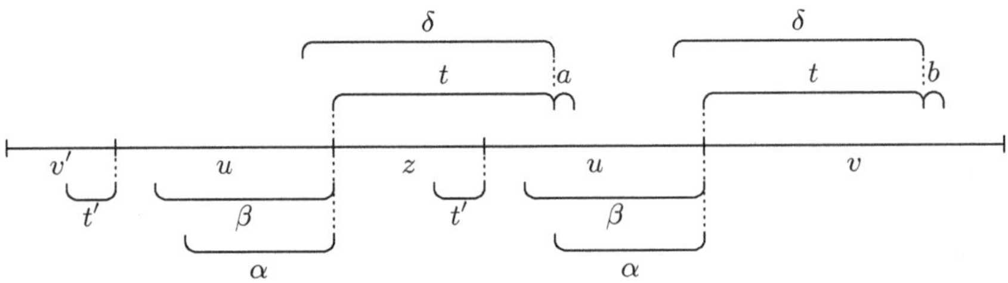

The example of long words where the period exceeds the length of the longest unbordered factors by Assous and Pouzet (see page 538) turns out to highlight the most interesting cases of this proof. We therefore use it as a running example throughout this section. The notation introduced above applied to a word of Assous and Pouzet is illustrated by the following figure. In this case t' is empty.

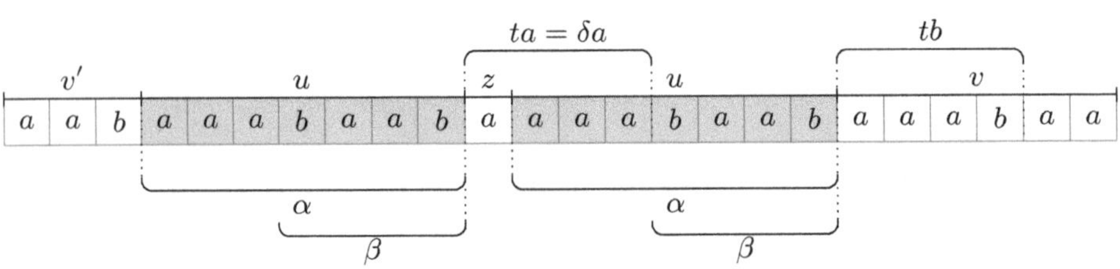

We can suppose w.l.o.g. that v' is as short as possible. This in particular implies the following claim.

Claim 3

$$|\alpha| \leq |u| \quad \textit{and} \quad |\beta| \leq |u| \ . \tag{1}$$

Proof. If α is longer than u, then the prefix $\hat{u}$ of α of length $\pi(\alpha)$ is unbordered by Remark 3. It is of length at least $|u|$, otherwise u is bordered. From $|u| = \tau_2(w)$ follows $|\hat{u}| = |u|$ since $\hat{u}$ occurs at least twice in w. We have a factorization $\hat{v}'\hat{u}\hat{z}\hat{u}\hat{v}$ of w where $\hat{v}' = v'u\alpha^{-1}$ and $|\hat{z}| = |z|$ and $\hat{v} = \hat{u}^{-1}\alpha v$; contradicting the minimality of $|v'|$. □

2.1 The First Factor

In this subsection we describe, using the factorization introduced above, a particular factor of w, which is likely to be unbordered and long; see the factor $uzuv_0d$ in the proof of Claim 5 below. The basic assumption of our proof, namely that there are no too long unbordered factors, will yield important additional restrictions on w.

Let γ denote the shorter of α and β, and let y_α and y_β denote the α- and β-suffix of uv for the rest of this proof. Moreover, let y be the shorter of y_α and y_β and let ξ be either α or β so that $y = y_\xi$. The following figure shall illustrate the considered setting by an example where $|\alpha| < |\beta|$ and $|y_\alpha| > |y_\beta|$, that is, we have $\gamma = \alpha$, $y = y_\beta$ and $\xi = \beta$.

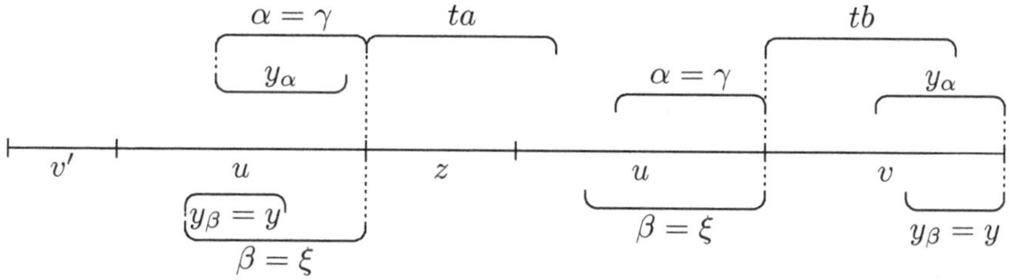

The same situation for our running example is depicted next.

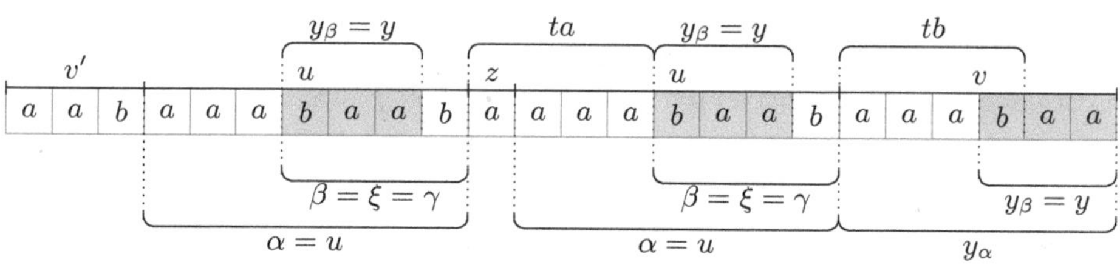

Claim 4. *If $v_0\gamma$ is a prefix of γv with $v_0 \neq \varepsilon$, then $uzu\gamma^{-1}v_0\gamma$ is unborderd.*

Proof. Suppose on the contrary that $uzu\gamma^{-1}v_0\gamma$ has a shortest border h. Note that h is, like every shortest border of a factor in w, not longer than $|u| = \tau_2(w)$. In fact $|h| < |u|$ since $|h| = |u|$ contradicts the maximality of $|z|$. If $|\gamma| < |h| < |u|$ then γ occurs more than once in u contradicting Remark 2. And finally, if $|h| \leq |\gamma|$ then u is bordered by h since then $h \leq_s \gamma \leq_s u$; a contradiction which concludes the proof. □

We shall now consider the ξ-critical prefix of w in order to prove the following inequalities.

Claim 5

$$|v'| < |u| \quad \textit{and} \quad |v| < |u| \quad \textit{and} \quad |v| \leq |ty|.$$

Proof. Within this proof suppose w.l.o.g. that $|v'| \leq |v|$. Note, the assumption that v' is as short as possible does not harm generality.

The claim is trivial if $|y| \geq |v|$. We therefore suppose that the ξ-critical prefix of w can be written as $v'uzuv_0d$, where d is a letter. We let g denote the ξ-suffix of $v'uzuv_0$. Assume first that $gd = \xi$ as illustrated in the next figure.

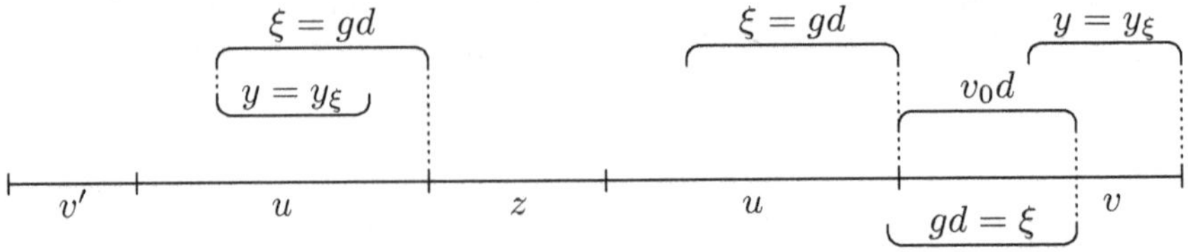

Then the word $uzuv_0d$ is unbordered, by Claim 4. From $|v| < |v_0d\xi|$ we obtain $|uzuv_0d| > |zuv|$. Therefore $|uzuv_0d| \geq |w|/2 + 1 > \frac{3}{7}|w| + 1$; a contradiction. This implies that gc is a prefix of ξ with $c \neq d$. Suppose $c \lhd^a d$ and consider βzuv_0d. Since $|\beta zuv_0d| > |zuv|$, it must be bordered, as above. Let hd be its shortest border.

Suppose $|h| \leq |g|$ as illustrated in the next figure.

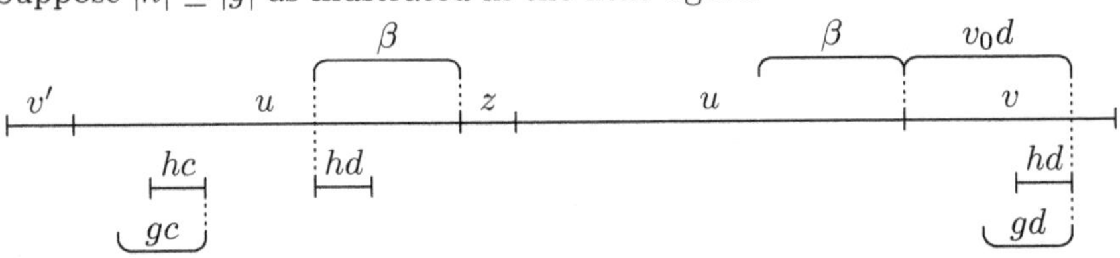

Then hd is a prefix of β and the occurrence of $hc \leq_s gc$ in ξ, and hence also in β, contradicts the maximality of β since $hd \blacktriangleleft_a hc$.

Suppose $|g| < |h| < |\beta|$ as illustrated in the next figure.

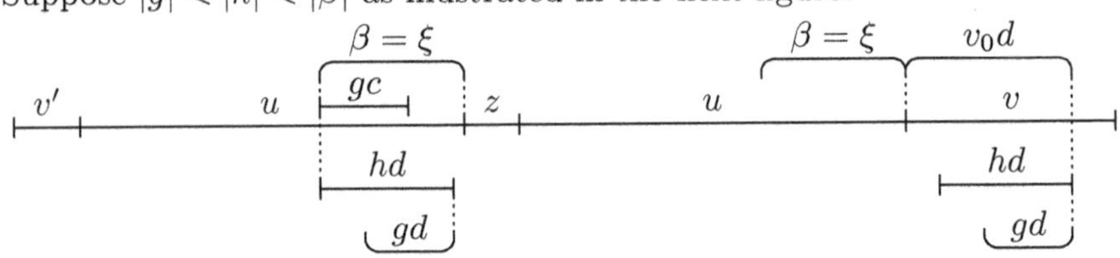

Then gd occurs in u and $\xi = \beta$ since $gd \blacktriangleleft_a gc$. Therefore h contradicts the assumption that g is the ξ-suffix of $v'uzuv_0$.

It remains that $|h| \geq |\beta|$ which implies $\beta \leq_p h$ as illustrated next.

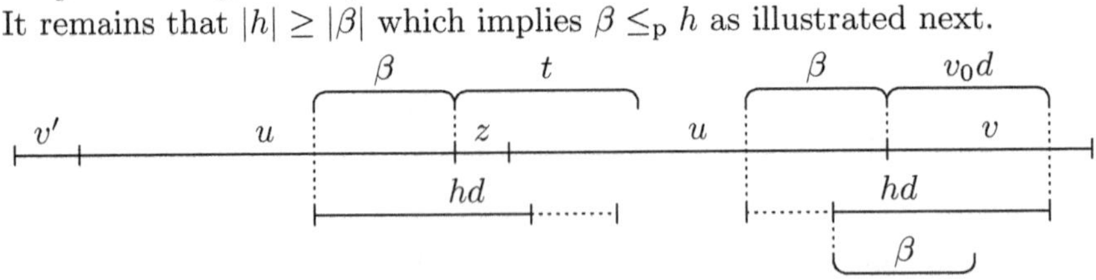

The choice of u implies $|h| < |u|$, whence either $h = \beta v_0$ or the word $uzuv_0h^{-1}\beta$ is unbordered, by Claim 4. From $|h| < |u|$ we have $|uzuv_0h^{-1}\beta| > |zuv| > |w|/2+1$. Therefore, $h = \beta v_0$, which implies $v_0d \leq_p t$, and $|v| \leq |ty|$. The remaining inequalities follow from $|\beta v| \leq |\beta v_0dy| = |hdy| < |u\beta|$, where the last inequality uses $|hd| \leq |u|$ and $|y| \leq |y_\beta| < |\beta|$. The possibility $d \lhd^a c$ is similar considering αzuv_0d. □

Suppose that $v \neq t$. Recall that tb is a prefix of v, and ta a prefix of zu. From $|v| \leq |ty|$, and from $|y| \leq |y_\alpha|$ we deduce that utb and y_α have in uv an overlap rb where r is nonempty. In other words, r is a suffix of ut such that $uv = utr^{-1}y_\alpha$.

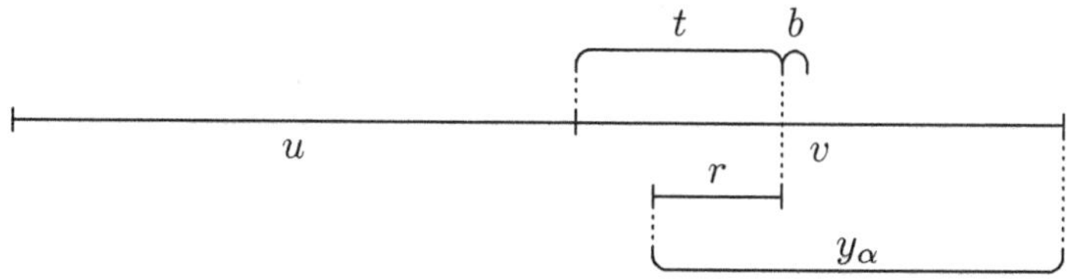

Since $|y_\alpha| < |\alpha|$, we have

$$|t| > |v| - |\alpha| + |r| \; . \tag{2}$$

The word rb is a prefix of α, and ra is a suffix of uta, which is a prefix of uzu. The maximality of α implies that ra is not a factor of $t'u$, and thus

$$|r| > |t| + |t'| - |z| \; . \tag{3}$$

2.2 The Second Factor

Let us now turn our attention to the word δ. In particular, we consider the factor $\delta t^{-1}zuv_\delta d$ as defined below in the proof of Claim 7.

The following claim points out that every factor of $t'uv$ is strictly less than δa w.r.t. $\lhd^a$. In particular, δa does not occur in $t'uv$.

Claim 6. *Let f be a factor of $t'uv$. Then $f \lhd^a \delta a$ and $f \neq \delta a$.*

Proof. Suppose on the contrary that there exists a factor f of $t'uv$ such that $\delta a \lhd^a f$. Note that the maximality of δ is contradicted, if f occurs in $t'ut$ or y_α. Therefore, we have that there exists a prefix $f'b$ of f such that $f' \leq_s t'ut$. But, we have $f'a \leq_s t'uta$, and hence, $f'a \lhd^a \delta a$. The contradiction follows now from $f'b \lhd^a f'a$. □

Let y_δ denote the δa-suffix of w. Note that

$$|y_\delta| < |v| - |t|, \tag{4}$$

since otherwise there is a suffix t_0 of $t'ut$ such t_0b is a prefix of y_δ, and t_0a is a suffix of $t'uta$ contradicting the maximality of δ.

Consider the following figure which already gives an illustration of the factor $\delta t^{-1}zuv_\delta d$ that will be defined below in the proof of Claim 7.

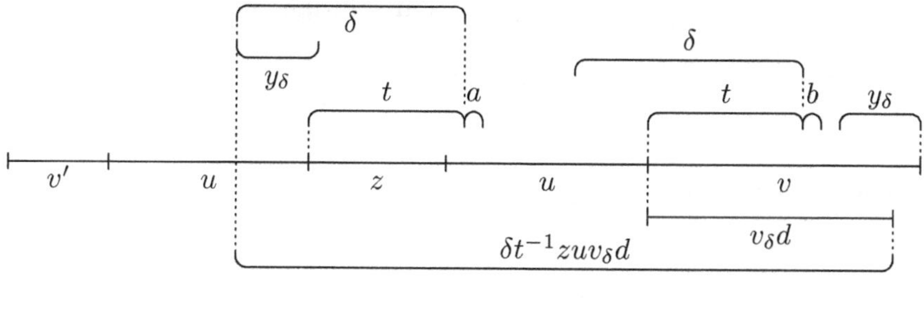

Our running example gives the following setting, with $d = b$.

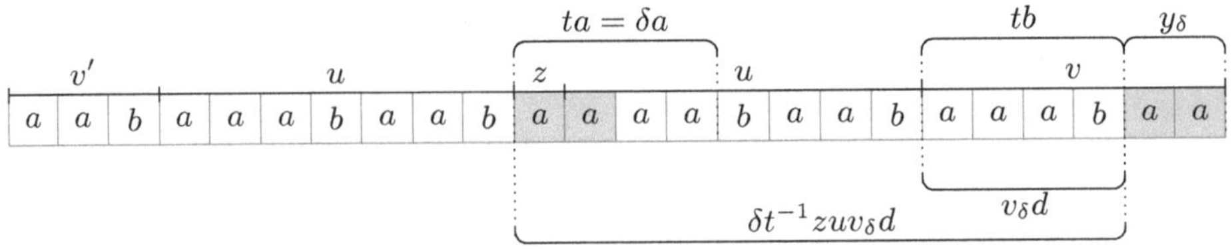

Claim 7

$$|\delta| - |t| + |z| + |u| + |v| - |y_\delta| < \frac{3}{7}|w| + 1 \tag{5}$$

Proof. Let $\delta t^{-1}zuv_\delta d$ be the δa-critical prefix of $v'uzuv$ and let g denote the δa-suffix of $\delta t^{-1}zuv_\delta$. This implies that $|v_\delta d| \geq |v| - |y_\delta|$ (in particular, $|v_\delta| \geq |t|$) and that gc is a prefix of δa and $c \neq d$. Note that $d \lhd^a c$, since all factors of uv are less than δa w.r.t. $\lhd^a$ by Claim 6.

Finally, we claim that $\delta t^{-1}zuv_\delta d$ is unbordered. Indeed, suppose on the contrary that there exists a shortest border hd. Since δa does not occur in uv and hd has to be shorter than u, we deduce that $|h| < |\delta|$. The maximality of g implies that $|h| \leq |g|$. But now hd is a suffix of gd whence hc is a factor of δa; a contradiction, since $hd \leq_{\mathrm{p}} \delta$ and $hd \lhd^a hc$. □

2.3 Case Analysis

In order to complete the proof we distinguish the following cases.

A Special Case. Consider first the special case where $t \neq v$ and z is empty and $u = \alpha$. It is not difficult to see that $uutb$ is unbordered.

Indeed, suppose that hb is the shortest border of $uutb$. The choice of u implies $|h| < |u|$, and since u is unbordered, we have $h \leq_{\mathrm{s}} t$. Now hb is a prefix of u and ha a factor of u; a contradiction to the maximality of α.

By $|uutb| \geq |w|/2 + 1 > \frac{3}{7}|w| + 1$ we can exclude this case.

Case 1. Let now either $t = v$ or $t' = v'$ but not both. By symmetry, we can suppose $t \neq v$ and $t' = v'$. Note that the assumption that v' is as short as possible does not harm the symmetry.

We are now going to show that the inequalities we have obtained in the previous subsections do not have a common solution. It is an exercise in the application of the simplex algorithm.

Inequality (3) can be transformed into

$$\mathrm{L}_1 := |r| - |t| - |t'| + |z| - 1 \geq 0. \tag{6}$$

The inequalities (4) and (5) imply,

$$|\delta| + |z| + |u| < \frac{3}{7}|w|,$$

which together with $|\delta| \geq |r|$ and $|w| = |v| + |v'| + 2|u| + |z|$ yield

$$\mathrm{L}_2 := 3|v'| + 3|v| - |u| - 4|z| - 7|r| > 0. \tag{7}$$

Similarly, if we use (5) with $|\delta| \geq |y_\delta|$, we obtain

$$\mathrm{L}_3 := 7|t| - 4|v| + 3|v'| - |u| - 4|z| + 7 > 0. \tag{8}$$

One can now check that under the assumption $|v'| = |t'|$ we have

$$28\,\mathrm{L}_1 + 4\,\mathrm{L}_2 + 3\,\mathrm{L}_3 + 7\,|v'ut| = -7,$$

a contradiction.

Case 2. Suppose $t \neq v$ and $t' \neq v'$. By the special case above, we can assume

$$\mathrm{L}_0 := |u| - |\alpha| + |z| - 1 \geq 0. \tag{9}$$

Inequality (2) can be transformed into

$$\mathrm{L}_4 := |t| + |\alpha| - |r| - |v| - 1 \geq 0, \tag{10}$$

The symmetry of v and v' yields, as a mirror variant of (8), the inequality

$$\mathrm{L}_5 := 7|t'| + 3|v| - 4|v'| - |u| - 4|z| + 7 > 0. \tag{11}$$

One can now check that

$$7\,\mathrm{L}_0 + 21\,\mathrm{L}_1 + 2\,\mathrm{L}_2 + 2\,\mathrm{L}_3 + 7\,\mathrm{L}_4 + 3\,\mathrm{L}_5 = 0,$$

again a contradiction.

Now, we have already proved that if $\tau(w) < \frac{3}{7}|w| + 1$, then v is a prefix of zu, and v' is a suffix of uz. It remains to consider this one more case.

Case 3. Let $t = v$ and $t' = v'$. Then $\pi(w) \leq |uz|$. Clearly, we can suppose that $\pi(w) > |u|$, since otherwise trivially $\pi(w) = \tau(w) = |u|$. Let $w = rs$ be a critical factorization of u. Then szr is unbordered of length $\pi(w)$, unless r is a prefix, and s is a suffix of z; see Remark 4. Suppose the latter possibility. Now, either one of the words uz and zu is unbordered of length $\pi(w)$ or u is both prefix and suffix of z. We are therefore left with the case $w = v'u^iz'u^jv$, with $i, j \geq 2$, where u is not a suffix of uz' and not a prefix of $z'u$. Moreover, v' is a suffix of u and v is a prefix of u. The assumption $\pi(w) > |u|$ now implies that z' is nonempty. Suppose, without loss of generality, $i \leq j$.

Similarly as above, we have that either $sz'u^{j-1}r$ or $z'u^j$ is unbordered. From $|u^jz'| < \frac{3}{7}|w| + 1$ we deduce

$$|v'v| > \left(\frac{4}{3}j - i\right)|u| + \frac{4}{3}|z'| - \frac{7}{3} \geq \left(\frac{4}{3}j - i\right)|u| - 1\,. \tag{12}$$

Case 3.1. $i = j$. If v' is a suffix of uz' and v a prefix of $z'u$, then we have $\pi(w) = \tau(w) = |z'u^j|$. Otherwise we obtain from Case 1 and Case 2 an unbordered factor of $v'uz'uv$ of length at least $\frac{3}{7}|v'uz'uv| + 1$. Moreover, this factor contains u as a factor, which can be substituted with u^j to obtain an unbordered factor of w of length at least $\frac{3}{7}|v'u^jz'u^jv| + 1$.

Case 3.2. $i < j$. Since $j \geq 3$, we obtain from (12) that $|v'v| \geq 2|u| - 1$; a contradiction with Claim 5.

This concludes the proof of Theorem 2.

References

1. Assous, R., Pouzet, M.: Une caractérisation des mots périodiques. Discrete Math. 25(1), 1–5 (1979)
2. Césari, Y., Vincent, M.: Une caractérisation des mots périodiques. C. R. Acad. Sci. Paris Sér. A 286, 1175–1177 (1978)
3. Crochemore, M., Perrin, D.: Two-way string-matching. J. ACM 38(3), 651–675 (1991)
4. Duval, J.-P.: Périodes et répétitions des mots du monoïde libre. Theoret. Comput. Sci. 9(1), 17–26 (1979)
5. Duval, J.-P.: Relationship between the period of a finite word and the length of its unbordered segments. Discrete Math. 40(1), 31–44 (1982)
6. Duval, J.-P., Harju, T., Nowotka, D.: Unbordered factors and Lyndon words. Discrete Math. 308(11), 2261–2264 (2008) (submitted in 2002)
7. Ehrenfeucht, A., Silberger, D.M.: Periodicity and unbordered segments of words. Discrete Math. 26(2), 101–109 (1979)
8. Harju, T., Nowotka, D.: Minimal Duval extensions. Internat. J. Found. Comput. Sci. 15(2), 349–354 (2004)
9. Harju, T., Nowotka, D.: Periodicity and unbordered words. In: Diekert, V., Habib, M. (eds.) STACS 2004. LNCS, vol. 2996, pp. 294–304. Springer, Heidelberg (2004)
10. Harju, T., Nowotka, D.: Periodicity and unbordered words: A proof of the extended Duval conjecture. J. ACM 54(4) (2007)
11. Holub, Š.: A proof of the extended Duval's conjecture. Theoret. Comput. Sci. 339(1), 61–67 (2005)
12. Mignosi, F., Zamboni, L.Q.: A note on a conjecture of Duval and Sturmian words. Theor. Inform. Appl. 36(1), 1–3 (2002)
13. Schützenberger, M.-P.: A property of finitely generated submonoids of free monoids. In: Algebraic theory of semigroups (Proc. Sixth Algebraic Conf., Szeged, 1976). Colloq. Math. Soc. János Bolyai, vol. 20, pp. 545–576. North-Holland, Amsterdam (1976)

Applications of Effective Probability Theory to Martin-Löf Randomness

Mathieu Hoyrup[1] and Cristóbal Rojas[2]

[1] LORIA - 615, rue du jardin botanique, BP 239
54506 Vandœuvre-lès-Nancy, France
hoyrup@loria.fr

[2] Institut de Mathématiques de Luminy, Campus de Luminy, Case 907
13288 Marseille Cedex 9, France
rojas@iml.univ-mrs.fr

Abstract. We pursue the study of the framework of *layerwise computability* introduced in a preceding paper and give three applications. (i) We prove a general version of Birkhoff's ergodic theorem for random points, where the transformation and the observable are supposed to be *effectively measurable* instead of *computable*. This result significantly improves V'yugin and Nandakumar's ones. (ii) We provide a general framework for deriving sharper theorems for random points, sensitive to the speed of convergence. This offers a systematic approach to obtain results in the spirit of Davie's ones. (iii) Proving an effective version of Prokhorov theorem, we positively answer a question recently raised by Fouché: can random Brownian paths reach any random number? All this shows that *layerwise computability* is a powerful framework to study Martin-Löf randomness, with a wide range of applications.

1 Introduction

Algorithmic randomness emerged as an early achievement of Kolmogorov's program to base probability theory on the theory of computing. Yet a framework allowing the combination of these two theories is still lacking: for instance, computable analysis is mainly concerned with effective versions of topological notions, and not probabilistic/measure-theoretic ones. For this reason, the study of algorithmic randomness has not reached its expected range of application: general probability theory. Let us recall the main contributions of algorithmic randomness to probability theory developed so far.

Theorems for random points. The main novelty brought by algorithmic randomness is that probabilistic laws can be strengthened in principle, holding at every random point and not only with probability one. Classical examples can be found in [1, 2, 3] for instance. When proving this kind of result the key hypothesis is the *computability* of the random variables involved. However, it is well-known that computability notions are the effective versions of topological ones (the computable functions are precisely the effectively continuous ones, the

S. Albers et al. (Eds.): ICALP 2009, Part I, LNCS 5555, pp. 549–561, 2009.

semi-decidable sets are precisely the effectively open sets, and so on). Hence the computability assumption on random variables is (i) inappropriate in principle, as probability theory is grounded on measure theory and not on topology; (ii) a priori too strong, as in the classical setting only properties as measurability, integrability are required. This leads to the following:

Problem 1. Theorems for random points should hold for "effectively measurable" objects and not only computable ones.

This problem has already been independently investigated in [4,5] where ergodic theorems for random points are proved for different types of "almost everywhere computable" functions. These works are, however, still far from catching the effective version of measurable functions. For instance in Birkhoff's ergodic theorem, nothing can be said about the mean sojourn time of algorithmically random points in fractal sets having effective constructions, as the *Smith-Volterra-Cantor* (or *fat Cantor*) set $A \subseteq [0,1]$, which is homeomorphic to the Cantor set and has Lebesgue measure $\frac{1}{2}$.

Information given by the randomness degree. A further contribution of algorithmic randomness to probability theory consists in making use of the "randomness degree" of a random point x to get additional information about the way x satisfies a given probabilistic law. For instance in [6], the speed of convergence in the Strong Law of Large Numbers is computed from the *compressibility coefficient*, or *deficiency of randomness* of each random sequence. This kind of result gives a much sharper insight into probabilistic phenomena and, we believe, new tools are needed in order to make this approach systematic and applicable on abstract spaces:

Problem 2. Having a general framework to get sharper theorems for random points, using the information given by the randomness degree.

Layerwise computability. In [7], working in the context of computable probability spaces (to which Martin-Löf randomness has been recently extended, see [8, 9]), effective versions of measure-theoretic notions were examined and another contribution of algorithmic randomness to probability theory was developed: the setting of a new framework for computability adapted to the probabilistic context. This was achieved by making a fundamental use of the existence of a universal Martin-Löf test to endow the space with what we call the *Martin-Löf layering*. In this new framework, which we call ***layerwise computability***, the *layerwise* versions of virtually all computability notions can be naturally defined. The contributions of this setting can be summarized in the following principle, supported by the main results in [7]:

Correspondence Principle (CP). Under effectivity assumptions, measure-theoretic notions correspond exactly to *layerwise* versions of topological ones.

Intuitively, this gives evidence that the *layering structure* grasps a large part of the probabilistic phenomena: each probabilistic notion, that by nature intimately depends on the underlying measure μ, can be expressed without referring to μ but only to its imprint on the space, captured by the *layering*. In this paper, elaborating on [7] and developing layerwise computability further, we give solutions to Problems 1 and 2. The CP is at the core of these solutions, that we briefly present now.

Solution to Problem 1. We prove general versions of theorems for random points and *effectively measurable* random variables, in particular Birkhoff's ergodic theorem. This is a significant improvement of [4,5] as it implies in particular a positive result for the Smith-Volterra-Cantor set. To prove these results we develop tools allowing to adapt the existent techniques (used in the *computable* context) to the *layerwise computable* context. Then, the results for *effectively measurable* objects follow from the CP. This strategy is very general and applicable in a wide range of situations.

Solution to Problem 2. As a further illustration of the CP we prove that under effectivity assumptions, *almost everywhere* convergence corresponds to the *layerwise* version of *uniform* convergence. This result gives evidence that the *layering* encodes information from which sharper results can be stated, providing a systematic approach to obtain results in the spirit of [6]. In particular, we use it to compute the speed of convergence of random points in both the Strong Law of Large Number and the Ergodic Theorem, in their general versions. The explicit connection between our framework and [6] is also given.

Our framework also enables us to give a simple answer to a question raised in [10] for algorithmically random Brownian motion (see Sect. 5.3).

In Sect. 2 we recall the background on computable probability spaces and Martin-Löf randomness and prove the effective version of a Prokhorov's result. In Sect. 3 we set the framework for *layerwise computability* and state the results relating it to effective measurability. In Sect. 4 we study the convergence of random variables from the effective point of view. We finish in Section 5 by applying all this machinery to obtain the general results announced above, giving solutions to Problems 1 and 2.

2 Preliminaries

2.1 Computable Probability Spaces

We work on the well-studied computable metric spaces (see [11]).

Definition 1. *A* **computable metric space** *is a triple* $(X, d, \mathcal{S})$ *where:*

1. *(X, d) is a separable metric space,*
2. *$\mathcal{S} = \{s_i : i \in \mathbb{N}\}$ is a countable dense subset of X with a fixed numbering,*
3. *$d(s_i, s_j)$ are uniformly computable real numbers.*

$\mathcal{S}$ is called the set of ***ideal points***. If $x \in X$ and $r > 0$, the metric ball $B(x, r)$ is defined as $\{y \in X : d(x, y) < r\}$. The set $\mathcal{B} := \{B(s, q) : s \in \mathcal{S}, q \in \mathbb{Q}, q > 0\}$ of ***ideal balls*** has a canonical numbering $\mathcal{B} = \{B_i : i \in \mathbb{N}\}$. An ***effectively open set*** is an open set U such that there is a r.e. set $E \subseteq \mathbb{N}$ with $U = \bigcup_{i \in E} B_i$. A compact set K is ***effectively compact*** if the set $\{\langle i_1,\dots,i_n\rangle : K \subseteq B_{i_1} \cup \dots \cup B_{i_n}\} \subseteq \mathbb{N}$ is r.e. Let $K \subset X$. A set V is ***effectively open in*** K if there is an effective open set U such that $V \cap K = U \cap K$. A set V is ***decidable in*** K if V and $X \setminus V$ are effectively open in K. A function $f : X \to Y$ is ***computable on*** K if the preimages of effectively open sets are effectively open in K, in a uniform way. A real function $f : X \to [-\infty, +\infty]$ is ***lower semi-computable*** if the sets $f^{-1}(q_i, +\infty)$ are uniformly effectively open, it is ***upper semi-computable*** if $-f$ is lower semi-computable. Any object that has some effectivity can be naturally encoded into a (possible more than one) integer, called its *Gödel number.*

Remark 1. Let K be effectively compact. It is not difficult to see that the complement $X \setminus K$ is an effective open set, uniformly in K, and that if U is an effective open set, then $K \setminus U$ is effectively compact, uniformly in U, K.

Several approaches to the computability of Borel probability measures have been proposed and happen to give the same notion, which can then be considered as a robust one.

Definition 2 (from [12,13,9]). *Let $(X, d, \mathcal{S})$ be a computable metric space. A Borel probability measure μ on X is* **computable** *if $\mu(B_{i_1} \cup \dots \cup B_{i_n})$ are lower semi-computable, uniformly in $i_1, \dots, i_n$.*

A **computable probability space** *is a pair (X, μ) where X is a computable metric space and μ is a computable Borel probability measure on X.*

Algorithmic Randomness. Martin-Löf randomness was first defined in [1] on the space of infinite symbolic sequences. Its generalization to abstract spaces has been investigated in [8, 9, 14, 15]. We follow the approaches [8, 9] developed on any computable probability space (X, μ).

Definition 3. *A* **Martin-Löf test (ML-test)** *V is a sequence of uniformly effective open sets V_n such that $\mu(V_n) < 2^{-n}$. A point x* **passes** *a ML-test V if $x \notin \bigcap_n V_n$. A point is* **Martin-Löf random (ML-random)** *if it passes all ML-tests. The set of ML-random points is denoted by ML_μ.*

Theorem 1 (adapted from [1]). *Every computable probability space (X, μ) admits a universal Martin-Löf test, i.e. a ML-test U such that for all $x \in X$, x is ML-random $\iff$ x passes the test U. Moreover, for each ML-test V there is a constant c (computable from any Gödel number of V) such that $V_{n+c} \subseteq U_n$ for all n.*

From now and beyond, we fix a particular universal ML-test U. One can assume w.l.o.g. that $U_{n+1} \subseteq U_n$. When the underlying space is complete, even if it is unbounded the finite character of probability measures makes the probabilistic phenomena concentrate in a small region. This is formally expressed by Prokhorov's theorem: on a complete separable metric every Borel probability measure is *tight.* We prove its effective version:

Theorem 2 (Effective Prokhorov theorem). *On a complete computable metric space, every computable Borel probability measure is effectively tight: the sets $K_n := X \setminus U_n$ are uniformly effective compact sets and $\mu(K_n) > 1 - 2^{-n}$.*

Effective Measurability. The following is an adaptation of [7] to complete spaces.

Definition 4. *A set A is* **effectively μ-measurable** *if there are uniformly effective compact sets C_n and open sets U_n such that $C_n \subseteq A \subseteq U_n$ and $\mu(U_n \setminus C_n) < 2^{-n}$. A function $f : X \to Y$ is* **effectively μ-measurable** *if there is a basis $\hat{\mathcal{B}} = \{\hat{B}_1, \hat{B}_2, \ldots\}$ of Y effectively equivalent to $\mathcal{B}$ such that $f^{-1}(\hat{B}_i)$ are uniformly effectively μ-measurable.*

In the original definition the sets C_n are complements of effective open sets. When the space is complete, requiring C_n to be effectively compact gives the same notion, using effective tightness (Thm. 2) and Rmk. 1. Any effective open set whose μ-measure is computable, as the complement of the fat Cantor set for the Lebesgue measure, is effectively μ-measurable. Conversely, the measure of any effectively μ-measurable set is computable. Decidable sets and computable functions are always effectively μ-measurable, whatever the computable measure μ may be. A set is effectively μ-measurable if and only if so is its indicator function.

3 Layerwise Computability

Now we enter in the main novelty of this article. With effective versions of measure-theoretic notions at our disposal, we can hope to solve Problem 1. However the notions developed so far are difficult to handle and rather heavy, see Def. 4. It was demonstrated in [7] that algorithmic randomness and the universal test offer an alternative elegant way of handling effective measurability notions.

Let (X, μ) be a *complete* computable probability space. It comes with a canonical universal ML-test U_n, with $U_{n+1} \subseteq U_n$. Hence the set of ML-random points is *layered* by an increasing sequence of uniformly effective compact sets: $\mathrm{ML} = \bigcup_n K_n$ (Thm. 2).

Definition 5 (Martin-Löf Layering). *Let (X, μ) be a computable probability space. We call the sequence $(K_n)_{n\in\mathbb{N}}$ the* **Martin-Löf layering** *of the space.*

Definition 6 (Layerwise computability notions)

1. *A set $A \subseteq X$ is* **layerwise semi-decidable** *if for all n, A is effectively open on K_n, uniformly in n, i.e. there are uniformly effective open sets U_n such that $A \cap K_n = U_n \cap K_n$,*
2. *A set $A \subseteq X$ is* **layerwise decidable** *if for all n, A is decidable on K_n, uniformly in n, i.e. both A and $X \setminus A$ are layerwise semi-decidable,*
3. *A function $f : (X, \mu) \to Y$ is* **layerwise computable** *if for all n, f is computable on K_n, uniformly in n, i.e. $f^{-1}(B_i)$ are uniformly layerwise semi-decidable (B_i are the canonical ideal open balls).*

More generally, every computability (or *effective topological*) notion has its layerwise counterpart. For instance, the layerwise counterpart of effective uniform convergence (i.e., in the sup norm) of functions will be examined in Sect. 4. In general, layerwise computability is not stable under composition, simply because layerwise computable functions may not preserve randomness. This can be overcome under measure-preservation:

Proposition 1 (from [7]). *Let (X, μ) be a computable probability space, $T : X \to X$ a layerwise computable function which preserves μ (i.e. $\mu \circ T^{-1} = \mu$) and $f : X \to Y$ a layerwise computable function. Then:*

1. *T preserves ML-randomness. Moreover there is a contant c such that $T(K_n) \subseteq K_{n+c}$ for all n.*
2. *$f \circ T$ is layerwise computable, uniformly in f and T.*

3.1 Relation with Effective Measurability

In [7], we prove the following equivalences:

Theorem 3. *Let $A \subseteq X$ be a set and $f : X \to Y$ a function.*

1. *A is effectively μ-measurable $\iff$ A is layerwise decidable,*
2. *$f : X \to Y$ is effectively μ-measurable $\iff$ f is layerwise computable.*

Therefore, under effectivity assumptions measure-theoretical notions are the layerwise versions of topological ones. Observe that the latter are expressed without referring to μ but only to the Martin-Löf layering. In other words, the layering catches the essential part of the probabilistic features. This is the first illustration of the Correspondence Principle (see Introduction).

3.2 Layerwise Tests

We now state the theorem which will allow to solve Problem 1: making theorems on random points hold under effective measurability assumptions. The surprising point is that weakening randomness tests to their layerwise versions leave them close to plain tests and does not spawn higher-order tests. This will enable us to strengthen many existing results.

Definition 7. *A* **layerwise Martin-Löf test** *A is a sequence of uniformly layerwise semi-decidable sets A_n such that $\mu(A_n) < 2^{-n}$. A* **layerwise integrable test** *is a layerwise lower semi-computable function $t : X \to [0, +\infty]$ such that $\int t \, d\mu < \infty$.*

Theorem 4. *Let U be a layerwise semi-decidable set, A a layerwise ML-test and t a layerwise integrable test.*

1. *If $\mu(U) = 1$ then $ML_\mu \subseteq U$.*
2. *If x is ML-random, then $x \notin \bigcap_n A_n$. Moreover, there is a constant c (computable from a Gödel number of the sequence A) such that $A_{n+c} \cap K_n = \emptyset$ for all n.*
3. *If x is ML-random, then $t(x) < \infty$. Moreover, there is a constant c such that $t < 2^{n+c}$ on K_n.*

4 Convergence of Random Variables

In [3] the following result for convergence of random variables on random points is stated: if computable functions converge almost everywhere in an effective way then they converge on ML-random points. Here we improve this in several ways:

- using layerwise tests, we weaken the hypothesis: the functions are now assumed to be effectively measurable only, which gives a solution to Problem 1,
- using the layering, we get information about the speed of convergence on random points, providing a solution to Problem 2,
- under effectivity assumptions, we get a characterization of a probabilistic notion (namely, almost everywhere convergence) as the layerwise version of a topological one (namely, uniform convergence), which further illustrate the Correspondence Principle, beyond Theorem 3.
- we give other results for random points under different types of assumptions on the convergence of the sequence.

Observe that what follows works on any computable probability space (algorithmic randomness was only developed on the Cantor space when [3] was written). Let $f_i : X \to \mathbb{R}$ be a sequence of random variables and f another random variable (expected to be the limit of f_i). Let $D_n(\delta) := \{x : \exists i \geq n, |f_i - f| > \delta\}$. It is a standard observation that the sequence f_i converge almost everywhere to f if and only if the measure of the sets $D_n(\delta)$ tends to zero, for each δ. This motivates the following:

Definition 8. *Functions f_n converge* **effectively almost everywhere (effectively a.e.)** *if $\mu(D_n(\delta))$ converge to 0, effectively from δ. In other words there is a computable function $n(\delta, \varepsilon)$ such that $\mu(D_{n(\delta,\varepsilon)}(\delta)) < \varepsilon$.*

As already said, V'yugin [3] proved that if f_n are uniformly *computable* functions that converge effectively a.e. then they converge at each ML-random point. Actually, the result also holds when the functions f_n are uniformly *effectively μ-measurable* and we can even go further. Let us first introduce the layerwise version of effective convergence for the uniform norm:

Definition 9. *Functions f_i* **converge layerwise effectively uniformly** *to f if for each k, the restrictions of f_i to K_k converge to the restriction of f to K_k for the uniform norm, effectively from k. In other words, there is a computable function $n(\delta, k)$ such that $\|f_i - f\|_{K_k} := \sup_{K_k} |f_i - f| \leq \delta$ for all $i \geq n(\delta, k)$.*

In the same way that uniform convergence implies pointwise convergence, such functions converge on each ML random point.

Proposition 2. *If f_i are uniformly layerwise computable functions that converge layerwise effectively uniformly to f then f is layerwise computable.*

As said above, effective a.e. convergence implies layerwise effective uniform convergence. Actually this is a characterization, which provides another illustration of the Correspondence Principle:

Theorem 5. *Let f_n be uniformly effectively μ-measurable functions. Then f_n converge effectively a.e. if and only if f_n converge layerwise effectively uniformly.*

At the same time, this result gives evidence that layerwise computability is a solution to Problems 1 and 2: this convergence for random points holds for *effectively μ-measurable* functions and not only *computable* ones, and the speed of convergence can be computed from the layer a random point belongs to.

Corollary 1. *If f_i are uniformly effectively μ-measurable functions that converge effectively a.e. to f, then f is effectively μ-measurable.*

The simplicity of this proof shall be very general, as soon as the Correspondence Principle holds: a result about effective measure-theoretical notions can be split into two parts (i) the layerwise version of the corresponding result for effective topological (i.e. computability) notions (as Prop. 2) and (ii) the characterizations of effective measure-theoretical notions as layerwise topological ones (as Thm. 3 and 5).

Non-effective Convergence. When the convergence a.e. is not effective we can still say something concerning random points.

Theorem 6. *Let f_n, f be uniformly layerwise computable functions, and c some (not necessarily computable) constant.*

- *If f_n converge a.e. to a constant c then* $\liminf f_n(x) \leq c \leq \limsup f_n(x)$ *for all $x \in ML$.*
- *If f_n converge a.e. to a layerwise computable function f, then* $\liminf f_n(x) \leq f(x) \leq \limsup f_n(x)$ *for all $x \in ML$.*

5 Applications

5.1 Ergodic Theorems for Effectively Measurable Functions

We now apply the tools developed so far to solve Problem 1 for Poincaré recurrence theorem and Birkhoff's ergodic theorem. The version of the latter theorem for random points has been proved by V'yugin [3] (i) on the Cantor space $X = \{0,1\}^{\mathbb{N}}$ and (ii) for a computable transformation $T : X \to X$ and a computable observable $f : X \to \mathbb{R}$. Point (i) is not a real restriction as the proof for general spaces remains unchanged. However the condition (ii) is an unnatural and *a priori* too strong restriction, as explained in the introduction. Moreover, on general spaces this restriction becomes much more important since the theorem cannot be applied to indicators of sets anymore (in connected spaces they cannot be computable as they are not continuous).

In [4], Birkhoff's ergodic theorem for random points is extended to include functions having some discontinuities at computable points. A further step is given in [5] where the result is proved to hold for the indicator functions of every (not necessarily constructive) set of continuity (i.e., a set whose boundary has measure zero). Yet, nothing can be said about some natural sets having effective constructions, like the ***Smith-Volterra-Cantor set*** (or *fat Cantor set*) whose Lebesgue measure is $\frac{1}{2}$ but has empty interior, and hence is not a continuity set.

We now give a definite solution by proving the ergodic theorems for *effectively measurable* functions. In particular, indicator functions of sets like the fat Cantor set fall in this class.

Theorem 7 (Poincaré recurrence theorem for random points). *Let (X, μ) be a computable probability space, $T : X \to X$ an effectively μ-measurable measure-preserving map and A a layerwise semi-decidable set with positive measure. Every ML-random point $x \in A$ falls infinitely often in A by iteration of T. If the system is moreover ergodic, then every ML-random point falls infinitely often in A by iteration of T.*

Theorem 8 (Ergodic theorem for random points). *Let (X, μ) be a computable probability space, $T : X \to X$ an effectively μ-measurable measure-preserving map and $f \in L^1(X, \mu)$ be an effectively μ-measurable observable. Then:*

(i) For every ML-random point x, the limit $\overline{f}(x) := \lim_{n\to\infty} \frac{1}{n} \sum_{i=0}^{n-1} f \circ T^i(x)$ exists.
(ii) If the system is moreover ergodic, then $\overline{f}(x) = \int f \, \mathrm{d}\mu$ for every ML-random x.

5.2 Layerwise Computable Speed of Convergence on Random Points

In this section we show how the framework developed so far provides a solution to Problem 2. Let us first recall some results established by Davie [6].

Davie's Results. To state them some background is needed first. On the Cantor space, implicitely endowed with the uniform measure λ, the *compressibility coefficient* or *deficiency of randomness* of a sequence ω is defined as $d_\lambda(\omega) = \sup_n\{n - H(\omega_{1:n})\}$ where $H(w)$ is the *Kolmogorov-Chaitin complexity* of the finite word w. A fundamental result from algorithmic randomness and information theory is that a sequence is ML-random w.r.t. λ if and only if $d_\lambda(\omega)$ is finite. Davie defines $K^c := \{\omega : d_\lambda(\omega) \leq c\}$ and proves:

Theorem 9 (Davie, 2001). *If A_i are uniformly effective open sets such that $\sum_i \mu(A_i)$ is a finite computable real number, then there is a computable function $n(c)$ such that for all $\omega \in K^c$ and all $m > n(c)$, $\omega \notin A_m$.*

Theorem 10 (Davie, 2001). *There is a computable function $n(c, \varepsilon)$ such that for all $\omega \in K^c$ and all $n > n(c, \varepsilon)$, $\left|\frac{S_n(\omega)}{n} - \frac{1}{2}\right| < \varepsilon$ where $S_n(\omega)$ is the number of ones in the prefix of ω of length n.*

The equivalence between the paradigm of effective measure theory (Martin-Löf's approach) and the paradigm of compressibility (Chaitin's approach) is a strong non-trivial result, partly based on the technical *coding theorem*. Davie's results follow this line as they relate the compressibility coefficient of a sequence to the way the sequence satisfies a probability law, and thus their proofs consist in a finer use of the coding theorem. In our framework, we stay on the side of effective measure theory. In this way, the relation between the layer a random point belongs to and the way it satisfy laws is much simpler to derive, as it is essentially already contained in the existing proofs. This provides a solution to Problem 2. At the same time, as layerwise computability provides a solution to Problem 1 too, our results hold for effectively measurable sets/functions.

As an illustration, we first state here the refined version of classical results in algorithmic randomness due to Solovay. The proofs are straightforward combinations of the usual proofs together with Thm. 4. Note that the first one is the generalization of Thm. 9 due to Davie.

Proposition 3 (Borel-Cantelli 1). *There is a computable function $n(c,p)$ such that if A_n are uniformly layerwise semi-decidable sets such that $\alpha := \sum_n \mu(A_n)$ is finite and computable, then there is a constant c, computable from a Gödel number of the sequence A_n and α, such that if $x \in K_p$ then $x \notin A_n$ for all $n \geq n(c,p)$.*

We can also get a weaker result when the sum is not computable.

Proposition 4 (Borel-Cantelli 2). *Let A_i be uniformly layerwise semi-decidable sets such that $\sum_i \mu(A_i) < \infty$. There is c, computable from a description of the sequence A_i, such that every x in K_n falls in the A_i's at most 2^{n+c} times.*

We can now easily prove:

Theorem 11 ((Very) Strong Law of Large Numbers.). *Let $X_i : (X,\mu) \to \mathbb{R}$ be i.i.d. effectively μ-measurable random variables such that $\int |X_i|^4 \, d\mu < +\infty$. Let $S_n := X_0 + \cdots + X_{n-1}$. Hence, there is a computable function $n(c,\varepsilon)$ such that if $x \in K_c$ then for all $n > n(c,\varepsilon)$, $\left|\frac{S_n(x)}{n} - \int X_0 \, d\mu\right| < \varepsilon$.*

Effective Convergence in Birkhoff's Theorem. The convergence of the Birkhoff averages is not effective in general. In [3], on the Cantor space V'yugin builds a computable probability measure which is invariant under the shift transformation, and such that the convergence of the averages of $\mathbf{1}_{[1]}$ is not effective. This measure is an infinite combination of ergodic measures and it is still an open question if a computable ergodic measure could be built for which the convergence is not effective.

However, in [16] it is shown that for a class of ergodic systems, the convergence in Birkhoff theorem is effective. Let us recall that a system is ergodic if and only if for any two integrable functions f and g, the quantity $\gamma_n(f,g) :=$

$\left|\frac{1}{n}\sum_{i<n}\int f\circ T^i.g\,\mathrm{d}\mu - \int f\,\mathrm{d}\mu\int g\,\mathrm{d}\mu\right|$ goes to 0. A system is said to be **$\ln^2$-*ergodic for* (f,g)**, if there is a constant $c_{f,g}>0$ such that $\gamma_n(f,g)\leq\frac{c_{f,g}}{(\ln(n))^2}$ for all $n\geq 2$.

Theorem 12 (Ergodic theorem for random points). *Let (X,μ) be a computable probability space, $T: X\to X$ be an effectively measurable measure-preserving map and $f\in L^1(X,\mu)$ be an effectively measurable function. If T is $\ln^2$-ergodic for (f,f), then there is a computable function $n(c,\varepsilon)$ such that if $x\in K_c$ then for all $n>n(c,\varepsilon)$, $\left|\frac{1}{n}\sum_{i<n}f(x)\circ T^i(x)-\int f\,\mathrm{d}\mu\right|<\varepsilon$.*

Relation between K_n and K^n. Let X be the Cantor space endowed with a computable Borel probability measure μ. The version of the *compressibility coefficient* or *deficiency of randomness* adapted to μ is $d_\mu(\omega):=\sup_n\{-\log\mu[\omega_{1:n}]-H(\omega_{1:n})\}$. This function is known to be the logarithm of a universal integrable μ-test, which means that for every integrable μ-test t there is a constant a such that $\log t\leq a+d_\mu$. On the other hand, every computable probability space admits a universal integrable test t_μ (see [8, 9]). Generalizing Davie, let us define $K^c:=\{x: t_\mu(x)\leq 2^c\}$. As $\mathrm{ML}_\mu=\bigcup_c K^c$, the sequence $(K^c)_{c\in\mathbb{N}}$ can be used as an alternative layering and underly alternative versions of Def. 6 and 9. Actually, this would lead to the same notions. Indeed, using classical results from algorithmic randomness and information theory (see [17,18]), it can be proved that there is a constant c such that $K_n\subseteq K^{n+c}$ and $K^n\subseteq K_{n+2\log n+c}$ for all n. Hence K^n are also uniformly effective compact sets and all layerwise computability notions relative to K^n are equivalent to the notions relative to K_n.

5.3 An Application to Brownian Motion

The study of Brownian motion from the algorithmic randomness point of view is carried out in [19,10]. Algorithmically random paths, called *complex oscillations* as they are defined in terms of Kolmogorov-Chaitin complexity, are the Martin-Löf random points of the computable probability space $(\mathcal{C}([0,1]),W)$, where $\mathcal{C}([0,1])$ is the space of continuous functions $x:[0,1]\to\mathbb{R}$ with the uniform norm and W is the Wiener probability measure. In [19] it is proved that if $t\in[0,1]$ is computable and x is a complex oscillation then $x(t)$ is not computable. At the end of [10] the following question is raised: can it be lower semi-computable?

We say that $y\in\mathbb{R}$ is λ-ML-random if $y=n+z$ where $n\in\mathbb{Z}$ and $z\in[0,1]$ is ML-random w.r.t. the Lebesgue measure λ on $[0,1]$. As noticed in [10], it is a corollary of [20] that $x(t)$ is actually λ-ML-random. But then can it be a Chaitin's Ω (which are lower semi-computable λ-ML-random reals)? The compactness of the layers (Thm. 2) enables us to give a positive answer. Indeed, Prop. 1 can be reinforced using Thm. 2:

Proposition 5. *Let (X,μ) and (Y,ν) be computable probability spaces such that X is complete. Let $T: X\to Y$ be a layerwise computable function which maps μ to ν. Then $T(ML_\mu)=ML_\nu$, i.e. T preserves randomness but it is also* onto.

Now, given a computable $t \in (0,1]$, the function $T_t : \mathcal{C}([0,1]) \to \mathbb{R}$ mapping x to $x(t)$ is computable. It pushes the Wiener measure W to a gaussian measure G. As G has bounded density w.r.t. the uniform measure and vice versa, ML_G is exactly the set of λ-ML-random reals. Hence,

Corollary 2. *Let x be a complex oscillation. For each computable $t \in (0,1]$, $x(t)$ is λ-ML-random. Moreover, given any λ-ML-random y and any non-zero computable t, there exists a complex oscillation x such that $x(t) = y$.*

References

1. Martin-Löf, P.: The definition of random sequences. Information and Control 9(6), 602–619 (1966)
2. Vovk, V.G.: The law of the iterated logarithm for random kolmogorov, or chaotic, sequences. Theory of Probability and Applications 32, 413–425 (1987)
3. V'yugin, V.V.: Effective convergence in probability and an ergodic theorem for individual random sequences. SIAM Theory of Probability and Its Applications 42(1), 39–50 (1997)
4. Nandakumar, S.: An effective ergodic theorem and some applications. In: STOC 2008: Proceedings of the 40th annual ACM symposium on Theory of computing, pp. 39–44. ACM Press, New York (2008)
5. Galatolo, S., Hoyrup, M., Rojas, C.: Effective symbolic dynamics, random points, statistical behavior, complexity and entropy. Inf. Comput. (in press, 2007)
6. Davie, G.: The Borel-Cantelli lemmas, probability laws and Kolmogorov complexity. Annals of Probability 29(4), 1426–1434 (2001)
7. Hoyrup, M., Rojas, C.: An application of Martin-Löf randomness to effective probability theory. In: Proceedings of CiE 2009. LNCS. Springer, Heidelberg (2009)
8. Gács, P.: Uniform test of algorithmic randomness over a general space. Theoretical Computer Science 341, 91–137 (2005)
9. Hoyrup, M., Rojas, C.: Computability of probability measures and Martin-Löf randomness over metric spaces. Inf. and Comput. (in press, 2009)
10. Fouché, W.L.: Dynamics of a generic brownian motion: Recursive aspects. Theor. Comput. Sci. 394(3), 175–186 (2008)
11. Weihrauch, K.: Computable Analysis. Springer, Berlin (2000)
12. Edalat, A.: When Scott is weak on the top. Mathematical Structures in Computer Science 7(5), 401–417 (1997)
13. Schröder, M.: Admissible representations of probability measures. Electronic Notes in Theoretical Computer Science 167, 61–78 (2007)
14. Zvonkin, A., Levin, L.: The complexity of finite objects and the development of the concepts of information and randomness by means of the theory of algorithms. Russian Mathematics Surveys 256, 83–124 (1970)
15. Hertling, P., Weihrauch, K.: Random elements in effective topological spaces with measure. Information and Computation 181(1), 32–56 (2003)
16. Galatolo, S., Hoyrup, M., Rojas, C.: A constructive Borel-Cantelli lemma. Constructing orbits with required statistical properties. Theor. Comput. Sci. (in press, 2007)

17. Li, M., Vitanyi, P.M.B.: An Introduction to Kolmogorov Complexity and Its Applications. Springer, Berlin (1993)
18. Gács, P.: Lecture notes on descriptional complexity and randomness, http://www.cs.bu.edu/faculty/gacs/papers/ait-notes.pdf
19. Fouché, W.L.: Arithmetical representations of brownian motion i. J. Symb. Log. 65(1), 421–442 (2000)
20. Kjos-Hanssen, B., Nerode, A.: The law of the iterated logarithm for algorithmically random brownian motion. In: Artemov, S., Nerode, A. (eds.) LFCS 2007. LNCS, vol. 4514, pp. 310–317. Springer, Heidelberg (2007)

An EPTAS for Scheduling Jobs on Uniform Processors: Using an MILP Relaxation with a Constant Number of Integral Variables*

Klaus Jansen

Institut für Informatik, Universität zu Kiel, Germany
kj@informatik.uni-kiel.de

Abstract. In this paper, we present an efficient polynomial time approximation scheme (EPTAS) for scheduling on uniform processors, i.e. finding a minimum length schedule for a set of n independent jobs on m processors with different speeds (a fundamental NP-hard scheduling problem). The previous best polynomial time approximation scheme (PTAS) by Hochbaum and Shmoys has a running time of $(n/\epsilon)^{O(1/\epsilon^2)}$. Our algorithm, based on a new mixed integer linear programming (MILP) formulation with a constant number of integral variables and an interesting rounding method, finds a schedule whose length is within a relative error ϵ of the optimum, and has running time $2^{O(1/\epsilon^2 \log(1/\epsilon)^3)} poly(n)$.

1 Introduction

We consider the following fundamental problem in scheduling theory. Suppose that we are given a set $\mathcal{J}$ of n independent jobs J_j with processing time p_j and a set $\mathcal{P}$ of m non-identical processors P_i that run at different speeds s_i. If job J_j is executed on processor P_i, the machine needs p_j/s_i time units to complete the job. The problem is to find an assignment $a : \mathcal{J} \to \mathcal{P}$ for the jobs to the processors that minimizes the total execution time, $\max_{i=1,\ldots,m} \sum_{J_j:a(J_j)=P_i} p_j/s_i$. This is the minimum time needed to complete the execution of all jobs on the processors. The problem is denoted $Q||C_{max}$ and it is also called the minimum makespan problem on uniform parallel processors. We may assume that the number m of processors is bounded by the number of jobs (otherwise select only the fastest n machines in $O(m)$ time). Furthermore, for simplicity we suppose that $s_1 \geq s_2 \geq \ldots \geq s_m$ (otherwise we have to sort the speed values).

Results. The problem for uniform (and also identical) processors has been demonstrated to be NP-hard [5,7] and the existence of a polynomial time algorithm for it would imply $P = NP$. Hochbaum and Shmoys [10,11] presented a family of polynomial time approximation algorithms $\{A_\epsilon | \epsilon > 0\}$ for scheduling on identical and uniform processors, where each algorithm A_ϵ generates a schedule of length $(1+\epsilon)OPT(I)$ for each instance I and has running time polynomial in the input size $|I|$. Such a family of algorithms is called a polynomial

* Research supported in part by EU research project AEOLUS, Algorithmic Principles for Building Efficient Overlay Computers, EU contract number 015964.

S. Albers et al. (Eds.): ICALP 2009, Part I, LNCS 5555, pp. 562–573, 2009.

time approximation scheme (PTAS). It is allowed that the running time of each algorithm A_ϵ is exponential in $1/\epsilon$. In fact, the running time of the PTAS for uniform processors by Hochbaum and Shmoys [11] is $(n/\epsilon)^{O(1/\epsilon^2)}$. If ϵ is small, then the running time of the algorithm can be very large. Two restricted classes of approximation schemes were defined that avoid this problem. An efficient polynomial time approximation scheme (EPTAS) is a PTAS with running time $f(1/\epsilon)poly(|I|)$ (for some function f), while a fully polynomial time approximation scheme (FPTAS) runs in time $poly(1/\epsilon, |I|)$ (polynomial in $1/\epsilon$ and the size $|I|$ of the instance). Since the scheduling problem on uniform (and also identical) processors is NP-hard in the strong sense (as it contains bin packing and 3-partition as special cases) [5], we cannot hope for an FPTAS. For identical processors, Hochbaum and Shmoys (see [9]) and Alon at el. [1] gave an EPTAS with running time $f(1/\epsilon) + O(n)$, where f is doubly exponential in $1/\epsilon$. The existence of an EPTAS for uniform processors is mentioned as an open problem by Epstein and Sgall [3]. Our main result is the following:

Theorem 1. *There is an EPTAS (a family of algorithms $\{A_\epsilon | \epsilon > 0\}$) which, given an instance I of $Q||C_{max}$ with n jobs and m processors with different speeds and a positive number $\epsilon > 0$, produces a schedule for the jobs of length $A_\epsilon(I) \leq (1+\epsilon)OPT(I)$. The running time of A_ϵ is $2^{O(1/\epsilon^2 \log(1/\epsilon)^3)} poly(n)$.*

Interestingly, the running time of our EPTAS is singly exponential in $1/\epsilon$.

Methods. We use the dual approximation method proposed by Hochbaum and Shmoys [11] to transform the scheduling problem into a bin packing problem with different bin sizes. Next, we structure the input by rounding bin sizes and processing times to values of the form $(1+\delta)^i$ and $\delta(1+\delta)^i$ with $i \in \mathbb{Z}$, respectively. After sorting the bins according to their sizes, $c_1 \geq \ldots \geq c_m$, we build three groups of bins: $\mathcal{B}_1$ with the largest K bins (where K is constant). Let G be the smallest index such that capacity $c_{K+G+1} \leq \gamma c_K$ where $\gamma < 1$ depends on ϵ (such an index G exists for $c_m \leq \gamma c_K$). In this case $\mathcal{B}_2$ is with the set of the next G largest bins where the maximum size $c_{max}(\mathcal{B}_2) = c_{K+1}$ divided by the minimum size $c_{min}(\mathcal{B}_2) = c_{K+G}$ is bounded by a constant $1/\gamma$ and $\mathcal{B}_3$ is the set with the remaining smaller bins of size smaller than γc_K. This generates a gap of constant size between the capacities of bins in $\mathcal{B}_1$ and $\mathcal{B}_3$. If the rate c_m/c_K (where c_m is the smallest bin size) is larger than the constant γ, then we obtain a simpler instance with only two groups $\mathcal{B}_1$ and $\mathcal{B}_2$ of bins. For $\mathcal{B}_1$ we compute all packings for the very large items (those which only fit there).

If there is a feasible packing, then we set up a mixed integer linear program (MILP), or an integer linear program (ILP) in the simpler case, to place the other items into the bins. The placement of a large item into the second group $\mathcal{B}_2$ is done via integral configuration variables (similar to the ILP formulation for bin packing by Fernandez de la Vega and Lueker [4]). We use fractional configuration variables for the placement of large items into $\mathcal{B}_3$. Furthermore, we use additional fractional variables to place small items into $\mathcal{B}_1$, $\mathcal{B}_2$, and $\mathcal{B}_3$. The MILP (and the ILP in the simpler case) has only a constant number of integral variables and, therefore, can be solved via the algorithm by Lenstra or

Kannan [13,14]. In order to avoid that the running time is doubly exponential in $1/\epsilon$, we use a recent result by Eisenbrand and Shmonin [2] about integer cones.

To apply their result we consider a system of equalities for the integral configuration variables and round the corresponding coefficients. Then each feasible solution of the modified MILP and ILP contains at most $O(1/\delta \log(1/\delta)^2)$ integral variables with values larger than zero. By choosing the strictly positive integral variables in the MILP and ILP, we are able to reduce the number of integral configuration variables from $2^{O(1/\delta \log(1/\delta))}$ to $O(1/\delta \log(1/\delta)^2)$. The number of choices is bounded by $2^{O(1/\delta^2 \log(1/\delta)^3)}$. Next, we consider a rounded version of the modified smaller MILP and ILP formulations in order to solve the corresponding LP feasibility problem more efficiently. Although we still have a huge number of variables, one can solve the LP feasibility problem for the MILP via the separation problem of the dual linear program and then using techniques from Grötschel, Lovasz and Schrijver [8].

Afterwards, we round the fractional variables in the MILP solution to integral values. In the first phase of the rounding we reduce the number of strictly positive fractional configuration variables for each block B_ℓ (that contains bins with similar capacities) from $2^{O(1/\delta \log(1/\delta))}$ to $O(1/\delta \log(1/\delta))$ using ideas from [12]. Afterwards we round down each such fractional variable to the next smaller integral value. In the second phase we transform a system of (in-)equalities for the other variables corresponding to the packing of the small items into a scheduling problem on unrelated machines. The fractional solution of the scheduling problem can be rounded into another solution with only few fractional values using ideas from [15]. The corresponding remaining fractional variables in the system of (in-)equalities are rounded down again to the next integral values. The effect of the rounding is that most of the items can be placed directly into the bins. Only a few of them cannot be placed this way, and here is where the K largest bins and the gap between $\mathcal{B}_1$ and $\mathcal{B}_3$ come into play. We prove that these items can be moved to the K largest bins by increasing their size only slightly.

Organization of the paper. In Section 2 we give definitions, notations and show how to structure the input and how to define the three bin groups. In Section 3 and 4 we consider the more general case with three groups of bins. In Section 3.1 we set up our MILP relaxation and in Section 3.2 we show how to solve it. Then in Section 4.1 we describe the rounding technique and in Section 4.2 we show how to pack the jobs via the rounded MILP solution. Here we bound also the total size of items that cannot be placed directly into bins. In the full paper we give additional details about how to solve and round the MILP.

2 Modifying the Input

First, we compute a 2-approximate solution using the algorithm by Gonzales et al. [6]. It generates a schedule of length $B(I) \leq 2OPT(I)$. Then we take the interval $[B(I)/2, B(I)]$ and use binary search to test values for the optimum or approximate schedule. In the following we choose a value $\delta < \epsilon$ such that $1/\delta$ is integral (we specify the value later). Notice that $OPT(I) \in [B(I)/2, B(I)]$

and that the length $(\delta/2)B(I) \leq \delta OPT(I)$. That implies that the interval $[B(I)/2, B(I)]$ can be divided into $1/\delta$ subintervals of length $\delta B(I)/2$ and that there is at least one subinterval $[B(I)/2+i(\delta/2)B(I), B(I)/2+(i+1)(\delta/2)B(I)]$ with $i \in \{0, \ldots, 1/\delta - 1\}$ that contains the optimum length $OPT(I)$. To find one of these intervals, we use a standard dual approximation method that for each value T either computes an approximate schedule of length $T(1+\alpha\delta)$ (where α is a constant) or shows that there is no schedule of length T. The scheduling problem can be transformed into a bin packing problem with variable bin sizes as described by Hochbaum and Shmoys [11]. For a given value T for the makespan we can generate m bins with capacities $c_i = T \cdot s_i$. Using the ordering of the speed values we have $c_1 \geq c_2 \geq \ldots \geq c_m$. The goal is now to find a packing for the jobs into these m bins. Let us round the processing time p_j of each job to the next pale $\bar{p}_j$ of the form $\delta(1+\delta)^{k_j}$ with $k_j \in \mathbb{Z}$, so $p_j \leq \bar{p}_j = \delta(1+\delta)^{k_j} \leq (1+\delta)p_j$. If we have a subset A of jobs with $\sum_{j \in A} p_j \leq c_i$, then the total increased processing time $\sum_{j \in A} \bar{p}_j$ is bounded by $c_i(1+\delta)$. Furthermore, we can round the enlarged capacities $c_i(1+\delta)$ to the next power c'_i of $(1+\delta)$. That implies $c_i(1+\delta) \leq c'_i = (1+\delta)^{\ell_i} \leq c_i(1+\delta)^2$ with $\ell_i \in \mathbb{Z}$. By normalization we may suppose that the minimum capacity $c'_{min} = \min_{i=1,\ldots,m} c'_i = 1$.

Lemma 1. *If there is a feasible packing of n jobs with processing times p_j into m bins with capacities $c_1 \geq \ldots \geq c_m$, then there is also a packing of n jobs with rounded processing times $\bar{p}_j = \delta(1+\delta)^{k_j} \leq (1+\delta)p_j$ into m bins with rounded bin capacities $c'_i = (1+\delta)^{\ell_i} \leq c_i(1+\delta)^2$.*

In the next step we divide the bins into different bin groups. The first bin group $\mathcal{B}_1$ consists of the K largest bins, where $K = O(1/\delta \log(1/\delta))$.

Lemma 2. *If $c_{max}(\mathcal{B})/c_{min}(\mathcal{B}) \leq C$ for some constant C and the set of capacities in $\mathcal{B}$ is $c(\mathcal{B}) = \{(1+\delta)^x, (1+\delta)^{x+1}, \ldots, (1+\delta)^y\}$ with $x, y \in \mathbb{Z}^+$ and $x < y$, then $|c(\mathcal{B})| \leq 2\log(C)/\delta + 1$ for any $\delta \in (0, 1/2]$.*

If $c_{min}(\mathcal{B})/c_{max}(\mathcal{B}') \geq C$ for two bin groups $\mathcal{B}$ and $\mathcal{B}'$ with $C > 1$, then there is a gap of size C between the capacities of the bins in the two groups. Depending on a constant $\gamma = \Theta(\delta^2)$ we have two bin or three bin groups:

Case 1: There is at least one bin with capacity at most $\gamma c'_K$. Let G be the smallest index such that $c'_{K+G+1} \leq \gamma c'_K$. This implies that $c'_{K+G} > \gamma c'_K$. In this case we have three groups of bins: $\mathcal{B}_1 = \{b_1, \ldots, b_K\}$, $\mathcal{B}_2 = \{b_{K+1}, \ldots, b_{K+G}\}$, and $\mathcal{B}_3 = \{b_{K+G+1}, \ldots, b_m\}$. Notice that $\mathcal{B}_2$ has a constant number of different capacities (using $c'_{K+1}/c'_{K+G} \leq c'_{K+1}/\gamma c'_K \leq 1/\gamma$). In addition we obtain a gap of at least $1/\gamma$ between the capacities in $\mathcal{B}_1$ and $\mathcal{B}_3$.

Case 2: All bins have capacity larger than $\gamma c'_K$. This implies that $c'_m > \gamma c'_K$. In this case we have only two groups of bins $\mathcal{B}_1 = \{b_1, \ldots, b_K\}$ and $\mathcal{B}_2 = \{b_{K+1}, \ldots, b_m\}$. In this case $\mathcal{B}_2$ has a constant number of different capacities (using $c'_{K+1}/c'_m \leq c'_{K+1}/\gamma c'_K \leq 1/\gamma$).

Let $\mathcal{B}'_1 = \{b_1, \ldots, b_{K'}\}$ be the subset of $\mathcal{B}_1$ with the bins that have capacity larger than $\delta/(K-1)c_{max}(\mathcal{B}_1)$. By a further modification of the bin packing instances we obtain the following result.

Lemma 3. *If there a solution for the original instance $(\mathcal{J}, \mathcal{M})$ of our scheduling problem with makespan T and corresponding bin sizes $c_1 \geq \ldots \geq c_m$, then there is a feasible packing for instance $(\mathcal{J}, \mathcal{B}_1' \cup \mathcal{B}_2 \cup \mathcal{B}_3)$ or instance $(\mathcal{J}, \mathcal{B}_1' \cup \mathcal{B}_2)$ with rounded bin capacities $\bar{c}_i = (1+\delta)^{\ell_i} \leq c_i(1+\delta)^3$ and rounded processing times $\bar{p}_j = \delta(1+\delta)^{k_j} \leq (1+\delta)p_j$. In addition we have one of the following four scenarios:*

1. *three bin groups $\mathcal{B}_1', \mathcal{B}_2, \mathcal{B}_3$ with gap of $1/\delta$ between $c_{min}(\mathcal{B}_1')$ and $c_{max}(\mathcal{B}_2)$ and gap of $1/\gamma$ between $c_{min}(\mathcal{B}_1)$ and $c_{max}(\mathcal{B}_3)$. Furthermore $\mathcal{B}_2$ has a constant number of different capacities.*
2. *three bin groups $\mathcal{B}_1', \mathcal{B}_2, \mathcal{B}_3$ with gap of $1/\gamma$ between $c_{min}(\mathcal{B}_1)$ and $c_{max}(\mathcal{B}_3)$ and constant number of different capacities in $\mathcal{B}_1' \cup \mathcal{B}_2$.*
3. *two bin groups $\mathcal{B}_1', \mathcal{B}_2$ with gap of $1/\delta$ between $c_{min}(\mathcal{B}_1')$ and $c_{max}(\mathcal{B}_2)$ and constant number of different capacities in $\mathcal{B}_2$.*
4. *two bin groups $\mathcal{B}_1', \mathcal{B}_2$ with a constant number of different capacities in $\mathcal{B}_1' \cup \mathcal{B}_2$.*

Notice that we have a set $\mathcal{J}_{tiny}$ of jobs with tiny processing time $\leq \delta \bar{c}_m$. Let S_{tiny} be the total size of tiny jobs, i.e. $S_{tiny} = \sum_{J_j \in \mathcal{J}_{tiny}} \bar{p}_j$. If there is a feasible schedule with makespan T, then the total processing time $\sum_{j \in \mathcal{J}} \bar{p}_j$ is smaller or equal to the total area of the corresponding bins $\sum_{i=1}^{m} \bar{c}_i$. If this inequality does not hold, then we can discard the choice with makespan T (in fact we have to increase the makespan in this case). Therefore, we can eliminate in a first step all tiny jobs. If there exists a packing for the other jobs into bins of size $\bar{c}_i$, then we can generate a feasible packing for all jobs into enlarged bins of size $\bar{c}_i(1+\delta)$. This can be done by a greedy algorithm that packs the tiny jobs into the free space left (by allowing to use an additional δ-fraction of the capacities). This works, since the processing time of each tiny job is at most $\delta \bar{c}_m \leq \delta \bar{c}_i$ for $i = 1, \ldots, m$ and the inequality above holds.

3 General Case with Three Groups

In this section we consider scenario 1 (with three bin groups). For the other simpler scenarios $2-4$ we refer to the full version of the paper. Let us suppose that $\mathcal{B}_1'$ contains $K' \leq K$ bins with capacities $\bar{c}_1 \geq \bar{c}_2 \geq \ldots \geq \bar{c}_{K'}$ where $\bar{c}_{K'} \geq \delta/((K-1)(\delta+1))\bar{c}_1$ and that $\mathcal{B}_2$ contains further bins $\bar{c}_{K+1} \geq \cdots \geq \bar{c}_m$ with $\bar{c}_{K+1} \leq \delta \bar{c}_{K'}$ (the other bins in $\mathcal{B}_1$ with smaller size can be neglected). This implies that we have a gap between $\mathcal{B}_1'$ and $\mathcal{B}_2$. In the first part of our algorithm we pre-assign the huge jobs with processing time larger than $\delta \bar{c}_{K'}$ to the first $K' \leq K$ machines. Using the properties above, there are at most $K'\bar{c}_1/(\delta \bar{c}_{K'}) \leq K(K-1)(1+\delta)/\delta^2$ many such jobs. If there are more jobs, then there is no feasible solution with makespan T and we are done. Here we use also the fact that $\bar{c}_{K+1} \leq \delta \bar{c}_{K'}$ and that, therefore, the huge jobs fit only on the first K' machines. Now we have to assign the huge jobs to the first K' machines. Since the number of machines $K' \leq K = O(1/\delta \log(1/\delta))$ and the number of jobs $H \leq K(K-1)(1+\delta)/\delta^2 \leq O(1/\delta^4 \log(1/\delta)^2)$ are both constant (where the values depend on

$1/\epsilon$), this can be done in constant time $f(1/\epsilon)$. In fact the number of possible assignments can be bounded by $(1/\delta \log(1/\delta))^{O(1/\delta^4 \log(1/\delta)^2)} \leq 2^{O(1/\delta^4 \log(1/\delta)^3)}$. Again, if there is no feasible assignment, then there is no corresponding schedule with makespan T. As an alternative we compute an approximate solution with accuracy ρ for the huge jobs. For this step we use the PTAS for scheduling on uniform machines [11]. The running time of the PTAS for uniform machines with a constant number of jobs (as above calculated) is $(1/\delta^4 \log(1/\delta)^2)^{O(1/\rho^2)} \leq 2^{O(1/\rho^2 \log(1/\delta))} = 2^{O(1/\delta^2 \log(1/\delta))}$ (using $\rho = \delta$). On the other hand, this increases the first K' bin capacities from $\bar{c}_i$ to $\bar{c}_i(1+\delta)$ for $i = 1, \ldots, K'$ and the makespan from $T(1+\delta)^3$ to $T(1+\delta)^4$. If there is no assignment for the huge jobs into the first K' enlarged bins of group $\mathcal{B}_1'$, then there is no schedule of length T and we are done. Otherwise we will find a feasible approximate assignment with free space $S_0 = \sum_{i=1}^{K'} \bar{c}_i(1+\delta) - \sum_{j=1}^{H} \bar{p}_j$ and use a mixed integer linear program (MILP) as described below.

3.1 The MILP Relaxation

Suppose that the set of different capacities in $\mathcal{B}_2$ and $\mathcal{B}_3$ is denoted by $\{\bar{c}(1), \ldots, \bar{c}(L)\}$ and $\{\bar{c}(L+1), \ldots, \bar{c}(L+N)\}$, respectively. Let $m_1, \ldots, m_{L+N}$ be the number of machines (bins) of size $\bar{c}(\ell) = (1+\delta)^{r_\ell}$ for $\ell = 1, \ldots, L+N$. The m_ℓ machines of the same speed form a block B_ℓ of bins with the same capacity $\bar{c}(\ell)$. We specify in the following the capacity of a bin group by $\bar{c}(\ell)$. Note that $\bar{c}(1) = \bar{c}_{K+1}$, $\bar{c}(L) = \bar{c}_{K+G}$, $\bar{c}(L+1) = \bar{c}_{K+G+1}$ and $\bar{c}(L+N) = \bar{c}_m$. Using our assumptions we have a constant number of different capacities in $\mathcal{B}_2$, i.e. $c_{max}(\mathcal{B}_2)/c_{min}(\mathcal{B}_2) \leq 1/\gamma$. In addition, there is a gap of $1/\gamma$ between $c_{min}(\mathcal{B}_1)$ and $c_{max}(\mathcal{B}_3)$. Furthermore, we have n_j jobs of size $\delta(1+\delta)^{k_j}$ for $j = 1, \ldots, P$ (all with processing time larger than $\delta\bar{c}(L+N) = \delta(1+\delta)^{r_{L+N}}$ and smaller or equal to $\delta\bar{c}_{K'}$). In the MILP below we use $C_1^{(\ell)}, \ldots, C_{h_\ell}^{(\ell)}$ as configurations or multisets with numbers $\delta(1+\delta)^j \in [\delta(1+\delta)^{r_\ell}, (1+\delta)^{r_\ell}]$ (these are large processing times corresponding to B_ℓ) where the total sum is bounded by $\bar{c}(\ell) = (1+\delta)^{r_\ell}$ (the capacity of the bins in block B_ℓ). Let $a(j, C_i^{(\ell)})$ be the number of occurrences of number $\delta(1+\delta)^j$ in configuration $C_i^{(\ell)}$ and let $size(C_i^{(\ell)}) = \sum_j a(j, C_i^{(\ell)})\delta(1+\delta)^j \leq \bar{c}(\ell)$ be the total sum of the numbers in $C_i^{(\ell)}$. In the MILP below we use an integral or fractional variable $x_i^{(\ell)}$ to indicate the length of the configuration $C_i^{(\ell)}$. For each job size $\delta(1+\delta)^{k_j} \leq (1+\delta)^{r_1}$, let a_j be the smallest index in $\{1, \ldots, L+N\}$ such that $\delta(1+\delta)^{k_j} \geq \delta(1+\delta)^{r_{a_j}}$. If there is no such index, then we have a tiny processing time $\delta(1+\delta)^{k_j} < \delta(1+\delta)^{r_{L+N}} = \delta\bar{c}_m$. These jobs are removed in the first step of our algorithm and will be added at the end. In addition for $j = 1, \ldots, P$ and $\ell = 0, \ldots, a_j - 1$ we use variables $y_{j,\ell}$ to indicate the number of jobs of size $\delta(1+\delta)^{k_j}$ to be placed as a small one in group B_ℓ with bin sizes $\bar{c}(\ell) = (1+\delta)^{r_\ell}$. B_0 represents here for simplicity the block with the largest K' bins. Suppose that the first P' (non-huge) job sizes $(1+\delta)^{k_j}$ are within $((1+\delta)^{r_1}, \delta c_{K'}]$. These job sizes do not fit into the bins in group $\mathcal{B}_2 \cup \mathcal{B}_3$. Therefore we use for these job sizes only one variable $y_{j,0} = n_j$ and set $a_j = 0$. We use now the following MILP:

$$\begin{array}{lll}
\sum_i x_i^{(\ell)} & \leq m_\ell & \text{for } \ell = 1, \ldots, L+N \\
\sum_{\ell,i} a(k_j, C_i^{(\ell)}) x_i^{(\ell)} + \sum_{\ell=0}^{a_j-1} y_{j,\ell} & = n_j & \text{for } j = P'+1, \ldots, P \\
\sum_i size(C_i^{(\ell)}) x_i^{(\ell)} + \sum_{j:\ell<a_j} y_{j,\ell}\delta(1+\delta)^{k_j} & \leq m_\ell \bar{c}(\ell) & \text{for } \ell = 1, \ldots, L+N \\
\sum_{j=1}^{P} y_{j,0}\delta(1+\delta)^{k_j} & \leq S_0 &
\end{array}$$

$$\begin{array}{lll}
x_i^{(\ell)} & integral \geq 0 & \text{for } \ell = 1, \ldots, L \text{ and } i = 1, \ldots, h_\ell \\
x_i^{(\ell)} & \geq 0 & \text{for } \ell = L+1, \ldots, L+N \text{ and } i = 1, \ldots, h_\ell \\
y_{j,\ell} & \geq 0 & \text{for } j = P'+1, \ldots, P \text{ and } \ell = 0, \ldots, a_j - 1 \\
y_{j,0} & = n_j & \text{for } j = 1, \ldots, P'
\end{array}$$

In the full version we show the following two results.

Lemma 4. *Each feasible packing for the jobs into the bins corresponds to a feasible solution of the MILP.*

Lemma 5. *The number of variables in the MILP is* $n^2 + n2^{O(1/\delta \log(1/\delta))}$*, the number of integral variables is at most* $2^{O(1/\delta \log(1/\delta))}$*, and the number of constraints (not counting the non-negativity constraints) is at most* $O(n)$*. Furthermore, the cardinalities of the sets* $P_\ell = |\{j \in \{P'+1, \ldots, P\} | \delta(1+\delta)^{k_j} \in (\delta(1+\delta)^{r_\ell}, (1+\delta)^{r_\ell}]\}|$ *(numbers of different large job sizes used in block* B_ℓ*) and the number* L *are bounded by* $O(1/\delta \log(1/\delta))$.

3.2 How to Solve the MILP?

The natural way to solve the MILP with a constant number of integral variables is to use the classical algorithm by Lenstra [14]. This would give running time $d^{O(d^2)} poly(s) = 2^{O(d^2 \log(d))} poly(s)$ where the dimension $d = 2^{O(1/\delta \log(1/\delta))}$ and s is the length of the input. Therefore, the running time can be bounded by $2^{2^{O(1/\delta^2 (\log(1/\delta))^2)}} poly(s)$ — doubly exponential in $1/\delta$. A better way is to use the algorithm by Kannan [13] with running time $d^{O(d)} poly(s)$ and to use a nice result by Eisenbrand and Shmonin [2] about integer cones $int - cone(X) = \{\lambda_1 x_1 + \ldots + \lambda_t x_t | t \geq 0; x_1, \ldots, x_t \in X; \lambda_1, \ldots, \lambda_t \in \mathbb{Z}_{\geq 0}\}$, where $X \subset \mathbb{R}^m$ is a finite set and m corresponds to the number of constraints.

Theorem 2. *[2] Let* $X \subset \mathbb{Z}^m$ *be a finite set of integer vectors and let* $b \in int - cone(X)$*. Then there exists a subset* $\tilde{X} \subset X$ *such that* $b \in int - cone(\tilde{X})$ *and* $|\tilde{X}| \leq 2m \log(4mM)$ *where* $M = max_{x \in X} \|x\|_\infty$.

In our context t corresponds to the number of integral variables, λ_i to the variables and x_i to the vectors with the coefficients of the variables. In the following we show how to apply this result to our integral variables $(x_i^{(\ell)})$. To do this we need integer coefficients with small size. In a second step we round the sizes or processing times of jobs to reduce the length s of the instance and to solve later the corresponding LP in the underlying algorithm of Kannan more efficiently. In the first step, each large size $\delta(1+\delta)^{k_j} \in C_i^{(\ell)}$ is rounded up to the next multiple of $\delta^2(1+\delta)^{r_\ell}$ for $\ell = 1,\ldots,L$. Let $round_\ell[\delta(1+\delta)^{k_j}]$ be the rounded number in block B_ℓ. This generates also modified configurations with total size at most $\bar{c}(\ell)(1+\delta) = (1+\delta)^{r_\ell+1}$ (since at most $1/\delta$ items in each configuration are rounded up). Notice that rounding up different numbers $\delta(1+\delta)^i \in (\delta(1+\delta)^{r_\ell}, (1+\delta)^{r_\ell}]$ up to $round_\ell[\delta(1+\delta)^i]$ generates different rounded numbers. Suppose by contradiction that two rounded numbers are equal $round_\ell[\delta(1+\delta)^{i+1}] = round_\ell[\delta(1+\delta)^i]$. Then the distance between the original numbers $\delta(1+\delta)^{i+1} - \delta(1+\delta)^i = \delta^2(1+\delta)^i$ is at most $\delta^2(1+\delta)^{r_\ell}$, and this is possible only if $i \le r_\ell$. But this is a contradiction, since i should be larger than r_ℓ. Let $\bar{C}_1^{(\ell)},\ldots,\bar{C}_{\bar{h}_\ell}^{(\ell)}$ be the sequence of all configuration or partitions of $\bar{c}(\ell)(1+\delta)$ into the rounded numbers $\{round_\ell[\delta(1+\delta)^{k_j}] | \delta(1+\delta)^{k_j} \in (\delta(1+\delta)^{r_\ell}, (1+\delta)^{r_\ell}]\}$ for $\ell = 1,\ldots,L$. Then, the equality for each job size $j \in \{P'+1,\ldots,P\}$ has now the form $\sum_{\ell,i:round[\delta(1+\delta)^{k_j}]\in\bar{C}_i^{(\ell)}} a(k_j,\bar{C}_i^{(\ell)})x_i^{(\ell)} + \sum_{\ell=0}^{a_j-1} y_{j,\ell} = n_j$ where $a(k_j,\bar{C}_i^{(\ell)})$ is the number of occurrences of the rounded value $round[\delta(1+\delta)^{k_j}]$ in $\bar{C}_i^{(\ell)}$. In addition we have as new constraints $\sum_i size(\bar{C}_i^{(\ell)})x_i^{(\ell)} + \sum_j y_{j,\ell}\delta(1+\delta)^{k_j} \le m_\ell\bar{c}(\ell)(1+\delta)$ for $\ell = 1,\ldots,L$ and $\sum_{j=1}^{P} y_{j,0}\delta(1+\delta)^{k_j} \le S_0$.

Clearly, if there is a feasible solution for the original MILP, then there is also a feasible solution for the MILP with modified coefficients. Now, the values $size(\bar{C}_i^{(\ell)})$ are multiples of $\delta^2(1+\delta)^{r_\ell}$ and bounded by $(1+\delta)^{r_\ell+1}$. By dividing the corresponding constraints by $\delta^2(1+\delta)^{r_\ell}$, the coefficients of $x_i^{(\ell)}$ are integral and bounded by $2/\delta^2$. We can prove now that each feasible integral solution of the modified MILP has at most $O(1/\delta(\log(1/\delta))^2)$ integral variables with values larger than zero. Let $P(\mathcal{B}_2)$ be indices of the large job sizes corresponding to a block $B_\ell \in \mathcal{B}_2$ (i.e. $P(\mathcal{B}_2) = \{j | \delta(1+\delta)^{k_j} \in (\delta(1+\delta)^{r_L}, (1+\delta)^{r_1}]\}$). The cardinality of $P(\mathcal{B}_2)$ can be bounded by $O(1/\delta\log(1/\delta))$. To prove the bound above for the number of integral variables consider for the $x_i^{(\ell)}$ variables of the blocks $B_\ell \in \mathcal{B}_2$ the following system of equalities:

$$\begin{array}{lll} \sum_i x_i^{(\ell)} & = \bar{m}_\ell & \text{for } \ell = 1,\ldots,L \\ \sum_{\ell,i} a(k_j,\bar{C}_i^{(\ell)})x_i^{(\ell)} & = \bar{n}_j & \text{for } j \in P(\mathcal{B}_2) \\ \sum_i \frac{size(\bar{C}_i^{(\ell)})}{\delta^2\bar{c}(\ell)} x_i^{(\ell)} & = Area(\ell,large) & \text{for } \ell = 1,\ldots,L, \end{array}$$

where the values $\bar{m}_\ell$, $\bar{n}_j$ and $Area(\ell,large)$ are given by the feasible solution. Then, the result by Eisenbrand and Shmonin [2] implies that there is an integral solution of this system with at most $2m\log(4mM) = 2(2L+|P(\mathcal{B}_2)|)\log(4(2L+|P(\mathcal{B}_2)|)2/\delta^2) \le O(1/\delta(\log(1/\delta))^2)$ many integral variables with value larger than zero (using $m = 2L + |P(\mathcal{B}_2)|$, and $L, |P(\mathcal{B}_2)| \le O(1/\delta\log(1/\delta))$ and $max_{x\in X}\|x\| \le 2/\delta^2$). Therefore, a feasible solution of the modified MILP contains only $O(1/\delta(\log(1/\delta))^2)$ integral nonzero variables $x_i^{(\ell)} > 0$ with $\ell \in \{1,\ldots, L\}$. For each choice with only $O(1/\delta(\log(1/\delta))^2)$ integral variables for $\mathcal{B}_2$, we set the remaining integral variables equal to 0 and obtain a smaller MILP instance. The number of choices or smaller MILP instances is bounded by $\binom{2^{O(1/\delta \log(1/\delta))}}{O(1/\delta(\log(1/\delta))^2)} \leq 2^{O(1/\delta^2 \log(1/\delta)^3)}$. In order to solve such MILP instances efficiently, we round the coefficients a second time (also in order to reduce the length s of the input). We round here the coefficients $size(\bar{C}_i^{(\ell)})$ for $\ell = 1, \ldots, L$, $size(C_i^{(\ell)})$ for $\ell = L+1, \ldots, L+N$ and the values $\delta(1+\delta)^{k_j}$ for each bin group B_ℓ to the next multiples of $\delta/(2n)(1+\delta)^{r_\ell}$, i.e. to $\bar{a}_{i,\ell}\delta/(2n)(1+\delta)^{r_\ell}$ and to $\bar{b}_{j,\ell}\delta/(2n)(1+\delta)^{r_\ell}$. In this process we have to enlarge the capacities of the bins to $\bar{c}'(\ell) = \bar{c}(\ell)(1+2\delta)$ for each bin group B_ℓ ($\ell = 1, \ldots, L+N$) and the capacities of the first K' bins to $\bar{c}'_i = \bar{c}_i(1+3\delta)$. We obtain modified MILP's with $O(1/\delta \log(1/\delta)^2)$ integral non-negative variables, $n^2 + n2^{O(1/\delta \log(1/\delta))}$ fractional non-negative variables, and the following constraints:

$$\begin{array}{lll}
\sum_i x_i^{(\ell)} & \leq m_\ell & \text{for } \ell = 1, \ldots, L+N \\
\sum_{\ell,i} a(k_j, \bar{C}_i^{(\ell)}) x_i^{(\ell)} + \sum_{\ell=0}^{a_j-1} y_{j,\ell} & = n_j & \text{for } j = P'+1, \ldots, P \\
\sum_i \bar{a}_{i,\ell} x_i^{(\ell)} + \sum_{j:\ell<a_j} \bar{b}_{j,\ell} y_{j,\ell} & \leq \lfloor \frac{m_\ell(1+2\delta)(2n)}{\delta} \rfloor & \text{for } \ell = 1, \ldots, L+N \\
\sum_{j=P'+1}^{P} \bar{b}_{j,0} y_{j,0} & \leq \lfloor \frac{\bar{S}_0^{new}(2n)}{\delta \bar{c}(1)} \rfloor &
\end{array}$$

where $\bar{S}_0^{(new)} = \sum_{i=1}^{K'} \bar{c}_i(1+\delta)^2 - \sum_{j=1}^{H} \bar{p}_j - \sum_{j=1}^{P'} n_j \delta(1+\delta)^{k_j}$. In the full version we give more details about the second rounding and show how to solve each modified MILP instance within $2^{O(1/\delta(\log(1/\delta))^3)} poly(n)$ time. Among all choices we obtain in $2^{O(1/\delta^2(\log(1/\delta))^3)} poly(n)$ time a feasible solution of one of the MILP instances (if there is a schedule with makespan at most T).

4 Generating an Approximate Schedule

4.1 Rounding the MILP Solution

First, we consider one block B_ℓ after another and round the $(x_i^{(\ell)})$ variables. For bins in B_ℓ that belong to the second group $\mathcal{B}_2$, the values of the variables $\bar{x}_i^{(\ell)}$ in our MILP solution are all integral. Let us study now a block B_ℓ that belongs to $\mathcal{B}_3$. Notice that again only a subset $P_\ell = \{j \in P | \delta(1+\delta)^{k_j} \in (\delta(1+\delta)^{r_\ell}, (1+\delta)^{r_\ell}]\}$ of job sizes that are large corresponding to B_ℓ have to be considered and that $|P_\ell|$ can be bounded as before by $\lfloor 2/\delta \log(1/\delta) \rfloor$. We denote with $\bar{m}_\ell$ the fractional number of bins assigned to B_ℓ, i.e. $\bar{m}_\ell = \sum_i \bar{x}_i^{(\ell)}$. In addition, let $n_j^{(\ell)}$ be the fractional number of jobs of size $\delta(1+\delta)^{k_j}$ assigned to block B_ℓ, i.e. $\sum_i a(k_j, \bar{C}_i^{(\ell)}) \bar{x}_i^{(\ell)} = n_j^{(\ell)}$. The total scaled area covered by the configurations is $Area(large, \ell) = \sum_i \bar{a}_{i,\ell} \bar{x}_i^{(\ell)}$ (using the rounded values $\bar{a}_{i,\ell}\delta/(2n)(1+\delta)^{r_\ell}$ for $size(C_i^{(\ell)})$; see also our full version). The generated solution $(\bar{x}_i^{(\ell)})$ of our MILP instance satisfies the following constraints:

$$\begin{array}{lll}
\sum_i x_i^{(\ell)} & = \bar{m}_\ell & \\
\sum_i a(k_j, \bar{C}_i^{(\ell)}) x_i^{(\ell)} & = n_j^{(\ell)} & \text{for } j \in P_\ell \\
\sum_i \bar{a}_{i,\ell} x_i^{(\ell)} & = Area(large, \ell) & \\
x_i^{(\ell)} & \geq 0 & \text{for } i = 1, \ldots, h_\ell.
\end{array}$$

This is a system with $|P_\ell| + 2 \leq O(1/\delta \log(1/\delta))$ linear equalities where all variables $x_i^{(\ell)}$ should be positive. The number q of variables $\bar{x}_i^{(\ell)} > 0$ is at most $2^{O(1/\delta \log(1/\delta))}$ for each block B_ℓ, but can be rounded down to a subset X_ℓ with at most $|P_\ell|+2$ configurations and corresponding variables $(\tilde{x}_i^{(\ell)})$ without violating the linear constraints (see our full version). For each configuration $C_i^{(\ell)} \in X_\ell$ we round down the fractional value $\tilde{x}_i^{(\ell)}$ to the next integral value $\lfloor \tilde{x}_i^{(\ell)} \rfloor \geq \tilde{x}_i^{(\ell)} - 1$. Then we use for the bin group B_ℓ exactly $\lfloor \tilde{x}_i^{(\ell)} \rfloor$ configurations of the type $C_i^{(\ell)}$ and place jobs of size $\delta(1+\delta)^{k_j}$ for $j \in P_\ell$ according to the multiset into the corresponding bins. For each configuration $C_i^{(\ell)} \in X_\ell$ with non-integral $\tilde{x}_i^{(\ell)}$ we need one additional bin to cover all $n_j^{(\ell)}$ jobs of size $\delta(1+\delta)^{k_j}$ for $j \in P_\ell$. For block B_ℓ let $\mathcal{J}_\ell$ be a collection with $\sum_{C_i^{(\ell)} \in X_\ell : \tilde{x}_i^{(\ell)} non-integral} a(k_j, C_i^{(\ell)})$ jobs of size $\delta(1+\delta)^{k_j}$ for $j \in P_\ell$.

The next step is to round the $(\bar{y}_{j,\ell})$ values of the MILP over all bin groups B_ℓ and job sizes $j \in P$. Since the values $\bar{y}_{j,\ell} = n_j$ are integral for each $j \leq P'$, we have only to round the other variables $\bar{y}_{j,\ell}$ with $j > P'$. Let N_j be the fractional total number of jobs assigned as a *small job* of size $\delta(1+\delta)^{k_j}$ to the blocks (i.e. $N_j = \sum_{\ell=0}^{a_j-1} \bar{y}_{j,\ell}$ for $j = P'+1, \ldots, P$). Let $Area(small, \ell)$ be the corresponding total scaled area of small jobs in bin group B_ℓ (i.e. $Area(small, \ell) = \sum_{j:\ell<a_j} \bar{b}_{j,\ell}\, \bar{y}_{j,\ell}$ for $\ell = 1, \ldots, L+N$). For group B_0 (these are the first K or K' bins) we denote with $Area(small, 0)$ the total scaled area of the assigned jobs $\sum_{j=P'+1}^{P} \bar{b}_{j,0}\, \bar{y}_{j,0}$. Then our values $(\bar{y}_{j,\ell})$ satisfy the following system of (in-) equalities:

$$\begin{array}{lll}
\sum_{\ell=0}^{a_j-1} y_{j,\ell} & = N_j & \text{for } j = P'+1, \ldots, P, \\
\sum_{j:\ell<a_j} \bar{b}_{j,\ell}\, y_{j,\ell} & = Area(small, \ell) & \text{for } \ell = 0, \ldots, L+N, \\
y_{j,\ell} & \geq 0 & \text{for } j = P'+1, \ldots, P \text{ and } \ell = 0, \ldots, L+N.
\end{array}$$

In the full version we show how to round the $(\bar{y}_{j,\ell})$ values such that there is at most one fractional variable $\tilde{y}_{j,\ell}$ for each group B_ℓ. Let $\mathcal{J}_B$ be a collection with one job of size $\delta(1+\delta)^{k_j}$ for each fractional variable $\tilde{y}_{j,\ell}$. These jobs are executed later as additional jobs on one of the machines in group B_ℓ. In the full version we show how to obtain the following result:

Lemma 6. *We can round a feasible solution $(\bar{x}, \bar{y})$ of the MILP (with at most $O(1/\delta \log(1/\delta)^2)$ integral variables $x_i^{(\ell)}$ for bins in $\mathcal{B}_2$) into another solution $(\tilde{x}, \tilde{y})$ with $\bar{c}'(\ell) = \bar{c}(\ell)(1+2\delta)$ such that it holds:*

$$\begin{array}{lll}
\sum_i \lfloor \tilde{x}_i^{(\ell)} \rfloor & \leq m_\ell & \text{for } \ell = 1, \ldots, L+N \\
\sum_{\ell,i} a(k_j, C_i^{(\ell)}) \lceil \tilde{x}_i^{(\ell)} \rceil + \sum_{\ell=0}^{a_j-1} \lceil \tilde{y}_{j,\ell} \rceil & \geq n_j & \text{for } j = P'+1, \ldots, P \\
\sum_i size(C_i^{(\ell)}) \lfloor \tilde{x}_i^{(\ell)} \rfloor + \sum_{j:\ell<a_j} \lfloor \tilde{y}_{j,\ell} \rfloor\, \delta(1+\delta)^{k_j} & \leq m_\ell \bar{c}'(\ell) & \text{for } \ell = 1, \ldots, L+N \\
\sum_{j=P'+1}^{P} \lfloor \tilde{y}_{j,0} \rfloor\, \delta(1+\delta)^{k_j} & \leq \bar{S}_0^{(new)}, &
\end{array}$$

where $\tilde{x}_i^{(\ell)} \geq 0$ and $\tilde{y}_{j,\ell} \geq 0$. Furthermore, $|\{i | x_i^{(\ell)} > 0\}| \leq O(1/\delta \log(1/\delta))$ for each block B_ℓ in $\mathcal{B}_3$. Furthermore, for each block B_ℓ with $\ell \in \{0, \ldots, L+N\}$ there is at most one fractional variable $\tilde{y}_{j,\ell}$.

4.2 Packing the Jobs via the Rounded MILP Solution

We place in a first phase the jobs as large ones according to the configurations and $\lfloor \tilde{x}_i^{(\ell)} \rfloor$ values and in a second phase the jobs as small ones according to the $\lfloor \tilde{y}_{j,\ell} \rfloor$ values in slightly enlarged bins. In the third phase we place the tiny jobs in $\mathcal{J}_{tiny}$ in the free space of the bins. This can be done due to the rounding phases and the area constraints. In the placement phases of the small and tiny jobs we have to enlarge the capacities $\bar{c}'(\ell) = \bar{c}(\ell)(1+2\delta)$ of the bins to $\bar{c}'(\ell)(1+\delta)$ for each bin in B_ℓ. In addition, we have to enlarge the capacities of the first K' bins to $\bar{c}'_i(1+\delta)$. After this step we place the set $\mathcal{J}_B$ on the machines: for each non-integral value $\tilde{y}_{j,\ell}$ we place a job of size $\delta(1+\delta)^{k_j}$ on one machine in group B_ℓ. Since for each group B_ℓ there is at most one job size $j \in \{P', \dots, P\}$ with $\tilde{y}_{j,\ell}$ non-integral and this size is small corresponding to the group B_ℓ, this increases the size of one bin in B_ℓ from $\bar{c}'(\ell)(1+\delta)$ to $\bar{c}'(\ell)(1+2\delta)$. Since we could have also one job size for group B_0, the size of one of the largest K bins is also increased to $\bar{c}'_i(1+2\delta)$.

Finally, we bound the total execution time of the non-placed jobs in $\cup_\ell \mathcal{J}_\ell$. For each block B_ℓ in $\mathcal{B}_3$ with $\ell \in \{L+1, \dots, L+N\}$ we obtain $|P_\ell| + 2 = \lfloor 2/\delta \log(1/\delta) \rfloor + 2$ additional bins of size $\bar{c}'(\ell) = \bar{c}(\ell)(1+2\delta) \le (1+\delta)^{r_\ell+2}$. These bins or the corresponding jobs are placed later on the first K machines. In this step we use also the machines $K'+1, \dots, K$. Let us specify $K := \lfloor 2/\delta \log(1/\delta) \rfloor + 2$. Now take one bin per group and estimate the total size of these bins among all groups $\ell = L+1, \dots, L+N$. Using the order $r_{L+1} > r_{L+2} > \dots > r_{L+N}$, the inequality $r_{(L+1)+\ell} \le r_{(L+1)} - \ell$, and the geometric series, we obtain $\sum_{\ell=L+1}^{L+N} \bar{c}(\ell) \le \sum_{\ell=0}^{N-1} (1+\delta)^{r_{L+1}-\ell} \le (1+\delta)^{r_{L+1}} \sum_{\ell=0}^{\infty} \frac{1}{(1+\delta)^\ell} = (1+\delta)^{r_{L+1}+1}/\delta$. Therefore, the sum of the enlarged bin sizes $\sum_{\ell=L+1}^{L+N} \bar{c}(\ell)(1+\delta)^2 \le (1+\delta)^{r_{(L+1)}+3}/\delta$. Now we have $(1+\delta)^{r_{L+1}+3}/\delta \le \delta c'_K$, if and only if $(1+\delta)^{r_{L+1}} \le \delta^2/(1+\delta)^3 c'_K$. Note that $(1+\delta)^{r_{L+1}} = c_{max}(\mathcal{B}_3)$, $c'_K = c_{min}(\mathcal{B}_1)$ and $c_{max}(\mathcal{B}_3) \le \gamma c_{min}(\mathcal{B}_1)$ using the gap construction in Section 2. The property above is satisfied for $\gamma \le \delta^2/(1+\delta)^3$. Therefore, we specify $\gamma := \delta^2/(1+\delta)^3$. In this case the sum of the capacities above is bounded by δ times the minimum capacity among bins in $\mathcal{B}_1$. In other words, we can take one bin per group B_ℓ among all groups in $\mathcal{B}_3$ and the corresponding jobs and place them on one of the K machines. Since the total size of these jobs is at most $\delta c'_K$, this enlarges the size of the i.th bin from $\bar{c}'_i(1+2\delta)$ to $\bar{c}'_i(1+2\delta) + \delta c'_K \le \bar{c}'_i(1+3\delta)$ for $i = 1, \dots, K$. In total we obtain the following result:

Lemma 7. *If there is a feasible solution of an MILP instance with bin capacities $\bar{c}(\ell)$ for blocks $B_\ell \in \mathcal{B}_2 \cup \mathcal{B}_3$ and capacities $\bar{c}_i$ for the first K largest bins, then the entire job set $\mathcal{J}$ can be packed into bins with enlarged capacities $\bar{c}(\ell)(1+2\delta)^2$ for blocks $B_\ell \in \mathcal{B}_2 \cup \mathcal{B}_3$ and enlarged capacities $\bar{c}_i(1+3\delta)^2$ for the first K bins.*

If there is a schedule with makespan at most T (and with corresponding bin sizes c_i), then the Lemma above implies a packing into bins of size at most $c_i(1+\delta)^3(1+3\delta)^2$ and a corresponding schedule of length $T(1+\delta)^3(1+3\delta)^2 \le OPT(1+\delta)^4(1+3\delta)^2 \le OPT(1+16\delta) \le OPT(1+\epsilon)$ for $\delta \le \epsilon/16$ and $\epsilon \le 1$.

We simply set $\delta = \frac{1}{\lceil 16/\epsilon \rceil}$ and obtain $\delta \leq \epsilon/16$, $\delta \geq \epsilon/(16+\epsilon) \geq \epsilon/(16+1)$ and that $1/\delta = \lceil 16/\epsilon \rceil$ is integral.

Acknowledgments. The author thanks Fritz Eisenbrand and Roberto Solis-Oba for many helpful discussions.

References

1. Alon, N., Azar, Y., Woeginger, G.J., Yadid, T.: Approximation schemes for scheduling on parallel machines. Journal on Scheduling 1, 55–66 (1998)
2. Eisenbrand, F., Shmonin, G.: Caratheodory bounds for integer cones. Operations Research Letters 34, 564–568 (2006)
3. Epstein, L., Sgall, J.: Approximation schemes for scheduling on uniformly related and identical parallel machines. Algorithmica 39, 43–57 (2004)
4. Fernandez de la Vega, W., Lueker, G.S.: Bin packing can be solved within $1+\epsilon$ in linear time. Combinatorica 1, 349–355 (1981)
5. Garey, M.R., Johnson, D.S.: Computers and Intractability: A Guide to the Theory of NP-completeness. W.H. Freeman, San Francisco (1979)
6. Gonzales, T., Ibarra, O.H., Sahni, S.: Bounds for LPT schedules on uniform processors. SIAM Journal on Computing 6, 155–166 (1977)
7. Graham, R.J., Lawler, E.L., Lenstra, J.K., Rinnooy Kan, A.H.G.: Optimization and approximation in deterministic sequencing and scheduling: a survey. Annals of Discrete Mathematics 5, 287–326 (1979)
8. Grötschel, M., Lovasz, L., Schrijver, A.: Geometric Algorithms and Combinatorial Optimization. Springer, Heidelberg (1987)
9. Hochbaum, D.S.: Various notions of approximations: good, better, best, and more. In: Hochbaum, D.S. (ed.) Approximation Algorithms for NP-Hard Problems, ch. 9, pp. 346–398. Prentice Hall, Englewood Cliffs (1997)
10. Hochbaum, D.S., Shmoys, D.B.: Using dual approximation algorithms for scheduling problems: practical and theoretical results. Journal of the ACM 34, 144–162 (1987)
11. Hochbaum, D.S., Shmoys, D.B.: A polynomial approximation scheme for scheduling on uniform processors: using the dual approximation approach. SIAM Journal on Computing 17, 539–551 (1988)
12. Jansen, K., Solis-Oba, R., Sviridenko, M.: Makespan minimization in job shops: a linear time approximation acheme. SIAM Journal on Discrete Mathematics 16, 288–300 (2003)
13. Kannan, R.: Minkowski's convex body theorem and integer programming. Mathematics of Operations Research 12, 415–440 (1987)
14. Lenstra, H.W.: Integer programming with a fixed number of variables. Mathematics of Operations Research 8, 538–548 (1983)
15. Plotkin, S.A., Shmoys, D.B., Tardos, E.: Fast approximation algorithms for fractional packing and covering problems. Mathematics of Operations Research 20, 257–301 (1995)

Popular Mixed Matchings*

Telikepalli Kavitha[1], Julián Mestre[2], and Meghana Nasre[1]

[1] Indian Institute of Science, Bangalore, India
{kavitha,meghana}@csa.iisc.ernet.in
[2] Max-Plank-Institut für Informatik, Saarbrücken, Germany
jmestre@mpi-inf.mpg.de

Abstract. We study the problem of matching applicants to jobs under one-sided preferences; that is, each applicant ranks a non-empty subset of jobs under an order of preference, possibly involving ties. A matching M is said to be *more popular* than T if the applicants that prefer M to T outnumber those that prefer T to M. A matching is said to be *popular* if there is no matching more popular than it. Equivalently, a matching M is popular if $\phi(M,T) \geq \phi(T,M)$ for all matchings T, where $\phi(X,Y)$ is the number of applicants that prefer X to Y.

Previously studied solution concepts based on the popularity criterion are either not guaranteed to exist for every instance (e.g., popular matchings) or are NP-hard to compute (e.g., least unpopular matchings). This paper addresses this issue by considering mixed matchings. A *mixed matching* is simply a probability distributions over matchings in the input graph. The function ϕ that compares two matchings generalizes in a natural manner to mixed matchings by taking expectation. A mixed matching P is popular if $\phi(P,Q) \geq \phi(Q,P)$ for all mixed matchings Q.

We show that popular mixed matchings *always* exist and we design polynomial time algorithms for finding them. Then we study their efficiency and give tight bounds on the price of anarchy and price of stability of the popular matching problem.

1 Introduction

We study the problem of matching a set of applicants $\mathcal{A}$ to a set of jobs $\mathcal{J}$ under one-sided preferences. More formally, the input consists of a bipartite graph $G = (\mathcal{A}, \mathcal{J}, E)$ and a rank function $r : E \to \mathbb{Z}$ that captures applicant preferences over the jobs. Given two jobs i and j in $\mathcal{J}$, an applicant a in $\mathcal{A}$ is said to *prefer* i to j if $r(a,i) < r(a,j)$; similarly, the applicant is *indifferent* between i and j if $r(a,i) = r(a,j)$.

For a given applicant a in $\mathcal{A}$, a's preference over jobs extends in a straightforward manner to matchings: Given matchings M and T, we say a prefers

* The second author was supported by an Alexander von Humboldt Fellowship. Part of this work was done when the first author visited MPI für Informatik through the DST-MPG partner group on Efficient Graph Algorithms, IISc Bangalore.

S. Albers et al. (Eds.): ICALP 2009, Part I, LNCS 5555, pp. 574–584, 2009.

matching M to T if a prefers $M(a)$ to $T(a)$, or if a is matched in M but not in T. Let $\phi(M,T)$ be the total number of applicants that prefer M to T:

$$\phi(M,T) = |\{a \in \mathcal{A} : a \text{ prefers } M \text{ to } T\}| . \tag{1}$$

We say M is *more popular* than T and write $M \succ T$, if $\phi(M,T) > \phi(T,M)$. The matching M is *popular* if there is no matching more popular than M.

Definition 1. *A matching M is popular if $\phi(M,T) \geq \phi(T,M)$ for every matching T.*

When dealing with a set of independent agents, it is always desirable that the solution concept used is stable. Popular matchings have this property in the sense that an applicant majority vote cannot force a migration to another matching.

Thus a popular matching seems a stable and desirable answer to the question of how to assign applicants to jobs bearing their preferences in mind. However, popular matchings do not provide a complete answer since there are instances that do not admit any popular matching [8]. Consider the instance $\mathcal{A} = \{a_1, a_2, a_3\}$, $\mathcal{J} = \{j_1, j_2, j_3\}$ where all applicants rank the jobs in the same way, say j_1 is better than j_2, which in turn is better than j_3.

a_1	j_1 j_2 j_3
a_2	j_1 j_2 j_3
a_3	j_1 j_2 j_3

The following three matchings $M_1 = \{(a_1, j_1), (a_2, j_2), (a_3, j_3)\}$, $M_2 = \{(a_2, j_1), (a_3, j_2), (a_1, j_3)\}$, and $M_3 = \{(a_3, j_1), (a_1, j_2), (a_2, j_3)\}$ demonstrate that the more-popular-than relation need not be acyclic since $M_1 \prec M_2 \prec M_3 \prec M_1$. In fact, it is easy to see that this instance admits no popular matching. There are, however, efficient algorithms for determining if a given instance admits a popular matching and computing such a matching provided one exists [3].

In an attempt to deal with the issue that not every instance admits a popular matching, McCutchen [15] proposed a measure that captures how unpopular a matching is. Let $\Delta(M,T)$ be the difference between the number of applicants who prefer T and the number of applicants who prefer M, that is,

$$\Delta(M,T) = \phi(T,M) - \phi(M,T). \tag{2}$$

He defined the *unpopularity margin* of M as $\max_T \Delta(M,T)$ and showed that computing a matching M minimizing this quantity is NP-hard.

Hence, we are faced with the unpleasant prospect of choosing between a solution concept that can be computed efficiently but may not exist, or one that always exists but cannot be computed efficiently. The main contribution of our paper is to introduce a new solution concept based on the popularity criterion that has the best characteristics of previous work. Namely, it is guaranteed to exist and can be computed efficiently.

Mixed matchings. Motivated by the notions of pure and mixed strategies from Game Theory we propose to study popular mixed matchings. A *mixed matching* P is a set $\{(M_1, p_1) \ldots, (M_k, p_k)\}$, where $\sum_{i=1}^{k} p_i = 1$ and for each $i = 1, \ldots, k$, M_i is a matching in G and $p_i \geq 0$. Thus a mixed matching is simply a probability distribution over matchings in G and a pure matching M can be thought of as the mixed matching $\{(M, 1)\}$.

The function $\phi(M, T)$ that allowed us to compare two pure matchings M and T generalizes to mixed matchings in a natural way. For mixed matchings $P = \{(M_1, p_1) \ldots, (M_k, p_k)\}$ and $Q = \{(T_1, q_1) \ldots, (T_l, q_l)\}$ we let $\phi(P, Q)$ be the expected number of applicants that prefer M to T where M and T are drawn from the probability distributions P and Q respectively; in other words,

$$\phi(P, Q) = \sum_{i=1}^{k} \sum_{j=1}^{l} p_i \, q_j \, \phi(M_i, T_j). \tag{3}$$

We are now ready to give the definition of *popular mixed matching.*

Definition 2. *A mixed matching P is popular if $\phi(P, Q) \geq \phi(Q, P)$ for all mixed matchings Q.*

For example, consider the instance above on jobs $\{j_1, j_2, j_3\}$ and applicants $\{a_1, a_2, a_3\}$ with identical preference lists that admits no popular (pure) matching. Consider the mixed matching

$$P = \{(M_1, 1/3), (M_2, 1/3), (M_3, 1/3)\}.$$

It is easy to see that we have $\phi(P, T) \geq \phi(T, P)$ for all matchings T in this graph, which in turn implies that $\phi(P, Q) \geq \phi(Q, P)$ for all mixed matchings Q in this graph. Thus the mixed matching P is popular in the instance above.

The rest of the paper is organized as follows. In Section 2 we prove that every instance admits a popular mixed matching by establishing a connection with a certain exponentially-large zero-sum game. In Section 3 we give efficient algorithms for solving these large game, which translate into efficient algorithms for computing a popular mixed matching. Our technique applies to a larger class of games and may be of independent interest. In Section 4, we give tight bounds on the efficiency of mixed popular matchings using the standard measures of price of anarchy and price of stability.

1.1 Related Work

Popular matchings were first studied by Gardenfors [8] in the context of the stable marriage problem where each side has preferences over members of the other side. When only one side has preferences, Abraham et al. [3] gave polynomial time algorithms to find a popular matching, or to report none exists. Their algorithm takes $O(m+n)$ time when the preference lists are strictly ordered and when the preference lists contain ties, they gave an $O(m\sqrt{n})$ time algorithm. Subsequently, Mahdian [11] showed that a popular matching exists with high probability, when preference lists are randomly constructed, and the number of jobs is a factor of $\alpha \approx 1.42$ larger than the number of applicants. In fact, he showed that a phase transition occurs at α; namely, if the ratio $\frac{|\mathcal{J}|}{|\mathcal{A}|}$ is smaller than α then with high probability popular matchings do not exist.

Manlove and Sng [14] generalized the algorithms of [3] to the case where each job has an associated *capacity*, the number of applicants that it can accommodate. Mestre [16] designed an efficient algorithm for the *weighted* popular matching problem, where each applicant is assigned a priority or weight and the definition of popularity takes into account these priorities.

McCutchen [15] proposed two quantities to measure the *unpopularity* of a matching and showed polynomial time algorithms to compute these measures for any fixed matching. He also showed that the problem of computing a matching that minimizes either of these measures is NP-hard. Huang et al. [9] gave algorithms to compute matchings with bounded values of these unpopularity measures in certain graphs.

The topic of mixed matchings has been studied extensively in the Economics literature; we are aware of [10,5,4,18,2,1,12,13]. This line of research is concerned with designing mechanisms that have a number of properties, such as envy-free, Pareto optimal, and strategy-proof. These properties can be defined in slightly different ways—the interested reader is referred to the article by Katta *et al.* [10] for an excellent overview of the various definitions commonly used. Roughly speaking, a mixed matching is envy-free if no applicant prefers to get other applicant's allocation to his own; it is Pareto optimal if it is not possible to improve someone's allocation without hurting someone else; and the mechanism is strategy proof if the applicants do not have an incentive to lie about their true preferences. Depending on how one defines an applicant's preference over mixed matchings it may [10,5] or may not [18] be possible to achieve simultaneously all these properties. Note that popular matchings take an orthogonal approach to this, placing emphasis not on individual applicants but on their aggregate or majority.

2 Existence of Popular Mixed Matchings

Recall that our input is a bipartite graph $G = (\mathcal{A} \cup \mathcal{J}, E)$ with $n = |\mathcal{A}| + |\mathcal{J}|$ vertices and $m = |E|$ edges, where each applicant $a \in \mathcal{A}$ has a preference list (can include ties) over its neighboring jobs. For ease of exposition we will introduce a unique last-resort job ℓ_a for every $a \in \mathcal{A}$, which we append at the end of a's preference list. This modification does not change the fact of whether the instance has a popular (mixed) matching or not, but it has the benefit that we can restrict our attention to applicant-complete assignments.

Theorem 1. *Every instance admits a popular mixed matching.*

Proof. We will model our problem as a two-person zero-sum game. The rows and columns of the payoff matrix S are indexed by all the possible matchings $M_1, \ldots, M_N$ in G. The (i, j)-th entry of the matrix S is $\Delta(M_i, M_j)$. Recall that

$\Delta(M_i, M_j)$ is the difference between the number of applicants who prefer M_j to M_i and the number of applicants who prefer M_i to M_j. A mixed strategy of the row player is a probability distribution $\pi = \langle p_1, \ldots, p_N \rangle$ over the rows of S; similarly, a mixed strategy of the column player is a probability distribution $\sigma = \langle q_1, \ldots, q_N \rangle$ over the columns of S.

The row player seeks a strategy π so that $\max_\sigma \Delta(\pi, \sigma)$ is minimized, where $\Delta(\pi, \sigma) = \sum_{i=1}^{N} \sum_{j}^{N} p_i \, q_j \, \Delta(M_i, M_j)$. Notice that π and σ can also be regarded as mixed matchings, in which case we can think of the row player as trying to find a mixed matching with least *expected* unpopularity margin.

It follows that the given instance admits a popular mixed matching if and only if

$$\min_{\pi} \max_{\sigma} \sum_{i=1}^{N} \sum_{j=1}^{N} p_i \, q_j \Delta(M_i, M_j) \leq 0.$$

It is easy to see that

$$\min_{\pi} \max_{\sigma} \sum_{i=1}^{N} \sum_{j=1}^{N} p_i \, q_j \Delta(M_i, M_j) \geq \min_{\pi} \sum_{i=1}^{N} p_i \, p_j \, \Delta(M_i, M_j) = 0 \qquad (4)$$

where the inequality follows from taking $\sigma = \pi$ and the equality from the fact that $\Delta(M_i, M_j) = -\Delta(M_j, M_i)$ for all i and j. Thus $\min_\pi \max_\sigma \Delta(\pi, \sigma)$ must be 0 if the instance admits a popular mixed matching.

The column player seeks a strategy σ so that $\min_\pi \Delta(\pi, \sigma)$ is maximized. By Von Neumann's Minimax Theorem [17] we have

$$\min_{\pi} \max_{\sigma} \sum_{i,j=1}^{N} p_i \, q_j \, \Delta(M_i, M_j) = \max_{\sigma} \min_{\pi} \sum_{i,j=1}^{N} p_i \, q_j \, \Delta(M_i, M_j). \qquad (5)$$

We can bound the right hand side of (5) analogous to (4) to get

$$\max_{\sigma} \min_{\pi} \sum_{i,j=1}^{N} p_i \, q_j \, \Delta(M_i, M_j) \leq \max_{\sigma} \sum_{i,j=1}^{N} q_i \, q_j \, \Delta(M_i, M_j) = 0. \qquad (6)$$

Combining Equations (4), (5), and (6) we get $\min_\pi \max_\sigma \Delta(\pi, \sigma) = 0$. Therefore every instance admits a popular mixed matching. □

3 Finding a Popular Mixed Matching

The proof of Theorem 1 implicitly provides an algorithm for computing a popular mixed matching. Namely, given a zero-sum game, its value and mixed strategies attaining that value can be computed using linear programming. Here we need to determine $p_1, \ldots, p_N$ adding up to 1, such that $\sum_{i=1}^{N} p_i \, \Delta(M_i, T) \leq 0$ for all pure matchings T and $p_i \geq 0$ for all i.

Unfortunately, the linear program (shown below) is too large to be solved efficiently.

$$\text{minimize } \tau \tag{LP1}$$

subject to

$$\begin{aligned} \textstyle\sum_i p_i\, \Delta(M_i, T) \leq \tau \quad \forall \text{ matchings } T \\ \textstyle\sum_i p_i = 1 \\ p_i \geq 0 \quad \forall i = 1, \ldots, N. \end{aligned}$$

Because the number of variables *and* the number of constraints in (LP1) is typically exponential, the ellipsoid algorithm cannot be applied directly. In order to do so, we must first reduce the number of variables. This can be achieved by working with *fractional matchings* instead of mixed matchings.

Let $\mathcal{X}$ be the set of fractional $\mathcal{A}$-complete matchings in G, that is,

$$\mathcal{X} = \left\{ x \in \mathbb{R}_+^m \;\middle|\; \begin{array}{l} \sum_{e \in E(a)} x(e) = 1 \text{ for } a \in \mathcal{A} \text{ and} \\ \sum_{e \in E(j)} x(e) \leq 1 \text{ for } j \in \mathcal{J}. \end{array} \right\} \tag{7}$$

where $E(u)$ is the set of edges incident on vertex u.

Let $\widehat{\mathcal{X}}$ be the set of extreme points of $\mathcal{X}$. Note that because $\mathcal{X}$ is integral we have $\widehat{\mathcal{X}} = \mathcal{X} \cap \{0, 1\}^m$, which corresponds to the set of integral $\mathcal{A}$-complete matchings in G.

Clearly, every mixed matching $P = \langle p_1, \ldots, p_N \rangle$ induces a fractional matching $\mathbf{x} = \sum_{M_i} p_i\, I_{M_i}$, where $I_{M_i} \in \{0, 1\}^m$ is the characteristic vector of the matching M_i. In turn, every $\mathbf{x} \in \mathcal{X}$ can be expressed as a convex combination of the extreme points of $\mathcal{X}$. This implies a many-to-one mapping between mixed and fractional matchings. Furthermore, given a fractional matching, we can build in polynomial time an equivalent mixed matching whose support is no larger than m using Carathéodory's Theorem [6].

The plan, therefore, is to replace the mixed matching in (LP1) with a fractional matching. In order to do so, we need to define the function $\Delta(\cdot)$ for fractional matchings $\mathbf{x}_1, \mathbf{x}_2 \in \mathcal{X}$:

$$\Delta(\mathbf{x}_1, \mathbf{x}_2) = \sum_{a \in \mathcal{A}} \sum_{\substack{(a,i) \in E(a) \\ (a,j) \in E(a)}} \mathbf{x}_1(a, i)\, \mathbf{x}_2(a, j)\, \mathsf{vote}_a(i, j). \tag{8}$$

where the term $\mathsf{vote}_a(i, j)$ captures a's willingness to switch from i to j:

$$\mathsf{vote}_a(i, j) = \begin{cases} -1 & \text{if } a \text{ prefers } i \text{ over } j, \\ 1 & \text{if } a \text{ prefers } j \text{ over } i, \\ 0 & \text{if } a \text{ is indifferent between } i \text{ and } j. \end{cases}$$

Lemma 1. *Let P and Q be two mixed matchings and let $\mathbf{x}_1$ and $\mathbf{x}_2$ be their corresponding fractional matchings. Then $\Delta(P, Q) = \Delta(\mathbf{x}_1, \mathbf{x}_2)$.*

Proof. P and Q are probability distributions $\langle p_1, \ldots, p_N \rangle$ and $\langle q_1, \ldots, q_N \rangle$ respectively, over matchings $M_1, \ldots, M_N$.

$$\begin{aligned}
\Delta(P, Q) &= \sum_{g,h=1}^{N} p_g \, q_h \, \Delta(M_g, M_h) \\
&= \sum_{g,h=1}^{N} p_g \, q_h \sum_{a \in \mathcal{A}} \mathsf{vote}_a(M_g(a), M_h(a)) \\
&= \sum_{g,h=1}^{N} p_g \, q_h \sum_{a \in \mathcal{A}} \sum_{\substack{(a,i) \in E(a) \\ (a,j) \in E(a)}} I_{M_g}(a,i) \, I_{M_h}(a,j) \, \mathsf{vote}_a(i,j) \\
&= \sum_{a \in \mathcal{A}} \sum_{\substack{(a,i) \in E(a) \\ (a,j) \in E(a)}} \sum_{g,h=1}^{N} p_g \, q_h \, I_{M_g}(a,i) \, I_{M_h}(a,j) \, \mathsf{vote}_a(i,j).
\end{aligned}$$

Regrouping the terms in the innermost sum we get

$$\begin{aligned}
\Delta(P, Q) &= \sum_{a \in \mathcal{A}} \sum_{\substack{(a,i) \in E(a) \\ (a,j) \in E(a)}} \left(\sum_{g=1}^{N} p_g \, I_{M_g}(a,i) \right) \left(\sum_{h=1}^{N} q_h \, I_{M_h}(a,j) \right) \mathsf{vote}_a(i,j) \\
&= \sum_{a \in \mathcal{A}} \sum_{\substack{(a,i) \in E(a) \\ (a,j) \in E(a)}} \mathbf{x}_1(a,i) \, \mathbf{x}_2(a,j) \, \mathsf{vote}_a(i,j) \\
&= \Delta(\mathbf{x}_1, \mathbf{x}_2).
\end{aligned}$$

□

It should be clear now that the following linear program is equivalent to (LP1).

$$\text{minimize } \tau \qquad \text{(LP2)}$$

subject to

$$\begin{aligned}
\Delta(\mathbf{x}, T) &\leq \tau \quad \forall \text{ matchings } T \\
\mathbf{x} &\in \mathcal{X}
\end{aligned}$$

Unlike (LP1), the linear program (LP2) has only $m + 1$ variables: τ and the coordinates $x_1, \ldots, x_m$ of $\mathbf{x}$. This new program can be solved in polynomial time using the ellipsoid method. To prove this we only need to design a polynomial-time separation oracle, which given an infeasible solution $(\mathbf{x}, \tau)$ returns a violated constraint of (LP2). It is trivial how to test $\mathbf{x} \in \mathcal{X}$ efficiently; so we only need to test whether there is a matching T such that $\Delta(\mathbf{x}, T) > \tau$. This is done by computing the *unpopularity margin* of $\mathbf{x}$.

Definition 3. *The unpopularity margin of* $\mathbf{x} \in \mathcal{X}$ *is* $\max_T \Delta(\mathbf{x}, T)$.

As mentioned earlier, McCutchen [15] studied the above measure for pure matchings. He gave a polynomial time algorithm based on min-cost flows to measure the unpopularity margin $\max_T \Delta(M, T)$ of a given matching M. We describe his algorithm below in the equivalent language of the maximum weight assignment problem and then show that it can be used to compute the unpopularity margin of a fractional matching also.

McCutchen's algorithm to determine the unpopularity margin of a matching T. Given a matching M we want to compute its unpopularity margin. The idea is to define a weight function w_M on the edges of G such that for any matching T, the weight of T under w_M equals $\Delta(M,T)$. Computing a unpopularity margin of M then reduces to solving the maximum weight assignment problem (G, w_M), which can be done in $O(n\,(m+n\log n))$ time [7].

The quantity $\Delta(M,T)$ can be expressed as a sum of individual votes $\sum_{a\in\mathcal{A}} \mathsf{vote}_a(T(a), M(a))$. Therefore, setting $w_M(a,j) = \mathsf{vote}_a(j, M(a))$ achieves the desired effect that $\Delta(M,T) = w_M(T)$.

McCutchen's method readily generalizes to fractional matchings. Let $\mathbf{x}$ be a fractional matching. We define the weight of an edge $(a,j) \in E$ as

$$w_{\mathbf{x}}(a,j) \;=\; \sum_{(a,i)\in E(a)} \mathbf{x}(a,i)\,\mathsf{vote}_a(i,j). \tag{9}$$

It is straightforward to verify that indeed $\Delta(\mathbf{x},T) = w_{\mathbf{x}}(T)$. Computing the unpopularity margin of a fractional matching then reduces to computing a maximum weight assignment.

The procedure for computing the unpopularity margin of a fractional matching provides the necessary oracle to apply the ellipsoid method to solve (LP2). Therefore we can find a popular mixed matching in polynomial time. While this settles the complexity of our problem, the scheme presented is not truly efficient as the ellipsoid algorithm is notoriously slow in practice. We will address this issue now by giving an alternative linear programming formulation whose size is only linear in the size of G.

Theorem 2. *There exists a linear programming formulation for finding a popular fractional matching with $m+n$ variables and constraints.*

Proof. Let us start by rewriting (LP2) as a mathematical program

$$\min_{\mathbf{x}\in\mathcal{X}}\ \max_{\mathbf{z}\in\widehat{\mathcal{X}}}\ w_{\mathbf{x}}(\mathbf{z}). \tag{10}$$

Because $\mathcal{X}$ is integral, the mathematical program (10) is in fact equivalent to the following program

$$\min_{\mathbf{x}\in\mathcal{X}}\ \max_{\mathbf{y}\in\mathcal{X}}\ w_{\mathbf{x}}(\mathbf{y}). \tag{11}$$

The main obstacle that we must overcome in order to formulate (11) as a pure linear program is the non-linear objective $w_{\mathbf{x}}(\mathbf{y})$. We can overcome this hurdle with the aid of linear programming duality. Consider the primal program $\max\,\{w_{\mathbf{x}}(\mathbf{y}) : \mathbf{y} \in \mathcal{X}\}$ and let $\mathcal{D}_{\mathbf{x}}$ denote the feasible region of its dual:

$$\mathcal{D}_{\mathbf{x}} = \left\{\boldsymbol{\alpha} \in \mathbb{R}^{|\mathcal{A}|},\ \boldsymbol{\beta} \in \mathbb{R}_{+}^{|\mathcal{J}|} \;:\; \alpha_a + \beta_j \ge w_{\mathbf{x}}(a,j) \text{ for each } (a,j) \in E\right\}.$$

By the Strong Duality Theorem we get

$$\max_{\mathbf{y}\in\mathcal{X}}\ w_{\mathbf{x}}(\mathbf{y}) \;=\; \min_{(\boldsymbol{\alpha},\boldsymbol{\beta})\in\mathcal{D}_{\mathbf{x}}} \sum_{a\in\mathcal{A}} \alpha_a + \sum_{j\in\mathcal{J}} \beta_j.$$

Therefore, (11) is equivalent to the following succinct linear program

$$\text{minimize} \sum_{a \in \mathcal{A}} \alpha_a + \sum_{j \in \mathcal{J}} \beta_j \tag{LP3}$$

subject to

$$\alpha_a + \beta_j \geq \textstyle\sum_{(a,i) \in E(a)} \mathbf{x}(a,i)\, \mathsf{vote}_a(i,j) \qquad \forall\, (a,j) \in E$$

$$\sum_{(a,j) \in E(a)} \mathbf{x}(a,j) = 1 \qquad \forall a \in \mathcal{A}$$

$$\sum_{(a,j) \in E(j)} \mathbf{x}(a,j) \leq 1 \qquad \forall j \in \mathcal{J}$$

$$\mathbf{x} \in \mathbb{R}_+^m$$

$$\boldsymbol{\beta} \in \mathbb{R}_+^{|\mathcal{J}|}$$

$$\boldsymbol{\alpha} \in \mathbb{R}^{|\mathcal{A}|}$$

Recall that (11) in turn is equivalent to (LP1). The new linear program has only $m + n$ variables and $m + n$ constraints. □

4 Efficiency of Popular Mixed Matchings

Now that we have settled the existence and computability of popular mixed matchings, the next logical step is to study how efficient this solution concept is. Recall that we inserted a last resort job for each applicant. Perhaps the most natural measure of efficiency is the actual cardinality of the matching, that is, the number of applicants that get a real job.

Of course, when we restrict our attention to popular matchings we may be ignoring larger matchings. The question we would like to investigate is how do popular mixed matchings compare to a maximum size matching that is not restricted by the popularity requirement. To answer this question we use the standard analytical tools used to measure the efficiency of Nash equilibria: the price of anarchy and stability.

Theorem 3. *The price of anarchy and the price of stability of the popular matching problem is 2.*

Due to lack of space, the proof of Theorem 3 is left for the journal version of the paper.

5 Concluding Remarks

In this paper we introduced the concept of popular mixed matchings and showed that they enjoy the best qualities of previously proposed solution concepts based in the popularity criterion; namely, they always exist and can be computed efficiently. For simplicity, we focused on the setting where all applicants are treated equally and there is a single copy of every job, but we note here that the same ideas apply to the more general setting where applicants have priorities or weights [16], multiple copies of the jobs are available [14], and two-sided preferences [8].

At the heart of our proofs is an interesting connection between popular matchings and the Nash equilibria of a certain zero-sum game with exponentially-many strategies. In general, finding these equilibria is equivalent to linear programming[1]; in our case, however, by exploiting problem-specific properties we could find an equivalent linear program whose size is polynomial on the size of the input graph. Our technique works for more general zero-sum games whose strategies correspond to the extreme points of a polytope $\mathcal{Q}$ and whose payoff matrix $S(\mathbf{x}, \mathbf{y})$ can be expressed as a bilinear function of fractional vectors $\mathbf{x}, \mathbf{y} \in \mathcal{Q}$.

We leave it as an open problem to design purely combinatorial algorithms for computing popular mixed matchings.

Acknowledgments. We thank Naveen Garg for helpful discussions, Khaled Elbassioni for suggesting the generalization to bilinear payoffs, and Debasis Mishra for asking us about popular fractional assignments.

References

1. Abdulkadiroğlu, T.S.A.: Ordinal efficiency and dominated sets of assignments. Journal of Economic Theory 112, 157–172 (2003)
2. Abdulkadiroğlu, A., Sönmez, T.: Random serial dictatorship and the core from random endowments in house allocation problems. Econometrica 66(3), 689–701 (1998)
3. Abraham, D.J., Irving, R.W., Kavitha, T., Mehlhorn, K.: Popular matchings. SIAM Journal on Computing 37(4), 1030–1045 (2007)
4. Bogomolnaia, A., Moulin, H.: A new solution to the random assignment problem. Journal of Economic Theory 100(2), 295–328 (2001)
5. Bogomolnaia, A., Moulin, H.: A simple random assignment problem with a unique solution. Economic Theory 75(3), 257–279 (2004)
6. Carathéodory, C.: Über den variabilitätsbereich der fourierschen konstanten von positiven harmonischen funktionen. Rend. Circ. Mat. Palermo 32, 193–217 (1911)
7. Edmonds, J., Karp, R.M.: Theoretical improvements in algorithmic efficiency for network flow problems. jacm 19(2), 248–264 (1972)
8. Gardenfors, P.: Match making: assignments based on bilateral preferences. Behavioural Sciences 20, 166–173 (1975)
9. Haung, C.-C., Kavitha, T., Michail, D., Nasre, M.: Bounded unpopularity matchings. In: Gudmundsson, J. (ed.) SWAT 2008. LNCS, vol. 5124, pp. 127–137. Springer, Heidelberg (2008)
10. Katta, A.-K., Sethuraman, J.: A solution to the random assignment problem on the full preference domain. Journal of Economic Theory 131(1), 231–250 (2005)

[1] Linear programming can be reduced to the problem of solving a zero-sum game of roughly the same size.

11. Mahdian, M.: Random popular matchings. In: Proceedings of the 8th ACM Conference on Electronic Commerce, pp. 238–242 (2006)
12. Manea, M.: A constructive proof of the ordinal efficiency welfare theorem. Journal of Economic Theory 141, 276–281 (2007)
13. Manea, M.: Asymptotic ordinal inefficiency of random serial dictatorship. Theoretical Economics (to appear, 2009)
14. Manlove, D., Sng, C.: Popular matchings in the capacitated house allocation problem. In: Azar, Y., Erlebach, T. (eds.) ESA 2006. LNCS, vol. 4168, pp. 492–503. Springer, Heidelberg (2006)
15. McCutchen, R.M.: The least-unpopularity-factor and least-unpopularity-margin criteria for matching problems with one-sided preferences. In: Laber, E.S., Bornstein, C., Nogueira, L.T., Faria, L. (eds.) LATIN 2008. LNCS, vol. 4957, pp. 593–604. Springer, Heidelberg (2008)
16. Mestre, J.: Weighted popular matchings. In: Bugliesi, M., Preneel, B., Sassone, V., Wegener, I. (eds.) ICALP 2006. LNCS, vol. 4051, pp. 715–726. Springer, Heidelberg (2006)
17. Neumann, J.V.: Zur theorie der gesellschaftsspiele. Math. Annalen 100, 295–320 (1928)
18. Zhou, L.: On a conjecture by Gale about one-sided matching problems. Journal of Economic Theory 52(1), 123–135 (1990)

Factoring Groups Efficiently

Neeraj Kayal[1] and Timur Nezhmetdinov[2,⋆]

[1] Microsoft Research Lab India,
196/36 2nd Main, Sadashivanagar Bangalore 560080 India
[2] Lehigh University, Pennsylvania, USA

Abstract. We give a polynomial time algorithm that computes a decomposition of a finite group G given in the form of its multiplication table. That is, given G, the algorithm outputs two subgroups A and B of G such that G is the direct product of A and B, if such a decomposition exists.

1 Introduction

1.1 Background

Groups are basic mathematical structures and a number of computer algebra systems do computations involving finite groups. One very successful area of research has been the design of algorithms that handle a given permutation group G. It is customary to specify G via a small set of generating permutations. Despite the succinctness of such representations, a substantial polynomial-time machinery has developed for computing with permutation groups [Luk]. A major stimulus for this activity was the application to the graph isomorphism problem (it is easily seen for example that the graph isomorphism problem reduces to the problem of computing the intersection of two permutation groups). Ensuing studies resulted in algorithms for deciphering the basic building blocks of the group making available constructive versions of standard theoretical tools [BKL79], [Luk87], [KT88], [Kan85a], [Kan85b], [Kan90], [BSL87]. But problems such as subgroup intersection and computing sylow subgroups are trivial for verbosely encoded groups - groups which are specified via their multiplication tables. On the other hand, the verbose representation begets its own set of problems. The central problem here is to design a polynomial-time algorithm that given two finite groups G_1 and G_2 decides whether they are isomorphic or not. Towards this end, the classification theorem for finite simple groups helps us by giving a polynomial time algorithm for isomorphism of finite simple groups. What prevents us in going from simple groups to arbitary groups is our lack of understanding of 'group products' - how to put together two groups to get a new one. So far, despite much effort [Mil78, AT04, Gal09] not much progress has been made in resolving the complexity of the general group isomorphism

⋆ This work was done while the second author was visiting DIMACS under REU program.

S. Albers et al. (Eds.): ICALP 2009, Part I, LNCS 5555, pp. 585–596, 2009.

problem. For example, even for groups whose derived series has length two, we do not know how to do isomorphism testing in NP$\cap$coNP. A less ambitious goal (a cowardly alternative?) is to develop algorithms that unravel the structure of a verbosely given group. In this work we devise an algorithm that accomplishes one such unravelling.

1.2 Direct Product of Groups

Given two groups A and B one of the most natural ways to form a new group is the direct product, denoted $A \times B$. As a set, the direct product group is the Cartesian product of A and B consisting of ordered pairs (a, b) and the group operation is component-wise.

$$(a_1, b_1) \cdot (a_2, b_2) = (a_1 \cdot a_2, b_1 \cdot b_2).$$

Given (the Cayley representation of) groups A and B, its trivial to compute (Cayley representation of) the group $G = A \times B$. In this article, we consider the inverse problem of factoring or decomposing a group G as a direct product of two of its subgroups. There are some very natural motivations for such a study. The fundamental theorem of finite abelian groups (Theorem 2) states that any finite abelian group can be written uniquely upto permutation as the direct product of cyclic groups of prime power order. This theorem means that the problem of finding an isomorphism between two given abelian groups [Kav07] is essentially the same as the problem of factoring an abelian group. In the general case, the Remak-Krull-Shcmidt theorem (Theorem 3) tells us that the factorization of a group as a direct product of indecomposable groups is "unique" in the sense that the isomorphism class of each of the components of the factorization is uniquely determined. This means that all such decompositions are structurally the same. This motivates us to devise an efficient algorithm which finds such a factorization. Computing the *direct factorization* of a group has also been found to be useful in practice. The computer algebra system GAP (Groups, Algorithms and Programming) implements an inefficient algorithm for this purpose [Hul].

1.3 Algorithm Outline

Notice that if we have an algorithm that efficiently computes *any nontrivial* factorization $G = A \times B$, we can also efficiently compute the complete factorization of G into indecomposable subgroups by recursing on A and B. Therefore we formulate our problem as follows: given a group G, find subgroups A and B such that $G = A \times B$ and both A and B are nontrivial subgroups of G. The algorithm is developed in stages, at each stage we solve a progressively harder version of the GroupDecomposition problem until we arrive at a complete solution to the problem. Each stage uses the solution of the previous stage as a subroutine.

- *G is abelian.* The proof of the fundamental theorem of finite abelian groups (Theorem 2) is constructive and gives a polynomial-time algorithm. This case has also been studied previously and a linear time algorithm is given in [CF07].

- *The subgroup A is known.* In this case we have to just find a B such that $G = A \times B$. We call it the GroupDivision problem and in section 4, an algorithm is devised in two substages.
  - B is abelian.
  - B is nonabelian.
- *A is unknown but an abelian A exists.* We call this the SemiAbelianGroupDecomposition problem and the algorithm is given in section 5.
- *A is unknown and all indecomposable direct factors of G are nonabelian.* In section 6 we describe the algorithm for this most general form of GroupDecomposition.

Let us now assume that we have an efficient algorithm for both GroupDivision and for SemiAbelianGroupDecomposition and outline the algorithm for solving GroupDecomposition using these subroutines. One of the main sources of difficulty in devising an efficient algorithm is that the decomposition of a group is not unique. Indeed, there can be superpolynomially many different decompositions of G. We fix a reference decomposition. We first analyze the different ways a group can decompose and come up with some invariants which do not depend on the particular decomposition at hand. Assume G is decomposable and let us fix a decomposition of G,

$$G = G_1 \times G_2 \times \ldots \times G_t.$$

with each G_i indecomposable. Let $Z_1 \stackrel{\text{def}}{=} \text{Cent}_G(G_2 \times \ldots \times G_t)$, where for any $A \subseteq G$, $\text{Cent}_G(A)$ denotes the subgroup of G consisting of all the elements of G that commutes with every element of A (it is called the centralizer of A in G). Our algorithm first computes Z_1 (the group Z_1 is invariant across different decompositions) and then uses this subgroup in order to solve GroupDecomposition. Notice that $Z_1 = G_1 \times \text{Cent}(G_2 \times \ldots \times G_t)$. By a repeated application of the subroutine SemiAbelianGroupDecomposition , we can obtain a decomposition of Z_1 into

$$Z_1 = H_1 \times Y,$$

where Y is an abelian group and H_1 has no abelian direct factors. An application of theorem 4 allows us to deduce that any such decomposition of Z_1 has the following properties:

1. H_1 is indecomposable and isomorphic to G_1.
2. $\exists Y_1 \trianglelefteq G$ such that $G = H_1 \times Y_1$.

Having obtained H_1, we obtain an appropriate Y_1 by invoking GroupDivision on (G, H_1) and thereby get a decomposition of G. We will now introduce some notation and then outline the procedure used to compute Z_1.

Notation. For a positive integer s, $[s]$ denotes the set $\{1, 2, \ldots, s\}$. We will denote the center of the group G by Z. For a set A of elements of G, $\langle A \rangle$ will denote the subgroup of G generated by the elements in A.

Computing Z_1. From the given group G, we construct a graph Γ_G which has the following properties:

1. The nodes of G correspond to conjugacy classes of G; however not all conjugacy classes of G are nodes of Γ_G. For a connected component Λ of Γ_G, let $\mathrm{Elts}(\Lambda) \subseteq G$ denote the set of all $g \in G$ that are members of some conjugacy class occuring in Λ.
2. (Proposition 3). If a decomposition of G contains t nonabelian indecomposable components then the number of connected components in Γ_G is at least t.
3. (Proposition 3). Let $G = G_1 \times \ldots \times G_t$, with each G_i indecomposable. Let Z be the center of G and let $\Lambda_1, \ldots, \Lambda_s$ be the set of connected components of Γ_G. Then there exists a partition

$$[s] = S_1 \uplus \ldots \uplus S_t$$

such that for any i,

$$G_i \pmod{Z} = \prod_{j \in S_i} \langle \mathrm{Elts}(\Lambda_j) \rangle \pmod{Z}.$$

4. (Proposition 5). The number of connected components $s \leq \log |G|$ and G/Z has a decomposition given by

$$G/Z = \langle \mathrm{Elts}(\Lambda_1) \rangle \pmod{Z} \times \ldots \times \langle \mathrm{Elts}(\Lambda_s) \rangle \pmod{Z}.$$

Now given only the group G and the constructed graph Γ_G, we do not the set $S_1 \subseteq [s]$ apriori. But $s \leq \log |G|$, so we can simply iterate over all possible sets S_1 in just $|G|$ iterations. Let us therefore assume that we have the appropriate S_1. Then the sought-after set Z_1 can be obtained as follows:

$$Z_1 \stackrel{\mathrm{def}}{=} \mathrm{Cent}_G(\bigcup_{j \notin S_1} \mathrm{Elts}(\Lambda_j)).$$

This completes the outline of the algorithm. Let us summarize the algorithm.

Algorithm 1. GroupDecomposition

Input. A group G in the form of a Cayley table.

1. Construct the conjugacy class graph Γ_G associated to the group G.
2. Compute the connected components $\Lambda_1, \ldots, \Lambda_s$ of Γ_G.
3. For each $S_1 \subseteq [s]$ do the following:
   (a) Let $Z_1 \stackrel{\mathrm{def}}{=} \mathrm{Cent}_G(\left\langle \bigcup_{j \notin S_1} \mathrm{Elts}(\Lambda_j) \right\rangle)$.
   (b) By repeated invocations to SemiAbelianGroupDecomposition determine $H_1, Y \trianglelefteq G$ such that $Z_1 = H_1 \times Y$ and H_1 has no abelian direct factors and Y is abelian.
   (c) Invoke GroupDivision on (G, H_1) to determine if there exists a $Y_1 \trianglelefteq G$ such that $G = H_1 \times Y_1$. If such a Y_1 is found then **output** (H_1, Y_1).
4. If no decomposition has been found, **output** *NO SUCH DECOMPOSITION.*

2 Preliminaries

2.1 Notation and Terminology

$\mathrm{Cent}(G)$ will denote the center of a group G and $|G|$ its size. For an element $a \in G$, we will denote $|\langle a \rangle|$ by $\mathrm{ord}(a)$. We will denote the conjugacy class of the element a by $\mathcal{C}_a$, i.e.

$$\mathcal{C}_a \stackrel{\text{def}}{=} \{g \cdot a \cdot g^{-1} \mid g \in G\} \subset G.$$

Let $A, B \subseteq G$. We write $A \leq G$ when A is a subgroup of G and $A \trianglelefteq G$ when A is a normal subgroup of G. $\mathrm{Cent}_G(A)$ will denote the subgroup of elements of G that commute with every element of A. i.e.

$$\mathrm{Cent}_G(A) \stackrel{\text{def}}{=} \{g \in G \mid a \cdot g = g \cdot a \quad \forall a \in A\}.$$

We will denote by $[A, B]$ the subgroup of G generated by the set of elements

$$\{a \cdot b \cdot a^{-1} \cdot b^{-1} | a \in A,\ b \in B\}.$$

We shall denote by $A \cdot B$ the set

$$\{a \cdot b \mid a \in A, b \in B\} \subseteq G.$$

We say that a group G is *decomposable* if there exist nontrivial subgroups A and B such that $G = A \times B$ and *indecomposable* otherwise. When A is a normal subroup of G we will denote by $B \pmod{A}$ the set of cosets $\{A \cdot b \mid b \in B\}$ of the quotient group G/A. We will say that a subgroup A of G is *a direct factor* of G if there exists another subgroup B of G such that $G = A \times B$ and we will call B a *direct complement* of A.

The canonical projection endomorphisms. When a group G has a decomposition

$$G = G_1 \times G_2 \times \ldots \times G_t$$

then associated with this decomposition is a set of endomorphisms $\pi_1, \ldots, \pi_t$ of G with

$$\pi_i : G \mapsto G_i, \quad \pi_i(g_1 \cdot g_2 \cdot \ldots \cdot g_t) = g_i.$$

where $g = g_1 \cdot g_2 \cdot \ldots \cdot g_t \in G$ $(\forall i \in [t] :\ g_i \in G_i)$ is an arbitary element of G. The π_i's we call the *canonical projection endomorphisms* of the above decomposition.

2.2 Background

Theorem 1. *(Expressing G as a direct product of A and B, cf. [Her75]) Let G be a finite group and A, B be subgroups of G. Then $G = A \times B$ if and only if the following three conditions hold:*

- *Both A and B are normal subgroups of G.*
- *$|G| = |A| \cdot |B|$.*
- *$A \bigcap B = \{e\}$.*

Theorem 2. *(The fundamental theorem of finite abelian groups, cf. [Her75]) Every finite abelian group G can be as the written product of cyclic groups of prime power order.*

Theorem 3. *(Remak-Krull-Schmidt, cf. [Hun74]) Let G be a finite group. If*

$$G = G_1 \times G_2 \times \ldots \times G_s$$

and

$$G = H_1 \times H_2 \times \ldots \times H_t$$

with each G_i, H_j indecomposable, then $s = t$ and after reindexing $G_i \cong H_i$ for every i and for each $r < t$,

$$G = G_1 \times \ldots \times G_r \times H_{r+1} \times \ldots \times H_t.$$

Notice that the uniqueness statement is stronger than simply saying that the indecomposable factors are determined upto isomorphism.

3 Invariants of Group Factorization

The main source of difficulty in devising an efficient algorithm for the decomposition of a group lies in the fact that the decomposition need not be unique. Let us therefore analyze what one decomposition should be in reference of another.

Lemma 1. *For a group G, suppose that $G = A \times B$. Then for a subset $C \subset G$,*

$G{=}C \times B \Longleftrightarrow C{=}\{\alpha \cdot \phi(\alpha) \mid \alpha \in A\}$, where $\phi : A \mapsto Cent(B)$ is a homomorphism.

Theorem 4. *(Characterization of the various decompositions of a group.) Let G be a finite group with*

$$G = G_1 \times G_2 \times \ldots \times G_t \tag{1}$$

with each G_i indecomposable. For $i \in [t]$, define M_i to be the normal subgroup of G as follows:

$$M_i \stackrel{\text{def}}{=} G_1 \times \ldots \times G_{i-1} \times G_{i+1} \times \ldots \times G_t,$$

so that $G = G_i \times M_i \forall i \in [t]$. If G has another decomposition

$$G = H_1 \times H_2 \times \ldots \times H_t \tag{2}$$

(the number of H_j's must equal t by Theorem 3) with each H_j indecomposable, then there exist t homomorphisms $\{\phi_r : G_r \mapsto Cent(M_r)\}_{r \in [t]}$ so that after reindexing, for each $r \in [t]$,

$$H_r = \{\alpha \cdot \phi_r(\alpha) \mid \alpha \in G_r,\ \phi_r(\alpha) \in Cent(M_r)\}$$

4 An Algorithm for GroupDivision

In this section we solve the group division problem which is used in step 3 of Algorithm I. Let us recall that GroupDivision is the following problem: given a group G and a normal subgroup $A \trianglelefteq G$, find a $B \trianglelefteq G$ such that $G = A \times B$, if such a decomposition exists. We will solve this problem itself in two stages. First, we devise an efficient algorithm assuming that the quotient group G/A is abelian and then use this as a subroutine in the algorithm for the general case.

4.1 When the Quotient Group G/A Is Abelian

In this case we can assume that $G = A \times B$ where B is abelian. Observe that in this case, for every coset $A \cdot g$ of A in G, we can pick an element $b \in A \cdot g$ such that $b \in \mathrm{Cent}(G)$ and $\mathrm{ord}_G(b) = \mathrm{ord}_{G/A}(A \cdot g)$. Also, the quotient group G/A is abelian and therefore using the abelian group decomposition algorithm, we can efficiently find a complete decomposition of G/A. So let

$$G/A = \langle A \cdot g_1 \rangle \times \ldots \times \langle A \cdot g_t \rangle$$

Now from each coset $A \cdot g_i$ we pick a representative element b_i such that $b_i \in \mathrm{Cent}(G)$ and $\mathrm{ord}_G(b_i) = \mathrm{ord}_{G/A}(A \cdot b_i)$. For any such set of b_i's, its an easy verification that $G = A \times \langle b_1 \rangle \times \ldots \times \langle b_t \rangle$.

4.2 When the Quotient Group G/A Is Nonabelian

We first give the algorithm and then prove its correctness.

Algorithm 2. GroupDivision
Input. A group G and a normal subgroup A of G.
Output. A subgroup C of G such that $G = A \times C$, if such a C exists.

1. Compute $T \stackrel{\text{def}}{=} \langle \{a \cdot g \cdot a^{-1} \cdot g^{-1} \mid a \in \mathrm{Cent}_G(A), g \in G\} \rangle$.
2. If T is not a normal subgroup of G then **output** *NO SUCH DECOMPOSITION.*
3. Compute $\tilde{G} \stackrel{\text{def}}{=} G/T$ and $\tilde{A} \stackrel{\text{def}}{=} \{T \cdot a \mid a \in A\} \trianglelefteq \tilde{G}$.
4. Verify that $T \bigcap A = \{e\}$. If not, output *NO SUCH DECOMPOSITION.* If yes, then we deduce that the canonical map $a \mapsto Ta$ is an isomorphism from A to $\tilde{A}$.
5. Using the abelian group division algorithm given above, determine if there exists a $\tilde{B} \trianglelefteq \tilde{G}$, *with* $\tilde{B}$ *abelian,* so that $\tilde{G} = \tilde{A} \times \tilde{B}$. If so, determine elements $Tg_1, Tg_2, \ldots, Tg_t \in G/T$ such that
$$\tilde{G} = \tilde{A} \times \langle Tg_1 \rangle \times \langle Tg_2 \rangle \times \ldots \times \langle Tg_t \rangle .$$
6. From each coset Tg_i, pick *any* representative element c_i. Compute $C \stackrel{\text{def}}{=} \langle T \bigcup \{c_1, \ldots, c_t\} \rangle \leq G$.
7. If $G = A \times C$ then **output** C else **output** *NO SUCH DECOMPOSITION.*

The algorithm clearly has polynomial running time and it remains for us to prove its correctness. To see whats going on in the algorithm above, let us assume that $G = A \times B$ and fix this decomposition of G. Its easy to verify that the subgroup T computed in step 1 is a normal subgroup of G and $T = [B, B]$. Also, $T = [B, B] \subseteq B$ and therefore $A \bigcap T$ must be $\{e\}$. This implies that the canonical mapping $a \mapsto T \cdot a$ is an isomorphism from A to $\tilde{A}$ This explains step 4 of the algorithm. Observe that the $\tilde{G}$ computed in step 3 has the decomposition

$$\tilde{G} = \tilde{A} \times (B/[B, B]).$$

But $B/[B, B]$ is an abelian group so we can use the previous algorithm and decompose $\tilde{G}$ into product of $\tilde{A}$ times a number of cyclic groups. By the end of step 6, we would have computed $c_1, \dots, c_t \in G$ such that

$$\tilde{G} = \tilde{A} \times \tilde{C}, \quad \text{where } \tilde{C} \stackrel{\text{def}}{=} \langle Tc_1 \rangle \times \langle Tc_2 \rangle \times \dots \times \langle Tc_t \rangle \leq G.$$

Proposition 1. *$C \trianglelefteq G$ and the elements of C and A together generate G. Furthermore, $C \bigcap A = \{e\}$.*

Summarizing, we have A and C are normal subgroups of G that span G and have a trivial intersection which means that $G = A \times C$, as required to prove the correctness of the algorithm.

5 An Algorithm for SemiAbelianGroupDecomposition

In this section, we solve the special case of GroupDecomposition when some of the indecomposable components of G are abelian groups. That is given G, we wish to find an abelian subgroup B and another subgroup A of G so that

$$G = A \times B, \quad \text{where } B \text{ is abelian.} \tag{3}$$

Since B is abelian, it has a decomposition into a direct product of cyclic groups. So let

$$B = \langle b_1 \rangle \times \dots \times \langle b_t \rangle .$$

so that G becomes

$$G = A \times \langle b_1 \rangle \times \dots \times \langle b_t \rangle .$$

Thus, if G has a decomposition of the form (3) then there exists a $b \in G$ such that $\langle b \rangle$ is a direct factor of G. Conversely, to find a decomposition of the form (3) it is sufficient to find a b such that $\langle b \rangle$ is a direct factor of G. Knowing B, we can find an appropriate direct complement of $\langle b \rangle$ efficiently using the algorithm for GroupDivision given previously. Lastly, given the group G, we find an appropriate b in polynomial-time by iterating over all the elements of G and using the algorithm for GroupDivision to verify whether $\langle b \rangle$ is a direct factor of G or not.

6 The Conjugacy Class Graph of a Group and Its Properties

Here we give the construction of the *conjugacy class graph* of a group. Consider a group G which has a decomposition

$$G = A \times B.$$

Fixing this decomposition, consider the conjugacy class $\mathcal{C}_g$ of an arbitary element $g = \alpha \cdot \beta \in G$, where $\alpha \in A, \beta \in B$. Observe that $\mathcal{C}_g = \mathcal{C}_\alpha \cdot \mathcal{C}_\beta$ and the elements of $\mathcal{C}_\alpha$ and $\mathcal{C}_\beta$ commute. More generally, we have

Observation 5. *If $G = G_1 \times \ldots \times G_t$ and $g = g_1 \cdot \ldots \cdot g_t$ is an arbitary element of G, with each $g_i \in G_i$ then*

$$\mathcal{C}_g = \mathcal{C}_{g_1} \cdot \mathcal{C}_{g_2} \cdot \ldots \cdot \mathcal{C}_{g_t}.$$

Furthermore for all $i \neq j$ each element of $\mathcal{C}_{g_i}$ commutes with every element of $\mathcal{C}_{g_j}$.

For the rest of this section, we fix the group G and a reference decomposition

$$G = G_1 \times G_2 \times \ldots \times G_t.$$

Let $\{\pi_i : G \mapsto G_i \mid i \in [t]\}$ be the set of canonical projection endomorphisms associated with the above decomposition. If any of the G_i's are abelian groups then we can obtain a decomposition of G using the algorithm for SemiAbelian-GroupDecomposition given in section 5. So henceforth we will assume that all the G_i's are nonabelian. Observation 5 above motivates the following definitions.

Definition 1. *We say that two conjugacy classes $\mathcal{C}_a$ and $\mathcal{C}_b$* commute *when for every $\alpha \in \mathcal{C}_a$ and $\beta \in \mathcal{C}_b$, α and β commute.*

Definition 2. *Call a conjugacy class* reducible *$\mathcal{C}_g$ if it is either a conjugacy class of an element from the center of G, or there exist two conjugacy classes $\mathcal{C}_a$ and $\mathcal{C}_b$ such that*

- *Neither a nor b belongs to the center of G.*
- *$\mathcal{C}_a$ and $\mathcal{C}_b$ commute.*
- *$\mathcal{C}_g = \mathcal{C}_a \cdot \mathcal{C}_b$*
- *$|\mathcal{C}_g| = |\mathcal{C}_a| \cdot |\mathcal{C}_b|$*

If a conjugacy class is not reducible, then call it irreducible.

Proposition 2. *If a conjugacy class $\mathcal{C}_g$ is irreducible then there exists a unique $i \in [t]$ such that $\pi_i(g) \notin Cent(G_i)$.*

Proof. If it happens that for all $i \in [t]$, $\pi_i(g) \in \mathrm{Cent}(G_i)$ then $g \in \mathrm{Cent}(G)$ so that the conjugacy class $\mathcal{C}_g$ is reducible by definition. If more than one $\pi_i(G)$ are noncentral elements then by observation 5, we would get that $\mathcal{C}_g$ is reducible.

The converse of this proposition is not true in general. The above proposition implies that corresponding to a conjugacy class $\mathcal{C}_g$, there exists a unique G_i such that $\pi_i(g) \notin \text{Cent}(G_i)$. Let us call this subgroup G_i the *indecomposable component associated to* the conjugacy class $\mathcal{C}_g$. Let us now define the *conjugacy class graph* Γ_G associated to a group G.

Definition 3. *The* graph *of a group G (denoted Γ_G) is a graph with irreducible conjugacy classes as nodes and such that a pair of nodes is connected by an edge iff the corresponding pair of conjugacy classes does* not *commute.*

The connected components of Γ_g can be computed efficiently and they give us information about the direct factors of G.

Proposition 3. *Let $\Lambda_1, \ldots, \Lambda_s$ be the connected components of Γ_G. Then $s \geq t$ and there is a partition*

$$[s] = S_1 \uplus S_2 \uplus \ldots \uplus S_t$$

such that $\langle \cup_{i \in S_j} Elts(\Lambda_i) \rangle \pmod{Z} = G_j \pmod{Z}$ for all j where $Z = Cent(G)$.

Proof. Let us consider two irreducible conjugacy classes $\mathcal{C}_g$ and $\mathcal{C}_h$. Let the indecomposable components associated with $\mathcal{C}_g$ and $\mathcal{C}_h$ be G_i and G_j respectively. Suppose that $i \neq j$. Then $\pi_j(g)$ and $\pi_i(h)$ are central elements of G so that every element of $\mathcal{C}_g$ commutes with every other element of $\mathcal{C}_h$. Thus there is no edge between the nodes corresponding to $\mathcal{C}_g$ and $\mathcal{C}_h$. This implies that if $\mathcal{C}_g$ and $\mathcal{C}_h$ are in the same connected component of Γ_G then the indecomposable components associated with $\mathcal{C}_g$ and $\mathcal{C}_h$ are the same. Each nonabelian component of G gives rise to at least one irreducible conjugacy class so that the number of connected components s of Γ_G is at least the number of indecomposable nonabelian components of G. The above argument shows that there exists a partition of $[s]$ into S_i such that $\langle \cup_{i \in S_j} \Lambda_i \rangle \pmod{Z} \subseteq G_j \pmod{Z}$ for all j. The inclusion is in fact an equality because all the irreducible conjugacy classes generate all the noncentral elements of G by construction.

In general it is not true that the number of connected components of Γ_G equals the number of indecomposable nonabelian groups in the factorization of G. The irreducible conjugacy classes of each of the G_i may be divided into more than one component. However we have the following:

Proposition 4. *If the center of group G is trivial, then the number of connected components of its graph is equal to the number of indecomposable groups in the factorization of G. Moreover, the subgroups generated by the conjugacy classes of each of the components are normal disjoint subgroups which together span G; thus we have the factorization of G.*

Using the proposition we can efficiently factor a group with a trivial center. When the center of the group is non-trivial it is no longer the case that each component of Γ_G generates one of the factors in the factorization of G. We would need to search through the partitions of the set of connected components to find the components associated to an indecomposable factor, say G_1. For that we need a bound on the number of components of the graph:

Proposition 5. *The number of connected components is upper bounded by* $\log|G|$.

7 Putting Everything together

We now have all the component steps of Algorithm I. So let us conclude with the proof of correctness of Algoritm I.

Theorem 6. *If the input to Algorithm I is a decomposable group G then it necessarily computes a nontrivial decomposition of G, otherwise it outputs* NO SUCH DECOMPOSITION. *Moreover, Algorithm I has running time polynomial in* $|G|$.

Proof. Clearly, if the group is indecomposable our algorithm outputs *NO SUCH DECOMPOSITION.* By Proposition 5, $s \leq \log|G|$ so that the number of iterations in step 3 is at most $|G|$. All the operations inside the loop (steps 3a to 3c) are polynomial-time computable so that the overall running time also poly($|G|$). It remains to show that if G is decomposable then our algorithm outputs a nontrivial decomposition. Let the given group G have a decomposition

$$G = G_1 \times \ldots \times G_t \tag{4}$$

with each G_i indecomposable. Let Z be the center of G. In the algorithm we iterate over all subsets of the connected components of Γ_G so let us assume that we have found the subset S_1 of indices of connected components corresponding to conjugacy class of elements of G_1. By Proposition 5 we have

$$G_2 \times G_3 \times \ldots \times G_t \pmod{Z} = \left\langle \cup_{j \notin S_1} \mathrm{Elts}(\Lambda_j) \right\rangle \pmod{Z}.$$

This means that in step 3a we would have computed

$$\begin{aligned} Z_1 &\stackrel{\mathrm{def}}{=} \mathrm{Cent}_G(\langle \cup_{j \notin S_1} \mathrm{Elts}(\Lambda_j) \rangle) \\ &= \mathrm{Cent}_G(G_2 \times G_3 \times \ldots \times G_t) \\ &= G_1 \times \mathrm{Cent}(G_2) \times \mathrm{Cent}(G_3) \times \ldots \times \mathrm{Cent}(G_t) \end{aligned}$$

Let us now consider the decomposition $Z_1 = H_1 \times Y$ obtained in step 3b of Algorithm I. By the Remak-Krull-Schmidt theorem (Theorem 3), all decompositions of Z_1 are isomorphic so that if H_1 is any direct factor of Z_1 which has no abelian direct factors then H_1 must be indecomposable and isomorphic to G_1. Furthermore by an application of theorem 4, we must have that H_1 must be of the form

$$H_1 = \{\alpha \cdot \phi(\alpha) \mid \alpha \in G_1,\ \phi(\alpha) \in (\mathrm{Cent}(G_2) \times \mathrm{Cent}(G_3) \times \ldots \times \mathrm{Cent}(G_t))\},$$

where $\phi : H_1 \mapsto \mathrm{Cent}(G_2) \times \mathrm{Cent}(G_3) \times \ldots \times \mathrm{Cent}(G_t)$ is a homomorphism. By lemma 1, we can replace G_1 by H_1 in the factorization (4) so that in fact

$$G = H_1 \times G_2 \times G_3 \times \ldots \times G_t.$$

In particular, this means that H_1 is a direct factor of G so that in step (3c), using the algorithm for GroupDivision , we necessarily recover a nontrivial factorization of G.

References

[AT04] Arvind, Toran: Solvable group isomorphism is (almost) in NP intersect coNP. In: Proceedings of the 19th Annual Conference on Computational Complexity, pp. 91–103 (2004)

[BKL79] Babai, L., Kantor, W.M., Luks, E.M.: Computational complexity and the classification of finite simple groups. In: Proceedings of the 24th FOCS, pp. 162–171 (1979)

[BSL87] Babai, L., Seress, A., Luks, E.M.: Permutation groups in nc. In: Proceedings of 19th STOC, pp. 409–420 (1987)

[CF07] Chen, L., Fu, B.: Linear and sublinear time algorithms for the basis of abelian groups. In: Electronic Colloquium on Computational Complexity, Technical report TR07-052 (2007)

[Gal09] Le Gall, F.: Efficient isomorphism testing for a class of group extensions. In: Proceedings of the annual Symposium on Theoretical Aspects of Computer Science (2009)

[Her75] Herstein, I.N.: Topics in Algebra, 2nd edn. John Wiley & Sons, New York (1975)

[Hul] Hulpke, A.: Gap project repository, `http://www.math.colostate.edu/~hulpke/gapproj/projects.htm`

[Hun74] Hungerford, T.W.: Algebra. Graduate Texts in Mathematics, vol. 73. Springer, New York (1974)

[Kan85a] Kantor, W.M.: Polynomial-time algorithms for finding elements of prime order and sylow subgroups. Journal of Algorithms 6, 478–514 (1985)

[Kan85b] Kantor, W.M.: Sylow's theorem in polynomial time. J. Comp. Syst. Sci. 30, 359–394 (1985)

[Kan90] Kantor, W.M.: Finding sylow normalizers in polynomial time. Journal of Algorithms 11, 523–563 (1990)

[Kav07] Kavitha, T.: Linear time algorithms for abelian group isomorphism and related problems. J. Comput. Syst. Sci. 73(6), 986–996 (2007)

[KT88] Kantor, W.M., Taylor, D.E.: Polynomial-time versions of sylow's theorem. Journal of Algorithms 9, 1–17 (1988)

[Luk] Luks, E.: Lectures on polynomial-time computation in groups, `http://ix.cs.uoregon.edu/~luks`

[Luk87] Luks, E.M.: Computing the composition factors of a permutation group in polynomial time. Combinatorica 7, 87–99 (1987)

[Mil78] Miller, G.: On the nlog n isomorphism technique. In: Proceedings of the tenth annual ACM symposium on Theory of computing (1978)

On Finding Dense Subgraphs*

Samir Khuller and Barna Saha

University of Maryland College Park
{samir,barna}@cs.umd.edu

Abstract. Given an undirected graph $G = (V, E)$, the density of a subgraph on vertex set S is defined as $d(S) = \frac{|E(S)|}{|S|}$, where $E(S)$ is the set of edges in the subgraph induced by nodes in S. Finding subgraphs of maximum density is a very well studied problem. One can also generalize this notion to directed graphs. For a directed graph one notion of density given by Kannan and Vinay [12] is as follows: given subsets S and T of vertices, the density of the subgraph is $d(S, T) = \frac{|E(S,T)|}{\sqrt{|S||T|}}$, where $E(S, T)$ is the set of edges going from S to T. Without any size constraints, a subgraph of maximum density can be found in polynomial time. When we require the subgraph to have a specified size, the problem of finding a maximum density subgraph becomes NP-hard. In this paper we focus on developing fast polynomial time algorithms for several variations of dense subgraph problems for both directed and undirected graphs. When there is no size bound, we extend the flow based technique for obtaining a densest subgraph in directed graphs and also give a linear time 2-approximation algorithm for it. When a size lower bound is specified for both directed and undirected cases, we show that the problem is NP-complete and give fast algorithms to find subgraphs within a factor 2 of the optimum density. We also show that solving the densest subgraph problem with an upper bound on size is as hard as solving the problem with an exact size constraint, within a constant factor.

1 Introduction

Given an undirected graph $G = (V, E)$, the density of a subgraph on vertex set S is defined as $d(S) = \frac{|E(S)|}{|S|}$, where $E(S)$ is the set of edges in the subgraph induced by S. The problem of finding a densest subgraph of a given graph G can be solved optimally in polynomial time, despite the fact that there are exponentially many subgraphs to consider [16,11]. In addition, Charikar [6] showed that we can find a 2 approximation to the densest subgraph problem in linear time using a very simple greedy algorithm (the greedy algorithm was previously studied by Asahiro et. al. [4]). This result is interesting because in many applications of analyzing social networks, web graphs etc., the size of the graph involved could be very large and so having a fast algorithm for finding an approximately dense subgraph is extremely useful. However when there is a size constraint specified - namely find a densest subgraph of exactly k vertices (*DkS*), the densest k subgraph problem becomes NP-hard [8,3]. When $k = \Theta(|V|)$, Asahiro et. al. [4] gave a constant factor approximation algorithm for the *DkS* problem. However for

* Research supported by NSF CCF 0728839 and a Google Research Award.

S. Albers et al. (Eds.): ICALP 2009, Part I, LNCS 5555, pp. 597–608, 2009.

general k, the algorithm developed by Feige, Kortsarz and Peleg [8] achieves the best approximation guarantee of $O(n^a)$, where $a < \frac{1}{3}$. Khot [13] showed that there does not exist any PTAS for the *DkS* problem under a reasonable complexity assumption. Closing the gap between the approximation factor and the hardness guarantee is an important open problem.

Recently, Andersen and Chellapilla [2] considered two variations of the problem of finding a densest k subgraph. The first problem, the densest at-least-k-subgraph problem (*DalkS*) asks for an induced subgraph of highest density among all subgraphs with at least k nodes. This relaxation makes *DalkS* significantly easier to approximate and Andersen et.al. gave a fast algorithm based on Charikar's greedy algorithm that guarantees a 3 approximation for the *DalkS* problem. In addition, Andersen [1] showed that this problem has a polynomial time 2 approximation, albeit with significantly higher running time. However it was left open as to whether or not this problem is NP-complete. The second problem studied was the densest at-most-k-subgraph problem (*DamkS*), which asks for an induced subgraph of highest density among all subgraphs with at most k nodes. For the *DamkS* problem, Andersen et.al. showed that if there exists an α approximation for *DamkS*, then there is a $\Theta(\alpha^2)$ approximation for the *DkS* problem, indicating that this problem is likely to be quite difficult as well.

For directed graphs, Kannan and Vinay [12] defined a suitable notion of density to detect highly connected subgraphs and provided a $\Theta(\log n)$ approximation algorithm for finding such dense components. Let $G = (V, E)$ be a directed graph and S and T be two subsets of nodes of V. Density corresponding to S and T is defined as $d(S,T) = \frac{|E(S,T)|}{\sqrt{|S||T|}}$, where $E(S,T)$ consists of the edges going from S to T. Charikar [6] showed that the problem can be solved in polynomial time by solving an LP using n^2 different values of a parameter. However a max-flow based technique similar to the one developed by Goldberg [11] for the densest subgraph problem in undirected graphs was not known for directed graphs. It was mentioned as one of the open problems in [6]. In addition to providing a polynomial time solution for the densest subgraph problem in directed graphs, Charikar also gave a 2 approximation algorithm which runs in $O(|V|^3 + |V|^2|E|)$ time.

Densest subgraph problems have received significant attention for detecting important substructures in massive graphs like web and different social networks. In a web graph, hubs (resource lists) and authorities (authoritative pages) on a topic are characterized by large number of links between them [15]. Finding dense subgraphs also acts as a useful primitive for discovering communities in web and social networks, for compressed representation of a graph and for spam detection [7,5,10]. Gibson et. al. [10] provided effective heuristics based on two-level fingerprints for finding large dense subgraphs in massive graphs. Their aim was to incorporate this step into web search engine for link spam control. Dourisboure gave a scalable method for identifying small dense communities in web graph [7]. Buehrer showed how large dense subgraphs can be useful in web graph compression and sub-sampling a graph [5]. In all these applications the underlying graph is massive and thus fast scalable algorithms for detecting dense subgraphs are required to be effective.

One of the main new insights in this paper is to illustrate the power of the flow based methods [11,16] to find dense subgraphs not only when there is no requirement on the size of the obtained subgraph, but also for cases when there is a constraint on the size of the obtained subgraph. Precisely our contributions are as follows:

1.1 Contributions

- For the densest subgraph problem without any size restrictions (Section 2):
  - We give a max-flow based polynomial time algorithm for solving the densest subgraph problem in directed graphs.
  - We give a linear time 2-approximation algorithm for the densest subgraph problem in directed graphs.
- For the densest at least k subgraph problem (Section 3):
  - We show that the densest at least k subgraph problem is NP-Hard.
  - For undirected graphs, we give a flow-based and LP based approximation algorithms, for the densest at least k subgraph problem. These run much faster than the polynomial time approximation algorithm of Andersen and deliver the same worst case approximation factor of 2.
  - We define the notion of densest at least k_1, k_2 subgraph problem for directed graphs and give a 2-approximation algorithm for it.
- Densest at most k subgraph problem (Section 4):
  - We show that approximating the densest at most k subgraph problem is as hard as the densest k subgraph problem within a constant factor, specifically an α approximation for *DamkS*, implies a 4α approximation for *DkS*.

2 Densest Subgraph without Any Size Restriction

In this section, first we give a max-flow based algorithm for the densest subgraph problem in directed graphs. For undirected graphs, Goldberg [11] developed a flow based algorithm, that finds a densest subgraph in polynomial time. However, for directed graphs, no flow based algorithm was known. Next we consider the greedy algorithm for densest subgraph in undirected graphs proposed by Charikar [6] and develop an extension of this algorithm to give a 2 approximation algorithm for finding a densest subgraph in directed graphs. This improves the running time from $O(|V|^3 + |V|^2|E|)$ to $O(|V| + |E|)$. We also give a very simple proof of 2-approximation for the greedy algorithm developed by Charikar [6] to obtain a densest subgraph in undirected graphs.

2.1 Max-flow Based Algorithm for Finding Densest Subgraphs in Directed Graphs

For a directed graph $G = (V, E)$, we wish to find two subsets of nodes S and T, such that $d(S,T) = \frac{|E(S,T)|}{\sqrt{|S||T|}}$ is maximized. Let us denote the optimum subsets of nodes by S^* and T^* respectively. To detect such subsets of nodes, we first guess the value of $\frac{|S^*|}{|T^*|}$ in the optimum solution. Since there are $|V|^2$ possible values, in $\Theta(|V|^2)$ time, it is possible to guess this ratio exactly[1]. Let this ratio be a. We create a bipartite graph

[1] If we want $(1 + \epsilon)$ approximation, only $O(\frac{\log |V|}{\epsilon})$ guessed values suffice.

$G' = (V_1, V_2, E)$, where $V_1 = V_2 = V$ and for every directed edge (i,j) in the original graph, we add an edge from vertex $i \in V_1$ to $j \in V_2$. We now wish to find $S \subseteq V_1$ and $T \subseteq V_2$, such that $\frac{|E(S,T)|}{\sqrt{|S||T|}}$ is maximized. We also know, $\frac{|S^*|}{|T^*|} = a$.

We add a source s and a sink t to $G' = (V_1, V_2, E)$. We guess the value of the optimum (maximum) density. Let our guessed value be g. The following edges with weights are then inserted into $G' = (V_1, V_2, E)$:

- We add an edge of weight m from source s to each vertex of V_1 and V_2, where $m = |E|$.
- We add an edge of weight $(m + \frac{g}{\sqrt{a}})$ from each vertex of V_1 to the sink t.
- We add an edge from each vertex j of V_2 to sink t of weight $m + \sqrt{a}g - 2d_j$, where d_j is the in-degree of j.
- All the edges going from V_1 to V_2 are given weight 0. For each edge going from V_1 to V_2, a reverse edge of weight 2 is added.

Now consider a s-t min-cut in this weighted graph. Since the cut $\{s\}, \{t, V_1, V_2\}$ has weight $m(|V_1| + |V_2|)$, the min-cut value is $\leq m(|V_1| + |V_2|)$. Now consider the cut $\{s, S \subseteq V_1, T \subseteq V_2\}$,$\{t, (V_1 \setminus S) \subseteq V_1, (V_2 \setminus T) \subseteq V_2\}$. The number of edges crossing the cut is,

$$m(|V_1|-|S|+|V_2|-|T|) + (m + \frac{g}{\sqrt{a}})|S| + \sum_{i \in T}(m + \sqrt{a}g - 2d_i) + \sum_{\substack{i \in T, j \in V_1 - S, \\ (j,i) \in E(G)}} 2$$

$$= m(|V_1| + |V_2|) + |S|\frac{g}{\sqrt{a}} + |T|\sqrt{a}g - 2|E(S,T)|$$

$$= m(|V_1| + |V_2|) + \frac{|S|}{\sqrt{a}}\left(g - \frac{|E(S,T)|}{|S|/\sqrt{a}}\right) + |T|\sqrt{a}\left(g - \frac{|E(S,T)|}{|T|\sqrt{a}}\right)$$

Let us denote the optimum density value by d_{OPT}. If $g < d_{OPT}$, then there exists S and T (corresponding to the optimum solution), such that both $\left(g - \frac{|E(S,T)|}{|S|/\sqrt{a}}\right)$ and $\left(g - \frac{E(S,T)}{|T|\sqrt{a}}\right)$ are negative. Therefore S and T are nonempty. If the guessed value $g > d_{OPT}$, let if possible S and T be non-empty. Let in this returned solution, $\frac{|S|}{|T|} = b$. We have,

$$\frac{|S|}{\sqrt{a}}\left(g - \frac{|E(S,T)|}{|S|/\sqrt{a}}\right) + |T|\sqrt{a}\left(g - \frac{E(S,T)}{|T|\sqrt{a}}\right)$$

$$= \sqrt{|S||T|}\frac{\sqrt{b}}{\sqrt{a}}\left(g - \frac{d(S,T)}{\sqrt{b}/\sqrt{a}}\right) + \sqrt{|S||T|}\frac{\sqrt{a}}{\sqrt{b}}\left(g - \frac{d(S,T)}{\sqrt{a}/\sqrt{b}}\right)$$

$$= \sqrt{|S||T|}\left(\left(\frac{\sqrt{b}}{\sqrt{a}} + \frac{\sqrt{a}}{\sqrt{b}}\right)g - 2d(S,T)\right) \tag{1}$$

Now, $\left(\frac{\sqrt{b}}{\sqrt{a}} + \frac{\sqrt{a}}{\sqrt{b}}\right) \geq 2 \; \forall$ reals a, b and we have, $g > d_{OPT} \geq d(S,T)$. Hence the value of (1) is > 0. Thus if S and T are non-empty, then this cut has value $>$

$m(|V_1| + |V_2|)$. Hence if $g > d_{OPT}$, min-cut is $(\{s\}, \{t, V_1, V_2\})$. If the guessed value $g = d_{OPT}$, then we get a cut of the same cost as the trivial min-cut, even by having S and T corresponding to S^* and T^* respectively. We can always ensure that we obtain a min-cut, which has the biggest size on the source side. Thus when the guessed value is correct, the optimum subsets S and T are obtained from the subsets of vertices of V_1 and V_2 that belong to the side of the cut that contains s. The algorithm detects the correct value of g using a binary search, similar to Goldberg's algorithm for finding a densest subgraph in undirected graphs [11]. Also it is easy to verify that, when the correct value of g is guessed, we have $b = a$. Using a parametric max-flow algorithm [9], the total time required is same as one flow computation within a constant factor.

2.2 2 Approximation Algorithm for the Densest Subgraph Problem in Undirected and Directed Graphs

We first consider Charikar's greedy algorithm [6] for densest subgraphs in undirected graphs. The greedy algorithm at each step chooses a vertex of minimum degree, deletes it and proceeds for $(n-1)$ steps, where $|V| = n$. At every step the density of the remaining subgraph is calculated and finally the one with maximum density is returned.

Algorithm 1. DENSEST-SUBGRAPH($G = (V, E)$)

$n \leftarrow |V|, H_n \leftarrow G$
for $i = n$ *to* 2
do $\begin{cases} \text{Let } v \text{ be a vertex in } H_i \text{ of minimum degree} \\ H_{i-1} \leftarrow H_i - \{v\} \end{cases}$
return (H_j, which has the maximum density among $H_i's, i = 1, 2, .., n$)

We show that the above greedy algorithm *Densest-Subgraph* achieves an approximation factor of 2 for undirected networks. This is not a new result. However our proof is simpler than the one given by Charikar. For directed graphs, Charikar developed a different greedy algorithm, that has a significantly higher time-complexity of $O(|V|^3 + |V|^2|E|)$. We show that the algorithm *Densest-Subgraph-Directed*, which is a generalization of the algorithm *Densest-Subgraph* detects a subgraph, with density within a factor of 2 of the optimum for directed graphs. This reduces the time complexity from $O(|V|^3 + |V|^2|E|)$ to $O(|V| + |E|)$.

Theorem 1. *The greedy algorithm* **Densest-Subgraph** *achieves a* 2*-approximation for the densest subgraph problem in undirected networks.*

Proof. Let $d_{OPT} = \lambda$. Observe that in an optimum solution, every vertex has degree $\geq \lambda$. Otherwise removing a vertex of degree $< \lambda$, will give a subgraph with higher density. Consider the iteration of the greedy algorithm when the first vertex of the optimum solution is removed. At this stage all the vertices in the remaining subgraph have degree $\geq \lambda$. If the number of vertices in the subgraph is s, then the total number of edges is $\geq \lambda s/2$, and the density is $\geq \lambda/2$. Since the greedy algorithm returns the subgraph with the highest density over all the iterations, it always returns a subgraph with density at least $\frac{1}{2}$ of the optimum. □

With a little work, one can make examples showing that the bound of 2 is tight for Charikar's algorithm (details omitted).

We now consider the case of directed graphs. In a directed graph, for each vertex we count its in-degree and out-degree separately. Let v_i be a vertex with minimum in-degree and v_o be a vertex with minimum out-degree. Then we say v_i has minimum degree, if the in-degree of v_i is at most the out-degree of v_o, else v_o is said to have the minimum degree. In the first case, the vertex with minimum degree belongs to the category IN. In the second case, it belongs to the category OUT. The greedy algorithm for directed graphs deletes the vertex with minimum degree and then depending on whether it is of category IN or OUT, either deletes all the incoming edges or all the outgoing edges incident on that vertex, respectively. If the vertex becomes a singleton, the vertex is deleted. To compute the density of the remaining graph after an iteration of *Densest-Subgraph-Directed*, any vertex that has nonzero out-degree is counted in the S side and all the vertices with non-zero in-degree are counted in the T side. Therefore the same vertex might appear both in S and T and will be counted once in S and once in T. We denote the optimum solution by (S^*, T^*).

Algorithm 2. DENSEST-SUBGRAPH-DIRECTED($G = (V, E)$)

$n \leftarrow |V|, H_{2n} \leftarrow G, i \leftarrow 2n$
while $H_i \neq \emptyset$
do
  Let v be a vertex in H_i of minimum degree
  if category(v) $= IN$
    then Delete all the incoming edges incident on v
    else Delete all the outgoing edges incident on v
  if v has no edges incident on it **then** Delete v
  Call the new graph $H_{i-1}, i \leftarrow i - 1$
return (H_j which has the maximum density among H_i's)

Define $\lambda_i = |E(S^*, T^*)| \left(1 - \sqrt{1 - \frac{1}{|T^*|}}\right)$ and $\lambda_o = |E(S^*, T^*)| \left(1 - \sqrt{1 - \frac{1}{|S^*|}}\right)$.

Lemma 1. *In an optimal solution, each vertex in S^*, has out-degree $\geq \lambda_o$ and each vertex in T^* has in-degree $\geq \lambda_i$.*

Proof. Suppose if possible, $\exists v \in S^*$ with out-degree $< \lambda_o$. Remove v from S^*. The density of the remaining subgraph is $> \frac{E(S^*, T^*) - \lambda_o}{\sqrt{(|S^*|-1)|T^*|}} = d_{OPT}$, which is not possible. Similarly, every vertex $v \in T^*$ has degree $\geq \lambda_i$. □

Theorem 2. *The greedy algorithm* **Densest-Subgraph-Directed** *achieves a 2 approximation for the densest subgraph problem in directed networks.*

Proof. Consider the iteration of the greedy algorithm, when the vertices in S have out-degree $\geq \lambda_o$ and the vertices in T have in-degree $\geq \lambda_i$. Let us call the set of vertices on the side of S and T by S' and T' respectively. Then the number of edges, $E' \geq |S'|\lambda_o$, and also $E' \geq |T'|\lambda_i$. Hence, the density $d(S', T') \geq \sqrt{\frac{|S'|\lambda_o |T'|\lambda_i}{|S'||T'|}} = \sqrt{\lambda_o \lambda_i}$. Substituting the values of λ_o and λ_i from Lemma 1, we get $d(S', T')^2 \geq$

$|E(S^*,T^*)|^2\left(1-\sqrt{1-\frac{1}{|S^*|}}\right)\left(1-\sqrt{1-\frac{1}{|T^*|}}\right)$. Now putting $|S^*| = \frac{1}{\sin^2\theta}$ and $|T^*|$ $= \frac{1}{\sin^2\alpha}$, we get $d(S',T') \geq \frac{|E(S^*,T^*)|}{\sqrt{|S^*||T^*|}} \frac{\sqrt{(1-cos\theta)(1-cos\alpha)}}{sin\theta sin\alpha} = \frac{d_{OPT}}{2cos\frac{\theta}{2}cos\frac{\alpha}{2}} \geq \frac{d_{OPT}}{2}$. □

3 Densest at Least *k* Subgraph Problem

For undirected graphs, the *DalkS* algorithm tries to find a subgraph of highest density among all subgraphs, that have size $\geq k$. We prove that the *DalkS* problem is NP-complete. and develop two algorithms; a combinatorial algorithm and one based on solving a linear programming formulation of the *DalkS* problem. Each algorithm achieves an approximation factor of 2. Finally we consider the *DalkS* problem in directed graphs, and give a 2-approximation algorithm for the problem.

Theorem 3. DalkS *is NP-Hard.*

Proof. We reduce the densest k subgraph problem (this problem is NP-hard [8,3]) to densest at least k subgraph problem. The entire proof can be found in [14]. □

We develop two algorithms for *DalkS* that both achieve an approximation factor of 2. We note that Andersen [1] proposed a 2 approximation algorithm, that requires n^3 max-flow computations. Even using the parametric flow computation [9] the running time is within a constant factor of n^2 flow computations. Whereas our first algorithm uses at most $max(1,(k-\gamma))$ flow computations using parametric flow algorithm and in general much less than that. Here γ is the size of the densest subgraph without any size constraint. The second algorithm is based on a linear programming formulation for *DalkS* and requires only a single solution of a LP.

3.1 Algorithm 1: Densest at Least *k* Subgraph

Let H^* denote the optimum subgraph and let d^* be the optimum density. The algorithm starts with the original graph G as G_0, and D_0 as $\emptyset$. In the ith iteration, the algorithm finds the densest subgraph H_i from G_{i-1} without any size constraint. If $|V(D_{i-1})| + |V(H_i)| \geq k$, the algorithm stops. Otherwise the algorithm adds H_i to D_{i-1} to obtain D_i. All the edges and the vertices of H_i are removed from G_{i-1}. For every vertex $v \in G_{i-1} \setminus H_i$, if v has l edges to the vertices in H_i, then in G_i a self loop of weight l is added to v. The algorithm then continues with G_i. When the algorithm stops, each subgraph D_i is padded with arbitrary vertices to make their size k. The algorithm then returns the D_j with maximum density.

Algorithm 3. DENSEST AT LEAST-K(G,k)

$D_0 \leftarrow \emptyset, G_0 \leftarrow G, i \leftarrow 1$
while $|V(D_i)| < k$
do $\begin{cases} H_i \leftarrow \text{maximum-density-subgraph}(G_{i-1}) \\ D_i \leftarrow D_{i-1} \cup H_i \\ G_i = \text{shrink}(G_{i-1}, H_i), i \leftarrow i+1 \end{cases}$
for each D_i
  do Add an arbitrary set of $max(k - |V(D_i)|, 0)$ vertices to it to form D_i'
return (D_j', which has the maximum density among the $D_i's$)

We prove that algorithm *Densest At least-k* achieves an approximation factor of 2.

Theorem 4. *The algorithm* **Densest At least-k** *achieves an approximation factor of* 2 *for the* DalkS *problem.*

Proof. If the number of iterations is 1, then H_1 is the maximum density subgraph of the original graph whose size is $\geq k$. Therefore $H^* = H_1$ and the algorithm returns it. Otherwise, say the algorithm iterates for $l \geq 2$ rounds. There can be two cases:

Case 1: *There exists a $l' < l$ such that $E(D_{l'-1}) \cap E(H^*) \leq \frac{E(H^*)}{2}$ and $E(D_{l'}) \cap E(H^*) \geq \frac{E(H^*)}{2}$.*

Case 2: *There exists no such $l' \leq l$.*

For case 2, we have for any $j \leq l-1$, $E(D_j) \cap E(H^*) \leq \frac{E(H^*)}{2}$. Therefore, $E(G_j) \cap E(H^*) \geq \frac{E(H^*)}{2}$. Consider $V' = V(G_j) \cap V(H^*)$. The density of the subgraph induced by V' in G_j is $\geq \frac{E(G_j) \cap E(H^*)}{|V'|} \geq \frac{E(H^*)}{2V(H^*)} = d^*/2$. Hence the density of H_l must be $\geq d^*/2$. So in case 1, for each $j \leq l$, the density of H_j is $\geq d^*/2$. Therefore the total number of edges in the subgraph D_l is $\geq \frac{d^* \sum_{j=1}^{l'} |V(H_j)|}{2}$, or the density of $D_{l'}$ is $\geq d^*/2$ and it has $\geq k$ vertices.

For case 1, the subgraph $D_{l'}$ has at least $E(H^*)/2$ edges and since $V(D_{l'}) \leq k$, the density of $D'_{l'}$ is $\geq \frac{d^*}{2}$.

Since the algorithm returns the subgraph D'_j with maximum density among all the D'_is, the returned subgraph has density at least $d^*/2$. □

There are example of graphs (see the extended version [14]) over which the approximation factor of $\frac{1}{2}$ is tight for algorithm *Densest At least-k Subgraph.*

3.2 Algorithm 2: Densest at Least k Subgraph

Next we give a LP based solution for the *DalkS* problem. Define a variable $x_{i,j}$ for every edge $(i,j) \in E(G)$ and a variable y_i for every vertex $i \in V(G)$. Consider now the following LP:

$$\text{maximize} \sum_{i,j} x_{i,j} \tag{2}$$

$$x_{i,j} \leq y_i \;, \forall (i,j) \in E(G); \; x_{i,j} \leq y_j \;, \forall (i,j) \in E(G)$$

$$\sum_i y_i = 1; \; y_i \leq \frac{1}{l} \;, \forall i \in V(G); \; x_{i,j}, y_i \geq 0 \;, \forall (i,j) \in E(G), \forall i \in V(G)$$

Here $l \geq k$ is the size of the optimum solution of the *DalkS* problem. Since there can be $n - k + 1$ possible sizes of the optimum solution, we can guess this value, putting different values for l. In Section 3.3 we show that by first running the algorithm *Densest-Subgraph* and then solving one single LP, we can guarantee a 2-approximation.

Lemma 2. *The optimum solution of LP (2) is greater than or equal to the optimum value of* DalkS.

Proof. Let the optimum solution for *DalkS* be obtained for a subgraph H having $l \geq k$ vertices and density λ. Consider a solution for the above LP, where each of the variables y_i corresponding to the vertices of H have value $\frac{1}{l}$. All the variable $x_{i,j}$ corresponding to the induced edges of H have value $\frac{1}{l}$. The solution is feasible, since it satisfies all the constraints of LP (2). The value of the objective function of the LP is $\sum_{(i,j) \in H} x_{i,j} = \frac{E(H)}{l} = \frac{E(H)}{V(H)} = \lambda$. Therefore the optimum value of the LP is $\geq \lambda$ □

Lemma 3. *If the value of the optimum solution of LP (2) is λ, a subgraph of size $\geq k$ with density $\geq \lambda/2$ can be constructed from that solution of LP (2).*

The proof can be found in the extended version [14]. The key idea is to show that there exists a value of $r \in [0, 1]$, such that if we consider the subgraph induced by the vertices with y value $\geq r$, then either it has density $\geq \lambda/2$ and size $\geq k$, or its size is $< k$, but it has more than half the number of edges the optimum solution has. In the later case, we can add arbitrary vertices to increase the size of the subgraph to k.

Theorem 5. *If the value of the optimum solution of LP (2) is λ, a subset S of vertices can be computed from the optimum solution of LP (2), such that*

$$d(G(S)) \geq \lambda \text{ and } |S| \geq k$$

Proof. Consider every possible subgraph by setting $r = y_i$ for all distinct values of y_i. By Lemma 3, there exists a value of r such that $|S(r)| \geq k$ and $\frac{|E(r)|}{|S(r)|} \geq v/2$, where v is an optimum solution of the LP. By Lemma 2 , $v \geq \lambda$ and hence the proof. □

The integrality gap of LP 2 is at least $\frac{5}{4}$. Also the approximation factor of $\frac{1}{2}$ is tight (see the extended version [14]).

3.3 Reducing the Number of LP Solutions

To reduce the number of LP solutions, we first run the algorithm *Densest-Subgraph*, consider the solutions over all the iterations that have $> k$ vertices and obtain the one with maximum density. We call this modified algorithm *Densest-Subgraph*$_{>k}$. We compare the obtained subgraph from *Densest-Subgraph*$_{>k}$ with the solution returned by the LP based algorithm with $l = k$. The final solution is the one which has higher density.

When the optimum solution for *DalkS* has exactly k vertices, Theorem 5 guarantees that we obtain a 2 approximation. Otherwise, the optimum subgraph has size $> k$. In this situation, the following lemma shows that the solution returned by *Densest-Subgraph*$_{>k}$ has density at least $\frac{1}{2}$ of the optimum solution of *DalkS*. Therefore using only a single solution of LP 2 along with the linear time algorithm *Densest-Subgraph*$_{>k}$, we can guarantee a 2 approximate solution for *DalkS*.

Lemma 4. *If the optimum subgraph of* DalkS *problem has size $> k$, then* **Densest-Subgraph**$_{>k}$ *returns a 2 approximate solution.*

Proof. Let the optimum density be λ. Since the size of the optimum solution is $> k$, if there exists any vertex in the optimum solution with degree $< \lambda$, then removing that we would get higher density and the size of the subgraph still remains $\geq k$. Hence all the vertices in the optimum solution has degree $\geq \lambda$. Now from Theorem 1, we get the required claim. □

3.4 Densest at Least k Subgraph Problem for Directed Graphs

Given a directed graph $G = (V,E)$ and integers k_1, k_2, the densest at least k directed subgraph *(DaLkDS) problem finds two subsets of nodes S and T containing at least k_1 and k_2 vertices respectively for which $\frac{E(S,T)}{\sqrt{|S||T|}}$ is maximized.*

In this section we give a 2 approximation algorithm for the *DaLkDS* problem. Since there are two parameters, k_1, k_2; we refer to this problem by *densest at least-k_1, k_2 problem* from now on.

3.5 Densest at Least k_1, k_2 Directed Subgraph Problem

Let S^*, T^* represent the optimum solution of *DaLkDS* and d^* represent the value of the density corresponding to S^*, T^*. Let the ratio $\frac{|S^*|}{|T^*|}$ be a. Since the possible values of a can be $\frac{i}{j}$, where $i \geq k_1, j \geq k_2$ and $i, j \leq |V|$, we can guess the value of a. We run the max-flow based algorithm of Section 2.1 (maximum-directed-density-subgraph) with the chosen a to obtain the densest directed subgraph without any size constraints. Instead of shrinking and removing the vertices and the edges in the densest directed subgraph, as in algorithm *Densest At least-k* for *DalkS*, we only remove the edges and maintain the vertices. We continue this procedure for the same choice of a, until at some round both the sizes of S and T thus obtained exceed k_1 and k_2 respectively. Let S_i and T_i be the partial subsets of vertices obtained up to the ith round. We append arbitrary vertices A and B to S_i and T_i to form S_i' and T_i' respectively, such that $|S_i'| \geq k_1$ and $|T_i'| \geq k_2$. The algorithm returns S_j', T_j', such that $d(S_j', T_j')$ is maximum over all the iterations.

Algorithm 4. DENSEST AT LEAST-$k_1, k_2(G, k_1, k_2, a)$

$S_0 \leftarrow \emptyset, T_0 \leftarrow \emptyset, G_0 \leftarrow G, i \leftarrow 1$
while $|S_{i-1}| < k_1$ or $|T_{i-1}| < k_2$
do $\begin{cases} H_i(S,T) \leftarrow \text{maximum-directed-density-subgraph}(G_{i-1}, a) \\ S_i \leftarrow S_{i-1} \cup H_i(S) \\ T_i \leftarrow T_{i-1} \cup H_i(T) \\ G_i = \text{shrink}(G_{i-1}, H_i), i \leftarrow i+1 \end{cases}$
for each S_i, T_i
do $\begin{cases} \text{Add arbitrary } max(k_1 - |S_i|, 0) \text{ vertices to } S_i \text{ to form } S_i' \\ \text{Add arbitrary } max(k_2 - |T_i|, 0) \text{ vertices to } T_i \text{ to form } T_i' \end{cases}$
return (S_j', T_j') which has maximum density among the (S_i', T_i')s

Theorem 6. *Algorithm* **Densest At least-k_1, k_2** *achieves an approximation factor of* 2 *for the* DaLkDS *problem.*

Proof. For a chosen a, algorithm *Densest At least-k_1, k_2* returns subsets $H_i(S)$ and $H_i(T)$ at iteration i, such that $\frac{|H_i(S)|}{|H_i(T)|} = a$. Suppose up to l_1th iteration, $|S_{l_1}| < k_1$ and $|T_{l_1}| < k_2$. Let $|S_{l_1+1}| \geq k_1$, but up to l_2th iteration, $|T_{l_2}| < k_2$. At iteration $l_2 + 1$, $|T_{l_2+1}| \geq k_2$. Now we consider the following cases,

Case 1: $|E(S_{l_1}, T_{l_1})| \geq |E(S^*, T^*)|/2$.

Case 2: $|E(S_{l_2}, T_{l_2}) \bigcap E(S^*, T^*)| \leq |E(S^*, T^*)|/2$.

Case 3: $\exists l', l_1 < l' \leq l_2$, *such that* $|E(S_{l'}, T_{l'})| > |E(S^*, T^*)|/2$ *and* $|E(S_{l'-1}, T_{l'-1})| \leq |E(S^*, T^*)|/2$.

These three cases are mutually exclusive and exhaustive. When case 1 occurs, we can append arbitrary vertices to S_{l_1} and T_{l_1} to make their sizes respectively k_1 and k_2. In that case $\frac{E(S'_{l_1}, T'_{l_1})}{\sqrt{|S'_{l_1}||T'_{l_1}|}} \geq \frac{E(S^*,T^*)}{2\sqrt{|S^*||T^*|}} = d^*/2$. When case 2 occurs, at iteration l_2 at least half of the edges of the optimum are still not covered. Since no vertices are ever deleted, choice of S^* and T^* maintains the ratio a and returns a density which is at least $\frac{d^*}{2}$. Then we have $\forall i = 1, 2, .., l_2$, $\frac{E(H_i(S), H_i(T))}{\sqrt{|H_i(S)||H_i(T)|}} \geq \frac{d^*}{2}$, or we have, $E(H_i(S), H_i(T)) \geq \sqrt{|H_i(S)||H_i(T)|}\frac{d^*}{2} = \sqrt{a}|H_i(T)|\frac{d^*}{2}$. Hence by summing over the iterations 1 to l_2+1 we get, $E(S_{l_2+1}, T_{l_2+1}) \geq \sqrt{a}\sum_{i=1}^{l_2+1}|H_i(T)|\frac{d^*}{2} \geq \frac{d^*}{2}\sqrt{a}|T_{l_2+1}| = \frac{d^*}{2}\sqrt{|T_{l_2+1}||S_{l_2+1}|}$. Hence we have, $\frac{E(S'_{l_2+1}, T'_{l_2+1})}{\sqrt{|S'_{l_2+1}||T'_{l_2+1}|}} = \frac{E(S_{l_2+1}, T_{l_2+1})}{\sqrt{|S_{l_2+1}||T_{l_2+1}|}} \geq \frac{d^*}{2}$.

When case 3 occurs, we have again $\forall i = 1, 2, .., l'$, $\frac{E(H_i(S), H_i(T))}{\sqrt{|H_i(S)||H_i(T)|}} \geq \frac{d^*}{2}$. Now following the same analysis as in case 2, $\frac{E(S_{l'}, T_{l'})}{\sqrt{|S_{l'}||T_{l'}|}} \geq \frac{d^*}{2}$. Since $|S'_{l'}| = |S_{l'}|$ might be much larger than k_1, the analysis as in case 1 cannot guarantee a 2-approximation. Let $X_1 = |\bigcup_{i=1}^{l'} H_i(S)| = |S_{l'}|$ and $X_2 = \sum_{i=1}^{l'}|H_i(S)|$. Similarly $Y_1 = |\bigcup_{i=1}^{l'} H_i(T)| = |T_{l'}|$ and $Y_2 = \sum_{i=1}^{l'}|H_i(T)|$. We have $Y_2 = \frac{X_2}{a} \geq \frac{X_1}{a} \geq \frac{k_1}{a} = k_2$. Also we have, $\frac{E(X_1,Y_1)}{\sqrt{|X_2||Y_2|}} \geq \frac{d^*}{2}$. So $\frac{E(X_1,Y_1)}{\sqrt{|S'_{l'}|}} = \frac{E(X_1,Y_1)}{\sqrt{|X_1|}} \geq \frac{E(X_1,Y_1)}{\sqrt{|X_2|}} \geq \sqrt{Y_2}\frac{d^*}{2} \geq \sqrt{k_2}\frac{d^*}{2}$. We add arbitrary vertices to Y_1 to make its size equal to k_2. Hence, $\frac{E(X_1,Y_1)}{\sqrt{|S'_{l'}|}} \geq \sqrt{|T'_{l'}|}\frac{d^*}{2}$. Also $\frac{E(X_1,Y_1)}{\sqrt{|T'_{l'}|}} = \frac{E(X_1,Y_1)}{\sqrt{k_2}} \geq \frac{E(X_1,Y_1)}{\sqrt{|Y_2|}} \geq \sqrt{|X_2|}\frac{d^*}{2} \geq \sqrt{|S'_{l'}|}\frac{d^*}{2}$. Multiplying we get, $\frac{E(S'_{l'},T'_{l'})^2}{\sqrt{|S'_{l'}||T'_{l'}|}} \geq \sqrt{|S'_{l'}||T'_{l'}|}\frac{d^{*2}}{4}$, or we have, $\frac{E(S'_{l'},T'_{l'})}{\sqrt{|S'_{l'}||T'_{l'}|}} \geq \frac{d^*}{2}$. □

For a particular value of a, using parametric max-flow, algorithm *Densest At least-*k_1, k_2 requires time of a single flow computation within a constant factor. However there are $|V|^2$ possible choices of a. An algorithm that does not need to guess the value of a, like *Densest- Subgraph-Directed*, is needed to reduce the time complexity, which is still open. Here we note that a LP based solution for this problem can be designed in the line of LP based algorithm for *DalkS* (details omitted).

4 Densest at Most *k* Subgraph Problem

Densest at most k subgraph problem (*DamkS*) tries to find a subgraph of highest density, whose size is at most k. Andersen et. al. [2] showed that an α approximation for *DamkS* implies a $\Theta(\alpha^2)$ approximation for the densest k subgraph problem. We prove that approximating *DamkS* is as hard as the *DkS* problem, within a constant factor. Precisely we prove the following theorem:

Theorem 7. *An α approximation algorithm for* DamkS *implies an 4α approximation algorithm for the densest k subgraph problem*

The proof can be found in an extended version [14].

5 Conclusion

In this paper, we have discussed different variations of the densest subgraph problems with and without size constraints. We have considered hardness issues related to these problems and have developed fast algorithms for them for both undirected and directed networks. All these problems can be generalized to weighted setting, with same time-complexity or sometimes with only a $\log|V|$ increase in running time. An interesting open question will be to design linear time algorithm with an approximation factor better than 2 for densest subgraph without any size constraint or to improve the approximation factor for *DalkS* problem. Obtaining faster algorithms for *densest at lease-*k_1, k_2 *subgraph* problem, or removing the requirement of guessing a in it or in the flow graph construction of maximum density directed subgraph will also be useful.

References

1. Andersen, R.: Finding large and small dense subgraphs. CoRR, abs/cs/0702032 (2007)
2. Andersen, R., Chellapilla, K.: Finding dense subgraphs with size bounds. In: Avrachenkov, K.E., Donato, D., Litvak, N. (eds.) WAW 2009. LNCS, vol. 5427, pp. 25–36. Springer, Heidelberg (2009)
3. Asahiro, Y., Hassin, R., Iwama, K.: Complexity of finding dense subgraphs. Discrete Appl. Math. 121(1-3), 15–26 (2002)
4. Asahiro, Y., Iwama, K., Tamaki, H., Tokuyama, T.: Greedily finding a dense subgraph. In: Karlsson, R., Lingas, A. (eds.) SWAT 1996. LNCS, vol. 1097, pp. 136–148. Springer, Heidelberg (1996)
5. Buehrer, G., Chellapilla, K.: A scalable pattern mining approach to web graph compression with communities. In: WSDM 2008, pp. 95–106 (2008)
6. Charikar, M.: Greedy approximation algorithms for finding dense components in a graph. In: Jansen, K., Khuller, S. (eds.) APPROX 2000. LNCS, vol. 1913, pp. 84–95. Springer, Heidelberg (2000)
7. Dourisboure, Y., Geraci, F., Pellegrini, M.: Extraction and classification of dense communities in the web. In: WWW 2007, pp. 461–470 (2007)
8. Feige, U., Kortsarz, G., Peleg, D.: The dense k-subgraph problem. Algorithmica 29, 410–421 (1997)
9. Gallo, G., Grigoriadis, M.D., Tarjan, R.E.: A fast parametric maximum flow algorithm and applications. SIAM J. Comput. 18(1), 30–55 (1989)
10. Gibson, D., Kumar, R., Tomkins, A.: Discovering large dense subgraphs in massive graphs. In: VLDB 2005, pp. 721–732 (2005)
11. Goldberg, A.V.: Finding a maximum density subgraph. Technical report (1984)
12. Kannan, R., Vinay, V.: Analyzing the structure of large graphs. Technical report (1999)
13. Khot, S.: Ruling out ptas for graph min-bisection, dense k-subgraph, and bipartite clique. SIAM J. Comput. 36(4), 1025–1071 (2006)
14. Khuller, S., Saha, B.: On finding dense subgraphs (2009), `http://www.cs.umd.edu/~samir/grant/ICALP09.pdf`
15. Kleinberg, J.M.: Authoritative sources in a hyperlinked environment. J. ACM 46(5), 604–632 (1999)
16. Lawler, E.: Combinatorial optimization - networks and matroids. Holt, Rinehart and Winston, New York (1976)

Learning Halfspaces with Malicious Noise

Adam R. Klivans[1], Philip M. Long[2], and Rocco A. Servedio[3]

[1] UT Austin
[2] Google
[3] Columbia University
klivans@cs.utexas.edu, plong@google.com, rocco@cs.columbia.edu

Abstract. We give new algorithms for learning halfspaces in the challenging *malicious noise* model, where an adversary may corrupt both the labels and the underlying distribution of examples. Our algorithms can tolerate malicious noise rates exponentially larger than previous work in terms of the dependence on the dimension n, and succeed for the fairly broad class of all isotropic log-concave distributions.

We give poly$(n, 1/\epsilon)$-time algorithms for solving the following problems to accuracy ϵ:

- Learning origin-centered halfspaces in $\mathbf{R}^n$ with respect to the uniform distribution on the unit ball with malicious noise rate $\eta = \Omega(\epsilon^2/\log(n/\epsilon))$. (The best previous result was $\Omega(\epsilon/(n\log(n/\epsilon))^{1/4})$.)
- Learning origin-centered halfspaces with respect to any isotropic log-concave distribution on $\mathbf{R}^n$ with malicious noise rate $\eta = \Omega(\epsilon^3/\log(n/\epsilon))$. This is the first efficient algorithm for learning under isotropic log-concave distributions in the presence of malicious noise.

We also give a poly$(n, 1/\epsilon)$-time algorithm for learning origin-centered halfspaces under any isotropic log-concave distribution on $\mathbf{R}^n$ in the presence of *adversarial label noise* at rate $\eta = \Omega(\epsilon^3/\log(1/\epsilon))$. In the adversarial label noise setting (or agnostic model), labels can be noisy, but not example points themselves. Previous results could handle $\eta = \Omega(\epsilon)$ but had running time exponential in an unspecified function of $1/\epsilon$.

Our analysis crucially exploits both concentration and anti-concentration properties of isotropic log-concave distributions. Our algorithms combine an iterative outlier removal procedure using Principal Component Analysis together with "smooth" boosting.

1 Introduction

A *halfspace* is a Boolean-valued function of the form $f = \text{sign}(\sum_{i=1}^n w_i x_i - \theta)$. Learning halfspaces in the presence of noisy data is a fundamental problem in machine learning. In addition to its practical relevance, the problem has connections to many well-studied topics such as kernel methods [26], cryptographic hardness of learning [15], hardness of approximation [6, 9], learning Boolean circuits [2], and additive/multiplicative update learning algorithms [17, 7].

Learning an unknown halfspace from correctly labeled (non-noisy) examples is one of the best-understood problems in learning theory, with work dating back to the famous Perceptron algorithm of the 1950s [21] and a range of efficient algorithms known for

S. Albers et al. (Eds.): ICALP 2009, Part I, LNCS 5555, pp. 609–621, 2009.

different settings [20, 16, 3, 18]. Much less is known, however, about the more difficult problem of learning halfspaces in the presence of noise.

Important progress was made by Blum *et al.* [2] who gave a polynomial-time algorithm for learning a halfspace under *classification noise*. In this model each label presented to the learner is flipped independently with some fixed probability; the noise does not affect the actual example points themselves, which are generated according to an arbitrary probability distribution over $\mathbf{R}^n$.

In the current paper we consider a much more challenging *malicious noise* model. In this model, introduced by Valiant [27] (see also [12]), there is an unknown target function f and distribution $\mathcal{D}$ over examples. Each time the learner receives an example, independently with probability $1-\eta$ it is drawn from $\mathcal{D}$ and labeled correctly according to f, but with probability η it is an arbitrary pair (x, y) which may be generated by an omniscient adversary. The parameter η is known as the "noise rate."

Malicious noise is a notoriously difficult model with few positive results. It was already shown in [12] that for essentially all concept classes, it is information-theoretically impossible to learn to accuracy $1-\epsilon$ if the noise rate η is greater than $\epsilon/(1+\epsilon)$. Indeed, known algorithms for learning halfspaces [25, 11] or even simpler target functions [19] with malicious noise typically make strong assumptions about the underlying distribution $\mathcal{D}$, and can learn to accuracy $1-\epsilon$ only for noise rates η much smaller than ϵ. We describe the most closely related work that we know of in Section 1.2.

In this paper we consider learning under the uniform distribution on the unit ball in $\mathbf{R}^n$, and more generally under any isotropic log-concave distribution. The latter is a fairly broad class of distributions that includes spherical Gaussians and uniform distributions over a wide range of convex sets. Our algorithms can learn from malicious noise rates that are quite high, as we now describe.

1.1 Main Results

Our first result is an algorithm for learning halfspaces in the malicious noise model with respect to the uniform distribution on the n-dimensional unit ball:

Theorem 1. *There is a* $\mathrm{poly}(n, 1/\epsilon)$*-time algorithm that learns origin-centered halfspaces to accuracy* $1-\epsilon$ *with respect to the uniform distribution on the unit ball in* n *dimensions in the presence of malicious noise at rate* $\eta = \Omega(\epsilon^2/\log(n/\epsilon))$.

Via a more sophisticated algorithm, we can learn in the presence of malicious noise under any isotropic log-concave distribution:

Theorem 2. *There is a* $\mathrm{poly}(n, 1/\epsilon)$*-time algorithm that learns origin-centered halfspaces to accuracy* $1-\epsilon$ *with respect to any isotropic log-concave distribution over* $\mathbf{R}^n$ *and can tolerate malicious noise at rate* $\eta = \Omega(\epsilon^3/\log(n/\epsilon))$.

We are not aware of any previous polynomial-time algorithms for learning under isotropic log-concave distributions in the presence of malicious noise.

Finally, we also consider a somewhat relaxed noise model known as *adversarial label noise*. In this model there is a fixed probability distribution P over $\mathbf{R}^n \times \{-1, 1\}$ (i.e. over labeled examples) for which a $1-\eta$ fraction of draws are labeled according to an unknown halfspace. The marginal distribution over $\mathbf{R}^n$ is assumed to be isotropic log-concave; so the idea is that an "adversary" chooses an η fraction of examples to mislabel, but unlike the malicious noise model she cannot change the (isotropic log-concave) distribution of the actual example points in $\mathbf{R}^n$. For this model we prove:

Theorem 3. *There is a* $\mathrm{poly}(n, 1/\epsilon)$*-time algorithm that learns origin-centered half-spaces to accuracy* $1-\epsilon$ *with respect to any isotropic log-concave distribution over* $\mathbf{R}^n$ *and can tolerate adversarial label noise at rate* $\eta = \Omega(\epsilon^3/\log(1/\epsilon))$.

1.2 Previous Work

Here is some of the most closely related previous work.

Malicious noise. General-purpose tools developed by Kearns and Li [12, 13] directly imply that halfspaces can be learned for any distribution over the domain in randomized poly(n,$1/\epsilon$) time with malicious noise at a rate $\Omega(\epsilon/n)$; the algorithm repeatedly picks a random subsample of the training data, hoping to miss all the noisy examples. Kannan (see [1]) devised a deterministic algorithm with a $\Omega(\epsilon/n)$ bound that repeatedly finds a group of $n+1$ examples that includes a noisy example, then removes the group. Kalai, et al. [11] showed that the poly(n,$1/\epsilon$)-time the averaging algorithm [24] tolerates noise at a rate $\Omega(\epsilon/\sqrt{n})$ when the distribution is uniform. They also described an improvement to $\tilde{\Omega}(\epsilon/n^{1/4})$ based on the observation that uniform examples will tend to be well-separated, so that pairs of examples that are too close to one another can be removed.

Adversarial label noise. Kalai, et al. showed that if the distribution over the instances is uniform over the unit ball, the averaging algorithm tolerates adversarial label noise at a rate $O(\epsilon/\sqrt{\log(1/\epsilon)})$ in poly(n,$1/\epsilon$) time. (In that paper, adversarial label noise was called "agnostic learning".) They also described an algorithm that fits low-degree polynomials that tolerates noise at a rate within an additive ϵ of the accuracy, but in $\mathrm{poly}\left(n^{1/\epsilon^4}\right)$ time; for log-concave distributions, their algorithm took $\mathrm{poly}\left(n^{d(1/\epsilon)}\right)$ time, for an unspecified function d. The latter algorithm does not require that the distribution is isotropic, as ours does.

Robust PCA. Independently of this work, Xu et al. [28] designed and analyzed an algorithm that performs principal component analysis when some of the examples are corrupted arbitrarily, as in the malicious noise model studied here.

1.3 Techniques

Outlier Removal. Consider first the simplest problem of learning an origin-centered halfspace with respect to the uniform distribution on the n-dimensional ball. A natural idea is to use a simple "averaging" algorithm that takes the vector average of the positive examples it receives and uses this as the normal vector of its hypothesis halfspace. Servedio [24] analyzed this algorithm for the random classification noise model, and Kalai *et al.* [11] extended the analysis to the adversarial label noise model.

Intuitively the "averaging" algorithm can only tolerate low malicious noise rates because the adversary can generate noisy examples which "pull" the average vector far from its true location. Our main insight is that the adversary does this most effectively when the noisy examples are coordinated to pull in roughly the same direction. We use a form of outlier detection based on Principal Component Analysis to detect such coordination. This is done by computing the direction $\mathbf{w}$ of maximal variance of the data set; if the variance in direction $\mathbf{w}$ is suspiciously large, we remove from the sample all points $\mathbf{x}$ for which $(\mathbf{w} \cdot \mathbf{x})^2$ is large. Our analysis shows that this causes many noisy examples, and only a few non-noisy examples, to be removed.

We repeat this process until the variance in every direction is not too large. (This cannot take too many stages since many noisy examples are removed in each stage.) While some noisy examples may remain, we show that their scattered effects cannot hurt the algorithm much.

Thus, in a nutshell, our overall algorithm for the uniform distribution is to first do outlier removal[1] by an iterated PCA-type procedure, and then simply run the averaging algorithm on the remaining "cleaned-up" data set.

Extending to Log-Concave Distributions via Smooth Boosting. We are able to show that the iterative outlier removal procedure described above is useful for isotropic log-concave distributions as well as the uniform distribution: if examples are removed in a given stage, then many of the removed examples are noisy and only a few are non-noisy (the analysis here uses concentration bounds for isotropic log-concave distributions). However, even if there were no noise in the data, the average of the positive examples under an isotropic log-concave distribution need not give a high-accuracy hypothesis. Thus the averaging algorithm alone will not suffice after outlier removal.

To get around this, we show that after outlier removal the average of the positive examples gives a (real-valued) *weak* hypothesis that has some nontrivial predictive accuracy. (Interestingly, the proof of this relies heavily on *anti*-concentration properties of isotropic log-concave distributions!) A natural approach is then to use a boosting algorithm to convert this weak learner into a strong learner. This is not entirely straightforward because boosting "skews" the distribution of examples; this has the undesirable effects of both increasing the effective malicious noise rate, and causing the distribution to no longer be isotropic log-concave. However, by using a "smooth" boosting algorithm [25] that skews the distribution as little as possible, we are able to control these undesirable effects and make the analysis go through. (The extra factor of ϵ in the bound of Theorem 2 compared with Theorem 1 comes from the fact that the boosting algorithm constructs "$1/\epsilon$-skewed" distributions.)

We note that our approach of using smooth boosting is reminiscent of [23, 25], but the current algorithm goes well beyond that earlier work. [23] did not consider a noisy scenario, and [25] only considered the averaging algorithm without any outlier removal as the weak learner (and thus could only handle quite low rates of malicious noise in our isotropic log-concave setting).

Finally, our results for learning under isotropic log-concave distributions with adversarial label noise are obtained using a similar approach. The algorithm here is in

[1] We note briefly that the sophisticated outlier removal techniques of [2, 5] do not seem to be useful in our setting; those works deal with a strong notion of outliers, which is such that no point on the unit ball can be an outlier if a significant fraction of points are uniformly distributed on the unit ball.

fact simpler than the malicious noise algorithm: since the adversarial label noise model does not allow the adversary to alter the distribution of the examples in $\mathbf{R}^n$, we can dispense with the outlier removal and simply use smooth boosting with the averaging algorithm as the weak learner. (This is why we get a slightly better quantitative bound in Theorem 3 than Theorem 2).

Organization. We present the simpler and more easily understood uniform distribution analysis first, proving Theorem 1 in Section 2. The proof of Theorem 2, which builds on the ideas of Theorem 1, is sketched in Section 3. In Section 1.2, we described some of the most closely related previous work. Because of space constraints the proof of Theorem 3 is omitted here and is given in the full version [14].

2 The Uniform Distribution and Malicious Noise

In this section we prove Theorem 1. As described above, our algorithm first does outlier removal using PCA and then applies the "averaging algorithm."

We may assume throughout that the noise rate η is smaller than some absolute constant, and that the dimension n is larger than some absolute constant.

2.1 The Algorithm: Removing Outliers and Averaging

Consider the following Algorithm A_{mu}:

1. Draw a sample S of $m = \mathrm{poly}(n/\epsilon)$ many examples from the malicious oracle.
2. Identify the direction $\mathbf{w} \in \mathbb{S}^{n-1}$ that maximizes

$$\sigma_{\mathbf{w}}^2 \stackrel{def}{=} \sum_{(\mathbf{x},y)\in S} (\mathbf{w}\cdot\mathbf{x})^2.$$

   If $\sigma_{\mathbf{w}}^2 < \frac{10m\log m}{n}$ then go to Step 4 otherwise go to Step 3.
3. Remove from S every example that has $(\mathbf{w}\cdot\mathbf{x})^2 \geq \frac{10\log m}{n}$. Go to Step 2.
4. For the examples S that remain let $\mathbf{v} = \frac{1}{|S|}\sum_{(\mathbf{x},y)\in S} y\mathbf{x}$ and output the linear classifier $h_{\mathbf{v}}$ defined by $h_{\mathbf{v}}(\mathbf{x}) = \mathrm{sgn}(\mathbf{v}\cdot x)$.

We first observe that Step 2 can be carried out in polynomial time:

Lemma 1. *There is a polynomial-time algorithm that, given a finite collection S of points in $\mathbf{R}^n$, outputs $\mathbf{w} \in \mathbb{S}^{n-1}$ that maximizes $\sum_{\mathbf{x}\in S}(\mathbf{w}\cdot\mathbf{x})^2$.*

Proof. By applying Lagrange multipliers, we can see that the optimal $\mathbf{w}$ is an eigenvector of $A = \sum_{\mathbf{x}\in S} xx^T$. Further, if λ is the eigenvalue of $\mathbf{w}$, then $\sum_{\mathbf{x}\in S}(\mathbf{w}\cdot\mathbf{x})^2 = \mathbf{w}^T A\mathbf{w} = \mathbf{w}^T(\lambda\mathbf{w}) = \lambda$. The eigenvector $\mathbf{w}$ with the largest eigenvalue can be found in polynomial time (see e.g. [10]). □

Before embarking on the analysis we establish a terminological convention. Much of our analysis deals with high-probability statements over the draw of the m-element sample S; it is straightforward but quite cumbersome to explicitly keep track of all of the failure probabilities. Thus we write "with high probability" (or "w.h.p.") in various places below as a shorthand for "with probability at least $1 - 1/\text{poly}(n/\epsilon)$." The interested reader can easily verify that an appropriate $\text{poly}(n/\epsilon)$ choice of m makes all the failure probabilities small enough so that the entire algorithm succeeds with probability at least $1/2$ as required.

2.2 Properties of the Clean Examples

In this subsection we establish properties of the clean examples that were sampled in Step 1 of A_{mu}. The first says that no direction has much more variance than the expected variance of $1/n$. Its proof, which uses standard tools from VC theory, is omitted due to space constraints.

Lemma 2. *W.h.p. over a random draw of ℓ clean examples S_{clean}, we have*

$$\max_{a \in \mathbb{S}^{n-1}} \left\{ \frac{1}{\ell} \sum_{(\mathbf{x},y) \in S_{\text{clean}}} (\mathbf{a} \cdot \mathbf{x})^2 \right\} \leq \frac{1}{n} + \sqrt{\frac{O(n) \log m}{\ell}}.$$

The next lemma says that in fact no direction has too many clean examples lying far out in that direction. Its proof, which uses Lemma 7 of [4], is omitted due to space constraints.

Lemma 3. *For any $\beta > 0$ and $\kappa > 1$, if S_{clean} is a random set of $\ell \geq \frac{O(1) \cdot n^2 \beta^2 e^{\beta^2 n/2}}{(1+\kappa)\ln(1+\kappa)}$ clean examples then w.h.p. we have*

$$\max_{a \in \mathbb{S}^{n-1}} \left\{ \frac{1}{\ell} \sum_{x \in S_{\text{clean}}} \mathbf{1}_{(a \cdot x)^2 > \beta^2} \right\} \leq (1+\kappa) e^{-\beta^2 n/2}.$$

2.3 What Is Removed

In this section, we provide bounds on the number of clean and dirty examples removed in Step 3.

The first bound is a Corollary of Lemma 3.

Corollary 1. *W.h.p. over the random draw of the m-element sample S, the number of clean examples removed during the any execution of Step 3 in A_{mu} is at most $6n \log m$.*

Proof. Since the noise rate η is sufficiently small, w.h.p. the number ℓ of clean examples is at least (say) $m/2$. We would like to apply Lemma 3 with $\kappa = 5\ell^4 n \log \ell$ and $\beta = \sqrt{\frac{10 \log m}{n}}$, and indeed we may do this because we have

$$\frac{O(1) \cdot n^2 \beta^2 e^{\beta^2 n/2}}{(1+\kappa)\ln(1+\kappa)} \leq \frac{O(1) \cdot n(\log m) m^5}{(1+\kappa)\ln(1+\kappa)} \leq O\left(\frac{m}{\log m}\right) \leq \frac{m}{2} \leq \ell$$

for n sufficiently large. Since clean points are only removed if they have $(\mathbf{a} \cdot \mathbf{x})^2 > \beta^2$, Lemma 3 gives us that the number of clean points removed is at most

$$m(1+\kappa) e^{-\beta^2 n/2} \leq 6m^5 n \log(\ell)/m^5 \leq 6n \log m.$$

□

The counterpart to Corollary 1 is the following lemma. It tells us that if examples are removed in Step 3, then there must be many *dirty* examples removed. It exploits the fact that Lemma 2 bounds the variance in *all* directions $\mathbf{a}$, so that it can be reused to reason about what happens in different executions of step 3.

Lemma 4. *W.h.p. over the random draw of S, whenever A_{mu} executes step 3, it removes at least $\frac{4m \log m}{n}$ noisy examples from S_{dirty}, the set of dirty examples in S.*

Proof. As stated earlier we may assume that $\eta \leq 1/4$. This implies that w.h.p. the fraction $\widehat{\eta}$ of noisy examples in the initial set S is at most $1/2$. Finally, Lemma 2 implies that $m = \tilde{\Omega}(n^2)$ suffices for it to be the case that w.h.p., for all $\mathbf{a} \in \mathbb{S}^{n-1}$, for the original multiset S_{clean} of clean examples drawn in step 1, we have

$$\sum_{(\mathbf{x},y)\in S_{\mathrm{clean}}} (\mathbf{a}\cdot\mathbf{x})^2 \leq \frac{2m}{n}. \tag{1}$$

We shall say that a random sample S that satisfies all these requirements is "reasonable". We will show that for any reasonable dataset, the number of noisy examples removed during the execution of step 3 of A_{mu} is at least $\frac{4m \log m}{n}$.

If we remove examples using direction $\mathbf{w}$ then it means $\sum_{(\mathbf{x},y)\in S}(\mathbf{w}\cdot\mathbf{x})^2 \geq \frac{10m \log m}{n}$. Since S is reasonable, by (1) the contribution to the sum from the clean examples that survived to the current stage is at most $2m/n$ so we must have

$$\sum_{(\mathbf{x},y)\in S_{\mathrm{dirty}}} (\mathbf{w}\cdot\mathbf{x})^2 \geq 10m\log(m)/n - 2m/n > 9m\log(m)/n.$$

Let us decompose S_{dirty} into $N \cup F$ where N ("near") consists of those points x s.t. $(\mathbf{w}\cdot\mathbf{x})^2 \leq 10\log(m)/n$ and F ("far") is the remaining points for which $(\mathbf{w}\cdot\mathbf{x})^2 > 10\log(m)/n$. Since $|N| \leq |S_{\mathrm{dirty}}| \leq \widehat{\eta}m$, (any dirty examples removed in earlier rounds will only reduce the size of S_{dirty}) we have $\sum_{(\mathbf{x},y)\in N}(\mathbf{w}\cdot\mathbf{x})^2 \leq (\widehat{\eta}m)10\log(m)/n$ and so

$$|F| \geq \sum_{(\mathbf{x},y)\in F} (\mathbf{w}\cdot\mathbf{x})^2 \geq 9m\log(m)/n - (\widehat{\eta}m)10\log(m)/n \geq 4m\log(m)/n$$

(the last line used the fact that $\widehat{\eta} < 1/2$). Since the points in F are removed in Step 3, the lemma is proved. □

2.4 Exploiting Limited Variance in Any Direction

In this section, we show that if all directional variances are small, then the algorithm's final hypothesis will have high accuracy.

We first recall a simple lemma which shows that a sample of "clean" examples results in a high-accuracy hypothesis for the averaging algorithm:

Lemma 5 ([24]). *Suppose $\mathbf{x}_1, \ldots, \mathbf{x}_m$ are chosen uniformly at random from $\mathbb{S}^{n-1}$, and a target weight vector $\mathbf{u} \in \mathbb{S}^{n-1}$ produces labels $y_1 = \mathrm{sign}(\mathbf{u}\cdot\mathbf{x}_1), \ldots, y_m = \mathrm{sign}(\mathbf{u}\cdot\mathbf{x}_m)$. Let $\mathbf{v} = \frac{1}{m}\sum_{t=1}^{m} y_t\mathbf{x}_t$. Then w.h.p. the component of $\mathbf{v}$ in the direction of $\mathbf{u}$ satisfies $\mathbf{u}\cdot\mathbf{v} = \Omega(\frac{1}{\sqrt{n}})$, while the rest of $\mathbf{v}$ satisfies $||\mathbf{v} - (\mathbf{u}\cdot\mathbf{v})\mathbf{u}|| = O(\sqrt{\log(n)/m})$.*

Now we can state Lemma 6.

Lemma 6. *Let $S = S_{\text{clean}} \cup S_{\text{dirty}}$ be the sample of m examples drawn from the noisy oracle $\text{EX}_\eta(f,\mathcal{U})$. Let*

- *S'_{clean} be those clean examples that were never removed during step 3 of A_{mu},*
- *S'_{dirty} be those dirty examples that were never removed during step 3 of A_{mu},*
- *$\eta' = \frac{|S'_{\text{dirty}}|}{|S'_{\text{clean}} \cup S'_{\text{dirty}}|}$, i.e. the fraction of dirty examples among the examples that survive step 3, and*
- *$\alpha = \frac{|S_{\text{clean}} - S'_{\text{clean}}|}{|S'_{\text{clean}} \cup S'_{\text{dirty}}|}$, the ratio of the number of clean points that were erroneously removed to the size of the final surviving data set.*

Let $S' \stackrel{def}{=} S'_{\text{clean}} \cup S'_{\text{dirty}}$. Suppose that , for every direction $\mathbf{w} \in \mathbb{S}^{n-1}$ we have

$$\sigma^2_{\mathbf{w}} \stackrel{def}{=} \sum_{(\mathbf{x},y)\in S'} (\mathbf{w}\cdot\mathbf{x})^2 \le \frac{10m\log m}{n}.$$

Then w.h.p. over the draw of S, the halfspace with normal vector $\mathbf{v} \stackrel{def}{=} \frac{1}{|S'|}\sum_{(\mathbf{x},y)\in S'} y\mathbf{x}$ has error rate

$$O\left(\sqrt{\eta'\log m} + \alpha\sqrt{n} + \sqrt{\frac{n\log n}{m}}\right).$$

Proof. The claimed bound is trivial unless $\eta' \le o(1)/\log m$ and $\alpha \le o(1)/\sqrt{n}$, so we shall freely use these bounds in what follows.

Let $\mathbf{u}$ be the unit length normal vector for the target halfspace. Let $\mathbf{v}_{\text{clean}}$ be the average of *all* the clean examples, $\mathbf{v}'_{\text{dirty}}$ be the average of the dirty (noisy) examples that were not deleted (i.e. the examples in S'_{dirty}), and $\mathbf{v}_{\text{del}}$ be the average of the clean examples that were deleted. Then

$$\begin{aligned}\mathbf{v} &= \frac{1}{|S'_{\text{clean}} \cup S'_{\text{dirty}}|}\sum_{(\mathbf{x},y)\in S'_{\text{clean}}\cup S'_{\text{dirty}}} y\mathbf{x}\\ &= \frac{1}{|S'_{\text{clean}} \cup S'_{\text{dirty}}|}\left(\left(\sum_{(\mathbf{x},y)\in S_{\text{clean}}} y\mathbf{x}\right) + \left(\sum_{(\mathbf{x},y)\in S'_{\text{dirty}}} y\mathbf{x}\right)\right.\\ &\qquad \left.- \left(\sum_{(\mathbf{x},y)\in S_{\text{clean}}-S'_{\text{clean}}} y\mathbf{x}\right)\right)\\ \mathbf{v} &= (1-\eta'+\alpha)\mathbf{v}_{\text{clean}} + \eta'\mathbf{v}'_{\text{dirty}} - \alpha\mathbf{v}_{\text{del}}.\end{aligned} \tag{2}$$

Let us begin by exploiting the bound on the variance in every direction to bound the length of $\mathbf{v}'_{\text{dirty}}$. For any $\mathbf{w} \in \mathbb{S}^{n-1}$ we know that

$$\sum_{(\mathbf{x},y)\in S'} (\mathbf{w}\cdot\mathbf{x})^2 \le \frac{10m\log m}{n}, \quad \text{and hence} \quad \sum_{(\mathbf{x},y)\in S'_{\text{dirty}}} (\mathbf{w}\cdot\mathbf{x})^2 \le \frac{10m\log m}{n}$$

since $S'_{\text{dirty}} \subseteq S'$. The Cauchy-Schwarz inequality now gives

$$\sum_{(\mathbf{x},y)\in S'_{\text{dirty}}} |\mathbf{w}\cdot\mathbf{x}| \le \sqrt{\frac{10m|S'_{\text{dirty}}|\log m}{n}}.$$

Taking $\mathbf{w}$ to be the unit vector in the direction of $\mathbf{v}'_{\text{dirty}}$, we have $||\mathbf{v}'_{\text{dirty}}|| =$

$$\mathbf{w}\cdot\mathbf{v}'_{\text{dirty}} = \mathbf{w}\cdot\frac{1}{|S'_{\text{dirty}}|}\sum_{(\mathbf{x},y)\in S'_{\text{dirty}}} y\mathbf{x} \le \frac{1}{|S'_{\text{dirty}}|}\sum_{(\mathbf{x},y)\in S'_{\text{dirty}}} |\mathbf{w}\cdot\mathbf{x}| \le \sqrt{\frac{10m\log m}{|S'_{\text{dirty}}|n}}. \tag{3}$$

Because the domain distribution is uniform, the error of $h_{\mathbf{v}}$ is proportional to the angle between $\mathbf{v}$ and $\mathbf{u}$, in particular,

$$\Pr[h_{\mathbf{v}} \ne f] = \frac{1}{\pi}\arctan\left(\frac{||\mathbf{v}-(\mathbf{v}\cdot\mathbf{u})\mathbf{u}||}{\mathbf{u}\cdot\mathbf{v}}\right) \le (1/\pi)\frac{||\mathbf{v}-(\mathbf{v}\cdot\mathbf{u})\mathbf{u}||}{\mathbf{u}\cdot\mathbf{v}}. \tag{4}$$

We have that $||\mathbf{v}-(\mathbf{v}\cdot\mathbf{u})\mathbf{u}||$ equals

$$\begin{aligned}&||(1-\eta'+\alpha)(\mathbf{v}_{\text{clean}}-(\mathbf{v}_{\text{clean}}\cdot\mathbf{u})\mathbf{u})+\eta'(\mathbf{v}'_{\text{dirty}}-(\mathbf{v}'_{\text{dirty}}\cdot\mathbf{u})\mathbf{u})-\alpha(\mathbf{v}_{\text{del}}-(\mathbf{v}_{\text{del}}\cdot\mathbf{u})\mathbf{u})||\\ &\quad\le 2||\mathbf{v}_{\text{clean}}-(\mathbf{v}_{\text{clean}}\cdot\mathbf{u})\mathbf{u}||+\eta'||\mathbf{v}'_{\text{dirty}}||+\alpha||\mathbf{v}_{\text{del}}||\end{aligned}$$

where we have used the triangle inequality and the fact that α, η are "small." Lemma 5 lets us bound the first term in the sum by $O(\sqrt{\log(n)/m})$, and the fact that $\mathbf{v}_{\text{del}}$ is an average of vectors of length 1 lets us bound the third by α. For the second term, Equation (3) gives us

$$\eta'||\mathbf{v}'_{\text{dirty}}|| \le \sqrt{\frac{10m(\eta')^2\log m}{|S'_{\text{dirty}}|n}} = \sqrt{\frac{10m\eta'\log m}{|S'|n}} \le \sqrt{\frac{20\eta'\log m}{n}},$$

where for the last equality we used $|S'| \ge m/2$ (which is an easy consequence of Corollary 1 and the fact that w.h.p. $|S_{\text{clean}}| \ge 3m/4$). We thus get

$$||\mathbf{v}-(\mathbf{v}\cdot\mathbf{u})\mathbf{u}|| \le O\left(\sqrt{\log(n)/m}\right)+\sqrt{20\eta'\log(m)/n}+\alpha. \tag{5}$$

Now we consider the denominator of (4). We have

$$\mathbf{u}\cdot\mathbf{v} = (1-\eta'+\alpha)(\mathbf{u}\cdot\mathbf{v}_{\text{clean}})+\eta'\mathbf{u}\cdot\mathbf{v}'_{\text{dirty}}-\alpha\mathbf{u}\cdot\mathbf{v}_{\text{del}}.$$

Similar to the above analysis, we again use Lemma 5 (but now the lower bound $\mathbf{u}\cdot\mathbf{v} \ge \Omega(1/\sqrt{n})$, Equation (3), and the fact that $||\mathbf{v}_{\text{del}}|| \le 1$. Since α and η' are "small," we get that there is an absolute constant c such that $\mathbf{u}\cdot\mathbf{v} \ge c/\sqrt{n}-\sqrt{20\eta'\log(m)/n}-\alpha$. Combining this with (5) and (4), we get

$$\Pr[h_{\mathbf{v}} \ne f] \le \frac{O\left(\sqrt{\frac{\log n}{m}}\right)+\sqrt{\frac{20\eta'\log m}{n}}+\alpha}{\frac{c}{\sqrt{n}}-\sqrt{\frac{20\eta'\log m}{n}}-\alpha} = O\left(\sqrt{\frac{n\log n}{m}}+\sqrt{\eta'\log m}+\alpha\sqrt{n}\right).$$

□

2.5 Proof of Theorem 1

By Corollary 1, with high probability, each outlier removal stage removes at most $6n \log m$ clean points.

Since each outlier removal stage removes at least $\frac{4m \log m}{n}$ noisy examples, there must be at most $O(n/(\log m))$ such stages. Consequently the total number of clean examples removed across all stages is $O(n^2)$. Since w.h.p. the initial number of clean examples is at least $m/2$, this means that the final data set (on which the averaging algorithm is run) contains at least $m/2 - O(n^2)$ clean examples, and hence at least $m/2 - O(n^2)$ examples in total. Consequently the value of α from Lemma 6 after the final outlier removal stage (the ratio of the total number of clean examples deleted, to the total number of surviving examples) is at most $\frac{2n^2}{m/2-n^2}$.

The standard Hoeffding bound implies that w.h.p. the actual fraction of noisy examples in the original sample S is at most $\eta + \sqrt{O(\log m)/m}$. It is easy to see that w.h.p. the fraction of dirty examples does not increase (since each stage of outlier removal removes more dirty points than clean points, for a suitably large poly(n/ϵ) value of m), and thus the fraction η' of dirty examples among the remaining examples after the final outlier removal stage is at most $\eta + \sqrt{O(\log m)/m}$. Applying Lemma 6, for a suitably large value $m = \text{poly}(n/\epsilon)$, we obtain $\Pr[h_{\mathbf{v}} \neq f] \leq O\left(\sqrt{\eta} \log m\right)$. Rearranging this bound, we can learn to accuracy ϵ even for $\eta = \Omega(\epsilon^2/\log(n/\epsilon))$. This completes the proof of the theorem. □

3 Isotropic Log-Concave Distributions and Malicious Noise

Our algorithm A_{mlc} that works for arbitrary log-concave distributions uses smooth boosting.

3.1 Smooth Boosting

A boosting algorithm uses a subroutine, called a *weak learner*, that is only guaranteed to output hypotheses with a non-negligible advantage over random guessing.[2] The boosting algorithm that we consider uses a *confidence-rated* weak learner [22], which predicts $\{-1, 1\}$ labels using continuous values in $[-1, 1]$. Formally, the *advantage* of a hypothesis h' with respect to a distribution $\mathcal{D}'$ is defined to be $\mathbf{E}_{x \sim \mathcal{D}'}[h'(x)f(x)]$, where f is the target function.

For the purposes of this paper, a boosting algorithm makes use of the weak learner, an example oracle (possibly corrupted with noise), a desired accuracy ϵ, and a bound γ on the advantage of the hypothesis output by the weak learner.

A boosting algorithm that is trying to learn an unknown target function f with respect to some distribution $\mathcal{D}$ repeatedly simulates a (possibly noisy) example oracle for f with respect to some other distribution $\mathcal{D}'$ calls a subroutine A_{weak} with respect to this oracle, receiving a *weak hypothesis*, which maps $\mathbf{R}^n$ to the continuous interval $[-1, 1]$.

[2] For simplicity of presentation we ignore the confidence parameter of the weak learner in our discussion; this can be handled in an entirely standard way.

After repeating this for some number of stages, the boosting algorithm combines the weak hypotheses generated during its various calls to the weak learner into a final aggregate hypothesis which it outputs.

Let $\mathcal{D}, \mathcal{D}'$ be two distributions over $\mathbf{R}^n$. We say that $\mathcal{D}'$ is $(1/\epsilon)$*-smooth with respect to* $\mathcal{D}$ if $\mathcal{D}(\mathbf{x}) \leq (1/\epsilon)\mathcal{D}'(\mathbf{x})$ for all $\mathbf{x} \in \mathbf{R}^n$.

The following lemma from [25] (similar results can be readily found elsewhere, see e.g. [8]) identifies the properties that we need from a boosting algorithm for our analysis.

Lemma 7 ([25]). *There is a boosting algorithm B and a polynomial p such that, for any $\epsilon, \gamma > 0$, the following properties hold. When learning a target function f using* $\mathrm{EX}_\eta(f, \mathcal{D})$, *we have: (a) If each call to A_{weak} takes time t, then B takes time $p(t, 1/\gamma, 1/\epsilon)$. (b) The weak learner is always called with an oracle* $\mathrm{EX}_{\eta'}(f, \mathcal{D}')$ *where $\mathcal{D}'$ is $(1/\epsilon)$-smooth with respect to $\mathcal{D}$ and $\eta' \leq \eta/\epsilon$. (c) Suppose that for each distribution* $\mathrm{EX}_{\eta'}(f, \mathcal{D}')$ *passed to A_{weak} by B, the output of A_{weak} has advantage γ. Then the final output h of B satisfies* $\Pr_{x \in \mathcal{D}}[h(x) \neq f(x)] \leq \epsilon$.

3.2 The Algorithm

Our algorithm for learning under isotropic log-concave distributions with malicious noise, Algorithm A_{mlc}, applies the smooth booster from Lemma 7 with the following weak learner, which we call Algorithm A_{mlcw}. (The value c_0 is an absolute constant that will emerge from our analysis.)

1. Draw $m = \mathrm{poly}(n/\epsilon)$ examples from the oracle $\mathrm{EX}_{\eta'}(f, \mathcal{D}')$.
2. Remove all those examples $(\mathbf{x}, y)$ for which $||\mathbf{x}|| > \sqrt{3n \log m}$.
3. Repeatedly
   - find a direction (unit vector) $\mathbf{w}$ that maximizes $\sum_{(\mathbf{x},y)\in S}(\mathbf{w} \cdot \mathbf{x})^2$ (see Lemma 1)
   - if $\sum_{(\mathbf{x},y)\in S}(\mathbf{w}\cdot\mathbf{x})^2 \leq c_0 m \log(n/\epsilon)$ then move on to Step 4, and otherwise
   - remove from S all examples $(\mathbf{x}, y)$ for which $(\mathbf{w} \cdot \mathbf{x})^2 > c_0 \log(n/\epsilon)$, and iterate again.
4. Let $\mathbf{v} = \frac{1}{|S|}\sum_{(\mathbf{x},y)\in S} y\mathbf{x}$, and return h defined by $h(\mathbf{x}) = \frac{\mathbf{v}\cdot\mathbf{x}}{3n \log m}$, if $|\mathbf{v} \cdot \mathbf{x}| \leq 3n \log m$, and $h(\mathbf{x}) = \mathrm{sgn}(\mathbf{v} \cdot \mathbf{x})$ otherwise.

Our main task is to analyze the weak learner. Given the following Lemma, Theorem 2 will be an immediate consequence of Lemma 7. The proof is omitted due to space constraints.

Lemma 8. *Suppose Algorithm A_{mlcw} is run using* $\mathrm{EX}_{\eta'}(f, \mathcal{D}')$ *where f is an origin-centered halfspace, $\mathcal{D}'$ is $(1/\epsilon)$-smooth w.r.t. an isotropic log-concave distribution $\mathcal{D}$, $\eta' \leq \eta/\epsilon$, and $\eta \leq \Omega(\epsilon^3/\log(n/\epsilon))$. Then w.h.p. the hypothesis h returned by A_{mlcw} has advantage $\Omega\left(\frac{\epsilon^2}{n \log(n/\epsilon)}\right)$.*

Proof Sketch. We exploit the fact that isotropic logconcave distributions are not *very* concentrated to show that clean examples tend to be classified correctly by a large margin. We then use concentration bounds to prove analogs of Lemmas 2 and 3, and put them together in a roughly similar way. □

References

[1] Arora, S., Babai, L., Stern, J., Sweedyk, Z.: The hardness of approximate optima in lattices, codes, and systems of linear equations. In: Proceedings of the 34th Annual Symposium on Foundations of Computer Science, pp. 724–733 (1993)

[2] Blum, A., Frieze, A., Kannan, R., Vempala, S.: A polynomial time algorithm for learning noisy linear threshold functions. Algorithmica 22(1/2), 35–52 (1997)

[3] Blumer, A., Ehrenfeucht, A., Haussler, D., Warmuth, M.: Learnability and the Vapnik-Chervonenkis dimension. Journal of the ACM 36(4), 929–965 (1989)

[4] Bshouty, N.H., Li, Y., Long, P.M.: Using the doubling dimension to analyze the generalization of learning algorithms. J. Comp. Sys. Sci. (to appear, 2009)

[5] Dunagan, J., Vempala, S.: Optimal outlier removal in high-dimensional spaces. J. Computer & System Sciences 68(2), 335–373 (2004)

[6] Feldman, V., Gopalan, P., Khot, S., Ponnuswami, A.: New results for learning noisy parities and halfspaces. In: Proc. FOCS, pp. 563–576 (2006)

[7] Freund, Y., Schapire, R.E.: Large margin classification using the perceptron algorithm. Machine Learning 37(3), 277–296 (1999)

[8] Gavinsky, D.: Optimally-smooth adaptive boosting and application to agnostic learning. Journal of Machine Learning Research 4, 101–117 (2003)

[9] Guruswami, V., Raghavendra, P.: Hardness of learning halfspaces with noise. In: Proc. FOCS, pp. 543–552. IEEE Computer Society Press, Los Alamitos (2006)

[10] Jolliffe, I.T.: Principal Component Analysis. Springer Series in Statistics (2002)

[11] Kalai, A., Klivans, A., Mansour, Y., Servedio, R.: Agnostically learning halfspaces. SIAM Journal on Computing 37(6), 1777–1805 (2008)

[12] Kearns, M., Li, M.: Learning in the presence of malicious errors. SIAM Journal on Computing 22(4), 807–837 (1993)

[13] Kearns, M., Schapire, R., Sellie, L.: Toward Efficient Agnostic Learning. Machine Learning 17(2/3), 115–141 (1994)

[14] Klivans, A., Long, P., Servedio, R.: Learning Halfspaces with Malicious Noise (2009), `http://www.cs.columbia.edu/~rocco/papers/icalp09malicious.html`

[15] Klivans, A., Sherstov, A.: Cryptographic hardness for learning intersections of halfspaces. In: Proc. FOCS, pp. 553–562 (2006)

[16] Littlestone, N.: Learning quickly when irrelevant attributes abound: a new linear-threshold algorithm. Machine Learning 2, 285–318 (1987)

[17] Littlestone, N.: Redundant noisy attributes, attribute errors, and linear-threshold learning using Winnow. In: Proc. COLT, pp. 147–156 (1991)

[18] Maass, W., Turan, G.: How fast can a threshold gate learn? In: Computational Learning Theory and Natural Learning Systems. Constraints and Prospects, vol. I, pp. 381–414. MIT Press, Cambridge (1994)

[19] Mansour, Y., Parnas, M.: Learning conjunctions with noise under product distributions. Information Processing Letters 68(4), 189–196 (1998)

[20] Novikoff, A.: On convergence proofs on perceptrons. In: Proceedings of the Symposium on Mathematical Theory of Automata, vol. XII, pp. 615–622 (1962)

[21] Rosenblatt, F.: The Perceptron: a probabilistic model for information storage and organization in the brain. Psychological Review 65, 386–407 (1958)

[22] Schapire, R., Singer, Y.: Improved boosting algorithms using confidence-rated predictions. Machine Learning 37, 297–336 (1999)

[23] Servedio, R.: PAC analogues of Perceptron and Winnow via boosting the margin. In: Proc. COLT, pp. 148–157 (2000)

[24] Servedio, R.: Efficient Algorithms in Computational Learning Theory. PhD thesis, Harvard University (2001)

[25] Servedio, R.: Smooth boosting and learning with malicious noise. Journal of Machine Learning Research 4, 633–648 (2003)

[26] Shawe-Taylor, J., Cristianini, N.: An introduction to support vector machines. Cambridge University Press, Cambridge (2000)

[27] Valiant, L.: Learning disjunctions of conjunctions. In: Proceedings of the Ninth International Joint Conference on Artificial Intelligence, pp. 560–566 (1985)

[28] Xu, H., Caramanis, C., Mannor, S.: Principal component analysis with contaminated data: The high dimensional case. In: JMLR (to appear, 2009)

General Scheme for Perfect Quantum Network Coding with Free Classical Communication

Hirotada Kobayashi[1,2], François Le Gall[2], Harumichi Nishimura[3], and Martin Rötteler[4]

[1] Principles of Informatics Research Division
National Institute of Informatics, Tokyo, Japan
[2] ERATO-SORST Quantum Computation and Information Project
Japan Science and Technology Agency, Tokyo, Japan
[3] School of Science, Osaka Prefecture University, Sakai, Japan
[4] NEC Laboratories America, Inc., Princeton, NJ, USA

Abstract. This paper considers the problem of efficiently transmitting quantum states through a network. It has been known for some time that without additional assumptions it is impossible to achieve this task *perfectly* in general — indeed, it is impossible even for the simple butterfly network. As additional resource we allow free classical communication between any pair of network nodes. It is shown that perfect quantum network coding is achievable in this model whenever classical network coding is possible over the same network when replacing all quantum capacities by classical capacities. More precisely, it is proved that perfect quantum network coding using free classical communication is possible over a network with k source-target pairs if there exists a classical linear (or even vector-linear) coding scheme over a finite ring. Our proof is constructive in that we give explicit quantum coding operations for each network node. This paper also gives an upper bound on the number of classical communication required in terms of k, the maximal fan-in of any network node, and the size of the network.

1 Introduction

Network coding was introduced by Ahlswede, Cai, Li and Yeung [1] to send multiple messages efficiently through a network. Usually, the network itself is given as a weighted, directed acyclic graph with the weights denoting the capacities of the edges. A typical example is the butterfly network in Fig. 1. In this example the task is to send one bit from s_1 to t_1 and another bit from s_2 to t_2, where each edge is a channel of unit capacity. It is obviously impossible to send two bits simultaneously by *routing* since the edge between n_1 and n_2 becomes a bottleneck. However, using *coding* at the nodes as shown in Fig. 1, it is feasible to send the two bits as desired. Two fundamental observations are in order: First, copying classical information is possible. In the example of the butterfly network, this is used for the operations performed at nodes s_1, s_2, and n_2. Second, and perhaps more importantly, information can be encoded. In the example, this is used at nodes n_1, t_1, and t_2 where the XOR operation (i.e.,

S. Albers et al. (Eds.): ICALP 2009, Part I, LNCS 5555, pp. 622–633, 2009.

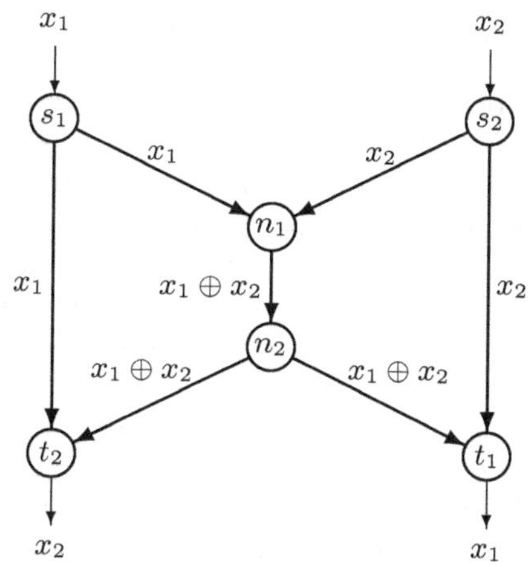

Fig. 1. The butterfly network and a classical linear coding protocol. The node s_1 (resp. s_2) has for input a bit x_1 (resp. x_2). The task is to send x_1 to t_1 and x_2 to t_2. The capacity of each edge is assumed to be one bit.

addition over the finite field $\mathbb{F}_2$) is applied. After the seminal result [1], network coding has been widely studied from both theoretical and experimental points of view and many applications have been found. A good resource is the network coding home page [11].

The multicast problem is a task that can be elegantly solved by network coding. In this problem, all messages at one source node must be sent to each of several target nodes. Ahlswede et al. [1] showed that the upper bound on the achievable rate given by the min-cut/max-flow condition is in fact always achievable. In other words, network coding allows one to send m messages if and only if the value of any cut between the source node and each target node is at least m. Li, Yeung and Cai [14] showed that such a rate is always achievable by *linear* coding over a sufficiently large finite field (in which the operation performed at each node is a linear combination over some finite field). Furthermore, this result was improved by Jaggi et al. [10] who showed that such encoding can be constructed in polynomial time with respect to the number of nodes. This implies that deciding whether a given multicast network has a (linear) network coding scheme can be solved in polynomial time. This contrasts with the general network coding problem for which Lehman and Lehman [12] showed that it is NP-hard to decide whether there exists a linear coding solution.

Another important subclass of network coding problems is the *k-pair problem* (also called the *multiple-unicast problem*). In this setting the network has k pairs of source/target nodes (s_i, t_i), and each source s_i wants to send a message x_i to the target t_i. Notice that Fig. 1 can be considered as a solution of a two-pair problem. It was shown by Dougherty and Zeger [4] that the solvability (resp. linear solvability) of any network coding problem can be reduced to the solvability (resp. linear solvability) of some instance of the k-pair problem. Combined with the Lehman-Lehman result, this implies that the linear solvability of the k-pair problem is NP-hard. Polynomial-time constructions of linear coding for fixed k have been investigated [8,19,20], but no complete answer has been obtained yet.

Recently, network coding has become a topic of research in quantum computation and information, giving rise to a theory of quantum network coding. The most basic setting is the following. The messages are quantum states, and the network is also quantum, i.e., each edge corresponds to a quantum channel. The question is whether quantum messages can be sent efficiently to the target nodes through the network (possibly using the idea of network coding). A very basic difficulty immediately arises as opposed to the classical case: quantum information cannot be copied [17]. Hence, multicasting quantum messages is impossible without imposing any extra conditions. One approach to work around this problem has been developed by Shi and Soljanin [18], who constructed a perfect multicasting scheme over families of quantum networks under the condition that the source owns many copies of quantum states. A more natural target may be the k-pair problem since here the number of inputs matches the number of outputs. For this problem, however, there are already a number of negative results known for the above basic setting. First, Hayashi et al. [7] showed that sending two qubits simultaneously and *perfectly (i.e., with fidelity one)* on the butterfly network is impossible. Leung, Oppenheim and Winter [13] extended this impossibility result to the case where the messages have to be sent in an asymptotically perfect way, and also to classes of networks other than the butterfly network. This means that some extra condition is needed to achieve perfect quantum network coding for the k-pair problem case as well.

Main results. The extra condition considered in this paper is to allow *free classical communication* to assist with sending quantum messages perfectly through the network. That is, any two nodes can communicate with each other through a classical channel which can be used freely (i.e., at no cost). Free classical communication as an extra resource often appears in quantum information theory, e.g., entanglement distillation and dilution (see Ref. [17]). Also, from a practical viewpoint, quantum communication is a very limited resource while classical communication is much easier to implement. Thus it would be desirable if the amount of quantum communication could be reduced using network coding with the help of classical communication. Another extra resource that may be considered (and rather popular in quantum information processing) is *entanglement*, such as shared EPR pairs. However, it has the weakness that, once used, quantum communication is needed to recreate it. Therefore, allowing free classical communication arguably is a comparatively mild additional resource for perfect quantum network coding.

The first result of this paper (Theorem 1) can be summarized as follows: if there exists a classical linear coding scheme over a ring R for a k-pair problem given by a graph $G = (V, E)$, then there exists a solution to the quantum k-pair problem over the same graph G if free classical communication is allowed. The idea to obtain this result is to perform a node-by-node simulation of the coding scheme solving the classical problem. For example, suppose that, in the classical coding scheme, a node v of G performs the map $(z_1, z_2) \mapsto f(z_1, z_2)$

where $f\colon R^2 \to R$ is some function. In the quantum case, this node will perform the quantum map $|z_1, z_2\rangle|0\rangle \mapsto |z_1, z_2\rangle|f(z_1, z_2)\rangle$. A basic observation is that the first two registers should be ideally "removed" in order to simulate properly the classical scheme. This task does not seem straightforward since the quantum state is in general a superposition of basis states, and this superposition has to be preserved so that the input state can be recovered at target nodes. Our key technique shows that this can be done if free classical communication is allowed, and if the classical scheme to be simulated is linear. More precisely, these registers are "removed" by measuring them in the Fourier basis associated with the additive group of R. An extra phase then appears, but it can be corrected *locally* at each target node as will be proved in Section 3. This requires (free) classical communication. In our construction it is sufficient to send at most $kM|V|$ elements of R, where M is the maximal fan-in of nodes of G.

A classical coding scheme is called *vector-linear* if the operation at each node is of the form $\sum_i A_i \boldsymbol{v}_i$, where the $\boldsymbol{v}_i$'s are the vectors input to the node and A_i is a matrix to apply to $\boldsymbol{v}_i$. The result above can be extended to the vector-linear coding case as well (Theorem 2). That is, if there exists a classical vector-linear coding scheme over a ring R for a k-pair problem given by a graph G, then there also exists a solution to the quantum k-pair problem over the same graph G (again if free classical communication is allowed). Notice that there are examples of graphs over which a vector-linear solution is known but no linear coding scheme exists (see Refs. [16,12]). There are also examples for which even vector-linear coding is not sufficient [3]. However, most of known networks solvable by network coding have vector-linear solutions, and hence our result is applicable quite widely (and actually is even applicable to the examples in Ref. [3]).

Related work. There are several papers studying quantum network coding on the k-pair problem in situations different from the most basic setting (perfect transmission of quantum states using only a quantum network of limited capacity). Hayashi et al. [7] and Iwama et al. [9] considered "approximate" transmission of qubits in the k-pair problem, and showed that transmission with fidelity larger than 1/2 is possible for a class of networks. Hayashi [6] showed how to achieve perfect transmission of two qubits on the butterfly network if two source nodes have prior entanglement, and if, at each edge, we can choose between sending two classical bits and sending one qubit. Leung, Oppenheim and Winter [13] considered various extra resources such as free forward/backward/two-way classical communication and entanglement, and investigated the lower/upper bounds of the rate of quantum network coding for their settings. The setting of the present paper is close to their model allowing free two-way classical communication. The difference is that Ref. [13] considered asymptotically perfect transmission while this paper focuses on perfect transmission. Also, Ref. [13] showed optimal rates for a few classes of networks while the present paper gives lower bounds for much wider classes of networks.

As mentioned in Ref. [13], free classical communication essentially makes the underlying directed graph of the quantum network undirected since quantum teleportation enables one to send a quantum message to the reverse direction of a directed edge. In this context, our result gives a lower bound of the rate of quantum network coding that might not be optimal even if its corresponding classical coding is optimal in the directed graph. However, even in the classical case, network coding over an undirected graph is much less known than that over a directed one. In the multicast network, the gap between the rates by network coding and by routing is known to be at most two [15], while there is an example for which the min-cut rate bound cannot be achieved by network coding [15,2]. Also notice that, in the k-pair problem, it is conjectured that fractional routing achieves the optimal rate for any undirected graph (see for example Ref. [5]). However, this conjecture has been proved only for very few families of networks, and remains one of the main open problems in the field of network coding.

2 The k-Pair Problem

The classical k-pair problem. We recall the statement of the k-pair problem in the classical case, and the definition of a solution to this problem. The reader is referred to, for example, Ref. [4] for further details.

An instance of a k-pair problem is a directed graph $G = (V, E)$ and k pairs of nodes $(s_1, t_1), \dots, (s_k, t_k)$. For $i \in \{1, \dots, k\}$, each x_i is given at the source s_i, and has to be sent to the target t_i through G under the condition that each edge has unit capacity. Let Σ be a finite set. A coding scheme over Σ is a choice of operations for all nodes in V: for each node $v \in V$ with fan-in m and fan-out n, the operation at v is written as n functions $f_{v,1}, \dots, f_{v,n}$, each from Σ^m to Σ, where the value $f_{v,i}(z_1, \dots, z_m)$ represents the message sent through the i-th outgoing edge of v when the inputs of the m incoming edges are $z_1, \dots, z_m$. A *solution* over Σ to an instance of the k-pair problem is a coding scheme over Σ that enables one to send simultaneously k messages x_i from s_i to t_i, for all $i \in \{1, \dots, k\}$. For example, the coding scheme in Fig. 1 is a solution over $\mathbb{F}_2$ to the two-pair problem associated with the butterfly graph.

For convenience, the following simple but very useful convention is assumed when describing a classical coding scheme. Each source s_i is supposed to have a "virtual" incoming edge from which it receives its input x_i. Also, each target t_i is supposed to have a "virtual" outgoing edge, where x_i must be output through. In this way, the source and target nodes perform coding operations on their inputs, and this convention enables us to ignore the distinction between source/target nodes and internal nodes. These conventions are illustrated in Fig. 1.

The quantum k-pair problem. We suppose that the reader is familiar with the basics of quantum information theory and refer to Ref. [17] for a good reference. In this paper we will consider d-dimensional quantum systems, i.e., quantum states that are normalized vectors in a complex Hilbert space of dimension d, for some positive integer d.

An instance of a quantum k-pair problem is, as in the classical case, a directed graph $G = (V, E)$ and k pairs of nodes $(s_1, t_1), \dots, (s_k, t_k)$. Let $\mathcal{H}$ be a Hilbert space. The goal is to send a quantum state $|\psi\rangle \in \mathcal{H}^{\otimes k}$ supported on the source nodes $s_1, \ldots, s_k$ (in this order) to the target nodes $t_1, \ldots, t_k$ (in this order). We consider the model where each edge of G can transmit one quantum state over $\mathcal{H}$. However, free classical communication is allowed between any two nodes of G. Let d be a positive integer. We say that an instance of the quantum k-pair problem is solvable over $\mathbb{C}^d$ if there exists a protocol solving this problem for $\mathcal{H} = \mathbb{C}^d$.

3 Perfect Quantum Network Coding

3.1 Linear Coding over Rings

This subsection considers instances of the k-pair problem for which there exists a solution using classical linear coding over rings, i.e., Σ is supposed to be a finite ring R (not necessarily commutative). A coding scheme is said *linear* over R if the functions $f_{v,i}$ corresponding to the encoding operations performed at each node $v \in V$ are linear.

The main result of this subsection is the following theorem.

Theorem 1. *Let $G = (V, E)$ be a directed graph and $(s_1, t_1), \ldots, (s_k, t_k)$ be k pairs of nodes. Let M be the maximal fan-in of nodes in G and R be a finite ring. Suppose that there exists a linear solution over R to the associated classical k-pair problem. Then the corresponding quantum k-pair problem is solvable over $\mathbb{C}^{|R|}$. Moreover, there exists a quantum protocol for this task that sends at most $kM|V|$ elements of R as free classical communication, i.e., at most $kM|V|\lceil \log_2 |R| \rceil$ bits of classical communication.*

The basic strategy for proving Theorem 1 is to perform a quantum simulation of the classical coding scheme. Before presenting the proof of this theorem, we need some preliminaries.

Let ϕ be a group isomorphism from the additive group of R to some abelian group $A = \mathbb{Z}_{r_1} \times \cdots \times \mathbb{Z}_{r_\ell}$ with $\Pi_{i=1}^{\ell} r_i = |R|$ (but ϕ is not necessarily a ring isomorphism). There are many possibilities for the choice of A and ϕ. One convenient choice is to take $\mathbb{Z}_{r_1} \times \cdots \times \mathbb{Z}_{r_\ell}$ to be the invariant factor decomposition of the additive group of R. For any $x \in R$ and $i \in \{1, \ldots, \ell\}$, let $\phi_i(x)$ denote the i-th coordinate of $\phi(x)$, i.e., an element of $\mathbb{Z}_{r_i}$. In the quantum setting, we suppose that each register contains a quantum state over $\mathcal{H} = \mathbb{C}^{|R|}$, and denote by $\{|z\rangle\}_{z \in R}$ an orthonormal basis of $\mathcal{H}$. We define a unitary operator W over the Hilbert space $\mathcal{H}$ as follows: for any $y \in R$, the operator W maps the basis state $|y\rangle$ to the state

$$\frac{1}{\sqrt{|R|}} \sum_{z \in R} \exp\Big(2\pi\iota \sum_{i=1}^{\ell} \frac{\phi_i(y) \cdot \phi_i(z)}{r_i}\Big) |z\rangle.$$

Here $\phi_i(y) \cdot \phi_i(z)$ denotes the product of $\phi_i(y)$ and $\phi_i(z)$, seen in the natural way as an element of the set $\{0, \ldots, r_i - 1\}$. Note that W is basically the quantum Fourier transform over the additive group of R.

Let m and n be two positive integers and $f_1, \ldots, f_n$ be n functions from R^m to R. Let $U_{f_1,\ldots,f_n}$ be the unitary operator over the Hilbert space $\mathcal{H}^{\otimes m} \otimes \mathcal{H}^{\otimes n}$ defined as follows: for any m elements $y_1, \ldots, y_m$ and any n elements $z_1, \ldots, z_n$ of R, the operator $U_{f_1,\ldots,f_n}$ maps the basis state $|y_1, \ldots, y_m\rangle|z_1, \ldots, z_n\rangle$ to the state

$$|y_1, \ldots, y_m\rangle|z_1 + f_1(y_1, \ldots, y_m), \ldots, z_n + f_n(y_1, \ldots, y_m)\rangle.$$

Now let us define the following quantum procedure ENCODING$(f_1, \ldots, f_n)$.

Procedure ENCODING$(f_1, \ldots, f_n)$

INPUT: quantum registers $\mathsf{Q}_1, \ldots, \mathsf{Q}_m$, each corresponding to $\mathcal{H}$
OUTPUT: quantum registers $\mathsf{Q}'_1, \ldots, \mathsf{Q}'_n$, each corresponding to $\mathcal{H}$, and elements $a_1, \ldots, a_m$ of R

1 Introduce n registers $\mathsf{Q}'_1, \ldots, \mathsf{Q}'_n$, each initialized to $|0_{\mathcal{H}}\rangle$.
2 Apply the operator $U_{f_1,\ldots,f_n}$ to $(\mathsf{Q}_1, \ldots, \mathsf{Q}_m, \mathsf{Q}'_1, \ldots, \mathsf{Q}'_n)$.
3 For each $i \in \{1, \ldots, m\}$, apply W to Q_i.
4 Measure the first m registers $\mathsf{Q}_1, \ldots, \mathsf{Q}_m$ in the computational basis. Let $a_1, \ldots, a_m \in R$ denote the outcomes of the measurements.
5 Output $\mathsf{Q}'_1, \ldots, \mathsf{Q}'_n$ and the m elements $a_1, \ldots, a_m$.

The behavior of this procedure on a basis state is described in the next proposition.

Proposition 1. *Suppose that the contents of the registers $\mathsf{Q}_1, \ldots, \mathsf{Q}_m$ is the state $|y_1, \ldots, y_m\rangle_{(\mathsf{Q}_1,\ldots,\mathsf{Q}_m)}$ for some elements $y_1, \ldots, y_m$ of R. Then the contents of the registers $\mathsf{Q}'_1, \ldots, \mathsf{Q}'_n$ after applying Procedure* ENCODING$(f_1, \ldots, f_n)$ *is a state of the form*

$$\exp\left(2\pi\iota g(y_1, \ldots, y_m)\right) |f_1(y_1, \ldots, y_m), \ldots, f_n(y_1, \ldots, y_m)\rangle_{(\mathsf{Q}'_1,\ldots,\mathsf{Q}'_n)},$$

where $g : R^m \to \mathbb{Q}$ is an additive group homomorphism determined by the measurement outcomes $a_1, \ldots, a_m$.

Proof. After Step 3, the resulting state is

$$\frac{1}{\sqrt{|R|^m}} \sum_{z_1,\ldots,z_m \in R} \exp\left(2\pi\iota \sum_{i=1}^{\ell} \sum_{j=1}^{m} \frac{\phi_i(y_j) \cdot \phi_i(z_j)}{r_i}\right) \times |z_1, \ldots, z_m\rangle_{(\mathsf{Q}_1,\ldots,\mathsf{Q}_m)} |f_1(y_1, \ldots, y_m), \ldots, f_n(y_1, \ldots, y_m)\rangle_{(\mathsf{Q}'_1,\ldots,\mathsf{Q}'_n)}.$$

At Step 4, if the measurement outcomes are $a_1, \ldots, a_m$, where each a_i is an element of R, then the state in $(\mathsf{Q}'_1, \ldots, \mathsf{Q}'_n)$ becomes

$$\exp\left(2\pi\iota \sum_{i=1}^{\ell} \sum_{j=1}^{m} \frac{\phi_i(a_j) \cdot \phi_i(y_j)}{r_i}\right) |f_1(y_1, \ldots, y_m), \ldots, f_n(y_1, \ldots, y_m)\rangle_{(\mathsf{Q}'_1,\ldots,\mathsf{Q}'_n)}.$$

This can be rewritten as

$$\exp\left(2\pi\iota g(y_1, \ldots, y_m)\right)|f_1(y_1, \ldots, y_m), \ldots, f_n(y_1, \ldots, y_m)\rangle_{(\mathsf{Q}'_1,\ldots,\mathsf{Q}'_n)},$$

where $g(y_1, \ldots, y_m) = \sum_{i=1}^{\ell} \sum_{j=1}^{m} \frac{\phi_i(a_j)\cdot\phi_i(y_j)}{r_i}$. Notice that g is an additive group homomorphism determined by the values of $a_1, \ldots, a_m$. □

Now we are ready to give the proof of Theorem 1.

Proof (of Theorem 1). Let $G = (V, E)$ be a graph on which there exists a linear solution over R to the classical k-pair problem associated with the pairs (s_i, t_i). For each node $v \in V$ with fan-in m and fan-out n, let $f_{v,1}, \ldots, f_{v,n}$ be the coding operations performed at node v in such a solution, where each $f_{v,i}$ is from R^m to R. Suppose that the input state of the quantum task is

$$|\psi_S\rangle_{(\mathsf{S}_1,\ldots,\mathsf{S}_k)} = \sum_{x_1,\ldots,x_k \in R} \alpha_{x_1,\ldots,x_k} |x_1\rangle_{\mathsf{S}_1} \otimes \cdots \otimes |x_k\rangle_{\mathsf{S}_k},$$

where the $\alpha_{x_1,\ldots,x_k}$'s are complex numbers such that $\sum_{x_1,\ldots,x_k\in R} |\alpha_{x_1,\ldots,x_k}|^2 = 1$. Here, for each $i \in \{1, \ldots, k\}$, S_i is a register owned by the node s_i.

The strategy is to simulate the solution to the associated classical task node by node. We shall show that the classical coding operation performed at a node with fan-in m can be simulated by sending km elements of R using free classical communication. The general bound $kM|V|$ claimed in the statement of the theorem then follows.

More precisely, let $v \in V$ be a node of G with fan-in m and fan-out n. The coding performed at node v is simulated as follows: the quantum procedure ENCODING$(f_{v,1}, \ldots, f_{v,n})$ is used on the m quantum registers input to v through its m incoming edges. The procedure outputs n registers and m elements $a_1, \ldots, a_m$ of R. Then all the elements $a_1, \ldots, a_m$ are sent to each target node (via free classical communication), and the n registers are sent along on the n outgoing edges of v. Such a simulation is done for all the nodes in V.

In what follows, we denote by $\mathcal{B}$ this strategy. We first describe the behavior of $\mathcal{B}$ when the input is a basis state.

Lemma 1. *Let $x_1, \ldots, x_k$ be k elements of R. Then the state after applying $\mathcal{B}$ to $|x_1\rangle_{\mathsf{S}_1} \otimes \cdots \otimes |x_k\rangle_{\mathsf{S}_k}$ is of the form*

$$e^{2\pi\iota h(x_1,\ldots,x_k)} |x_1\rangle_{\mathsf{T}_1} \otimes \cdots \otimes |x_k\rangle_{\mathsf{T}_k},$$

where $h \colon R^k \to \mathbb{Q}$ is an additive group homomorphism depending only on the outcomes of the measurements done during the procedure. Here, for each $i \in \{1, \ldots, k\}$, the register T_i is owned by the target node t_i.

Proof (of Lemma 1). Since the classical coding scheme is linear, Proposition 1 ensures that, at any step of the protocol, the quantum state of the system is of the form

$$\beta |y_1\rangle_{\mathsf{Q}_1} \otimes \cdots \otimes |y_D\rangle_{\mathsf{Q}_D} \quad (1)$$

for some positive integer D (depending on the step of the protocol) and some phase β (depending on the step of the protocol, the outcomes of the measurements done, and the values $x_1, \ldots, x_k$), such that each $y_i \in R$ can be written as a linear combination of the x_j's, in a way that corresponds to the associated classical coding scheme. Here each Q_i is a register owned by some node of the graph G.

Let us get back to the simulation at node v described above to work out the general form of the phase β. Suppose that the current state of the quantum system is given by Eq. (1). Suppose, without loss of generality, that the coding at node v is done on the first m registers. In other words, the simulation performed at node v is done over the state $\beta|y_1,\ldots,y_m\rangle_{(\mathsf{Q}_1,\ldots,\mathsf{Q}_m)} \otimes |y_{m+1},\ldots,y_D\rangle_{(\mathsf{Q}_{m+1},\ldots,\mathsf{Q}_D)}$ where each $y_i = \sum_{j=1}^{k} \gamma_{i,j}x_j$ for some constants $\gamma_{i,j} \in R$ (depending on the step of the protocol). Then, from Proposition 1, the simulation done at this step (using the procedure ENCODING$(f_{v,1},\ldots,f_{v,n})$) can be seen as transforming this state into the state

$$\beta e^{(2\pi\iota h_v(x_1,\ldots,x_k))}|f_{v,1}(y_1,\ldots,y_m),\ldots,f_{v,n}(y_1,\ldots,y_m)\rangle_{(\mathsf{Q}'_1,\ldots,\mathsf{Q}'_n)} \otimes|y_{m+1},\ldots,y_D\rangle_{(\mathsf{Q}_{m+1},\ldots,\mathsf{Q}_D)},$$

where, if g denotes the function in the statement of Proposition 1,

$$h_v(x_1,\ldots,x_k) = g\left(\sum_{j=1}^{k}\gamma_{1,j}x_j,\ldots,\sum_{j=1}^{k}\gamma_{m,j}x_j\right).$$

Since g is a group homomorphism, the function $h_v\colon R^k \to \mathbb{Q}$ is a group homomorphism also.

From the observation that the classical coding scheme solves the associated classical k-pair problem, we conclude that the state after applying $\mathcal{B}$ can be written as

$$e^{2\pi\iota\sum_{v\in V} h_v(x_1,\ldots,x_k)}|x_1\rangle_{\mathsf{T}_1}\otimes\cdots\otimes|x_k\rangle_{\mathsf{T}_k}.$$

The claimed form is obtained by defining the function h as $h(x_1,\ldots,x_k) = \sum_{v\in V} h_v(x_1,\ldots,x_k)$. Notice that h is determined only by the values of the measurements (the constants $\gamma_{i,j}$ are fixed by the choice of the classical coding scheme). □

Now we proceed with the proof of Theorem 1. Lemma 1 implies that the state after applying $\mathcal{B}$ must be of the form

$$\sum_{x_1,\ldots,x_k\in R}\alpha_{x_1,\ldots,x_k}e^{2\pi\iota h(x_1,\ldots,x_k)}|x_1\rangle_{\mathsf{T}_1}\otimes\cdots\otimes|x_k\rangle_{\mathsf{T}_k},$$

where, for each $i \in \{1,\ldots,k\}$, the register T_i is owned by the target node t_i. Also, Lemma 1 guarantees that each target node t_i knows the function h since the values of all the measurement have been sent to it. Since h is an additive group homomorphism, it can be written as $h(x_1,\ldots,x_k) = h_1(x_1)+\cdots+h_k(x_k)$, where the function $h_i\colon R \to \mathbb{Q}$ maps x_i to $h(0,\ldots,0,x_i,0,\ldots,0)$, for each $i \in \{1,\ldots,k\}$.

Now for each $i \in \{1, \ldots, k\}$ the target node t_i applies the map Y_i to its register, where Y_i is defined as

$$Y_i\colon |x\rangle_{\mathsf{T}_i} \mapsto e^{-2\pi\iota h_i(x)}|x\rangle_{\mathsf{T}_i},$$

for any $x \in R$. This step corrects the phases and the resulting state is $|\psi_S\rangle_{(\mathsf{T}_1,\ldots,\mathsf{T}_k)}$. This concludes the proof of Theorem 1. □

In Theorem 1, for the clarity of the proof, we gave the bound $kM|V|\lceil\log_2|R|\rceil$ of the number of classical bits to be sent. For concrete networks, this bound can be improved significantly: (i) at each node, the measurement outcomes $a_1, \ldots, a_m$ have only to be sent to the target nodes t_j such that $\gamma_{i,j} \neq 0$ for some index $i \in \{1, \ldots, m\}$, and (ii) any node performing only a copy operation does not require any free classical communication to be simulated quantumly. Furthermore, we can reduce the amount of classical communication to $O(1)$ in the subclass of k-pair problems considered in Ref. [8], as shown in the following corollary.

Corollary 1. *Suppose that there exists a classical linear coding scheme over a constant-size finite field $\mathbb{F}$ that solves a k-pair problem for a fixed constant k. Then the corresponding quantum k-pair problem is solvable over $\mathbb{C}^{|\mathbb{F}|}$ by sending at most $O(1)$ elements of $\mathbb{F}$ as free classical communication.*

Proof. Iwama et al. [8] showed that if k and $|\mathbb{F}|$ are constant, we can find a classical linear coding scheme such that the total number of non-trivial linear operations is a constant (only depending on k and $|\mathbb{F}|$). Then the corollary follows from Theorem 1. □

3.2 Vector-Linear Coding

This subsection shows how to simulate classical vector-linear coding over any ring R (possibly not commutative). This is one of the most general settings considered in the literature, see for example Ref. [3].

Let R be a finite ring. Let Σ be the R-module R^q for some positive integer q. Informally, this module is the analogue of the usual vector space $\mathbb{F}^q$ of dimension q over a finite field $\mathbb{F}$, but here $\mathbb{F}$ is replaced by the ring R. A coding scheme is said *q-vector-linear* over R if, for each function $f_{v,i}\colon (R^q)^m \to R^q$ corresponding to the i-th encoding operation performed at the node $v \in V$, there exist matrices $B_j^{(v,i)}$ of size $q \times q$ over R such that $f_{v,i}(\boldsymbol{y}_1, \ldots, \boldsymbol{y}_m) = \sum_{j=1}^m B_j^{(v,i)}\boldsymbol{y}_j$ for all $(\boldsymbol{y}_1, \ldots, \boldsymbol{y}_m) \in (R^q)^m$.

Again let ϕ be a group isomorphism from R to an abelian group $\mathbb{Z}_{r_1} \times \cdots \times \mathbb{Z}_{r_\ell}$, and, for any $x \in R$ and $i \in \{1, \ldots, \ell\}$, denote by $\phi_i(x)$ the i-th coordinate of $\phi(x)$, i.e., an element of $\mathbb{Z}_{r_i}$. Given an element $\boldsymbol{x} = (x_1, \ldots, x_q) \in R^q$, let $\psi_i(\boldsymbol{x})$ denote the element $(\phi_i(x_1), \ldots, \phi_i(x_q))$ in $\mathbb{Z}_{r_i}^q$ corresponding to the projections of all the coordinates of $\boldsymbol{x}$ to $\mathbb{Z}_{r_i}$.

The following theorem is proved in a manner similar to Theorem 1.

Theorem 2. *Let $G = (V, E)$ be a directed graph and $(s_1, t_1), \ldots, (s_k, t_k)$ be k pairs of nodes. Let M be the maximal fan-in of nodes in G, R be a finite ring* *and q be a positive integer. Suppose that there exists a q-vector-linear solution over R to the associated classical k-pair problem. Then the corresponding quantum k-pair problem is solvable over $\mathbb{C}^{|R|^q}$. Moreover, there exists a quantum protocol that sends at most $kM|V|$ elements of R^q as free classical communication.*

Proof. All the results of Subsection 3.1 hold similarly by using the following Fourier transform W' instead of W. The unitary operator W' is defined over the Hilbert space $\mathcal{H} = \mathbb{C}^{|R|^q}$ by its action on the basis states of $\mathcal{H}$: for any $\boldsymbol{y} \in R^q$, the operator W' maps the state $|\boldsymbol{y}\rangle$ to the state

$$\frac{1}{\sqrt{|R|^q}} \sum_{\boldsymbol{z} \in R^q} \exp\Big(2\pi\iota \sum_{i=1}^{\ell} \frac{\psi_i(\boldsymbol{y}) \cdot \psi_i(\boldsymbol{z})}{r_i}\Big) |\boldsymbol{z}\rangle,$$

where $\psi_i(\boldsymbol{y}) \cdot \psi_i(\boldsymbol{z})$ denotes the inner product of the vectors $\psi_i(\boldsymbol{y})$ and $\psi_i(\boldsymbol{z})$. If we denote by A the abelian group of R, then W' is basically the quantum Fourier transform over the abelian group A^q. □

4 Concluding Remarks

This paper has presented a protocol to achieve perfect quantum network coding with free classical communication. The proposed protocol works for all k-pair problems that can be solved by linear or by vector-linear coding over any finite ring, encompassing a broad class of networks that have been studied classically.

There are still several open problems. A natural question is whether perfect quantum network coding (with free classical communication) is possible for any instance of the k-pair problem solvable classically. Another open problem is a converse of the results of this paper, i. e., to determine whether there exists an undirected network such that quantum coding is possible (with free classical communication) but classical coding is not possible.

Acknowledgement. HK is partially supported by the Grant-in-Aid for Scientific Research (B) Nos. 18300002 and 21300002 of the Ministry of Education, Culture, Sports, Science and Technology of Japan. HN is partially supported by the Grant-in-Aid for Young Scientists (B) No. 19700011 of the Ministry of Education, Culture, Sports, Science and Technology of Japan.

References

1. Ahlswede, R., Cai, N., Li, S.-Y.R., Yeung, R.W.: Network information flow. IEEE Transactions on Information Theory 46(4), 1204–1216 (2000)
2. Al-Bashabsheh, A., Yongacoglu, A.: On the capacity bound of undirected networks. arXiv.org e-Print archive, arXiv:0804.4455 (2008)
3. Dougherty, R., Freiling, C.F., Zeger, K.: Insufficiency of linear coding in network information flow. IEEE Transactions on Information Theory 51(8), 2745–2759 (2005)

4. Dougherty, R., Zeger, K.: Nonreversibility and equivalent constructions of multiple-unicast networks. IEEE Transactions on Information Theory 52(11), 5067–5077 (2006)
5. Harvey, N., Kleinberg, R., Lehman, A.: On the capacity of information networks. IEEE Transactions on Information Theory 52(6), 2410–2424 (2006)
6. Hayashi, M.: Prior entanglement between senders enables perfect quantum network coding with modification. Physical Review A 76(4), 040301(R) (2007)
7. Hayashi, M., Iwama, K., Nishimura, H., Raymond, R., Yamashita, S.: Quantum network coding. In: Thomas, W., Weil, P. (eds.) STACS 2007. LNCS, vol. 4393, pp. 610–621. Springer, Heidelberg (2007)
8. Iwama, K., Nishimura, H., Paterson, M., Raymond, R., Yamashita, S.: Polynomial-time construction of linear network coding. In: Aceto, L., Damgård, I., Goldberg, L.A., Halldórsson, M.M., Ingólfsdóttir, A., Walukiewicz, I. (eds.) ICALP 2008, Part I. LNCS, vol. 5125, pp. 271–282. Springer, Heidelberg (2008)
9. Iwama, K., Nishimura, H., Raymond, R., Yamashita, S.: Quantum network coding for general graphs. arXiv.org e-Print archive, quant-ph/0611039 (2006)
10. Jaggi, S., Sanders, P., Chou, P.A., Effros, M., Egner, S., Jain, K., Tolhuizen, L.: Polynomial time algorithms for multicast network code construction. IEEE Transactions on Information Theory 51(6), 1973–1982 (2005)
11. Koetter, R.: Network coding home page, `http://tesla.csl.uiuc.edu/~koetter/NWC/`
12. Lehman, A., Lehman, E.: Complexity classification of network information flow problems. In: Proceedings of the 15th ACM-SIAM Symposium on Discrete Algorithms, pp. 142–150 (2004)
13. Leung, D., Oppenheim, J., Winter, A.: Quantum network communication — the butterfly and beyond. arXiv.org e-Print archive, quant-ph/0608223 (2006)
14. Li, S.-Y.R., Yeung, R.W., Cai, N.: Linear network coding. IEEE Transactions on Information Theory 49(2), 371–381 (2003)
15. Li, Z., Li, B., Lau, L.C.: A constant bound on throughput improvement of multicast network coding in undirected networks. IEEE Transactions on Information Theory 55(3), 1016–1026 (2009)
16. Médard, M., Effros, M., Ho, T., Karger, D.: On coding for non-multicast networks. In: Proceedings of the 41st Annual Allerton Conference on Communication, Control and Computing (2003)
17. Nielsen, M.A., Chuang, I.L.: Quantum Computation and Quantum Information. Cambridge University Press, Cambridge (2000)
18. Shi, Y., Soljanin, E.: On multicast in quantum networks. In: Proceedings of the 40th Annual Conference on Information Sciences and Systems, pp. 871–876 (2006)
19. Wang, C.-C., Shroff, N.B.: Beyond the butterfly – a graph-theoretic characterization of the feasibility of network coding with two simple unicast sessions. In: Proceedings of the IEEE International Symposium on Information Theory, pp. 121–125 (2007)
20. Wang, C.-C., Shroff, N.B.: Intersession network coding for two simple multicast sessions. In: Proceedings of the 45th Annual Allerton Conference on Communication, Control and Computing (2007)

Greedy Δ-Approximation Algorithm for Covering with Arbitrary Constraints and Submodular Cost

Christos Koufogiannakis and Neal E. Young*

Department of Computer Science, University of California, Riverside
{ckou,neal}@cs.ucr.edu

Abstract. This paper describes a greedy Δ-approximation algorithm for MONOTONE COVERING, a generalization of many fundamental NP-hard covering problems. The approximation ratio Δ is the maximum number of variables on which any constraint depends. (For example, for vertex cover, Δ is 2.) The algorithm unifies, generalizes, and improves many previous algorithms for fundamental covering problems such as vertex cover, set cover, facilities location, and integer and mixed-integer covering linear programs with upper bound on the variables.

The algorithm is also the first Δ-competitive algorithm for *online* monotone covering, which generalizes online versions of the above-mentioned covering problems as well as many fundamental online paging and caching problems. As such it also generalizes many classical online algorithms, including LRU, FIFO, FWF, BALANCE, GREEDY-DUAL, GREEDY-DUAL SIZE (a.k.a. LANDLORD), and algorithms for connection caching, where Δ is the cache size. It also gives new Δ-competitive algorithms for *upgradable* variants of these problems, which model choosing the caching strategy *and* an appropriate hardware configuration (cache size, CPU, bus, network, etc.).

1 Introduction

The classification of general techniques is an important research program within the field of approximation algorithms. What are the scopes of, and the relationships between, the various algorithm-design techniques such as the primal-dual method, the local-ratio method [5], and randomized rounding? Within this research program, an important question is which problems admit optimal and fast *greedy* approximation algorithms, and by what techniques [25,11]?

We give here a single online greedy Δ-approximation algorithm for a combinatorially rich class of *monotone* covering problems, including many classical covering problems as well as online paging and caching problems. The approximation ratio, Δ, is the maximum number of variables on which any constraint depends. (For VERTEX COVER, $\Delta = 2$.)

For some problems in the class, no greedy (or other) Δ-approximation algorithms were known. For others, previous greedy Δ-approximation algorithms were known, but with non-trivial and seemingly problem-specific analyses. For VERTEX COVER and SET

* Partially supported by NSF awards CNS-0626912, CCF-0729071.

S. Albers et al. (Eds.): ICALP 2009, Part I, LNCS 5555, pp. 634–652, 2009.

problem	approximation ratio	method	where	comment	
VERTEX COVER	$2 - \ln\ln \hat{\Delta}/\ln \hat{\Delta}$	local ratio	[28]	see also [32,7,44,27,29,21,37]	
SET COVER	Δ	greedy; LP	[6]; [31,32]	$\Delta = \max_i \lvert\{j : A_{ij} > 0\}\rvert$	$\star$
CIP, 0/1-variables	$\max_i \sum_j A_{ij}$	greedy	[10,26]		$\star$
CIP	Δ	ellipsoid	[15,46]	KC-ineq., high-degree-poly time	$\star$
MONOTONE COVER	Δ	greedy	[our §2]	$\min\{c(x) : x \in S\ (\forall S \in \mathcal{C})\}$	new
CMIP	Δ	greedy	[our §3]	near-linear-time implementation	new
FACILITY LOCATION	Δ	greedy	[our §4]	linear-time implementation	new
PROBABILISTIC CMIP	Δ	greedy	[our §4]	quadratic-time implementation	new
online problem	competitive ratio	deterministic online			
PAGING	$k = \Delta$	potential function	[48,47]	e.g. LRU, FIFO, FWF, Harmonic	$\star$
CONNECTION CACHING	$O(k)$	reduction to paging	[18,1]		$\star$
WEIGHTED CACHING	k	primal-dual	[52,47]	e.g. Harmonic, Greedy-Dual	$\star$
FILE CACHING	k	primal-dual	[53,14]	e.g. Greedy-Dual-Size, Landlord	$\star$
UNW. SET COVER	$O(\log(\Delta)\log(n/\mathsf{opt}))$	primal-dual	[13]	unweighted	
CLP	$O(\log n)$	fractional	[13]	$\min\{c \cdot x : Ax \geq b; x \leq u\}$,	
MONOTONE COVER	Δ	potential function	[our §2]	includes the above and CMIP...	new
UPGRADABLE CACHING	$d + k$	reduction to mono. cover	[our §5]	d components, k files in cache	new

Fig. 1. Some Δ-approximation covering algorithms and deterministic online algorithms. "$\star$" = generalized or strengthened here.

COVER, in the early 1980's, Hochbaum gave an algorithm that rounds a solution to the standard LP relaxation [33]; Bar-Yehuda and Even gave a linear-time greedy algorithm [6]. A few years later, for SET MULTICOVER, Hall and Hochbaum gave a quadratic-time primal-dual algorithm [26]. In the late 1990's, Bertsimas and Vohra generalized all of these results with a quadratic-time primal-dual algorithm for covering integer programs (CIP), restricted to $\{0, 1\}$-variables and integer constraint matrix A, and with approximation ratio $\max_i \sum_j A_{ij} \geq \Delta$ [10]. Most recently, in 2000, Carr et al. gave the first (and only previous) Δ-approximation for general CIP with $\{0, 1\}$ variables [15].[1] They state (without details) that their result extends to allow general upper bounds on the variables (restricting $x_j \in \{0, 1, 2, \ldots, u_j\}$). In 2009 (independently of this work), [46] gives details of an extension to CIP with general upper bounds on the variables. Both [15] and [46] use exponentially many valid "Knapsack Cover" (KC) inequalities to reduce the integrality gap to Δ. Their algorithms solve the LP using the ellipsoid method, so the running time is a high-degree polynomial.

Online paging and caching algorithms are also (online) monotone covering problems, as they can be formulated as online SET COVER [2]. These problems also have a rich history (see Fig. 1, and [12]).

All of the classical covering problems above (vertex cover, set cover, mixed integer linear programs with variable upper bounds (CMIP) and others (facility location, probabilistic variants of these problems, etc.), as well as online variants (paging, weighted caching, file caching, (generalized) connection caching, etc.) are special cases of what we call **monotone covering**. Formally, a monotone covering instance is specified by a collection $\mathcal{C} \subset 2^{\mathbb{R}_+}$ of constraints and a non-negative, non-decreasing,

[1] The standard LP relaxation has an arbitrarily large integrality gap (e.g. $\min\{x_1 : 10x_1 + 10x_2 \geq 11; x_2 \leq 1\}$ has gap 10).

submodular[2] objective function, $c : \mathbb{R}_+^n \to \mathbb{R}_+$. The problem is to compute $\min\{c(x) : x \in \mathbb{R}_+^n, (\forall S \in \mathcal{C})\ x \in S\}$. Each constraint $S \in \mathcal{C}$ must be monotone (i.e., closed upwards), but can be non-convex.

Monotone covering allows each variable to take values throughout $\mathbb{R}_+$, but can still model problems with restricted variable domains. For example, formulate vertex cover as $\min\{\sum_v c_v x_v : x \in \mathbb{R}_+^V, (\forall (u,w) \in E)\ \lfloor x_u \rfloor + \lfloor x_w \rfloor \geq 1\}$. Given any 2-approximate solution x to this formulation (which allows $x_u \in \mathbb{R}_+$), rounding each x_u down to its floor gives a 2-approximate integer solution. Generally, to model problems where each variable x_j should take values in some closed set $U_j \subset \mathbb{R}_+$ (e.g. $U_j = \{0,1\}$ or $U_j = \mathbb{Z}_+$), one allows $x \in \mathbb{R}_+^n$ but replaces each monotone constraint $x \in S$ by the monotone constraint $x \in \mu^{-1}(S)$, where $\mu^{-1}(S) = \{x : \mu(x) \in S\}$ and $\mu_j(x) = \max\{z \in U_j, z \leq x_j\}$. If $x \in \mathbb{R}_+^n$ is any Δ-approximate solution to the modified problem, then $\mu(x)$ will be a Δ-approximate solution respecting the variable domains. (For vertex cover each $U_j = \mathbb{Z}_+$ so $\mu_j(x) = \lfloor x_j \rfloor$.)[3]

Section 2 describes our greedy Δ-approximation algorithm (Alg. 1) for monotone covering. It is roughly the following: *consider the constraints in any order; to satisfy a constraint, raise each variable in the constraint continuously and simultaneously, at rate inversely proportional to its cost. At termination, round x down to $\mu(x)$ if appropriate.*

The proof of the approximation ratio is relatively simple: with each step, the cost incurred by the algorithm is at most Δ times the reduction in the *residual cost* — the minimum possible cost to augment the current x to feasibility. The algorithm is online (as described below), and admits distributed implementations (see [39]).

The running time depends on the implementation, which is problem specific, but can be fast. Section 2 describes linear-time implementations for vertex cover, set cover, and (non-metric) facility location. Section 3 describes a nearly linear-time implementation for covering mixed integer linear programs with variable upper bounds (CMIP). (In contrast, the only previous Δ-approximation algorithm (for CIP, a slight restriction of CMIP) uses the ellipsoid method; its running time is a high-degree polynomial [15].) Section 4 describes an extension to a *probabilistic* (two-stage) variant of monotone covering, which naturally has submodular cost. The implementation for this case takes time $O(N\widehat{\Delta}\log\Delta)$, where N is the number of non-zeros in the constraint matrix and $\widehat{\Delta}$ is the maximum number of constraints in which any variable appears. (For comparison, [30] gives a $\ln(n)$-approximation algorithm for the special case of probabilistic set

[2] Formally, $c(x)+c(y) \geq c(x \wedge y) + c(x \vee y)$, where $x \wedge y$ (and $x \vee y$) are the component-wise minimum (and maximum) of x and y. Intuitively, there is no positive synergy between the variables: let $\partial_j c(x)$ denote the rate at which increasing x_j would increase $c(x)$; then, increasing x_i (for $i \neq j$) does not increase $\partial_j c(x)$. Any separable function $c(x) = \sum_j c_j(x_j)$ is submodular, the product $c(x) = \prod_j x_j$ is not. The maximum $c(x) = \max_j x_j$ is submodular, the minimum $c(x) = \min_j x_j$ is not.

[3] In this setting, if the cost is defined only on the restricted domain, it should be extended to $\mathbb{R}_+^n$ for the algorithm. One way is to take the cost of $x \in \mathbb{R}_+^n$ to be the expected cost of $\hat{x}$, where $\hat{x}_j$ is rounded up or down to its nearest elements a, b in U_j such that $a \leq x_j \leq b$: take $\hat{x}_j = b$ with probability $\frac{b-x_j}{b-a}$, otherwise take $\hat{x}_j = a$. If a or b doesn't exist, let $\hat{x}_j$ be the one that does.

cover; the algorithm is based on submodular-function minimization [45], resulting in high-degree-polynomial run-time.[4])

Section 5 discusses *online* monotone covering. Following [13], an online algorithm must maintain a current x; as constraints $S \in \mathcal{C}$ are revealed one by one, the algorithm must increase coordinates of x to satisfy $x \in S$. The algorithm can't decrease coordinates of x. An algorithm is Δ-competitive if $c(x)$ is at most Δ times the minimum cost of any solution x^* that meets all the constraints.

The greedy algorithm (Alg. 1) is an online algorithm. Thus, it gives Δ-competitive algorithms for online versions of all of the covering problems mentioned above. It also generalizes many classical deterministic online algorithms for paging and caching, including LRU, FIFO, FWF for paging [48], Balance and Greedy Dual for weighted caching [16,52], Landlord [53], a.k.a. Greedy Dual Size [14], for file caching, and algorithms for connection caching [18,19,20,1]. The competitive ratio Δ is the cache size, commonly denoted k, or, in the case of file caching, the maximum number of files ever held in cache — at most k or $k + 1$, depending on the specification. This is the best possible competitive ratio for deterministic online algorithms for these problems.

Section 5 also illustrates the generality of online monotone covering by describing a $(k + d)$-competitive algorithm for a new class of **upgradable** caching problems. In upgradable caching, the online algorithm chooses not only which pages to evict, but also how to configure and upgrade the relevant hardware components (determining such parameters as the cache size, CPU, bus, and network speeds, etc.) In the competitive ratio, d is the number of configurable hardware parameters. We know of no previous results for upgradable caching, although the classical online rent-or-buy (a.k.a. ski rental) problem [36] and its "multislope" generalization [41] have the basic characteristic (paying a fixed cost now can reduce many later costs; these are special cases of online monotone covering with $\Delta = 2$).

Section 6 describes a natural randomized generalization of Alg. 1, with more flexibility in incrementing the variables. This yields a *stateless* online algorithm, generalizing the Harmonic k-server algorithm (as it specializes for paging and weighted caching [47]) and Pitt's weighted vertex-cover algorithm [4].

Section 7 concludes by discussing the relation of the analysis here to the primal-dual and local-ratio methods. As a rule of thumb, greedy approximation algorithms can generally be analysed naturally via the primal-dual method, and sometimes even more naturally via the local-ratio method. The results here extend many primal-dual and local-ratio results. We conjecture that it is possible, but unwieldy, to recast the analysis here via primal-dual. It can be recast as a local-ratio analysis, but in a non-traditional form.

For distributed implementations of Alg. 1 running in $O(\log^2 n)$ rounds (or $O(\log n)$ for $\Delta = 2$), see [39].

We assume throughout that the reader is familiar with classical covering problems [51,34] as well as classical online paging and caching problems and algorithms [12].

---

[4] [30] also mentions a 2-approximation for probabilistic vertex cover, without details.

Alternatives to Δ-Approximation: log-Approximations, Randomized Online Algorithms. In spite of extensive work, no $(2-\varepsilon)$-approximation algorithm for constant $\varepsilon > 0$ is yet known for vertex cover [28,32,7,44,27,29,21,37]. For small Δ, it seems that Δ-approximation may be the best possible in polynomial time.

As an alternative when Δ is large, many covering problems considered here also admit $O(\log \widehat{\Delta})$-approximation algorithms, where $\widehat{\Delta}$ is the maximum number of constraints in which any variable occurs. Examples include a greedy algorithm for set cover [35,42,17] (1975) and greedy $O(\log \max_j \sum_i A_{ij})$-approximation algorithms for CIP with $\{0,1\}$-variables and integer A [22,24] (1982). Srinivasan gave $O(\log \widehat{\Delta})$-approximation algorithms for general CIP without variable upper bounds [49,50] (2000); these were extended to CIP with variable upper bounds by Kolliopoulos et al. [38] (2005). (The latter algorithm solves the CIP relaxation with KC inequalities, then randomly rounds the solution.) The class of $O(\log(\hat{\Delta}))$-approximation algorithms for general CIP is not yet fully understood; these algorithms could yet be subsumed by a single fast greedy algorithm.

For most online problems here, no *deterministic* online algorithm can be better than Δ-competitive. But many online problems admit better-than-Δ-competitive *randomized* algorithms. Examples include rent-or-buy [36,40], paging [23,43], weighted caching [2,14], connection caching [18], and file caching [3]. Some cases of online monotone covering (e.g. vertex cover) are unlikely to have better-than-Δ-competitive randomized algorithms. It would interesting to classify which cases admit better-than-Δ-competitive randomized online algorithms.

greedy algorithm for monotone covering (monotone constraints $\mathcal{C}$, submodular objective c) alg. 1
output: feasible $x \in S$ ($\forall S \in \mathcal{C}$), Δ-approximately minimizing $c(x)$ (see Thm. 1)
1. Let $x \leftarrow \mathbf{0}$. ... *$\Delta = \max_{S\in\mathcal{C}} |\mathsf{vars}(S)|$ is the max # of vars any constraint depends on*
2. While $\exists\, S \in \mathcal{C}$ such that $x \notin S$, do **step**(x, S) for any S such that $x \notin S$.
3. Return x. ... *or $\mu(x)$ in the case of restricted variable domains; see the introduction.*

subroutine step$_c(x, S)$: ... *makes progress towards satisfying $x \in S$.*
1. Choose a scalar *step size* $\beta \geq 0$. ... *choose β subject to restriction in Thm. 1.*
2. For $j \in \mathsf{vars}(S)$, let $x'_j \in \mathbb{R}_+ \cup \{\infty\}$ be the maximum such that raising x_j to x'_j would raise $c(x)$ by at most β.
3. For $j \in \mathsf{vars}(S)$, let $x_j \leftarrow x'_j$. ... *if c is linear, then $x'_j = x_j + \beta/c_j$ for $j \in \mathsf{vars}(S)$.*

2 The Greedy Algorithm for Monotone Covering (Alg. 1)

Fix an instance of monotone covering. Let $\mathsf{vars}(S)$ denote the variables in x that constraint $x \in S$ depends on, so that $\Delta = \max_{S\in\mathcal{C}} |\mathsf{vars}(S)|$.

The algorithm (Alg. 1) starts with $x = \mathbf{0}$, then repeats the following step until all constraints are satisfied: *choose any unmet constraint and a* step size $\beta > 0$; *for each variable x_j that the constraint depends on ($j \in \mathsf{vars}(S)$), raise that variable so as to increase the cost $c(x)$ by at most β.* (The step increases the total cost by at most $\Delta\beta$.)

The algorithm returns x (or, if variable domains are restricted as described in the introduction, $\mu(x)$).

The algorithm returns a Δ-approximation, as long as each step size β is *at most* the minimum cost to optimally augment x to satisfy S, that is, $\min\{c(\hat{x}) - c(x) : \hat{x} \in S, \hat{x} \geq x\}$. Denote this cost $\mathsf{distance}_c(x, S)$. Also, let $\mathsf{residual}_c(x)$ be the *residual cost* of x — the minimum cost to augment x to full feasibility, i.e., $\mathsf{distance}_c(x, \cap_{S \in \mathcal{C}} S)$.

Theorem 1. *For monotone covering, the greedy algorithm (Alg. 1) returns a Δ-approximate solution as long as it chooses step size $\beta \leq \mathsf{distance}_c(x, S)$ in each step (and eventually terminates).*

Proof. First, a rough intuition. Each step starts with $x \notin S$. Since the optimal solution x^* is in S and S is monotone, there must be *at least one* $k \in \mathsf{vars}(S)$ such that $x_k < x^*_k$. By raising all x_j for $j \in \mathsf{vars}(S)$, the algorithm makes progress "covering" at least that coordinate x^*_k of x^*. Provided the step increases x_k to $x'_k \leq x^*_k$, the cost incurred can be charged to a corresponding portion of the cost of x^*_k (intuitively, to the cost of the part of x^*_k in the interval $[x_k, x'_k]$; formally, to the *decrease in the residual cost* from increasing x_k, provably at least β). Since the step increases $c(x)$ by at most $\beta\Delta$, and results in a charge to $c(x^*)$ of at least β, this proves the Δ-approximation.

Here is the formal proof. By inspection (using that c is submodular) each step of the algorithm increases $c(x)$ by at most $\beta|\mathsf{vars}(S)| \leq \beta\Delta$. We show that $\mathsf{residual}(x)$ decreases by at least β, so the invariant $c(x)/\Delta + \mathsf{residual}(x) \leq \mathsf{opt}$ holds, proving the theorem.

Let x and x', respectively, be x before and after a given step. Let feasible $x^* \geq x$ be an optimal augmentation of x to full feasibility, so $c(x^*) - c(x) = \mathsf{residual}(x)$. Let $x \wedge y$ (resp. $x \vee y$) denote the component-wise minimum (resp. maximum) of x and y. By the submodularity of c, $c(x') + c(x^*) \geq c(x' \vee x^*) + c(x' \wedge x^*)$. (Equality holds if c is separable (e.g. linear).)

Rewriting gives $[c(x^*) - c(x)] - [c(x' \vee x^*) - c(x')] \geq c(x' \wedge x^*) - c(x)$.

The first bracketed term is $\mathsf{residual}(x)$. The second is at least $\mathsf{residual}(x')$, because $x^* \vee x' \geq x'$ is feasible. Thus,

$$\mathsf{residual}(x) - \mathsf{residual}(x') \geq c(x' \wedge x^*) - c(x). \tag{1}$$

To complete the proof, we show the right-hand side of (1) is at least β.

Case 1. Suppose $x'_k < x^*_k$ for some $k \in \mathsf{vars}(S)$. (In this case it must be that increasing x_k to x'_k costs β.)

Let y be x with just x_k raised to x'_k. Then $c(x' \wedge x^*) \geq c(y) = c(x) + \beta$.

Case 2. Otherwise $x' \wedge x^* \in S$, because $x^* \in S$ and $x'_j \geq x^*_j$ for all $j \in \mathsf{vars}(S)$. Also $x' \wedge x^* \geq x$.

Thus, the right-hand side of (1) is at least $\mathsf{distance}_c(x, S)$. By assumption this is at least β. □

Choosing the step size, β. In a sense, the algorithm reduces the given problem to a sequence of subproblems, each of which requires computing a lower bound on $\mathsf{distance}(x, S)$ for the current x and a given unmet constraint S. To completely specify the algorithm, one must specify how to choose β in each step.

Thm. 1 allows β to be small. At a minimum, $\mathsf{distance}(x, S) > 0$ when $x \notin S$, so one can take β to be infinitesimal. Then Alg. 1 raises x_j for $j \in \mathsf{vars}(S)$ continuously at rate inversely proportional to $\partial c(x)/\partial x_j$ (at most until $x \in S$).

Another, generic, choice is to take β just large enough to satisfy $x \in S$. This also satisfies the theorem:

Observation 1. *Let β be the minimum step size so that* $\mathsf{step}(x, S)$ *brings x into S. Then* $\beta \leq \mathsf{distance}_c(x, S)$.

Thm. 1 can also allow β to be *more* than large enough to satisfy the constraint. Consider $\min\{x_1 + 2x_2 : x \in S\}$ where $S = \{x : x_1 + x_2 \geq 1\}$. Start with $x = \mathbf{0}$. Then $\mathsf{distance}(x, S) = 1$. The theorem allows $\beta = 1$. A single step with $\beta = 1$ gives $x_1 = 1$ and $x_2 = 1/2$, so that $x_1 + x_2 = 3/2 > 1$.

Generally, one has to choose β small enough to satisfy the theorem, but large enough so that the algorithm doesn't take too many steps. The computational complexity of doing this has to be addressed on a per-application basis. Consider a simple subset-sum example: $\min\{c \cdot x : x \in S\}$ where the single constraint S contains $x \geq 0$ such that $\sum_j c_j \min(1, \lfloor x_j \rfloor) \geq 1$. Computing $\mathsf{distance}(\mathbf{0}, S)$ is NP-hard, but it is easy to compute a useful β, for example $\beta = \min_{j:x_j<1} c_j(1 - x_j)$. With this choice, the algorithm will satisfy S within Δ steps.

As a warm-up, here are linear-time implementations for facility location, set cover, and vertex cover.

Theorem 2. *For (non-metric) facility location, set cover, and vertex cover, the greedy Δ-approximation algorithm (Alg. 1) has a linear-time implementation. For facility location Δ is the maximum number of facilities that might serve any given customer.*

Proof. Formulate facility location as minimizing the submodular objective $\sum_j f_j \max_i x_{ij} + \sum_{ij} d_{ij} x_{ij}$ subject to, for each customer i, $\sum_{j \in N(i)} \lfloor x_{ij} \rfloor \geq 1$ (where $j \in N(i)$ if customer i can use facility j).[5]

The implementation starts with all $x_{ij} = 0$. It considers the customers i in any order. For each it does the following: let $\beta = \min_{j \in N(i)}[d_{ij} + f_j(1 - \max_{i'} x_{i'j})]$ (the minimum cost to raise x_{ij} to 1 for any $j \in N(i)$). Then, for each $j \in N(i)$, raise x_{ij} by $\min[\beta/d_{ij}, (\beta + f_j \max_{i'} x_{i'j})/(d_{ij} + f_j)]$ (just enough to increase the cost by β). By maintaining, for each facility j, $\max_{i'} x_{i'j}$, the above can be done in linear time, $O(\sum_i |N(i)|)$.

Vertex cover and set cover are the special cases when $d_{ij} = 0$. □

3 Nearly Linear-Time Implementation for Covering Mixed Integer Linear Programs

Theorem 3. *For CMIP (covering mixed integer linear programs with upper bounds), the greedy algorithm (Alg. 1) can be implemented to return a Δ-approximation in*

[5] The standard ILP is not a covering ILP due to constraints $x_{ij} \leq y_j$. The standard reduction to set cover increases Δ exponentially.

subroutine stepsize$_c(x, S(I, A_i, u, b_i))$ **(for CMIP)** alg. 2

1. Order $I = (j_1, j_2, \ldots, j_k)$ by decreasing A_{ij}. ... *So* $A_{ij_1} \geq A_{ij_2} \geq \cdots \geq A_{ij_k}$.
   Let $J = J(x, S)$ contain the minimal prefix of I such that $x \notin S(J, A_i, u, b_i)$.
   Let S' denote the relaxed constraint $S(J, A_i, u, b_i)$.
2. Let $U = U(x, S) = \{j : x_j \geq u_j; A_{ij} > 0\}$ contain the variables that have hit their upper bounds.
3. Let $\beta_J = \min_{j \in J-U}(1 - x_j + \lfloor x_j \rfloor)c_j$ be the minimum cost to increase any floored term in S'.
4. Let $\beta_{\overline{J}} = \min_{j \in \overline{J}-U} c_j b'_i / A_{ij}$, where b'_i is the slack (b_i minus the value of the left-hand side of S'), be the minimum cost to increase the sum of fractional terms in S' to satisfy S'.
5. Return $\beta = \min\{\beta_J, \beta_{\overline{J}}\}$.

$O(N \log \Delta)$ time, where Δ is the maximum number of non-zeroes in any constraint and N is the total number of non-zeroes in the constraint matrix.

Proof (sketch). Fix any CMIP instance $\min\{c \cdot x : x \in \mathbb{R}^n_+; Ax \geq b; x \leq u; x_j \in \mathbb{Z}\ (j \in I)\}$.

Model each constraint $A_i x \geq b_i$ using a monotone constraint $S \in \mathcal{C}$ of the form

$$\sum_{j \in I} A_{ij} \lfloor \min(x_j, u_j) \rfloor + \sum_{j \in \overline{I}} A_{ij} \min(x_j, u_j) \ \geq \ b_i \qquad S(I, A_i, u, b_i)$$

where set I contains the indexes of the integer variables.

Given such a constraint S and an $x \notin S$, the subroutine stepsize(x, S) (Alg. 2) computes a step size β satisfying Thm. 1 as follows. Let S', J, U, β_J, $\beta_{\overline{J}}$, and β be as in Alg. 2. That is, $S' = S(J, A_i, u, b_i)$ is the relaxation of $S(I, A_i, u, b_i)$ obtained by relaxing the floors in S (in order of increasing A_{ij}) as much as possible, while maintaining $x \notin S'$; $J \subseteq I$ contains the indices j of variables whose floors are not relaxed. Increasing x to satisfy S' requires (at least) either: (i) increasing $\sum_{j \in J-U} A_{ij} \lfloor x_j \rfloor$, at cost at least β_J, or (ii) increasing $\sum_{j \in \overline{J}-U} A_{ij} x_j$ by at least the slack b'_i of the constraint S', at cost at least $\beta_{\overline{J}}$. Thus, distance$(x, S) \geq$ distance$(x, S') \geq \min\{\beta_J, \beta_{\overline{J}}\} = \beta$. This choice satisfies Thm. 1, so the algorithm returns a Δ-approximate solution.

Lemma 1. *For any S, Alg. 1 calls* step(x, S) *with $\beta =$* stepsize(x, S) *(from Alg. 2) at most* $2|\text{vars}(S)|$ *times.*

Proof (sketch). Let j be the index of the variable x_j that determines β in the algorithm (β_J in case (i) of the previous proof, or $\beta_{\overline{J}}$ in case (ii)). The step increases x_j by β / c_j. This may bring x_j to (or above) its upper bound u_j. If not, then, in case (i), the left-hand side of S' increases by at least A_{ij}, which, by the minimality of $J(x)$ and the ordering of I, is enough to satisfy S'. Or, in case (ii), the left-hand side increases by the slack b'_i (also enough to satisfy S'). Thus the step either the increases the set $U(x)$ or satisfies S', increasing the set $J(x)$. □

The naive implementations of stepsize() and step() run in time $O(|\text{vars}(S)|)$ (after the A_{ij}'s within each constraint are sorted in preprocessing). By the lemma, with this implementation, the total time for the algorithm is $O(\sum_S |\text{vars}(S)|^2) \leq O(N\Delta)$. By a careful heap-based implementation, this time can be reduced to $O(N \log \Delta)$ (proof omitted). □

4 (Two-Stage) Probabilistic Monotone Covering

An instance of *probabilistic* monotone covering is specified by an instance $(c, \mathcal{C})$ of monotone covering, along with *activation* probabilities p_S for each constraint $S \in \mathcal{C}$ and a non-decreasing, submodular *first-stage* objective W. The first stage requires the algorithm to commit to a vector $x^S \in S$ for each $S \in \mathcal{C}$. In the second stage, the algorithm must pay to satisfy the activated constraints, where each constraint S is independently activated with probability p_S. The algorithm pays $c(\hat{x})$, where $\hat{x}$ is the minimal vector such that $\hat{x} \geq x^S$ for each active S ($\hat{x}_j = \max\{x^S_j : S \text{ active}\}$). The problem is to choose the first-stage vectors to minimize the first-stage cost $W(x^S : S \in \mathcal{C})$ plus the *expected* second-stage cost, $E[c(\hat{x})]$. This (expected) cost is submodular as long as c is.

Observation 2. *Probabilistic monotone covering reduces to monotone covering.*

Probabilistic CMIP is the special case where W is linear and the pair $(c, \mathcal{C})$ define a CMIP.

For example, consider a two-stage probabilistic facilities location problem specified by first-stage costs f^1, d^1, an activation probability p_i for each customer i, and second-stage costs f^2, c^2. The algorithm assigns to each customer i a facility $j(i) \in N(i)$ (those that can serve i), by setting $x_{ij(i)} = 1$ (satisfying constraints $\sum_{j \in N(i)} \lfloor x_{ij} \rfloor \geq 1$), then paying the first-stage cost $\sum_j f^1_j \max_i x_{ij} + \sum_{ij} d^1_{ij} x_{ij}$. Then, each customer i is activated with probability p_i. Facilities assigned to activated customers are opened by setting $\hat{x}_{ij} = 1$ if $x_{ij} = 1$ and i is active. The algorithm then pays the second-stage cost $\sum_j f^2_j \max_i \hat{x}_{ij} + \sum_{ij} d^2_{ij} \hat{x}_{ij}$. The algorithm should minimize its total expected payment. The degree $\Delta = \max_i |N(i)|$ is the maximum number of facilities that any given customer is eligible to use.

Theorem 4. *For probabilistic CMIP,*

(a) The greedy Δ-approximation algorithm can be implemented to run in $O(N\widehat{\Delta} \log \Delta)$ time, where $\widehat{\Delta}$ is the maximum number of constraints per variable and $N = \sum_{S \in \mathcal{C}} |\mathsf{vars}(S)|$ is the input size.

(b) When $p = 1$, it can be implemented to run in time $O(N \log \Delta)$ (generalizes CMIP and facilities location).

Proof (sketch). Let $X = (x^S)_{S \in \mathcal{C}}$ be the matrix formed by the first-stage vectors. Let random variable $\hat{x}$ be as described in the problem definition ($\hat{x}_j = \max\{\hat{x}^S_j : S \text{ active}\}$), so the problem is to choose X subject to $x^S \in S$ for each S to minimize $C(X) = W \cdot X + E[c \cdot \hat{x}]$. This function is submodular, increasing, and continuous in X.

To satisfy Thm. 1, the subroutine $\mathsf{step}(X, S)$ must compute the step size β to be at most $\mathsf{distance}(X, S)$ (the minimum possible increase in $C(X)$ required to satisfy S).

For a given X and S, have $\mathsf{step}(X, S)$ compute β as follows. For a given X, the rate at which increasing x_j^S would increase $C(X)$ is

$$c'_j = w_j^S + c_j \Pr[x_j^S = \hat{x}_j] = w_j^S + c_j p_S \prod \{1 - p_R : x_j^R > x_j^S, j \in \mathsf{vars}(R)\}.$$

This rate does not change until x_j^S reaches $t_j = \min\{x_j^R : x_j^R > x_j^S, j \in \mathsf{vars}(R)\}$.

Take $\beta = \min(\beta_t, \mathsf{stepsize}_{c'}(x^S, S))$, where $\beta_t = \min\{(t_j - x_j^S)c'_j : j \in \mathsf{vars}(S)\}$ is the minimum cost to bring any x_j^S to its threshold, and $\mathsf{stepsize}()$ is the subroutine from Section 3, using the (linear) cost vector c' defined above. This β is a valid lower bound on $\mathsf{distance}(X, S)$, because β_t is a lower bound on the cost to bring any x_j^S to its next threshold, while $\mathsf{stepsize}_{c'}(x^S, S)$ is a lower bound on the cost to satisfy S without bringing any x_j^S to its threshold.

If $\mathsf{step}(X, S)$ uses this β, the number of steps to satisfy S is at most $O(|\mathsf{vars}(S)|\widehat{\Delta})$. Each step either (i) makes some x_j^S reach its next threshold (and each x_j^S crosses at most $\widehat{\Delta}$ thresholds), or (ii) increases the number of "floored" variables or increases the number of variables at their upper bounds (which by the analysis of $\mathsf{stepsize}()$ from Section 3, can happen at most $2|\mathsf{vars}(S)|$ times). Thus, the total number of steps is $O(\sum_S |\mathsf{vars}(S)|\widehat{\Delta})$, that is, $O(N\widehat{\Delta})$. (Implementation details needed to achieve amortized time $O(\log \Delta)$ per step are omitted.) This completes the proof sketch for part (a).

For part (b) of the theorem, note that in this case the product in the equation for $C_j^S(X)$ is 1 if $x_j^S = \max_R x_j^R$ and 0 otherwise. Each variable has at most one threshold to reach, so the number of calls to $\mathsf{step}(X, S)$ is reduced to $O(|\mathsf{vars}(S)|)$. This allows an implementation in total time $O(N \log \Delta)$. □

5 Online Monotone Covering and Caching with Upgradable Hardware

Recall that in online monotone covering, each constraint $S \in \mathcal{C}$ is revealed one at a time; an online algorithm must raise variables in x to bring x into the given S, without knowing the remaining constraints. Alg. 1 (with, say, $\mathsf{step}(x, S)$ taking β just large enough to bring $x \in S$; see Observation 1) can do this, so it yields a Δ-competitive online algorithm.[6]

Corollary 1. *The greedy algorithm (Alg. 1) gives a Δ-competitive online monotone covering algorithm.*

Example: generalized connection caching. As discussed in the introduction (following the formulation of weighted caching as online set cover from [2]) this result naturally generalizes a number of known results for paging, weighted caching, file caching,

[6] If the cost function is linear, in responding to S this algorithm needs to know S and the values of variables in S and their cost coefficients. For general submodular costs, the algorithm may need to know not only S, but *all* variables' values and the whole cost function.

connection caching, etc. To give just one example, consider connection caching. A request sequence r is given online. Each request $r_t = (u_t, w_t)$ activates the connection (u_t, w_t) (if not already activated) between nodes u_t and w_t. If either node has more than k active connections, then one of them other than r_t (say r_s) must be closed at cost $\mathsf{cost}(r_s)$. Model this problem as follows. Let variable x_t indicate whether connection r_t is closed before the next request to r_t after time t, so the total cost is $\sum_t \mathsf{cost}(r_t)x_t$. For each node u and each time t, for any $(k+1)$-subset $Q \subseteq \{r_s : s \le t; u \in r_s\}$, at least one connection $r_s \in Q - \{r_t\}$ (where s is the time of the most recent request to r_s) must have been closed, so the following constraint[7] is met: $\sum_{r_s \in Q - \{r_t\}} \lfloor x_s \rfloor \ge 1$.

Corollary 1 gives the following k-competitive algorithm for online connection caching. When a connection request (u, w) occurs at time t, the connection is activated and x_t is set to 0. If a node, say u, has more than k active connections, the current x violates the constraint above for the set Q containing u's active connections. Node u applies the $\mathsf{step}()$ subroutine for this constraint: it raises x_s for all the connections $r_s \in Q - \{r_t\}$ at rate $1/\mathsf{cost}(r_s)$ simultaneously, until some x_s reaches 1. It closes any such connection r_s.

Remark on $k/(k-h+1)$-competitiveness. The classic ratio of $k/(k-h+1)$ (versus opt with cache size $h \le k$) can be reproduced in such a setting as follows. For any set Q as described above, opt must meet the stronger constraint $\sum_{r_s \in Q - \{r_t\}} \lfloor x_s \rfloor \ge k-h+1$. In this scenario, the proof of Thm. 1 extends to show a ratio of $k/(k-h+1)$ (use that the variables are $\{0,1\}$, so there are at least $k-h+1$ variables x_j such that $x_j < x_j^*$).

Upgradable online problems. Standard online caching problems model only the caching strategy. In practice other parameters (e.g., the size of the cache, the speed of the CPU, bus, network, etc.) must also be chosen well. In *upgradable* caching, the algorithm chooses not only the caching strategy, but also the hardware configuration. The hardware configuration is assumed to be determined by how much has been spent on each of some d components. The configuration is modeled by a vector $y \in \mathbb{R}_+^d$, where y_i has been spent so far on component i.

In response to each request, the algorithm can upgrade the hardware by increasing the y_i's. Then, if the requested item r_t is not in cache, it is brought in. Then items in cache must be selected for eviction until the set Q of items remaining in cache is cachable, as determined by some specified predicate $\mathsf{cachable}_t(Q, y)$. The cost of evicting an item r_s is specified by a function $\mathsf{cost}(r_s, y)$.

The $\mathsf{cachable}()$ predicate and $\mathsf{cost}()$ function can be specified arbitrarily, subject to the following restrictions. Predicate $\mathsf{cachable}_t(Q, y)$ must be non-decreasing in y (upgrading the hardware doesn't cause a cachable set to become uncachable) and non-increasing with Q (any subset of a cachable set is cachable). The function $\mathsf{cost}(r_s, y)$ must be non-increasing in y (upgrading the hardware doesn't increase the eviction cost of any item). To model (standard, non-upgradable) file caching, take $\mathsf{cachable}_t(Q, y)$ to be true if $\sum_{r_s \in Q} \mathsf{size}(r_s) \le k$.

---

[7] This presentation assumes that the last request must stay in cache. If not, don't subtract $\{r_t\}$ from Q in the constraints. The competitive ratio goes from k to $k+1$.

In general, the adversary is free to constrain the cache contents at each step t in *any* way that depends on t and the hardware configuration, as long as upgrading the cache or removing items does not make a cachable set uncachable. Likewise, the cost of evicting any item can be determined by the adversary in *any* way that depends on the item and the hardware configuration, as long as upgrading the configuration does not increase any eviction cost. This gives a great deal of flexibility in comparison to the standard model. For example, the adversary could insist (among other constraints) that no set containing both of two (presumably conflicting) files can be cached. Or, upgrading the hardware could reduce the eviction cost of some items arbitrarily, even to zero.

The optimal cost is achieved by choosing an optimal hardware configuration at the start, then handling all caching decisions optimally. To be competitive, an algorithm must also choose a good hardware configuration: an algorithm is Δ-competitive if its total cost (eviction cost plus final hardware configuration cost, $\sum_i y_i$) is at most Δ times the optimum. (Naturally, when the algorithm evicts an item, it pays the eviction cost in its *current* hardware configuration. Later upgrades do not reduce earlier costs.)

Next we describe how to model the upgradable problem via online monotone covering with degree $\Delta = k + d$, where k is the maximum number of files ever held in cache and d is the number of hardware components. This gives a simple $(k + d)$-competitive online algorithm for upgradable caching.

Theorem 5. *Upgradable caching has a $(d + k)$-competitive online algorithm, where d is the number of upgradable components and k is the maximum number of files that can be held in the cache.*

Proof (sketch). Let variable y_i for $i = 1, \dots, d$ denote the amount invested in component i, so that the vector y gives the current hardware configuration. Let x_t be the cost (if any) incurred for evicting the tth requested item r_t at any time before its next request. The total final cost is $\sum_i y_i + \sum_t x_t$. At time t, if some subset $Q \subseteq \{r_s : s \leq t\}$ of the items is not cachable, then at least one item $r_s \in Q - \{r_t\}$ (where s is the time of the most recent request to r_s) must have been evicted, so the following constraint is met:

$$\mathsf{cachable}_t(Q, y) \text{ or } \textstyle\sum_{r_s \in Q - \{r_t\}} \lfloor x_s / \mathsf{cost}(r_s, y) \rfloor \geq 1. \qquad S_t(Q)$$

The restrictions on cachable and cost ensure that this constraint is monotone in x and y.

The greedy algorithm initializes $y = \mathbf{0}$, $x = \mathbf{0}$ and $Q = \emptyset$. It caches the subset Q of requested items r_s with $x_s < \mathsf{cost}(r_s, y)$. To respond to request r_t (which adds r_t to the cache if not present), the algorithm raises each y_i and each x_s for r_s in $Q - \{r_t\}$ at unit rate. It evicts any r_s with $x_s \geq \mathsf{cost}(r_s, y)$, until $\mathsf{cachable}_t(Q, y)$ holds for the cached set Q. The degree[8] Δ is the maximum size of $Q - \{r_t\}$, plus d for y. □

This result generalizes easily to "upgradable" monotone caching, where investing in some d components can relax constraints or reduce costs.

Restricting groups of items (such as segments within files). The `http` protocol allows retrieval of segments of files. To model this in this setting, consider each file f

[8] The algorithm enforces just *some* constraints $S_t(Q)$; Δ is defined w.r.t. the problem defined by those constraints.

subroutine rstep$_c(x, S)$	alg. 3
1. Fix an arbitrary probability $p_j \in [0, 1]$ for each $j \in \mathsf{vars}(S)$.	*... taking each* $p_j = 1$ *gives Alg. 1*
2. Choose a scalar step size $\beta \geq 0$.	
3. For $j \in \mathsf{vars}(S)$ with $p_j > 0$, let X_j be the max. s.t. raising x_j to X_j would raise $c(x)$ by $\leq \beta/p_j$.	
4. For $j \in \mathsf{vars}(S)$ with $p_j > 0$, with probability p_j, let $x_j \leftarrow X_j$.	*... these events can be dependent if desired!*

subroutine stateless-rstep$_c(x, S, U)$: $\cdots$ *do rstep, and keep each* x_j *in its (countable) domain* $U_j \cdots$	alg. 4
1. For $j \in \mathsf{vars}(S)$, let $X_j = \min\{z \in U_j; z > x_j\}$ (or $X_j = x_j$ if the minimum is undefined).	
2. Let α_j be the increase in $c(x)$ that would result from increasing just x_j to X_j.	
3. Do $\mathsf{rstep}_c(x, S)$, choosing any $\beta \in (0, \min_j \alpha_j]$ and $p_j = \beta/\alpha_j$ (or $p_j = 0$ if $X_j = x_j$).	

as a group of arbitrary segments (e.g. bytes or pages). Let x_t be the *number of segments* of file r_t evicted before its next request. Let $c(x_t)$ be the cost to retrieve the cheapest x_t segments of the file, so the total cost is $\sum_t c(x_t)$. Then, for example, to say that the cache can hold at most k segments total, add constraints of the form (for appropriate subsets Q of requests) $\sum_{s \in Q} \mathsf{size}(r_s) - \lfloor x_s \rfloor \leq k$ (where $\mathsf{size}(r_s)$ is the number of segments in r_s). When the greedy algorithm increases x_s to x'_s, the online algorithm evicts segments $\lfloor x_s \rfloor + 1$ through $\lfloor x'_s \rfloor$ of file r_s (assuming segments are ordered by cheapest retrieval).

Generally, any monotone restriction that is a function of just the *number* of segments evicted from each file (as opposed to which specific segments are evicted), can be modeled. (For example, *"evict at least 3 segments of* r_s *or at least 4 segments from* r_t*"*: $\lfloor x_s/3 \rfloor + \lfloor x_t/4 \rfloor \geq 1$.) Although the caching constraints constrain file segments, the competitive ratio will be the maximum number of files (as opposed to segments) referred to in any constraint.

6 Randomized Variant of Alg. 1 and Stateless Online Algorithm

This section describes a randomized, online generalization of Alg. 1. It has more flexibility than Alg. 1 in how it increases variables. This can be useful, for example, in distributed settings, in dealing with numerical precision issues, and in obtaining *stateless* online algorithms (an example follows).

The algorithm is Alg. 1, modified to call subroutine $\mathsf{rstep}_c(x, S)$ (shown in Alg. 3) instead of $\mathsf{step}_c(x, S)$. The subroutine has more flexibility in incrementing x. Its step-size requirement is a bit more complicated.

Theorem 6. *For monotone covering suppose the randomized greedy algorithm terminates, and, in each step,* β *is at most* $\min\{E[c(x \uparrow_p \hat{x}) - c(x)] : \hat{x} \geq x; \hat{x} \in S\}$, *where* $x \uparrow_p \hat{x}$ *is a random vector obtained from* x *by raising* x_j *to* $\hat{x}_j$ *with probability* p_j *for each* $j \in \mathsf{vars}(S)$. *Then the algorithm returns a* Δ*-approximate solution in expectation.*

If the objective $c(x)$ is linear, the required upper bound on β above simplifies to $\mathsf{distance}_{c'}(x, S)$ where $c'_j = p_j c_j$.

Proof (sketch). We claim that, in each step, the expected increase in $c(x)$ is at most Δ times the expected decrease in $\mathsf{residual}(x)$. This implies (by the optional stopping theorem) that $E[c(x_{\text{final}})] \leq \Delta \times \mathsf{residual}(\mathbf{0})$, proving the theorem.

Fix any step starting with a given x. Let (r.v.) x' be x after the step. Fix feasible $x^* \geq x$ s.t. $\mathsf{residual}(x) = c(x^*) - c(x)$. Inequality (1) holds; to prove the claim we show $E_{x'}[c(x' \wedge x^*) - c(x)] \geq \beta$. Since $x^* \geq x$ and $x' = x \uparrow_p X$, this is equivalent to $E[c(x \uparrow_p X) - c(x)] \geq \beta$.

(**Case 1.**) Suppose $X_k < x_k^*$ for some $k \in \mathsf{vars}(S)$ with $p_k > 0$. Let y be obtained from x by raising just x_k to X_k. Then with probability p_k or more, $c(x \uparrow_p X) \geq c(y) \geq c(x) + \beta/p_k$. Thus the expectation is at least β.

(**Case 2.**) Otherwise, $X_j \geq x_j^*$ for all j with $p_j > 0$. Then $E[c(x \uparrow_p X) - c(x)] \geq E[c(x \uparrow_p x^*) - c(x)]$. Since $x^* \geq x$ and $x^* \in S$, this is at least β by the assumption on β. □

A stateless online algorithm. As described in the introduction, when the variables have restricted domains ($x_j \in U_j$), Alg. 1 constructs x and then "rounds" x down to $\mu(x)$. In the online setting, Alg. 1 maintains x as constraints are revealed; meanwhile, it uses $\mu(x)$ as its current online solution. In this sense, it is not *stateless*. A stateless algorithm can maintain only one online solution, each variable of which should stay in its restricted domain.

Next we use Thm. 6 to give a stateless online algorithm. The algorithm generalizes the Harmonic k-server algorithm as it specializes for paging and caching [47], and Pitt's weighted vertex cover algorithm [4]. Given an unsatisfied constraint S, the algorithm increases each x_j for $j \in \mathsf{vars}(S)$ to its next largest allowed value, with probability inversely proportional to the resulting increase in cost. (The algorithm can be tuned to increase just one, or more than one, x_j. It repeats the step until the constraint is satisfied.)

Formally, the stateless algorithm is the randomized algorithm from Thm. 6, but with the subroutine $\mathsf{rstep}_c(x, S)$ replaced by **stateless-rstep**$_c(x, S, U)$ (in Alg. 4), which executes $\mathsf{rstep}_c(x, S)$ in a particular way. (A¡ technicality: if $0 \notin U_j$, then x_j should be initialized to $\min U_j$ instead of 0. This does not affect the approximation ratio.)

Theorem 7. *For monotone covering with discrete variable domains as described above, there is a stateless randomized online Δ-approximation algorithm.*

Proof (sketch). By inspection **stateless-rstep**$_c(x, S, U)$ maintains each $x_j \in U_j$.

We show that **stateless-rstep**$_c(x, S, U)$ performs $\mathsf{rstep}_c(x, S)$ in a way that satisfies the requirement on β in Thm. 6. Let $\hat{x}$ be as in the proof of Thm. 6, with the added restriction that each $\hat{x}_j \in U_j$. Since $\hat{x} \in S$ but $x \notin S$, there is a $k \in \mathsf{vars}(S)$ with $\hat{x}_k > x_k$. Since $\hat{x}_k \in U_k$, the choice of X_k ensures $\hat{x}_k \geq X_k$. Let y be obtained from x by raising x_k to X_k. Then, $E[c(x \uparrow_p \hat{x}) - c(x)] \geq p_k[c(y) - c(x)] = p_k\alpha_k = \beta$, satisfying Thm. 6. □

7 Relation to Primal-Dual and Local-Ratio Methods

Primal-Dual. Here we speculate about how Thm. 1 might be cast as a primal-dual analysis. Given a vector v, consider its "shadow" $s(v) = \{x : \exists_j x_j \geq v_j\}$. Any monotone set S is the intersection of the shadows of its boundary points: $S = \bigcap_{v \in \partial S} s(v)$. Thus, any monotone covering instance can be recast to use only shadow sets for constraints. Any shadow set $s(v)$ is of the form $s(v) = \{x : \sum_j \lfloor x_j/v_j \rfloor \geq 1\}$, a form similar to that of the CMIP constraints $S(I, A_i, u, b_i, d)$ in Section 3. We conjecture that the Knapsack Cover (KC) inequalities from [15] for CIP can be generalized to give valid inequalities with integrality gap Δ for constraints of this form. (Indeed, the result in Section 3 easily extends to handle such constraints.) This could yield an appropriate relaxation on which a primal-dual analysis could be based.

For even simple instances, generating a Δ-approximate primal-dual pair for the greedy algorithm here requires a "tail-recursive" dual solution implicit in some local-ratio analyses [9], as opposed to the typical forward-greedy dual solution.[9] Even if the above program (extended to non-linear cost functions!) can be carried out, it seems likely to lead to a less intuitive proof than that of Thm. 1.

Local-Ratio. The local-ratio method has most commonly been applied to problems with variables x_j taking values in $\{0, 1\}$ and with linear objective function $c \cdot x$ (see [7,4,9,5]; for one exception, see [8]). In these cases, each step of the algorithm is typically interpreted as modifying the problem by repeatedly *reducing* selected objective function weights c_j by some β. At the end, the x, where x_j is raised from 0 to 1 if $c_j = 0$, gives the solution. At each step, the weights to lower are chosen so that the change must decrease OPT's cost by at least β, while increasing the cost for the algorithm's solution by at most $\Delta\beta$. This guarantees a Δ-approximate solution.

In contrast, recall that Alg. 1 raises selected x_j's fractionally by β/c_j. At the end, x_j is rounded down to $\lfloor x_j \rfloor$. Each step costs $\beta\Delta$, but reduces the *residual cost* by at least β.

For problems with variables x_j taking values in $\{0, 1\}$ and with linear objective function $c{\cdot}x$, Alg. 1 can be given the following straightforward local-ratio interpretation. Instead of raising x_j by β/c_j, reduce c_j by β. At the end, instead of setting x_j to $\lfloor x_j \rfloor$, set $x_j = 1$ if $c_j = 0$. With this reinterpretation, a standard local-ratio analysis applies.

To understand the relation between the two interpretations, let c' denote the modified weights in the above reinterpretation. The reinterpreted algorithm maintains the following invariants: Each modified weight c'_j stays equal to $c_j(1 - x_j)$ (for c and x in the original interpretation; this is the cost to raise x_j the rest of the way to 1). Also, the residual cost $\mathsf{residual}(x)$ in the original interpretation equals (in the reinterpreted algorithm) the minimum cost to solve the original problem but with weights c'.

This local-ratio reinterpretation is straightforward and intuitive for problems with $\{0, 1\}$ variables and a linear objective. But for problems whose variables take values in more general domains, it does not extend cleanly. For example, suppose a variable x_j takes values in $\{0, 1, 2, \ldots, u\}$. The algorithm cannot afford to reduce the weight c_j,

[9] For example, consider $\min\{x_1 + x_2 + x_3 : x_1 + x_2 \geq 1, x_1 + x_3 \geq 2\}$. If the greedy algorithm does the constraints in *either* order and chooses β maximally, it gives a solution of cost 4. In the dual $\max\{y_{12} + 2y_{13} : y_{12} + y_{13} \leq 1\}$, the only way to generate a solution of cost 2 is to set $y_{13} = 1$ and $y_{12} = 0$. If the primal constraint for y_{12} is considered first, y_{12} cannot be assigned a non-zero value. Instead, one should consider the dual variables for constraints for which steps were done, in the *reverse* order of those steps, raising each until a constraint is tight.

and then, at termination, set x_j to u for j with $c_j = 0$ (this can lose a factor of u in the approximation). Instead, one has to reinterpret the modified weight c'_j as a vector of weights $c'_j : \{1, \ldots, u\} \to \mathbb{R}_+$ where $c'_j(i)$ is the cost to raise x_j from $\max\{x_j, i-1\}$ to $\min\{x_j, i\}$ (initially $c'_j(i) = c_j$). When the original algorithm lowers x_j by β/c_j, reinterpret this as leaving x_j at zero, but lowering the non-zero $c'_j(i)$ with minimum i by β. At the end, take x_j to be the maximum i such that $c'_j(i) = 0$. We show next that this approach is doable (if less intuitive) for monotone covering.

At a high level, the local-ratio method requires only that the objective be decomposed into "locally approximable" objectives. The common weight-reduction presentation of local ratio described above gives one decomposition, but others have been used. A local-ratio analysis for an integer programming problem with non-$\{0,1\}$ variable domains, based on something like $\mathsf{residual}(x)$, is used in [8]. Here, the following decomposition (different than [8]) works:

Lemma 2. *Any algorithm returns a Δ-approximate solution x provided there exist $\{c^t\}$ and r such that*

(a) for any x, $c(x) = c(\mathbf{0}) + r(x) + \sum_{t=1}^{T} c^t(x)$,
(b) for all t, and any x and feasible x^*, $c^t(x) \leq c^t(x^*)\Delta$,
(c) the algorithm returns x such that $r(x) = 0$.

Proof. Let x^* be an optimal solution. Applying properties (a) and (c), then (b), then (a),

$$c(x) \;=\; c(\mathbf{0}) + \textstyle\sum_{t=1}^{T} c^t(x) \;\leq\; c(\mathbf{0})\Delta + \sum_{t=1}^{T} c^t(x^*)\Delta \;+\; r(x^*)\Delta \;=\; c(x^*)\Delta.$$
□

Next we describe how to use the proof of Thm. 1 (based on residual cost) to generate such a decomposition.

Let $\mathsf{distance}(x, y) = c(x \vee y) - c(x)$ (the cost to raise x to dominate y).

For any x, define $c^t(x) = \mathsf{distance}(x^{t-1}, x) - \mathsf{distance}(x^t, x)$, where x^t is Alg. 1's x after t calls to $\mathsf{step}()$.

Define $r(x) = \mathsf{distance}(x^T, x)$, where x^T is the algorithm's solution.

For linear c note $c^t(x) = \sum_j c_j \big|[0, x_j] \cap [x_j^{t-1}, x_j^t]\big|$, the cost for x "between" x^{t-1} and x^t.

Lemma 3. *These c^t and r have properties (a-c) from Lemma 2, so the algorithm gives a Δ-approximation.*

Proof. Part (a) holds because the sum in (a) telescopes to $\mathsf{distance}(\mathbf{0}, x) - \mathsf{distance}(x^T, x) = c(x) - c(\mathbf{0}) - r(x)$.

Part (c) holds because the algorithm returns x^T, and $r(x^T) = \mathsf{distance}(x^T, x^T) = 0$.

For (b), consider the tth call to $\mathsf{step}()$. Let β be as in that call.

The triangle inequality holds for $\mathsf{distance}()$, so, for any $\hat{x}$, $c^t(\hat{x}) \leq \mathsf{distance}_c(x^{t-1}, x^t) = c(x^t) - c(x^{t-1})$.

As proved in the proof of Thm. 1, $c(x^t) - c(x^{t-1})$ is at most $\beta\Delta$.

Also in the proof of Thm. 1, it is argued that $\beta \leq \mathsf{distance}(x^{t-1}, \cap_{S \in \mathcal{C}} S) - \mathsf{distance}(x^t, \cap_{S \in \mathcal{C}} S)$.

By inspection that argument holds for any $x^* \in \cap_{S \in C} S$, giving $\beta \leq \text{distance}(x^{t-1}, x^*) - \text{distance}(x^t, x^*)$.

The latter quantity is $c^t(x^*)$. Thus, $c^t(\hat{x}) \leq \beta\Delta \leq c^t(x^*)\Delta$. □

References

1. Albers, S.: On generalized connection caching. Theory of Computing Systems 35(3), 251–267 (2002)
2. Bansal, N., Buchbinder, N., Naor, J.: A primal-dual randomized algorithm for weighted paging. In: The forty-third IEEE symposium on Foundations Of Computer Science, pp. 507–517 (2007)
3. Bansal, N., Buchbinder, N., Naor, J.S.: Randomized competitive algorithms for generalized caching. In: The fourtieth ACM Symposium on Theory Of Computing, pp. 235–244. ACM, New York (2008)
4. Bar-Yehuda, R.: One for the price of two: A unified approach for approximating covering problems. Algorithmica 27(2), 131–144 (2000)
5. Bar-Yehuda, R., Bendel, K., Freund, A., Rawitz, D.: Local ratio: a unified framework for approximation algorithms. ACM Computing Surveys 36(4), 422–463 (2004)
6. Bar-Yehuda, R., Even, S.: A linear-time approximation algorithm for the weighted vertex cover problem. J. Algorithms 2(2), 198–203 (1981)
7. Bar-Yehuda, R., Even, S.: A local-ratio theorem for approximating the weighted vertex cover problem. Annals of Discrete Mathematics 25(27-46), 50 (1985)
8. Bar-Yehuda, R., Rawitz, D.: Efficient algorithms for integer programs with two variables per constraint 1. Algorithmica 29(4), 595–609 (2001)
9. Bar-Yehuda, R., Rawitz, D.: On the equivalence between the primal-dual schema and the local-ratio technique. SIAM Journal on Discrete Mathematics 19(3), 762–797 (2005)
10. Bertsimas, D., Vohra, R.: Rounding algorithms for covering problems. Mathematical Programming: Series A and B 80(1), 63–89 (1998)
11. Borodin, A., Cashman, D., Magen, A.: How well can primal-dual and local-ratio algorithms perform? In: Caires, L., Italiano, G.F., Monteiro, L., Palamidessi, C., Yung, M. (eds.) ICALP 2005. LNCS, vol. 3580, pp. 943–955. Springer, Heidelberg (2005)
12. Borodin, A., El-Yaniv, R.: Online computation and competitive analysis. Cambridge University Press, New York (1998)
13. Buchbinder, N., Naor, J.: Online primal-dual algorithms for covering and packing problems. In: Brodal, G.S., Leonardi, S. (eds.) ESA 2005. LNCS, vol. 3669, pp. 689–701. Springer, Heidelberg (2005)
14. Cao, P., Irani, S.: Cost-aware www proxy caching algorithms. In: the USENIX Symposium on Internet Technologies and Systems on USENIX Symposium on Internet Technologies and Systems table of contents, pp. 18–18 (1997)
15. Carr, R.D., Fleischer, L.K., Leung, V.J., Phillips, C.A.: Strengthening integrality gaps for capacitated network design and covering problems. In: The eleventh ACM-SIAM Symposium On Discrete Algorithms, Philadelphia, PA, USA, pp. 106–115. Society for Industrial and Applied Mathematics (2000)
16. Chrobak, M., Karloff, H., Payne, T., Vishwanathan, S.: New results on server problems. SIAM J. Discrete Math. 4(2), 172–181 (1991)
17. Chvatal, V.: A greedy heuristic for the set-covering problem. Mathematics of Operations Research 4, 233–235 (1979)
18. Cohen, E., Kaplan, H., Zwick, U.: Connection caching. In: The thirty-first ACM Symposium on Theory Of Computing, pp. 612–621 (1999)

19. Cohen, E., Kaplan, H., Zwick, U.: Connection caching under various models of communication. In: The twelfth ACM Symposium on Parallel Algorithms and Architectures, pp. 54–63 (2000)
20. Cohen, E., Kaplan, H., Zwick, U.: Connection caching: Model and algorithms. Journal of Computer and System Sciences 67(1), 92–126 (2003)
21. Dinur, I., Safra, S.: On the hardness of approximating minimum vertex cover. Annals of Mathematics, 162 (2005)
22. Dobson, G.: Worst-case analysis of greedy heuristics for integer programming with nonnegative data. Mathematics of Operations Research 7(4), 515–531 (1982)
23. Fiat, A., Karp, R.M., Luby, M., McGeoch, L.A., Sleator, D.D., Young, N.E.: Competitive paging algorithms. J. Algorithms 12, 685–699 (1991)
24. Fisher, M.L., Wolsey, L.A.: On the Greedy Heuristic for Continuous Covering and Packing Problems. SIAM Journal on Algebraic and Discrete Methods 3, 584–591 (1982)
25. Gonzales, T. (ed.): Approximation Algorithms and Metaheuristics, ch 4. (Greedy Methods). Tylor and Francis Books, CRC Press (2007)
26. Hall, N.G., Hochbaum, D.S.: A fast approximation algorithm for the multicovering problem. Discrete Applied Mathematics 15(1), 35–40 (1986)
27. Halldorsson, M.M., Radhakrishnan, J.: Greed is good: Approximating independent sets in sparse and bounded-degree graphs. In: The twenty-sixth ACM Symposium on Theory Of Computing, pp. 439–448 (1994)
28. Halperin, E.: Improved approximation algorithm for the vertex cover problem in graphs and hypergraphs. SIAM Journal on Computing 31(5), 1608–1623 (2002)
29. Hastad, J.: Some optimal inapproximability results. Journal of the ACM 48(4), 798–859 (2001)
30. Hayrapetyan, A., Swamy, C., Tardos, E.: Network design for information networks. In: The sixteenth ACM-SIAM Symposium On Discrete Algorithms, pp. 933–942 (2005)
31. Hochbaum, D.S.: Approximation algorithms for the set covering and vertex cover problems. SIAM Journal on Computing 11, 555–556 (1982)
32. Hochbaum, D.S.: Efficient bounds for the stable set, vertex cover, and set packing problems. Discrete Applied Mathematics 6, 243–254 (1983)
33. Hochbaum, D.S.: Approximation algorithms for the set covering and vertex cover problems. SIAM Journal on Computing 11, 555 (1982)
34. Hochbaum, D.S.: Approximation algorithms for NP-hard problems. PWS Publishing Co., Boston (1996)
35. Johnson, D.S.: Approximation algorithms for combinatorial problems. In: The fifth ACM Symposium On Theory Of Computing, vol. 25, pp. 38–49 (1973)
36. Karlin, A.R., Manasse, M.S., Rudolph, L., Sleator, D.D.: Competitive snoopy caching. Algorithmica 3, 77–119 (1988)
37. Khot, S., Regev, O.: Vertex cover might be hard to approximate to within 2-ε. Journal of Computer and System Sciences 74, 335–349 (2008)
38. Kolliopoulos, S.G., Young, N.E.: Approximation algorithms for covering/packing integer programs. Journal of Computer and System Sciences 71(4), 495–505 (2005)
39. Koufogiannakis, C., Young, N.E.: Distributed and parallel algorithms for weighted vertex cover and other covering problems. In: The twenty-eighth ACM symposium Principles of Distributed Computing (2009)
40. Lotker, Z., Patt-Shamir, B., Rawitz, D.: Rent, lease or buy: Randomized algorithms for multislope ski rental. In: The twenty fifth Symposium on Theoretical Aspects of Computer Science, pp. 503–514 (2008)

41. Lotker, Z., Patt-Shamir, B., Rawitz, D.: Rent, lease or buy: Randomized algorithms for multislope ski rental. In: Albers, S., Weil, P. (eds.) STACS, Internationales Begegnungs- und Forschungszentrum fuer Informatik (IBFI), Schloss Dagstuhl, Germany. Dagstuhl Seminar Proceedings, vol. 08001, pp. 503–514 (2008)
42. Lovasz, L.: On the ratio of optimal integral and fractional covers. Discrete Math. 13, 383–390 (1975)
43. McGeoch, L.A., Sleator, D.D.: A strongly competitive randomized paging algorithm. Algorithmica 6(1), 816–825 (1991)
44. Monien, B., Speckenmeyer, E.: Ramsey numbers and an approximation algorithm for the vertex cover problem. Acta Informatica 22, 115–123 (1985)
45. Orlin, J.B.: A faster strongly polynomial time algorithm for submodular function minimization. In: Fischetti, M., Williamson, D.P. (eds.) IPCO 2007. LNCS, vol. 4513, pp. 240–251. Springer, Heidelberg (2007)
46. Pritchard, D.: Approximability of sparse integer programs. Technical report, arxiv.org (2009), http://arxiv.org/abs/0904.0859
47. Raghavan, P., Snir, M.: Memory versus randomization in on-line algorithms. IBM Journal of Research and Development 38(6), 683–707 (1994)
48. Sleator, D.D., Tarjan, R.E.: Amortized efficiency of list update and paging rules. Communications of the ACM 28(2), 202–208 (1985)
49. Srinivasan, A.: Improved approximation guarantees for packing and covering integer programs. SIAM Journal on Computing 29, 648–670 (1999)
50. Srinivasan, A.: New approaches to covering and packing problems. In: The twelveth ACM-SIAM Symposium On Discrete Algorithms, pp. 567–576 (2001)
51. Vazirani, V.V.: Approximation algorithms. Springer, Heidelberg (2001)
52. Young, N.E.: The k-server dual and loose competitiveness for paging. Algorithmica 11, 525–541 (1994)
53. Young, N.E.: On-line file caching. Algorithmica 33(3), 371–383 (2002)

Limits and Applications of Group Algebras for Parameterized Problems

Ioannis Koutis[1,⋆] and Ryan Williams[2,⋆⋆]

[1] Computer Science Department, Carnegie Mellon University, Pittsburgh, PA, USA
jkoutis@cs.cmu.edu

[2] School of Mathematics, Institute for Advanced Study, Princeton, NJ, USA
ryanw@ias.edu

Abstract. The algebraic framework introduced in [Koutis, Proc. of the 35th ICALP 2008] reduces several combinatorial problems in parameterized complexity to the problem of detecting multilinear degree-k monomials in polynomials presented as circuits. The best known (randomized) algorithm for this problem requires only $O^*(2^k)$ time and oracle access to an arithmetic circuit, i.e. the ability to evaluate the circuit on elements from a suitable group algebra. This algorithm has been used to obtain the best known algorithms for several parameterized problems. In this paper we use communication complexity to show that the $O^*(2^k)$ algorithm is essentially *optimal* within this *evaluation oracle* framework. On the positive side, we give new applications of the method: finding a copy of a given tree on k nodes, a spanning tree with at least k leaves, a minimum set of nodes that dominate at least t nodes, and an m-dimensional k-matching. In each case we achieve a faster algorithm than what was known. We also apply the algebraic method to problems in exact counting. Among other results, we show that a combination of dynamic programming and a variation of the algebraic method can break the trivial upper bounds for exact parameterized counting in fairly general settings.

1 Introduction

The algebraic framework introduced in [14] reduces several parameterized problems including the k-path problem, the m-set k-packing problem, and several graph packing problems to the following ground problem:

The k-monomial detection problem. Given a (commutative) arithmetic circuit C over a set of variables X, decide whether the polynomial $P(X)$ represented by C contains a multilinear monomial of degree k.

In other words, construing $P(X)$ as a sum of products, we wish to know if the sum contains a degree-k monomial with no squares. Here, an arithmetic circuit is a directed acyclic graph with a single sink, sources labelled by variables from a set X, and multiplication and addition gate labels at all other nodes. The decision algorithm given in [14] solves the *odd* k-monomial detection problem where the additional assumption that $P(X)$ contains an odd number of multilinear

⋆ Research partially supported by NSF Grant CCF-0635257.

⋆⋆ Research partially supported by NSF Grant CCF-0832797.

S. Albers et al. (Eds.): ICALP 2009, Part I, LNCS 5555, pp. 653–664, 2009.

monomials is made. The technique used in [14] to reduce the parameterized packing problems to the odd k-monomial detection problem was extended and completed in [19] where the general k-term problem was essentially reduced to the odd problem by working over a larger group algebra.

The algebraic method of [14,19] is based on choosing a commutative algebra $\mathcal{A}$ and randomized assignments $X \to \mathcal{A}$ such that (i) squares (and by commutativity, non-multilinear monomials) evaluate to the zero element of $\mathcal{A}$, and (ii) some multilinear monomial does not evaluate to zero, with good probability. For instance, the algorithm presented in [19] uses certain assignments from the group algebra $GF(2^{3+\log_2 k})[\mathbb{Z}_2^k]$ (for the definition of a group algebra, see the Appendix). Elements in $GF(2^\ell)[\mathbb{Z}_2^k]$ can be described in space $O(\ell 2^k)$ and the complexity of addition and multiplication in the algebra is $O^*(\ell 2^k)$.[1] This gives an $O^*(2^k t)$ algorithm for k-monomial detection, where t is the number of gates in C. The algorithm requires only *oracle access* to an "extended version" C' of C, and it can be implemented as one evaluation of C' over $GF(2^\ell)[\mathbb{Z}_2^k]$, or as $O^*(\ell 2^k)$ evaluations of C' over small polynomials in $\mathbb{Z}[x]$.

The technique yields the fastest known parameterized algorithms for the k-path problem, the m-set k-packing problem (packing k sets, each of size m), and more. In Section 2 we also show how to obtain faster algorithms for finding a copy of a given tree on k nodes, a spanning tree with k leaves, a minimum set of nodes that dominate at least t nodes in a graph, and an m-dimensional k-matching, all by simple reductions to k-monomial detection. Some of these algorithms match the best known upper bounds for the original NP-hard problems, i.e. the Hamiltonian path problem, and the k-set (n/k)-packing problem, in the sense that further improvement on the parameterized algorithms will imply further progress on the original problems. Thus, an intriguing and natural question is whether k-monomial detection can be solved faster, by evaluating circuits over a more exotic $\mathcal{A}$, with a significant reduction in the lengths of descriptions for elements in $\mathcal{A}$ (much less than 2^k), and the related complexity of arithmetic in $\mathcal{A}$. If so, this would have tremendous implications in exact algorithms.

The answer is, unfortunately, negative. In Section 3 we use communication complexity to show that for any commutative algebra $\mathcal{A}$ used to evaluate the circuit C, the lengths of elements in $\mathcal{A}$ must be at least $\Omega(2^k/k)$ in order to perform k-monomial detection. For all of the applications listed above, commutativity of multiplication is required.[2] Thus the $O^*(2^k)$ algorithm for k-monomial detection is optimal in some sense, and further progress on the relevant parameterized problems are only via algorithms that exploit specific properties of circuits for k-path and set packing, or perform different kinds of operations altogether. Much research in complexity theory focuses on understanding the limits of restricted algorithmic frameworks. Such cases include the study of deterministic

[1] Throughout the paper the $O^*()$ notation hides factors polynomial in the instance size n and the parameter k.

[2] We state our result for commutative algebras, for the sake of clarity and coherence with our algorithmic results that all use commutativity. However, our lower bound holds in the non-commutative setting as well, under the appropriate definition.

oracle-based algorithms for the computation of convex volumes [7], of randomized oracle-based volume computation algorithms [16], of local minimum search algorithms for black-box functions [1], or in the area of parameterized kernelization [5]. Apart from the obvious potential of saving unneeded efforts on upper bounds, this type of research often reveals surprising connections.

Although the algebraic method is superior to older approaches such as color coding [3] and the divide-and-color approaches [11,6] for *decision* problems, its potential has not been explored in the context of the related *counting* problem:

Exact multilinear k-monomial counting. Given a (commutative) arithmetic circuit C describing an n-variate polynomial P over $\mathbb{R}$, compute the sum of the coefficients of the degree-k multilinear monomials in P.

A simple enumeration-based algorithm for the problem runs in $O^*(2^k\binom{n}{k})$ time. Given the #W[1] hardness of exact counting of k-paths [8] (a special case), an $O(f(k)poly(n))$ algorithm for the counting problem is unlikely to exist. However, certain recent results indicate that improvements may be possible. Björklund et al. gave an $O^*\left(\binom{n}{k/2}\right)$ algorithm for computing the number of k-paths [4]. Using rectangular permanents, Vassilevska and Williams gave algorithms for counting small subgraphs with large independent sets [18]. In Section 4 we show that a *combination* of enumeration-based algorithm and the algebraic method is able to break the naive $\binom{n}{k}$ upper bound for this problem when the polynomial P is a product of two polynomials P_1, P_2 each of which have total degree less than k. One application of this is an $O^*(n^{mk/2})$ algorithm for counting k-packings of m-sets over a universe of size n.

2 Faster Parameterized Algorithms

To further demonstrate the power of k-monomial detection, we show how it can be used to obtain faster randomized algorithms for the following problems:

k-Tree. Given a tree T on k nodes and a graph G on n nodes, decide if there is a (not necessarily induced) copy of T in G.

k-Leaf Spanning Tree. Given an undirected graph G on n nodes, decide if G contains a spanning tree with at least k leaves.

t-Dominating Set. Given a graph $G = (V, E)$, find a minimum set of nodes S that dominate at least t nodes in the graph. That is, $|S \cup N(S)| \geq t$ where $N(S) = \{v \mid (u, v) \in E, u \in S\}$.

m-Dimensional k-Matching. Given mutually disjoint sets U_i, for $i = 1, \ldots, m$, and a collection $\mathcal{C}$ of m-tuples from $U_1 \times \ldots \times U_m$, decide whether $\mathcal{C}$ contains a sub-collection of k mutually disjoint m-tuples.

Each of these can be solved by formulating them as k-monomial detection instances in some way. To the best of our knowledge, the only other algorithm we know for k-Tree follows from the color-coding method [3], running in $O^*((2e)^k)$ time. For k-Leaf Spanning Tree, the best known algorithm is deterministic and runs in $O^*(4^k)$ time [10]. The best known (randomized) algorithm for t-Dominating Set runs in time $O^*((4+\varepsilon)^t)$ [12]. The best known (randomized) algorithm for the k m-Dimensional Matching Problem runs in time $O^*(2^{mk})$ [14]. In addition, in all these problems, given an algorithm for the decision version of the problem, standard reductions can be used to recover an algorithm for the search version, with the same exponential dependence on the parameter.

Before we proceed to showing the reductions we consider the following generalization of the results in [14,19].

Lemma 1. *Let $P(X, z)$ be a polynomial represented by a commutative arithmetic circuit C. The existence of a term of the form $z^t Q(X)$ in $P(X, z)$, where $Q(X)$ is a multilinear monomial of degree at most k, can be decided in time $O^*(2^k t^2 |C|)$ and space $O(t|C|)$.*

The idea of the proof is to assign values $\mathcal{X}$ from the appropriate algebra $\mathcal{A}$ to the variables X, in exactly the same fashion as in [19], evaluate the polynomial $P(\mathcal{X}, z)$ symbolically, then read off the answer from the coefficient $Q(\mathcal{X}) \in \mathcal{A}$ of z^t. The details will appear in the full version of the paper.

Theorem 1. *The k-Tree problem can be solved in $O^*(2^k)$ time.*

Proof. Suppose T is a k-node tree we wish to find in G. Let the nodes of T be $\{1, \dots, k\}$, and let the nodes of G be $\{1, \dots, n\}$. We define an arithmetic circuit $C_{T,i,j}(x_1, \dots, x_n)$ inductively, for all $i \in [k]$ and $j \in [n]$.

- If $|V(T)| = 1$, simply define $C_{T,i,j} := x_j$.
- If $|V(T)| > 1$, let $T_{i,1}, \dots, T_{i,\ell}$ be the connected subtrees of T remaining after node i is removed from T. For all $t = 1, \dots, \ell$, let $n_{i,t} \in [k]$ be the (unique) node in $T_{i,t}$ that is a neighbor of i in T. Define

$$C_{T,i,j} := \prod_{t=1}^{\ell} \left(\sum_{j' : (j,j') \in E(G)} x_j \cdot C_{T_{i,t}, n_{i,t}, j'} \right).$$

Observe that the size of $C_{T,i,j}(x_1, \dots, x_n)$ is at most $O(|V(T)| \cdot |E(G)|)$. The polynomial $C_{T,i,j}$ enumerates those homomorphisms that map nodes of T into nodes of G, such that node i in T is mapped to node j in G. Each monomial represents the range of some homomorphism. More precisely, the monomial $x_{j_1} \cdots x_{j_k}$ is present in the polynomial if and only if there is a homomorphism where the nodes of T are mapped to the vertices $\{j_1, \dots, j_k\} \subseteq V$ of G. These mappings are all homomorphisms, since $i' \in V(T)$ can only be mapped to $j' \in V(G)$ if $(i, i') \in E(T)$, $(j, j') \in E(G)$, and $i \in V(T)$ is already mapped to $j \in V(G)$.

Now consider the sum-product expansion of $Q = \sum_{j \in V(G), i \in V(T)} C_{T,i,j}$. Q is a sum over all connected subgraphs S of G on k vertices, where each monomial corresponds to the range of some homomorphism from T into S. Note Q also includes homomorphisms that map distinct nodes of T to a common node in G. However, those homomorphisms correspond precisely to monomials with *squares* in them. That is, the multilinear monomials of Q correspond to homomorphisms that map all nodes in T to distinct nodes in G, i.e. an isomorphic copy of T in G. Therefore, by calling k-monomial detection on the arithmetic circuit defined by Q, we can detect whether G contains a copy of T in $O^*(2^k)$ time. □

Theorem 2. *The k-Leaf Spanning Tree problem can be solved in $O^*(2^k)$ time.*

Proof. To find a spanning tree with at least k leaves, first observe that it suffices to find a connected subgraph S with at least k nodes of degree one. To get a spanning tree for G, compute a spanning tree T over G with the nodes and edges of S removed. Connect T and S by an edge, and take the spanning tree to be the union of these two subgraphs (removing any extraneous edges that may form cycles). This was observed in [10].

It suffices to show how to detect if G has a connected subgraph with at least k degree one nodes. Let the vertices of G be $\{1, \ldots, n\}$. We define an arithmetic circuit $C_{k,t,i}(x_1, \ldots, x_n, z)$ inductively.

- Define $C_{1,1,i} := x_i \cdot z$. If $t < k$ or $k \leq 0$, define $C_{k,t,i} := 1$.
- For $k > 1$, define

$$C_{k,t,i} := \sum_{t'=1}^{t} \sum_{k'=1}^{k} \sum_{j:(i,j)\in E} (C_{k',t',j} \cdot C_{k-k',t-t',i}).$$

By induction, one can prove that the monomials of $C_{k,t,i}$ correspond to the connected subgraphs on t nodes which include node i of G and have up to k degree ≤ 1 nodes. Each such subgraph has a monomial corresponding to its leaves. A multilinear monomial of degree k in the coefficient of z^k implies that there is a subgraph rooted at i that has k nodes of degree one. Testing whether the circuit $Q = \sum_{t=k+1}^{n} \sum_{i=1}^{n} C_{k,t,i}$ has a multilinear monomial in the coefficient of z^k determines if G has a connected subgraph with at least k degree-one nodes. By Lemma 1 this can be done. Finally, note that the size of Q as an arithmetic circuit is poly(n, k). □

Theorem 3. *The t-Dominating Set problem can be solved in $O^*(2^t)$ time.*

Proof. Let the vertex set be $V = \{1, \ldots, n\}$ and $X = \{x_1, \ldots, x_n\}$, where x_i corresponds to the vertex i. Consider the polynomial

$$P(X, z) = \left(\sum_{i \in V} \left((1 + z \cdot x_i) \cdot \prod_{j\ :\ (i,j)\in E} (1 + z \cdot x_j) \right) \right)^k,$$

where z is an extra indeterminate. P is a sum of monomials in which each monomial of the form $z^t x_{i_1} \cdots x_{i_t}$ for *distinct* i_j appears if and only if the t nodes $\{i_1, \ldots, i_t\}$ are dominated by a set of at most k nodes. In addition, every other term of the form $z^t Q(X)$ contains a square since $Q(X)$ has total degree t. Then, the proof follows from Lemma 1 and trials with increasing values of k. □

Theorem 4. *The m-Dimensional k-Matching problem can be solved in time $O^*(2^{(m-1)k})$.*

Proof (Sketch). Encode each element u in $U = \bigcup_{i=2}^{m} U_i$ by a variable $x_u \in X$. Encode each m-tuple $t = (u_1, \ldots, u_m) \in \mathcal{C} \subseteq U_1 \times \cdots \times U_m$ by the monomial

$M_t = \prod_{i=2}^{m} x_{u_i}$. Assume $U_1 = \{u_{1,1}, \ldots, u_{1,n}\}$, and let $T_j \subseteq \mathcal{C}$ denote the subset of m-tuples whose first coordinate is $u_{1,j}$. Consider the polynomial

$$P(X, z) = \prod_{j=1}^{n} \left(1 + \sum_{t \in T_j} (z \cdot M_t) \right),$$

where z is an extra indeterminate. The coefficient of z^k is a polynomial $Q(X)$ which contains a multilinear $((m-1)k)$-term if and only if $\mathcal{C}$ contains a k-matching. The proof follows from Lemma 1. □

3 Lower Bound for Multilinear Detection

We now investigate whether the approach to finding k-paths in [14,19] can be improved upon further. One burning question from this work is whether is it possible to determine if an arbitrary arithmetic circuit C has a multilinear k-monomial in much less time than $O(2^k|C|)$, by evaluating C over a more exotic algebraic structure. We shall prove that this is *not* the case. For any commutative $\mathcal{A}$, we show that if it can be used to detect multilinear monomials in an arithmetic circuit (even randomly), then $|\mathcal{A}| \geq 2^{\Omega(2^k/\sqrt{k})}$. This implies a lower bound of $\Omega(\frac{2^k}{\sqrt{k}}|C|)$. That is, the group algebras used in [14,19] (which have $|\mathcal{A}| \leq 2^{O(2^k \log k)}$) are essentially optimal for their purpose: commutativity of $\mathcal{A}$ is required for all the applications of k-monomial detection that we have shown.

At a high level, our proof uses hypothetical fast multilinear monomial detection in order to design communication protocols that are too efficient to exist. Let us informally recall some notions from communication complexity. Let $f : \{0,1\}^n \times \{0,1\}^n \to \{0,1\}$. Suppose two parties wish to compute $f(x, y)$, but one party is given x (the x-party), the other is given y (the y-party). The parties must communicate bits about their inputs in order to compute f. A *simultaneous public-coin protocol for f on n-bits* is specified by a pair of functions (g_1, g_2), and works as follows on all n-bit strings. Initially, the x-party and y-party see a common string z chosen at random from a distribution independent of x and y (to maximize the parties' capabilities, we assume the distribution is uniform). The x-party computes $g_1(x, z)$, the y-party computes $g_2(y, z)$, and both send their answers to a third party. We require that the third party holding the two messages can compute f (with high probability), on all x and y of length n. The *randomized public-coin simultaneous communication complexity of* (g_1, g_2) is the maximum length $L(n)$ of a message sent in the protocol (g_1, g_2) over all n-bit strings. The corresponding communication complexity of f is the minimum $L(n)$ achieved by any n-bit protocol (g_1, g_2) for f.

The set disjointness function $DISJ_n(x, y)$ on $x, y \in \{0,1\}^n$ is defined to be $\vee_{i=1}^{n}(x_i \wedge y_i)$. We utilize the following fact:

Theorem 5 ([15],p.79). *The randomized public-coin communication complexity of $DISJ_n$ is $\Omega(n)$.*

Theorem 6. *Let $\mathcal{A}$ be a commutative semiring. Suppose there is a distribution $\mathcal{D}$ on elements from $\mathcal{A}$, an element $e \in \mathcal{A}$, and constants $d_1, d_2 \in [0,1]$, $d_1 < d_2$ such that for every circuit $C(x_1, \ldots, x_n)$ representing a degree-k polynomial:*

- *C has a multilinear monomial*
  $\Longrightarrow \Pr_{(e_1,\ldots,e_n)\in\mathcal{D}^n}[C(e_1,\ldots,e_n) = e] \le d_1$
- *C does not have a multilinear monomial*
  $\Longrightarrow \Pr_{(e_1,\ldots,e_n)\in\mathcal{D}^n}[C(e_1,\ldots,e_n) = e] \ge d_2$.

Then $|\mathcal{A}| \ge 2^{\Omega(2^k/\sqrt{k})}$. Furthermore, for every k there is a circuit C with $n = k$ for which this lower bound holds.

In other words, the group algebra utilized in [19] is optimal within poly(k) factors: any other algebra could only yield a slightly better asymptotic upper bound. As a consequence it is not possible to solve Hamilton Path much faster than $O(2^n)$ by solving k-monomial detection over a more interesting algebra.

Proof. Using multilinear detection, we design a protocol for $DISJ_N$. Let $N > 0$ and $k = n = \log N + \frac{1}{2}\log\log N + c$ where $c > 0$ is sufficiently large. Without loss of generality, assume k is even. Let $\mathcal{S} = \{S_1, \ldots, S_\ell\}$ be an intersecting set system over $[k]$ such that $|S_i| = k/2$, for all i, and $\ell \ge N$. That is, we have $S_i \cap S_j \ne \varnothing$ for all i, j. For example, we may take $\mathcal{S} = \{S \subseteq [k] \mid |S| = k/2 \;\&\; 1 \in S\}$. Observe that for this collection, when c is large enough we have

$$|\mathcal{S}| = \binom{k-1}{k/2} \ge \Omega\left(\frac{2^k}{\sqrt{k}}\right) = \Omega\left(\frac{2^{\log N + \frac{1}{2}\log\log N + c}}{\sqrt{\log N + \frac{1}{2}\log\log N + c}}\right) \ge N$$

by Stirling's inequality. Hence $\ell \ge N$.

We now give a protocol for set disjointness, assuming the existence of a good $\mathcal{A}$. Let $a, b \in \{0,1\}^N$. For all $i = 1, \ldots, \ell$, define the monomials

$$P_i = \prod_{j \in S_i} x_j \quad \text{and} \quad Q_i = \prod_{j \notin S_i} x_j.$$

Now define an arithmetic circuit

$$C(x_1, \ldots, x_k) = \left(\sum_{i=1}^{n} a_i P_i\right) \cdot \left(\sum_{i=1}^{n} b_i Q_i\right).$$

Note that C represents a homogeneous polynomial of degree k. We claim that C has a square-free monomial if and only if $DISJ_N(a,b) = 1$. This follows from the fact that the monomial $P_i \cdot Q_j$ has a square if and only if $i \ne j$.[3]

[3] Note this is the portion of the argument where commutativity of multiplication is used. In the non-commutative setting, the condition is that the monomial $P_i \cdot Q_j$ contains two instances of the same variable in its product. This condition is precisely the same as that required for all the algorithmic applications.

Let $C_a = \sum_{i=1}^{n} a_i P_i$ and $C_b = \sum_{i=1}^{n} b_i Q_i$. To get a communication protocol for set disjointness, the x-party uses public randomness to obtain $e_i \in \mathcal{A}$ for each x_i, computes $v = C_a(e_1, \ldots, e_k) \in \mathcal{A}$, and sends an $O(\log|\mathcal{A}|)$-bit string corresponding to v. Similarly, the y-party obtains all $e_i \in \mathcal{A}$, evaluates $w = C_b(e_1, \ldots, e_k)$, and sends w. The third party outputs *disjoint* if and only if $e = v \cdot w$ (where $e \in \mathcal{A}$ has the properties of the theorem's hypothesis).

Under the hypotheses of the theorem (and repeating the protocol $O(1)$ times to obtain a good estimate of $\Pr_{(e_1,\ldots,e_n)\in\mathcal{D}^n}[C(e_1, \ldots, e_n) = e]$), the above is a correct public-coin protocol for $DISJ_n$. It follows from Theorem 5 that $\log|\mathcal{A}| \geq cN$, for some constant $c > 0$. Finally, note that $N \geq \Omega\left(\frac{2^k}{\sqrt{k}}\right)$. □

4 Exact Parameterized Multilinear Counting

We will base our algorithm on the following lemma.

Lemma 2. *Let $P(x_1, \ldots, x_n)$ be a commutative polynomial with coefficients from $\mathbb{R}$, given by an arithmetic circuit C. For all $\{i_1, \ldots, i_k\} \subseteq [n]$, the coefficient of the monomial $x_{i_1} \cdots x_{i_k}$ in P can be computed in time $O^*(2^k)$. It follows that the restriction of P to its multilinear monomials of degree k can be computed in time $O^*(2^k\binom{n}{k})$.*

Proof (Sketch). Without loss of generality, we can assume that all gates of C have two inputs. Consider the coefficient of $x_1 \cdots x_k$. Note that it is easy to find the sub-circuit C' that computes $P' = P(x_1, \ldots, x_k, 0, \ldots, 0)$. We will view C' as a circuit acting on polynomials. Let $\mathcal{R}(Q)$ denote the operation that restricts the polynomial Q to its multilinear terms. The key idea is to place an $\mathcal{R}$-gate in the output of every multiplication and addition gate of C'. Note that this does not change the coefficient of $x_1 \cdots x_k$ in the output of C', because all the terms that get deleted produce only non-multilinear terms in the rest of the computation.

Note now that the inputs of all multiplication and addition gates in C' contain only multilinear terms. There are at most 2^k such terms. It follows that the complexity of computing $\mathcal{R}(P_1 + P_2)$ is $O^*(2^k)$. It remains to show that the complexity of computing $\mathcal{R}(P_1 \cdot P_2)$ is also $O^*(2^k)$. First observe that the operation can be written as the sum of at most k^2 operations of the form $\mathcal{R}(Q_1 \cdot Q_2)$ where Q_1, Q_2 are k-variate *homogeneous* polynomials consisting of those multilinear terms in P_1, P_2 that have degrees *exactly* d and $t - d$ respectively, for $t \in [1, k]$ and $d \in [1, t]$. Let $Q_1'(z)$ and $Q_2'(z)$ be the polynomials resulting by the substitution $x_i \to z^{2^{i-1}}$ in Q_1, Q_2. The degrees of Q_1' and Q_2' are at most $2^k - 1$, hence $Q_1' \cdot Q_2'$ can be computed in $O^*(2^k)$ via the Fast Fourier Transform. Now observe that $\mathcal{R}(Q_1 \cdot Q_2)$ can be recovered by the restriction of $Q_1' \cdot Q_2'$ to its terms of the form z^j where the binary form of j contains exactly t bits. □

This lemma gives an upper bound for the exact multilinear k-monomial counting problem. We show that this upper bound can be improved upon for polynomials of the form $P \cdot Q$, where the multilinear terms of P, Q both have total degree smaller than k. We need the following (properly adapted) theorem from the theory of error correcting codes [17][Chap. 5, Theorem 8], also used in the deterministic construction of k-wise probability spaces [2][Proposition 6.5].

Theorem 7. *There is a 0-1 matrix M with n columns and m rows with $2^m \leq 2(n+1)^{\lfloor k/2 \rfloor}$, such that every k columns of M are linearly independent over $\mathbb{Z}_2$. Moreover, the matrix can be constructed in $O(n^{k/2+1})$ time.*

We are now ready to prove the key theorem of this section. We use the group algebra $\mathbb{R}[\mathbb{Z}_2^k]$.

Theorem 8. *Let $P \cdot Q \in \mathbb{R}[X]$ be an n-variate polynomial over $\mathbb{R}$, such that P and Q contain only multilinear monomials of degree at most d. The sum of the coefficients of the multilinear k-terms in $P \cdot Q$, can be computed in time $O^*(2^d\binom{n}{d} + n^{\lfloor k/2 \rfloor})$.*

Proof. Applying the algorithm of Lemma 2 to the circuits for P and Q, we can assume that P, Q contain only multilinear terms. Furthermore, assume (for now) that all multilinear terms of $P \cdot Q$ have degree exactly k. Let m be the number of rows of the matrix M in Theorem 7. Note $m \approx \frac{k}{2} \log n$. The algorithm is as follows. Let $A : X \to \mathbb{R}[\mathbb{Z}_2^m]$ be the assignment $x_i \mapsto (\mathbf{v_0} + M_i)$, where $M_i \in \mathbb{Z}_2^m$ is the ith column of M, and $\mathbf{v_0} \in \mathbb{Z}_2^m$ is the zero vector. Let $\bar{A}$ be the assignment $x_i \mapsto (\mathbf{v_0} - M_i)$. Compute $\Pi = P(A) \cdot Q(\bar{A})$ over $\mathbb{R}[\mathbb{Z}_2^m]$, and return the coefficient of $\mathbf{v_0}$ in Π.

Let us first consider the algorithm's correctness. Every term in $P \cdot Q$ is a multiple of the form $t_P(A) \cdot t_Q(\bar{A})$ where t_P, t_Q are multilinear monomials in P and Q respectively. If t_P and t_Q both contain a variable x_i, then $t_P(A) \cdot t_Q(\bar{A})$ is a multiple of $(\mathbf{v_0} + M_i)(\mathbf{v_0} - M_i)$ which is equal to 0, since $\mathbf{v_0}^2 = M_i^2 = \mathbf{v_0}$. If $t_P t_Q$ is multilinear, then because columns of M_i are independent and by Lemma 2.2 of [14], the coefficient of $\mathbf{v_0}$ in $t_P(A) \cdot t_Q(\bar{A})$ equals 1. This proves correctness.

Now we concentrate on the time complexity. We first note that we can compute each product $t(A)$ inductively, as follows. Assume we have evaluated $t'(A)$ where t' consists of the first d' factors of t. By Lemma 2.2 of [14] the product $t'(A)$ is a sum of $2^{d'}$ distinct vectors from $\mathbb{Z}_2^m$. Hence, we can store $t'(A)$ in a sparse form, where only the identities of $2^{d'}$ vectors are kept. The identity of each of the 2^m vectors of $\mathbb{Z}_2^m$ requires just m bits. Given this sparse storage, a sparse form product of the form $t'(A)(\mathbf{v_0} + M_i)$ can be evaluated fairly simply in time $O^*(2^{d'})$. Hence, $t(A)$ can be evaluated in sparse form in time $\sum_{i=1}^{d} 2^i = O(2^d)$. For the sum $P(A)$, we keep a single vector of size 2^m, which we update every time a term $t(A)$ is computed. The discussion is very similar for the assignment $\bar{A}$. Hence, $P(A)$ and $Q(A)$ can be computed in time $O(2^d\binom{n}{d})$, as there are at most $\binom{n}{d}$ monomials in P and Q.

Finally, we want $P(A) \cdot Q(\bar{A})$. Since $P(A)$ and $Q(\bar{A})$ are both elements of $\mathbb{R}[\mathbb{Z}_2^m]$, their multiplication can be performed via a Fast Fourier Transform method in time $O^*(2^m)$ [19]. Finally note that if $P \cdot Q$ contains multilinear terms of degree different than k, these can be treated by the introduction of a free indeterminate z, as in Lemma 1. □

To show an application of Theorem 8 we consider the m-set k-packing problem, over a universe of n elements and an instance consisting of t sets. It is clear that the number of k-packings can be counted exactly in time $O^*(\binom{t}{k})$. For large enough t, we can do better. The "algebraization" of this problem gives a simple polynomial I which is a product of sums [13], satisfying the requirements of Theorem 8. This gives the following corollary.

Corollary 1. *The number of m-set k-packings can be counted in $O^*(n^{\lceil mk/2 \rceil})$.*

It is also interesting that the above construction can be used to compute the parity of the coefficient without restrictions in the type of the circuit, as stated in the following theorem whose proof is given in the full version of the paper.

Theorem 9. *Let C be a circuit describing a degree-k polynomial P over $\mathbb{Z}$. The parity of the sum of the coefficients of the multilinear monomials in P can be computed in time $O^*(n^{k/2})$ and poly($|C|$) space.*

5 Final Remarks

Using the techniques of this paper the following theorem can also be shown.

Theorem 10. *Let A be a $k \times n$ matrix with entries from a commutative semiring S. The permanent of A can be computed in time and space $O^*(2^k)$, assuming a unit cost for the addition and multiplication operations over S. If $S = \mathbb{R}$, the space complexity can be reduced to poly(n).*

To the best of our knowledge the only previous related results were an $O^*(3^k)$ time, $O^*(2^k)$ space algorithm in [9], and more recently an $O^*(4^k)$ time poly(n) space algorithm for matrices over arbitrary commutative semirings [18]. The proof of this result along with other proofs omitted in this paper will appear in the full version of the paper. A number of interesting questions remain open:

- Theorem 7 can be used to get a weak derandomization of our earlier results, and solve more sophisticated problems. To what extent can we improve on the construction of M in this result?
- Is there an algorithm for multilinear k-monomial counting that improves over the naive upper bounds for *general* circuits?
- Can the algebraic method be applied to improve the time complexity for *approximate* counting? For this question, color coding is still the best known approach.

References

1. Aldous, D.: Minimization algorithms and random walk on the d-cube. Annals of Probability 2, 403–413 (1983)
2. Alon, N., Babai, L., Itai, A.: A fast and simple randomized parallel algorithm for the maximal independent set problem. J. Algorithms 7(4), 567–583 (1986)

3. Alon, N., Yuster, R., Zwick, U.: Color coding. Journal of the ACM 42(4), 844–856 (1995)
4. Björklund, A., Husfeldt, T., Kaski, P., Koivisto, M.: The fast intersection transform with applications to counting paths. CoRR, abs/0809.2489 (2008)
5. Bodlaender, H.L., Downey, R.G., Fellows, M.R., Hermelin, D.: On problems without polynomial kernels (extended abstract). In: Aceto, L., Damgård, I., Goldberg, L.A., Halldórsson, M.M., Ingólfsdóttir, A., Walukiewicz, I. (eds.) ICALP 2008, Part I. LNCS, vol. 5125, pp. 563–574. Springer, Heidelberg (2008)
6. Chen, J., Lu, S., Sze, S.-H., Zhang, F.: Improved algorithms for path, matching, and packing problems. In: Proc. 18th ACM-SIAM Symposium on Discrete Algorithms (SODA 2007), pp. 298–307 (2007)
7. Elekes, G.: A geometric inequality and the complexity of computing volume. Discrete & Computational Geometry 1, 289–292 (1986)
8. Flum, J., Grohe, M.: The parameterized complexity of counting problems. In: FOCS 2002: Proceedings of the 43rd Symposium on Foundations of Computer Science, Washington, DC, USA, p. 538. IEEE Computer Society, Los Alamitos (2002)
9. Kawabata, T., Tarui, J.: On complexity of computing the permanent of a rectangular matrix. IEICE Trans. Fundamentals, E82-A
10. Kneis, J., Langer, A., Rossmanith, P.: A new algorithm for finding trees with many leaves. In: Hong, S.-H., Nagamochi, H., Fukunaga, T. (eds.) ISAAC 2008. LNCS, vol. 5369, pp. 270–281. Springer, Heidelberg (2008)
11. Kneis, J., Mölle, D., Richter, S., Rossmanith, P.: Divide-and-color. In: WG: Graph-Theoretic Concepts in Computer Science, 32nd International Workshop, pp. 58–67 (2006)
12. Kneis, J., Mölle, D., Rossmanith, P.: Partial vs. complete domination: t-dominating set. In: van Leeuwen, J., Italiano, G.F., van der Hoek, W., Meinel, C., Sack, H., Plášil, F. (eds.) SOFSEM 2007. LNCS, vol. 4362, pp. 367–376. Springer, Heidelberg (2007)
13. Koutis, I.: A faster parameterized algorithm for set packing. Information Processing Letters 94(1), 4–7 (2005)
14. Koutis, I.: Faster algebraic algorithms for path and packing problems. In: Aceto, L., Damgård, I., Goldberg, L.A., Halldórsson, M.M., Ingólfsdóttir, A., Walukiewicz, I. (eds.) ICALP 2008, Part I. LNCS, vol. 5125, pp. 575–586. Springer, Heidelberg (2008)
15. Kushilevitz, E., Nisan, N.: Communication Complexity. Cambridge Univ. Press, Cambridge (1996)
16. Rademacher, L., Vempala, S.: Dispersion of mass and the complexity of randomized geometric algorithms. In: FOCS '06: Proceedings of the 47th Annual IEEE Symposium on Foundations of Computer Science, Washington, DC, USA, pp. 729–738. IEEE Computer Society, Los Alamitos (2006)
17. Sloane, N.J., MacWilliams, F.J.: The Theory of Error-Correcting Codes. North-Holland, Amsterdam (1983)
18. Vassilevska, V., Williams, R.: Finding, minimizing, and counting weighted subgraphs. In: STOC (to appear, 2009)
19. Williams, R.: Finding paths of length k in $O^*(2^k)$. Information Processing Letters 109(6), 301–338 (2009)

A Appendix Group Algebras of $\mathbb{Z}_2^k$

Let $\mathbb{Z}_2^k$ be the group consisting of k-dimensional 0-1 vectors with the group multiplication being entry-wise addition modulo 2. The group algebra $R[\mathbb{Z}_2^k]$, where R is a ring, is the set of all linear combinations of the form $\sum_{v\in\mathbb{Z}_2^k} a_v v$ where $a_v \in K$. The addition operator of $R[\mathbb{Z}_2^k]$ is defined by

$$\sum_{v\in\mathbb{Z}_2^k} a_v v + \sum_{v\in\mathbb{Z}_2^k} b_v v = \sum_{v\in\mathbb{Z}_2^k} (a_v + b_v)v.$$

Multiplication by a scalar $\alpha \in R$ is defined by

$$\alpha \sum_{v\in\mathbb{Z}_2^k} a_v v = \sum_{v\in\mathbb{Z}_2^k} (\alpha a_v)v.$$

The multiplication operator of $R[\mathbb{Z}_2^k]$ is defined by

$$\left(\sum_{v\in\mathbb{Z}_2^k} a_v v\right) \cdot \left(\sum_{u\in\mathbb{Z}_2^k} b_u u\right) = \sum_{v,u\in\mathbb{Z}_2^k} (a_v b_u)(uv).$$

It can be verified that $R[\mathbb{Z}_2^k]$ is commutative, provided that R is commutative.

Sleep with Guilt and Work Faster to Minimize Flow Plus Energy

Tak-Wah Lam[1,⋆], Lap-Kei Lee[1], Hing-Fung Ting[1], Isaac K.K. To[2,⋆⋆], and Prudence W.H. Wong[2,⋆⋆]

[1] Department of Computer Science, University of Hong Kong
{twlam,lklee,hfting}@cs.hku.hk
[2] Department of Computer Science, University of Liverpool
{isaacto,pwong}@liverpool.ac.uk

Abstract. In this paper we extend the study of flow-energy scheduling to a model that allows both sleep management and speed scaling. Our main result is a sleep management algorithm called IdleLonger, which works online for a processor with one or multiple levels of sleep states. The design of IdleLonger is interesting; among others, it may force the processor to idle or even sleep even though new jobs have already arrived. IdleLonger works in both clairvoyant and non-clairvoyant settings. We show how to adapt two existing speed scaling algorithms AJC [15] (clairvoyant) and LAPS [9] (non-clairvoyant) to the new model. The adapted algorithms, when coupled with IdleLonger, are shown to be $O(1)$-competitive clairvoyant and non-clairvoyant algorithms for minimizing flow plus energy on a processor that allows sleep management and speed scaling.

The above results are based on the traditional model with no limit on processor speed. If the processor has a maximum speed, the problem becomes more difficult as the processor, once overslept, cannot rely on unlimited extra speed to catch up the delay. Nevertheless, we are able to enhance IdleLonger and AJC so that they remain $O(1)$-competitive for flow plus energy under the bounded speed model. Non-clairvoyant scheduling in the bounded speed model is left as an open problem.

1 Introduction

Speed scaling, flow and energy. Energy consumption has become a major issue in the design of microprocessors, especially for battery-operated devices. Many modern processors support dynamic speed scaling to reduce energy usage. Recently there is a lot of theory research on online job scheduling taking speed scaling and energy usage into consideration (see [10] for a survey). The challenge arises from the conflicting objectives of providing good quality of service and conserving energy. Among others, the study of minimizing flow time plus energy has attracted much attention [1,3,4,5,9,14,15]. The results to date are based on a speed scaling model in which a processor, when running at speed s, consumes energy at the rate of s^α, where α is typically 3 (the cube-root rule [7]). Most

⋆ T.W. Lam is partially supported by HKU Grant 7176104.
⋆⋆ I.K.K. To & P.W.H. Wong are partially supported by EPSRC Grant EP/E028276/1.

S. Albers et al. (Eds.): ICALP 2009, Part I, LNCS 5555, pp. 665–676, 2009.

research assumes the traditional *infinite speed model* [16] where any speed is allowed; some considers the more realistic *bounded speed model* [8], which imposes a maximum processor speed T.

Total flow time is a commonly used QoS measure for job scheduling. The flow time (or simply flow) of a job is the time elapsed since the job arrives until it is completed. In the online setting, jobs with arbitrary sizes arrive at unpredictable times. They are to be run on a processor which allows preemption without penalty. To understand the tradeoff between flow and energy, Albers and Fujiwara [1] initiated the study of minimizing a linear combination of total flow and total energy. The intuition is that, from an economic viewpoint, users are willing to pay a certain (say, ρ) units of energy to reduce one unit of time. By changing the units of time and energy, one can further assume that $\rho = 1$ and thus would like to optimize flow plus energy.

Under the infinite speed model, Albers and Fujiwara [1] considered jobs of unit size, and their work was extended to jobs of arbitrary sizes by Bansal, Pruhs and Stein [5]. The BPS algorithm scales the speed as a function of unfinished work and is $O((\frac{\alpha}{\ln \alpha})^2)$-competitive for minimizing flow plus energy. Bansal et al. [3] later adapted the BPS algorithm to the bounded speed model; the competitive ratio remains $O((\frac{\alpha}{\ln \alpha})^2)$ if the online algorithm is given extra speed. Recently, Lam et al. [15] gave a new algorithm AJC whose speed function depends on the number of unfinished jobs. AJC avoids the extra speed requirement and improves the competitive ratio to $O(\frac{\alpha}{\ln \alpha})$. Recall that α is typically equal to 3. Then the competitive ratios of BPS are estimated to be 7.9 and 11.9, and AJC 3.25 and 4 under the infinite and bounded speed model, respectively. More recently, Bansal et al. [4] further showed that AJC can be adapted to be 3-competitive, independent of α. All these results assume clairvoyance, i.e., the size of a job is known when the job arrives.

The non-clairvoyant setting, where job size is known only when the job is completed, is practically important. Chan et al. [9] have recently given a non-clairvoyant speed scaling algorithm LAPS that is $O(\alpha^3)$-competitive for total flow plus energy in the infinite speed model.

Sleep management. In earlier days, energy reduction was mostly achieved by allowing a processor to enter a low-power *sleep* state, yet waking up requires extra energy. In the (embedded systems) literature, there are different energy-efficient strategies to bring a processor to sleep during a period of zero load [6]. This is an online problem, usually referred to as *dynamic power management.* The input is the length of the period, known only when the period ends. There are several interesting results with competitive analysis (e.g., [2, 11, 13]). In its simplest form, the problem assumes the processor is in either the *awake* state or the *sleep* state. The awake state always requires a static power $\sigma > 0$. To have zero energy usage, the processor must enter the sleep state, but a wake-up back to the awake state requires $\omega > 0$ energy. In general, there can be multiple intermediate sleep states, which demand some static power but less wake-up energy.

It is natural to study job scheduling on a processor that allows both sleep states and speed scaling. More specifically, a processor in the awake state can run at any speed $s \geq 0$ and consumes energy at the rate $s^\alpha + \sigma$, where $\sigma > 0$ is static power and s^α is the dynamic power[1]. Here job scheduling requires two components: a *sleep management algorithm* to determine when to sleep or work, and a *speed scaling algorithm* to determine which job and at what speed to run. Notice that sleep management here is not the same as in dynamic power management; in particular, the length of a sleep or idle period is part of the optimization (rather than the input). Adding a sleep state actually changes the nature of speed scaling. Assume no sleep state, running a job slower is a natural way to save energy. Now one can also save energy by sleeping more and working faster later. It is even more complicated when flow is concerned. Prolonging a sleeping period by delaying job execution can save energy, yet it also incurs extra flow. Striking a balance is not trivial. In the theory literature, the only relevant work is by Irani et al. [12]; they studied deadline scheduling on a processor with one sleep state and infinite speed scaling. They showed an $O(1)$-competitive algorithm to minimize the energy for meeting the deadlines of all jobs.

Our contributions. This paper initiates the study of flow-energy scheduling that exploits both speed scaling and multiple sleep states. We give a sleep management algorithm called IdleLonger, which works for a processor with one or multiple levels of sleep states. IdleLonger works in both clairvoyant and non-clairvoyant settings. We adapt the clairvoyant speed scaling algorithm AJC [15] and the non-clairvoyant algorithm LAPS [9] to take the static power σ into consideration. Under the infinite speed model, these adapted algorithms together with IdleLonger are shown to be $O(1)$-competitive for minimizing flow plus energy in the clairvoyant and non-clairvoyant settings, respectively. More precisely, the ratios are $O(\frac{\alpha}{\ln \alpha})$ and $O(\alpha^3)$ (recall that α is a constant).

For the bounded speed model, the problem becomes more difficult since the processor, once overslept, cannot rely on unlimited extra speed to catch up the delay. Nevertheless, we are able to enhance IdleLonger and AJC to observe the maximum processor speed. They remain $O(1)$-competitive for flow plus energy under the bounded speed model.

Sleep management algorithm IdleLonger. When the processor is sleeping, it is natural to delay waking up until sufficient jobs have arrived. The non-trivial case is when the processor is idle (i.e., awake but at zero speed), IdleLonger has to determine when to start working again or go to sleep. At first glance, if some new jobs arrive while the processor is idle, the processor should run the jobs immediately so as to avoid extra flow. Yet this would allow the adversary to easily keep the processor awake, and it is difficult to achieve $O(1)$-competitiveness. In an idle period, IdleLonger considers the (static) energy and flow accumulated during the period as two competing quantities. Only if the flow exceeds the energy, IdleLonger would start to work. Otherwise, IdleLonger will remain idle until the energy reaches to a certain level; then the processor goes to sleep even in the presence of jobs.

[1] Static power is dissipated due to leakage current and is independent of processor speed, and dynamic power is due to dynamic switching loss and increases with the speed.

Analysis framework. Apparently, a sleep management algorithm and a speed scaling algorithm would affect each other; analyzing their relationship and their total cost could be a complicated task. Interestingly, the results of this paper stem from the fact that we can isolate the analysis of these algorithms. We divide the total cost (flow plus energy) into two parts, *working cost* (incurred while working on jobs) and *inactive cost* (incurred at other times). We upper bound the inactive cost of IdleLonger independent of the speed scaling algorithm. For the working cost, although it does depend on both algorithms, our potential analysis of the speed scaling algorithms reveals that the dependency on the sleep management algorithm is limited to a simple quantity called *inactive flow*, which is the flow part of the inactive cost. Intuitively, large inactive flow means many jobs are delayed due to prolonged sleep, and hence the processor has to work faster later to catch up, incurring a higher working cost. It is easy to minimize inactive flow at the sacrifice of the energy part of the inactive cost. IdleLonger is designed to maintain a good balance between them. In conclusion, coupling IdleLonger with AJC and LAPS, we obtain competitive algorithms for flow plus energy.

Organization of the paper. Section 1.1 defines the model formally. Sections 2 and 3 focus on the infinite speed model and discuss the sleep management algorithm IdleLonger and two speed scaling algorithms. Finally, Section 4 presents our results on the bounded speed model.

1.1 Model and Notations

The input is a sequence of jobs arriving online. We denote the release time and work requirement (or size) of a job J as $r(J)$ and $w(J)$, respectively.

Speed and power. We first consider the setting with one sleep state. At any time, a processor is in either the *awake* state or the *sleep* state. In the former, the processor can run at any speed $s \geq 0$ and demands power in the form $s^\alpha + \sigma$, where $\alpha > 1$ and $\sigma > 0$ are constants. We call s^α the dynamic power and σ the static power. In the sleep state, the speed is zero and the power is zero. State transition requires energy; without loss of generality, we assume a transition from the sleep state to the awake state requires an amount ω of energy, and the reverse takes zero energy. To simplify our work, we assume state transition takes no time.

Next we consider the setting with $m > 1$ levels of sleep. A processor is in either the *awake* state or the *sleep-i* state, where $1 \leq i \leq m$. The awake state is the same as before, demanding static power σ and dynamic power s^α. For convenience, we let $\sigma_0 = \sigma$. The sleep-m state is the only "real" sleep state, which has static power $\sigma_m = 0$; other sleep-i states have decreasing positive static power σ_i such that $\sigma_0 > \sigma_1 > \sigma_2 > \cdots > \sigma_{m-1} > \sigma_m = 0$. We denote the wake-up energy from the sleep-i state to the awake state as ω_i. Note that $\omega_m > \omega_{m-1} > \cdots > \omega_1 > 0$.

It is useful to differentiate two types of awake state: with zero speed and with positive speed. The former is called *idle* state and the latter is *working* state.

Flow and energy. Consider any schedule of jobs. The flow $F(J)$ of a job J is the time elapsed since it arrives and until it is completed. The total flow is

$F = \sum_J F(J)$. Note that $F = \int_0^\infty n(t)\, \mathrm{d}t$, where $n(t)$ is the number of unfinished jobs at time t. Based on this view, we divide F into two parts: F_{w} is the flow incurred during time intervals of working state, and F_{i} for idle or sleep state. The energy usage is also divided into three parts: W denotes the energy due to wake-up transitions, E_{i} is the idling energy (static power consumption in the idle or intermediate sleep state), and E_{w} is the working energy (static and dynamic power consumption in the working state). Our objective is to minimize the *total cost* $G = F_{\mathrm{w}} + F_{\mathrm{i}} + E_{\mathrm{i}} + E_{\mathrm{w}} + W$. We call $F_{\mathrm{w}} + E_{\mathrm{w}}$ the *working cost*, and $F_{\mathrm{i}} + E_{\mathrm{i}} + W$ the *inactive cost*.

2 Sleep Management Algorithm IdleLonger

This section presents a sleep management algorithm called IdleLonger that determines when the processor should sleep, idle, and work (with speed > 0). IdleLonger can be coupled with any *speed scaling algorithm*, which specifies which job and at what speed the processor should run when the processor is working. As a warm-up, we first consider the case with a single sleep state. Afterwards, we consider the general case of multiple sleep states.

In this section, we derive an upper bound of the inactive cost of IdleLonger independent of the choice of the speed scaling algorithm. Section 3 will present two speed scaling algorithms for the clairvoyant and non-clairvoyant settings, respectively, and analyze their working costs when they are coupled with IdleLonger. In conclusion, putting IdleLonger and each of these two speed scaling algorithms together, we can show that both the inactive cost and working cost are $O(1)$ times of the total cost of the optimal offline algorithm OPT.

2.1 Sleep Management Algorithm for a Single Sleep State

When the processor is in the working state and sleep state, it is relatively simple to determine the next transition. In the former, the processor keeps on working as long as there is an unfinished job; otherwise switch to the idle state. In the sleep state, we avoid waking up immediately after a new job arrives as this requires energy. It is natural to wait until the new jobs have accumulated enough flow, say, at least the wake-up energy ω, then we let the processor to switch to working state direct. Below we refer the flow accumulated due to new jobs over a period of idle or sleep state as the *inactive flow* of that period.

When the processor is in idle state, it is non-trivial when to switch to the sleep or working state. Intuitively, the processor should not stay in idle state too long, because it consumes energy (at the rate of σ) but does not get any work done. Yet to avoid frequent wake-up in future, the processor should not sleep immediately. Instead the processor should wait for possible job arrival and sleep only after the idling energy (i.e., σ times the length of idling interval) reaches the wake-up energy ω. When a new job arrives in the idle state, a naive idea is to let the processor switch to the working state to process the job immediately; this avoids accumulating inactive flow. Yet this turns out to be a bad strategy as it becomes too difficult to sleep; e.g., the adversary can use some tiny jobs sporadically, then the processor would never accumulate enough idling energy to sleep.

It is perhaps counter-intuitive that IdleLonger always prefers to idle a bit longer, and it can switch to the sleep state even in the presence of unfinished jobs. The idea is to consider the inactive flow and idling energy at the same time. Note that when an idling period gets longer, both the inactive flow and idling energy increase, but at different rates. We imagine that these two quantities are competing with each other.

> The processor switches from the idle state to the working state once the inactive flow catches up with the idling energy. If the idling energy has exceeded ω before the inactive flow catches up with the idling energy, the processor switches to the sleep state.

Below is a summary of the above discussion. For simplicity, IdleLonger is written in a way that it is being executed continuously. In practice, we can rewrite the algorithm such that the execution is driven by discrete events like job arrival, job completion and wake-up.

Algorithm 1. IdleLonger(A): A is any speed scaling algorithm

At any time t, let $n(t)$ be the number of unfinished jobs at t.
In working state: If $n(t) > 0$, keep working on jobs according to the algorithm A; else (i.e., $n(t) = 0$), switch to idle state.
In idle state: Let $t' \leq t$ be the last time in working state ($t' = 0$ if undefined). If the inactive flow over $[t', t]$ equals $(t - t')\sigma$, then switch to working state;
Else if $(t - t')\sigma = \omega$, switch to sleep state.
In sleep state: Let $t' \leq t$ be the last time in working state ($t' = 0$ if undefined). If the inactive flow over $[t', t]$ equals ω, switch to working state.

Below we upper bound the inactive cost of IdleLonger (the working cost will be dealt with in Section 3). It is useful to define three types of time intervals. An $\mathrm{I_w}$-interval is a maximal interval in idling state with a transition to the working state at the end, and similarly an $\mathrm{I_s}$-interval for that with a transition to the sleep state. Furthermore, an $\mathrm{IS_w}$-interval is a maximal interval comprising an $\mathrm{I_s}$-interval, a sleeping interval, and finally a wake-up transition. As the processor starts in the sleep state, we allow the first $\mathrm{IS_w}$-interval containing no $\mathrm{I_s}$-interval.

Consider a schedule of IdleLonger(A). Recall that the inactive cost is composed of W (wake-up energy), F_i (inactive flow), and E_i (idling energy). We further divide E_i into two types: $\boldsymbol{E}_{\mathbf{iw}}$ is the idling energy incurred in all $\mathrm{I_w}$-intervals, and $\boldsymbol{E}_{\mathbf{is}}$ for all $\mathrm{I_s}$-intervals.

By the definition of IdleLonger, we have the following property.

Property 1. (i) $F_\mathrm{i} \leq W + E_\mathrm{iw}$, and (ii) $E_\mathrm{is} = W$.

Therefore, the inactive cost of IdleLonger, defined as $W + F_\mathrm{i} + E_\mathrm{iw} + E_\mathrm{is}$, is at most $3W + 2E_\mathrm{iw}$. The non-trivial part is to upper bound W and E_iw. Our main result is stated below. For the optimal offline algorithm OPT, we divide its total cost G^* into two parts: W^* is the total wake-up energy, and $C^* = G^* - W^*$ (i.e., the total flow plus the working and idling energy).

Theorem 1. $W + E_{\mathrm{iw}} \le C^* + 2W^*$.

Corollary 1. *The inactive cost of* IdleLonger *is at most* $3C^* + 6W^*$.

The rest of this section is devoted to proving Theorem 1. Note that W is the wake-up energy consumed at the end of all $\mathrm{IS_w}$-intervals, and E_{iw} is the idling energy of all $\mathrm{I_w}$-intervals. All these intervals are disjoint. Below we show a charging scheme such that, for each $\mathrm{IS_w}$-interval, we charge OPT a cost at least ω, and for each $\mathrm{I_w}$-interval, we charge OPT at least the idling energy of this interval. Thus, the total charge to OPT is at least $W + E_{\mathrm{iw}}$. On the other hand, we argue that the total charge is at most $C^* + 2W^*$. Therefore, $W + E_{\mathrm{iw}} \le C^* + 2W^*$.

The charging scheme for an $\mathrm{IS_w}$-interval $[t_1, t_2]$ is as follows. The target is at least ω.

Case 1. If OPT switches from or to the sleep state in $[t_1, t_2]$, we charge OPT the cost ω of the first wake-up in $[t_1, t_2]$ (if it exists) or of the last wake-up before t_1.

Case 2. If OPT is awake throughout $[t_1, t_2]$, we charge OPT the static energy $(t_2 - t_1)\sigma$. Note that in an $\mathrm{IS_w}$-interval, IdleLonger has an idle-sleep transition, and hence $(t_2 - t_1)\sigma > \omega$.

Case 3. If OPT is sleeping throughout $[t_1, t_2]$, we charge OPT the inactive flow (i.e., the flow incurred by new jobs) over $[t_1, t_2]$. In this case, OPT and IdleLonger have the same amount of inactive flow during $[t_1, t_2]$, which equals ω (because IdleLonger wakes up at t_2).

For an $\mathrm{I_w}$-interval, we use the above charging scheme again. The definition of $\mathrm{I_w}$-interval allows the scheme to guarantee a charge of $(t_2 - t_1)\sigma$ instead of ω. Specifically, as an $\mathrm{I_w}$-interval ends with an idle-working transition, the inactive flow accumulated in $[t_1, t_2]$ is $(t_2 - t_1)\sigma$, and the latter cannot exceed ω. Therefore, the charge of Case 1, which equals ω, is at least $(t_2 - t_1)\sigma$. Case 2 charges exactly $(t_2 - t_1)\sigma$. For Case 3, we charge OPT the inactive flow during $[t_1, t_2]$. Note that OPT and IdleLonger accumulate the same inactive flow, which is $(t_2 - t_1)\sigma$.

Summing over all $\mathrm{I_w}$- and $\mathrm{IS_w}$-intervals, we have charged OPT at least $W + E_{\mathrm{iw}}$. On the other hand, since all these intervals are disjoint, in Cases 2 and 3, the charge comes from non-overlapping flow and energy of C^*. In Case 1, each OPT's wake-up from the sleep state is charged for ω at most twice, thus the total charge is at most $2W^*$. In conclusion, $W + E_{\mathrm{iw}} \le C^* + 2W^*$.

2.2 Sleep Management Algorithm for $m \ge 2$ Levels of Sleep States

We extend the previous sleep management algorithm to allow intermediate sleep states, which demand less idling (static) energy than the idling state, and also less wake-up energy than the final sleep state (i.e., sleep-m state). We treat the sleep-m state as the only sleep state in the single-level setting, and adapt the transition rules of the idling state for the intermediate sleep states. The key idea is again to compare inactive flow against idling energy continuously. To ease our discussion, we treat the idle state as the sleep-0 state with wake-up energy $\omega_0 = 0$.

Algorithm 2. IdleLonger(A): A is any speed scaling algorithm

At any time t, let $n(t)$ be the number of unfinished jobs at t.
In working state: If $n(t) > 0$, keep working on the jobs according to the algorithm A; else if $n(t) = 0$, switch to idle state.
In sleep-j state, where $0 \leq j \leq m-1$: Let $t' \leq t$ be the last time in the working state, and let t'', where $t' \leq t'' \leq t$, be the last time switching from sleep-$(j-1)$ state to sleep-j state. If the inactive flow over $[t', t]$ equals $(t - t'')\sigma_j + \omega_j$, then wake up to the working state;
Else if $(t - t'')\sigma_j = (\omega_{j+1} - \omega_j)$, switch to sleep-$(j+1)$ state.
In sleep-m state: Let $t' \leq t$ be the last time in the working state. If the inactive flow over $[t', t]$ equals ω_m, then wake up to the working state.

When we analyze the multi-level algorithm, the definition of W (total wake-up cost) and F_{i} (total inactive flow) remain the same, but E_{is} and E_{iw} have to be generalized. Below we refer a maximal interval during which the processor is in a particular sleep-j state, where $0 \leq j \leq m$, as a *sleep interval* or more specifically, a sleep-j interval. Note that all sleep intervals, except sleep-m intervals, demand idling (static) energy. We denote $\boldsymbol{E_{\mathrm{iw}}}$ as the idling energy for all sleep intervals that end with a wake-up transition, and $\boldsymbol{E_{\mathrm{is}}}$ the idling energy of all sleep intervals ending with a (deeper) sleep transition.

IdleLonger imposes a rigid structure of sleep intervals. Define $\ell_j = (\omega_{j+1} - \omega_j)/\sigma_j$. A sleep-$j$ interval can appear only after a sequence of lower level sleep intervals, which starts with an sleep-0 interval of length ℓ_0, followed by a sleep-1 interval of length ℓ_1, ..., and finally a sleep-$(j-1)$ interval of length ℓ_{j-1}. Consider a maximum sequence of such sleep intervals that ends with a transition to the working state. We call the entire time interval enclosed by this sequence an $\mathrm{IS_w}[j]$-interval for some $0 \leq j \leq m$ if the deepest (also the last) sleep subinterval is of level j. It is useful to observe the following lemma about an $\mathrm{IS_w}[j]$-interval. Its proof is left in the full paper.

Lemma 1. *Consider any* $\mathrm{IS_w}[j]$*-interval* $[t_1, t_2]$, *where* $0 \leq j \leq m$. *Assume that the last sleep-j (sub)interval is of length ℓ. Then,* $\omega_j + \ell\sigma_j \leq \omega_k + (t_2 - t_1)\sigma_k$ *for any* $0 \leq k \leq m$.

It is not hard to see that the rigid sleeping structure of IdleLonger allows us to maintain Property 1 as before. That is, (i) $F_{\mathrm{i}} \leq W + E_{\mathrm{iw}}$, and (ii) $E_{\mathrm{is}} = W$. Thus, the inactive cost, which is equal to $F_{\mathrm{i}} + E_{\mathrm{iw}} + E_{\mathrm{is}} + W$, is still at most $3W + 2E_{\mathrm{iw}}$. In the rest of this section, we prove that W and E_{iw} have the same upper bound as before.

Theorem 2. *In the setting of $m \geq 2$ sleep states,* $W + E_{\mathrm{iw}} \leq C^* + 2W^*$.

To account for W and E_{iw}, it suffices to look at all $\mathrm{IS_w}[j]$-intervals, where $0 \leq j \leq m$. For each $\mathrm{IS_w}[j]$-interval, we show how to charge OPT a cost $\omega_j + \ell\sigma_j$, where ℓ is length of the deepest sleep subinterval (it is useful to recall that $\omega_0 = 0$ and $\sigma_m = 0$). Then we argue that the total cost charged is at least $W + E_{\mathrm{iw}}$ and at most $C^* + 2W^*$.

Without loss of generality, we can assume that in a maximal interval $[r_1, r_2]$ that OPT is not working, if OPT has ever slept (in sleep-1 or deeper sleep state), then $[r_1, r_2]$ contains only one sleep transition, which occurs at r_1, and the processor remains in the same sleep state until r_2.

Charging scheme. Consider any $\mathrm{IS_w}[j]$-interval $[t_1, t_2]$, where $0 \leq j \leq m$. Let ℓ be the length of the sleep-j (sub)interval in this interval.

Case 1. If OPT has ever switched from or to the sleep-1 or deeper sleep state in $[t_1, t_2]$, let $k \geq 1$ be the deepest sleep level involved in the entire interval. Note that OPT uses static energy at least $(t_2 - t_1)\sigma_k$ during $[t_1, t_2]$. We charge OPT the sum of $(t_2 - t_1)\sigma_k$ and ω_k (in view of a wake-up from sleep-k state inside $[t_1, t_2]$ or after t_2; if there is no-wake up after t_2, then we charge OPT the first wake-up). By Lemma 1, this charge is at least $\omega_j + \ell\sigma_j$.

Case 2. If OPT is working or idle throughout $[t_1, t_2]$, we charge OPT the static energy $(t_2 - t_1)\sigma_0$, which, by Lemma 1, is at least $\omega_j + \ell\sigma_j$.

Case 3. If OPT is sleeping (at any level except zero) throughout $[t_1, t_2]$, we charge OPT the inactive flow over $[t_1, t_2]$. Note that OPT has the same amount of inactive flow as IdleLonger. By definition of a wake-up transition in IdleLonger, the inactive flow equals $\omega_j + \ell\sigma_j$.

Since $\mathrm{IS_w}[j]$-intervals are all disjoint, the flow and idling (static) energy charged to OPT by Cases 1, 2 and 3 come from different parts of C^*. For Case 1, each of OPT's wake-up from a sleep state is charged at most twice. Thus, $W + E_{\mathrm{iw}} \leq C^* + 2W^*$, completing the proof of Theorem 2.

3 Clairvoyant and Non-clairvoyant Speed Scaling

We consider both the clairvoyant and non-clairvoyant settings; in the latter, job size is only known when the job completes. IdleLonger can actually work in both settings as its decision does not depend on the job size. When there is no sleep state and the power function is in the form of s^α, [15] and [9] gave respectively a clairvoyant speed scaling algorithm AJC and a non-clairvoyant speed scaling algorithm LAPS that are $O(1)$-competitive for flow plus energy. These algorithms always run the jobs at a speed proportional to $n(t)^{1/\alpha}$, where $n(t)$ is the number of unfinished jobs at time t. In the sleep setting, when the processor is working, it requires at least the static power σ; if σ is large, running at a speed comparable to $n(t)^{1/\alpha}$ would be too slow to be cost effective as the dynamic power could be way smaller than σ. Indeed AJC and LAPS have unbounded competitive ratios no matter what sleep management algorithm is used. This section shows how to analyze the following simple adaptations of AJC and LAPS to the sleep setting, and upper bound their working costs in terms of OPT's total cost.

Clairvoyant algorithm SAJC. At any time t, SAJC runs the job with the shortest remaining work at the speed $(n(t) + \sigma)^{1/\alpha}$.

Non-clairvoyant algorithm SLS. At any time t, SLS runs at speed $(1 + \frac{3}{\alpha})(n(t) + \sigma)^{1/\alpha}$, and runs the $\lceil (\frac{1}{2\alpha})n(t) \rceil$ unfinished jobs with the latest release times by splitting the speed equally among these jobs.

The analysis of SAJC (resp. SLS) is valid no matter what (non-clairvoyant) sleep management algorithm Slp is being used together. Ideally we want to upper bound the working cost of SAJC and SLS solely in terms of the total cost of OPT, yet this is not possible as the working cost also depends on Slp. Below we give an analysis in which the dependency on Slp is bounded by the inactive flow incurred by Slp. More specifically, let G_{w} and F_{i} be respectively the working cost and the inactive flow of Slp(SAJC) (resp. Slp(SLS)). Again, we use C^* to denote the total cost of OPT minus the wake-up energy; the latter is denoted by W^*. We will show that $G_{\mathrm{w}} = O(C^* + F_{\mathrm{i}})$.

Let us look at a simple case. If Slp always switches to working state whenever there are unfinished jobs, then $F_{\mathrm{i}} = 0$. In this case we can easily adapt the analysis of [15] (resp. [9]) to bound G_{w} in terms of C^* only. However, the inactive cost of Slp may be unbounded in this case. On the other hand, consider a sleep management algorithm that prefers to wait for more jobs before waking up to work (e.g., IdleLonger). Then SAJC (resp. SLS) would start at a higher speed and G_{w} can be much larger than C^*. Roughly speaking, the excess is due to the fact that the online algorithm is sleeping while OPT is working. Note that the cost to catch up the work lagged behind increases at a rate depending on $n(t)$. This motivates us to bound the excess in terms of F_{i}. Based on this idea, we can adapt the potential analyses of AJC [15] and LAPS [9] to show Theorem 3 below, where $\beta = 2/(1 - \frac{\alpha-1}{\alpha^{\alpha/(\alpha-1)}})$ and $\gamma = (1 + (1 + \frac{3}{\alpha})^{\alpha}) \le 1 + e^3$. Together with the results on inactive cost of IdleLonger (Property 1 and Corollary 1), it implies that the clairvoyant algorithm IdleLonger(SAJC) is $(2\beta+2)$-competitive for flow plus energy, and the non-clairvoyant algorithm IdleLonger(SLS) is $((4\alpha^3 + \alpha)\gamma + 3)$-competitive for flow plus energy. Detailed proofs will be given in the full paper.

Theorem 3. (i) *With respect to* Slp(SAJC), $G_{\mathrm{w}} \le \beta C^* + (\beta - 2)F_{\mathrm{i}}$. (ii) *With respect to* Slp(SLS), $G_{\mathrm{w}} \le (4\alpha^3 + 1)\gamma C^* + (\alpha - 1)\gamma F_{\mathrm{i}}$.

Corollary 2. *In the setting of single sleep state or multiple sleep states,*

(i) *the total cost of the clairvoyant algorithm* IdleLonger(SAJC) *is at most* $(2\beta + 2)$ *times of the total cost of OPT, and*
(ii) *the total cost of the non-clairvoyant algorithm* IdleLonger(SLS) *is at most* $((4\alpha^3 + \alpha)\gamma + 3)$ *times of the total cost of OPT.*

4 Bounded Speed Model

This section extends the sleep management algorithm IdleLonger and the clairvoyant speed scaling algorithm SAJC to the bounded speed model. We consider the setting where the processor speed is upper bounded by a constant $T > 0$, and there are $m \ge 1$ levels of sleep states. We show that the total cost (comprising inactive and working cost) of IdleLonger(SAJC) is $O(1)$ times of the optimal offline algorithm OPT.

Adaptation. In the bounded speed model, IdleLonger (see Section 2) still works and the inactive cost is $O(1)$ times of OPT's total cost. However, IdleLonger often allows a long sleep, then a speed scaling algorithm, without the capability to speed up arbitrarily, cannot always catch up the progress of OPT and may have unbounded working cost. Thus, we adapt IdleLonger to wake up earlier, especially when too many new jobs have arrived. To this end, we add one more wake-up condition to IdleLonger. Recall that $\sigma(=\sigma_0)$ is the static power in the working state.

> In the sleep-j state, where $0 \leq j \leq m$, if the number of unfinished jobs exceeds σ, the processor wakes up to the working state.

Recall that SAJC runs at the speed $(n(t)+\sigma)^{1/\alpha}$, where $n(t)$ is the number of unfinished jobs at time t. To adapt SAJC to the bounded speed model, we simply cap the speed at T. I.e., at any time t, the processor runs at the speed $\min\{(n(t)+\sigma)^{1/\alpha}, T\}$.

Inactive cost of IdleLonger. The rigid structure of sleep intervals remains the same as before, and the inactive cost is still at most $3W + 2E_{\rm iw}$, where W is the wake-up energy and $E_{\rm iw}$ is the idling energy incurred in those idling or intermediate sleep intervals that end with a wake-up transition (see Section 2 for details). However, due to the additional wake-up rule, IdleLonger has a slightly worse bound on W plus $E_{\rm iw}$. Our main result is stated in Theorem 4. Again, W^* denotes the wake-up energy of OPT, and C^* is the total cost of OPT minus W^*.

Theorem 4. (i) $W + E_{\rm iw} \leq C^* + 3W^*$. (ii) *The inactive cost of* IdleLonger *is at most* $3C^* + 9W^*$.

To prove Theorem 4(i), we extend the charging scheme in Section 2.2 to show that for each $\mathrm{IS_w}[j]$-interval, OPT can be charged with a cost at least $\omega_j + \ell\sigma_j$, where ℓ is the length of the deepest sleep subinterval (recall that $\omega_0 = 0$, $\sigma_0 = \sigma$ and $\sigma_m = 0$). The three cases of the old charging scheme remain the same, except that Case 3 is restricted to $\mathrm{IS_w}[j]$-intervals where IdleLonger wakes up at the end due to excessive inactive flow. We supplement Case 3 with a new scheme (Case 4) to handle $\mathrm{IS_w}[j]$-intervals with wake-ups due to more than σ unfinished jobs.

Charging scheme – Case 4. If OPT is sleeping (at any level except zero) throughout an $\mathrm{IS_w}[j]$-interval $[t_1, t_2]$, and IdleLonger has accumulated more than σ unfinished jobs at t_2, we consider two scenarios to charge OPT, depending on $n_o(t_1)$, the number of unfinished jobs in OPT at t_1.

(a) Suppose $n_o(t_1) \geq \sigma_0$. We charge OPT the inactive flow of these jobs over $[t_1, t_2]$, which is at least $(t_2 - t_1)\sigma_0$. By Lemma 1, this charge is at least $\omega_j + \ell\sigma_j$.

(b) Suppose $n_o(t_1) < \sigma_0$. Note that OPT stays in a sleep-k state, for some $k \geq 1$, in the entire interval and uses static energy $(t_2 - t_1)\sigma_k$ during $[t_1, t_2]$. We charge OPT the sum of $(t_2 - t_1)\sigma_k$ and ω_k (in view of OPT's first wake-up after t_2, which must exist because new jobs have arrived within $[t_1, t_2]$). By Lemma 1, this charge is at least $\omega_j + \ell\sigma_j$.

In conclusion, we are able to charge OPT, for each $\mathrm{IS_w}[j]$-interval, a cost at least $\omega_j + \ell\sigma_j$. Therefore, the sum of the charges to all $\mathrm{IS_w}[j]$-intervals is at least $W + E_{\mathrm{iw}}$. On the other hand, recall that Case 1 has a total charge at most $2W^*$. Case 2, 3 and 4(a) have a total charge at most C^*. We can argue that OPT is charged by Case (4b) with a cost at most W^*; details are left in the full paper. Then we have $W + E_{\mathrm{iw}} \leq C^* + 3W^*$. And Theorem 4(ii) follows directly. In the full paper, we will adapt the potential analysis of AJC [15] and show that the working cost of SAJC is still $O(\frac{\alpha}{\ln \alpha})$ times OPT's total cost. Therefore, IdleLonger(SAJC) remains $O(\frac{\alpha}{\ln \alpha})$-competitive for flow plus energy.

References

1. Albers, S., Fujiwara, H.: Energy-efficient algorithms for flow time minimization. TALG 3(4) (2007)
2. Augustine, J., Irani, S., Swany, C.: Optimal power-down strategies. In: FOCS, pp. 530–539 (2004)
3. Bansal, N., Chan, H.L., Lam, T.W., Lee, L.K.: Scheduling for speed bounded processors. In: Aceto, L., Damgård, I., Goldberg, L.A., Halldórsson, M.M., Ingólfsdóttir, A., Walukiewicz, I. (eds.) ICALP 2008, Part I. LNCS, vol. 5125, pp. 409–420. Springer, Heidelberg (2008)
4. Bansal, N., Chan, H.L., Pruhs, K.: Speed scaling with an arbitrary power function. In: SODA, pp. 693–701 (2009)
5. Bansal, N., Pruhs, K., Stein, C.: Speed scaling for weighted flow time. In: SODA, pp. 805–813 (2007)
6. Benini, L., Bogliolo, A., de Micheli, G.: A survey of design techniques for system-level dynamic power management. IEEE Transactions on VLSI Systems 8(3), 299–316 (2000)
7. Brooks, D., Bose, P., Schuster, S., Jacobson, H., Kudva, P., Buyuktosunoglu, A., Wellman, J., Zyuban, V., Gupta, M., Cook, P.: Power-aware microarchitecture: Design and modeling challenges for next-generation microprocessors. IEEE Micro 20(6), 26–44 (2000)
8. Chan, H.L., Chan, W.T., Lam, T.W., Lee, L.K., Mak, K.S., Wong, P.W.H.: Energy efficient online deadline scheduling. In: SODA, pp. 795–804 (2007)
9. Chan, H.L., Edmonds, J., Lam, T.W., Lee, L.K., Marchetti-Spaccamela, A., Pruhs, K.: Nonclairvoyant speed scaling for flow and energy. In: STACS, pp. 255–264 (2009)
10. Irani, S., Pruhs, K.: Algorithmic problems in power management. SIGACT News 32(2), 63–76 (2005)
11. Irani, S., Shukla, S., Gupta, R.: Online strategies for dynamic power management in systems with multiple power-saving states. ACM Tran. Embeded Comp. Sys. 2(3), 325–346 (2003)
12. Irani, S., Shukla, S., Gupta, R.: Algorithms for power savings. TALG 3(4) (2007)
13. Karlin, A., Manasse, M., McGeoch, L., Owicki, S.: Competitive randomized algorithms for non-uniform problems. In: SODA, pp. 301–309 (1990)
14. Lam, T.W., Lee, L.K., To, I.K.K., Wong, P.W.H.: Competitive non-migratory scheduling for flow time and energy. In: SPAA, pp. 256–264 (2008)
15. Lam, T.W., Lee, L.K., To, I.K.K., Wong, P.W.H.: Speed scaling functions for flow time scheduling based on active job count. In: Halperin, D., Mehlhorn, K. (eds.) ESA 2008. LNCS, vol. 5193, pp. 647–659. Springer, Heidelberg (2008)
16. Yao, F., Demers, A., Shenker, S.: A scheduling model for reduced CPU energy. In: FOCS, pp. 374–382 (1995)

Improved Bounds for Flow Shop Scheduling

Monaldo Mastrolilli and Ola Svensson

IDSIA – Switzerland
{monaldo,ola}@idsia.ch

Abstract. We resolve an open question raised by Feige & Scheideler by showing that the best known approximation algorithm for flow shops is essentially tight with respect to the used lower bound on the optimal makespan. We also obtain a nearly tight hardness result for the general version of flow shops, where jobs are not required to be processed on each machine.

Similar results hold true when the objective is to minimize the sum of completion times.

1 Introduction

In the *flow shop scheduling problem* we have a set of n jobs that must be processed on a given set of m machines that are located in a fixed order. Each job j consists of a sequence of m operations, where the i-th operation must be processed during $p_{ij} \in \mathbb{Z}^+$ time units without interruption on the i-th machine. A feasible schedule is one in which each operation is scheduled only after all operations preceding it in its job have been completed, and each machine processes at most one operation at a time. A natural generalization of the flow shop problem is to not require jobs to be processed on all machines, i.e., a job still requests the machines in compliance with their fixed order but may skip some of them. We will refer to this more general version as *generalized flow shops* or *flow shops with jumps*. Generalized flow shop (and thus flow shop) scheduling is a special case of the *acyclic job shop* scheduling, which only requires that within each job all operations are performed on different machines, which in turn is a special case of the general *job shop* scheduling, where each job may have many operations on a single machine.

For any given schedule, let C_j be the completion time of the last operation of job j. We consider the natural and typically considered objectives of minimizing the *makespan*, $C_{\max} = \max_j C_j$, and the *sum of weighted completion times*, $\sum w_j C_j$, where w_j are positive integers. The goal is to find a feasible schedule which minimizes the considered objective function. In the notation of Graham et al. [6] the flow shop scheduling problem is denoted as $F||\gamma$, where γ denotes the objective function to be minimized. We will sometimes abbreviate the generalized flow shop problem by $F|jumps|\gamma$.

1.1 Literature Review

Flow shops have long been identified as having a number of important practical applications and have been widely studied since the late 50's (the reader is

S. Albers et al. (Eds.): ICALP 2009, Part I, LNCS 5555, pp. 677–688, 2009.

referred to the survey papers of Lawler et al. [9] and of Chen, Potts & Woeginger [2]). To find a schedule that minimizes the makespan, or one that minimizes the sum of completion times, was proved to be strongly NP-hard in the 70's, even for severely restricted instances (see e.g. [2]).

From then, many approximation methods have been proposed. Since the quality of an approximation algorithm is measured by comparing the returned solution value with a polynomial time computable lower bound on the optimal value, the goodness of the latter is very important. For a given instance, let C^*_{max} denote the minimum makespan taken over all possible feasible schedules. If D denotes the length of the longest job (the dilation), and C denotes the time units requested by all jobs on the most loaded machine (the congestion), then $lb = max[C, D]$ is a known trivial lower bound on C^*_{max}. To the best of our knowledge, no significant stronger lower bound is known on C^*_{max}, and all the proposed approximation algorithms for flow shops (but also for the more general job shop, acyclic job shop and the more constrained case of permutation flow shops) have been analyzed with respect to this lower bound (see, e.g., [10,11,4,17,5,13]).

Even though the trivial lower bound might seem weak a surprising result by Leighton, Maggs & Rao [10] says that for acyclic job shops, if all operations are of unit length, then $C^*_{max} = \Theta(lb)$. If we allow operations of any length, then Feige & Scheideler [4] showed that $C^*_{max} = O(lb \cdot \log lb \log\log lb)$. They also showed their analysis to be nearly tight by providing acyclic job shop instances with $C^*_{max} = \Omega(lb \cdot \log lb/ \log\log lb)$. The proofs of the upper bounds in [10,4] are nonconstructive and make repeated use of (a general version) of Lovasz local lemma. Algorithmic versions appeared in [1,3]. Recently, the authors showed that the best known approximation algorithm for acyclic job shops is basically tight [12]. More specifically, it was shown that for every $\epsilon > 0$, the (acyclic) job shop problem cannot be approximated within ratio $O(\log^{1-\epsilon} lb)$, unless NP has quasi-polynomial Las-Vegas algorithms.

In contrast to acyclic job shops, the strength of the lower bound lb for flow shop scheduling is not well understood, and tight results are only known for some variants. A notable example is given by the *permutation flow shop* problem, that is a flow shop problem with the additional constraint that each machine processes all the jobs in the same order. Potts, Shmoys & Williamson [14] gave a family of permutation flow shop instances with $C^*_{max} = \Omega(lb \cdot \sqrt{\min[m, n]})$. This lower bound was recently showed to be tight, by Nagarajan & Sviridenko [13], who gave an approximation algorithm that returns a permutation schedule with makespan $O(lb \cdot \sqrt{\min[m, n]})$.

Feige & Scheideler's upper bound for acyclic jobs [4] is also the best upper bound for the special case of flow shops. As flow shops have more structure than acyclic job shops and no flow shop instances with $C^*_{max} = \omega(lb)$ were known, one could hope for a significant better upper bound for flow shops. The existence of such a bound was raised as an open question in [4]. Unfortunately our recent inapproximability results for acyclic job shops do not apply to flow shops, since in [12] our construction builds upon the lower bound construction for acyclic job shop, which does not seem to generalize to flow shop [4]. The only known inapproximability result is due to Williamson et al. [18], and states that when the number of machines and jobs are part of the input, it is NP-hard to approximate $F||C_{\max}$ with unit time operations, and at most three operations per job, within a ratio better than 5/4. It is a long standing open problem if the above algorithms $F||C_{\max}$, are tight or even nearly tight (see, e.g. "Open problem 6 " in [16]).

A similar situation holds for the objective $\sum w_j C_j$. Queyranne & Sviridenko [15] showed that an approximation algorithm for the above mentioned problems that produces a schedule with makespan a factor $O(\rho)$ away from the lower bound lb can be used to obtain a $O(\rho)$-approximation algorithms for other objectives, including the sum of weighted completion times. The only known inapproximability result is by Hoogeveen, Schuurman & Woeginger [7], who showed that $F||\sum C_j$ is NP-hard to approximate within a ratio better than $1+\epsilon$ for some small $\epsilon > 0$.

1.2 Our Results

In this paper, we show that the best known upper bound [4] is essentially the best possible, by proving the existence of instances of flow shop scheduling for which the shortest feasible schedule is of length $\Omega(lb \cdot \log lb / \log \log lb)$. This resolves (negatively) the aforementioned open question by Feige & Scheideler [4].

Theorem 1. *There are flow shop instances for which any schedule has makespan* $\Omega(lb \cdot \log lb / \log \log lb)$.

If we do not require a job to be processed on *all* machines, i.e. generalized flow shops, we prove that it is hard to improve the approximation guarantee. Theorem 2 shows that generalized flow shops, with the objective to either minimize makespan or sum of completion times, have no constant approximation algorithm unless $P = NP$.

Theorem 2. *For all sufficiently large constants* K*, it is NP-hard to distinguish between generalized flow shop instances that have a schedule with makespan* $2K \cdot lb$ *and those that have no solution that schedules more than half of the jobs within* $(1/8)K^{\frac{1}{25}(\log K)} \cdot lb$ *time units. Moreover this hardness result holds for generalized flow shop instances with bounded number of operations per job, that only depends on* K.

By using a similar reduction, but using a stronger assumption, we give a hardness result that essentially shows that the current approximation algorithms for generalized flow shops, with both makespan and sum of weighted completion times objective, are tight.

Theorem 3. *Let* $\epsilon > 0$ *be an arbitrarily small constant. There is no* $O\left((\log lb)^{1-\epsilon}\right)$*-approximation algorithm for* $F|jumps|C_{max}$ *or* $F|jumps|\sum C_j$*, unless* $NP \subseteq ZTIME(2^{O(\log n)^{O(1/\epsilon)}})$.

No results of this kind were known for a flow shop problem. Moreover, this paper extends and significantly simplifies the recent hardness results by the authors for the acyclic job shop problem [12].

In summary, the consequences of our results are among others that in order to improve the approximation guarantee for flow shops, it is necessary to (i) improve the used lower bound on the optimal makespan and (ii) use the fact that a job needs to be processed on *all* machines.

2 Job and Flow Shops Instances with Large Makespan

We first exhibit an instance of general job scheduling for which it is relatively simple to show that any optimal schedule is of length $\Omega(lb \cdot \log lb)$. The construction builds upon the idea of jobs of different "frequencies", by Feige & Scheideler [4], but we will introduce some important differences that will be decisive for the flow shop case. The resulting instance slightly improves[1] on the bound by Feige & Scheideler [4], who showed the existence of job shop instances with optimal makespan $\Omega(lb \cdot \log lb / \log\log lb)$.

The construction of flow shop instances with "large" makespan is more complicated, as each job is required to have exactly one operation for every machine, and all jobs are required to go through all the machines in the same order. The main idea is to start with the aforementioned job shop construction, which has very cyclic jobs, i.e., jobs have many operations on a single machine. The flow shop instance is then obtained by "copying" the job shop instance several times and, instead of having cyclic jobs, we let the i-th long operation of a job to be processed in the i-th copy of the original job shop instance. Finally, we insert additional zero-length operations to obtain a flow shop instance. We show that the resulting instance has optimal length $\Omega(lb \cdot \log lb / \log\log lb)$.

2.1 Job Shops with Large Makespan

Construction. For any integer $d \geq 1$ consider the job shop instance with d machines $m_1, \ldots, m_d$ and d jobs $j_1, \ldots, j_d$. We say that job j_i has *frequency* i, which means that it has 3^i so-called *long-operations* on machine m_i, each one of them requires 3^{d-i} time units. And between any two consecutive long-operations, job j_i has *short-operations* that requires 0 time units on the machines $m_1, \ldots, m_{i-1}$. Note that the length of all jobs and the load on all machines are 3^d, which we denote by lb. For a small example see Figure 1.

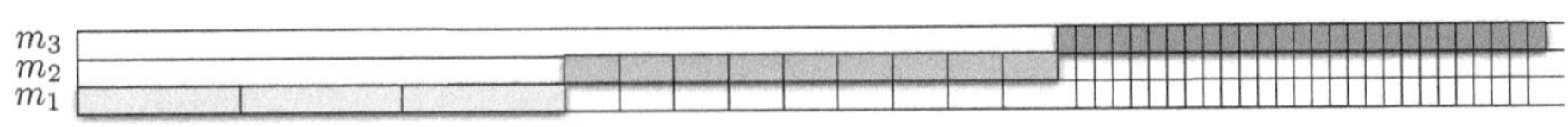

Fig. 1. An example of the construction for job shop with $d = 3$

[1] However, in their construction all operations of a job have the same length which is not the case for our construction.

Analysis. Fix an arbitrarily feasible schedule for the jobs. We shall show that the length of the schedule must be $\Omega(lb \cdot \log lb)$.

Lemma 1. *For $i, j : 1 \leq i < j \leq d$, the number of time units during which both j_i and j_j perform operations is at most $\frac{lb}{3^{j-i}}$.*

Proof. During the execution of a long-operation of j_i (that requires 3^{d-i} time units), job j_j can complete at most one long-operation that requires 3^{d-j} time units (since its short-operation on machine m_i has to wait). As j_i has 3^i long-operations, the two jobs can perform operations at the same time during at most $3^i \cdot 3^{d-j} = \frac{3^d}{3^{j-i}} = \frac{lb}{3^{j-i}}$ time units. □

It follows that, for each $i = 1, \ldots, d$, at most a fraction $1/3 + 1/3^2 + \cdots + 1/3^i \leq 1/3 + 1/3^2 + \cdots + 1/3^d \leq \frac{1}{3-1} = 1/2$ of the time spent for long-operations of a job j_i is performed at the same time as long-operations of jobs with lower frequency. Hence a feasible schedule has makespan at least $d \cdot lb/2$. As $d = \Omega(\log lb)$ (recall that $lb = 3^d$), the optimal makespan of the constructed job shop instance is $\Omega(lb \cdot \log lb)$.

2.2 Flow Shops with Large Makespan

Construction. For sufficiently large integers d and r, consider the flow shop instance defined as follows:

- There are r^{2d} groups of machines[2], denoted by $M_1, M_2, \ldots, M_{r^{2d}}$. Each group M_g consists of d machines $m_{g,1}, m_{g,2}, \ldots, m_{g,d}$ (one for each frequency). Finally the machines are ordered in such a way so that $m_{g,i}$ is before $m_{h,j}$ if either (i) $g < h$ or (ii) $g = h$ and $i > j$. The latter case will ensure that, within each group of machines, long-operations of jobs with high frequency will be scheduled before long-operations of jobs with low frequency, a fact that is used to prove Lemma 3.
- For each frequency $f = 1, \ldots, d$, there are $r^{2(d-f)}$ groups of jobs, denoted by $J_1^f, J_2^f, \ldots, J_{r^{2(d-f)}}^f$. Each group J_g^f consists of r^{2f} copies, referred to as $j_{g,1}^f, j_{g,2}^f, \ldots, j_{g,r^{2f}}^f$, of the job that must be processed during $r^{2(d-f)}$ time units on the machines

$$m_{a+1,f}, m_{a+2,f}, \ldots, m_{a+r^{2f},f} \text{ where } a = (g-1) \cdot r^{2f}$$

  and during 0 time units on all the other machines that are required to create a flow shop instance. Let J^f be the set of jobs that correspond to frequency f, i.e., $J^f = \{j_{g,a}^f : 1 \leq g \leq r^{2(d-f)}, 1 \leq a \leq r^{2f}\}$.

Note that the length of all jobs and the load on all machines are r^{2d}, which we denote by lb. Moreover, the total number of machines and number of jobs are both $r^{2d} \cdot d$. In the subsequent we will call the operations that require more than 0 time units *long-operations* and the operations that only require 0 time units *short-operations*. For an example of the construction see Figure 2.

[2] These groups of machines "correspond" to copies of the job shop instance in subsection 2.1.

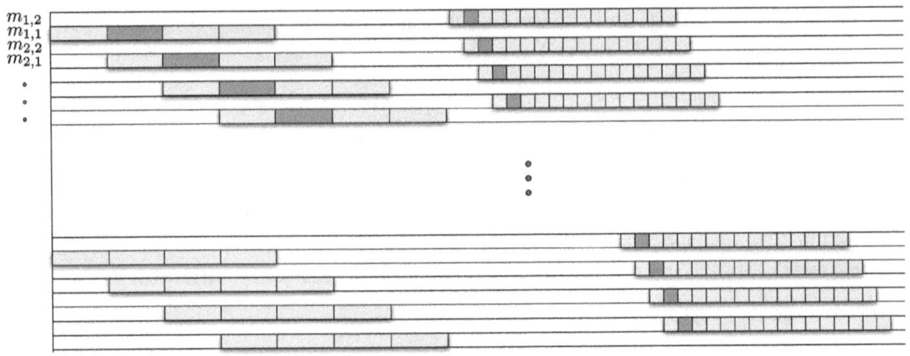

Fig. 2. An example of the construction for flow shop scheduling with $r = d = 2$. Only long-operations on the first 4 and last 4 groups of machines are depicted. The long-operations of one job of each frequency are highlighted in dark gray.

Analysis. We shall show that the length of the schedule must be $\Omega(lb \cdot \min[r, d])$. As $lb = r^{2d}$, instances constructed with $r = d$ has optimal makespan $\Omega(lb \cdot \log lb / \log\log lb)$.

Fix an arbitrarily feasible schedule for the jobs. We start by showing a useful property. For a job j, let $d_j(i)$ denote the delay between job j's i-th and $i+1$-th long-operations, i.e., the time units between the end of job j's i-th long-operation and the start of its $i+1$-th long-operation (let $d_j(i) = \infty$ for the last long-operation). We say that the i-th long-operation of a job j of frequency f is *good* if $d_j(i) \leq \frac{r^2}{4} \cdot r^{2(d-f)}$.

Lemma 2. *If the schedule has makespan less than $r \cdot lb$ then the fraction of good long-operations of each job is at least $\left(1 - \frac{4}{r}\right)$.*

Proof. Assume that the considered schedule has makespan less than $r \cdot lb$. Suppose toward contradiction that there exists a job j of frequency f so that j has at least $\frac{4}{r} r^{2f}$ long-operations that are not good. But then the length of j is at least $\frac{4}{r} r^{2f} \cdot \frac{r^2}{4} \cdot r^{2(d-f)} = r \cdot r^{2d} = r \cdot lb$, which contradicts that the makespan of the considered schedule is less than $r \cdot lb$. □

We continue by analyzing the schedule with the assumption that its makespan is less than $r \cdot lb$ (otherwise we are done). In each group M_g of machines we will associate a set $T_{g,f}$ of time intervals with each frequency $f = 1, \dots, d$. The set $T_{g,f}$ contains the time intervals corresponding to the *first half* of all good long-operations scheduled on the machine $m_{g,f}$.

Lemma 3. *Let $k, \ell : 1 \leq k < \ell \leq d$ be two different frequencies. Then the sets $T_{g,k}$ and $T_{g,\ell}$, for all $g : 1 \leq g \leq r^{2d}$, contain disjoint time intervals.*

Proof. Suppose toward contradiction that there exist time intervals $t_k \in T_{g,k}$ and $t_\ell \in T_{g,\ell}$ that overlap, i.e., $t_k \cap t_\ell \neq \emptyset$. Note that t_k and t_ℓ correspond to good long-operations of jobs of frequencies k and ℓ, respectively. Let us say that the good long-operation corresponding to t_ℓ is the a-th operation of some job j. As t_ℓ and t_k overlap, the a-th long-operation of j must overlap the first half of the long-operation corresponding to t_k. As job j has a short operation on machine $m_{g,k}$ after its long-operation on machine $m_{g,\ell}$ (recall that machines are ordered $m_{g,d}, m_{g,d-1}, \ldots, m_{g,1}$ and $\ell > k$), job j's $(a+1)$-th operation must be delayed by at least $r^{2(d-k)}/2 - r^{2(d-\ell)}$ time units and thus $d_j(a) > r^{2(d-k)}/2 - r^{2(d-\ell)} > \frac{r^2}{4} r^{2(d-\ell)}$, which contradicts that the a-th long-operation of job j is good. □

Let $L(T_{g,f})$ denote the total time units covered by the time intervals in $T_{g,f}$. We continue by showing that there exists a g such that $\sum_{f=1}^{d} L(T_{g,f}) \geq \frac{lb}{4} \cdot d$. With this in place, it is easy to see that any schedule has makespan $\Omega(d \cdot lb)$ since all the time intervals $\{T_{g,f} : f = 1, \ldots, d\}$ are disjoint (Lemma 3).

Lemma 4. *There exists a* $g \in \{1, \ldots, r^{2d}\}$ *such that*

$$\sum_{f=1}^{d} L(T_{g,f}) \geq \frac{lb}{4} \cdot d$$

Proof. As $\sum_{f=1}^{d} L(T_{g,f})$ adds up the time units required by the *first half* of each good long-operation scheduled on a machine in M_g, the claim follows by showing that there exist one group of machines M_g from $\{M_1, M_2, \ldots, M_{r^{2d}}\}$ so that the total time units required by the good long-operations on the machines in M_g is at least $\frac{lb \cdot d}{2}$.

By lemma 2 we have that the good long-operations of each job requires at least $lb \cdot \left(1 - \frac{4}{r}\right)$ time units. Since the total number of jobs is $r^{2d}d$ the total time units required by all good long-operations is at least $lb \cdot \left(1 - \frac{4}{r}\right) \cdot r^{2d}d$. As there are r^{2d} many groups of machines, a simple averaging argument guarantees that in at least one group of machines, say M_g, the total time units required by the good long-operations on the machines in M_g is at least $lb \cdot \left(1 - \frac{4}{r}\right) d > lb \cdot d/2$. □

3 Hardness of Generalized Flow Shops

Theorem 2 and Theorem 3 are proved by presenting a gap-preserving reduction Γ from the graph coloring problem to the generalized flow shop problem. Γ has two parameters r and d. Given an n-vertex graph G whose vertices are partitioned into d independent sets, it computes in time polynomial in n and r^d, a generalized flow shop instance $S(r, d)$ where all jobs have the same length r^{2d} and all machines the same load r^{2d}. Hence, $lb = r^{2d}$. Instance $S(r, d)$ has a set of r^{2d} jobs and and a set of r^{2d} machines for each vertex in G. The total number of jobs and the total number of machines are thus both $r^{2d}n$. Moreover, each job has at most $(\Delta + 1)r^{2d}$ operations. By using jobs of different frequencies, as done in the gap construction, we have the property that "many" of the jobs corresponding to adjacent vertices cannot be scheduled in parallel in any feasible schedule. On the other hand, by letting jobs skip those machines corresponding to non-adjacent vertices, jobs corresponding to an independent set in G *can* be scheduled in parallel (their operations can overlap in time) in a feasible schedule. This ensures that the following completeness and soundness hold for the resulting generalized flow shop instance $S(r, d)$.

- Completeness case: If G can be colored using L colors then $C^*_{max} \leq lb \cdot 2L$;
- Soundness case: For any $L \leq r$. Given a schedule where at least half the jobs finish within $lb \cdot L$ time units, we can, in time polynomial in n and r^d, find an independent set of G of size $n/(8L)$.

In Section 3.1 we describe the gap-preserving reduction Γ. With this construction in place, Theorem 2 easily follows by using a result by Khot [8], that states that it is NP-hard to color a K-colorable graph with $K^{\frac{1}{25}(\log k)}$ colors, for sufficiently large constants K. The result was obtained by presenting a polynomial time reduction that takes as input a SAT formula ϕ together with a sufficiently large constant K, and outputs an n-vertex graph G with degree at most $2^{K^{O(\log K)}}$. Moreover, (completeness) if ϕ is satisfiable then graph G can be colored using K colors and (soundness) if ϕ is not satisfiable then graph G has no independent set containing $n/K^{\frac{1}{25}(\log K)}$ vertices (see Section 6 in [8]). Note that the soundness case implies that any feasible coloring of the graph uses at least $K^{\frac{1}{25}(\log K)}$ colors. We let $\mathcal{G}[c, i]$ be the family of graphs that either can be colored using c colors or have no independent set containing a fraction i of the vertices. To summarize, for sufficiently large K and $\Delta = 2^{K^{O(\log K)}}$, it is NP-hard to decide if an n-vertex graph G in $\mathcal{G}[K, 1/K^{\frac{1}{25}(\log K)}]$ with bounded degree Δ has

$$\chi(G) \leq K \text{ or } \alpha(G) \leq \frac{n}{K^{\frac{1}{25}(\log K)}}$$

where $\chi(G)$ and $\alpha(G)$ denote the chromatic number and the size of a maximum independent set of G, respectively. As the vertices of a graph with bounded degree Δ can, in polynomial time, be partitioned into $\Delta + 1$ independent sets, we can use Γ with parameters $d = \Delta+1$ and $r = K^{\frac{1}{25}(\log K)}$ (r is chosen such that the condition $L \leq r$ in the soundness case of Γ is satisfied for $L = K^{\frac{1}{25}(\log K)}/8$). It follows, by the completeness case and soundness case of Γ, that it is NP-hard to distinguish if the obtained scheduling instance has a schedule with makespan at most $lb \cdot 2K$, or no solution schedules more than half of the jobs within $lb \cdot K^{\frac{1}{25}(\log K)}/8$ time units. Moreover, each job has at most $(\Delta+1)r^{2d}$ operations, which is a constant that only depends on K.

The proof of Theorem 3 is similar to the proof of Theorem 2 with the exception that the graphs have no longer bounded degree. Due to space limits the proof is omitted; it follows the one provided by the authors for the acyclic job shop problem [12].

3.1 Construction

Here, we present the reduction Γ for the general flow shop problem where jobs are allowed to skip machines. Given an n-vertex graph $G = (V, E)$ whose vertices are partitioned into d independent sets, we create a generalized flow shop instance $S(r,d)$, where r and d are the parameters of the reduction. Let $I_1, I_2, \ldots I_d$ denote the independent sets that form a partition of V.

$S(r,d)$ is very similar to the gap instance described in Section 2.2. The main difference is that in $S(r,d)$ distinct jobs can be scheduled in parallel if their corresponding vertices in G are not adjacent. This is obtained by letting a job to skip those machines corresponding to non-adjacent vertices. (The gap instance of Section 2.2 can be seen as the result of the following reduction when the graph G is a complete graph with d nodes). For convenience, we give the complete description with the necessary changes.

- There are r^{2d} groups of machines, denoted by $M_1, M_2, \ldots, M_{r^{2d}}$. Each group M_g consists of n machines $\{m_{g,v} : v \in V\}$ (one for each vertex in G). Finally the machines are ordered in such a way so that $m_{g,u}$ is before $m_{h,v}$ if either (i) $g < h$ or (ii) $g = h$ and $u \in I_k, v \in I_\ell$ with $k > \ell$. The latter case will ensure that, within each group of machines, long-operations of jobs with high frequency will be scheduled before long-operations of jobs with low frequency, a fact that is used to prove Lemma 8.
- For each $f : 1 \leq f \leq d$ and for each vertex $v \in I_f$ there are $r^{2(d-f)}$ groups of jobs, denoted by $J_1^v, J_2^v, \ldots, J_{r^{2(d-f)}}^v$. Each group J_g^v consists of r^{2f} copies, referred to as $j_{g,1}^v, j_{g,2}^v, \ldots, j_{g,r^{2f}}^v$, of the job that must be processed during $r^{2(d-f)}$ time units on the machines

$$m_{a+1,v}, m_{a+2,v}, \ldots, m_{a+r^{2f},v} \text{ where } a = (g-1) \cdot r^{2f}$$

  and during 0 time units on machines corresponding to adjacent vertices, i.e., $\{m_{a,u} : 1 \leq a \leq r^{2d}, \{u,v\} \in E\}$ in an order such that it results in a generalized flow shop instance. Let J^v be the jobs that correspond to the vertex v, i.e., $J^v = \{j_{g,i}^v : 1 \leq g \leq r^{2(d-f)}, 1 \leq i \leq r^{2f}\}$.

Note that the length of all jobs and the load on all machines are r^{2d}, which we denote by lb. The total number of machines and total number of jobs are both $r^{2d} \cdot n$. Moreover, each job has at most $(\Delta+1)r^{2d}$ operations. In the subsequent we will call the operations that require more than 0 time units *long-operations* and the operations that only require 0 time units *short-operations*. For an example of the construction see Figure 3.

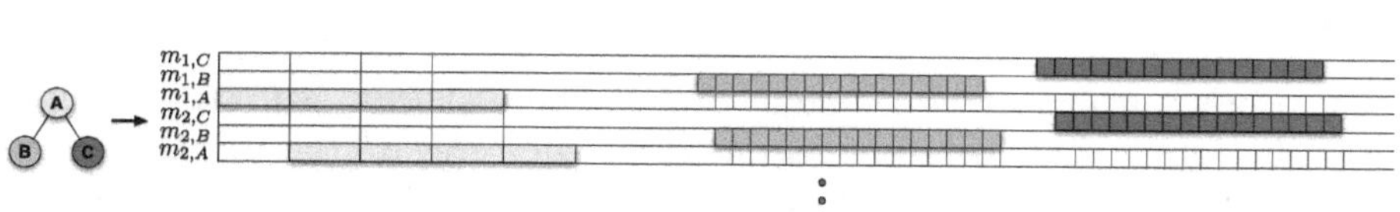

Fig. 3. An example of the reduction with $r = 2, d = 2, I_1 = \{A\}$ and $I_2 = \{B, C\}$. Only the two first out of $r^{2d} = 16$ groups of machines are depicted with the jobs corresponding to A,B, and C to the left, center, and right respectively.

Completeness. We prove that if the graph G can be colored with "few" colors then there is a relatively "short" schedule to the general flow shop instance.

Lemma 5. *There is a schedule of $S(r,d)$ with makespan $2lb \cdot \chi(G)$.*

Proof. We start by showing that all jobs corresponding to non-adjacent vertices can be scheduled within $2 \cdot lb$ time units.

Claim. Let IS be an independent set of G. Then all the jobs $\bigcup_{v \in IS} J^v$ can be scheduled within $2 \cdot lb$ time units.

Proof of Claim. Consider the schedule defined by scheduling the jobs corresponding to each vertex $v \in IS$ as follows. Let I_f be the independent set with $v \in I_f$. A job $j^v_{g,i}$ corresponding to vertex v is then scheduled without interruption starting from time $r^{2(d-f)} \cdot (i-1)$.

The schedule has makespan at most $2 \cdot lb$ since a job is started at latest at time $r^{2(d-f)} \cdot (r^{2f} - 1) < lb$ and requires lb time units in total.

To see that the schedule is feasible, observe that no short-operations of the jobs in $\bigcup_{v \in IS} J^v$ need to be processed on the same machines as the long-operations of the jobs in $\bigcup_{v \in IS} J^v$ (this follows from the construction and that the jobs correspond to non-adjacent vertices). Moreover, two jobs $j^v_{g,i}, j^{v'}_{g',i'}$, with either $g \neq g'$ or $v \neq v'$, have no two long-operations that must be processed on the same machine. Hence, the only jobs that might delay each other are jobs belonging to the same vertex v and the same group g, but these jobs are started with appropriate delays (depending on the frequency of the job). □

We partition set V into $\chi(G)$ independent subsets $V_1, V_2, \ldots, V_{\chi(G)}$. By the above lemma, the jobs corresponding to each of these independent sets can be scheduled within $2 \cdot lb$ time units. We can thus schedule the jobs in $\chi(G)$-"blocks", one block of length $2 \cdot lb$ for each independent set. The total length of this schedule is $2lb \cdot \chi(G)$. □

Soundness. We prove that, given a schedule where many jobs are completed "early", we can, in polynomial time, find a "big" independent set of G.

Lemma 6. *For any $L \leq r$. Given a schedule of $S(r,d)$ where at least half the jobs finish within $lb \cdot L$ time units , we can, in time polynomial in n and r^d, find an independent set of G of size at least $n/(8L)$.*

Proof. Fix an arbitrarily schedule of $S(r,d)$ where at least half the jobs finish within $lb \cdot L$ time units. In the subsequent we will disregard the jobs that do not finish within $lb \cdot L$ time units throughout the analysis. Note that the remaining jobs are at least $r^{2d}n/2$ many. As for the gap construction (see Section 2.2), we say that the i-th long-operation of a job j of frequency f is *good* if the delay $d_j(i)$ between job j's i-th and $i+1$-th long-operations is at most $\frac{r^2}{4} \cdot r^{2(d-f)}$. In each group M_g of machines we will associate a set $T_{g,v}$ of time intervals with each vertex $v \in V$. The set $T_{g,v}$ contains the time intervals corresponding to the *first half* of all good long-operations scheduled on the machine $m_{g,v}$. We also let $L(T_{g,v})$ denote the total time units covered by the time intervals in $T_{g,v}$. Scheduling instance $S(r,d)$ has a similar structure as the gap instances created in Section 2.2 and has similar properties. By using the fact that all jobs (that were not disregarded) have completion time at most $L \cdot lb$ which is by assumption at most $r \cdot lb$, Lemma 7 follows from the same arguments as Lemma 2.

Lemma 7. *The fraction of good long-operations of each job is at least* $\left(1 - \frac{4}{r}\right)$.

Consider a group M_g of machines and two jobs corresponding to adjacent vertices that have long-operations on machines in M_g. Recall that jobs corresponding to adjacent vertices have different frequencies. By the ordering of the machines, we are guaranteed that the job of higher frequency has, after its long-operation on a machine in M_g, a short-operation on the machine in M_g where the job of lower frequency has its long-operation. The following lemma now follows by observing, as in the proof of Lemma 3, that the long-operation of the high frequency job can only be good if it is *not* scheduled in parallel with the first half of the long-operation of the low frequency job.

Lemma 8. *Let* $u \in I_k$ *and* $v \in I_l$ *be two adjacent vertices in* G *with* $k > l$. *Then the sets* $T_{g,u}$ *and* $T_{g,v}$, *for all* $g : 1 \leq g \leq r^{2d}$, *contain disjoint time intervals.*

Finally, Lemma 9 is proved in the very same way as Lemma 4. Their different inequalities arise because in the gap instance we had $d \cdot r^{2d}$ jobs and here we are considering at least $r^{2d}n/2$ jobs that were not disregarded.

Lemma 9. *There exists a* $g \in \{1, \ldots, r^{2d}\}$ *such that*

$$\sum_{v \in V} L(T_{g,v}) \geq \frac{lb \cdot n}{8}.$$

We conclude by a simple averaging argument. Set g so that $\sum_{v \in V} L(T_{g,v})$ is at least $\frac{lb \cdot n}{8}$, such a g is guaranteed to exist by the lemma above. As all jobs that were not disregarded finish within $L \cdot lb$ time units, at least $\frac{lb \cdot n}{8}/(L \cdot lb) = \frac{n}{8L}$ time intervals must overlap at some point during the first $L \cdot lb$ time units of the schedule, and, since they overlap, they correspond to different vertices that form an independent set in G (Lemma 8). Moreover, we can find such a point in the schedule by, for example, considering all different blocks and, in each block, verify the start and end points of the time intervals. □

Acknowledgments

This research is supported by the Swiss National Science Foundation project "Approximation Algorithms for Machine scheduling Through Theory and Experiments III" Project N. 200020-122110/1.

References

1. Beck, J.: An algorithmic approach to the lovasz local lemma. Random Structures and Algorithms 2(4), 343–365 (1991)
2. Chen, B., Potts, C., Woeginger, G.: A review of machine scheduling: Complexity, algorithms and approximability. Handbook of Combinatorial Optimization 3, 21–169 (1998)
3. Czumaj, A., Scheideler, C.: A new algorithm approach to the general lovász local lemma with applications to scheduling and satisfiability problems (extended abstract). In: STOC, pp. 38–47 (2000)
4. Feige, U., Scheideler, C.: Improved bounds for acyclic job shop scheduling. Combinatorica 22(3), 361–399 (2002)
5. Goldberg, L., Paterson, M., Srinivasan, A., Sweedyk, E.: Better approximation guarantees for job-shop scheduling. SIAM Journal on Discrete Mathematics 14(1), 67–92 (2001)
6. Graham, R., Lawler, E., Lenstra, J., Kan, A.R.: Optimization and approximation in deterministic sequencing and scheduling: A survey. In: Annals of Discrete Mathematics, vol. 5, pp. 287–326. North-Holland, Amsterdam (1979)
7. Hoogeveen, H., Schuurman, P., Woeginger, G.J.: Non-approximability results for scheduling problems with minsum criteria. INFORMS Journal on Computing 13(2), 157–168 (2001)
8. Khot, S.: Improved inapproximability results for maxclique, chromatic number and approximate graph coloring. In: FOCS, pp. 600–609 (2001)
9. Lawler, E., Lenstra, J., Kan, A.R., Shmoys, D.: Sequencing and scheduling: Algorithms and complexity. Handbook in Operations Research and Management Science 4, 445–522 (1993)
10. Leighton, F.T., Maggs, B.M., Rao, S.B.: Packet routing and job-shop scheduling in O(congestion + dilation) steps. Combinatorica 14(2), 167–186 (1994)
11. Leighton, F.T., Maggs, B.M., Richa, A.W.: Fast algorithms for finding O(congestion + dilation) packet routing schedules. Combinatorica 19, 375–401 (1999)
12. Mastrolilli, M., Svensson, O.: (Acyclic) jobshops are hard to approximate. In: FOCS, pp. 583–592 (2008)
13. Nagarajan, V., Sviridenko, M.: Tight bounds for permutation flow shop scheduling. In: Lodi, A., Panconesi, A., Rinaldi, G. (eds.) IPCO 2008. LNCS, vol. 5035, pp. 154–168. Springer, Heidelberg (2008)
14. Potts, C., Shmoys, D., Williamson, D.: Permutation vs. nonpermutation flow shop schedules. Operations Research Letters 10, 281–284 (1991)
15. Queyranne, M., Sviridenko, M.: Approximation algorithms for shop scheduling problems with minsum objective. Journal of Scheduling 5(4), 287–305 (2002)
16. Schuurman, P., Woeginger, G.J.: Polynomial time approximation algorithms for machine scheduling: ten open problems. Journal of Scheduling 2(5), 203–213 (1999)
17. Shmoys, D., Stein, C., Wein, J.: Improved approximation algorithms for shop scheduling problems. SIAM Journal on Computing 23, 617–632 (1994)
18. Williamson, D., Hall, L., Hoogeveen, J., Hurkens, C., Lenstra, J., Sevastianov, S., Shmoys, D.: Short shop schedules. Operations Research 45, 288–294 (1997)

A 3/2-Approximation Algorithm for General Stable Marriage

Eric McDermid*

Department of Computing Science, University of Glasgow, Glasgow G12 8QQ, UK
mcdermid@dcs.gla.ac.uk

Abstract. In an instance of the stable marriage problem with ties and incomplete preference lists, stable matchings can have different sizes. It is APX-hard to compute a maximum cardinality stable matching, but there have recently been proposed polynomial-time approximation algorithms, with constant performance guarantees for both the general version of this problem, and for several special cases. Our contribution is to describe a $\frac{3}{2}$-approximation algorithm for the general version of this problem, improving upon the recent $\frac{5}{3}$-approximation algorithm of Király. Interest in such algorithms arises because of the problem's application to centralized matching schemes, the best known of which involve the assignment of graduating medical students to hospitals in various countries.

1 Introduction

1.1 Classical Stable Marriage

An instance I of the classical *stable marriage problem* (SM) involves a set of n men and n women, each of whom provides a linearly ordered preference list of all the members of the opposite sex. A *matching* M is a one-to-one correspondence between the men and women of I. A (man,woman) pair (m, w) is said to be a *blocking pair* for a matching M if m and w are not matched to each other in M, but m prefers w to his partner in M and w prefers m to her partner in M. A matching that admits no blocking pairs is called a *stable matching.* Every SM instance admits a stable matching (in general there may be exponentially many [13,7]) and such a matching can be found in $O(n^2)$ time by the well-known Gale/Shapley algorithm [2].

The stable marriage problem has received much attention from researchers because of the problem's widespread practical applications to centralized matching schemes. Several nations, including the US, Canada, and Japan use centralized matching algorithms to assign graduating medical students to their first hospital jobs, based on the medical students' and hospitals' preferences for one another (see for example [17,16,18]). Larger-scale schemes than these exist, for example in Hungary, the annual assignment of over 100,000 first-year students to universities is handled by a centralized matching algorithm [1].

* Supported by Engineering and Physical Sciences Research Council grant EP/E011993/1.

S. Albers et al. (Eds.): ICALP 2009, Part I, LNCS 5555, pp. 689–700, 2009.

1.2 Stable Marriage with Ties and Incomplete Lists

Primarily because of the scale of such applications, it is unrealistic to assume that the agents involved will be able to produce preference lists ranking every member of the opposite set. Indeed, in all of the above-mentioned applications, the agents submit incomplete preference lists. So one natural generalization of the stable marriage problem is to allow *incomplete* preference lists, in which the men and women rank a subset of the members of the opposite set. As in the case of SM, a stable matching always exists, and can be found by a straightforward generalization of the Gale/Shapley algorithm [4]. While a stable matching may not match everyone, every stable matching matches precisely the same set of men and women [3] – thus all stable matchings have the same size. An alternative generalization of SM arises when we relax the requirement that an agent's preference list be strictly ordered, thus allowing *ties* to appear in the preference lists. Such a relaxation is realistic – consider, for example, a hospital that must otherwise attempt to produce a genuinely strict ranking of hundreds of medical students. In the case of ties it is still computationally easy to find a perfect stable matching by arbitrarily breaking the ties and running the Gale/Shapley algorithm.

When *both* ties and incomplete lists are allowed, the stable matchings for the instance may have different sizes, and, in the worst case, an arbitrary stable matching can be as little as one-half the size of a maximum cardinality stable matching [15]. Since a prime objective of most centralized matching schemes is to match as many agents as possible, the obvious question is the following: "Given an instance of the stable marriage problem with ties and incomplete lists (SMTI), how can we efficiently compute a maximum cardinality stable matching?". Unfortunately, it is NP-hard to do so [15], and hard to approximate [19,5].

1.3 Approximation Results for MAX-SMTI

A straightforward approximation algorithm for MAX-SMTI arises by breaking the ties of the given instance and computing an arbitrary stable matching [15]. A stable matching must always be a maximal matching, hence a 2-approximation is easy. A number of approximation algorithms with a performance guarantee better than 2 have appeared in the literature for both the general version of this problem [9,10,11] and for several special cases [5,6,8].

In a very recent paper, Király [12] provided two simple and elegant algorithms which effectively superseded all previously known approximation algorithms for MAX-SMTI, (save only the randomized algorithm given for the very special case studied in [6]). Király's first algorithm provides a $\frac{3}{2}$-approximation for the restricted case of MAX-SMTI in which ties are allowed to appear only on the women's side (this is the only restriction). The second algorithm provides a $\frac{5}{3}$-approximation for general MAX-SMTI, in which the preference lists and ties of the men and women are completely unrestricted.

```
set every man to be unmatched, unpromoted, and unexhausted;
while (∃m: m is unmatched and (m is unpromoted or m is unexhausted))
    if (m is exhausted)
        promote m and set m to be unexhausted;
        reactivate m so that he begins proposing again from the start of his list;
    w ← next woman on m's preference list;                /* m proposes to w */
    if (w is unmatched)
        M ← M ∪ (m, w);                                    /* w accepts m */
    else if (w prefers m to her partner m')
        M ← M ∪ {(m, w)} − {(m', w)};          /* w rejects m' and accepts m */
        set m' to be exhausted if w is the last woman on his list;
    else                                                   /* w rejects m */
        set m to be exhausted if w is the last woman on his list;
```

Fig. 1. Király's algorithm

1.4 Our Contribution

Our contribution in this paper is to provide a three phase $\frac{3}{2}$-approximation algorithm for general MAX-SMTI, improving upon Király's general performance guarantee of $\frac{5}{3}$. Our work builds from some ideas used in Király's first algorithm (henceforth *Király's algorithm*) in the sense that one of the three phases uses a generalization of that algorithm.

2 Background

2.1 Király's Algorithm

For completeness of presentation, we describe a version of Király's algorithm using the concept of *promotion* from a tie rather than that of extra score used by Király [12]. For convenience, we will henceforth refer to the entries of an agent's preference list as a series of ties, where each tie is of size ≥ 1. As input to this algorithm, the men of the instance have strictly ordered preference lists and the women have no restriction on the nature of the ties in their preference lists. The idea behind the algorithm is to iteratively allow men to make proposals to the women on their preference lists, as in the Gale/Shapley algorithm, but with an additional feature. The change here is that a man m who is unmatched after proposing to every woman on his preference list – we use the term *exhausted* to describe such a man – is given one "second chance" in which m is promoted ahead of each tie in which he appears, and is then allowed to propose to each woman on his list a second time.

At the start of the algorithm, each man is set to be *unmatched*, *unpromoted*, and *unexhausted*. The main body of the algorithm is a while loop, which continues as long as there exists a man m who is (i) unmatched and (ii) either unpromoted or unexhausted (or both). If m is exhausted, m is set to be promoted. The operation of promoting m involves examining each woman w who finds m acceptable, and, if m is in a tie of size at least 2 on w's list, m is promoted immediately ahead of this tie on w's preference list. Furthermore, m is set to be unexhausted and is *reactivated*, meaning he will now begin again making proposals to women starting from the beginning of his preference list. The algorithm proceeds by m proposing to the top woman w on his preference list to whom he has not yet proposed (or to whom he has proposed only once, if he has been reactivated). When a man m proposes to a woman w, she rejects her current partner (if any) and accepts m if m is a *strict* improvement for her, taking into account any promotions that may have occurred. Otherwise, she retains her current partner and rejects m. On rejection, a man becomes (or remains) unmatched. When a man has been rejected by every woman on his list, he is set to be exhausted.

When Király's algorithm ends, each man is either (i) matched (possibly having been previously promoted as well), or (ii) promoted, exhausted, and unmatched. A pseudocode description of Király's algorithm is given in Figure 1.

2.2 Gallai-Edmonds Decomposition Theorem

Phase 2 of our approximation algorithm uses a classical result regarding bipartite matchings known as the *Gallai-Edmonds decomposition theorem*. In this subsection we review the parts of this theorem that we will need in the forthcoming sections. To this end, let $G = (U \cup V, E)$ be a bipartite graph and M a maximum cardinality matching of G. With respect to M, we partition the vertex set of G in the following way. A vertex v is said to be *odd* (respectively, *even*) if there exists an odd (respectively, even) length alternating path from some unmatched vertex to v. A vertex v is said to be *unreachable* if there is no alternating path to v beginning at some unmatched vertex. The following Gallai-Edmonds decomposition theorem provides an important characterization of the set of maximum cardinality matchings of G with respect to this vertex partition [14].

Theorem 1. *Let $G = (U \cup V, E)$ be a bipartite graph and M be a maximum cardinality matching for G. Let $\mathcal{E}$, $\mathcal{O}$, and $\mathcal{U}$ be the set of even, odd, and unreachable vertices as defined above with respect to G and M. Then,*

(1) $\mathcal{E}$, $\mathcal{O}$, and $\mathcal{U}$ are pairwise disjoint. Every maximum matching for G partitions the vertex set of G into the same sets of even, odd, and unreachable vertices.
(2) In any maximum-cardinality matching of G, every vertex in $\mathcal{O}$ is matched with some vertex in $\mathcal{E}$, and every vertex in $\mathcal{U}$ is matched with another vertex in $\mathcal{U}$. The size of a maximum-cardinality matching is $|\mathcal{O}| + |\mathcal{U}|/2$.
(3) There is no edge in G connecting a vertex in $\mathcal{E}$ with another vertex in $\mathcal{E}$ or a vertex in $\mathcal{U}$.

We note that the Gallai-Edmonds decomposition of a bipartite graph can be obtained as a by-product of a maximum cardinality matching algorithm. Having reviewed this important theorem, we proceed to the discussion of our approximation algorithm.

3 The Approximation Algorithm

Our approximation algorithm consists of 3 phases. A pseudocode description is given in Figures 2 and 3. In the first phase, we use an approach somewhat

```
M ← ∅;
set all men to be unmatched, unpromoted, unexhausted, and unstalled;
Phase 1:
while (∃m: m is unmatched and unstalled and (m is unpromoted or m is unexhausted))
    if (m is exhausted)
        promote m and set m to be unexhausted;
        reactivate m so that he begins proposing again from the start of his list;
    t ← m's current tie;
    if (|t| ≥ 2)
        if (t contains exactly one unmatched woman w)
            promote w ahead of t;
        else if (t contains no unmatched woman)
            break t arbitrarily;
        else
            set m to be stalled;
    else
        w ← only woman in t;                              /* m proposes to w */
        if (w is unmatched)
            M ← M ∪ (m, w);                               /* w accepts m */
            unstall the appropriate men, if any;
        else if (w prefers m to her partner m')
            M ← M ∪ {(m, w)} − {(m', w)};     /* w rejects m' and accepts m */
            set m' to be exhausted if w is the last woman on his list;
        else                                              /* w rejects m */
            set m to be exhausted if w is the last woman on his list;
if (the set S of stalled men is empty)
    return M;
else
    go to phase 2;
```

Fig. 2. Phase 1 of the approximation algorithm

similar to the Király algorithm, adapted to take into account the ties in the men's preference lists. In this phase men again may become promoted, exhausted, and matched, but may also enter a different state in which they become *stalled*. The meaning of this state will become clear in the description of phase 1 below. Phase 1 takes as input a matching M, and the set of women and men. Prior to calling the phase 1 algorithm for the first time, each man is set to be unmatched, unpromoted, unexhausted, and unstalled, and the matching M is initialized to be empty.

3.1 Phase 1

In the first phase of the algorithm, the men iteratively make proposals to the women on their preference lists in a similar way to the Király algorithm. The main body of this phase is again a while loop, which continues as long as there exists a man m who is (i) unmatched and unstalled and (ii) unpromoted or unexhausted (or both). If m is exhausted, he is promoted, set to be unexhausted, and is *reactivated*, precisely as described in the Király algorithm. Next, we let

Phase 2:
Construct the phase-2 graph $G = (U \cup V, E)$;
$N \leftarrow$ maximum cardinality matching in G;
identify the sets $\mathcal{E}$, $\mathcal{O}$, and $\mathcal{U}$;
$N' \leftarrow$ subset of N obtained by removing all pairs
    (m, w) such that $m \in \mathcal{O}$ and $w \in \mathcal{E}$;
if $(N' = \emptyset)$
    go to phase 3;
else
    for $(m, w) \in N'$
        promote w ahead of m's current tie; /* m proposes to w */
        $M \leftarrow M \cup (m, w)$;
        set m to be unstalled;
    unstall all men in U who are unmatched in N;
    go to phase 1;
Phase 3:
for $(m, w) \in N$ /* m proposes to w */
    $M \leftarrow M \cup (m, w)$;
return M;

Fig. 3. Phases 2 and 3 of the approximation algorithm

t denote the first tie on m's preference list containing a woman w to whom m has not yet proposed (or to whom he has proposed only once, if he has been reactivated). We refer to t as m's *current tie*. The algorithm then proceeds based on the following cases concerning t. The first case is if the size of t is at least 2. If t also contains exactly one unmatched woman w, w is promoted ahead of t on m's preference list. If instead t contains no unmatched women, t is broken arbitrarily on m's preference list, creating a total order of these women to replace t on his list. Otherwise, t must contain at least 2 unmatched women, and m is set to be stalled. The second case is if the size of t is exactly one. In this case m proposes to w, the only woman in t. When a man m proposes to a woman w, she accepts if m is a strict improvement for her, taking into account any promotions that have been made. Otherwise, she rejects m. When an unmatched woman w becomes matched, the men who, as a result, have now just one unmatched woman in their current tie are unstalled. As before, when a man has been rejected by every woman on his preference list, he is set to be exhausted.

The primary task of phase 1 ends with the termination of this while loop. At this point in the execution of the approximation algorithm every man m is in exactly one of three categories: (i) m is matched to an acceptable woman (and possibly is promoted as well), or (ii) m is exhausted, promoted and unmatched, having been rejected by every woman on his preference list despite his promotion, or (iii) m is stalled.

If the set S of stalled men is empty, the algorithm returns the current matching and halts. Otherwise, we proceed to phase 2.

3.2 Phase 2

Phase 2 of the algorithm takes as input a matching M, along with the set of men and women. The goal of the algorithm in this phase is to attempt to match a certain subset of the stalled men. We construct a bipartite graph $G = (U \cup V, E)$ with U being the set of men in S and V being the set of unmatched women appearing in the current tie of some man in S. We refer to these men and women and the vertices of G representing them interchangeably. The set of edges are those (man,woman) pairs (m, w) such that $w \in V$ appears in m's current tie. We call this graph the *phase-2 graph*. The algorithm then computes a maximum cardinality matching N in G.

We proceed by removing selected pairs from N in the following way. We identify the sets $\mathcal{E}$, $\mathcal{O}$, and $\mathcal{U}$ of vertices as described according to the Gallai-Edmonds decomposition theorem in Section 2.2. All pairs in N consisting of a man $m \in \mathcal{O}$ and a woman $w \in \mathcal{E}$ are removed from N, yielding a new matching $N' \subseteq N$. One of the crucial properties of N' (proved in Lemma 2) is that, for each man m who is matched in N', if $w_1, w_2, \dots w_t$ are the unmatched (in M) women in m's current tie, then $w_1, w_2, \dots, w_t$ are also matched in N'. This important property of N' is key to the establishment of the performance guarantee of the algorithm.

If N' is empty, we proceed to phase 3. Otherwise, for every pair $(m, w) \in N'$, w is promoted locally ahead of m's current tie. Man m then proposes to w, who accepts because she is unmatched in M, and this pair is added to M. All the men matched in N' are now set to be unstalled.

At this point in phase 2, the assignment of any man not in S has remained unchanged, as the matching has changed only by matching previously unmatched women to men in S. However, the situation of the men who were in S at the beginning of phase 2 has, of course, changed. We claim (proven in Lemma 2) that those men m remaining in S fall into one of two categories: (i) m was matched in N, is not matched in N', and still has at least 2 unmatched women in his current tie, or (ii) m was unmatched in N and every woman in his current tie is now matched in M. The men in (ii) are set to be unstalled, and the algorithm returns to phase 1.

3.3 Phase 3

Phase 3 takes as input the current matching M along with the matching N constructed in the execution of phase 2 which passed control to phase 3. The algorithm arrives at phase 3 if and only if the matching N' of phase 2 is empty. We will show (in Lemma 1) that this implies that N matches every man in S. The algorithm terminates after the man in each pair in N proposes to his partner in N – all of these women are single – and these pairs are added to M. The current matching M is returned.

4 Correctness

We next establish a few key properties of the algorithm, and verify certain claims made in the description of the pseudocode.

Lemma 1. *Let S denote the set of stalled men at the start of an arbitrary execution of phase 2 of the approximation algorithm. If the matching N' constructed in this call is empty, then the corresponding maximum cardinality matching N matches every man in S.*

Proof. Suppose N' is empty, and that a man m is unmatched by N. Let w be an arbitrary neighbor of m in G. Since m is not matched in N, m is even (i.e. $m \in \mathcal{E}$), and therefore w is odd ($w \in \mathcal{O}$). Since N is maximal, w was matched to a man m' in N, who therefore must also be even. But this implies the pair (m', w) could not have been removed from N, as it consists of an even man and an odd woman.

Corollary 1. *Phase 3 of the approximation algorithm finds a matching that matches every man who was in S in the preceding execution of phase 2.*

Lemma 2. *Let m be a stalled man in the set S with current tie t at the start of an arbitrary execution of phase 2. Then, exactly one of the following is true of m when that execution of phase 2 ends (i.e., the instant before either goto statement in phase 2 is invoked).*

(1) m was matched in N', so m is now matched in M to a woman in t, and every woman in t is matched in M.
(2) m was matched in N but not in N', m's current tie is still t, and there are at least two women in t who are still unmatched in M.
(3) m was unmatched in N, m's current tie is still t, and every woman in his current tie is now matched in M.

Proof. (1) Suppose m was matched in N'. Then, m was an even or unreachable vertex with respect to M. Therefore, all neighbors of m in G are either odd or unreachable, and could not have been deleted from N, for only even women are removed from N. It follows that all of m's neighbors are in N', and therefore they all receive proposals in this execution of phase 2, and are matched in M.

(2) If instead m is matched to a woman w in N but is unmatched in N', then m was removed from N because he is an odd vertex. We establish the claim by showing there is another even woman $w' \neq w$ who is adjacent to m and is unmatched in N' as well. To see this, consider the path of odd length that makes m an odd vertex. This path cannot reach him via his partner in the matching, for alternate edges in that path would have to be edges in the matching. Hence the first edge in the path would be in the matching (since the last edge is), contradicting the fact that the starting vertex in the path must be unmatched. Therefore this path must reach him from another neighboring vertex w', which must be even. This woman is unmatched in N', for she can only be matched to an odd man in N or unmatched in N.

(3) Finally, if m is unmatched by N he is an even vertex. All women in his current tie are therefore odd vertices, are matched in N because N is maximal, and could not have been removed from N. Therefore, these women are all matched in N' and all receive proposals in this execution of phase 2, and hence are matched in M.

Lemma 3. *On termination of the approximation algorithm, any man who remains unmatched has been promoted, and has been rejected by every woman on his list even after becoming promoted.*

Proof. The execution of the algorithm can only halt in one of two places. The first place is at the end of phase 1, on the condition that there are no stalled men. This implies that every unmatched man is promoted and has still been rejected by every woman on his list. The other point at which the algorithm may terminate is in phase 3. Now, control reaches phase 3 only if, in phase 2, it is discovered that N' is empty, implying that N matches every man in S by Corollary 1. Notice that when this happens nothing is done in phase 2 to modify the assignment of any agent, rather phase 2 simply passes control to phase 3, which matches every man in S. Hence, the unmatched men are those who were unmatched after the final call to phase 1, and, as described above, they must have become exhausted while promoted.

Lemma 4. *Suppose a woman w becomes matched to a man m at some point in the execution of the approximation algorithm. Then, w only rejects m if she accepts a proposal from a man ranked at least as highly as m on w's (original) preference list.*

Proof. Matched women can only change their partner in one place in the approximation algorithm, and that is when receiving a proposal from a man they strictly prefer, possibly after promotions, to their current partner in phase 1. This new suitor must be ranked at least as highly as w's current partner on w's original preference list.

Lemma 5. *The matching M returned at the end of the approximation algorithm is a stable matching.*

Proof. Suppose that (m, w) blocks M. The essence of the approximation algorithm from a man's point of view is a left-to-right sweep of his preference list in which, if necessary, he becomes promoted and again makes another left-to-right sweep of his preference list. Hence, for m to prefer w, he must have proposed to her at least once, whether it be in phase 1 or phase 2 (he cannot have proposed to her in phase 3, for otherwise they would be matched in M). The fact that w has rejected m along with Lemma 4 implies that w does not prefer m to her current partner in M, and hence (m, w) do not block M.

Lemma 6. *The approximation algorithm runs in $O(n^{3/2}L)$ time, where n is the sum of the men and women and L is the sum of the lengths of the preference lists.*

Proof. The algorithm essentially constitutes one or two partial or complete left to right sweeps of the men's preference lists, interleaved with calls to phase 2. The total number of calls to phase 2 is bounded by the number of men, as each call to phase 2 either strictly increases the size of M or passes control to phase 3, in which phase the algorithm terminates. Any one execution of phase 2 requires a total of $O(\sqrt{|V|}|E|) = O(\sqrt{n}L)$ time, as the construction of N is the dominant step of phase 2. In the worst case $\Omega(n)$ calls could be made to phase 2, each of which computes a matching N of size $\Omega(n)$ but a matching N' of size $O(1)$. These successive calls to phase 2 would clearly dominate the complexity, yielding a bound of $O(n^{3/2}L)$.

5 The Performance Guarantee

Let M be the stable matching returned by the approximation algorithm and let M_{opt} denote an optimal stable matching for a given instance of MAX-SMTI. Consider the symmetric difference $M \oplus M_{opt}$ of these two matchings. The components of the underlying graph of $M \oplus M_{opt}$ consist of alternating cycles and paths. Each cycle component in $M \oplus M_{opt}$ is of even length, so the ratio of M-edges to M_{opt}-edges in these components is one. For an alternating path component, the ratio of M_{opt}-edges to M-edges is always at most 3/2 except for a component which is a path of length 3 with its endpoints in M_{opt}. Therefore, if we can establish that $M \oplus M_{opt}$ contains no such path, we will have shown the ratio of M_{opt}-edges to M-edges in each component is at most 3/2, establishing that the approximation algorithm is a $\frac{3}{2}$-approximation algorithm. To this end, suppose for a contradiction that $P_4 = w' - m - w - m'$ exists in $M \oplus M_{opt}$ such that (m, w) is the M-edge, and (m, w') and (m', w) are the M_{opt}-edges. We begin with several facts with easy proofs.

Fact 1. *The man m' in P_4 must be exhausted and promoted. This follows from Lemma 3 and the fact that m' is unmatched in M.*

Fact 2. *The man m in P_4 was never promoted by the approximation algorithm. This is because he has an unmatched woman w' on his preference list, and thus could never have become exhausted, for w' can never have received a proposal.*

Fact 3. *Woman w in P_4 strictly prefers m to m' in her original preference list. If w strictly prefers m' to m, then (m', w) is a blocking pair for M, a contradiction. If, instead, she were indifferent between these two men, she could not have rejected m', who, by Lemma 3, must have proposed to w at some point after being promoted. But at that moment w was matched to m or someone ranked lower, and m was never promoted.*

Fact 4. *Man m in P_4 is indifferent between w and w' in his original preference list. If he strictly prefers w' to w, then (m, w') blocks M. But by the previous fact, m forms a blocking pair with w in M_{opt} if he strictly prefers w to w'.*

Lemma 7. *Man m' proposed to woman w prior to the end of the final execution of phase 1 and was rejected by her. Hence, w is matched prior to any potential call to phase 3.*

Proof. Every man who participates in phase 3 becomes matched, hence m' did not participate in phase 3, and since no matched man becomes unmatched during phase 3 or phase 2, man m' was unmatched at the end of the final call to phase 1. By Lemma 3, m' proposed to w even after becoming promoted, but since he is single in M, she must have rejected him. Hence, w is matched to someone at least as good as m' at the final call to phase 1.

Now, we arrive at the contradiction. Since matched women never become unmatched, man m always had woman w' unmatched on his preference list, and by Fact 4 she is tied with w. In which phase of the algorithm can m have become matched to w? It cannot have been in phase 1, for the phase 1 algorithm does not allow him to propose to w, regardless of whether or not w is matched, because of w' being tied with w and unmatched. But m cannot become matched to w in some call to phase 2 either, for the fact that w' is unmatched at the end of the algorithm implies she could never be in N' at any call to phase 2. By Lemma 2, this implies that the pair (m, w) would have to be deleted in the creation of N' as well. Thus, we conclude that m became matched to w in phase 3. However, men only become matched to unmatched women in phase 3, implying that w is single at the start of phase 3, a contradiction to Lemma 7.

Theorem 2. *The polynomial-time approximation algorithm outputs a stable matching at least $\frac{2}{3}$ the size of an optimal stable matching.*

6 Conclusion

We have presented a polynomial-time approximation algorithm for general MAX-SMTI with a performance guarantee of 3/2, improving the previously best known algorithm for this problem. There is an example, due to Yanagisawa [20], which shows this analysis is tight. We note that our approximation algorithm also extends to the many-to-one generalization of SMTI (the Hospitals/Residents setting) by a technique involving "cloning" [8], with the same performance guarantee of 3/2.

Acknowledgements. I am indebted to Robert Irving and David Manlove for reading several previous drafts of this paper, and providing invaluable feedback, which greatly improved the presentation. In particular, the proofs of Lemmas 3, 5, and 7, and the description of the pseudocode were much improved by their suggestions. Thanks to Robert Irving for suggesting the use of the Gallai-Edmonds decomposition theorem, which significantly shortened the proofs of Lemmas 1 and 2. Thanks also to Peter Biró and Christine Cheng for helpful comments on an earlier draft.

References

1. Biró, P.: Student Admissions in Hungary as Gale and Shapley Envisaged. University of Glasgow, Computing Science Department Research Report, TR-2008-291 (2008)
2. Gale, D., Shapley, L.: College admissions and the stability of marriage. American Mathematical Monthly 69, 9–15 (1962)

3. Gale, D., Sotomayor, M.: Some remarks on the stable matching problem. Discrete Applied Mathematics 11, 223–232 (1985)
4. Gusfield, D., Irving, R.W.: The Stable Marriage Problem: Structure and Algorithms. MIT Press, Cambridge (1989)
5. Halldórsson, M., Iwama, K., Miyazaki, S., Yanagisawa, H.: Improved approximation results for the stable marriage problem. ACM Trans. Algorithms 3(3), 30 (2007)
6. Halldórsson, M., Iwama, K., Miyazaki, S., Yanagisawa, H.: Randomized approximation of the stable marriage problem. Theoretical Computer Science 325(3), 439–465 (2004)
7. Irving, R.W., Leather, P.: The complexity of counting stable marriages. SIAM Journal on Computing 15(3), 655–667 (1986)
8. Irving, R.W., Manlove, D.F.: Approximation algorithms for hard variants of the stable marriage and hospitals/residents problems. Journal of Combinatorial Optimization 16, 279–292 (2008)
9. Iwama, K., Miyazaki, S., Okamoto, K.: A $\left(2 - c\frac{\log n}{n}\right)$-approximation algorithm for the stable marriage problem. In: Hagerup, T., Katajainen, J. (eds.) SWAT 2004. LNCS, vol. 3111, pp. 349–361. Springer, Heidelberg (2004)
10. Iwama, K., Miyazaki, S., Yamauchi, N.: A $(2 - c(1/\sqrt(N)))$-Approximation Algorithm for the Stable Marriage Problem. Algorithmica 51(3), 342–356 (2008)
11. Iwama, K., Miyazaki, S., Yamauchi, N.: A 1.875–approximation algorithm for the stable marriage problem. In: 18th ACM/SIAM Symposium on Discrete Algorithms, pp. 288–297 (2007)
12. Király, Z.: Better and simpler approximation algorithms for the stable marriage problem. In: Halperin, D., Mehlhorn, K. (eds.) Esa 2008. LNCS, vol. 5193, pp. 623–634. Springer, Heidelberg (2008)
13. Knuth, D.E.: Mariages Stables. Les Presses de L'Université de Montréal (1976)
14. Lovász, L., Plummer, M.D.: Matching Theory. Annals of Discrete Mathematics, vol. 29. North-Holland, Amsterdam (1986)
15. Manlove, D.F., Irving, R.W., Iwama, K., Miyazaki, S., Morita, Y.: Hard variants of stable marriage. Theoretical Computer Science 276(1-2), 261–279 (2002)
16. Canadian Resident Matching Service, `http://www.carms.ca/jsp/main.jsp`
17. National Resident Matching Program, `http://www.nrmp.org/about_nrmp/how.html`
18. Scottish Foundation Allocation Scheme, `http://www.nes.scot.nhs.uk/sfas/`
19. Yanagisawa, H.: Approximation algorithms for stable marriage problems, PhD thesis, Kyoto University, Graduate School of Informatics (2007)
20. Yanagisawa, H.: Personal communication (2008)

Limiting Negations in Formulas*

Hiroki Morizumi**

Graduate School of Information Sciences, Tohoku University,
Sendai 980-8579, Japan
morizumi@ecei.tohoku.ac.jp

Abstract. Negation-limited circuits have been studied as a circuit model between general circuits and monotone circuits. In this paper, we consider limiting negations in formulas. The minimum number of NOT gates in a Boolean circuit computing a Boolean function f is called the inversion complexity of f. In 1958, Markov determined the inversion complexity of every Boolean function and particularly proved that $\lceil \log_2(n+1) \rceil$ NOT gates are sufficient to compute any Boolean function on n variables. We determine the inversion complexity of every Boolean function in formulas, i.e., the minimum number of NOT gates (NOT operators) in a Boolean formula computing (representing) a Boolean function, and particularly prove that $\lceil n/2 \rceil$ NOT gates are sufficient to compute any Boolean function on n variables. Moreover we show that if there is a polynomial-size formula computing a Boolean function f, then there is a polynomial-size formula computing f with at most $\lceil n/2 \rceil$ NOT gates. We consider also the inversion complexity in formulas of negation normal form and prove that the inversion complexity is at most polynomials of n for every Boolean function on n variables.

1 Introduction

When we consider Boolean circuits with a limited number of NOT gates, there is a basic question: Can a given Boolean function be computed by a circuit with a limited number of NOT gates? This question has been answered by Markov [11] in 1958. The *inversion complexity* of a Boolean function f is the minimum number of NOT gates required to construct a Boolean circuit computing f, and Markov completely determined the inversion complexity of every Boolean function f. In particular, it has been shown that $\lceil \log_2(n+1) \rceil$ NOT gates are sufficient to compute any Boolean function.

After more than 30 years from the result of Markov, Santha and Wilson [17] investigated the inversion complexity in *constant depth circuits* and showed that on the restriction $\lceil \log_2(n+1) \rceil$ NOT gates are not sufficient to compute a Boolean function. The result has been extended to *bounded depth circuits* by Sung and Tanaka [19]. Recently we completely determined the inversion complexity of

* Some results of this paper have appeared in the preliminary version [12].

** Present Affiliation: Graduate School of Informatics, Kyoto University, Kyoto 606-8501, Japan (morizumi@kuis.kyoto-u.ac.jp).

S. Albers et al. (Eds.): ICALP 2009, Part I, LNCS 5555, pp. 701–712, 2009.

every Boolean function in *non-deterministic circuits*, and particularly proved that one NOT gate is sufficient to compute any Boolean function if we can use an arbitrary number of guess inputs [13].

A Boolean circuit whose gates have fan-out one is called a *formula*. Formulas are one of well-studied circuit models in circuit complexity theory [6,21]. Note that a Boolean circuit whose gates have fan-out one corresponds to a Boolean formula. In this paper, we consider the inversion complexity in formulas, which corresponds to the minimum number of NOT operators in a Boolean formula representing a Boolean function. As far as we know, there is no result for the inversion complexity in formulas. We completely determine the inversion complexity of every Boolean function in formulas, and particularly prove that $\lceil n/2 \rceil$ NOT gates are sufficient to compute any Boolean function on n variables. (Section 3)

The best known lower bound on the size of formulas computing an explicit Boolean function is $\Omega(n^{3-o(1)})$ by Håstad [9], while exponential lower bounds have been known on the size of *monotone* formulas (actually even monotone circuits [1,4,16]). This situation is similar to the present situation on the size of circuits. More precisely, while exponential lower bounds have been known on the size of monotone circuits, no superlinear lower bound has been known so far on the size of circuits. The study of negation-limited circuits [2,3,5,10,18] is motivated by the situation: what happens if a limited number of NOT gates are allowed? By the reason above, we consider negation-limited formulas and show that if there is a polynomial-size formula computing a Boolean function f, then there is a polynomial-size formula computing f with at most $\lceil n/2 \rceil$ NOT gates. (Section 4) The fact means that when one try to prove a superpolynomial size lower bound for formulas, it is enough to prove a superpolynomial size lower bound for formulas with at most $\lceil n/2 \rceil$ NOT gates. To prove a superpolynomial lower bound on the size of formulas computing an explicit function is one of the most challenging open problems in computational complexity theory.

We consider also the inversion complexity in *negation normal form (NNF)* formulas, which are formulas with a restriction that all NOT operators are used in literals. The restriction does not affect the size of formulas since DeMorgan's laws convert an arbitrary formula to an NNF formula with the same size. On the other hand, the restriction strongly affects the inversion complexity. (An example is in Section 5.) We prove that the inversion complexity in NNF formulas is at most polynomials of n for every Boolean function on n variables. (Section 5)

2 Preliminaries

2.1 Definitions

A *circuit* is an acyclic Boolean circuit which consists of AND gates of fan-in two, OR gates of fan-in two and NOT gates. A *formula* is a circuit whose gates have fan-out one. A *negation normal form (NNF)* formula is a formula with a restriction that NOT gates are connected only from inputs. (Formulas (with no restriction) and NNF formulas often are not distinguished if we are interested in the size, since the restriction does not affect the size. In this paper we distinguish formulas and NNF formulas since the restriction strongly affects the number of NOT gates.) A *monotone* formula is a formula which consists of only AND gates and OR gates. The size of a formula is defined to be the number of inputs in the formula. (Each of inputs corresponding to a same variable is distinguished.) By the definition, the size of a formula is precisely one more than the number of AND gates and OR gates in the formula. We denote by $L(f)$ the size of the smallest formula computing a function f. We denote by $not(C)$ the number of NOT gates in a formula C.

Let x and x' be Boolean vectors in $\{0,1\}^n$. $x \leq x'$ means $x_i \leq x'_i$ for all $1 \leq i \leq n$. $x < x'$ means $x \leq x'$ and $x_i < x'_i$ for some i. A Boolean function f is called *monotone* iff $f(x) \leq f(x')$ whenever $x \leq x'$.

The threshold-k-function $T^n_k(x_1, \ldots, x_n)$ is 1 iff $\Sigma^n_{i=1} x_i \geq k$. The k-th slice f^k $(0 \leq k \leq n)$ of a Boolean function f is defined by

$$f^k(x_1, \ldots, x_n) = \begin{cases} 0 & \text{for } \Sigma^n_{i=1} x_i < k; \\ f(x_1, \ldots, x_n) & \text{for } \Sigma^n_{i=1} x_i = k; \\ 1 & \text{for } \Sigma^n_{i=1} x_i > k. \end{cases}$$

By the definition, the threshold-k-function T^n_k and the k-th slice f^k are monotone for all $0 \leq k \leq n$.

2.2 Markov's Theorem

We denote the inversion complexity of a Boolean function f in circuits by $I(f)$. Markov gave the tight bound of $I(f)$ for every Boolean function f. A *chain* is an increasing sequence $x^1 < x^2 < \cdots < x^k$ of Boolean vectors in $\{0,1\}^n$. The *decrease* $d_X(f)$ of a Boolean function f on a chain X is the number of indices i such that $f(x^i) \not\leq f(x^{i+1})$ for $1 \leq i \leq k-1$. The *decrease* $d(f)$ of f is the maximum of $d_X(f)$ over all increasing sequences X.

Theorem 1 (Markov[11]). *For every Boolean function f,*

$$I(f) = \lceil \log_2(d(f) + 1) \rceil.$$

In Theorem 1, the Boolean function f can also be a multi-output function.

3 Inversion Complexity in Formulas

3.1 Result

We denote by $I_F(f)$ the inversion complexity of a Boolean function f in formulas. In discussion about formulas, we consider only single-output Boolean functions. We completely determine the inversion complexity of every Boolean function in formulas.

Theorem 2. *For every Boolean function f,*

$$I_F(f) = d(f).$$

Since $d(f) \leq \lceil n/2 \rceil$ for every n-input Boolean function f by the definition of $d(f)$, $I_F(f) \leq \lceil n/2 \rceil$ for every f. $I_F(f)$ is exponentially larger than $I(f)$, i.e., inversion complexity in circuits as shown in the following corollary:

Corollary 1. *For every Boolean function f,*

$$I(f) = \lceil \log_2(I_F(f) + 1) \rceil.$$

Proof. It is proved by Theorem 1 and Theorem 2. □

In the rest of this section, we prove Theorem 2.

3.2 Upper Bound

We prove $I_F(f) \leq d(f)$. We use a similar argument to the one which is used to prove Theorem 1 [8].

Proof (the upper bound of $I_F(f)$). We use induction on $d(f)$.
Base: $d(f) = 0$. Then f is monotone and $I_F(f) = 0$.
Induction Step: Suppose

$$I_F(f') \leq d(f')$$

for every Boolean function f' such that $d(f') \leq d(f) - 1$.

First we separate f to two functions f_1 and f_2 as follows. See Fig. 1. Let S be the set of all vectors $x \in \{0,1\}^n$ such that for every chain X starting with x, $d_X(f) = 0$. We define f_1 and f_2 as follows:

$$f_1(x) = \begin{cases} f(x) & \text{if } x \in S; \\ 0 & \text{otherwise,} \end{cases}$$

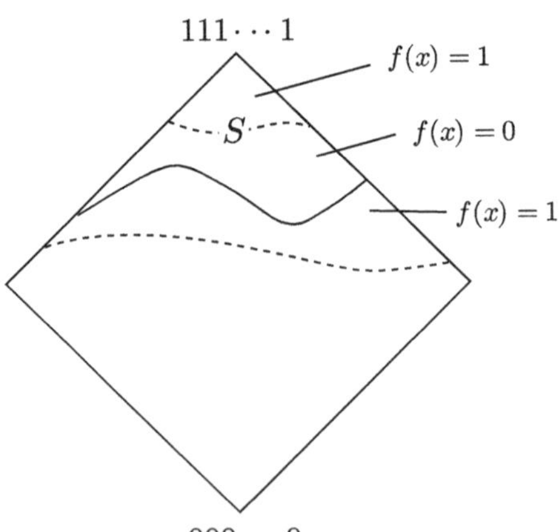

Fig. 1. The separation of f

and

$$f_2(x) = \begin{cases} 1 & \text{if } x \in S; \\ f(x) & \text{otherwise.} \end{cases}$$

We define f_t as follows:

$$f_t(x) = \begin{cases} 1 & \text{if } x \in S; \\ 0 & \text{otherwise.} \end{cases}$$

By the definitions of f_1 and S,

$$d(f_1) = 0. \tag{1}$$

Next we show that $d(f_t) = 0$. Let x and x' be Boolean vectors in $\{0,1\}^n$ such that $x \leq x'$. Suppose that $f_t(x') = 0$, i.e., $x' \notin S$. Since $x' \notin S$, there is a chain X' starting with x' and such that $d_{X'}(f) \geq 1$. Then for a chain X which starts with x and includes X', $d_X(f) \geq 1$. Therefore $x \notin S$. Thus if $f_t(x') = 0$, then $f_t(x) = 0$, which means

$$d(f_t) = 0. \tag{2}$$

Finally we show that

$$d(f_2) \leq d(f) - 1. \tag{3}$$

We assume that $d(f_2) > d(f) - 1$. Since $f_2(x) = 1$ for $x \in S$, there is a chain X_1 ending in a vector $x' \notin S$ and such that $d_{X_1}(f_2) > d(f) - 1$. Since x' is not in S, there is a chain X_2 starting with x' and such that $d_{X_2}(f) \geq 1$. Let X' be the chain which is obtained by connecting X_1 and X_2. Then,

$$\begin{aligned} d_X(f) &= d_{X_1}(f) + d_{X_2}(f) \\ &= d_{X_1}(f_2) + d_{X_2}(f) \\ &> (d(f) - 1) + 1 = d(f). \end{aligned}$$

Thus a contradiction happens.

By the supposition and Eq. (1) to (3), there are a formula C_2 computing f_2 such that $not(C_2) \leq d(f_2) \leq d(f) - 1$ and formulas C_1 and C_t computing f_1 and f_t respectively such that $not(C_1) = not(C_t) = 0$. We construct a formula C computing f from C_1, C_2 and C_t as follows:

$$f_1 \vee (f_2 \wedge \neg f_t).$$

The number of NOT gates in C is

$$\begin{aligned} not(C) &= not(C_1) + not(C_2) + not(C_t) + 1 \\ &\leq d(f). \end{aligned}$$

We show that C computes f for each of the following two cases.

Case 1: The input x is in S.

Then $f_1(x) = f(x)$ and $f_t(x) = 1$. Therefore

$$f_1 \vee (f_2 \wedge \neg f_t) = f_1 = f.$$

Case 2: The input x is not in S.

Then $f_1(x) = 0$, $f_2(x) = f(x)$ and $f_t(x) = 0$. Therefore

$$f_1 \vee (f_2 \wedge \neg f_t) = f_2 = f.$$

Thus the formula C computes f and has at most $d(f)$ NOT gates. Therefore $I_F(f) \leq d(f)$. □

3.3 Lower Bound

We prove $I_F(f) \geq d(f)$. If the input of a NOT gate N is 0 and the output is 1, then we call the state of N *up*. If otherwise, we call the state *down*. We denote by $not_d(C, x)$ the number of NOT gates whose states are down in a formula C given x as the input of C.

Proof (the lower bound of $I_F(f)$). Let C be a formula computing f. Let X be an increasing sequence $x^1 < x^2 < \cdots < x^k$ of Boolean vectors in $\{0,1\}^n$ such that $d_X(f) = d(f)$.

Lemma 1. *Let x and x' be Boolean vectors in $\{0,1\}^n$ such that $x < x'$, $f(x) = 1$ and $f(x') = 0$. Then,*

$$not_d(C, x') - not_d(C, x) \geq 1.$$

Proof. We change the input of C from x to x'. Let $N_1, N_2, \ldots, N_m$ be all NOT gates which change from down state to up state at the time. Since $x < x'$, each N_i for $1 \leq i \leq m$ is connected from N_i' which changes from up state to down state by a path including no NOT gate. Since the output of C changes from 1 to 0, the output of C is also connected from N_o' which changes from up state to down state by a path including no NOT gate. $N_1', N_2', \ldots, N_m'$ and N_o' are distinguished from each other, since C is a formula. Thus the number of NOT gates whose states are down increases by at least one. □

Lemma 2. *Let x and x' be Boolean vectors in $\{0,1\}^n$ such that $x < x'$. Then,*

$$not_d(C, x') - not_d(C, x) \geq 0.$$

Proof. We can use a similar argument to the one of Lemma 1 except we do not consider N_o'. □

Since on X the number of indices i such that $f(x^i) = 1$ and $f(x^{i+1}) = 0$ is at least $d(f)$, by Lemma 1 and 2,

$$not_d(C, x^k) - not_d(C, x^1) \geq d(f).$$

Thus C includes at least $d(f)$ NOT gates. □

4 The Size of Negation-Limited Formulas

4.1 Result

In this section, we show that a relation between the size of negation-limited formulas and general formulas. We denote by $L_r(f)$ the size of the smallest formula with at most r NOT gates computing f. In Section 3, we showed that $I_F(f) \leq \lceil n/2 \rceil$ for any function f. In this section, we show that an arbitrary formula can be converted to a formula with at most $\lceil n/2 \rceil$ NOT gates and the size increases only polynomially.

Theorem 3. *For every n-input Boolean function f,*

$$L_{\lceil n/2 \rceil}(f) \leq L(f) \cdot O(n^{6.3}).$$

Theorem 3 implies that when one try to prove a superpolynomial lower bound for $L(f)$, it is enough to prove a superpolynomial lower bound for $L_{\lceil n/2 \rceil}(f)$. In the rest of this section, we prove Theorem 3.

4.2 Proof

Similar relations to Theorem 3 are known in circuits [5,7,8,14]. (We will mention it again in Sec. 6.) All of these proofs are based on a construction of negation-limited inverters. In formulas, we cannot use a similar proof method by the obstacle that the fan-out of gates in a formula is bounded to one.

We denote the vector $(x_1, \ldots, x_{i-1}, x_{i+1}, \ldots, x_n)$ by $(x)_i$. It is known that the monotone function $T_k^{n-1}((x)_i)$ operates as $\neg x_i$ on some conditions, which fact relates to monotone circuit complexity for slice functions. (See, e.g., Sec. 6.13 of [21].) We use a similar idea as the following lemma.

Lemma 3. *$T_k^{n-1}((x)_i) = \neg x_i$ if $\Sigma_{j=1}^n x_j = k$.*

Proof. If $x_i = 0$, then $(\Sigma_{j=1}^n x_j) - x_i = k$. Therefore $T_k^{n-1}((x)_i) = 1 = \neg x_i$. If $x_i = 1$, then $(\Sigma_{j=1}^n x_j) - x_i = k - 1$. Therefore $T_k^{n-1}((x)_i) = 0 = \neg x_i$. □

We call a Boolean function g a *pseudo k-th slice* of f iff $g(x) = f(x)$ for all x such that $\Sigma_{j=1}^n x_j = k$. By Lemma 3, $T_k^{n-1}((x)_i)$ is a pseudo k-th slice of $\neg x_i$. (Actually $T_k^{n-1}((x)_i)$ is the k-th slice of $\neg x_i$, though it is not needed for the proof of Theorem 3.)

Proof (Theorem 3). We suppose that n is an even number and omit the case that n is an odd number, which is easier than the case that n is even. We construct a formula C which computes f and includes $n/2$ NOT gates and has size $L(f) \cdot O(n^{6.3})$.

First we construct n monotone formulas C_k's $(0 \leq k \leq n-1)$ computing a pseudo k-th slice of f if k is an even number and computing a pseudo k-th slice of $\neg f$ if k is an odd number by the following way. See also Fig. 2.

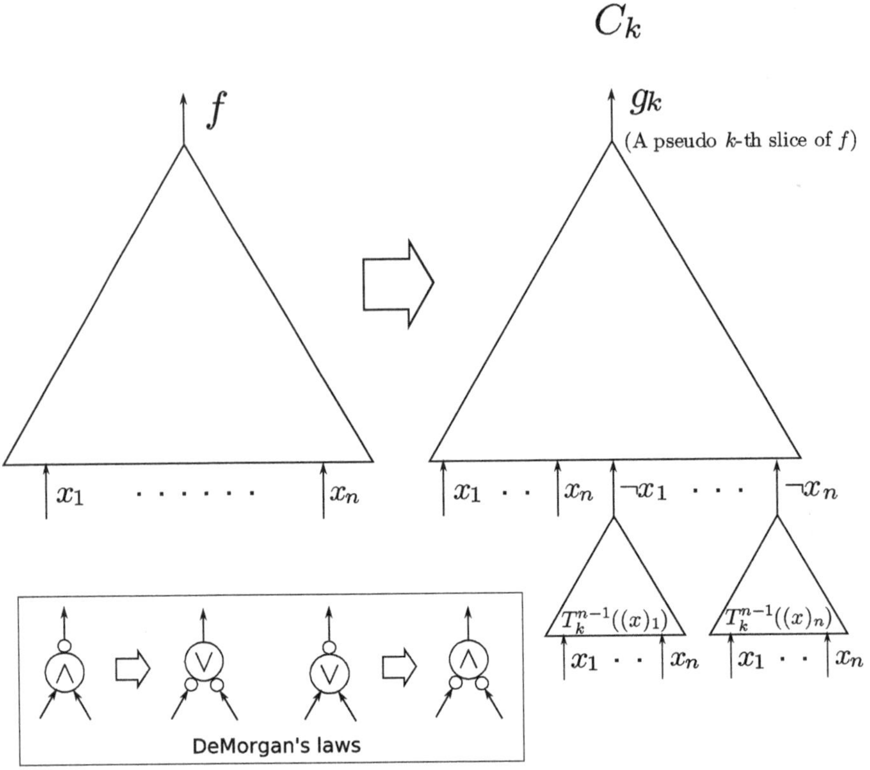

Fig. 2. The construction of C_k (k is an even number)

1. Construct a formula computing f if k is even and computing $\neg f$ if k is odd whose size is $L(f)$.
2. Move all NOT gates to the input side by DeMorgan's laws.
3. Replace each $\neg x_i$ (precisely, each x_i which connect to a NOT gate and the NOT gate) to a formula computing a pseudo k-th slice of $\neg x_i$.

In the third step, we use $T_k^{n-1}((x)_i)$ as a pseudo k-th slice of x_i. Since $T_k^{n-1}((x)_i)$ is monotone, we can construct a monotone formula computing $T_k^{n-1}((x)_i)$. Thus the obtained formula C_k has no NOT gate. If an input x given to C_k satisfies $\Sigma_{j=1}^{n} x_j = k$, then pseudo k-th slices of $\neg x_i$ equal $\neg x_i$ and C_k computes f if k is even and $\neg f$ if k is odd. Therefore C_k computes a pseudo k-th slice of f or $\neg f$. The size of each C_k's is $L(f) \cdot O(n^{5.3})$ for all $0 \leq k \leq n-1$, since each $\neg x_i$ was replaced to a monotone formula computing $T_k^{n-1}((x)_i)$ and the known best upper bound on the size of monotone formulas computing the threshold functions is $O(n^{5.3})$ [20].

Let g_k be the function computed by C_k for $0 \leq k \leq n-1$. As shown above, g_k is a pseudo k-th slice of f (resp. $\neg f$) if k is even (resp. odd). We construct a formula C as follows:

$$\bigvee_{i=0}^{n/2} h_i$$

where

$$h_i = T^n_{2i} \wedge (g_{2i} \vee T^n_{2i+1}) \wedge \neg((g_{2i+1} \wedge T^n_{2i+1}) \vee T^n_{2i+2}) \text{ for } 0 \le i \le n/2 - 1,$$

$$h_{n/2} = \begin{cases} T^n_n & \text{if } f(1,1,\ldots,1) = 1; \\ 0 & \text{otherwise.} \end{cases}$$

Since we construct formulas computing threshold functions (T^n_i $(0 \le i \le n+1)$) with no NOT gate, $not(C) = n/2$. The size of C is $L(f) \cdot O(n^{6.3})$, since the size of each C_k's is $L(f) \cdot (n^{5.3})$ for all $0 \le k \le n-1$ and at most $O(n)$ monotone formulas computing the threshold functions, which can be constructed with size $O(n^{5.3})$ [20], are used.

We show that C computes f, i.e., $\vee_{i=0}^{n/2} h_i = f$. To show it, it is enough that we prove the following equation:

$$h_i = \begin{cases} f & \text{if } \Sigma^n_{j=1} x_j = 2i \text{ or } 2i+1; \\ 0 & \text{otherwise,} \end{cases} \quad \text{for } 0 \le i \le n/2.$$

Case 1: $0 \le i \le n/2 - 1$.

If $\Sigma^n_{j=1} x_j < 2i$, then $T^n_{2i}(x) = 0$. Therefore $h_i = 0$.

If $\Sigma^n_{j=1} x_j = 2i$, then $T^n_{2i}(x) = 1$, $T^n_{2i+1}(x) = 0$ and $T^n_{2i+2}(x) = 0$. Therefore $h_i = g_{2i}$. Since g_{2i} is a pseudo $2i$-th slice of f, $h_i = f$.

If $\Sigma^n_{j=1} x_j = 2i+1$, then $T^n_{2i}(x) = 1$, $T^n_{2i+1}(x) = 1$ and $T^n_{2i+2}(x) = 0$. Therefore $h_i = \neg g_{2i+1}$. Since g_{2i+1} is a pseudo $(2i+1)$-th slice of $\neg f$, $h_i = f$.

If $\Sigma^n_{j=1} x_j > 2i+1$, then $T^n_{2i+2}(x) = 1$. Therefore $h_i = 0$.

Case 2: $i = n/2$.

If $\Sigma^n_{j=1} x_j < 2i = n$, then $T^n_n(x) = 0$. Therefore $h_i = h_{n/2} = 0$.

If $\Sigma^n_{j=1} x_j = 2i = n$, then $T^n_n(x) = 1$. Therefore $h_i = h_{n/2} = f(x)$.

The case that $\Sigma^n_{j=1} x_j > 2i = n$ does not happen since $\Sigma^n_{j=1} x_j \le n$. □

5 Inversion Complexity in NNF Formulas

5.1 Result

In this section, we consider the inversion complexity in NNF formulas. The restriction that formulas are NNF strongly affects the inversion complexity. Consider $\neg T^n_{\lceil n/2 \rceil}$ which is a function such that it equals the negation of the majority function for example. The inversion complexity of $\neg T^n_{\lceil n/2 \rceil}$ in formulas is 1, since $T^n_{\lceil n/2 \rceil}$ is a monotone function. On the other hand, the inversion complexity of $\neg T^n_{\lceil n/2 \rceil}$ in NNF formulas is at least n by the following simple observation. Suppose an input such that $x_i = 0$ and $\Sigma^n_{j=1} x_j = \lceil n/2 \rceil - 1$. If we change x_i from 0 to 1, the output changes from 1 to 0. Therefore there is at least one input of $\neg x_i$ in any NNF formula computing $\neg T^n_{\lceil n/2 \rceil}$ for all $1 \le i \le n$.

We prove that the inversion complexity in NNF formulas is at most polynomials of n for every Boolean function on n variables. Note that the minimum size of formulas is exponential for almost all Boolean functions by the counting argument. (See, e.g., Sec. 4.3 of [21].) We denote by $I_{NNF}(f)$ the inversion complexity of a Boolean function f in NNF formulas.

Theorem 4. *For every n-input Boolean function f,*

$$I_{NNF}(f) \leq O(n^{5.57}).$$

In the rest of this section, we prove Theorem 4.

5.2 Proof

The proof of Theorem 4 is based on the following lemma:

Lemma 4. *For every n-input Boolean function f,*

$$I_{NNF}(f) \leq \Sigma_{i=0}^{n} I_{NNF}(\neg T_i^n).$$

Proof. We construct an NNF formula C computing f using $\Sigma_{i=0}^{n} I_{NNF}(\neg T_i^n)$ NOT gates as follows:

$$\wedge_{i=0}^{n}(f^i \vee \neg T_i^n).$$

In the construction, we use monotone formulas computing f^i's (the i-th slices of f) and NNF formulas computing $\neg T_i^n$'s with $I_{NNF}(\neg T_i^n)$ NOT gates for all $0 \leq i \leq n$. Since f^i is monotone for all i, one can construct a monotone formula computing f^i. By the construction above, the obtained formula C is an NNF formula and has $\Sigma_{i=0}^{n} I_{NNF}(\neg T_i^n)$ NOT gates.

We prove that C computes f, i.e., $\wedge_{i=0}^{n}(f^i \vee \neg T_i^n) = f$. Let s be the number such that $s = \Sigma_{j=1}^{n} x_j$. If $i < s$, then $(f^i \vee \neg T_i^n) = 1$ since $f^i(x) = 1$. If $i > s$, then $(f^i \vee \neg T_i^n) = 1$ since $T_i^n(x) = 0$. Therefore $\wedge_{i=0}^{n}(f^i \vee \neg T_i^n) = (f^s \vee \neg T_s^n)$. Since $f^s(x) = f(x)$ and $T_s^n(x) = 1$, $(f^s \vee \neg T_s^n) = f$. Thus C computes f. □

Theorem 4 is easily obtained from Lemma 4.

Proof (Theorem 4). A formula computing $\neg T_i^n$ can be converted to an NNF formula computing $\neg T_i^n$ with the same size by DeMorgan's laws. The number of NOT gates in a formula is less than or equal the size of the formula. Thus

$$I_{NNF}(\neg T_i^n) \leq L(\neg T_i^n).$$

The known best upper bound on the size of formulas computing the threshold functions T_i^n for $0 \leq i \leq n$ is $O(n^{4.57})$ [15]. By the upper bound and Lemma 4,

$$\begin{aligned} I_{NNF}(f) &\leq \Sigma_{i=0}^{n} I_{NNF}(\neg T_i^n) \\ &\leq (n+1) \cdot L(\neg T_i^n) \\ &\leq (n+1) \cdot O(n^{4.57}) \\ &\leq O(n^{5.57}). \end{aligned}$$

□

6 Concluding Remarks

In this paper, we studied negation-limited formulas. In Section 4, we gave a relation between the size of general formulas and the size of negation-limited formulas. Also for circuits, similar relations are known, which are obtained from a construction of negation-limited inverters. We denote by $size(f)$ the size of the smallest circuit computing a function f and denote by $size_r(f)$ the size of the smallest circuit with at most r NOT gates computing f.

Proposition 1. *([5]) For every function f,*

$$size_{\lceil \log(n+1) \rceil}(f) \le 2size(f) + O(n \log n).$$

Proposition 2. *([14]) For every function f,*

$$size_{\log^{1+o(1)} n}(f) \le 2size(f) + O(n).$$

These propositions imply that when one try to prove a $\omega(n \log n)$ (resp. superlinear) lower bound for $size(f)$, it is enough to prove a $\omega(n \log n)$ (resp. superlinear) lower bound for $size_{\lceil \log(n+1) \rceil}(f)$ (resp. $size_{\log^{1+o(1)} n}(f)$). The attempt has not given a lower bound for $size(f)$ so far. One of the reasons is that $\neg x_1, \ldots, \neg x_n$ can be generated by an inverter with the limited number of NOT gates and it is also difficult to prove a size lower bound for circuits which have $x_1, \ldots, x_n, \neg x_1 \ldots, \neg x_n$ as the input and consist of AND gates and OR gates. In formulas, negation-limited inverters are not efficient since fan-out of all gates is bounded to 1. Although Theorem 3 may not be useful to prove a size lower bound for formulas, we expect that one can know the effect of NOT gates in formulas better through Theorem 3.

In Section 5, we proved a polynomial upper bound of the inversion complexity of every Boolean function in NNF formulas. Lower bounds of it and the determination of the exact value remain open. It seems to be not so easy even to determine the inversion complexity of $\neg T^n_{\lceil n/2 \rceil}$, i.e., the negation of the majority function. It is easily obtained that $I_{NNF}(f) \le L(f)$ since a formula C which computes f and has size $L(f)$ can be converted an NNF formula C' computing f and $not(C')$ is at most the size of C', i.e., $L(f)$. Therefore it seems to be a difficult problem to prove a large lower bound on $I_{NNF}(f)$ for an explicit function f since it also means a large lower bound on $L(f)$. In the proof of Theorem 4, we use a trivial upper bound as the upper bound of $I_{NNF}(\neg T^n_i)$. If we can prove a better upper bound of $I_{NNF}(\neg T^n_i)$, Theorem 4 is improved.

Acknowledgement

The author would like to thank Kazuyuki Amano for helpful comments.

References

1. Alon, N., Boppana, R.B.: The monotone circuit complexity of Boolean functions. Combinatorica 7(1), 1–22 (1987)
2. Amano, K., Maruoka, A.: A superpolynomial lower bound for a circuit computing the clique function with at most $(1/6)\log\log n$ negation gates. SIAM J. Comput. 35(1), 201–216 (2005)
3. Amano, K., Maruoka, A., Tarui, J.: On the negation-limited circuit complexity of merging. Discrete Applied Mathematics 126(1), 3–8 (2003)
4. Andreev, A.E.: On a method for obtaining lower bounds for the complexity of individual monotone functions. Sov. Math. Doklady 31(3), 530–534 (1985)
5. Beals, R., Nishino, T., Tanaka, K.: On the complexity of negation-limited Boolean networks. SIAM J. Comput. 27(5), 1334–1347 (1998)
6. Boppana, R., Sipser, M.: The complexity of finite functions. In: Leeuwen, J.v. (ed.) Handbook of Theoretical Computer Science, vol. A: Algorithms and Complexity, pp. 757–804. Elsevier Science Publishers, Amsterdam (1990)
7. Fischer, M.J.: The complexity of negation-limited networks - a brief survey. In: Brakhage, H. (ed.) GI-Fachtagung 1975. LNCS, vol. 33, pp. 71–82. Springer, Heidelberg (1975)
8. Fischer, M.J.: Lectures on network complexity, Technical Report 1104, CS Department, Yale University (1974) (revised 1996)
9. Håstad, J.: The shrinkage exponent of de Morgan formulas is 2. SIAM J. Comput. 27(1), 48–64 (1998)
10. Jukna, S.: On the minimum number of negations leading to super-polynomial savings. Inf. Process. Lett. 89(2), 71–74 (2004)
11. Markov, A.A.: On the inversion complexity of a system of functions. J. ACM 5(4), 331–334 (1958)
12. Morizumi, H.: A note on the inversion complexity of Boolean functions in Boolean formulas. CoRR (Computing Research Repository), arXiv:0811.0699 (2008)
13. Morizumi, H.: Limiting negations in non-deterministic circuits. Theoret. Comput. Sci. (accepted)
14. Morizumi, H., Suzuki, G.: Negation-limited inverters of linear size. In: Hong, S.-H., Nagamochi, H., Fukunaga, T. (eds.) ISAAC 2008. LNCS, vol. 5369, pp. 605–614. Springer, Heidelberg (2008)
15. Paterson, M.S., Pippenger, N., Zwick, U.: Optimal carry save networks. In: Boolean Function Complexity. London Mathematical Society Lecture Note Series, vol. 169, pp. 174–201. Cambridge Univ. Press, Cambridge (1992)
16. Razborov, A.A.: Lower bounds on the monotone complexity of some Boolean functions. Sov. Math. Doklady 31, 354–357 (1985)
17. Santha, M., Wilson, C.: Limiting negations in constant depth circuits. SIAM J. Comput. 22(2), 294–302 (1993)
18. Sung, S., Tanaka, K.: An exponential gap with the removal of one negation gate. Inf. Process. Lett. 82(3), 155–157 (2002)
19. Sung, S., Tanaka, K.: Limiting negations in bounded-depth circuits: an extension of Markov's theorem. In: Ibaraki, T., Katoh, N., Ono, H. (eds.) ISAAC 2003. LNCS, vol. 2906, pp. 108–116. Springer, Heidelberg (2003)
20. Valiant, L.G.: Short monotone formulae for the majority function. J. Algorithms 5(3), 363–366 (1984)
21. Wegener, I.: The Complexity of Boolean Functions. Teubner/Wiley (1987)

Fast Polynomial-Space Algorithms Using Möbius Inversion: Improving on Steiner Tree and Related Problems*

Jesper Nederlof

Department of Informatics, University of Bergen, N-5020 Bergen, Norway
jesper.nederlof@ii.uib.no

Abstract. Given a graph with n vertices, k terminals and bounded integer weights on the edges, we compute the minimum STEINER TREE in $\mathcal{O}^*(2^k)$ time and polynomial space, where the $\mathcal{O}^*$ notation omits $\mathtt{poly}(n,k)$ factors. Among our results are also polynomial-space $\mathcal{O}^*(2^n)$ algorithms for several $\mathcal{NP}$-complete spanning tree and partition problems.

The previous fastest known algorithms for these problems use the technique of dynamic programming among subsets, and require exponential space. We introduce the concept of branching walks and extend the Inclusion-Exclusion algorithm of Karp for counting Hamiltonian paths. Moreover, we show that our algorithms can also be obtained by applying Möbius inversion on the recurrences used for the dynamic programming algorithms.

1 Introduction

One of the most widely used techniques for achieving moderately exponential time algorithms for $\mathcal{NP}$-hard problems is dynamic programming among subsets, but unfortunately an exponential storage requirement seems to be inherent to this technique. As mentioned by Woeginger [20] this requirement makes them useless in practice. Therefore polynomial-space exact algorithms have already been studied for several $\mathcal{NP}$-hard problems [6,7,11,16,17,20]. Hence, from both a theoretical and a practical perspective it is desirable to identify those dynamic programming algorithms that can be improved to require polynomial space, preferably maintaining the best known upper bound on the running time. In this paper we improve several algorithms in this way.

In 2006, Björklund et al. [6] drew new attention to the principle of Inclusion-Exclusion: they gave $\mathcal{O}^*(2^n)$-time algorithms[1] for several set partition problems, the most prominent one being k-COLORING. They also mention a simple adjustment to their algorithm to achieve an $\mathcal{O}^*(2.24^n)$-time algorithm with polynomial space for k-COLORING. Also related to this are the $\mathcal{O}^*(2^n)$-time polynomial-space algorithms for #HAMILTONIAN PATH by Karp [16] and (implicitly) Kohn et al. [17], and for #PERFECT MATCHING by Björklund and Husfeldt [2].

* This work is supported by the Research Council of Norway.

[1] The $\mathcal{O}^*$ notation omits polynomial factors, and n denotes the number of nodes of the graph.

S. Albers et al. (Eds.): ICALP 2009, Part I, LNCS 5555, pp. 713–725, 2009.

Table 1. An overview of problems that can be solved using polynomial space and with the given time bound using Inclusion-Exclusion. For the problems indicated with a *, we provide the first polynomial-space algorithm with the given running time.

	Problem	References
$\mathcal{O}^*(2^k)$	Steiner tree*[2]	[3,9,11,12]
$\mathcal{O}^*(2^n)$	Degree constrained spanning tree*	[13]
	Max internal spanning tree*	[10]
	#c-Spanning forests*	[4,15]
	Cover polynomial*	[4]
	#Hamiltonian path	[16,17]
	#Perfect matching	[2]
$\mathcal{O}^*(2.24^n)$	k-Coloring	[6,7]

We show that some dynamic programming algorithms can be improved to obtain polynomial-space algorithms with the same worst-case running time. Our algorithms heavily rely on the work of Björklund et al. [2,3,4,5,6]. The results can be read from Table 1: we add five problems to the list of polynomial space Inclusion-Exclusion algorithms (note that this list is not exhaustive).

Steiner Tree is one of the most well-studied $\mathcal{NP}$-complete problems. The Dreyfus-Wagner [9] dynamic programming algorithm has been the fastest exact algorithm for over 30 years. However, recently Björklund et al. [3] gave an $\mathcal{O}^*(2^k)$-time algorithm[2] for the variant with bounded integer weights, and Fuchs et al. [12] gave an $\mathcal{O}^*(c^k)$-time algorithm for the general case, for any $c > 2$. Both algorithms use $\Omega(2^k)$ space. In [11], Fomin et al. initiated the study of polynomial space algorithms for Steiner Tree. They gave polynomial space algorithms with running time bounded by $\mathcal{O}(5.96^k n^{\mathcal{O}(\log k)})$ and $\mathcal{O}(1.60^n)$ where n is the number of nodes in the graph. They pose the question whether Steiner Tree is fixed parameter tractable with respect to k when there is a polynomial space restriction. We answer this question affirmatively by providing an algorithm that runs in $\mathcal{O}^*(2^k)$ time and meets the restriction. Using the techniques of [11], this also leads to a polynomial-space $\mathcal{O}^*(1.3533^n)$-time algorithm.

The Max Internal Spanning Tree (MIST) and Degree Constrained Spanning Tree (DCST, also called Min-Max Degree Spanning Tree) problems are natural generalizations of Hamiltonian path. In [10], Fernau et al. ask if there exists an $\mathcal{O}^*(2^n)$-time algorithm to solve MIST. In [13], Gaspers et al. ask if there exists an $\mathcal{O}^*(2^n)$-time algorithm solving DCST. We answer both questions by giving polynomial-space algorithms with this running time.

The Cover polynomial of a directed graph introduced by Graham and Chung [8] generalizes all problems that can be solved using two operations named deletion and contraction of edges, and is designed to be the directed analogue of the Tutte polynomial. We improve the $\mathcal{O}^*(3^n)$-time polynomial-space algorithm of Björklund et al. [4] to an $\mathcal{O}^*(2^n)$-time polynomial-space algorithm. We also give the same improvement for #c-Spanning forests, which is one particular

[2] k stands for the number of terminals.

case of the Tutte polynomial. For more information about the Tutte polynomial we refer to [4].

The paper is organised as follows: in Section 2 we recall the principle of Inclusion-Exclusion and the well-known Hamiltonian path algorithm. After this we provide a natural extension by introducing the concept of *branching walks* and give the resulting algorithms. In the next section we show how the Inclusion-Exclusion algorithms can be obtained from dynamic programming algorithms by taking the *zeta transform* of the associated recurrences. After this, we give a more structural approach to the subset products introduced in [3] and provide applications.

We use the following definitions: for any set S, 2^S is the *power set* of S, i.e. the set consisting of all subsets of S. For a boolean expression b, $[b]$ stands for 1 if b is `true`, and for 0 if b is `false`. Let $G = (V, E)$ be a (directed) graph. Throughout the paper, we use $|V| = n$. The graph *induced by* X, where $X \subseteq V$, is the graph $G[X]$ with nodeset X and all edges in E only adjacent to nodes in X. For $v \in V$, $N(V)$ are all nodes adjacent to v. A *walk of length* k in G is a tuple $W = (v_1, \ldots, v_{k+1}) \in V^{k+1}$ such that $(v_i, v_{i+1}) \in E$ for each $0 \leq i \leq k$. W is from v if $v_1 = v$, and W is *cyclic* if $v_1 = v_{k+1}$. Let $G' = (V', E')$ also be a graph, a *homomorphism from* G *to* G' is a function $\phi : V \to V'$ such that $(u, v) \in E$ implies $(\phi(u), \phi(v)) \in E'$.

2 Inclusion-Exclusion Formulations

Let us start this section by stating the principle of Inclusion-Exclusion. The following theorem can be found in many textbooks on discrete mathematics. For a proof see for example Bax [1].

Theorem 1 (Folklore). *Let U be a set and $A_1, \ldots, A_n \subseteq U$. With the convention $\bigcap_{i \in \emptyset} \overline{A_i} = U$, the following holds:*

$$\Big| \bigcap_{i \in \{1,\ldots,n\}} A_i \Big| = \sum_{X \subseteq \{1,\ldots,n\}} (-1)^{|X|} \Big| \bigcap_{i \in X} \overline{A_i} \Big| \tag{1}$$

In this paper, we call any application of the above theorem an *IE-formulation*. In this context we will refer to the set U as the *universe*, and to $A_1, \ldots, A_n$ as the *requirements*. Moreover, we call the task of computing $|\bigcap_{i \in X} \overline{A_i}|$ for an arbitrary $X \subseteq \{1, \ldots, n\}$ the *simplified problem*. Note that if the simplified problem can be computed in polynomial time, there exists an $\mathcal{O}^*(2^n)$-time polynomial-space algorithm that evaluates Equation 1.

We mention that it is also possible to break Theorem 1 down into smaller steps, i.e. we choose a subset of $\{1, \ldots, n\}$ and apply the theorem. This has recently been used for a faster exact algorithm for Dominating Set in van Rooij et al. [19]; see also Bax [1]. Other methods to speed up IE-formulations are given by Björklund et al. [2,5] and are surveyed in [18].

We continue this section by giving some IE-formulations. The first one is well-known, but illustrative and it will be extended in the next subsections.

2.1 Hamiltonian Paths

Given a graph $G = (V, E)$, a Hamiltonian path is a walk that contains each node exactly once[3]. The #Hamiltonian Path problem is to count the number of Hamiltonian paths. The following IE-formulation is due to Karp [16]: define the universe U as all walks of length $n-1$ in G, and define A_v as all walks of length $n-1$ that contain node v, for each $v \in V$. With these definitions, the left-hand side of Equation 1, $|\bigcap_{v\in V} A_v|$, is the number of Hamiltonian paths in G. Now it remains to show how to solve the simplified problem: given $R \subseteq V$ and $s \in R$, let $w_k(s, R)$ be the number of walks from s of length k in $G[R]$. Then $w_k(s, R)$ admits the following recurrence:

$$w_k(s,R) = \begin{cases} 1 & \text{if } k = 0 \\ \sum_{t \in N(s)\cap R} w_{k-1}(t,R) & \text{otherwise} \end{cases} \tag{2}$$

Notice that $w_k(s, R)$ is $\mathcal{O}(n^n)$, hence the number of bits needed to represent this value is polynomially bounded, and that

$$|\bigcap_{v\in X} \overline{A_v}| = \sum_{s\in V\setminus X} w_{n-1}(s, V\setminus X)$$

Hence, the simplified problem can be solved in polynomial time using dynamic programming on Equation 2 (the parameter R is fixed but is added for clearness). Thus it takes $\mathcal{O}^*(2^n)$ time and polynomial space to evaluate Equation 1.

2.2 Steiner Tree

Now we are ready for our first new result. Assume a graph $G = (V, E)$ and weight function $w : E \to \mathbb{Z}_+$ are given. The Steiner Tree problem is the following: given a set of *terminals* $K \subseteq V$ and an integer c, does there exist a subtree $T = (V', E')$ of G such that $K \subseteq V'$ and $\sum_{e\in E'} w(e) \le c$. In this section we will give an extension of the results in the previous section to obtain a new IE-formulation for Steiner Tree with *unit weights*, meaning $w(e) = 1$ for every edge $e \in E$. We introduce the following definition:

Definition 2. *A* branching walk *B in $G = (V, E)$ is a pair (T_B, ϕ) where $T_B = (V_B, E_B)$ is an ordered tree and $\phi : V_B \to V$ is a homomorphism from T_B to G.* The length *of B, denoted with $|B|$, is $|E_B|$. For a node $s \in V$, B is* from *s if the root of T_B is mapped to s by ϕ. If a branching walk is said to be* unordered, *T_B is an unordered tree.*

We will use $\phi(V_B) = \{\phi(u)|u \in V_B\}$ and $\phi(E_B) = \{(\phi(u), \phi(v))|(u, v) \in E_B\}$, hence $\phi(V_B) \subseteq V$ and $\phi(E_B) \subseteq E_B$. A branching walk is a natural generalization of a walk: notice that a branching walk is a walk if T_B is a path. The definition is particularly useful in combination with the following lemma:

[3] This is slightly different from the usual definition, since a path corresponds to two walks in both directions.

Lemma 3. *Let $s \in K$. There exists a subtree $T = (V', E')$ of G such that $K \subseteq V'$ and $|E'| \leq c$ if and only if there exists a branching walk $B = (T_B = (V_B, E_B), \phi)$ from s such that $K \subseteq \phi(V_B)$ and $|B| \leq c$.*

Proof. For the first part, choose $T_B = T$ and $\phi : V_B \to V' = V_B$ to be the identity function. For the second part, we can take T to be a spanning tree of the graph $(\phi(V_B), \phi(E_B))$, and it has the required properties. □

Consider the following IE-formulation: let $s \in K$ be an arbitrarily chosen terminal, and define the universe U as all branching walks from s of length c. For each $v \in K$, define a requirement A_v that consists of all elements of U that contain terminal v (i.e. $v \in \phi(V_B)$). It follows that the left-hand side of Equation 1, $|\bigcap_{v \in X} A_v|$, is the number of branching walks that contain all terminals. Using Lemma 3 this is larger than 0 if and only if the instance of STEINER TREE is a yes-instance.

It remains to show how the simplified problem can be solved. For $R \subseteq K$, let $R' = (V \setminus K) \cup R$, and define $b_j^R(s)$ as the number of branching walks from s of length j in $G[R']$, where $s \in R'$. Note that the simplified problem is to compute

$$|\bigcap_{v \in X} \overline{A_v}| = b_c^{K \setminus X}(s)$$

for a given set $X \subseteq K$ of terminals. Now $b_c^R(s)$ can be computed in polynomial time using the following lemma:

Lemma 4. *Let $R \subseteq K$ and $s \in R'$, then*

$$b_j^R(s) = \begin{cases} 1 & \text{if } j = 0 \quad (3a) \\ \sum_{t \in N(s) \cap R'} \sum_{j_1 + j_2 = j-1} b_{j_1}^R(t)\, b_{j_2}^R(s) & \text{otherwise} \quad (3b) \end{cases}$$

Proof. There is one branching walk of length 0, $B = (T_B, \phi)$, from s with T_B being a single node and ϕ mapping this single node to s, hence Case 3a. If the length $j = |E_B|$ is larger than 0, take the first child c_1 of the root of T_B. Notice that $(s, \phi(c_1))$ has to be in E; therefore, $t \in N(s) \cap R'$. Now any tree T_B consists of an edge from the root r to c_1, and two trees rooted at r and c_1. Hence, B also consists a two branching walks, one from t and one from s. The lengths of these branching walks have to sum up to $j - 1$ since the edge (r, c_1) already contributes 1 to the length of B. Now it remains to sum over all possibilities of distributions of the length, and hence Case 3b also holds. □

From Equation 3 it follows that for each $j > 0$ and $s \in R'$, $b_j^R(s)$ is $\mathcal{O}((nj)^j)$. Hence, the number of bits needed to represent $b_j^R(s)$ is polynomially bounded.

Theorem 5. *The* STEINER TREE *problem with unit weights can be solved in $\mathcal{O}^*(2^k)$ time and polynomial space, where k is the number of terminals.*

Proof. Due to Lemma 3 the considered IE-formulation solves STEINER TREE, and we can use dynamic programming on Equation 3b to compute the simplified problem in polynomial time. □

The following result is a direct consequence of Theorem 5 and the considerations of Section 4 in [11]:

Corollary 6. STEINER TREE *with unit weights can be solved in* $\mathcal{O}^*(1.3533^n)$ *time using polynomial space.*

2.3 Further IE-Formulations

In this section we give IE-formulations for two problems that lead to algorithms running in $\mathcal{O}^*(2^n)$-time and using polynomial space. In the following assume we are given a graph $G = (V, E)$ and an integer c. In both IE-formulations we define A_v, for each $v \in V$, to be all elements of U that contain node v. The universe U itself will be more tailor-made for both problems.

Degree constrained spanning tree. The DEGREE CONSTRAINED SPANNING TREE problem, also called MIN-MAX DEGREE SPANNING TREE, asks whether G has a spanning tree with maximum degree at most c. Define U as all branching walks (T_B, ϕ) of length $n-1$ such that T_B has maximum degree at most c. For $R \subseteq V$, define $d_j^R(g, s)$ as the number of branching walks (T_B, ϕ) from s of length j in $G[R]$ such that the degree of the root of T_B is at most g. The simplified problem is to compute

$$|\bigcap_{v \in X} \overline{A_v}| = \sum_{s \in V \setminus X} d_{n-1}^{V \setminus X}(c, s)$$

and it can be computed in polynomial time with dynamic programming using:

$$d_j^R(g, s) = \begin{cases} [g \geq 0] & \text{if } j = 0 \\ \sum_{t \in N(s) \cap R} \sum_{j_1 + j_2 = j-1} d_{j_1}^R(c-1, t)\, d_{j_2}^R(g-1, s) & \text{otherwise} \end{cases}$$

To see that the equation holds, notice that $d_0^R(g, s) = [g \geq 0]$ by definition and if $j > 0$, we count combinations of two branching walks: in the branching walk from t we are allowed to choose $c-1$ neighbors, and in the one from s we are allowed to use one neighbor less than before.

Max internal spanning tree. The MAX INTERNAL SPANNING TREE asks whether G has a spanning tree with at least c internal nodes (i.e. nodes with degree at least 2). Define the universe U as all branching walks (T_B, ϕ) of length $n-1$ such that T_B has at most $n-(c+1)$ leaves. For $R \subseteq V$, define $m_{g,j}^R(s)$ as the number of branching walks in $G[R]$ of length j from s having at most g leaves.

$$m_{g,j}^R(s) = \begin{cases} [g \leq 1] & \text{if } j = 0 \\ \sum_{t \in N(s) \cap R} \sum_{g_1 + g_2 = g} \sum_{j_1 + j_2 = j-1} m_{g_1, j_1}^R(t)\, m_{g_2, j_2}^R(s) & \text{otherwise} \end{cases}$$

In the equation we count all possible distributions of the length and the number of leaves in the branching walks from s and t. Now the simplified problem is to compute $\sum_{s \in V \setminus X} m^{V \setminus X}_{n-(c+1),j}(s)$. To see that the IE-formulation solves the problem, note there exists $B \in \bigcap_{v \in V} A_v$ if and only if there exists $(T_B, \phi) \in \bigcap_{v \in V} A_v$ such that the root of T_B has degree 1. And in T_B, the number of internal nodes is the number of non-leaves minus 1.

Theorem 7. Degree Constrained Spanning Tree *and* Max Internal Spanning Tree *can be solved in* $\mathcal{O}^*(2^n)$ *time and polynomial space.*

Proof. The discussed IE-formulations solve the problems due to Lemma 3 and the above considerations, and the simplified problems can be solved in polynomial time with dynamic programming on the stated recurrences. □

3 Möbius Inversion

In this section we study an algebraic equivalent of Inclusion-Exclusion, called Möbius inversion. Basically, it consists of the following two transforms:

Definition 8. *Given a function* $f : 2^V \to \mathbb{Z}_+$ *and* $Y \subseteq V$, *the* zeta transform $\zeta f(Y)$ *and the* Möbius transform $\mu f(Y)$, *are defined as:*

$$\zeta f(Y) = \sum_{X \subseteq Y} f(X) \qquad \mu f(Y) = \sum_{X \subseteq Y} (-1)^{|Y \setminus X|} f(X)$$

Now the principle of Möbius inversion can be formulated as the following theorem (this is already folklore, but we make the equivalence relation more clear with a proof):

Theorem 9 (Folklore). *The Möbius transform is the inverse of the zeta transform; that is, for every* $Y \subseteq V$, $f(Y) = \mu\zeta f(Y)$.

Proof. Define U and $A_v \subseteq U$ for $v \in V$ such that for $X \subseteq V$ we have:

$$f(X) = |\bigcap_{v \in X} A_v \setminus \bigcup_{v \in V \setminus X} A_v|$$

This can be done by defining $U = \{e_i^X | X \subseteq V, 1 \leq i \leq f(X)\}$ such that $e_i^X \in A_v$ if and only if $v \in X$. Now $\zeta f(Y) = |\bigcap_{i \in V \setminus Y} \overline{A_i}|$, and hence $\mu\zeta f(Y)$ is equal to the right-hand of Equation 1. Since $f(Y)$ is equal to the left-hand side of Equation 1, the result follows from Theorem 1. □

So intuitively, using the terminology of the previous section, any IE-formulation is equivalent to applying Möbius inversion and the simplified problem is to compute $\zeta f(V \setminus X)$. Now we will reobtain the IE-formulation of Karp [16] discussed in Subsection 2.1 by applying Möbius inversion to the classical dynamic programming approach.

Hamiltonian path revisited. Let us again consider #HAMILTONIAN PATH. Let $h(s,R)$ be the number of Hamiltonian paths from s in $G[R \cup s]$. Recall the dynamic programming algorithm of Held and Karp [14]:

$$h(s,R) = \begin{cases} 1 & \text{if } R = \emptyset \\ \sum_{t \in N(s) \cap R} h(t, R \setminus t) & \text{otherwise} \end{cases}$$

We start by adding a parameter k, which is the length of the Hamiltonian paths we are counting. Although this seems superfluous because we know that each Hamiltonian path in $G[R \cup s]$ has length $|R|$, it gives us some needed flexibility.

$$h_k(s,R) = \begin{cases} [R = \emptyset] & \text{if } k = 0 \\ \sum_{t \in N(s) \cap R} h_{k-1}(t, R \setminus t) & \text{otherwise} \end{cases}$$

Now $h_{n-1}(s, V \setminus s)$ is the number of Hamiltonian paths from s. Consider the following slightly different function

$$h'_k(s,R) = \begin{cases} [R = \emptyset] & \text{if k=0} \quad (4a) \\ \sum_{t \in N(s) \cap R} h'_{k-1}(t, R \setminus t) + h'_{k-1}(t, R) & \text{otherwise} \quad (4b) \end{cases}$$

Notice that $h'_{|R|}(s,R) = h_{|R|}(s,|R|)$, since the term $h'_{k-1}(t,R)$ added in Case 4b is 0 if $k \leq |R|$. As a next step, we take the zeta transform on both sides of Equation 4. For Case 4a, we have $\zeta h'_0(s,R) = 1$, and for Case 4b:

$$\begin{aligned} \zeta h'_k(s,R) &= \sum_{X \subseteq R} \sum_{t \in N(s) \cap X} h'_{k-1}(t, X \setminus t) + h'_{k-1}(t, X) \\ &= \sum_{t \in N(s) \cap R} \sum_{t \in X \subseteq R} h'_{k-1}(t, X \setminus t) + h'_{k-1}(t, X) \\ &= \sum_{t \in N(s) \cap R} \zeta h'_{k-1}(t, R) \end{aligned}$$

It is immediate that $\zeta h'_k(t,R) = w_k(s,R)$, and we obtained the IE-formulation of Subsection 2.1.

3.1 Subset Products

An application for which Möbius inversion is particularly suited is the computation of subset products, introduced by Björklund et al. We will use the following:

Definition 10 ([3]). *Given two functions* $f, g : 2^V \to \mathbb{Z}_+$, *the* cover product $(f *_c g)(Y)$, *for* $Y \subseteq V$ *is defined as:*

$$(f *_c g)(Y) = \sum_{A \cup B = Y} f(A) g(B)$$

Assuming f and g can be evaluated in polynomial time, the naive way to compute $(f *_c g)(Y)$ would take $\mathcal{O}^*(3^n)$ time. In [3], Björklund et al. implicitly use the following theorem in order to obtain an $\mathcal{O}^*(2^n)$ algorithm:

Theorem 11 ([3]). *Given two functions* $f, g : 2^V \to \mathbb{Z}_+$, *the following holds for* $Y \subseteq V$:

$$\zeta(f *_c g)(Y) = (\zeta f(Y))\ (\zeta g(Y))$$

Proof. Consider the following rewriting:

$$\zeta(f *_c g)(Y) = \sum_{X \subseteq Y} \sum_{A \cup B = X} f(A) g(B) = \Big(\sum_{A \subseteq Y} f(A)\Big)\Big(\sum_{B \subseteq Y} g(B)\Big)$$

The first equality follows by definition. For the second equality notice that for each $A, B \subseteq Y$, there exists exactly one $X \subseteq Y$ such that $A \cup B = X$, hence we can sum over each combination of two subsets A and B. Now the theorem follows from the definition of zeta transform. □

Steiner Tree revisited. Let us again consider Steiner Tree. Recall we denote R' for $(V \setminus K) \cup R$. Our starting point is an adjusted version of the famous Dreyfus-Wagner recurrence [3,9]: for $R \subseteq K$, integer c and $t \in V$ we are going to define $s_c(t, R)$ such that it will be larger than 0 if and only if there exists a subtree $T = (V', E')$ of the graph $G[R' \cup t]$ such that $R \cup t \subseteq V'$ and $\sum_{e \in E'} w(e) \leq c$. For $c \leq 0$, we have $s_c(t, R) = [c = 0 \wedge R = \emptyset]$, and for $c > 0$ define:

$$s_c(t, R) = \sum_{u \in N(t) \cap R'} g_{c - w(t,u)}(t, u, R \setminus u)$$

$$g_c(t, u, R) = \sum_{c_1 + c_2 = c} \sum_{A \cup B = R} s_{c_1}(u, A)\, s_{c_2}(t, B)$$

We use a slightly different variant s'_c of s_c. Define $s'_0(t, R)$ to be $s_0(t, R)$, and for $c > 0$:

$$s'_c(t, R) = \sum_{u \in N(t) \setminus K} g(R) + \sum_{u \in N(t) \cap R} g(R) + g(R \setminus u)$$

where we shorthand $g'_{c - w(t,u)}(t, u, R)$ with $g(R)$, and the definition of g' is obtained by replacing s with s' in the definition of g (hence s' does not depend on s). Note that $s'_c(t, R) > 0$ if and only if $s_c(t, R) > 0$, since $0 \leq g(R) \leq g(R \setminus u)$. Now we take the zeta transform of both s'_c and g'_c:

$$\begin{aligned}
\zeta s'_c(t,R) &= \sum_{X\subseteq R}\Big(\sum_{u\in N(t)\setminus K} g(X) + \sum_{u\in N(t)\cap X} g(X) + g(X\setminus u)\Big) \\
&= \sum_{u\in N(t)\setminus K} \zeta g(R) + \sum_{u\in N(t)\cap R}\ \sum_{u\in X\subseteq R} g(X) + g(X\setminus u) \\
&= \sum_{u\in N(t)\setminus K} \zeta g(R) + \sum_{u\in N(t)\cap R} \zeta g(R) \\
&= \sum_{u\in N(t)\cap R'} \zeta g(R)
\end{aligned}$$

$$\begin{aligned}
\zeta g'_c(t,u,R) &= \sum_{X\subseteq R}\ \sum_{c_1+c_2=c}\ \sum_{A\cup B=X} s'_{c_1}(u,A)\, s'_{c_2}(t,B) \\
&= \sum_{c_1+c_2=c} \zeta(s'_{c_1}(u) *_c s'_{c_2}(t))(R) \\
&= \sum_{c_1+c_2=c} \zeta s'_{c_1}(u,R)\, \zeta s'_{c_2}(t,R)
\end{aligned}$$

Combining both derivations gives us

$$\zeta s'_c(t,R) = \sum_{u\in N(t)\cap R'}\ \sum_{c_1+c_2=c-w(t,u)} \zeta s'_{c_1}(u,R)\, \zeta s'_{c_2}(t,R)$$

comparing this with Equation 3, we see that $\zeta s'_c(t,R) = b_c^R(t)$ in the special case of unit weights. And the following result also follows:

Theorem 12. *The* STEINER TREE *problem with bounded integer weights can be solved in $\mathcal{O}^*(2^k)$ and polynomial space.*

3.2 Further Applications

In this subsection we give some other applications of the methods considered in the previous subsection, continuing the work of Björklund et al. [3,4].

Cover Polynomial. We use $x^{\underline{i}}$ for the falling factorial $\frac{x!}{(x-i)!}$. A Hamiltonian cycle of a graph is a cyclic walk that contains all nodes exactly once. The cover polynomial of a directed graph $D = (V, A)$ can be defined as (see also [4,8]):

$$\sum_{i,j} C_V(i,j) x^{\underline{i}} y^{\underline{j}}$$

where $C_V(i,j)$ can be interpreted as the number of ways to partition V into i directed paths and j directed cycles of D. Since paths and cycles with l edges contain $l+1$ and l nodes respectively, the sum of the lengths of the paths and cycles in such a partition will be $n-i$. Moreover, if V is covered, the path and cycles are disjoint because of this size restriction. This allows us to define C_Y using the cover product, such that C_V matches the above interpretation:

$$C_Y(i,j) = \frac{1}{i!j!} \sum_{l_1+\ldots l_{i+j}=n-i} (h_{l_1} *_c \ldots *_c h_{l_i} \; *_c \; c_{l_{i+1}} *_c \ldots *_c c_{l_{i+j}})(Y)$$

where $h_l(Y)$ and $c_l(Y)$ are the number of Hamiltonian paths and Hamiltonian cycles of length l in $D[Y]$, respectively (note that like before we use the redundant parameter l, for obtaining the efficient computable zeta transform). Recall Equation 4 and note we can replace $h_l(Y)$ with $h'_l(Y) = \sum_{s \in V} h'_l(s,Y)$, and $\zeta h'_l(Y)$ is the number of walks of length l in $D[Y]$. We mention that one can in a analogue way replace $c_t(Y)$ with $c'_t(Y)$ such that $\zeta c'_l(Y)$ is the number of cyclic walks of length l in $D[Y]$. Now we apply Theorem 11 on the cover products and obtain:

$$\zeta C_Y(i,j) = \frac{1}{i!j!} \sum_{l_1+\ldots l_{i+j}=n-i} \Big(\prod_{t=1}^{i} \zeta h'_{l_t}(Y)\Big) \Big(\prod_{t=i+1}^{i+j} \zeta c'_{l_t}(Y)\Big)$$

which can be computed in polynomial using standard dynamic programming, since $\zeta h'_l(Y) = \sum_{s \in Y} w_k(s,Y)$ and $\zeta c'_l(Y)$ also can.

#c-Spanning forests. A c-spanning forest of $G = (V, E)$ is an acyclic subgraph of G with exactly c connected components. Denote $\tau(c)$ for the number of c-spanning forests of G. Assume an ordering $\prec$ on the nodeset V is given. For $Y \subseteq V$, define $\hat{b}_l(Y)$ as the number of unordered branching walks $(T_B = (V_B, E_B), \phi)$ in G such that $Y \subseteq \phi(V_B)$ and $\phi(r)$ is minimum among $\phi(V_B)$, where r is the root of T_B (recall from Subsection 2.2 that $\phi(V_B) = \{\phi(u)|u \in V_B\}$). Now we can write $\tau(c)$ as follows:

Lemma 13

$$\tau(c) = \frac{1}{c!} \sum_{l_1+\ldots+l_c=n-c} (\hat{b}_{l_1} *_c \ldots *_c \hat{b}_{l_c})(V) \tag{5}$$

Proof. A set of c branching walks of total length $n - c$ can only cover V if it induces a c-spanning forest. Every tree in this spanning forest corresponds to one unordered branching walk from the minimum node it contains. Hence obtain the equality. □

Now we can use Möbius inversion and Theorem 11 on Equation 5 to obtain

$$\tau(c) = \mu\Big(\frac{1}{c!} \sum_{l_1+\ldots+l_c=n-c} \prod_{i=1}^{c} (\zeta \hat{b}_{l_1})(V)\Big)$$

and it remains to show how to compute $\zeta \hat{b}_{l_1}(R)$ for $R \subseteq V$. For $s \in R$, define $\hat{b}^R_{j,q}(s)$ as the number of unordered branching walks (T_B, ϕ) from s of length j

in $G[R]$ such that no child of the root of T_B is mapped to one of the first $q-1$ neighbors of s in $G[R]$ with respect to the ordering $\prec$. Notice that:

$$\hat{b}^R_{j,q}(s) = \begin{cases} 0 & \text{if } q > |N(s) \cap R| \\ 1 & \text{else if } j = 0 \\ b^R_{j,q+1}(s) + \sum\limits_{j_1+j_2=j-1} \hat{b}^R_{j_1,1}(N^s_q)\, \hat{b}^R_{j_2,q+1}(s) & \text{otherwise} \end{cases}$$

where N^s_q is the qth-first element of the set $N(s) \cap R$ with respect to the ordering $\prec$. Now $\hat{b}_l(R) = \sum_{s \in R} b^{R_s}_{l,1}(s)$, where R_s stands for the set of all elements e in R such that $s \prec e$.

Theorem 14. Cover Polynomial *and* #c-Spanning forests *can be solved in* $\mathcal{O}^*(2^n)$ *time and polynomial space.*

4 Conclusion

We studied applications where the zeta transform is computable in polynomial time. As mentioned in the introduction, our algorithms considerably improve on dynamic programming in practice: in addition to improving the space requirement, our algorithms can potentially be made faster in practice when combined with techniques from [1,19]. We want to mention that applying Möbius inversion to a problem is not straightforward: first one has to come up with a function with the wanted properties, in order to succesfully apply Möbius inversion.

To support finding more applications, it is interesting whether more subset products with similar nice properties can be found, for some examples, we also refer to [3].

Acknowledgements. The author wants to thank his advisor Pinar Heggernes, Daniel Lokshtanov and the anonymous referees for their valuable support and insightful remarks on this paper.

References

1. Bax, E.T.: Recurrence-Based Reductions for Inclusion and Exclusion Algorithms Applied to #P Problems (1996)
2. Björklund, A., Husfeldt, T.: Exact algorithms for exact satisfiability and number of perfect matchings. In: Bugliesi, M., Preneel, B., Sassone, V., Wegener, I. (eds.) ICALP 2006. LNCS, vol. 4051, pp. 548–559. Springer, Heidelberg (2006)
3. Björklund, A., Husfeldt, T., Kaski, P., Koivisto, M.: Fourier meets möbius: fast subset convolution. In: STOC, pp. 67–74 (2007)
4. Björklund, A., Husfeldt, T., Kaski, P., Koivisto, M.: Computing the tutte polynomial in vertex-exponential time. In: FOCS, pp. 677–686 (2008)
5. Björklund, A., Husfeldt, T., Kaski, P., Koivisto, M.: Trimmed moebius inversion and graphs of bounded degree. In: STACS, pp. 85–96 (2008)

6. Björklund, A., Husfeldt, T., Koivisto, M.: Set partitioning via inclusion–exclusion. SIAM Journal on Computing, special issue dedicated to selected papers from FOCS 2006, 575–582 (2006)
7. Bodlaender, H.L., Kratsch, D.: An exact algorithm for graph coloring with polynomial memory. Technical Report UU-CS-2006-015, Department of Information and Computing Sciences, Utrecht University (2006)
8. Chung, F.R.K., Graham, R.L.: On the cover polynomial of a digraph. J. Comb. Theory, Ser. B 65(2), 273–290 (1995)
9. Dreyfus, S., Wagner, R.: The Steiner problem in graphs. Networks 1, 195–207 (1972)
10. Fernau, H., Raible, D., Gaspers, S., Stepanov, A.A.: Exact exponential time algorithms for max internal spanning tree. CoRR, abs/0811.1875 (2008)
11. Fomin, F.V., Grandoni, F., Kratsch, D.: Faster steiner tree computation in polynomial-space. In: Halperin, D., Mehlhorn, K. (eds.) Esa 2008. LNCS, vol. 5193, pp. 430–441. Springer, Heidelberg (2008)
12. Fuchs, B., Kern, W., Molle, D., Richter, S., Rossmanith, P., Wang, X.: Dynamic programming for minimum Steiner trees. Theory of Computing Systems 41(3), 493–500 (2007)
13. Gaspers, S., Saurabh, S., Stepanov, A.A.: A moderately exponential time algorithm for full degree spanning tree. In: Agrawal, M., Du, D.-Z., Duan, Z., Li, A. (eds.) TAMC 2008. LNCS, vol. 4978, pp. 479–489. Springer, Heidelberg (2008)
14. Held, M., Karp, R.M.: A dynamic programming approach to sequencing problems. Journal of the Society for Industrial and Applied Mathematics 10(1), 196–210 (1962)
15. Jaeger, F., Vertigan, D.L., Welsh, D.J.A.: On the computational complexity of the jones and tutte polynomials. Mathematical Proceedings of the Cambridge Philosophical Society 108(01), 35–53 (1990)
16. Karp, R.M.: Dynamic programming meets the principle of inclusion and exclusion. Oper. Res. Lett. 1, 49–51 (1982)
17. Kohn, S., Gottlieb, A., Kohn, M.: A generating function approach to the traveling salesman problem. In: ACM 1977: Proceedings of the 1977 annual conference, pp. 294–300. ACM, New York (1977)
18. Nederlof, J.: Inclusion exclusion for hard problems. Master's thesis, Utrecht University (August 2008)
19. van Rooij, J.M.M., Nederlof, J., van Dijk, T.C.: Inclusion/exclusion meets measure and conquer: Exact algorithms for counting dominating sets. Technical Report UU-CS-2008-043, Utrecht, The Netherlands (2008)
20. Woeginger, G.J.: Space and time complexity of exact algorithms: Some open problems. In: Downey, R.G., Fellows, M.R., Dehne, F. (eds.) IWPEC 2004, vol. 3162, pp. 281–290. Springer, Heidelberg (2004)

Superhighness and Strong Jump Traceability

André Nies*

University of Auckland

Abstract. Let A be a c.e. set. Then A is strongly jump traceable if and only if A is Turing below each superhigh Martin-Löf random set. The proof combines priority with measure theoretic arguments.

1 Introduction

A lowness property of a set $A \subseteq \mathbb{N}$ specifies a sense in which A is computationally weak.

(I) Usually this means that A has limited strength when used as an oracle. An example is superlowness, $A' \leq_{tt} \emptyset'$. Further examples are given by traceability properties of A. Such a property specifies how to effectively approximate the values of certain functions (partial) computable in A. For instance, A is *jump traceable* [1] if $J^A(n)\downarrow$ implies $J^A(n) \in T_n$, for some uniformly c.e. sequence $(T_n)_{n\in\mathbb{N}}$ of computably bounded size. Here J is the jump functional: If $X \subseteq \mathbb{N}$, we write $J^X(n)$ for $\Phi_n^X(n)$.

(II) A further way to be computationally weak is to be easy to compute. A lowness property of this kind specifies a sense in which many oracles compute A. For instance, consider the property to be a base for ML-randomness, introduced in [2]. Here the class of oracles computing A is large enough to admit a set that is ML-random relative to A. By [3] this property coincides with the type (I) lowness property of being low for ML-randomness.

As our main result, we show a surprising further coincidence of a type (I) and a type (II) lowness property for c.e. sets. The type (I) property is strong jump traceability, introduced in [4], and studied in more depth in [5]. We say that a computable function $h\colon \mathbb{N} \to \mathbb{N} \setminus \{0\}$ is an *order function* if h is nondecreasing and unbounded.

Definition 1. *$A \subseteq \mathbb{N}$ is* strongly jump traceable *(s.j.t.) if for each order function h, there is a uniformly c.e. sequence $(T_n)_{n\in\mathbb{N}}$ such that $\forall n\, |T_n| \leq h(n)$ and $\forall n\, [J^A(n)\downarrow \rightarrow\ J^A(n) \in T_n]$.*

Figueira, Nies and Stephan [4] built a promptly simple set that is strongly jump traceable. Cholak, Downey and Greenberg [5] showed that the strongly jump traceable c.e. sets form a proper subideal of the K-trivial c.e. sets under Turing reducibility.

We say that a set $Y \subseteq \mathbb{N}$ is *superhigh* if $\emptyset'' \leq_{tt} Y'$. This notion was first studied by Mohrherr [6] for c.e. sets. For background and results on superhighness see

* The author was partially supported by the Marsden Fund of New Zealand, grant no. 03-UOA-130.

S. Albers et al. (Eds.): ICALP 2009, Part I, LNCS 5555, pp. 726–737, 2009.

[7,8]. The type (II) property is to be Turing below each superhigh ML-random set. Thus our main result is that *a c.e. set A is strongly jump traceable* if and only if *A is Turing below each superhigh Martin-Löf random set.*

The property to be Turing below each superhigh ML-random set can be put into a more general context. For a class $\mathcal{H} \subseteq 2^\omega$, we define the corresponding diamond class

$$\mathcal{H}^\diamond = \{A\colon\ A \text{ is c.e. } \&\ \forall Y \in \mathcal{H} \cap \mathsf{MLR}\,[A \leq_T Y]\}.$$

Here MLR is the class of ML-random sets. Note that $\mathcal{H}^\diamond$ determines an ideal in the c.e. Turing degrees. By a result of Hirschfeldt and Miller (see [7, 5.3.15]), for each null Σ^0_3 class, the corresponding diamond class contains a promptly simple set A. Their construction of A is via a non-adaptive cost-function construction (see [7, Section 5.3] for details on cost functions). That is, the cost function can be given in advance. This means that the construction can be viewed as injury-free. In contrast, the direct construction of a promptly simple strongly jump traceable set in [4] varies Post's construction of a low simple set, and therefore has injury.

In [9] a result similar to our main result was obtained when $\mathcal{H}$ is the class of superlow sets Y (namely, $Y' \leq_{tt} \emptyset'$). Earlier, Hirschfeldt and Nies had obtained such a coincidence for the class $\mathcal{H}$ of ω-c.e. sets Y (namely, $Y' \leq_{tt} \emptyset'$).

In all cases, to show that a c.e. strongly jump traceable set A is in the required diamond class, one finds an appropriate collection of benign cost functions; this key concept was introduced by Greenberg and Nies [10]. The set A obeys each benign cost function by the main result of [10]. This implies that A is in the diamond class.

It is harder to prove the converse inclusion: each c.e. set in $\mathcal{H}^\diamond$ is s.j.t. Suppose an order function h is given. For one thing, similar to the proof of the analogous inclusion in [9], we use a variant of the golden run method introduced in [12]. One wants to restrict the changes of A to the extent that A is strongly jump traceable. To this end, one attempts to define a "naughty set" $Y \in \mathcal{H} \cap \mathsf{MLR}$. It exploits the changes of A in order to avoid being Turing above A. The number of levels in the golden run construction is infinite, with the e-th level based on the Turing functional Φ_e. If the golden run fails to exist at level e then $A \neq \Phi_e^Y$. If this is so for all e then $A \not\leq_T Y$, contrary to the hypothesis that $A \in \mathcal{H}^\diamond$. Hence a golden run must exist. Since it is golden it successfully builds the required trace for J^A with bound h.

A further ingredient of our proof stems from ideas that started in Kurtz [13] and were elaborated further, for instance, in Nies [12,14]: mixing priority arguments and measure theoretic arguments. In contrast, the proof in [9] is not measure theoretic. (Indeed, they prove, more generally, that for *each* non-empty Π^0_1 class P, each c.e. set Turing below every superlow member of P must be strongly jump traceable. This stronger statement has no analog for superhighness, for instance because all members of P could be computable.) Here we need to make the naughty set Y superhigh. This is done by coding $\emptyset''$ (see [7, 3.3.2]) in the style of Kučera, but not quite into Y: the coding strings change due to the activity of the tracing procedures. The number of times they change is computably bounded. So the coding yields $\emptyset'' \leq_{tt} Y'$.

Notation. Suppose f is a unary function and $\widetilde{f}$ is binary. We write

$$\forall n\, f(n) = \lim_s^{\mathrm{comp}} \widetilde{f}(n,s)$$

if there is a computable function $g : \mathbb{N} \to \mathbb{N}$ such that for all n, the set

$$\{s > 0\colon\ \widetilde{f}(n,s) \neq \widetilde{f}(n,s-1)\}$$

has cardinality less than $g(n)$, and $\lim_s \widetilde{f}(n,s) = f(n)$.

We let $X' = \{n\colon\ J^X(n)\downarrow\}$, and $X'_t = \{n\colon\ J^X_t(n)\downarrow\}$. We use Knuth's bracket notation in sums. For instance, $\sum_n n^{-2}\, [\![n \text{ is odd}]\!]$ denotes $1+1/9+1/25+\ldots = \pi^2/8$.

A forthcoming paper by Greenberg, Hirschfeldt and Nies (Characterizing the s.j.t. sets via randomness) contains a new proof of Theorem 2 using the language of "golden pairs". This makes it possible to cut some parameters.

2 Benign Cost Functions and Shigh$^\diamond$

Note that a function f is d.n.c. relative to $\emptyset'$ if $\forall x\, \neg f(x) = J^{\emptyset'}(x)$. Let P be the $\Pi^0_1(\emptyset')$ class of $\{0,1\}$-valued functions that are d.n.c. relative to $\emptyset'$. The PA sets form a null class (see, for instance, [7, 8.5.12]). Relativizing this to $\emptyset'$, we obtain that the class $\{Z\colon\ \exists f \leq_T Z \oplus \emptyset' [f \in P]\}$ is null. Then, since $\mathrm{GL}_1 = \{Z\colon\ Z' \equiv_T Z \oplus \emptyset'\}$ is conull, the following class, suggested by Simpson, is also null:

$$\mathcal{H} = \{Z\colon\ \exists f \leq_{tt} Z' [f \in P]\}. \tag{1}$$

This class clearly contains Shigh because $\emptyset''$ truth-table computes a function that is d.n.c. relative to $\emptyset'$. Since $\mathcal{H}$ is Σ^0_3, by a result of Hirschfeldt and Miller (see [7, 5.3.15]) the class $\mathcal{H}^\diamond$ contains a promptly simple set. We strengthen this:

Theorem 1. *Let A be a c.e. set that is strongly jump traceable. Then $A \in \mathcal{H}^\diamond$.*

Proof. In [10] a cost function c is defined to be *benign* if there is a computable function g with the following property: if $x_0 < \ldots < x_n$ and $c(x_i, x_{i+1}) \geq 2^{-e}$ for each i, then $n \leq g(e)$. For each truth table reduction Γ we define a benign cost function c such that for each Δ^0_2 set A, and each ML-random set Y,

$$A \text{ obeys } c \text{ and } \Gamma^{Y'} \text{ is } \{0,1\}\text{-valued d.n.c. relative to } \emptyset' \Rightarrow A \leq_T Y.$$

Let (I_e) be the sequence of consecutive intervals of $\mathbb{N}$ of length e. Thus $\min I_e = e(e+1)/2$. We define a function $\alpha \leq_T \emptyset'$. We are given a partial computable function p and (via the Recursion Theorem) think of p as a reduction function for α, namely, p is total, increasing, and $\forall x\ \alpha(x) \simeq J^{\emptyset'}(p(x))$.

At stage s of the construction we define the approximation $\alpha_s(x)$. Suppose $x \in I_e$. If $p(y)$ is undefined at stage s for some $y \in I_e$ let $\alpha_s(x) = 0$. Otherwise, let

$$\mathcal{C}_{e,s} = \{Y\colon\ \exists t_{\,v \leq t \leq s} \forall x \in I_e\, [1 - \alpha_t(x) = \Gamma(Y'_t, p(x))]\}, \tag{2}$$

where $v \le s$ is greatest such that $v = 0$ or $\alpha_v \restriction I_e \neq \alpha_{v-1} \restriction I_e$. (Thus, $\mathcal{C}_{e,s}$ is the set of oracles Y such that Y' computes α correctly at some stage t after the last change of $\alpha \restriction_{I_e}$.)

Construction of α.
Stage $s > 0$. For each $e < s$, if $\lambda\mathcal{C}_{e,s-1} \le 2^{-e+1}$ let $\alpha_s \restriction I_e = \alpha_{s-1} \restriction I_e$. Otherwise change $\alpha \restriction I_e$: define $\alpha_s \restriction I_e$ in such a way that $\lambda\mathcal{C}_{e,s} \le 2^{-e}$.

Claim. *$\alpha(x) = \lim_s \alpha_s(x)$ exists for each x.*

We use a measure theoretic fact suggested by Hirschfeldt in a related context (see [7, 1.9.15]). Suppose $N, e \in \mathbb{N}$, and for $1 \le i \le N$, the class $\mathcal{B}_i$ is measurable and $\lambda\mathcal{B}_i \ge 2^{-e}$. If $N > k2^e$ then there is a set $F \subseteq \{1, \ldots, N\}$ such that $|F| = k+1$ and $\bigcap_{i \in F} \mathcal{B}_i \neq \emptyset$.

Suppose now that $0 = v_0 < v_1 < \ldots < v_N$ are consecutive stages at which $\alpha \restriction I_e$ changes. Thus $p \restriction I_e$ is defined. Then $\lambda\mathcal{B}_i \ge 2^{-e}$ for each $i \le N$, where

$$\mathcal{B}_i = \{Y\colon\ Y'_{v_{i+1}} \restriction_k \neq Y'_{v_i} \restriction_k\},$$

and $k = \mathsf{use}\ \Gamma(\max p(I_e))$, because $\lambda\mathcal{C}_e$ increased by at least 2^{-e} from v_i to v_{i+1}. Note that the intersection of any $k+1$ of the $\mathcal{B}_i$ is empty. Thus $N \le 2^e k$ by the measure theoretic fact. ◇

Since α is Δ^0_2, by the Recursion Theorem, we can now assume that p is a reduction function for α. Then in fact we have a computable bound g on the number of changes of $\alpha \restriction I_e$ given by $g(e) = 2^e \mathsf{use}\ \Gamma(\max p(I_e))$.

To complete the proof, let A be a c.e. set that is strongly jump traceable. We define a cost function c by $c(x,s) = 2^{-x}$ for each $x \ge s$; if $x < s$, and $e \le x$ is least such that $e = x$ or $\alpha_s \restriction I_e \neq \alpha_{s-1} \restriction I_e$, let

$$c(x,s) = \max(c(x,s-1), 2^{-e}).$$

Note that the cost function c is benign as defined in [10]: if $x_0 < \ldots < x_n$ and $c(x_i, x_{i+1}) \ge 2^{-e}$ for each i, then $\alpha_s \restriction I_e \neq \alpha_{s-1} \restriction I_e$ for some s such that $x_i < s \le x_{i+1}$. Hence $n \le g(e)$ where g is defined after the claim.

By [10] fix a computable enumeration $(A_s)_{s\in\mathbb{N}}$ of A that obeys c. (The rest of the argument actually works for a computable approximation $(A_s)_{s\in\mathbb{N}}$ of a Δ^0_2 set A.)

We build a Solovay test $\mathcal{G}$ as follows: when $A_{t-1}(x) \neq A_t(x)$, we put $\mathcal{C}_{e,t}$ defined in (2) into $\mathcal{G}$ where e is largest such that $\alpha \restriction I_e$ has been stable from x to t. Then $2^{-e} \le c(x,t)$. Since $\lambda\mathcal{C}_{e,t} \le 2^{-e+1} \le 2c(x,t)$ and the computable approximation of A obeys c, $\mathcal{G}$ is indeed a Solovay test.

Choose s_0 such that $\sigma \not\prec Y$ for each $[\sigma]$ enumerated into $\mathcal{G}$ after stage s_0. To show $A \le_T Y$, given an input $y \ge s_0$, using Y as an oracle, compute $s > y$ such that $\alpha_s(x) = \Gamma(Y'_s; x)$ for each $x < y$. Then $A_s(y) = A(y)$: if $A_u(y) \neq A_{u-1}(y)$ for $u > s$, let $e \le y$ be largest such that $\alpha \restriction I_e$ has been stable from y to u. Then by stage $s > y$ the set Y is in $\mathcal{C}_{e,s} \subseteq \mathcal{C}_{e,t}$, so we put Y into $\mathcal{G}$ at stage u, contradiction.

In the following we give a direct construction of a null Σ^0_3 class containing the superhigh sets. Note that the class $\mathcal{H}$ defined in (1) is such a class. However, the proof below uses techniques of independent interest. For instance, they might be of use to resolve the open question whether superhighness itself is a Σ^0_3 property.

Proposition 1. *There is a null Σ^0_3 class containing the superhigh sets.*

Proof. For each truth-table reduction Φ, we uniformly define a null Π^0_2 class $\mathcal{S}_\Phi$ such that $\emptyset'' = \Phi(Y') \rightarrow Y \in \mathcal{S}_\Phi$.

We build a Δ^0_2 set D_Φ. Then, by the Recursion Theorem we have a truth-table reduction Γ_Φ such that $\emptyset'' = \Phi(Y') \rightarrow D_\Phi = \Gamma(Y')$. We define D_Φ in such a way that $\mathcal{S}_\Phi = \{Y\colon D_\Phi = \Gamma(Y')\}$ is null. Also, $\mathcal{S}_\Phi$ is Π^0_2 because

$$Y \in \mathcal{S}_\Phi \leftrightarrow \forall w \, \forall i > w \exists s > i \, D_\Phi(w, s) = \Gamma(Y'_s; w).$$

Claim. *For each string σ, the real number $r_\sigma = \lambda\{Z\colon \sigma \prec Z'\}$ is the difference of left-c.e. reals uniformly in σ* (see [7, 1.8.15]).

To see this, note that for each finite set F the class $\mathcal{C}_F = \{Z\colon F \subseteq Z'\}$ is uniformly Σ^0_1. Let $F(\sigma) = \{j < |\sigma|\colon \sigma(j) = 1\}$, then

$$r_\sigma = \lambda(\mathcal{C}_{F(\sigma)} - \bigcup_{r<|\sigma| \& \sigma(r)=0} \mathcal{C}_{\{r\} \cup F(\sigma)}).$$

This proves the claim. Now, for each τ let $b_\tau = \lambda\{Z\colon \tau \prec \Gamma(Z')\}$. Then $b_\tau = \sum_\sigma r_\sigma \, [\![\tau = \Gamma^\sigma]\!]$ is uniformly difference left-c.e.

One can define the Δ^0_2 set $D = D_\Phi$ in such a way that $2b_{D\restriction n+1} \leq b_{D\restriction n}$ for each n. Then $2^{-n} \geq \lambda\{Y\colon D_\Phi \restriction_n = \Gamma(Y') \restriction_n\}$ for each n, so $\mathcal{S}_\Phi$ is null.

3 Each Set in Shigh$^\diamond$ Is Strongly Jump Traceable

Theorem 2. *Let A be a c.e. set that is Turing below all ML-random superhigh sets. Then A is strongly jump traceable.*

Proof. Let h be an order function. We will define a ML-random superhigh set Z such that $A \leq_T Z$ implies that A is jump traceable via bound h. In fact for an arbitrary given set G we can define Z such that $G \leq_{tt} Z'$. If also $G \geq_{tt} \emptyset''$, then Z is superhigh.

Preliminaries. Let λ denote the uniform measure on Cantor space. We will need a lower bound on the measure of a non-empty Π^0_1 class of ML-random sets. This bound is given uniformly in an index for the class (Kučera; see [7, 3.3.3]). Let $Q_0 \subseteq \mathsf{MLR}$ be the complement $2^\omega - \mathcal{R}_1$ of the second component of the standard universal ML-test.

Lemma 1. *Given an effective listing $(P^v)_{v\in\mathbb{N}}$ of Π^0_1 classes, $P^v \subseteq Q_0$, there is a constant c_0 such that $\lambda P^v \leq 2^{-K(v)-c_0} \rightarrow P^v = \emptyset$.*

We assume an indexing of all the Π^0_1 classes. Given an index for a Π^0_1 class P we have an effective approximation $P = \bigcap_t P_t$ where P_t is a clopen set ([7, Section 1.8]).

The basic set-up. For each e, a procedure R^e (with further parameters to be discussed later) builds a c.e. trace $(T_x)_{x\in\mathbb{N}}$ with bound h. Either for almost all x,

$J^A(x)\downarrow$ implies $J^A(x) \in T_x$, or R^e shows that $A \neq \Phi_e^Z$. Since Z is superhigh, the first alternative must hold for some e.

When a new computation $w = J^A(x)\downarrow$ with use u appears, R^e activates a sub-procedure S_x^e. This sub-procedure waits for evidence that $A\restriction_u$ is stable before putting w into the trace set T_x. By first waiting long enough, it makes sure that an $A\restriction_u$ change after this tracing can happen for at most $h(x)$ times, so that $|T_x| \leq h(x)$. S_x^e also calls an instance of the next procedure R^{e+1}. Thus, during the construction we can have many runs of each of the procedures R^e and S_x^e.

The environment of a procedure. Each R^e has as further parameters a Π_1^0 class P and a number $r \in \mathbb{N}$. It assumes that $Z \in P$ and $2^{-r} < \lambda P$. Each S_x^e activated by $R^e(P, r)$ will specify an appropriate subclass $Q \subseteq P$ and a number $q \in \mathbb{N}$, and call $R^{e+1}(Q, q)$.

Initially we call $R^0(Q_0, 2)$.

The two phases of S_x^e. A procedure S_x^e alternates between Phases I, and II. When changing phases it returns control to R^e. In our first approximation to describing the construction, once a computation $w = J^A(x)\downarrow$ with use u appears, S_x^e enters Phase I. It considers the Σ_1^0 class $C = \{Z\colon \Phi_e^Z\restriction_u = A\restriction_u\}$. It calls $R^{e+1}(Q, q)$ where $Q = P - C$ and q is obtained by Lemma 1. If it stays here then, because $Z \in Q$, its outcome is that $\Phi_e^Z \neq A$.

For a threshold δ depending only on r and x, once $\lambda(P_s \cap C_s) > \delta$ at stage s it lets $D = C_s$ and puts w into T_x. Now the outcome is that $J^A(x)$ has been traced. So S_x^e can return and stay inactive unless $A\restriction_u$ changes.

Once $A\restriction_u$ has changed, S_x^e enters Phase II by calling $R^{e+1}(Q, q)$ where now $Q = P \cap D$ and q is obtained by Lemma 1. Its outcome is again that $\Phi_e^Z \neq A$, this time because $\Phi_e^Z\restriction_u$ is the previous value of $A\restriction_u$ (here we use that A is c.e.).

If, later on, $P \cap D$ becomes empty, then S_x^e returns. It is now turned back to the beginning and may start again in Phase I when a new computation $J^A(x)$ appears. Note that P has now lost a measure of δ. So S_x^e can go back to Phase I for at most $1/\delta$ times.

The golden run. For some e we want a run of R^e such that each sub-procedure S_x^e it calls returns. For then, the c.e. trace $(T_x)_{x\in\mathbb{N}}$ this run of R^e builds is a trace for J^A. If no such run R^e exists then each run of R^e eventually calls some S_x^e which does not return, and therefore permanently runs a procedure R^{e+1}. If $Z \in \bigcap P_e$ where P_e is the parameter of the final run of a procedure R^e, then $A \not\leq_T Z$. So we have a contradiction if we can define a set $Z \in \bigcap_e P_e$ such that $G \leq_{\mathrm{tt}} Z'$.

Ensuring that $G \leq_{\mathrm{tt}} Z'$. For this we have to introduce new parameters into the procedures S_x^e.

Note that $G \leq_{\mathrm{tt}} Z'$ iff there is a binary function $f \leq_T Z$ such that $\forall x\, G\restriction_x = \lim_s^{\mathrm{comp}} f(x, s)$ (namely, the number of changes is computably bounded). We will define Z such that Z' encodes G. We use a variant of Kučera's method to code into ML-random sets. We define strings $z_\gamma = \lim_s^{\mathrm{comp}} z_{\gamma,s}$ and let $Z = \bigcup_{\gamma \prec G} z_\gamma$. The strings $z_{\gamma,s}$ are given effectively, and for each s they are pairwise

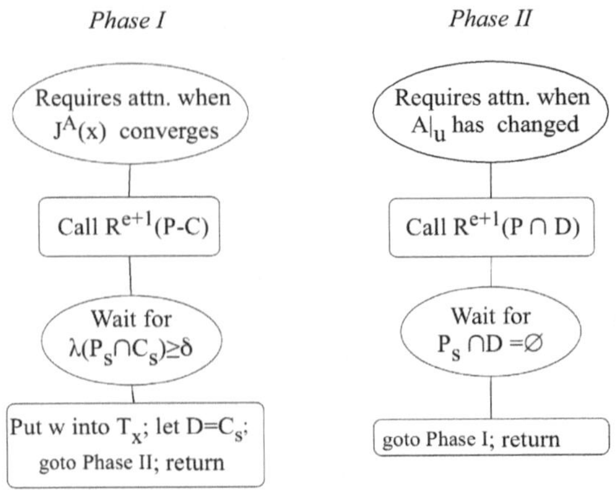

Fig. 1. Diagram for the procedure S^e_x

incomparable. Then we let $f(x, s) = \gamma$ if $|\gamma| = x$ and $z_{\gamma,s} \prec Z$, and $f(x, s) = \varnothing$ if there is no such γ.

Firstly, we review Kučera's coding into a member of a Π^0_1-class P of positive measure. For a string x let $\lambda(P|x) = 2^{|x|}\lambda(P \cap [x])$.

Lemma 2 (Kučera; see [7], 3.3.1). *Suppose that P is a Π^0_1 class, x is a string, and $\lambda(P|x) \geq 2^{-l}$ where $l \in \mathbb{N}$. Then there are at least two strings $w \succeq x$ of length $|x|+l+1$ such that $\lambda(P|w) > 2^{-l-1}$. We let w_0 be the leftmost and w_1 be the rightmost such string.*

In the following we code a string β into a string y_β on a Π^0_1 class P.

Definition 2. Given a Π^0_1 class P, a string z such that $P \subseteq [z]$, and $r \in \mathbb{N}$ such that $2^{-r} < \lambda P$, we define a string

$$y_\beta = \mathsf{kuc}(P, r, z, \beta)$$

as follows: $y_\varnothing = z$; if $x = y_\beta$ has been defined, let $l = r + |\beta|$, and let $y_{\beta\frown b} = w_b$ for $b \in \{0, 1\}$, where the strings w_b are defined as in Lemma 2.

Note that for each β we have $\lambda(P \mid y_\beta) \geq 2^{-r-|\beta|}$ and

$$|y_\beta| \leq |z| + |\beta|(r + |\beta| + 1). \tag{3}$$

At stage s we have the approximation $y_{\beta,s} = \mathsf{kuc}(P_s \cap [z], r, z, \beta)$. While $y_{\beta,s}$ is stable, the string w_b in the recursive definition above changes at most 2^l times. Thus, inductively, $y_{\beta,s}$ changes at most $2^{|\beta|(r+|\beta|+1)}$ times.

For each e, η we may have a version of R^e denoted $R^{e,\eta}(P, r, z_\eta)$. It assumes that η has already been coded into the initial segment z_η of Z, and works within $P \subseteq [z_\eta]$. It calls procedures $S^{e,\eta\alpha}_x(P, r, z_\eta)$ for certain x, α. In this case we let $z_{\eta\alpha} = y_\alpha = \mathsf{kuc}(P, r, z_\eta, \alpha)$.

For each x, once $J^A(x) \downarrow$, $R^{e,\eta}$ wishes to run $S^{e,\eta\alpha}_x$ for all α of a certain length m defined in (5) below, which increases with $h(x)$. Thus, as x increases, more and more bits beyond η are coded into Z. The trace set T_x will contain all the numbers enumerated by procedures $S_x^{e,\eta\alpha}$ where $|\alpha| = m$. We ensure that m is small enough so that $|T_x| \leq h(x)$. To summarize, a typical sequences of calls of procedures is

$$R^{e,\eta} \rightarrow S_x^{e,\eta\alpha} \rightarrow R^{e+1,\eta\alpha}.$$

Formal details. Some ML-random set $Y \not\geq_T \emptyset'$ is superhigh by pseudo jump inversion as in [7, 6.3.14]. Since $A \leq_T Y$ and A is c.e., A is a base for ML-randomness; see [7, 5.1.18]. Thus A is superlow. Hence there is an order function g and a computable enumeration of A such that $J^A(x)[s]$ becomes undefined for at most $g(x)$ times.

We build a sequence of Π^0_1 classes $(P^n)_{n\in\mathbb{N}}$ as in Lemma 1. If $n = \langle e, \gamma, x, i\rangle$, then since $K(n) \leq^+ 2\log\langle e, \gamma\rangle + 2\log x + 2\log i$, we have

$$P^{\langle e,\gamma,x,i\rangle} \neq \emptyset \;\Rightarrow\; \lambda P^{\langle e,\gamma,x,i\rangle} \geq 2^{-q} \tag{4}$$

where $q = 2\log\langle e, \gamma\rangle + 2\log x + 2\log i + c$ for some fixed $c \in \mathbb{N}$. By the Recursion Theorem we may assume that we know c in advance.

The construction starts off by calling $R^{0,\varnothing}(Q_0, 3, \varnothing)$.

Procedure $R^{e,\eta}(P, r, z)$, where $z \in 2^{<\omega}$, $P \subseteq \mathsf{MLR} \cap [z]$ is a Π^0_1 class and $r \in \mathbb{N}$. This procedure enumerates a c.e. trace $(T_x)_{x\in\mathbb{N}}$. (It assumes that $2^{-r} < \lambda P$.)

For each string α of length at most the stage number s, see whether some procedure $S_x^{e,\eta\alpha}(P)$ requires attention, or is at (b) or (e), and no procedure $S_y^{e,\eta\beta}(P)$ for $\beta \prec \alpha$ satisfies the same condition. If so, choose x least for α and activate $S_x^{e,\eta\alpha}(P)$. (This suspends any runs $S_z^{e,\rho}$ for $\eta\alpha \preceq \rho$. Such a run may be resumed later.)

Procedure $S_x^{e,\eta\alpha}(P, r, z)$, where $|\alpha|$ is the greatest $m > 0$ such that, if $n = m(r + m + 1)$, we have

$$2^{|\eta\alpha|}2^{2n+r+2} \leq h(x). \tag{5}$$

There only is such a procedure if x is so large that m exists.

Let $y_{\alpha,s} = \mathsf{kuc}(P_s, \alpha, r, z)$. Let

$$\delta = 2^{-|y_{\alpha,s}|-m-r-1}.$$

(Comment: $S_x^{e,\eta\alpha}(P, r, z)$ cannot change $y_{\alpha,s}$. It only changes "by itself" as P_s gets smaller. This makes the procedure go back to the beginning. So in the following we can assume y_α is stable.)

Phase I.

(a) $S_x^{e,\eta\alpha}$ *requires attention* if $w = J^A(x)\downarrow$ with use u. Let

$$C = [y_\alpha] \cap \{Z\colon \Phi_e^Z\restriction_u = A\restriction_u\},$$

a Σ^0_1 class. Let $C_s = [y_{\alpha,s}] \cap \{Z\colon \Phi_e^Z\restriction_u = A\restriction_u [s]\}$ be its approximation at stage s, which is clopen.

(b) WHILE $\lambda(P_s \cap C_s) < \delta$ run in case $e < s$ the procedure

$$R^{e+1,\eta\alpha}(Q, q, y_{\alpha,s});$$

here Q is the Π^0_1 class $P \cap [y_{\alpha,s}] - C$, and

$$q = 2\log\langle e, \eta\alpha\rangle + 2\log x + 2\log i + c,$$

where i is the number of times $S^{e,\eta\alpha}_x$ has called $R^{e+1,\eta\alpha}$ (the constant c was defined after (4) at the beginning of the formal construction). Then $2^{-q} < \lambda Q$ unless $Q = \emptyset$. Meanwhile, if $y_{\alpha,s} \neq y_{\alpha,s-1}$ put w into T_x, cancel all sub-runs, GOTO (a), and RETURN. Otherwise, if $A_s \upharpoonright_u \neq A_{s-1} \upharpoonright_u$ cancel all sub-runs, GOTO (a) and RETURN.
(Comment: if the run $S^{e,\eta\alpha}_x$ stays at (b) and $Z \in Q$, then $A \upharpoonright_u = \Phi^Z_e \upharpoonright_u$ fails, so we have defeated Φ_e.)

(c) Put w into T_x, let $D = C_s$, GOTO (d), and RETURN. (Thus, the next time we call $S^{e,\eta\alpha}_x(P)$ it will be in Phase II.)

Phase II.

(d) $S^{e,\eta\alpha}_x$ requires attention again if $A \upharpoonright_u$ has changed.

(e) WHILE $P_s \cap D \neq \emptyset$ RUN in case $e < s$

$$R^{e+1}(P \cap D, q, y_{\alpha,s})$$

where $q \in \mathbb{N}$ is defined as in (b). Meanwhile, if $y_{\alpha,s} \neq y_{\alpha,s-1}$ cancel all sub-runs, GOTO (a), and RETURN.
(Comment: if the run $S^{e,\eta\alpha}_x$ stays at (e) and $Z \in Q$ then again $A \upharpoonright_u = \Phi^Z_e \upharpoonright_u$ fails, this time because $Z \in D$ and $\Phi^Z_e \upharpoonright_u$ is an old version of $A \upharpoonright_u$.)

(f) GOTO (a) and RETURN.

Verification. The function g was defined at the beginning of the formal proof. First we compute bounds on how often a particular run $S^{e,\eta\alpha}_x$ does certain things.

Claim 1. *Consider a run $S^{e,\eta\alpha}_x(P, r, z)$ called by $R^{e,\eta}(P, r, z)$. As in the construction, let $m = |\alpha|$ and $n = m(r + m + 1)$.*

(i) While $y_{\alpha,s}$ does not change, the run passes (f) for at most 2^{m+r+1} times.
(ii) The run enumerates at most 2^{2n+r+2} elements into T_x.
(iii) It calls a run $R^{e+1,\eta\alpha}$ at (b) or (e) for at most $2^{n+1}g(x)$ times.

To prove (i), as before let $\delta = 2^{-|y_\alpha|-m-r-1}$. Note that each time the run passes (f), the class $P \cap [y_\alpha]$ loses $\lambda D \geq \delta$ in measure. This can repeat itself at most 2^{m+r+1} times. (This argument allows for the case that the run of $S^{e,\eta\alpha}_x$ is suspended due to the run of some $S^{e,\eta\beta}_z$ for $\beta \prec \alpha$. If $S^{e,\eta\beta}_z$ finishes then $S^{e,\eta\alpha}_x$, with the same parameters, continues from the same point on where it was when it was suspended.)
(ii) There are at most 2^n values for y_α during a run of $S^{e,\eta\alpha}_x$ by the remarks after Definition 2. Therefore this run enumerates at most $2^n 2^{n+r+1} + 2^n$ elements into T_x where at most 2^n elements are enumerated when y_α changes.

(iii): for each value y_α there are at most $2g(x)$ calls, namely, at most two for each computation $J^A(x)$ (g is defined at the beginning of the formal proof). ◇

Note that $|T_x| \leq h(x)$ by (ii) of Claim 1 and (5).

Strings $z_{\gamma,s}$, $\gamma \in 2^{<\omega}$ are used to code the given set G into Z'. Let $z_{\varnothing,s} = \varnothing$.

- If $z_{\eta,s}$ has been defined and $R^{e,\eta}(P, r, z_{\eta,s})$ is running at stage s, then for all β such that no procedure $S^{e,\eta\alpha}$ is running for any $\alpha \prec \beta$, let $z_{\eta\beta,s} = \mathsf{kuc}(P, r, z_{\eta,s}, \beta)$.
- If α is maximal under the prefix relation such that $z_{\eta\alpha,s}$ is now defined, it must be the case that $R^{e+1,\eta\alpha}(Q, q, z_{\eta\alpha})$ runs. So we may continue the recursive definition. Note that $|\alpha| > 0$ by the condition that $m > 0$ in (5).

Claim 2. *For each γ, $z_\gamma = \lim_s z_{\gamma,s}$ exists, with the number of changes computably bounded in γ.*

We say that a run of $S_x^{e,\rho}$ is a *k-run* if $|\rho| \leq k$. For each number parameter p we will let $\overline{p}(k, v)$ denote a computable upper bound for p computed from k, v. Such a function is always chosen nondecreasing in each argument.

To prove Claim 2, we think of k as fixed and define by simultaneous recursion on $v \leq k$ computable functions $\overline{r}(k, v), \overline{x}(k, v), \overline{b}(k, v), \overline{c}(k, v)$ with the following properties:

(i) *$\overline{r}(k, v)$ bounds r in any call $R^{e,\eta}(Q, r)$ where $|\eta| \leq k$ and $e \leq v$.*
(ii) *$\overline{x}(k, v)$ bounds the largest x such that some k-run $S_x^{e,\eta\alpha}$ is started where $e \leq v$.*
(iii) *For each x, $\overline{b}(k, v)$ bounds the number of times a k-run $S_x^{e,\eta\alpha}$ for $e \leq v$ requires attention.*
(iv) *For each x, $\overline{c}(k, v)$ bounds the number of times a run $R^{e+1,\eta\alpha}$ is started by some k-run $S_x^{e,\eta\alpha}$ for $e \leq v$.*

Fix γ such that $|\gamma| = k$. In the following we may assume that $\eta\alpha \preceq \gamma$, because then the actual bounds can be obtained by multiplying with 2^k.

Suppose now $k \geq v \geq 0$ and we have defined the bounds in (i)–(iv) for $v-1$ in case $v > 0$. We define the bounds for v and verify (i)–(iv). We may assume $e = v$, because then the required bounds are obtained by adding the bounds for $k, v-1$ to the bounds now obtained for $e = v$.

(i) First suppose that $v = 0$. Then $\eta = \varnothing$, so let $\overline{r}(k, 0) = 3$. If $v > 0$, we define a sequence of Π^0_1 classes as in Lemma 1: if for the i-th time a run $S_x^{e-1,\rho}$ calls a run $R^{e,\rho}(Q, q)$ we let $P^{\langle e,\rho,x,i\rangle} = Q$. By the inductive hypothesis (iii) and (iv) for $v-1$ we have a bound $\overline{i}(v, x)$ on the largest i such that a class $P^{\langle v,\eta\alpha,x,i\rangle}$ is defined (when $S_x^{v-1,\eta}$ in (b) or (e) starts a run $R^{v,\eta}$). Thus let $\overline{r}(k, v) = 2\log\langle v, \gamma\rangle + 2\log\overline{x}(k, v-1) + 2\log\overline{i}(v, \overline{x}(k, v-1)) + c$.

To prove (ii) and (iii), suppose $R^{e,\eta}(Q, r)$ calls $S_x^{e,\eta\alpha}$. Let $m = |\alpha|$ and $n = m(r + m + 1)$. Then $n \leq k(\overline{r}(k, v) + k + 1)$.

(ii) We have $h(x) < 2^{k+2k(\overline{r}(k,v)+k+1)+3}$ because m is chosen maximal in (5). Since h is an order function, this gives the desired computable bound $\overline{x}(k, v)$ on x.

(iii) By Claim 1(i), for each value of y_α, the run can pass (f) for at most $2^{k+\overline{r}(k,v)+1}$ times. Further, it can require attention $2^n + g(\overline{x}(k,v))$ more times because y_α changes or because $J^A(x)$ changes. This allows us to define $\overline{b}(k,v)$.
(iv) By Claim 1(iv) a run $R^{v+1,\eta\alpha}$ is started for at most $\overline{b}(k,v)2^{k+1}g(\overline{x}(k,v))$ times.

This completes the recursive definition of the four functions. Now, to obtain Claim 2, fix γ. One reason that z_γ changes is that (A) some run $S_y^{e,\rho}$ for $\rho \preceq \gamma$, calls $R^{e+1,\rho}$ in (e). This run is a k-run for $k = |\gamma|$. By (ii) and (iii), the number of times this happens is computably bounded by $\overline{b}(k,k)\overline{x}(k,k)$. While it does not happen, z_γ can also change because (B) for some $\eta\alpha \preceq \gamma$ as in the construction, y_α changes because some P_s, which defines y_α, decreases. Since there is a computable bound $\overline{l}(k)$ on the length of z_γ by (i) of this claim and (3), while the first reason does not apply, this can happen for at most $2^{\overline{l}(k)}$ times. Thus in total z_γ changes for at most $\overline{b}(k,k)\overline{x}(k,k)2^{\overline{l}(k)}$ times. $\diamond$

Now let $Z = \bigcup_{\gamma \prec G} z_\gamma$. By Claim 2 we have $G \leq_{tt} Z'$.

Claim 3. (Golden Run Lemma) *For some $\eta \prec G$ and some e, there is a run $R^{e,\eta}(P,r)$ (called a golden run) that is not cancelled such that, each time it calls a run $S_x^{e,\eta\alpha}$ where $\eta\alpha \prec G$, that run returns.*

Assume the claim fails. We verify the following for each e.

(i) There is a run $R^{e,\eta}$ that is not cancelled; further, $S_x^{e,\eta\alpha}(P)$ is running for some x, where $\eta\alpha \prec G$, and eventually does not return.
(ii) $A \neq \Phi_e^Z$.

(i) We use induction. For $e = 0$ clearly the single run of $R^{0,\varnothing}$ is not cancelled. Suppose now that a run of $R^{e,\eta}$ is not cancelled. Since we assume the claim fails, some run $S_x^{e,\eta\alpha}$, $\eta\alpha \prec G$, eventually does not return. From then on the computation $J^A(x)$ it is based on and y_α are stable. So the run calls $R^{e+1,\eta\alpha}$ and that run is not cancelled.
(ii) Suppose the run $S_x^{e,\eta\alpha}(P,r,z)$ that does not return has been called at stage s. Suppose further it now stays at (b) or (e), after having called $R^{e,\eta\alpha}(Q,q,y_\alpha)$. Since $y_{\eta\alpha}$ is stable by stage s, we have $Z \in Q$. Hence $A \neq \Phi_e^Z$ by the comments in (b) or (e). $\diamond$

Let $(T_x)_{x\in\mathbb{N}}$ be the c.e. trace enumerated by this golden run.

Claim 4. *$(T_x)_{x\in\mathbb{N}}$ is a trace for J^A with bound h.*

As remarked after Claim 1, we have $|T_x| \leq h(x)$. Suppose x is so large that m in (5) exists. Suppose further that the final value of $w = J^A(x)$ appears at stage t. Let $\eta\alpha \prec G$ such that $|\alpha| = m$.

As the run is golden and by Claim 1(i), eventually no procedure $S_y^{e,\eta\beta}(P)$ for $\beta \prec \alpha$ is at (b) or (e). Thus, from some stage $s > t$ on, the run $S_x^{e,\eta\alpha}$ is not suspended. If y_α has not settled by stage s then w goes into T_x. Else $\lambda(P \mid y_{\alpha,s}) > 2^{-r-|\alpha|}$. Since $S_x^{e,\eta\alpha}$ returns each time it is called, the run is at

(a) at some stage after t. Also, $P_s \cap C_s$ must reach the size $\delta = 2^{-|y_\alpha|-|\alpha|-r-1}$ required for putting w into T_x.

As a consequence, we can separate highness properties within the ML-random sets. See [7, Def. 8.4.13] for the weak reducibility $\leq_{JT}$, and [10] for the highness property "$\emptyset'$ is c.e. traceable by Y". Note that JT-hardness implies both this highness property and superhighness.

Corollary 1. *There is a ML-random superhigh Δ^0_3 set Z such that $\emptyset'$ is not c.e. traceable by Z. In particular, Z is not JT-hard.*

Proof. By [7, Lemma 8.5.19] there is a benign cost function c such that each c.e. set A that obeys c is Turing below each ML-random set Y such that $\emptyset'$ is c.e. traceable by Y. By [7, Exercise 8.5.8] there is an order function h such that some c.e. set A obeys c but is not jump traceable with bound h. Then by the proof of Theorem 2 there is a ML-random superhigh set $Z \leq_T \emptyset''$ such that $A \not\leq_T Z$. Hence Z is not JT-hard.

References

1. Nies, A.: Reals which compute little. In: Logic Colloquium 2002. Lecture Notes in Logic, pp. 260–274. Springer, Heidelberg (2002)
2. Kučera, A.: On relative randomness. Ann. Pure Appl. Logic 63, 61–67 (1993)
3. Hirschfeldt, D., Nies, A., Stephan, F.: Using random sets as oracles. J. Lond. Math. Soc (2) 75(3), 610–622 (2007)
4. Figueira, S., Nies, A., Stephan, F.: Lowness properties and approximations of the jump. Ann. Pure Appl. Logic 152, 51–66 (2008)
5. Cholak, P., Downey, R., Greenberg, N.: Strongly jump-traceability I: the computably enumerable case. Adv. in Math. 217, 2045–2074 (2008)
6. Mohrherr, J.: A refinement of low_n and high_n for the r.e. degrees. Z. Math. Logik Grundlag. Math. 32(1), 5–12 (1986)
7. Nies, A.: Computability and Randomness. Oxford University Press, Oxford (2009); Oxford Logic Guides, pp. xv + 443
8. Kjos-Hanssen, B., Nies, A.: Superhighness. Notre Dame J. Formal Logic (to appear)
9. Greenberg, N., Hirschfeldt, D., Nies, A.: Characterizing the strongly jump traceable sets via randomness (to appear)
10. Greenberg, N., Nies, A.: Benign cost functions and lowness properties (to appear)
11. Downey, R., Hirschfeldt, D., Nies, A., Stephan, F.: Trivial reals. In: Proceedings of the 7th and 8th Asian Logic Conferences, pp. 103–131. Singapore Univ. Press, Singapore (2003)
12. Nies, A.: Lowness properties and randomness. Adv. in Math. 197, 274–305 (2005)
13. Kurtz, S.: Randomness and genericity in the degrees of unsolvability. Ph.D. Dissertation, University of Illinois, Urbana (1981)
14. Nies, A.: Non-cupping and randomness. Proc. Amer. Math. Soc. 135(3), 837–844 (2007)

Amortized Communication Complexity of Distributions

Jérémie Roland and Mario Szegedy

[1] NEC Laboratories America
jroland@nec-labs.com
[2] Rutgers University
szegedy@cs.rutgers.edu

Abstract. Consider the following general communication problem: Alice and Bob have to simulate a probabilistic function $\mathbf{p}$, that with every $(x, y) \in \mathcal{X} \times \mathcal{Y}$ associates a probability distribution on $\mathcal{A} \times \mathcal{B}$. The two parties, upon receiving inputs x and y, need to output $a \in \mathcal{A}$, $b \in \mathcal{B}$ in such a manner that the (a, b) pair is distributed according to $\mathbf{p}(x, y)$. They share randomness, and have access to a channel that allows two-way communication. Our main focus is an instance of the above problem coming from the well known EPR experiment in quantum physics. In this paper, we are concerned with the amount of communication required to simulate the EPR experiment when it is repeated in parallel a large number of times, giving rise to a notion of amortized communication complexity.

In the 3-dimensional case, Toner and Bacon showed that this problem could be solved using on average 0.85 bits of communication per repetition [1]. We show that their approach cannot go below 0.414 bits, and we give a fundamentally different technique, relying on the reverse Shannon theorem, which allows us to reduce the amortized communication to 0.28 bits for dimension 3, and 0.410 bits for arbitrary dimension. We also give a lower bound of 0.13 bits for this problem (valid for one-way protocols), and conjecture that this could be improved to match the upper bounds. In our investigation we find interesting connections to a number of different problems in communication complexity, in particular to [2]. The results contained herein are entirely classical and no knowledge of the quantum phenomenon is assumed.

1 Communication Complexity of Distributions

Communication complexity has been an amazingly potent tool for studying lower bounds for circuits, branching programs, VLSI and streaming data. Lately it is also used to quantify non-local nature of quantum systems.

Recall that in the original version of the model [3] Alice and Bob jointly evaluate a Boolean predicate $f(x, y)$ $(x \in \mathcal{X},\ y \in \mathcal{Y})$ through exchanging messages. Throughout, we will be concerned with the following generalization of the model:

Let $\mathcal{X}$ and $\mathcal{A}$ be the sets of inputs and possible outputs for Alice, and $\mathcal{Y}$ and $\mathcal{B}$ be the sets of inputs and possible outputs for Bob.

Task: A task $\mathbf{p}$ is specified by a function $\mathbf{p} : \mathcal{X} \times \mathcal{Y} \to \text{Distrib}(\mathcal{A} \times \mathcal{B})$, where $\text{Distrib}(\mathcal{A} \times \mathcal{B})$ is the set of all probability distributions on $\mathcal{A} \times \mathcal{B}$.

S. Albers et al. (Eds.): ICALP 2009, Part I, LNCS 5555, pp. 738–749, 2009.

Alice and Bob meet the specification $\mathbf{p}$ if upon receiving $x \in \mathcal{X}$ and $y \in \mathcal{Y}$ their output pair (a, b) is distributed according to $\mathbf{p}(x, y)$. Task $\mathbf{p}$ is completely described by the probabilities

$$\mathbf{p}(a, b||x, y) \stackrel{def}{=} \text{the probability of } (a, b) \text{ under distribution } \mathbf{p}(x, y).$$

Alice and Bob share randomness from a common source Λ, i.e. in addition to their input they both receive λ, where $\lambda \in \Lambda$ is picked randomly.

Even with unlimited computational power Alice and Bob usually need to communicate to produce the desired output. The exact rules concerning the communication are critical for our analysis of very low communication problems. Under the wrong definition, Alice may signal to Bob simply by her choice of sending or not sending a bit. To exclude this, we postulate that Alice and Bob are either in send-mode or in receive-mode or in output mode. The communication runs in rounds. After each round the players get into a new mode, which is a function of the player's input, the shared random string λ, and the messages received so far by the player. *A protocol must satisfy that in each round either of two cases happens:* 1. one player is in send-mode and the other is in receive mode; 2. both players are in output mode. No other combination is permitted. Note that if the parties needed to make random choices, we could add them to the shared randomness, Λ. Thus we assume that the protocol is deterministic for any fixed λ.

Protocol: A protocol P for a given simulation task $\mathbf{p}$ is a probability distribution $p(\lambda)$ over deterministic protocols P_λ, each solving a task $\mathbf{p}_\lambda$, such that $\mathbf{p} = \sum_\lambda p(\lambda)\mathbf{p}_\lambda$.

Note that since any $\lambda \in \Lambda$ corresponds to a deterministic protocol, we may extend Λ to the set of all possible deterministic communication protocols with inputs in $\mathcal{X} \times \mathcal{Y}$ and outputs in $\mathcal{A} \times \mathcal{B}$ (we would just set $p(\lambda) = 0$ for all deterministic protocols that never occur when executing the shared randomness protocol P).

LHV: Let Λ_0 be the set of all deterministic protocols that do not use any communication. A task $\mathbf{p}$ is in LHV if it may be simulated using a distribution over protocols in Λ_0 only, that is, if there is a zero (classical) communication protocol for it.

LHV stands for Local Hidden Variable referring to $\lambda \in \Lambda$, which is the only source of correlation between Alice and Bob. Note that these correlations do not violate locality because we assume that the parties receive the "hidden" λ when they are not yet spatially separated.

Fix the input and output sets $\mathcal{X}, \mathcal{Y}, \mathcal{A}, \mathcal{B}$ for the rest of this paragraph.

Bell inequality: A Bell inequality is an inequality of the form

$$\sum_{x,y,a,b} B_{xyab}\, \mathbf{p}(a, b||x, y) \leq B_0, \tag{1}$$

which holds for all $\mathbf{p} \in$ LHV.

Notice that the left hand side of Eq. (1) is a linear functional, which we will shortly denote as $B(\mathbf{p})$. A task $\mathbf{p}$ is in LHV if and only if it satisfies all Bell inequalities.

In other words, LHV is a convex set. We are now interested in tasks outside LHV, which may be identified by the fact that they violate some Bell inequality. This means that such a task $\mathbf{p}$ may not be simulated using shared randomness only, and that some additional communication is required. Let P be a communication protocol simulating $\mathbf{p}(a,b||x,y)$ using shared randomness Λ, $M(x,y,\lambda)$ be the transcript of the messages on input x and y when the shared randomness is fixed to λ, and $|M|$ be the length in bits of this transcript. We define the worst-case cost $C_w(P)$ as the maximal number of bits communicated between Alice and Bob in any particular execution of the protocol, that is, $C_w(P) = \max_{x,y,\lambda} |M(x,y,\lambda)|$, where the maximum is over inputs $(x,y) \in \mathcal{X} \times \mathcal{Y}$ and shared randomness $\lambda \in \Lambda$ such that $p(\lambda) \neq 0$. We then define the worst-case communication complexity as $C_w(\mathbf{p}) = \min_P C_w(P)$. In this paper, we are more interested in the average cost:

Average cost: Given a distribution D on $\mathcal{X} \times \mathcal{Y}$, the average cost $C^D(P)$ is the expected number of bits communicated between Alice and Bob, where the expectation is taken over the shared randomness $\lambda \in \Lambda$ and the inputs $(x,y) \in D$,

$$C^D(P) = \sum_{\lambda \in \Lambda} p(\lambda) \sum_{(x,y) \in \mathcal{X} \times \mathcal{Y}} D(x,y)\,|M(x,y,\lambda)|. \tag{2}$$

Average communication complexity: $C(\mathbf{p}) = \max_D C^D(\mathbf{p})$, where $C^D(\mathbf{p}) = \min_P C^D(P)$ is the *distributional* average communication complexity for fixed input distribution D, the minimum being taken over all protocols P implementing $\mathbf{p}$, and the maximum over all distributions D on $\mathcal{X} \times \mathcal{Y}$.

We emphasize that even when we are concerned with the average case complexity, P needs to meet the specification for every input pair $(x,y) \in \mathcal{X} \times \mathcal{Y}$.

Example. The CHSH correlations ($\mathbf{p}_\mu$): Let us define the task $\mathbf{p}_\mu$ for $0 \leq \mu \leq 1$ as follows: $\mathcal{X} = \mathcal{Y} = \{0,1\}$, $\mathcal{A} = \mathcal{B} = \{1,-1\}$ and

$$\mathbf{p}_\mu(a,b||x,y) = \frac{1 + \mu\, ab\,(-1)^{x\cdot y}}{4}.$$

The task is defined in such a way that for all $(x,y) \in \mathcal{X} \times \mathcal{Y}$, the relation $ab = (-1)^{x\cdot y}$ between the inputs and the outputs has to be satisfied with probability $\frac{1+\mu}{2}$. It is not hard to show that $\mathbf{p}_\mu$ can be implemented classically with zero communication only for $0 \leq \mu \leq 1/2$. In particular, for $\mu > 1/2$, $\mathbf{p}_\mu$ violates the so-called CHSH Bell inequality [4]:

$$\sum_{x,y,a,b} ab\,(-1)^{x\cdot y} \mathbf{p}(a,b||x,y) \leq 2,$$

so that in a classical world, this task requires communication to be implemented. However, if Alice and Bob are separated in space, but they share a pair of entangled qubits, in the quantum world they can solve $\mathbf{p}_{1/\sqrt{2}}$ *with no communication* whatsoever. This is because quantum correlations may violate Bell inequalities, and therefore have a non-local character, as was first shown by Bell [5].

The EPR-Bohm experiment ($\mathbf{p}_{\dim=d}$): The inputs to Alice and Bob are unit vectors $\boldsymbol{x}$ and $\boldsymbol{y}$ from the d-dimensional sphere $\mathbb{S}_{d-1}$. The output is again an element of $\{1,-1\}$, a for Alice, and b for Bob, with the specification

$$\mathbf{p}_{\dim=d}(a,b||\boldsymbol{x},\boldsymbol{y}) = \frac{1 - ab\,\boldsymbol{x}\cdot\boldsymbol{y}}{4}. \tag{3}$$

$\mathbf{p}_{\dim=d}$ arises from the EPR-Bohm experiment [6,7], and can be solved in the quantum world with zero communication.

The communication complexity of $\mathbf{p}_{\dim=d}$, and quantum distributions in general, has been studied in a series of paper [8,9,10,11]. $\mathbf{p}_{\dim=3}$ is particularly interesting because it corresponds to Bohm's original version of the experiment, involving a maximally entangled qubit pair, the most simple quantum system that captures the essential properties of entanglement. The best known protocol for $\mathbf{p}_{\dim=3}$ was presented by Toner and Bacon, and uses one bit of communication [1]. This was shown to be optimal, even for the average complexity, by Barrett, Kent and Pironio [12], hence, $C_w(\mathbf{p}_{\dim=3}) = C(\mathbf{p}_{\dim=3}) = 1$ bit.

The higher dimensional case $\mathbf{p} = \mathbf{p}_{\dim=d}$ has been studied by Degorre, Laplante and Roland [13], who proved that the average communication complexity scaled at most as $C(\mathbf{p}_{\dim=d}) = O(\log d)$. This was significantly improved by Regev and Toner [14], who showed that bounded worst-case communication was sufficient, by providing an explicit 2-bit protocol, so that $C_w(\mathbf{p}_{\dim=d}) \leq 2$. When considering average communication, they could improve their protocol to 1.82 bits.

The amortized communication cost of simulating $\mathbf{p}_{\dim=d}$ (and its powers) will be the focus of this paper, and will be described in more details in the next section.

Finiteness: In this example $\mathcal{X}$, $\mathcal{Y}$, and probability space Λ are infinite, equipped with some measure. The communication still needs to be bounded. Note that as in the finite case, each bit communicated in the protocol by a given party can be described by a measurable function that goes from this party's input, the shared randomness and the bits communicated so far into $\{0,1\}$. Even though the domain of these functions is infinite, the parties can compute them for free because they are computationally unbounded. We also need to modify formula (2) by replacing the sum with an integral.

2 Amortization

We now consider the task $\mathbf{p}^{\otimes n}$, given by the n-fold parallelization of $\mathbf{p}$,

$$\mathbf{p}^{\otimes n}(\boldsymbol{a},\boldsymbol{b}||\boldsymbol{x},\boldsymbol{y}) = \prod_{i=1}^{n} \mathbf{p}(a_i,b_i||x_i,y_i).$$

We then define the following communication complexity.

Amortized communication complexity: $C_\infty(\mathbf{p}) = \max_D C_\infty^D(\mathbf{p})$, where $C_\infty^D(\mathbf{p}) = \lim_{n\to\infty} C^{D^{\otimes n}}(\mathbf{p}^{\otimes n})/n$ is the *distributional* amortized communication complexity, and $D^{\otimes n}(\boldsymbol{x},\boldsymbol{y}) = \prod_{i=1}^{n} D(x_i,y_i)$.

2.1 Entropic Complexity

Let P be a communication protocol simulating $\mathbf{p}(a,b||x,y)$ using shared randomness Λ, and let D be the input distribution.

Entropic cost $C_H^D(P)$: Conditional entropy $H(M|\Lambda)$ of the transcript M of the messages communicated between Alice and Bob, given the shared randomness $\lambda \in \Lambda$.

We also define the corresponding (distributional and non-distributional) entropic communication complexities for a task $\mathbf{p}$ as $C_H^D(\mathbf{p}) = \min_P C_H^D(P)$ and $C_H(\mathbf{p}) = \max_D C_H^D(\mathbf{p})$, where the minimum is taken over all protocols P implementing $\mathbf{p}$, and the maximum is taken over all distributions D on $\mathcal{X} \times \mathcal{Y}$.

2.2 The Input Distribution

As first observed by Yao [15], von Neumann's minmax principle [16] implies the following statement.

Theorem 1. *Let C_* be any of C, C_∞, C_H. We have $C_*(\mathbf{p}) = \min_P \max_D C_*^D(P)$.*

Note that for a fixed protocol P, the maximum over distributions D is achieved for a given input couple $(x,y) \in \mathcal{X} \times \mathcal{Y}$.

For specific tasks, symmetries allow to make assumptions on the hardest distribution, which attains $C_*(\mathbf{p}) = \max_D \min_P C_*^D(P)$. In particular, for the CHSH problem $\mathbf{p}_\mu$, we can show that the uniform distribution is the hardest distribution.

Claim 1. *Let C_* be any of C, C_∞, C_H. Then, $C_*(\mathbf{p}_\mu) = C_*^U(\mathbf{p}_\mu)$, where U is the uniform distribution on $\{0,1\}^2$.*

Similarly, for the EPR-Bohm problem $\mathbf{p}_{\dim=d}$, we may assume that the hardest distribution has uniform marginals (this observation is due to Toner and Bacon [1]). For an input distribution D, we will denote D_A and D_B the marginal distributions of x and y, respectively.

Claim 2. *Let C_* be any of C, C_∞, C_H. Then, there exists a distribution U on $\mathbb{S}_{d-1} \times \mathbb{S}_{d-1}$ with uniform marginals U_A and U_B such that $C_*^U(\mathbf{p}_{\dim=d}) = \max_D C_*^D(\mathbf{p}_{\dim=d})$.*

Note that for $\mathbf{p}_{\dim=d}$, we can only show that the marginals of the hardest distribution are uniform, not that the hardest distribution itself is uniform. However, let us note that when restricting to one-way communication protocols, the communication complexity only depends on the marginal distribution for the player sending the messages. For any notion of communication complexity, we add a superscript $\rightarrow$ when we only consider protocols restricted to one-way communication.

Claim 3. *Let C_* be any of C, C_∞, C_H, C_I, and let D, D' be two distributions on $\mathcal{X} \times \mathcal{Y}$ having the same marginal distributions for x, that is, $D_A = D'_A$. Then, $C_*^{\rightarrow,D}(\mathbf{p}) = C_*^{\rightarrow,D'}(\mathbf{p})$.*

2.3 Relation between the Communication Complexities

We will use as an intermediate step the following cost for communication protocols using *private* randomness only, first introduced by Chakrabarti *et al.* [17]:

Information cost $C_I^D(P)$: Mutual information $I(XY:M)$ between the inputs X,Y and the transcript M of the messages communicated between Alice and Bob.

As for the other complexities, we also define the information complexities $C_I^D(\mathbf{p}) = \min_P C_I^D(P)$ and $C_I(\mathbf{p}) = \min_P \max_D C_I^D(P)$, where the minimum is taken over all *private* randomness protocols P implementing $\mathbf{p}$ (note that in the presence of shared randomness, there always exists a protocol P such that $C_I^D(P) = 0$, so this quantity would not be relevant). Chakrabarti *et al.* have shown that this complexity satisfies the following direct sum property.

Theorem 2 ([17,2]). *If $D = D_A \otimes D_B$ is a product distribution, then $C_I^{D^{\otimes n}}(\mathbf{p}^{\otimes n}) = n\, C_I^D(\mathbf{p})$.*

In the case of one-way communication, the complexity only depends on the marginal distribution, so we have the following corollary.

Corollary 3. *For one-way communication, $C_I^{\rightarrow,D^{\otimes n}}(\mathbf{p}^{\otimes n}) = n\, C_I^{\rightarrow,D}(\mathbf{p})$.*

Finally, we will also use the reverse Shannon theorem [18], which in our notations may be stated as follows:

Theorem 4 ([18]). *Let $\mathbf{p} : \mathcal{X} \rightarrow \mathrm{Distrib}(\mathcal{B})$ be a simulation task with no output on Alice's side, and no input on Bob's side. Then, $C_\infty^{\rightarrow,D}(\mathbf{p}) \leq I(X:B)$.*

Our statement is slightly different from [18] but may be proved using the same construction. Informally, it says that a communication channel $X \rightarrow B$ may be simulated, in the limit of a large number of repetitions, using (free) shared randomness and one-way communication at most $I(X:B)$ per repetition.

Proposition 5. *The communication complexities satisfy the following relations: $C_\infty^D(\mathbf{p}) \leq C_I^D(\mathbf{p}) \leq C_H^D(\mathbf{p}) \leq C^D(\mathbf{p}) \leq C_w(\mathbf{p})$. For a product input distribution $D = D_A \otimes D_B$, we also have $C_\infty^{D_A\otimes D_B}(\mathbf{p}) = C_I^{D_A\otimes D_B}(\mathbf{p})$. Similarly, for any input distribution D but restricting to one-way communication protocols, $C_\infty^{\rightarrow,D}(\mathbf{p}) = C_I^{\rightarrow,D}(\mathbf{p})$.*

Proof (Sketch). These relations are based on fundamental propositions of information theory, such as Shannon's source coding theorem. Let us focus on the less obvious relations, involving $C_I(\mathbf{p})$.

[$C_I^D(\mathbf{p}) \leq C_H^D(\mathbf{p})$]. Let $C_H^D(\mathbf{p}) = H(M|\Lambda)$ be achieved by a protocol P with shared randomness Λ, where M is the transcript of the messages communicated during the protocol. Let us build a protocol P' using private randomness only. In this protocol, only Alice knows the random string Λ, and her first action is to send Λ to Bob. From there, the players proceed as in protocol P. Since the transcript of P' is the concatenation of Λ and M, we have $C_I^D(P') = I(XY:M\Lambda)$. From the facts that $I(XY:\Lambda) = 0$ (since the shared randomness is independent from the inputs), and $H(M|XY\Lambda) = 0$ (since the messages depend deterministically on the inputs and the randomness), it is straightforward to check that $I(XY:M\Lambda) = H(M|\Lambda)$.

[$C_\infty^D(\mathbf{p}) \le C_I^D(\mathbf{p})$]. Let $C_I^D(\mathbf{p}) = I(XY : M)$ be achieved by a protocol P without shared randomness, where M is the transcript of the messages communicated during the protocol P. These messages are alternatingly sent by Alice to Bob and vice-versa. Let us denote by M_k the kth message and $M_{[k]}$ the restriction of the transcript to the first k messages. We may express the information complexity of $\mathbf{p}$ as:

$$C_I^D(\mathbf{p}) = \sum_{k=1}^{t} I(XY : M_k | M_{[k-1]}),$$

where t is the maximal number of rounds of the protocol (possibly infinite). Let us focus on the kth message M_k, and suppose it is sent by Alice to Bob. In a particular execution of the protocol, the partial transcript $M_{[k-1]}$ will be fixed to some string m, which is at this point known to both Alice and Bob, so that

$$I(XY : M_k | M_{[k-1]}) = \sum_m p(m)\, I(X : M_k | M_{[k-1]} = m),$$

where $p(m) = \Pr[M_{[k-1]} = m]$, and we have used the fact that M_k only depends on X and $M_{[k-1]}$, and not on Y. Alice now needs to send M_k to Bob, which only depends on X when we condition on $M_{[k-1]}$, so she actually needs to simulate a communication channel $X \longrightarrow M_k$. Since the partial transcript $M_{[k-1]} = m$ happens with probability $p(m)$, this particular channel will have to be simulated on average $n \cdot p(m)$ times when repeating the protocol n times, and the reverse Shannon theorem (Theorem 4) ensures that as n goes to infinity, this simulation may be achieved using shared randomness and communication $I(X : M_k | M_{[k-1]} = m)$ per repetition. By compressing similarly each successive message, and averaging over all possible transcripts, we get that $C_\infty^D(\mathbf{p}) \le I(XY : M) = C_I^D(\mathbf{p})$.

[$C_I^{D_A \otimes D_B}(\mathbf{p}) \le C_\infty^{D_A \otimes D_B}(\mathbf{p})$]. For a product input distribution $D = D_A \otimes D_B$, Theorem 2 implies that $C_I^D(\mathbf{p}) = C_I^{D^{\otimes n}}(\mathbf{p}^{\otimes n})/n$. Moreover, since $C_I^{D^{\otimes n}}(\mathbf{p}^{\otimes n}) \le C^{D^{\otimes n}}(\mathbf{p}^{\otimes n})$, we obtain $C_I^D(\mathbf{p}) \le C^{D^{\otimes n}}(\mathbf{p}^{\otimes n})/n$ and, in the limit $n \to \infty$, $C_I^D(\mathbf{p}) \le C_\infty^D(\mathbf{p})$. Similarly, Corollary 3 implies that for any input distribution but one-way communication, $C_I^{\to,D}(\mathbf{p}) \le C_\infty^{\to,D}(\mathbf{p})$.

3 Lower Bound on the Entropic Complexity

The previous best upper bound on the amortized communication complexity of $\mathbf{p}_{\text{dim}=3}$ is due to Toner and Bacon, who proved that $C_\infty(\mathbf{p}_{\text{dim}=3}) \le \text{Si}(\pi)/(\pi \ln 2) \approx 0.85$ bits [1], where $\text{Si}(x)$ is the sine integral function. Indeed, they showed that in their one-bit protocol, the conditional entropy of the messages given the shared randomness (what we defined as the entropic cost) is only 0.85 bits, so that one can use Shannon's source coding theorem to compress the communication from 1 bit to 0.85 bits. In this section we prove that the entropic complexity of $\mathbf{p}_{\text{dim}=d}$ is at least 0.414 bits for any $d \ge 2$, which shows that to reduce the communication further, a new technique is required. Such a technique will be presented in the next section.

We prove the lower bound on the entropic complexity by adapting a method proposed by Pironio for lower bounds on the average communication complexity [19]. The idea behind the following theorem is that a task $\mathbf{p}$ outside LHV may violate a Bell inequality, so that it will require to use deterministic protocols P_λ for $\lambda \notin \Lambda_0$, simulating tasks $\mathbf{p}_\lambda$ outside LHV, with some probability. More precisely, we consider for each deterministic protocol a "violation per entropy" ratio. To achieve the same violation as the task $\mathbf{p}$ using as little communication as possible (where the communication is counted as the entropy of the messages), one should use a distribution over deterministic protocols that have a large violation per entropy ratio. In particular, the deterministic protocol having the largest ratio gives a lower bound on the entropic communication complexity of $\mathbf{p}$.

Theorem 6. *Let B be a linear functional over the set of tasks, which defines a Bell inequality $B(\mathbf{p}_\lambda) \leq B_0$ satisfied for all $\lambda \in \Lambda_0$, but violated by a simulation task $\mathbf{p}$, that is, $B(\mathbf{p}) > B_0$. Then, the entropic communication complexity of $\mathbf{p}$ is lower bounded as follows:*

$$C_H^D(\mathbf{p}) \geq \frac{B(\mathbf{p}) - B_0}{B(\mathbf{p}_{\lambda^*}) - B_0}\, C_H^D(P_{\lambda^*}),$$

where P_{λ_*} is a deterministic protocol for a task $\mathbf{p}_{\lambda^*}$ such that

$$\frac{B(\mathbf{p}_{\lambda^*}) - B_0}{C_H^D(P_{\lambda^*})} = \max_{\lambda \notin \Lambda_0} \frac{B(\mathbf{p}_\lambda) - B_0}{C_H^D(P_\lambda)}.$$

This may be proved along the lines of the proof of Proposition 1 in [19], which gives a similar statement for the average communication complexity. We may now completely determine the entropic communication complexity of $\mathbf{p}_\mu$:

Theorem 7. *For any $1/2 \leq \mu \leq 1$ we have, $C_H(\mathbf{p}_\mu) = 2\mu - 1$.*

Note that for $0 \leq \mu \leq 1/2$, we trivially have $C_H(\mathbf{p}_\mu) = 0$.

Proof. The lower bound comes from the previous theorem. This is then shown to be tight by giving an explicit protocol. Since we have shown in Claim 1 that $C_H(\mathbf{p}_\mu) = C_H^U(\mathbf{p}_\mu)$, it suffices to consider the uniform input distribution.

[$C_H^U(\mathbf{p}_\mu) \geq 2\mu - 1$]. We use the CHSH inequality [4], which is defined by a linear functional B acting on a task $\mathbf{p}$ as:

$$B(\mathbf{p}) = \sum_{x,y,a,b} ab\,(-1)^{x \cdot y} \mathbf{p}(a, b || x, y).$$

It is straightforward to check that $B(\mathbf{p}) \leq 2$ for all $\mathbf{p}$ in LHV, so we set $B_0 = 2$. For the simulation task $\mathbf{p}_\mu$, we have $B(\mathbf{p}_\mu) = 4\mu$, so that the inequality is violated as soon as $\mu > 1/2$. Moreover, $\max_{P_\lambda}(B(\mathbf{p}_\lambda) - B_0)/C_H^U(P_\lambda)$ is attained by a protocol P_{λ^*} where one player sends his input to the other, such that $C_H^U(P_{\lambda^*}) = 1$ and $B(\mathbf{p}_{\lambda^*}) = 4$. We then obtain

$$C_H^U(\mathbf{p}_\mu) \geq \frac{4\mu - 2}{4 - 2} 1 = 2\mu - 1.$$

[$C_H^U(\mathbf{p}_\mu) \leq 2\mu - 1$]. Let us consider the extreme cases $\mu = 1/2$ and $\mu = 1$. For $\mu = 1/2$, there exists a shared randomness protocol $P_{1/2}$ without any communication ($\mathbf{p}_{1/2}$ is in LHV), therefore satisfying $C_H^U(P_{1/2}) = 0$. On the other hand, for $\mu = 1$, there exists a protocol P_1 with one bit of communication (one of the player sends his input to the other), that is, $C_H^U(P_1) = 1$. It is also straightforward to show that $\mathbf{p}_\mu = (2 - 2\mu)\mathbf{p}_{1/2} + (2\mu - 1)\mathbf{p}_1$, so that for implementing $\mathbf{p}_\mu$, it suffices to use the protocol $P_{1/2}$ with probability $(2 - 2\mu)$ and the protocol P_1 with probability $(2\mu - 1)$. By linearity, the obtained protocol has entropic cost $2\mu - 1$.

Using a reduction from $\mathbf{p}_{1/\sqrt{2}}$ to $\mathbf{p}_{\mathrm{dim}=d}$, Theorem 7 implies as a corollary a lower bound on the entropic complexity of $\mathbf{p}_{\mathrm{dim}=d}$.

Claim 4. *Let C_* be any of C, C_∞, C_H. Then, $C_*(\mathbf{p}_{\mathrm{dim}=d}) \geq C_*(\mathbf{p}_{1/\sqrt{2}})$ for any $d \geq 2$.*

Proof. The key observation is that the task $\mathbf{p}_{1/\sqrt{2}}$ for uniformly distributed inputs is equivalent to the task $\mathbf{p}_{\mathrm{dim}=d}$ for a special distribution $\tilde{D}$, where the inputs are uniform over two vectors $\{\boldsymbol{x}_0, \boldsymbol{x}_1\}$ for Alice and two vectors $\{\boldsymbol{y}_0, \boldsymbol{y}_1\}$ for Bob, laid out such that $\boldsymbol{x}_i \cdot \boldsymbol{y}_j = (-1)^{i \cdot j}/\sqrt{2}$. We then have $C_*(\mathbf{p}_{\mathrm{dim}=d}) \geq C_*^{\tilde{D}}(\mathbf{p}_{\mathrm{dim}=d}) = C_*^U(\mathbf{p}_{1/\sqrt{2}})$, which concludes the proof since we have shown that $C_*^U(\mathbf{p}_\mu) = C_*(\mathbf{p}_\mu)$.

Corollary 8. $C_H(\mathbf{p}_{\mathrm{dim}=d}) \geq \sqrt{2} - 1 \approx 0.414$ *bits.*

This lower bound means that for parallel repetitions of the problem, if we simply compress the messages using Shannon's source coding theorem, we may not reduce the communication further than 0.414 bits. We show in the next section that we can beat this lower bound by using another technique, based on the reverse Shannon theorem.

4 A New Protocol

In this section, we show how to reduce the communication for parallel repetitions of the problem of simulating $\mathbf{p}_{\mathrm{dim}=d}$, beating the lower bound on the entropic communication complexity derived in the previous section. We use a result due to Degorre *et al.*, which shows that the problem reduces to a distributed sampling task:

Theorem 9 ([20,13]). *Let $\boldsymbol{x}$ and $\boldsymbol{y}$ be Alice's and Bob's inputs. If Alice and Bob share a random variable $\boldsymbol{v} \in \mathbb{S}_{d-1}$ distributed according to a probability measure*

$$\rho(\boldsymbol{v}||\boldsymbol{x}) = \frac{|\boldsymbol{x} \cdot \boldsymbol{v}|}{\mathcal{R}_d},$$

where $\mathcal{R}_d = \int_{\mathbb{S}_{d-1}} |\boldsymbol{x} \cdot \boldsymbol{v}|\, d\boldsymbol{v}$, then they are able to simulate $\mathbf{p}_{\mathrm{dim}=d}$ without any further resource.

This observation leads to an apparently very bad communication protocol for $\mathbf{p}_{\mathrm{dim}=d}$ with private randomness only: using her input and private randomness, Alice locally samples $\boldsymbol{v}$ according to the distribution $\rho(\boldsymbol{v}||\boldsymbol{x})$, and then communicates $\boldsymbol{v}$ to Bob. This would require infinite communication, but the point is that the information cost of this protocol would actually be not only finite, but also rather low, so that for parallel repetitions of the problem, and with the help of shared randomness, we may significantly reduce the communication using the reverse Shannon theorem. In particular, we prove the following upper bound:

Theorem 10. *The amortized communication complexity of* $\mathbf{p}_{\text{dim}=d}$ *satisfies*

$$C_\infty(\mathbf{p}_{\text{dim}=d}) \le \begin{cases} \frac{1}{\ln 2}\left[\ln \frac{(d-1)A_d}{A_{d-1}} - \sum_{k=0}^{\frac{d}{2}-1} \frac{1}{2k+1}\right] & \text{for } d \text{ even,} \\ \frac{1}{\ln 2}\left[\ln \frac{(d-1)A_d}{2A_{d-1}} - \sum_{k=1}^{\frac{d-1}{2}} \frac{1}{2k}\right] & \text{for } d \text{ odd,} \end{cases} \tag{4}$$

where $A_d = \int_{\mathbb{S}_{d-1}} d\boldsymbol{v}$ *is the surface area of the d-dimensional sphere.*

In particular, we have

1. $C_\infty(\mathbf{p}_{\text{dim}=2}) \le \frac{1}{\ln 2}(\ln \pi - 1) \approx 0.21$ bits,
2. $C_\infty(\mathbf{p}_{\text{dim}=3}) \le 1 - \frac{1}{2\ln 2} \approx 0.28$ bits,
3. $C_\infty(\mathbf{p}_{\text{dim}=d}) \le \frac{1}{2\ln 2}(\ln \pi - \gamma) \approx 0.410$ bits for arbitrary d, where γ is the Euler-Mascheroni constant.

Proof. Let P be the following private randomness protocol for $\mathbf{p}_{\text{dim}=d}$: using her input together with private randomness, Alice samples a random variable V according to the distribution $\rho(\boldsymbol{v}||\boldsymbol{x})$ defined above and communicates the obtained sample $\boldsymbol{v}$ to Bob. By Theorem 9, they are then able to solve task $\mathbf{p}_{\text{dim}=d}$. More precisely, it suffices for the players to set their respective outputs as $a = \text{sgn}(\boldsymbol{x} \cdot \boldsymbol{v})$ and $b = \text{sgn}(\boldsymbol{y} \cdot \boldsymbol{v})$, where $\text{sgn}(x) = 1$ if $x \ge 0$, and -1 otherwise [20,13].

We have shown in the previous section that the reverse Shannon theorem implies that $C_\infty^D(\mathbf{p}_{\text{dim}=d}) \le C_I^D(\mathbf{p}_{\text{dim}=d})$ (Proposition 5) and also that the hardest distribution for $\mathbf{p}_{\text{dim}=d}$ has uniform marginals (Claim 2), so it suffices to compute the information cost $C_I^D(P) = I(X:V)$ for a distribution D with uniformly distributed $\boldsymbol{x}$. The computation of $I(X:V)$ will be given in the full version of the paper, and yields Eq. (4).

For completeness, let us note that using the same technique, we can prove the following upper bound on the amortized communication complexity of $\mathbf{p}_\mu$:

Theorem 11. *For any* $1/2 \le \mu \le 1$, *we have* $C_\infty(\mathbf{p}_\mu) \le 1 - H[\mu]$, *where* $H[\mu] = \mu \log \frac{1}{\mu} + (1-\mu)\log \frac{1}{1-\mu}$.

Proof. Let v be a random bit correlated with x, such that $p(x = v) = \mu$. The channel defined by the Markov process $X \to V$ is then a binary symmetric channel, with channel capacity $1 - H[\mu]$. It is straightforward to show that if Alice may use such a channel to communicate information about her input x to Bob, it is sufficient to simulate $\mathbf{p}_\mu$. Indeed, it suffices for Alice and Bob to output

$$a = (-1)^{(x\oplus v\oplus 1)\cdot\lambda_0}\,(-1)^{(x\oplus v)\cdot\lambda_1},$$
$$b = (-1)^{\lambda_0}\,(-1)^{y\cdot v},$$

where λ_0, λ_1 are shared unbiased random bits. The reverse Shannon theorem then ensures that asymptotically, the channel $X \to V$ may be simulated using on average $1 - H[\mu]$ bits per repetition.

In the next section, we will show that this protocol is optimal, at least when the players are restricted to one-way communication.

5 The Difference Method

Amortized lower bounds are notoriously hard to prove. Examples include the Shannon capacity of graphs [21] and the parallel repetition theorem of Raz [22]. In some lucky cases the situation is better. Quantum values of XOR games [23] and the communication complexity of correlation [2] are examples, where mathematics seems to be in our favor. Incidentally, both topics have relevance to lower bounding amortized communication complexity. We develop a new method we call the *difference method*, which so far we could apply only in the one-way communication context. Note, however, that all efficient protocols we know for this problem are one-way.

Theorem 12. *For any* $1/2 \leq \mu \leq 1$*, we have* $C_\infty^{\rightarrow}(\mathbf{p}_\mu) \geq 1 - H[\mu]$.

This matches the upper bound of Theorem 11, showing that the above protocol is optimal, at least for one-way communication.

Proof. Since the hardest distribution is the uniform distribution U (Claim 1), we have $C_\infty^{\rightarrow}(\mathbf{p}_\mu) = C_\infty^{\rightarrow,U}(\mathbf{p}_\mu)$. Moreover, Proposition 5 implies that $C_\infty^{\rightarrow,U}(\mathbf{p}_\mu) = C_I^{\rightarrow,U}(\mathbf{p}_\mu)$, so it suffices to show that for any protocol for $\mathbf{p}_\mu$, $I(X:M) \geq 1 - H[\mu]$, where x is an unbiased random bit. The idea of the proof is to reduce the problem to the communication complexity problem of a correlation *à la* Harsha *et al.* [2] and then, following their approach, use the mutual information between Alice's input and Bob's output to bound the communication. To reduce to [2], 1. We have to get rid of Bob's input; 2. We have to get rid of Alice's output. If we fix Bob's input and omit Alice's output, we get nothing. Nevertheless, since the communication is one-way, when Bob receives Alice's message, he can just compute the output on any input he wants to. We show that if we run the protocol with a random input x on Alice's side, take both $y = 0$ and $y' = 1$ as inputs on Bob's side, and receive outputs b and b', respectively from Bob, then the product $b \cdot b'$ will contain a lot of information about Alice's input, x.

Observe that $(a \cdot b) \cdot (a \cdot b') = b \cdot b'$. The specification of $\mathbf{p}_\mu$ tells us that $a \cdot b$ should take 1 with probability $(1+\mu)/2$ and -1 with probability $(1-\mu)/2$. Also, $a \cdot b'$ should take $(-1)^x$ with probability $(1+\mu)/2$ and $(-1)^{x+1}$ with probability $(1-\mu)/2$. The union bound gives that the probability that $b \cdot b' = (-1)^x$ is at least μ. This shows that the mutual information $I(X:E)$ between x and $e = b \cdot b'$ is at least $1 - H[\mu]$, where we have used the fact that $H(X) = 1$ (since X is an unbiased random bit). The data processing inequality on the Markov chain $X \rightarrow M \rightarrow E$ then implies that $I(X:M) \geq I(X:E)$.

For $\mu = 1/\sqrt{2}$, we have $C_\infty^{\rightarrow}(\mathbf{p}_{1/\sqrt{2}}) \geq 1 - H[1/\sqrt{2}]$, and in turn, using the reduction from Claim 4, we obtain the following lower bound on $C_\infty^{\rightarrow}(\mathbf{p}_{\mathrm{dim}=d})$ as a corollary:

Corollary 13. $C_\infty^{\rightarrow}(\mathbf{p}_{\mathrm{dim}=d}) \geq 1 - H[1/\sqrt{2}] \approx 0.13$ *bits for any* $d \geq 2$.

Acknowledgements

J. Roland would like to thank Julien Degorre and Sophie Laplante for many interesting discussions. Part of this work was done while J. Roland was affiliated with U.C. Berkeley and FNRS Belgium.

References

1. Toner, B.F., Bacon, D.: Communication Cost of Simulating Bell Correlations. Phys. Rev. Lett. 91, 187904 (2003)
2. Harsha, P., Jain, R., McAllester, D., Radhakrishnan, J.: The communication complexity of correlation. In: Proc. 22nd CCC, pp. 10–23 (2007)
3. Yao, A.C.C.: Some complexity questions related to distributive computing. In: Proc. 11th STOC, pp. 209–213 (1979)
4. Clauser, J.F., Horne, M.A., Shimony, A., Holt, R.A.: Proposed Experiment to Test Local Hidden-Variable Theories. Phys. Rev. Lett. 23, 880–884 (1969)
5. Bell, J.S.: On the Einstein Podolsky Rosen paradox. Physics 1, 195 (1964)
6. Einstein, A., Podolsky, B., Rosen, N.: Can quantum-mechanical description of physical reality be considered complete? Phys. Rev. 47, 777–780 (1935)
7. Bohm, D., Aharonov, Y.: Discussion of Experimental Proof for the Paradox of Einstein, Rosen, and Podolsky. Phys. Rev. 108, 1070–1076 (1957)
8. Maudlin, T.: Bell's inequality, information transmission, and prism models. In: Biennal Meeting of the Philosophy of Science Association, pp. 404–417 (1992)
9. Brassard, G., Cleve, R., Tapp, A.: Cost of Exactly Simulating Quantum Entanglement with Classical Communication. Phys. Rev. Lett. 83, 1874–1877 (1999)
10. Steiner, M.: Towards quantifying non-local information transfer: finite-bit non-locality. Phys. Lett. A 270, 239–244 (2000)
11. Cerf, N.J., Gisin, N., Massar, S.: Classical Teleportation of a Quantum Bit. Phys. Rev. Lett. 84, 2521–2524 (2000)
12. Barrett, J., Kent, A., Pironio, S.: Maximally nonlocal and monogamous quantum correlations. Phys. Rev. Lett. 97(17), 170409 (2006)
13. Degorre, J., Laplante, S., Roland, J.: Classical simulation of traceless binary observables on any bipartite quantum state. Phys. Rev. A 75, 012309 (2006)
14. Regev, O., Toner, B.: Simulating quantum correlations with finite communication. In: Proc. 48th FOCS, pp. 384–394 (2007)
15. Yao, A.C.C.: Probabilistic computations: Toward a unified measure of complexity. In: Proc. 18th FOCS, pp. 222–227 (1977)
16. von Neumann, J.: Zur Theorie der Gesellschaftsspiele. Math. Ann. 100(1), 295–320 (1928)
17. Chakrabarti, A., Shi, Y., Wirth, A., Yao, A.: Informational complexity and the direct sum problem for simultaneous message complexity. In: Proc. 42nd FOCS, pp. 270–278 (2001)
18. Bennett, C., Shor, P., Smolin, J., Thapliyal, A.: Entanglement-assisted capacity of a quantum channel and the reverse Shannon theorem. IEEE Trans. Inf. Theor. 48(10), 26–37 (2002)
19. Pironio, S.: Violations of Bell inequalities as lower bounds on the communication cost of nonlocal correlations. Phys. Rev. A 68(6), 062102 (2003)
20. Degorre, J., Laplante, S., Roland, J.: Simulating quantum correlations as a distributed sampling problem. Phys. Rev. A 72, 062314 (2005)
21. Lovász, L.: On the Shannon capacity of a graph. IEEE Trans. Inf. Theor. 25(1), 1–7 (1979)
22. Raz, R.: A parallel repetition theorem. SIAM J. Comput. 27(3), 763–803 (1998)
23. Cleve, R., Slofstra, W., Unger, F., Upadhyay, S.: Perfect parallel repetition theorem for quantum XOR proof systems. In: Proc. 22nd CCC, pp. 109–114 (2007)

The Number of Symbol Comparisons in QuickSort and QuickSelect

Brigitte Vallée[1], Julien Clément[1],
James Allen Fill[2], and Philippe Flajolet[3]

[1] GREYC, CNRS and University of Caen, 14032 Caen Cedex, France
[2] Applied Mathematics and Statistics, The Johns Hopkins University, Baltimore, MD 21218-2682, USA
[3] Algorithms Project, INRIA-Rocquencourt, 78153 Le Chesnay, France

Abstract. We revisit the classical `QuickSort` and `QuickSelect` algorithms, under a complexity model that fully takes into account the elementary comparisons between symbols composing the records to be processed. Our probabilistic models belong to a broad category of information sources that encompasses memoryless (i.e., independent-symbols) and Markov sources, as well as many unbounded-correlation sources. We establish that, under our conditions, the average-case complexity of `QuickSort` is $O(n \log^2 n)$ [rather than $O(n \log n)$, classically], whereas that of `QuickSelect` remains $O(n)$. Explicit expressions for the implied constants are provided by our combinatorial–analytic methods.

Introduction

Every student of a basic algorithms course is taught that, on average, the complexity of Quicksort is $O(n \log n)$, that of binary search is $O(\log n)$, and that of radix-exchange sort is $O(n \log n)$; see for instance [13,16]. Such statements are based on specific assumptions—that the comparison of data items (for the first two) and the comparison of symbols (for the third one) have unit cost—and they have the obvious merit of offering an easy-to-grasp picture of the complexity landscape. However, as noted by Sedgewick, these simplifying assumptions suffer from limitations: they do not make possible a precise assessment of the relative merits of algorithms and data structures that resort to different methods (e.g., comparison-based versus radix-based sorting) in a way that would satisfy the requirements of either information theory or algorithms engineering. Indeed, computation is not reduced to its simplest terms, namely, the manipulation of totally elementary symbols, such as bits, bytes, characters. Furthermore, such simplified analyses say little about a great many application contexts, in databases or natural language processing, for instance, where information is highly “non-atomic”, in the sense that it does not plainly reduce to a single machine word.

First, we observe that, for commonly used data models, the mean costs S_n and K_n of *any* algorithm under the symbol-comparison and the key-comparison model, respectively, are connected by the universal relation $S_n = K_n \cdot O(\log n)$. (This results from the fact that at most $O(\log n)$ symbols suffice, with high probability, to distinguish n keys; cf. the analysis of the height of digital trees, also known as “*tries*”, in [3].) The surprise is that there are cases where this

S. Albers et al. (Eds.): ICALP 2009, Part I, LNCS 5555, pp. 750–763, 2009.

upper bound is tight, as in `QuickSort`; others where both costs are of the same order, as in `QuickSelect`. In this work, we show that the expected cost of `QuickSort` is $O(n \log^2 n)$, *not* $O(n \log n)$, when *all* elementary operations—symbol comparisons—are taken into account. By contrast, the cost of `QuickSelect` turns out to be $O(n)$, in both the old and the new world, albeit, of course, with different implied constants. Our results constitute broad extensions of earlier ones by Fill and Janson (`Quicksort`, [7]) and Fill and Nakama (`QuickSelect`, [8]).

The sources that we deal with include memoryless ("Bernoulli") sources with possibly *non-uniform* independent symbols and Markov sources, as well as many non-Markovian sources with unbounded memory from the past. A central idea is the modelling of the source via its *fundamental probabilities*, namely the probabilities that a word of the source begins with a given prefix. Our approach otherwise relies on methods from analytic combinatorics [10], as well as on information theory and the theory of dynamical systems. It relates to earlier analyses of digital trees ("tries") by Clément, Flajolet, and Vallée [3,19].

Results. We consider a totally ordered *alphabet* Σ. What we call a probabilistic *source* produces infinite words on the alphabet Σ (see Definition 1). The set of *keys* (words, data items) is then Σ^∞, endowed with the strict lexicographic order denoted '$\prec$'. Our main objects of study are two algorithms, applied to n keys assumed to be *independently drawn* from the same source $\mathcal{S}$, namely, the standard sorting algorithm `QuickSort`(n) and the algorithm `QuickSelect`(m, n), which selects the mth smallest element. In the latter case, we shall mostly focus our attention on situations where the rank m is proportional to n, being of the form $m = \alpha n$, so that the algorithm determines the αth quantile; it will then be denoted by `QuickQuant`$_\alpha(n)$. We also consider the cases where rank m equals 1 (which we call `QuickMin`), equals n (`QuickMax`), or is randomly chosen in $\{1, \ldots, n\}$ (`QuickRand`).

Our main results involve constants that depend on the source $\mathcal{S}$ (and possibly on the real α); these are displayed in Figures 1–3 and described in Section 1. They specialize nicely for a binary source $\mathcal{B}$ (under which keys are compared via their binary expansions, with uniform independent bits), in which case they admit pleasant expressions that simplify and extend those of Fill and Nakama [8] and Grabner and Prodinger [11] and lend themselves to precise numerical evaluations (Figure 2). The conditions Λ–tamed, Π–tamed, and "periodic" are made explicit in Subsection 2.1. Conditions of applicability to classical source models are discussed in Section 3.

Theorem 1. *(i) For a Λ–tamed source $\mathcal{S}$, the mean number T_n of symbol comparisons of* `QuickSort`(n) *involves the entropy $h(\mathcal{S})$ of the source:*

$$T_n = \frac{1}{h(\mathcal{S})}\, n \log^2 n + a(\mathcal{S})\, n \log n + b(\mathcal{S})\, n + o(n), \quad \text{for some } a(\mathcal{S}), b(\mathcal{S}) \in \mathbb{R}. \quad (1)$$

(ii) For a periodic source, a term $nP(\log n)$ is to be added to the estimate (1), *where $P(u)$ is a (computable) continuous periodic function.*

Theorem 2. *For a Π-tamed source $\mathcal{S}$, one has the following, with $\delta = \delta(\mathcal{S}) > 0$.*
(i) The mean number of symbol comparisons $Q_n^{(\alpha)}$ for `QuickQuant`$_\alpha$ *satisfies*

$$Q_n^{(\alpha)} = \rho_{\mathcal{S}}(\alpha)n + O(n^{1-\delta}). \tag{2}$$

(ii) The mean number of symbol comparisons, $M_n^{(-)}$ for `QuickMin` (n) *and $M_n^{(+)}$ for* `QuickMax`(n)*, satisfies with $\epsilon = \pm$,*

$$M_n^{(\epsilon)} = \rho_{\mathcal{S}}^{(\epsilon)} n + O(n^{1-\delta}), \qquad \textit{with} \quad \rho_{\mathcal{S}}^{(+)} = \rho_{\mathcal{S}}(1), \qquad \rho_{\mathcal{S}}^{(-)} = \rho_{\mathcal{S}}(0). \tag{3}$$

(iii) The mean number M_n of symbol comparisons performed by `QuickRand`(n) *satisfies*

$$M_n = \gamma_{\mathcal{S}}\, n + O(n^{1-\delta}), \qquad \textit{with} \quad \gamma_{\mathcal{S}} = \int_0^1 \rho(\alpha) d\alpha. \tag{4}$$

These estimates are to be compared to the classical ones (see, e.g., [13, p.634] and [12], for extensions), relative to the *number of comparisons* in `QuickQuant`$_\alpha$:

$$K_n^{(\alpha)} = \kappa(\alpha)n + O(\log n), \qquad \kappa(\alpha) := 2(1 - \alpha\log(\alpha) - (1-\alpha)\log(1-\alpha)). \tag{5}$$

General strategy. We operate under a general model of source, parametrized by the unit interval $\mathcal{I}$. Our strategy comprises three main steps. The first two are essentially *algebraic*, while the last one relies on complex *analysis*.

Step (a)**.** We first show (Proposition 1) that the mean number S_n of symbol comparisons performed by a (general) algorithm $\mathcal{A}(n)$ applied to n words from a source $\mathcal{S}$ can be expressed in terms of two objects: (i) the *density* ϕ_n of the algorithm, which uniquely depends on the algorithm and provides a measure of the mean number of key-comparisons performed near a specific point; (ii) the family of *fundamental triangles*, which solely depends on the source and describes the location of pairs of words which share a common prefix.

Step (b)**.** This step is devoted to computing the density of the algorithm. We first deal with the *Poisson model*, where the number of keys, instead of being fixed, follows a Poisson law of parameter Z. We provide an expression of the Poissonized density relative to each algorithm, from which we deduce the mean number of comparisons under this model. Then, simple algebra yield expressions relative to the model where the number n of keys is fixed.

Step (c)**.** The exact representations of the mean costs is an alternating sum which involves two kinds of quantities, the size n of the set of data to be sorted (which tends to infinity), and the fundamental probabilities (which tend to 0). We approach the corresponding asymptotic analysis by means of *complex integral representations* of the Nörlund–Rice type. For each algorithm–source pair, a series of Dirichlet type encapsulates both the properties of the source and the characteristics of the algorithm—this is the *mixed Dirichlet series*, denoted by $\varpi(s)$, whose *singularity structure in the complex plane* is proved to condition our final asymptotic estimates.

1 Algebraic Analysis

The developments of this section are essentially formal (algebraic) and *exact*; that is, no approximation is involved at this stage.

1.1 A General Source Model

Throughout this paper, a totally ordered (finite or denumerable) alphabet Σ of "symbols" or "letters" is fixed.

Entropy constant [Quicksort, Theorem 1]: $h(\mathcal{S}) := \lim_{k\to\infty} \big[-\frac{1}{k} \sum_{w\in\Sigma^k} p_w \log p_w \big]$.

Quantile constant $\rho_{\mathcal{S}}(\alpha)$ [QuickQuant, Theorem 2, Part *(i)*]

$$\rho_{\mathcal{S}}(\alpha) := \sum_{w\in\Sigma^\star} p_w \, L\left(\frac{|\alpha-\mu_w|}{p_w}\right).$$

Here, $\mu_w := (1/2)\,(p_w^{(+)} + p_w^{(-)})$ involves the probabilities $p_w^{(+)}$ and $p_w^{(-)}$ defined in Subsection 1.1; the function L is given by $L(y) := 2[1+H(y)]$, where $H(y)$ is expressed by $y^+ := (1/2)+y$, $y^- := (1/2)-y$, and

$$H(y) := \begin{cases} -(y^+\log y^+ + \, y^- \log y^-), & \text{if } 0 \le y < 1/2 \\ 0, & \text{if } y = 1/2 \\ y^-(\log y^+ - \log|y^-|), & \text{if } y > 1/2. \end{cases} \tag{6}$$

— **Min/max constants** [QuickMin, QuickMax, Theorem 2, Part *(ii)*]:

$$\rho_{\mathcal{S}}^{(\epsilon)} = \sum_{w\in\Sigma^\star} p_w \big[1 - \frac{p_w^{(\epsilon)}}{p_w} \log\big(1+\frac{p_w}{p_w^{(\epsilon)}}\big)\big].$$

— **Random selection constant** [QuickRand, Theorem 2, Part *(iii)*]:

$$\gamma_{\mathcal{S}} = \sum_{w\in\Sigma^\star} p_w^2 \big[2 + \frac{1}{p_w} + \sum_{\epsilon=\pm} \big[\log\big(1+\frac{p_w^{(\epsilon)}}{p_w}\big) - \big(\frac{p_w^{(\epsilon)}}{p_w}\big)^2 \log\big(1+\frac{p_w}{p_w^{(\epsilon)}}\big)\big]\big].$$

Fig. 1. The main constants of Theorems 1 and 2, relatively to a general source $(\mathcal{S})$

$h(\mathcal{B}) = \log 2$ [entropy]

$$\rho_{\mathcal{B}}^{(\epsilon)} = 4 + 2\sum_{\ell\ge 0} \frac{1}{2^\ell} \sum_{k=1}^{2^\ell-1} \big[1 - k\log\big(1+\frac{1}{k}\big)\big] \doteq 5.27937\,82410\,80958.$$

$$\gamma_{\mathcal{B}} = \frac{14}{3} + 2\sum_{\ell=0}^{\infty} \frac{1}{2^{2\ell}} \sum_{k=1}^{2^\ell-1} \big[k+1+\log(k+1) - k^2\log\big(1+\frac{1}{k}\big)\big] \doteq 8.20730\,88638.$$

Fig. 2. The constants relative to a binary source

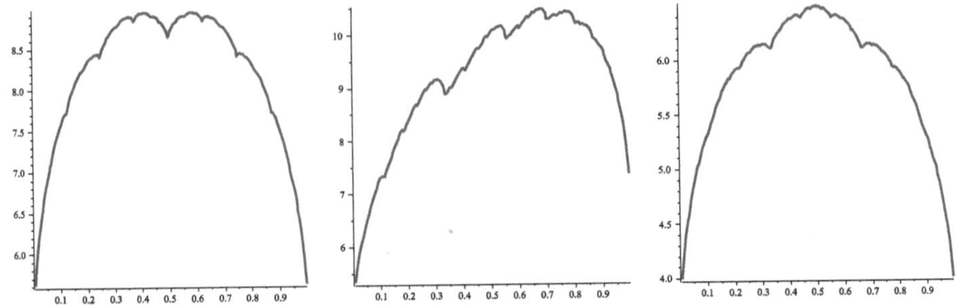

Fig. 3. Plot of the function $\rho_{\mathcal{S}}(\alpha)$ for $\alpha \in [0,1]$ and the three sources: Bern($\frac{1}{2}, \frac{1}{2}$), Bern($\frac{1}{3}, \frac{2}{3}$), and Bern($\frac{1}{3}, \frac{1}{3}, \frac{1}{3}$). The curves illustrate the fractal character of the constants involved in `QuickSelect`. (Compare with $\kappa(\alpha)$ in Eq. (5), which appears as limit of $\rho_{\mathcal{S}}(\alpha)$, for an unbiased r–ary source $\mathcal{S}$, when $r \to \infty$.)

Definition 1. *A* probabilistic source, *which produces infinite words of Σ^∞, is specified by the set $\{p_w, w \in \Sigma^\star\}$ of* fundamental probabilities p_w, *where p_w is the probability that an infinite word begins with the finite prefix w. It is furthermore assumed that $\pi_k := \sup\{p_w : w \in \Sigma^k\}$ tends to 0, as $k \to \infty$.*

For any prefix $w \in \Sigma^\star$, we denote by $p_w^{(-)}, p_w^{(+)}, p_w$ the probabilities that a word produced by the source begins with a prefix w' of the same length as w, which satisfies $w' \prec w$, $w' \succ w$, or $w' = w$, respectively. Since the sum of these three probabilities equals 1, this defines two real numbers $a_w, b_w \in [0,1]$ for which

$$a_w = p_w^{(-)}, \quad 1 - b_w = p_w^{(+)}, \quad b_w - a_w = p_w.$$

Given an infinite word $X \in \Sigma^\infty$, denote by w_k its prefix of length k. The sequence (a_{w_k}) is increasing, the sequence (b_{w_k}) is decreasing, and $b_{w_k} - a_{w_k}$ tends to 0. Thus a unique real $\pi(X) \in [0,1]$ is defined as common limit of (a_{w_k}) and (b_{w_k}). Conversely, a mapping $M : [0,1] \to \Sigma^\infty$ associates, to a number u of the interval $\mathcal{I} := [0,1]$, a word $M(u) := (m_1(u),\ m_2(u),\ m_3(u), \ldots) \in \Sigma^\infty$. In this way, the lexicographic order on words ('$\prec$') is compatible with the natural order on the interval $\mathcal{I}$; namely, $M(t) \prec M(u)$ if and only if $t < u$. Then, the fundamental interval $\mathcal{I}_w := [a_w, b_w]$ is the set of reals u for which $M(u)$ begins with the prefix w. Its length equals p_w.

Our analyses involve the two Dirichlet series of fundamental probabilities,

$$\Lambda(s) := \sum_{w \in \Sigma^\star} p_w^{-s}, \qquad \Pi(s) := \sum_{k \geq 0} \pi_k^{-s}; \tag{6}$$

these are central to our treatment. They satisfy, for $s < 0$, the relations $\Pi(s) \leq \Lambda(s) \leq \Pi(s+1)$. The series $\Lambda(s)$ always has a singularity at $s = -1$, and $\Pi(s)$ always has a singularity at $s = 0$, The regularity properties of the source can be expressed in terms of Λ near $s = -1$, and Π near 0, as already shown in [19].

An important subclass is formed by *dynamical sources*, which are closely related to dynamical systems on the interval; see Figure 4 and [19]. One starts with a partition $\{\mathcal{I}_\sigma\}$ indexed by symbols $\sigma \in \Sigma$, a coding map $\tau : \mathcal{I} \to \Sigma$

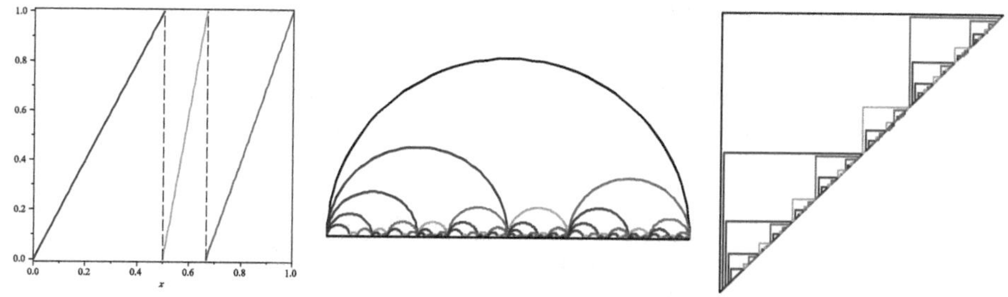

Fig. 4. *Left*: the shift transformation T of a ternary Bernoulli source with $\mathbb{P}(0) = 1/2$, $\mathbb{P}(1) = 1/6$, $\mathbb{P}(2) = 1/3$. *Middle*: fundamental intervals $\mathcal{I}_w$ materialized by semi-circles. *Right*: the corresponding fundamental triangles $\mathcal{T}_w$.

which equals σ on $\mathcal{I}_\sigma$, and a shift map $T : \mathcal{I} \to \mathcal{I}$ whose restriction to each $\mathcal{I}_\sigma$ is increasing, invertible, and of class $\mathcal{C}^2$. Then the word $M(u)$ is the word that encodes the trajectory $(u, Tu, T^2u, \ldots)$ via the coding map τ, namely, $M(u) := (\tau(u), \tau(Tu), \tau(T^2u), \ldots)$. All memoryless (Bernoulli) sources and all Markov chain sources belong to the general framework of Definition 1: they correspond to a piecewise linear shift, under this angle. For instance, the standard binary system is obtained by $T(x) = \{2x\}$ ($\{\cdot\}$ is the fractional part). Dynamical sources with a non-linear shift allow for correlations that depend on the entire past (e.g., sources related to continued fractions obtained by $T(x) = \{1/x\}$).

1.2 The Algorithm $\texttt{QuickVal}_\alpha$

We consider an algorithm that is dual of `QuickSelect`: it takes as input a set of words $\mathcal{X}$ and a *value* V, and returns the *rank* of V inside the set $\mathcal{X} \cup \{V\}$. This algorithm is of independent interest and is easily implemented as a variant of `QuickSelect` by resorting to the usual partitioning phase, then doing a comparison between the *value* of the pivot and the input *value* V (rather than a comparison between their ranks). We call this modified `QuickSelect` algorithm $\texttt{QuickVal}_\alpha$ when it is used to seek the rank of the data item with value $M(\alpha)$. The main idea here is that the behaviors of $\texttt{QuickVal}_\alpha(n)$ and $\texttt{QuickQuant}_\alpha(n)$ should be asymptotically similar. Indeed, the α*–quantile* of a random set of words of large enough cardinality must be, with high probability, close to the word $M(\alpha)$.

1.3 An Expression for the Mean Number of Symbol Comparisons

The first object of our analysis is the *density* of the algorithm $\mathcal{A}(n)$, which measures the number of key–comparisons performed by the algorithm. The second object is the collection of *fundamental triangles* (see Figure 4).

Definition 2. *The* density *of an algorithm $\mathcal{A}(n)$ which compares n words from the same probabilistic source $\mathcal{S}$ is defined as follows:*

$\phi_n(u,t)\,du\,dt$:= *"the mean number of key–comparisons performed by* $\mathcal{A}(n)$ *between words* $M(u'), M(t')$*, with* $u' \in [u, u+du]$ *and* $t' \in [t, t+dt]$*."*

For each $w \in \Sigma^\star$*, the* fundamental triangle *of prefix* w*, denoted by* $\mathcal{T}_w$*, is the triangle built on the fundamental interval* $\mathcal{I}_w := [a_w, b_w]$ *corresponding to* w*; i.e.,* $\mathcal{T}_w := \{(u,t) : a_w \le u \le t \le b_w\}$*. We set* $\mathcal{T} \equiv \mathcal{T}_\epsilon := \{(u,t) : 0 \le u \le t \le 1\}$*.*

Our first result establishes a relation between the mean number S_n of symbol-comparisons, the density ϕ_n, and the fundamental triangles $\mathcal{T}_w$:

Proposition 1. *Fix a source on alphabet* Σ*, with fundamental triangles* $\mathcal{T}_w$*. For any integrable function* g *on the unit triangle* $\mathcal{T}$*, define the* integral transform

$$\mathcal{J}[g] := \sum_{w\in\Sigma^\star} \int_{\mathcal{T}_w} g(u,t)\,du\,dt.$$

Then the mean number S_n *of symbol comparisons performed by* $\mathcal{A}(n)$ *is equal to*

$$S_n = \mathcal{J}[\phi_n] = \sum_{w\in\Sigma^\star} \int_{\mathcal{T}_w} \phi_n(u,t)\,du\,dt,$$

where ϕ_n *is the density of algorithm* $\mathcal{A}(n)$*.*

Proof. The *coincidence function* $\gamma(u,t)$ is the length of the largest common prefix of $M(u)$ and $M(t)$, namely, $\gamma(u,t) := \max\{\ell : m_j(u) = m_j(t),\ \forall j \le \ell\}$. Then, the number of symbol comparisons needed to compare two words $M(u)$ and $M(t)$, is $\gamma(u,t)+1$ and the mean number S_n of symbol comparisons performed by $\mathcal{A}(n)$ satisfies

$$S_n = \int_{\mathcal{T}} [\gamma(u,t)+1]\,\phi_n(u,t)\,du\,dt,$$

where $\phi_n(u,t)$ is the density of the algorithm. Two useful identities are

$$\sum_{\ell\ge 0} (\ell+1)\mathbf{1}_{[\gamma=\ell]} = \sum_{\ell\ge 0} \mathbf{1}_{[\gamma\ge\ell]} \qquad [\gamma \ge \ell] = \bigcup_{w\in\Sigma^\ell} (\mathcal{I}_w \times \mathcal{I}_w)\,.$$

The first one holds for *any* integer-valued random variable γ ($\mathbf{1}_A$ is the indicator of A). The second one follows from the definitions of the coincidence γ and of the fundamental intervals $\mathcal{I}_w$. Finally, the domain $\mathcal{T} \cap [\gamma \ge \ell]$ is a union of fundamental triangles. □

1.4 Computation of Poissonized Densities

Under the *Poisson model* Poi_Z of parameter Z, the number N of keys is no longer fixed but rather follows a Poisson law of parameter Z:

$$\mathbb{P}[N = n] = e^{-Z}\frac{Z^n}{n!}$$

Then, the expected value $\widetilde{S}_Z$ in this model, namely,

$$\widetilde{S}_Z := e^{-Z} \sum_{n\geq 0} \frac{Z^n}{n!} S_n = e^{-Z} \sum_{n\geq 0} \frac{Z^n}{n!} \mathcal{J}[\phi_n] = \mathcal{J}\left[e^{-Z} \sum_{n\geq 0} \frac{Z^n}{n!} \phi_n\right] = \mathcal{J}[\widetilde{\phi}_Z],$$

involves the Poissonized density $\widetilde{\phi}_Z$, defined as

$$\widetilde{\phi}_Z(u,t) := e^{-Z} \sum_{n\geq 0} \frac{Z^n}{n!} \phi_n(u,t).$$

The following statement shows that the Poissonized densities relative to `QuickSort` and `QuickVal` admit simple expressions, which in turn entail nice expressions for the mean value $\widetilde{S}_Z$ in this Poisson model, via the equality $\widetilde{S}_Z = \mathcal{J}[\widetilde{\phi}_Z]$.

Proposition 2. *Set* $f_1(\theta) := e^{-\theta} - 1 + \theta$. *The Poissonized densities of* `QuickSort` *and* $\texttt{QuickVal}_\alpha$ *satisfy*

$$\begin{aligned} \widetilde{g}_Z(u,t) &= 2(t-u)^{-2} f_1(Z(t-u)), \\ \widetilde{g}_Z^{(\alpha)}(u,t) &= 2\left(\max(\alpha,t) - \min(\alpha,u)\right)^{-2} f_1\left(Z(\max(\alpha,t) - \min(\alpha,u)\right). \end{aligned}$$

The mean number of comparisons of `QuickSort` *and* $\texttt{QuickVal}_\alpha$ *in the Poisson model satisfy*

$$\begin{aligned} \widetilde{S}_Z &= 2\mathcal{J}[(t-u)^{-2} \cdot f_1(Z(t-u))] \\ \widetilde{S}_Z^{(\alpha)} &= 2\mathcal{J}\left[(\max(\alpha,t) - \min(\alpha,u))^{-2} \cdot f_1(Z(\max(\alpha,t) - \min(\alpha,u))\right]. \end{aligned}$$

Proof. The probability that $M(u')$ and $M(t')$ are both keys for some $u' \in [u, u + du]$ and $t' \in [t, t+dt]$ is $Z^2 du\, dt$. Denote by $[x,y] := [x(u,t), y(u,t)]$ the smallest closed interval that contains $\{u,t,\alpha\}$ ($\texttt{QuickVal}_\alpha$) or $\{u,t\}$ (`QuickSort`). Conditionally, given that $M(u)$ and $M(t)$ are both keys, $M(u)$ and $M(t)$ are compared if and only if $M(u)$ or $M(t)$ is chosen as the pivot amongst the set $\mathcal{M} := \{M(z);\ z \in [x,y]\}$. The cardinality of the "good" set $\{M(u), M(t)\}$ is 2, while the cardinality of $\mathcal{M}$ equals $2 + N[x,y]$, where $N[x,y]$ is the number of keys strictly between $M(x)$ and $M(y)$. Then, for any fixed set of words, the probability that $M(u)$ and $M(t)$ are compared is $2/(2 + N[x(u,t), y(u,t)])$. To evaluate the mean value of this ratio in the Poisson model, we remark that, if we draw Poi (Z) i.i.d. random variables uniformly distributed over $[0,1]$, the number $N(\lambda)$ of those that fall in an interval of (Lebesgue) measure λ is Poi (λZ) distributed, so that

$$\mathbb{E}\left[\frac{2}{N(\lambda)+2}\right] = \sum_{k\geq 0} \frac{2}{k+2} e^{-\lambda Z} \frac{(\lambda Z)^k}{k!} = \frac{2}{\lambda^2 Z^2} f_1(\lambda Z). \qquad \square$$

1.5 Exact Formulae for S_n

We now return to the model of prime interest, where the number of keys is a fixed number n.

Proposition 3. *Assume that $\Lambda(s)$ of (6) converges at $s = -2$. Then the mean values S_n, associated with* `QuickSort` *and* `QuickVal`$_\alpha$ *can be expressed as*

$$S_n = \sum_{k=2}^{n} \binom{n}{k} (-1)^k \varpi(-k), \qquad \textit{for } n \geq 2,$$

where $\varpi(s)$ is a series of Dirichlet type given, respectively, by

$$\varpi(s) = 2\mathcal{J}[(t-u)^{-(2+s)}], \qquad \varpi(s) = 2\mathcal{J}[(\max(\alpha, t) - \min(\alpha, u))^{-(2+s)}].$$

The two functions $\varpi(s)$ are defined for $\Re s \leq -2$; they depend on the algorithm and *the source and are called the* mixed Dirichlet series.

For `QuickSort`, the series $\varpi(s)$ admits a simple form in terms of $\Lambda(s)$:

$$\varpi(s) = \frac{\Lambda(s)}{s(s+1)}, \qquad S_n = 2\sum_{k=2}^{n} \frac{(-1)^k}{k(k-1)} \binom{n}{k} \sum_{w \in \Sigma^\star} p_w^k. \tag{7}$$

For `QuickVal`, similar but more complicated forms can be obtained.

Proof (Prop. 3). Consider any sequence S_n whose Poissonized version is of the form

$$\widetilde{S}_Z \equiv e^{-Z} \sum_{n \geq 0} S_n \frac{Z^n}{n!} = \int_D \lambda(x)\, f_1(Z\mu(x))\, dx, \tag{8}$$

for some domain $D \subset \mathbb{R}^d$ and some weights $\lambda(x), \mu(x) \geq 0$, with, additionally, $\mu(x) \leq 1$. By expanding f_1, then exchanging the order of summation and integration, one obtains

$$\widetilde{S}_Z = \sum_{k=2}^{\infty} (-1)^k \varpi(-k) \frac{Z^k}{k!}, \qquad \text{where} \quad \varpi(-k) := \int_D \lambda(x)\, \mu(x)^k\, dx. \tag{9}$$

Analytically, the form (9) is justified as soon as the integral defining $\varpi(-2)$ is convergent. Then, since S_n is related to $\widetilde{S}_Z$ via the relation $S_n = n![Z^n](e^Z \widetilde{S}_Z)$, it can be recovered by a binomial convolution. □

2 Asymptotic Analysis

For asymptotic purposes, the *singular structure* of the involved functions, most notably the mixed Dirichlet series $\varpi(s)$, is essential. With tameness assumptions described in Subsection 2.1, it becomes possible to develop *complex integral representations* (Subsection 2.2), themselves strongly tied to Mellin transforms [9,18]. We can finally conclude with the proofs of Theorems 1 and 2 in Subsection 2.3.

2.1 Tamed Sources

We now introduce important regularity properties of a source, via its Dirichlet series $\Lambda(s), \Pi(s)$, as defined in (6). Recall that $\Lambda(s)$ always has a singularity at $s = -1$, and $\Pi(s)$ always has a singularity at $s = 0$.

Definition 3. *Consider a source $\mathcal{S}$, with $\Lambda(s), \Pi(s)$ its Dirichlet series.*

(*a*) *The source $\mathcal{S}$ is Π–*tamed *if the sequence π_k satisfies $\pi_k \le Ak^{-\gamma}$ for some $A > 0, \gamma > 1$. In this case, $\Pi(s)$ is analytic for $\Re(s) < -1/\gamma$.*

(*b*) *The source $\mathcal{S}$ is Λ–*tamed *if $\Lambda(s)$ is meromorphic in a "hyperbolic domain"*

$$\mathcal{F} = \left\{ s \;\middle|\; \Re s \le -1 + \frac{c}{(1 + |\Im s|)^d} \right\} \qquad (c, d > 0), \tag{10}$$

is of polynomial growth as $|s| \to \infty$ in $\mathcal{F}$, and has only a pole at $s = -1$, of order 1, with a residue equal to the inverse of the entropy $h(\mathcal{S})$.

(*c*) *The source $\mathcal{S}$ is Λ–*strongly tamed *if $\Lambda(s)$ is meromorphic in a half-plane*

$$\mathcal{F} = \{\Re(s) \le \sigma_1\}, \tag{11}$$

for some $\sigma_1 > -1$, is of polynomial growth as $|s| \to \infty$ in $\mathcal{F}$, and has only a pole at $s = -1$, of order 1, with a residue equal to the inverse of the entropy $h(\mathcal{S})$.

(*d*) *The source $\mathcal{S}$ is* periodic *if $\Lambda(s)$ is analytic in $\Re(s) < -1$, has a pole of order 1 at $s = -1$, with a residue equal to the inverse of the entropy $h(\mathcal{S})$, and admits an imaginary period $i\Omega$, that is, $\omega(s) = \omega(s + i\Omega)$.*

Essentially all "reasonable" sources are Π– tamed and "most" classical sources are Λ–tamed; see Section 3 for a discussion. The properties of $\varpi(s)$ turn out to be related to properties of the source via the Dirichlet series $\Lambda(s), \Pi(s)$.

Proposition 4. *The mixed series $\varpi(s)$ of* `QuickVal`$_\alpha$ *relative to a Π–tamed source with exponent γ is analytic and bounded in $\Re(s) \le -1+\delta$ with $\delta < 1-1/\gamma$. The mixed series $\varpi(s)$ of* `QuickSort` *is meromorphic and of polynomial growth in a hyperbolic domain* (10) *when the source is Λ–tamed, and in a vertical strip* (11) *when the source is Λ–strongly tamed.*

2.2 General Asymptotic Estimates

The sequence of numerical values $\varpi(-k)$ lifts into an analytic function $\varpi(s)$, whose singularities essentially determine the asymptotic behaviour of `QuickSort` and `QuickSelect`.

Proposition 5. *Let (S_n) be a numerical sequence which can be written as*

$$S_n = \sum_{k=2}^{n} \binom{n}{k} (-1)^k \varpi(-k), \qquad \textit{for } n \ge 2.$$

(i) If the function $\varpi(s)$ is analytic in $\Re(s) < C$, with $-2 < C < -1$, and is of polynomial growth with order at most r, then the sequence S_n admits a Nörlund–Rice representation, *for $n > r+1$ and any $-2 < d < C$:*

$$S_n = \frac{1}{2i\pi} \int_{d-i\infty}^{d+i\infty} \varpi(s) \frac{n!}{s(s+1)\cdots(s+n)}\, ds. \tag{12}$$

(ii) If, in addition, $\varpi(s)$ is meromorphic in $\Re(s) < D$ for some $D > -1$, has only a pole (of order $k_0 \geq 0$) at $s = -1$ and is of polynomial growth in $\Re(s) < D$ as $|s| \to +\infty$, then S_n satisfies $S_n = A_n + O(n^{1-\delta})$, for some $\delta > 0$, with

$$A_n = -\text{Res}\left(\frac{n!\,\varpi(s)}{s(s+1)\cdots(s+n)}; s = -1\right) = n\left(\sum_{k=0}^{k_0} a_k \log^k n\right).$$

2.3 Application to the Analysis of the Three Algorithms

We conclude our discussion of algorithms `QuickSort` and `QuickVal`, then that of `QuickQuant`.

— *Analysis of* `QuickSort`. In this case, the Dirichlet series of interest is

$$\frac{\varpi(s)}{s+1} = \frac{2}{s+1}\mathcal{J}[(t-u)^{-(2+s)}] = 2\frac{\Lambda(s)}{s(s+1)^2}\,. \tag{13}$$

For a Λ–tamed source, there is a triple pole at $s = -1$: a pole of order 1 arising from $\Lambda(s)$ and a pole of order 2 brought by the factor $1/(s+1)^2$. For a periodic source, the other poles located on $\Re s = -1$ provide the periodic term $nP(\log n)$, in the form of a Fourier series. This concludes the proof of Theorem 1.

— *Analysis of* $\texttt{QuickVal}_\alpha$. In this case, the Dirichlet series of interest is

$$\frac{\varpi(s)}{s+1} = \frac{2}{s+1}\mathcal{J}[(\max(\alpha,t) - \min(\alpha,u))^{-(2+s)}]\,.$$

Proposition 4 entails that $\varpi(s)$ is analytic at $s = -1$. Then, the integrand in (12) has a simple pole at $s = -1$, brought by the factor $1/(s+1)$ and Proposition 5 applies as soon as the source $\mathcal{S}$ is Π–tamed. Thus, for $\delta < 1 - (1/\gamma)$:

$$V_n^{(\alpha)} = \rho_{\mathcal{S}}(\alpha)n + O(n^{1-\delta}). \tag{14}$$

The three possible expressions of the function $(u,t) \mapsto \max(\alpha,t) - \min(\alpha,u)$ on the unit triangle give rise to three intervals of definition for the function H defined in Figure 1 (respectively, $]-\infty,-1/2]$, $[-1/2,+1/2]$, $[1/2,\infty[$).

— *Analysis of* $\texttt{QuickQuant}_\alpha$. The main chain of arguments connecting the asymptotic behaviors of $\texttt{QuickVal}_\alpha(n)$ and $\texttt{QuickQuant}_\alpha(n)$ is the following.
(*a*) The algorithms are asymptotically "similar enough": If $X_1,\ldots,X_n$ are n i.i.d. random variables uniform over $[0,1]$, then the α–*quantile* of set X is with high probability close to α. For instance, it is at distance at most $(\log^2 n)/\sqrt{n}$ from α with an exponentially small probability (about $\exp(-\log^2 n)$).
(*b*) The function $\alpha \mapsto \rho_{\mathcal{S}}(\alpha)$ is Hölder with exponent $c = \min(1/2, 1-(1/\gamma))$.
(*c*) The error term in the expansion (14) is uniform with respect to α.

3 Sources, `QuickSort`, and `QuickSelect`

We can now recapitulate and examine the situation of classical sources with respect to the tameness and periodicity conditions of Definition 3. *All* the sources listed below are Π–tamed, so that Theorem 2 (for `QuickQuant`) is systematically applicable to them. The finer properties of being Λ–tamed, Λ–strongly tamed, or periodic only intervene in Theorem 1, for `Quicksort`.

Memoryless sources and Markov chains. The memoryless (or Bernoulli) sources correspond to an alphabet of some cardinality $r \geq 2$, where symbols are taken independently, with p_j the probability of the jth symbol. The Dirichlet series is

$$\Lambda(s) = (1 - p_1^{-s} - \cdots - p_r^{-s})^{-1}$$

Despite their simplicity, memoryless sources are *never* Λ–strongly tamed. They can be classified into three subcategories (Markov chain sources obey a similar trichotomy).

(*i*) A memoryless source is said to be *Diophantine* if at least one ratio of the form $(\log p_i)/(\log p_j)$ is irrational and poorly approximable by rationals. *All Diophantine sources are* Λ*-tamed in the sense of Definition 3(b);* i.e., the existence of a hyperbolic region (10) can be established. Note that, in a measure-theoretic sense, *almost all memoryless sources are Diophantine.*
(*ii*) A memoryless source is periodic when *all* ratios $(\log p_i)/(\log p_1)$ are rational (this includes cases where $p_1 = \cdots = p_r$, in particular, the model of uniform random bits). The function $\Lambda(s)$ then admits an imaginary period, and its poles are regularly spaced on $\Re(s) = -1$. This entails the presence of an additional periodic term in the asymptotic expansion of the cost of `QuickSort`, which corresponds to complex poles of $\Lambda(s)$; this is Case (*ii*) of Theorem 1.
(*iii*) Finally, *Liouvillean sources* are determined by the fact that the ratios are not all rational, but *all* the irrational ratios are very well approximated by rationals (i.e., have infinite irrationality measure); for their treatment, we can appeal to the Tauberian theorems of Delange [6]

Dynamical sources were described in Section 1. They are principally determined by a *shift* transformation T. As shown by Vallée [19] (see also [3]), the Dirichlet series of fundamental intervals can be analysed by means of *transfer operators*, specifically the ones of *secant type*. We discuss here the much researched case where T is a non-linear expanding map of the interval. (Such a source may be periodic only if the shift T is conjugated to a piecewise affine map of the type previously discussed.) One can then adapt deep results of Dolgopyat [4], to the secant operators. In this way, it appears that a dynamical source is strongly tamed as soon as the branches of the shift (i.e., the restrictions of T to the intervals $\mathcal{I}_\sigma$) are not "too often" of the "same form"—we say that such sources are of *type* $D1$. Such is the case for continued fraction sources, which satisfy a "Riemann hypothesis" [1,2]. The adaptation of other results of Dolgopyat [5] provides natural instances of tamed sources with a hyperbolic strip: this occurs as soon as the branches all have the same geometric form, but not the same arithmetic features—we say that such sources are of type $D2$.

Theorem 3. *The memoryless and Markov chain sources that are Diophantine are Λ-tamed. Dynamical sources of type D2 are Λ-tamed and dynamical sources of type D1 are Λ-strongly tamed. The estimates of Theorem 1 and 2 are then applicable to all these sources, and, in the case of Λ-strongly tamed sources, the error term of* `QuickSort` *in* (1) *can be improved to $O(n^{1-\delta})$, for some $\delta > 0$.*

It is also possible to study "intermittent sources" in the style of what is done for the subtractive Euclidean algorithm [20, §2,6]: a higher order pole of $\Lambda(s)$ then arises, leading to complexities of the form $n \log^3 n$ for `QuickSort`, whereas `QuickQuant` remains of order n.

Acknowledgements. The authors are thankful to Také Nakama for helpful discussions. Research for Clément, Flajolet, and Vallée was supported in part by the SADA Project of the ANR (French National Research Agency). Research for Fill was supported by The Johns Hopkins University's Acheson J. Duncan Fund for the Advancement of Research in Statistics. Thanks finally to Jérémie Lumbroso for a careful reading of the manuscript.

References

1. Baladi, V., Vallée, B.: Euclidean algorithms are Gaussian. Journal of Number Theory 110, 331–386 (2005)
2. Baladi, V., Vallée, B.: Exponential decay of correlations for surface semi-flows without finite Markov partitions. In: Proc. AMS 133, vol. 3, pp. 865–874 (2005)
3. Clément, J., Flajolet, P., Vallée, B.: Dynamical sources in information theory: A general analysis of trie structures. Algorithmica 29(1/2), 307–369 (2001)
4. Dolgopyat, D.: On decay of correlations in Anosov flows. Annals of Mathematics 147, 357–390 (1998)
5. Dolgopyat, D.: Prevalence of rapid mixing (I). Ergodic Theory and Dynamical Systems 18, 1097–1114 (1998)
6. Delange, H.: Généralisation du théorème de Ikehara. Annales scientifiques de l'École Normale Supérieure Sér. 3, 71(3), 213–142 (1954)
7. Fill, J.A., Janson, S.: The number of bit comparisons used by Quicksort: An average-case analysis. In: Proceedings of the ACM-SIAM Symposium on Discrete Algorithms (SODA 2004), pp. 293–300 (2001)
8. Fill, J.A., Nakama, T.: Analysis of the expected number of bit comparisons required by Quickselect. Algorithmica, ArXiv:0706.2437 (to appear), Extended abstract in Proceedings ANALCO, pp. 249–256. SIAM Press (2008)
9. Flajolet, P., Sedgewick, R.: Mellin transforms and asymptotics: finite differences and Rice's integrals. Theoretical Comp. Sc. 144, 101–124 (1995)
10. Flajolet, P., Sedgewick, R.: Analytic Combinatorics. Cambridge University Press, Cambridge (2009); available electronically from the authors' home pages
11. Grabner, P., Prodinger, H.: On a constant arising in the analysis of bit comparisons in Quickselect. Quaestiones Mathematicae 31, 303–306 (2008)
12. Kirschenhofer, P., Prodinger, H., Martínez, C.: Analysis of Hoare's FIND Algorithm. Random Structures & Algorithms 10, 143–156 (1997)
13. Knuth, D.E.: The Art of Computer Programming, 2nd edn. Sorting and Searching, vol. 3. Addison-Wesley, Reading (1998)

14. Nörlund, N.E.: Leçons sur les équations linéaires aux différences finies. Gauthier-Villars, Paris (1929)
15. Nörlund, N.E.: Vorlesungen über Differenzenrechnung. Chelsea Publishing Company, New York (1954)
16. Sedgewick, R.: Quicksort. Garland Pub. Co., New York (1980); Reprint of Ph.D. thesis, Stanford University (1975)
17. Sedgewick, R.: Algorithms in C, Parts 1–4, 3rd edn., pp. 1–4. Addison-Wesley, Reading (1998)
18. Szpankowski, W.: Average-Case Analysis of Algorithms on Sequences. John Wiley, Chichester (2001)
19. Vallée, B.: Dynamical sources in information theory: Fundamental intervals and word prefixes. Algorithmica 29(1/2), 262–306 (2001)
20. Vallée, B.: Euclidean dynamics. Discrete and Continuous Dynamical Systems 15(1), 281–352 (2006)

Computing the Girth of a Planar Graph in $O(n \log n)$ Time

Oren Weimann[1] and Raphael Yuster[2]

[1] Massachusetts Institute of Technology, Cambridge, MA 02139, USA
oweimann@mit.edu
[2] Department of Mathematics, University of Haifa, Haifa, Israel
raphy@math.haifa.ac.il

Abstract. We give an $O(n \log n)$ algorithm for computing the girth (shortest cycle) of an undirected n-vertex planar graph. Our solution extends to any graph of bounded genus. This improves upon the best previously known algorithms for this problem.

1 Introduction

The *girth* of a graph is the length of its shortest cycle, or infinity if the graph does not contain any cycles. In addition to being a basic combinatorial characteristic of graphs, the girth has tight connections to many other graph properties. The connection between the girth of a graph and its chromatic number was studied by Erdős [13], Lovasz [19], Bollobás [4], and Cook [6]. Other important graph properties related to the girth include the minimum or average degree of the vertices, the diameter, the connectivity, the maximum genus, and the existence of certain type of minors (see Diestel's book [8] for a review of results).

The problem of computing the girth of a graph is among the most natural and easily stated algorithmic graph problems. Itai and Rodeh [17] were the first to suggest an efficient algorithm to compute the girth. They presented an $O(nm)$-time algorithm for a graph of n vertices and m edges, and an $O(n^2)$-time algorithm if an additive error of one is allowed. Monien [20] showed that finding the shortest cycle of even length is easier and can be done in $O(n^2\alpha(n))$ time, where $\alpha(n)$ is the inverse Ackermann function. Yuster and Zwick improved this to a pure $O(n^2)$ time algorithm [25]. Vazirani and Yannakakis [24] and Robertson *et al.* [23] studied the connection between such even-length cycles and Pfaffian orientations. Finding a cycle of a given size has also been extensively studied (see Alon *et al.* [1,2] for references).

For the case of planar graphs, Eppstein [11] proved that the girth can be found in $O(n)$ time provided it is bounded by some constant. His result extended that of Itai and Rodeh [17] and of Papadimitriou and Yannakakis [21] who proved this for girth bounded by 3. For the general case, when the girth is not bounded by a constant, Djidjev [9] presented and algorithm that computes the girth in $O(n^{5/4} \log n)$ time. Djidjev's solution uses dynamic data structure for shortest paths [10], as well as a clever use of hammock decompositions [14]. Djidjev's

S. Albers et al. (Eds.): ICALP 2009, Part I, LNCS 5555, pp. 764–773, 2009.

algorithm is the fastest algorithm that solves this problem *directly*. However, there is another, *indirect* approach to solve the girth problem in planar graphs. It is a known fact that cuts in an embedded planar graph correspond to cycles in the *dual* plane graph. Furthermore, minimum cuts correspond to shortest cycles in the dual plane graph. Chalermsook *et al.* [5] gave an $O(n \log^2 n)$ time algorithm for the minimum-cut problem in planar graphs. This algorithm can be used to solve the girth problem in planar graphs with positive edge weights in the same time, by reducing it (in linear time) to the min-cut problem in planar graphs. We note that this reduction introduces (not necessarily constant) weights in the dual graph even if the original graph was unweighted.

In this paper, we give an $O(n \log n)$-time algorithm for finding the girth of a planar graph. Apart from being faster than Djidjev's $O(n^{5/4} \log n)$ algorithm and from the $O(n \log^2 n)$ minimum-cut based algorithm, the structure of our algorithm is different – it is a simple divide-and-conquer, and we require no dynamic data structures. In addition, just like in Djidjev's case, our result extends from planar graphs to graphs of bounded genus. Unlike the minimum-cut based algorithm, or Djidjev's algorithms, we *do not* need to find an embedding of the graph in the plane (notice that any minimum-cut based algorithm must first embed the graph since it needs to construct the dual graph).

The rest of the paper is organized as follows. In Section 2 we recall some definitions and facts about planar graphs and bounded genus graphs. Sections 3 contains the description and proof of the algorithm for planar graphs, and Section 4 describes the generalization to bounded genus graphs. The final section contains some concluding remarks.

2 Preliminaries

A *planar embedding* of a graph assigns each vertex to a distinct point on the sphere, and assigns each edge to a simple curve between the points corresponding to its endpoints, with the property that the curves intersect only at their endpoints. A graph G is planar if it has a planar embedding. Consider the set of points on the sphere that are not assigned to any vertex or edge; each connected component of this set is a *face* of the embedding. A planar embedding on the sphere translates to a planar embedding in the plane where a chosen face becomes the outer face. If all the vertices of G lie on a single face, G is said to be *outerplanar* (or 1-outerplanar). G is *k-outerplanar* if the deletion of the vertices on the outer face results in a $(k-1)$-outerplanar graph.

The genus of a graph is the minimum number of handles that must be added to a sphere so that the graph can be embedded in the resulting surface with no crossing edges. A planar graph therefore has genus 0. Euler's formula states that a graph embedded on a surface of genus g with n vertices, m edges, and f faces, satisfies

$$n - m + f = 2 - 2g \ . \qquad (1)$$

A separator is a set of vertices whose removal leaves no connected component of more than $2n/3$ vertices. If G is a planar graph, then it has a separator of

$O(\sqrt{n})$ vertices [18], and if G has genus $g > 0$, then it has a separator of $O(\sqrt{gn})$ vertices [15]. The corresponding separators can be found in $O(n+g)$ time. Every k-outerplanar graph has a separator of size $O(k)$ [3,22] that can be found in $O(n)$ time.

3 The Algorithm for Planar Graphs

In this section we prove the following main theorem of our paper.

Theorem 1. *The girth of an undirected n-vertex planar graph can be computed in $O(n \log n)$ time.*

Given an embedded planar graph G, the size of each face of G is clearly an upper bound on G's girth. Notice however that the shortest cycle is not necessarily a face. Djidjev's algorithm [9], begins by computing, in $O(n)$ time, the size h of the minimal face of G. It then uses h to decide which of two procedures to apply in order to compute the girth. One procedure is used if h is below some specific threshold, and another if it is above. Our algorithm begins by computing, in $O(n)$ time, an *upper bound*, h, for the minimal face size of any embedding. We therefore avoid the need to compute an embedding explicitly. Unlike Djidjev's algorithm, our algorithm is a single divide-and-conquer procedure whose running-time is independent of h.

Our general idea is to cover G with $O(n/k)$ overlapping k-outerplanar graphs where $k = 2h$. The cover is constructed so that the smallest cycle in G is entirely contained within one of these k-outerplanar graph. This means that we can compute G's girth by independently computing the girth of each k-outerplanar graph. We use a simple algorithm on each k-outerplanar graph that exploits the fact that it has an $O(k)$ separator. We next describe this algorithm. In order to use it later we need the algorithm to work even if G's edges have positive edge lengths (and we seek the shortest, rather than smallest, cycle).

k-Outerplanar Graphs with Nonnegative Edge-Lengths

Given a k-outerplanar graph G with n vertices and nonnegative edge-lengths we describe an algorithm that computes G's shortest cycle in $O(kn \log n)$ time. The algorithm first constructs the $O(k)$-sized separator, and is then applied recursively on each of the connected components resulting from the removal of the separator. The recursive calls find G's shortest cycle in the case that it does not pass through any of the separator vertices. We are therefore left with finding G's shortest cycle in the case that it includes one or more of the separator vertices.

To do this, we first run a single-source shortest path algorithm from every separator vertex. Henzinger*et al.* [16] gave an $O(n)$-time algorithm for planar graphs with nonnegative edge-lengths that computes the distances from a given source v to all vertices of G. Therefore, in $O(kn)$ time, we can construct the shortest-path tree from every separator vertex. Suppose that the shortest cycle of G passes through some separator vertex v. The following lemma states an important connection between this cycle and the shortest-path tree from v to all vertices of G.

Lemma 1. *Let G be a connected graph with positive edge-lengths. If a vertex v lies on a shortest cycle, and if T is a shortest paths tree from v then there is a shortest cycle that passes through v and has exactly one edge not in T.*

Proof. Suppose that the shortest cycle of G is of length s. Among all cycles of length s that pass through v, let C be the one with the least number of edges not in T. Assume for contradiction that this number is $k \geq 2$. A vertex $u \neq v$ on C partitions C into two parts C_1 and C_2 that are the two v-to-u simple paths in C. Since $k \geq 2$, there exists a vertex u so that both C_1 and C_2 contain an edge that is not in T. This is illustrated in Fig. 1

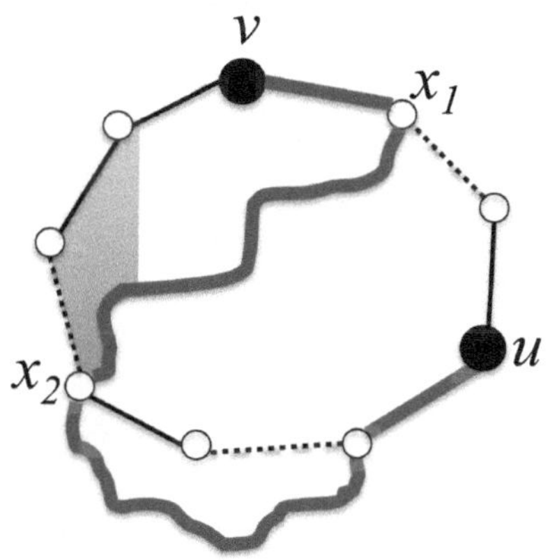

Fig. 1. A cycle passing through a vertex v. The solid edges belong to the shortest-paths tree T and the dashed edges do not. The path P in bold (red) is the shortest path from v to u. The shaded area is a shorter cycle formed by a prefix of P and a part of the original cycle.

Denote by P the path in T from the root v to the vertex u. Suppose that the only vertices that P and C_1 (resp. C_2) share are u and v. Then the cycle C' formed by P together with C_1 (resp. C_2) is of length at most s. This is because P is a shortest v-to-u path and thus not longer than C_2 (resp. C_1). However, C' has less than k edges that are not in T since all the edges of P are in T. This contradicts the fact that C is the shortest cycle with the least number of edges not in T.

Therefore, P must share some vertex $x_1 \notin \{u, v\}$ with C_1 and some vertex $x_2 \notin \{u, v\}$ with C_2. Without loss of generality we assume that x_1 appears before x_2 in P and that x_2 is the first vertex of P in $C_2 - \{u, v\}$. The prefix of P that ends in x_2, together with the part of C_2 between v and x_2, form a cycle C'. However, again, C' has less than k edges that are not in T, and this contradicts the fact that C is the shortest cycle with the least number of edges not in T. □

The above lemma suggests the following $O(n)$-time procedure to find the shortest cycle in case it passes through v. Let T be the shortest-path tree rooted at v, and let $d_v(x)$ denote the length of the shortest path from v to x. For each edge (x, y) not in T whose length is $\ell(x, y)$ we look at $d_v(x)+d_v(y)+\ell(x, y)$ and take the minimum of this sum over all edges $(x, y) \notin T$.

Suppose the shortest cycle is of length s. Notice that if v is indeed part of a shortest cycle then by Lemma 1 we are guaranteed to find it using the above process. If on the other hand no shortest cycle passes through v then the value we get from this process is not smaller than s. This is because every value $d_v(x) + d_v(y) + \ell(x, y)$ that we consider corresponds to either an actual cycle or a cycle attached to a path (in the case where the shortest paths to x and to y share a common prefix). The $O(n)$ time complexity follows from the shortest paths algorithm of Henzinger*et al.* [16] and from the fact that the number of edges in G is $O(n)$ and each edge is checked in $O(1)$.

We have thus established that in $O(kn)$ time we can find the shortest cycle in the case that it passes through a separator vertex. If the removal of the separator results in $t \geq 2$ connected components, then the total time-complexity of all the recursive calls is therefore

$$T(n) = T(n_1) + T(n_2) + \cdots + T(n_t) + O(kn),$$

$$\text{where } \sum_{i=1}^{t} n_i \leq n \text{ and every } n_i \leq 2n/3 \ .$$

The solution to this recurrence is $T(n) = O(kn \log n)$ (for the standard analysis of such recurrences see, e.g. [7]).

This concludes our description of the k-outerplanar $O(kn \log n)$-time algorithm. Notice that this algorithm works even if the graph is directed. Indeed, suppose we want to compute the shortest directed cycle containing the separator vertex v. We start by deleting all the edges incoming to v. We then apply the Henzinger*et al.* algorithm (that works also for directed graphs) from source v. Let $d_v(x)$ denote the length of the shortest v-to-x path. We scan all edges (x, v) that we deleted before and take the minimum of $d_v(x) + \ell(x, v)$.

Since any planar graph G has a separator of size $\sqrt{|G|}$, our algorithm for directed planar graphs runs in time

$$T(G) = T(G_1) + T(G_2) + \cdots + T(G_t) + O(|G|^{3/2}),$$

$$\text{where } \sum_{i=1}^{t} |G_i| \leq |G| \text{ and every } |G_i| \leq 2|G|/3 \ .$$

This gives a total of $O(n^{3/2})$ time. We next show that in the undirected case this can be improved to $O(n \log n)$ by dividing the planar graph into many k-outerplanar graphs.

Covering the Graph by k-Outerplanar Graphs

Before we can cover G by k-outerplanar graphs, we will need to modify G in order to make sure that each edge of G is incident with a vertex whose degree is *at least* 3. We note that this is opposite of Djidjev's algorithm [9] which makes sure that the maximum degree is 3. We may assume, of course, that G is 2-connected as otherwise we can run the algorithm on each 2-connected component separately. In particular, this implies that G has no vertices of degree 0 or degree 1. We may also assume that G is not a simple cycle as this case is trivial. We apply the following contraction to G repeatedly. We remove every vertex u of degree 2 whose two neighbors v_1, v_2 are not connected and add an edge (v_1, v_2) whose length is the sum of lengths of the edges (u, v_1) and (u, v_2). Once this contraction process ends we obtain a graph G' with the property that the girth of G is equal to the length of the shortest cycle in G'. Therefore, it suffices to compute the shortest cycle of G'. Notice that if h is the minimum face length of any embedding of G, then the number of edges on a shortest cycle of G' is also bounded by h. This is because the girth of G is bounded by h and we have only contracted edges to get from G to G'. The following lemma states two important properties of G'.

Lemma 2. *In order to compute the girth of an n-vertex planar graph G, for which some embedding has minimum face length h, it suffices to compute the shortest cycle of the planar graph G', which has nonnegative edge-lengths and $O(n/h)$ vertices.*

Proof. Fix an embedding of G with minimal face length h. We will prove that the graph G' obtained by the above process has $O(n/h)$ vertices. We denote m as the number of edges in G, F denotes the set of all faces of G, $|x|$ denotes the size of a face $x \in F$ and f denotes the number of faces of G. Notice that the transformation from G to G' does not change the total number of faces. We will show first that $f = O(n/h)$. Since any edge of G belongs to two faces, then

$$2m = \sum_{x \in F} |x| \geq \sum_{x \in F} h = fh .$$

In any planar graph $m \leq 3n - 6$ so we get that $f \leq 2m/h \leq 6n/h = O(n/h)$.

Let m' and n' denote the number of edges and vertices of G'. We need to show that $n' = O(n/h)$, or, equivalently, that $m' = O(n/h)$. We denote T as the set of vertices of G with degree at least 3 and set $t = |T|$. As the set of vertices with degree 2 in G' is an independent set, we have that

$$\sum_{v \in T} deg(v) \geq m' .$$

On the other hand,

$$\sum_{v \in T} deg(v) + 2(n' - t) = 2m' .$$

By Euler's formula we know that

$$m' = n' + f - 2 \leq n' + 6n/h .$$

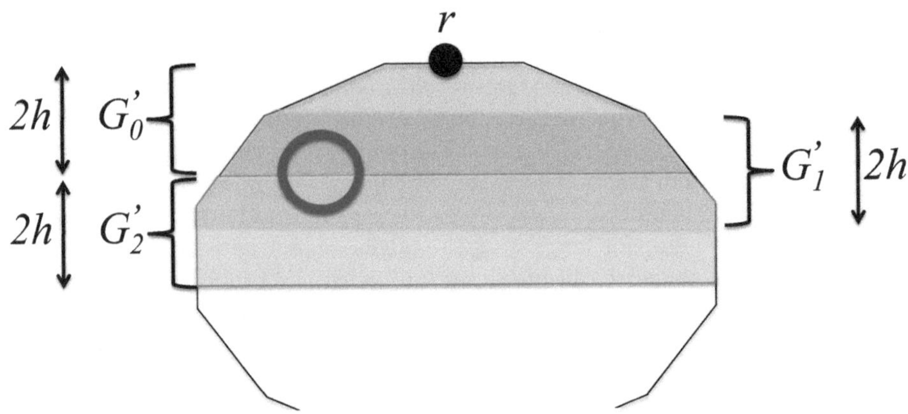

Fig. 2. A decomposition of a graph G' into overlapping $2h$-outerplanar graphs according to a breadth-first search of G' that starts in an arbitrary vertex r. The shortest cycle is guaranteed to be completely contained within one of these $2h$-outerplanar graphs, in this case in G'_1.

It follows that

$$\sum_{v \in T} (deg(v) - 2) = 2(m' - n') = 2f - 4 \leq 12n/h \ .$$

Since $deg(v) \geq 3$ for each $v \in T$ we have that

$$\frac{1}{3} \sum_{v \in T} deg(v) \leq 12n/h \ .$$

Consequently,

$$m' \leq 36n/h$$

as required. □

Lemma 2 actually provides a way to compute an upper bound h for the minimum face length of any embedding of G. We simply construct G' resulting in n' vertices, and set $h = \min\{n, \lfloor 36n/n' \rfloor\}$.

Now that we can work with a graph G' that has only $O(n/h)$ vertices we can finally describe how to cover G' by k-outerplanar graphs. Consider a breadth-first search of G' that starts in an arbitrary vertex r (and can be done in linear-time). Define G'_i as the graph induced by the vertices whose distance from r is between $ki/2$ and $k + ki/2$ for $k = 2h$ and $i = 0, 1, \ldots, \frac{2(n-k)}{k}$. In this way, every G'_i overlaps with at most two other graphs, G'_{i-1} and G'_{i+1}. This is depicted in Fig. 2. It is easy to verify that every G'_i is indeed a $(k+1)$-outerplanar graph. Furthermore, recall that the shortest cycle in G' has at most h edges. Therefore, it must be entirely contained within a single G'_i. This is because we chose k to be $2h$ and the overlap between two adjacent G'_i's to be $k/2$.

Finally, we run our k-outerplanar graph algorithm on every G'_i separately to find its shortest cycle. We then return the shortest cycle among these cycles. The time complexity is thus

$$\sum_i O(k|G'_i| \log |G'_i|) \leq O(2h \log n) \cdot \sum_i |G'_i| \ .$$

The $O(n \log n)$ total time complexity is achieved by noticing that every vertex in G' appears in at most three G'_i's therefore $\sum_i |G'_i| = O(|G'|)$, which is equal to $O(n/h)$ by Lemma 2. This completes the proof of Theorem 1. □

4 Extension for Bounded Genus Graphs

In this section we show how to adjust the proof of Theorem 1 so that it extends to graphs with bounded genus. We therefore obtain the following theorem.

Theorem 2. *For every fixed positive integer g, the girth of an undirected n-vertex graph whose genus is at most g can be computed in $O(n \log n)$ time.*

We outline the adjustments to the proof of Theorem 1 that are required in order to obtain Theorem 2. Regarding the shortest paths algorithm of Henzinger *et al.* [16] that we use, as pointed out in [16], separators of size $O(n^{1-\varepsilon})$ suffice for the application of their algorithm, provided that the separator can be found in linear time. Thus their algorithm remains $O(n)$ when applied to graphs with bounded genus.

In the proof of Theorem 1 the $(k+1)$-outerplanar graphs are just obtained by taking $k+1$ consecutive layers of a breadth-first search from a given vertex. We then use the fact that such graphs have $O(k)$-sized separators, and such separators are guaranteed to exist in subgraphs of these $(k+1)$-outerplanar graphs, as subgraphs of $(k+1)$-outerplanar graphs are also $(k+1)$-outerplanar. In other words, we simply use the fact that k-outerplanar graphs have *tree-width* $O(k)$. Now, suppose we perform breadth-first search in a genus g graph, and let G'_i be obtained by taking the $k+1$ consecutive layers s through $s+k$ of that search. We would like to claim that G'_i is analogous to a $(k+1)$-outerplanar graph in a "genus g" setting. Consider any embedding of the graph on a genus g surface. We can contract all vertices in layers above s to a single vertex z. The resulting graph is a minor of the original graph, thus the genus does not increase. Notice that now the diameter of G'_i becomes $O(k)$ and the genus remains at most g. A result of Eppstein [12] shows that graphs with bounded genus g have separators, as well as tree-width, of the same order as the diameter. It follows that G'_i has tree-width $O(k)$ as well, so the same analysis as in the case of k-outerplanar graphs holds in the bounded genus setting.

Another point of minor difference is in Euler's formula when applied in the proof of Lemma 2. Instead of using the fact that in planar graphs we have $m \leq 3n - 6$ we use the fact that in genus g graphs we have $m \leq 3n - 6 + 6g$. As g is bounded we still have $f = O(n/h)$ as in the planar case. Similarly, instead of using $m' = n' + f - 2$ which holds in the planar case we use $m' = n' + f - 2 + 2g$ and since g is bounded this still gives us that $m' = O(n/h)$ as in the planar case.

We have therefore shown that Theorem 1 can be adjusted to apply to the bounded genus setting, thereby proving Theorem 2. □

5 Concluding Remarks and Open Problems

We have presented the fastest algorithm for computing the girth of an undirected planar graph and bounded genus graph. Our algorithm runs in $O(n \log n)$ time, improving the previous best algorithms. It would be interesting to extend this algorithm to undirected graphs with arbitrary positive *real* edge weights. It would also be interesting to find an $o(n^{3/2})$ algorithm for directed planar graphs.

References

1. Alon, N., Yuster, R., Zwick, U.: Color-coding. Journal of the ACM 42(4), 844–856 (1995)
2. Alon, N., Yuster, R., Zwick, U.: Finding and counting given length cycles. Algorithmica 17, 354–364 (1997)
3. Bodlaender, H.: A partial k-arboretum of graphs with bounded treewidth. Theor. Comput. Sci. 209(1-2), 1–45 (1998)
4. Bollobás, B.: Chromatic number, girth and maximal degree. Discrete Mathematics 24, 311–314 (1978)
5. Chalermsook, P., Fakcharoenphol, J., Nanongkai, D.: A deterministic near-linear time algorithm for finding minimum cuts in planar graphs. In: SODA, pp. 828–829 (2004)
6. Cook, R.J.: Chromatic number and girth. Periodica Mathematica Hungarica 6, 103–107 (1975)
7. Cormen, T.H., Leiserson, C.E., Rivest, R.L., Stein, C.: Introduction to Algorithms, 2nd edn. The MIT Press and McGraw-Hill Book Company (2001)
8. Diestel, R.: Graph theory, 2nd edn. Springer, Heidelberg (2000)
9. Djidjev, H.N.: Computing the girth of a planar graph. In: Welzl, E., Montanari, U., Rolim, J.D.P. (eds.) ICALP 2000. LNCS, vol. 1853, pp. 821–831. Springer, Heidelberg (2000)
10. Djidjev, H., Pantziou, G.E., Zaroliagis, C.D.: Improved algorithms for dynamic shortest paths. Algorithmica 28(4), 367–389 (2000)
11. Eppstein, D.: Subgraph isomorphism in planar graphs and related problems. Journal of Graph Algorithms and Applications 3(3), 1–27 (1999)
12. Eppstein, D.: Diameter and treewidth in minor-closed graph families. Algorithmica 27(3), 275–291 (2000)
13. Erdős, P.: Graph theory and probability. Canadian Journal of Math 11, 34–38 (1959)
14. Frederickson, G.N.: Planar graph decomposition and all pairs shortest paths. Journal of the ACM 38(1), 162–204 (1991)
15. Gilbert, J.R., Hutchinson, J.P., Tarjan, R.E.: A separator theorem for graphs of bounded genus. Journal of Algorithms 5(3) (1984)
16. Henzinger, M.R., Klein, P.N., Rao, S., Subramanian, S.: Faster shortest-path algorithms for planar graphs. Journal Comp. Sys. Sci. 55(1), 3–23 (1997)

17. Itai, A., Rodeh, M.: Finding a minimum circuit in a graph. SIAM J. Comput. 7(4), 413–423 (1978)
18. Lipton, R.J., Tarjan, R.E.: A separator theorem for planar graphs. SIAM J. Appl. Math. 36, 177–189 (1979)
19. Lovasz, L.: On chromatic number of finite set systems. Acta Math. Acad. Sci. Hun. 19, 59–67 (1968)
20. Monien, B.: The complexity of determining a shortest cycle of even length. Computing 31, 355–369 (1983)
21. Papadimitriou, C.H., Yannakakis, M.: The clique problem for planar graphs. Information Processing Letters 13, 131–133 (1981)
22. Robertson, N., Seymour, P.D.: Graph minors. iii. planar tree-width. J. Comb. Theory, Ser. B 36(1), 49–64 (1984)
23. Robertson, N., Seymour, P.D., Thomas, R.: Permanents, pfaffian orientations, and even directed circuits. Ann. of Math. 150(3), 929–975 (1999)
24. Vazirani, V.V., Yannakakis, M.: Pfaffian orientations, 0/1 permanents, and even cycles in directed graphs. In: Lepistö, T., Salomaa, A. (eds.) ICALP 1988. LNCS, vol. 317, pp. 667–681. Springer, Heidelberg (1988)
25. Yuster, R., Zwick, U.: Finding even cycles even faster. SIAM J. Discrete Math. 10(2), 209–222 (1997)

Elimination Graphs

Yuli Ye and Allan Borodin

Department of Computer Science,
University of Toronto,
Toronto, Ontario, Canada M5S 3G4
{y3ye,bor}@cs.toronto.edu

Abstract. A graph is chordal if it does not contain any induced cycle of size greater than three. An alternative characterization of chordal graphs is via a perfect elimination ordering, which is an ordering of the vertices such that, for each vertex v, the neighbors of v that occur later than v in the ordering form a clique. Akcoglu et al [1] define an extension of chordal graphs whereby the neighbors of v that occur later than v in the elimination order have at most k independent vertices. We refer to such graphs as sequentially k-independent graphs. We study properties of such families of graphs, and we show that several natural classes of graphs are sequentially k-independent for small k. In particular, any intersection graph of translates of a convex object in a two dimensional plane is a sequentially 3-independent graph; furthermore, any planar graph is a sequentially 3-independent graph. For any fixed constant k, we develop simple, polynomial time approximation algorithms for sequentially k-independent graphs with respect to several well-studied NP-complete problems.

1 Introduction

We assume familiarity with standard graph theory terminology; all graphs in this paper will be finite, simple, connected and undirected. Let $G = (V, E)$ be a graph of n vertices and m edges. If $X \subseteq V$, the subgraph of G induced by X is denoted by $G[X]$. For a particular vertex $v_i \in V$, let $d(v_i)$ denote its degree and $N(v_i)$ denote the set of neighbors of v_i. We use $\alpha(G)$ to denote the size of a maximum independent set of G. Given an ordering of vertices $v_1, v_2, \ldots, v_n$, we let $V_i = \{v_i, \ldots, v_n\}$. A graph is **chordal** if it does not contain any induced cycle of size greater than three. An alternative characterization of chordal graphs is via a perfect elimination ordering.

Definition 1. *A* **perfect elimination ordering** *is an ordering of vertices v_1, v_2, …, v_n such that for any v_i, $1 \leq i \leq n$, $\alpha(G[N(v_i) \cap V_i]) = 1$.*

A natural extension to this perfect elimination ordering is to relax the size of the maximum independent set. Surprisingly, this extension seems to have only been relatively recently proposed in Akcoglu et al [1] and not studied subsequently.

Definition 2. *A* **k-independence ordering** *is an ordering of vertices v_1, v_2, …, v_n such that for any v_i, $1 \leq i \leq n$, $\alpha(G[N(v_i) \cap V_i]) \leq k$. The minimum*

S. Albers et al. (Eds.): ICALP 2009, Part I, LNCS 5555, pp. 774–785, 2009.

of such k over all orderings is called the **sequential independence number**[1], *which we denote as* $\lambda(G)$.

This extension of a perfect elimination ordering leads to a natural generalization of chordal graphs.

Definition 3. *A graph G is* **sequentially** k**-independent** *if* $\lambda(G) \leq k$.

For several natural classes of graphs, the sequential independence number is bounded by a small constant.

- **Chordal graphs:** since a chordal graph admits a perfect elimination ordering, chordal graphs are sequentially 1-independent graphs. All sub-classes of chordal graphs are then clearly sequentially 1-independent graphs; for example, interval graphs and trees.
- **Graphs with a bounded average degree on every induced subgraph:** it is not hard to see that $\lambda(G)$ is bounded above by the maximum average degree over all induced subgraphs of G, though this bound is usually not tight.
- **Claw-free graphs:** if a graph is k-claw-free, then it is a sequentially $(k-1)$-independent graph. For example, line graphs are sequentially 2-independent graphs. Note that the converse is not always true since for example, a k-ary tree is not k-claw-free, but it is a sequentially 1-independent graph.
- **Graphs with a constant tree-width:** since the tree-width of G can be viewed as the smallest k such that G is a partial k-tree, it is not hard to see graphs with a tree-width k are sequentially k-independent graphs.
- **Intersection graphs of geometric objects:** disk graphs and unit disk graphs are sequentially 5 and 3-independent graphs, respectively. Marathe et al [26] show that simple heuristics achieve good approximations for various problems for unit disk graphs. We extend most of their results in this paper, and explain a connection between the unit disk graphs and planar graphs as observed in their paper.

Akcoglu et al [1] show that the (weighted) maximum indepedent set (MIS) problem has a simple k-approximation algorithm for any sequentially k-independent graph. We call attention to two interesting examples.

- **The interval scheduling problem (ISP) and the job interval scheduling problem (JISP):** For a given set of (weighted) intervals on the real line, the goal is trying to schedule a set of intervals of maximum size (or total weight in the weighted case) without any overlapping. There are simple algorithms that solve ISP optimally in both the unweighted and weighted cases. This is not a surprise since if we order intervals according to non-decreasing finishing times, then it is a perfect elimination ordering and the underlying intersection graph of ISP is a chordal graph. The job interval scheduling problem is an extension of ISP and has been extensively studied in the literature, for example, see [4][15][31]. In JISP, each interval belongs to a job. A job can be scheduled

[1] Akcoglu et al [1] refer to this as the directed local independence number.

onto one and only one of its intervals. The objective is to find the maximum number (or total weight in the weighted case) of jobs that can be scheduled without conflicts. Using the same ordering by finishing time, the intersection graph for the JISP problem is a sequentially 2-independent graph.

- **The axis parallel rectangles problem:** This problem is studied by Berman and DasGupta in [9] and is motivated by applications to non-overlapping local alignment problems in computational molecular biology. The input is a set of axis parallel rectangles such that, for each axis, the projection of a rectangle on this axis does not enclose that of another. The goal is to select a subset of independent (non-overlapping projection on both axes) rectangles with maximum cardinality (or total weight in the weighted case). It is not hard to see that sorting the rectangles by (say) their rightmost x-coordinate yields a 3-independence ordering; hence the underlying graph is a sequentially 3-independent graph. This also extends to D dimension, where the underlying graph is a sequentially $(2D-1)$-independent graph.

Both of the JISP and local alignment problems are MAX SNP-hard although the current NP-hardness inapproximations are very weak. The existence of local ratio approximation algorithms in Bar-Noy et al [3] and Berman and DasGupta in [9] for the above two problems was our initial motivation for investigating how the intersection graph structure underlies the success of those algorithms. In fact, such "elimination structure" occurs in many natural graph classes and extends in greater generality as we shall see.

2 Chordal Graphs and Their Generalizations

The study of chordal graphs can be traced back to Hajnal and Surányi [21] in the late 1950s. Fulkerson and Gross [18] characterized chordal graphs in terms of perfect elimination orderings; this was also observed by Rose [29] in 1970. Following this, Rose, Tarjan and Lueker [30] introduced the first linear-time algorithm for producing a perfect elimination ordering, known as the lexicographic breadth-first search (LBFS); and later, Tarjan [32] gave an even simpler algorithm known as the maximum cardinality search (MCS).

Many generalizations of chordal graphs have been proposed and studied, for example as in [14][22][25]. One very close extension of chordal graphs related to sequentially k-independent graphs is defined by Jamison and Mulder [25], in which they use the minimum vertex clique cover number instead of the maximum independent set number for the neighborhood property. Since for $k \geq 3$, it is NP-hard to determine whether or not a graph has a k vertex clique cover, it seems unlikely that we can recognize such graphs in polynomial time for $k \geq 3$.

Another relevant (but not comparable) class of graphs are the k-interval graphs as defined by Trotter and Harary [33], and independently by Griggs and West [20]. Butman, Hermelin, Lewenstein and Rawitz [12] studied minimum vertex cover, minimum dominating set and maximum clique on k-interval graphs and were able to obtain approximation algorithms for each of problems. One disadvantage of such a graph class is that recognizing a k-interval graph for $k \geq 2$

is NP-hard [35] and some of the algorithms on k-interval graphs G require a k-interval representation of G. We note that a complete bipartite graph $K_{n,n}$ has interval number $\frac{n^2+1}{2n}$ [20] and sequential independence number n. Furthermore, the interval number of chordal graphs can be arbitrarily large [34], so it follows that the class of k-interval graphs and the class of sequentially k-independent graphs are incomparable.

3 Properties of Sequentially Independent Graphs

In this section, we study general properties of sequentially independent graphs. We first give the following basic lemma.

Lemma 1. *Any induced subgraph of a sequentially k-independent graph is a sequentially k-independent graph.*

Lemma 1 ensures that we can test if a graph is sequentially k-independent by repeatedly removing a vertex whose neighbors in the remaining graph have independent set size at most k, until there is no vertex remaining or for every vertex v in the remaining graph G, $\alpha(G[N(v)]) > k$. This k-elimination process, if successful (*i.e.*, no vertices remain at the end of the process), also constructs a k-independence ordering. Note that at each step, we can check for each vertex v to see if $\alpha(G[N(v)]) > k$ in $O(k^2 n^{k+1})$ time by enumerating all subsets of size $k+1$. Therefore this k-elimination process terminates in $O(k^2 n^{k+3})$ time. By the observation of Itai and Rodeh [24], and results in [17], we can improve the time complexities of a 2-elimination process to $O(n^{4.376})$, a 3-elimination process to $O(n^{5.334})$ and a 4-elimination process to $O(n^{6.220})$. We can further improve the time complexity of the k-elimination process by the following theorem.

Theorem 1. *A sequentially k-independent graph can be recognized in $O(k^2 n^{k+2})$ time, and a k-independence ordering of a sequentially k-independent graph can be constructed in $O(k^2 n^{k+2})$ time.*

Note that by a reduction to the independent set problem, finding the sequential independence number is complete for $W[1]$, hence it is unlikely to have a fixed parameter tractable solution. But this does not exclude the possibility to improve the current complexity bound for a small k. It is interesting to note that recognizing a chordal graph, *i.e.*, a sequentially 1-independent graph, can be done in linear time using LBFS or MCS, while our generic algorithm runs in time $O(n^3)$. It might be possible to improve on the bounds above by using different techniques. Note that we are mostly interested in sequentially k-independent graphs with small k's, because in practice, we either know a-priori that a graph is a sequentially k-independent graph with some small constant k, or we want to test whether or not it is the case. In many specific cases like JISP and non-overlapping local alignment, the complexity of computing a k-independence ordering can be reduced to $O(n \log n)$. It is not hard to show that the sequential independence number is bounded by some graph parameters.

Theorem 2. *A graph G with n vertices and m edges has sequential independence number no more than* $\min\{\lfloor \frac{n}{2} \rfloor, \lfloor \sqrt{m} \rfloor, \lfloor \frac{\sqrt{1+4[\binom{n}{2}-m]}+1}{2} \rfloor\}$.

4 Natural Classes of Sequentially Independent Graphs

In Section 1, we saw several known examples of sequential independent graphs. In this section, we show two more natural classes of graphs that fit our definition.

4.1 Translates of a Convex Object

We consider the intersection graph of a set of translates of a convex object in the two dimensional plane, *i.e.*, each translate is represented by a vertex; two vertices are adjacent if two associated translates are overlapping.

Theorem 3. *The intersection graph of translates of a convex object in a two dimensional plane is a sequentially* 3*-independent graph.*

The bound in Theorem 3 is in fact tight. If we allow different sizes, but still the same shape and orientation, the resulting intersection graph is sequentially 5-independent by always looking at the translate with the smallest size. It follows immediately that disk graphs and unit disk graphs are sequentially 5 and 3-independent graphs respectively. We conjecture that Theorem 3 extends to higher dimensions as follows: the intersection graph of translates of a convex object in D dimensional space is a sequentially $(2D-1)$-independent graph.

4.2 Planar Graphs

We first show relatively trivial result for arbitrary genus. Since the average vertex degree of a graph G with genus g is less than $6 + \frac{12(g-1)}{n}$, where n is the number of vertices in G and any induced subgraph of G has genus no more than g, the following theorem is immediate.

Theorem 4. *A graph G with n vertices and genus g has sequential independence number no more than* $\lfloor 6 + \frac{12(g-1)}{n} \rfloor$.

Theorem 4 implies that any planar graph is a sequentially 5-independent graph, but this is not tight.

Theorem 5. *Any planar graph is a sequentially* 3*-independent graph.*

Proof. We prove the contrapositive. Suppose G is a planar graph but it is not a sequentially 3-independent graph, then at a certain step of the 3-elimination process, we will end up with an induced subgraph G^* of G such that for any vertex v in G^*, we have $\alpha(G^*[N(v)]) > 3$. We look at a planar embedding of G^*. It is not hard to show, using a counting argument, that the Euler characteristic of G^* is non-positive, hence a contradiction. Therefore any planar graph is sequentially 3-independent. □

5 Algorithmic Aspects of Sequentially Independent Graphs

In this section, we show several algorithmic results for sequentially k-independent graphs. We do not explicitly state the time complexity for those algorithms since (with the exception of the vertex cover problem) they depend on the complexity for constructing a k-independence ordering. But by Theorem 1, all algorithms discussed here run in polynomial time when k is a fixed constant.

It is well known that many NP-complete problems can be solved optimally when restricted to chordal graphs [19][36]. We show that if we restrict the graph to be a sequentially k-independent graph, we get simple approximation algorithms for several NP-complete problems. In fact, the structure of sequentially k-independent graphs gives a unified treatment for many previous results. Note that for both minimization problem and maximization problems, we always consider approximation ratios to be greater or equal to one.

5.1 Weighted Maximum m-Colorable Subgraph

The weighted maximum independent set problem is NP-complete for general graphs and NP-hard to approximate within $n^{1-\epsilon}$, but polynomial time solvable for chordal graphs. Akcoglu et al [1] show that the local ratio technique ([5][8]) achieves a k-approximation for the weighted maximum independent set on sequentially k-independent graphs. The local ratio technique is usually described as a recursive algorithm. As in Berman and DasGupta [8], we view it as an iterative algorithm, modeled as stack algorithms in [10]. For some graph classes such as those defined by the JISP and axis paralell rectangles problems mentioned in section 1, the orderings in the stack algorithm satisfy the locally defined orderings in priority algorithms [11] and hence we obtain greedy (in the unweighted case) or greedy-like stack algorithms.

The interval scheduling problem is often extended to scheduling on m machines. For identical machines, the graph-theoretic formulation of this problem leads to the following natural generalization of the MIS problem. Given a graph $G = (V, E)$ with a positive weight $w(v)$ for each vertex v, an m-colorable subgraph of G is an induced subgraph $G[V']$ on a subset V' of V such that $G[V']$ is m-colorable. A maximum m-colorable subgraph is an m-colorable subgraph with maximum number (or total weight in the weighted case) of the vertices. For chordal graphs, the unweighted case of the problem is polynomial time solvable for any fixed m, but NP-complete otherwise. Chakaravarthy and Roy [13] recently showed that for chordal graphs, the problem is NP-hard and has a simple 2-approximation algorithm for the weighted case. Here we strengthen this result.

Theorem 6. *For all fixed constant $k \geq 1$, there is a polynomial time algorithm that achieves a $(k + 1 - \frac{1}{m})$-approximation for weighted maximum m-colorable subgraph if G is a sequentially k-independent graph.*

We first describe the algorithm. We use almost the same idea for the weighted maximum independent set except that we now use a stack S_c for each color class

c. At each step i, the algorithm considers the vertex v_i in the k-independence ordering, and computes the updated weight

$$\bar{w}(v_i) = w(v_i) - \sum_{v_j \in S_c \cap N(v_i)} \bar{w}(v_j)$$

for each color class c. If $\bar{w}(v_i)$ is non-positive for every color class, then reject v_i without coloring it. Otherwise, find a color class with the largest $\bar{w}(v_i)$ value and assign v_i with that color. We next give the following permutation lemma.

Let M be an m by m square matrix, and $\sigma \in \Sigma$ be a permutation on $\{1, 2, \ldots, n\}$. and σ_i be the ith element in the permutation.

Lemma 2. *There exists a permutation σ such that*

$$\sum_i M_{i\sigma_i} \leq \frac{1}{m} \sum_{i,j} M_{ij}.$$

Let $c_1, c_2, \ldots, c_m$ be the color classes and $S_1, S_2, \ldots, S_m$ be the sets of vertices that have been put onto the stacks at the end of the push phase. Let S be the union of all the stacks. Now we prove the theorem.

Proof. It is not hard to see that the algorithm achieves at least the total weight $W = \sum_{v_t \in S} \bar{w}(v_t)$ of all stacks. The goal is to show that the weight of the optimal solution will be at most $(k + 1 - \frac{1}{m}) \cdot W$. Let A be the output of the algorithm and O be the optimal solution. For each given vertex v_i in O, let c_{oi} be its color class in O, and c_{si} be its color class in S if it is accepted into one of the stacks. Let S^i_{oi} be the content of the stack in color class c_{oi} when the algorithm considers v_i, then we have three cases:

1. If v_i is rejected during the push phase of the algorithm then we have

$$w(v_i) \leq \sum_{v_j \in S^i_{oi} \cap N(v_i)} \bar{w}(v_j).$$

   In this case, we can view that $w(v_i)$ is charged to all $\bar{w}(v_j)$ with $v_j \in S^i_{oi} \cap N(v_i)$, each of which appears in the same color class c_{oi} and is charged at most k times coming from the same color class.
2. If v_i is accepted into the same color class during the push phase of the algorithm then we have

$$w(v_i) = \bar{w}(v_i) + \sum_{v_j \in S^i_{oi} \cap N(v_i)} \bar{w}(v_j).$$

   In this case, we can view that $w(v_i)$ is charged to $\bar{w}(v_i)$ and all $\bar{w}(v_j)$ with $v_j \in S^i_{oi} \cap N(v_i)$. Note that they all appear in the same color class c_{oi}; $\bar{w}(v_i)$ is charged at most once and each $\bar{w}(v_j)$ is charged at most k times coming from the same color class.
3. If v_i is accepted into a different color class during the push phase of the algorithm then we have

$$w(v_i) \le \bar{w}(v_i) + \sum_{v_j \in S_{oi}^i \cap N(v_i)} \bar{w}(v_j).$$

In this case, we can view that $w(v_i)$ is charged to $\bar{w}(v_i)$ and all $\bar{w}(v_j)$ with $v_j \in S_{oi}^i \cap N(v_i)$. Note that each $\bar{w}(v_j)$ appears in the same color class c_{oi} and is charged at most k times coming from the same color class. However $\bar{w}(v_i)$ in this case is in a different color class c_{si} and is charged at most once coming from a different color class.

If we sum up for all $v_i \in O$, we have

$$\sum_{v_i \in O} w(v_i) \le \sum_{v_i \in S \cap O \wedge c_{oi} \neq c_{si}} \bar{w}(v_i) + k \sum_{i=1}^{m} \sum_{v_t \in S_i} \bar{w}(v_t).$$

The inequality holds when we sum up for all $v_i \in O$, since the number of charges coming from the same color class can be at most k; the number of charges coming from a different color class can be at most one, which only appears when v_i is accepted into a stack of a different color class (comparing to the optimal) during the push phase of the algorithm. Therefore we have the extra term $\sum_{v_i \in S \cap O \wedge c_{oi} \neq c_{si}} \bar{w}(v_i)$. Now if we can permute the color classes of the optimal solution so that for any $v_i \in S \cap O$, $c_{oi} = c_{si}$, then the term $\sum_{v_i \in S \cap O \wedge c_{oi} \neq c_{si}} \bar{w}(v_i)$ disappears and we achieve a k-approximation. But it might be the case that no matter how we permute the color classes of the optimal solution, we always have some $v_i \in S \cap O$ with $c_{oi} \neq c_{si}$. We construct the weight matrix M in the following way. An *assignment* $c_i \to c_j$ is to assign the color class c_i of O to the color class c_j of S. A vertex is *misplaced* with respect to this assignment $c_i \to c_j$ if it is in $S \cap O$ and its color class is c_i in O, but is not c_j in S. We then set M_{ij} be total updated weight of misplaced vertices with respect to the assignment $c_i \to c_j$. Note that the total weight of the matrix is $(m-1)\sum_{v_i \in S \cap O} \bar{w}(v_i)$, and applying Lemma 2, there exists a permutation of the color class in O such that

$$\sum_{v_i \in S \cap O \wedge c_{oi} \neq c_{si}} \bar{w}(v_i) \le \frac{m-1}{m} \sum_{v_i \in S \cap O} \bar{w}(v_i) \le \frac{m-1}{m} \sum_{v_i \in A} w(v_i).$$

Therefore, we have

$$\sum_{v_i \in O} w(v_i) \le \frac{m-1}{m} \sum_{v_i \in A} w(v_i) + k \sum_{i=1}^{m} \sum_{v_t \in S_i} \bar{w}(v_t) \le (k + 1 - \frac{1}{m}) \sum_{v_i \in A} w(v_i).$$

□

5.2 Minimum Vertex Cover

Minimum vertex cover is one of the most celebrated problems for approximation algorithms, because there exist simple 2-approximation algorithms, yet for general graphs no known algorithm can achieve approximation ratio $2 - \epsilon$ for any fixed $\epsilon > 0$. In this section, we show a $(2 - \frac{1}{k})$-approximation algorithm for minimum vertex cover on sequentially k-independent graphs. The algorithm shares the same spirit of the result of Bar-Yehuda and Even [6] for the $\frac{5}{3}$-approximation of vertex cover for planar graphs, and can be viewed as a generalization of that algorithm. The algorithm first removes all triangles in the graph, and then finds a maximum matching in local neighborhood of a minimum degree vertex. Baker's PTAS algorithm [2] for minimum vertex cover on planar graphs depends on a planar embedding and would not be considered as a simple combinatorial algorithm.

Theorem 7. *There is a polynomial time algorithm that achieves a $(2 - \frac{1}{k})$-approximation for minimum vertex cover if G is a sequentially k-independent graph. Furthermore, for triangle free graphs sequentially k-independent graphs with $k > 1$, the algorithm is a greedy algorithm (in the sense of [11]).*

The running time of this algorithm is dominated by the time to remove all triangles which can be done in $n \times n$ matrix multiplication time $O(n^{\omega}) \approx O(n^{2.376})$. We can further improve the ratio to $2 - \frac{2}{k+1}$ using a result of Hochbaum [23]. This yields a $\frac{3}{2}$-approximation for planar graphs.

Theorem 8. *There is a polynomial time algorithm that achieves a $(2 - \frac{2}{k+1})$-approximation for minimum vertex cover if G is a sequentially k-independent graph.*

5.3 Minimum Vertex Coloring

For chordal graphs, a greedy algorithm on the reverse ordering of any perfect elimination ordering gives an optimal coloring. For sequentially k-independent graphs, the same greedy algorithm achieves a k-approximation.

Theorem 9. *For all fixed constant $k \geq 1$, there is a polynomial time algorithm that achieves a k-approximation for the minimal vertex coloring if G is a sequentially k-independent graph.*

6 Sequential Neighborhood Properties

We view properties such as "chromatic number at most k" as global properties, and properties like "k-claw-freeness" as (universal) neighborhood properties. In contrast, we refer to "bounded sequential independence number" as a sequential neighborhood property. Such properties give rise to a general development of "elimination graphs".

Definition 4. *A graph G has a* **(universal) neighborhood property** *with respect to graph property P if for all vertices $v_1, v_2, \ldots, v_n$, P holds on $G[N(v_i)]$. The set of all graphs satisfying such a neighborhood property is denoted as $\hat{G}(P)$. A graph G has a* **sequential neighborhood property** *with respect to graph property P if there exists an ordering of vertices $v_1, v_2, \ldots, v_n$ such that for any v_i, $1 \leq i \leq n$, P holds on $G[N(v_i) \cap V_i]$. The set of all graphs satisfying such a sequential neighborhood property is denoted as $\tilde{G}(P)$.*

If the property P is closed on induced subgraphs, then $\hat{G}(P)$ is a sub-family of $\tilde{G}(P)$, and both $\hat{G}(P)$ and $\tilde{G}(P)$ can be recognized in polynomial time provided

P can be tested in polynomial time. Using this notation, we can express many known graph classes as well as defining many new interesting graph classes. For illustration, we consider the following five graph properties:

1. $|V|_k$: the size of the vertex set is no more than k.
2. VCC_k: the minimum vertex clique cover size is no more than k.
3. IS_k: the maximum independent set size is no more than k.
4. IDS_k: the minimum independent dominating set size is no more than k.
5. DS_k: the minimum dominating set size is no more than k.

Using these defined properties, the set of sequentially k-independent graphs discussed in this paper is $\tilde{G}(IS_k)$, the set of k-claw-free graphs is $\hat{G}(IS_{k-1})$, and the set of trees is $\tilde{G}(|V|_1)$. The JISP graphs are in $\tilde{G}(VCC_2)$ as well as in $\tilde{G}(IS_2)$, and graphs with tree-width no more than k are in $\tilde{G}(|V|_k)$. Jamison and Mulder's generalization of chordal graphs is $\tilde{G}(VCC_k)$. Since $|V|_k \Rightarrow VCC_k \Rightarrow IS_k \Rightarrow IDS_k \Rightarrow DS_k$, , we have $\hat{G}(|V|_k) \subset \hat{G}(VCC_k) \subset \hat{G}(IS_k) \subset \hat{G}(IDS_k) \subset \hat{G}(DS_k)$. Similarly, we have $\tilde{G}(|V|_k) \subset \tilde{G}(VCC_k) \subset \tilde{G}(IS_k) \subset \tilde{G}(IDS_k) \subset \tilde{G}(DS_k)$. We can construct examples to show that all the above inclusions are proper.

7 Conclusion and Open Questions

We considered a generalization of chordal graphs due to Akcoglu et al [1] based on a specific type of elimination ordering. We showed that several natural classes of graphs have a small sequential independence number, and gave a unified approach for several optimization problems when such structure is present. Since independence naturally extends to hypergraphs, we think our results also extend to hypergraphs.

There are many open questions. The first and perhaps the most important issue is to improve the time complexity to recognize a sequentially k-independent graph for small constants k.

Our second question is on the algorithmic aspects of sequentially k-independent graphs. We have studied the weighted maximum m-colorable subgraph, minimum vertex cover, and minimum vertex coloring problems. What can be said about other basic graph problems such as weighted minimum vertex cover and minimum edge coloring? Several other problems can be solved optimally in polynomial time for chordal graphs such as maximum clique and minimum clique cover. Can we $O(k)$-approximate such problems for sequentially k-independent graphs? A particularly interesting problem is minimum independent dominating set. The unweighted case can be solved optimally in polynomial time for chordal graphs, but the weighted case is NP-complete even for chordal graphs. Can we $O(k)$-approximate the independent dominating set problem for sequentially k-independent graphs?

Finally, we think there is a correspondence between algorithm paradigms and problem structures. We have seen multiple evidences for simple algorithms based on local decisions. The most notable one is the matroid and greedoid structures in correspondence to greedy algorithms, which have been studied extensively in the literature. One less studied class is claw-free graphs. In [7], Berman gives a $\frac{d}{2}$-approximation for maximum weight independent set in d-claw free graphs. We also note that Edmond's weighted matching algorithm [16] can also be extended to claw free graphs [27][28]. Both of these algorithms are local search based algorithms. Finally, as illustrated in this paper, various problems solved (or approximated) by the local ratio technique seem to be connected with the graphs with small sequential independence number. We believe there is a unified view of all these algorithms.

Acknowledgment

We thank Derek Corneil and the referees for their helpful comments. This research is supported by the Natural Sciences and Engineering Research Council of Canada and the Department of Computer Science, University of Toronto.

References

1. Akcoglu, K., Aspnes, J., DasGupta, B., Kao, M.-Y.: Opportunity Cost Algorithms for Combinatorial Auctions. Applied Optimization 74: Computational Methods in Decision-Making, Economics and Finance, 455-479 (2002)
2. Baker, B.S.: Approximation algorithms for NP-complete problems on planar graphs. Journal of the Association for Computing Machinery 41(1), 153–180 (1994)
3. Bar-Noy, A., Bar-Yehuda, R., Freund, A., Naor, J., Schieber, B.: A unified approach to approximating resource allocation and scheduling. JACM 48(5), 1069–1090 (2001)
4. Bar-Noy, A., Guha, S., Naor, J., Schieber, B.: Approximating the throughput of multiple machines in real-time scheduling. SIAM Journal on Computing 31(2), 331–352 (2001)
5. Bar-Yehuda, R., Bendel, A., Freund, A., Rawitz, D.: Local ratio: A unied framework for approxmation algorithms in memoriam: Shimon Even 1935-2004. Computing Surveys 36, 422–463 (2004)
6. Bar-Yehuda, R., Even, S.: On approximating a vertex cover for planar graphs. In: Proceedings of the Fourteenth Annual ACM Symposium on Theory of Computing, pp. 303–309 (1982)
7. Berman, P.: A $\frac{d}{2}$ approximation for maximum weight independent set in d-claw free graphs. Nord. J. Comput. 7(3), 178–184 (2000)
8. Berman, P., DasGupta, B.: Improvements in throughout maximization for real-time scheduling. In: STOC, pp. 680–687 (2000)
9. Berman, P., DasGupta, B.: A simple approximation algorithm for nonoverlapping local alignments (Weighted Independent Sets of Axis Parallel Rectangles). Biocomputing 1, 129–138 (2002)
10. Borodin, A., Cashman, D., Magen, A.: How well can primal-dual and local-ratio algorithms perform? In: Caires, L., Italiano, G.F., Monteiro, L., Palamidessi, C., Yung, M. (eds.) ICALP 2005. LNCS, vol. 3580, pp. 943–955. Springer, Heidelberg (2005)
11. Borodin, A., Nielsen, M., Rackoff, C.: (Incremental) priority algorithms. In: SODA, pp. 752–761 (2002)
12. Butman, A., Hermelin, D., Lewenstein, M., Rawitz, D.: Optimization problems in multiple-interval graphs. In: Proceedings of SODA 2007, pp. 268–277 (2007)

13. Chakaravarthy, V.T., Roy, S.: Approximating maximum weight K-colorable subgraphs in chordal graphs. Information Processing Letters (to appear)
14. Chmeiss, A., Jégou, P.: A generalization of chordal graphs and the maximum clique problem. Information Processing Letters 62, 61–66 (1997)
15. Chuzhoy, J., Ostrovsky, R., Rabani, Y.: Approximation algorithms for the job interval selection problem and related scheduling problems. Mathematics of Operations Research 31(4), 730–738 (2006)
16. Edmonds, J.: Paths, trees, and flowers. Canad. J. Math. 17, 449–467 (1965)
17. Eisenbrand, F., Grandoni, F.: On the complexity of fixed parameter clique and dominating set. Theoretical Computer Science 326(1-3), 57–67 (2004)
18. Fulkerson, D.R., Gross, O.A.: Incidence matrices and interval graphs. Pacific J. Math. 15, 835–855 (1965)
19. Gavril, F.: Algorithms for minimum coloring, maximum clique, minimum covering by cliques, and maximun independent set of a chordal graph. SIAM Journal on Computing 1(2), 180–187 (1972)
20. Griggs, J.R., West, D.B.: Extremal values of the interval number of a graph. SIAM Journal on Algebraic and Discrete Methods 1(1), 1–14 (1980)
21. Hajnal, A., Surányi, J.: Über die Auflösung von Graphen in Voltständige Teilgraphen. Annales Universitatis Sci. Budapestiensis 1, 113–121 (1958)
22. Hayward, R.: Weakly triangulated graphs. J. Comb. Theory 39, 200–208 (1985)
23. Hochbaum, D.: Efficient bounds for the stable set, vertex cover, and set packing problems. Discrete Applied Mathematics 6, 243–254 (1983)
24. Itai, A., Rodeh, M.: Finding a minimum circuit in a graph. SIAM Journal of Computing 7, 413–423 (1978)
25. Jamison, R.E., Mulder, H.M.: Tolerance intersection graphs on binary trees with constant tolerance 3. Discrete Math. 215, 115–131 (2000)
26. Marathe, M.V., Breu, H., Hunt III, H.B., Ravi, S.S., Rosenkrantz, D.J.: Simple Heuristics for Unit Disk Graphs. Networks 25, 59–68 (1995)
27. Minty, G.J.: On maximal independent sets of vertices in claw-free graphs. Journal of Combinatorial Theory B-28, 284–304 (1980)
28. Nakamura, D., Tamura, A.: A revision of Minty's algorithm for finding a maximum weighted stable set of a claw-free graph. Journal of the Operations Research Society of Japan 44(2), 194–204 (2001)
29. Rose, D.J.: Triangulated graphs and the elimination process. J. Math. Anal. Appl. 32, 597–609 (1970)
30. Rose, D.J., Tarjan, R.E., Lueker, G.S.: Algorithmic aspects of vertexelimination on graphs. SIAM J. Cornput. 5, 266–283 (1976)
31. Spieksma, F.C.R.: On the approximability of an interval scheduling problem. Journal of Scheduling 2, 215–227 (1999)
32. Tarjan, R.E.: Maximum cardinality search and chordal graphs. Lecture Notes CS 259 (unpublished, 1976)
33. Trotter Jr., W.T., Harary, F.: On double and multiple interval graphs. Journal of Graph Theory 3(3), 205–211 (1979)
34. West, D.B.: Parameters of partial orders and graphs: Packing, covering and representation. In: Proceedings of the Conference on Graphs and Order, 1984, Banff, Canada, pp. 267–350. Reidel (1985)
35. West, D.B., Shmoys, D.B.: Recognizing graphs with fixed interval number is NP-complete. Discrete Applied Mathematics 8, 295–305 (1984)
36. Yannakakis, M., Gavril, F.: The maximum k-colorable subgraph problem for chordal graphs. Information Processing Letters 24(2), 133–137 (1987)

Author Index

FSC
www.fsc.org
MIX
Papier | Fördert
gute Waldnutzung
FSC® C083411

Kolibri 360 GmbH
Ferdinand-Jühlke-Straße 7
99095 Erfurt, Deutschland
produktsicherheit@kolibri360.de